UNIVERSITY CHEMISTRY
Frontiers and Foundations from a Global and Molecular Perspective

James G. Anderson

The MIT Press
Cambridge, Massachusetts
London, England

The MIT Press would like to thank the anonymous peer reviewers who provided comments on drafts of this book. The generous work of academic experts is essential for establishing the authority and quality of our publications. We acknowledge with gratitude the contributions of these otherwise uncredited readers.

This book was set in Minion Pro by New Best-set Typesetters Ltd. Printed and bound in the United States of America.

Library of Congress Cataloging-in-Publication Data

Names: Anderson, James G., author.
Title: University chemistry : frontiers and foundations from a global and molecular perspective / James G. Anderson, Harvard University.
Description: Cambridge : The MIT Press, 2022. | Includes bibliographical references and index.
Identifiers: LCCN 2021030650 | ISBN 9780262542654 (paperback)
Subjects: LCSH: Chemistry.
Classification: LCC QD33 .A498 2022 | DDC 540—dc23
LC record available at https://lccn.loc.gov/2021030650

10 9 8 7 6 5 4 3 2 1

CONTENTS

The Periodic Table of the Elements

by Robert Carson, version 1.4

Legend

atomic mass
or most stable mass number
1st ionization energy in kJ/mol

chemical symbol

name

electron configuration

55.845
762.5 1.83
Fe
26
[Ar] 3d⁶ 4s²

atomic number
electronegativity

oxidation states
most common are bold

+6
+5
+4
+3
+2
+1
-2

radioactive elements have masses in parenthesis

Categories
- alkali metals
- alkaline metals
- other metals
- transition metals
- lanthanoids
- actinoids
- metalloids
- nonmetals
- halogens
- noble gases
- unknown elements

electron configuration blocks
s, d, p, f

notes
- as of yet, elements 113-118 have no official name designated by the IUPAC.
- 1 kJ/mol ≈ 96.485 eV.
- all elements are implied to have an oxidation state of zero.

group	1	2	3	4	5	6	7	8	9	10	11	12	13	14	15	16	17	18
period 1	H 1																	He 2
2	Li 3	Be 4											B 5	C 6	N 7	O 8	F 9	Ne 10
3	Na 11	Mg 12											Al 13	Si 14	P 15	S 16	Cl 17	Ar 18
4	K 19	Ca 20	Sc 21	Ti 22	V 23	Cr 24	Mn 25	Fe 26	Co 27	Ni 28	Cu 29	Zn 30	Ga 31	Ge 32	As 33	Se 34	Br 35	Kr 36
5	Rb 37	Sr 38	Y 39	Zr 40	Nb 41	Mo 42	Tc 43	Ru 44	Rh 45	Pd 46	Ag 47	Cd 48	In 49	Sn 50	Sb 51	Te 52	I 53	Xe 54
6	Cs 55	Ba 56	Lu 71	Hf 72	Ta 73	W 74	Re 75	Os 76	Ir 77	Pt 78	Au 79	Hg 80	Tl 81	Pb 82	Bi 83	Po 84	At 85	Rn 86
7	Fr 87	Ra 88	Lr 103	Rf 104	Db 105	Sg 106	Bh 107	Hs 108	Mt 109	Ds 110	Rg 111	Cn 112	Uut 113	Uuq 114	Uup 115	Uuh 116	Uus 117	Uuo 118

Lanthanoids
La 57 | Ce 58 | Pr 59 | Nd 60 | Pm 61 | Sm 62 | Eu 63 | Gd 64 | Tb 65 | Dy 66 | Ho 67 | Er 68 | Tm 69 | Yb 70

Actinoids
Ac 89 | Th 90 | Pa 91 | U 92 | Np 93 | Pu 94 | Am 95 | Cm 96 | Bk 97 | Cf 98 | Es 99 | Fm 100 | Md 101 | No 102

The Elements (Atomic Numbers and Atomic Masses)

Name	Symbol	Atomic Number	Atomic Mass*	Name	Symbol	Atomic Number	Atomic Mass*
Actinium	Ac	89	(227)	Neodymium	Nd	60	144.2
Aluminum	Al	13	26.98	Neon	Ne	10	20.18
Americium	Am	95	(243)	Neptunium	Np	93	(244)
Antimony	Sb	51	121.8	Nickel	Ni	28	58.70
Argon	Ar	18	39.95	Niobium	Nb	41	92.91
Arsenic	As	33	74.92	Nitrogen	N	7	14.01
Astatine	At	85	(210)	Nobelium	No	102	(253)
Barium	Ba	56	137.3	Osmium	Os	76	190.2
Berkelium	Bk	97	(247)	Oxygen	O	8	16.00
Beryllium	Be	4	9.012	Palladium	Pd	46	106.4
Bismuth	Bi	83	209.0	Phosphorus	P	15	30.97
Bohrium	Bh	107	(267)	Platinum	Pt	78	195.1
Boron	B	5	10.81	Plutonium	Pu	94	(242)
Bromine	Br	35	79.90	Polonium	Po	84	(209)
Cadmium	Cd	48	112.4	Potassium	K	19	39.10
Calcium	Ca	20	40.08	Praseodymium	Pr	59	140.9
Californium	Cf	98	(249)	Promethium	Pm	61	(145)
Carbon	C	6	12.01	Protactinium	Pa	91	(231)
Cerium	Ce	58	140.1	Radium	Ra	88	(226)
Cesium	Cs	55	132.9	Radon	Rn	86	(222)
Chlorine	Cl	17	35.45	Rhenium	Re	75	186.2
Chromium	Cr	24	52.00	Rhodium	Rh	45	102.9
Cobalt	Co	27	58.93	Roentgenium	Rg	111	(272)
Copper	Cu	29	63.55	Rubidium	Rb	37	85.47
Curium	Cm	96	(247)	Ruthenium	Ru	44	101.1
Darmstadtium	Ds	110	(281)	Rutherfordium	Rf	104	(263)
Dubnium	Db	105	(262)	Samarium	Sm	62	150.4
Dysprosium	Dy	66	162.5	Scandium	Sc	21	44.96
Einsteinium	Es	99	(254)	Seaborgium	Sg	106	(266)
Erbium	Er	68	167.3	Selenium	Se	34	78.96
Europium	Eu	63	152.0	Silicon	Si	14	28.09
Fermium	Fm	100	(253)	Silver	Ag	47	107.9
Fluorine	F	9	19.00	Sodium	Na	11	22.99
Francium	Fr	87	(223)	Strontium	Sr	38	87.62
Gadolinium	Gd	64	157.3	Sulfur	S	16	32.07
Gallium	Ga	31	69.72	Tantalum	Ta	73	180.9
Germanium	Ge	32	72.61	Technetium	Tc	43	(98)
Gold	Au	79	197.0	Tellurium	Te	52	127.6
Hafnium	Hf	72	178.5	Terbium	Tb	65	158.9
Hassium	Hs	108	(227)	Thallium	Tl	81	204.4
Helium	He	2	4.003	Thorium	Th	90	232.0
Holmium	Ho	67	164.9	Thulium	Tm	69	168.9
Hydrogen	H	1	1.008	Tin	Sn	50	118.7
Indium	In	49	114.8	Titanium	Ti	22	47.88
Iodine	I	53	126.9	Tungsten	W	74	183.9
Iridium	Ir	77	192.2	Uranium	U	92	238.0
Iron	Fe	26	55.85	Vanadium	V	23	50.94
Krypton	Kr	36	83.80	Xenon	Xe	54	131.3
Lanthanum	La	57	138.9	Ytterbium	Yb	70	173.0
Lawrencium	Lr	103	(257)	Yttrium	Y	39	88.91
Lead	Pb	82	207.2	Zinc	Zn	30	65.41
Lithium	Li	3	6.941	Zirconium	Zr	40	91.22
Lutetium	Lu	71	175.0				
Magnesium	Mg	12	24.31				
Manganese	Mn	25	54.94				
Meitnerium	Mt	109	(268)				
Mendelevium	Md	101	(256)				
Mercury	Hg	80	200.6				
Molybdenum	Mo	42	95.94				

See https://www.nist.gov/pml/atomic-weights-and-isotopic-compositions-relative-atomic-masses

> The physical sciences hold the key to solving problems of immense importance to this and subsequent generations. These challenges demand answers based upon scientific and technical advances initiated now, but extending decades into the future. To provide the concepts and context, this text recognizes a central premise:
>
> "If I learned anything in my forty one years of teaching, it is that the best way to transmit knowledge and stimulate thought is to teach from the top down. Begin by posing large problems, questions, and concepts of the highest significance and then peel off layers of causation as currently understood. Do not teach from the bottom up."
>
> —E. O. Wilson

Preface

It is the responsibility of universities to look forward in time to consider what graduates of today will face in the decades ahead, and to then reform curricula to provide the required foundation for those graduates. Current university graduates face coming to terms with a number of questions: What technical forces are shaping the modern world? Where are the frontiers of innovation and what implications do those advances hold for professional endeavors, not just in technology, but also in international economics, government, ethics, public health, law, and education? Which public policy strategies are founded on sound scientific and technological understanding and which are not? Which country's economic and societal structures are responsive to emerging advances and which are not?

Yet a growing compendium of research has shown that university graduates in the United States are, to a remarkable degree, lacking both fundamental knowledge of the physical sciences and the associated judgment with respect to consequences for society and for the nation's future. The growing number of studies and documented interviews with university students across the country in their junior and senior years as well as on commencement day, is a testament to an educational strategy in significant need of reconsideration.

Research has shown that a significant contributor to this lack of an effective scientific foundation required for active engagement in a rapidly changing international arena, is that introductory chemistry and physics are taught as isolated subjects. Courses taught as isolated course material, without a larger and compelling context, create an effectively *exclusive* rather than *inclusive* message—it becomes a process of irreversible elimination that has shed individuals of immense creativity and talent from the sciences. But of perhaps greater importance, this exclusion has left a chasm between undergraduates and the sciences and it is clear that this separation projects forward in time onto the structure of society following graduation that has served neither science nor society well.

There are other important considerations calling for a reassessment of the strategy behind the teaching of university introductory chemistry. The first is the recognition that the physical sciences hold the key to solving problems of immense importance to this and subsequent generations—for example, the challenge of producing the energy required to meet the rapidly escalating demands of a global society that will approach a population of 10 billion by 2050 with a rapidly increasing standard of living in developing countries. This is a situation unique in

The question is: how do we develop an innovative university-level strategy leading to the proactive engagement of science and technology in the core objectives of the nation's future? A critical part of the answer lies in the structure and objectives of introductory physical science courses. This is so because introductory chemistry and physics courses, if taught as subjects isolated from each other and from the larger compelling challenges we face, serve to irreversibly eliminate students from the physical sciences early and thereby from a contextual understanding of the central role of science in our future. This has served neither science nor society well.

All university graduates today, independent of chosen field, face coming to terms with a number of questions:

1.
What technical forces are shaping the modern world?

2.
Where are the frontiers of innovation and what implications do those advances hold for professional endeavors …

3.
not just in technology, but also in international economics, government, ethics, public health, law and education?

4.
Which public policy strategies are founded on sound scientific and technological understanding and which are not?

human history. It required 7000 human generations to reach a global population of 2 billion from the emergence of modern humans 160,000 years ago. In the span of a single human lifetime, 1945 to 2045, world population will increase from 2 billion to approximately 10 billion people. To supply energy to this global society *today* is an $8 trillion/year enterprise. To cover increases in both population and per capita income, the scale of the problem and the attendant implications test the limits of human intuition, even well informed human intuition. For example, our global society must build the *equivalent* of two large fossil fuel burning power plants *per day* between now and 2050 to meet nominal demand. While constraints on how primary energy is generated expose the consequences of such a decision, the construction of two large fossil fuel burning power plants per day for the next forty years sets the scale of the increasing demand for primary energy generation globally.

National security, competitive economic considerations, and human health should dominate public policy debates. But public policy debates are dominated in large part by a lack of essential information and perspective that should be an integral part of a modern university education. Yet the numbers reveal that university graduates are part of the problem. Society must contend with the collision between global energy demand and the constraints on policy choices. This is a challenge that demands answers based upon scientific and technical advances initiated now, but extending over decades into the future. This in turn engages a new vision for chemistry that requires a command of thermodynamics, electrochemistry, quantum mechanics, molecular bonding, kinetics, catalysis, materials, and biological processes at the molecular level.

Another perspective has engendered a reassessment of university introductory chemistry courses, a perspective that has emerged from a very different line of reasoning. Specifically, research done into how students learn scientific principles and develop quantitative reasoning skills. The currently practiced strategy in introductory chemistry is to present lectures and text material that covers the basic formalism and theory followed by assigned problem sets. Solid evidence has shown that there are two basic failures with the "formalism first" approach to teaching. First, it results in "disembodied knowledge." While the instructor delivering lectures sees the landscape clearly, students cannot attach the knowledge delivered to them to a context or, as is frequently the case, to their aspirations for the future. As a result this formalism becomes a sea of largely meaningless symbols and facts to memorize. Research has shown that knowledge obtained in this way is filed away in the brain in a separate compartment and building links to that compartment after the fact is much harder and less effective than if it had been filed correctly from the start.

The most common symptom of the problem is that students are unable to recognize when these ideas do and do not apply when confronted with new situations that they have not explicitly studied, a problem exacerbated by the wide use of standardized tests so prevalent in scientific secondary educational infrastructure today.

The second failure of the "formalism first" approach involves student attitudes about the subject and how it is best learned and the closely related issue of interest in the subject. If the student first sees the subject as a collection of abstract formalism, it gets classified as "abstract meaningless stuff and the only way to learn it is by rote memorization." Studies of student attitudes about science show that instructors are seldom successful at undoing initial damage that is created by first introducing abstract formalism. This lack of a robust set of connections to the student's past knowledge and experience and to their future aspirations also results in much lower long-term retention of learning.

In contrast, if new conceptual material is preceded by a compelling problem to solve, and then in the context of solving that problem, or set of problems, a formalism emerges, there is a proactive engagement of the concepts with the larger contextual vision. This provides an individual seeking to understand a scientific principle with a much more robust and useful knowledge base.

So it is the union of these two perspectives: (1) the critical role played by the physical sciences in the solution of unprecedented societal need and objectives, and (2) compelling evidence for how scientific principles and quantitative reasoning skills are mastered at the university level that establishes the structure of this text. The objective of the text is to (1) establish the imperative defining the central role that the principles of chemistry and physics play in the unfolding global challenges, (2) develop the core concepts of, among others: energy and energy transformations, thermodynamics, chemical equilibria, acid/base and redox reactivity, electrochemistry, quantum mechanics, molecular bonding, kinetics, catalysis, materials, and nuclear, and (3) establish the direct link between those core concepts and the larger global context. To accomplish this, the structure of the text consists of three primary elements. Each chapter opens with a "Framework" section that establishes the larger context defining why the scientific concepts treated in that chapter are critically important to science and technology's role in addressing rapidly emerging challenges at the global scale. Second, the chapter "Core" that addresses the quantitative concepts central to the modern physical sciences. The third element in the chapter structure are the "Case Studies" that link the concepts in the chapter Core to the larger global context. Those Case Studies fall into three categories: (1) Case Studies designed to build quantitative reasoning skills, (2) Case Studies designed to build a technology backbone, and (3) Case Studies designed to build a global energy backbone to supply the dramatic expansion in the need for primary energy generation.

Thus, as the adjoining figure illustrates in graphical form, the Case Studies link the central concepts in contemporary physical science and technology into the global concepts of human health, technology leadership, economic choices, national security, energy production and distribution, climate consequences of energy production choices, and international negotiations, to pick a few examples.

This development of the context and the concepts together in the structure of the text does not imply, however, that these changes are done at the expense of rigor in the development of scientific principles. The depth of the development of both the context and the scientific principles are gauged to provide the student with the preparation needed to compete successfully today and in the future in the international scientific and societal arena.

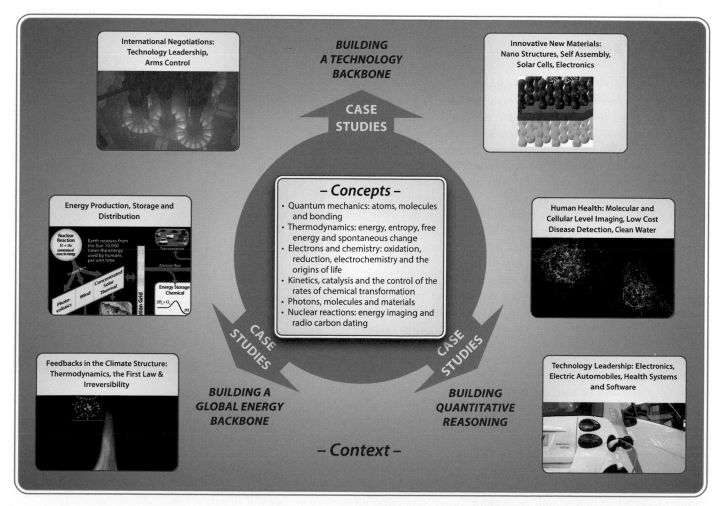

Acknowledgments

Restructuring introductory university chemistry involves directly addressing decades of tradition that no matter how entrenched, none-the-less provide remarkable opportunities for instigating changes that project the subject into a position of markedly greater importance for both students and faculty. I would first like to thank the Harvard University Department of Chemistry and Chemical Biology for its unflinching support over many years leading to the reconstruction of introductory chemistry at the university. In particular Ted Betley, Charles Lieber, Adam Cohen, David Evans, Roy Gordon, Rick Heller, Dudley Herschbach, Eric Jacobsen, Dan Nocera, George Whitesides, Sunney Xie, and Xiaowei Zhuang to name a few. Second, I would like to thank my close colleagues in the teaching of the course for which this text was developed and used over the past six years: Gregory Tucci, Tim Kaxiras, Sirinya Matchacheep, and Lu Wang. These four individuals made the year-in-year-out teaching and development of this course a profoundly enjoyable endeavor.

I would like to acknowledge extremely important discussions with Carl Wieman during the development of the text structure. I also acknowledge the courage of two individuals who taught the course at the University of Wisconsin during and through the evolution of the text—Fleming Crim and Randall Goldsmith, both of whom made important suggestions and contributed decidedly to the refinement of the text. Innumerable associates provided editing and proof reading, among them graduate students and postdocs in my research group, teaching fellows in the course, and Kenneth H. McDonald.

Ralph Cicerone highlighted the text as President of the National Academy of Sciences and Mark Cardillo repeatedly noted the development of the course in coverage from the Dreyfus Foundation. Both early and subsequently, Dan Kaveney and David J. C. MacKay provided key advice on the options for making the text available to a wide readership.

Of primary importance, I owe a great debt of thanks to Robert Stanhope for converting illegible handwriting, much of it done on aircraft, and scribbled diagrams over the past six years, into the complete text as it now appears. Also to Joyce Makrides for conversion of handwriting to text files in the early phases of the book's development.

Finally, I wish to thank my wife Shirine for putting up over these years with the evenings and weekends lost to the development of new chapters and case studies, often just days before posting to the course website!

About the Author

Jim Anderson is the Philip S. Weld Professor in the Departments of Chemistry and Chemical Biology, Earth and Planetary Sciences and the School of Engineering and Applied Sciences, Harvard University. He was Chairman, Department of Chemistry and Chemical Biology, Harvard University, 1998–2001. He was elected to the National Academy of Sciences in 1992, the American Philosophical Society in 1998, the American Academy of Arts and Sciences in 1985, a Fellow of the American Association for the Advancement of Science in 1986, a Fellow of the American Geophysical Union in 1989.

The Anderson research group addresses three domains in the physical sciences: (1) chemical reactivity viewed from the microscopic perspective of electron structure, molecular orbitals, and reactivities of radical-radical and radical-molecule systems; (2) chemical catalysis sustained by free radical chain reactions that dictate the macroscopic rate of chemical transformation in Earth's stratosphere and troposphere; and (3) mechanistic links between chemistry, radiation, and dynamics that control climate.

He received the 2021 Dreyfus Prize in the Chemical Sciences, the 2019 Alumnus Summa Laude Dignatus from the University of Washington, the 2017 Lichtenberg Medal from the Göttingen Academy of Sciences and Humanities, the 2016 Polanyi Medal of the British Royal Academy of Chemistry for work on free radical kinetics, the 2016 Benton Medal for Public Service by the University of Chicago, and the 2012 Smithsonian American Ingenuity Award in the Physical Sciences. He received the E.O. Lawrence Award in Environmental Science and Technology; the American Chemical Society's Gustavus John Esselen Award for Chemistry in the Public Interest; the United Nations Vienna Convention Award for Protection of the Ozone Layer; Harvard University's Ledlie Prize for Most Valuable Contribution to Science by a Member of the Faculty; and the American Chemical Society's National Award for Creative Advances in Environmental Science and Technology. He has testified on numerous occasions before both Senate and House committees on national energy and climate issues.

ENERGY

CONCEPTUAL FOUNDATION AND THE LAWS THAT GOVERN ITS TRANSFORMATIONS

1

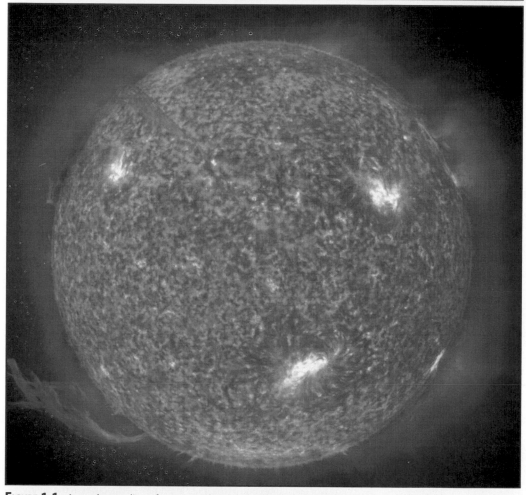

FIGURE 1.1 An understanding of energy at the global and molecular level requires an understanding of how nature seamlessly transforms energy among and between various categories. The Sun represents an important case in point because at the core of the Sun the extremely high pressure and high temperature (100 million K) sustain nuclear fusion reactions converting hydrogen nuclei to helium nuclei generating some 10^{34} joules of energy a year. This energy release moves to the surface of the Sun by the transport of hot material and by the transport of electromagnetic radiation outward to the Sun's surface. The temperature of the visible disk of the Sun, called the photosphere, resides at 5800 K, emitting visible light (electromagnetic radiation). That solar radiation in the visible is emitted outward into the blackness of Space. The Earth intercepts approximately one part in 10^{10} of that solar radiation, which is 10,000 times the total energy consumption rate of the global economy. The solar radiation received by the Earth is transformed into the kinetic energy of the motion of the atmosphere and the oceans, into the building of organisms through photosynthesis, into the power that sustains humanity, and into the heat that warms the planet.

Setting the Context 1: Energy Scale in Joules

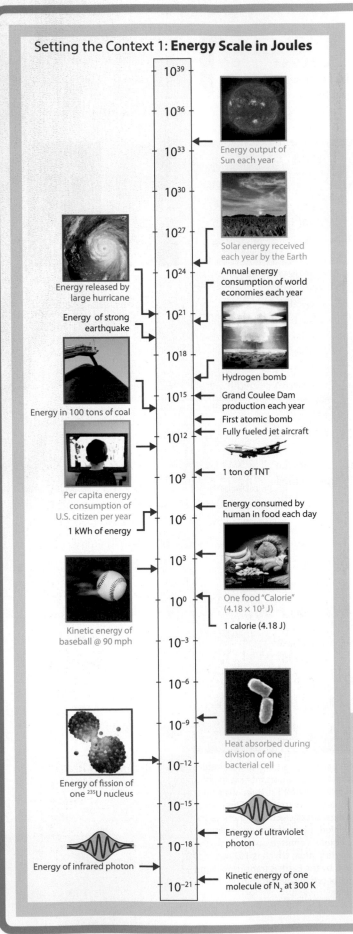

Framework

Why do we care about understanding the concept of energy? What is energy anyway? Why does the understanding of energy at the molecular level dominate nearly every scientific discussion of chemical, physical, or biological change? Why does the subject of energy at the global level dominate nearly every discussion surrounding international negotiations, political debates, and investment strategies?

There are important answers to these questions that emerge from different lines of reasoning. One perspective is that until a *scientific* understanding of energy emerged in the 18th and 19th century, little progress in scientific understanding in its larger context was achieved. More recently, an understanding of energy and the control of energy at the molecular level underpins developments in drug delivery for cancer treatments, self-assembly of complex molecular architectures, innovative new materials for nanoelectronics, low cost high efficiency solar cells for electricity production from sunlight, low cost disease detection, low cost water purification and desalinization, and innovative electrodes for production of hydrogen from sunlight for developing nations.

Numbers are important. We receive 10,000 times the amount of energy from the Sun, shown in Figure 1.1, per unit time than is consumed by all mankind. The human endeavor of energy extraction, distribution and use is today an \$8 trillion per year enterprise, dwarfing its next closest competitor. Developing a 1% niche in this industry is an \$80 billion per year opportunity. Energy dominates consideration of economic decisions and economic forecasts. Another perspective is that supplying energy to the global population is the single most important prerequisite for human health, food supplies, education, and civilized existence. As change accelerates in the global arena, the future increasingly belongs to those in command of an understanding of energy and energy consequences at the global and molecular level. A key reason for this is that picturing what occurs at the submicroscopic or molecular level directly clarifies the cause of changes that occur at the macroscopic, observable scale. Alternatively stated, the lack of an understanding of the principles that underlie the concept of energy will increasingly

FIGURE 1.2 Becoming familiar with the orders of magnitude that comprise the range of energies occurring in both nature and in the human endeavor constitutes one of the most important perspectives in the union of science and society. We will constantly be returning to the calculation and analysis of "how much energy" is associated with global consumption, with the output of the Sun, with the interaction of photons of light with molecules, with the energy release from a nuclear reaction or a chemical reaction, and with the kinetic energy of a single molecule. While we will subsequently become familiar with how to calculate the energy in each category of interest, and how to relate each of these energy categories, we present here a range of energies from the microscopic to the macroscopic as a framework to build an intuition concerning how the world works—in the universally adopted unit of joules, so named for James Prescott Joule whom we will meet in the chapter Core.

become an impairment to professional success as globalization becomes increasingly important.

An important consideration at the intersection of science and the role science now plays in societal objectives is that learning and understanding are accelerated by an imperative. We reside today at an unprecedented point in human history for which the physical sciences hold the key to developing a strategy for global scale decisions of critical importance to this and subsequent generations. This increasingly powerful union of science and society places the physical sciences not only in a position of opportunity, but also in a position of responsibility for establishing a rational foundation for progress. Current university graduates face coming to terms with a number of questions: What technical forces are shaping the modern world? What are the most pressing problems this and subsequent generations will face? Where are the frontiers of innovation and what implications do they hold for professional endeavors in science, technology, international economics, government, ethics, public health, law, and education?

From the emergence of modern humans as a species 160,000 years ago, it required 7000 human generations to reach a global population of 2 billion in the middle of the 20th century. In the span of a single human lifetime, 1945 to 2045, world population will increase by a factor of five to approximately 10 billion. To recognize the scale of human demand for energy in the face of this population increase, combined with the expected growth in the standard of living for the developing economies in Asia, Africa, and South America, our global society must build the *equivalent* of two large fossil fuel burning power plants *per day* between now and 2050; alternatively, a nuclear power plant *every day* between now and 2050. This sets the scale of the challenges—but what are the consequences?

As we will see, when carefully considered, pressure applied to the structures of global society by increasing global population and per capita standard of living transform the union between the core concepts in the physical sciences and the structure of an effective university education. Specifically the union is transformed by the objective of linking an understanding of the scientific and technical fundamentals into an integrated university education that does not, at an early stage, split the sciences from the broader context of societal challenges and opportunities. Operating effectively within the international arena without a grasp of the scientific and technical fundamentals that define a path forward will, for current university graduates, become increasingly difficult.

An important component of an ability to understand the scientific and technical fundamentals is a grasp of the orders of magnitude associated with energy scales and power scales. Figure 1.2 begins the development of a foundation defining the orders of magnitude of energy associated with important examples. These examples run from the energy output of the Sun per year (10^{34} joules) to the kinetic energy of a single molecule of N_2 at room temperature (10^{-21} joules). Figure 1.3 spans the orders of magnitude of the associated power scales. The energy scales and power scales deserve particular attention and we will calculate each entry as the text unfolds and so we will repeatedly refer back to these scales as we move forward. These numbers matter, and an important degree of insight emerges from knowing the ratio of key examples such as the ratio of solar energy received by the Earth to the energy used by the global economy. The importance of the relationship between energy and power on the one hand, and the calculation of the magnitude associated with important categories of energy and power on the other is the subject of Case Studies 1.1, 1.2, and 1.3 at the chapter end. The calculations incorporated within these three Case Studies, highlighted as sidebars to the right, are essential for establishing quantitative reasoning and for building a foundation in the link between science and technology.

A key structural feature of the approach adopted here is the relationship between *concepts* intrinsic to the scientific fundamentals and *context* that constitutes

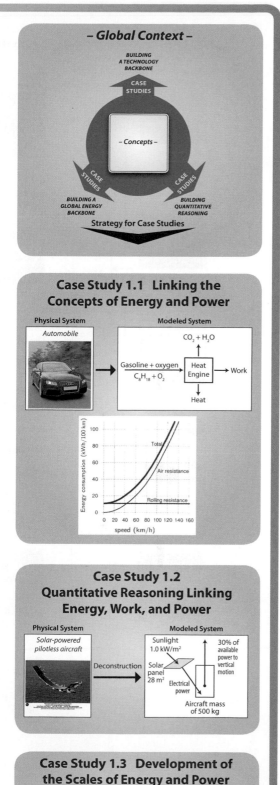

Case Study 1.1 Linking the Concepts of Energy and Power

Case Study 1.2 Quantitative Reasoning Linking Energy, Work, and Power

Case Study 1.3 Development of the Scales of Energy and Power

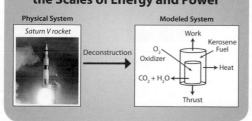

Setting the Context 2: **Power Scale in Watts**

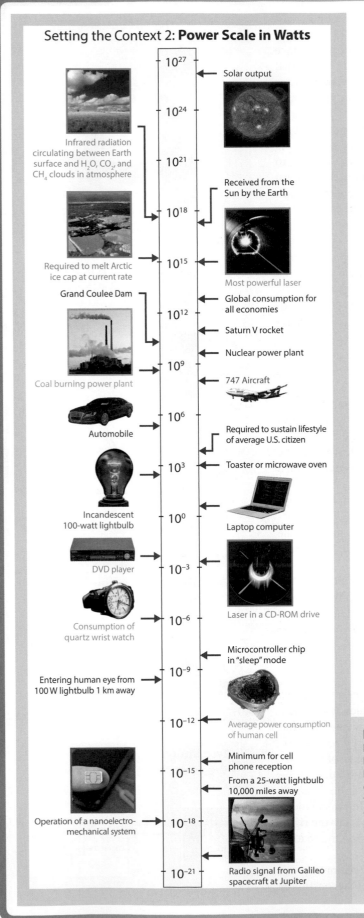

the global demand for innovative solutions serving both personal and societal objectives. Figure 1.4 presents this strategy in graphical form, placing the *concepts* that form the foundation of the modern physical sciences at the center of the diagram: quantum mechanics, thermodynamics, electrochemistry, kinetics, catalysis, photochemistry, materials, and nuclear chemistry. The outer circle of Figure 1.4 summarizes the *context* that requires an understanding of the concepts in order to participate proactively in the affairs of our global society as the future unfolds: energy production and storage, innovative new materials, feedbacks in the climate structure, human health, technology leadership, national security, and international negotiations. Indeed, a working knowledge of the concepts at the core of Figure 1.4 and the union of those concepts with the context laid out in Figure 1.4 is necessary for informed participation in a modern scientific discourse and in a modern democracy.

One other aspect of Figure 1.4 that requires note is the *input*, displayed at the bottom center of the diagram. The rapid increase in global population, particularly in Asia and Africa, as we approach the middle of the century is forcing dynamic change linking the concepts to the context. Because this global population increase is occurring primarily in the developing world, where per capita energy consumption is a small fraction of what it is in the developed countries, the global demand for energy will increase *far* more rapidly than will population. For example, the per capita consumption of energy in the US is nearly 50 times the per capita energy consumption of nations in central Africa, yet a major fraction of world population growth will occur in Africa.

We can trace these connections between a given *concept* that lies within the inner circle of Figure 1.4 and the surrounding *context* that drives the imperative for progress by asking some questions: Why is an understanding of thermodynamics so important to our future? Within the domain of thermodynamics lie the answers to critical, practical questions: why is a gasoline-powered automobile only 15% efficient in converting the energy contained in the chemical bonds of octane, C_8H_{18}, to the motion of the vehicle? Why does it require only 15 kWh of energy to propel an electric car 100 km when it requires 125 kWh of chemical energy to propel a gasoline automobile the same 100 km? From thermodynamics emerges the dual role of energy and entropy that together quantitatively establish the

FIGURE 1.3   While energy determines the amount of work that can be done, power determines how fast or how quickly that energy can be converted from one form of energy to another. Thus the distinction between energy and power becomes one of paramount importance in any discussion of energy transfer. Power is quite simply the amount of energy transferred per unit time so power = energy/time and the universal unit of power is the watt, named after James Watt, such that 1 watt = 1 joule/second. In the case of an electrical device for example, a 100 watt light bulb uses 100 joules of energy in one second. Frequently for mechanical devices such as a car, we define its power in terms of horsepower (hp). But 1000 watts, or a kilowatt (kW), is just 1.3 hp so a 200 horsepower car is a 154 kW car. Examine this power scale carefully—just as with energy, it provides an important perspective!

criteria for spontaneous change. Spontaneous change is what drives all chemical reactions; the chemical transformation from reactants to products. The release of *free energy* in a chemical reaction unifies the concepts of energy and entropy, but what is *free energy* and how specifically is it defined?

A further study of thermodynamics over the course of this text will demonstrate that *all forms of energy*, whether associated with a living cell, an electromechanical device, or the Earth's climate system, are inter-convertible in accordance with the First Law of Thermodynamics. The First Law of Thermodynamics states that *energy is neither created nor destroyed, but energy can be converted from one form to another.* In practice this means we must *quantitatively* account for how and where energy has moved within a system or between a system and its surroundings. While the First Law appears on the face of it to be rather self evident,

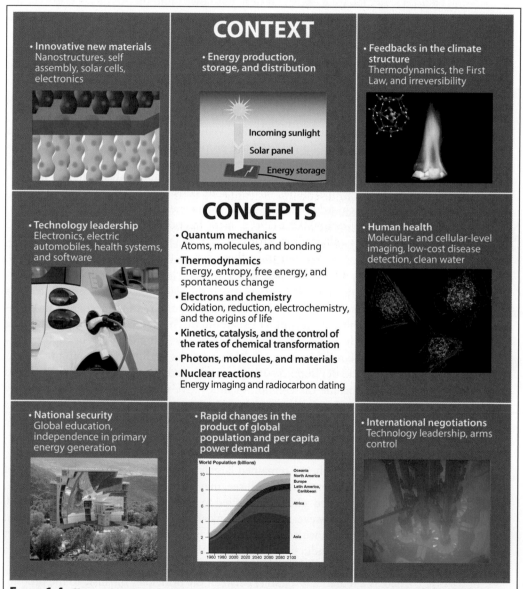

FIGURE 1.4 The combination of rapidly developing innovation and rapidly increasing world population places the physical sciences in a position of primary responsibility for the solution to key societal objectives. The *concepts* central to chemistry and the physical sciences are summarized in the center of the diagram. Mastering the conceptual core of the physical sciences is essential for understanding the larger *context* of how those principles are connected to immediate challenges including global energy production, innovative new materials, economic leadership, human health, feedbacks in the climate system, international negotiations, and natural security.

it is one of the most universally applicable and powerful laws in science. We will see why.

For example, an important axiom that can be drawn directly from the First Law of Thermodynamics is: *it is the net flow of heat into the reservoirs of the climate system that define the course of events as we move into the future.* It is the irreversible changes to the climate structure that result from the *retention* of that heat that matters most to society, not simply "global warming." An important example is the floating ice encompassing the Arctic Ocean that has remained in place for the past 3 million years. However, in the last 30 years, 75 percent of the permanent ice has been lost from this system initiating potent feedbacks that accelerate the removal of the remaining ice. It is now possible to pass unencumbered from the Pacific to the Atlantic Ocean in summer—the "northwest passage" that drove exploration from the 15th century on. The numbers matter: this loss of ice volume means that a *net* 5×10^{21} joules of energy, as heat, has flowed into the Arctic Ice Cap over the past 30 years. Yet the energy per unit time required to melt this Arctic ice is but one part in 50,000 of the infrared energy circulating between the Earth's surface and the water vapor, carbon dioxide, methane, and cloud structures in the atmosphere. In order to place this in context, we must understand energy scales and the relative orders of magnitude associated with categories of energy. We must also develop an understanding of feedbacks and the role they play in physical and biological systems.

Next consider quantum mechanics. Why is quantum mechanics so important to our future? Recognition of the wave properties of the electron initiates fundamental insight into the structure of the hydrogen atom and multielectron atomic systems. In addition, tracing the scientific origin of quantum mechanics gives us an opportunity to discuss the Scientific Method in concrete terms—this is the subject of Case Study 1.4.

We must come to understand the scientific foundation that underpins multielectron atoms before we can explore rapid developments in materials research that extend from the fabrication of tubular geometries that capture sunlight, to nanotechnology for increasingly small and efficient computer chips, to the increasing sophistication of the internet and the electronic devices linked to it. As we will see as the text unfolds, the direct impact of quantum mechanics on modern society establishes the union of quantum mechanics with key challenges at the global scale, such as the harvesting of solar energy. This energy can be captured using an array of renewable energy gathering devices leading to an economic revolution in the race to develop new energy storage and distribution systems. Business opportunities include the creation of innovative new materials for high energy density batteries, new materials for the conversion of sunlight to renewable fuels for supplying energy to both developed and developing countries, and innovative systems capable of converting those fuels directly to electricity or for heating, water purification, and transportation—all essential for basic human health.

The concept of energy remained largely misunderstood throughout most of recorded history. Energy, as we will see, may be defined as the ability to do *work*, which is easily grasped when it is applied to automobile engines, electric motors, water pumps, etc. However, when we think about energy associated with chemical reactions within cells, with transformations in chemical bonds, with the transport of electrons along conduction pathways in the cell, the concept of work becomes more abstract and, at the outset, less intuitive.

There is a remarkable parallel between the principles intrinsic to the production, storage and use of energy within the global economy—indeed the management of the world's resources—and the production, storage, and use of energy at the cellular scale in living organisms. Cells require energy to function—to execute the myriad tasks that define the distinction between the living and the inanimate—including the synthesis of glucose (fuel) from carbon dioxide and water.

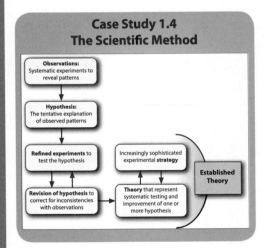

**Case Study 1.4
The Scientific Method**

The work of the cell spans the range from the contraction of muscles, to the mental recognition of patterns, to the replication of DNA.

A key part of our objective in this text is the development of quantitative reasoning. Quantitative reasoning is developed in large part through a practice of deconstruction and then reconstruction. This involves the deconstruction of a problem into its component parts that can then be addressed by identifying the scientific principle or principles that will facilitate a solution to the problem. This is in stark contrast to the process of simply sticking numbers into an equation—a tactic rewarded in standardized tests, but antithetical to the practice and the application of science. *Models* of a system or a problem are a great aid in this process of deconstruction into component parts. The reconstruction is then a process by which the parts of that simplified model are reassembled.

As we develop in this text both the central concepts in modern physical science and the close coupling of those fundamental concepts to the larger global context, a key aspect that places these considerations in quantitative context is the calculation of energy consumption by the global economic endeavor. While this is, on the face of it, a complicated task, there are ways of addressing calculations of total global energy consumption that are both accurate and straightforward. This is the subject of Case Study 1.5.

A final note on the chapter structure used in this text: each chapter is initiated by a Framework section that sets the context for the Core of the chapter; the Core of the chapter in turn develops the concepts at the foundation of modern chemistry, and of modern physical sciences more broadly. Following the chapter Core is a set of Case Studies that use the fundamentals developed in the Core to develop quantitative reasoning skills and to expand upon the connection between the chapter's (and previous chapters') concepts and the larger context of the subject. However, this text structure will not compromise rigor in the development of scientific principles. The depth of the development of the scientific foundation is gauged to provide the preparation needed to compete successfully today and in the future in the international arena—scientific and societal.

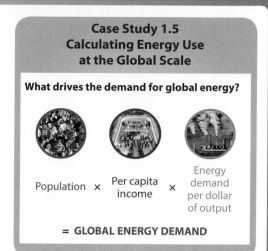

**Case Study 1.5
Calculating Energy Use
at the Global Scale**

What drives the demand for global energy?

Population × Per capita income × Energy demand per dollar of output

= GLOBAL ENERGY DEMAND

Chapter Core

Road Map for Chapter 1

In the sections that follow, we develop first an understanding of the concept of energy by tracing its scientific development—a treatment that demonstrates how elusive the concept of energy turned out to be. As we will see, in order to understand energy we must develop a facility for understanding how nature transforms various types of energy among and between various important categories of energy. The Chapter 1 Road Map is displayed in Figure 1.5.

Road Map to Core Concepts	
Understanding the Concept of Energy: A Prerequisite for Scientific Progress (**p. 9**)	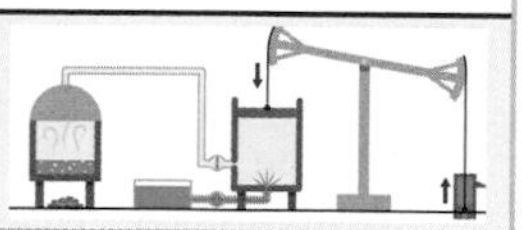
Energy, Work, and Newton's Laws: Conservation of Mechanical Energy (**p. 12**)	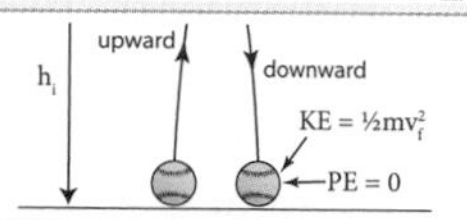
Conservation of Energy: Tracking the Flow of Energy from Macroscopic Mechanical Energy to Microscopic Molecular Motion (**p. 15**)	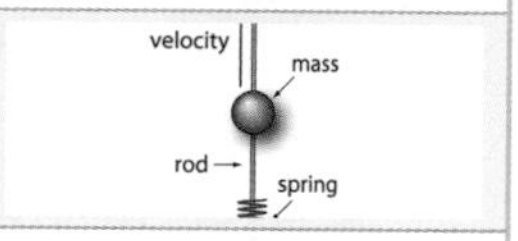
Potential Energy Diagram and the Reaction Coordinate Diagram (**p. 20**)	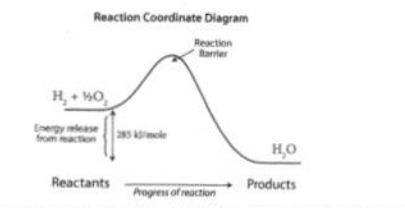
The Mechanical Equivalent of Heat and the Concept of Heat Capacity (**p. 22**)	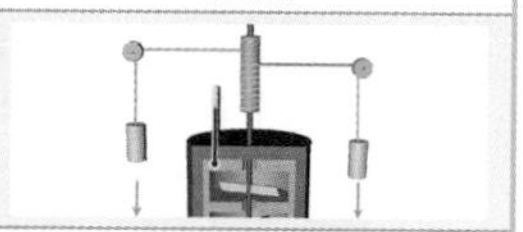
Kinetic Theory Interpretation of Temperature (**p. 26**)	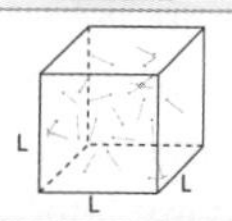
Role of Electromagnetic Radiation in the Transformation of Energy Among Various Categories of Energy (**p. 30**)	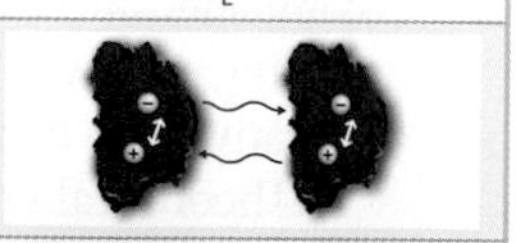
Blackbody Radiation and the Stefan–Boltzmann Law (**p. 34**)	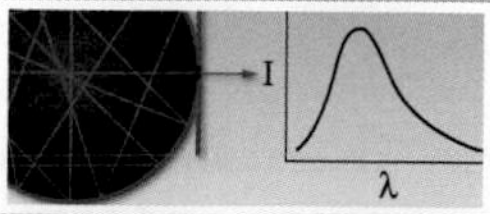
Energy and Power: A Critically Important Distinction (**p. 38**)	

FIGURE 1.5   Summary of the major concepts developed in the chapter core.

Emergence of Energy as a Scientific Concept

Energy is transformed continuously between its various forms in ways that can range from the obvious—for example, when a battery supplies voltage to an electric motor, thereby converting chemical energy to the kinetic energy of motion—to the more subtle—for example, when sunlight (electromagnetic radiation) induces the formation of organic material through the process of photosynthesis.

Sidebar 1.1—James Watt

The initial motivation for all of the pioneers of steam, most notably Thomas Newcomen (1663–1729) and James Watt (1738–1819), was to develop an efficient means to pump water from underground mines. In this, Watt's primary achievement was to improve on the earlier innovation by Thomas Newcomen. But where Newcomen failed to profit much from his invention, Watt was remarkably successful. He was recognized early in life as an accomplished inventor and entrepreneur. He received more lasting fame in later life when he was elected, at age 47 (in 1785), to the Royal Society of London. Fifteen years after his death in 1819, a statue honoring him was installed in Westminster Abbey.

James Watt

The inscription on Watt's statue reads as follows:

"The King, His Ministers, and many of the Nobles and Commoners of the Realm raised this monument to JAMES WATT who, directing the force of an original Genius, early exercised in philosophic research, to the improvement of the Steam Engine, enlarged the resources of his Country, increased the power of Man, and rose to an eminent place among the most illustrious followers of science and the real benefactors of the World."

Andrew Carnegie began his biography of Watt (Carnegie, 1905) as follows: "James Watt, born in Greenock, January 19, 1736, had the advantage, so highly prized in Scotland, of being of good kith and kin." Watt's father, James, was a merchant who supplied nautical goods and mathematical instruments. He was known for his skill in fabricating delicate instruments, a skill inherited by his son. His grandfather, Thomas, identified as a "Professor of Mathematics" (Marsden, 2002), taught navigation and surveying."

Recognizing various *forms* of energy and understanding the scientific principles involved in the *transformations* that take place between these energy categories is fundamental to scientific understanding at any level of inquiry. Energy is found in the form of molecular motion, electrical energy, nuclear energy, electromagnetic energy, kinetic energy of macroscopic objects, gravitational energy, etc.

The emergence of energy as a well defined scientific concept has a tortured past. Aristotle (381–322 BC) was the first to grapple with the many connotations of the concept of energy and to attempt to make sense of its many complexities. He first referred to it in his *Metaphysics* by joining **in** (εν) and **work** (εργον) to form ενεργεα or "energea" that he linked with "entelechia" or "complete reality." The verb energea thus came to signify motion, action, work, and/or change. But little progress was made on this concept of energy through the Roman period, the Middle Ages, and the Renaissance.

Aristotle

A number of legendary scientific pioneers had a rickety and flawed conceptual grasp of energy. At the approach of the mid-eighteenth century, the term energy became indistinguishable from power and force. As excerpted from Vaclav Smil's 2006 treatise, *Energy*, the foundation of the concept of energy traced the following course:

In 1748, David Hume (1711–1776) complained, in *An Enquiry Concerning Human Understanding*, that "There are no ideas, which occur in metaphysics, more obscure and uncertain, than those of *power, force, energy* or *necessary connexion*, of which it is every moment necessary for us to treat in all our disquisitions."

In 1807, in a lecture at the Royal Institution, Thomas Young (1773–1829) defined energy as the product of the mass of a body and the square of its velocity, thus offering an inaccurate formula (the mass should be halved) and restricting the term only to kinetic (mechanical) energy. Three decades later the seventh edition of the Encyclopedia Britannica (completed in 1842) offered only a very brief and unscientific entry, describing energy as "the power, virtue, or efficacy of a thing. It is also used figuratively, to denote emphasis in speech."

Theoretical energy studies reached a satisfactory (though not a perfect) coherence and clarity before the end of the nineteenth century when, after generations of hesitant progress, the great outburst of Western intellectual and inventive activity laid down the firm foundations of modern science and soon afterwards developed many of its more sophisticated concepts. The ground work for these advances began in the seventeenth century, and advanced considerably during the course of the eighteenth, when it was aided by the adoption both of Isaac Newton's (1642–1727) comprehensive view of physics and by engineering experiments, particularly those associated with James Watt's (1736–1819) improvements of steam engines. As the saying goes, science owes more to the steam engine than the steam engine to science.

During the early part of the nineteenth century a key contribution to the multifaceted origins of modern understanding of energy were the theoretical deductions of a young French engineer, Sadi Carnot (1796–1832), who set down the universal principles applicable to producing kinetic energy from heat and defined the maximum efficiency of an ideal (reversible) heat engine. Shortly afterwards, Justus von Liebig (1803–1873), one of the founders of modern chemistry and science-based agriculture, offered a basically correct interpretation of human and animal metabolism, by ascribing the generation of carbon dioxide and water to the oxidation of foods or feeds.

The formulation of one of the most fundamental laws of modern physics had its origin in a voyage to Java made in 1840 by a young German physician, Julius Robert von Mayer (1814–1879), as ship's doctor. The blood of patients he bled there (the practice of bleeding as a cure for many ailments persisted well into the nineteenth century) appeared much brighter than the blood of patients in Germany.

Mayer had an explanation ready: blood in the tropics does not have to be as oxidized as blood in temperate regions, because less energy is needed for body metabolism in warm places. But his answer led him to another key question. If less heat is lost in the tropics due to radiation, how about the heat lost as a result of physical work (that is, expenditure of mechanical energy), which clearly warms its surroundings, whether done in Europe or tropical Asia? Unless we put forward some mysterious origin, that heat, too, must come from the oxidation of blood—and hence heat and work must be equivalent and convertible at a fixed rate. And so began the formulation of the law of the conservation of energy. In 1842 Mayer published the first quantitative estimate of the equivalence, and three years later extended the idea of energy conservation to all natural phenomena, including electricity, light, and magnetism, and gave details of his calculation based on an experiment with gas flow between two insulated cylinders.

The correct value for the equivalence of heat and mechanical energy was found by the English physicist James Prescott Joule (1818–1889), after he conducted a large number of careful experiments. Joule used very sensitive thermometers to measure the temperature of water being churned by an assembly of revolving vanes driven by descending weights: this arrangement made it possible to measure fairly accurately the mechanical energy invested in the churning process. In

James Prescott Joule

1847 Joule's painstaking experiments yielded a result that turned out to be less than one percent of the actual value. *The law of conservation of energy—that energy can be neither created nor destroyed—*is now commonly known as the first law of thermodynamics.

In 1850 the German theoretical physicist Rudolf Clausius (1822–1888) published his first paper on the mechanical theory of heat, in which he proved that *the maximum performance obtainable from an engine using the Carnot cycle depends solely on the temperatures of the heat reservoirs, not on the nature of the working*

Sidebar 1.2—Development of the Watt Steam Engine: John B. Fenn, *Engines, Energy, and Entropy*

In Newcomen's engine, steam under pressure from a boiler pushed a piston to the top of a vertical cylinder as shown in the diagram below. The steam was then condensed by injecting water into the cylinder. The resulting vacuum allowed atmospheric pressure to push the piston to the bottom of the cylinder.

Newcomen's engines were used almost exclusively for pumping water. When power was needed to turn other machinery, a Newcomen engine would pump water to a level from which it could fall through a water wheel! (This practice has had a reincarnation. Nowadays, some electric generating power plants use off-peak power to pump water to elevated reservoirs. During peak demand, the water runs downhill through hydraulic turbines whose output supplements the capacity of the steam plant.) In spite of its wide use, the Newcomen engine left much to be desired. It was large, ungainly, and had an extravagant appetite for fuel.

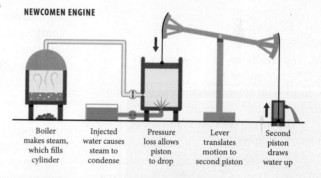

NEWCOMEN ENGINE

| Boiler makes steam, which fills cylinder | Injected water causes steam to condense | Pressure loss allows piston to drop | Lever translates motion to second piston | Second piston draws water up |

In 1763 the Scottish instrument maker James Watt became disturbed by the Newcomen engine's great waste of heat through the alternate warming and cooling of the cylinder. He introduced the idea of a separate condenser that would remain cool yet communicate with the cylinder by a valve opened at a propitious point in the cycle (see figure right).

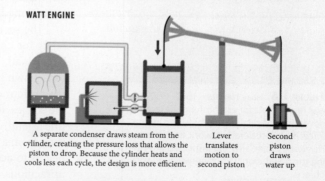

WATT ENGINE

| A separate condenser draws steam from the cylinder, creating the pressure loss that allows the piston to drop. Because the cylinder heats and cools less each cycle, the design is more efficient. | Lever translates motion to second piston | Second piston draws water up |

Thus the cylinder could be kept warm all the time—in fact, could be insulated to keep from losing heat to the surroundings. Watt's condenser greatly cut fuel consumption and increased efficiency. It marked the real beginning of the steam engine age, which we are still very much in.

substance, and that there can never be a positive heat flow from a colder to a hotter body. Clausius continued to refine this fundamental idea and in his 1864 paper he coined the term entropy—from the Greek τροπη (transformation)—to *measure the degree of disorder in a closed system.* Clausius also crisply formulated the second law of thermodynamics: entropy of the universe tends to maximum. In practical terms this means that in a closed system (one without any external supply of energy) the availability of useful energy can only decline. A lump of coal is a high-quality, highly ordered (low entropy) form of energy; its combustion will produce heat, a dispersed, low-quality, disordered (high entropy) form of energy. The sequence is irreversible: diffused heat (and emitted combustion gases) cannot ever be reconstituted as a lump of coal. Heat thus occupies a unique position in the hierarchy of energies: all other forms of energy can be completely converted to it, but its conversion into other forms can never be complete, as only a portion of the initial input ends up in the new form. [*End of Smil's essay*]

Energy, Work, and Newton's Laws

An analysis of the developments leading to a coherent scientific definition of energy revealed the importance of studying how energy is *transformed*. It is for just this reason that we place considerable emphasis in this chapter on both the *transformations* that occur between different types or categories of energy and the *changes* that can be initiated or sustained by the *expenditure* of energy at both the macroscopic and microscopic (molecular) level.

So reaching a quantitative understanding of what energy *is* required a broad recognition of the many forms energy could take, the subtle ways energy could be transformed in nature, and of paramount importance—the fact that energy is a conserved quantity. It cannot be created nor destroyed. While we regard the conservation of energy as an unassailable fact from our first exposure to science, the fact that it *is* conserved changes our discourse and our strategy of thinking about energy. In particular the focus immediately turns to the question: if I can't quantitatively account for the energy change associated with a given physical or chemical transformation, then where did that unaccounted-for energy go? It must have gone somewhere—where is it? This is not unlike tracking money through a complex web of international transactions. Money is not lost, it is ingeniously transformed.

The secrets surrounding how energy serves as the singularly important currency for changes in natural systems began to be revealed through the work of James Watt, Thomas Young, and others as described above. But it was Isaac Newton, shown in Figure 1.6, who, well before Watt's and Young's contributions, made the profound connections that began to expose the true nature of energy. While the stories about Newton watching an apple fall from a tree are woven into scientific legend, consider what is actually occurring when you observe such an event. It is far from a trivial transformation. The apple, suspended at rest above the ground, begins to accelerate downward, gaining kinetic energy as it falls. This is a rather miraculous, yet seamless, interchange of energy from one form (gravitational potential energy) to another form (kinetic energy). When the apple hits the ground, it comes to rest. Has the energy disappeared? As we will see, the answer is no. But Newton was initially concerned with a different question. Suppose, he reasoned, we reverse the process by returning that apple to the branch from which it fell. Newton recognized that to do so he would have to apply a *force* against gravity, over a *distance*, from the ground to the point from which the apple fell. That quan-

FIGURE 1.6 Isaac Newton (1642–1727) was responsible for establishing the laws governing the relationship between inertia, force, mass, acceleration, work, and energy. He would later clarify the principles governing the gravitational force between celestial objects.

tity, he reasoned (the force times the distance) must quantitatively represent the same "thing" as the kinetic energy contained in the apple as it hit the ground. That is, the work required to raise the apple in the Earth's gravitational field was quantitatively the same as the kinetic energy of motion possessed by the apple at the moment it hit the ground. The common, and apparently exchangeable, quantity involved in this physical transformation, was *energy*. Except now this exchangeable currency, energy, could be quantitatively determined by measuring the force and the distance and multiplying them to obtain the *work* required to place the apple at its initial position.

It is obvious to us when we do physical work that it takes more work to push a 100 lb box 200 feet than it does to push it 100 feet (and that a 50 lb box takes less work to push an equal distance than a 100 lb box). But we still do not have a quantitative definition of *force*, even though we can reliably measure distance.

What Newton recognized was that the apple was accelerated downward by a force—the force of gravity—but more generally, that a force is necessary to instigate a *change* in the state of motion, either the speed or the direction of an object's motion. An object changes speed or direction *only* when acted on by a force. Newton referred to the tendency for a body to *retain* its current state of motion as the property of *inertia*: if at rest, a body remains at rest. If in motion, a body remains in motion *unless* acted on by an external force. This recognition lead to **Newton's First Law of Motion**: *Every body persists in its state of rest or of motion in a straight line unless it is compelled to change that state by forces impressed upon it.*

It was well understood in Newton's time that speed was quantitatively defined as the change in distance divided by the change in time:

Speed = change in distance / change in time

In addition, it was also understood that the concept of velocity involved both *speed* and *direction* and that acceleration was defined as the change in velocity divided by the change in time:

acceleration = change in velocity / change in time

where the velocity of an object is defined by both its speed *and* its direction.

Check Yourself 1

Isaac Newton developed the foundations of physics by his First Law that clarified the concept of *inertia*, and his Second Law that related force to mass and acceleration, F = ma. Newton also established the law governing the gravitational force between two masses m_1 and m_2.

Once Newton recognized the profound importance of inertia, which clarified the role of force as the agent of *change*, and moreover that the change in the current state of motion was acceleration, he had "only" to understand the relationship between *force* and *acceleration*. It was clear that a more massive body required a greater force to achieve the same acceleration, so mass must play a role in quantitatively linking force and acceleration. He also recognized that different masses accelerated equally in the Earth's gravitational field that created a force proportional to mass. From that he concluded that the proportionality between force and acceleration was simply the object's mass (as opposed to the mass squared, the square root of the mass, or something else).

This formed the basis for **Newton's Second Law of Motion**: *The acceleration, a, of a body is proportional to the force, F, applied to that body.*

$$F \propto a$$

1. Newton expressed the gravitational force between two masses as

    $$F = G\frac{m_1 m_2}{r^2}$$

    where G is the gravitational constant equal to 6.67×10^{-11} m³/kg·sec², m_1 and m_2 are the two masses, and r is the distance between the centers of the two objects. If you are at a distance from the Earth equal to the distance from the Earth to the moon, calculate the force the Earth exerts on you from that distance.

2. When we write the expression that the force on an object at the Earth's surface is F = mg, where g is the "acceleration of gravity" based on Newton's law for gravity, what is the functional form of g in terms of G, r, and the masses?

3. Calculate the value of g at the Earth's surface.

Sidebar 1.3—Velocity of Molecules in a Gas

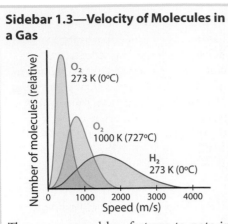

There are several key features to note in this distribution of molecular kinetic energies. The first is that the distribution is a distinct function of temperature. The distribution of molecular speeds for molecular oxygen at 273 K peaks (reaches a maximum) at a velocity of about 450 m/sec, whereas the O_2 velocity distribution at 1000 K peaks at nearly 1000 m/sec. The number of molecules at very low velocities drops to zero, as does the number at high velocities. Of particular importance, however, is that as the temperature increases, the number of molecules at the high velocity (and thus high kinetic energy) end of the distribution increases dramatically. Consider, for example, the fraction of O_2 molecules with molecular velocities equal to or greater than 1000 m/sec at 273 K (~1%) versus at 1000 K (~30%). The importance of molecular mass on the velocity distribution is also dramatic. The molecular mass of O_2 ($32 \times 1.66 \times 10^{-27}$ kg = 5.3×10^{-26} kg) is sixteen times that of H_2 ($2 \times 1.66 \times 10^{-27}$ kg = 3.3×10^{-27} kg), but the velocity corresponding to the maximum fraction of H_2 molecules at 273 K is some four times that of O_2.

Average speed for gases at 300 K	
Gas	Speed (cm/sec)
H	27×10^4
He	14×10^4
N	7.3×10^4
O	6.8×10^4
H_2O	6.4×10^4
N_2	5.2×10^4
O_2	4.8×10^4

Furthermore, the proportionality constant between the force and the acceleration is the mass, m, of the body so that

$$F = ma$$

But it is *energy* that is the elusive entity. What has all this force, mass, and acceleration have to do with energy? The answer is that work, which is force times distance, represents the key concept linking force and energy. Specifically, if we suspend a mass m in a gravitational field, it will experience a force F = ma = mg where g is the acceleration of gravity. If we raise that mass to a height h from the ground, we will do an amount of work

$$w = \text{force} \times \text{distance} = mgh \qquad (1.1)$$

If we release that mass, it will accelerate downward, converting that invested work to kinetic energy such that just before it reaches the ground, it will have kinetic energy, $\frac{1}{2} mv^2$, equal to that invested work, mgh, so we can equate the work done with the kinetic energy:

$$\tfrac{1}{2} mv^2 = mgh$$

Alternatively, we can run repeated experiments, for example with a baseball as shown in Figure 1.7, where we send the mass vertically with different velocities and measure the height the mass reaches. Each time, $\frac{1}{2} mv^2 = mgh$, and the mass returns to Earth with the same *downward* velocity that it had *upward* velocity (except for small losses due to air drag that we will discuss subsequently). Thus, if we designate mgh as the *potential* energy, PE, and $\frac{1}{2} mv^2$ as the kinetic energy, KE, we recognize that this quantity PE + KE remains constant. It is neither created nor destroyed. So if we define PE + KE as the total energy E_T, we have

$$E_T = PE + KE = \text{constant}.$$

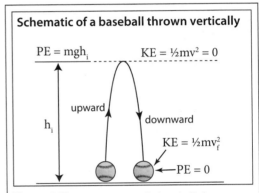

FIGURE 1.7 The trajectory of a baseball, thrown vertically from the ground, exemplifies the exchange between potential energy and kinetic energy. The baseball leaves the ground with initial vertical velocity of v_i and initial kinetic energy of $\frac{1}{2} mv_i^2$. It rises vertically until it acquires an amount of potential energy, mgh, equal to its initial kinetic energy. It then falls toward the ground, converting potential energy to kinetic energy until it reaches the ground with downward velocity opposite in direction but equal in magnitude to v_i.

Or if $(E_T)_{ground} = (PE + KE)_{ground}$ is the sum of potential energy and kinetic energy at the ground and if $(E_T)_{max} = (PE + KE)_{max}$ is the sum of potential energy plus kinetic energy at the maximum height the mass reaches, then, because the conservation of energy dictates that $\Delta E_T = (E_T)_{ground} - (E_T)_{max} = 0$, we have $(E_T)_{ground} = (E_T)_{max}$ so $(mgh + \frac{1}{2} mv^2)_{ground} = (mgh + \frac{1}{2} mv^2)_{max}$. If we begin initially with the mass moving vertically upward at the ground with velocity v_i at h = 0, then $(mgh)_{ground} = 0$ and $(\frac{1}{2} mv^2)_{ground} = \frac{1}{2} mv_i^2$. At the maximum height, the final velocity, $v_f = 0$, and with the mass at height, h, $(mgh + \frac{1}{2} mv_f^2)_{max} = mgh + 0$ so our conservation of energy expression gives us

$$\tfrac{1}{2} mv_i^2 = mgh.$$

Thus the *height*, h, reached by the mass with initial vertical velocity v_i is

$$h = v^2/2g$$

It is important to keep this principle of conservation of energy clearly in mind, which we can do for any point in the trajectory of the mass using an *energy chart* to keep quantitative track. This is displayed in Figure 1.8.

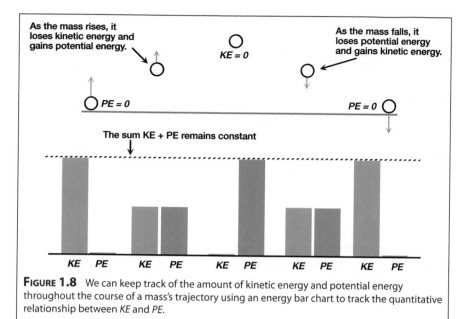

FIGURE 1.8   We can keep track of the amount of kinetic energy and potential energy throughout the course of a mass's trajectory using an energy bar chart to track the quantitative relationship between *KE* and *PE*.

So it was the link among inertia, force, acceleration, and work that exposed the central principle of *energy* and of *energy conservation*.

Check Yourself 2

If you dive off a cliff 30 meters above the water, how fast will you be going when you reach the water?

There is another important relationship that emerges from Newton's laws. Since F = ma it follows from the definition of acceleration a = $\Delta v/\Delta t$ that

$$F = m\frac{\Delta v}{\Delta t}$$

But for a body of given mass m, the *momentum* of that body is P = mv and thus $\Delta P = m\Delta v$

$$F = m\left(\frac{\Delta v}{\Delta t}\right) = \frac{\Delta P}{\Delta t} \qquad (1.2)$$

Conservation of Energy: Tracking the Flow of Energy

This brings us back to the question: when Newton's apple struck the ground, what happened to the kinetic energy of the apple's motion? Did the energy disappear? If we trust the law of conservation of energy, then it must go *somewhere*. It must be accounted for.

To answer this question, we must build a more specific *model* of how energy is transformed from the macroscopic motion of a body to the microscopic motion of the molecules that comprise it. For the purpose of clarification, let's designate

So let's work a quick problem linking the conversion of potential energy to kinetic energy. Suppose we want to calculate the velocity of a person who dives off a cliff into a lake, river, or ocean. Suppose the ledge is 30 meters above the water; what will be the speed of the diver at the moment the diver enters the water?

That is, if you dive from a cliff 30 meters above the water, how fast will you be moving when you reach the surface?

Initial Total Energy = Final Total Energy

$$\tfrac{1}{2}mv_i^2 + mgh_i = \tfrac{1}{2}mv_f^2 + mgh_f$$

$$\tfrac{1}{2}\cancel{mv_i^2}_{0} + mgh_i = \tfrac{1}{2}mv_f^2 + \cancel{mgh_f}_{0}$$

So

$$mgh_i = \tfrac{1}{2}mv_f^2$$

The mass cancels $gh_i = \tfrac{1}{2}v_f^2$
And we have

$$v_f = \sqrt{2gh}$$

If you dive from a cliff 30 meters above the water, how fast will you be moving when you reach the surface?

$$v_f = \sqrt{2gh} = \sqrt{2\left(9.8\ \text{m/sec}^2\right)30\ \text{m}}$$

$$= 24\ \text{meters/sec}$$

That's about 50 mph!

the sum of the macroscopic kinetic plus potential energy of a mass, m, as the macroscopic *mechanical energy* of the system, E_{mech} such that now

$$E_{mech} = KE + PE \qquad\qquad (1.3)$$

Now let's suppose we have a mass moving up and down in a gravitational field on a *frictionless* rod with a spring at the bottom, such that at the bottom end of the downward motion, the mass compresses the spring, then rebounds with the same kinetic energy it had when it struck the spring. This apparatus is shown in Figure 1.9. In the absence of friction between the mass and the rod, and in the absence of any energy dissipation in the spring, the motion of the mass obeys the conservation of total energy law developed above, namely that the change in mechanical energy is the final mechanical energy subtracted from the initial mechanical energy:

$$\Delta E_{mech} = (E_{mech})_f - (E_{mech})_i = 0 \qquad (1.4)$$

The conservation of energy equation requires that the mass m on the rod will continue to oscillate indefinitely. Suppose, however, that the motion of the mass on the rod is no longer frictionless and that there is some energy dissipation in the spring. In this case, our conservation of total energy law, *as expressed by the equation* $\Delta E_{mech} = \Delta(KE + PE) = 0$, will clearly be violated as the friction will remove mechanical energy from the system (mass, rod, spring in a gravitational field). The vertical oscillation of the mass on the rod will decay in amplitude until the mass comes to rest on the top of the spring—motionless. At that point, direct observation has demonstrated that rather than $\Delta E_{mech} = 0$, in fact $\Delta E_{mech} = (E_{mech})_i$, the initial mechanical energy of the system. The conservation of energy equation requires that $\Delta E_{mech} = (E_{mech})_i - (E_{mech})_f$ so if $(E_{mech})_f = 0$, then $\Delta E_{mech} = (E_{mech})_i - 0 = (E_{mech})_i$ and so $\Delta E_{mech} = (E_{mech})_i$. But if we trust in the principle of energy conservation, then we must *find* that lost energy, because there is no longer any macroscopic mechanical energy associated with the system.

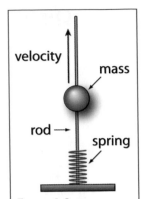

velocity

mass

rod →

spring

FIGURE 1.9 The apparatus consisting of a mass, rod and spring is a system that can be used to track quantitatively the conversion of macroscopic mechanical energy to microscopic thermal energy contained in the atoms that compose the system.

1. The Fahrenheit scale (°F) was introduced by Gabriel Fahrenheit (1686–1736) so that 0°F corresponded to a typical cold day in winter and 100°F to the *approximate* temperature of the human body. When this temperature scale is referenced to the freezing and boiling point of water of 1 atmosphere pressure their values are 32°F and 212°F respectively. The Fahrenheit scale was adopted into the system of British units and as a result, while lacking a degree of scientific logic, was projected into common discourse throughout the 19th and 20th century to the present day.

2. Not long after the introduction of the Fahrenheit scale, Anders Celsius (1701–1744) introduced the Celsius temperature scale based upon 0°C as the freezing point of water and 100°C as the boiling point of water at one atmosphere pressure. The Fahrenheit scale has 180°F between the boiling and freezing point of water, while the Celsius scale has 100°C as the temperature difference between the same two points. Thus, one degree Celsius is 180/100 = 1.8 times as large as one degree Fahrenheit.

3. Approximately 100 years after the introduction of the Fahrenheit and Celsius scales, William Thomson (later Lord Kelvin) proposed that a scientific scale of temperature must be adopted that had an *absolute zero* such that at that temperature atoms or molecules would have zero energy associated with their motion. Thomson defined this *Kelvin* temperature scale against the Celsius scale such that

   $$T(K) = T(°C) + 273.15$$

   using, among other considerations, the properties of gases.

Check Yourself 3—Temperature Scales

Temperature scales that have remained in common use today share three characteristics in common. They all designate zero degrees to be a physically useful reference, they all span a temperature range convenient for a given application, and they can all be interconverted. The temperature scales in common use are the Fahrenheit scale, the Celsius scale, and the Kelvin scale. How do these three temperature scales differ and what was their origin?

So what happened to the energy lost by the oscillating mass due to friction? The answer is that it went into increasing the *temperature* of the oscillating mass, the rod, and the spring itself by virtue of the conversion of the macroscopic kinetic energy of the oscillating mass, m, to the microscopic motion of the atoms and molecules that comprise the system. But if we view the system to be the mass, rod, and spring, then we can express this increase in the microscopic energy of the molecules that make up the system as a new term, ΔU_{therm}, where U_{therm} is the *thermal* energy of the molecules in the mass, rod, and spring. Then our conservation of energy equation for the mass, rod, spring system is:

$$-\Delta E_{mech} = \Delta U_{therm}$$

The minus sign represents the fact that the macroscopic mechanical energy *lost* by the oscillating mass is *gained* by the molecular motion of the atoms in the ball, rod, and spring—the thermal energy. Thus we can write

$$\Delta U_{therm} = -\Delta E_{mech} = -\Delta(KE + PE)$$

because the increase in thermal energy of the mass, rod, spring system equals the mechanical energy lost from the macroscopic motion of the mass on the rod. This gives us a more general, and thus more powerful, statement of the conservation of energy that was expressed in equations 1.3 and 1.4:

$$\Delta KE + \Delta PE + \Delta U_{therm} = \Delta E_{mech} + \Delta U_{therm} = 0 \qquad \textbf{(1.5)}$$

Once we accept as true the hypothesis that energy is neither created nor destroyed, then the issue becomes simply one of finding *where the energy has gone* and to represent that as a *new term* in the energy conservation equation. In the case of equation 1.5, we "discovered" that the mechanical energy, E_{mech}, of the *macroscopic* mass oscillating on the rod was being converted to the *microscopic* motion of the molecules, U_{therm}, that comprise the rod, mass, and spring, thereby raising the temperature of the system.

Can we prevent the oscillating mass from coming to rest and express this in our conservation of energy equation 1.5? Yes, if we recognize that the frictional dissipation of energy is, in fact, a "nonconservative" force (a force that removes macroscopic mechanical energy, and in so doing converts it to microscopic energy of molecular motion) and that the removal of mechanical energy, ΔE_{mech}, *could* be made up by an amount of work, W_{nc}, added in each cycle to counter the *nonconservative* (frictional) force. In this formulation, the product of the frictional force times the distance (the amplitude of the motion of the mass along the rod) is equal to the mechanical energy that would be lost if this work, W_{nc}, were not expended to sustain, unattenuated, the motion of the mass on the rod. Then we can write our *conservation of energy equation* as an equality between the change in mechanical energy, ΔE_{mech}, and the work done, W_{nc}, needed to make up for the loss of mechanical energy resulting from the nonconservative frictional force. For if we choose not to insert into the system an amount of work, W_{nc}, to counter the nonconservative (frictional) force, the decay of mechanical energy ($\Delta E_{mech} = \Delta(KE + PE)$) will occur such that $-\Delta E_{mech} = W_{nc}$. While we can think of a number of ways of keeping the mass oscillating vertically on the rod at the same amplitude (a small kick downward at the top of the trajectory, a small kick upward at the bottom of the trajectory, etc.), we can write

$$-\Delta E_{mech} = -\Delta(KE + PE) = +W_{nc}$$

Because the increase in thermal energy of the mass, rod, spring system equals the mechanical energy lost from the motion of the mass on the rod, we have:

$$\Delta(KE + PE) + \Delta U_{therm} =$$
$$\Delta E_{mech} + \Delta U_{therm} = -W_{nc} + \Delta U_{therm} = 0$$

Focusing specifically on the relationship between the work done to counter the nonconservative force of friction, W_{nc}, and the change (increase) in the internal energy of the microscopic energy of the molecules that comprise the mass, rod, and spring, ΔU_{therm}, we have:

$$-W_{nc} + \Delta U_{therm} = 0 \qquad \text{or} \qquad W_{nc} = \Delta U_{therm}$$

This expresses the fact that the work done to sustain the macroscopic kinetic energy of the oscillating mass is converted to the microscopic kinetic energy of the molecules that comprise the system.

We now have a formulation that can answer the question: What happened to the kinetic energy of Newton's falling apple when it struck the ground? Just as with the moving mass on the rod, the answer lies in the conversion of *macroscopic* motion of a body (the apple) to the *microscopic* energy contained in the motion of molecules that make up a solid body. In particular, when the apple strikes the ground, the kinetic energy of its downward motion is converted to the *increase* in the velocity of the molecules that make up the ground at the point of impact as well as the velocity of the molecules that make up the apple at the point of impact with the ground. This is manifest in the increase in *temperature* of the collision "zone" between the apple and the ground displayed graphically in Figure 1.10.

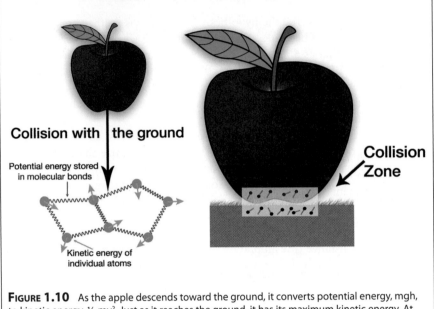

FIGURE 1.10   As the apple descends toward the ground, it converts potential energy, mgh, to kinetic energy, ½ mv². Just as it reaches the ground, it has its maximum kinetic energy. At the point of collision, the macroscopic kinetic energy of the apple is converted to the microscopic energy of the molecules in the collision zone, increasing the velocity of those molecules and thereby increasing their temperature.

1. If the apple fell from a height of 5 meters, how fast was the apple moving when it struck the ground?
2. What happened to the energy contained in the descending apple when it struck the ground?
3. If this kinetic energy of motion initially went into a region just at the interface of the apple and the ground, the "collision zone," shown in Figure 1.10, did the temperature of that zone increase or decrease?
4. If (1) the temperature of the material in that collision zone increases in direct proportion to the amount of energy deposited in the zone, (2) the amount of energy required to raise the temperature 1 K per unit mass of the apple and ground within the zone is 4 kJ/kg, and (3) the mass of the collision zone is 0.05 kg, what is the change in temperature of the collision zone after the apple hits the ground?

Check Yourself 4

When Newton's apple struck the ground, it instantly lost its kinetic energy of motion, ½ mv².

Energy at the Molecular Level: Microscopic and Macroscopic Forms of Energy

There are other revelations contained in this formulation of the conservation of energy. One is the distinction between the macroscopic (mechanical) energy, E_{mech}, of the moving mass of the ball and the microscopic energy contained in the motion of the atoms that comprise the system, U_{therm}. This distinction will turn out to be very important to our understanding of the First Law of Thermodynamics. Another important concept contained in the formulation of equation 1.3 is that we can *amend* our Conservation of Energy equation to handle increasingly complex systems *by identifying the categories of energy involved in the transformations of energy that occur in the system under consideration.* Our example was the recognition that inclusion of nonconservative (frictional) forces in a system comprised of a moving mass on a rod leads to the inclusion of the thermal energy term, ΔU_{therm}, representing the increase in microscopic energy of the system resulting from the increased velocity of the individual atoms contained in the ball, rod, and spring.

Consider the energy contained in molecular motion within a vessel containing air (80% N_2, 20% O_2) at room temperature. We know that the kinetic energy of motion of a *single molecule* of nitrogen can be simply written as

$$KE_{N_2} = \tfrac{1}{2} m v_{N_2}^2$$

where m is the mass of N_2 ((28 amu) $\times$ (1.66 $\times$ 10^{-27} kg/amu) = 4.65 $\times$ 10^{-26} kg) and v_{N_2} is the velocity of the nitrogen molecule. We will learn how to calculate that molecular velocity directly in Chapter 12, but what we do know is that the molecular velocity must be *at least* as great as the speed of sound (340 m/sec) because sound is transmitted through air by molecular-molecular collisions. The average velocity of a nitrogen molecule at room temperature exceeds the speed of sound somewhat and is actually about 500 m/sec. With a mass per molecule of 4.65 $\times$ 10^{-26} kg, the kinetic energy of a single nitrogen molecule is

$$KE = \tfrac{1}{2}mv^2 = \tfrac{1}{2}(4.65 \times 10^{-26} \text{ kg})(5.2 \times 10^2 \text{ m/sec})^2$$
$$= 6.3 \times 10^{-21} \text{ kg m}^2/\text{sec}^2 = 6.3 \times 10^{-21} \text{ joules/molecule}$$

This is an important number—the kinetic energy contained in a single nitrogen molecule at room temperature—because it defines, in a practical sense, one of the smallest increments of energy in nature. Check back to the energy scale in Figure 1.2.

Check Yourself 5—The Mole

We will make constant use of the concept of a *mole* of a chemical substance and will examine the concept of a mole at some length in the next chapter. But a mole is simply a way of counting the large number of molecules in a given macroscopic object, one we can hold in our hand or weigh on a laboratory scale.

This amount of kinetic energy per molecule of air leads to an important calculation: the amount of kinetic energy contained in the thermal motion of the atoms or molecules of a macroscopic body at rest. If we consider a soccer ball and calculate the kinetic energy of the ensemble of air molecules contained inside that soccer ball at rest, we have, assuming a volume for the soccer ball of 4 liters, the kinetic energy of molecules in the soccer ball:

$$KE_{molecules} = \left(6.3 \times 10^{-21} \text{ joules/molecule}\right)\left(6 \times 10^{23} \text{ molecules/mole}\right)\left(4 \text{ liters}\right)\left(\frac{1 \text{ mole}}{22.4 \text{ liters}}\right)$$

$$= 6.2 \times 10^2 \text{ joules} = 620 \text{ joules}$$

This then represents the *microscopic kinetic energy* of the molecular motion of the air inside the soccer ball when the soccer ball is at rest.

The *macroscopic kinetic energy* of the soccer ball constitutes an illustrative example comparing the kinetic energy of the molecular motion of the gas inside an object, with the kinetic energy of the ball's motion. If the ball is kicked (hard!) and reaches a velocity of 25 meters/sec (53 mph), the kinetic energy of the ball (mass 0.45 kg) would be

$$KE_{ball} = \tfrac{1}{2}mv^2 = \tfrac{1}{2}(0.45 \text{ kg})(25 \text{ m/sec})^2$$

Just as we refer to eggs conveniently by the "dozen," so too do we refer to molecules conveniently by the "mole." The only difference is that there are a very large number of molecules in a mole. There are, to a reasonable level of accuracy, 6.022 $\times$ 10^{23} molecules in a mole of a substance. An easy way to recall the magnitude of a mole, also called *Avogadro's Number*, is to remember that Avogadro's birth time was 6 in the morning of October 23! It is also very useful to remember that at a temperature of 273K and a pressure of one atmosphere, a mole of gas has a volume of 22.4 liters.

1. If each nitrogen molecule at room temperature has a kinetic energy of 6 $\times$ 10^{-21} joules, how much kinetic energy is contained in a mole of nitrogen molecules?

2. If each nitrogen molecule weighs 4.65 $\times$ 10^{-26} kg, what is the mass of a mole of nitrogen molecules?

3. If the energy release from the fission of a single ^{235}U nucleus is 3.2 $\times$ 10^{-11} joules/nucleus, what is the energy release from a mole of ^{235}U nuclei?

4. If the energy release from the combustion of one hydrogen molecule with oxygen to form water
$$H_2 + \tfrac{1}{2} O_2 \rightarrow H_2O$$
is 4.8 $\times$ 10^{-19} joules, how much energy is released in the combustion of a mole of H_2 molecules?

5. What is the *ratio* of the amount of energy produced by the fission of a mole of ^{235}U versus the combustion of a mole of H_2 molecules?

$$= 141 \text{ kg m}^2/\text{sec}^2 = 141 \text{ joules}$$

That is significantly *less* than the kinetic energy of the microscopic molecular motion within the ball when the ball is at rest!

So we recognize from this quick calculation that, first, molecules at room temperature are moving at *very* high velocity, some 500 m/sec (1000 mph), which is comparable to the muzzle velocity of a high power rifle; and, second, that the kinetic energy of molecular motion often exceeds the kinetic energy of the macroscopic motion of the body containing the molecules. We compare the velocities of some important objects in Sidebar 1.4.

Exchange of Kinetic and Potential Energy on a Surface

We can also represent the exchange of kinetic and potential energy by investigating a rolling sphere as it progresses over a surface that contains a series of hills and valleys as displayed in Figure 1.11.

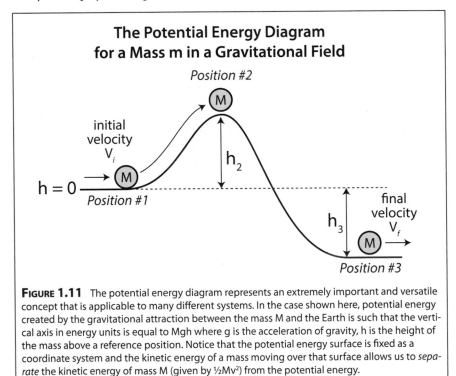

The Potential Energy Diagram for a Mass m in a Gravitational Field

FIGURE 1.11 The potential energy diagram represents an extremely important and versatile concept that is applicable to many different systems. In the case shown here, potential energy created by the gravitational attraction between the mass M and the Earth is such that the vertical axis in energy units is equal to Mgh where g is the acceleration of gravity, h is the height of the mass above a reference position. Notice that the potential energy surface is fixed as a coordinate system and the kinetic energy of a mass moving over that surface allows us to *separate* the kinetic energy of mass M (given by ½Mv²) from the potential energy.

As the sphere rolls past position #1 in Figure 1.11, it has an initial kinetic energy $\frac{1}{2} mv_i^2$. As it moves toward position #2, it is losing kinetic energy as it "climbs" the hill toward position #2 because it is converting kinetic energy into potential energy such that the sum $E_{mech} = KE + PE$ remains *constant* (where we assume the motion of the ball over the surface is frictionless.) When it reaches position #2, it has *gained* an amount of potential energy, given by mgh_2, but it has *lost* an amount of kinetic energy given by $-mgh_2$. If the amount of initial kinetic energy, $\frac{1}{2} mv_i^2$, is *less* than mgh_2, then the rolling sphere will reverse direction and pass again through position #1 moving in the opposite direction, but in the absence of friction, with a velocity equal in magnitude to that of the initial condition at position #1. However, if the initial kinetic energy *exceeds* mgh_2, the sphere will pass over the crest of the hill and will convert potential energy to kinetic energy as it proceeds to the right, down the hill toward position #3 such that it will have *added* additional kinetic energy. The additional kinetic energy is equal to the potential energy lost in moving from position #1 to position #3, an amount mgh₃. Figure 1.11 represents a simple but extremely important concept called a **Potential Energy Surface (PES)**. In this example, the potential energy surface graphically represents a vertical displacement in a gravitational field that determines the potential energy (mgh) at any point along the surface referenced to a specific position. It also, interestingly, serves to effectively *separate* the *kinetic energy* ($\frac{1}{2}$ mv²) from the *potential energy.* This separation of the kinetic energy from the potential energy on a PES is a characteristic that will become increasingly important as more complicated systems are considered.

This idea of a potential energy surface that defines the amount of kinetic energy required to surmount a barrier has an important and very useful analogy when applied to a chemical reaction. We know that when we mix hydrogen gas and oxygen, the mixture of H_2 and O_2 will coexist without reaction indefinitely if left alone. We also know that if we touch a match to the mixture, it will explode, releasing considerable energy. We also know that when H_2 and O_2 react, there is only one product: water or H_2O. All of this information can be summarized in a single diagram, called a *Reaction Coordinate Diagram*. Figure 1.12 is a reaction coordinate diagram for H_2 reacting with O_2 to form H_2O. The reaction coordinate diagram has several important features. First, the diagram *separates* reactants (H_2 and O_2) from the products (H_2O) of the chemical reaction. The chemical reaction thus progresses from left to right in Figure 1.12. Second, note that the reactants, H_2 and O_2, are *separated* from the reaction product, H_2O, by a barrier—not unlike the barrier in the potential energy diagram Figure 1.11. That barrier represents the energy that must be invested to initiate the chemical reaction. This barrier is the reason we must use a match or a spark to ignite the hydrogen–oxygen mixture. This is also the reason your automobile has an ignition system and "spark plugs." A carefully timed spark is what ignites the mixture of gasoline and oxygen in the cylinders of the automobile's engine.

The reason a barrier to the reaction exists is that the outer electrons in H_2 and O_2 repel each other (like electrical charges repel) and the colliding H_2 and O_2 molecules must have sufficient kinetic energy to overcome the potential energy associated with this repulsive force to place the nuclei in close proximity to form new chemical bonds. Third, the net energy *release* resulting from the chemical reaction is given by the *difference* in energy between the hydrogen and oxygen *reactants* and the H_2O *products.* Fourth, the reaction coordinate diagram keeps track of the number of hydrogen atoms (two on the reactant side of the barrier, two on the product side

Sidebar 1.4—Comparing High Speed Objects

Examination of the figure in the sidebar 1.3 on page 14 that displays the velocity distribution of molecules of various masses at different temperatures raises the question of how those speeds compare with other objects such as a high power rifle bullet, a commercial jet aircraft, or a supersonic fighter aircraft. Inspection of the figure on molecular speeds reveals that an O_2 molecule at room temperature has, on average, a speed of approximately 450 m/s. That corresponds to 935 miles/hr. A commercial jet travels at approximately 550 mph, so that an oxygen molecule at room temperature has a speed nearly twice that of a commercial jet aircraft. Supersonic aircraft, such as the SR-71 high altitude spy plane, travel at Mach 3.5, three and one-half times the speed of sound. Since the speed of sound is 770 mph (344 m/s, 1230 km/hr, 1130 ft/s), the SR-71 has a speed of 2695 mph (1204 m/s, 4340 km/hr, 3955 ft/s). Thus the SR-71 has a velocity approximately 2.5 times the *average* speed of the O_2 molecule at room temperature. Notice, however, that because the O_2 molecules have a *distribution* of speeds, some of the molecules have speeds in excess of 1000 m/s, and thus some of the O_2 molecules in this room have speeds *in excess of the Mach 3.5 jet aircraft*. What about the rifle bullet? The speed of rifle bullets ranges from 200 m/s (22 caliber rim-fire cartridge) to 1500 m/s (large charge center-fire cartridge), so the average speed of an O_2 molecule at room temperature is about twice that of a "normal" rifle bullet and about one-third that of a high power rifle bullet. Notice, from inspection of the figure in sidebar 1.3, that even at room temperature the H atom has an average speed that exceeds that of the Mach 3.5 jet aircraft *and* that of the high power rifle bullet! We will examine the relationship between the temperature, mass, and velocity of molecules in a subsequent discussion.

SR-71 aircraft with full afterburner

Bullet leaving revolver

of the barrier) and of the oxygen atoms (one on the reactant side, one on the product side). The reason the number of atoms of each element must "balance" (must be equal) on the reactant side and the product side is that a chemical reaction cannot produce or destroy atoms. A chemical reaction can only change that bonding *structure* of product molecules with respect to reactant molecules.

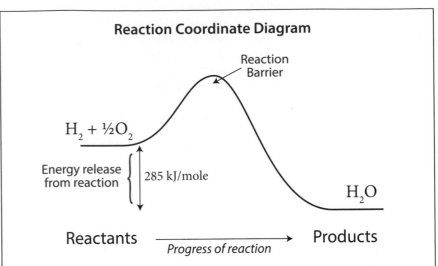

FIGURE 1.12   The potential energy diagram for a chemical reaction, in this case between hydrogen and oxygen, represents the relative energy between reactants ($H_2 + O_2$) and products (H_2O), so that the energy difference represents the energy released in the rearrangement of the chemical bonds. In addition to the energy difference between reactants and products, the energy "barrier" between reactants and products represents a potential energy "hill" created by the electron-electron repulsion of the valence electrons in H_2 and O_2 that must be surmounted in order to execute the chemical reaction.

1. Suppose we have a mass of 1 kg rolling in the absence of friction at a speed of 10 m/sec at position 1. Suppose the height of position 2, h_2 in Figure 1.11, is 3 meters and that position 3 is 2 meters below position 1. Sketch the potential energy surface for this system, assuming that the potential energy is zero at position 1.

2. If the total (mechanical) energy of the system is $E_{mech} = KE + PE$, what is the total energy of the system?

3. What is the speed of the mass at position 2? What is the speed of the mass at position 3?

4. Now construct a *quantitative* bar energy chart at position 1, position 2, and position 3.

5. Is the potential energy at position 3 positive or negative? If it is positive, what is the total energy of the system at position 3? If it is negative, is it physically reasonable to have a negative potential energy? Discuss.

Before proceeding further with the development of the scientific understanding of *energy*, it is important to examine how science and the techniques of scientific inquiry are developed in parallel with the emergence of the conceptual framework of *energy*. This story of how the scientific method and the emergence of modern concepts of energy evolved constitutes a cornerstone in scientific thought.

Check Yourself 6—Energy Bar Chart and a Mass Rolling on a Potential Energy Surface

To keep track of the quantitative relationship between the kinetic energy, potential energy and total energy of the mass moving in the potential energy diagram of Figure 1.11, we employ the energy bar chart introduced in Figure 1.8.

The Mechanical Equivalent of Heat

Scientific discovery profits in spectacular fashion from the serendipitous: the discovery of penicillin, radioactivity, x-rays, carbon nanostructures, etc. It is also of critical importance to realize that science progresses by the intuitive formulation of theory, which is *then* submitted to the testing of that theory by innovative experimental strategies. This is the Scientific Method and it is treated explicitly in Case Study 1.4 at the end of the chapter. Science advances less by plodding through logical protocol than by imagination that leaps forward, then to be eventually linked back by innovative testing of hypotheses.

We know from direct experience that when two objects at different temperatures are put in contact, heat spontaneously flows from the hotter one to the colder one. The flow of thermal energy is spontaneous; it serves to equalize the temperature of the two bodies in contact. While some aspects of the flow of thermal energy seem self evident, society and science have long sought a more coherent understanding of the relationship between heat, the transformation of energy, and, in fact, of temperature itself. From the Greeks through the Roman period to the Renaissance and beyond, the explanation of the flow of heat engaged a discussion of the movement of a *fluid* substance called *caloric*.

It was believed that a body at a high temperature contained a great deal of "caloric" and one at low temperature contained just a small amount of caloric. When the two bodies were brought in contact, the body high in caloric lost a fraction to the other until both bodies reached the same temperature. The concept of *heat as a substance* whose total amount remained constant, eventually could not withstand experimental interrogation. The first really conclusive evidence that heat could not be a *substance* was articulated by Benjamin Thompson (1753–1814), an American from Woburn, Massachusetts, who later became Count Rumford of Bavaria. In a paper read before the Royal Society of England in 1798 he wrote,

> I … am persuaded, that a habit of keeping the eyes open to everything that is going on in the ordinary course of the business of life has oftener led, as it were by accident, or in the playful excursions of the imagination … to useful doubts and sensible schemes for investigation and improvement, than all the more intense meditations of philosophers, in the hours expressly set apart for study. It was by accident that I was led to make the experiments of which I am about to give an account.

Rumford made his discovery while supervising the boring of military cannons for the Bavarian government. To prevent overheating, the bore of the cannon was kept full of water. The water was constantly replenished as it boiled away during the boring process. It was accepted that caloric had to be supplied to water to boil it. The continuous production of caloric was explained by the hypothesis that when a substance was more finely subdivided, as in boring, its capacity for retaining caloric became smaller, and that the caloric released in this way was what caused the water to boil. Rumford noticed, however, that the water boiled away even when his boring tools became so dull that they were no longer cutting or subdividing matter. He writes, after ruling out by experiment all possible caloric interpretations,

> … in reasoning on this subject, we must not forget to consider that most remarkable circumstance, that the source of Heat generated by friction, in these Experiments, appeared evidently to be *inexhaustible* … it appears to me to be extremely difficult, if not quite impossible, to form any distinct idea of any thing capable of being excited and communicated in the manner the Heat was excited and communicated in these Experiments, except it be *MOTION*.

What Rumford noticed, quite simply, was that as the cutting blade grew dull, little metal was cut but the amount of water boiled away remained largely unchanged. What remained unchanged was the *motion* of the cutting blade. Thus Rumford concluded that the heat was generated by the motion of the cutting head, not the cutting of the metal to release "caloric."

As we noted above, the idea that heat is related to energy was pursued by a number of scientists in the 1800s. But it was an English brewer, James Prescott Joule (1818–1889), who, by a series of brilliantly simple experiments, established

Benjamin Thompson

the key quantitative link between the release of mechanical energy of a macroscopic body (specifically the release of potential energy, mgh, of a mass falling in a gravitational field) and the increase in temperature of a system comprised of a water bath and copper paddles (displayed in panel (a) of Figure 1.13). The experimental apparatus that Joule designed transferred mechanical potential energy released as the masses fell through a gravitational field to a water bath via the friction between the paddle blades and the water. That experimental system provided the physical coupling between the macroscopic mechanical energy release, mgh, and the increase in the temperature of the apparatus, ΔT_{app}. Conservation of energy states that energy cannot be created nor destroyed. The release of mechanical potential energy could, with Joule's apparatus, only go into the water and the copper that comprised the apparatus as the heat added, q_{added}, to that system. Thus, q_{added}, the *transfer of energy*, in the units of energy, must equal the release of potential energy, mgh, from the falling mass in the gravitational field. Thus, in energy units, q_{added} must equal mgh. But moreover, when the amount of potential energy release was doubled, the observed increase in the temperature of the paddle wheel water apparatus doubled as well. This simple proportionality between the amount of potential energy release and the increase in temperature ($\Delta T_{app} \propto mgh$), combined with $q_{added} = mgh$, meant that the heat added, the potential energy release, and the increase in temperature of the apparatus could be equated: $mgh = q_{added} = C\Delta T_{app}$, where C is termed the *heat capacity* of the object into which heat is added. Thus the heat capacity C is just the proportionality factor between q_{added} and ΔT_{app}. Through repeated experiments, Joule determined that a given amount of work was always equivalent to the same amount of heat added to a system. This is known as the **mechanical equivalent of heat.**

Check Yourself 7—Joule and the Relationship between the Heat Added and the Increase in Temperature

James Prescott Joule made his living by brewing beer in England in the mid 19th century. He made his mark on science by quantitatively linking heat and work using a thermometer, a paddle wheel, and a pulley-coupled mass as shown in Figure 1.13.

1. Joule demonstrated that the *increase* in temperature of the paddle wheel and water system doubled when the distance the mass was allowed to fall doubled. If the energy required to raise the water-paddle wheel system 1 degree Celsius was 4×10^4 joules and the mass that descended in the gravitational field that turned the paddle wheel was 10 kg, how far would the mass have to fall to raise the temperature 1°C ?

2. Since Joule lived in town and carried out his experiments there, could he have carried out the experiment using the quantities cited in part 1? Suggest how he might have designed the experiment since he determined the *mechanical equivalent of heat* to an accuracy of 1%.

Joule's work linking the release of mechanical energy with the amount of heat released into a chamber of water, thereby raising the temperature of the water, set in place two critically important concepts. First, Joule established a *quantitative scale* for measuring the amount of energy transferred. It is for this reason we use a scale for energy with units of the joule. At the time of Joule's work, the calorie was defined as the amount of heat required to raise the temperature of 1 gram of water 1°C. In the International System of Units (the SI system of units) the energy scale is defined in terms of the joule (lower case joules is used as the energy unit) which is equal to 1 kg m²/s². The conversion between joules (J) and calories (cal) is that 4.184 J = 1 cal, which could be directly determined by Joule's original system. Second, if mechanical energy was equivalent to the energy appearing as heat—the "mechanical equivalent of heat"—then what about other categories of energy being quantitatively converted to heat? The answer is that we can use an array of different energy types to achieve the same transfer of heat to the same system of paddles and water to measure the change in temperature. This is displayed in Figure 1.13 panels (a) through (d).

Panel (a) displays Joule's original technique for quantitatively linking the release of mechanical energy to the production of heat. Panel (b) uses the same apparatus, but instead of allowing the masses to fall and the paddles to turn, sunlight is collected and imaged on the bottom of the water-paddle wheel system. The change in temperature, ΔT, with the same physical system, allows the amount

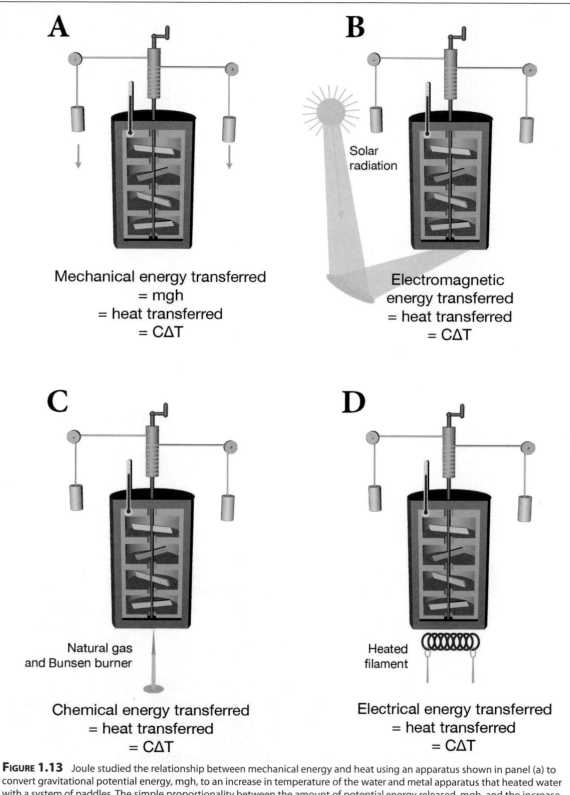

A

Mechanical energy transferred
= mgh
= heat transferred
= CΔT

B

Solar radiation

Electromagnetic
energy transferred
= heat transferred
= CΔT

C

Natural gas
and Bunsen burner

Chemical energy transferred
= heat transferred
= CΔT

D

Heated
filament

Electrical energy transferred
= heat transferred
= CΔT

FIGURE 1.13   Joule studied the relationship between mechanical energy and heat using an apparatus shown in panel (a) to convert gravitational potential energy, mgh, to an increase in temperature of the water and metal apparatus that heated water with a system of paddles. The simple proportionality between the amount of potential energy released, mgh, and the increase in temperature, ΔT, established the relationship between the heat added, q_{added} = mgh, and the increase in temperature such that q_{added} = CΔT. Panel (b) replaces the release of potential energy as a source of heat with sunlight thereby quantitatively linking electromagnetic radiation to mechanical energy. Panels (c) and (d) establish the quantitative relationship to chemical energy and electrical energy.

Solution: First they realized that they must convert the 600 Calories to the correct units to quantitatively compare their gain in potential energy, $mg\Delta h$, where Δh is the vertical height of the hike, with Calorie units. They realize that 1 food Calorie equals 10^3 cal, so 600 food Calories equals 600 kcal, which equals 6.0×10^5 cal.

However, there are 4.18 J/cal so

$(6.0 \times 10^5 \text{ cal}) \, 4.18 \text{ J/cal} = 2.5 \times 10^6$ J.

Thus, the gain in potential energy, $mg\Delta h$, which is equal to the work, W, expended to climb the mountain is

$W = mg\Delta h = 2.5 \times 10^6 \text{ J} = 2.5 \times 10^6$ kg m^2/sec^2

Thus with m = 80 kg
any g = 9.8 m/sec^2

$$\Delta h = \frac{2.5 \times 10^6 \text{ kg m}^2/\text{sec}^2}{(80 \text{ kg})(9.8 \text{ m/sec}^2)}$$

$= 3.2 \times 10^3$ m = 3.2 km

This would be a large mountain indeed, since 3.2 km equals 10.6 thousand feet or a mountain nearly 11,000 feet high.

However, the human body does not convert chemical (i.e. food) energy to mechanical (potential) energy with 100 percent efficiency. Just as with a gasoline (heat) engine, the human body converts about 20 percent of the chemical energy to mechanical energy. Thus, the students would need to climb a mountain (0.20) 3.2 km = 640 meters high or equivalently 2,120 feet high.

of *energy contained in sunlight* to be measured and placed on the same absolute scale as mechanical energy. Panel (c) replaces sunlight with a Bunsen burner that reacts natural gas, CH_4, with oxygen. Thus the energy released from the chemical reaction leads to an increase in temperature of the system, ΔT, providing a *quantitative measure of chemical energy* placed on the same absolute scale as mechanical energy and electromagnetic radiation (sunlight). Panel (d) uses a heated filament supplied by electricity to add energy to the same water-paddle wheel apparatus. By again observing the increase in temperature, ΔT, electrical energy can be placed on the same energy scale as mechanical energy, electromagnetic energy, and chemical energy.

The "equivalent" in the terminology *mechanical equivalent of heat* thus applies far more broadly to other categories of energy. This insight unified and clarified the concept of energy, moving science into the modern age.

Check Yourself 8—Calculating the Mechanical Equivalent of Heat for a Human

One day a pair of students each purchased a large ice cream cone that contained 600 Calories. They decided to offset that caloric intake by taking an afternoon hike. The first thing they did was calculate how high up the mountain they would need to climb to expend the 600 Calories contained in the ice cream. Both students weighed approximately 80 kg. To what height up the mountain would they need to hike?

Heat and Temperature at the Molecular Level

This simple proportionality between the heat added to a body, q_{added}, and the temperature of the body increase, ΔT_{body}, begs the question: what is going on at the *molecular* level within the body as the heat is being added? We can gain some insight into this important question by reexamining part of Figure 1.10, the schematic of the network of masses and springs that represent the atoms and the bonds between them that comprise the macroscopic body, as displayed in Figure 1.14.

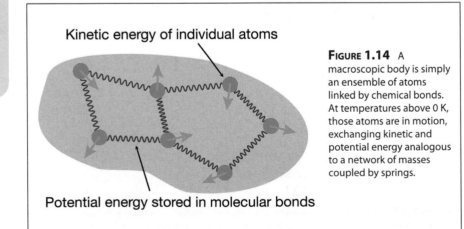

Kinetic energy of individual atoms

Potential energy stored in molecular bonds

FIGURE 1.14 A macroscopic body is simply an ensemble of atoms linked by chemical bonds. At temperatures above 0 K, those atoms are in motion, exchanging kinetic and potential energy analogous to a network of masses coupled by springs.

One aspect of Joule's studies of the mechanical equivalent of heat that was so transformational was the idea that the mechanical energy of a macroscopic body could be quantitatively converted to the microscopic motion of atoms within the body into which energy, as heat, had flowed. But what was also transformational was that for a given object (such as the paddle wheel-water systems of Figure 1.13) when the amount of mechanical energy converted to heat was doubled, the temperature increase, ΔT, also doubled. Therefore, if q_{added} is the amount of heat added to the body, then $\Delta T_{body} \propto q_{added}$. Different materials have a different quantitative relationship between the magnitude of the temperature increase, ΔT_{body}, and the amount of heat added, but for the same material and the same amount of material, the simple proportionality $\Delta T_{body} \propto q_{added}$ always holds. For a given material, as the amount of the material into which that heat is added increases, so too must the amount of heat added increase to achieve the same temperature increase, ΔT_{body}. We are acquainted with this in practice. If you double the amount of water in a pot on the stove, you must add twice as much heat to bring it from room temperature, 20°C, to the boiling point, 100°C.

As we saw in Joule's experiments, to express this relationship between ΔT_{body} and the heat added to achieve that increase in temperature, the proportionality $\Delta T_{body} \propto q$ is replaced by the equality

$$q_{added} = C \Delta T_{body}$$

The heat capacity of that particular macroscopic body, C, is the proportionality factor between q_{added} and ΔT_{body}. For example the heat capacity, C, of the paddle wheel-water system of Figure 1.13 is unique to that system. If the system is doubled in size, then the heat capacity of that new system is twice as large as the original one.

This simple proportionality between the heat capacity, C_i, of a given amount of material, i, and the amount of material expressed as its mass, m_i, is given by

$$C_i = c_i \, m_i$$

where c_i is the *specific heat* of substance i, which is the heat capacity *per unit mass* of a specific material. For example, the specific heat of copper is 0.38 J/°C·g so the amount of heat, q_{added}, required to raise the temperature of 100 grams of copper 10°C is

$$q_{added} = c_{copper} \, m_{copper} \, \Delta T = (0.38 \text{ J/°C·g})(100 \text{ g})(10°C) = 3.8 \times 10^2 \text{ joules}$$

The specific heat of liquid water is 4.184 J/°C·g so the amount of heat required to raise the temperature of 100 g of liquid water 10°C is

$$q_{added} = c_{H_2O(\ell)} \, m_{H_2O(\ell)} \, \Delta T = (4.184 \text{ J/°C·g})(100 \text{ g})(10°C)$$
$$= 4.2 \times 10^3 \text{ joules}$$

The key point is that the temperature of a macroscopic body is a quantitative measure of the sum of the kinetic energy and potential energy of the *ensemble of atoms* that comprise the mass of the body. This is the *thermal energy* of the body. Notice the key role played by the law of energy conservation at play here. Joule demonstrated that when a mass falls twice as far in a gravitational field, twice as much heat is added to a body, inducing a temperature change, ΔT, that always turned out to be twice as large. The only way energy can be stored in, for example, a block of copper, is by virtue of the kinetic energy of motion of the copper atoms and the potential energy in the bonds between those copper atoms. Just as with our mass, rod, and spring example where there is a constant exchange of kinetic and potential energy, in the case of kinetic and potential energy at the molecular level, there is nowhere else for that energy to go once it is stored as thermal energy by virtue of the energy contained in the motion of the atoms that comprise the body. This means that there is a simple chain of logic leading from the release of mechanical potential energy of an observable macroscopic body falling through a gravitational field releasing potential energy, mgh, to the transfer of heat into the body leading to a temperature increase ΔT_{body} of the body, as represented schematically in Figure 1.15. But the ultimate recipient of that mechanical energy, mgh, is the microscopic motion of the atoms in the body.

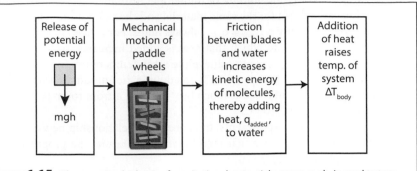

FIGURE 1.15 The organized release of gravitational potential energy, mgh, is used to turn the paddles connected to a central shaft that transfers that energy through friction to the surrounding water. The addition of that energy to the molecules of water raises the temperature of the system by an amount, ΔT_{body}. That increase in temperature of the water is a result of the increased kinetic energy of the molecules of water and of the copper atoms in the apparatus.

There is a compelling logic that emerges from the simple proportionality between the increase in temperature, ΔT, and the energy added to a given mass of material such as a block of iron or a given number of molecules of a gas. Because that given mass of material must contain a fixed number of atoms or molecules, it follows that the *average* energy, ε, of an atom or molecule that comprises that given mass of material must be proportional to the temperature of that mass of material. Thus we know that

$$\varepsilon \propto T$$

However, while we know ε is proportional to T, we would like to know the proportionality constant that would allow us to *equate* ε and T. While we will explore this in quantitative detail in Chapter 4, we can achieve some important insights from what we already know.

From our analysis of Newton's Laws, we know that when a particle of mass m collides in the x direction with a wall, as shown in Figure 1.16, and rebounds, the change in momentum will be $\Delta P_x = m(-v_x) - mv_x = -2mv_x$. If a *single* molecule is contained in a box with a dimension L on each edge, with no other molecules present, it will collide with the opposite wall and return in a time period $\Delta t = 2L/v_x$. The resultant force delivered to the wall by this sequence of collisions would be, given our expression for the impulse from equation 1.2, and taking the absolute value of $\Delta P_x / \Delta t$ because the force is positive:

$$F = \left| \frac{\Delta P_x}{\Delta t} \right| = \frac{2mv_x}{2L/v_x} = \frac{2}{L}\left(\frac{1}{2}mv_x^2 \right)$$

The resulting pressure exerted on the wall can be calculated from the relationship between force, F, and pressure, p, where

$$p = \frac{F}{A}$$

with A representing the area over which the force is applied.

Therefore, in the case of our box of dimension L,

$$p = \frac{F}{A} = \frac{F}{L^2} = \frac{2}{L^3}\left(\frac{1}{2}mv_x^2 \right) = \frac{2}{V}\left(\frac{1}{2}mv_x^2 \right)$$

This results in the expression

$$pV = 2\left(\frac{1}{2}mv_x^2 \right)$$

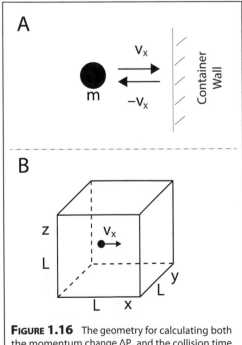

FIGURE 1.16 The geometry for calculating both the momentum change ΔP_x and the collision time interval Δt.

However, a gas is characterized by molecular motion with velocities independent of direction, so

$$\overline{v_x^2} = \overline{v_y^2} = \overline{v_z^2} = \frac{1}{3}\overline{v^2}$$

This gives us the important result that

$$pV = 2\left(\frac{1}{2}m\overline{v_x^2}\right) = 2\left(\frac{1}{2}m\left(\frac{1}{3}\overline{v^2}\right)\right) = \frac{2}{3}\left(\frac{1}{2}m\overline{v^2}\right)$$

Finally, $\frac{1}{2}m\overline{v^2}$ is the average kinetic energy of each molecule, which we designate as ε, and so

for N total molecules we have $pV = \frac{2}{3}N\varepsilon$.

However, the Perfect Gas Law gives us the relationship

$$pV = nRT$$

and so, equating our two expressions for pV we have

$$nRT = 2/3 \ N\varepsilon$$

With R=8.3 J/mole-K expressing the gas constant in units of joules per *mole* per K, and with k_B, the Boltzmann constant in units of joules per *molecule* per K, we have $nR = k_B N$ and

$$\frac{3}{2}\left(\frac{nR}{N}\right)T = \frac{3}{2}k_B T = \varepsilon$$

This gives us the *proportionality constant* between the average kinetic energy per molecule, ε, and the temperature T, in Kelvin:

$$\varepsilon = \frac{3}{2}k_B T$$

As we will see, this is an extremely versatile expression relating the average energy per atom or molecule, ε, to the observed temperature of any ensemble of atoms or molecules that comprise a macroscopic object.

While $\varepsilon = \frac{3}{2}k_B T$ defines the proportionality factor between ε and T, it is the *functional* form of the relationship between ε and T *and* the simplicity of that relationship that is the most important consideration here. It is also important to examine the *magnitude* of the average energy associated with each atom or molecule in a body that determines its temperature. The value of the Boltzmann constant is

$$k_B = 1.3807 \times 10^{-23} \ J/K \cdot molecule$$

The amount of energy $k_B T$ is a very important quantity and is referred to as the *thermal energy* representing the *average energy* of an atom or molecule that makes up a macroscopic body at temperature T. Thermal energy is the ultimate benefactor of the cascade of organized forms of energy (kinetic energy of a moving body, chemical energy, nuclear energy, etc.) to increasingly disorganized forms of energy.

Solution:

1. The average translational kinetic energy per molecule is

   $\varepsilon = \frac{3}{2} k_B T = \frac{3}{2} (1.38 \times 10^{-23}$ J/K·molecule$)$ $(273$ K$) = 5.6 \times 10^{-21}$ J/molecule

   One mole of gas contains 6.02×10^{23} molecules, N_A, so the kinetic energy of a mole of gas molecules at 273 K is:

   $N_A \varepsilon = (5.6 \times 10^{-21}$ J/molecule$)(6.02 \times 10^{23}$ molecules$) = 3.4 \times 10^3$ J

2. A remarkable aspect of the expression $\varepsilon = \frac{3}{2} k_B T$ for the average translational kinetic energy of a gas molecule is that it contains no reference to the mass of the atom or molecule. The reason is that when molecules collide, no matter what their relative masses are, the collisions act to equalize the translational kinetic energy, $\frac{1}{2} mv^2$, of all atoms or molecules in the gas mixture. Thus $\varepsilon = \frac{3}{2} k_B T$ contains no reference to the molecular mass.

3. The reason that a mole of argon atoms in a 1 liter vessel exerts the same pressure as a mole of helium atoms comes directly from the fact that the average kinetic energy per atom in the gas mixture is the same for both argon and helium. Thus $\frac{1}{2} m_{Ar} v_{Ar}^2 = \frac{1}{2} m_{He} v_{He}^2$ and so:
   $$m_{Ar} v_{Ar}^2 = m_{He} v_{He}^2$$

   But the force exerted per molecule is, from our discussion on page 28, given by the *impulse*

   $F_{Ar} = \Delta P_{Ar}/\Delta t$ for argon, and
   $F_{He} = \Delta P_{He}/\Delta t$ for helium

   Also from page 28 we have

   $\Delta P_{Ar} = -2m_{Ar}(v_x)_{Ar}$ and $\Delta t = 2L/(v_x)_{Ar}$

   For helium:
   $\Delta P_{He} = -2m_{He}(v_x)_{He}$ and $\Delta t = 2L/(v_x)_{He}$

   So the pressure (force divided by area) exerted by each atom in the vessel is

   $p_{Ar} = F_{Ar}/A = |\Delta P_{Ar}/\Delta t|/A = 2m_{Ar}(v_x)_{Ar}/(2L/(v_x)_{Ar})/A = m_{Ar}(v_x)_{Ar}^2/(LA) = m_{Ar}(v_x)_{Ar}^2/V = m_{Ar}(\frac{1}{3}v_{Ar}^2)/V$

   and $p_{He} = F_{He}/A = |\Delta P_{He}/\Delta t|/A = 2m_{He}(v_x)_{He}/(2L/(v_x)_{He})/A = m_{He}(v_x)_{He}^2/(LA) = m_{He}(v_x)_{He}^2/V = m_{He}(\frac{1}{3}v_{He}^2)/V$

   But if from $\varepsilon = \frac{3}{2} k_B T$ we have $m_{Ar} v_{Ar}^2 = m_{He} v_{He}^2$, then the pressure exerted by one molecule or one mole of each gas (at the same temperature) will be equal.

Check Yourself 9

The expression for the average kinetic energy, ε, of a molecule or atom, $\varepsilon = \frac{3}{2} k_B T$ is a remarkably versatile relation—with it we can solve a wide range of important problems.

1. What is the total translational energy of the molecules in 1 mole of a gas at STP?
2. Why was it unnecessary to specify the specific molecule in question 1?
3. Explain why 1 mole of argon (molecular weight 40 amu) in a one liter flask at 273 K exerts the same pressure as a mole of helium (molecular weight 4 amu) in a one liter flask at 273 K.

To grasp the scale of energy contained in a single molecule at room temperature, we calculate (1) the thermal energy of a molecule at room temperature and (2) the number of molecules required to comprise a system containing 10J of thermal energy at 20°C. Because the thermal energy of a molecule is $\frac{3}{2} k_B T$ the thermal energy of a single molecule is $\frac{3}{2} k_B T = (\frac{3}{2}) (1.38 \times 10^{-23}$ J/K$) (293K) = 6.0 \times 10^{-21}$ J. The thermal energy of n molecules is $n \frac{3}{2} k_B T$, then with

$$\frac{3}{2} nk_B T = 10 \text{ J}$$

it follows that

$$n = \frac{10 \text{ J}}{\frac{3}{2} k_b T} = \frac{10 \text{ J}}{(1.38 \times 10^{-23} \text{ J/K}) \, 293 \text{ K}} = 1.7 \times 10^{21} \, molecules$$

As noted above, the thermal energy, $\frac{3}{2} k_B T$, of a single molecule is, in a practical sense, the smallest amount of energy we will consider, and thus it constitutes the lower limit on our energy scale in Figure 1.2.

Energy Transformations: The Central Role of Electromagnetic Radiation

A primary objective in this chapter is to elevate the concept of energy to a prominent position within the conceptual structure of the physical sciences. It is, as we are coming to see, impossible to develop an understanding of energy without a keen sense of the various categories of energy types and particularly *transformations* among and between types of energy. The four panels in Figure 1.13 are a case in point. Consider panel (b) in Figure 1.13. This panel captures the process wherein a beam of sunlight is focused on the base of the same paddle wheel-water system that Joule used to establish the mechanical equivalent of heat. Using the same thermometer, Joule could easily (and perhaps did!) measure the amount of energy delivered to the base of his system by that beam of sunlight. But what happens specifically when this sunlight, electromagnetic radiation, falls on the metal surface of Joule's system? We know the light "disappears" and the heat "appears." How are these two events coupled—recognizing that energy is a conserved quantity?

While we will develop increasingly the details of the answer to this question in subsequent chapters, there are important conclusions that we can draw between the motion of molecules at the microscopic level, and the absorption of electromagnetic radiation. Because the temperature of the system illuminated by electromagnetic radiation increases, we know that the atoms or molecules composing the system must have undergone an increase in average energy in the form of the kinetic energy of motion of the atoms that make up the body. The fact that matter is made up of charged particles—electrons and protons—is a key part of the answer. First, we explore (a) the character of electromagnetic radiation and (b) that light (electromagnetic radiation) is both *absorbed* by material composed of charged particles (as all material is) and *emitted* by material comprised of charged particles in *motion*.

The transformation between electromagnetic radiation and solid surfaces.

In any discussion concerning transformations among and between various forms of energy (kinetic, gravitational, chemical, electromagnetic, thermal, etc.), electromagnetic energy occupies a central role. Light from the Sun, the wireless connection to your computer, cooking with microwaves in an oven, all involve the transfer of energy and the conversion of energy to one form or another using electromagnetic radiation. What is rather surprising is that the structure of the electromagnetic wave is common to all types of radiation that propagate either through vacuum or through materials that are to some degree transparent, depending upon the molecular level interaction between the electromagnetic radiation and the atoms that comprise the particular material. It was a young James Clerk Maxwell (1831–1879) in 1865 who theorized that electromagnetic radiation was comprised of *mutually perpendicular oscillating electric and magnetic fields* and that the direction of oscillation of these two fields was in turn perpendicular to the direction of propagation of those fields. The geometry of the electromagnetic wave is captured in Figure 1.17. The diagram identifies the electric and magnetic field components, as well as the direction of propagation.

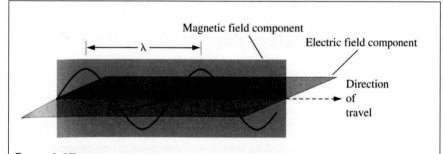
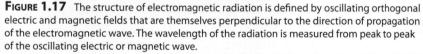

FIGURE 1.17 The structure of electromagnetic radiation is defined by oscillating orthogonal electric and magnetic fields that are themselves perpendicular to the direction of propagation of the electromagnetic wave. The wavelength of the radiation is measured from peak to peak of the oscillating electric or magnetic wave.

Electromagnetic radiation is characterized first and foremost by the *wavelength* of the radiation, λ, and by the velocity of the propagation, the magnitude of which is the speed of light, c. Another characterization of electromagnetic energy is the *frequency* of that radiation. The wavelength, λ, is defined to be the distance between two consecutive peaks (or troughs) in a wave, as displayed in Figure 1.17. The frequency, ν, is defined as the number of cycles (waves) per second that pass a given position (or point) in space. Because all electromagnetic radiation travels with the speed of light, shortwave radiation is of higher frequency than longwave radiation. A very simple but important equation links wavelength, λ, and frequency, ν, because the product of the wavelength (in meters) and the frequency (in sec⁻¹ or hertz) equals the speed of light:

$$\lambda \nu = c$$

Figure 1.18 maps out the electromagnetic spectrum extending from gamma rays (a product of nuclear fusion that we will discuss in Chapter 13 on nuclear chemistry) at the high energy end to radio waves at the low energy end of the spectrum. What is important to garner from Figure 1.18 is the relationship between the *wavelength* of the electromagnetic radiation and the *size* (the scale) of known entities whose physical dimensions are comparable to those *wavelengths*. For example, x-rays have a wavelength equivalent to the size of individual atoms, ultraviolet light has a wavelength comparable to the size of a virus, visible light to that of a bacterium, microwaves to that of the width of a human finger and radio waves to the height of a human being.

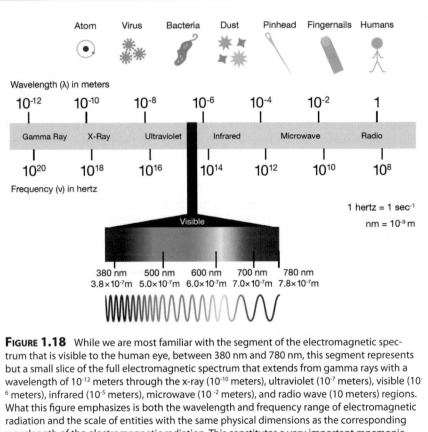

FIGURE 1.18 While we are most familiar with the segment of the electromagnetic spectrum that is visible to the human eye, between 380 nm and 780 nm, this segment represents but a small slice of the full electromagnetic spectrum that extends from gamma rays with a wavelength of 10⁻¹² meters through the x-ray (10⁻¹⁰ meters), ultraviolet (10⁻⁷ meters), visible (10⁻⁶ meters), infrared (10⁻⁵ meters), microwave (10⁻² meters), and radio wave (10 meters) regions. What this figure emphasizes is both the wavelength and frequency range of electromagnetic radiation and the scale of entities with the same physical dimensions as the corresponding wavelength of the electromagnetic radiation. This constitutes a very important mnemonic device for linking the type of electromagnetic radiation to well known objects.

We will refer repeatedly to the basis of electromagnetic radiation throughout the course when we examine *transformations* between different types of energy, when we consider the propagation of radiation through different materials, and when we consider the molecular level interaction between light and the electrons in the various energy levels of atoms and molecules. Thus Figure 1.18 deserves particular attention. Given that matter is composed of charged particles—electrons and protons—in general when the electric field of the electromagnetic radiation impinges on the surface of a material, that electric field applies a force to the charged particles.

If an object (for example an electron or proton) has a charge q, and is placed in an electric field, E, the *force* on that charge is given by

$$F = q\,E$$

Thus, when light falls on the surface of a material, the force on a negative charge (for example an electron) is

$$F = -e\,E$$

and the force on a positive charge will be

$$F = +\,q\,E$$

Thus as the electromagnetic wave passes into the material, a force will be applied to these charge centers that alternately forces the charges apart and then forces them back together, as displayed schematically in Figure 1.19.

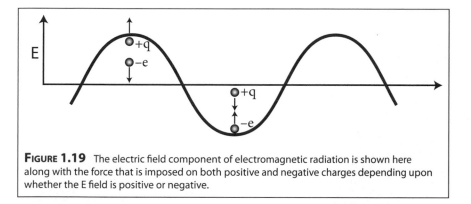

FIGURE 1.19   The electric field component of electromagnetic radiation is shown here along with the force that is imposed on both positive and negative charges depending upon whether the E field is positive or negative.

This is what physically couples the *electric field* in the radiation to the *kinetic energy* of the atoms in the material, increasing the kinetic energy of motion of the atoms that comprise the material. We can sketch, as shown in Figure 1.20, the atomic level interaction between the electric field of the electromagnetic radiation and the charges that comprise the material illuminated by that radiation.

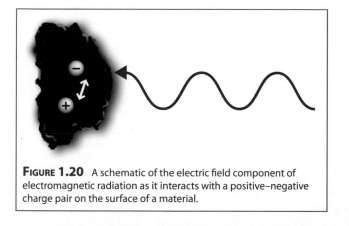

FIGURE 1.20   A schematic of the electric field component of electromagnetic radiation as it interacts with a positive–negative charge pair on the surface of a material.

Another critically important fact that emerged from Maxwell's (and others') work was that an *accelerating charge radiates electromagnetic radiation.* Moreover, a positive and negative charge pair, oscillating with respect to their relative distance of separation, can *generate* electromagnetic radiation, as shown in Figure 1.21. That oscillation of positive and negative charge centers is sustained by the thermal motion of the atoms and molecules that comprise the material. Thus we can capture the *emission* of electromagnetic radiation by the oscillating charges in the material in the analogous diagram to the one detailing the absorption of electromagnetic radiation:

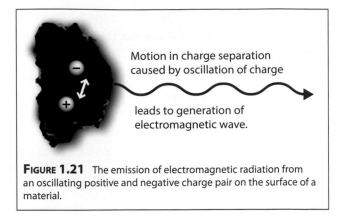

FIGURE 1.21 The emission of electromagnetic radiation from an oscillating positive and negative charge pair on the surface of a material.

The conclusion is that all materials with a temperature above (absolute) zero on the Kelvin scale emit electromagnetic radiation. This establishes a very important, and dynamic, exchange of electromagnetic radiation between any two bodies, as shown in Figure 1.22. Electromagnetic radiation emitted by one body is absorbed by the other and vice versa as depicted here:

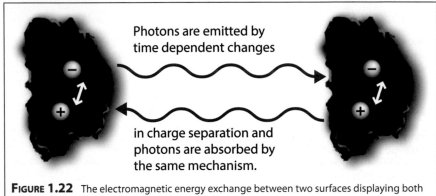
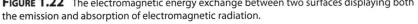

FIGURE 1.22 The electromagnetic energy exchange between two surfaces displaying both the emission and absorption of electromagnetic radiation.

This underscores the reciprocity inherent in the emission and absorption of electromagnetic radiation: light is emitted by the time dependent motion of charges that comprise matter, and light is absorbed by inducing changes on the charge displacement of the positive and negative charges within matter.

Because of this ubiquitous emission and absorption of radiation between all bodies that possess thermal energy, any discussion of the transformation of energy between various categories of energy must explicitly consider the transfer of electromagnetic radiation.

Notice that in Figure 1.13, organized forms of energy—potential energy of a falling body, energy contained in molecular bonds, electrons moving in a wire—are converted to the microscopic kinetic energy of molecules, thereby raising the temperature of a body (e.g., the water-paddle wheel system of Joule in Figure 1.13). But the surfaces of bodies are always emitting electromagnetic radiation. For bodies with temperatures near room temperature, this emitted electromagnetic radiation falls in the infrared spectral range displayed in Figure 1.18. When you are sitting in a room, you are radiating infrared radiation outward, and the walls of the room are radiating infrared radiation into you. You consume food that is oxidized to provide the thermal energy, heat, to make up the difference between what you radiate into your surroundings and what your surroundings radiate into you. Thus, when you are sitting in class, you are bathed in infrared radiation from the walls, ceiling, floor, and from those sitting around you. As we will see, the Earth also radiates infrared radiation into the blackness of space. But in this case the infrared radiation from the outer envelope of the atmosphere is emitted outward but not returned. On the other hand, the exchange of infrared radiation between the Earth's *surface* and the water vapor, clouds, CO_2, and other gas in the atmosphere controls the flow of energy within the climate structure. As we will see, it is the flow of energy within the climate structure of the Earth that determines the response of the system to increasing levels of CO_2, not simply "global warming."

Increasingly, quantitative studies of the emission of electromagnetic radiation from the surface of materials as the 20th century approached began to reveal an increasingly coherent picture of the character of this emitted radiation. For example, an iron plate, which could be heated to temperatures of over 1000 K, or titanium, to temperatures approaching 2000 K, provided direct observations of the intensity of radiation as a function of wavelength displayed in Figure 1.23.

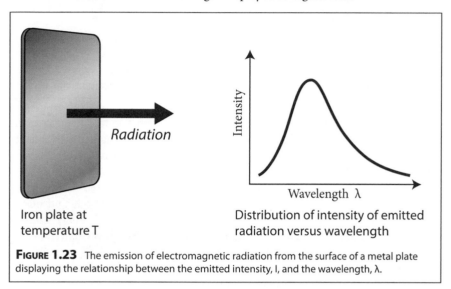

Iron plate at temperature T

Distribution of intensity of emitted radiation versus wavelength

FIGURE 1.23   The emission of electromagnetic radiation from the surface of a metal plate displaying the relationship between the emitted intensity, I, and the wavelength, λ.

Figure 1.23 shows the intensity of radiation, I, as a function of wavelength, λ, from a plate of iron. When the emission from a plate of polished nickel or a vat of water was examined, the details of the shape of the curve of I vs. λ at a given temperature changed somewhat, but the characteristic shape remained: at very small λ the intensity approached zero, then rose to a maximum as the wavelength increased, then fell back to zero at large λ.

But what was also apparent was that as the temperature of the plate was raised, the total energy emitted per unit of time from that surface of the metal plate increased *very* rapidly with increasing temperature.

While the quantitative details of emitted intensity varied somewhat as a function of wavelength depending on the metal or the condition of the surface of the metal, these differences could be eliminated entirely by forming a cavity of the material with a small hole to allow the emitted radiations to escape. The use of this "blackbody cavity," as it came to be called, eliminated any variables specific to the material and placed those studies on a path to quantitative consistency. A diagram of such a cavity is shown in Figure 1.24.

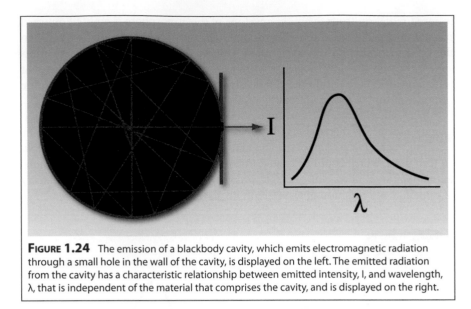

Figure 1.24 The emission of a blackbody cavity, which emits electromagnetic radiation through a small hole in the wall of the cavity, is displayed on the left. The emitted radiation from the cavity has a characteristic relationship between emitted intensity, I, and wavelength, λ, that is independent of the material that comprises the cavity, and is displayed on the right.

So the curve on the right of the diagram above, defining the intensity of energy flow, I, as a function of wavelength, λ, emitted from the aperture of a cavity, or "blackbody," characterizes the radiation emitted from the cavity wall resulting from multiple reflections within the cavity.

Figure 1.25 displays how the relationship between the emitted intensity I vs. λ changes with temperature. For the case of a blackbody at 6000 K and one at 3000 K, as the temperature is increased, the *peak* of the blackbody curve shifts to shorter wavelengths. Note also that the *area* under the curve of I versus λ increases dramatically. You experience this in many venues: As the heating element of a stove gets *hotter*, sequentially the color shifts from invisible infrared radiation that you can feel with your hand to deep red that you can see with your eyes, then to orange at the highest temperature. The amount of heat flow you sense increases dramatically in the transition to higher temperature. Another example is that as the battery of your car begins to fail, the filaments in the headlights dim and turn increasingly red from their normal bright whitish-yellow.

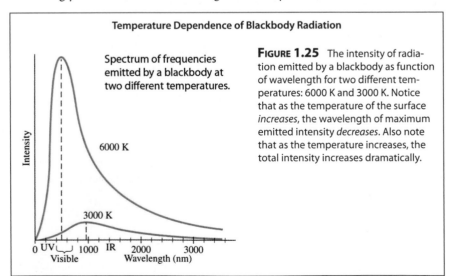

Temperature Dependence of Blackbody Radiation

Spectrum of frequencies emitted by a blackbody at two different temperatures.

Figure 1.25 The intensity of radiation emitted by a blackbody as function of wavelength for two different temperatures: 6000 K and 3000 K. Notice that as the temperature of the surface *increases*, the wavelength of maximum emitted intensity *decreases*. Also note that as the temperature increases, the total intensity increases dramatically.

The total amount of energy emitted by a blackbody is expressed quantitatively by a simple equation, termed the Stefan–Boltzmann Law, which states that the rate at which energy leaves a body at temperature T is given by

$$\Delta E_{rad}/\Delta t = (\text{Area of Body})\sigma T^4 = A\sigma T^4$$

where σ is the Stefan–Boltzmann constant,

$$\sigma = 5.67 \times 10^{-8} \text{ W/(m}^2 \text{ K}^4),$$

and A is the area of the emitting surface.

What is remarkable is that the complicated motion of atoms or molecules within the emitting material reduces to such a simple expression for the total energy emitted by the surface per unit time.

The Stefan–Boltzmann Law is one of the most powerful and widely applicable laws governing the transformation of energy from the thermally induced fluctuations of charge in solids to the generation of electromagnetic radiation. It relates the energy flow, per unit area, from a surface in units of joules/m²-sec, to the temperature, T, of that emitting surface. Thus, each area increment, A, of a surface will emit energy at a rate of F, given by

$$F = A\sigma T^4$$

as shown in Figure 1.26.

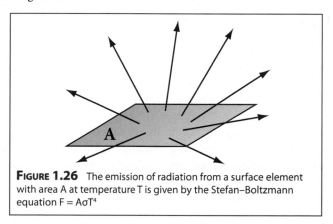

FIGURE 1.26 The emission of radiation from a surface element with area A at temperature T is given by the Stefan–Boltzmann equation $F = A\sigma T^4$

This makes it very easy to calculate the rate at which energy is emitted from any solid body at temperature, T. For example a sphere of radius r at temperature T will emit an amount of energy per unit time into the space that surrounds it given by the simple expression $F = A\sigma T^4$, as shown in Figure 1.27:

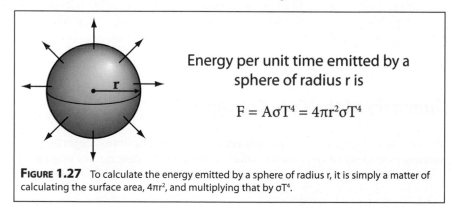

Energy per unit time emitted by a sphere of radius r is

$$F = A\sigma T^4 = 4\pi r^2 \sigma T^4$$

FIGURE 1.27 To calculate the energy emitted by a sphere of radius r, it is simply a matter of calculating the surface area, $4\pi r^2$, and multiplying that by σT^4.

The Stefan–Boltzmann law puts in place a simple expression that can be used to calculate the energy radiated from your body, and to relate that quantity to the number of Calories of food energy that would be required if infrared energy was not trapped by your clothing and were infrared radiation not emitted *into* you by the walls surrounding you. This is developed quantitatively in Check Yourself 10 and Case Study 1.3.

1. Using the remarkably simple expression for the flow of radiant energy from a body at temperature, T, of $F = A\sigma T^4$ where A is the area of the body emitting electromagnetic radiation and σ is the Stefan–Boltzmann constant, calculate the energy, emitted by the human body in the infrared. The first step is to "model" the complex structure of the human body as a simple geometric structure that will yield a reasonable value for the area, A, in the Stefan–Boltzmann equation.

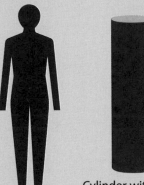

Human   Cylinder with same surface area A

2. Calculate the flow of energy per unit time from this "model" human at a temperature of 37°C. How many joules per day are radiated in the infrared (IR)?

3. You know that there are 4.18 joules in a calorie, and that there are 10^3 calories in a food "Calorie" (a capital C denotes the food Calories listed on products you purchase and consume). How many Calories must you consume per day to sustain this energy loss by infrared radiation from your body?

4. How does this compare with your normal intake of food?

5. How do you reconcile the energy emitted from your body in the IR each day and the amount of food you consume?

Check Yourself 10: Stefan–Boltzmann & the Consumption of Food

The Stefan–Boltzmann law provides remarkable insight into how the exchange of energy actually works in nature. We know that the human body is sustained by the food we eat and that our body temperature must be held within a *very* few degrees of 38°C for us to function—a little too hot and we have a fever (or heat stroke), a little too cold and we suffer from hypothermia.

Energy and Power: A Very Important Distinction

In any discussion of energy, the question invariably arises: how fast is the energy transferred or transformed in a given process? This issue of the *rate* at which energy is transferred is addressed by the concept of *power*, which is defined as:

$$\text{Power} = \text{Energy/Time}$$

We can write this in equation form as $P_w = \Delta E/\Delta t$ where ΔE is the amount of energy transformed or transferred in time Δt. Thus the power, P_w, represents the rate at which energy is transferred and, with energy measured in joules, power has the units of joules/sec. The standard unit of power is the *watt*, which is joules/sec, so 1 watt is 1 joule/sec. The watt was named after James Watt, who facilitated the introduction of the steam engine into the industrial revolution.

But we know from our discussion of work, force, and distance that the energy transferred, ΔE, is equal to the work done, which in turn is equal to the force applied times the distance, say Δx, over which that force was applied. Therefore, we can write $\Delta E = \Delta w = F\Delta x$ where ΔE is the energy transferred, Δw is the work done by the force F applied over a distance Δx. Thus, if we insert this expression into our definition of power we can write

$$P_w = \Delta E/\Delta t = \Delta w/\Delta t = F\,\Delta x/\Delta t = Fv$$

where v is the velocity of the object to which the force is applied.

This topic is of sufficient importance that an entire Case Study is devoted to developing both a quantitative and an intuitive understanding of the distinction between energy and power—Case Study 1.1.

Summary of the Core Concepts

We close the chapter core with a sequence of brief descriptions categorizing the summary concepts of the chapter before moving to the case studies linking the concepts to the larger context addressed by the text.

Summary Concepts

Understanding the Concept of Energy: A Prerequisite for Scientific Progress

Until energy was understood in a rigorous scientific sense, modern scientific inquiry could not and did not progress. The central role that energy plays in the instigation of spontaneous *change* within physical, chemical, and biological systems places it at the forefront of any effort to comprehend the natural world. In the course of that intensive inquiry into the nature of energy, it became apparent that while nature works ceaselessly to achieve a state of lowest energy, nature also works ceaselessly to achieve a state of maximum disorganization. This inexorable drive by nature to seek the state of lowest energy *and* the state of maximum disorganization establishes the concepts of both energy and entropy in the determination of spontaneous change. It is for this reason that we place inordinate focus on energy in this opening chapter.

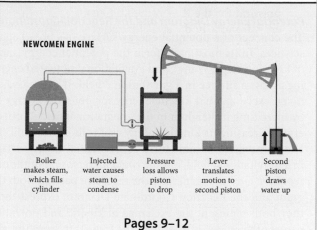

NEWCOMEN ENGINE

| Boiler makes steam, which fills cylinder | Injected water causes steam to condense | Pressure loss allows piston to drop | Lever translates motion to second piston | Second piston draws water up |

Pages 9–12

Energy, Work, and Newton's Laws: Conservation of Mechanical Energy

The evolution of energy as a scientific concept emerged from the foundation of Newton's Laws of motion. Newton's First Law clarified the importance of *inertia* and states that every body persists in its state of rest or of motion in a straight line unless it is compelled to change that state by forces impressed upon it. This clarified the role of force as the agent of change, and that change was the *acceleration* imparted to the body by the external force. Newton's Second Law defined the relationship between force and acceleration: the acceleration, a, of a body is proportional to the force applied. Moreover, the force is equal to the product of the mass of the body and acceleration, F = ma. This relationship between force and acceleration provided the relationship between work and energy, where work is the product of force times distance. The sum of kinetic energy, $KE = \frac{1}{2}mv^2$, and potential energy, $PE = mgh$, is equal to the total mechanical energy $E_T = PE + KE = mgh + \frac{1}{2}mv^2$ of the macroscopic system.

Schematic of a baseball thrown vertically

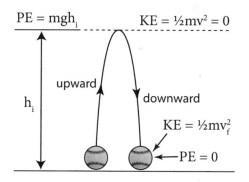

$$PE = mgh_i \qquad\qquad KE = \frac{1}{2}mv^2 = 0$$

h_i

upward

downward

$$KE = \frac{1}{2}mv_f^2$$

$$PE = 0$$

Pages 12–15

Conservation of Energy: Tracking the Flow of Energy from Macroscopic Mechanical Energy to Microscopic Molecular Motion

With the incorporation of nonconservative frictional forces in mechanical systems, the principle of the conservation of energy requires that we discover how the loss of the sum of kinetic and potential energy, $E_T = PE + KE$, results in the increase of energy in some *other form* such that energy is neither created nor destroyed. The recipient of the energy lost (as mechanical energy of the macroscopic bodies in a system that includes friction) is the kinetic energy of the individual atoms that comprise the body. The motion of atoms and molecules is expressed as the thermal energy, U_{therm}, that is the recipient of the decrease in macroscopic mechanical energy. The conservation of energy is expressed as the fact that the sum of macroscopic mechanical energy, E_{mech}, and the microscopic thermal energy, U_{therm}, is a constant: $E_{mech} + U_{therm} = $ constant.

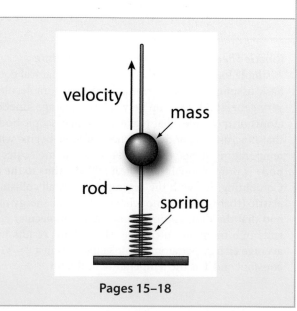

velocity

mass

rod →

spring

Pages 15–18

Summary Concepts

Potential Energy Diagram and the Reaction Coordinate Diagram

The concept of a potential energy surface has broad applicability in the physical sciences. In its most basic form the potential energy surface tracks the exchange of kinetic energy, $\frac{1}{2} mv^2$, and gravitational potential energy, mgh, of a mass moving across a surface in a gravitational field. If the mass has kinetic energy, $\frac{1}{2} mv^2$, it can exchange that kinetic energy for potential energy if the moving mass rises by increasing its height h in the gravitational field. It can also release potential energy, increasing its kinetic energy as it rolls down a hill. The Reaction Coordinate Diagram takes the same form as a Potential Energy Diagram for a mass moving in a gravitational field, but the *coulomb repulsion* of the outer electrons creates the *repulsive force* that generates the repulsive barrier. Both the Potential Energy Diagram and the Reaction Coordinate Diagram express the conservation of energy, they both represent the exchange of kinetic and potential energy, and they both express the release of energy in the transition from initial to final conditions.

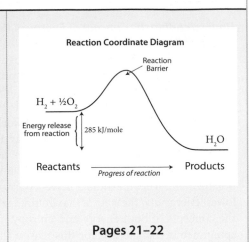

Reaction Coordinate Diagram

Reaction Barrier

$H_2 + \frac{1}{2}O_2$

Energy release from reaction   285 kJ/mole

Reactants   *Progress of reaction*   Products

H_2O

Pages 21–22

The Mechanical Equivalent of Heat and the Concept of Heat Capacity

Following the contributions of Newton it was well understood that energy could be exchanged between potential and kinetic forms, but it could not be created nor destroyed. The concept of heat was far more subtle. The idea that heat was a *substance* was finally dispelled by Benjamin Thompson (Count Rumford) who correctly identified heat with the process of the *transfer of energy*. But it was Joule who, by a series of brilliantly simple experiments, established the key *quantitative* link between the release of mechanical energy of a macroscopic body and the increase in temperature of a system that transferred that mechanical potential energy to a water bath via the friction between the paddle blades and the water. That experimental system provided the physical coupling between the macroscopic mechanical energy release, mgh, and the increase in the temperature of the water bath, ΔT_{bath}. Thus q_{added} must equal the release of potential energy, mgh, from the falling mass in the gravitational field. This simple proportionality between the amount of potential energy release and the increase in temperature ($\Delta T_{bath} \propto$ mgh), combined with $q_{added} = mgh$, meant that the heat added, the potential energy release, and the increase in temperature of the apparatus could be equated: mgh = $q_{added} = C\Delta T_{bath}$ where C is termed the *heat capacity* of the object into which heat is added.

Mechanical energy transferred
= mgh
= heat transferred
= CΔT

Pages 22–26

Kinetic Theory Interpretation of Temperature

A simple logic that emerges from the mechanical equivalent of heat coupled with basic assumptions concerning the theory of molecular motion in gases provides a very general relationship between the average kinetic energy, ε, of the individual atoms or molecules that make up a macroscopic body and the temperature, T, of that macroscopic body. The chain of logic begins with the observed fact that the temperature change, ΔT, of an ensemble of molecules that comprise a macroscopic body is proportional to the heat added—thus to the energy added—to that body. Combining this with the assumption that all collisions at the molecular level are elastic (that is they do not change the total energy of the ensemble of molecules) and that the momentum change in any molecular collision with a vessel wall results in a momentum change $\Delta P = -2mv$, results in the determination that the average energy per atom or molecule, ε, is just $\varepsilon = \frac{3}{2} k_B T$ where k_B is Boltzmann's constant and T is the temperature in Kelvin.

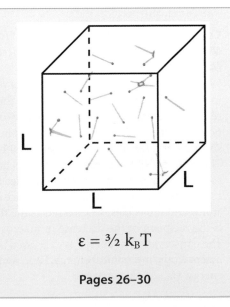

L   L   L

$$\varepsilon = \frac{3}{2} k_B T$$

Pages 26–30

Summary Concepts

Role of Electromagnetic Radiation in the Transformation of Energy Among Various Categories of Energy

The concept of energy cannot be grasped in a scientific sense without an understanding of how energy is transformed from one category to another. The basic truth of this is revealed in the work of Newton, Count Rumford, Joule, and others. Joule recognized very early in his experiments that he must carefully insulate his paddle wheel-water system in order to get accurate results. Joule recognized that he could not let energy of any form escape from his experimental system or his results would be inaccurate. Yet the addition of heat, it was recognized, increased the average speed of the atoms that comprised the object into which the heat flowed. From the work of Maxwell it was recognized that oscillating charges emit electromagnetic radiation, and it was clear that electromagnetic radiation falling on an object transferred heat to that body—by necessity increasing the average speed of the atoms that comprised the object. Careful experiments revealed that objects at room temperature and above emit radiation in the infrared region of the spectrum by virtue of the oscillatory motion of charges that comprise the material.

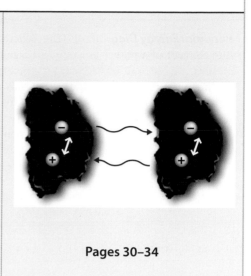

Pages 30–34

Blackbody Radiation and the Stefan–Boltzmann Law

Intensive studies of the radiation emitted by various materials, particularly metals, revealed that while the mechanism by which light was emitted by the surface of high temperature objects was complicated, the flow of energy from most surfaces obeyed a remarkably simple expression. Specifically the rate at which energy was emitted per unit area depended only on the temperature of the surface raised to the fourth power, with only minor modifications for different emitting materials. This lead to the Stefan–Boltzmann law that states that the flux of energy, F, emitted from a surface per unit time is simply

$$F = A\, \sigma\, T^4$$

where A is the area from which the radiation is emitted, T is the temperature of the surface, and σ is the Stefan–Boltzmann constant, $\sigma = 5.67 \times 10^{-8}$ joules/sec·m²·K⁴. Experiments demonstrated that if a material was formed as a cavity with a small opening for observing the escaped radiation, that the relationship between the emitted intensity of radiation and the wavelength dependence of the radiation was dependent only on temperature and not on the material itself. The *shape* of the intensity curve *I* vs. λ was to revolutionize modern science as Case Study 1.4 makes clear.

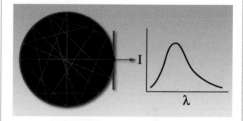

Pages 34–38

Energy and Power: A Critically Important Distinction

With the development of the concept of energy comes the immediate question, how rapidly is that energy used, or more to the point, how rapidly is that energy *transformed*? The answer to that question is quantitatively defined by the concept of power, which is the rate at which energy is transformed:

$$P_w = \text{Power} = \text{Energy Transformed} / \text{time}$$

Typically an amount of energy, ΔE, is transformed in time, Δt, and the power, P_w, is given by

$$P_w = \Delta E / \Delta t$$

The unit of power is the watt, and one watt is equal to 1 joule/sec.

Saturn V rocket

Page 38

CASE STUDY 1.1   Quantitatively Linking the Concepts of Energy and Power: Some Important Examples

KEY CONCEPTS:

"There is a fact, or if you wish, a *law*, governing all natural phenomena that are known to date. There is no known exception to this law—it is exact so far as we know. The law is called the *conservation of energy.*"

Richard Feynman

1. The Scientific Concept of Energy

A scientific understanding of energy constitutes the foundation of virtually all of modern scientific thought. The scientific concept of energy began with an understanding of (1) the relationship between force, work, and energy, and grew to encompass (2) the concept of the conservation of energy: that energy is neither created nor destroyed.

It was Newton who recognized that in order to impart potential energy to a mass in a gravitational field, work had to be done to raise the mass a given distance, h. In so doing, work, a force equal to the acceleration of gravity, g, times the mass, m, of the object, multiplied times the distance, h, such that $W = F \cdot d = mgh$ quantitatively defined the potential energy of the mass. Newton also recognized that if that mass were released to fall through the gravitational field, the kinetic energy *gained* as the body accelerated downward was equal to the potential energy *lost*. This was a clear example of the conservation of energy, in this case the mechanical energy $E_M = mgh + \frac{1}{2}mv^2 =$ constant and therefore $\Delta E_M = 0$.

It was Joule who quantitatively related mechanical energy to heat setting in place the concept of the "mechanical equivalent of heat." This not only set the absolute scale for energy (the "joule"), but also brought the recognition that energy occurs in many different forms (many different categories), but that all forms of energy can be placed on the same energy scale and thus share the same unit of measure—the "joule" equal to $1 \text{ kg } (m/s)^2$. Realizing that energy occurs in many different forms: mechanical, electrical, chemical, nuclear, electromagnetic, thermal, etc., the principle of the conservation of energy—that energy is neither created nor destroyed—requires that the definition of total energy must encompass all forms of energy that can be transformed from one form to another within the system under consideration. Thus $E_T = E_{mech} + E_{chem} + E_{nuc} + E_{EM} + E_{therm} + \ldots$. So while energy can be readily exchanged among and between these categories, total energy, E_T does not change so $\Delta E_T = 0$.

2. Relationship between Energy and Power

In a vast majority of systems, it is the *rate* at which energy is converted from one form or category of energy to another that is of great importance. For example, the *rate* at which chemical energy contained in gasoline is converted to mechanical energy—the acceleration of an automobile—determines the delivered "horsepower" of that mechanical system. Power is defined as the rate at which energy is converted from one form to another, and so quantitatively **Power** = $\Delta E / \Delta t$, the energy conversion per unit time. It is therefore of great importance to carefully distinguish between energy and power. While the SI unit of energy is the joule (J), the unit of power is the watt (W), which is 1 J/sec.

3. Modeling of a System to Calculate Energy and Power

In this Case Study we will develop the ability not only to quantitatively distinguish between energy and power, and to calculate energy and power for a number of important systems, but also to "model" a system by breaking a physical system into the minimum number of subunits so that both the energy and the power can be calculated for the functional operation of that system.

CASES:

Case 1: An Electric Light Bulb

It is important for an analysis of each system we consider to "model" that system as a translation of the physical system as it appears upon visual inspection into a *schematic* that captures the essential features of the device under consideration. We do that in Figure CS1.1a for the light bulb wherein we convert the bulb into a sketch that captures the flow of electrons into a filament that is heated to a high temperature producing *visible light* from the filament and *heat* that emerges in the infrared. The electrons flowing out of the filament are returned to the power station to complete the circuit.

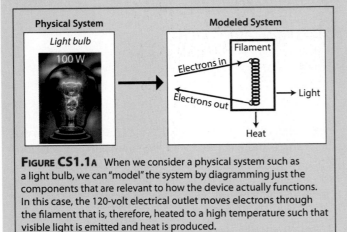

FIGURE CS1.1A When we consider a physical system such as a light bulb, we can "model" the system by diagramming just the components that are relevant to how the device actually functions. In this case, the 120-volt electrical outlet moves electrons through the filament that is, therefore, heated to a high temperature such that visible light is emitted and heat is produced.

But to "power" this light bulb you must purchase energy from the "power company," and your monthly bill will arrive with a statement of the cost per unit of energy and a charge for each unit of that energy. You pay for the energy used, independent of the rate at which you use it. But the bill will read an amount of energy in kilowatt-hours (designated kWh). What does this mean? Inspection of the unit kWh reveals that this is power (kilowatts) multiplied by time (hours) so the quantity is (Power)(time) = Energy. Therefore, the bill is indeed for the purchase of units of energy. But how large is the energy unit: kWh? One kilowatt is 10^3 watts or 10^3 J/sec. One hour is 3.6×10^3 seconds so

$$1 \text{ kWh} = (10^3 \text{ joules/sec}) \, 3.6 \times 10^3 \text{ sec} = 3.6 \times 10^6 \text{ joules}$$

Thus the kWh unit is 3.6×10^6 times larger than the joule, and it is a unit far more appropriate than the joule for measuring energy involving *macroscopic* systems. The other advantage of the kWh is that it stands as a constant reminder that energy is power multiplied by time—just as power is energy divided by time!

So if our 100 watt light bulb is on for a period of one hour, it will use an amount of energy equal to

$$(100 \text{ watts}) 1 \text{ hour} = (0.1 \text{ kilowatts}) \, 1 \text{ hour} = 0.1 \text{ kWh}$$

If you live in New York this will cost (electricity is 17¢/kWh) 1.7 cents, if you live in Washington State it will cost 0.7¢, and if you live in California, 1.4¢. If you left that single 100 watt light bulb on for a year, it would cost $149 in New York, $75 in Washington State, and $123 in California.

Case 2: A Gas Furnace

For the case of a gas furnace, the energy contained in the bond structure of natural gas, CH_4, and molecular oxygen, O_2, *relative* to the energy contained in the bond structure of the products, water and carbon dioxide, is released as heat in the "burner" of the gas furnace. In a typical home furnace, this combustion of natural gas produces 20,000 joules/sec or 20 kW of power. The 20 kW of power produced by the combustion of natural gas in the furnace appears as *thermal energy* or heat, to be used to warm the house, heat hot water, etc. If the furnace is "too small" for the house, it means that the heat loss from the house through its exterior is "too fast" to allow a furnace of a given *power* (energy per unit time) to maintain the house at a comfortable temperature. We can model the gas furnace by converting the physical system as it appears on visual inspection into a schematic that captures the essential components of the combustion of natural gas (methane or CH_4) in oxygen to produce a flame that converts the molecular structure of methane and oxygen to the molecular structure of water and carbon dioxide. The link between the physical presence of the furnace and the schematic sketch that models the key features of the energy conversion is shown in Figure CS1.1b.

Natural gas is typically purchased from the gas company in units of the "therm" that corresponds to 100 cubic feet of gas delivered at a temperature of 20°C at a pressure of 14.7 pounds/square inch (one atmosphere pressure). The energy content of this amount of natural gas is 1.06×10^8 joules or 2.9×10^1 kWh, which can be rounded to 30 kWh for the purpose of comparison. For example, natural gas sold on the commodities market at the beginning of 2013 for $0.35/therm so that in terms of the cost per unit of energy, this corresponds to 30 kWh/$0.35 or 1.2 cents per kWh, markedly less than the cost of an equivalent amount of energy delivered to your house in the form of electricity! On the other hand, a liquid fuel such as heating oil, with an energy content of approximately 10 kWh/liter, costs about $1 a liter. Thus, 30 kWh of energy from fuel oil costs approximately $3, whereas the same amount of energy from natural gas costs $0.35—a difference of nearly a factor of ten.

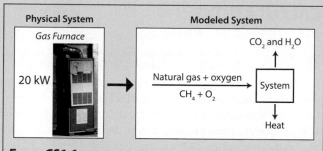

FIGURE CS1.1B   A gas furnace employs a burner that simply controls the mixing of natural gas (primarily methane, CH_4) with oxygen to form carbon dioxide, CO_2, and water, H_2O. The energy contained in the bonding structure of CH_4 and O_2 relative to the energy contained in the bonding structure of CO_2 and H_2O is released as heat. That heat is then distributed to warm the house. The essentials of this conversion of chemical energy to heat, which we will explore in a subsequent chapter, are represented in the simplified model in the right-hand panel.

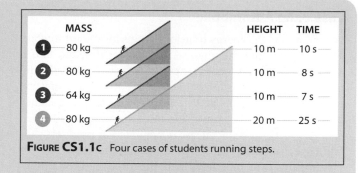

FIGURE CS1.1C   Four cases of students running steps.

Case 3: A Human Being at Rest

How much *power* is required to sustain a human being? We can calculate this quickly by recognizing that the caloric content of the food we eat goes primarily to maintaining our body temperature of 38°C. This is true unless we engage in significant exercise. To calculate the power required to sustain us we simply need to divide the (caloric) energy contained in what we eat per day by the number of seconds in a day. A typical person consumes about 2000 Calories per day. The Calorie unit (with a capital C) is equal to 10^3 regular calories, so 2000 Calories is 2×10^6 calories. There are 4.2 joules in a calorie.

We can model the case of energy transformations for a human being by again distilling the human organism into a schematic of the energy conversion processes. The intake of food is represented by the C_xH_y hydrocarbon molecule, "food." When the hydrocarbon is burned by reacting it with oxygen carried in the blood stream by the hemoglobin molecule, the reaction produces the heat to sustain a body temperature of 38°C with the chemical products of combustion exhaled as CO_2 and H_2O.

Case 4: Human Being Running Steps

We can demonstrate the distinction between energy and power while at the same time reviewing the relationship between work and potential energy by considering the following problem. Four students run up the stairs in the time shown. In the following calculational exercises, we will determine the energy and power for each case. Those cases are displayed in Figure CS1.1c.

Case 5: Human Being on a Bicycle

If we put the human being on a bicycle, how much power does a person typically deliver to the bicycle to propel it? The answer to this question depends upon a number of factors: how fast the bike is ridden, the physical training and conditioning of the rider, how steep the hill, how much the bike and rider weigh, how much head wind, etc. But here we get considerable help from the equation relating power, energy per unit time, work per unit time, force, and velocity. Specifically:

$$P = \Delta E/\Delta t = \Delta W/\Delta t = F\,\Delta x/\Delta t = Fv$$

Let's assume for this calculation that the bike is ridden on the flat (we can extend this quite easily to riding up hill). We can again convert the physical reality of the bike and rider into a model of the system that includes the input of the hydrocarbon fuel—the food eaten by the rider—the work performed to move the bicycle forward, the heat generated within the rider's body and by the frictional losses associated with the motion of the bicycle, and the products of respiration, CO_2 and H_2O.

Thus, for a given speed, v, of say 20 km/hr (5.6 m/sec), the force required to sustain this speed is made up of the rolling friction of the bicycle and the resistance resulting from moving through the atmosphere. The rolling resistance of the bicycle is independent of speed, but the air drag is proportional to the *square* of the speed. Those two factors are plotted in Figure CS1.1d for a typical bicycle. We will use this relationship between energy consumption and speed in the next section.

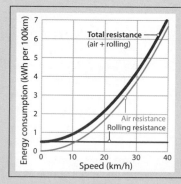

FIGURE CS1.1D   Representation of the energy consumption of a bicycle per 100 km assuming the drag area of the cyclist is 0.75 m² and a cyclist plus bike mass of 90 kg. In subsequent case studies, we will derive how to calculate rolling resistance and air resistance (drag) from first principles.

Case 6: An Automobile

We can model the energy transformations for the automobile by again distilling the physical presence of the auto into a diagram that represents the input of octane (C_8H_{18}), the intake of O_2 for combustion of the fuel, the production of work to propel the automobile, the release of heat as a by-product of inefficiency in the conversion of chemical energy to mechanical work, and the release of combustion products CO_2 and H_2O from the exhaust pipe of the automobile.

For this case, on a level road, the force required to sustain a given speed (let's choose 100 km/hr) must be sufficient to overcome both the rolling friction of the tires and the air drag. For a typical automobile of weight 1500 kg and frontal area 2.25 m², the rolling resistance and air drag are displayed in Figure CS1.1e in terms of the energy consumption per unit of distance. This provides the necessary information to calculate the required power.

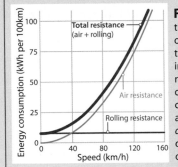

FIGURE CS1.1E Representation of the energy consumption of an automobile per 100 km traveled. The calculation takes into account the rolling resistance of the car and the air drag resulting from pushing the car through the atmosphere. It is assumed that the product of the *drag coefficient* and *frontal area* of the car is 1 m², and the mass of the car is 1500 kg.

CALCULATIONS:

Problem 1

Consider Case 3. Sketch the link between the physical system (for the case of a human) and the modeled system, but with the "fuel" represented by the hydrocarbon molecule (food) C_xH_y. Calculate power required to sustain a human at rest if the person consumes 2000 calories per day.

Problem 2

Consider Case 4. Calculate the energy required for each example (a to d), then calculate the power expended. Now calculate the energy required in example (d) and compare that with the daily energy consumption of a human being at rest. Rank, from largest to smallest, the power outputs, P_a to P_d, in Figure CS1.1c.

Problem 3

Consider Case 5. Calculate the power to sustain a speed of 20 km/hr in units of kW and horsepower.

Problem 4

Consider the analysis of an automobile powered by an "internal combustion" engine—an engine that uses gasoline or diesel as a fuel.

First construct a diagram of the modeled system that captures the key aspects of the automobile that convert the chemical energy of the fuel to mechanical work. Next calculate the power demand to sustain a cruise speed of 100 km/hr on a flat road in both horsepower and kW. For a typical automobile, the drag generated by the transmission constitutes a loss of approximately 15%.

CASE STUDY 1.2 Linking Energy, Work, and Power in Contemporary Systems

KEY CONCEPTS:

1. Conversion of Energy among Different Types of Energy

The law of the conservation of energy requires that we keep quantitative track of all forms or types of energy contained in the system under consideration. Thus when we express the "conservation of energy," this means that the *total* energy of the system, E_T, does not change such that $\Delta E_T = 0$. It is essential that we include all categories of energy in our calculation of that total energy, so that for example the total energy of a system that contains mechanical energy, chemical energy, nuclear energy, electromagnetic energy, and thermal energy can be expressed as $E_T = E_{mech} + E_{chem} + E_{nuc} + E_{EM} + E_{therm}$. Then while we know that $\Delta E_T = 0$, we also know that we can accomplish work by converting one form of energy to another, say electromagnetic energy (e.g. sunlight) to mechanical energy to fly a solar powered aircraft. Or we can convert chemical energy (e.g. gasoline) to mechanical energy to propel an automobile.

As we will see in our development of thermodynamics, electrochemistry, quantum mechanics of molecular bonding, kinetics, and nuclear chemistry in subsequent chapters, modern technology is developing ever more sophisticated ways of interconverting energy to provide the energy needed to accomplish a particular task or to sustain the human endeavor more broadly, in increasingly sustainable ways. This Case Study emphasizes (a) the pathways available for interconverting light to electricity, (b) the remarkable efficiency and low cost of electricity versus gasoline for transportation, as well as (c) the ability to generate a major fraction of U.S. energy needs from solar power.

2. Placing Different Types of Energy on the Same Absolute Scale

We know from our discussion of the mechanical equivalent of heat that once we have "calibrated" the amount of potential energy in joules required to heat a mass, m, with heat capacity, C, through a change in temperature, ΔT, we can immediately measure the amount of energy contained in a stream of photons from the Sun, or the amount of energy released in burning a certain amount of natural gas. The diagram at the right, from Figure 1.13, shows the example of how the energy in a beam of sunlight can be determined directly by measuring the increase in temperature using the same system (similar to the one used by James Prescott Joule) used to measure the energy release from a mass, m, falling a distance, Δh, through a gravitational field where $\Delta E_{mech} = mg\Delta h = q_{heat} = C\Delta T$.

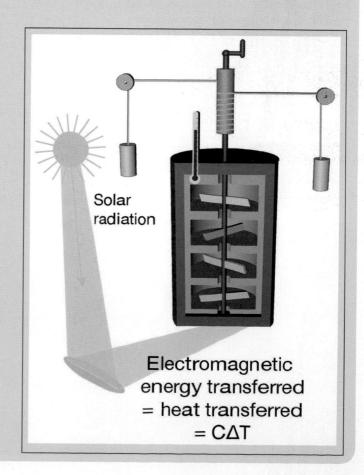

Solar radiation

Electromagnetic energy transferred = heat transferred = CΔT

CASES:

The objective of this case study is to establish a foundation for problem solving that involves the analysis of larger, coupled problems. These require the breaking down of that problem into its component parts, linking those parts to the appropriate physical and chemical models that then allow the scientific principles to be applied that can solve each component of the problem. This is then followed by reconstructing the components of the problem in order to solve the overall problem.

Consider the following problems.

Case 1:

The development of new materials that are stronger per unit mass than steel, and even aluminum, has ushered in a new era of lightweight composites that are opening new opportunities in transportation systems and flight systems. For example lightweight, solar-powered pilotless aircraft can fly for as long as the lighting conditions permit. They are being considered for remote surveillance, communications, and reconnaissance uses. In addition, as we will see in subsequent chapters, new materials allowing much greater energy storage per unit mass in batteries provide the opportunity to sustain flight for solar-powered aircraft through the night providing continuous flight for weeks.

We can model such an aircraft by representing the physical system as a schematic as we did in Case Study 1.1. An example is shown here:

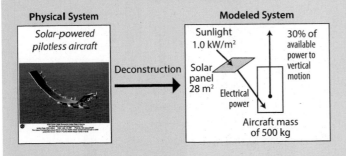

Case 2:

Chevy Volt

Chevrolet has recently introduced the Chevy Volt, a new type of electric car that can extend its range using a gasoline engine to charge the on-board lithium ion batteries for long-range driving. For short trips, the Chevy Volt is powered solely by electricity. It has a set of rechargeable batteries (16 kWh capacity) that provide electricity to the car's motor. The Chevy Volt uses 70% of the battery output before it switches to the gas engine, thus providing an electric motor-only range of about 40 miles.

The Chevrolet Volt represents the new class of "plug-in hybrids" distinguished by the fact that they can drive for extended distances without using the on-board gasoline engine.

Case 3:

The United States is endowed with a remarkable mix of wind power, solar power, geothermal power, and hydroelectric power. While we will analyze each of these we consider here the case of solar power and we consider two types: photovoltaics (PVs) and concentrated solar thermal. Photovoltaics (shown here) produce electricity directly from sunlight with an efficiency of 15–20%.

Concentrated solar thermal uses large linear parabolic reflectors to concentrate the solar energy onto a transparent tube at the focal point that raises the fluid temperature to > 300°C. That high temperature fluid is then used to drive a steam turbine to generate electricity. Concentrated high temperature solar thermal typically achieves efficiencies of ~30%.

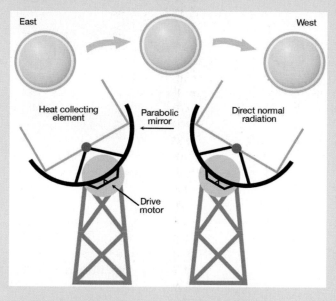

We can then use yearly average observed solar power received at specific locations to calculate the area required to generate any given amount of electrical power.

Displayed here is a sampling of yearly average sunshine in units of W/m².

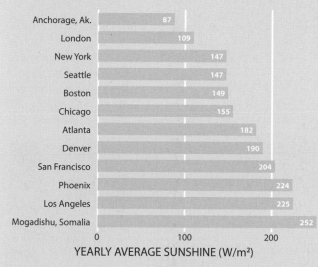

YEARLY AVERAGE SUNSHINE (W/m²)

PROBLEMS:

Problem 1

Consider Case 1:

- Consider a 500 kg solar-powered aircraft carrying 28 m² of solar cells with 15% efficiency. The solar constant (power available from the Sun per unit area) is 1.0 kW/m². Assume that 30% of engine power is spent on vertical climb; the rest goes into horizontal propulsion. Determine the maximum vertical speed of this aircraft.

- If the vertical component of the wind velocity fluctuates about zero with an amplitude of +/- 0.2 m/s, can the solar aircraft take off safely on its own? Compare the result with the rate of climb of a conventional two-seat Cessna 150 airplane, 3.4 m/s.

Problem 2

Consider Case 2:

- On average, Americans drive about 29 miles a day. Given that the electricity prices are about 11 cents per kWh, what is the average cost of electricity (in dollars) to run this electric car for a year?

- For the Chevy Cavalier, which gets 26 miles/gallon, with gas costing $4.00/gallon, for a year of driving, what is the cost of fuel? If gasoline costs $3.00/gallon?

- The Volt uses 11.2 kWh of energy to go 40 miles. If the efficiency of the electricity generating power plant is 50% for conversion of chemical energy of the fuel to electricity and 5% of the electrical energy is lost getting the electricity from the generating plant to the home for charging the battery of the Volt, how much chemical energy is required to drive the Volt 40 miles?

- How much chemical energy is required to drive the Cavalier 40 miles? (Hint: the energy content per liter for gasoline is approximately 10 kWh/liter, which is discussed in detail in Case Study 1.3.)

Problem 3

Consider Case 3:

The United States uses approximately 3 terawatts (TW) of power to sustain its economic system.

- If one-third of this total power consumption is generated by concentrated solar thermal, what collective area is required if we assume 30% efficiency and 200 W/m²?

- The area of the state of Arizona is 2.9×10^5 km² or 2.95×10^{11} m². What fraction of the area of the state is required?

- If we generate 1 TW of electrical power by PVs, what fraction of the state of Arizona is required?

CASE STUDY 1.3   Development of the Scales of Energy and Power

KEY CONCEPTS:

1. Spanning the Range of Orders of Magnitude in Energy and Power

It is often said that what distinguishes a scientist from a non-scientist is the ability to estimate, or directly calculate, the order of magnitude of the energy (or power where appropriate) associated with important natural phenomena and important systems that are central to the global economic structure. This ability to calculate orders of magnitude in energy or power requires first that we can calculate the *range* of the orders of magnitude spanned by these phenomena, and then that we can successfully fill in the intermediate orders of magnitude for important intervening cases so as to construct a coherent sequence.

To this end we first analyze the upper limit of our range of energy and power by considering the Sun, we analyze the lower bound of the energy scale by considering a single nitrogen molecule at room temperature, and important cases in between.

2. Energy Emitted by the Sun

The Stefan–Boltzmann law provides a remarkably simple way to calculate the *rate* of energy emission (thus the power) produced by the Sun. The Stefan–Boltzmann law is given on the top of page 37 wherein it states that the energy emitted per unit of time from a surface at temperature T is $\Delta E/\Delta t =$ (area of the emitting surface) $\sigma T^4 = A \sigma T^4$ where σ is the Stefan–Boltzmann constant: $\sigma = 5.67 \times 10^{-8}$ J/s·m²·K⁴ or, because 1 J/s =1 W, $\sigma = 5.67 \times 10^{-8}$ W/m²·K⁴. An examination of the units of the Stefan–Boltzmann constant reveals an important fact: when the Stefan–Boltzmann constant is multiplied by the area of the emitting surface times the fourth power of the temperature in Kelvin, the *power* (energy per unit time) emitted by the surface is directly calculated. Thus we can, for example, immediately calculate the power emitted by the Sun if we know its surface temperature (5800 K) and its area; or we can calculate the power emitted by the Earth to space (temperature of the outer atmosphere 255 K) or the power emitted by the human body (311 K).

3. Energy of a Single Nitrogen Molecule

One of the most important and versatile expressions in the physical sciences is the equation, derived on page 29, relating the average kinetic energy per molecule, ε, to the temperature in Kelvin, namely $\varepsilon = 3/2\, k_B T$ where k_B is the Boltzmann constant equal to 1.38×10^{-23} J/K·molecule. There are a number of important aspects to this expression. First is the

remarkable simplicity of the expression itself. Second, note that it does not contain the mass of the molecule. This is a manifestation of the "equipartition of energy"—the principle that all molecules in collisional contact and at equilibrium have the same amount of energy.

4. Energy Contained in One Liter of Liquid Fuel

A vast proportion of the systems that sustain modern society—automobiles, aircraft, ships, rockets, etc.—derive the energy required to operate from liquid fuels. While those liquid fuels differ somewhat in the details of how much energy is contained in a liter, they are all reasonably close to an easy number to remember: one liter of liquid fuel contains ~10 kWh of energy when reacted with oxygen (i.e. burned) to form $CO_2 + H_2O$.

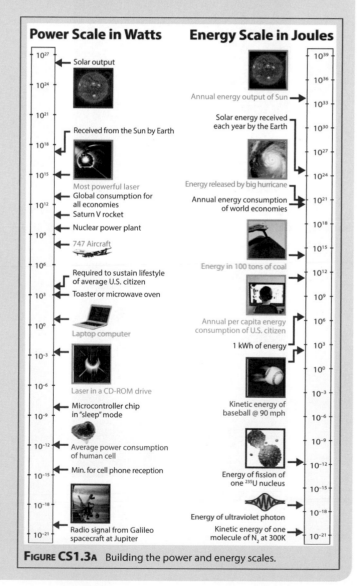

FIGURE CS1.3A   Building the power and energy scales.

CASES:

Case 1: The Sun

While we will develop the subject of nuclear reactions more fully in Chapter 13, we note here that energy is generated at the core of stars such as our Sun by the conversion of hydrogen to helium in a sequence of *nuclear reactions* (i.e. reactions that change the structure of the nucleus). These reactions are classed *fusion* reactions because they *join* lighter nuclei to form heavier nuclei, thereby converting mass to energy according to Einstein's equation $E = mc^2$. The fusion reactions occur at the core of the Sun, at temperatures in excess of 1×10^8 K, 100 million K. As a result, that energy is radiated from the Sun's surface, which approximates a blackbody surface.

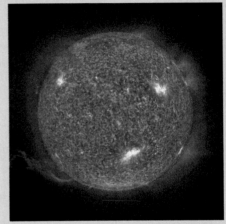

We can take the physical entity of the Sun displayed above, and model it in order to calculate the energy per unit time emitted to space. That schematic of that model is

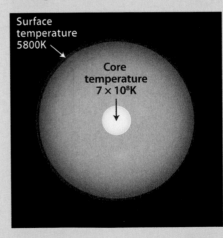

Surface temperature 5800K

Core temperature 7×10^8K

The radius of the Sun is 7.0×10^8 meters.

Case 2: Energy of a Single Nitrogen Molecule

Review the derivation of the average kinetic energy of a molecule in a gas that is presented on pages 28 and 29. That simple but powerful microscopic (i.e. molecule by molecule) derivation has important consequences at the macroscopic level.

A simple logic that emerges from the mechanical equivalent of heat coupled with basic assumptions concerning the theory of molecular motion in gases provides a very general relationship between the average kinetic energy, ε, of the individual atoms or molecules that make up a macroscopic body and the temperature, T, of that macroscopic body. The chain of logic begins with the observed fact that the temperature change, ΔT, of an ensemble of molecules that comprise a macroscopic body is proportional to the heat added—thus to the energy added—to that body. Combining this with the assumption that all collisions at the molecular level are elastic (that is, they do not change the total energy of the ensemble of molecules) and that the momentum change in any molecular collision with a vessel wall results in a momentum change ΔP = –2mv, results in the determination that the average energy per atom or molecule, ε, is just $\varepsilon = \frac{3}{2} k_B T$ where k_B is Boltzmann's constant and T is the temperature in Kelvin.

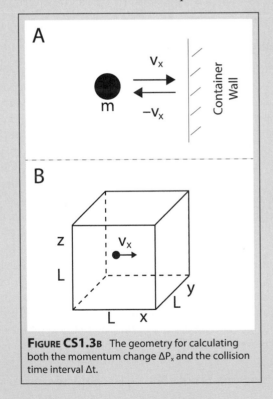

FIGURE CS1.3B The geometry for calculating both the momentum change ΔP_x and the collision time interval Δt.

Case 3: Energy Content of Fossil Fuels

As it turns out, we are greatly aided in the calculation of energy scales by the fact that 80% of the global primary energy production is from fossil fuels, and fossil fuels are all hydrocarbon based. While the structure of those fuels differs—natural gas is mostly methane, CH_4; gasoline is mostly octane, C_8H_{18}; and coal is a complicated structure, of crosslinked carbon and hydrogen (see Case Study 2.3 and 2.5)—all fossil fuels are composed primarily of carbon-hydrogen bonds so to first order, their *energy content per unit mass* will be roughly equivalent. This is also true of the hydrocarbons we eat such as peanut butter, etc. Thus we can calculate by direct analogy the energy content of those substances per unit mass.

If we examine the label of chunky peanut butter container we see that it contains approximately 1700 Calories per cup, and so a cup of peanut butter has

$$\left(\frac{1700\ \text{Cal}}{260\text{g}}\right)\left(\frac{1000\text{g}}{\text{kg}}\right)\left(\frac{10^3\text{cal}}{\text{Cal}}\right) 4.2\ \text{joules}/\text{cal} \approx$$

$$30\ \text{MJ/kg} = \frac{30 \times 10^6 \text{J/kg}}{3.6 \times 10^6 \text{J/kWh}} = 8\ \text{kWh/kg}$$

We also know that oil, gasoline, and peanut butter float on water, and water has a density of 1 kg/liter, so we can estimate that peanut butter has a density of 0.9 kg/liter so it has an energy content *per liter* of

$$(8\ \text{kWh/kg})(0.9\ \text{kg/liter}) = 7.2\ \text{kWh/liter}$$

This represents a typical energy-per-liter figure for many hydrocarbons. Gasoline has a somewhat higher energy content per liter, 9.3 kWh/ℓ, which we will approximate as 10 kWh/ℓ.

CALCULATIONS:

Problem 1

Consider Case 1:

- The radius of the Sun is 7.0×10^8 meters. If the surface temperature of the Sun is 5800 K, how many joules per second does the Sun emit to space?

- How many joules of energy does the Sun produce in a year?

- The Earth is 1.5×10^{11} meters from the Sun and has a radius of 6.4×10^6 meters. How much energy does the Earth receive from the Sun each year?

Problem 2

Consider Case 2:

- Calculate the average kinetic energy of a gas phase **nitrogen** molecule (N_2) at room temperature (295 K) and at 1000 K. What is the average velocity of that N_2 molecule at 295 and 1000 K?

- Calculate the average kinetic energy of a gas phase **mercury** atom at room temperature and at 1000 K. What is the average velocity of that Hg atom at 295 and 1000 K?

Problem 3

Consider Case 3:

An Automobile: Energy content of fuel tank, kinetic energy of motion, and power.

The first step is to transpose the physical system of the automobile into a model that captures the elements of the system related to the objective of the calculation. This is done in the diagram:

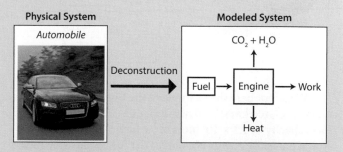

How much energy is stored in the fuel tank if the tank capacity is 60 liters (15.2 gallons)?

If the energy contained in the fuel tank is converted to mechanical energy to move the automobile with 100% efficiency and the force required to move the auto at 100 km/hr is 800 kg m/sec², how far will the auto go on a full tank?

In practice, the car can only travel 500 km before running out of fuel. What happened? What happened is that we calculated the distance traveled using the *assumption* that 100% of the energy contained in the chemical energy of the fuel was converted to the work needed to move the car at 100 km/hr. In fact, an automobile typically converts only about 15% of the energy contained in the fuel into mechanical energy to move the vehicle. Thus we must multiply our distance traveled by the efficiency with which the chemical energy of the fuel is converted to mechanical energy in order to properly calculate the distance traveled on a full tank of fuel. If the conversion of chemical energy to the mechanical energy to propel the car is only 15 percent, how far will the car go on a full tank of gas? Where was the energy lost? Most of the energy was lost in the conversion of chemical energy to mechanical energy in the engine itself—a subject that we will study in Chapter 4. The remainder was lost to friction in the transmission and differential of the car's drive train. The bottom line, however, is that 85% of the chemical energy is wasted; only 15% is available to move the car! This has, as we will see, important consequences when we set an energy strategy for the future.

Finally, if the automobile is moving at a speed of 100 km/hr, how much kinetic energy does the automobile have and what fraction is that of the chemical energy stored in a full fuel tank?

Problem 4

A Jet Liner
Again the first step is to transpose the physical system of the jet aircraft into a model that captures the elements of the system related to the objective of the calculation.

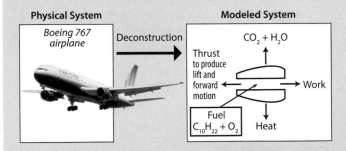

We adopt here the example of the Boeing 767 that has a range of 11,400 km (6,150 nautical miles) (767-400ER) and a fuel capacity of 1.1×10^5 liters of jet fuel.

What is the energy content of the fuel in the fully fueled aircraft in units of joules and of kWh?

If the aircraft operates at a cruise speed of 850 km/hr (530 mph) and flies 10,000 km, what is the average power delivered to propel the aircraft if the jet engine efficiency is 30%?

Problem 5

We have seen that the power output of the Sun is enormous, 3.9×10^{26} W. Even so, solar energy accounts for less than 2% of the total world energy consumption. However, there are situations when solar power is the only viable source of energy. For instance, the Mars rovers Spirit and Opportunity run exclusively off of solar energy collected by their solar panels. They convert the absorbed solar radiation into electricity that is stored in their two lithium batteries, allowing them to operate at night or when the sun is low in the sky.

a. Mars is somewhat smaller than the Earth and further from the Sun. The average distance of Mars from the Sun is about 2.3×10^8 km and its radius is 3397 km. How much power does Mars receive from the Sun?

b. The rovers are equipped with high efficiency solar cells that can operate with over 30% efficiency. The rovers are 1.6 m long and 2.3 m wide, and half of that area is covered by solar panels. From this information, estimate the amount of power that can be generated when the solar panels of a rover are fully illuminated.

TAKING THE NEXT STEP

CASE STUDY 1.4 Application of the Scientific Method—The "UV Catastrophe"

Science has two fundamental objectives: To explain natural phenomena and to use the understanding of such phenomena to predict future events.

We begin with the Greeks; Aristotle brought *deductive reasoning* to the forefront of disciplined thought. *Deduction* is the process of reasoning in which a conclusion follows necessarily from the stated premise (i.e., inferences by reasoning from the general to the specific). A simple example of deduction in mathematics is that *if* a = b and b = c, *then* a = c. While *deduction* is a category of logic, it is not sufficient for the advancement of scientific knowledge. Aristotle recognized this—he put forward the assumption that nature was built from four fundamental substances: earth, fire, air, and water. All substances, all materials, were thereby formed from combinations of these four components.

Some two thousand years later, the structure of the scientific method began to explicitly emerge in the 1600s with the work of Galileo, Francis Bacon, and Robert Boyle. In the eighteenth and nineteenth centuries David Hume, Thomas Young, Isaac Newton, and Sadi Carnot advanced the sophistication of the scientific method, as outlined in the Chapter Core. A central tenet of the scientific method is that careful, repeatable *observations* of natural phenomena or events constitutes the initiation of a chain of logic that in turn develops a strategy for a concise, coherent and testable theoretical structure to develop, that in turn builds a scientific framework.

The approach uses the logic of *inductive reasoning* wherein, with no initial assumptions, a series of observations have been made such that a pattern begins to develop leading to a generalization: a law that explains the ensemble of observations. One of the most famous examples of the scientific method—and the process of reasoning by *induction*—was the work by Copernicus (1473–1543) in which he made an extensive series of astronomical observations that proved that Earth (along with the other planets) revolved around the Sun rather than the belief of the day that all astronomical bodies rotated about Earth.

But the *scientific method* represents a dynamic exchange between these observations that provide a pattern for behavior of natural processes and the development of a theory—or a *model*—that seeks to articulate the specifics of a given natural phenomenon. The dynamic component of the scientific method lies in the creation of a *hypothesis*, emergent from the observations, that provides a tentative explanation for the pattern contained in those observa-

tions. The purpose of the hypothesis in the practice of the scientific method is both to capture an explanation of the phenomenon in concise form *and* to clarify what observations, and experiments, would be most powerful for testing the veracity of the hypothesis. It is commonly recognized that the hypothesis may well require modification resulting from the development of more rigorous experimental techniques.

Building on this dynamic exchange between observation and hypothesis is the development of a *theory* that represents *both* (1) a union of hypotheses to more fully represent laws guiding the structure and function of natural phenomena and (2) a foundation for prediction of other phenomena and/or for what will happen in the future. This linkage is represented in Figure CS1.4a.

While the scientific method is a construct that must be adapted to various categories of the scientific endeavor, it is nonetheless a process of immense importance today, just as it was four centuries ago as the techniques of modern science emerged on a global scale. In fact, there are examples in contemporary science where the lack of vigorous appreciation for the scientific method has led to serious mistakes associated with the national research endeavor in fields as disparate as chemical biology, climate, and materials research.

A particularly important example of how the scientific method serves to guide scientific progress is represented by the example of Max Planck, who developed the hypothesis

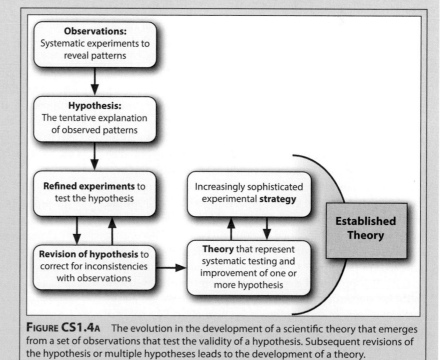

FIGURE CS1.4A The evolution in the development of a scientific theory that emerges from a set of observations that test the validity of a hypothesis. Subsequent revisions of the hypothesis or multiple hypotheses leads to the development of a theory.

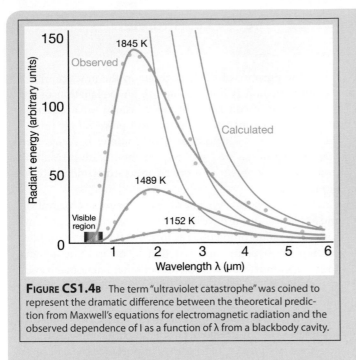

Figure CS1.4b The term "ultraviolet catastrophe" was coined to represent the dramatic difference between the theoretical prediction from Maxwell's equations for electromagnetic radiation and the observed dependence of I as a function of λ from a blackbody cavity.

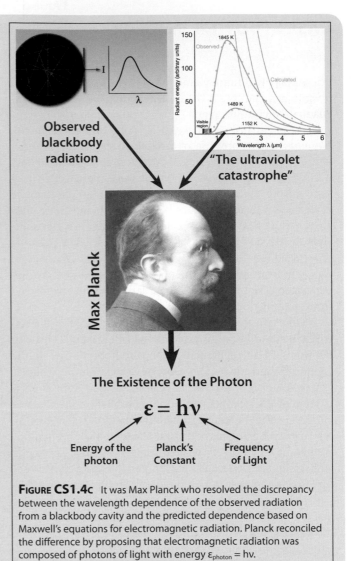

Figure CS1.4c It was Max Planck who resolved the discrepancy between the wavelength dependence of the observed radiation from a blackbody cavity and the predicted dependence based on Maxwell's equations for electromagnetic radiation. Planck reconciled the difference by proposing that electromagnetic radiation was composed of photons of light with energy $\varepsilon_{photon} = h\nu$.

that electromagnetic radiation (light) consisted of discrete packets of energy—or quanta of light. The reason that this example is so important is that the triumph of 19th century physics was the concise development of the mathematical equations describing the propagation and behavior of electromagnetic radiation—Maxwell's Equation, named after James Clark Maxwell. These equations described electromagnetic radiation in terms of electromagnetic waves. The wave nature of electromagnetic radiation was a universally accepted cornerstone of physics as the nineteenth century gave way to the twentieth.

Yet the *wavelength dependence* of the intensity of radiation emitted by a blackbody held key information that was to revolutionize the physical sciences. It was the *shape* of the intensity, I, versus wavelength displayed in Figure CS1.4b that presented a very serious problem for Maxwell's theory of electromagnetism. In particular, Maxwell's wave formulation of electromagnetic radiation predicts that the energy density of radiation within the blackbody cavity should be proportional to $1/\lambda^4$. If that were true, the intensity of radiation emitted by the black body would increase without limit as $\lambda \to 0$. This in turn meant that rather than glowing red, a blackbody should emit violet light. As it gets hotter, it would simply become a brighter and brighter *blue*. The *prediction* of the wavelength dependence of emitted radiation according to Maxwell's formulation of electromagnetic theory is superimposed on the *observed* wavelength dependence in Figure CS1.4b. This profound departure between the observed and predicted dependence of the intensity emitted by the blackbody cavity was so serious that at the turn of the 20th century it was universally referred to as the "ultraviolet catastrophe."

It was Max Planck, a young German physicist, who proposed in 1900 a solution that was to be the first step in the revolution of modern chemistry and physics. Planck proposed that the radiant energy of light comes in discrete packets or bundles called **quanta** each of which possesses electromagnetic energy ε proportional to the frequency, ν, of the light such that:

$$\varepsilon = h\nu$$

The proportionality constant h, now called **Planck's constant**, has a value $h = 6.626 \times 10^{-34}$ J·sec.

What was transformational about this hypothesis put forward by Planck in 1900, as summarized in Figure CS1.4c, was that it implied a wave-particle duality for electromagnetic radiation. That is, light possesses both the properties of waves as described by Maxwell's theory, *and* the properties of particles. This *wave-particle duality* formed the first cornerstone in the formulation of quantum mechanics that constitutes the foundation of modern theories of chemical bonding and in the understanding of the interaction of these photons with the structure of atoms and molecules.

What is remarkable is that Planck himself was not convinced that his own hypothesis was correct—it was such a radical idea that it had little chance of being accepted by the physics community. Yet by inductive reasoning based on the observed dependence of intensity on wavelength emitted from blackbodies combined with the simple representation of the quanta of energy as $\varepsilon = h\nu$, the observed wavelength dependence of the intensity of blackbody radiation exactly matched the *calculated* wavelength dependence of blackbody radiation as displayed in Figure CS1.4c. It was a number of years before additional evidence was gathered (the "photo-electric effect") that lead to the acceptance of Planck's hypothesis that electromagnetic radiation is comprised of discrete packets (quanta) of radiation. It represents a triumph of the scientific method.

CASE STUDY 1.5 Calculating Global Energy and Power Consumption Now and in the Future

KEY CONCEPTS:

1. Conservation of Energy

In any calculation of energy or power consumption, particularly for an array of energy transformations intrinsic to the global economic endeavor, it is important to review again the Law of the Conservation of Energy. One way of reviewing and rethinking this critically important law is to consider the numerous ways the law is articulated. We can gain insight into the law by considering carefully the perspective offered by the different formulations or expressions of the law.

Example 1:

Richard Feynman has famously said: "It is just a strange fact that we can calculate some number, and when we finish watching nature go through her tricks and calculate the number again, it is the same. It is important to realize that in science today, we have no knowledge of what energy *is*. There is a fact, or if you wish a law, governing all natural phenomena that are known to date. There is no known exception to this law—it is exact as far as we know. The law is called the *conservation of energy*."

Example 2:

Realizing that energy occurs in many different forms: mechanical, electrical, chemical, nuclear, electromagnetic, thermal, etc., the principle of the conservation of energy—that energy is neither created nor destroyed—requires that the definition of total energy encompass all forms of energy that can be transformed from one form to another within the system under consideration. Thus $E_T = E_{mech} + E_{chem} + E_{nuc} + E_{EM} + E_{therm}\ldots$. So while energy can be readily exchanged among and between these categories, total energy, E_T does not change so $\Delta E_T = 0$.

That is, while a particular form of energy, say chemical energy contained in the bond structure of gasoline (C_8H_{18}) and oxygen (O_2) *relative* to the chemical energy of CO_2 and H_2O is used to propel an automobile, that change in chemical energy, ΔE_{chem}, appears as a change in mechanical energy, ΔE_{mech}, and as heat, ΔE_{therm}. But if we carefully add all the ΔEs together, they sum exactly to zero.

Example 3:

One of the first formulations of the law of the conservation of energy was by William Thomson (Lord Kelvin) in the latter half of the 19th century: "Heat and work are both forms of energy. In any process, energy can be converted from one form to another, but it is never created nor destroyed."

2. Calculating Global Scale Energy and Power Consumption

One of the most important capabilities of the scientific approach to important problems is the ability to represent a vast number of individual events in a simple but accurate formulation. A key example is the calculation of global primary energy consumption by all humans on Earth—a daunting task. Primary energy consumption is the energy extracted from its initial source such as natural gas, gasoline, coal, nuclear, hydroelectric, wind, solar, etc. The analysis begins with the reasonable assumption that the total global primary energy consumption (TGEC) is proportional to global population (Pop). The more difficult question is how to relate population to energy consumption. We can measure per capita income (PCI) in the majority of countries, as well as the energy consumption of each country, and we can estimate the energy required to produce a given dollar of gross domestic product termed the primary energy intensity (PEI). Thus TGEC is just the product of population times per capita income times the primary energy intensity. Therefore TGEC = (Pop)(PCI)(PEI).

CASES:

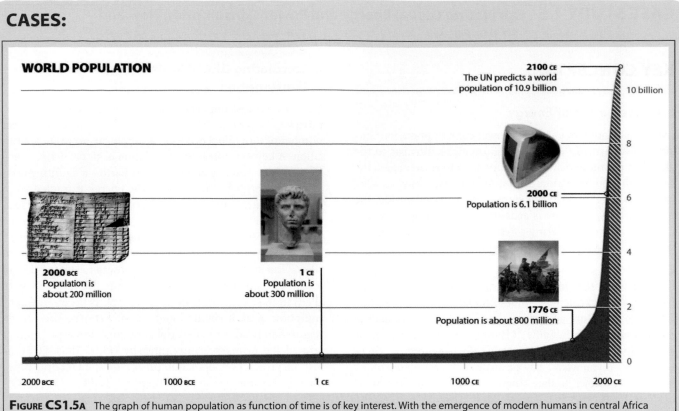

WORLD POPULATION

2000 BCE
Population is
about 200 million

1 CE
Population is
about 300 million

1776 CE
Population is about 800 million

2000 CE
Population is 6.1 billion

2100 CE
The UN predicts a world
population of 10.9 billion

10 billion

8

6

4

2

0

2000 BCE   1000 BCE   1 CE   1000 CE   2000 CE

FIGURE CS1.5A The graph of human population as function of time is of key interest. With the emergence of modern humans in central Africa approximately 160,000 years ago, it required 7000 generations to reach a global population of 2 billion in the middle of the 20th century. Then within one human lifetime, population is forecast to increase by 8 billion people to 10 billion by 2050.

Case 1: Global Population

Modern humans emerged from central Africa some 160,000 years ago. Remarkably, it required approximately 158,000 years for the global population of humans to reach 250 million—this occurred during the height of the Roman Empire in 1 AD. This global population remained at 250 million until the Renaissance and remarkably global population did not reach 1 billion until 25 years after the American Revolution that occured in 1776. At the time of the Second World War global population had reached approximately 2.3 billion, primarily because agricultural production had increased markedly because of the use of fixed nitrogen as a fertilizer. Thus from the emergence of modern humans to the mid-twentieth century it had required some 7000 generations to reach a global population of two billion. In sharp contrast, over the span of a single human lifetime, 1945 to 2045, human population is forecast to increase by 8 billion, to approximately 10 billion people. This is summarized in Figure CS1.5a.

Case 2: Per Capita Income

While population increase receives the most important attention as the driver of increased energy use, it is actually increases in per capita income that is the primary driver. The reason for this is that it is the developing countries (e.g., In-

dia, central African nations, etc.) that are projected to have the greatest increase in population, but also have the lowest per capita income that will, when modernization takes place as it has in China, have the largest increase in per capita income. The concentration of global population growth in Asia is obvious from the recent U.N. projection displayed in Figure CS1.5b.

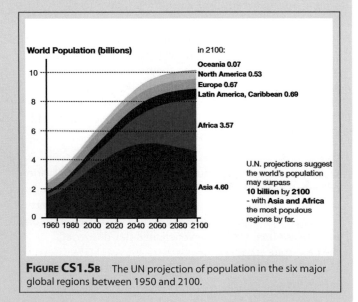

World Population (billions)

in 2100:

Oceania 0.07
North America 0.53
Europe 0.67
Latin America, Caribbean 0.69

Africa 3.57

Asia 4.60

U.N. projections suggest the world's population may surpass **10 billion** by **2100** - with **Asia and Africa** the most populous regions by far.

10

8

6

4

2

0

1960 1980 2000 2020 2040 2060 2080 2100

FIGURE CS1.5B The UN projection of population in the six major global regions between 1950 and 2100.

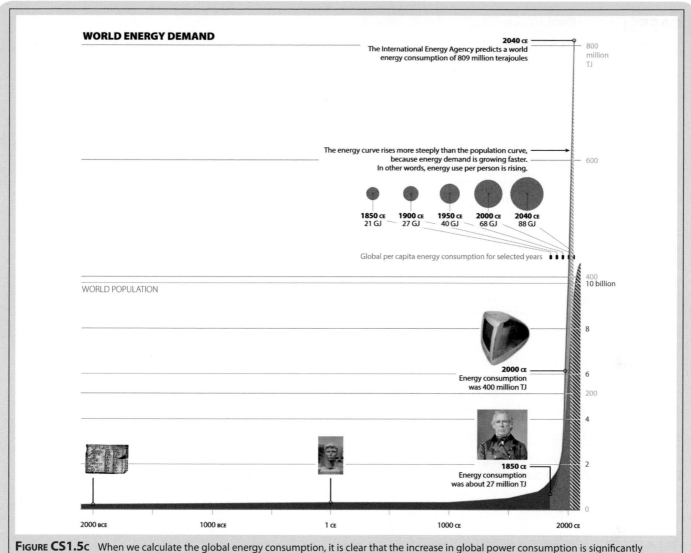

FIGURE CS1.5c When we calculate the global energy consumption, it is clear that the increase in global power consumption is significantly greater than the increase in global population.

It is instructive, therefore, to superimpose the increase in global energy consumption on the previously shown graph of global population to emphasize how much more rapidly global energy consumption is increasing than is global population. This remarkable contrast is displayed graphically in Figure CS1.5c.

CALCULATIONS:

Problem 1

The per capita income, globally averaged, was $5,500 per annum in 2014. Calculate the global energy consumption in 2014 if the PEI globally averaged is 1.5×10^7 J/$GDP.

Problem 2

Calculate world energy demand for 2050 if per capita income is $8,000/p·year in 2050. Assume PEI does not change.

We have therefore calculated global energy demand per year in 2014 and 2050. It is important to also calculate the global *power* demand because for many applications, it is the *rate* of energy consumption that is most important. Power is, of course, represented in units of J/sec or watts. Given that we have calculated global energy demand in J/yr, we need simply convert to J/sec to determine global power demand in watts.

Problem 3

Calculate global power demand in 2014.

Problem 4

Calculate global power demand in 2050.

When we calculate the global power demand in 2050 and compare it with the global energy demand in 2014, two factors are important. One is the population increase from 7 billion to 10 billion—an increase of 43%. The other is the increase in per capita power requirement that is proportional to per capita income. This increase is from $5500 in 2014 to $8000 in 2050, an increase of 46 percent. An illustrative calculation that makes clear the scale of the buildup in global energy production to meet this demand is to calculate the rate at which natural gas burning power plants must be constructed to supply the increased demand. If we assume each new power plant provides an additional 500 MW, then this figure, in combination with our calculated increase in global energy demand between 2014 and 2050, provides the required information to calculate the total number of new power plants needed. If we then calculate the number required per year between 2014 and 2050, we can calculate the number of new power plants required per day!

Problem 5

Calculate the number of coal burning power plants that must be built per year between now and 2050 to meet increase in demand.

ATOMIC AND MOLECULAR STRUCTURE

ENERGY FROM CHEMICAL BONDS

2

> "It was quite the most incredible event that has ever happened to me in my life. It was almost as incredible as if you fired a 15-inch shell at a piece of tissue paper and it came back and hit you."
>
> —Sir Ernest Rutherford, 1936
>
> Quoted in Abraham Pais, *Inward Bound* (1986), 189, from E. N. da C. Andrade, *Rutherford and the Nature of the Atom* (1964), 111.

FIGURE 2.1 The image of campers sitting around a campfire contains within it a remarkable number of important transformations occurring between different categories of energy. The combustion of wood (cellulose) releases chemical energy that is transmitted to the campers as electromagnetic radiation (light) in both the visible and infrared part of the spectrum. That electromagnetic radiation is absorbed by the campers as heat, which augments the heat generated within their bodies by the oxidation of glucose. The body of each camper emits infrared radiation, which is absorbed and reradiated back to them by the clothes they wear.

Framework

This unprecedented point in human history holds both remarkable opportunity and levels of risk that are dependent upon how rapidly leading nations of the world comprehend and then adjust their strategic choices at the intersection between energy generation and the global consequences thereof. China has been the pacesetter in the construction of new coal burning power plants—approximately two 500 MW plants per week in the latter half of the first decade of this century. In 2006, China passed the United States as the largest global emitter of CO_2 from fossil fuel combustion. But with an economic growth of ~9% per year, China's CO_2 emission will be double that of the US by 2014. This 9% growth rate each year leads to an exponential growth rate in

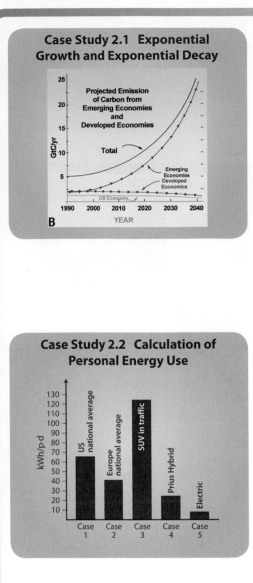

Case Study 2.1   Exponential Growth and Exponential Decay

Case Study 2.2   Calculation of Personal Energy Use

carbon emission. Case Study 2.1 examines the mathematics of exponential growth and decay. However with nearly four times the population of the US, China consumes approximately one-third the amount of energy per person as does the US. To sustain every individual in the US requires approximately 10,000 watts (10 kilowatts or 10kW) of power; in China 3000 watts per person, in India 800 watts per person, and in central Africa 350 watts per person. Yet in the midst of remarkable economic expansion China recognized that the rate of glacial loss in the Tibetan plateau may well place the water supplies for Asia in serious question within a generation. In response, the leadership of China has moved rapidly and aggressively on two fronts. First, they have decided the real battle of the next decade is how rapidly they can enter and then lead the new technology of alternative, renewable energy generation through building and selling wind and solar technology internationally. Second, they have moved to build extensive wind farms, solar thermal, and geothermal systems in order to curb dependence on foreign petroleum as the number of automobiles sold in China escalates at a rate unseen in the history of the industry.

It is also important to calculate the energy consumption of an individual to quantitatively assess the importance of each contribution energy makes to sustaining an individual's living conditions: home, food, transportation, purchased goods, etc. That analysis is the subject of Case Study 2.2.

As we will see in this chapter, we cannot comprehend the fundamentals that underpin the transformation that is occurring in the rearrangement of energy generation, storage, and distribution within global economic structures without a thorough grounding in atomic and molecular structure and thermochemistry. Why is this so? The reason is that thermochemistry provides answers to an array of key questions. For example, how is the energy contained in the chemical bond extracted to produce useful work required to sustain society? Why is virtually all of the energy contained in a chemical bond lost as heat under some circumstances, but is effectively channeled to build new and complex molecules in other instances? How is the chemical energy of one reaction coupled to subsequent chemical reactions leading to the formation of a desired chemical product? Why is chemical catalysis, the technique of speeding up a chemical reaction by controlling the mechanism of the reaction, so important for controlling the flow of energy in chemical systems? Why are gasoline engines 20% efficient, diesel engines 30% efficient, and electric motors 95% efficient? How does photosynthesis guide energy pathways through membranes without losing that chemical energy to heat—a form of energy inappropriate to the synthesis of new bond structures used to build and fuel organic systems?

Nations that prosper will be those tuned to the pathways that are innovative in the development of techniques for primary energy generation. Strategic choices for primary energy generation will serve any nation's ability to compete economically in the world market, achieve national security through energy independence, and properly forecast changes in the Earth's climate structure. The climate structure in turn controls water supplies from glacial and snow pack runoff, sea level, severe storm frequency and intensity, rainfall patterns, temperatures, and thereby social infrastructure.

To address these questions, we first consider the processes involved in the familiar example of hydrocarbon combustion and human comfort—a campfire surrounded by a group of organisms—humans, as displayed in Figure 2.1; humans that are themselves sustained by the oxidation of hydrocarbons that they have consumed as "food." While this example provides a model that may appear trivial in contrast to the scale of global energy, it provides a prototype for a series of important processes in atomic and molecular structure and in thermochemistry. It also provides an important context before turning to the development of atomic and molecular structure of matter.

A picture of campers sitting around a fire at night contains, within a single image, some very important concepts fundamental to the study of molecular structure and the release of energy from a chemical bond. It also serves to emphasize the transformations in that energy that couple the components of the system together. First, there is the release of chemical energy by virtue of the combustion of wood (lignin and cellulose):

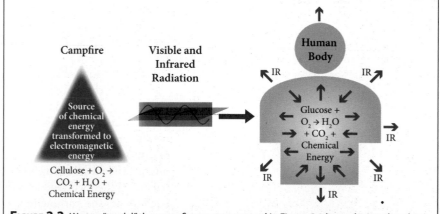 + oxygen → water + carbon dioxide + chemical energy.

The energy released in the reaction increases the temperature of tiny particles of carbon and/or "ash" that collectively appear as a flame because of the blackbody emission from the solid surfaces of the particles. The particles are hot enough that blackbody radiation occurs in the visible region of the spectrum and is thus detected by the human eye, but also clearly sensed as warmth by the campers sitting around the fire. Notice, however, that the flame transitions to red at the edges and then disappears because the particles cool, shifting to the red end of the visible spectrum and then, with further cooling, emitting radiation in the infrared that is invisible to the human eye. The chemical energy released in the combustion of wood is transformed to electromagnetic radiation (visible and infrared) and then absorbed by the campers in proximity to the fire. The campers, in turn, are sustained by the oxidation of glucose formed from what they have had for dinner, releasing chemical energy that maintains their body temperature at 38°C (311 K). The clothes, worn by the campers, absorb the infrared radiation from their bodies and then reradiate that infrared back into their bodies. The clothes also eliminate convective loss of energy to the surrounding air (energy transport via the macroscopic motion of the air). The rate of energy transport through the campers' clothes depends upon the *difference* in temperature between the inside and outside of their garments—the greater the temperature difference, the higher the rate of energy loss through the garments. The electromagnetic energy (visible and infrared) produced by the fire directly transmits energy to their bodies, through exposure to their faces, but also reduces the *temperature difference* between the outer and inner surface of their clothes, reducing the flow of energy outward from their bodies through their clothes.

We can model the systems contained in the image of the campers sitting around the campfire, Figure 2.1, by capturing the sources of energy, the transformations of energy taking place, as well as the barriers to energy flow. This "model" of the system is shown in Figure 2.2. Notice that in order to maintain a comfortable temperature of 38°C, the campers will select a distance from the fire and a thickness of outer garments to comfortably stabilize their body temperature at 38°C.

While we can represent respiration (the oxidation of organic material to form CO_2 and H_2O) as the burning of lignin or cellulose in the fire, what is the comparable mechanism for the oxidation of hydrocarbons that "fuels" the human beings in this picture?

FIGURE 2.2 We can "model" the campfire scene captured in Figure 2.1 by replacing the physical objects (fire, people, clothes, etc.) by simplified "subsystems" that capture, in this case, just the chemical transformations, the boundaries defining each of the subsystems from their surroundings, and the energy transformations taking place that link the subsystems together.

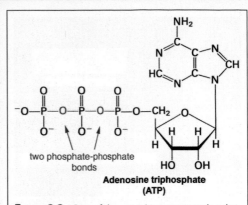

FIGURE 2.3 One of the most important molecules used in nature to control the flow and the coupling of energy within living organisms is adenosine triphosphate, or ATP. The containment of available energy for biological systems is in the double bond of the phosphate groups linked to the adenosine structure. ATP will be discussed in detail in Chapter 7.

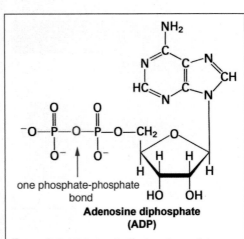

FIGURE 2.4 With the elimination of one of the high energy phosphate bonds, ATP is converted to adenosine diphosphate or ADP, which, as a consequence, contains less chemical energy than does ATP. Living organisms universally use the conversion of ATP to ADP to provide energy for life.

Cells have the capacity to extract energy from molecules (primarily sugars) through a coupled series of reactions that release energy in a controlled sequence that can either release heat directly or control that energy to be used for muscular control, nerve function, etc. As we have seen, heat is a highly disorganized form of energy that is not available for the controlled formation of specific chemical bonds. In almost all organisms this transfer of energy along chains of coupled reactions is accomplished by the molecule adenosine triphosphate or ATP. The structure of ATP is shown in Figure 2.3. The energy content of ATP that is readily transferred is in the phosphate subunit of the molecule, highlighted in red in that sidebar. The conversion of ATP to adenosine diphosphate, or ADP shown in Figure 2.4, occurs in the reaction with water

$$ATP \xrightarrow{H_2O} ADP + energy$$

and energy is released in going from two phosphate-phosphate bonds to the single phosphate-phosphate bond. We will develop the full complement of reactions involving ATP and ADP as the chapters unfold. We turn first to how we measure the release of energy from any chemical bond.

We are at liberty to select a number of examples in developing the relationship between the atomic and molecular structure of matter and the release of energy from transformations in the bonding structure of molecules, but we will select the combustion of octane, C_8H_{18}, as our primary focus here. The reasons for this are many, but octane, or gasoline as it is commonly referred to, is the prototypical liquid hydrocarbon fuel. It not only fuels cars, trucks, buses, etc., but it is also a very good prototype for diesel fuel, jet engine fuel, heating oil, and nearly every other liquid form of petroleum.

This chapter sets in place the fundamentals of the atomic and molecular structure of matter. As we will see, from the revelation that all matter is composed of discrete indivisible particles that could not be created nor destroyed emerged a series of key scientific developments beginning with the discovery of the electron and the discovery of the nucleus. Those discoveries set the stage for theories of molecular bonding. When we write a chemical reaction, the fundamental existence of atoms as the conserved (neither created nor destroyed) indivisible constituents of matter is explicitly recognized. Thus when we analyze the combustion of gasoline in an automobile engine, we must first represent the molecular structure of the gasoline molecule. We have already noted that octane can be represented by the molecular formula C_8H_{18}. While gasoline is a mixture of hydrocarbon molecules ranging from 6 to 12 carbon atoms, a reasonable *average* composition for gasoline is in fact this molecular formula for octane, C_8H_{18}.

When octane is combusted in air, the octane molecule reacts with O_2 to form the products carbon dioxide and water, releasing energy because the CO_2 and H_2O products are more stable (at a lower energy) than C_8H_{18} and O_2. Because atoms are neither created nor destroyed in a chemical reaction, we must have an equal number of carbon atoms, hydrogen atoms, and oxygen atoms on the reactant side as on the product side of a chemical reaction. While we will develop the methods for balancing chemical reactions in this chapter, we write down the balanced chemical equation for the burning (a.k.a. combustion or oxidation) of octane as

$$2C_8H_{18} + 25O_2 \rightarrow 16CO_2 + 18H_2O + energy$$

The question is: how much energy does gasoline actually contain relative to the products CO_2 and H_2O? How could we measure quantitatively the amount of energy contained in, say, a liter of gasoline? In a fully fueled jet aircraft? A full tank of gas in an automobile? The answers to these questions are important and, as we will see, quite surprising in practice.

So let's sort out how we might get a reasonably accurate answer to the question of how much energy is contained in a liter of gasoline even before we precisely link atomic and molecular structure to thermochemistry in the next two chapters. First, we know from experience that if we allow energy to flow into a block of steel, for example, its temperature will rise. If we place a block of steel on a stove, the flow of heat from the combustion of gas (natural gas, CH_4) will enter the steel block and its temperature will rise. In fact, as we saw in Chapter 1, extensive experiments have demonstrated that there is a very simple relationship between the amount of energy passing as heat into a mass of material, and the *change* in temperature of that material.

$$\Delta T \propto \left[\begin{array}{c} \text{amount of energy entering} \\ \text{the mass of material} \end{array} \right]$$

Because the amount of energy entering a mass, m, of material is directly proportional to the *change* in temperature of that body, we can write the expression

$$\left[\begin{array}{c} \text{Amount of energy} \\ \text{entering the mass, m} \end{array} \right] = C\Delta T$$

where C, the "heat capacity" of the body, is a constant specific to the material into which the energy is flowing, and to the *amount* of material—the mass of the body into which the energy is flowing.

This is very important in practice because it means that we can measure the energy contained in the reaction of, say, a liter of gasoline, by capturing that energy and simply measuring the increase in temperature of the material into which the energy released in the reaction has flowed. We could get a reasonably accurate measure of the energy contained in the molecular structure of gasoline relative to the products of the reaction, CO_2 and H_2O, simply by burning the liter of gasoline in a steel box, first measuring the temperature of the box just before we ignited the liter of fuel and then measuring the temperature of the box just after the gasoline was consumed in combustion. We could improve our measurement by insulating the box so no energy was lost by the emission of infrared radiation or by conduction to the surrounding air from the box during the combustion of the gasoline. In addition we could add the gasoline fairly slowly so the flame burned at a controlled rate such that the steel box did not explode from the rapid release of energy. A simple experiment to obtain a good estimate of the energy contained in gasoline is shown in Figure 2.5.

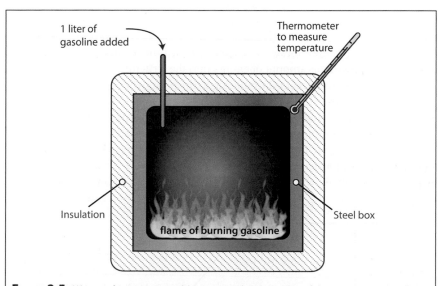

Thermometer to measure temperature

1 liter of gasoline added

Insulation

flame of burning gasoline

Steel box

FIGURE 2.5 We can obtain a reasonably accurate determination of the energy contained in the bond structure of gasoline (C_8H_{18}) using a simple experiment that combusts gasoline in a steel box wrapped with insulation to prevent the escape of heat. The observation of the increase in temperature of the steel box in combination with the equation $q = C\Delta T$ quantitatively establishes the energy released in the chemical reaction $2C_8H_{18} + 25O_2 \rightarrow 16CO_2 + 18H_2O$.

Experiments have demonstrated that for steel, every joule of energy added will increase the temperature by approximately 2°C for every gram of steel contained in the body of material. To be precise, if we construct a steel box that weighs 100 kilograms, it will require 46kJ of energy to raise the temperature of the box by 1°C. So let's run our experiment to measure the energy contained in a liter of gasoline.

When we begin the experiment the temperature of the box is 20°C, room temperature. We start the flow of gasoline into the box and ignite it, allowing it to slowly burn and we watch the temperature of the box. Because metals conduct heat very quickly (as we will consider in detail in the chapters 8, 9, and 10) the entire box will rapidly reach approximately the same temperature as the gasoline combusts, but very little energy will escape because of the insulation.

When we measure the temperature of the 100 kg (220 lb!) steel box after the 1 liter of gasoline has burned, we find that the temperature is approximately 760°C! Thus, because 46 kJ of energy is required to raise the 100kg steel box by 1°C, the total energy release from 1 liter of gasoline is

$$\left[\begin{array}{c} \text{Energy release from} \\ \text{1 liter of gasoline} \end{array}\right] = (46 \text{ kJ/°C})\Delta T = (46 \text{ kJ/°C})(740°C) = 3.4 \times 10^4 \text{ kJ}.$$

This is a very important observation. It says that gasoline contains approximately 3×10^7 joules of energy per liter. A liter of gasoline weights 0.80 kilograms so gasoline contains 3.4×10^7 joules/0.80kg or approximately 4.3×10^7 joules/kg. Just to put this in perspective:

- Gasoline delivers 15 times the energy per unit mass as TNT.
- Gasoline has 1000 times as much energy as an equal weight of flashlight batteries and 100 times as much energy as an equal weight of computer batteries.
- Liquid hydrogen has 4.5 times less energy per liter than does gasoline.

When the World Trade Center was attacked in 2001, it wasn't the impact of the aircraft that did the damage. The plane weighed 130 tons (1.2×10^5 kg) and was moving at 300 miles per hour. The kinetic energy of the aircraft was ½ mv² = ½ $(1.2 \times 10^5$ kg$)(1.3 \times 10^2$ m/sec$)^2 = 1 \times 10^9$ joules. The (nearly) fully fueled aircraft carried 60 tons (5.4×10^4 kg) of fuel, which, from our measurement of the energy contained in a kg of fuel (4×10^7 joules/kg), means the energy content of the fuel in the aircraft was approximately 2.2×10^{12} joules of energy. Thus the energy contained in the fuel of the aircraft was 2200 times the kinetic energy of the aircraft in flight. The release of this amount of energy into the structure of the building raised the temperature of the steel in the building to the point where the structure melted and began to collapse. The mass of the upper floors begin to accelerate downward such that the underlying structure could not arrest the downward velocity of the upper part of the building and the entire structure collapsed.

We have, with our insulated steel box into which we slowly added a liter of gasoline that combusted, successfully measured the energy content of octane when reacted with air, which we can draw in an energy diagram:

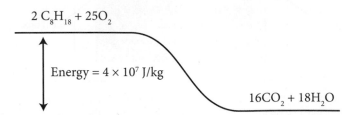

In the process of combusting gasoline, we have released all the energy as heat that went into elevating the temperature of the steel box. We can sketch this in Figure 2.6 to emphasize that we combined gasoline with oxygen producing car-

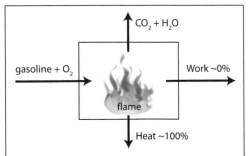

FIGURE 2.6 When we burn gasoline in a container open to the atmosphere, all the energy contained in the chemical bonds of gasoline is released as heat and as work done in the resulting expansion of heated air. No useful mechanical work is extracted.

bon dioxide and water. Energy was released primarily as heat. Expansion of gas resulting from the formation of CO_2 and H_2O molecules did some work on the surroundings, but as we will see, very little compared with the release of heat. Thus virtually all of the energy released in the combustion of gasoline went to heat. This is also summarized in Figure 2.6. Given that fossil fuels provide 80% of all primary energy, where do fossil fuels come from? How were fossil fuels formed? This is an important topic that is treated in Case Study 2.3.

When we use gasoline to power an automobile rather than combusting gasoline in an open steel box, the only difference is that the engine of the car converts *some* of the energy contained in the bonding structure of octane to work, moving the car forward. This forward motion is a result of a series of small explosions within the cylinders of the engine that turns the crankshaft of the engine that turns the wheels of the car. A diagram that expresses the combustion of gasoline in oxygen, to form CO_2 and H_2O with the release of energy because of the chemical reaction, is shown in Figure 2.7 for the case of an automobile. In the case of even a modern high-efficiency gasoline-powered car, Figure 2.7 properly reflects the fact that only about 15% of the energy released from the rearrangement of chemical bonds of gasoline and oxygen to produce carbon dioxide and water is available to power the automobile: 85% of the energy is lost as heat!

Becoming familiar with the macroscopic release of energy from chemical bond rearrangement turns out to be very important both for understanding chemical reactions and for understanding the structure of global energy that guides our choices now and in the future. Before we consider how to break the problem down to the molecular level, we examine one more aspect of energy release from the burning of hydrocarbons.

First, if you open your electric bill you will immediately notice that you are charged a certain amount for each kilowatt-hour of energy that you consume. Of course a watt (joule/sec) is a unit of *power;* that is, energy per unit time. A kilowatt is 1000 watts. A kilowatt-hour (or kWh) is then 1000 watts times the length of time that 1000 watts is used, in this case 1 hr or 3.6×10^3 seconds. Power times time is energy. Thus, you are charged on your electric bill for the amount of *energy* that you use in a month; the number of kWh you have consumed. The amount you pay for electricity varies significantly in different regions of the country. Energy in units of kWhs is generally used for the billing of electricity use. However, the kWh unit of energy is very important for understanding energy production, use, and storage for any *macroscopic* or global scale issue involving energy. The reasons for this are numerous, but one is that the unit of the *joule*, which we will use for all molecular scale calculations, is a *very* small number. A good example is that just one liter of gasoline contains 3.4×10^7 joules of energy. A kWh of energy contains $(10^3 \text{ watts})(3.6 \times 10^3 \text{ sec/hour}) = (10^3 \text{ joules/sec})(3.6 \times 10^3 \text{ sec/hr}) = 3.6 \times 10^6$ joules. Thus a liter of gasoline contains $3.4 \times 10^7 \text{J}/3.6 \times 10^6 \text{J/kWh}$ or 9.4 kWh of energy. Diesel fuel contains 9.9 kWh of chemical energy per liter and jet fuel contains about 9.5 kWh of chemical energy per liter. As we will see, this unit of energy is extremely versatile for executing very illustrative calculations on the macroscopic or global scale. In order to facilitate these calculations we will often approximate the energy context of liquid fuels as 10 kWh/liter.

But we are also interested in the molecular scale. So because we have succeeded in determining the amount of energy contained in one liter of gasoline, from the density of gasoline we can calculate that gasoline contains 4.2×10^7 joules per kg, and we know gasoline can be approximated by the molecular formula C_8H_{18}. How do we calculate the energy contained in a *single* C_8H_{18} molecule when reacted with O_2 to form water and carbon dioxide? *All* we need to know is the number of molecules of C_8H_{18} in a kg.

A single number provides the link between the *macroscopic* measurement of the energy contained in a kilogram of octane and the energy release from a single

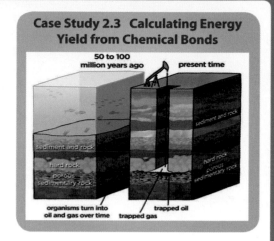

Case Study 2.3 Calculating Energy Yield from Chemical Bonds

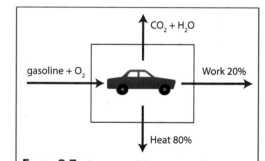

FIGURE 2.7   An automobile contains a "heat engine" capable of converting the energy contained in the chemical bonds of the fuel into work that moves the car through the conversion of chemical energy to mechanical energy. Surprisingly, even the most efficient modern gasoline engines only convert about 20% of that chemical energy of the fuel to work. The other 80% is lost to heat.

molecule of C_8H_{18}, and that is *Avogadro's number*. Avogadro's number tells you the number of molecules contained in a certain mass of a substance—a mass equal to the *molar mass* of that substance. The molar mass of a substance, as we will discuss in some detail in the chapter, is simply equal to mass of the molecule obtained by adding the atomic mass of each atom in the molecule to obtain the total. Octane, C_8H_{18}, contains 8 atoms of carbon at 12.0 g/mole each and 18 atoms of hydrogen at 1.0 g/mole each for a total of $8 \times 12 + 18 \times 1 = 114$g/mole for the *molar mass* of octane. Avogadro's number is thus the number of octane molecules in 114g of C_8H_{18}—and Avogadro's number is a *very* large number: 6.02×10^{23} molecules. Our liter of gasoline weighs 0.76 kg so, because 114 g of C_8H_{18} contains 6.02×10^{23} C_8H_{18} molecules, 0.76 kg of octane contains

$$\left(\frac{0.76 \text{ kg}}{0.114 \text{ kg}} \right) 6.02 \times 10^{23} = 4.0 \times 10^{24} \text{ molecules of } C_8H_{18}$$

But that 0.76 kg of C_8H_{18} yields 3.2×10^7 joules of energy when combusted so on a *per molecule* basis, that corresponds to

$$\frac{3.2 \times 10^7 \text{ joules}/0.76 \text{ kg}}{4.0 \times 10^{24} \text{ molecules}/0.76 \text{ kg}} = 8.5 \times 10^{-18} \text{ J/molecule}$$

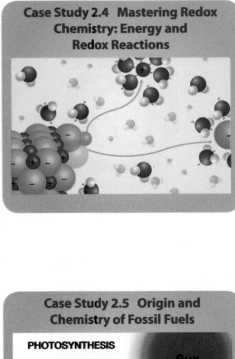

Case Study 2.4 Mastering Redox Chemistry: Energy and Redox Reactions

This is a very small number because Avogadro's number is so large! Because of this, chemistry has developed a convenient unit — the *mole*. Quite simply, by international agreement, a mole of the isotope carbon 12, ^{12}C, weighs 12.000 grams. All other elements are scaled to the mass of carbon 12 in the statement of their atomic weight, which is listed in the periodic table. As this chapter will describe, because elements occur in nature with *differing* numbers of neutrons in the nucleus for a given number of protons (the number of protons uniquely determines the identity of the element), thereby creating *isotopes* of the same element, the atomic weight reflects the weighted average of those isotopes in the naturally occurring element. So, for convenience the *mole* is used to count atoms and molecules just as the dozen is used to count eggs.

Thus, at the macroscopic level, our 8×10^{-18} J/ C_8H_{18} molecule becomes

$$(8 \times 10^{-18} \text{ J/molecule})(6.02 \times 10^{23}/\text{mole}) = 4.8 \times 10^6 \text{ J/mole} = 4.8 \times 10^3 \text{ kJ/mole}$$

While our example of the oxidation of C_8H_{18} by molecular oxygen represents an important example of an oxidation reaction, Case Study 2.4 introduces the subject of oxidation-reduction reactions as an important class of reactions critical to the development of life on the planet and to the evolution of life.

Because of the importance of fossil fuels to the generation of primary energy globally, Case Study 2.5 reviews the rapidly changing picture of fossil fuel resources globally.

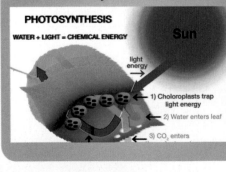

Case Study 2.5 Origin and Chemistry of Fossil Fuels

This Framework has established the context for the importance of calculating the energy contained in chemical bonds whether at the molecular level through the use of Avogadro's number and the mole or at the macroscopic level from direct experiments such as the combustion of gasoline in air as we have worked out from our experiment with the steel box. We turn then to an examination of the atomic and molecular structure of matter that establishes the foundation for understanding the chemical behavior of all substances as well as how energy release from molecular bond rearrangement can be harnessed to achieve a remarkable range of objectives.

Chapter Core

Road Map for Chapter 2

The entire structure of modern chemistry is built upon the atomic view of matter—the idea that all matter is constructed from discrete, indivisible particles known as atoms that cannot be created nor destroyed in a chemical reaction. The development of the Atomic Theory of Matter began a series of developments in chemistry that included:

Road Map to Core Concepts	
1. Atomic View of Matter	
2. Discovery of the Electron and the Nucleus	
3. Atomic Mass, Mass Number, and Atomic Symbols: Isotopes	
4. Molecular Structure	
5. Stoichiometry, Avogadro's Number, and Molar Mass	
6. Balancing Chemical Reactions	
7. Oxidation-Reduction Reactions	

These are the concepts that constitute the foundation for all modern studies of chemistry.

The Atomic View of Matter

Virtually without exception, our major conceptual advances in the physical sciences have sprung from an observation or series of observations that refute the community-held views of the day, replacing those views with a radically different perspective. This was the case when the concept of a material substance called *caloric* was replaced by the concept that heat was not a substance but rather the *transfer* of thermal energy; it was the case when Copernicus demonstrated that the Earth did not sit at the center of the solar system. While we take it for granted today that material is built from atoms and the bonds they form with other atoms to form molecules, throughout a vast fraction of human history, matter was viewed as continuous. The idea that the physical material around us was divisible into specific, identifiable entities was rejected in scientific thinking until the beginning

FIGURE 2.8 Antoine Lavoisier, the father of modern chemistry, in his laboratory in France. Lavoisier lived from 1743 to 1794, and was beheaded in the French Revolution as a member of the French aristocracy at the age of 50.

FIGURE 2.9 Joseph Proust (1754–1826), who established the Law of Definite Proportions.

FIGURE 2.10 John Dalton (1766–1844), who developed the Atomic Theory of Matter released in 1808.

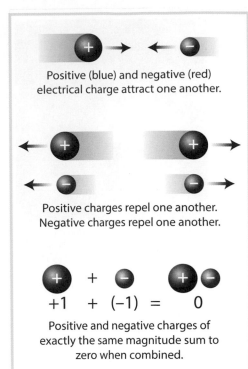

Positive (blue) and negative (red)
electrical charge attract one another.

Positive charges repel one another.
Negative charges repel one another.

+1 + (−1) =   0

Positive and negative charges of
exactly the same magnitude sum to
zero when combined.

FIGURE 2.11 J. J. Thomson (1856–1940), who
discovered the existence of the electron in his
laboratory with a cathode ray tube.

of the 19th century. In fact there were leading scientists that rejected the notion into the early part of the 20th century. A sequence of three discoveries, articulated in the form of scientific laws, changed the very foundation of how the structure of materials was viewed in science.

The first law that led to a new atomic theory of matter was formulated in 1789 by Antoine Lavoisier, shown in Figure 2.8. Lavoisier is often called the father of modern chemistry because he was the first to recognize, isolate, and name oxygen in 1778 and hydrogen in 1783. Lavoisier discovered that while matter may change its form, shape, or chemical characteristics, its mass always remains *constant*. In other words, mass is *conserved* in such transformations. This realization led Lavoisier to formulate his *Law of the Conservation of Mass* in 1789: *In a chemical reaction, matter is neither created nor destroyed.*

It is interesting to point out that it was Lavoisier's deft discoveries of oxygen and hydrogen as gases produced and/or consumed by chemical reactions that led him to include these important gases in his analysis of both reactant and product masses involved in a chemical reaction, opening the pathway to an understanding of mass conservation. Lavoisier brought the application of high accuracy measurements of mass to the practice of chemistry, initializing a new era of analytical chemistry that remains an important part of chemistry today. Lavoisier was a member of the French aristocracy and was beheaded in the French Revolution on the 8th of May 1794 at the age of 50.

In 1799 another French chemist, Joseph Proust, shown in Figure 2.9, convinced many scientific skeptics that the elements that make up a given compound were always present in fixed proportions. Establishing this fact lead to his formulation of the *Law of Definite Proportions*, also known as the *Law of Definite Composition*, which states that: *Independent of its source, a particular compound is composed of the same elements in the same fraction by mass.*

The *fraction* by *mass* that a given element contributes is obtained by dividing the mass of each element by the total mass of the compound.

Five years later, in 1804, John Dalton, an English physicist and chemist shown in Figure 2.10, published his *law of Multiple Proportions*, which stated that: *When two elements, A and B, react to form two different compounds, the masses of B that combine with a fixed mass of A can be expressed as a ratio of whole numbers.*

Thus when two elements react to form two different compounds, they do so in a ratio of (small) whole numbers. It was clear from an analysis of Dalton's laboratory notebooks from that time, and from his discussions presented at the Manchester (England) Literacy and Philosophical Society, that he was already strongly of the opinion that matter was comprised of discrete, irreducible particles, or "atoms."

The year modern chemistry was born was 1808 when John Dalton released his treatise *A New System of Chemical Philosophy*. In this work Dalton drew together the *Law of the Conservation of Mass*, the *Law of Definite Proportions*, and the *Law of Multiple Proportions* to create what came to be known as *The Atomic Theory* of *Matter*. In his formulations, Dalton presented four postulates that he deemed to be universally true:

1. All matter consists of discrete, indivisible particles known as *atoms* that cannot be created or destroyed.

2. Atoms of a given element are distinct from those of any other element and cannot be changed into atoms of another element, and a chemical reaction seems only to change the way those atoms are grouped together.

3. All atoms of a given element possess the same mass and the same chemical properties that are distinct from that of any other element.

4. Atoms of one element can combine with atoms of other elements to form chemical compounds; a given compound always has the same ratio of those atoms.

We can immediately recognize the lineage of Dalton's atomic theory from the laws of mass conservation, definite proportions, and multiple proportions. We can also recognize the short comings—the "indivisible particles" and the notion that an "element cannot change into another." However, while we will improve upon the atomic theory of Dalton's, his theory revolutionized chemistry as both an intellectual pursuit and a practical contributor to society.

It was nearly one hundred years before the *structure* of the atom was interrogated—an investigation that brought a new perspective to the structure of the atom itself.

Discovery of the Electron

The fact that the new Dalton model of matter revolutionized the study of both physics and chemistry did not mean that the theory was complete. For example the Dalton model could not successfully explain why atoms could bond together to form molecules, why carbon and oxygen could form either carbon monoxide, CO, or carbon dioxide, CO_2.

It was the union of the studies of electricity and chemical composition that opened key new insights in the late 19th century. The English physicist J. J. Thomson, shown in Figure 2.11, working with evacuated tubes to which a high voltage could be applied, discovered the fact that these "atoms" were not indivisible. In fact they contained negatively charged particles that could be separated from the positive component of the atom by a strong electric field. An example of the original device used in these studies is shown in Figure 2.12—it was termed a *cathode ray tube*. Cathode ray tubes, as Figure 2.12 shows, consist of two electrodes, one negative—the cathode, and one positive—the anode, separated by a few centimeters of space that can be evacuated and filled with small amounts of any number of different gases. With the application of high voltage between these electrodes, a beam of light appears between the electrodes and a "current" flows through the tube. These glowing discharges were a fascination of the day, but it was J. J. Thomson who systematically studied those discharges using specifically designed tubes and using both electric and magnetic fields to study the properties of these charged entities that traversed the tube when voltage was applied. Thomson, through a series of increasingly decisive experiments, determined that those "cathode rays" were streams of particles that (1) traveled in straight trajectories, (2) were independent of the element that comprised the cathode, and (3) carried a negative charge. By studying the curvature of the trajectories of these subatomic particles, Thomson established the *ratio* of the particle's charge to the particle's mass—the "charge-to-mass-ratio"—demonstrating that those particles were 2000 times less massive than the hydrogen atom.

This discovery rewrote the atomic model—the atom was not indivisible, it contained negative particles and positive particles. The new subatomic particle discovered by Thomson was called an *electron*. As is quite often the case, many in the scientific community reacted with bemused disbelief. But additional experiments that tested Thomson's hypothesis proved him correct. The scientific method, in the fullness of time, is the ultimate adjudicator of disputes among scientists.

The cathode ray tube constitutes the basis for both the display tube on a television set shown in Figure 2.13 (now replaced by a flat screen system that employ, for example, liquid crystal displays—LCDs, that employ light emitting diodes—LEDs) and the "neon" signs that are simply evacuated tubes with a high voltage applied containing various mixtures of the rare gases (helium, neon, argon, etc.) that are ubiquitous in the advertising community.

While Thomson discovered the electron and measured the charge to mass

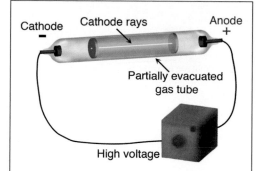

FIGURE 2.12 An early simple cathode ray tube consisting of an evacuated tube, a "cathode" that was negatively charged, and an "anode" that was positively charged. High voltage applied between the anode and cathode generated a stream of electrons flowing from the cathode to the anode.

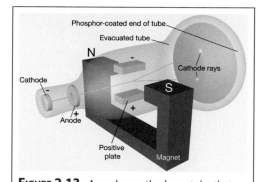

FIGURE 2.13 A modern cathode ray tube that provides precise control over the electron trajectory using both electric and magnetic fields to direct electrons to specific locations on the phosphor-coated "screen" at the cathode ray tube's terminus.

FIGURE 2.14 Robert Millikan (1868–1953), developer of the oil drop experiment that provided the most accurate measurements of the charge on the electron.

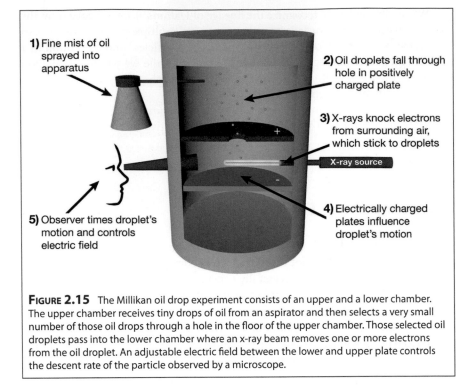

1) Fine mist of oil sprayed into apparatus

2) Oil droplets fall through hole in positively charged plate

3) X-rays knock electrons from surrounding air, which stick to droplets

X-ray source

4) Electrically charged plates influence droplet's motion

5) Observer times droplet's motion and controls electric field

FIGURE 2.15 The Millikan oil drop experiment consists of an upper and a lower chamber. The upper chamber receives tiny drops of oil from an aspirator and then selects a very small number of those oil drops through a hole in the floor of the upper chamber. Those selected oil droplets pass into the lower chamber where an x-ray beam removes one or more electrons from the oil droplet. An adjustable electric field between the lower and upper plate controls the descent rate of the particle observed by a microscope.

ratio of the electron, it was the American physicist Robert Millikan, Figure 2.14, working at the University of Chicago, who determined the charge on the electron. This was extremely important, for not only did it place an absolute value on the electron charge, it also, in combination with Thomson's measurement of the charge-to-mass ratio of the electron, provided a determination of the mass of the electron itself. While some experiments in science are clean and decisive, Millikan's experiment, called the "Millikan oil drop experiment," was anything but. Here is how it worked.

Millikan built an experimental apparatus, shown in Figure 2.15, that aspirated a mist of oil droplets into an upper chamber. A few of those tiny oil droplets passed through an opening into a second lower chamber where they passed through a collimated beam of x-rays (high energy electromagnetic radiation, Figure 2.15) that detached one or more electrons from one or more of the oil droplets. The lower chamber was comprised of an upper plate that was positively charged and a lower plate that was negatively charged. By adjusting the voltage between the two plates, the vertical velocity of the oil droplets could be controlled; in fact a given droplet could be suspended such that the force on the oil droplet resulting from the imposed electric field just balanced the gravitational force resulting from the mass of the oil droplet within the Earth's gravitational field.

The experimental complement, as shown in Figure 2.15, was complete with the addition of a microscope that provided the identification of a single oil droplet that could be manipulated with the electric field to reveal both its mass by measuring its fall velocity (cross checked by observing its size and knowing the density of the oil) and its charge by determining the electric field necessary to suspend it motionless in the gravitational field. Millikan's strategy was to observe a large number of cases, which would include a range of integral numbers of electrons removed by the x-ray beam, then solve for the minimum incremental charge.

This work was completed in 1909, yielding a charge on the electron of -1.602×10^{-19} coulombs or -1.602×10^{-19} C. This value is within 1% of the value accepted today after decades of refinement.

When combined with Thomson's measurement of -5.686×10^{-12} kg/C for the mass-to-charge ratio of the electron using his modified cathode ray tube, it was then possible to calculate the mass of the electron:

$$\text{Mass of Electron} = (\text{mass/charge})_{elec} \times (\text{charge})_{elec}$$
$$= (-5.686 \times 10^{-12} \text{ kg/C})(-1.602 \times 10^{-19} \text{ C})$$
$$= 9.109 \times 10^{-31} \text{ kg}$$

As we will see in the chapters that follow, chemical behavior is dominated first and foremost by *electrons*! And this control of chemical reactivity by electrons is in large measure a result of its charge, coupled with the exceedingly small mass of this component of atomic structure.

Check Yourself 1

On a dry day you often notice that after walking on a rug you build up a "static charge" that is released as a spark when you touch another person.

1. If you accumulate a charge of -25 μC (microcoulombs), how many excess electrons have you accumulated?

2. How many electrons are required to produce a charge of -1.0 coulombs?

3. What is the mass of -1.0 coulombs of electrons?

Discovery of the Atomic Nucleus

While Thomson's discovery of the electron was irrefutable, and in conjunction with Millikan's work demonstrated that the electron was but a tiny fraction of the mass of the lightest atom—hydrogen—the scientific world was confronted with a major mystery. What was the structure of the atom? The first answer to this was a new model of the atom, a model put forth by Thomson himself. That model was the "plum-pudding" model; an atomic structure characterized by a continuous, positively charged sphere (the pudding) punctuated by an occasional electron.

But Thomson was intent on testing his model of the atomic structure. After all, if he was wrong he wanted to be the first one to know. A powerful motivator leading to good science!

Thus Ernest Rutherford, a young New Zealander, shown in Figure 2.16, who had worked with Thomson and was a proponent of the "plum-pudding" model, decided to use an important new discovery—radioactivity—as a new experimental tool. Henri Becquerel and Marie Curie, shown in Figure 2.17, discoverers of radioactivity, had identified three distinct forms of radioactivity: alpha (α) particles, beta (β) particles, and gamma (γ) rays. Those will be treated in some detail in Chapter 13.

FIGURE 2.16 Ernest Rutherford (1871–1937), shown in the laboratory with his apparatus used to discover the nuclear architecture of atomic structure.

FIGURE 2.17 Madame Marie Curie (1867–1934), codiscoverer of radio activity, pioneered the study of isotopes and won two Nobel prizes.

The α particles had the attributes Rutherford needed—they were massive (relative to the electron) and they were emitted from radioactive elements at high velocity. In setting out to confirm Thomson's model, Rutherford built an experimental apparatus that was remarkably simple but powerful in its decisive ability to test this new model of atomic structure. As noted, that model of atomic structure was built on the notion of a continuous sphere of positive charge containing the majority of mass that has electrons of equal total negative charge distributed through the sphere. That model is graphically depicted in the left panel of Figure 2.18. The apparatus that Rutherford built to verify the plum-pudding model is displayed in Figure 2.19, and consists of (1) a radioactive *source* centered within a lead block that serves to collimate a beam of α particles emitted from the source, (2) a thin target of gold only a few molecules in thickness, and (3) a circumferential target that detects α particles scattered from the target and identifies the angle of the scattered α particles.

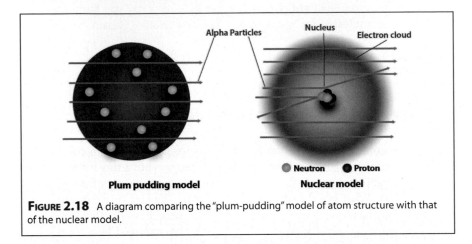

FIGURE 2.18 A diagram comparing the "plum-pudding" model of atom structure with that of the nuclear model.

In the opening phases of the experiment in 1909, Rutherford believed he had confirmed Thomson's plum-pudding model. Virtually all the detected α particles displayed little or no deflection from a straight trajectory. But as the data set evolved to include more and more observations, a remarkable and at first unbelievable conclusion emerged. Approximately one in 20,000 α particles was scattered back toward the source!

As the data base grew it became increasingly clear that approximately one α particle in 20,000 possessed a trajectory that carried the particle backwards in the direction of the source, as shown in the right-hand panel in Figure 2.18 and in Figure 2.19. This was an unexpected result for Rutherford because it refuted the plum-pudding model he had intended to verify with his experimental strategy. Rutherford's now famous comment was that the results were "about as credible as if you had fired a 15-inch shell at a piece of tissue paper and it came back and hit you." But continuing experiments proved these observations to be correct. An entirely new model of atomic structure was required and Rutherford provided that new model. He reasoned that the mass association with the positive charge which constituted the dominant fraction of the mass present, must be a highly concentrated point mass. This implied a structure of the atom with a positively charged central "nucleus" that was largely empty space within which very small negatively charged electrons resided. The contrast between the "plum-pudding" model and the "nuclear model" is displayed in Figure 2.18.

Rutherford thus proposed a *nuclear theory* of the atom with three primary components:

1. A nucleus that contains the positive charge and virtually all the mass of the atom.

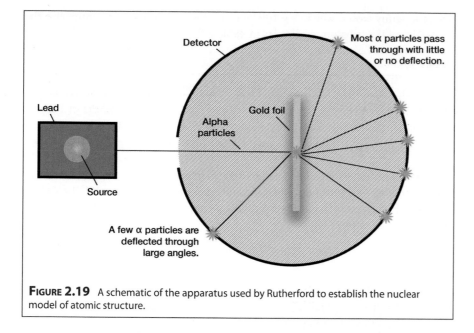

FIGURE 2.19 A schematic of the apparatus used by Rutherford to establish the nuclear model of atomic structure.

2. Virtually the entire volume of the atom is empty space with extremely small electrons dispersed within that volume.

3. The number of positively charged particles in the nucleus (called protons) is equal to the number of negatively charged electrons.

Yet again the world of atomic structure was fundamentally and irreversibly transformed, this time by Rutherford's nuclear model. However, it was immediately clear that the mass of the nucleus was not simply an ensemble of positively charged particles because while helium had two electrons, it had a mass four times that of hydrogen. It was Rutherford working with a student, James Chadwick, who demonstrated two decades later, that the unaccounted for mass was explained by the existence of another subatomic particle. That particle resided in the nucleus and was of approximately the same mass as the proton, but carried no intrinsic charge. The particle was called the *neutron*.

We can now summarize the view of the atom with a graphic, shown in Figure 2.20, depicting the nucleus residing at the center of an electron cloud. The diameter of the electron cloud is approximately 10^{-10} meters, but the nucleus is 5 orders of magnitude smaller in diameter, approximately 10^{-15} meters. The nucleus contains protons that possess a positive charge and neutrons that contain no charge. The mass of the proton is 1.67262×10^{-27} kg and the mass of the neutron is 1.67493×10^{-27} kg. While the nucleus contains 99.97% of the atom's mass, it occupies one quadrillionth the volume. The density (mass/volume) of the nucleus is thus remarkable by any standard we are normally acquainted with. The density of the nucleus can be calculated from the diameter and the mass and is approximately 10^{18} kg/m³. Lead, which we regard as a high-density material, has a density of approximately 1×10^4 kg/m³, which is 10^{14} times less dense than

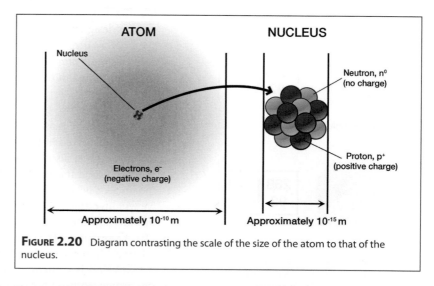

FIGURE 2.20 Diagram contrasting the scale of the size of the atom to that of the nucleus.

nuclear material. There are many interesting comparisons. A nucleus the size of a period in this text would weigh approximately 100,000 kg (100 tons); a coin the size of a penny would weigh 3×10^{12} kg (6 billion tons!). We can summarize the properties of the subatomic particles in Table 2.1.

TABLE 2.1				
	Mass (kg)	Mass (amu)	Charge (relative)	Charge (C)
Proton	1.67262×10^{-27}	1.00727	+1	$+1.60218 \times 10^{-19}$
Neutron	1.67493×10^{-27}	1.00866	0	0
Electron	0.00091×10^{-27}	0.00055	−1	-1.60218×10^{-19}

It is a common procedure to assign masses to protons, neutrons, and atoms in terms of *atomic mass units* or *amu*. One amu is equal to 1.66054×10^{-27} kg so that the mass of an element in amu is essentially a count of the number of protons and neutrons in the nucleus.

1. How many electrons would it take to equal the mass of a proton?

2. A carbon atom has six protons and six neutrons. How many electrons would it take to equal the mass of the carbon nucleus?

3. Considering Figure 2.20, what is the ratio of the volume of the atom to the volume of the nucleus?

Check Yourself 2
Rutherford's discovery of the structure of the atom through his discovery of the nucleus and the size and (subsequently) the mass of the nucleus established important relationships with the electron.

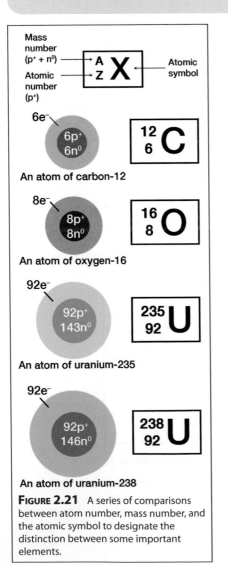

FIGURE 2.21 A series of comparisons between atom number, mass number, and the atomic symbol to designate the distinction between some important elements.

Atomic Number, Mass Number, and Atomic Symbol

The *atomic number* of a given element is equal to the number of *protons* in the nucleus of that element, and is thus equal to the number of *electrons* the element possesses. The key point is that what makes an element distinct, what defines its *chemical properties*, is the number of electrons it possesses in its electrically neutral configuration. That is, when the number of electrons and protons are equal. Thus the number of protons defines the element. All atoms of a particular element have the same *atomic number* and that atomic number is equal to the number of protons present in the nucleus. Each element thus has a different *atomic number* from every other element. As we will see it is the number of electrons an element has that determines the chemical characteristics of an atom, so therefore each element is designated by its atomic number. The atomic number is designated by the symbol Z as displayed in Figure 2.21.

This *mass number* of an atom is equal to the total number of protons and neutrons contained in the nucleus. The *mass number* is designated by the symbol A, also in Figure 2.21.

This gives rise to the convention of designating each element by its atomic symbol with the mass number as a superscript and the atomic number as a subscript as displayed in Figure 2.21. All carbon atoms have 6 protons. The common form of carbon also has 6 neutrons so we write this form of carbon as $^{12}_{6}\text{C}$. The most common form of helium has 2 protons and 2 neutrons so we write $^{4}_{2}\text{He}$. As the nucleus gets larger, the number of protons and neutrons is no longer equal as we will investigate in Chapter 13. For example, the stable form of uranium has 92 protons and 146 neutrons and is thus designated $^{238}_{92}\text{U}$.

A word about the *chemical* symbols. In a majority of cases, there is an obvious relationship between the chemical symbol and the name of the elements. Oxygen is O, carbon is C, neon is Ne, bromine is Br, etc. However, just to make things interesting, some of the chemical symbols are drawn from their Latin names. The symbol for tin is Sn from the Latin *stannum*; the symbol for sodium is Na from the Latin *natrium*.

Isotopes

While all atoms of a given element have the same number of *protons*, the same number of *electrons*, and the same *chemical symbol*, all atoms of a given element do not necessarily have the same number of *neutrons*. Thus all atoms of a given element do not necessarily have the same *mass*. Atoms with the same number of protons but with a different number of neutrons are called *isotopes* of that element. Isotopes are ubiquitous and they are profoundly important to nearly all branches of science from medical research to nuclear energy to paleobiology.

Perhaps the most famous example of the distinction between isotopes in light elements is that of carbon. There are three important isotopes of carbon. Carbon 12 is the common form, with six protons and six neutrons, carbon 13 has six protons and seven neutrons, and carbon 14 has six protons and eight neutrons. As we will examine in more detail in Chapter 13, carbon 12 and carbon 13, written $^{12}_{6}C$ and $^{13}_{6}C$, are *stable* isotopes. That is, they remain a carbon 12 or a carbon 13 and do not "decay" into other nuclei by the loss of a neutron. Carbon 14, written $^{14}_{6}C$, is radioactive and decays to stable nitrogen $^{14}_{7}N$ with the emission of particles from the nuclei, as we will discuss in Chapter 13. Carbon 14 is formed in the upper atmosphere by the collision of a stable nitrogen 14 and a high-energy neutron. The carbon 14 so formed is incorporated continuously into living organisms, built from the CO_2 present in the atmosphere. When the organism dies, carbon 14 is no longer incorporated, and carbon 14 decays with the emission of radioactivity such that with time, the number of radioactive events per unit mass, within the organism that has died, decreases. This provides a *clock* timing the delay from the death of the organism to the present. The method, termed *radiocarbon dating*, developed by Willard Libby, shown in Figure 2.22, at the University of Chicago in 1947, has revolutionized our ability to determine the age of artifacts obtained from ancient sites. The details of the

FIGURE 2.22   Willard F. Libby (1908–1980).

method will be presented in Chapter 13 on nuclear chemistry. The use of isotopes has spread to virtually every branch of modern science, and is a subject we will revisit often in the course.

Check Yourself 3

Problem: The electronics industry uses silicon (Si) in large quantities. Moreover, computers engage state-of-the-art electronics so Si is a dominant element in modern industry and society. Analysis demonstrates that Si occurs in three naturally occurring isotopes: ^{30}Si, ^{29}Si, and ^{28}Si. Given these three isotopes, determine the number of electrons, neutrons, and protons in each of these isotopes.

Solution: First, recognize that the mass number, A, is given by the superscript in each of the silicon isotope designations. Next, recall that the mass number is the sum of the neutrons and protons in the nucleus of each isotope.

When we consider the periodic table, provided with this text, we note that the atomic number, Z, of Si provides us with the number of electrons and protons that is common to all isotopes of Si. Thus with an atomic number Z = 14 for Si, we can calculate the number of protons, neutrons and electrons.

1) ^{30}Si has 14 electrons,
   30 – 14 = 16 neutrons and 14 protons.

2) ^{29}Si has 14 electrons,
   29 – 14 = 15 neutrons and 14 protons.

3) ^{28}Si has 14 electrons,
   28 – 14 = 14 neutrons and 14 protons.

Exercise: How many electrons, protons and neutrons are in

(1) $^{131}_{53}X$ (2) $^{41}_{20}R$ (3) $^{11}_{5}Q$?

Identify each of the elements represented by X, R, and Q.

Molecular Structure

When we move from a discussion of the elements to a discussion of the molecules and compounds constructed from those atoms such as the enzyme shown in Figure 2.23, which can convert CO_2 to C_2O_4, we enter into a world of profound variety and nearly limitless options. When we consider the variety of chemical properties

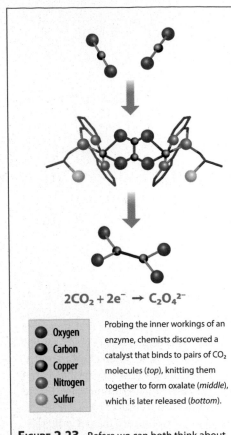

$$2CO_2 + 2e^- \rightarrow C_2O_4{}^{2-}$$

● Oxygen	Probing the inner workings of an
● Carbon	enzyme, chemists discovered a
● Copper	catalyst that binds to pairs of CO_2
● Nitrogen	molecules (*top*), knitting them
● Sulfur	together to form oxalate (*middle*),
	which is later released (*bottom*).

FIGURE 2.23 Before we can both think about and converse about molecular structures, we must name the molecular structures. Shown here is the conversion of CO_2 to the anion $C_2O_4{}^{2-}$, the ethanedioate anion, which results in the extraction of CO_2 produced by fossil fuel combustion.

"Chlorine is a deadly poison gas employed on European battlefields in World War I. Sodium is a corrosive metal which burns upon contact with water. Together they make a placid and unpoisonous material, table salt. Why each of these substances has the properties it does is a subject called chemistry."

—Carl Sagan

exhibited by the array of chemical compounds available in nature *relative* to the range of properties of the elements themselves, it is as though we were to compare the range of options in the expression of ideas using a fully developed language compared to what is available with just the letters used to create the "alphabet" of that language. So it is with chemistry.

While the remarkable variety in chemical properties available to us represents a powerful opportunity, we must first acquire a working knowledge of how these compounds are constructed and how they are named. We note this close relationship between the *structure* of compounds and the strategy for *naming* them as an important general point. It is also true that many of the conventions used in the naming of compounds are based on historical evolution, not on a logical protocol. This is unfortunate, but such is the case with any language.

The change in chemical and physical properties that occurs when elements combine to form a compound comprised of those same atoms is remarkable indeed. When we ignite a mixture of hydrogen and oxygen, we begin with the most commonly occurring form of pure hydrogen (diatomic hydrogen gas, H_2) and react that H_2 with the most commonly occurring form of pure oxygen (diatomic oxygen gas, O_2), the product is a liquid at room temperature; H_2O. Water is, as we will see repeatedly in this text, the single most important molecule to life on this planet and bear virtually no relationship in any chemical or physical sense to the gases, H_2 and O_2, that react to produce it. There are other striking examples. We combine, as displayed in Figure 2.24, the most commonly occurring pure form of chlorine, the diatomic gas Cl_2, which is a highly toxic gas, with the most commonly occurring form of sodium, the solid metal that bursts into flames when placed in water, and the product is a crystalline structure sodium chloride, NaCl. Thus we combine two toxic, highly reactive elements to form a compound that resides on every dinner table and is used universally with the food we eat. How can this be?

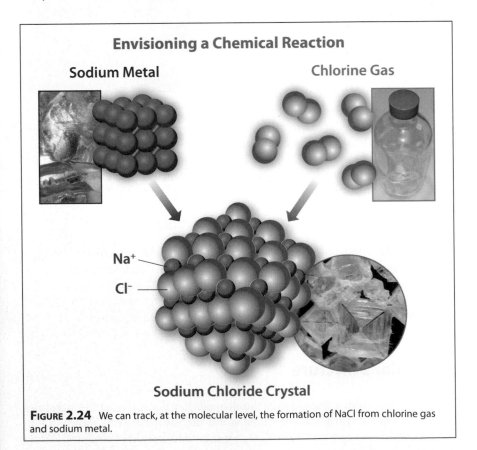

FIGURE 2.24 We can track, at the molecular level, the formation of NaCl from chlorine gas and sodium metal.

2.18

While we are concerned with virtually hundreds of thousands of compounds, we note that only a small number of elements actually occur free in nature. The noble gases—helium (He), neon (Ne), argon (Ar), krypton (Kr), xenon, (Xe), and radon (Rn)—are so inert that they do not react with other elements to form compounds but exist in air as separate atoms. As we have seen, hydrogen, oxygen, and nitrogen occur in their most common elemental states as H_2, O_2, and N_2. Sulfur occurs as S_8. Carbon occurs in extensive, nearly pure, deposits as coal. A limited number of metals exist in the Earth's crust in pure elemental form, most notably copper (Cu), silver (Ag), gold (Au), and platinum (Pt).

In the formation of chemical bonds, it is the electrons that form the bond to which we now turn our attention. In the formation of any chemical bond, the electrons position themselves so as to achieve the strongest bond—that is, to minimize the potential energy of the combination of electrons and protons. The electrons are said to "delocalize" from their position around the free, unbound atom in order to achieve the bond associated with the formation of the chemical compound created from its constituent atoms.

When we consider the formation of a chemical bond, we begin with two positively charged nuclei and a swarm of negatively charged electrons. By what miracle could that mix of attracting and repelling entities ever form a stable molecule? It is in fact remarkable indeed. We began with two nuclei, both of which strongly repel, and we move them to within a *very small* distance so, by Coulomb's law (which states that the force between two charges is proportional to the product of the charges divided by the *square* of the distance between them), the repulsive force between them is very *large*. We then place the electrons within the surrounding volume, and while there are an equal number of positive and negative charges, there is no guarantee that the mixture of electrons and nuclei constitutes a stable structure. In contemplating the positioning of the electrons to provide for the possibility of a stable bond, we recognize that if the electrons were positioned preferentially *between* the nuclei, two objectives would be achieved. First, the presence of the electrons would, by virtue of their negative charge, attract the positive charge of the nuclei toward the center of the chemical bond. Second, the presence of the cloud of negative electrons between the nuclei would "shield" the positive charge of the first nucleus from the positive charge of the second nucleus, thus reducing the repulsion between them.

It turns out, though it is by no means obvious, that this preferential placement of electrons between the nuclei is just, by a very small margin, adequate to form a stable union among the nuclei and the swarm of electrons.

This *sharing* of electrons between two nuclei was the idea behind the development of the first systematic theory of the chemical bond by G. N. Lewis, who published his ideas beginning in 1916. Applied to a hydrogen molecule, H_2, the Lewis theory postulated that a hydrogen molecule is formed when two atoms come together with the sharing of the electron donated by each atom to the bond:

$$H\overset{\bm\cdot}{} + \cdot H \longrightarrow H\!:\!H$$

The diagram on the right, with the electron pair shared by the newly formed H_2 molecule, is called a *Lewis structure* and we can represent that chemical bond in any number of ways as shown below:

H:H   H—H

While this chemical bond that binds two hydrogen atoms to form an H_2 molecule is the simplest bond in chemistry, it still requires significant computing power to quantitatively and accurately calculate the bond strength because the strength of the bond depends upon the small difference between large numbers: the mutual

G. N. Lewis (1875–1946)

repulsion of the electrons, the mutual repulsion of the protons, and the attraction of the electron–proton interaction.

This electron dot Lewis diagram represents the classic electron pair bond defining the sharing of electrons between the nuclei that seek to fill the available electron positions, so as to achieve an electron configuration that emulates the closed shell structure of the stable noble gases. In the case of H_2, the electron pair constitutes a "closed shell" that emulates the first noble gas, helium. This type of *shared electron* bond is called a *covalent* bond, and it is the most common category of bonding in chemistry. But hydrogen and helium occupy only the first row in the periodic table. For elements in the second and third rows of the periodic table, *observational* evidence demonstrates that the requirement of a "closed shell" requires an octet (8) of valence, or outer shell, electrons. For example a hydrogen atom, with one electron, will form a chemical bond with atomic chlorine, with seven electrons, to form the HCl molecule (hydrogen chloride) with eight electrons forming an "octet" of electrons about the Cl atom and a pair of shared electrons between H and Cl.

$$\text{H} \cdot + \cdot \ddot{\underset{..}{\text{Cl}}} : \longrightarrow \text{H} : \ddot{\underset{..}{\text{Cl}}} :$$

However, there is an important pattern in the periodic table wherein, as we move to the right-hand side, the propensity for the atoms to attract electrons *increases*. The ability of an atom of an element to draw electrons to it is called *electronegativity*. The greater the attractive power of an atom to draw electrons to it in a chemical bond, the greater is its *electronegativity*. Thus in the formation of the bond between H atoms (from the left-hand side of the periodic table) and Cl atoms (from the right-hand side), electrons are preferentially drawn toward the chlorine atom in HCl even though the bonding electrons are still viewed as "shared."

There are important cases, however, where the sharing becomes sufficiently unequal that one of the atoms in the bond effectively removes an electron from the less electronegative atom forming a positive ion, a *cation*, and a negative ion, an *anion*. This is the case for a metal, sodium, bonding to a chlorine atom. We can represent this case with a diagram.

$$\text{Na} \cdot + \cdot \ddot{\underset{..}{\text{Cl}}} : \longrightarrow \text{Na}^+ \; : \ddot{\underset{..}{\text{Cl}}} :^-$$

Sodium, in the act of losing an outer "valence" electron, is left with an inner electron "core" like that of the noble gas neon. Chlorine, with the addition of a single electron from sodium, acquires an electron configuration of argon, another noble gas.

We thus have two ions, Na^+ and Cl^-, that attract one another directly as two opposite charges with a force simply calculated from Coulomb's law. This type of bond is called an *ionic bond*.

While all chemical bonds are formed from the rearrangement of electron positions around the nuclei of the atoms that comprise the molecules within any compound, it is common practice to classify chemical bonds according to two categories:

1. The case where an electron is almost completely transferred from one atom to another. This is called an *ionic* bond because the element *donating* the electron becomes a *positive* ion, or *cation*. The atom that *receives* the electron becomes a *negative* ion, or *anion*.

2. The case where electrons belonging to the atoms involved in a chemical bond are *shared* between the atoms that comprise the bond. This is termed a *covalent* bond.

Development of the theory behind chemical bond formation occupies Chapters 11 and 12, but it is important to develop here the relationship between the structure of bonds and naming of compounds associated with their bond structures.

Chemical Formulas

In the representation of the bonding structure of chemical compounds there is a hierarchy of complexity employed that is determined by the degree of detail required to transmit the chemical information in question. We will consider here both the convention for molecules and molecular compounds and for ions and ionic compounds. We begin with formulas for molecular compounds. The simplest and most commonly used representation is the *chemical formula* that indicates the elements present in the compound and the number of atoms or ions of each element. The most common example of a molecular formula for a molecular compound is for water, H_2O. The second most common example of the *molecular formula* is, arguably, for carbon dioxide, CO_2. There are two points to observe with the molecular formula. First, the subscript indicates the number of atoms of each element present in the molecule. Second, notice that the least electronegative atom is listed first and the more electronegative atom is listed second. The same is true for carbon tetrachloride, CCl_4.

Chemical Formulas and Molecular Models

The next level of complexity representing molecular bond formation introduces the *structural formula*, which employs lines representing the covalent bonds that link the atoms together in the molecular structure. For example we write the *structural formula* for water as H – O – H, for carbon dioxide O – C – O, and for hydrogen peroxide H – O – O – H. However, many molecules are not inherently linear. In fact neither water nor hydrogen peroxide are linear. Thus, it is typical to represent the geometric structure of a molecule more fully in the structural formula by including an approximate representation of bond geometry in the structural formula. Thus the structural formula for water becomes

$$H-O-H \longrightarrow \quad H \diagup \overset{O}{\diagdown} H$$

indicating the bent geometry of this important molecule.

The structural formula for hydrogen peroxide becomes

$$H-O-O-H \longrightarrow \quad H \diagup O-O \diagup^{H}$$

The structural formula quite often also includes the existence of double bonds so the structural formula for carbon dioxide becomes

$$O-C-O \longrightarrow O=C=O$$

indicating a *linear geometry* but with two double bonds, one each between the carbon center and the oxygen end members. As we will see, many important molecules have increasingly complex geometries that while not fully captured by the structural formula are nevertheless approximated. Two important examples are methane:

$$CH_4 \qquad \longrightarrow \qquad \begin{matrix} & H & \\ & | & \\ H- & C & -H \\ & | & \\ & H & \end{matrix}$$

molecular formula   *structural formula*

and ammonia:

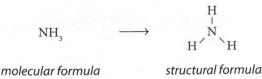

molecular formula   *structural formula*

An important molecule that demonstrates the way a structural formula captures both geometric structure and the distinction between single and double bonds is benzene:

C_6H_6 $\longrightarrow$

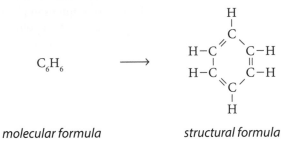

molecular formula   *structural formula*

Molecular Models

The next level of sophistication is the representation of the *three-dimensional* structure of a molecule using either a *ball-and-stick model* or a *space-filling model*. This ability to capture the 3D geometry of a molecule becomes increasingly important as we develop the relationship between structure and function in chemical reactions.

An important example is methane, CH_4, shown in Figure 2.25, a molecule involved in every issue from global energy ("natural gas" is largely CH_4) to molecular synthesis (CH_4 is the starting material for many polymers). We can use methane to compare and contrast the molecular formula, structural formula, ball-and-stick model, and space-filling model as shown in Figure 2.25.

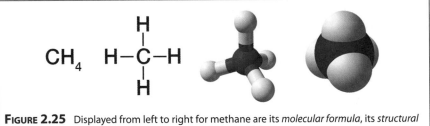

FIGURE 2.25 Displayed from left to right for methane are its *molecular formula*, its *structural formula*, its *ball-and-stick model*, and its *space-filling model*.

In stepping from the structural formula of methane to the ball-and-stick model, the *tetrahedral* structure of the molecule is revealed. In the transition from the ball-and-stick model to the space-filling model, the volume actually occupied by the bonding electrons is revealed. With the advent of increasingly sophisticated calculations of molecular structure, the space-filling model can be further refined by the use of color to represent the degree of delocalization of electron density in the bonding structure. Throughout the chapters in the book we will select from the various options for representing molecules depending on the required detail. Table 2.2 summarizes the comparison between the name of the compound, the molecular formula, the structural formula, and the ball-and-stick model for a few important examples.

Table 2.2

Benzene, Acetylene, Glucose, and Ammonia

Name of Compound	Molecular Formula	Structural Formula	Ball-and-Stick Model
Benzene	C_6H_6	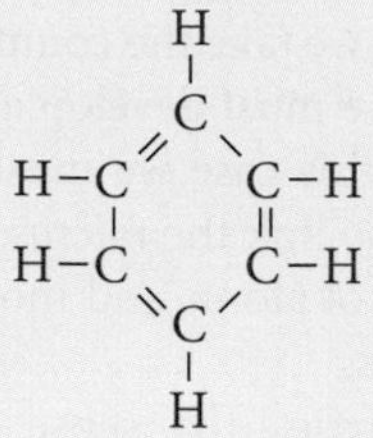	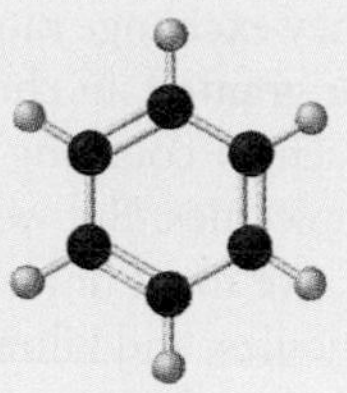
Acetylene	C_2H_2	H−C≡C−H	
Glucose	$C_6H_{12}O_6$	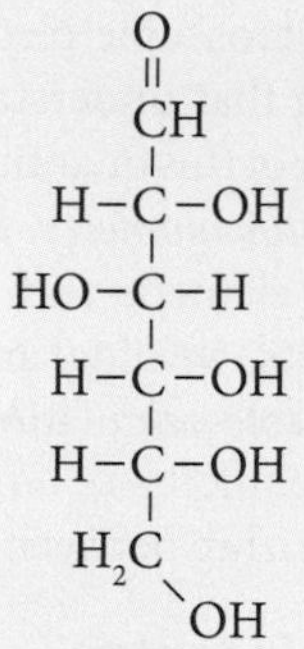	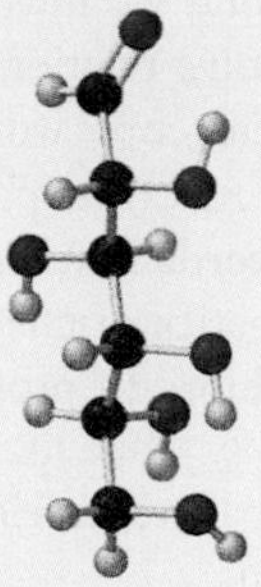
Ammonia	NH_3	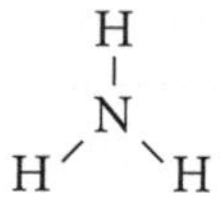	

For the case of an ionic substance that contains a net positive or negative charge, that net charge is indicated with a bracket and a superscript indicating the net charge. Displayed here are the ball-and-stick models for the anions phosphate and carbonate.

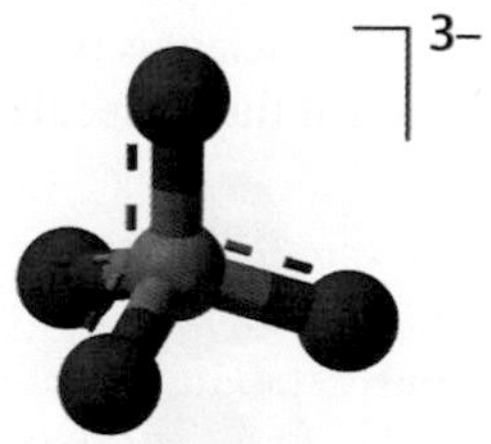

Phosphate ion, PO_4^{3-}

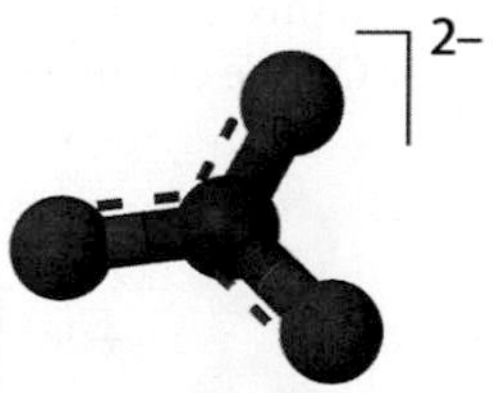

Carbonate ion, CO_3^{2-}

Stoichiometry

Counting matters. We practice this every day in our transactions with money where we routinely give or receive increments of currency counted to the penny. Individuals, corporations, governments—they all use "money" as a transaction tool for counting that extends internationally from country to country through currency exchange rates. While we take the counting of dollars, euros, yuans, yen, etc. for granted, in chemistry we must develop a consistent, universally accepted approach to counting—but in this case we must count electrons, nuclei, atoms, and molecules. But we must also link the *microscopic* world with the *macroscopic* world. We must link the world of atoms and molecules with the world of grams, liters, joules, and kilowatt-hours.

Undeniably, chemistry carries the sense of an intense and disciplined marketplace with spontaneous trading, with electrons—most notably valence electrons—serving as the exchanged and rearranged currency in the market of chemical change.

In order to execute calculations in the domain of chemistry, first we must be aware that matter is composed of specific combinations of atoms and molecules and second we must link this microscopic picture of matter through established *quantitative* relationships linking that understanding with useful calculations in our macroscopic world. The study of this quantitative link between the microscopic and the macroscopic is termed *stoichiometry*. Stoichiometry is derived from the Greek words *stoicheon*, meaning element, and *metron*, meaning measure. It is the mathematics of chemical amounts, weights, and measures, and a facility with it opens doors to profoundly valuable calculations in the marketplace of modern society.

The Framework of Stoichiometry

To set up our examination of stoichiometry, we review some fundamentals. From the most basic perspective, a chemical reaction is a microscopic collision between individual atoms or molecules and some integral number of particles are converted to some other integral number of particles. The microscopic transformation is represented in a concise form by a *chemical equation*:

$$2H_2 + O_2 \rightarrow 2H_2O$$

which represents the combination of two hydrogen molecules with one oxygen molecule to yield two water molecules. The stoichiometric coefficients are the factors by which each of the reactants and each of the products are multiplied so as to ensure that the number of atoms of each element are the same on both sides of the chemical reaction. In this case the stoichiometric coefficient is two for both H_2 and H_2O and is one for O_2. We can model this reaction at the molecular level:

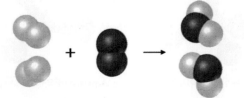

While on the face of it, the link between the *chemical equation* and the graphic representing one oxygen molecule combining with two hydrogen molecules to yield two water molecules is a straightforward visualization, it is a rearrangement with nothing *created* nor *destroyed*. Four hydrogen atoms and two oxygen atoms go in, four hydrogen atoms and two oxygen atoms emerge from the rearrangement.

But to be more specific (because in chemistry we *always* focus on the fate of valence electrons) four hydrogen nuclei, two oxygen nuclei, and twenty electrons enter the chemical transformation, and the *same* four hydrogen nuclei, two oxygen nuclei, and twenty electrons emerge. The number of electrons is strictly conserved; the number and identity of the nuclei is strictly conserved. While we will see, in our studies of nuclear chemistry in Chapter 13, that mass and energy can be transformed into one another in *nuclear* rearrangements, in *molecular* rearrangements the total mass is conserved in all chemical reactions. From this fact follows the law of *conservation of mass* for a chemical reaction. It is the law of the microscopic chemical domain and it constitutes the law that allows us to simply count atoms involved in a chemical reaction and from that count calculate the mass associated with that transformation. But of course grams do not react with grams, atoms and molecules do, so first we need a consistent way to count atoms.

Atomic Mass

We seek to bridge between the microscopic and the macroscopic, so in addition to stoichiometry (where we count atoms on both sides of a chemical reaction and adjust the stoichiometric coefficients to balance the number of atoms on each side of the chemical equation), we need to establish the *relative atomic mass* of the elements in the periodic table. The ^{12}C isotope of carbon contains six protons, six neutrons, and six electrons. By international agreement, ^{12}C is taken as the *reference* isotope and its mass is *defined* as 12.000 atomic mass units, or amu. The masses of all other atoms are then measured relative to ^{12}C to provide the dimensionless values of atomic mass displayed in the periodic table. Check this out—it is important to recognize where those numbers come from. For example, an atom of hydrogen (relative mass = 1.00794) is approximately 1/12 the mass of ^{12}C, and oxygen (relative mass = 15.9994) is approximately 16/12 as massive as ^{12}C. From this periodic table of atomic mass relative to ^{12}C, we can compute the molecular mass by simply adding up the appropriate atomic masses. For our covalently bonded H_2O molecules this gives 1.00794 + 1.00794 + 15.9994 = 18.0153.

A key point to note is that the atomic masses listed in the periodic table are *weighted averages* of the individual isotopes of a given element in its natural abundance. Carbon, for instance, is itself an isotopic mixture. In its natural abundance on Earth, carbon is 98.9% ^{12}C and 1.1% ^{13}C such that its average relative mass works out to be 12.01. A very interesting case is chlorine, because it has a large fractural separation between its two dominant isotopes. The lighter isotope, ^{35}Cl, has a mass of 34.9689 and the heavier isotope, ^{37}Cl, has a mass of 36.9659. But the isotopic fraction of ^{35}Cl is 75.77% and that of ^{37}Cl 24.23% so the average relative mass is

$$(0.7577)(34.9689) + (0.2423)(36.9659) = 35.45$$

This is recorded as the single number that appears in the periodic table.

Check Yourself 4

Problem: Remarkably, the known isotopes of silver are 46 in number. However, only 2 occur naturally: ^{107}Ag and ^{109}Ag. Using mass spectrometer data provided in the table below, calculate the *atomic mass* of Ag.

Isotope	Abundance (%)	Mass (amu)
^{109}Ag	48.16	108.90476
^{107}Ag	51.84	106.90509

Approach: This problem has the objective of highlighting how to calculate the single number, the "atomic mass" of the element that appears as a single number in the periodic table. To calculate the atomic mass we must know both the measured mass of each isotope in amu and the fractional abundance. We must then multiply the fractional abundance of each isotopic mass by each isotopic mass. The atomic mass is then the sum of the isotopic portions.

Solution: First we calculate the contribution to the atomic mass from each isotope:

Contribution of atomic mass from ^{109}Ag
= [Abundance(g)][Mass(amu)]
= (0.4816)(108.90476 amu) = 52.45 amu

Contribution of atomic mass from ^{107}Ag
= (0.5184)(106.90509 amu) = 55.42 amu

Thus the *atomic mass* of Ag is
= 52.45 amu + 55.42 amu = 107.87 amu

Check: To check our calculations in this case we can refer to the periodic table, which gives the *atomic mass* of silver as 107.87 amu. Our calculation is correct!

Avogadro's Number

The requirement for a practical quantitative link between the microscopic and macroscopic is that it must allow us to discuss a *manageable* amount of material with a conveniently large but *fixed* number of atoms or molecules. By international agreement this is taken to be the number of atoms in 12.000 grams of ^{12}C. This 12.000 grams of ^{12}C contains Avogadro's number (N_A) of carbon-12 atoms. N_A is a measurable quantity, and it is a *very* large number indeed: 6.02×10^{23}. The fact that it takes that many atoms to constitute a macroscopically realizable mass of a substance is a stark reminder of how small individual atoms are. Avogadro's number of species, N_A, is called a *mole* of that substance and it is typically abbreviated as *mol*. From our mole of ^{12}C atoms in 12.000g of isotopically pure carbon-12, we generalize to mean that a mole of *anything* represents 6.02×10^{23} (Avogadro's number) of those "objects." We can specify a mole of fluorine atoms, a mole of glucose molecules ($C_6H_{12}O_6$), a mole of bricks or a mole of automobiles. The mole is a collective unit just as is a "dozen" or a "score."

> **Approach:** We are given (1) the density of aluminum, which is the mass per unit volume, (2) the element, Al, that has a molar mass of 26.98 g/mol. We know the volume of a sphere is $V = 4/3\pi r^3$ and that there are Avogadro's number of atoms in a mole: 6.022×10^{23}. Thus we have what we need to solve the problem because:
>
> 1. The number of moles of aluminum is
>
> $$\frac{8.55 \times 10^{22} \text{ atoms}}{6.022 \times 10^{23} \text{ atoms}} = 1.42 \text{ mol}$$
>
> 2. The mass of aluminum
> $(0.142 \text{ mol})(26.98 \text{ g Al/mol Al}) = 3.83 \text{ g}$
>
> 3. The volume of aluminum
> $(3.83 \text{ g})(1 \text{ cm}^3/2.70 \text{ g}) = 1.419 \text{ cm}^3$
>
> 4. The relationship between volume and radius
>
> $$V = \frac{4}{3}\pi r^3 \quad \text{so}$$
>
> $$r = \sqrt[3]{\frac{3V}{4\pi}} = \sqrt[3]{\frac{3\left(1.419 \text{ cm}^3\right)}{4\pi}} = 0.697 \text{ cm}$$

Check Yourself 5

It is of considerable importance to be able to intercalculate the number of atoms in a body, the density of the body, and the volume of the body. It serves to illustrate the very large number of atoms in any body with a volume large enough to be seen with the human eye. So suppose we have the density of aluminum, 2.70 g/cm³, and we are given the fact that a particular sphere of aluminum contains 8.55×10^{22} aluminum atoms.

Problem: Calculate the radius of that aluminum sphere.

Molar Mass

We seek to link the counting of atoms or of molecules at the molecular level with the observed mass of substances in the macroscopic world. To do this we define the *molar mass*, M, of *any* species as the mass of one mole of that species with, by international agreement, the molar mass of $^{12}_6C$ is 12.000 g/mol. The *molar mass* for all other atoms scales in direct proportion. The molar mass of hydrogen atoms is 1.00794 g/mol, the molar mass of oxygen atoms is 15.9994 g/mol, and the molar mass of H_2O is 18.0153 g/mol.

But this is an important step—an important link between the microscopic and the macroscopic. Because now, equipped with the concept of the *molar mass*, we can quantitatively link what we weigh on a normal scale that gives us the mass in grams to the *number* of atoms or molecules on the scale. As a result we know exactly how to select measured quantities of a material and match that to the number of atoms or molecules in a given chemical transformation.

Let's look again at the reaction of H_2 with O_2 to form water. This reaction is interesting for a number of reasons, not the least of which is that we can demonstrate it in a chemistry lecture hall *and* it will become an increasingly important means for storing and releasing energy. So just as two *molecules* of H_2 react with one *molecule* of O_2 to form two *molecules* of water, so too do two *moles* of H_2 react with one *mole* of O_2 to form two *moles* of water.

Suppose we execute this reaction by first placing 1 mole of H_2 in a balloon at standard temperature and pressure (that balloon would have a volume of 22.4 liters!). Then we touched a match to the balloon. There would ensue, in a fraction of a second, an explosion with a flash of light (electromagnetic radiation) and a large sphere of water vapor.

From our new stoichiometric perspective, here is what has happened. Beginning with 6.02×10^{23} molecules of hydrogen, H_2, just as the reaction between hydrogen and oxygen was initiated with the high temperature of the match flame, molecules of hydrogen emerged from the balloon into a "sea" of oxygen molecules. One-by-one the H_2 molecules reacted with O_2 in the room air such that 6.02×10^{23} molecules of H_2O were produced. In the process one-half mole ($\frac{1}{2} \times 6.02 \times 10^{23}$ = 3.01×10^{23} molecules) of O_2 was extracted from the air in the room. A superbly accurate counting machine this stoichiometry is! In a fraction of a second 6.02×10^{23} H_2 molecules went into the chemical reaction, 6.02×10^{23} H_2O molecules emerged, and 3.01×10^{23} O_2 molecules were removed from the air in the room. Stoichiometry is locked by molecular structure, molecular structure is locked by the bonding of atoms.

Notice that with the conversion of 6.02×10^{23} molecules of H_2, the reaction stops abruptly. While there is still a sea of oxygen in the room, there is no more H_2 to engage the chemical transformation of hydrogen and oxygen to water. Thus H_2 is the reactant in short supply, it is the scarce commodity, and it is termed the *limiting reactant* or *limiting reagent*. Note that we have not, in our explosion of the hydrogen balloon, carefully controlled the fraction of reactants, yet the fraction of hydrogen molecules was precisely mated to the fraction of oxygen molecules to produce the correct number of H_2O molecules. This example of the *limiting reagent* is very important and will recur repeatedly.

Check Yourself 6

We add 4.86 g of metallic zinc (Zn) to 75.0 mL of 1.49 M (1.49 moles/liter) hydrochloric acid, HCl, solution, and the reaction

$$Zn + 2HCl \rightarrow ZnCl_2 + H_2$$

proceeds to completion. Find the number of liters of hydrogen gas produced at STP, and the molar concentration of zinc chloride.

Solution: We do not know at the start whether those amounts of reagents conform to the balanced chemical equation, so we calculate

$$\text{mol Zn} = (4.86 \text{ g})(1 \text{ mol}/65.39 \text{ g})$$
$$= 0.0743 \text{ mol Zn}$$

$$\text{mol HCl} = (0.0750 \text{ L})(1.49 \text{ mol/L})$$
$$= 0.1118 \text{ mol HCl}$$

Because the ratio mol HCl/mol Zn < 2, where 2 is the ratio of stoichiometric coefficients, the amount of HCl present is insufficient to consume all the Zn, and therefore HCl is the limiting reagent. (We could have reached this same conclusion by using the balanced chemical equation to compute the ratio of combining masses, and comparing that to the mass ratio actually present.) The amounts of $ZnCl_2$ and H_2 formed are thus determined by the amount of HCl. For H_2,

$$L \ H_2 = 0.1118 \text{ mol HCl } (1 \text{ mol } H_2/2 \text{ mol HCl})(22.4 \text{ L/mol } H_2) = 1.25 \text{ L}$$

and for $ZnCl_2$,

$$\text{mol } ZnCl_2 = 0.1118 \text{ mol HCl } (1 \text{ mol } ZnCL_2/2 \text{ mol HCl}) = 0.0559 \text{ mol}$$

If we assume negligible change in solution volume during reaction, then M $ZnCl_2$ = 0.0559 mol/0.0750 L = 0.745 M. If the reaction had been the result of mixing two solutions, the volumes would be (approximately) additive, and the molarity of the products would be obtained using the total solution volume.

Balancing Chemical Equations

Just as nature "balances" the fraction of atoms contained in reactants to match that of the atoms contained in the products, so too must we do the same when we write a chemical equation expressing that chemical reaction. By balancing a chemical equation we simply mean that the number of atoms contained in the reactants must equal the number of those same atoms in the products. As noted before, the coefficients required to do this are termed the "stoichiometry coefficients" and for the reaction of hydrogen and oxygen to form water:

$$2H_2 + O_2 \rightarrow 2H_2O$$

the stoichiometric coefficient of the H_2 is two, that for O_2 is one, and the stoichiometric coefficient for H_2O is two. A key point in determining the stoichiometric coefficients is to recognize the difference between (1) the *subscript* in a chemical formula that establishes the identity of the molecule in the chemical equation and

(2) the stoichiometric coefficient itself. The subscript determines the molecular structure of the reactants and products. The stoichiometric coefficients determine the *amount* of each molecule engaged in the chemical transformation.

In order to demonstrate the process of balancing a chemical equation consider the combustion (burning) of natural gas (CH_4) using a Bunsen burner. When we "light" the Bunsen burner in a room containing oxygen we know that the *reactants* are CH_4 and O_2. We also know, by virtue of a long history of experiments, that the products formed in the burning of CH_4 in O_2 are carbon dioxide and water. This is shown graphically in Figure 2.26. The *molecular structures* involved as reactants and as products of this reaction have been established by direct observation. Based on this we write the reactants on the left-hand side of the chemical equation and the products on the right-hand side of the equation.

$$CH_4 + O_2 \rightarrow CO_2 + H_2O$$
$$\uparrow$$

chemical equation, but an
unbalanced
chemical equation

This is an important step—it specifically identifies all the reactants, all the products, and the specific molecular formula associated with each. In balancing a chemical reaction (that is in determining the stoichiometric coefficients that determine the *relative* amounts of each reactant molecule (or atom) and of each product molecule), we *cannot* change the subscripts in each chemical formula. We can only change the coefficient determining the amount of each molecule that takes part in the chemical reaction. For methane combustion, when we count the number of atoms on the reactant (left) side of the equation, we have four hydrogen atoms, one carbon atom, and two oxygen atoms. On the product (right) side of the equation, we have two hydrogen atoms, three oxygen atoms, and one carbon atom. Thus as written, the chemical equation implies that we are destroying two hydrogen atoms and producing one oxygen atom. We are thus violating the law of conservation of mass.

We begin the process of balancing our chemical reaction by noting that a single carbon atom appears in only one molecule on the reactant side and only one molecule on the product side; CH_4 and CO_2, respectively. Thus, the stoichiometric coefficient must be, in the final balanced chemical equation, the *same* for both CH_4 and CO_2. We begin by setting the stoichiometric coefficient for each equal to one. But CH_4 has four hydrogen atoms, so in order to balance the hydrogen atoms on the left- and right-hand sides of the equation, we multiply the hydrogen containing molecule (H_2O) on the right by the stoichiometric coefficient 2:

$$CH_4 + O_2 \rightarrow CO_2 + 2H_2O$$
$$\uparrow$$

still **unbalanced**

Now the number of hydrogen atoms on the left- and right-hand sides balance, the number of carbon atoms on the left- and right-hand sides balance, *but* we have four oxygen atoms on the right-hand (product) side, but only two oxygen atoms on the left-hand (reactant) side of the chemical equation. If we multiply the number of oxygen molecules in the left-hand side by a factor of two, we have

$$CH_4 + 2O_2 \rightarrow CO_2 + 2H_2O$$

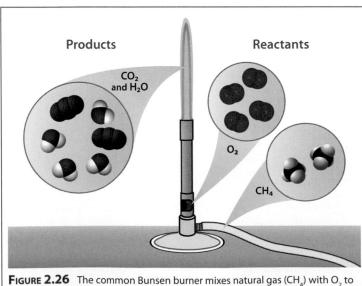

Products   Reactants

CO_2
and H_2O

O_2

CH_4

FIGURE 2.26 The common Bunsen burner mixes natural gas (CH_4) with O_2 to form CO_2 and H_2O, releasing heat as a result of the chemical reaction.

We now have one carbon atom, four hydrogen atoms, and four oxygen atoms on the left-hand (reactant) side of the chemical reaction and one carbon atom, four hydrogen atoms, and four oxygen atoms on the right-hand (product) side. The chemical reaction is balanced.

Given that 80% of the world's energy is produced by the combustion of natural gas, petroleum and coal, it is important to examine this important class of reactions—the combustion of carbon/hydrogen based fuels—in some detail. To this end, we will develop the technique of balancing chemical equations using two important cases: (1) methyl alcohol, which can be used as a liquid fuel for transportation and (2) octane, C_8H_{18}, which is the primary constituent of gasoline. We will use a four step process applied to each of these examples in Table 2.3. It is important to work through the examples in Table 2.3 to become familiar with balancing chemical reactions.

TABLE 2.3

Balancing Chemical Equations: A Stepwise Approach	Write a balanced chemical equation for the combustion of methanol, CH_3OH, in air	Write a balanced chemical equation for the combustion of gasoline, which is comprised primarily of octane, C_8H_{18}
1. Write an initial chemical equation that specifies the chemical formula for the reactants on the left side and the products on the right side.	As a general rule, whenever a compound containing C, H, and O is combusted in air, it reacts with O_2 to form carbon dioxide and water; thus the unbalanced skeletal equation is: $$CH_3OH + O_2 \rightarrow CO_2 + H_2O \text{ (unbalanced)}$$	Hydrocarbon compounds, those containing carbon and hydrogen, no matter how complex, burn in air to produce CO_2 and H_2O. This covers the combustion chemistry of some 3 million compounds! Thus our unbalanced skeletal equation is: $$C_8H_{18} + O_2 \rightarrow H_2O + CO_2 \text{ (unbalanced)}$$
2. Balance the atoms that occur in more complex substances first.	Carbon atoms are balanced on both sides of the chemical equation. CH_3OH has 4 hydrogen atoms, so we place a stoichiometric coefficient of 2 in front of H_2O on the right-hand side: $$CH_3OH + O_2 \rightarrow CO_2 + 2H_2O \text{ (unbalanced)}$$	To balance carbon atoms, we apply a stoichiometric coefficient of 8 to CO_2 on the right-hand side: $$C_8H_{18} + O_2 \rightarrow H_2O + 8CO_2 \text{ (unbalanced)}$$ To balance the hydrogen atoms we apply a stoichiometric coefficient of 9 to H_2O on the right side: $$C_8H_{18} + O_2 \rightarrow 9H_2O + 8CO_2 \text{ (unbalanced)}$$
3. Balance atoms that occur as free elements on either side of the equation last.	This balances H atoms, but we have 4 atoms on the right and 3 on the left. If we place a stoichiometric coefficient of 3/2 in front of the O_2 on the left, we will balance O atoms on both sides of the chemical equation: $$CH_3OH + 3/2 \, O_2 \rightarrow CO_2 + 2H_2O$$	We have now balanced the carbon and hydrogen atoms, but there are 25 oxygen atoms on the right-hand side so we apply a stoichiometric coefficient of 25/2 to O_2 on the left-hand side: $$C_8H_{18} + 25/2 \, O_2 \rightarrow 9H_2O + 8CO_2$$ The equation is now balanced for all C, H and O atoms
4. If the balanced equation contains coefficient fractions, clear these by multiplying the entire equation by the denominator of the fraction.	Although the equation is now balanced, and is strictly speaking an acceptable chemical equation, the convention is to clear fractions. Thus we multiply each stoichiometry coefficient by a factor of 2: $$2CH_3OH + 3O_2 \rightarrow 2CO_2 + 4H_2O$$	We can clear the fractions in the stoichiometric coefficients by multiplying by a factor of 2: $$2C_8H_{18} + 25O_2 \rightarrow 18H_2O + 16CO_2$$

Solution: First, based on the meanings of the prefixes micro and milli, convert from microliters to liters, and then to milliliters. At this point, bring in density as a conversion factor to obtain the mass in grams. The remaining factors use molar mass to convert mass to the number of moles of substance and, finally, the Avogadro constant to convert number of moles to the number of molecules. In summary, the conversion pathway is $\mu L \rightarrow L \rightarrow mL \rightarrow g \rightarrow mol \rightarrow molecules$.

$$? \text{ molecules } C_2H_6S =$$

$$1.0\mu L \times \frac{1\times10^{-6}L}{1\mu L} \times \frac{1000mL}{1L} \times \frac{0.84 \text{ g } C_2H_6S}{1 \text{ mL}}$$

$$\times \frac{1 \text{ mol } C_2H_6S}{62.1 \text{ g } C_2H_6S} \times \frac{6.022\times10^{23} \text{ molecules } C_2H_6S}{1 \text{ mol } C_2H_6S}$$

$$= 8.1 \times 10^{18} \text{ molecules } C_2H_6S$$

Practice Example A:
Gold has a density of 19.32 g/cm³. A piece of gold foil is 2.50 cm on each side and 0.100 mm thick. How many atoms of gold are in this piece of gold foil?

Practice Example B:
If the 1.0 μL sample of liquid ethyl mercaptan is allowed to evaporate and distribute itself throughout a chemistry lecture room with dimensions 62 ft × 35 ft × 14 ft, will the odor of the vapor be detectable in the room? The limit of human detectability is 9×10^{-4} μmol/m³.

Check Yourself 7
Natural gas that leaks into buildings and homes can be dangerous for two reasons. First, a spark from any electrical device can ignite a mixture of CH_4 and O_2, leading to a large explosion. Second, CH_4 can replace O_2 in a room, reducing the amount of available oxygen to breathe. To alert inhabitants to a gas leak, very small amounts of substance is added to the (odorless) CH_4 before it is placed in the pipeline. The volatile liquid ethyl mercaptan, C_2H_6S, is one of the most odoriferous substances known. It is sometimes added to natural gas to make gas leaks detectable. How many C_2H_6S molecules are contained in a 1.0 μL sample? The specific gravity of ethyl mercaptan is 0.826.

While we have now discussed a number of cases of hydrocarbon combustion in O_2 that, in virtually all cases, produces H_2O and CO_2 as a product, let's examine the reverse reaction: that of photosynthesis. Photosynthesis, as we have already discussed, uses energy from solar photons to create a remarkable variety of organic structures from the simple ingredients H_2O and CO_2, both taken directly from the atmosphere through the leaf structure of the plant. Let's use our knowledge of moles, molar mass, and balanced chemical equations to solve a *problem*.

Problem:
Plants convert carbon dioxide and water into glucose ($C_6H_{12}O_6$) and molecular oxygen. Through laboratory measurements you establish that a particular plant your are studying consumes 37.8 g CO_2 in a period of a week. What mass of glucose can the plant synthesize from the CO_2 extracted directly from the atmosphere?

Solution:
First we must write the balanced chemical equation for the reaction:

$$CO_2 + H_2O \rightarrow O_2 + C_6H_{12}O_6$$

Balancing the carbon atoms:

$$6CO_2 + H_2O \rightarrow O_2 + C_6H_{12}O_6$$

Then the hydrogen atoms:

$$6 CO_2 + 6H_2O \rightarrow O_2 + C_6H_{12}O_6$$

Finally we balance the oxygen atoms:

$$6CO_2 + 6H_2O \rightarrow 6O_2 + C_6H_{12}O_6$$

Next we recognize that, if we supplied the plant with water throughout the period, there was sufficient H_2O to react with the (essentially limitless) supply of CO_2 taken from the atmosphere.

Now the problem becomes one of interconverting the mass of

the reactants and or products to moles. A systematic approach to the problem is to first "model" the problem by visualizing what is occurring.

Examine What Is Given and What Is the Objective:
- Information Provided: 37.8 grams of CO_2 consumed
- Objective: Calculate the *Mass* of $C_6H_{12}O_6$

What we know from our model:
1. We can relate the number of moles of CO_2 required (given the availability of adequate H_2O) for the production of 1 mole of glucose.
2. We can calculate the molar mass of both CO_2 and $C_6H_{12}O_6$

Set a Strategy:
Using our balanced chemical equation in the model, we can calculate the number of *moles* of CO_2 given the number of *grams* of CO_2 and then, through the stoichiometric coefficients in our model, we can calculate the *ratio* of moles of glucose to moles of carbon dioxide. Once we have the number of moles of $C_6H_{12}O_6$, we can calculate the mass of glucose produced by calculating the molar mass of $C_6H_{12}O_6$ and then multiplying by the number of moles of glucose.

- Molar mass of CO_2 = 12.01 + 2(16.00) = 44.01 g/mol
- Number of moles of CO_2 = 37.8 g CO_2/44.01 g CO_2/mole
$$= 0.859 \text{ moles } CO_2$$
- Ratio of number of moles of $C_6H_{12}O_6$ to number of moles of CO_2: From stoichiometric coefficients in our model: 1 mole $C_6H_{12}O_6$ to 6 moles CO_2
- Number of moles of $C_6H_{12}O_6$: 1/6 (0.859 moles CO_2) = 0.143 moles glucose
- Mass of glucose: multiply the molar mass times the number of moles of glucose
- Molar mass of glucose: 6(12.01) + 12(1.01) + 6(16.00) = 180.18 grams $C_6H_{12}O_6$/mole
- Mass of glucose from 37.8 g CO_2 is then: (0.143 moles $C_6H_{12}O_6$)(180.18 grams $C_6H_2O_6$/mole) = 25.77 grams

Check How Reasonable the Results Are in the Context of the Model:
Note first that the calculated mass of the glucose produced is less than the mass of CO_2 consumed. Examination of the model for this process shows that each carbon atom in glucose has *one* oxygen atom in the molecular structure, while CO_2, the carbon containing reactant, has two oxygen atoms associated with each carbon atom. The photosynthesis reaction has substituted hydrogen atoms (which are much less massive than oxygen) for oxygen atoms. Thus the carbon containing product should contain less mass than the carbon containing reactant. It does.

Oxidation-Reduction Reactions

Chemical reactions are all about electron rearrangement—the formation of molecular bonds involving the outer most electrons in an atom and then the rearrangement of those electrons to form new molecular structures as a result of a chemical reaction. Oxidation-reduction reactions or *redox reactions* are reactions in which electrons are *transferred* from one reactant atom to another in the process of a chemical reaction taking place to form products—new chemical structures. Redox reactions constitute one of the most important classes of reactions, and as we develop a picture of modern chemical research and its intimate link to solving global energy problems, there will be repeated analysis of this category of chemical reaction. Thus we consider redox reactions here in some detail. As we will see, while redox reactions often involve oxygen, they need not. However, we will use combustion reactions, reactions that involve hydrocarbons reacting with oxygen to produce water and carbon dioxide, to introduce redox chemistry.

We have identified the fact that for a *redox reaction*, the key event is the net movement of one or more electrons. Any *loss* of electrons from a given atom is referred to as *oxidation*; a designation that emerged because it was in fact combustion of substances (hydrocarbons) in air that lead to an understanding of electron transfer in chemical reactions. But electrons do not simply *drop off* an atom or molecule. Electrons are *pulled* off by another atomic or molecular structure that has a greater ability to draw electrons to it. We will examine this tendency to extract electrons in subsequent chapters. For every *oxidation* event where an electron is given up, another atom or molecule has succeeded in forcefully extracting that electron or electrons into its valence shell. Thus while *oxidation* is the loss of an electron, *reduction* is the gaining of an electron. The processes of oxidation and reduction are locked together; what one member of a chemical reaction loses, the other gains because electrons are neither created nor destroyed in a chemical reaction. In the competitive market place of chemical reactions, electrons are transferred but not lost. The redox transfer of an electron is thus a gain of charge by one entity and a loss of charge by another in pair-wise organization. To keep a close accounting of this electron transfer, this redox process, we use a convenient device known as an *oxidation* state. An *atom's oxidation state* (we keep track of electron exchange atom-by-atom) is a positive or negative number that designates that atom's electronic charge *relative to a conveniently assigned normal or standard condition*. The oxidation state for an atom in elemental form is zero. Without outside influence (such as what happens in the formation of a chemical bond) all atoms are in full possession of their electrons. When these atoms in elemental form engage in the formation of chemical bonds, it is necessary to develop a procedure for assigning oxidation numbers. The rules for assigning oxidation states are laid out in Table 2.4. This table is a hierarchy of rules for assigning oxidation states, and we will follow the hierarchy laid out in Table 2.4 every time we assign oxidation numbers in a chemical reaction.

When assigning oxidation numbers, which must be done in general for thousands of compounds, the oxidation number of a given atom (or element) depends upon the other elements in the compound. That is why there is a hierarchy of rules, particularly for the nonmetals that represent a very large range in their ability to extract electrons from other atoms. But there are important patterns to this giving and receiving of electrons (or more appropriately to this stealing and surrendering of electrons). The group 1A and 2A elements are unable to retain electrons in vir-

TABLE 2.4

Rules of oxidation states	Example
1. As noted, the oxidation state of an atom in a free element is zero	Fe oxidation state: 0 Cl in Cl_2 oxidation state: 0
2. The oxidation state of a monoatomic ion is equal to its charge	Na^+ oxidation state: +1 F^- oxidation state: -1
3. Because electrons are neither produced nor destroyed in a redox reaction: • For a neutral molecule or molecular formula, the sum of the oxidation states is zero • For an ion, the sum of the oxidation states is equal to the charge of the ion	H_2O: $2 \times$ (H oxidation state) $+ 1 \times$ (O oxidation state) $= 0$ PO_4^{3-}: $1 \times$ (P oxidation state) $+ 4 \times$ (O oxidation state) $= -3$
4. Metals, *when they form compounds*, have positive oxidation states • Group 1A metals always have an oxidation state of +1 • Group 2A metals always have and oxidation state of +2	NaCl Na has a +1 oxidation state MgF_2 Mg has a +2 oxidation state
5. When *nonmetals* form compounds, those nonmetals are assigned values according to the table displayed here. Within this table, there is an ordering of importance with the most important rule beginning at the top	**Oxidation States of Nonmetals** | Nonmetal	Oxidation State	Example
---	---	---
Fluorine	– 1	MgF_2 – 1 ox state
Hydrogen	+ 1	H_2O + 1 ox state
Oxygen	– 2	CO_2 – 2 ox state
Group 7A	– 1	CCl_4 – 1 ox state
Group 6A	– 2	H_2S – 2 ox state
Group 5A	– 3	NH_3 – 3 ox state	

tually all cases, so they have oxidation states of +1 and +2 respectively. As for those elements that "steal" electrons, it is important to give priority to the elements highest on the list in Table 2.4. Finally, when assigning oxidation states to elements not covered by rules 4 and 5, deduce their oxidation state by first assigning oxidation state to all other elements in the molecular structure and then using Rule 3. For example, in the reaction of elemental carbon with oxygen, the reaction is

$$C + 2O \rightarrow CO_2$$

The oxidation state of O is –2, so with two oxygen atoms, and the fact that the total of the oxidation states for a neutral molecule must be 0, carbon has an oxidation state of +4.

Check Yourself 8

Determining the oxidation state of each element in a molecule requires practice—in particular the engagement of rules presented in Table 2.4. As an example, determine the oxidation state of each of the underlined elements in the following molecules:

(1) $\underline{Fe}_3O_4$ (2) $Na\underline{H}$ (3) $\underline{Al}_2O_3$

(4) $\underline{P}_4$ (5) $\underline{Mn}O_4{}^-$ (6) $H_2\underline{O}_2$

Oxidation and Reduction in Combustion Reactions

Given that combustion reactions are an important category of redox reactions, lets consider what happens to the electrons associated with the combustion of natural gas. We first write the balanced chemical equation for CH_4 combustion:

$$CH_4 + 2O_2 \rightarrow 2H_2O + CO_2$$

Next we assign oxidation numbers to *every* atom in the reaction, both reactants on the left and products on the right.

From Table 2.4, there is no fluorine, so we assign an oxidation state of +1 for hydrogen, and then set the sum of oxidation numbers for those compounds containing hydrogen equal to zero:

CH_4 so C has an oxidation number of –4

H_2O so O has an oxidation number of –2

Oxygen is the next member below hydrogen in Table 2.4. Thus for

CO_2 carbon has an oxidation number of +4

In this reaction, therefore, we can assign oxidation numbers

Solution:

(1) $\underline{Fe}_3O_4$
Inspection of Table 2.4 defining the order of assigning oxidation states of nonmetals, we note that Fe_3O_4 does not contain fluorine, F; does not contain hydrogen, and next in order in oxygen. Thus we assign each O atom an oxidation state of –2. Thus the four oxygen atoms contribute a total of –8. To balance the oxidation state of oxygen, each Fe atom must contribute +8/3 such that the net for the (neutral) molecule Fe_3O_4 is $4(-2) + [3(+ \text{8/3})] = 0$.

(2) $Na\underline{H}$
Rule 4 states that the oxidation state for Group 1A metals, of which Na is an example, always have an oxidation state of +1. Therefore, because Rule 3 states that "for a neutral molecule or molecular formula, the sum of the oxidation states must be zero," the oxidation state of H must be –1.

(3) $\underline{Al}_2O_3$
As with example (1) above, given that this molecule has no fluorine or hydrogen, the rule for nonmetals applies and O atoms have an oxidation state of –2. Thus the three oxygen atoms contribute an oxidation state of –6 to the molecule, and to balance, each Al atom must have an oxidation state of +3.

(4) $\underline{P}_4$
This is the formula for elemental phosphorus. For an atom within a free element, the oxidation state must be zero as reflected in Rule 1 of Table 2.4. This must also be the case because the oxidation state of an atom defines its relative ability to extract or donate electron density by virtue of the formation of the molecular bond. If all atoms in a molecule are the same, then there is no relative advantage or disadvantage between atoms in a molecule, so the oxidation state must be zero!

(5) $\underline{Mn}O_4{}^-$
This is the permanganate ion—a very common ion in chemistry. We note first that the sum of oxidation states for all atoms in the molecule must be –1, the net change on the ion. This is Rule 2 in Table 2.4. Referring next to assignment of oxidation states to the nonmetals in the structure, each oxygen atom has an oxidation state of –2 (there are no F atoms or H atoms), so oxygen contributes an oxidation of $(4)(-2) = -8$ to the molecule. This must be balanced by a Mn oxidation state of +7 such that $(4)(-2) + (+7) = -1$ to satisfy Rule 2 in Table 2.4.

(6) $H_2\underline{O}_2$
In this case we apply the assignment of oxidation states of elements within a molecular structure in order: There are no fluorine atoms. There are two H atoms so each H atom is assigned an oxidation state of +1. Thus to balance the oxidations states of all atoms in the molecule, each O atom must have an oxidation state of –1.

to all atoms in the balanced chemical equation

$$CH_4 + 2O_2 \rightarrow CO_2 + 2H_2O$$

In the reaction overall, we see that carbon has been *oxidized* (its oxidation number changed from −4 to +4) and oxygen has been reduced (its oxidation number changed from 0 to −2). Thus, indeed, the combination of hydrocarbons in air is an oxidation-reduction reaction—a *redox* reaction. Most importantly, we have used *oxidation numbers* to watch over and quantitatively evaluate the *movement of electron charge* in a chemical reaction.

Redox Reactions Where Oxygen Is Not Involved

While reactions with oxygen were largely responsible for elucidating the importance of electron transfer or electron movement in chemical reactions, and thus the term *oxidation-reduction* (redox) reactions became common to the language of chemistry, the presence of oxygen is not necessary for oxidation-reduction to take place. All that is required is the *transfer* of electrons to take place in the reaction.

Therefore, *redox* reactions include any reaction wherein there is a *change in the oxidation state* of atoms in going from reactants to products. In going from reactants to products:

- The oxidizing agent removes electrons from another substance (i.e oxidizes another substance), and in the process is itself reduced.
- The reducing agent donates an electron to another substance and is itself oxidized.

While we will return repeatedly to oxidation-reduction processes, we consider here two examples of redox reactions that do not involve oxygen.

Movement of electron charge occurs in the formation of both covalent (as we have just seen) and ionic compounds. Consider first the example of hydrogen reacting with chlorine to produce HCl:

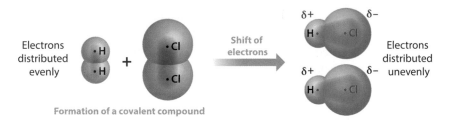

Formation of a covalent compound

In this case we can assign oxidation numbers using our rules:

$$H_2 + Cl_2 \rightarrow 2HCl$$

and we can see explicitly in the figure above that hydrogen has transferred electron density to Cl in the HCl bond.

If we react zinc metal with iron cations in solution:

$$Zn(s) + Fe^{2+}(aq) \rightarrow Zn^{2+}(aq) + Fe(s)$$

there is an explicit transfer of electrons from zinc to the iron cation, forming neutral iron atoms. In this case $Zn(s)$ is oxidized and $Fe^{2+}(aq)$ is reduced.

The central importance of redox reactions to chemical reactivity is covered in Case Study 2.1.

Summary Concepts

1. Atomic View of Matter

Modern chemistry is built upon the fundamental concept that matter is comprised of discrete entities called atoms. The atomic theory of matter, articulated in 1808 by John Dalton engaged four postulates:

- All matter consists of discrete indivisible particles as *atoms* that cannot be created or destroyed.
- Atoms of a specific element are distinct from any other element, and while a chemical reaction can rearrange those atoms into different molecular structures, a chemical reaction cannot change one element into another.
- All atoms of a particular element possess the chemical properties that are distinct from all other elements.
- A given chemical compound always has the same ratio of atoms that comprise the specific molecular structure.

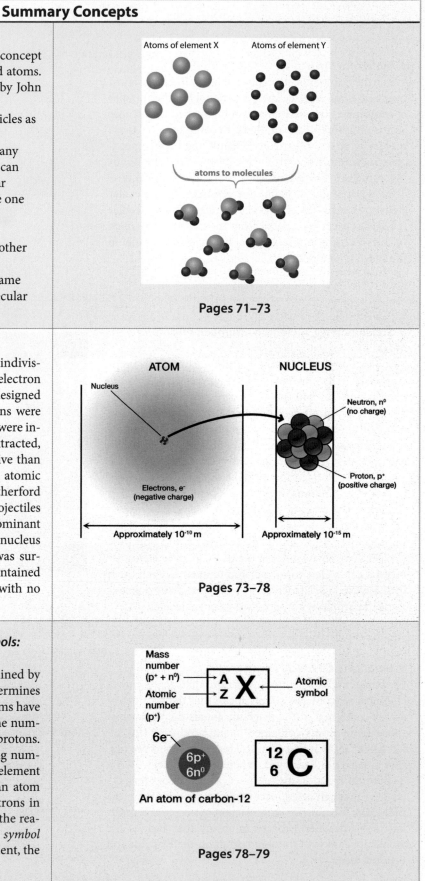

Pages 71–73

2. Discovery of the Electron and the Nucleus

The view that all matter in nature is comprised of indivisible *atoms* was dispelled with the discovery of the electron by J. J. Thompson using an array of carefully designed cathode ray tubes that demonstrated that electrons were (1) particles that carried a negative charge and (2) were independent of the element from which they were extracted, and (3) were approximately 2000 times less massive than the hydrogen atom. The discovery rewrote the atomic model. The atomic nucleus was discovered by Rutherford using $\propto$ particles (helium nuclei) as subatomic projectiles fired at a thin gold film demonstrating that the dominant fraction of the atoms' mass was contained in a tiny nucleus at the center of the atom and that the nucleus was surrounded by a cloud of electrons. The nucleus contained both protons with positive charge and neutrons with no net charge.

Pages 73–78

3. Atomic Mass, Mass Number, and Atomic Symbols: Isotopes

The unique identification of an element is determined by the number of protons in the nucleus, which determines the *atomic number* of a given element. Because atoms have no net charge in their naturally occurring state, the number of electrons in an atom equals the number of protons. But because atoms can have nuclei with a differing number of protons and neutrons, atoms of the same element can have a different mass. The *mass number* of an atom is equal to the *total* number of protons and neutrons in the nucleus. This differing number of neutrons is the reason elements have different *isotopes*. The *atomic symbol* identifies a specific isotope by identifying the element, the atomic number and the mass number.

Pages 78–79

Summary Concepts

4. Molecular Structure

While the number of elements that occur in nature is just over one hundred (see the periodic table), the number of molecular structures that can be formed from those atoms numbers in the millions. Keeping track of those molecular structures involves identifying the geometrical structure of the particular molecule *and* the bonding structure associated with molecular architecture. This requirement to define the geometry of molecular structures leads to chemical formulas and molecular models of ascending detail from the *molecular formula* to the *structural formula* to the *ball and stick* model to the *space-filling* model.

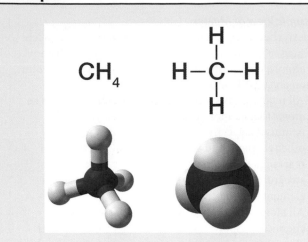

Pages 79–85

5. Stoichiometry, Avogadro's Number, and Molar Mass

Stoichiometry in chemistry is all about counting—keeping track of the number of atoms (and electrons) in the reacting species and the number of atoms in the product species. Atoms cannot be created nor destroyed in a chemical reaction. It is also important, in this counting of atoms, to relate the number of atoms or molecules that are present in a sample that we typically employ in the macroscopic world. To make this conversion from the number of atoms or molecules contained in a given sample to the mass or volume of that sample we use an agreed upon count—Avogadro's number, which is the number of $^{12}_{6}C$ atoms in 12.000 grams of isotopically pure carbon. Avogadro's number is 6.02×10^{23}. The *molecular mass* is calculated from the atomic mass of each atom in the molecular structure. A *mole* of any substance is Avogadro's number of those atoms or molecules.

Pages 85–89

6. Balancing Chemical Reactions

By balancing a chemical equation that represents a chemical reaction we simply mean that the number of atoms contained in the reactants equals the number of those same atoms contained in the products. This reflects the fact that in a chemical reaction, atoms are neither created nor destroyed. When, for example, we write a chemical reaction for octane, C_8H_{18}, with oxygen, O_2, we step through a sequence of properly identifying the molecular formula for each reactant molecule and product molecule, then proceed to ensure that each of the atoms appears with the same count on the reactant and product side.

Step 1: Identify the correct molecular formulas
$$C_8H_{18} + O_2 \rightarrow CO_2 + H_2O$$
Step 2: Balance carbon atoms
$$C_8H_{18} + O_2 \rightarrow 8\,CO_2 + H_2O$$
Step 3: Balance hydrogen atoms
$$C_8H_{18} + O_2 \rightarrow 8\,CO_2 + 9\,H_2O$$
Step 4: Balance oxygen atoms
$$C_8H_{18} + 25/2\,O_2 \rightarrow 8\,CO_2 + 9\,H_2O$$
Step 5: Clear Fraction
$$2\,C_8H_{18} + 25\,O_2 \rightarrow 16\,CO_2 + 18\,H_2O$$

Pages 89–93

Summary Concepts

7. Oxidation-Reduction Reaction

Oxidation-reduction reactions, or *redox reactions*, are reactions in which electrons are transferred or delocalized from one reactant atom to another in the process of a chemical reaction taking place to form products. Redox reactions constitute an extremely important class of chemical reaction, and are considered in some depth in Case Study 2.1. The redox transfer of an electron is a gain of charge by one entity and a loss of charge by another in pair-wise organization because electrons are neither created nor destroyed in a chemical reaction. To keep specific count of the electrons transferred or shifted in a chemical reaction we use the convention of an *oxidation state*. An atom's oxidation state is a positive or negative number that designates the atom's electronic charge relative to a conveniently assigned standard state. The rules assigning oxidation states are reviewed at right.

Rules of Oxidation States

1. As noted, the oxidation state of an atom in a free element is zero

2. The oxidation state of a monoatomic ion is equal to its charge

3. Because electrons are neither produced nor destroyed in a redox reaction:
   - For a neutral molecule or molecular formula, the sum of the oxidation states is zero.
   - For an ion the sum of the oxidation states is equal to the charge of the ion.

4. Metals, *when they form compounds*, have positive oxidation states
   - Group 1A metals always have an oxidation state of +1
   - Group 2A metals always have an oxidation state of +2

5. When *nonmetals* form compounds, those nonmetals are assigned values according to the table displayed here. Within this table, there is an ordering of importance with the most important rule beginning at the top

Oxidation States of Nonmetals

Nonmetal	Oxidation State	Example
Fluorine	− 1	MgF_2 − 1 ox state
Hydrogen	+ 1	H_2O + 1 ox state
Oxygen	− 2	CO_2 − 2 ox state
Group 7A	− 1	CCl_4 − 1 ox state
Group 6A	− 2	H_2S − 2 ox state
Group 5A	− 3	NH_3 − 3 ox state

Pages 93–96

CASE STUDY 2.1 Exponential Growth, Exponential Decay, and the "Law of 70"

KEY CONCEPTS:

As is frequently the case, when a quantity, Q, grows at a constant *percentage* rate it exhibits what is termed *exponential growth*. The reason is that when a quantity Q grows at a constant percentage rate, it means that the growth rate, $\Delta Q/\Delta t$, is *proportional* to the amount of that quantity, Q, so

$$\Delta Q / \Delta t \propto Q .$$

The proportionality constant, k, then establishes the equality between the growth rate, $\Delta Q/\Delta t$ and the quantity Q:

$$\Delta Q / \Delta t = kQ$$

Thus processes that adhere to this simple fractional growth rate follow the "law" that:

$$\Delta Q/Q = \text{constant fraction per unit time}$$

What is remarkable is that this expression applies to processes as diverse as the growth rate of a bank account at a fixed interest rate, radioactive decay, chemical reaction rates, growth rates of CO_2 emission from developing countries, molecular absorption of photons, and growth rates of gross domestic product (GDP) of a nation's economy.

Taking the most common example of the growth of a bank account at a fixed interest rate, supposing the savings account balance is Q and the interest is a fixed 5% per annum, we would have $\Delta Q/Q = 5\%$ per year. Then $[(\Delta Q/Q)/\Delta t] = 0.05 = k$ when t is measured in years and k is a constant.

Then: $\Delta Q/Q = k\Delta t$

Note: if the quantity Q *decays* at a fixed rate rather than *grows*, the equation is simply

$$-\Delta Q/Q = k\Delta t$$

In the limit of $\Delta Q \to 0$, we can write the differential equation:

$$\lim_{\Delta Q \to 0} \Delta Q / Q = dQ / Q = k\, dt \qquad (1)$$

While this differential equation expresses these mathematical relationships between the rate of increase of Q and the quantity Q, what we would like to solve for is the quantity Q at any time, t, which we will designate as Q_t, and the quantity Q initially present, Q_0.

We can solve for Q_t in terms of Q_0 by integrating Eq. (1) from the initial time, t = 0, and some final time, t = t:

$$\int_{initial}^{final} dQ / Q = \int_{t=0}^{t=t} k\,dt$$

The left-hand side of the expression is

$$\int_{Q_0}^{Q_t} dQ / Q = \ln\left(Q_t / Q_0 \right)$$

The right-hand side is, because k is a constant and can be removed from the integral,

$$\int_{t=0}^{t} k\,dt = k \int_{t=0}^{t} dt = kt$$

Thus we have

$$\ln\left(Q_t / Q_0 \right) = kt \qquad (2)$$

    or

$$Q_t = Q_0 e^{kt} \qquad (3)$$

We can plot Q as a function of t as shown in Figure CS2.1a

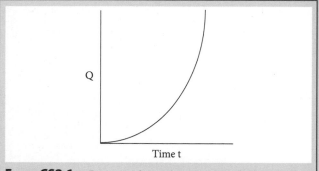

FIGURE CS2.1A Exponential growth, when plotted graphically, has the characteristic shape wherein the quantity Q that, for example, depends on time, t, increases slowly at small values of t but then increases very rapidly with increasing t.

It is the dependence of Q_t on time that establishes the designation "exponential growth."

This *constant percentage rate* increase in Q is characteristic of (1) money residing in a bank account at a fixed rate of interest, (2) economic growth for a nation with a constant percentage rate of economic growth, (3) global growth in energy consumption, just to note a few examples. A key attribute of exponential growth is that the "doubling time," the period of time required for Q to double, t_2, is a constant, independent of Q. We can see this immediately by setting $Q_t = 2Q_0$ in Eq. (2) and t = t_2 giving

$$\ln\left(Q_t / Q_0 \right) = \ln(2) = kt_2$$

Solving for t_2 we have

$$t_2 = \ln(2) / k = 0.693 / k$$

In practice, this is a very convenient expression because the *percentage growth rate*, p, is just 100 times the growth rate, k, so

$$p = 100k$$

and thus

$$t_2 = \frac{0.693}{k} = \frac{69.3}{p} \approx \frac{70}{p}$$

It turns out that this simple expression has remarkable practical applications. This is the "rule of 70," which tells us the number of years required for a doubling of Q for a given percentage increase per year. Thus if Q increases at a constant 7% per year, that quantity will double in

$$t_2 = \frac{70}{p} = \frac{70}{7} = 10 \text{ years}$$

For example, in recent years the economy of China has been growing at 9% per year. Thus the economy of China will double in size every 7.8 years.

The identical mathematics is applicable to exponential decay.

$$\frac{dQ}{Q} = -kdt$$

$$Q = Q_0 e^{-kt}$$

The exponential decay curve is displayed in Figure CS2.1b.

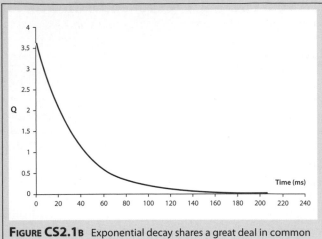

FIGURE CS2.1B Exponential decay shares a great deal in common mathematically with exponential growth except in an exponential decay, the quantity Q decreases initially, very rapidly, and then the rate of decrease diminishes with increasing time.

The "half life," $t_{1/2}$, when $Q_t/Q_0 = \frac{1}{2}$

$$t_{1/2} = \frac{0.693}{k} = \frac{70}{p}$$

So 7% decay per year yields a reduction by ½ in $t_{1/2} = \frac{70}{7} =$ 10 years. Both of these expressions, exponential growth and exponential decay, will turn out to be immensely valuable.

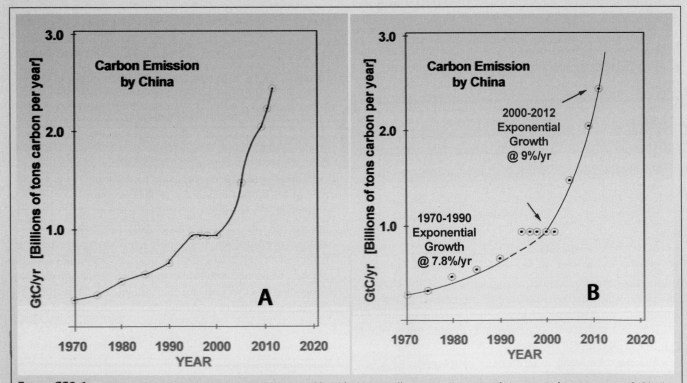

FIGURE CS2.1C Carbon dioxide emission from the combustion of fossil fuel is typically measured in units of gigatons carbon per year, or GtC/yr. In order to convert the mass of CO_2 released from combustion to the mass of carbon released, we need to multiply by the ratio of molecular weights. The molecular weight of CO_2 is 44 amu and that of carbon, 12 amu. The release rate of CO_2 from fossil fuel combustion in China has risen exponentially since 1970. Between 1970 and 1990, growth was 7.8% per year. From 2000 to 2012, 9% per year.

2.40

Exponential growth is used for investments, energy consumption rates, population growth, economic growth, molecular concentration growth, etc. Exponential decay is used for rates of disappearance of chemical species undergoing reactions (Chapter 12), radioactive decay (Chapter 13), absorption of photons (Chapter 11), disappearance of ice structure, etc.

As an important global example of exponential growth, we consider the important case of the increase in carbon emission from China as its economy expands at unprecedented rates. Figure CS2.1c displays the increase in carbon emission from China between the years 1970 and 2012. In 1970, China emitted 0.27 billion tons of *carbon* in the form of CO_2. If we wish to calculate the mass of CO_2 this represents, we must multiply this figure by the ratio of the molecular weight of CO_2 (44 amu) to the molecular weight of carbon (12 amu). Thus 0.27 billion tons of carbons per year (usually written as 0.27 GtC/yr) corresponds to (44/12) 0.27 GtC/yr = 1.0 Gt CO_2/yr. The reason emissions are usually expressed in terms of Gt of carbon rather than of CO_2 is that other carbon containing compounds such as methane, CH_4, are emitted and a vast majority of those carbon compounds are oxidized to form CO_2 in the atmosphere.

An inspection of Figure CS2.1c, panel A, reveals the remarkable rate of increase in CO_2 emission from the combustion of fossil fuel in China as the Chinese economy expands.

Panel B of Figure CS2.1c displays the fitting of the data points with two exponential growth curves. One covers the period 1970 to 1990, for which the fractional increase in CO_2 emissions was 7.8% per year. The second period from 2000 to 2012 is fit by another exponential representing a fractional growth rate of 9.0% per year. This example sets the stage for examining the mathematical properties of exponential growth and decay.

Problem 1

Part A:
Carbon emission worldwide increased at an annual rate of 4% between 1990 and 2010. Beginning in 1990, how long would it take to double global carbon emission?

Part B:
If global carbon emission between 2020 and 2040 increases by 9.2% per year, how many years will it take to double the amount of carbon released each year? In fact, at current growth rates in the developing economies, this *would be* the projected increase per year of carbon release if fossil fuels continue to be the principle source of primary energy.

Part C:
It is important to compare the carbon release rate of the US and China in both the short term and long term. In Figure

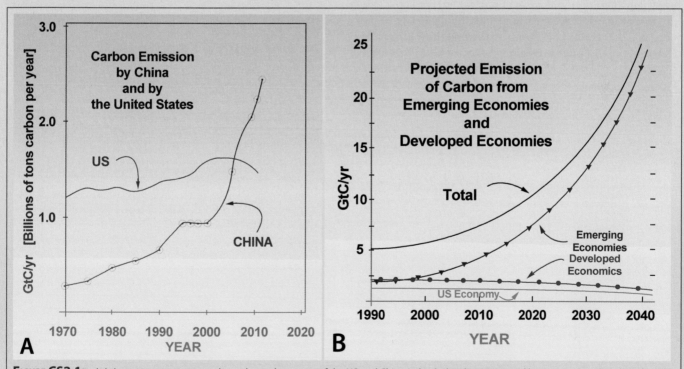

FIGURE CS2.1D It is important to compare the carbon release rate of the US and China in both the short term and long term. Panel A plots the CO_2 release rate in units of GtC/yr for the US and China between 1970 and 2012. China passed the US in carbon emission in 2006. Panel B displays the longer term impact of the exponential growth in carbon release from the developing economies (primarily in China, India, and South America) and from the developed countries. The impact of exponential growth is stark.

CS2.1d, panel A, we plot the superposition of carbon release in the US and carbon release in China from 1970 to 2010. From an inspection of Figure CS2.1d, panel A, we can see that in 2006, carbon emission from China equaled that in the US. If China's emission rate is increasing at 9% per year and the US emission rate does not change, when will the emission rate from China be double that of the US?

Problem 2

Suppose we are studying the growth in the number of bacteria in a biochemistry experiment. If there are initially 100 bacteria and the bacteria exhibit an exponential growth rate of $k = 0.05$:

(a) After a period of 10 hours, what will be the population of bacteria?

(b) What is the doubling time, t_2, for these bacteria?

Problem 3

Considering again a study of bacterial growth, if there are initially 6,000 bacteria, and after a period of 1 hour, it is determined that there are 6,800 bacteria, what period of time is required to reach a population of 10,000 bacteria?

Problem 4

A key consideration in the use of nuclear energy is the half-life, $t_{1/2}$, of key isotopes. If $t_{1/2}$ of plutonium 239 is 24,000 years, and 100 grams of ^{239}Pu are in a sample, how long will it take until only 10 grams remain?

CASE STUDY 2.2   Calculation of Personal Energy Use and an Analysis of Options

The multiple ways that we, as individuals, use energy makes a quantitative analysis of the problem potentially complex. However, it is remarkably instructive to break the problem of personal energy use into specific categories such as sustaining a home, driving an automobile, flying, purchased goods, etc. Without this analysis by category, and the options available to the individual within each of these categories, it becomes difficult if not impossible to understand global energy use. Thus we pursue a quantitative understanding of personal energy use in some detail in this important Case Study.

One of the most important, informative and innovative books written on the subject of global energy technology and energy use is *Sustainable Energy—Without the Hot Air* by the late David J. C. MacKay of the University of Cambridge. One of the many innovations and insights offered in this remarkable book is the careful analysis of the most effective units applied to different aspects of the calculation of both personal energy use and global energy use. For our purposes in this Case Study on personal energy use, we adopt the unit of kWh/person·day so that we can quantitatively analyze each of the ways we, as individuals, use energy. This allows us to decisively understand what aspects of our daily routine use specific amounts of energy—large and small—but moreover it informs us how our daily decisions impact our energy "footprint."

Thus we proceed to analyze each of the important categories of personal energy use employing this energy unit of kWh/person·day consistently across the key categories.

Category 1: Personal Energy Use for an Automobile

The category of personal energy use in driving an automobile is both an important introduction into how to use the units of kWh per person per day (kWh/p·d) as well as an introduction to the large range of energy amounts spanning the range from a large SUV in traffic to an electric automobile (or electric truck).

1.1. The gas- or diesel-fueled case

In order to calculate the energy consumed per day for a gas or diesel powered automobile, we need: (1) the distance traveled per day, (2) the energy per unit of fuel, (3) distance per unit of fuel.

1. The average distance traveled per day in the United States is 50 km/day.

2. The energy per unit of fuel is 10 kWh/l.

3. The distance per unit of fuel depends on the vehicle. We consider these examples of vehicles:

a. The national average fuel economy in the US is 18 miles/gallon or 7.8 km/liter.

b. The average fuel economy in Europe is 12 km/liter.

c. A large sport utility vehicle (SUV) in traffic in the US is 4.3 km/liter.

Thus using our equation for the energy use per person per day:

$$\begin{bmatrix} \text{Energy used per} \\ \text{day per person} \end{bmatrix} =$$

$$\left[\left(\begin{array}{c} \text{distance traveled} \\ \text{per day} \end{array} \right) \middle/ \left(\begin{array}{c} \text{distance per} \\ \text{unit of fuel} \end{array} \right) \right] \times \left(\begin{array}{c} \text{energy per} \\ \text{unit of fuel} \end{array} \right)$$

For case (a) the US national average fuel economy of 7.8 km/liter:

[Energy used per person per day] =
[(50 km/day)/(7.8 km/l)] × [10 kWh/l] = 64 kWh/p·d

For case (b) the average fuel economy in Europe of 12 km/l:

[Energy used per person per day] =
[(50 km/day)/(12 km/l)] × [10 kWh/l] = 42 kWh/p·d

For case (c) the SUV in traffic for which 4.3 km/l:

[Energy used per person per day] =
[(50 km/day)/(4.3 km/l)] × [10 kWh/l] = 116 kWh/p·d

Next we consider the case of an electric car.

1.2. Personal energy use per day for an electric car

The Tesla Model S is a large passenger car that uses 15 kWh of electrical energy per 100 km. For an electric car our calculation of energy use per person per day is

[Energy used per person per day] =
[50 km/day] × [15 kWh/100 km] = 7.5 kWh/p·d

1.3. Summary of the energy used per person per day for the four vehicle types

A summary can best be compared graphically. See Figure CS2.2a. The differences are remarkable both among the three types of internal combustion vehicles that range from 42 kWh/p·d for the vehicle with European average fuel economy to 116 kWh/p·d for the SUV in traffic, as well as that of the electric vehicle.

Remarkably the electric vehicle uses just 7.5 kWh/p·d. When we study the thermodynamics of the internal combustion engine in Chapter 3 and then electrochemistry in Chapter 7, we will see why there is this remarkable difference.

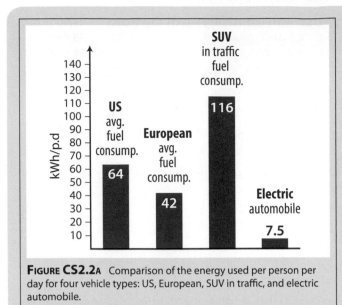

Figure CS2.2a Comparison of the energy used per person per day for four vehicle types: US, European, SUV in traffic, and electric automobile.

Category 2: Heating, Cooling, and Electrical Use to Sustain a Household

In this analysis we view the house as a system as displayed in Figure CS2.2b. We consider two types of energy use to sustain a household. The first is the energy required to heat the home with natural gas. The second is the electricity used to provide lighting, operate computers, television, etc. Just as was the case for the analysis of the automobile types, the energy use per day per person can range over a remarkable difference. However we will use typical numbers in order to judge what fraction of a person's energy consumption is taken up by the home in which they live.

If we consider, for example, a typical home in the northeastern section of the US, a dominant energy use is to heat the home. In the southwest of the US the dominant energy use is to cool the home. We consider here the case of the northeastern US.

2.1. Heating at Home
The average home in the northeastern US uses 5×10^3 m³/yr of natural gas. This converts to $(5 \times 10^3$ m³/yr$)(0.8$ kg/m³$)$ = 4×10^3 kg/yr of natural gas usage. However, natural gas (CH_4) provides 13.5 kWh/kg of heat given the fact that modern natural gas furnaces are ~95% efficient. Thus our average home will use $(13.5$ kWh/kg$)(4 \times 10^3$ kg/yr$)$ = 5.4×10^4 kWh/yr. Converting this to a per-day energy consumption of $(5.4 \times 10^4$ kWh/yr$)(3.65 \times 10^2$ d/yr$)$ = 1.5×10^2 kWh/d. If we assume a family of four living in the house, this translates to $(1.5 \times 10^2$ kWh/d$)/(4$ persons$)$ = 37 kWh/p·d.

2.2. Electricity Use at Home
Analysis has demonstrated that the average home in the US consumes 10,000 kWh of electrical power each year to oper-

ate lighting, electronics, and other personal devices. This corresponds to $(10{,}000$ kWh/yr$)$ / $(365$ d/yr$)$ = 27 kWh/d. For a household with four members, this translates to 6.8 kWh/p·d which we approximate as 7 kWh/p·d.

2.3. Personal Energy Use at Work
In order to account for personal energy consumption at work, a contribution that includes heating, cooling, lighting, electronics, etc., analysis has demonstrated that this adds approximately 15 kWh/d per employee averaged across the US.

2.4. Summary of Personal Energy Use for Both Home and Workplace
When we combine the contributions to our personal energy budget, we thus have

Home heating	Home elect.	Workplace
37 kWh/p·d +	7 kWh/p·d +	15 kWh/p·d

= 59 kWh/p·d

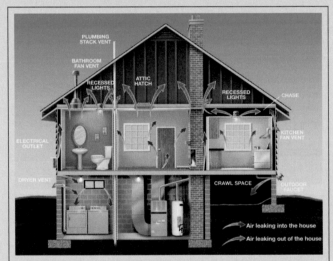

Figure CS2.2b The house viewed as a system provides a direct way to analyze the kWh/p·d demand in your personal energy budget. The house system requires heating, cooling, lighting, support for electronics, etc.

Category 3: Personal Energy Use Associated with Flying

FIGURE CS2.2c   A Boeing 767 is an aircraft used routinely for extended flights, whether transcontinental or for intercontinental flights.

Flying in commercial airliners, displayed in Figure CS2.2c, represents an example of just how important this aspect of your personal energy budget can be if you fly frequently. In our quantitative analysis we consider first a single round trip flight between New York and San Francisco, a distance of approximately 10,000 km. A Boeing 767 that has a passenger capacity of 360 uses 2×10^5 liters of jet fuel for each leg of the round trip. Thus, using our fact that each liter of fuel contains 10 kWh of chemical energy, we have

$$\begin{bmatrix} \text{Energy consumed for} \\ \text{round trip per person} \end{bmatrix}$$

$$= 2 \times \left[\left(\begin{array}{c} \text{fuel volume consumed} \\ \text{each leg} \end{array} \right) \Big/ \left(\begin{array}{c} \text{number of} \\ \text{passengers} \end{array} \right) \right] \times \left(\begin{array}{c} \text{energy per} \\ \text{unit volume} \end{array} \right)$$

$$= 2 \times [(200{,}000 \text{ liters})/(360 \text{ passengers})] \times 10 \text{ kWh}/\ell$$

$$= 11 \times 10^3 \text{ kWh/passenger}$$

$$= 2 \times [(2 \times 10^5 \, \ell)/(360 \text{ pass.})] \times 10 \text{ kWh}/\ell$$

$$= 1.1 \times 10^4 \text{ kWh/p}$$

If we assume that this round trip flight between New York and San Francisco is the only flight an individual takes in a year, then we have 1.1×10^4 kWh/p/365 d = 30 kWh/p·d.

When we compare this contribution to our personal energy budget, it is remarkable that a single round trip transcontinental flight is nearly half the energy consumption per person, per day needed for a *year* of driving an automobile with average US fuel consumption. Also that single round trip flight from New York to San Francisco equals nearly half the yearly energy consumption required to sustain our home and workplace energy demand!

However, a single round trip flight across the US is a rather modest amount of flying. If we analyze the calculation for an individual who flies frequently on business, the resulting energy demand becomes remarkable. Suppose we take a typical yearly itinerary:

1	Round trip NY to Beijing	24,000 km
6	Coast-to-coast, U.S.	58,000 km
10	NY to Chicago equivalent	38,000 km
10	Boston to Washington or equivalent	13,000 km
		133,000 km

Given that we just calculated that the energy consumed for a single 10,000 km round trip flight was 30 kWh/p·d, the above itinerary is 13.3 times as far, so this constitutes (13.3)30 kWh/p·d = 400 kWh/p·d. This figure of 400 kWh/p·d for a frequent flyer turns out to be the dominant contribution to an individual's personal budget. This remarkable contrast is displayed in Figure CS2.2d.

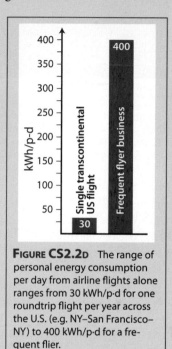

FIGURE CS2.2d   The range of personal energy consumption per day from airline flights alone ranges from 30 kWh/p·d for one roundtrip flight per year across the U.S. (e.g. NY–San Francisco–NY) to 400 kWh/p·d for a frequent flier.

Category 4: Contribution of Purchased Goods to an Individual's Personal Energy Budget

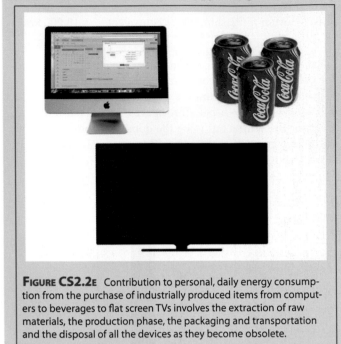

FIGURE CS2.2E Contribution to personal, daily energy consumption from the purchase of industrially produced items from computers to beverages to flat screen TVs involves the extraction of raw materials, the production phase, the packaging and transportation and the disposal of all the devices as they become obsolete.

The flow of manufactured goods (examples of which are displayed in Figure CS2.2e), particularly in the developed world, engages an intricate but interesting web of processes and phases that together constitute a surprisingly important contribution to our personal energy budget. To address this network of steps leading to the fabrication, delivery and sale of manufactured goods, we break the sequence into four phases:

1. Extraction and delivery of the raw materials required for manufacture.

2. Production of the devices or systems themselves.

3. Application of the goods for their intended purpose.

4. Removal and disposal.

We consider an array of the most important examples.

4.1 Automobiles and Roads

The production of a new car consumes approximately 80,000 kWh of energy for the first two phases above: extraction of raw materials and the manufacture of the automobile. In the language of production, this figure of 80,000 kWh refers to the *embedded* or *embodied energy*. For the automobile with a typical lifetime of 15 years, the embodied energy calculates to an average energy cost of 80,000 kWh/(15 yr)(365 d/yr) or 15 kWh/p·d.

For the use phase, we have done the calculation in Category 1 above where we determined that for an automobile with average US fuel economy, 62 kWh/p·d is an appropriate figure, for an SUV in traffic 116 kWh/p·d and for an electric automobile 7.5 kWh/p·d. Note, therefore, that the embodied energy in the production of the car is considerably less than the energy required to operate the automobile unless it is an electric car where the operating energy budget is about *one-half* the embodied energy.

However, we also need to consider the energy required to build and maintain the roads on which the automobiles are operated as displayed in Figure CS2.2f. In the US, roadways require on average 35,000 kWh of energy *per meter* of roadway. If we assume an average lifetime of 40 years for the road, this adds approximately 2 kWh/p·d to or personal energy budget.

FIGURE CS2.2F When we account for the energy to produce an automobile, we must also account for the cost of building the roads on which the automobiles are driven.

4.2 Houses

The contribution of house construction, displayed in Figure CS2.2g, to our personal energy budget is one of the most surprising numbers in this analysis. If we assume that the typical house lasts for 100 years, this works out to be approximately 2.3 kWh/day. If the house is occupied by on average 2.3 individuals, this gives us just **1 kWh/p·d**. Given that it requires some 44 kWh/p·d to heat, cool and electrify the home, the actual construction of the home is a very small part of the budget!

FIGURE CS2.2G The cost of constructing a house is, on the face of it, an expensive undertaking. However, because the lifetime of most houses is in the order of 100 years, the contribution to the energy budget per person per day is quite modest.

4.3 Electronics and Computers

The embodied energy of a typical computer, displayed in Figure CS2.2h, is 2000 kWh, but a typical computer lasts but two to three years, so these devices including television sets contribute approximately 2.5 kWh/p·d to our personal energy budget.

FIGURE CS2.2H Computers, because they are quite energy intensive to manufacture and they become obsolete in a very few years, are a significant contribution to the energy budget.

4.4 Summary of the Contribution to Our Personal Energy Budget by the Purchase of Automobiles, Home Construction, Roads, Computers, Drink Containers, Newspapers, Batteries, etc.

The sum of all these contributions gives an additional 30 kWh/p-d to our energy budget.

4.5 Delivery of Products to Market

One of the many interesting considerations in calculating the contribution to our personal energy budget is the transportation cost of delivery. To establish the cost of delivery to market, we first summarize the large range of cost per ton-km in Figure CS2.2i for each mode of transport: ship, rail, truck, or aircraft. The unit ton-km (t-km) is the energy required to move 1000 kg = 1 ton a distance of 1 km. For each of our transportation modes, the approximate figures are:

1. Shipping   0.1 kWh/t·km
2. Rail   0.1 kWh/t·km
3. Trucking   1.1 kWh/t·km
4. Aircraft   1.65 kWh/t·km

The advantage of this approach is that we can now simply add up all the ton-km figures for each mode of transportation to calculate the overall contribution to our personal energy budget.

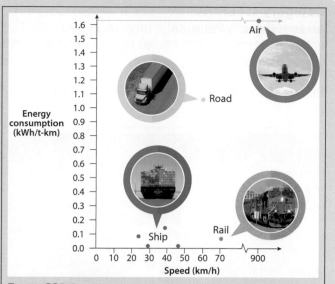

FIGURE CS2.2I The relationship between energy consumption in kWh/t-km and the speed of transport for the major carriers reveals a number of conclusions. Hauling freight by truck is more than an order of magnitude more energy intensive than shipping by rail. Shipping by air is only 50% more energy intensive than shipping by truck. Transport by water is remarkably efficient, but that efficiency depends on hull design leading to the scatter of points.

4.5.1 Transport by Rail

As an example in 2016, 2.7×10^{12} t-km were carried by rail service as shown in Figure CS2.2j. Thus with an energy intensity of 0.1 kWh/t-km for rail, we have

$$\begin{bmatrix} \text{Energy consumption} \\ \text{per person per day} \end{bmatrix} = \frac{\left[\left(2.7 \times 10^{12}\,\text{t-km}\right)\left(0.1\,\text{kWh/t-km}\right)\right]}{\left[\left(365\,\text{days}\right)\left(3.2 \times 10^{8}\,\text{people}\right)\right]}$$

$$= 2.4\,\text{kWh/p·d}$$

FIGURE CS2.2J Freight hauling by train is not only extremely energy efficient, it can also be done using electricity—a fact that will become increasingly important.

4.5.2 Transport by Road

As an example, in 2016 2.0×10^{12} t-km were transported within the US by truck, displayed in Figure CS2.2k. Given the corresponding energy intensity of 1 kWh/t-km, this computes to 2.0×10^{12} kWh of expended energy per year. For 365 days and 320 million people in the US, this gives 17 kWh/p·d.

Figure CS2.2k While freight shipping by truck is convenient, it is also very energy intensive.

4.5.3 Shipping of Imports

United States ports supported 9.1×10^{12} t-km of shipped goods in 2018; with a corresponding energy intensity of 0.1 kWh/t-km we have:

$$\frac{\left(9.1 \times 10^{12}\,\text{t-km}\right)\left(0.1\,\text{kWh/t-km}\right)}{\left(365\,\text{days}\right)\left(3.2 \times 10^{8}\,\text{p}\right)} = 7.8\,\text{kWh/p·d}$$

4.5.4 Air Freight of Imports

Into the US in 2018, 4×10^{10} t-km of goods were transported with an energy intensity of 1.65 kWh/t-km. So this contributes

$$\frac{\left(4 \times 10^{10}\,\text{t-km}\right)\left(1.65\,\text{kWh/t-km}\right)}{\left(365\,\text{days}\right)\left(3.2 \times 10^{8}\,\text{p}\right)} = 0.6\,\text{kWh/p·d}$$

4.6 Embodied Energy of Imported Goods

We turn next to the embodied energy of *imports* to the U.S. This includes the flat screen TVs, computers, DVD players, refrigerators, air conditioners, lawn mowers, clothing, cameras, microwave ovens, printers, copiers, cars, etc. brought into the U.S. from abroad, increasingly from China. An analysis of the embodied energy content of those devices demonstrates that 10 kWh/kg is a reasonable estimate. With 9.1×10^{11} kg of imports per year to the U.S., this translates to 80 kWh/p·d of embodied energy.

From this figure we must subtract the contribution from automobile imports, because we have already calculated the embodied cost of U.S. auto purchases. Total U.S. auto sales in the second half of the first decade of this century were between 8 and 9 million cars, SUVs, and pickup trucks. Of this, approximately half were built overseas and shipped to the U.S. by sea. Thus we must subtract from the total imported embodied energy figure of 80 kWh/p·d, one-half the embodied energy per person per day that we calculated for the automobile (14 kWh/p·d). Thus, subtracting 7 kWh/p·d from 80 kWh/p·d gives us 73 kWh/p·d for the embodied energy of all imported items to the U.S.

Thus the embodied energy for purchased products, including transportation works out to be approximately:

Embodied Energy of Domestic Consumer Goods:

Automobiles and roads	16 kWh/p·d
House	1 kWh/p·d
Computers	2.5 kWh/p·d
Drink Containers	2 kWh/p·d
Packaging	6 kWh/p·d
Newspapers, Magazines & Junk Mail	0.3 kWh/p·d

Transportation of Consumer Goods:

Transportation within U.S.	
Rail	2.5 kWh/p·d
Truck	18 kWh/p·d
Air Freight	0.5 kWh/p·d
Shipping to U.S. ports	8 kWh/p·d

Embodied Energy of Imports:

Embodied Energy of Imports	73 kWh/p·d
	―――――
	130 kWh/p·d

Analysis of the Range of Contributions to Our Personal Energy Budget

One of the most illuminating ways to employ the calculations of energy use per person per day is to explore the range over which the combination of factors that comprise a person's energy budget factor into the overall personal energy budget. For example, we could see the remarkable range of energy use per person per day for the case of operating an automobile. If we drive a gas powered SUV in traffic this demands 116 kWh/p·d. If we drive an electric automobile the figure drops by a factor of nearly 16 to 7.5 kWh/p·d!

So if we summarize the ranges for the other contributions, we can immediately infer how lifestyle or life choices affect the totality of our personal energy budget. In Category 2, energy to sustain a household and our workplace, we arrived at a figure of 59 kWh/p·d for an average free standing house. If we occupy a large house, as many people in the United States now do, the figure can easily be double that, or approximately 120 kWh/p·d. On the other hand, if we live in an apartment or attached townhouse the figure can be half that of a free standing house, or approximately 30 kWh/p·d.

We can thus construct a graphic similar to Figure CS2.2a but for sustaining a household as displayed here in Figure CS2.2l.

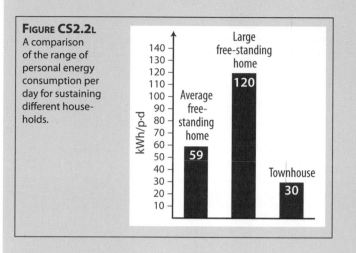

FIGURE CS2.2L
A comparison of the range of personal energy consumption per day for sustaining different house-holds.

Similarly the distance we fly each year for pleasure or business markedly affects our personal energy budget as summarized in Figure CS2.2m. As we calculated under Category 3, if we take a single round trip flight from New York to San Francisco and back, this constitutes an additional 30 kWh/p·d to our individual energy budget. If on the other hand a person flies extensively for business, this can easily increase the contribution of flying to 400 kWh/p·d.

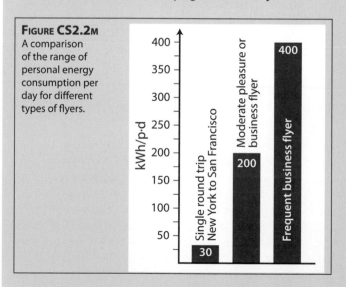

FIGURE CS2.2M
A comparison of the range of personal energy consumption per day for different types of flyers.

With respect to the consumption of consumer goods, again there is a large range of energy use per person per day depending on an individual's inclination or financial capacity to purchase goods as reviewed in Figure CS2.2n. Thus we calculated that on average in the US the combination of embodied energy of domestic and imported goods in com-

bination with required transportation gave a figure of 138 kWh/p·d. If we approximate this as 140 kWh/p·d, then a person with a more modest propensity to shop would draw perhaps half of that or 70 kWh/p·d. A person with more extravagant shopping habits could easily raise to average figure to twice the average or 280 kWh/p·d. This provides another important range for our consideration.

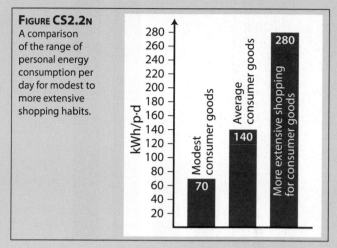

FIGURE CS2.2N
A comparison of the range of personal energy consumption per day for modest to more extensive shopping habits.

With this analysis of the range of personal energy consumption, we can then quickly relate lifestyle choices with personal energy consumption. Thus we consider four cases as shown in Figure CS2.2o:

Case 1: lives in a townhouse, drives an electric car, has modest shopping goals and flies one round trip excursion between the East Coast and West Coast. Personal energy budget: 138 kWh/p·d.

Case 2: lives in a detached house, drives a gas automobile with average fuel economy, purchases the average number of consumer goods and flies the equivalent of two round trip excursions between the East and West Coasts. Personal energy budget: 324 kWh/p·d.

Case 3: lives in a large detached house, drives an SUV in traffic, purchases twice the average number of consumer goods and engages in moderate pleasure/business air travel. Personal energy budget: 716 kWh/p·d.

Case 4: lives in a large detached house, drives an SUV in traffic, purchases twice the average amount of consumer goods and travels extensively for business and/or pleasure. Personal energy budget: 916 kWh/p·d.

We can summarize these four cases graphically in Figure CS2.2o.

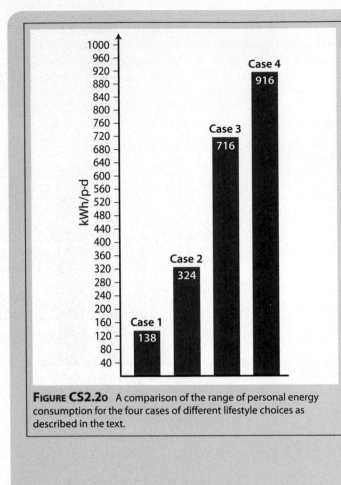

Figure CS2.2o A comparison of the range of personal energy consumption for the four cases of different lifestyle choices as described in the text.

2.50

CASE STUDY 2.3 Calculating Energy Yield from Chemical Bonds: Chemistry of Fossil Fuels

KEY CONCEPTS:

We can initiate this discussion of the extraction of chemical energy from the combustion of fossil fuels by starting from the example of the combustion of molecular hydrogen in air—wherein the air supplies the O_2 required to convert the H_2 and O_2 to water:

$$2 H_2 + O_2 \rightarrow 2 H_2O$$

First, we know from experience that a balloon filled with hydrogen without outside intervention will not explode. Nor will hydrogen released into the air explode spontaneously. The reason is that, as displayed in Figure 1.12, there is an *energy barrier* that must be surmounted before the reaction is initiated. If we touch a flame to a balloon (or a spark to a mixture of gasoline and air), the high temperature of the flame will impart very high kinetic energy to a limited number of H_2 and O_2 molecules and the barrier to reaction will be surmounted, releasing an amount of energy equal to the enthalpy difference between the reactants (H_2 and O_2) and the products (H_2O). The chemical energy of H_2 results from its ability to react with O_2 to form molecules of significantly lower enthalpy of formation, H_2O. Two molecules of H_2O are formed, each with a pair of O-H bonds. This suggests that we can develop a strong intuitive sense for the energy release in a given chemical reaction by comparing the bond "strength" (the energy required to convert a chemical bond into its two atomic constituents) represented by some *average* bond dissociation enthalpy for that particular pair of atoms. This would work if, and only if, O-H bond strengths, C-O bond strengths, C-H bond strengths, etc. were reasonably equivalent *in different molecules*. A priori there is no reason to believe that this would be true. But, importantly, it turns out to be a reasonably accurate approximation for reasons we will develop in Chapters 10 and 11.

For our purposes here, we note that the H-H bond is strong as a result of the close proximity of the valence electrons to the nuclei. However, O-H bonds are even stronger because oxygen atoms have a higher nuclear charge and the valence electrons "see" significantly higher charge on the O atom nucleus. This draws electrons preferentially toward the O atom in a bond. As a result O atoms are said to have higher *electronegativity* than H atoms—a concept we will develop fully in Chapters 9 and 10. The O-H bond enthalpy is not exactly the same in H_2O as in ethanol, C_2H_5OH, because the strength of the O-H bond is affected to some degree by the presence of the other bonds in the molecular structure. But the differences are not large. This allows us to create a table of *average* bond energies for atom-atom pairs as displayed in Table CS2.3a.

TABLE CS2.3A

Bond	Enthalpy (kJ mol^{-1})	Bond	Enthalpy (kJ mol^{-1})
H—H	432	C≡O	1071
O=O	494	C—C	347
O—H	460	C=C	611
C—H	410	C⋯C^a	519
C—O	360	N=O	623
C=O	799	N≡N	941

^{a}Aromatic, 1.5 bond order.

Examination of Table CS2.3a reveals at least five important facts:

1. Virtually all the important bonding pairs in the structure of fuels are included in the table.
2. With the greater electronegativity of oxygen atoms over carbon atoms, the O-H bond is stronger than the C-H bond.
3. Double bonds are stronger than single bonds between the same two atoms.
4. Triple bonds are stronger than double bonds.
5. The strongest bond is the triple bond of CO followed by the triple bond of N_2.

Finally, it is the combination of relatively weak O = O bonds (developed in Chapter 11 and 12) and strong O-H and C = O bonds that makes combustion reactions so potent as chemical energy sources.

Comparison of Fuel Energies

We can use the short list of average bond dissociation enthalpies in Table CS2.3a to calculate the combustion enthalpies for a wide range of fuels. We can begin by tabulating the combustion energies of several important categories of fuel types: hydrogen, methane, petroleum, coal, ethanol, and carbohydrates. This is presented in Table CS2.3b in energy units of kJ in terms of the reaction enthalpies, energy release per mol of O_2, per mol of fuel, per gram of fuel, and in terms of the moles of CO_2 formed per 1000 kJ released.

TABLE CS2.3B

	Energy Content (kJ)				
	Reaction Enthalpy	Per mol O$_2$	Per mol Fuel	Per g Fuel	Mol CO$_2$/ 1000 kJ
Hydrogen $2 H_2 + O_2 \rightarrow 2 H_2O$	482	482	241	120	0
Methane $CH_4 + 2 O_2 \rightarrow CO_2 + 2 H_2O$	810	405	810	51.6	1.2
Petroleum $2(-CH_2-) + 3 O_2$ $\rightarrow 2 CO_2 + 2 H_2O$	1220	407	610	43.6	1.6
Coal $4(-CH-) + 5 O_2$ $\rightarrow 4 CO_2 + 2 H_2O$	2046	409	512	39.3	2.0
Ethanol $C_2H_5OH + 3 O_2$ $\rightarrow 2 CO_2 + 3 H_2O$	1257	419	1257	27.3	1.6
Carbohydrate $(-CHOH-) + O_2$ $\rightarrow CO_2 + H_2O$	447	447	447	14.9	2.2

There are several very important observations to be made in the investigation of Table CS2.3b. *First*, the reaction enthalpies span a wide range for the reaction as written. The enthalpies are not directly comparable because the number of moles of chemical bond transformations involved vary considerably. For example, the reaction enthalpy for CH_4 combustion is considerably *larger* than for hydrogen combustion (810 kJ vs. 482 kJ). However, viewed on a per-mole basis, the energy released per mole of O_2 is *smaller* for CH_4 (405 kJ) than for H_2 (482 kJ). On the other hand, one mole of CH_4 has a higher energy content than a mole of H_2 (810 kJ vs. 241 kJ). Thus because one mole of any gas occupies the same volume (22.4 liters at STP), a cubic meter of CH_4 contains over *three times* the energy of a cubic meter of H_2. The reason is that there are four C–H bonds in methane that are converted to C=O and H–O bonds on combustion. There is a single H–H bond in H_2 that is converted to O–H bonds.

However, there are different figures of merit for different fuel uses. For example, if it is weight that matters most, then H_2 is the best choice. From Table CS2.3b hydrogen combustion yields 120 kJ per gram of fuel vs. the next best, methane, at 51.6 kJ of enthalpy release per gram of fuel. For this reason rockets use liquid hydrogen as a fuel because the fuel weight is a large fraction of the system weight during first-stage burn.

Second, Table CS2.3b shows the schematic reaction for both petroleum and coal. This distinguishes representative composition and bonding arrangements, reducing the analysis to interpretable bonding pair atoms. For example, petroleum is largely comprised of saturated hydrocarbons and thus we consider the relevant combustion reaction of a –CH$_2$– group in a hydrocarbon chain:

$$2(-CH_2-) + 3 O_2 \rightarrow 2 CO_2 + 2 H_2O$$

In this analysis we include the C–C bond energy *once* per CH$_2$ group because each of the group's two C-C bonds joins to a neighbor and this avoids double counting in the polymer chain. Thus as written above the reaction releases 1220 kJ:

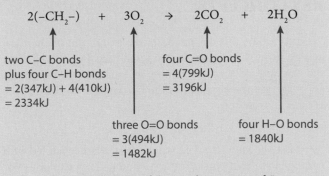

Bond enthalpies of the products = 5036 kJ
Bond enthalpies of the reactants = 3816 kJ

Therefore, enthalpy release in reaction of two (–CH$_2$–) groups = 1220 kJ. On a per-mole of O_2 basis, the energy release is 407 kJ (1220/3 = 407 kJ), which is nearly the same as that of methane. On a per-gram basis, petroleum produces 43.6 kJ, somewhat less than the 51.6 kJ for methane.

Third, the H/C ratio of a fuel controls the energy content per gram of fuel. The H/C ratio of saturated hydrocarbons is > 2:1 because of the CH$_3$ groups at the terminus of the (branched) chain molecules that make up the fuel. On the other hand, petroleum has a significant fraction of aromatic molecules with H/C ratios < 2:1. As displayed in Figure CS2.5g, the molecular structure of coal is largely built from aromatic rings with an H/C ratio of 1:1 (or slightly less). Thus our proto molecule in combustion is

$$4(-CH-) + 5O_2 \rightarrow 4CO_2 + 2H_2O$$

The C atom in an aromatic structure is bonded to neighboring carbon atoms by bonds with an order of 1.2. As was the case with our analysis, the energy accounting counts only one C⋯C bond per (–CH–) unit. Thus the energy release for the above combustion reaction as written is

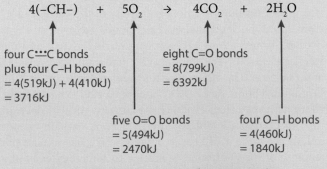

Bond enthalpies for the products = 8232 kJ
Bond enthalpies for the reactants = 6186 kJ

The enthalpy release is thus 2046 kJ for the coal reaction as written, which is 409 kJ per mole of O_2 and 39.3 kJ/g of fuel. In practice, coal contains both minerals and water so in practice typical values for "hard coal" (bituminous or anthracite) are 29–33 kJ/g and for "soft coal" (lignite) typically 17–21 kJ/g.

Problem 1

Let's consider gasoline, which is mostly octane, C_8H_{18}. Octane's structure is shown below:

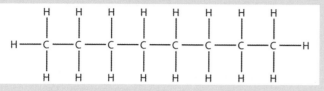

a. Based on the list of average bond dissociation enthalpies in Table CS2.3a, calculate the combustion enthalpy for octane in terms of kJ per mol of reaction:

$$2\,C_8H_{18} + 25\,O_2 \rightarrow 16\,CO_2 + 18\,H_2O$$

b. Experimental data for the combustion enthalpy for octane is 10920 kJ per mol of reaction, and that equals 5460 kJ per mol of octane. What is the energy content of octane in terms of kJ per gram of fuel (kJ/g)? What is the H/C ratio of octane? The H/C ratio of the crude oil is < 2:1. Do you expect the energy content per gram of crude oil higher or lower than that of gasoline?

c. What is the energy content of octane in terms of kWh per liter fuel (kWh/l)? Do you think it is a good approximation to use 10 kWh/l as the energy content per liter for gasoline as discussed in Case Study 1.3? (The density of octane is 703 g/l).

Problem 2

While we will consider biofuels in some detail in Case Study 13.3, let's consider ethanol, the structure of which is shown below:

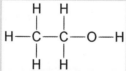

a. Based on the list of average bond dissociation enthalpies in Table CS2.3a, calculate the combustion enthalpy for ethanol in terms of (i) kJ per mol of reaction, and (ii) kJ per gram of fuel (kJ/g).

$$C_2H_5OH + 3\,O_2 \rightarrow 2\,CO_2 + 3\,H_2O$$

b. Why does ethanol produce significantly less energy per unit mass than does petroleum at ~44 kJ/g?

Problem 3

a. Based on the list of average bond dissociation enthalpies in Table CS2.3a, calculate the combustion (or respiration) energy release for carbohydrates in terms of (i) kJ per mol of O_2, and (ii) kJ per gram of fuel (kJ/g):

$$(\text{-CHOH-}) + O_2 \rightarrow CO_2 + H_2O$$

b. Explain qualitatively why the energy *per gram* for carbohydrates is only about one-third of that for hydrocarbons. (Within the context of nutrition, that is also why fats, which contain primarily hydrocarbons, have many more calories per gram than do carbohydrates.)

CASE STUDY 2.4 Mastering Redox Chemistry: Energy and Redox Reactions

KEY CONCEPTS:

In order to sustain life on the planet, it is important to develop an understanding of how life evolved and from where the energy came that allowed it to prosper. To develop an understanding of the coupling between energy at the molecular level and the development of self sustaining organisms it is essential that we explore the importance of water as a medium that supports all life forms on Earth and the relationship between water and molecular oxygen in the chemical balance of life.

To this end we begin with the fact that *life is sustained by, is powered by, redox reactions.* Nature has discovered how to use the *energy released* when electrons are *transferred* to or from atoms when *reactant* molecules are converted to another molecular structure, i.e. *product* molecules. So let's review oxidation-reduction reactions. When we treat an electron "transfer" in an oxidation-reduction reaction, we seek a convenient way to register the movement of electron density in that chemical reaction. The accounting method is to employ the "oxidation state" of each atom involved in the reaction. In brief, an atom's oxidation state is a positive or negative number designed to reflect electron density shifts local to that particular atom *relative* to some standard state. That standard state is an atom in elemental form—the form that represents the electron distribution it would have in the absence of another atom that may (or may not) have a greater ability to draw electron density to it when engaged in a chemical bond. We can demonstrate both the redistribution of electron charge, the oxidation state, and the energy release

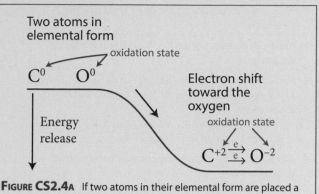

FIGURE CS2.4A If two atoms in their elemental form are placed a distance apart equal to the internuclear distance of a corresponding molecule and then the electrons are allowed to delocalize toward the atom with the greater ability to attract electrons to it in a chemical bond, the energy of the molecule so formed drops. This changes the oxidation state from *zero* for the elemental form to a new oxidation state in the molecule formed as shown.

resulting from that redistribution of electron charge in the following diagram for carbon monoxide, Figure CS2.4a. If we begin with the carbon and oxygen a distance apart equal to the bond length in CO, but with the atoms each in their elemental form (including the distribution of electron density about each atom), they will both have an oxidation state of *zero*, by definition. If we now allow the relative ability of each atom to attract electron density to it in a chemical bond, electron density will move toward the oxygen, will "delocalize" toward the oxygen, releasing energy as the electron density shifts toward the oxygen, thereby reducing the energy of the ensemble of atoms, in this case the two atoms in CO.

The oxidation state formalism takes a subtle shift in electron density and turns it into an all or nothing statement of electron "ownership." Just as in an election, if candidate A receives just 51% of the vote and candidate B receives 49% of the votes, candidate A is declared the winner. So it is with a chemical bond. In the case of CO, electron density shifts toward the oxygen and it is deemed the "winner" of the electrons and is awarded an oxidation number of –2. The electron is conserved, so the carbon atom is deemed the loser (of electron density) and awarded an oxidation number of +2.

It requires a large computer to accurately calculate the distribution of electron charge in a molecule. Oxidation numbers are an approximation that provide remarkable insight into how electrons move, how they delocalize, in a chemical reaction. Because of the conservation of electrons, the sum of the oxidation numbers assigned to atoms in a molecular structure

YELLOWSTONE PARK A thermal pool in Yellowstone National Park contains beautifully colored mats of bacterial communities that include *thermophilic cyanobacteria*. One of many examples of the adaptability of bacteria to challenging conditions!

TABLE CS2.4A

Rules of oxidation states
1. As noted, the oxidation state of an atom in a free element is zero
2. The oxidation state of a monoatomic ion is equal to its charge
3. Because electrons are neither produced nor destroyed in a redox reaction, the *sum of the oxidation states* of all atoms in: • A neutral molecule or molecular formula is 0   • An ion is equal to the charge of the ion
4. Metals, when they form compounds, have positive oxidation states   • Group 1A metals always have an oxidation state of +1   • Group 2A metals always have an oxidation state of +2
5. When *nonmetals* form compounds, those nonmetals are assigned values according to the table displayed here. Within this table, there is an ordering of importance with the most important rule beginning at the top

non-metals

Oxidation States of Nonmetals

Nonmetal	Oxidation State	Example
Fluorine	− 1	MgF_2 − 1 ox state
Hydrogen	+ 1	H_2O + 1 ox state
Oxygen	− 2	CO_2 − 2 ox state
Group 7A	− 1	CCl_4 − 1 ox state
Group 6A	− 2	H_2S − 2 ox state
Group 5A	− 3	NH_3 − 3 ox state

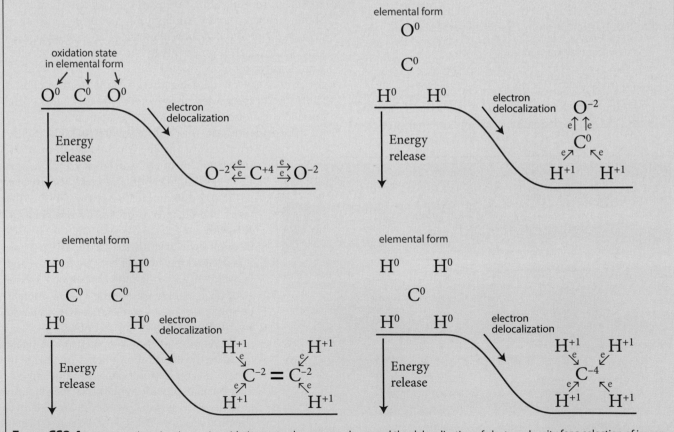

FIGURE CS2.4B We can relate the change in oxidation state, the energy release and the delocalization of electron density for a selection of important cases: CO_2, H_2CO, C_2H_4, and CH_4.

must equal the net charge on the molecule or ion. In lieu of a large computer calculation, the assignment of oxidation numbers follows a hierarchical pattern based on the competitive ability of individual atoms to draw electron density to them in a chemical bond. This set of rules that establish the hierarchy was presented in Table 2.4, and is reviewed here in Table CS2.4a.

We can sketch some important examples by displaying the transition from the elemental state with oxidation state zero to the oxidation state of the molecule after bond formation, by designating of the direction of electron delocalization and the resulting release of energy as shown in Figure CS2.4b.

Note that the double bond between the carbon atoms in C_2H_2, ethane, is not associated with any electron delocalization. Thus, to summarize, *oxidation states* provide an approximate way, in the absence of a full computer calculation, to describe the shift or delocalization of electron density in a molecule or polyatomic ion. The process of assigning oxidation:

- Assumes that the more electronegative atoms takes (or "wins") the whole electron as if a shared electron bond (a covalent bond) were really an ionic bond.
- Employs an arrow to mark the shift of electron density relative to the atom in its elemental state.
- Each arrow designating this shift in electron density is designated by an increase of 1 for the atom that is oxidized (at the tail of the arrow) and a decrease of 1 for the atom that is reduced (at the arrow tip).
- Bonds between like atoms, shown without arrows, do not result in an electron shift—an electron delocalization—and thus do not engender a change in oxidation state.

The idea of an oxidation-reduction reaction, or redox reaction, emerged in the developing understanding of chemical reactions in the study of rapid, intense combustion of hydrocarbon fuels in the presence of oxygen. A very familiar example is the burning of natural gas in oxygen:

$$CH_4 + 2O_2 \rightarrow 2H_2O + CO_2$$

We can immediately assign oxidation states to each of the atoms in the reactant molecules and in the product molecules using Table CS2.4a:

$$C^{-4}H_4^{+1} + 2O_2^0 \rightarrow 2H_2^{+1}O^{-2} + C^{+4}O_2^{-2}$$

The reaction is a redox reaction because the carbon is *oxidized* from C^{-4} to C^{+4} and the oxygen is *reduced* from O^0 to O^{-2}.

But redox reactions are not always fast and intense. For example the iron used for building bridges or constructing automobiles is invariably involved in a redox reaction where Fe in its elemental state reacts with oxygen (mediated by the presence of water) to produce Fe_2O_3—the process of corrosion.

$$2Fe + 3/2\ O_2 \rightarrow Fe_2O_3$$

In this case both *reactants* are in their elemental state with an oxidation state of zero. The product (oxygen with a metal) has the following oxidation state:

$$2Fe^0 + 3/2\ O_2^0 \rightarrow Fe_2^{+3}O_3^{-2}$$

While the reaction releases considerable energy in the form of heat, the reaction takes place over an extended period of time (often years) such that the heat flows from the reaction so slowly that it is imperceptible.

With this background on redox reactions, we proceed to a discussion of how organisms, particularly bacteria, use redox reactions to survive. If energy can be extracted, it can be used to sustain vital processes that, when taken together, constitutes a living organism. The key is to channel the energy captured from redox reactions and then use that energy to instigate such objectives as muscular contraction, molecular synthesis, response to light or sound, or keeping the organism at a fixed temperature. Organisms have evolved the ability to design membranes from protein building blocks to use energy from redox reactions to sustain a remarkable range of life forms. Before we can understand such systems, we must advance our understanding of oxidation-reduction process for they are the reactions that form the foundation, the motive energy, behind life's most fundamental processes.

What is remarkable is that life forms that evolved to the point where they could synthesize molecular structures from H_2O and CO_2, photosynthesis:

$$H_2O + CO_2 + sunlight \rightarrow CH_2O + O_2$$

were evolved from life forms that had to operate using redox reactions that did not involve oxygen. There are, today, regions of the world's oceans, lakes and estuaries where oxygen, O_2, is not available. These "dead zones" are usually the result of introducing molecules that react with and remove molecular oxygen from the water system, returning that system to the conditions that were present prior to 3.5 billion years ago when, as we will see, oxygen levels in the atmosphere began to rise to levels comparable to O_2 levels found today. Our atmosphere began to approach current levels of O_2 approximately 1 billion years ago. So we use this transition between aerobic (oxygen containing) and anaerobic (oxygen depleted) condition to examine more fully how redox reactions work and how we can harness an understanding of redox chemistry to create new opportunities.

When oxygen is available (an aerobic environment), the dominant biological redox process is respiration:

$$CH_2O + O_2 \rightarrow CO_2 + H_2O + energy$$

Question:

Demonstrate explicitly why this reaction is a redox reaction.

Answer:

We know from the chapter that a redox reaction must involve the *movement* of one or more electrons from one reagent to or toward another. But more precisely, we must keep track of the electrons assigned to each *atom* in both reactants and products in order to determine whether electrons have been transferred even though that "transfer" may simply be the delocalization or movement to or from the atom's center. So let's do this accounting explicitly for the reactants and products in this reaction. To do this we use the Rules for Oxidation States in Table CS2.2a.

All the atoms in our respiration reaction are nonmetals, so we assign oxidation states, according to the order presented in the table: to hydrogen first, oxygen second, and then apply the requirement that the oxidation states sum to zero for each of our (neutral) molecules (Rule 3).

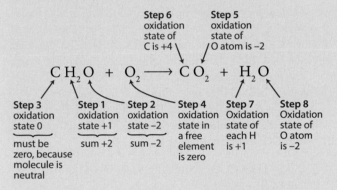

Thus, when we examine the *movement* of electrons on an atom-by-atom basis in the transition from reactant to product

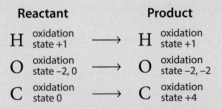

By virtue of the carbon atom from a molecular structure involving two hydrogen atoms and an oxygen atom (CH_2O) with resulting oxidation state of zero, to a molecular structure involving two oxygen atoms (CO_2), the carbon atom donated 4 electrons and was thus oxidized. In the process, O was reduced because it transitioned from an oxidation state of zero in O_2, to an oxidation state of -2 in CO_2. *All* higher life forms obtain their energy by respiration. The abundance of free oxygen makes life easy because an immense amount of energy can be harvested with this strong oxidizing agent available.

Notice that it is the *combination* of the organism's ability to *channel* the energy released from the *rearrangement* of electron density in the redox reaction that constitutes the key transformation. This is done by using sophisticated membranes that use the energy release to carry out processes vital to life (molecular synthesis, muscle contraction, etc.) that constitutes this remarkable union between electron rearrangement and life. Both membrane architecture and redox reactions are required. We focus here on developing an understanding of redox processes. We will examine the role of membranes in subsequent Case Studies.

What is universal is that redox reactions provide the energy wherein reactants go to products and energy is released:

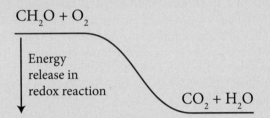

However, for more than 1 billion years, life forms were primarily the domain of bacteria, principally because bacteria evolved to employ redox reactions in the *absence* of oxygen. In fact bacteria evolved to become the masters of the game of extracting energy from an array of redox reactions in the absence of oxygen.

Question:

Why is the chemistry of oxygen and of water so central to redox reactions and to the development of life forms?

Answer:

We will repeatedly examine the implications of electron *movement* in chemical transformations that result from the differing ability of atoms to draw electrons to them resulting in a lower energy configuration. Water is the medium within which life takes place—whether at the ocean bottom in vents that spawned primitive life forms 3.5 billion years ago or within complex cells of advanced life forms. Water's central role in all this emerges from two primary traits: (1) the ability of the oxygen to extract electron density from hydrogen in a chemical bond and (2) the shape of the water molecule. It is bent with electron density localized on the ("electron rich") oxygen and with the "electron poor" hydrogens extending outward as shown in Figure CS2.4c.

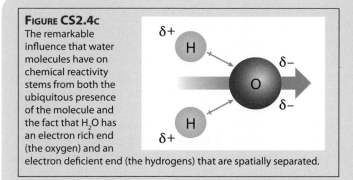

FIGURE CS2.4c The remarkable influence that water molecules have on chemical reactivity stems from both the ubiquitous presence of the molecule and the fact that H_2O has an electron rich end (the oxygen) and an electron deficient end (the hydrogens) that are spatially separated.

While we will fully develop the principles that determine this structure in subsequent chapters, we will concentrate here on the consequences of this combination of *shape* and *charge distribution* intrinsic to the water molecule.

When an ionic solid—a solid comprised of atoms held together by ionic bonds as described on page 2.17—is immersed in water, the *negative* (oxygen) end of H_2O attracts the positive (cation) of the ionic solid while the *positive* (hydrogen) end of H_2O attracts the negative (anion) of the ionic solid. This becomes a battle between the ionic bonding between the cation and anion in the solid and the Coulomb attraction between the polar water molecules and the *separated* cations and anions in the water. We can depict this graphically in Figure CS2.4d, which displays the release of the *cation* that is then surrounded by numerous water molecules with their negative (oxygen) ends turned toward the central cation.

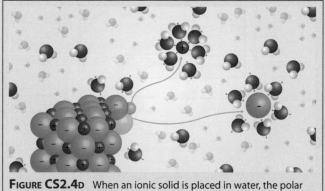

FIGURE CS2.4d When an ionic solid is placed in water, the polar nature of water is such that the anions are extracted from the solid by the positive (hydrogen) end of water and the cations are attracted to the negative (oxygen) end of the water molecule.

Also shown is the corresponding anion that, when released from the ionic solid, is surrounded by the accessible positive (hydrogen) ends of *multiple* H_2O solvent molecules.

Why does the ionic solid decide to break up into cations and anions? Because, in so doing, energy is released. We can sketch this on an energy diagram shown in Figure CS2.4e.

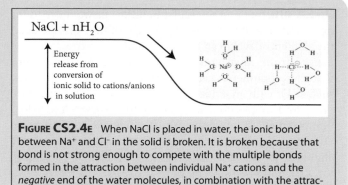

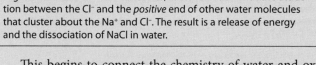

FIGURE CS2.4e When NaCl is placed in water, the ionic bond between Na^+ and Cl^- in the solid is broken. It is broken because that bond is not strong enough to compete with the multiple bonds formed in the attraction between individual Na^+ cations and the *negative* end of the water molecules, in combination with the attraction between the Cl^- and the *positive* end of other water molecules that cluster about the Na^+ and Cl^-. The result is a release of energy and the dissociation of NaCl in water.

This begins to connect the chemistry of water and oxygen with the concept of redox reactions and then to the mechanisms by which redox reactions power life.

Elements can exist in different oxidation states depending upon how many electrons are added to or removed from their valence shells. In an aqueous medium, the amount of energy released—the stability of a given oxidation state—depends on the properties of water as described above. For Group 1 and Group 2 metals, for example sodium and magnesium, there are 1 and 2 weakly bound electrons respectively in the valence shell so invariably if energy is released in the solution (the addition of the ionic solid to water), Na^+ and Mg^{2+} are formed from ionic solids that dissociate in water. All metals form cations (positive ions) in water. As we move to the right in the periodic table, electrons are bound more tightly and we can have multiple possibilities for oxidation states. An important example is iron, for which either Fe^{2+} or Fe^{3+} can exist in water.

Thus we have a general picture of metals, M, donating electrons to the surrounding aqueous medium as shown in Figure CS2.4f.

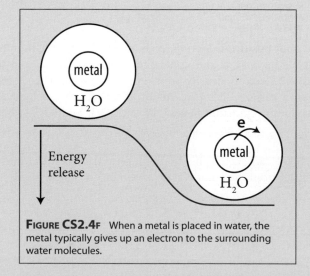

FIGURE CS2.4f When a metal is placed in water, the metal typically gives up an electron to the surrounding water molecules.

As we will see in greater detail in the development of

atomic structure and molecular bonding, as we move to the nonmetals that occupy the upper right of the periodic table, the elements develop an increasing ability to extract electron density from other atoms in a chemical bond. This means that these nonmetals can acquire negative oxidation states, such as O_2 with an oxidation state –2 in water. It is standard procedure to use Roman numerals to designate the oxidation number that systematically distinguishes the oxidation number from the actual charge. The lowest oxidation state attainable by fluorine, oxygen, nitrogen, and carbon are –I, –II, –III, and –IV respectively.

Identifying Redox Reactions

PROBLEM Use oxidation numbers to decide which of the following are redox reactions:
(a) $CaO(s) + CO_2(g) \longrightarrow CaCO_3(s)$
(b) $4KNO_3(s) \xrightarrow{\Delta} 2K_2O(s) + 2N_2(g) + 5O_2(g)$
(c) $NaHSO_4(aq) + NaOH(aq) \longrightarrow Na_2SO_4(aq) + H_2O(l)$
PLAN To determine whether a reaction is an oxidation-reduction process, we use Table 4.3 to assign each atom an O.N. and see if it changes as the reactants become products.
SOLUTION

$$\overset{+2}{\underset{-2}{|}}\ \overset{+4}{\underset{-2}{|}}\ \overset{+4}{\underset{+2\ -2}{|}}$$
(a) $CaO(s) + CO_2(g) \longrightarrow CaCO_3(s)$

Because each atom in the product has the same O.N. that it had in the reactants, we conclude that this is *not* a redox reaction.

O.N. decreased: *reduction*
$$\overset{+5}{\underset{+1\ -2}{|}}\ \overset{+1\ -2}{|}\ \overset{0}{|}\ \overset{0}{|}$$
(b) $4KNO_3(s) \xrightarrow{\Delta} 2K_2O(s) + 2N_2(g) + 5O_2(g)$
O.N. increased: *oxidation*

In this case, the O.N. of N changes from +5 to 0, and the O.N. of O changes from −2 to 0, so this *is* a redox reaction.

$$\overset{+6}{\underset{+1+1\ -2}{|}}\ \overset{+1}{\underset{+1\ -2}{|}}\ \overset{+6}{\underset{+1\ -2}{|}}\ \overset{+1\ -2}{|}$$
(c) $NaHSO_4(aq) + NaOH(aq) \longrightarrow Na_2SO_4(aq) + H_2O(l)$

The O.N. values do not change, so this is *not* a redox reaction.
COMMENT The reaction in part (c) is an acid-base reaction in which HSO_4^- transfers an H^+ to OH^- to form H_2O. In the net ionic equation for a strong acid–strong base reaction,

$$\overset{+1}{|}\ \overset{-2}{\underset{+1}{|}}\ \overset{+1\ -2}{|}$$
$$H^+(aq) + OH^-(aq) \longrightarrow H_2O(l)$$

we see that the O.N. values remain the same on both sides of the equation. Therefore, an acid-base reaction is **not** a redox reaction.

FOLLOW-UP PROBLEM Use oxidation numbers to decide which, if any, of the following equations represents a redox reaction:
(a) $NCl_3(l) + 3H_2O(l) \longrightarrow NH_3(aq) + 3HOCl(aq)$
(b) $AgNO_3(aq) + NH_4I(aq) \longrightarrow AgI(s) + NH_4NO_3(aq)$
(c) $2H_2S(g) + 3O_2(g) \longrightarrow 2SO_2(g) + 2H_2O(g)$

But positive oxidation levels are also accessible to nonmetals because of the very large propensity for oxygen to extract electron density from other elements. This is the reason carbon, nitrogen, sulfur, and chlorine are in their maximum oxidation states +IV, +V, +VI, and +VII when surrounded by an oxide: CO_2 (or CO_3^{2-}), NO_3^{2-}, SO_4^{2-}, and ClO_4^-.

We can sketch this by noting the transfer of electron density from the nonmetal to oxygen in a chemical bond:

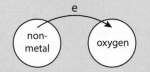

What is important to recognize is that in the assignment

of oxidation states, the designation represents a movement, a shift, a "delocalization" away from the nonmetal to oxygen, not a complete transfer of an electron or electrons. As a result, the actual charges on the nonmetal in a nonmetal to oxygen bond are much less than the +4, +5, +6, +7 in CO_2, NO_3^{2-}, SO_4^{2-}, ClO_4^-, respectively. However, the reason we keep careful track of the oxidation state of an atom in a molecular structure and why we keep track of how that oxidation state *changes* in a chemical reaction is that *energy* is at stake. A change in oxidation state virtually without exception implies a change in *energy*. This is why living organisms evolved to become masters of manipulation when it comes to oxidation states and thus of redox reactions.

As we will see, this is why nitrogen is so important for plant growth. Eight electrons must be removed from *N* in the conversion of NH_3 to NO_3^- and this provides an "energy ladder" by which organisms use the oxidation states of nitrogen to sustain themselves. We summarize the terminology for redox reactions in Figure CS2.4g.

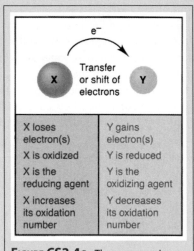

FIGURE CS2.4G There are a series of terms applied when an electron is *either* transferred completely or shifts partially from a donating species X to a receiving species Y. The four major examples are presented here.

CASE STUDY 2.5 Origin and Chemistry of Fossil Fuels

KEY CONCEPTS:

Fossil fuels today constitute 80% of primary energy generation globally. Figure CS2.5a displays graphically the formation of coal, petroleum, and natural gas, which are the three fossil fuels we will focus on here.

The question is, how did these fossil fuels form, what is their energy content, and how are they utilized in practice? As we develop an understanding of how energy is generated and used from fossil fuels, and the role they play in the planet's past, present, and future, it is important to establish how the Earth's carbon cycle functions. We will then turn to the origin of the major fossil fuels: petroleum, natural gas, and coal. This sets the foundation for investigating quantitatively the release of chemical energy from fossil fuel combustion.

I. The Global Carbon Cycle

The genesis of fossil fuels begins with the production of organic material from photosynthesis as we first explored in the Framework for Chapter 1. From the perspective of fuel production, only about 0.5% of the energy received from the Sun at the Earth's surface is converted by photosynthesis to chemical energy in the form of carbohydrates. As pointed out in Chapter 1, the overall reaction representing photosynthesis can be given by $6CO_2 + 6H_2O \rightarrow C_6H_{12}O_6 + 6O_2$ as displayed in Figure CS2.5b. In this expression $C_6H_{12}O_6$ represents the basic structure of the carbohydrates produced. Carbohydrates are so named because their basic chemical structure, $(CH_2O)_n$, contains two atoms of hydrogen and one atom of oxygen for each carbon atom in the structure.

The products, $C_6H_{12}O_6$ and O_2, are at a higher energy than the reactants CO_2 and H_2O by virtue of the capture of energy from the sun. The amount of energy contained in the products of the photosynthesis reaction is ~450 kJ/mole of carbon—an amount of energy that can be released as heat by combustion or an amount of energy that can be channeled by biological processes to build or sustain a living organism. As a rule, green plants use about one-half the energy contained in their carbohydrate production for their own energy needs.

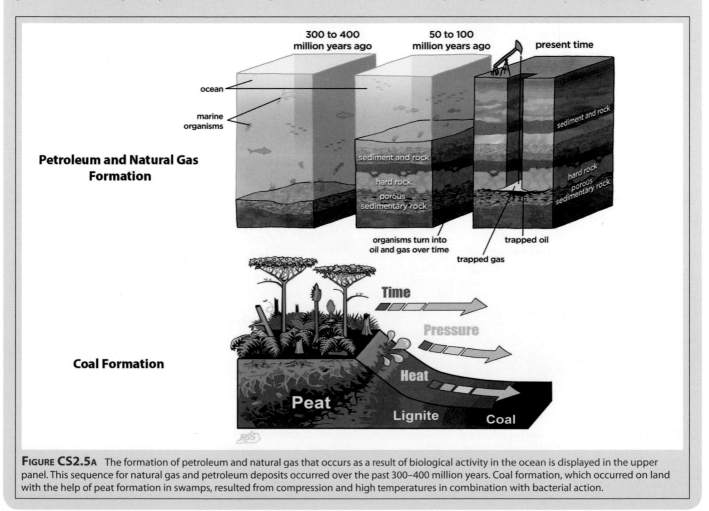

300 to 400 million years ago

50 to 100 million years ago

present time

ocean

marine organisms

Petroleum and Natural Gas Formation

sediment and rock

sediment and rock

hard rock

hard rock

porous sedimentary rock

porous sedimentary rock

organisms turn into oil and gas over time

trapped oil

trapped gas

Time

Pressure

Coal Formation

Heat

Peat

Lignite

Coal

FIGURE CS2.5A   The formation of petroleum and natural gas that occurs as a result of biological activity in the ocean is displayed in the upper panel. This sequence for natural gas and petroleum deposits occurred over the past 300–400 million years. Coal formation, which occurred on land with the help of peat formation in swamps, resulted from compression and high temperatures in combination with bacterial action.

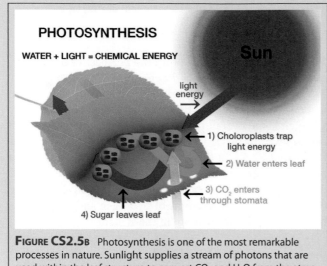

PHOTOSYNTHESIS

WATER + LIGHT = CHEMICAL ENERGY

Sun

light energy

1) Choloroplasts trap light energy

2) Water enters leaf

3) CO_2 enters through stomata

4) Sugar leaves leaf

FIGURE CS2.5B Photosynthesis is one of the most remarkable processes in nature. Sunlight supplies a stream of photons that are used within the leaf structure to convert CO_2 and H_2O from the atmosphere into complex organic structures that build the plant structure as well as produce carbohydrates that provide the plant with food.

The remainder is converted to other biological molecules or is invested in the growth of plant tissue. It is the energy contained in the growth of plant tissue that supplies the basis for fossil fuel production *or* (as we will see in Case Study 13.1) the production of biofuels. In either case the production of plant tissue is termed *net primary productivity*.

Over the past 350 million years the processes of photosynthesis and respiration

$$6CO_2 + 6H_2O \underset{\text{respiration}}{\overset{\text{photosynthesis}}{\rightleftarrows}} C_6H_{12}O_6 + 6O_2$$

have been closely balanced, and the exchange of carbon between CO_2 in the atmosphere and that contained in plant tissue constitutes a nearly closed system of carbon exchange. This cycle is displayed schematically in Figure CS2.5c. While this diagram appears to be complex, the figure's caption defines important aspects that are essential for an understanding of climate coupling.

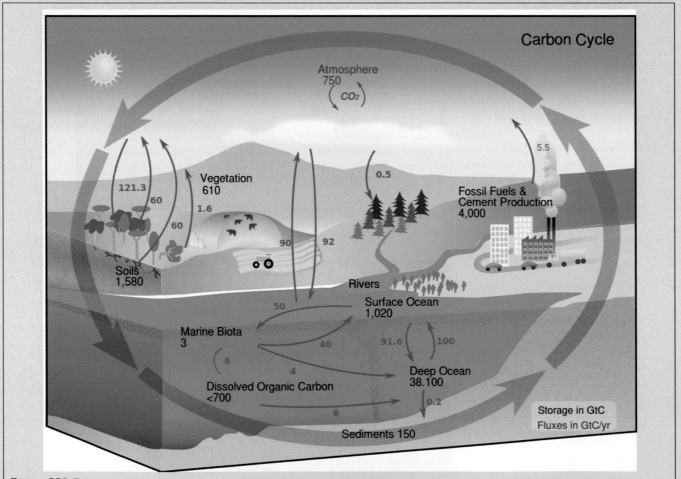

FIGURE CS2.5C The carbon budget of the Earth system is critical to the unfolding relationship between energy and the consequences of fossil fuel combustion as a source of primary energy at the global level. The carbon budget is represented graphically here in terms of (a) the reservoirs of carbon (e.g. the atmosphere, surface ocean, marine biota, etc.) in units of GtC and (b) the fluxes indicated by the arrows that link the reservoirs where the units are in GtC/yr. The distinction between those quantities that refer to natural reservoirs and fluxes vs. those affected by human activity should be carefully noted.

However, and very importantly, approximately 1 part in 10^4 of the organic matter produced by photosynthesis is buried such that it is no longer in contact with the O_2 that is also produced by photosynthesis. Over the millions of years that have transpired since photosynthesis developed at large scale on the Earth, a large amount of reduced carbon has built up in the soils and outer-crust structure of the planet. Some of the buried carbon accumulated in significant deposits that were subject to elevated pressures and temperatures and in some cases an assortment of bacterial action that transformed their molecular structure.

As we will develop further in subsequent Chapters (see for example Case Study 12.2) the build up of oxygen in the atmosphere resulted from the burial of the organic material, removing it as a reactant for respiration—thus leaving an excess of O_2 in the atmosphere that reached approximately 21% of the atmosphere about 400 million years ago.

This brings us to an interesting question. An estimate of the energy content of recoverable fossil fuels is ~5×10^{22} joules of energy. If this fossil fuel was burned in the course of providing primary energy to the global population, how much oxygen would remain in the atmosphere?

Will We Run Out of Oxygen?

Q. Some people are concerned that we will use up earth's O_2 supply if we burn all the fossil fuel. Are they right to worry?

A. (1) To decide this question, first calculate how many moles of O_2 would be used up, given that the energy available in recoverable fossil fuels is estimated to be 5×10^{19} kJ and 407 kJ of energy are released per mole of O_2 on average, when fossil fuel is burned (this value differs from the 450 kJ absorbed in many carbohydrate reactions, because fossil fuels differ in composition from carbohydrates):

Dividing 5×10^{19} kJ by 407 kJ mol^{-1} of O_2

gives 1.2×10^{17} mol of O_2.

(2) Next, calculate what fraction of the atmosphere's O_2 this represents, using the following facts: The atmosphere weighs 1000 g for each square centimeter (cm^2) of earth's surface, and is 21% O_2 by weight; the radius of earth (r) is 6.4×10^6 meters (m).

Since each square centimeter (cm^2) of the earth's surface accounts for 1000 g of air, or 210 g of O_2, we need to know the surface area. This value can be obtained from the radius using the formula for the area of a sphere:

$$a = (r^2) \times 4\pi$$

r is given as 6.4×10^6 m, which we need to convert to centimeters by multiplying by 100 cm m^{-1}. Then we multiply the area (in cm^2) by the weight of O_2 per cm^2, and finally divide by the molecular weight of O_2 (32 g mol^{-1}) to find the number of moles:

$$\text{mol } O_2 = (6.4 \times 10^6 \text{ m} \times 100 \text{ cm/m})^2 \times$$
$$4\pi \, (210 \text{ g/cm}^2 \, (32 \text{ g mol}^{-1}))$$
$$= 3.4 \times 10^{19} \text{ mol}$$

This value is 283 times greater than the answer in part (1).

II. Origins of Petroleum and Natural Gas

New discoveries of natural gas and petroleum reserves have altered the calculation of fossil fuel availability markedly since 2010. This has brought a focus back on how those fuels evolved over the past 300 million years.

The key distinction is that natural gas and petroleum deposits are of *marine origin*. The oceans are a vast biological "engine" with a production of approximately 400 billion tons of reduced carbon *annually* via photosynthesis in surface waters. A major fraction of this biogenic carbon is recycled to the atmosphere as displayed in Figure CS2.3c. However, a small fraction settles to the bottom where there is no oxygen available for respiration. This biological detritus becomes entombed in a matrix of porous clay and/or sandstone. Anaerobic bacteria digest the organic matter releasing methane, ammonia (NH_3) and water. The molecules most difficult to digest are the hydrocarbon-based lipids. Remarkably, the hydrocarbons in recovered petroleum have both molecular structure and carbon number distributions that closely resemble those found in the lipids of living organisms. Interestingly, all petroleum deposits contain derivatives of the hydrocarbon hopane ($C_{30}H_{52}$) that is commonly found in the membrane structure of bacteria. The structure of bacteriohopanetetrol is shown in Figure CS2.5d.

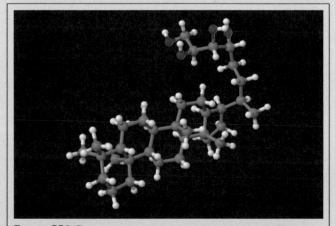

FIGURE CS2.5D The molecular structure of hopane is displayed here in 3 dimensions where carbon atoms are grey, hydrogen atoms are white, and oxygen atoms are red.

With the continuing burial of the mixed organic-inorganic matrix, both the temperature and the pressure rise. The combination of increasing pressure and temperature terminates the bacterial action because the bacteria cannot survive those conditions. However, the pressure and temperature are sufficiently high to induce the chemical rearrangement of those molecular structures. The result is that large quantities of methane are released, along with a mixture of other light hydrocarbons. These gases, capable of diffusing through some porous rock structures, accumulate in pockets *under impermeable* rock structures. The remaining heavy organic compounds mixed with water migrate through the rock structures until the increasing pressure squeezes the water from the emulsified mixture leaving the oil trapped in the porous layers of the rock.

This process of accumulation following bacterial action and increasing formation of natural gas and petroleum deposits has taken place over hundreds of millions of years as detailed in Figure CS2.5e.

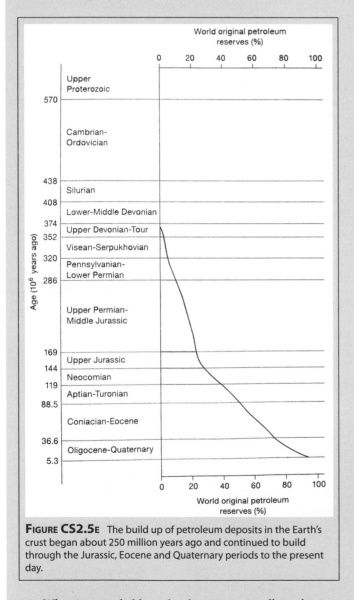

FIGURE CS2.5E The build up of petroleum deposits in the Earth's crust began about 250 million years ago and continued to build through the Jurassic, Eocene and Quaternary periods to the present day.

What is remarkable is that humans are well on the way to extracting a significant fraction of those deposits over a period of just one and a half centuries.

As we will see in the next chapter, a great deal more methane is contained in permafrost deposits and undersea deposits as clathrates than is contained in geological deposits of petroleum and natural gas. Methane clathrates are found in ocean sediments at combined depths of 300 meters or more as well as under the Arctic permafrost. As was the case with the geological deposits, this clathrate CH_4 is formed by the anaerobic decomposition of organic matter by bacterial action. Current estimates are that the chemical energy content of these methane clathrates distributed world-wide is at least *twice* that of all coal, petroleum and natural gas deposits *combined*. The containment of this methane in clathrates and CO_2 in permafrost represents one of the most potent feedbacks in the climate structure because increasing temperatures and melt regions of the Arctic will release both methane and carbon dioxide from these reservoirs.

Finally, a fraction of the hydrocarbon in the Earth's crust is abiotic in origin. At great depths, the heat and pressure can convert carbonate rock ($CaCO_3$) from oxidized to reduced forms of carbon—specifically CH_4. In the biogenic (via photosynthesis) organic compounds, the ^{13}C isotope is depleted relative to ^{12}C because the heavier isotope reacts more slowly in biochemical reactions including photosynthesis. The result is that methane collected from great depth tends to have increased $^{13}C/^{12}C$ ratios when compared with biochemical sources.

III. Origins of Coal

A primary distinction between coal and the marine origins of petroleum and natural gas is that coal is of *terrestrial origin*. Coal deposits are a result of plant matter provided from large, thickly wooded swamps that dominated the planet between 250 and 50 million years ago. During that period from the Middle Jurassic to the Eocene (see Figure CS2.5e), the Earth's climate was warm and moist. As we discussed in the Framework to this Chapter, woody plants are made up mainly of lignin and cellulose. While aerobic bacteria rapidly oxidize cellulose to CO_2 and H_2O after the plant dies, lignin is much more resistant to bacterial action. The structure of the lignin polymer is displayed in the upper panel of Figure CS2.5f. The lignin polymer, which gives trees a remarkable strength-to-weight ratio, is a complex three-dimensional structure built upon multiple benzene rings. The primary building units are coniferyl and sinapyl alcohols from coniferous and deciduous plants respectively.

In these warm swamps, the lignin accumulates underwater, slowly compacting into peat. Over millions of years the peat layers of these primeval swamps are transformed to deposits of coal. The motion of tectonic plates buried these deposits and subjected them to high pressures and temperatures for extended periods. This high pressure, high temperature process gradually eliminated the oxygen content via the expulsion of CO_2 and H_2O from these deposits. In addition,

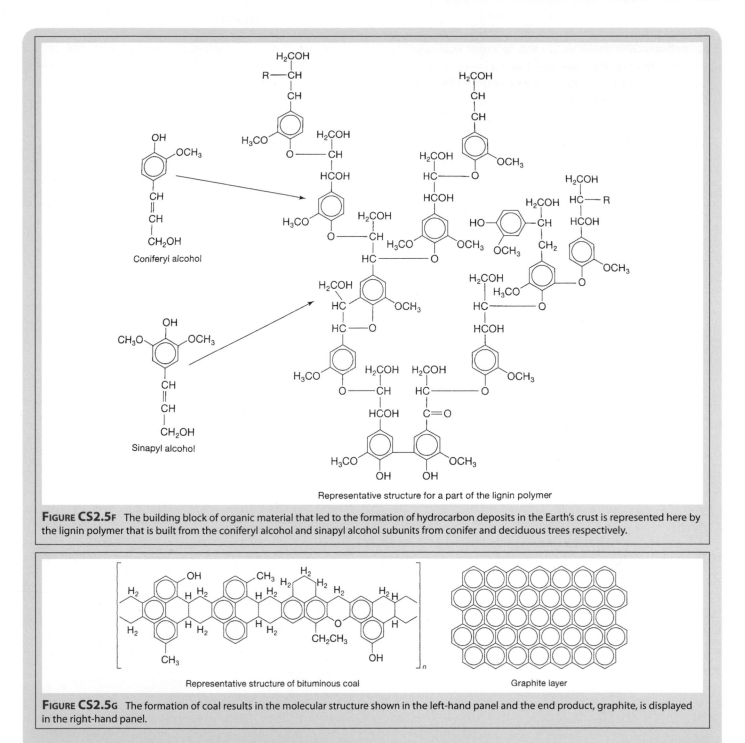

Representative structure for a part of the lignin polymer

FIGURE CS2.5F The building block of organic material that led to the formation of hydrocarbon deposits in the Earth's crust is represented here by the lignin polymer that is built from the coniferyl alcohol and sinapyl alcohol subunits from conifer and deciduous trees respectively.

Representative structure of bituminous coal   Graphite layer

FIGURE CS2.5G The formation of coal results in the molecular structure shown in the left-hand panel and the end product, graphite, is displayed in the right-hand panel.

the aromatic (hydrocarbon ring) structures condensed, resulting in a closely packed structure of hydrocarbon rings with a very high carbon content as displayed in the left-hand panel of Figure CS2.5g. Were this metamorphosis to continue; the final result would be the structure of graphite shown in the right-hand panel of Figure CS2.5g.

Problem 1

Based on the list of average bond dissociation enthalpies in Table CS2.3a, calculate the energy needed for the photosynthesis reaction in terms of kJ per mol of carbon.

$$6\,CO_2 + 6\,H_2O \rightarrow C_6H_{12}O_6 + 6\,O_2.$$

The structure of $C_6H_{12}O_6$ is

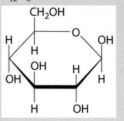

Thermochemistry
Development of the
First Law of Thermodynamics

3

Estimated U.S. Energy Consumption in 2020: 92.9 Quads

Lawrence Livermore
National Laboratory

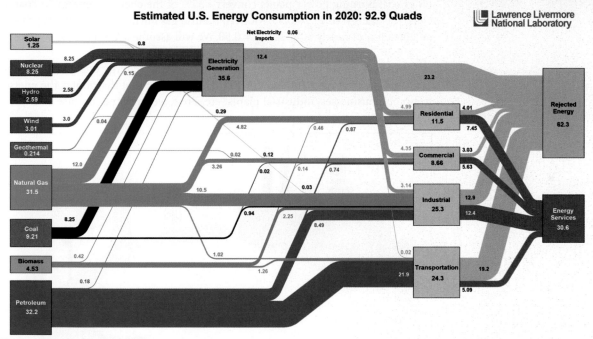

Figure 3.1 The Lawrence Livermore National Laboratory developed a graphical representation of energy flow through the residential, commercial, industrial and transportation sectors of the United States that links primary energy generation on the far left of the diagram through its uses to the end result. This energy flow diagram provides key insight and is published online each year by the U.S. Department of Energy: https://flowcharts.llnl.gov.

Framework

Why does the First Law of Thermodynamics play such a central role in developing the link between scientific concepts and real world context? For example, why is approximately 65% of primary energy input wasted on the national scale when primary energy sources, shown on the left-hand side of Figure 3.1, are converted to forms that deliver energy input to residential, commercial, industrial, and transportation use? Why do the principles of thermodynamics establish the fact that only 20% of the chemical energy contained in gasoline is converted to usable work to propel an automobile? This represents a remarkable loss of primary energy captured in the flow of energy from the petroleum input (in green in Figure 3.1) through the transportation sector to "Rejected Energy" on the right-hand side of

the diagram. Why are carbon based fuels, gasoline and diesel, inherently limited in their efficiency for producing mechanical work? Why are technological developments associated with converting photons to electrons to realize new methods of primary energy generation critically important for opening pathways to innovative solutions for economic development and for sustaining emerging economies? Achieving an understanding of the laws of thermodynamics, it turns out, constitutes a pivotal part of setting national and international energy policy as we move into the decades ahead.

On the face of it, the First Law of Thermodynamics is disarmingly simple. It states that energy is neither created nor destroyed (when mass energy $E = mc^2$ is included). The fact that energy is a conserved quantity immediately suggests that in any physical or chemical process we must keep track of that energy. This means quantitatively accounting for the origin of that energy in any physical or chemical transformation, as well as accounting for where the energy goes. But the fact that energy is neither created nor destroyed turns out to be a powerful constraint leading to remarkable insight into how processes work at both the molecular and global scale.

We set the context for our study of the First Law by examining an illustrative example of energy flow within our economic structure—the sequence of events from the combustion of coal in an electricity power plant to the delivery of that power to our living space. Consider first the efficiency of a power plant using the expression for efficiency, ε, given by ε = (what you get)/(what you pay for) which in this case is ε = (electrical energy produced)/(chemical energy purchased). Typically the energy that provides the high temperature steam that drives the turbines used to generate electric power comes either from coal or natural gas. On average, older coal burning power plants convert ~35% of the chemical energy in coal to electrical energy, so ε_{coal} = 0.35. Modern natural gas burning power plants achieve an efficiency of nearly 50%, so ε_{gas} = 0.50. We will assume a US national average of ε_{ave} = 0.38.

Now let's consider what happens to that electrical energy once it has left the coal-fired generating plant and enters the power grid that distributes the energy to homes, businesses, industrial plants, etc. First, we examine the case of electric

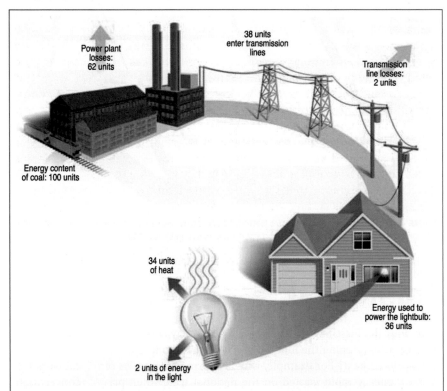

FIGURE 3.2 An interesting case of tracking the effective delivery of energy from its source—the chemical energy content of a fossil fuel such as coal—to its end use is that of tracing the generation of electrical energy in a coal burning power plant through to the production of visible light from an incandescent lightbulb. If we begin with 100 units of chemical energy in coal and drive a turbine to produce electricity, typically 62 of the 100 units of chemical energy (from the combustion of coal) is lost as heat. Thus 38 units of electrical energy is distributed to the electrical grid that delivers the electrical power to consumers. The transmission of electrical energy typically results in the lose of about two units of electrical energy, thus delivering 36 units of electrical energy to the house. Of the 36 units of electrical energy delivered to the incandescent light bulb, only two units of light, light visible to the human eye, is delivered by the lightbulb; the rest is emitted by the lightbulb as heat.

lighting for your home, office, dorm room, etc. Using a telling graphic from *What You Need to Know About Energy*, published by the US National Academy of Sciences (2008), we can quantitatively trace the path from energy generation to light output (Figure 3.2). With a (typical) 38% efficiency in initial generation, we lose 62 units of the initial 100 energy units contained in the chemical bonds of combusted coal relative to the product CO_2 and H_2O. Typically, between 2 units and 8 units are lost in the electrical transmission lines, as shown in Figure 3.2. We will assume the more efficient of these such that 36 units of electrical energy of the original 100 units of chemical energy are delivered to the home. However, with the use of an incandescent bulb, 34 units of that delivered electrical energy are emitted as *heat* and only 2 units are emitted as usable, that is visible, light. Thus, only 2% of the original chemical energy contained in the coal combusted in the power plant actually results in useful energy! Thermodynamics will allow us to calculate each of these contributions.

There are a number of key conclusions to be drawn. First, the initial step of fossil fuel combustion in the power plant is very inefficient—between 60 and 70% of the primary chemical energy is lost before any useful energy is generated. Second, it is that first step that releases the large amounts of CO_2 (as well as soot, nitrates, sulfates, mercury, etc., depending on the type of fossil fuel) into the atmosphere. Third, one unit of energy saved at the usage end translates into 10 to 20 units at the production end. Thus, conservation is very important. Fourth, if energy were produced at the site of its use (for example, photovoltaics to collect energy to supply lighting or air conditioning), the demand for primary power generation using coal or natural gas would drop dramatically. It is also of fundamental importance to recognize that if renewable forms of energy such as solar or wind energy replaced fossil fuel as the primary energy source, the total energy required would decrease by more than 60%! Finally, note that the conversion of electrical energy to mechanical energy by the electric motor is > 90% efficient—contrast that with a gasoline engine that is < 20% efficient, which is a key point for the next generation of automobiles, trucks, and buses.

In order to understand and apply the First Law, we must develop a clear understanding of the distinction between temperature, heat, work, and the thermal energy contained within the system under study. As we already know, work involves a force operating over a distance—it is inherently a *macroscopic* concept. Heat, in stark contrast, involves the transfer of energy at the molecular level by the motion of molecules and collisions between molecules at the *microscopic* level.

Key questions need answers. How does the release of energy from a chemical reaction actually produce mechanical work? How does a heat engine work and what controls and ultimately limits its efficiency? What is heat and how is heat defined in a scientific sense? How is the flow of heat measured quantitatively? These issues are developed in the chapter core and then treated quantitatively in Case Study 3.1.

While the principles of thermodynamics illuminate the quantitative foundations for how energy is transformed to provide what civilization needs to sustain health, prosperity, and stability, what does thermodynamics have to say about the consequences of particular choices for primary fuels? It turns out that thermodynamics not only informs us about how chemical energy release powers the planet, it also informs us as to how the flow of heat into the subsystem of the climate set in motion feedbacks that initiate irreversible changes in the Earth's climate structure.

An important example, treated quantitatively in Case Study 3.2, is the directly observed disappearance of the Arctic Ice Cap. In the last 35 years, 75% of the Arctic Ice Cap has melted as a result of the flow of heat into that system. With the melting of Arctic ice comes the inflow of warm water from lower latitudes and the inflow of warm air into the Arctic Basin, inflow of air that has not been cooled by passing over cold fields of snow and ice. And finally comes the inflow of heat from

Case Study 3.1 The Carnot Cycle and Heat Engine Efficiency

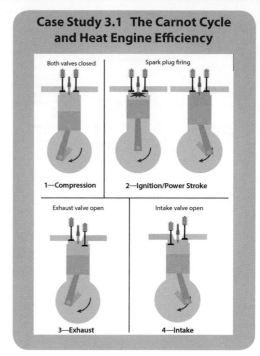

Case Study 3.2 Thermodynamics, Heat Pumps, and the Personal Energy Budget

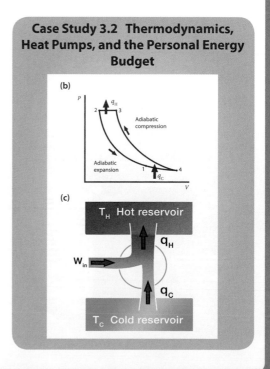

Case Study 3.3 Geothermal Energy

EGS Development Sequence

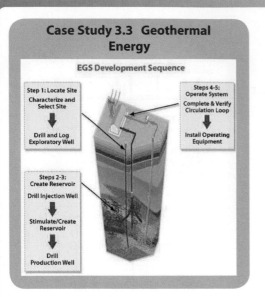

Case Study 3.4 The "50 Questions": Segment No. 1

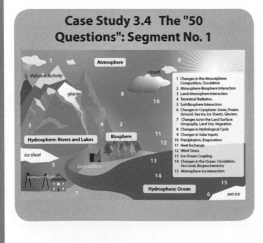

sunlight absorbed in the summer by the now-open Arctic Ocean rather than being reflected to space by the white surface of Arctic ice cover that has been in place for the past 100,000 years. It is the irreversible *retention* of this heat that sets the path forward for the planet, not simply "global warming." These feedbacks mean that the second half of the Arctic Ice Cap will disappear more rapidly than the first half. It is now believed that the Arctic will be free of permanent ice by 2025 or before, where "permanent ice" refers to the ice volume that remains at the end of the summer melt season.

The laws of thermodynamics hold additional insights that are essential for understanding unfolding challenges for society. For example, global warming is most often expressed in terms of increases in the global *mean* temperature over a particular period of time. But 70% of the globe is covered by oceans that are, on average, 3500 meters deep. Water, as we will see, has a large *heat capacity* per unit volume, which means that a great deal of heat flows into the global system for a small change in temperature, suppressing the *observed* increase in average global temperature. This draws a stark contrast between the concepts of heat and temperature—a distinction that only thermodynamics can clarify.

Other remarkable insights emerge from an understanding of thermodynamics. Why, for example, do you deliver twice as much heat to your house by burning natural gas in an electricity generating power plant and then using that electricity to power a heat pump in your house rather than burning that natural gas directly in a furnace *in your home*? The laws of thermodynamics tell us why, and Case Study 3.3 demonstrates this critically important point explicitly.

We know that over 80% of the primary energy used to sustain the global economy is extracted from fossil fuels—primarily coal, petroleum, and natural gas. The energy is extracted by converting carbon–hydrogen bonds and carbon–carbon bonds to H_2O and CO_2. But why are H_2O and CO_2 the universal product of fossil fuel combustion? And how do we quantitatively couple calculations of this energy release into work that can drive turbines for electrical generators, internal combustion engines for automobiles and trucks, and jet engines for aircraft?

An understanding of the devices that are capable of transforming chemical energy to work, first discussed in Chapter 1, relies on the extensive use of physical *models* that capture the essential elements of a complicated mechanical device in the simplest possible way. In so doing, the physical model of a thermodynamic engine becomes amenable to mathematical analysis. It is this process of constructing a simplified model of an otherwise complicated "heat engine" and then linking that model to the mathematics required to analyze the system quantitatively that makes thermodynamics a powerful foundation for linking science and technology in the modern world. It is also the basis for advancing our development of quantitative reasoning as developed in Case Study 3.3 using the intriguing distinction between *heat engines* and *heat pumps*.

Case Study 3.4 continues the pattern of using one Case Study in each chapter to develop quantitative reasoning. An important part of the development of quantitative reasoning involves developing the quantitative answers to important global scale questions related to global scale energy and power. Thus Case Study 3.4 introduces the first segment of what we term the "50 Questions"—fifty quantitative questions and answers that are critical to understanding the evolving challenges that lie ahead for all of us.

It is also of critical importance to use advances in our understanding of the role of thermodynamics in the developing global technology in order to explore how limitations imposed by thermodynamics can be used to elucidate alternative methods for primary energy generation. It is for this reason that new methods for energy generation using high-temperature geothermal techniques are featured in Case Study 3.5.

Chapter Core

Road Map for Chapter 3

In the sections that follow, we develop the principles central to thermochemistry with a focus on the development of the First Law of Thermodynamics. This requires a clear differentiation between *work* and *heat*. This is the context for the following Core Concepts that are addressed in order in this chapter.

Road Map to Core Concepts
Constructing a Model: A System and its Surroundings
Work Done On or By a System
Concept of Internal Energy
State Variables in Thermodynamics
Work Produced by a Chemical Reaction
Development of the First Law of Thermodynamics
Heat, Heat Capacity, and the Bomb Calorimeter
Enthalpy: A State Variable for Thermodynamic Changes at Constant Pressure
Standard Enthalpies of Formation
Standard Heats of Reaction
Hess's Law
Processes That Occur on a pV Surface
Linking the Thermodynamic Machine, the pV Diagram and the Energy Bar Chart
Spontaneous Change, Irreversibility, and Disequilibrium
Thermodynamics of Phase Transitions

Development of the First Law of Thermodynamics

Relationship Between the Energy Change of the System, of the Surroundings, and of the Universe

As we develop the First Law of Thermodynamics, it will become increasingly important to clearly distinguish between the *system* and the *surroundings*. The reason is that we must constantly refer to the heat, q, or the work, w, exchanged *between* two clearly defined domains. Figure 3.3 clarifies this distinction and emphasizes that the combination of the *system* and the *surroundings* constitutes all of matter. In thermodynamics, this sum of the system and the surroundings is referred to as the *universe* to emphasize the totality of matter as represented in Figure 3.3.

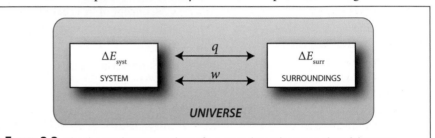
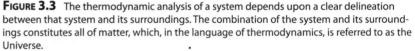

FIGURE 3.3 The thermodynamic analysis of a system depends upon a clear delineation between that system and its surroundings. The combination of the system and its surroundings constitutes all of matter, which, in the language of thermodynamics, is referred to as the Universe.

It is important when solving problems in thermodynamics to carefully define the boundary between the system and the surroundings. A key result of the conservation of energy is that the change in energy of the system, ΔE_{syst}, is equal to but opposite in sign to the energy change of the surroundings, ΔE_{surr}. Therefore

$$\Delta E_{system} = -\Delta E_{surr}$$

However, conservation of energy also dictates that energy is neither created nor destroyed in the totality of the universe, so

$$\Delta E_{univ} = 0$$

and because

$$\Delta E_{univ} = \Delta E_{syst} + \Delta E_{surr}$$

we have, in turn, the important relationship between ΔE_{univ}, ΔE_{syst}, ΔE_{surr}, and the conservation of energy.

$$\Delta E_{univ} = \Delta E_{syst} + \Delta E_{surr} = \Delta E_{syst} - \Delta E_{syst} = 0$$

The reason this conservation of energy expression involving ΔE_{univ} is so important is that we can often draw conclusions about what must happen within the totality of the universe and independently determine ΔE_{syst}, thereby determining, ΔE_{surr} by direct calculation. In general, then, if we know any two of the quantities ΔE_{univ}, ΔE_{syst}, or ΔE_{surr}, we can solve for the third. This approach turns out to provide considerable insight into an array of important problems.

Energy Exchange between a System and Its Surroundings

In Chapter 1 we developed an energy equation that represented both the total energy of a system of macroscopic objects and microscopic atoms and molecules, and represented the total energy of that system, E_{syst}. The total energy of the system, E_{syst}, represents the sum of the kinetic energy, KE, and potential energy, PE, of the macroscopic bodies that together constitute the mechanical energy, E_{mech}, of the system, and thermal energy, U_{therm}, that represents the microscopic energy of molecular motion that comprise the system such that

$$E_{syst} = E_{mech} + U_{therm} \qquad (3.1)$$

We also constructed a model of the system by identifying and separating the *system* from the *surroundings*. This model took into account both the exchange of kinetic and potential energy, which constitutes the mechanical (macroscopic) energy of the system, $E_{mech} = KE + PE$, and the conversion of that mechanical energy to thermal energy represented by the microscopic motion of atoms and molecules in the system, U_{therm}.

We can represent the model of the *system* by the specific case of an oscillating mass, moving on a rod with a spring system, wherein the friction between the mass and the rod (and dissipation within the spring) lead to an inexorable change in the mechanical energy of the system, converting mechanical energy to thermal energy until the mass ceases to oscillate on the rod as described in the sidebar.

But we can generalize the model to include the capability to externally supply mechanical energy to the oscillating mass on the rod to make up for the conversion of mechanical energy ($KE + PE$) to thermal energy caused by friction between the mass and the rod such that

$$\Delta E_{syst} = \Delta E_{mech} + \Delta U_{therm} = w_{ext}$$

where w_{ext} is the work externally applied to the system by the *surroundings*. This could be accomplished, for example, by adding to the downward velocity of the mass on each cycle to increase its kinetic energy, and thus its mechanical energy, by an amount equal to the energy dissipated in each cycle by friction.

We can represent that by a model that includes the work done on the system by an external force and we can also represent the possibility that the system can do work on the surroundings in that same model, as shown in Figure 3.4.

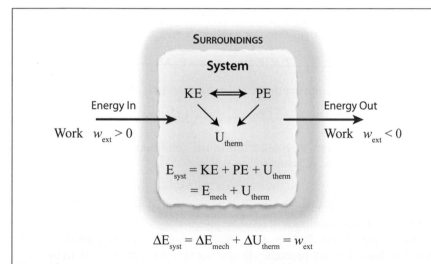

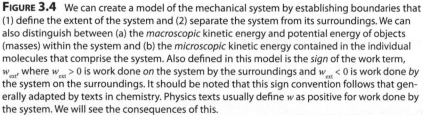

FIGURE 3.4 We can create a model of the mechanical system by establishing boundaries that (1) define the extent of the system and (2) separate the system from its surroundings. We can also distinguish between (a) the *macroscopic* kinetic energy and potential energy of objects (masses) within the system and (b) the *microscopic* kinetic energy contained in the individual molecules that comprise the system. Also defined in this model is the *sign* of the work term, w_{ext}, where $w_{ext} > 0$ is work done *on* the system by the surroundings and $w_{ext} < 0$ is work done *by* the system on the surroundings. It should be noted that this sign convention follows that generally adapted by texts in chemistry. Physics texts usually define w as positive for work done by the system. We will see the consequences of this.

Constructing a Model: A *System* and its *Surroundings*

The word system stems (as do many terms in thermodynamics) from Greek words meaning "to bring together" or "to combine." In the discussion of thermodynamic processes we must constantly grapple with the problem of keeping track of the flow of thermal energy (heat) through a definite surface and the mechanical work done on a specific ensemble of (macroscopic) objects. Thus, by nature, we define a system by its *boundary* that separates the *system*—that part of the physical world upon which we focus our attention—from the rest of the world, the surroundings. This is displayed in the adjoining figure.

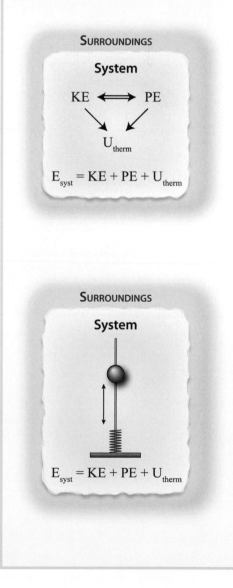

While we were concerned in Chapter 1 in our discussion of the conservation of energy for *isolated* systems, we turn now to the issue of how energy is transferred between the *surroundings* and the *system*.

In the study of chemical thermodynamics—thermochemistry—we are repeatedly confronted with how work is exchanged between a system and its surroundings. This turns out to be true whether we are quantitatively analyzing the heat and work produced by a chemical reaction or whether we are analyzing the work done by a "heat engine" such as an automobile engine that burns gasoline to generate kinetic energy.

A particularly important example of how work is exchanged between a system and its surroundings involves the work done by a piston that can move in response to a change in pressure within an otherwise closed vessel. This is shown diagrammatically in Figure 3.5.

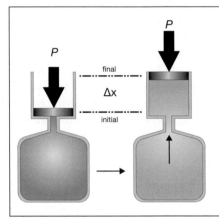

FIGURE 3.5 One of the most useful expressions for the work done by the expansion of a working substance is to convert the product of (Force)(distance) that is equal to the work done to (Force/area)(area)(distance of piston motion) and then write (Force/area) as pressure and (area)(distance) as the change in volume ΔV. Thus at constant pressure, the work is $w = -p\Delta V$. This is displayed at left as the gas contained in a beaker with a movable piston of area, A, that moves through a displacement, Δx.

In this case, the piston moves a distance Δx against the constant external pressure, P. We know that work is equal to the product of a force times a distance, so in our case (Figure 3.5):

$$w = (\text{Force})(\text{Distance}) = F \cdot \Delta x$$

However, the force, F, on the piston is the pressure, P, which is defined as the force per unit area times the area. Thus, for a piston of area A:

$$F = (F/\text{area})\,\text{area} = P \cdot A$$

and the work done in moving the piston a distance Δx is:

$$w = F \cdot \Delta x = P\,A\,\Delta x$$

The volume change in going from initial conditions to final conditions is $\Delta V = A\Delta x$ so the work, w, is just:

$$w = P\,A\,\Delta x = P\Delta V$$

This is a *very* useful expression for the work done by virtue of a change in volume, ΔV, under conditions of *constant pressure*.

Notice the units of this equation. With the pressure in atmospheres and the volume in liters, it is not immediately obvious that the units of work in this equation are energy (i.e., joules), as we know they must be from the First Law of Thermodynamics. However, pressure is force per unit area or mass times acceleration, divided by length squared. Volume is length cubed, so the product is:

$$\left[\frac{(\text{mass})\ \text{acceleration}}{\ell^2}\right]\ell^3 = \left[(\text{mass})\ \text{acceleration}\right]\ell^1 = \left[(\text{mass})\frac{\ell}{\sec^2}\right]\ell = (\text{mass})(\text{vel})^2$$

The units of energy indeed!

Check Yourself 1

An important example of the pressure—volume work is the inflation of a balloon under conditions of 1.00 atm pressure. Suppose for example we begin with a deflated balloon of volume 0.05 L and we inflate the balloon to 2.00 L. Calculate the amount of work required in joules.

Solution:

First identify the *system* and the *surroundings*. For this case, we take the system to be the balloon. Thus given $V_1 = 0.05$ L and $V_2 = 2.00$ L, and the pressure is a constant p = 1.00 atm, we calculate the work, w.

The work, w, can be directly calculated as
$$w = -p\Delta V = -p(V_2 - V_1)$$

Thus $V_2 - V_1 = (2.00 - 0.05)$ L = 1.95 L
p = 1.0 atm

so $w = -p(V_2 - V_1) = -(1.00 \text{ atm})(1.95 \text{ L})$
= −1.95 L-atm

Next we need to convert the units of L-atm to joules. Using the conversion of 101.3 J/L-atm we have:

$(-1.95 \text{ L-atm}) \times (101.3 \text{ J/L-atm}) = -197.5 \text{ J}$

It is always important to carefully note the sign of the answer. The system, the balloon, has done work on the surroundings, thus the sign on w should be negative—and it is.

We say, for example, that the *surroundings* do work *on* the *system* if the piston compresses the gas by moving downward. Alternatively, the system does work on the surroundings if the piston moves vertically upward as it does in Figure 3.5. So our model of the system in Figure 3.4, wherein $w_{ext} > 0$ represents net work done *on* the system, where energy is transferred *from* the surroundings to the system. Conversely, if work is done *on* the surroundings *by* the system, $w_{ext} < 0$.

But is the *mechanical interaction* of a force operating over a distance that constitutes the external work, w_{ext}, done on our system the *only* way to transfer energy to the system from the surroundings? Suppose we place the system on a stove and ignite the burner? There will quite rapidly be an increase in U_{therm} such that $\Delta U_{therm} > 0$ because the temperature of the system has increased. But we have done *no work* on the system, so $w_{ext} = 0$; however:

$$\Delta E_{syst} = \Delta E_{mech} + \Delta U_{therm} = w_{ext}$$

is clearly not zero; i.e.,

$$\Delta E_{syst} = \Delta E_{mech} + \Delta U_{therm} \neq 0.$$

So we have clearly violated our Energy Equation.

The problem is that while we have properly accounted for the exchange of energy by *mechanical interaction* between the system and the surroundings, we have not accounted for the energy transferred between the system and the surroundings by *thermal interaction*, by the transfer of energy at the microscopic level.

Therein lies the origin of the concept of *heat*. Heat is the energy transferred from the surroundings to a system by thermal interaction and it requires an additional term in our Energy Equation. As developed in Chapter 1, heat, the thermal (microscopic) energy transferred from a system to the surroundings, is designated by the symbol, q, such that our Energy Equation becomes

$$\Delta E_{syst} = \Delta E_{mech} + \Delta U_{thermal} = w + q$$

This places the transfer of energy from a system to its surroundings by *microscopic interaction*, q, in a position of parity with the transfer of energy by *macroscopic interaction*, w.

There is another perspective on the equation $\Delta E_{syst} = w + q$. Since w represents the *macroscopic* transfer of energy to the system from the surroundings and q represents the *microscopic* transfer of energy to the system from the surroundings, and energy must be either macroscopic or microscopic, what other way could there be to transfer energy? So if energy is neither created nor destroyed, the equation must be true from the perspective of pure logic.

The Concept of Internal Energy

We turn now to a more careful consideration of how we define the energy of the system, E_{syst}. Consideration of our equation $E_{syst} = E_{mech} + U_{therm}$ representing the energy of the system as the sum of *mechanical energy* (kinetic plus potential energy of the macroscopic objects within the system) and the *thermal energy* (associated with atomic and molecular-scale kinetic and potential energy) immediately raises the question: What about the energy contained in the bonds of the molecules that might release energy if converted to another chemical compound?

We considered an explicit example of this in Chapter 2 when we combusted octane, C_8H_{18}, releasing chemical energy to raise the temperature of the steel box that contained the burning gasoline as displayed in the reaction coordinate diagram Figure 3.6.

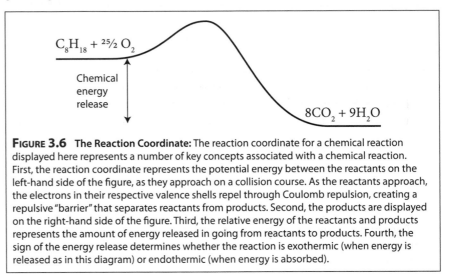

$C_8H_{18} + {}^{25}\!/_2 \, O_2$

Chemical energy release

$8CO_2 + 9H_2O$

FIGURE 3.6   The Reaction Coordinate: The reaction coordinate for a chemical reaction displayed here represents a number of key concepts associated with a chemical reaction. First, the reaction coordinate represents the potential energy between the reactants on the left-hand side of the figure, as they approach on a collision course. As the reactants approach, the electrons in their respective valence shells repel through Coulomb repulsion, creating a repulsive "barrier" that separates reactants from products. Second, the products are displayed on the right-hand side of the figure. Third, the relative energy of the reactants and products represents the amount of energy released in going from reactants to products. Fourth, the sign of the energy release determines whether the reaction is exothermic (when energy is released as in this diagram) or endothermic (when energy is absorbed).

Indeed, as we saw with the combustion of octane, *chemical energy* contained in the bonding structure of chemical reactants relative to the energy contained in the bonding structure of products is very important when considering the total energy of a system. So, too, is the *nuclear energy* stored in the atomic nuclei that is released in either fission or fusion nuclear reactions that we will study in Chapter 13. That energy can be of primary importance when adding up the energy contained in the atoms and molecules of a *system*.

Recognition of the importance of energy transfer between the system and the surroundings by *microscopic* processes leads to a major modification of our *Energy Equation*, with the addition of the heat term, *q*. So too does a more careful consideration of the forms of *microscopic energy* contained *within* the system—a consideration that leads to a reformulation of the *Energy Equation* yet again. Specifically, energy contained in chemical and nuclear energy categories must be added to the kinetic energy of molecular motion at the microscopic level to fully define the internal energy of a system.

Consideration of all sources of *microscopic* energy taken together is particularly important for *thermochemistry*, which is why it is given a specific symbol, the *internal energy, U*. The internal energy of a system is defined as the sum of all contributions to the system energy resulting from the atoms and molecules that comprise that system:

$$U_{syst} = U_{therm} + U_{chem} + U_{nuclear} + \ldots$$

We are now in a position to expand our definition of the *internal energy* of the system to include the mechanical energy, E_{mech}, of the objects contained in the system, *as well as* the thermal energy of molecular motion, U_{therm}, the chemical internal energy, U_{chem}, and the nuclear energy, $U_{nuclear}$, contained in the nuclei of the atoms that comprise the system. Thus, with this more complete accounting for the categories of internal energy, we have

$$E_{syst} = E_{mech} + U_{syst} = E_{mech} + U_{therm} + U_{chem} + U_{nuclear}$$

We will focus in our development of the First Law of Thermodynamics on the internal energy of the system, U_{syst}, defined as the sum of the *thermal energy*, U_{therm}, contained in the atoms and molecules of the system and the *chemical* energy, U_{chem}, contained in the bonding structure of the molecules that comprise the system relative to the bonding structure of molecules that could be formed as a result of some chemical transformation of those molecules. We will add the term U_{nuc} when we consider nuclear reactions in Chapter 13, but will ignore it for now, because the processes we consider through the end of Chapter 12 involve only U_{therm} and U_{chem}.

Internal energy of a system, U_{syst}, is the sum of energies for all of the individual particles (electrons, protons, neutrons, etc. that comprise the atoms and, via chemical bonds, the molecules) in a sample of matter that constitutes the system. Therefore, included in this *total internal energy inventory* is

- translational kinetic energy of molecules in the gas, liquid, and solid phase;
- kinetic energy associated with molecular *rotations* and *vibrations*, which includes the potential energy of the bond that is exchanged with the kinetic energy of vibration; and
- energy stored in the chemical bonds (ionic and covalent) that can be released in a chemical reaction converting reactants to products within the system.

Next, we further simplify our equation for the energy of the system, E_{syst}, by excluding the *mechanical energy*, E_{mech}, possessed by macroscopic objects within the system that have their own (macroscopic) kinetic and/or potential energy. This does not compromise our development of the First Law of Thermodynamics because the contribution to the total energy of the system by the mechanical energy of moving objects within the system can easily be included. For *thermochemistry*, then, we focus on the internal energy, U_{syst}, such that

$$U_{syst} = U_{therm} + U_{chem}$$

State Variables in Thermodynamics

Thermodynamics is an inherently quantitative branch of science and thus the measurement of quantities is particularly important. So too is the character of the quantities that we measure and record in thermodynamic studies. One important category of measurements involves quantities that determine the *state* of a system independent of how that system reached that state prior to the point in time when the observation was made. While we haven't explicitly identified quantities that define the state of a system, independent of the history of how that state was achieved, we have already examined important examples. One is the potential energy of a mass in a gravitational field. If a mass is suspended a height, h, in a gravitational field, by virtue of that position alone, it has a potential energy equal to mgh. But the work required to deliver that mass to a height h depends on how that mass attained that height and thereby the potential energy, mgh. That mass might have been raised to that height by a (nearly) frictionless pulley. Or it may have been pulled up the incline on a rolling device or dragged up over a rough surface. In each case, the *work* required would be quantitatively very different but

State Variables

We are constantly seeking to express dynamic and changing systems in terms of constants that remain invariant even though they may capture billions of individual events. The *state* of our *system* is defined by a specific set of quantities that establish its properties. Important examples include composition, temperature, pressure, volume, mass, etc. A distinguishing characteristic of a *state variable*, particularly for thermodynamics, is that *a state variable is independent of the path taken to achieve that state.* Such is the case for temperature, pressure, internal energy, etc. In contrast, heat and work are not state variables as they depend very much on the path taken.

the potential energy would be the same *independent of the path taken*. In this case the potential energy defines, independent of path, the *state* of the system and it is, in the language of thermodynamics, a state variable. Work, in stark contrast, depends very much on the path taken so work, w, is not a state variable. Temperature is another important example of a state variable because temperature is determined entirely by the velocity of the atoms and molecules that make up the object in question. If that body is a block of copper, and that block of copper is at a temperature of 20°C, that block may have cooled from 100°C, adding heat to its *surroundings* in the process. Or it may have warmed from 0°C extracting heat from its surroundings. It would be impossible to tell from examining the velocity of the atoms that make up the block of copper which path was taken. Yet the velocity of the atoms that comprise that copper block uniquely defines the temperature. Temperature is thus a *state* variable; the heat emitted or absorbed is not a state variable.

Consider the implications of defining a *state variable* when applied to a chemical reaction. Suppose, as Figure 3.7 demonstrates, we pour sulfuric acid, H_2SO_4, into a beaker containing zinc filings. As we will discover in Chapter 6, metal is attacked by a strong acid, forming a salt and hydrogen gas, H_2 in a reaction that releases energy:

$$Zn(s) + H_2SO_4(aq) \rightarrow H_2(g) + ZnSO_4(aq) + energy$$

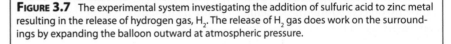

FIGURE 3.7　The experimental system investigating the addition of sulfuric acid to zinc metal resulting in the release of hydrogen gas, H_2. The release of H_2 gas does work on the surroundings by expanding the balloon outward at atmospheric pressure.

The "apparatus" shown in Figure 3.7 includes a thermometer and a sidearm to which a balloon is attached. As the reaction proceeds, the balloon inflates and the temperature of the system increases. The balloon, as it inflates, constitutes work done by the system on the surroundings. The increase in temperature results in heat released by the chemical reaction. A key question here is: how is the boundary between the system and the surrounding best defined? One reasonable choice would be to define the *system* as the chemicals: $Zn(s)$, $H_2SO_4(aq)$, $H_2(g)$, $ZnSO_4(aq)$. Everything else would then be defined as the surroundings—beaker, balloon, stopper, thermometer, and everything else in the universe. With this definition, the *energy* from the system (the chemical reaction) is transferred to the surroundings as work done to inflate the balloon and the heat transferred that increases the temperature of the beaker. As the thermal energy flows outward from the beaker, the bench top, the air in the room, etc. increase in temperature.

Alternatively, we could define the system to be the chemicals, the beaker, the stopper, and the balloon. And we could insulate the system such that no heat is transferred from the system to the surroundings. Then q = 0 and the system only does work on the surroundings by virtue of the increased volume of the balloon. The work can be calculated directly from w = –pΔV where p is equal to one atmosphere and ΔV is the change in volume of the balloon. We can represent what has occurred on an energy scale by considering the schematic in Figure 3.8. Before the

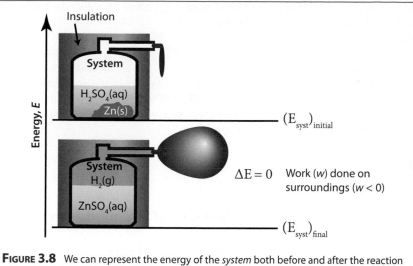

FIGURE 3.8 We can represent the energy of the *system* both before and after the reaction takes place. The energy of the *system* decreases because the *system* has done work on the *surroundings* by virtue of the fact that the expansion of the balloon against the pressure of 1 atm has done an amount of work equal to $-p_{atm}\Delta V$.

reaction has proceeded (but *just* after the sulfuric acid has been added to the zinc) the energy of the system is $(E_{syst})_{init} = (U_{therm})_{init} + (U_{chem})_{init}$. After the reaction has progressed to completion and the balloon has fully expanded, the final energy of the system, $(E_{syst})_{final} = (U_{therm})_{final} + (U_{chem})_{final}$ will be less than $(E_{syst})_{init}$ because work has been done by the system on the surroundings. If, as is shown in Figure 3.9, the balloon is replaced by a piston, the displacement, the volume change, ΔV, is then $A\Delta x$, where A is the piston area. The work done by the system is then $-p\Delta V$ where p is the pressure of one atmosphere. On an energy diagram we thus contrast the initial and final states of the system and the change in energy of the system:

$$\Delta E_{syst} = (E_{syst})_{final} - (E_{syst})_{initial} = w = -p\Delta V$$

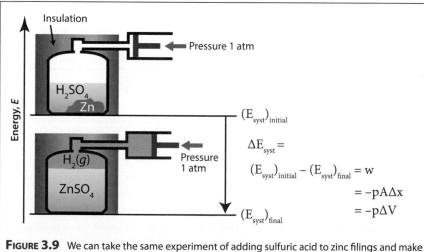

FIGURE 3.9 We can take the same experiment of adding sulfuric acid to zinc filings and make it quantitative by replacing the balloon by a piston that allows the direct determination of the displacement, Δx, facilitating the calculation of the work, $w = -p_{atm}A\Delta x = -p_{atm}\Delta V$.

where w is the work done by the system in its surroundings, $-pA\Delta x$. Notice, in particular, that the *energy* of the system has *decreased* in going from the initial to the final state. Conservation of energy:

$$\Delta E_{univ} = \Delta E_{syst} + \Delta E_{surr} = 0$$

Solution:

First you must find the limiting reagent. Zinc turns out to be limiting. The number of moles of hydrogen gas produced is then equal to the moles of Zn consumed:

$$n_{gas} = n_{H_2} = \left(14.66 \text{ g Zn}\right)\left(\frac{1 \text{ mol Zn}}{65.39 \text{ g Zn}}\right)\left(\frac{1 \text{ mol H}_2}{1 \text{ mol Zn}}\right)$$

$$= 0.2242 \text{ mol}$$

Next we note that the reaction took place at constant temperature (25°C) so T = constant and, from the Perfect Gas Law pV = nRT:

$$\Delta(pV) = \Delta(nRT)$$

So, for constant pressure:

$$p\Delta V = RT\Delta n$$

Thus w= $-p\Delta V = -\Delta nRT$

and the work is then:

$$w = -(0.2242 \text{ mol})(0.08206 \text{ L atm K}^{-1} \text{ mol}^{-1})(298 \text{ K})$$

$$= -5.483 \text{ L atm} = (-5.483 \text{ L atm})(101.3 \text{J/L atm})$$

$$= -555.5 \text{ J}$$

Note that $\Delta n_{gas} > 0$ implies w < 0; that is, work is done by the system on the surroundings. Note also that the calculation does not presuppose the presence of a container (the balloon) to catch the gas. An open reaction flask still does work, the evidence being the bubbles of gas evolving from the reaction mixture, but it is not useful work. A reaction such as this was used on a much larger scale by Jacques Charles to fill the early hydrogen balloons for manned flight.

tells us that because $\Delta E_{syst} < 0$, then ΔE_{surr} must be > 0 so the energy of the surroundings has increased.

Check Yourself 2

For the reaction $Zn(s) + H_2SO_4(aq) \rightarrow H_2(g) + ZnSO_4(aq)$ depicted in Figure 3.7, 300 mL of 1.00 M H_2SO_4 was added to 14.66 g Zn metal. Calculate the work, assuming the reaction went to completion, and the temperature of the expanding gas was 25°C.

Introduction to the First Law of Thermodynamics

The term *thermodynamics* suggests both (1) a study of an active transformation, dynamics, and (2) the central role played by heat as well as other forms of energy. The implication of dynamical transformation or exchange suggests that we must quantitatively define the distinction between a *system* under study and the *surroundings* within which the system resides. Thus we consider how we define how energy is *exchanged* between the system and its surroundings.

What is remarkable about the First Law of Thermodynamics is that it considers the potentially complicated issue of energy exchange between a system and its surroundings, and simplifies the quantitative treatment of the problem using pure logic.

As we developed in the first section of the chapter, work is the product of force times a distance of physical displacement. Work is thus inherently a *macroscopic* quantity. Heat, in sharp contrast, is the energy transferred between the system and its surroundings by strictly thermal interaction via molecular interaction. Heat is therefore inherently a *microscopic* quantity.

We can express the First Law of Thermodynamics as follows: heat, q, and work, w, are the *only* means by which energy is transferred between the system and its surroundings. We can express this in equation form by focusing on the change in internal energy of the system, ΔU_{syst}. Thus, the change in internal energy of the system is

$$\Delta U_{syst} = q + w \tag{3.2}$$

This is the statement of the First Law of Thermodynamics in equation form, but notice that the left-hand side of the equation refers only to the system while the right-hand side of the equation defines the exchange of microscopic energy (*q*) and macroscopic energy (*w*) *across* the boundary between the system and the surroundings.

An important aspect of the First Law is that it explicitly represents a statement of the conservation of energy, because by equating the change in internal energy of the system to the sum of the heat and work quantitatively exchanged *between* the system and the surroundings, energy is neither created nor destroyed.

The First Law of Thermodynamics also engages the pure logic that if energy is exchanged between the system and its surroundings, it must be by either the *macroscopic* energy transfer or the *microscopic* energy transfer or a combination of both: following on the logic that energy must either be macroscopic or microscopic.

It is also important at this stage in the development of the First Law of Thermodynamics to emphasize the following points:

(a) work and heat are not *contained within* the thermodynamic system—work and heat exist only as the forms of energy *transferred* between the system and its surroundings,

(b) internal energy, U_{syst}, is the *only* form of energy contained within the thermodynamic system,

(c) if the system is isolated from its surroundings, then $\Delta U_{syst} = 0$.

Development of the *First Law of Thermodynamics*, $\Delta U_{syst} = q + w$, as a quantitative tool for the solution of important problems is dependent upon establishing important conventions that clarify the *sign* of both the work term (w) and the heat term (q) in our expression above for the First Law. For any system, we adopt the convention that

- The heat, q, *entering* the system has a *positive* sign such that the heat *absorbed*, $q > 0$.
- Heat leaving the system representing energy transfer *from* the system *to* the surroundings has a negative sign such that $q < 0$.
- Work done *by* the *surroundings on* the *system* is positive, $w > 0$.
- Work done *on* the *surroundings* by the system is negative, $w < 0$.

This convention can be summarized, as displayed in Figure 3.10, by recognizing that energy entering the system has a positive sign; energy leaving the system has a negative sign.

Check Yourself 3

For the reaction system described in Check Yourself 2, the temperature was found to increase from 22°C to a maximum of 47°C, and then it began to fall again as the reaction subsided. Calculate the heat and the heat per mole of Zn.

We turn first to the question of how the internal energy *change* (ΔU_{syst}) of a system is instigated by thermal energy transfer (heat) by using a device called a bomb calorimeter, shown schematically in Figure 3.11.

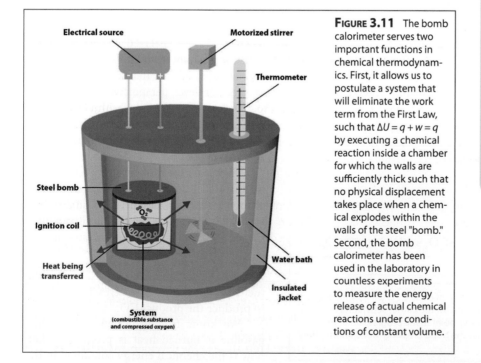

FIGURE 3.10 The definition of a system and its surroundings is fundamental to the quantitative accounting of the heat, q, added to the system from the surroundings and the work, w, done on the system by the surroundings. While the heat, q, is transferred by molecular level (microscopic) processes, work, w, is transferred by mechanical displacement wherein a force acts over a distance. Thus the movable piston in the diagram.

Solution:

Using the idealized two-stage analysis and $q = -mc\Delta T$; for 300 mL of solution, $m \approx 300$ g. We have

$$q = -(300 \text{ g})(4.18 \text{ J g}^{-1}\text{K}^{-1})(47 - 22)\text{K} = -31 \text{ kJ}$$

Note that, although the temperatures are given in Celsius, only their difference is required, and thus no conversion to the Kelvin scale is necessary, since 1°C = 1 K. From the first example, 0.2242 mol Zn reacted, giving

$$q/n = -31 \text{ kJ}/0.2242 \text{ mol} = -138 \text{ kJ/mol}$$

You should realize that the quoted temperature rise was not as high as it could have been, since the temperature probe shown in Figure 3.7 is detecting heat coming through the flask wall while the reaction is still in progress; the escaping heat is not available to raise the temperature of the system itself.

FIGURE 3.11 The bomb calorimeter serves two important functions in chemical thermodynamics. First, it allows us to postulate a system that will eliminate the work term from the First Law, such that $\Delta U = q + w = q$ by executing a chemical reaction inside a chamber for which the walls are sufficiently thick such that no physical displacement takes place when a chemical explodes within the walls of the steel "bomb." Second, the bomb calorimeter has been used in the laboratory in countless experiments to measure the energy release of actual chemical reactions under conditions of constant volume.

The purpose of the bomb calorimeter, which has been used extensively in chemical research, is to (1) create a practical physical model of a *system* and its *surroundings*, and (2) remove the work (w) term from the First Law of Thermodynamics such that the change in internal energy of the system (ΔU_{syst}) is equal to the heat term (q) alone: $\Delta U_{\text{syst}} = q + \cancel{w} = q$. This is done first by building a bomb with rigid walls such that the work term, which is the product of a force times a physical displacement (Force × displacement), is driven to zero by eliminating any *deflection* in the wall such that the physical displacement of the "bomb" is zero no matter what occurs within the bomb that contains the chemical reaction under study. Second, the bomb is loaded with chemical reactants that can be ignited externally, usually by a filament "flashed" by a pulse of electric current, as shown in Figure 3.11. Prior to detonation of the reactants, the initial internal energy, U_i, of the system is the sum of (1) the thermal energy of molecular motion, U_{therm}, and (2) the chemical energy associated with the chemical bonds of the reactants, U_{chem}. Thus, we can write

$$U_i = (U_{\text{therm}})_i + (U_{\text{chem}})_i .$$

After the chemical reaction has been initiated by the electric spark, and the system has returned to *equilibrium* such that there is *no temperature difference* between the bomb and the water bath, the system will have an internal energy, U_f is the sum of $(U_{\text{therm}})_f$, the thermal energy of the bomb, and $(U_{\text{chem}})_f$, the chemical energy contained within the bomb. $(U_{\text{chem}})_f$ is presumably zero because all the material combusted in the bomb has been converted to products thereby releasing the available chemical energy. The *change* in internal energy of the system is then

$$\Delta U_{\text{syst}} = U_f - U_i .$$

Employing our statement of the First Law of Thermodynamics $\Delta U_{\text{syst}} = q_{\text{rxn}} + w$, where q_{rxn} is the energy release of the chemical reaction transferred as thermal energy to the water bath, bomb, metal enclosure, etc. by virtue of the temperature difference between the bomb and the water bath. The work, w, done by the bomb on the water bath is zero because the bomb is designed such that the wall deflection is zero as noted above. Thus, we are left with the simple equation:

$$\Delta U_{\text{syst}} = U_f - U_i = q_{\text{rxn}} + \cancel{w}^{0} = q_{\text{rxn}} = q_V \qquad \textbf{(3.3)}$$

wherein the quantity q_V is the heat (thermal energy) transferred at *constant volume*.

But, while the bomb calorimeter has successfully *removed* the work term (w) from quantitative consideration in our statement of the First Law, two questions immediately emerge: First, from a molecular level perspective, how was the *chemical energy*, U_{chem}, converted to *thermal energy*, U_{therm}, within the bomb calorimeter? Second, how do we *measure* the energy released in the chemical reaction, q_{rxn}, at constant volume?

To answer the first question, let's assume we inserted into the "bomb" an amount of chemical reactant, octane, C_8H_{18} (gasoline), and sufficient O_2 that the chemical reaction

$$2\ C_8H_{18} + 25\ O_2 \rightarrow 16\ CO_2 + 18\ H_2O$$

can proceed to completion such that all of the fuel (octane) is consumed (reacted with O_2) to produce the product CO_2 and H_2O.

We know from experience that when gasoline is burned heat is produced. Then how is the chemical energy contained in the

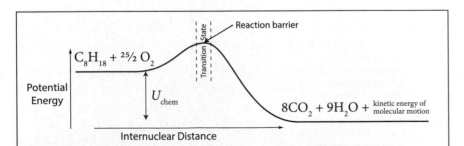

FIGURE 3.12 The potential energy surface for the reaction of octane (gasoline) with oxygen producing CO_2 and H_2O is displayed here. The highest point on the barrier separating reactants and products is termed the "transition state" for it is at this point in the progression of the reaction along the path from reactants to products that the bond breaking–bond reformation takes place.

bonds of C_8H_{18} and O_2 as they are converted to CO_2 and H_2O actually released? The answer to this question is aided by referring to our potential energy surface for the reaction. This potential energy surface or "reaction coordinate" shown in Figure 3.12 (introduced in Chapter 1) is a plot of potential energy on the vertical axis and internuclear distance on the horizontal axis. The reactants, C_8H_{18} and O_2, are shown on the left side of the diagram. As they approach, their intermolecular distance begins to decrease and the electron-electron repulsion begins to increase thereby increasing the potential energy until the "reaction barrier," displayed in Figure 3.12, is surmounted. As the chemical bonds rearrange at the "transition state" shown in Figure 3.12, the newly formed product molecules at the energy barrier then move rapidly to products, converting the *potential energy* that they (the products) possess at the instant of their formation, to *kinetic energy* of molecular motion as they move down the potential energy surface. The products of the reaction, CO_2 and H_2O, representing the *new bond structure* of the products, are shown on the right at a lower potential energy than that of the reactants. The products CO_2 and H_2O "explode" away from the point of formation, carrying with them a large amount of translational energy, vibrational energy, and rotational energy. These new molecules (CO_2 and H_2O) contain, by virtue of the chemical energy released in the reaction, an extremely large amount of kinetic energy (translation, vibration, rotation). Those newly formed CO_2 and H_2O molecules then collide repeatedly with the molecules around them within the bomb of the calorimeter, transferring their kinetic energy to the other molecules. This exchange of kinetic energy via molecule-molecule collision continues until the energy is partitioned among the energy modes (translation, vibration, rotation) of all molecules equally; thus establishing a new *temperature* for the ensemble of molecules contained in the bomb of the calorimeter.

The mixture of high kinetic energy molecules contained within the walls of the bomb segment of the device shown in Figure 3.11 then begin transferring kinetic energy to the entire system, and increasing the temperature of the surrounding combination of water, steel jacket, stirring system, and thermometer. This may be represented at the molecular level as shown in Figure 3.13.

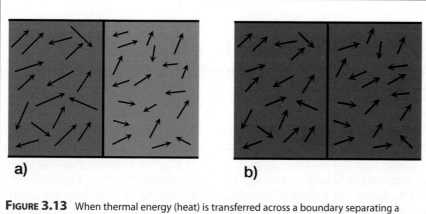

a)   b)

FIGURE 3.13 When thermal energy (heat) is transferred across a boundary separating a high temperature body (the left side of panel **a**) from a low temperature body (the right side of panel **a**) the molecules moving with higher kinetic energy in the hot body collide with the slower moving molecules within the low temperature body. This process continues until energy is equally partitioned in the two bodies and the system has achieved thermal equilibrium as shown in panel **b**. The temperatures of the system of two bodies lies between the original temperatures of the hot and cold bodies.

In the figure above, the hot gas, with molecules moving at high velocity, within the bomb shown on the left, panel (a), collide with the cold molecules of the walls of the calorimeter, transferring their kinetic energy, collision by collision, to the entire system.

As a result, the high kinetic energy contained in the molecules that comprise the walls of the bomb vessel and the gas contained in the bomb is transferred to the water bath in a process identical to that displayed in Figure 3.13. This transfer continues until the bomb vessel and the water have reached the same temperature and the *net* transfer of kinetic energy ceases. Note that while the net transfer of energy has ceased, there remains a dynamic exchange of kinetic energy on a *molecule-by-molecule* level. The interior of the calorimeter has now reached a state of *thermal* equilibrium where the *temperature difference* between the molecules within the bomb and within the water, steel container, etc. is now zero.

So how is the amount of energy released in the chemical reaction actually measured? While the internal energy of the (calorimeter) system,

$$U_{syst} = U_{chem} + U_{therm},$$

has remained unchanged, there has been a net conversion of chemical internal energy, U_{chem}, to thermal internal energy, U_{therm}. But we cannot, in any practical sense, accurately account for U_{therm} by adding the kinetic energy of translation, rotation, and vibration for *each* molecule to calculate a total. We must employ another strategy. That strategy is to measure a *change*, specifically the increase in *temperature*, of the bomb and the surrounding water bath, which we can easily do with the thermometer shown in Figure 3.11. But how is the temperature going to quantitatively define q_{rxn} produced in the chemical reaction as it appears in the First Law, given in Equation 3.2?

We recognize that by building the wall of the bomb sufficiently strong that no work is done by the bomb on the surrounding water such that

$$\Delta U_{syst} = q_{rxn} + w = q_{rxn} = q_v$$

where q_v is the "heat of reaction" at constant volume. But how do we calculate q_v from the temperature measurement of the water in the calorimeter? We answered this question in principle when we examined the experiments of James Joule in Chapter 1. Let's now consider the answer to the question in more detail.

Heat and Heat Capacity: How Thermal Energy Transfer (Heat) Is Calculated from a Temperature Change

We consider two cubes, diagramed in Figure 3.14, each containing a gram of water. It was established by the work of James Prescott Joule (and others) that the amount of energy required to raise 1 gram of water by 1°C was 1 calorie of energy; more appropriately in SI units, it requires 4.18 joules of energy to raise 1 gram of water by 1°C. So we run a series of experiments that involve (1) establishing an initial temperature for cube A and cube B and then (2) placing the cubes together until they are of *equal temperature*—i.e., such that they have reached *thermal equilibrium*. We know that for each 1°C a block *increases* in temperature, 4.18 joules of energy flowed into that 1 gram of water. We also know that for each 1°C a cube *decreases* in temperature, 4.18 joules of energy flowed out of that 1 gram of water.

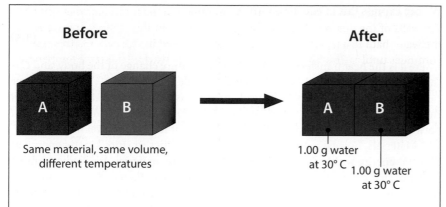

Before

A   B

Same material, same volume,
different temperatures

After

A   B

1.00 g water
at 30° C
1.00 g water
at 30° C

FIGURE 3.14   When two cubes of the same material with the same volume are at different temperatures, heat will flow from the warmer to the cooler cube in such a way that we can quantitatively deduce the amount of heat that flows between the two bodies. A series of such experiments, represented in Table 3.1, provides important insight into the relationship between heat flow and temperature change.

We can then run a series of experiments wherein we measure the initial temperatures of block A and of block B, and then record the final temperature of the two in contact, calculating each time the energy gained (or lost) by block A and the energy lost (or gained) by block B. We run a series of experiments and record the data:

TABLE 3.1

Run	Initial Temperature of A (°C)	Initial Temperature of B (°C)	Final Temperature of A & B (°C)	Heat, A (joules)	Heat, B (joules)
1	10.0	40.0	25.0	+62.7	−62.7
2	20.0	40.0	30.0	+41.8	−41.8
3	30.0	40.0	35.0	+20.9	−20.9
4	40.0	40.0	40.0	0.0	0.0
5	10.0	50.0	30.0	+83.6	−83.6
6	20.0	50.0	35.0	+62.7	−62.7
7	30.0	50.0	40.0	+41.8	−41.8

Four important facts emerge from our inspection of the data:

a)   When cube A and B have different temperatures, the cooler object always gains heat and the warmer object always loses it. We conclude that kinetic energy of molecular motion always flows from the warmer object to the cooler object. We express this spontaneous transfer of internal thermal energy, U_{therm}, as heat, q, transferred from cube B to cube A.

b)   When the initial temperatures of A and B are equal, no heat is exchanged between the two cubes.

c)   When the masses of A and B are the same, and when they are comprised of the same material, the final temperature is the *average* of the two initial temperatures.

d)   When the temperature change that an object undergoes is *doubled*, so too is the amount of heat exchanged *doubled*.

This fourth observation, (d), is particularly important. It says that the heat gained or lost by an object is directly proportional to the temperature change that it undergoes.

Heat Capacity of the World's Oceans

While we routinely discuss the heat capacity of objects in the laboratory or in industrial settings, it is becoming increasingly important to grasp the scale of the heat capacity of objects on the global scale because quantitatively analyzing the flow of thermal energy (heat) into those systems defines the trajectory upon which we are moving as the addition of CO_2 to the atmosphere traps increasing amounts of infrared radiation. We begin first by calculating the heat capacity of the world's oceans.

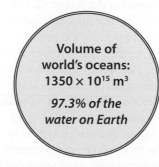

Volume of world's oceans: 1350×10^{15} m³

97.3% of the water on Earth

With a volume of 1350×10^{15} m³ this equals 1.35×10^{24} cm³. The density of water is 1g/cm³ so the mass of the world's oceans is 1.35×10^{24} g. The specific heat of water is 4.2 J/g-°C so the heat capacity of the ocean is:

$$C_{ocean} = mC_{H_2O} = (4.2 \text{ J/g-°C}) \, 1.35 \times 10^{24} \text{ g}$$
$$= 5.7 \times 10^{24} \text{ J/°C}$$

We can round this number off to achieve the easily remembered number of 6000 ZJ/°C as the heat capacity of the world's oceans.

We can express this in equation form by writing

$$q = C\Delta T \qquad \text{(3.4)}$$

where q is the heat transferred, ΔT is the temperature change of the body, and C is the "heat capacity" of the object. Notice that while "heat capacity" is a term universally used in thermodynamics, it treads dangerously close to violating the concept that heat is not a substance to be stored but rather is the thermal energy *transferred* between a system and its surroundings by virtue of molecular level interaction.

Materials, however, possess very different abilities to store kinetic energy as thermal energy within the bonds that comprise the material. We know from experience that a gram of water has a higher heat capacity than a gram of wood. But we already know that the thermal energy component of a body's internal energy, U_{therm}, is composed of the sum of the kinetic energy (translational, rotational, vibrational) of the individual molecules that comprise the material. Thus, a material with more molecules per unit volume will quite probably have a higher heat capacity. But it is also clear that the bonding structure and the vibrational and rotational modes that characterize the material will play a role in the heat capacity of a body as well because heat can flow into these modes of molecular motion.

We can verify by experiment that the heat capacity, C, in Equation 3.4 is directly proportional to the mass of the object under consideration. That is,

$$C = mc$$

where C is the heat capacity of the *object*, m is the mass of that object, and c is the *specific heat* (in joules/g-°C) of the material that comprises the object. It is the *specific heat* that we use in all thermodynamic calculations involving various substances, because we wish to tabulate the number of joules/g-°C *specific* to a given substance. Then it is simply a matter of multiplying that tabulated number by the mass of the object to determine the heat capacity of the object.

We can then write our quantitative formulation of the thermal energy transferred as heat, q, in the form

$$q = (\text{mass})(\text{specific heat})(\text{difference in temperature}) = mc\Delta T$$

Notice that the units of c are joules/g-°C such that when c is multiplied by the mass of the object in grams and the temperature *change* in °C or Kelvin, that q is given in joules. Some important examples are given in Table 3.2.

Volume of Other Water Containing Systems

While the world's oceans contain a vast majority of the available water (97.3%) at the Earth's surface, the volume of water in other systems is of great importance for analyzing changes in the Earth's climate. We consider here five examples:

1. The glaciers and polar ice. Combining the volume of ice contained in the Antarctic, the Arctic, and the major glacial systems of the continents, the volume of water contained in the ice is 2.1% of the total volume of water at the Earth's surface. That is 29×10^{15} m³ of water or 29×10^{21} cm³ of water. The volume of water tied up in the Greenland glacial system is 2.9×10^{21} cm³, or approximately 10% of the total volume of water tied up in the world's ice/glacial system.

2. The underground aquifers contain 8.4×10^{21} cm³ of water or 0.6% of the total.

3. The lakes and rivers of the world contain 0.2×10^{21} cm³ of water or 0.01%.

4. The atmosphere contains 0.013×10^{21} cm³ of water or 0.001% of the total.

5. The biosphere contains 0.006×10^{21} cm³ of water.

TABLE 3.2	
Specific Heats	
Substance	Specific Heat, J g⁻¹ °C⁻¹ (25 °C)
Carbon (graphite)	0.711
Copper	0.387
Ethyl alcohol	2.45
Gold	0.129
Granite	0.803
Iron	0.4498
Lead	0.128
Olive oil	2.0
Silver	0.235
Water (liquid)	4.18

Note that liquids have greater specific heats than solids. This results from the greater number of degrees of freedom (rotation, vibration, translation) of the molecules in the liquid, relative to those in the solid.

Application of the Bomb Calorimeter

Development of the First Law of Thermodynamics provides the quantitative foundation for the determination of the energy released from the chemical reaction using the bomb calorimeter. In particular, we will use the bomb calorimeter to measure energy release in the combustion of gasoline, C_8H_{18}, recognizing that because the bomb calorimeter eliminates any physical expansion of the system, no work is done during the course of the reaction, and thus $\Delta U_{syst} = q + w = q_{rxn}$ where $q_{rxn} = q_V =$ heat release at constant volume for the chemical reaction.

We define (a) the *system* as the chemical reaction and (b) the *surroundings* as all the components of the calorimeter: the calorimeter case, the water contained within the calorimeter, the cup holding the reactants, the reaction chamber, the thermometer, and the steering mechanism.

The heat gained by the calorimeter, q_{cal}, is then equal to the heat gain of each of the components of the bomb calorimeter such that:

$$q_{cal} = q_{case} + q_{water} + q_{cup} + q_{chamber} + q_{therm} + q_{stir}$$

Moreover, $-q_{cal} = q_{rxn}$, because the heat produced by the chemical reaction (the system), q_{rxn}, is equal but opposite in sign to the heat gained by the calorimeter (the surroundings), q_{cal}.

But as an experimental system, the heat capacity, C, of the entire calorimeter system can be determined independently, either at the individual component level or by submersion of the entire calorimeter in a bath of known temperature and volume of water, and then measuring the final temperature of the calorimeter plus water bath just as we did in generating Table 3.1. Then, for any subsequent measurement of $q_{rxn,}$ we can use

$$q_{calor} = (\text{heat capacity of calorimeter})\Delta T = C_{calor} \Delta T$$

Now we are ready to determine the heat of reaction, q_{rxn}, of octane using our bomb calorimeter. We have determined through careful and repeated studies that the *heat capacity* of the calorimeter is 5.62 kJ/°C. The combustion of 1 gram of C_8H_{18} in the bomb of the calorimeter causes the temperature to increase from 22.50°C to 31.08°C. What is the heat of combustion (heat of reaction) of octane expressed in kilojoules per mole?

Step 1:

Calculate q_{calor} by multiplying the heat capacity of the calorimeter (5.62 kJ/°C) by the observed increase in temperature (8.58°C):

$$q_{calor} = (8.58°C)(5.62 \text{ kJ/°C}) = 48.2 \text{ kJ}$$

Thus

$$q_{rxn} = -q_{calor} = -48.2 \text{ kJ}$$

Therefore the heat of combustion of octane *per gram* is

$$q_{rxn} = -48.2 \text{ kJ/1 gram} = -48.2 \text{ kJ/g}$$

Step 2:

To calculate the heat of combustion *per mole* of octane, recognize that the molecular weight of octane (C_8H_{18}) is

$$(8 \times 12) + (18 \times 1) = 114 \text{ g/mole}$$

So *per mole* we have

$$q_{rxn} = q_V = (-48.2 \text{ kJ/g}) \times 114 \text{ g/mole} = -5.50 \times 10^3 \text{ kJ/mole}$$

Solution:

The solution to the problem involves three steps:

Step 1:

Recognize that $q_{cal} = (C_{cal}) \Delta T$

and that $q_{rxn} = -q_{cal}$

and that $\Delta T = 27.69°C - 23.50°C = 4.19°C$

Step 2:

Calculate q_{rxn} of the reaction:

$q_{cal} = (C_{cal}) \Delta T = (7.45 \text{ kJ/°C})(4.19°C) = 31.2 \text{ kJ}$

Step 3:

Calculate $\Delta U_{syst} = q_{rxn} = \Delta U_{chem}$ in kJ/mol of $C_6H_{12}O_6$

$\Delta U_{chem} = -31.2 \text{ kJ/2 mol } C_6H_{12}O_6$
$= -15.6 \text{ kJ/mol } C_6H_{12}O_6$

Verify the sign:

Since heat *leaves* the system, the sign should be *negative*, as it is.

Check Yourself 4

Suppose 2.00 mol of glucose ($C_6H_{12}O_6$) is reacted in a bomb calorimeter and the temperature of the calorimeter increases from 23.50°C to 27.69°C.

The heat capacity of the calorimeter is $C_{cal} = 7.45 \text{ kJ/°C}$.

Calculate: $\Delta U_{syst} = \Delta U_{chem}$ in kJ/mol of $C_6H_{12}O_6$.

Enthalpy

To this point, we have dealt with the change in chemical internal energy, ΔU_{chem}, only under conditions of constant volume in our bomb calorimeter, for which

$$\Delta U_{syst} = \Delta U_{therm} + \Delta U_{chem} = q + \not{w} = q_v$$

where $w = -p\Delta V = 0$ because the rigid walls of the bomb prevented any change in volume, $\Delta V = 0$. So the release of chemical energy upon combustion appeared as an increase in thermal energy at constant volume, q_v. However, a great many chemical reactions occur under conditions of *constant pressure* and not *constant volume*.

This immediately raises the question: When we measure the heat of reaction at constant volume, q_v, how does that compare quantitatively with the heat of reaction measured at constant pressure?

We know, because the system will do work *on its surroundings* ($w < 0$) at constant pressure (*because the volume will increase*), that the heat of reaction at constant pressure, q_p, will be greater (less negative) than q_v because

$$\Delta U = q_v = w + q_p$$

and $w < 0$ because when the *system* does work on the *surrounding*, $w < 0$. But by how much?

The relation $q_v = q_p + w$ is the key starting point from which we can deduce the answer to our question. First, we know that, for a process at *constant volume*, $\Delta U_{syst} = q_v$ and we know that $w = -p\Delta V$ so we can write

$$\Delta U_{syst} = q_v = q_p + w = q_p - p\Delta V$$

Thus, $q_p = \Delta U_{syst} + p\Delta V$. But now we recognize that U_{syst}, p, and V are all *state variables* because their values are *each independent* of the path taken to reach that given state.

Suppose now we define a new *state* variable

$$H = U + pV$$

Then the *change* in that state variable

$$\begin{aligned}
\Delta H &= H_f - H_i \\
&= (U_f + p_f V_f) - (U_i + p_i V_i) \\
&= (U_f - U_i) + (p_f V_f - p_i V_i) \\
&= \Delta U + \Delta(pV)
\end{aligned}$$

If the process is carried out at *constant temperature and pressure*, then $\Delta(pV) = p\Delta V$ and

$$\Delta H = \Delta U + p\Delta V = (q_p - p\Delta V) + p\Delta V$$
$$= q_p$$

where q_p is the heat released at constant pressure.

This new state variable, H, is called the *enthalpy* and $\Delta H = q_p$ is the enthalpy change for a chemical process at constant pressure, which is the thermal energy (heat) produced by the chemical reaction at constant pressure.

Because we live, by and large, in a *constant pressure* world, the state variable enthalpy is a variable of great importance. In fact, ΔH released in a chemical reaction is routinely referred to as the *energy* release resulting from the change in bond structure in going from reactants to products in a chemical reaction. The enthalpy change, ΔH, is listed in the appendix of all chemistry texts and all thermochemistry data sources as the quantitative measure of the relative energy contained in the bonds of molecules. A table of ΔH constitutes Appendix B of this text. *If we lived in a constant volume world, those tables in the appendix of textbooks would list ΔU, not ΔH!*

It is important to consider the magnitude of the enthalpy change of a chemical reaction, ΔH, and the $p\Delta V$ work term associated with a given reaction. What is their relative magnitude?

Suppose we react two moles of carbon monoxide in the gas phase, $CO(g)$, with a mole of $O_2(g)$ to form two moles of carbon dioxide in the gas phase:

$$2CO(g) + O_2 \rightarrow 2CO_2(g)$$

If the reaction is run at constant pressure, we discover that 566.0 kJ of energy is *released* such that

$$q_p = -566.0 \text{ kJ}$$

and since $\Delta H_{rxn} = q_p$, it follows that

$$\Delta H_{rxn} = -566.0 \text{ kJ}$$

To evaluate the pressure-work term, $p\Delta V$, we write, from the Perfect Gas Law $pV = nRT$, so at constant pressure

$$p\Delta V = \Delta nRT$$

where Δn is the change in the number of moles of gas and T is the temperature of the gas mixture, which we keep constant.

But the change in the number of moles is just $\Delta n = n_f - n_i = 2 - 3 = -1$ moles. Thus,

$$p\Delta V = RT\Delta n = (8.3 \times 10^{-3} \text{ kJ/mole-K})(298 \text{ K})(-1) = -2.5 \text{ kJ}$$

This is an important and fairly general result that in *most* cases the $-p\Delta V$ term is small compared with the change in enthalpy, ΔH, of a reaction, and thus the thermal energy (heat) produced in a chemical reaction at constant pressure is approximately equal to that produced at constant volume.

To summarize:

$$\Delta H = q_p$$
$$\Delta U = q_v$$
$$\Delta U = \Delta H - p\Delta V$$

and because a large fraction of chemical reactions occur at constant pressure, it is the enthalpy, ΔH, that appears most frequently.

Reactions in the Liquid Phase: An Important Example of Reactions at Constant Pressure

Adenosine triphosphate (ATP) is used in the cell for the formation of proteins. In the reaction, ATP is hydrolyzed to form adenosine diphosphate (ADP) and phosphate (HPO_4^{2-}) in the reaction

$$ATP^{4-} + H_2O \rightarrow ADP^{3-} + HPO_4^{2-} + H^+$$

Problem:
If 10 grams of ATP hydrolyzed to ADP and HPO_4^{2-} in 50 grams of water at constant pressure in a calorimeter, it is observed that the temperature of the water increases by 2.1°C. What is q of the reaction? What is ΔH for the reaction?

Step 1:
We know that the temperature increased so the reaction is exothermic and q_{rxn} is thus negative.

$$q = (50 \text{ g})(4.18 \text{ J/g-°C})(2.1 \text{ K}) = 439 \text{ J}$$

Thus $q_{rxn} = -439$ J as heat was released by the reaction.

The molecular weight of ATP is 573 g/mole.

So: $\Delta H_{rxn} = (-439 \text{ J}/10\text{g})(573 \text{ g/mole}) = -25 \text{ kJ/mole}$

Constant Pressure Calorimetry

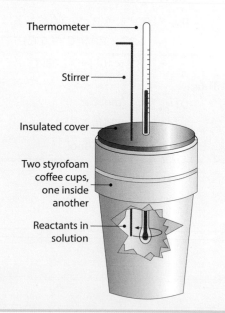

Thermometer

Stirrer

Insulated cover

Two styrofoam coffee cups, one inside another

Reactants in solution

Step 1:
Recognize that for this case the *system* is the reaction

$$Ca + 2\,HCl \rightarrow H_2 + CaCl_2$$

and that the *surroundings* are the solution within which the reaction takes place. Moreover, q_{rxn} is taking place at constant pressure, so $q_{rxn} = -q_{cal}$ and that $\Delta H_{rxn} = q_{rxn}/mols\ Ca$.

Step 2:
Calculate
(1) the mass of the solution,
(2) the heat capacity of the calorimeter,
(3) the temperature increase, and
(4) the number of moles of calcium reacted.

The mass of the solution:
$$m_{sol} = (150\ mL)(1.00\ g/mL)$$
$$= 1.50 \times 10^2 g$$

The heat capacity of the calorimeter:
$$C_{cal} = (4.18\ J/g \cdot °C\ )(1.50 \times 10^2\ g)$$
$$= 6.27 \times 10^2\ J/°C$$

The temperature increase:
$$\Delta T = 29.1°C - 24.5°C = 4.6°C$$

The number of moles of Ca:
$$(0.250\ g)/(40.1\ g/mol) = 6.2 \times 10^{-3}\ mol$$

Step 3:
Determine the enthalpy change for the reaction, ΔH_{rxn}, in units of kJ/mol Ca:

$$q_{rxn} = -q_{cal} = -q = -C_{cal}\Delta T$$
$$= -(6.27 \times 10^2\ J/°C)(4.6°C)$$
$$= -28.8 \times 10^2\ J = -2.88 kJ$$

Thus
$$\Delta H_{rxn} = q_{rxn}/mol\ Ca$$
$$= -2.88 kJ/6.2 \times 10^{-3}\ mol$$
$$= -464.5\ kJ/mol$$

Check Yourself 5—Measuring ΔH_{rxn} in a Coffee-Cup Container

The "Coffee Cup Calorimeter" displayed on the previous page, is very important for measuring the enthalpy release of a reaction in the liquid phase at constant pressure.

Problem: Determine the enthalpy change, ΔH_{rxn}, for the reaction of calcium with hydrochloric acid if 0.250 g of calcium is reacted with sufficient HCl to make 150 mL of solution in the calorimeter. The temperature rises from 24.5°C to 29.1°C as a result of the chemical reaction. The density of the solution is 1.00 g/mL. The specific heat capacity of the solution is 4.18 J/g · °C.

Standard Enthalpies of Formation

In the application of thermochemistry to a broad range of important calculations we need a convention by which the enthalpy change for a given reaction, called the *enthalpy of reaction*, ΔH_R, can be readily calculated. The convention is to define the *standard enthalpy of formation*, ΔH_f°, to specific molecular species, and then *tabulate* those values of ΔH_f° for each of the molecular species. Because enthalpy is a state function, we are concerned only with *changes* in enthalpy ΔH, so the absolute scale is not important in such a tabulation. In order to set the scale for *standard enthalpies of formation*, the convention is to assign enthalpy values of zero to elements in their standard states. Specifically, the enthalpy of formation, ΔH_f°, is defined as zero for O_2, H_2, N_2, and C(graphite) in their standard states at one atmosphere pressure, and 25°C. This is shown graphically in Figure 3.15, wherein $\Delta H_f^\circ = 0$ sets the scale for enthalpies of formation for a broad range of molecular species, both positive (energy *required* to form a molecular structure from its elements in their standard state) and negative (energy released in the formation of the species from their standard states). A great deal of experimental work over time has gone into the determination of the enthalpies of formation for hundreds of compounds—information that is now available in tables, specifically Appendix B of this text. A selection of important examples is shown in Table 3.3.

One of the most important applications for the enthalpies of formation ΔH_f°, is to determine whether a reaction is *thermodynamically allowed*. That is, will energy be released in the reaction such that the reaction is spontaneous or is energy required to carry reactants to products wherein the reaction is thermodynamically forbidden and thus requires the external input of energy in order to proceed?

To demonstrate, we ask whether the reaction of methane (natural gas) with the hydroxyl radical (OH) will proceed from reactants to products.

$$CH_4 + OH \rightarrow CH_3 + H_2O$$

To answer this question, we combine the concepts of the *standard enthalpy of formation*, ΔH_f°, with the concept of *standard enthalpy of reaction* ΔH_R such that

$$\Delta H_R^\circ = \sum_{Products} \Delta H_f^\circ - \sum_{Reactants} \Delta H_f^\circ$$

Note that because, in thermochemistry, we have defined the condition wherein heat flows *from* the system *to* the surrounding as $q < 0$ (Figure 3.10), then if energy is released in a chemical reaction, $\Delta H_R^\circ < 0$.

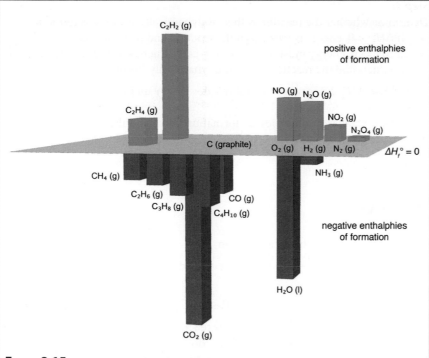

FIGURE 3.15 Each compound, each molecule, has an enthalpy of formation, ΔH_f°, that is referenced to the enthalpies of formation of the elements in their standard state. Shown here on the plane of $\Delta H_f^\circ = 0$ are the examples $O_2(g)$, $H_2(g)$, $N_2(g)$, and C(graphite). A number of important enthalpies of formation, both positive and negative, are displayed relative to the plane of $\Delta H_f^\circ = 0$.

Step 1:

Look up the enthalpy of formation in a tabulation—Appendix B in this text—for each of the products and reactants:

$$\Delta H_f^\circ(CH_3) = +146.7 \text{ kJ/mole}$$
$$\Delta H_f^\circ(H_2O) = -285.8 \text{ kJ/mole}$$
$$\Delta H_f^\circ(CH_4) = -74.5 \text{ kJ/mole}$$
$$\Delta H_f^\circ(OH) = +37.2 \text{ kJ/mole}$$

Step 2:

Calculate $\sum\limits_{\text{Products}} \Delta H_f^\circ$ and $\sum\limits_{\text{Reactants}} \Delta H_f^\circ$:

$$\sum_{\text{Products}} \Delta H_f^\circ = (\text{number moles } CH_3)\Delta H_f^\circ(CH_3) + (\text{number moles } H_2O)\Delta H_f^\circ(H_2O)$$
$$= 1(+146.7 \text{ kJ/mole}) + 1(-285.8 \text{ kJ/mole})$$
$$= -139.1 \text{ kJ/mole}$$

$$\sum_{\text{Reactants}} \Delta H_f^\circ = (\text{number moles } CH_4)\Delta H_f^\circ(CH_4) + (\text{number moles } OH)\Delta H_f^\circ(OH)$$
$$= 1(-74.5 \text{ kJ/mole}) + 1(+37.2 \text{ kJ/mole})$$
$$= -37.3 \text{ kJ/mole}$$

Step 3:

Calculate the enthalpy change of the reaction under standard conditions (1 atm and 25°C):

$$\Delta H_R^\circ = \sum_{\text{Products}} \Delta H_f^\circ - \sum_{\text{Reactants}} \Delta H_f^\circ = -139.1 \text{ kJ/mole} - (-37.3 \text{ kJ/mole})$$
$$= -101.8 \text{ kJ/mole}$$

TABLE 3.3 Standard enthalpies of formation for some common compounds

Compound	$\Delta H^\circ_{f, 298}$ kJ/mol
$H_2O(l)$	−285.83
$H_2O(g)$	−241.82
$CO(g)$	−110.52
$CO_2(g)$	−393.51
$CH_4(g)$	−74.81
$C_2H_2(g)$	226.73
$C_2H_4(g)$	52.30
$C_2H_6(g)$	−84.68
$CH_3OH(l)$	−238.66
$C_2H_5OH(l)$	−277.69
$C_6H_6(l)$	49.028
$C_6H_6(g)$	82.93
$C_6H_{12}O_6(s)$	−1260
$I-C_8H_{18}(l)$	−208.2
$SiO_2(s)$	−910.94
$NH_3(g)$	−46.11
$NO(g)$	90.25
$NO_2(g)$	33.18
$O_3(g)$	142.7
$H_2S(g)$	−20.63
$SO_2(g)$	−296.81
$HCl(g)$	−92.31
$NaCl(s)$	−411.15
$NH_4Cl(s)$	−314.4
$NaHCO_3(s)$	−950.81
$Na_2CO_3(s)$	−1130.68
$MgO(s)$	−601.7
$CaO(s)$	−635.09
$CaCO_3(s)$	−1206.92
$Fe_2O_3(s)$	−824.2
$Al_2O_3(s)$	−1675.7

Step 4:
Determine whether the reaction is thermodynamically allowed or forbidden:
- If $\Delta H^\circ_R < 0$, energy is *released* in the reaction and it is "allowed."
- If $\Delta H^\circ_R > 0$, energy must be supplied to carry the reaction from reactants to products and the reaction is thermodynamically "forbidden."

In our case $\Delta H^\circ_R < 0$, *so the reaction releases energy and is thermodynamically allowed.*

Again, standard enthalpies of formation for a number of important compounds are given in Table 3.3. Can you recognize patterns in ΔH_f?

Hess's Law

Consider an arbitrary chemical reaction

$$\text{Reactants} \rightarrow \text{Products}$$

or in short hand

$$R \rightarrow P$$

If we wish to calculate $\Delta H_{R \rightarrow P}$ for this reaction, and we know nothing (thermodynamically!) about the reaction, but we do know the enthalpy change for reactants, R, forming an intermediate, I, and the intermediate forming the product, P,

$$R \rightarrow I \qquad \Delta H_{R \rightarrow I}$$

and

$$I \rightarrow P \qquad \Delta H_{I \rightarrow P}$$

then because H is a *state function* and thus ΔH is *independent* of the path taken, we can write

$$\Delta H_{R \rightarrow P} = \Delta H_{R \rightarrow I} + \Delta H_{I \rightarrow P}$$

This is the foundation for Hess's Law, which states that:

If a process occurs in steps—even if the steps are hypothetical—then the enthalpy change for the overall process is the same as the sum of the enthalpy changes of the individual steps.

There is an illustrative example of Hess's Law in the analogy of the potential energy of a mass in a gravitational field. Suppose we wish to calculate the potential energy (mgh) of a mass, m, at a given floor of a large apartment building—say at position x in Figure 3.16. Because potential energy is a state variable (it does not depend on the path taken to reach position x) we can calculate the potential energy via a number of different paths. We could raise the mass through path 1 in Figure 3.16 directly. Or, we could raise the mass through path 2 to a higher floor, then *subtract* the potential energy released in going from the top of path 2 to point x. Through either path we would arrive at the *same value* for the potential energy at point x.

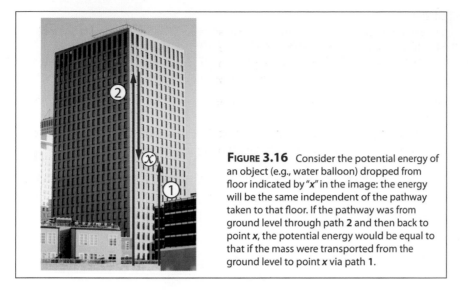

FIGURE 3.16 Consider the potential energy of an object (e.g., water balloon) dropped from floor indicated by "*x*" in the image: the energy will be the same independent of the pathway taken to that floor. If the pathway was from ground level through path **2** and then back to point *x*, the potential energy would be equal to that if the mass were transported from the ground level to point *x* via path **1**.

In the execution of calculations using Hess's Law, there are three rules (each of which results from the fact that enthalpy is a state variable) that are worth reviewing:

1. Enthalpy change is directly proportional to the amounts of substances in a system.

$$N_2(g) + O_2(g) \rightarrow 2NO(g) \qquad \Delta H = 180.5 \text{ kJ}$$
$$\tfrac{1}{2}\,N_2(g) + \tfrac{1}{2}\,O_2(g) \rightarrow NO(g) \qquad \Delta H = \tfrac{1}{2}(180.5 \text{ kJ})$$
$$= 90.25 \text{ KJ}$$

2. ΔH changes sign when the process is reversed.

$$NO(g) \rightarrow \tfrac{1}{2}\,N_2(g) + \tfrac{1}{2}\,O_2(g) \qquad \Delta H = -90.25 \text{ kJ}$$

3. If a process occurs in steps, the enthalpy change for the overall process is the sum of the enthalpy changes for the individual steps. Suppose we need to know ΔH for the reaction:

$$\tfrac{1}{2}\,N_2(g) + O_2(g) \rightarrow NO_2(g) \qquad \Delta H = ?$$

But we are given the enthalpy change for two *different* reactions, say,

$$\tfrac{1}{2}\,N_2(g) + \tfrac{1}{2}\,O_2(g) \rightarrow NO(g) \qquad \Delta H = 90.3 \text{ kJ}$$
$$NO(g) + \tfrac{1}{2}\,O_2(g) \rightarrow NO_2(g) \qquad \Delta H = -57.1 \text{ kJ}$$

First we recognize that when those two reactions are added together, they add to yield the reaction for which we wish to calculate the enthalpy of reaction:

$$\tfrac{1}{2}\,N_2(g) + \tfrac{1}{2}\,O_2(g) \rightarrow NO(g)$$
$$NO(g) + \tfrac{1}{2}\,O_2(g) \rightarrow NO_2(g)$$
$$\overline{\tfrac{1}{2}\,N_2(g) + O_2(g) \rightarrow NO_2(g)}$$

But when we *add* chemical reactions, we *add* the corresponding enthalpies of reaction:

$$\tfrac{1}{2}\,N_2(g) + \tfrac{1}{2}\,O_2(g) \rightarrow NO(g) \qquad \Delta H = 90.3 \text{ kJ}$$
$$NO(g) + \tfrac{1}{2}\,O_2(g) \rightarrow NO_2(g) \qquad \Delta H = -57.1 \text{ kJ}$$
$$\overline{\tfrac{1}{2}\,N_2(g) + O_2(g) \rightarrow NO_2(g) \qquad \Delta H = 90.3 \text{ kJ} - 57.1 \text{ kJ}}$$
$$= 33.18 \text{ kJ}$$

Thus ΔH for the net reaction is:

$$\tfrac{1}{2}\,N_2(g) + O_2(g) \rightarrow NO_2(g) \qquad \Delta H = 33.18 \text{ kJ}$$

Step 1:

Recognize that if the reactions for which ΔH is known can be re-arranged such that if added together, the net reaction is a reaction under consideration, then ΔH for

$$3C(s) + 4H_2(g) = C_3H_8(g)$$

can be calculated from the entropies of the individual reactions for which ΔH is known. If we multiply reaction [1] by a factor of 2 such that:

$$4H_2(g) + 2O_2(g) \rightarrow 4H_2O(g)$$

and we reverse reaction [2] such that:

$$3CO_2(g) + 4H_2O(g) \rightarrow C_3H_8(g) + 5O_2(g)$$

and finally, if we multiply reaction [3] by a factor of 3:

$$3C(s) + 3O_2(g) \rightarrow 3CO_2(g)$$

We can now add these three reactions together to yield the net desired reaction:

[1'] $4H_2(g) + 2O_2(g) \rightarrow 4H_2O(g)$

[2'] $3CO_2(g) + 4H_2O(g) \rightarrow C_3H_8(g) + 5O_2(g)$

[3'] $3C(s) + 3O_2(g) \rightarrow 3CO_2(g)$

___

net $3C(s) + 4H_2(g) \rightarrow C_3H_8(g)$

Step 2:

Next we need to determine ΔH for each of the three reactions, which when taken together, yield the net reaction.

The first reaction above, reaction [1'], is reaction [1] multiplied by a factor of two. Thus ΔH for reaction [1'] is ΔH = 2(−483.6 kJ) = −967.2 kJ.

The second reaction above, reaction [2'], is the reverse of reaction [2]. Thus ΔH for reaction [2'] is ΔH = +2043 kJ.

The third reaction above, reaction [3'], is reaction [3] multiplied by a factor of three. Thus ΔH for reaction [3'] is ΔH = 3(−393.5 kJ) = −1180.5 kJ.

Step 3:

Assemble the reactions for which the recalculated values of ΔH are now available to yield the net overall reaction and sum the values of ΔH for the individual reactions to calculate ΔH for the net reaction:

[1'] $4H_2(g) + 2O_2(g) \rightarrow 4H_2O(g)$ ΔH = −967.2 kJ

[2'] $3CO_2(g) + 4H_2O(g) \rightarrow C_3H_8(g) + 5O_2(g)$ ΔH = +2043 kJ

[3'] $3C(g) + 3O_2(g) = 3CO_2(g)$ ΔH = −1180.5 kJ

___

net $3C(s) + 4H_2(g) \rightarrow C_3H_8(g)$ ΔH = −104.7 kJ

Check Yourself 6—Hess's Law

Problem: Find the enthalpy change, ΔH_{rxn}, for the reaction of elemental carbon with hydrogen gas to form the product propane in the gas phase.

$$3C(s) + 4H_2(g) \rightarrow C_3H_8(g)$$

Using the following reaction enthalpies:

[1] $2H_2(g) + O_2(g) \rightarrow 2H_2O(g)$ ΔH = −483.6 kJ

[2] $C_3H_8(g) + 5O_2(g) \rightarrow 3CO_2(g) + 4H_2O(g)$ ΔH = −2043 kJ

[3] $C(s) + O_2(g) \rightarrow CO_2(g)$ ΔH = −393.5 kJ

Pressure-Volume Work and the First Law

We have now, by virtue of the development of the concept of *enthalpy*, where

$$\Delta U = \Delta U_{therm} + \Delta U_{chem} = q_v = w + q_p \text{ and}$$

$$\Delta H = \Delta U + \Delta(pV) = q_p \text{ at constant temperature and pressure,}$$

developed a powerful and practical way of measuring the energy release from a chemical reaction under conditions for which no work is done (constant volume $\Delta U_{syst} = q + \cancel{w} = q_v$) and under conditions of constant pressure where work is done ($\Delta H = \Delta U + \Delta(pV) = q_p$).

We also know that, while the transfer of thermal energy (heat) by the microscopic collision of molecules is a complicated process, the quantitative measure of that thermal energy (heat) transfer can be easily determined by measuring the increase in temperature, ΔT, using the mass, m, and specific heat, c, such that

$$q = mc\Delta T$$

But in order to quantitatively determine the amount of work that can be extracted from the energy release in a chemical reaction we need to address the *macroscopic* transfer of energy from the system to the surroundings (or vice versa) in a more thorough way. We have already developed the important concept that *work, w,* is energy transferred to a system from the surroundings by a force operating over a distance. It is a distinctly *macroscopic* concept representing the *mechanical interaction* of the system operating on the surroundings, or visa versa. Figure 3.5 displayed a very important prototype system, or model, using a piston acting on a closed volume.

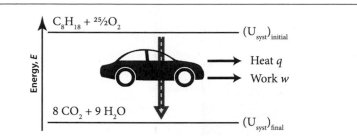

FIGURE 3.17 We recognize from our examples of the combustion of gasoline (octane) in an open steel box in Chapter 2 that all of the available energy contained in the chemical bonds of C_8H_{18} and O_2 relative to CO_2 and H_2O can be released simply as heat, q. When that same octane is combusted in an automobile engine, some of the energy is released as heat (80%) and some as work (20%). The net result is that the energy of the system (the chemical system comprised of C_8H_{18} and O_2) *decreases* in going to its final state (CO_2 and H_2O) and q and w are delivered to the surroundings.

Of considerable importance was the fact that for such a system,

$$w = \text{Force} \times \text{Distance} = (\text{Force/Area})(\text{Area})(\text{Distance})$$
$$= F \cdot \Delta x = (F/A)(A)(\Delta x)$$
$$= (\text{Pressure})(\text{Area})(\text{Distance})$$
$$= p \cdot A \cdot \Delta x = -p\Delta V \qquad \textbf{(3.5)}$$

where $A \cdot \Delta x$ is equal to the change in volume, and P is the pressure applied to the top of this piston (usually by the surrounding atmosphere). Also of considerable importance is that because ΔV is *positive*, work is done by the system on the surroundings and therefore $w < 0$.

We are now familiar with the ability of chemical reactions to change the internal energy, ΔU_{syst}, of a system and then to have that kinetic energy (of the molecules to which that chemical energy was imparted) transferred to another component of the system as heat, q, by virtue of a temperature difference (as was the case for the calorimeter). But we are also familiar with the concept that the combustion of octane (gasoline) can do work on its surroundings, because that is exactly what happens when we drive a car: gasoline is fed to the engine, the automobile moves under your command, carbon dioxide and water pour out the exhaust pipe, and the engine produces heat (as well as work) as a by-product of the combustion process. The conversion of octane and molecular oxygen to carbon dioxide and water in an automobile engine is sketched on an energy scale in Figure 3.17. Understanding how chemical energy is converted to *work* constitutes the foundation upon which the global energy structure is built because 80% of our primary energy generation comes from the combustion of fossil fuels. So we now turn to the question of how work is quantitatively integrated into the First Law.

First, we capture the expression of our First Law of Thermodynamics with a *thermodynamic energy model*, wherein the system and the surroundings are designated by a physical boundary; we identify changes in the internal energy of the system as $\Delta U_{syst} = q + w$ where $U_{syst} = U_{therm} + U_{chem}$. We also separate the *energy in* ($w > 0$ and/or $q > 0$) from the *energy out* ($w < 0$ and/or $q < 0$), as displayed in Figure 3.18.

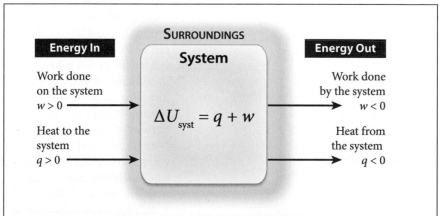

FIGURE 3.18 Sign Convention in Chemical Thermodynamics: Specification of the system that establishes the boundaries within which the change in internal energy of that system ΔU_{syst} is defined and the sign convention that heat *into* the system *from* the surroundings is positive, $q > 0$, and work done on the system by the surroundings is positive, $w > 0$. Energy removed from the system to the surroundings corresponds to work done *by* the system, $w < 0$, and/or heat flow *from* the system to the surroundings, $q < 0$.

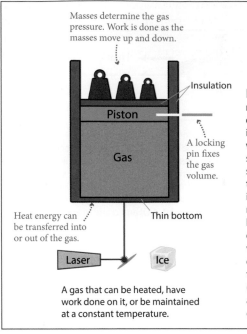

Masses determine the gas pressure. Work is done as the masses move up and down.

Insulation

Piston

Gas

A locking pin fixes the gas volume.

Heat energy can be transferred into or out of the gas.

Thin bottom

Laser   Ice

A gas that can be heated, have work done on it, or be maintained at a constant temperature.

FIGURE 3.19 A thermodynamic machine capable of dissecting the distinction between the change in internal energy of the system, ΔE_{syst}, as well as heat added or removed from that system and/or work done on or by the system. This physical manifestation of the First Law of Thermodynamics includes (1) a piston that can either move freely in the cylinder or be locked by a pin insertion, (2) insulation eliminating heat flow from the cylinder walls or piston, (3) adjustable masses to control pressure, (4) a "working medium" that is a perfect gas for which $pV = nRT$, (5) a *source* of thermal energy (a laser), and (6) a *sink* of thermal energy (a cube of ice).

We emphasize (repeat) two points:

- The First Law doesn't concern itself with (i.e., tell us anything about) the absolute magnitude of U_{syst}, only how heat and work *change* the internal energy of the system, ΔU_{syst}.

- The system's internal energy is not the only thing that changes. We can, by virtue of heat and/or work done on or by the system, change the pressure, volume, or temperature of the system. The First Law tells us only about the *change* in the system's internal energy, ΔU_{syst}, and we must use other laws, such as the Perfect Gas Law, to link changes in pressure, volume, molar concentration, temperature, etc., to one another. We will see that the First Law *in combination with* the Perfect Gas Law ($pV = nRT$) constitutes a potent diagnostic approach to understanding the conversion among and between various categories of energy.

We turn, first, to the question of how to devise a system that can, when coupled to the First Law of Thermodynamics, dissect the heat term at constant volume, q_v. We adopt a machine, displayed in Figure 3.19, that is comprised of a piston that contains a volume of gas in a cylinder constructed such that the piston can be locked in place with a pin (to operate at *constant volume*). With the pin extracted the piston is free to move for measurements made at *constant pressure* and, in addition, objects of various masses can be added or removed from the top of the piston to increase or decrease the pressure of the gas within the piston-cylinder volume. While the walls of the cylinder and the top of the piston are *insulated*, the bottom of the cylinder is a thin wall that can be heated by a laser (or a Bunsen burner) or cooled by a device such as a block of ice. Chemicals can be added to the volume contained within the piston/cylinder system such that chemical energy can be released into or removed from the volume. This is a very versatile machine with which to study the First Law of Thermodynamics. We will use it repeatedly. It is the *physical manifestation* of the equations we will use to represent the processes *mathematically*.

Returning to our expression, for the work done *on* a system by virtue of a piston moving such that the volume changes by ΔV at a constant pressure P, we saw that the work, w, done on the system is given by

$$w = -p\Delta V$$

We can use our machine to make extremely important observations, measurements, and deductions that are of far reaching significance. We begin by examining the case where the piston of our machine has a mass placed on the top of the piston to create a pressure inside the vessel of 2.5 atmospheres, as shown in Figure 3.20.

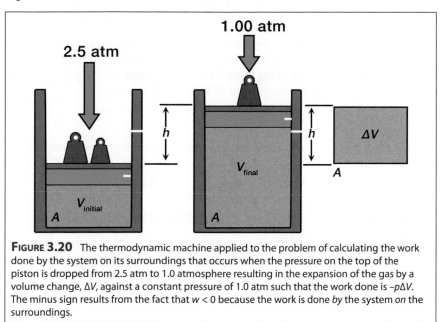

FIGURE 3.20 The thermodynamic machine applied to the problem of calculating the work done by the system on its surroundings that occurs when the pressure on the top of the piston is dropped from 2.5 atm to 1.0 atmosphere resulting in the expansion of the gas by a volume change, ΔV, against a constant pressure of 1.0 atm such that the work done is $-p\Delta V$. The minus sign results from the fact that $w < 0$ because the work is done *by* the system *on* the surroundings.

Since 1 atmosphere pressure is 1.01×10^5 Pascal (or 14.1 pounds per square inch!) we would need to add a mass equal to the pressure *increase* (2.5 atm–1.00 atm) times the area of the piston because pressure, p, is equal to the force, F, divided by the area, A, of the piston.

To calculate the amount of work done, consider a specific case:

Problem:

Calculate the pressure-volume ($-p \cdot \Delta V$) work for the case of 0.10 mol He at an initial pressure of 2.5 atm that expands against a constant pressure of 1 atm. How much work, in joules, is done during the expansion if the temperature is fixed at 298 K?

Solution:

Step 1:

Employ the Perfect Gas Law,

$$pV = nRT,$$

where p is the pressure in atmospheres, V is the volume in liters, n is the number of moles of gas, T is the temperature in K, and R is the gas constant = 0.0821 L-atm $mol^{-1}K^{-1}$.

Step 2:

Calculate the initial volume at 2.5 atm for the 0.1 mole of He by solving the Gas Law for volume:

$$V_i = nRT/p = (0.1 \text{ mol})(0.0821 \text{ L-atm mol}^{-1}K^{-1}) \, 298/2.5 \text{ atm}$$
$$= 0.979 \text{ L}$$

so the initial volume is

$$V_i = 0.979 \text{ L}$$

Step 3:
Calculate the volume after expansion has taken place to achieve the final volume. We use the same equation, but the final pressure is 1 atm:

$$V_f = nRT/p = (0.1 \text{ mol})(0.0821 \text{ L-atm-mol}^{-1}\text{K}^{-1})\ 298/1.00 \text{ atm}$$
$$= 2.45 \text{ L}$$

Step 4:
To determine the work done by the system on the surroundings recognize that the removal of the mass from the piston top (or pulling the pin from the piston to allow it to move upward against the pressure of 1 atm) is equal to the product of the external pressure, p_{ext}, times the change in volume, ΔV, so that

$$w = -p_{ext}\Delta V$$

because the work is done *by* the system on the surrounding, introducing the minus sign. Thus,

$$w = -(1.0 \text{ atm})(2.45 \text{ L} - 0.978 \text{ L})$$
$$= -1.47 \text{ L-atm}$$

But the problem asks (as it should!) for the answer in energy units of joules. The conversion factor for joules is

$$8.315 \text{ J/mol-K} = 0.0821 \text{ L-atm/mol-K}$$

or

$$101.3 \text{ J/L-atm}$$

Therefore

$$w = -(1.47 \text{ L-atm})(101 \text{ J/L-atm})$$
$$= -1.5 \times 10^2 \text{ J}$$

Notice that the sign is negative because the system (piston, cylinder) has done work *on* the surroundings—see Figure 3.20.

Isochoric, Isobaric, and Isothermal Processes

Several key points emerge from this rather simple problem. First, work is a mechanical exchange of energy by macroscopic forces acting between the system and the surroundings over some displacement that is manifest as a change of volume. Second, this calculation of work always involves transformations on a diagram of pressure versus volume. Such a plot is displayed for three important cases in the sidebar entitled *Processes that Occur on a* pV *Surface.*

Case (a), when the pressure of the system changes but the volume does not. This was, in fact, the case in point for our calorimeter experiment when the "bomb" of the calorimeter was designed with rigid walls such that there was no deflection following combustion of octane and thus, since work is force times displacement,

$$w = F \cdot d = -p\Delta V = 0.$$

This is termed an *isochoric* process. Case (b) corresponds to the case we have just worked out, wherein the helium gas expanded from 0.978 L to 2.45 L under a constant pressure of 1 atm. This is called an *isobaric*, constant pressure, process. Case (c) is for the very important case in which both the volume and the pressure change, but the *temperature* does not. This is termed an *isothermal* process.

We can use our thermodynamic machine, pictured in Figure 3.19, to examine the behavior of all three cases, and we begin with isochoric cooling.

Isochoric Processes

Problem:

Using our thermochemical machine of Figure 3.19, design and execute a process that will decrease the pressure in the gas cylinder *without changing the volume*. Show how this process is executed by tracing the path on a pV diagram.

Solution:

Step 1:

First recognize that an isochoric process is one wherein the volume does not change through the course of events that define a specific trajectory on the pV diagram. To accomplish this with our thermodynamic machine, we recognize that we must first insert the pin into the piston through the cylinder wall such that the volume is set at the initial condition: Volume = $V_i = V_f$, and it remains at that volume.

Step 2:

If we are to decrease the pressure in the cylinder without changing the volume and without initiating a chemical reaction (that could add or remove molecules from the volume within the cylinder), then the internal energy of the molecules within the cylinder must be *reduced* by removing energy from the *thermal* component of the internal energy. This can only be done by transferring kinetic energy of molecular motion, ΔU_{therm}, of molecules within the cylinder to the surroundings in the form of heat, q, causing the gas temperature and pressure to *decrease*. To execute the extraction of thermal energy (heat) *from* the gas within the cylinder, we place the machine on a block of ice.

Step 3:

Remove the cylinder from the ice when the desired pressure is reached.

Step 4:

Adjust the mass sitting on top of the piston such that the final mass balances the new gas pressure. Notice that this act of balancing the inside pressure with the outside mass (force) must be done without allowing the piston to move; otherwise, work will be done either *on* the system if the mass *removed* was insufficient (so too much mass remained on the piston) or *by* the system on the surroundings if too little mass remained on the piston when the pin is pulled.

Step 5:

Remove the locking pin and verify that the piston does not move.

This entire sequence is reviewed on the sidebar on the next page. The isochoric process representing the "cycle" we have just executed is shown on a pV diagram in Figure 3.21.

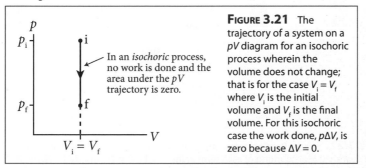

In an *isochoric* process, no work is done and the area under the pV trajectory is zero.

FIGURE 3.21 The trajectory of a system on a pV diagram for an isochoric process wherein the volume does not change; that is for the case $V_i = V_f$ where V_i is the initial volume and V_f is the final volume. For this isochoric case the work done, $p\Delta V$, is zero because $\Delta V = 0$.

There are a growing number of quantities that we must keep track of when we link the First Law to physical systems, and a growing number of quantities that *change*, even for a rather simple cycle such as the isochoric sequence that we just traced. This will become increasingly true as we explore isobaric and isothermal

Processes that Occur on a pV Surface

There are three simple thermodynamic processes that take place on a pV diagram that we will use repeatedly. The first is an *isochoric* process that occurs at a fixed volume as shown in panel (a). Because the area under the pV curve is zero, the work done is also zero. An *isobaric* process occurs at constant pressure as shown in panel (b) and thus the work is simply $-p\Delta V$ where the minus sign results from the fact that work is done by the system on the surrounding, the area within the box defined by the boundaries V_i and V_f, is (ΔV) and the pressure is p. An *isothermal* process is one that occurs at constant temperature as shown in panel (c). To calculate the work done in this case we must break the progression from V_i to V_f into small segments, calculate the work for each, $p\Delta V$, and then add them up to find the total:

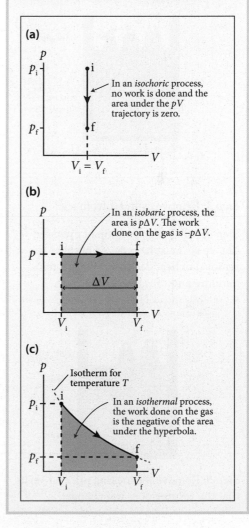

(a) In an *isochoric* process, no work is done and the area under the pV trajectory is zero.

(b) In an *isobaric* process, the area is $p\Delta V$. The work done on the gas is $-p\Delta V$.

(c) Isotherm for temperature T. In an *isothermal* process, the work done on the gas is the negative of the area under the hyperbola.

Thermodynamic Machine and an Isochoric Process

Step 1: An isochoric process occurs at a fixed volume, so we insert the pin to lock the piston in place.

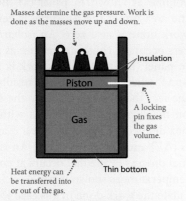

Masses determine the gas pressure. Work is done as the masses move up and down.

Insulation

Piston

Gas

A locking pin fixes the gas volume.

Heat energy can be transferred into or out of the gas.

Thin bottom

Step 2: Place base of cylinder on ice block to extract thermal energy.

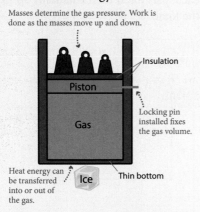

Masses determine the gas pressure. Work is done as the masses move up and down.

Insulation

Piston

Gas

Locking pin installed fixes the gas volume.

Heat energy can be transferred into or out of the gas.

Ice

Thin bottom

Step 3: Remove cylinder from ice.

Step 4: Adjust mass on top of cylinder. The mass must be reduced because the removal of heat decreased the gas temperature and thus the gas pressure.

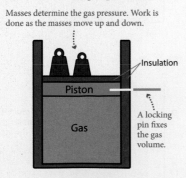

Masses determine the gas pressure. Work is done as the masses move up and down.

Insulation

Piston

Gas

A locking pin fixes the gas volume.

Step 5: Remove the locking pin and verify that the volume does not change.

processes, so we seek a consistent *format* with which we can break down and dissect such thermochemical cycles.

First, we recognize the need to identify the categories of energy and of the *transfer* of energy as work or heat (for example U_{therm}, U_{chem}, etc.) into which the total energy of the system is partitioned before and after a step on the pV diagram is executed.

Second, we recognize that because the total energy of the system plus surroundings remains constant through the process (the First Law!), the *sum* of the *initial* internal energy of the system plus the work (w) and heat (q) exchanged with the surroundings, must equal the final internal energy of the system. But the fraction of internal energy tied up in each category of internal energy may change.

Third, we identify explicitly that we must consider:
- The initial and final internal energy of the system that is comprised of the thermal energy, U_{therm}, and the chemical energy, U_{chem}.
- The work, w, and heat, q, which represent, respectively, (a) the *macroscopic* exchange of energy *to* or *from* the system and (b) the *microscopic* exchange of energy *to* or *from* the system.

We can represent the sequence on our pV diagram using a bar chart shown in Figure 3.22 for the process that quantitatively captures each of the quantities *and* keeps track of the *sign* of the *change*. We consider this bar chart in combination with the pV diagram. As the gas in the volume of our *isochoric* process decreased in temperature when the base of the cylinder was cooled, the initial thermal energy, $U_{therm\,i}$, decreased as heat, q, was removed *from* the system, $q < 0$. The chemical energy, U_{chem}, remained unchanged through the course of the process because no chemical reaction took place. Similarly, the work done *on* or *by* the piston/cylinder system is equal to zero because we locked the piston in place with the pin. Thus, the structure of our bar chart first identifies the *initial internal* energy, $(U_{therm})_i$ and $(U_{chem})_i$, and the magnitude and sign of the work and heat terms. The *final* internal energy, $(U_{therm})_f$ and $(U_{chem})_f$, at the end of the process is accounted for quantitatively by the bar graphs on the righthand side of the figure. Thus the bar chart takes the form displayed in Figure 3.22 for the isochoric process displayed on the pV diagram of Figure 3.21.

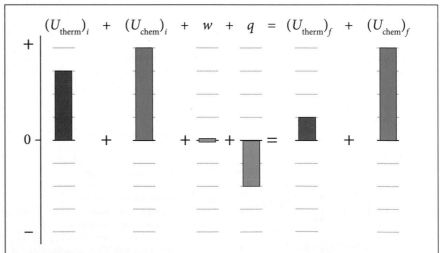

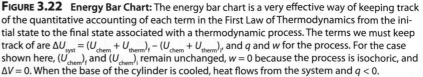

FIGURE 3.22 Energy Bar Chart: The energy bar chart is a very effective way of keeping track of the quantitative accounting of each term in the First Law of Thermodynamics from the initial state to the final state associated with a thermodynamic process. The terms we must keep track of are $\Delta U_{syst} = (U_{chem} + U_{therm})_f - (U_{chem} + U_{therm})_i$ and q and w for the process. For the case shown here, $(U_{chem})_f$ and $(U_{chem})_i$ remain unchanged, $w = 0$ because the process is isochoric, and $\Delta V = 0$. When the base of the cylinder is cooled, heat flows from the system and $q < 0$.

Notice that it is the *combination* of the thermochemical machine, the pV diagram, and the energy bar chart that systematically describes what is occurring in a thermochemical process. Becoming comfortable with this triad (machine, pV diagram, and energy bar chart) will greatly aid understanding thermochemical processes.

The Triad for Dissecting Thermochemical Processes: The Pressure–Volume Machine, the Pressure–Volume Diagram, and the Initial → Final Bar Chart

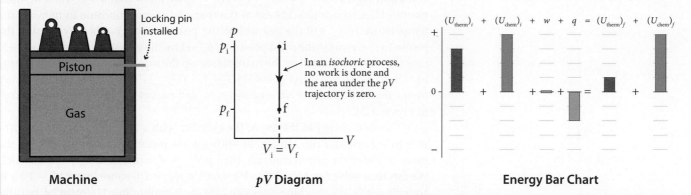

| **Machine** | **pV Diagram** | **Energy Bar Chart** |

The analysis of thermochemical processes requires the visualization of three primary elements: (1) the physical system that we depict here as the "machine," (2) the pressure-volume graph that maps out the trajectories of the thermodynamic change in moving from the initial condition to the final condition for a given thermodynamic step, and (3) the energy bar chart that provides a quantitative accounting of the terms in the First Law of Thermodynamics (U_{therm}, U_{chem}, w, and q) for the initial conditions and the final conditions for the given thermodynamic step. In the case shown here we are analyzing an *isochoric* process for which the volume of the working substance, the gas, does not change in going from the initial to the final condition. Thus in the machine, a pin locks the position in place so the volume cannot change. As heat flows from the gas, the pressure drops with the temperature at a fixed volume, from initial $p_i = nRT_i/V$ to the final $p_f = nRT_f/V$ conditions.

Isobaric Processes

We turn next to the case of an *isobaric* process shown in Figure 3.23 on the pressure-volume surface. For the isobaric process, as we proceed from the *initial* point on the pressure-volume plot, at p_iV_i, to the final point on the pressure-volume plot, at p_fV_f, the *pressure remains the same*. Thus, while $p_i = p_f$, the volume increases from V_i to V_f, and $V_f > V_i$, as shown in Figure 3.23.

One of the most powerful and versatile features of the pressure-volume (pV) diagram is that it provides a simple, straightforward means for calculating the work done for any trajectory across the pressure-volume diagram, independent of the functional form of that trajectory. To demonstrate this, we begin with the simplest case—this isobaric case that occurs at constant pressure. In this particular case, we know that the work is given by

$$w = -p\Delta V = p(V_f - V_i)$$

where V_f is the final volume and V_i is the initial volume. We can sketch this on a pV diagram as shown in Figure 3.23 and, moreover, see by inspection that the *shaded* area under the line representing the trajectory on the pressure-volume diagram is *quantitatively* equal to the work done in moving from V_i to V_f at pressure p_0, specifically $p_0\Delta V$.

But we must also keep track of what occurs physically in an isobaric process with our thermodynamic machine and then track the *initial* and *final* internal energies and the work and heat terms using our energy bar graph. Given that this is an isobaric process, we know that the *pressure* must be constant so that (a) the mass placed on top of the cylinder must not change during the expansion; (b) the

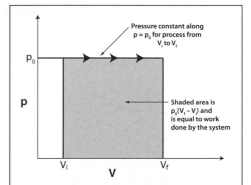

FIGURE 3.23 The trajectory of a system on a pressure-volume diagram for an *isobaric* process, wherein the pressure remains constant but the volume increases from V_i to V_f, is shown here. The area of the shaded region under the trajectory from V_i to V_f is equal to the work done by the system at pressure p during the constant pressure expansion. The shaded area is also easy to calculate in this case; it is just $p_0(V_f - V_i)$.

piston must move upward, so the pin must be removed; and (c) the machine must do work *on* the surroundings so that $w < 0$ by our convention defined in Figure 3.18.

By the Perfect Gas Law we know that $p_iV_i = nRT_i$ and $p_fV_f = nRT_f$. Because $p_i = p_f = p_0$ we can write $V_i = (nR/p_0)T_i$ and $V_f = (nR/p_0)T_f$, and so $\Delta V = V_f - V_i = (nR/p_0)(T_f - T_i)$. Therefore, because $\Delta V > 0$, $T_f > T_i$ and $(U_{therm})_f > (U_{therm})_i$. Since U_{chem} is unchanged, $(U_{chem})_i = (U_{chem})_f$ and thus with $(U_{therm})_f > (U_{therm})_i$ and with $w < 0$, and knowing that $w + q = (U_{therm})_f - (U_{therm})_i$, we know that $q > 0$. Thus we must use our laser source to *add heat* to the thermodynamic machine to increase the temperature (U_{therm}) of the gas within the piston-cylinder—thereby causing the piston to rise against the fixed pressure (p_0) set by the pressure of the atmosphere plus the pressure created by the mass placed on the top of the piston. In our energy bar chart, $q > 0$, $w < 0$ and $q + w = (U_{therm})_f - (U_{therm})_i$. This triad of the thermodynamic machine, pressure-volume diagram, and energy bar chart are summarized in Figure 3.24.

If we add energy to the gas in the cylinder with a laser and we keep the pressure in the cylinder constant, p_0, by allowing the piston to move, but keeping the mass on top of the piston constant, then $p_iV_i = p_0V_i = nRT_i$ and $p_fV_f = p_0V_f = nRT_f$. We can then solve for $\Delta V = (V_f - V_i) = (nR/p_0)(T_f - T_i)$, so we know $(T_f - T_i) > 0$; so with $w < 0$ (the machine does work on the surroundings), q must be positive, as shown in the bar chart of Figure 3.24. Again it is important to carefully think through what is occurring physically, how the process maps out on a pressure-volume plot and the sign and magnitude of q and w with the energy bar chart.

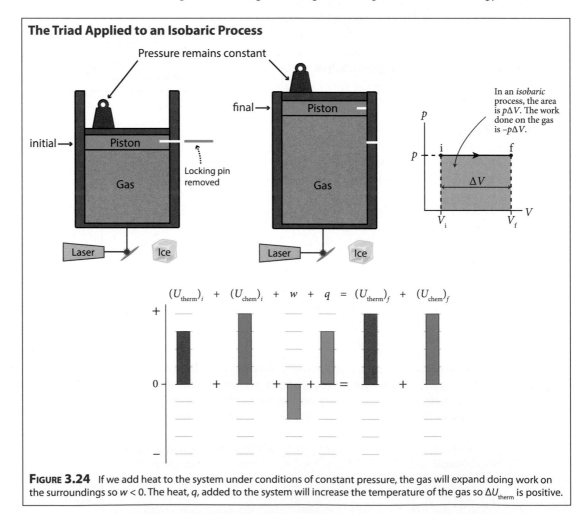

FIGURE 3.24 If we add heat to the system under conditions of constant pressure, the gas will expand doing work on the surroundings so $w < 0$. The heat, q, added to the system will increase the temperature of the gas so ΔU_{therm} is positive.

Isothermal Processes

We consider next the behavior of an *isothermal* process, a thermodynamic process from which the temperature does not change. First, we recognize, from the Perfect Gas Law, $pV = nRT$, that if the *temperature* of a contained volume of gas does not change, then pV is a *constant* for the process. The question is, can we design a process that will increase the volume in the cylinder of our thermodynamic machine without changing the temperature? If so, what does the process look like on a pV diagram?

To answer this question, we first step back to consider the feasibility of such a process. If the piston does work on the surroundings by virtue of moving upward, then $w < 0$, but if we heat the bottom of the cylinder, then $q > 0$ and $\Delta U_{syst} = w + q$ can indeed equal zero, as it must because there is no change in temperature and no chemical reaction takes place within the volume:

$$\Delta U_{syst} = \Delta U_{therm} + \Delta U_{chem} = 0$$

Examining each term again, we see that $\Delta U_{chem} = 0$, because no chemical reaction is taking place. With $\Delta U_{chem} = 0$ so, too, must $\Delta U_{therm} = 0$ because if the temperature does not change (isothermal), then the thermal energy component of the internal energy change must be equal to zero. Thus, we proceed with a strategy:

1. Employ a laser to transfer energy to the base of the cylinder. Energy will be added to the gas in the cylinder by microscopic processes wherein the electromagnetic energy of the laser beam is absorbed by the metal base of the thermochemical machine, transferring the kinetic energy of motion in the metal base to the gas molecules in the volume, resulting in the expansion of the gas.

2. The product of the pressure and the volume, pV, must remain constant in an isothermal process, so we must continuously remove mass from the piston top to balance the *increase* in volume with the *decrease* in pressure. Note in particular that the energy, q, added to the system by the laser goes *entirely* to the work, w, performed by the system and that while $q > 0$, $w < 0$. The work is negative.

3. When the desired final volume is reached, we switch off the laser and graph the process on a pV diagram, tracking the progression from the initial pressure and volume, $p_i V_i$, to the final pressure and volume, $p_f V_f$. We can systematically track, using the bar chart, the initial and final values of U_{therm} and U_{chem}, the changes in internal energy, ΔU_{therm} and ΔU_{chem}, as well as the *macroscopic* energy transfer *by* the system *to* the surroundings, $w < 0$, and the microscopic energy transfer to the system, $q > 0$, by virtue of the laser.

What may be counterintuitive in this isothermal process is that heat has been added to the system, but the temperature did not increase! This can only be accomplished by using a thermochemical machine that is capable of converting the microscopic energy transferred to the system into work; q is exactly equal

Work done when the pressure is constant

One of the most powerful and versatile features of the pressure-volume (pV) diagram is that it provides a simple, straight-forward means for calculating the work done for any trajectory across the pV graph, independent of the functional form of that trajectory. To demonstrate this, we begin with the simplest case—that of a process that occurs at constant pressure. In this particular case, we know that the work is given by:

$$w = -p(V_f - V_i) = -p\Delta V$$

Where V_f is the final volume and V_i is the initial volume. We can sketch this on a pV diagram as shown below and moreover, we can see by inspection that the shaded area under the line representing the trajectory on the pV surface is specifically equal to the work done in moving from V_i to V_f at pressure p.

But we can put ourselves in a position to generalize this calculation for more complicated cases by noting that we can break this trajectory from V_i to V_f into a number of smaller increments dV_1, dV_2, ... dV_n. We can add up each of these increments to again calculate the area under the trajectory from V_i to V_f as shown in the sketch below.

As we will see, the point of breaking this simple case into incremental steps will prove to be most useful when considering more complicated functional forms on the pV diagram. What remains unchanged, however complicated the functional form becomes, is that the work done is always the area under the curve on the pV plot.

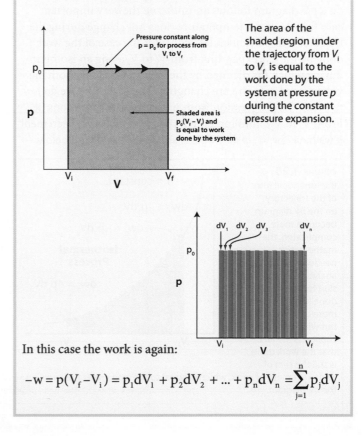

Pressure constant along $p = p_0$ for process from V_i to V_f

Shaded area is $p_0(V_f - V_i)$ and is equal to work done by the system

The area of the shaded region under the trajectory from V_i to V_f is equal to the work done by the system at pressure p during the constant pressure expansion.

In this case the work is again:

$$-w = p(V_f - V_i) = p_1 dV_1 + p_2 dV_2 + ... + p_n dV_n = \sum_{j=1}^{n} p_j dV_j$$

The Most Versatile Integral in Science

It is remarkable that a single mathematical relationship in integral calculus is all that is required to solve the most important problems in radioactive decay, chemical kinetics, thermodynamics, light absorption by molecules, exponential growth of global energy demand, etc. That integral is the disarmingly simple expression:

$$\int_{x_1}^{x_2} dx/x = \ln(x_2/x_1)$$

where $\ln(x_2/x_1)$ is the logarithm to the base e of the ratio x_2/x_1. We will use this integral throughout this course.

This equation follows from:

$$\int dx/x = \ln(x)$$

$$\int_{x_1}^{x_2} dx/x = \ln(x_2) - \ln(x_1)$$

$$= \ln(x_2/x_1)$$

to the macroscopic energy (work, w) executed by the system on the surroundings. The First Law of Thermodynamics (the conservation of energy) tells us that

$$\Delta U = \Delta U_{therm} + \Delta U_{chem} = q + w$$

and that since (in this case) $\Delta U_{chem} = 0$, and $\Delta U_{therm} = 0$ because the process is isothermal,

$$\Delta U_{therm} + \Delta U_{chem} = q + w = 0$$

so $q = -w$.

We can calculate the work done in this isothermal process by breaking the progression from the initial condition to the final condition into increments as shown in Figure 3.25. The work is then the sum of each of the increments $p_i \Delta V_i$ in Figure 3.25. Alternatively we can replace the summation of the small increments by an integral as shown in the sidebar such that

$$w = -nRT \int_{v_i}^{v_f} dV/V = -nRT \ln\left(V_f/V_i\right)$$

The union of what is occurring physically with the trajectory on the pressure-volume diagram and the energy bar chart that accounts for q and w is displayed as the triad in Figure 3.26.

Work on a pV Trajectory for an Isothermal Process

Suppose we now consider the case for which the trajectory on a pV diagram follows an *isotherm*: the very important case for which the temperature does not change during the trajectory. In particular, let's consider the case of the work done by the gas expanding from V_i to V_f along an isotherm during which (as dictated by the Perfect Gas Law) both the volume *and* pressure are changing. The first thing we do is to break the progression from initial to final conditions (ie. From V_i to V_f) into increments such that for each increment j, we have $dw_j = -p_j dV_j$. This is shown in the sketch below.

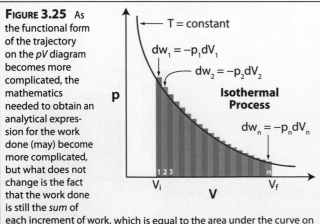

FIGURE 3.25 As the functional form of the trajectory on the pV diagram becomes more complicated, the mathematics needed to obtain an analytical expression for the work done (may) become more complicated, but what does not change is the fact that the work done is still the *sum* of each increment of work, which is equal to the area under the curve on the pV diagram.

For this case of the *isothermal* process, while the *pressure* is changing constantly as we move along the isothermal curve on the pV plot, the temperature remains fixed so we can use the Perfect Gas Law $pV = nRT$ to substitute for the pressure at each increment along the trajectory and write:

$$w = -\sum_{j=1}^{n} p_j dV_j = \sum_{j=1}^{n} nRT dV_j/V_j$$

However, as the sidebar highlighting "the most important integral in science" points out, as we go to the limit of small increments in dV_j, we can replace the illustrative but tedious sum by the *integral* expression which, through the power of calculus, allows us to write the solution in a simple "closed" form:

$$w = -\sum_{j=1}^{n} p_j dV_j = -\sum_{j=1}^{n} nRT dV_j/V_j$$

$$= -nRT \int_{V_i}^{V_f} dV/V = -nRT \ln V_f/V_i$$

That is, the work done in the isothermal expansion from V_i to V_f is simply equal to nRT times the natural logarithm of the ratio V_f/V_i.

The Triad Applied to an Isothermal Process

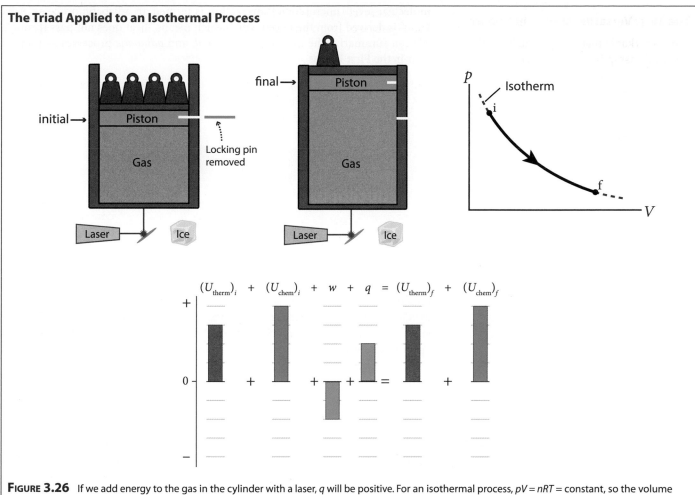

FIGURE 3.26   If we add energy to the gas in the cylinder with a laser, q will be positive. For an isothermal process, $pV = nRT =$ constant, so the volume must increase to keep the product pV constant. This means we must continuously remove mass from the top of the piston. This balancing act keeps U_{therm} constant throughout the process. Because U_{chem} is also constant (no chemical reaction within the cylinder) we know from $\Delta U_{syst} = q + w$ that the thermal energy added, q, must be offset by the same amount of work done such that $q = -w$ for the process. This is captured by the isothermal trajectory in the pV diagram and the quantitative accounting in the bar chart.

Adiabatic Processes

We have thus far investigated three important thermodynamic processes:

- *isochoric*, for which $w = 0$ because there is no macroscopic displacement, and thus $\Delta V = 0$;
- *isobaric*, for which the pressure does not change, and thus $\Delta p = 0$;
- *isothermal*, for which the temperature, and thus the thermal component of the internal energy, remains constant, $\Delta U_{therm} = 0$, through the course of the process.

But we also recognize that, for a process that does not involve a chemical transformation, $\Delta U_{chem} = 0$, the three quantities that appear in the First Law are ΔU_{therm}, w, and q, such that

$$\Delta U_{therm} = q + w$$

But if $\Delta U_{therm} = 0$ is an isothermal process and $w = 0$ is an isochoric process, what is a process for which $q = 0$? A process in which no energy is transferred by molecular level (microscopic) interaction is termed an adiabatic process. "Adiabatic" is derived from the Greek word which means "heat does not pass through." We can summarize the *isochoric, isothermal,* and *adiabatic* processes as they appear in the First Law as

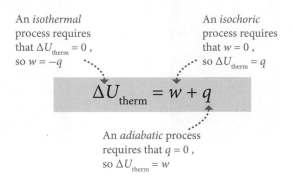

An *isothermal* process requires that $\Delta U_{\text{therm}} = 0$, so $w = -q$

An *isochoric* process requires that $w = 0$, so $\Delta U_{\text{therm}} = q$

$$\Delta U_{\text{therm}} = w + q$$

An *adiabatic* process requires that $q = 0$, so $\Delta U_{\text{therm}} = w$

While an adiabatic process may seem highly idealized because microscopic kinetic energy (thermal energy) always flows from warm bodies to cool bodies, adiabatic processes can in fact be closely emulated by employing very effective *insulation* to limit q to values far less than either w or ΔU_{therm} in the First Law such that q can be quantitatively ignored. Another situation that can be treated as approximately adiabatic is a process that occurs in a very short period compared with any other process in the system. A prime example of the latter is the operation of a gasoline or diesel engine in which the piston stroke occurs in such a short period of time that very little heat is transferred from the combustion zone during a single stroke. It turns out that adiabatic processes are very important in thermodynamic systems.

So we have, for an adiabatic process, a situation very common to critically important cycles in both nature and in the generation of useful work from combustion. For an adiabatic process, we also have a rather simple form of the First Law:

$$\Delta U_{\text{therm}} = w$$

because $q = 0$!

Compressing a gas adiabatically, for which w > 0, increases the thermal energy of the gas so $\Delta U_{\text{therm}} > 0$, resulting in an increase in the temperature of the gas. Therefore, an adiabatic compression of a gas *raises* its temperature and conversely an adiabatic expansion of a gas *lowers* its temperature. Thus, by virtue of an adiabatic process, the temperature of a gas can be raised *without using heat*. It is also important to recognize that the *work* expended on (by) a gas in an adiabatic process goes entirely to heating (cooling) the gas because $q = 0$.

Adiabatic processes are some of the most important we will study. Though the First Law for adiabatic processes may appear quite simple, $\Delta U_{\text{therm}} = w$, the *quantitative manipulation* of adiabatic processes on the *pV* diagram engages an array of very important concepts, and some slightly more interesting mathematics.

First, we have developed the concept of specific heat, wherein we deduced by *direct experimental studies* on an array of solids and liquids that we could write

$$q = mc\Delta T$$

where q was the thermal energy (heat) transferred to a solid or liquid, c was the specific heat of the material, m was the mass of the material, and ΔT was the temperature change (in °C or K). We now recognize that, in such cases, $w = 0$ in the application of the First Law because, to a high degree of precision, the volume of the material does not change because both solids and liquids are incompressible; so there can be little or no macroscopic displacement to produce (or take up) work.

Second, we recognize that substances can undergo a change in phase—that is, a solid may be converted to a liquid, a liquid to a gas, etc. by raising the temperature. But when the medium under consideration is a gas, it becomes *compressible* such that for the same number of moles of a gas, the volume can increase or decrease in response to changing pressure as defined by the Perfect Gas Law, $pV = nRT$. When we attempt to write a simple proportionality between the heat added, q, and the change in temperature, ΔT, such that

$$q = \text{mc}\Delta T$$

we discover something very important: If we execute a process at *constant volume* (isochoric) between two isotherms on our pV diagram (path 1 in Figure 3.27), the *amount* of *thermal energy transferred* to the system will not be equal to the case of moving between the same two isotherms along the path of constant pressure (isobaric) indicated by path 2 in Figure 3.27.

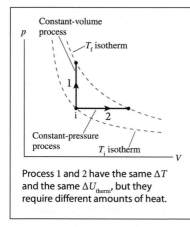

Process 1 and 2 have the same ΔT and the same ΔU_{therm}, but they require different amounts of heat.

FIGURE 3.27   A plot of two trajectories in the pV graph, each of which carries the system from one isotherm, T_i, to a second isotherm T_f. Along segment 1 from T_i to T_f, the process is isochoric so $\Delta V = 0$ and no work is done. In this case the heat added to the system is $q = nc_v\Delta T$, where c_v is the molar heat capacity at constant volume. Along segment 2 between the same two isotherms, the volume increases but the pressure remains constant, so work is done by the system on the surroundings, and the heat added to the system is $q = nc_p\Delta T$, where c_p is the molar heat capacity at constant pressure. Note that because ΔU_{syst} is the same for both path 1 and path 2, $c_p > c_v$ because work had to be done along the path of constant pressure.

We recognize why this is so from our analysis of isochoric and isobaric processes with our thermochemical machine, our pV diagrams, and our bar graphs. Specifically, while *no work is done along path 1* in moving between two isotherms, *work is done along path 2*.

Recognition of these two points requires some thought when we treat compressible fluids (in this case, gases) with the First Law. We highlight the issue of the path dependence of the heat required to engender the same change in temperature by defining two distinct quantities: the specific heat of a gas at constant volume, c_v, corresponding to path 1 in Figure 3.27, and the specific heat of a gas at constant pressure, c_p, corresponding to path 2 in Figure 3.27. It is common practice to write specific heats for gases in terms of the *molar* specific heat rather than the specific heat per unit mass, as is typically done for solids and liquids. Thus, we write

$$q = nc_v\Delta T$$

for the temperature change at *constant **volume*** for n moles of a gas to which an amount of thermal energy, q, has been added. In an analogous way, we write

$$q = nc_p\Delta T$$

for the temperature change at *constant **pressure*** for n moles of a gas to which an amount of thermal energy, q, has been added. Respectively, c_v is the *molar specific heat at constant volume* and c_p is the *molar specific heat at constant pressure*.

It is quite informative when the molar specific heats of gases are compared, particularly when we divide gases into monatomic gases and diatomic gases, as displayed in Table 3.4.

TABLE 3.4 Molar specific heats of gases (J/mol K)			
Gas	c_p	c_v	$c_p - c_v$
MONATOMIC GASES			
He	20.8	12.5	8.3
Ne	20.8	12.5	8.3
Ar	20.8	12.5	8.3
DIATOMIC GASES			
H_2	28.7	20.4	8.3
N_2	29.1	20.8	8.3
O_2	29.2	20.9	8.3

What is most obvious is that the molar heat capacities at constant pressure are virtually identical for monatomic gases, as they are for molar heat capacities at constant volume. By necessity, then, the *difference* between c_p and c_v is the same for all monatomic gases. But inspection of diatomic gases reveals that, while there are small differences in c_p between various diatomics *and* there are small differences in c_v between various diatomics, the difference between c_p and c_v, i.e., $c_p - c_v$, is virtually identical for diatomics and in fact is the same as $c_p - c_v$ for monatomic gases.

There are important ideas that lie behind this lack of variation between atomic/molecular values for c_p and for c_v and behind the identical values for the *difference* between c_p and c_v.

First, ΔU_{therm} (which is the change in the (molecular level) microscopic energy—the thermal energy—of a gas) is the same no matter what path is followed in going through a temperature difference $\Delta T = T_f - T_i$. This is the *definition* of a *state variable*, a quantity that is path independent and a quantity that defines the state of the system.

Second, the First Law, $\Delta U_{therm} = q + w$, states that a gas cannot distinguish between the transfer of microscopic (molecular level) energy and macroscopic (mechanical) energy.

Therefore, no matter what path is taken that results in the change in thermal energy of the gas, ΔU_{therm}, the temperature change, ΔT, will be the same.

So let's consider first path 1 in Figure 3.27, which occurs along a path such that no work can be done on or by the system:

$$\left(\Delta U_{therm}\right)_1 = \cancel{w} + q = 0 + q_v = nc_v\Delta T$$

Along path 2 in Figure 3.27, the work done (by the system) is $-p\Delta V$, so

$$(\Delta U_{therm})_2 = w + q = -p\Delta V + q_p = -p\Delta V + nc_p\Delta T$$

But the fact that $(\Delta U_{therm})_1 = (\Delta U_{therm})_2$ allows us to equate the two expressions such that

$$nc_v\Delta T = -p\Delta V + nc_p\Delta T$$

But we also know that, by virtue of the Perfect Gas Law,

$$pV = nRT$$

and therefore that

$$\Delta(pV) = \Delta(nRT)$$

For a constant pressure process,

$$\Delta(pV) = p_f V_f - p_i V_i$$

but $p_f = p_i = p$, so

$$\Delta(pV) = pV_f - pV_i = p(V_f - V_i) = p\Delta V$$

and therefore that

$$p\Delta V = nR\Delta T$$

This expression can be substituted directly into our equation $(\Delta U_{therm})_1 = (\Delta U_{therm})_2$, which is

$$nc_V \Delta T = -p\Delta V + nc_p \Delta T$$

to yield

$$nc_V \Delta T = -nR\Delta T + nc_p \Delta T$$

However, $n\Delta T$ cancels, yielding

$$c_V = -R + c_p$$

or

$$c_p - c_V = R$$

This is remarkable. It is exactly what we see in the experimentally determined data of Table 3.4. The above expression is a universally useful result. It emerged, remember, from the union of the First Law and the Perfect Gas Law.

While we are now equipped to quantitatively tackle an *adiabatic* process (a process for which $q = 0$), it is of considerable interest to step back and assess what we have just accomplished.

We demonstrated that

$$\Delta U_{therm} = nc_V \Delta T$$

for a *constant volume process*. But we also now recognize that ΔU_{therm} is identical for all processes that carry us between the same two isotherms (T_i and T_f such that $\Delta T = T_f - T_i$ is the same). Therefore,

$$\Delta U_{therm} = nc_V \Delta T$$

is the same *for any ideal gas process*! But we opened this discussion by pointing out that the path mattered; namely $q = nc_p \Delta T$ for a constant pressure process and $q = nc_V \Delta T$ for a constant volume process. How do we reconcile this (potential) contradiction? The reconcilliation lies in the fact that while ΔU_{therm} is *path independent* because U_{therm} is a state variable, the heat, q, *depends on the path*.

If we write the First Law as

$$q = \Delta U_{therm} - w,$$

then we see the resolution to the apparent contradiction. In our constant volume (isochoric) path, $w = 0$ by definition, so the amount of microscopic energy (heat) transferred is used exclusively to change the thermal energy of the system. In sharp contrast, in a process at constant pressure (isobaric), a quantity of microscopic energy (heat) transferred to the system leaves the system as work as a result of the expanding gas ($\Delta V > 0$ so $-p\Delta V < 0$). Thus from

$$q = \Delta U_{therm} - w = \Delta U_{therm} + p\Delta V$$

Showing pV^γ = constant for Adiabatic Process

We examine an adiabatic ($q = 0$) process for which a small increment of work, dw, done by a gas causes an incremental change in the thermal energy, dU_{therm}, such that (with $dq = 0$)

$$dU_{therm} = dw$$

But because we can write

$$dU_{therm} = nc_V dT$$

for any process for which the working medium is a perfect gas, we also know that $dw = -pdV$. Thus we have

$$dU_{therm} = -pdV$$

And the ideal gas law gives us

$$p = \frac{nRT}{V}$$

so by substitution

$$dU_{therm} = nc_V dT = -pdV = -nRT\frac{dV}{V}$$

or

$$\frac{dT}{T} = -\frac{R}{c_V}\frac{dV}{V}$$

But

$$\frac{R}{c_V} = \frac{c_p - c_V}{c_V} = \frac{c_p}{c_V} - 1 = \gamma - 1$$

where, as is commonly employed, $\gamma = c_p/c_V$. Thus we can integrate

$$\frac{dT}{T} = -\frac{R}{c_V}\frac{dV}{V}$$

from initial to final conditions to obtain

$$\int_{T_i}^{T_f} dT/T = -\int_{V_i}^{V_f} (\gamma - 1)dV/V$$

$$\ln\left(T_f/T_i\right) = -\ln\left(V_f/V_i\right)^{\gamma-1}$$

Note: we used the log identity $a \ln x = \ln(x)^a$ and $\log a - \log b = \log(a/b)$.

So then we have

$$\ln\left(T_f/T_i\right) = \ln\left(V_i/V_f\right)^{\gamma-1}$$

or

$$T_f V_f^{\gamma-1} = T_i V_i^{\gamma-1}$$

and, from the ideal gas law $T = pV/nR$, we have

$$p_f V_f^\gamma = p_i V_i^\gamma$$

for an *adiabatic* process.

the process that takes place along path 1 (isochoric) requires *less heat* than that along path 2 (isobaric). Thus, while U_{therm} is a state variable and does not depend upon the path by which the system achieved that state, both heat, q, and work, w, *depend on the path taken*.

We are now in a position to address the characteristics of an adiabatic process ($q = 0$) on a pressure-volume diagram. We recall that since $\Delta U_{therm} = w + q$, when $q = 0$, then any work done on or by the system is equal to the change in internal thermal energy, ΔU_{therm}, such that

$$\Delta U_{therm} = nc_V\Delta T = w$$

We have seen in our numerous examples investigating trajectories of processes on a pV diagram that a gas follows a hyperbola during an isothermal process (pV = constant). We now need to explore the trajectory on a pressure-volume diagram for an *adiabatic* process.

If we define the ratio of the molar heat capacity at constant pressure, c_p, to the molar heat capacity at constant volume, c_V, as γ, then

$$\gamma = c_p/c_V$$

This ratio, γ, is a very important quantity in thermodynamics.

An *adiabatic* process is one that satisfies the relationship

$$pV^\gamma = \text{constant}$$

or

$$p_f V_f^\gamma = p_i V_i^\gamma$$

How do we know this? It is proven in the sidebar on the previous page! Notice that the adiabatic pV relation bears a strong mathematical resemblance to the isothermal case for which pV = constant.

Figure 3.28 maps out trajectories on a pressure-volume diagram for adiabatic processes. These trajectories are called *adiabats* and, because $\gamma = c_p/c_V > 1$, these trajectories are steeper than the corresponding trajectories for an isothermal process.

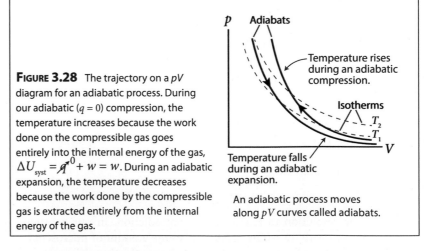

Figure 3.28 The trajectory on a pV diagram for an adiabatic process. During our adiabatic ($q = 0$) compression, the temperature increases because the work done on the compressible gas goes entirely into the internal energy of the gas, $\Delta U_{syst} = q^0 + w = w$. During an adiabatic expansion, the temperature decreases because the work done by the compressible gas is extracted entirely from the internal energy of the gas.

An understanding of adiabatic and isothermal processes turns out to be critically important for understanding heat engines and heat pumps. Examples of heat engines include internal combustion gasoline engines for transportations as well as steam engines and steam turbines. Steam turbines are the thermodynamic method by which the combustion of coal and natural gas are converted to electric-

---

Why $\Delta U_{therm} = nc_V\Delta T$ Holds Even When the Process Does Not Occur at Constant Volume

When we consider two processes that carry us from an initial position **i** on a pV diagram along two paths, **path 1** at constant volume and **path 2** at constant pressure, we know we can write $q_1 = nc_V\Delta T$ for path 1 and $q_2 = nc_p\Delta T$ for path 2.

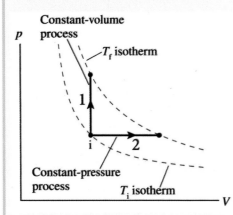

But because both process 1 and process 2 end on the same isotherm, both have the same ΔT so, *for a perfect gas*, both must have the same U_{therm} as well as the same ΔU_{therm}! Thus, the *very* versatile equation

$$\Delta U_{therm} = nc_V\Delta T \text{ (for an ideal gas)}$$

is true *even when the process does not occur at constant volume*. This turns out to be one of the most versatile expressions in thermodynamics.

ity. Case Study 3.1 investigates heat engines in the detail required to understand important elements in the global energy picture. The operation of heat pumps, the topic of Case Study 3.2, engages the thermodynamic principles used for refrigeration but equally important, the method by which homes and buildings can be heated or cooled *far* more efficiently than is currently done in the US. As we will see, because the conversion of the chemical energy contained in hydrocarbons is so *inefficiently* converted to work in a heat engine, and because heat pumps can dramatically reduce the energy required to heat homes and buildings, a significant part of the global energy strategy for the future rests in the hands of these principles of thermodynamics. We turn, therefore, to develop the key relationships between spontaneous change, reversibility vs. irreversibility, and equilibrium in thermodynamic systems designed to convert heat to work or work to heat. As we will see, it is the Carnot cycle that ties these concepts together quantitatively, and the Carnot cycle is the central focus of Case Study 3.1.

Phase Changes and the Thermodynamics of Melting, Vaporization, and Sublimation

To this point we have considered the First Law of Thermodynamics with respect to heat, q, and work, w, for a single phase: solid, liquid, or gas. Many important processes involve the thermodynamics of changes from solid to liquid and from liquid to gas. We note those phase changes here in the context of enthalpy changes, ΔH, associated with those phase changes. We focus on enthalpy changes because the vast majority of phase transitions take place at *constant pressure*, which the thermodynamic variable enthalpy was specifically created to address. We summarize the terminology of phase changes in Figure 3.29, which demonstrates the phase transitions from solid to liquid to gas and visa versa on an enthalpy scale.

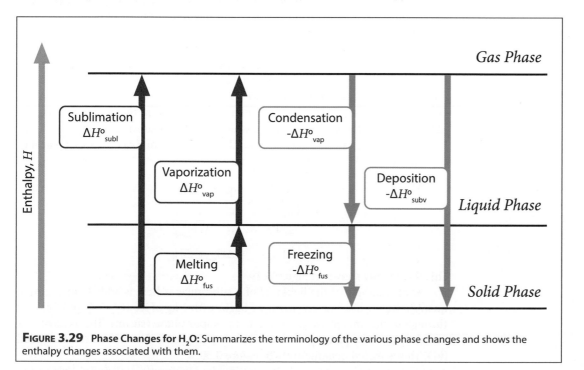

FIGURE 3.29 **Phase Changes for H_2O:** Summarizes the terminology of the various phase changes and shows the enthalpy changes associated with them.

The most common phase transitions, melting (solid to liquid) and vaporization (liquid to gas), are associated with specific enthalpy changes ΔH as follows:

(a) For a pure substance (such as water), each phase change has a specific enthalpy change per mole at the temperature of that phase change. The enthalpy change for the phase transition from liquid to gas is referred to as the *heat* of *vaporization*, $\Delta H°_{vap}$. For water

$$H_2O(\ell) \rightarrow H_2O(g)$$

$$\Delta H = \Delta H°_{vap} = 40.7 \text{ kJ/mol @ } 100°C$$

The reverse process, the condensation from vapor to liquid, is equal in magnitude but opposite in sign:

$$H_2O(g) \rightarrow H_2O(\ell)$$

$$\Delta H = -\Delta H°_{vap} = -40.7 \text{ kJ/mol @ } 100°C$$

(b) The enthalpy change for the phase transition from solid to liquid is the *heat of fusion*

$$H_2O(s) \rightarrow H_2O(\ell)$$

$$\Delta H = \Delta H°_{fus} = 6.02 \text{ kJ/mol @ } 0°C$$

The reverse process, the freezing from liquid to solid, is equal in magnitude but opposite in sign:

$$H_2O(\ell) \rightarrow H_2O(s)$$

$$\Delta H = -\Delta H°_{fus} = -6.02 \text{ kJ/mol @ } 0°C$$

Note that the amount of energy per mol to transition from solid to liquid is *much less* than the energy required to transition from liquid to vapor. The reason is that in going from solid to liquid it is necessary merely to provide the energy to allow molecules to slide past one another whereas to vaporize a substance requires that all the *intermolecular* bonds be broken to allow complete separation of the molecules.

A less common phase transition is sublimation. Sublimation is the direct phase transition from solid to vapor and is familiar in the case of solid carbon dioxide (dry ice) that "disappears" without passing through the liquid phase (thus the name!). Sublimation is also familiar in the winter in dry, cold climates where snowfall is followed by the direct loss of snow at temperatures below freezing without the appearance of liquid water. Sublimation is also used in the preparation of "freeze-dried" foods. The reverse of sublimation is termed *deposition* and is used extensively in the production of electronic circuits and for sophisticated films that control the absorption and transmission of electromagnetic radiation. The enthalpy change, ΔH, for sublimation is referred to as the *heat of sublimation* ($\Delta H°_{subl}$) and is the enthalpy change when 1 mol of a substance sublimes. From Hess's law we can calculate $\Delta H°_{subl}$ from the heat of fusion ($\Delta H°_{fus}$) and the heat of vaporization ($\Delta H°_{vap}$) by recognizing that

$$\begin{array}{ll} \text{Solid} \rightarrow \text{liquid} & \Delta H°_{fus} \\ \underline{\text{Liquid} \rightarrow \text{gas}} & \underline{\Delta H°_{vap}} \\ \text{Solid} \rightarrow \text{gas} & \Delta H°_{subl} \end{array}$$

$$\text{and } \Delta H_{subl} = \Delta H_{fus} + \Delta H_{vap}$$

This is explicitly shown in Figure 3.29.

As an example of the behavior of a substance as heat is added, we examine quantitatively what happens when we begin with 1 kg of ice at −40°C and follow it through to the final phase transition to the vapor phase (steam). This is most conveniently done by plotting the temperature (°C) versus the amount of heat added (kJ). This is traced quantitatively in Figure 3.30.

3.46

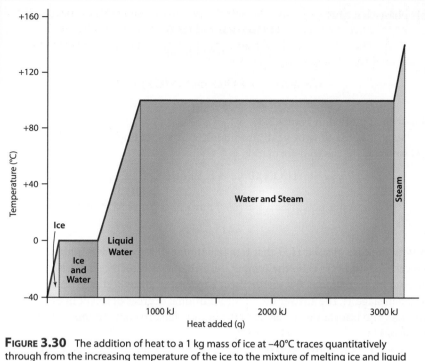

FIGURE 3.30 The addition of heat to a 1 kg mass of ice at −40°C traces quantitatively through from the increasing temperature of the ice to the mixture of melting ice and liquid water. Once all the ice has melted, the liquid water begins to increase in temperature as further heat is added. As the boiling point of water is reached, added heat goes into the formation of steam and the temperature ceases to increase until all the water is converted to steam. The final temperature is 140°C.

Stage 1 on the far left of Figure 3.30 is the addition of heat to ice at −40°C. As heat enters the structure of ice, the kinetic energy of the H_2O molecules in the ice structure increases via the physically constrained rotational and vibrational motion of the molecules. The change is

$$H_2O(s) \ (-40°C) \rightarrow H_2O(s)(0°C)$$

The relationship between the heat added, q, and the change in temperature is just

$$q = n \times c_{H_2O(s)} \times \Delta T$$

where n is the number of moles of H_2O, $c_{H_2O(s)}$ is the molar heat capacity of solid H_2O, and ΔT is the temperature change in °C (or K). Thus we must first calculate the number of moles of ice in 1 kg of ice. H_2O has a molar mass of 18.02 g/mol so 1 kg of H_2O corresponds to

$$\frac{1000 \text{ g}}{18.02 \text{ g/mol}} = 55.5 \text{ mol } H_2O$$

The molar heat capacity of solid H_2O is 37.6 J/mole · °C so for a temperature rise from −40°C to 0°C,

$$q = (55.5 \text{ mol})(37.6 \text{ J/mol} \cdot °C)(40°C)$$

$$= 83.5 \text{ kJ}$$

Because the increase in temperature is directly proportional to the amount of heat added, Stage 1 in our graph of temperature vs. heat added is linear as displayed in Figure 3.30.

Stage 2 is the addition of heat to melt solid H_2O to liquid H_2O. The change associated with melting occurs at one temperature, 0°C, as the inflow of thermal energy (heat) converts the crystal structure of solid H_2O to liquid where the H_2O molecules are free to slide past one another. The enthalpy change associated with this phase change is

$$H_2O(s) \rightarrow H_2O(\ell) \text{ @ } 0°C$$

$$\Delta H°_{fus} = 6.02 \text{ kJ/mol}$$

and

$$q = n(\Delta H°_{fus}) = (55.5 \text{ mol})(6.02 \text{ kJ/mol})$$

$$= 334.1 \text{ kJ}$$

This stage is plotted in Figure 3.30.

Stage 3 is the flow of heat into liquid H_2O, increasing its temperature from 0°C to 100°C, so the transformation is

$$H_2O(\ell)(0°C) \rightarrow H_2O(\ell)(100°C)$$

so

$$q = n \times c_{H_2O(\ell)} \times \Delta T = (55.5 \text{ mol})(75.4 \text{ J/mol °C})(100°C)$$

$$= 418.5 \text{ kJ}$$

As with the case of solid H_2O that remains in one phase as heat is added, so too is the case for liquid H_2O so the temperature change is directly proportional to the amount of heat added. Why is the heat capacity in stage 3 greater than the heat added in stage 1? The answer lies in the additional degrees of freedom of vibration, rotation, and translation available in the liquid phase in comparison with the solid phase of H_2O.

Stage 4 is the conversion of liquid H_2O to vapor phase H_2O. As heat is added in this phase transition, the temperature remains constant as the H_2O-H_2O bonds are broken in liquid phase H_2O, releasing individual H_2O molecules into the gas phase. The transformation is thus

$$H_2O(\ell) \rightarrow H_2O(g) \text{ @ } 100°C$$

$$\Delta H = \Delta H°_{vap}$$

and

$$q = n \times (\Delta H°_{vap}) = (55.5 \text{ mol}) \, 40.75 \text{ kJ/mol} = 2258.9 \text{ kJ}$$

Note that this is a *very* large amount of energy. In the transition from solid H_2O at –40°C to vapor phase H_2O (steam) at 100°C, this is by far the largest increment of enthalpy. This is revealed graphically in Figure 3.30.

Stage 5 is the increase in temperature of the steam from 100°C to 140°C

$$H_2O(g)(100°C) \rightarrow H_2O(g)(140°C)$$

so

$$q = n \times c_{H_2O(g)} \times \Delta T = (55.5 \text{ mol})(33.1 \text{ J/mol °C}) \, 40°C$$

$$= 73.5 \text{ kJ}$$

This step-by-step calculation is simply a reflection of Hess's Law for we have transitioned from 1 kg of ice at –40°C to 1 kg of steam at +140°C in five steps and have added a total amount of heat equal to the sum of the heat added in each stage. This totals to 3168.5 kJ, the dominant fraction going to the vaporization of liquid water to steam.

The key points to take away from this sequence are

1. Within a given phase, the addition of heat results in a proportional increase in temperature. The amount of heat gained or lost depends on the amount of the substance and on the molar heat capacity.

2. For the case of a transition from one phase to another, the addition of heat results in no change in temperature and the heat added is equal to the number of moles of material that has undergone the phase change times the *enthalpy* change, ΔH, for that phase transition.

3. By far the greatest amount of energy in going from the solid to the vapor phase occurs in the phase transition from liquid to vapor.

The amount of heat absorbed, for *no* increase in temperature, is very important for understanding the thermodynamic behavior of systems and is given a name— *latent heat*. For example, the *latent heat* released in the condensation of water vapor to liquid water is $-\Delta H°_{vap}$. As it turns out, this latent heat release is of central importance to the thermodynamics of, for example, hurricanes and other severe storm events as well as to the forecasting of weather and climate. It is also of critical importance for an understanding of irreversible changes to the Earth's climate system to recognize that it requires very *little* energy to melt the polar ice caps because ΔH_{fus} is so small. Thus as more infrared radiation is trapped by the release of CO_2 in the combustion of fossil fuels, more moisture enters the atmosphere as the oceans warm, increasing the intensity of severe storms. Yet this trapping of infrared radiation accelerates the irreversible loss of the polar ice caps.

To appraise the energy release of a storm system, consider the following problem.

Problem:

How much energy is released in a 2 cm rainfall over an area 10 by 10 km? If 1 ton of TNT releases 4.18×10^6 kJ of energy, how many tons of TNT would be equivalent to the energy released in rainfall?

Solution:

We know that 2259 kJ of energy are released by condensing 1 L of water, so we can apply that conversion to the larger volume of water in the rainfall. That volume is found by multiplying the depth times the area, remembering that 1 mL = 1 cm³, and that 1 km = 10^3 m = 10^5 cm:

$$\text{Volume of rain} = 2 \text{ cm} \times (10 \times 10^5 \text{ cm})^2 = 2 \times 10^{12} \text{ cm}^3$$
$$= (2 \times 10^{12} \text{ cm}^3)(1 \text{ ml/cm}^3)(1\text{L}/10^3 \text{ ml})$$
$$= 2 \times 10^9 \text{ L}$$

$$\text{Energy from rain} = 2 \times 10^9 \text{ L} \times 2259 \text{ kJ/L} = 4.52 \times 10^{12} \text{ kJ}$$

And since 1 ton of TNT releases 4.18×10^6 kJ of energy, we can divide this factor into the energy from rainfall to obtain the TNT equivalents.

$$\text{Equivalent TNT} = 4.52 \times 10^{12} \text{ kJ}/4.18 \times 10^6 \text{ kJ per ton TNT} = 1.1 \times 10^6 \text{ tons TNT.}$$

This gives some feeling for the scale of energy in storms! A modest rainfall releases the equivalent of about a million tons of TNT. Compare the latent heat releases in a hurricane with that of an atomic bomb. Which is larger?

We turn now to summarize the key concepts in this chapter.

Summary Concepts

1. Constructing a Model: A System and its Surroundings

The word system stems (as do many terms in thermodynamics) from the Greek words meaning "to bring together" or "to combine." In the discussion of thermodynamic processes we must constantly grapple with the problem of keeping track of the flow of thermal energy (heat) through a definite surface, or mechanical work done on a specific ensemble of (macroscopic) objects. Thus, by nature, we define a system by its *boundary* that separates the *system*—that part of the physical world upon which we focus our attention—from the rest of the world, the surroundings. This is displayed in the adjoining figure.

SURROUNDINGS

System

Energy In
Work done on the system $w > 0$

Heat to the system $q > 0$

$$\Delta U_{syst} = q + w$$

Energy Out
Work done by the system $w < 0$

Heat from the system $q < 0$

Pages 134–135

2. Work Done On or By a System

A particularly important example of how work is exchanged between a system and its surroundings involves the work done by a piston that can move in response to a change in pressure within an otherwise closed vessel. For example, as shown to the right, the release of hydrogen gas (H_2) in the chemical reaction of an acid (HCl) with a metal (Zn) releases H_2, increasing the pressure within the vessel, forcing the piston to move against the external (atmospheric) pressure. The work, w, is the product of force times distance: $w = -F\Delta x$. However, the force $F = (F/area)\ area = p \cdot A$ because pressure is force per unit area. If the piston has area A, then $w = -pA\Delta x$ and the change in volume is $\Delta V = A\Delta x$ and the work done at constant pressure $w = -pA\Delta x = -p\Delta V$.

Pages 136–137

3. Concept of Internal Energy

The internal energy, U_{syst}, of a system is generally taken to be independent of the mechanical energy, E_{mech}, of a system where E_{mech} refers to the kinetic and potential energy of the *macroscopic* objects in a system. The internal energy of a system, in contrast, is equal to the sum of the thermal energy of the microscopic (molecular) motion of the material that comprises the system and the chemical energy contained in the bonding structure of reactant molecules relative to the chemical energy contained in the bonding structure of product molecules such that $U_{syst} = U_{therm} + U_{chem}$.

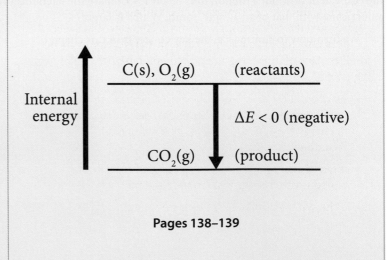

Internal energy

C(s), O_2(g)   (reactants)

$\Delta E < 0$ (negative)

CO_2(g)   (product)

Pages 138–139

Summary Concepts

4. State Variables in Thermodynamics

We are constantly seeking to express dynamic and changing systems in terms of constants that remain invariant even though they may capture billions of individual events. The *state* of a thermodynamic *system* is defined quantitatively by a specific set of quantities that establish its properties. Important examples include composition, temperature, pressure, volume, mass, etc. A distinguishing characteristic of a *state variable*, particularly for thermodynamics, is that *a state variable is independent of the path taken to achieve that state*. Such is the case for temperature, pressure, internal energy, etc. In contrast, heat and work are not state variables as they depend very much on the path taken.

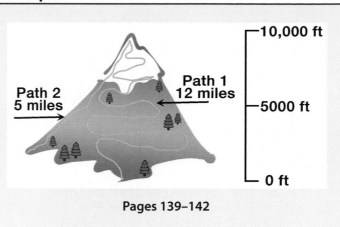

Pages 139–142

5. Work Produced by a Chemical Reaction

Chemical reactions can produce work by producing a gas that increases the pressure within a vessel or by increasing the temperature of a gas through the release of energy in the conversion of reactants to products. For example, as shown to the right, the release of hydrogen gas, H_2, in the chemical reaction of an acid (H_2SO_4) with a metal (Zn) increases the pressure within a vessel, forcing the piston to move against the external atmospheric pressure yielding work $w = -p\Delta V$.

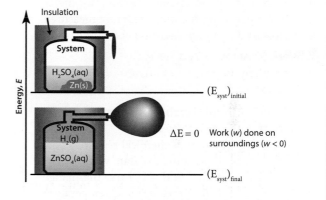

Pages 141–142

6. Development of the First Law of Thermodynamics

As the word *thermodynamics* implies, we are concerned first and foremost with setting in place laws that succinctly define the means by which a system *exchanges* energy with its surroundings. This is the domain of the First Law of Thermodynamics, and it is a disarmingly powerful law, as we have seen.

It is also important at this stage in the development of the First Law of Thermodynamics, to emphasize the following points:

(a) work and heat are not *contained within* the thermodynamic system—work and heat exists only as the forms of energy *transferred* between the system and its surroundings;

(b) internal energy, U_{syst}, is the only form of energy contained within the thermodynamic system;

(c) if the system is isolated from its surroundings, then $\Delta U_{syst} = 0$.

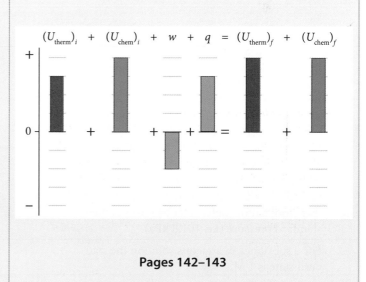

$$(U_{therm})_i + (U_{chem})_i + w + q = (U_{therm})_f + (U_{chem})_f$$

Pages 142–143

Summary Concepts

7. Heat, Heat Capacity, and the Bomb Calorimeter

Measurements of the energy release from a chemical reaction are typically done by confining the reaction in a vessel such that no work is done during the course of the chemical reaction. The First Law then becomes $\Delta U_{chem} = q + w = q + 0 = q_{rxn}$ because no work is done (the containment vessel does not expand or contract) and the heat release is then equal to the energy release in going from reactants to products. The heat produced in the reaction, q_{rxn}, is determined by measuring the increase in the temperature of the so-called bomb calorimeter shown at right using the independently measured heat capacity, C_{cal}, of the calorimeter and the equation $q_{rxn} = C_{cal}\Delta T$.

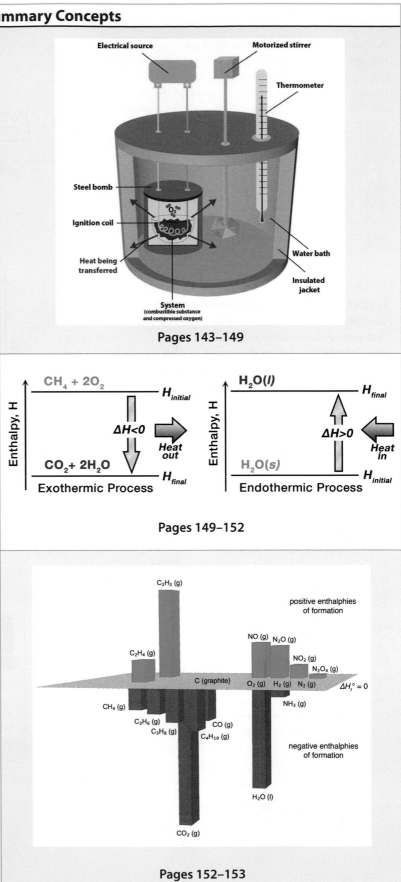

Pages 143–149

8. Enthalpy: A State Variable for Thermodynamic Changes at Constant Pressure

Because a large fraction of chemical reactions take place at a constant pressure, the state variable, enthalpy, is defined such that $H = U + pV$. The change in enthalpy at constant pressure and temperature is then $\Delta H = \Delta U + \Delta(pV) = \Delta U + p\Delta V$. But $\Delta U = q_p - p\Delta V$ so $\Delta H = \Delta U + p\Delta V = q_p - p\Delta V + p\Delta V = q_p$. Thus the energy in a chemical reaction at constant pressure is just the heat release at constant pressure, q_p.

Pages 149–152

9. Standard Enthalpies of Formation

Thermodynamic variables depend, to a degree, on the conditions under which they are measured. This has lead to the universally agreed upon definition of the *Standard State*, which specifies conditions and concentrations.

1. For a gas, the standard state is 1 atm but has more specifically been refined to be 100.0 kPa = 1.00 bar whereas 1 atm is 101.3 kPa = 1.013 bar. All standard tables in this text and most thermodynamic tables refer to 1 atm as the standard pressure (i.e. 101.3 kPa).

2. For a substance in aqueous solution the standard state is 1 M.

3. For a pure substance (element or compound) the standard state is the most stable form of the substance at 1 atm and 25°C (298 K).

Pages 152–153

Summary Concepts

10. Standard Heats of Reaction

The Standard Heat of Reaction, ΔH_{rxn} or ΔH_R, is the sum of the standard heats of formation of the *products* of the reaction minus the sum of the standard heats of formation of the *reactants* $\Delta H^{\circ}_R = \Sigma m \, \Delta H^{\circ}_{f(products)} - \Sigma n \, \Delta H^{\circ}_{f(reactants)}$ where m and n are the molar amounts of the products and reactants.

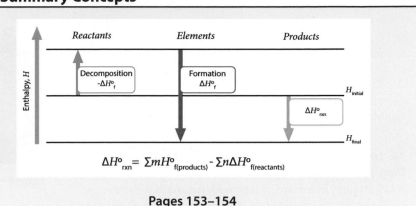

$$\Delta H^{\circ}_{rxn} = \Sigma m H^{\circ}_{f(products)} - \Sigma n \Delta H^{\circ}_{f(reactants)}$$

Pages 153–154

11. Hess's Law

One of the reasons that enthalpy is such an important thermodynamic variable is that a very large number of heats of reaction, ΔH, can be calculated from a small number of heats of formation, ΔH°_f. This is a consequence of the fact that enthalpy is a state variable. Hess's law states that: If a process occurs in steps, even if those steps are hypothetical, the enthalpy change for the overall process is the sum of the enthalpy changes for each of the individual steps.

Hess's Law

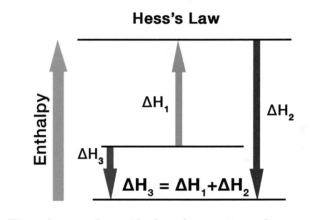

The change in enthalpy for a stepwise process is the sum of the enthalpy changes of the steps.

Pages 154–156

12. Processes That Occur on a pV Surface

There are four thermodynamic processes that take place on a pV diagram that we will use repeatedly. The first is an isochoric process that occurs at a fixed volume as shown in panel (a). Because the area under the pV curve is zero, the work done is also zero. An isobaric process occurs at constant pressure as shown in panel (b) and thus the work is simply $p\Delta V$, the area within the box defined by the boundaries V_i and V_f, (ΔV), and the pressure p. An isothermal process is one that occurs at constant temperature as shown in panel (c). To calculate the work done in this case we must break the progression from V_i to V_f into small segments, calculate the work for each, $p\Delta V$, and then add them to find the total. Finally, the adiabatic process for which $q = 0$ is shown in panel (d).

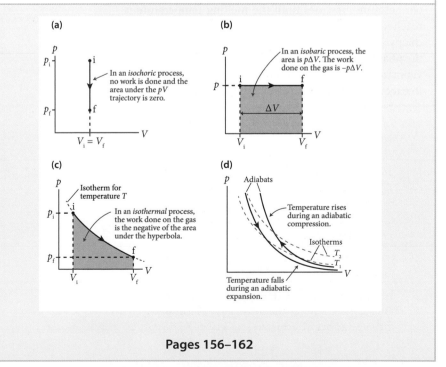

Pages 156–162

Summary Concepts

13. Linking the Thermodynamic Machine, the pV Diagram, and the Energy Bar Chart

A key strategy in mastering thermodynamics is to develop the ability to couple what is occurring in the physical world with what occurs on the plot of pressure vs. volume and to link those two perspectives with the First Law of Thermodynamics using the energy bar chart.

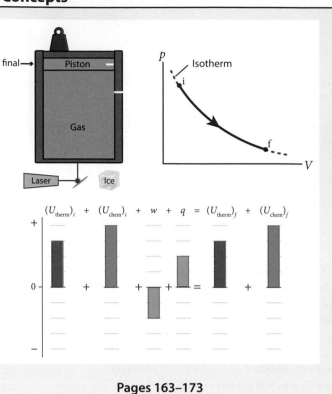

$$(U_{therm})_i \; + \; (U_{chem})_i \; + \; w \; + \; q \; = \; (U_{therm})_f \; + \; (U_{chem})_f$$

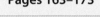

Pages 163–173

14. Thermodynamics of Phase Transitions

Molecules in a liquid or solid are held together by intermolecular attraction. When a solid is heated, its temperature increases until the melting temperature is reached. The *enthalpy of fusion*, ΔH_{fus}, is the molar enthalpy defining the amount of heat required to execute the phase transition from solid to liquid. The enthalpy of vaporization, ΔH_{vap}, defines the amount of heat required to execute the phase transition from liquid to gas.

Pages 173–177

CASE STUDY 3.1 The Carnot Cycle: The Critical Link Between Spontaneity, Reversibility, and Heat Engine Efficiency

KEY CONCEPTS:

Thermodynamics can answer some critically important questions: Why is 80% of the chemical energy contained in gasoline lost to heat and thus unavailable to propel an automobile? Why, even if coal is burned to produce electricity and that electricity is used to power an automobile, is only half as much chemical energy is expended to propel an electric automobile when compared to a gasoline powered automobile? Why is a diesel engine significantly more efficient than a gasoline engine? Why is a "heat pump" remarkably efficient for heating or cooling your home?

The answers to these, and many other questions, emerge from a powerful extension of the first law of thermodynamics. By virtue of early and remarkable insight by a young French physical chemist Nicolas Léonard Sadi Carnot who, through intellectual insight as discussed in the chapter, related the maximum amount of work that could be done by a reciprocating engine to the heat added in each cycle of the engine. That heat added could be supplied by the combustion of gasoline, the addition of steam, or the combustion of coal dust.

For example, the engine in a gasoline powered automobile, as displayed in Figure CS3.1a, engages four steps in a cycle (thus the terminology "4-cycle engine"). These steps are:

1. A compression stroke wherein the mixture of air and gasoline vapor is reduced in volume and thereby increased in pressure by the upward movement of a piston in a cylinder.

2. The power stroke initiated as the piston reaches the top of its stroke at which point the *volume* of fuel vapor and air is a minimum. At that point, an electrical "spark" across the gap of the sparkplug initiates the combustion of the fuel-air mixture. This detonation of the fuel-air releases the chemical energy as heat—a precipitous increase in temperature within the gases contained in the cylinder. This release of heat from the chemical reaction increases the *temperature* of the gases in the cylinder, which produces a requisite increase in *pressure* from the gas law $P = nRT/V$. This precipitous increase in pressure forces the piston downward, applying force through the connecting rod to the crankshaft. The crankshaft rotation supplies the torque to the driveshaft linking the crankshaft rotation through the transmission and differential to the drive wheels of the automobile.

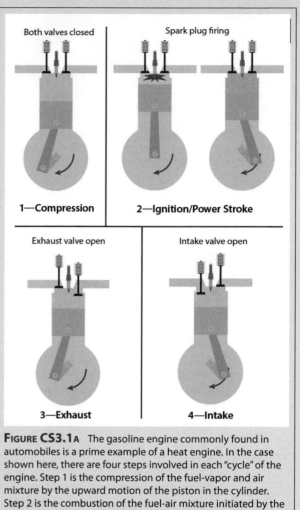

FIGURE CS3.1A The gasoline engine commonly found in automobiles is a prime example of a heat engine. In the case shown here, there are four steps involved in each "cycle" of the engine. Step 1 is the compression of the fuel-vapor and air mixture by the upward motion of the piston in the cylinder. Step 2 is the combustion of the fuel-air mixture initiated by the spark timed to occur when the piston has reached the top of the stroke that results in the power stroke. Step 3 cleans the cylinder of the combustion products. Step 4 draws a fresh charge of fuel-air into the cylinder.

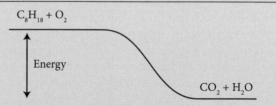

3. At the bottom of the power stroke, a valve at the top of the cylinder opens, allowing the gases that are now composed of N_2, CO_2, H_2O, and any remaining O_2 (along with NO, CO, and other combustion products) to escape into the exhaust manifold that is connected to the exhaust pipe. The upward motion of the piston clears the cylinder of exhaust gases.

4. As the piston reaches the top of its stroke, the exhaust valve closes. As the piston moves downward, the "intake"

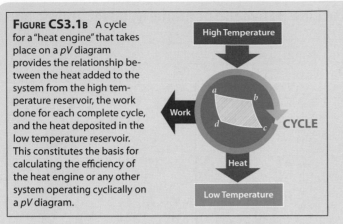

FIGURE CS3.1B A cycle for a "heat engine" that takes place on a *pV* diagram provides the relationship between the heat added to the system from the high temperature reservoir, the work done for each complete cycle, and the heat deposited in the low temperature reservoir. This constitutes the basis for calculating the efficiency of the heat engine or any other system operating cyclically on a *pV* diagram.

valve opens, allowing a fresh mixture of fuel vapor and air to be drawn into the cylinder. As the piston reaches the bottom of its stroke, the intake valve closes and the power stroke is initiated as the piston begins to move up, compressing the fuel-air mixture. This returns us to step 1, above, in the sequence.

It is, therefore this four-step sequence that constitutes a full cycle of a "four cycle" gasoline engine. While this cycle is completed thousands of times per minute as you drive your car, the thermodynamic analysis is applied to a *single* sequence of the four steps that constitute a complete cycle.

Strategically what we are doing is quantitatively analyzing the net work done by the reciprocating engine represented by the *cycle* on the *pV* diagram as shown in Figure CS3.1b.

When we proceed in a clockwise direction around the cycle, work is produced as heat flows from the high temperature reservoir to the low temperature reservoir. The *efficiency* of the cycle is equal to the *ratio* of work produced to heat added, W_{net}/q_{added}, for the cycle where (upper case) W_{net} is the net work done for the cycle and q_{added} is the heat added for the cycle from the high temperature reservoir.

Our objective, therefore, is to calculate the ratio W_{net}/q_{added} where W_{net} represents the net work for the full cycle and q_{added} is the heat added to the system at the high temperature reservoir for the full cycle.

A key "question" of great practical importance is: why is a diesel engine so much more efficient than a gasoline engine? To answer this question, we review how to determine the work done in a cycle on the *pV* diagram.

The Carnot Cycle and the Critical Link Between Spontaneity and Reversibility

A young French engineer by the name of Sadi Carnot published a paper in 1824 with the title "On the Motive Power of Heat." While it was one of a very few papers Carnot published for which we have a record (he died of cholera at age 36 and was buried with many of his manuscripts), it is a hallmark of human intelligence. What Carnot realized was that

all engines (primarily steam engines at the time) are simply mechanical devices that convert heat (from steam or fuel) into work in a repeating cycle. The engine performs by simply repeating the same sequence over and over again. Carnot reasoned that if he could bring logic to a full cycle of the engine, he could discover the theoretical limit to the engine's *efficiency*, ε, which could be expressed as

$$\varepsilon = \frac{\text{work done by the engine in a cycle}}{\text{heat added in a cycle}}$$

To this end, Carnot constructed a closed cycle on the *pV* diagram such that:

1. The cycle returned to its original position on a *pV* diagram.

2. The cycle was constructed of four stages or legs.

3. Each of those stages was carried out *reversibly*, which meant that each leg was comprised of a large number of *small* increments. The reason each stage had to progress reversibly was that, as we will see, this defines the *maximum* efficiency of the cycle for producing work from a given amount of heat, where the heat is supplied by burning fuel.

4. The four reversible stages of the "Carnot cycle" consisted of a pair of isothermal trajectories and a pair of adiabatic trajectories on a *pV* diagram shown in Figure CS3.1c.

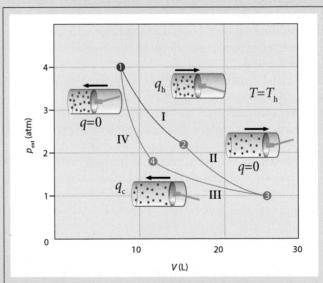

FIGURE CS3.1c The Carnot cycle is comprised of four reversible paths on a *pV* diagram. Path I and III are reversible isothermal legs for which $pV = $ constant. Path II and IV are reversible adiabatic legs for which $pV^\gamma = $ constant with $\gamma = c_p/c_v = 5/3$. Heat q_h is taken up from a high temperature reservoir at temperature T_h during path I. Heat q_c is discharged into the cold reservoir in path III at temperature T_c. The net work done is given by the area enclosed by the cycle on the *pV* plot.

We consider each leg of the closed cycle in order:

Leg I

The first segment is an *isothermal expansion* with the addition of heat q_h at high temperature, T_h. From the First Law of Thermodynamics,

$$\Delta U_I = q_I + w_I = q_h + w_I$$

Because the first segment, I in Figure CS3.1c, is isothermal, it follows that $\Delta U_I = 0$ and thus $q_h = -w_I$. Because heat enters the system, q_h is positive, and the gas expands and does work.

This first isothermal segment in the Carnot cycle appears to be a simple transition on a pV diagram, but in fact, it contains key aspects linking spontaneous change with the concept of reversibility and the concept of equilibrium in a thermodynamic system. The conceptual links develop as follows.

We know that for the case of constant pressure, where we will take that pressure to be the external pressure, p_{ext}, the work is just

$$w = -p_{ext}\Delta V$$

To make this quantitative, we choose the example of one mole of an ideal gas at 273 K with an initial pressure $p_1 = 2.0$ atm and an initial volume $V_1 = 11.2$ ℓ. The gas expands *isothermally* (as our first stage in the Carnot cycle) to a final state at 273 K, $p_2 = 1.0$ atm, and $V_2 = 22.4$ ℓ. Now comes the critical distinction linking spontaneity with irreversibility. *If* $p_{ext} < p$ (where p is the pressure within the cylinder), the gas will expand *spontaneously* to the final state with $p = p_{ext} = 1.0$ atm. We can represent this course of action on a pV diagram (as we did in the sidebar on page 172) displayed in the top panel of Figure CS3.1e on the following page by two legs linking state 1 and state 2. The first leg is the drop in pressure from $p_{ext} = 2.0$ atm to $p_{ext} = 1.0$ atm at the fixed volume of 11.2 ℓ. The next leg is the expansion from 11.2 ℓ to 22.4 ℓ at the fixed pressure of $p_{ext} = 1.0$ atm. This constitutes the trajectory displayed in the upper panel of Figure CS3.1e. The work done *by* the gas, $-p_{ext}\Delta V$, is simply the area under the curve of p_{ext} versus volume with $p_{ext} = 1.0$ atm.

However, we know that the work done by the expanding gas is just equal to the *area* under the curve of p vs. V as delineated in this chapter. This fact leads us to four very important observations:

(a) If we proceed from State 1 to State 2 in small increments, as shown in the *lower panel* in Figure CS3.1e, the *area* under the curve connecting the initial state with the final state *increases* significantly in comparison with the irreversible path. In fact, in the limit of these pressure *steps* approaching zero (limit $\Delta p \to 0$) $\Delta p_{limit} = dp$, the trajectory will approach a smooth curve resulting in the achievement of the *maximum* area under the curve in the pV plot in the lower panel of Figure CS3.1e. This would yield the *maximum* amount of work

by the system given the initial and final states.

(b) If we chose to incrementally increase the pressure, p_{ext}, at any point in this sequence of small pressure changes we could step reversibly back up the trajectory in the lower panel in Figure CS3.1e, ultimately returning to the initial state.

(c) If we precipitously drop the pressure from 2.0 atm to 1.0 atm, following the trajectory in the upper panel of Figure CS3.1e, the system will expand inexorably to the final state. The amount of work, w, done by the system would then be considerably *less* than if we followed the incremental path depicted in the upper trajectory of Figure CS3.1e. The expansion under these conditions would be *spontaneous*, but would not be reversible. We cannot work our way backwards along the lower path, retracing our steps to the initial state following the trajectory in the upper panel of Figure CS3.1e backwards.

(d) We thus have created an important distinction wherein an *irreversible path* is *spontaneous* and a *reversible path* is *nonspontaneous*. In fact, in the limit of very small increases or decreases in p_{ext}, the system has no propensity to change its state and thus it is in equilibrium. We have, therefore, pairs of contrasting conditions that turn out to be very important for understanding thermodynamic systems. Figure CS3.1d displays these closely related concepts by linking opposites across the diagram and equivalents down the diagram.

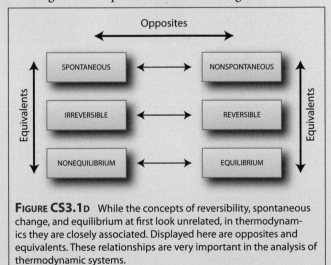

FIGURE CS3.1D While the concepts of reversibility, spontaneous change, and equilibrium at first look unrelated, in thermodynamics they are closely associated. Displayed here are opposites and equivalents. These relationships are very important in the analysis of thermodynamic systems.

On the face of it, the nonspontaneous, reversible, equilibrium case seems to be of little value for any real system. However, as Figure CS3.1e graphically displays, the reversible case is the one for which the *work extracted* is a *maximum*. This is the reason the reversible limiting case is so important—it puts in place *quantitatively* the maximum amount of work that can be produced by the system in this first leg of the cycle.

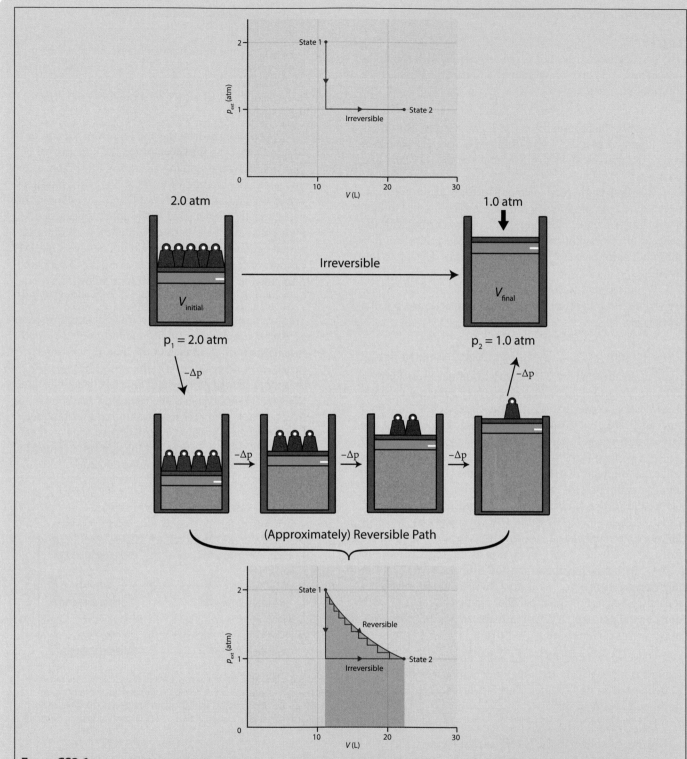

Figure CS3.1E Irreversible and nearly reversible isothermal expansions illustrated by gas in a cylinder pushing a piston against the atmosphere. For the irreversible path, the disk weights, which compress the gas to a pressure greater than atmospheric, are removed all at once, whereas for the nearly reversible path the weights are removed one at a time. The latter path takes longer, but produces more work. A closer approximation to a reversible expansion would be achieved by replacing the weights with a tall beaker of sand, and removing the sand one grain at a time. The expansion could be reversed by replacing a grain of sand. The pV diagram at the bottom of the figure illustrates these two paths. The work done by the system (the gas) is the area under the P versus V curve in each case. The darker blue area gives the extra work gained by running the process reversibly. The stair-step curve shows the approximation to the reversible limit obtained by removing the weights one at a time. To construct the pV diagram, 1 mol of gas at 273 K is assumed to behave ideally. The cylinder would have to be in good thermal contact with an ice bath to maintain constant temperature.

We can calculate the work for the reversible path in Figure CS3.1e just as we did in Figure 3.31:

$$-w_{rev} = \int_{V_1}^{V_2} P dV = \int_{V_1}^{V_2} \frac{nRT}{V} dV = nRT \int_{V_1}^{V_2} dV / V$$

$$= nRT \ln\left(V_2 / V_1\right)$$

This is, in fact, a mathematical expression for the area under the curve in the lower panel of Figure CS3.1e—the quantitative expression for the *reversible* work done in the first leg of the Carnot cycle shown in Figure CS3.1c.

We can repeat the calculation for the *irreversible* path in the upper panel of Figure CS3.1d. It is just

$$-w_{irrev} = p_2 \Delta V = p_2(V_2 - V_1) = nRT(1 - V_1/V_2)$$

With R = 8.31 J/K–mol, and the values of V_1 and V_2 from the previous page we have

$$-w_{rev} = 1570 \text{ joules}$$

$$-w_{irrev} = 1130 \text{ joules}$$

The ratio is thus

$$\frac{w_{rev}}{w_{irrev}} = \frac{1570}{1130} = \frac{\ln(2)}{1/2} = 1.39$$

so the reversible path represents an amount of work 39% greater than the irreversible path. But here is a key point. For a reversible *compression*, all that is involved is a *reversal in sign* because we work incrementally back up the reversible curve and the work done *on* the system in compression is equal in magnitude but opposite in sign to the work done reversibly *by* the system in expansion. The irreversible path is a different matter. If we wish to recompress gas from V_2 = 24.2 liters at constant P_{ext}, we would need 2.0 atm pressure and twice as much work would be done *on* the gas as was done *by* the gas in the expansion step. We can summarize two important conclusions:

1. Reversible expansion *yields* maximum work and reversible compression *requires* minimum work.

2. Irreversible expansion *yields* less than maximum work and irreversible compression *requires* greater than minimum work.

Now we are ready to consider the remaining three legs of the Carnot cycle displayed in Figure CS3.1c.

Leg II

The second leg is an adiabatic expansion, so q = 0. Because of the continued expansion, work is being done, but with q = 0, no heat is added and the internal energy and temperature must decrease:

$$\Delta U_{II} = nc_v \Delta T_{II} = w_{II}$$

where all quantities are *negative*. If the final temperature is T_c, then $\Delta T_{II} = T_h - T_c$, because in Leg I we followed an isothermal trajectory along $T = T_h$.

Leg III

At this point in the Carnot cycle, all the work has been extracted and it is now necessary to find our way back to the initial state so the cycle can begin again from the same initial point in the *pV* diagram. This can be accomplished in two steps, one isothermal and the second adiabatic. Leg III is an *isothermal compression* in which the heat q_c is produced by the compression and deposited in the cold reservoir at temperature T_c. Because the trajectory on the *pV* diagram is isothermal, $\Delta U_{III} = 0$ and $q_c = -w_{III}$ with $q_c < 0$ so heat flows *out* of the system. The amount of work is, from our expression for isothermal work, just the now familiar expression

$$-w_{III} = nRT_c \ln(V_4/V_3).$$

Leg IV

The final leg brings us back to the initial state along an *adiabatic* trajectory so q again is zero and

$$\Delta U_{IV} = nc_v \Delta T_{IV} = w_{IV}$$

where all quantities are positive and

$$\Delta T_{IV} = T_c - T_h$$

What distinguishes the Carnot cycle is that each leg is carried out reversibly between temperatures T_h and T_c. Importantly, this ensures the *maximum* work in expansion and the *minimum* work in compression. This was exactly what Carnot wanted because his objective was to determine the theoretical *upper limit* for the efficiency of the heat engine in converting heat to work.

That maximum theoretical efficiency, ε, is then simply the net work extracted for the given amount of heat added, q_h, so

$$\varepsilon = \frac{-\left(w_I + w_{II} + w_{III} + w_{IV}\right)}{q_h}$$

First we note that the work terms, $nc_v \Delta T$ from our analysis of Leg II and Leg IV, exactly cancel so

$$\varepsilon = \frac{-\left(w_I + w_{III}\right)}{q_h}$$

But also for the isothermal stages legs I and III that $-w = q$ so

$$\varepsilon = \frac{q_h + q_c}{q_h}$$

Gathering the terms for each of the four legs:

$$w_I = -nRT_h \ln(V_2/V_1)$$

$$w_{II} = nc_v (T_h - T_c)$$

$$w_{III} = -nRT_c \ln(V_4/V_3)$$

$$w_{IV} = nc_v(T_c - T_h)$$

We have

$$\varepsilon = \frac{nRT_h \ln\left(V_2/V_1\right) + nRT_c \ln\left(V_4/V_3\right)}{nRT_h \ln\left(V_2/V_1\right)}$$

Again because w_{II} and w_{IV} are equal but opposite in sign, they cancel. But Leg II allows us to simplify this expression for the efficiency, ε, because

$$dU = dq + dw \quad \text{and if} \quad dq = 0$$
$$dU = dw$$

But $dU = nc_v dT$ and

$$dw = -pdV = -\frac{nRT}{V}dV$$

so with $nc_v\, dT = -nRT\, dV/V$, we have

$$nc_v \int_{T_h}^{T_c} \frac{dT}{T} = -nR \int_{V_2}^{V_3} \frac{dV}{V}$$

and

$$nc_v \ln(T_c/T_h) = -nR \ln(V_3/V_2)$$

Leg IV yields the same result with the temperature limits reversed and the volume ratio V_1/V_4 such that

$$nc_v \ln(T_h/T_c) = -nR \ln(V_1/V_4)$$

Because $\ln(x_2/x_1) = -\ln(x_1/x_2)$ we have

$$nc_v \ln\left(T_c/T_h\right) = -nc_v \ln\left(T_h/T_c\right) = nR\ln\left(V_3/V_2\right)$$
$$= -nR\ln\left(V_1/V_4\right)$$

or, with cancellation $V_3/V_2 = V_4/V_1$ or $V_2/V_1 = V_3/V_4$ and

$$\varepsilon = \frac{nRT_h \ln\left(V_2/V_1\right) + nRT_c \ln\left(V_4/V_3\right)}{nRT_h \ln\left(V_2/V_1\right)}$$
$$= \frac{T_h \ln\left(V_2/V_1\right) + T_c \ln\left(V_4/V_3\right)}{T_h \ln\left(V_2/V_1\right)}$$
$$= \frac{T_h - T_c}{T_h}$$

Thus our initially complex expression for the maximum theoretical efficiency of a heat engine collapses sequentially through a series of cancellations. We review the sequence as follows:

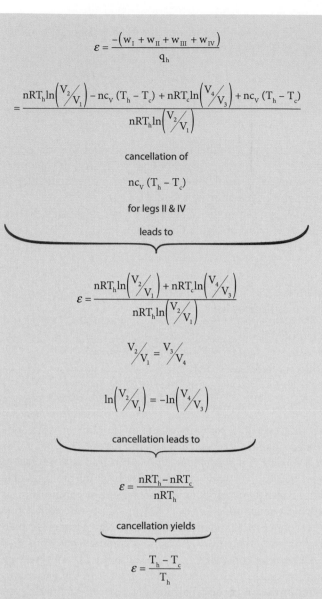

$$\varepsilon = \frac{-\left(w_I + w_{II} + w_{III} + w_{IV}\right)}{q_h}$$

$$= \frac{nRT_h\ln\left(V_2/V_1\right) - nc_v(T_h - T_c) + nRT_c\ln\left(V_4/V_3\right) + nc_v(T_h - T_c)}{nRT_h\ln\left(V_2/V_1\right)}$$

cancellation of

$$nc_v(T_h - T_c)$$

for legs II & IV

leads to

$$\varepsilon = \frac{nRT_h\ln\left(V_2/V_1\right) + nRT_c\ln\left(V_4/V_3\right)}{nRT_h\ln\left(V_2/V_1\right)}$$

$$V_2/V_1 = V_3/V_4$$

$$\ln\left(V_2/V_1\right) = -\ln\left(V_4/V_3\right)$$

cancellation leads to

$$\varepsilon = \frac{nRT_h - nRT_c}{nRT_h}$$

cancellation yields

$$\varepsilon = \frac{T_h - T_c}{T_h}$$

The conclusion is remarkable and strikingly simple. It says, quantitatively, that the efficiency of the Carnot engine depends only on the temperatures of the hot and cold reservoirs and not on any of the details of the heat engine itself. Because the Carnot cycle is based upon reversible pathways it represents the *maximum* efficiency possible and thus a heat engine cannot be 100% efficient as long as T_h and T_c are finite.

We can now, using this expression for the maximum efficiency of a heat engine in terms of the hot and cold reservoir temperatures, T_h and T_c respectively, summarize the efficiency of the Carnot cycle in terms of either temperature or the heat added using equation 3.3 so

$$\varepsilon_{Carnot} = \frac{T_h - T_c}{T_h} = \frac{q_h + q_c}{q_h} \tag{3.3}$$

As we will see in the next chapter, this link between heat and temperature in the Carnot cycle and the related relationships between reversibility, spontaneity, and equilibrium leads to the quantitative foundation for the new state variable, *entropy*.

Problem 1

If an automobile of mass 1200 kg employs a gasoline (C_8H_{18}) engine that we model here as a heat engine for which the Carnot Cycle is applicable, calculate the height in meters that can be obtained if the automobile is ascending a slope and has 1 gallon of fuel available. Assume the exhaust temperature at the exhaust valve is 760°C and the engine cylinder temperature at the top of the stroke is 2200°C.

Problem 2

A Carnot engine operates between two temperature reservoirs maintained at 200°C and 20°C, respectively. If the desired output of the engine is 15 kW, determine (i) the heat transferred per second from the high-temperature reservoir and (ii) the heat transferred per second to the low-temperature reservoir.

CASE STUDY 3.2 Thermodynamics, Heat Pumps, and the Personal Energy Budget

KEY CONCEPTS:

Heat Pumps

Residing in most kitchens is a device that moves heat from one place to another, decreasing the temperature in one place and increasing it in another. It's the refrigerator, and its thermodynamic cycle is shown in Figure CS3.2a. Heat is transferred from a cold reservoir at temperature T_C by virtue of the input of work, W_{in}, expended to move that heat to the high temperature reservoir at temperature T_H. If you pull the refrigerator away from the wall, you will notice that the back of the refrigerator is quite warm relative to the surrounding air and dramatically warmer than the air in the refrigerator or freezer. The fact that the refrigerator is plugged into the wall and that the compressor can be heard operating immediately suggests that *work* (mechanical energy) is being expended to move heat *from* the interior of the refrigerator *into* the surrounding room.

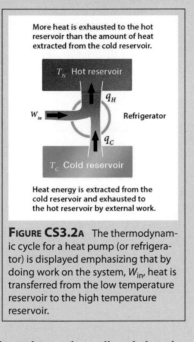

More heat is exhausted to the hot reservoir than the amount of heat extracted from the cold reservoir.

Heat energy is extracted from the cold reservoir and exhausted to the hot reservoir by external work.

FIGURE CS3.2A The thermodynamic cycle for a heat pump (or refrigerator) is displayed emphasizing that by doing work on the system, W_{in}, heat is transferred from the low temperature reservoir to the high temperature reservoir.

The first question is, what is the efficiency of "moving heat around?" Herein lies the power of the Carnot Cycle! We invested all the effort in Chapter 3 and Case Study 3.1 to answer this question for a "heat engine" whose purpose is to take heat at high temperature to do work and then exhaust the remaining heat to a low temperature reservoir. Are things different for a "cold engine" that moves heat from a cold reservoir to a warm one? An inspection of the diagram for a heat engine when compared to the diagram for a heat pump (Figure CS3.2b) reveals that they are virtually identical. The primary difference is that the heat engine, moving *clockwise* around the Carnot cycle, takes in heat at high temperature, produces work, and expels heat to a low temperature reservoir. If the cycle runs backwards, it requires work but, by virtue of that work, heat is transferred from a low temperature reservoir to a high temperature reservoir.

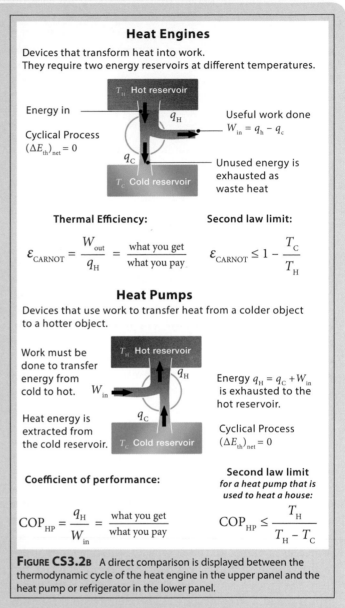

Heat Engines

Devices that transform heat into work.
They require two energy reservoirs at different temperatures.

Energy in

Cyclical Process $(\Delta E_{th})_{net} = 0$

Useful work done $W_{in} = q_h - q_c$

Unused energy is exhausted as waste heat

Thermal Efficiency:

$$\varepsilon_{CARNOT} = \frac{W_{out}}{q_H} = \frac{\text{what you get}}{\text{what you pay}}$$

Second law limit:

$$\varepsilon_{CARNOT} \leq 1 - \frac{T_C}{T_H}$$

Heat Pumps

Devices that use work to transfer heat from a colder object to a hotter object.

Work must be done to transfer energy from cold to hot.

Heat energy is extracted from the cold reservoir.

Energy $q_H = q_C + W_{in}$ is exhausted to the hot reservoir.

Cyclical Process $(\Delta E_{th})_{net} = 0$

Coefficient of performance:

$$COP_{HP} = \frac{q_H}{W_{in}} = \frac{\text{what you get}}{\text{what you pay}}$$

Second law limit:
for a heat pump that is used to heat a house:

$$COP_{HP} \leq \frac{T_H}{T_H - T_C}$$

FIGURE CS3.2B A direct comparison is displayed between the thermodynamic cycle of the heat engine in the upper panel and the heat pump or refrigerator in the lower panel.

So what is the theoretical efficiency of such a "heat pump" that moves heat from a low temperature reservoir (e.g. the outside at temperature T_C) to a high temperature reservoir (e.g. the inside of the house at temperature T_H)? We can determine this, as is done in the sidebar, by a minor modification of our full treatment in Case Study 3.1, and write the efficiency of the heat pump, ε_{HP}, as

$$\varepsilon_{HP} = \frac{T_H}{T_H - T_C}$$

where T_C is the temperature of the cold reservoir, the outside temperature, and T_H is the high temperature reservoir, the inside of the house.

To see what this means, let's assume the outside temperature is at the freezing point, 0°C. But T_1 must be the temperature absolute, in Kelvin, so T_C = 273 K. If we wish the inside temperature to be 20°C, then T_H = 293 K. Thus our *theoretical maximum coefficient of performance* (COP, as it is commonly referred to for heat pumps), for simply moving heat from reservoir 1 (outside) at temperature T_1 (273 K) to reservoir 2 (inside the house) at temperature T_2, (293 K) is

$$\text{COP}_{HP} = \frac{T_2}{T_2 - T_1} = \frac{293 \text{ K}}{20 \text{ K}} \approx 15$$

This is a remarkable result, because it says that we can move 15 times as much thermal energy (i.e. heat) as the amount of work expended (i.e. electricity). The "Carnot efficiency" is the theoretical maximum—*in practice* the actual efficiency that can be achieved by a heat pump is between 4 and 5 depending on the sophistication of the equipment. But this, it turns out, is very important. It says that even at the lower end of the efficiency range, for every kilowatt of electrical power delivered to the heat pump, *four to five* kilowatts of heat are delivered into the house. In fact, a run-of-the-mill heat pump will deliver 4 kWh of heat to the house for every kWh of electrical power delivered to the heat pump. State-of-the-art systems will deliver 5 kWh of heat for every 1 kWh of electrical energy. To emphasize how important this is, consider the following example. If we use electricity to power a space heater, we will put 1 kWh of heat into the house for each kWh of electricity we purchase. This is an efficiency of 100%. If we use a heat pump to do the same thing, from a standard issue heat pump 4 kWh of heat enter the house for each kWh of power put into the heat pump—an efficiency of 400%!

There are typically two types of heat pumps. One variety uses a heat exchanger operating in air such that the outside air is cooled (heat is extracted) by transferring heat from the atmosphere to the inside of the house. This is called an "air source" heat pump. The other type uses cooling loops buried in the ground, cooling the ground outside by drawing heat from the ground and moving it to the inside of the house. This is called a "ground source" heat pump with the outside unit consisting of a box that sits outside the house and the heating/cooling unit that is inside the house. These devices can operate in either direction: heating in winter and cooling in summer.

We are now equipped with the information needed to calculate the impact of the heat pump on energy efficiency for the heating of buildings and homes. The first step is to track the implications of using a heat pump to warm houses, buildings, factories, etc. by comparing the heat delivered relative to the energy invested. Let's assume we begin with 100 units of energy delivered by the combustion of natural gas for two examples:

Example 1:

A condensing boiler that is 90% efficient. We can diagram this in a single step as displayed in Figure CS3.2c.

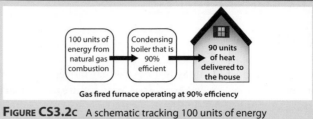
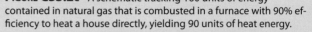

Gas fired furnace operating at 90% efficiency

FIGURE CS3.2c A schematic tracking 100 units of energy contained in natural gas that is combusted in a furnace with 90% efficiency to heat a house directly, yielding 90 units of heat energy.

The result is that of the 100 units of energy invested, 90 units of heat are delivered to the interior of the house or building.

Example 2:

The generation of electricity in a modern electrical power generating plant fueled by natural gas is approximately 50% of the chemical energy contained in the natural gas (CH_4). If we assume a transmission loss of 5% in the power delivery grid, this results in (0.95)(0.50) 100 units = 47.5 units of electrical energy delivered to the house or building. A diagram of the sequence from the combustion of natural gas to the electrical energy delivered to the house is displayed in Figure CS3.2d. A coefficient of performance by the heat pump of 4 is assumed.

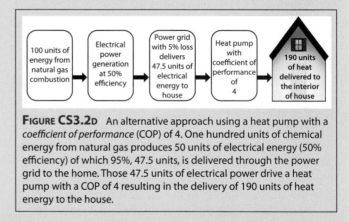

FIGURE CS3.2d An alternative approach using a heat pump with a *coefficient of performance* (COP) of 4. One hundred units of chemical energy from natural gas produces 50 units of electrical energy (50% efficiency) of which 95%, 47.5 units, is delivered through the power grid to the home. Those 47.5 units of electrical power drive a heat pump with a COP of 4 resulting in the delivery of 190 units of heat energy to the house.

The comparison of these two examples displayed in Figures CS3.2c and Figure CS3.2d is of great importance to any energy strategy. Notice that while the heat pump is supplied with electrical power *after* suffering a loss of 52% of the chemical energy contained in the natural gas combusted to drive the electrical power generation, the system still delivered more than *twice* the amount of heat to the building than the heat delivered by the direct combustion of natural gas in a furnace of 90% efficiency within the building. While this may be quite counter intuitive, it is a direct result of the laws of thermodynamics. But also note that the heat pump is driven by electrical energy. If that electrical energy is derived from solar or wind energy, the total energy required decreases by ×2.

To demonstrate the thermodynamic principles of the heat pump, we will examine quantitatively what happens when we use the heat engine run backwards such that work is done *on* the system to transfer heat from a cold reservoir to a hot reservoir (rather than the Carnot cycle wherein heat from a high temperature reservoir is used to *produce* work and exhaust the remaining heat to a low temperature reservoir). In order to examine the sequence of events in progressing counterclockwise around the thermodynamic cycle, we use three tightly linked diagrams:

- A schematic of the mechanical devices that actually executes the conversion of work to move the heat *from* a low temperature reservoir to a high temperature reservoir;
- The *pV* diagram that tracks each leg of the thermodynamic cycle; and
- The diagram tracking the flow of energy, both heat and work, that occurs in a given cycle of the system.

These three diagrams are displayed in Figure CS3.2e.

Just as the case of the Carnot cycle, we take each leg of the thermodynamic cycle and analyze it sequentially. Beginning with point 4 in diagram CS3.2e(b) we progress counterclockwise to point 3 along an adiabatic compression leg that decreases the volume and increases the pressure. It requires work to do this, work that is done by the compressor shown in Figure CS3.2e(a). It is this segment of the cycle that receives the work expended, W_{in}, and it is this segment for which you pay the electrical power company for the electricity to drive the electric motor that turns the shaft on the compressor.

But here is the first key point. The heat pump can extract heat from the cold reservoir (the outside of the house) *if and only if* the gas temperature of the low temperature heat exchanger is at a temperature *lower* than the outside temperature. The reason, of course, is that heat flows only from a high temperature reservoir to a low temperature reservoir. Thus this heat pump must achieve a temperature in its low temperature heat exchanger that is *lower* than the outside temperature. We will see how that is achieved as we progress around the cycle. So as we progress up the adiabatic compression segment from point 4 to point 3, the volume decreases, the pressure increases and the *temperature* increases. This high temperature gas is then passed through a high temperature heat exchanger where the gas cools in an isobaric (constant pressure) segment of the thermodynamic cycle, transferring heat to the high temperature reservoir—the house we are interested in heating. But here is the second key point. The heat pump can transfer heat q_H to the house if and only if the temperature along segment $3 \rightarrow 2$ is *higher* than the temperature of the house interior. The reason again is that heat only flows from high temperature to low temperature.

So here is the point. The heat pump must establish a low temperature in its low temperature heat exchanger that is colder, at a *lower* temperature, than the low temperature reservoir (the outside of the house) such that heat is induced to flow *into* the low temperature heat exchanger. Then the heat pump must do work *on* the gas to raise its temperature to a temperature *greater* than the high temperature reservoir, the interior of the house, so as to induce heat to flow *into* the house.

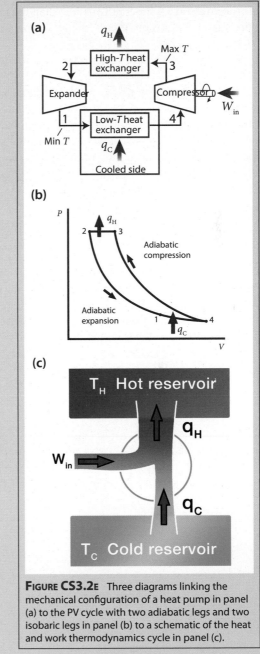

FIGURE CS3.2E Three diagrams linking the mechanical configuration of a heat pump in panel (a) to the PV cycle with two adiabatic legs and two isobaric legs in panel (b) to a schematic of the heat and work thermodynamics cycle in panel (c).

So, following the isobaric segment from 3 to 2, the gas cools as heat flows into the house. At point 2, the gas is allowed to *expand* adiabatically as the pressure drops and the gas expands. It is allowed to expand such that its temperature at part 1 is considerably colder than when it began at point 4. As the final segment is completed from point 1 to point 4, the gas warms isobarically as heat flows from the cold temperature reservoir *into* the gas from outside the house as it returns to the starting point at 4.

The requirement that a heat pump can only extract heat $q_{OUTSIDE}$ at a low temperature and deliver q_{INSIDE} at a higher temperature by establishing a reservoir at a temperature *less* than the outside temperature and then provide a temperature inside the house at a temperature *greater* than the inside temperature places important thermodynamic constraints on the system.

In order to work this heat pump cycle quantitatively, consider the following.

Returning to our opening discussion of the common household refrigerator, we note that the refrigerator operates the same way a heat pump that heats our home operates. The only difference is that when calculating the coefficient of performance (COP) for a refrigerator, recall that

$$COP = \frac{\text{what you get}}{\text{what you pay}}$$

so for the case of a refrigerator, "what you get" is the heat *extracted* from the low temperature reservoir, q_C. Thus, for the refrigerator

$$(COP)_{REF} = \frac{T_C}{T_H - T_C}$$

whereas for the heat pump that warms our home, "what you get" is the heat delivered to the house, q_H, so

$$(COP)_{HP} = \frac{T_H}{T_H - T_C}$$

Problem 1

Given that it takes about 1.0×10^3 kWh per month for heating a one-bedroom apartment in winter, consider the following three heating options, and for each option, calculate (1) the money one needs to pay for the heating and (2) the total amount of energy expended in terms of natural gas combustion. Note that the electricity below is from a modern electrical power generating plant fueled by natural gas with 50% efficiency, and we assume a transmission loss of 5% in the power delivery.

a. A condensing boiler: 90% efficiency; the natural gas price is $0.04 / kWh.

b. An electric space heater: 100% efficiency; the electricity price is $0.11 / kWh.

c. A heat pump: the coefficient of performance is 4; the electricity price is $0.11 / kWh.

Please show your calculation process and put the results in the following table:

	Price/ month	Total energy consumption/month
a) condensing boiler		
b) space heater		
c) heat pump		

3.66

CASE STUDY 3.3 High Temperature Geothermal Energy

KEY CONCEPTS:

High temperature geothermal energy refers to the extraction of heat contained within the materials of the Earth's crust. This thermal energy is supplied by (1) natural radioactive decay of uranium, potassium, and thorium and (2) the outward flow of heat from the Earth's core-mantle system that resulted from the accretion process during the formation of the planet. The dominant heat source, however, is radioactive decay. Figure CS3.3a displays the Earth's thermal structure and the scale of the methods which vary from deep high-temperature systems for direct steam extraction for electricity generation, to the injection of water into high temperature rocks, to intermediate temperature for direct heating of houses and offices, to the use of shallow thermal reservoir for heat pump systems.

While geothermal energy was long viewed as a fringe technique restricted to very specific regions characterized by local volcanic or hot spring sources—termed hydrothermal sources—research in recent years has revealed the remarkable potential of geothermal sources capable of supplying a very significant fraction of the energy demands of the US. High temperature geothermal energy has some very powerful advantages over many other competing sources of primary energy:

1. It can be designed to produce very low emission of infrared active compounds that force the increasing trapping of heat in the climate system.
2. Very large amounts of energy can be extracted with very limited impact on the local environment.
3. The supply of energy is reliable, continuous, and capable of a very high fraction of full-capacity production in combination with
4. The ability to provide electric power generation from high temperature sources but with additional capability of providing active heating of buildings and houses with lower temperature sources.
5. When used in conjunction with heat pumps, low temperature (0–50°C) sources can dramatically reduce heating/cooling costs for buildings and homes as described in detail in Case Study 3.2.

The engagement of geothermal energy generation has already been employed as major national energy sources in the Philippines, Ireland, and El Salvador. The United States, which is richly endowed with geothermal source regions, is currently the world leader in delivered geothermal energy capacity of 3 GW with 80% of that capacity in California alone.

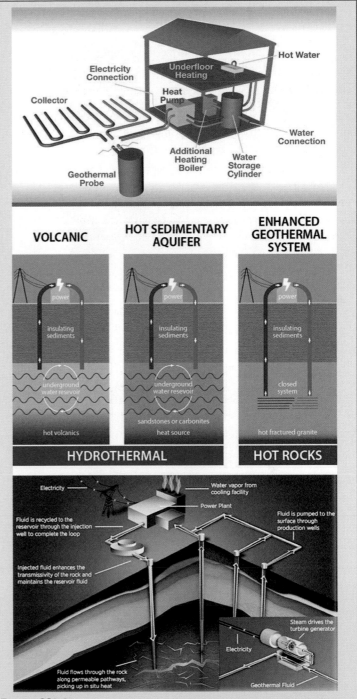

FIGURE CS3.3a Geothermal energy is a term that refers to the extraction of thermal energy from the Earth from a number of different depths and a number of different temperatures and energy sources. The upper panel displays a system for extracting heat from shallow depths that is used in conjunction with a heat pump for heating or cooling a single house or building. The center panel displays an array of heat sources from hydrothermal to enhanced geothermal that injects water into a deep reservoir of high temperature rock. The bottom panel displays a modern system for implementing enhanced geothermal systems (EGS) as described in the text.

I. Scale of Global Geothermal Resource

If we limit our analysis to the upper 10 km of the Earth's crust (a depth that engages the range of currently practical drilling depths) that shell curtains 50,000 times more energy than all the oil and gas reserves in the world. This is an important number.

We can graphically summarize the temperature structure as a function of depth down to 10 km as shown in Figure CS3.3b. The regions with the highest underground temperatures are in areas with active or geologically young volcanoes. In the US those regions occur primarily in the Western half, but at greater depths the zones of high temperature become more broadly dispersed geographically.

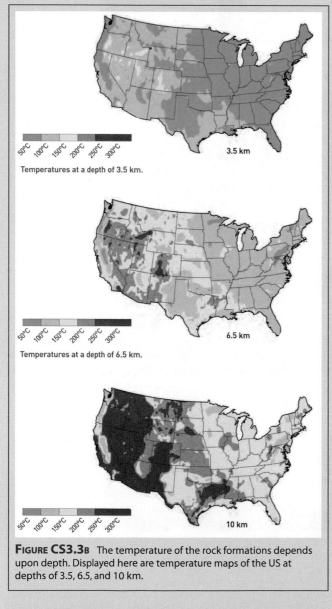

Temperatures at a depth of 3.5 km.

Temperatures at a depth of 6.5 km.

FIGURE CS3.3B The temperature of the rock formations depends upon depth. Displayed here are temperature maps of the US at depths of 3.5, 6.5, and 10 km.

There are three primary categories of geothermal energy:

1. *Hydrothermal*, which refers to the direct availability of very high temperature water or steam (depending on the pressure containment).
2. *Dry rock* high temperature zones that require the introduction of water and extraction of high-temperature water and/or steam.
3. Moderate temperature regions, often closer to the surface, that are useful for direct-heating purposes at depths of 10 to a few hundred feet below the surface.
4. Ambient temperature systems that, while appearing of little value for heating, actually still contain vast amounts of thermal energy because they are, even at the freezing point of water, still at 273 K and thus a very modest amount of work is required to raise the temperature to 20°C or 293 K as discussed in Case Study 3.2.

While category (1) above (hydrothermal) is the most obvious candidate for exploitation as a source of electricity power generation, by far the greatest potential capacity resides within the domain of category (2) which is the "heat mining" technique termed Enhanced Geothermal Systems (EGS) that uses the injection of water into hot, porous rock with the re-capture/extraction of that water in a closed cycle. This technology makes available >500 GW of electricity generating capacity within the US alone in the coming few decades. Thus EGS constitutes a capacity of greater than half the electricity generation demand of the U.S. But more than that, as renewable energy (solar, wind and high temperature geothermal) replaces the low efficiency primary energy from fossil fuels, 500 GW of energy from high temperature geothermal can produce nearly half the *total* energy demmand of the US.

II. Methods of Geothermal Energy Capture

1. Hydrothermal Convection.

As noted in the introduction, the most obvious and easily exploited form of geothermal results when water seeps into the Earth's crust, creating pockets of high-temperature water that can be accessed by directly drilling and extracting steam to drive a conventional electricity generator's turbine system as shown in the left panel of Figure CS3.3c.

There are three designs for those direct extraction power plants. In its simplest design, displayed in the left panel of Figure CS3.3c, the steam is fed directly to the turbine system then into a condenser that collects the water and recycles it to the reservoir from which it was extracted. In a second design, hot water in the liquid phase under high pressure is "flashed" into steam by releasing that pressure just prior to passing the steam into a turbine. This is displayed in the middle panel. A third approach is one step more complicated—it is the "binary" or two-stage approach wherein the hot water extracted under high pressure is passed through a heat exchanger where a second liquid is heated. The second liquid is usually selected to have a lower boiling temperature than water and that second liquid vaporizes to steam that then drives the electric power-generating turbine.

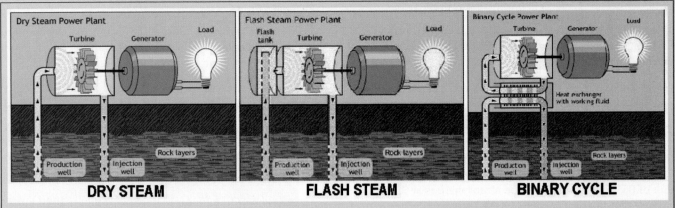

FIGURE CS3.3c There are typically three different methods for driving a steam turbine using hydrothermal convection. The first and simplest, displayed in the left-hand panel, injects the dry steam from the underground source directly into the steam turbine generator. The middle panel displays the system that carries high-pressure steam from the production well and releases it into a chamber at lower pressure, converting super-heated water to steam. The panel on the right-hand side displays the arrangement using a second fluid with a lower boiling point and a heat exchanger prior to injection into the turbine.

The selection of the design depends upon the characteristics of the hydrothermal source. If the water comes from the geothermal well directly as steam it can be used in the simplest design shown in the left panel of Figure CS3.3c. If the geothermal well produces water at high enough pressure and temperature it can be "flashed" for use in the system displayed in the middle panel. If the water temperature is not high enough to use as high-pressure water vapor, the heat exchanger can be used to properly tailor the vapor pressure of a selected second substance as shown in the right panel.

The largest hydrothermal systems in the US are found in northern California at the "Geysers" facility with a net delivery capacity of 725 MW that is similar in output to a nuclear reactor or to 300 2.5 MW wind turbines. That facility, shown in Figure CS3.3d, meets nearly 60 percent of the average electricity demands for the California north coast region that extends from the Golden Gate Bridge to the Oregon border.

FIGURE CS3.3d A large hydrothermal system that powers a significant fraction of homes along the northern California coast is the "Geysers" facility.

2. "Direct Use" Geothermal Heat.

Hot spring water is used to deliver heat directly to buildings, homes, greenhouses, fruit drying, fish drying, fish farms, and spas in Oregon, Idaho, Virginia, and Georgia. This technique constitutes a major heat source in Reykjavik, Iceland where a population of 115,000 is supplied with heat from hot water piped in from 25 km away. Iceland now gets > 50% of its *total* primary energy from geothermal sources.

3. Ground Source Heat Pumps.

The thermodynamics of heat pumps is analyzed in detail in Case Study 3.2 and that analysis reveals the remarkable leverage afforded by using electricity to drive a compressor system that extracts heat at one (lower) temperature does a limited amount of work and expels heat at a higher temperature. Modern systems can produce 500 kWh of heat energy for every 100 kWh of electrical energy used. More than 600,000 ground source heat pumps supply both heating and cooling to US homes and the number is rising quickly because payback time for system installation has decreased to less than five years in most cases, a 20% annual return on investment!

4. Enhanced Geothermal Systems.

Given the many inherent advantages of geothermal energy for the US, the question becomes one of potential capacity, and this is where the rapidly emerging technology of capturing heat in dry areas that possess high temperature porous rock becomes critically important. This "heat mining," typically referred to as Enhanced Geothermal Systems (EGS), thus becomes the primary technology capable of expanding geothermal sources such that it becomes a major player in the primary energy generation categories in the US. It is the subject of EGS to which we now turn.

Problem 1

Where do the resources for geothermal energy exist in the US? What is the potential for electricity generation from geothermal energy in the US?

Problem 2

A power utility company desires to use the hot groundwater from a hot spring to power a heat engine. If the groundwater is at 95 °C, estimate the maximum power output if a mass flux of 0.2 kg/s is possible. The atmosphere is at 20 °C. Assuming that the heat capacity of the hot groundwater is 75.4 J/mol °C, and that all the heat released from the hot spring when it cools from 95 °C to atmosphere temperature is transferred to the heat engine.

Problem 3: Questions around Harvard

a. There are revolving doors at the entrance of the Harvard Science Center. What is the heat transfer (in terms of joules) every time a swing door is used?

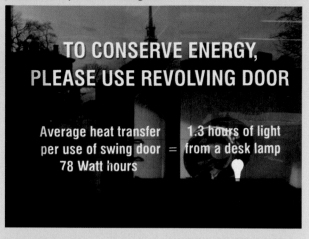

b. At Harvard, replacing the current heating systems with heat pumps is one of the ongoing Green Initiatives. In the Harvard campus map "Sustainability" layer shown below, several locations (e.g., Quad Athletic Center) are using ground source heat pumps. If a heat pump (with a coefficient of performance = 4) were used to heat the Science Center, how much electrical energy (in terms of joules) would we need to provide the pump in order to compensate for the amount of heat lost per use of the swing door?

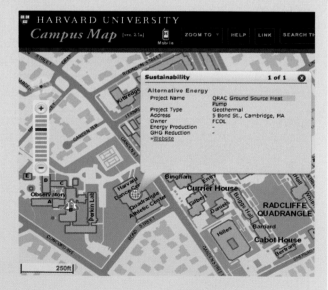

3.70

CASE STUDY 3.4   Linking Global Scale Calculations: The "50 Questions in Global Scale Energy and Power" Part I.

There are quantitative relationships involving global energy and power calculations that are essential for wise stewardship of national security, international relations, and informed public policy decisions in a modern democracy. The most important of these constitute the "50 Questions" segment of the course that we will explore as the text unfolds. Revisiting these key questions will be a recurring theme of these Case Studies.

We begin with the first 10 of these question categories, which serve as a review of the material covered in the first three chapters of the text. It is important to work out each of the calculations and to begin examining each calculation in the context of global energy and power.

Category 1—Power and energy from the Sun

What is the power output of the Sun in watts? How much energy does the Sun produce per year in joules? In kWh? What fraction of that energy is intercepted by the Earth? Thus, how many kWh of energy fall on the Earth in a year?

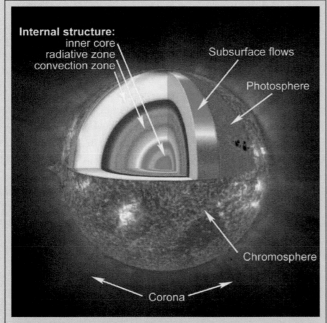

FIGURE CS3.4A While the Sun is a complicated system that converts the release of nuclear energy from the fusion of hydrogen to produce helium and then transports that energy to the Sun's surface, we can easily calculate the power produced from the Sun just by knowing its surface temperature.

Category 2—Power and energy from entering the climate system

How many kWh are absorbed into the Earth's climate system from the Sun each year? How many kWh of energy circulate between (a) the Earth's surface and (b) the clouds, water, and carbon dioxide in the atmosphere? How do you calculate this quantity?

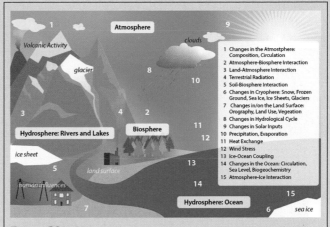

FIGURE CS3.4B The climate system consists of all physical, chemical, and biological subsystems of the terrestrial and ocean structures.

Category 3—Global power and energy consumption in the world economy

What was the global energy demand in 2018 in kWh per year? How is it calculated? What is the corresponding power consumption in watts? What will the approximate global energy demand be in 2050 in joules? In kWh? How is it calculated? What will the approximate global power demand in watts be in 2050 and how is it calculated? What is the ratio of energy received from the sun in a year to the energy consumed by the global economy in the same period?

FIGURE CS3.4C While the global energy consumption is difficult to calculate on a system-by-system basis, it has been determined the global energy consumption depends primarily on just these quantities: population, per capita income, and the amount of energy required for each dollar of gross domestic product.

Category 4—Increase in global energy demand expressed in terms of fossil fuel burning power plants

If the increase in energy demand between now and 2050 were to be supplied by the construction of coal burning power plants (~500MW each), how many of those plants would have to be constructed per week between now and 2050? How many 1 GW nuclear plants would be required?

Figure CS3.4D Nuclear power plants typically produce between 1 and 1.5 gigawatts of power.

Category 5—Energy per year to melt the Arctic Ice Cap

What fraction of the permanent ice in the Arctic Ice Cap has been lost in the past 30 years? How much energy is required each year to melt the Arctic Ice Cap at its current rate of disappearance?

What is the ratio of (a) the energy per year required to melt the Arctic Ice Cap to (b) the energy circulating between the Earth's surface and the clouds, water vapor, and CO_2 in the atmosphere? What is the ratio of (a) global energy consumption by the world economies to (b) energy required per year to melt the Arctic Ice Cap?

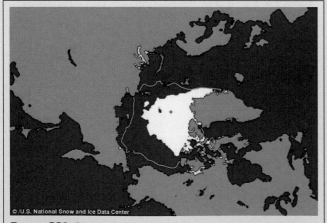

Figure CS3.4E The energy to melt the Arctic Ice Cap is surprisingly small.

Category 6—Petroleum imports to the US

What are the five leading nations from which we import petroleum? What percentage of US oil consumption is imported? Is that fraction increasing or decreasing in 2019? How many barrels of oil does the US import each year? At $100/bbl, how much does this add to our trade deficit each year? What fraction of our trade deficit is this in 2019? If a tax of $20/bbl were placed on US oil imports, how much tax revenue would that raise?

Figure CS3.4F A major contributor to the balance of payments deficit in the US results directly from the purchase of petroleum outside our borders.

Category 7—Comparison of energy magnitudes

What is the ratio of the energy contained in 100 tons of coal to the energy contained in the energy generated by the Grand Coulee Dam in a year? What is the ratio of the energy contained in the fuel of a fully fueled jet liner to the kinetic energy of the airliner in flight? What is the ratio of the energy contained in the gas tank of an automobile to the kinetic energy of the car at 100 km/hr? What is the power output of the first stage of the Saturn V rocket that launched men to the moon?

Figure CS3.4G It is important to compare the ratio of major items in the scales of energy and power.

Category 8—Personal energy budget

On average in the US, how much energy per person per day is expended to drive automobiles? To heat homes? To purchase "stuff"? How many kWh of energy per day, averaged over the year, did you consume in flying in the last year? What is the ratio of (a) energy expended by the average US citizen to purchase "stuff" to (b) energy expended to heat/cool their home?

What is the ratio of (a) the energy consumption per person per day to build a house to (b) energy consumed to deliver the newspaper and junk mail to the same house? What is the energy required per person per day for a standard US diet vs. a vegan diet?

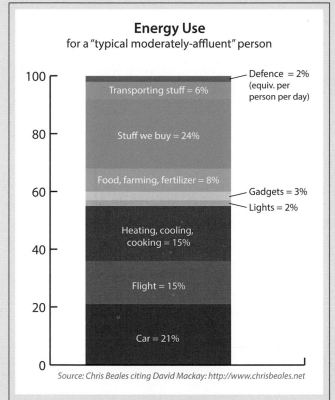
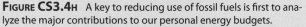

Figure CS3.4H A key to reducing use of fossil fuels is first to analyze the major contributions to our personal energy budgets.

Category 9—Origin of fossil fuels

How were the deposits of coal formed? How were the deposits of petroleum formed? If we burn all known fossil fuel reserves, by what fraction will the oxygen level of the atmosphere decrease? Why is the energy content per kg of CH_4 higher than that of coal? Why is the carbon dioxide emission from natural gas dramatically less per unit of energy produced than from coal?

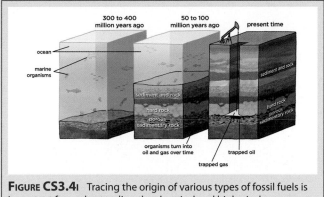

Figure CS3.4I Tracing the origin of various types of fossil fuels is important for understanding the chemical and biological processes involved in fossil fuel deposits.

Category 10—Sea level rise

How many meters of sea level rise are contained in the Greenland glacial system? In the Arctic floating ice? In the West Antarctic ice shelf?

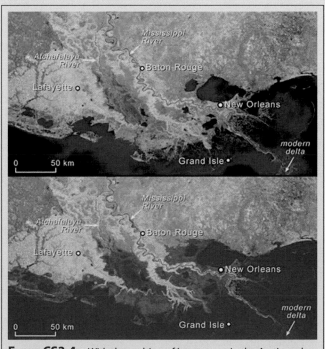

Figure CS3.4J With the melting of ice systems in the Arctic and Antarctic comes the increasing risk of large changes in sea level.

ENTROPY

AND THE SECOND LAW OF THERMODYNAMICS

4

> "Since a given system can never of its own accord go over into another equally probable state but into a more probable one, it is likewise impossible to construct a system of bodies that after traversing various states returns periodically to its original state, that is a perpetual motion machine."
>
> —Ludwig Eduard Boltzmann

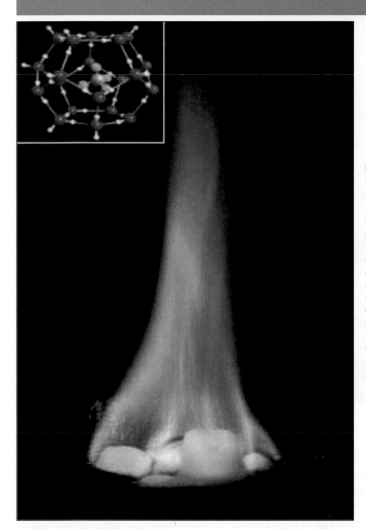

FIGURE 4.1 The ability of water molecules to organize, through hydrogen bonding, around nonpolar molecules, such as methane, accounts for the existence of hydrates, or more specifically clathrates. In those structures the water molecules are packed around the nonpolar organizing center (CH_4) by forming ordered arrays. What is remarkable is that methane clathrates store more than 10,000 gigatons of carbon in permafrost and ocean floor deposits. Those clathrates burn when ignited as shown in the figure. Recalling that approximately 8 gigatons of carbon is released to the atmosphere each year by the burning of fossil fuels, CH_4 clathrates hold more than 1000 times the yearly release of carbon via fossil fuel combustion. A global inventory of methane clathrates is currently underway, but we know that there are between two and five times as much fuel energy stored as methane in clathrates than in all fossil fuel reserves of coal, petroleum, and natural gas.

Framework

Nature is driven by *spontaneous* processes—processes that proceed without external intervention. With our advancing understanding of energy, with the insight brought by the First Law of Thermodynamics, and with our ability to track energy transformations from highly *organized* forms of energy (mechanical energy, chemical energy, electromagnetic energy) that inexorably cascade to disorganized thermal energy, comes a sharpened recognition that there are other factors that drive spontaneous processes in nature. In particular, we begin to recognize, as our understanding of energy advances, that a great many processes that occur in nature are not driven toward a state of lower energy, as is the case for an exothermic chemical reaction or a mass falling in a gravitational field. As we will see, spontaneous processes are guided by *both* the release of energy and by the tendency to seek a configuration of increasing disorganization.

This relationship between energy release, the tendency toward increasing disorganization, and the spontaneous formation of unique structures in nature is exemplified by the existence of mixed hydrocarbon-water compounds in nature. Figure 4.1 captures a rather remarkable event—what you see is a sample of ice removed from the ocean floor at depths of greater than ~300 meters. But the ice is burning!

It burns because within the ice structure, methane CH_4 is trapped within a cage of water molecules as shown in the upper left inset of the figure. This ice-methane structure is called a *methane clathrate* or *methane hydrate*. A remarkable amount of methane is contained in the clathrates that exist at low temperature and elevated pressure in permafrost and ocean floor deposits. The methane is formed from microbial decomposition of organic matter in soils and in sediments. Exploration to determine the sizes of global deposits of these methane clathrates is currently in progress, but it is known that *at least* twice as much fuel energy is contained in those structures as in the total of the Earth's oil, gas, and coal deposits combined. The principal methane clathrate deposits in the U.S. occur in the Gulf of Mexico, off the west and east coast, and on the north slope of Alaska. As an example of the scale of these methane clathrate deposits in the U.S., the methane contained in those deposits in the Gulf of Mexico *alone* is estimated to be $610 \times 10^{12}\,m^3$. The yearly consumption of methane (natural gas) in the U.S. is $650 \times 10^9\,m^3$. Thus the Gulf of Mexico deposits alone would supply the U.S. with natural gas at current consumption rates for $610 \times 10^{12}\,m^3/(650 \times 10^9\,m^3/yr)$ = 940 years. Research into how methane as a fuel could be extracted from those deposits is in progress but there is potential risk because methane is a greenhouse gas (it absorbs infrared radiation at a wavelength of 7.6 μm, 7.6×10^{-6} meters). Because those clathrates are stable only at low temperature, they pose a significant threat to climate stability. If just 0.5% of the permafrost structures in Alaska and northern Siberia alone were to melt each year, resulting from the loss of the Arctic Ice Cap, approximately 8 Gt C (gigatons carbon) would be added to the atmosphere *each year*. This would more than double the amount of carbon added to the atmosphere each year from the combustion of fossil fuel worldwide. Figure 4.2 displays the importance of methane clathrate and carbon dioxide permafrost melt rates per year of just 0.5% of the reservoir stored in the permafrost of Alaska and Siberia.

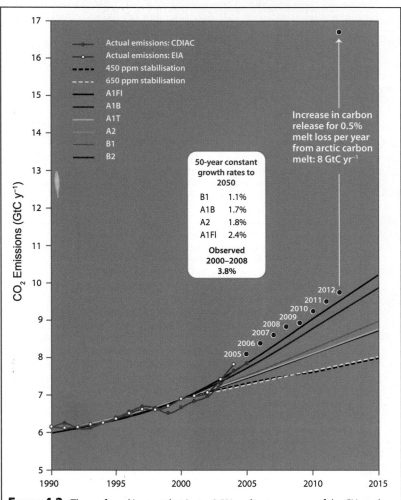

FIGURE 4.2 The profound impact that just a 0.5% melt rate per year of the CH_4 and CO_2 containing permafrost in Alaska and Siberia would have on the carbon added to the atmosphere by all fossil fuel burning is displayed. Of particular importance: just 0.5% of the methane and carbon dioxide contained in the upper 3 meters of the soil in northern Alaska and Siberia alone is equal to the annual relase of carbon dioxide from the burning of fossil fuels globally.

While methane storage in these clathrate structures will play an increasingly controversial role in the global debate at the intersection of energy and climate, at the molecular level the formation of these structures is an illustration of how *spontaneous* processes behave in nature. Specifically, the ability of water molecules to hydrogen bond to one another, as displayed in Figure 4.3, accounts for the existence of these crystalline clathrates.

The most common naturally occurring gas hydrate structure is composed of water molecules hydrogen bonded and organized around the methane molecules shown in Figure 4.4. The unit cell of the methane clathrate contains 8 methane molecules caged within 46 water molecules. Because methane is trapped within the crystal structure defined by the 46 cage water molecules, methane gas in hydrate form is significantly compressed. For example, one liter of methane clathrate solid contains, on average, 168 liters of methane gas at standard temperature and pressure.

What is important as we develop the concept of spontaneous processes in our study of thermodynamics is recognizing that the presence of methane within those bonding structures of water constitutes a *spontaneous* transformation from free methane gas and pure ice (with the structure shown in Figure 4.3) to the methane clathrate structure, shown in Figure 4.4. By filling voids in the ice structure, the "guest" methane molecules stabilize the clathrate hydrogen-bonded crystal structure. These methane clathrates are stable to temperatures considerably higher than the melting point of "pure" ice. Methane clathrates melt at 18°C at atmospheric pressure. This was, incidentally, a serious problem in natural gas distribution systems because the methane clathrates would clog large supply pipelines. The problem was eliminated by removing water vapor from the methane before pumping it into the supply network.

The ability of nature to *spontaneously* synthesize intricate structures that are of immense practical importance to society begins to beg an array of important questions that, on the face of it, appear to be unrelated.

Why does heat flow from regions of high temperature to regions of low temperature? Why do organized forms of energy, for example, the energy contained in the bonding structures of hydrocarbons, spontaneously and inexorably transition to highly disorganized forms of energy such as heat? Why, when tracking the energy content of coal from an electricity generator power plant, do we extract but 2 percent of that energy as useful light from an incandescent light bulb in our living room? How would the energy structure of the U.S. change if we transitioned to electricity-powered vehicles from internal combustion vehicles? Why would it cost an equivalent amount of less than 1 dollar a gallon if we used electricity rather than gasoline to power our automobiles?

We set the context for our study of the Second Law of Thermodynamics by examining an illustrative example of energy flow within our economic structure—the sequence of events from the combustion of coal in an electricity power plant to the delivery of that power to our living space. We consider first the efficiency of a power plant using our knowledge of the maximum efficiency of a heat engine converting an amount of heat, Q, to usable work, W. We know from our study of the Carnot cycle in Chapter 3 that the maximum efficiency of a system that converts heat to work is given by

$$(\text{Eff})_{max} = W/Q = (T_1 - T_2)/T_1 = 1 - T_2/T_1$$

where T_1 is the high temperature reservoir and T_2 is the low

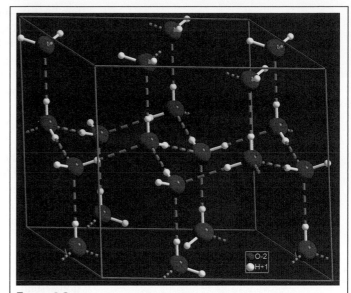

FIGURE 4.3 The structure of ice is shown with the hydrogen bonding displayed between hydrogen atoms(the small solid circles) and the electronegative oxygen centers (the larger open circles) that constitutes the crystal structure. The planes of oxygen atoms in the crystal structure are differentiated by the shading of the open circles.

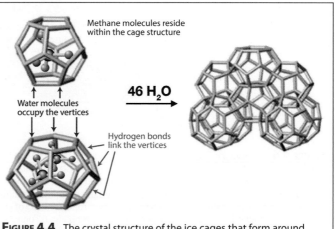

FIGURE 4.4 The crystal structure of the ice cages that form around methane molecules to form the methane clathrate, the unit cell of which consists of 8 methane molecules and 46 water molecules, is shown here as the union between two distinct cage structures.

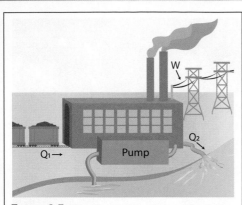

Figure 4.5 The chemical bonds in coal provide the heat input, Q_1. The power plant produces an amount of work, W, that is delivered to the grid. At the power plant, Q_2 is waste heat that must be absorbed by the surroundings.

temperature reservoir. Thus, because the actual efficiency is always less than the maximum given by the Carnot cycle, we have Eff < $W/Q_1 = (T_1 - T_2)/T_1$. Referring to Figure 4.5 to distinguish between the heat in, Q_1, and heat out, Q_2, we can "model" our electrical power plant, wherein the heat, Q_1 is supplied by coal and the work, W, is the electrical energy delivered to the power grid. Then we can rewrite the above equation as

$$W/Q_1 \leq 1 - T_2/T_1$$

Because the high temperature reservoir, T_1, is limited by the maximum temperature of the initial plumbing carrying the steam from the boiler to the turbine, modern coal fired electrical power plants are limited to $T_1 \approx 900$ K. The low temperature reservoir is typically $T_2 \approx 350$ K. Thus the *maximum efficiency* for conversion of highly organized energy to produce heat (Q_1) that is converted to work, W, in the form of electrical energy is

$$W/Q_1 = 1 - 350 \text{ K}/900 \text{ K} \approx 0.60$$

In practice this *maximum possible* efficiency is, through losses in the generation system, reduced to about 38%. Thus, 62% of the energy content of the coal is wasted—it is lost as (highly disorganized) heat and dumped into the surrounding environment, indicated by Q_2 in Figure 4.5.

Now let's consider what happens to that electrical energy once it has left the coal-fired generating plant and enters the power grid that distributes the energy to homes, businesses, industrial plants, etc. First, we examine the case of electric lighting for your home, office, dorm room, etc. Using a telling graphic from *What You Need to Know About Energy* (NAS Press, 2008), we can quantitatively trace the path from energy generation to light output (Figure 4.6). With a (typical) 38% efficiency in initial generation, we lose 62 units of the initial 100 energy units contained in the chemical bonds of the coal. Typically, between 2 units and 8 units are lost in the transmission lines, as shown in Figure 4.6. We will assume the more efficient of these such that 36 units of the original 100 units are delivered to the home. However, with the use of an incandescent bulb, 34 units are emitted as heat and only 2 units are emitted as usable light. Thus, only 2% of the original energy contained in the coal actually results in useful energy!

If, instead of an incandescent lightbulb, a compact fluorescent bulb is used, then 10 units of that energy are available for useful lighting—a factor of five improvement in energy efficiency. This is why all incandescent lightbulbs should be eliminated from houses and buildings as quickly as possible. If, on the other hand, we use light emitting diodes (LEDs), a device we will investigate in Chapter 11, we use only one-*tenth* the energy to produce the same amount of useful light that is emitted by an incandescent light. Thus, if LEDs replaced incandescent

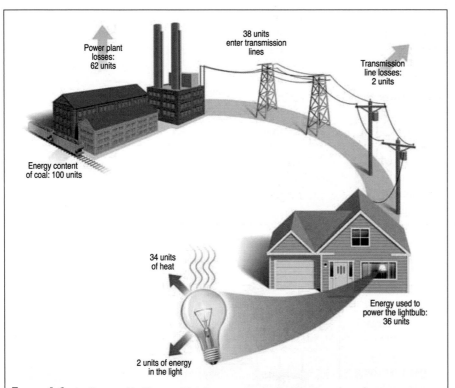

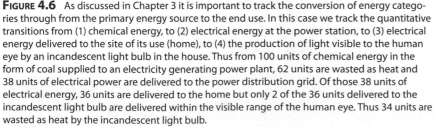

Figure 4.6 As discussed in Chapter 3 it is important to track the conversion of energy categories through from the primary energy source to the end use. In this case we track the quantitative transitions from (1) chemical energy, to (2) electrical energy at the power station, to (3) electrical energy delivered to the site of its use (home), to (4) the production of light visible to the human eye by an incandescent light bulb in the house. Thus from 100 units of chemical energy in the form of coal supplied to an electricity generating power plant, 62 units are wasted as heat and 38 units of electrical power are delivered to the power distribution grid. Of those 38 units of electrical energy, 36 units are delivered to the home but only 2 of the 36 units delivered to the incandescent light bulb are delivered within the visible range of the human eye. Thus 34 units are wasted as heat by the incandescent light bulb.

bulbs, 20 units of light energy would be produced instead of 2 units of light energy produced by incandescent lighting.

A second important example involves the sequence of events when that electricity is used to drive a pumping system, as is the case in factories, agriculture, heating, etc., across the industrial and domestic spectrum. This sequence is captured in Figure 4.7, a recent graphic by the Rocky Mountain Institute (Amory Lovin lecture, Harvard University, 2008). Again, beginning with 100 units of energy contained in the coal burned to produce electricity in a power plant, 70 units are lost in the initial generation, yielding 30 units of electrical energy delivered to the grid. It is common to see differences in power plant efficiency, notable in comparing Figure 4.6 and Figure 4.7, because different power plants have significantly different efficiencies. Transmission and distribution losses remove 9%. Electrical motor losses remove 10% more, drive train losses 2% more, pump system losses 25% more, throttle and pipe losses 33% and 20%, respectively; so that only ~9% of the original 100 units of energy actually survive to execute the objective of the overall system.

There are a number of key conclusions to be drawn. First, the initial step is very inefficient—between 60 and 70% of the primary energy is lost before any useful energy is generated. Second, it is that first step that releases the large amounts of CO_2 (as well as soot, nitrates, sulfates, mercury, etc.) into the atmosphere. Third, one unit of energy saved at the usage end translates into 10 to 20 units at the production end. Thus, conservation is very important. Fourth, if energy were produced at the site of its use (for example, photovoltaics to collect energy to supply lighting or air conditioning) the demand for primary power generation using coal would drop dramatically. Finally, note that the conversion of electrical energy to mechanical energy by the electric motor is 90% efficient—contrast that with a gasoline engine that is ≤ 20% efficient, which is a key point for the next generation of automobiles, trucks, and buses.

One of the most dramatic and immediate ways that the US can direct its energy policy toward the twin objectives of (1) reducing carbon dioxide emissions and (2) reducing dependence on foreign oil sources, is to recognize the profound advantages in shifting automobiles from a reliance on the century-old internal combustion engine to electrically powered transportation.

To place this problem in context, we begin by noting that the US *imports* between 12 and 13 million barrels of oil *per day*. This constitutes approximately 60% of domestic petroleum consumption. The price per barrel for imported oil ranged

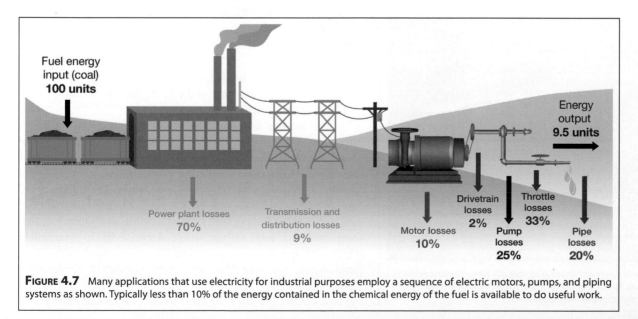

FIGURE 4.7 Many applications that use electricity for industrial purposes employ a sequence of electric motors, pumps, and piping systems as shown. Typically less than 10% of the energy contained in the chemical energy of the fuel is available to do useful work.

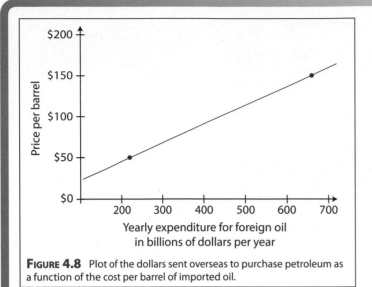

FIGURE 4.8 Plot of the dollars sent overseas to purchase petroleum as a function of the cost per barrel of imported oil.

between \$50 and \$150 per barrel in 2008 and settled between \$80 and \$100 per barrel in 2010. As shown in Figure 4.8, which plots the yearly U.S. expenditure for the purchase of foreign oil against the price per barrel, the U.S. spent, on a per yearly basis, approximately 450 billion dollars to purchase petroleum. This is a major contribution to our balance of payments each year.

Combustion of oil in 2006 accounted for 44% of US CO_2 emissions, with 33% of the total CO_2 emissions resulting from transportation, which includes cars, trucks, trains, ships, and aircraft. Gasoline powered cars and light trucks were responsible for 40% of total oil consumption, with diesel powered vehicles accounting for an additional 10%.

What steps would be required to eliminate the need to import foreign oil? Because we import 60% of the oil that we use, if we could reduce our dependence on oil by that amount we would eliminate foreign imports. We can immediately recognize these important factors in sizing up this problem:

1. Vehicles use 70% of all petroleum in this country.

2. The internal combustion engine extracts only 20% of the energy contained in liquid fuels that can be made available to propel the car—losing 80% to heat.

3. If we include drive train loss, idling loss, loss to accessories, etc., only 13% of the fuel energy is actually expended to move the car.

4. 75% of the fuel use is related to the weight of the vehicle.

Thus there are two very large and very effective steps that can (and must) be taken. The first is to move immediately to much lighter vehicles—but not at the expense of safety, comfort, nor performance. This is achieved through the use of carbon-fiber technology to reduce the vehicular weight from 2000–2500 kg down to 500 kg. An example of such an automobile is shown in Figure 4.9.

The second large and important step is to replace the internal combustion engine with electric propulsion. This transition can be effected very quickly by moving to the technology of hybrid electric vehicles (HEV) that can be, and are, built in a number of different combinations of joint electric-internal combustion propulsion options. These combinations are typically classified as HEV 0, HEV 20, HEV 40, and HEV 60 options, representing the distance the vehicle can cover without using the onboard gasoline or diesel engine—in this case, 0, 20, 40, or 60 miles, respectively.

The average distance driven per day, per car in the US is about 30 miles. If HEV 60 autos constituted the fleet of cars in the US, the result would be that the gasoline (or diesel) assisted travel would be satisfied with just 10% of the current petroleum usage. Replacing 90% of gasoline consumption by electrical propulsion would be equivalent to raising the fleet's average fuel efficiency from the current 17 mpg to close to 150 mpg. If this were realized, total oil use would decrease by approximately 35%, cutting need for imported oil by approximately 60%. General Motors has invested a considerable amount in the development of the Chevy Volt, which is an HEV 40 vehicle, and Toyota is introducing the HEV 40 Prius in 2011. Both the Chevy Volt

FIGURE 4.9 A dominant contributor to low efficiency for automobiles is weight and drag. New carbon fiber construction and aerodynamic design can radically improve the situation.

and the Toyota "plug in" Prius use the new lithium ion battery technology that we will examine in Chapter 7. There are also an array of all-electric powered automobiles moving toward production that have a range of > 200 miles between chargings of the battery.

By switching to the HEV or all electric technology, an immediate reduction in operating cost would be realized. The reason is that, as we have seen, the internal combustion engine wastes some 85% of the energy content of the fuel, while an electric motor is 90% efficient—wasting only 10% of the energy delivered to it.

The chemical energy content of a gallon of gasoline is (9.4 kWh/L) × (3.8 L/gallon) = 35.7 kWh/gallon. But because an electric motor is 0.90/0.15 = 6 times as efficient as a gasoline automobile, we need only 35.7 kWh/gallon divided by 6 = 5.95 kWh, which we will approximate as 6.0 kWh of electricity, to obtain the same driving range as that produced by a gallon of gasoline. In the U.S., the cost for 1 kWh of electricity ranges from 6–7 cents in a number of western and mid-western states, to 19–20 cents in the northeast. At 7 cents per kWh, it would take approximately 42 cents to drive the distance covered by a gallon of gasoline. At 20 cents per kWh, it would cost $1.20 to drive the distance covered by a gallon of gasoline. Thus, in addition to dramatically reducing the need for imported oil, there is an immediate advantage of reducing the per-mile cost of driving. Of course, the objective of eliminating CO_2 emissions from the transportation sector involves generating the primary energy supplied to the power grid without the use of fossil fuels.

This brings us back to the critical importance of determining what drives the vast range of spontaneous processes in nature. The coherent, scientifically rigorous framework that *quantitatively* defines spontaneous processes in nature constitutes an objective of great importance. To this end, we recognize that it is the *spontaneous* nature of change that is the compelling concept that must be understood, and that while nature seeks to find pathways to states of lower energy, other factors come powerfully into play.

For instance, when red hot lava fed from a volcano reaches the cool waters of the ocean, we know instinctively what will happen: thermal energy, contained in the hot lava, flows into the ocean until the lava has cooled to the temperature of the water around it. We know that when we release carbonation from a bottle, the gas rushes out of the vessel into its surroundings. In these cases, is the final state at *lower energy* than the initial state, or are other factors controlling the process?

In pursuit of a scientific understanding of spontaneous processes, it becomes important to develop the ability to distinguish between spontaneous and nonspontaneous processes. Examples of nonspontaneous processes include the construction of a stone wall, the working of a chemistry problem set, and decomposition of water by electrolysis into hydrogen and oxygen. These nonspontaneous events can occur only as long as there is external intervention. But nonspontaneous changes have another characteristic in common: They can occur only when accompanied by, or coupled to, some process that is spontaneous. The stone mason and the chemistry student are sustained by biochemical reactions that are themselves spontaneous. The electrolysis of water requires a source of electrons driven through the electrolysis cell by chemical or mechanical change.

The observation that all *nonspontaneous* events occur at the expense of spontaneous processes leads us, through *inductive* reasoning, to an important conclusion: All that happens can be traced either directly or indirectly to spontaneous change. This, in turn, implies that we must progress from a position of understanding energy, to a position of understanding the genesis and characteristics of spontaneous change in nature, and by *deduction*, there must be at least one other compelling tendency, independent of energy, that drives spontaneous change.

Consider the processes shown in the figure below: the corrosion of iron, the combustion of octane, and the melting of ice. The first two examples are easily

Case Study 4.1 Entropy, Free Energy, and the Maximum Amount of Work that can be Extracted from a Fuel Cell

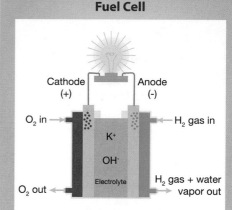

Case Study 4.2 The Most Probable Distribution Is the Boltzmann Distribution

Case Study 4.3 Methods for Energy Storage

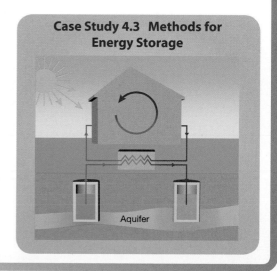

Expansion of an ideal gas into a vaccuum

(a) Initial condition

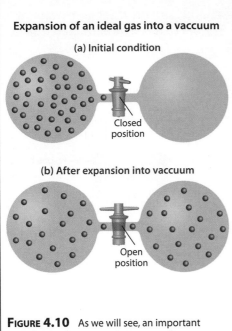

Closed position

(b) After expansion into vaccuum

Open position

FIGURE 4.10 As we will see, an important example of the increase in entropy of a spontaneous process is the expansion of a gas from an initial condition of, for example, 1 liter at 1.0 atm pressure to 2 liters at 0.5 atm pressure. This can be achieved by opening a valve between two bulbs, each with a volume of 1 liter, wherein one of the bulbs initially holds 1 atm of the gas, while the other is a vacuum.

understood in terms of exothermicity—the release of chemical energy resulting from the tendency of exothermic change to proceed spontaneously.

Spontaneous Change: what controls it?
Spontaneous change: a change that occurs without outside intervention.

On the basis of observed processes in nature, it is very tempting to conclude that spontaneous processes are driven exclusively by pathways leading to the release of energy such that proceeding in the direction of decreasing energy is the motive force behind spontaneous change. As we saw in Chapter 3, changes that lower the potential energy (mechanical or chemical) of a system can be referred to as exothermic. Within this view of spontaneous processes, we could deduce a law: *Exothermic changes have a tendency to proceed spontaneously.* Indeed, for centuries this was held as a fundamental tenet defining spontaneous changes in nature.

But what about the melting of arctic ice? We know from experience that ice melts spontaneously when in contact with a temperature reservoir just a fraction of a degree higher in temperature than the melting point of ice. Now we have a conundrum, for while we know that ice melts spontaneously, we also know that in the process of melting, thermal energy must be flowing *into* the ice as it melts. The transfer of thermal energy (heat) leaves the liquid water from the melted ice at a *higher internal energy* than that of the original ice. Thus, the process of melting is an endothermic process! Yet it is spontaneous. This revelation opens an awareness of many other spontaneous processes that are endothermic: Expansion of a gas into a vacuum (Figure 4.10); the "cold pack" chemical envelopes used for medical purposes; the evaporation of ocean water into the atmosphere; the formation of a solution created by the intermixing of substances in the liquid phase.

Emergent in these examples is a pattern of molecular-level processes that exhibit the characteristic that systems tend to move *toward* a state of *higher probability* and away from a state of low probability, even if they must move "uphill" in energy. This fact immediately engages us in a treatment of what defines the probability of a given state, or configuration, and it requires us to consider how microscopic (molecular) level processes are quantitatively linked to macroscopic systems that are changed by this propensity of molecular systems to seek states of higher probability.

Chapter Core

Road Map for Chapter 4

We turn now to the development of a quantitative basis for defining spontaneous processes in nature beginning with the development of the relationship between disorder and the tendency of systems to inexorably seek a state of higher probability. This development begins at the molecular level and leads to the concept of *entropy* that, in partnership with *energy*, constitutes the foundation for quantitatively defining criteria for determining whether a process is spontaneous or nonspontaneous. We will see that this concept of *entropy*, which can be understood at both the molecular (microscopic) level and the macroscopic level, sets in place the foundation for the *Second Law of Thermodynamics* that states that *whenever a spontaneous event takes place in the universe, the total entropy of the universe must increase*. What is remarkable is that this disarmingly simple statement leads directly to the formulation of the most important thermodynamic variable—the Gibbs free energy. The remainder of the chapter and significant parts of chapters 5, 6, and 7 that treat on chemical equilibrium, acid-base reactions, and electrochemistry, are built upon developing a thorough understanding of Gibbs free energy.

Road Map to Core Concepts	
1. Development of the Concept of Entropy Beginning with the Microscopic Perspective	Ten Molecules
2. Boltzmann and the Microscopic Formulation of Entropy	
3. Qualitative Prediction of Entropy Change: Establishing the Sign of ΔS	
4. Quantitative Treatment of Entropy: Calculating ΔS for a System	
5. Joining the Macroscopic and Microscopic: Calculation of Entropy Change, ΔS	
6. Second Law of Thermodynamics, Gibbs Free Energy, and Spontaneous Change	
7. Absolute Value for Entropy: Third Law of Thermodynamics	
8. Calculation of Entropy Change for a Chemical Reaction	
9. Calculation of Gibbs Free Energy for a Reaction	

Determination of Probability at the Molecular Level

We examine the issue of probability at the molecular level from two complementary perspectives.

Perspective #1:

Suppose we examine the behavior of molecules in a two-chamber apparatus, shown in Figure 4.11.

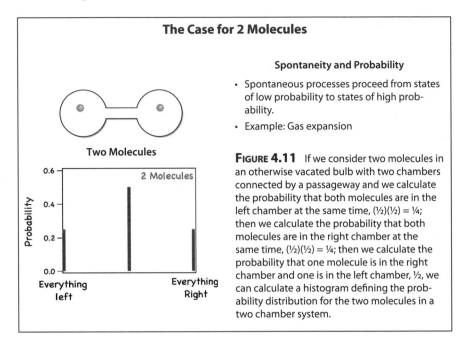

The Case for 2 Molecules

Spontaneity and Probability

- Spontaneous processes proceed from states of low probability to states of high probability.
- Example: Gas expansion

Figure 4.11 If we consider two molecules in an otherwise vacated bulb with two chambers connected by a passageway and we calculate the probability that both molecules are in the left chamber at the same time, (½)(½) = ¼; then we calculate the probability that both molecules are in the right chamber at the same time, (½)(½) = ¼; then we calculate the probability that one molecule is in the right chamber and one is in the left chamber, ½, we can calculate a histogram defining the probability distribution for the two molecules in a two chamber system.

We begin this (important!) thought experiment with the chambers evacuated except for two indistinguishable gas-phase molecules that we observe over a period of time during which we, at certain intervals, count the number of times we observe the available configurations, which are (i) both molecules in the left-hand chamber, (ii) both molecules in the right-hand chamber, (iii) one molecule in the left-hand chamber and one molecule in the right-hand chamber. We can draw a histogram that summarizes our findings. The probability of finding one molecule in the left-hand chamber is ½, just as the probability of getting "heads" in a coin toss is ½. The probability of finding both molecules in the left-hand chamber is (½)(½) = (¼), just as the probability of tossing two "heads" in a row is (½)(½) = (¼). But the same logic applies to the right-hand chamber; the probability of finding both molecules in the right-hand chamber is (½)(½) = (¼). The probability of finding one molecule in the left chamber and one molecule in the right chamber is ½. Thus, we construct our histogram accordingly. Notice that for our case of two identical molecules, there are three ($n + 1$) possible configurations and that there is a higher probability of finding the molecules evenly distributed between the two chambers.

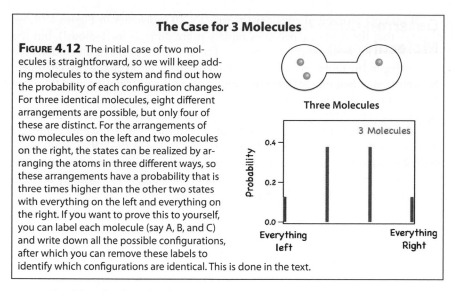

The Case for 3 Molecules

Figure 4.12 The initial case of two molecules is straightforward, so we will keep adding molecules to the system and find out how the probability of each configuration changes. For three identical molecules, eight different arrangements are possible, but only four of these are distinct. For the arrangements of two molecules on the left and two molecules on the right, the states can be realized by arranging the atoms in three different ways, so these arrangements have a probability that is three times higher than the other two states with everything on the left and everything on the right. If you want to prove this to yourself, you can label each molecule (say A, B, and C) and write down all the possible configurations, after which you can remove these labels to identify which configurations are identical. This is done in the text.

Let's add a third molecule as shown in Figure 4.12. But now let's label the molecules as A, B, and C. And, in addition, let's explicitly list out the possible arrangements:

1. All in the left chamber: (ABC)–(   ) $(\frac{1}{2})(\frac{1}{2})(\frac{1}{2}) = \frac{1}{8}$

2. All in the right chamber: (   )–(ABC) $(\frac{1}{2})(\frac{1}{2})(\frac{1}{2}) = \frac{1}{8}$

3. Two in the left and one in the right:

         (AB)–(C) $(\frac{1}{2})(\frac{1}{2})(\frac{1}{2}) = \frac{1}{8}$
         (AC)–(B) $(\frac{1}{2})(\frac{1}{2})(\frac{1}{2}) = \frac{1}{8}$
         (BC)–(A) $(\frac{1}{2})(\frac{1}{2})(\frac{1}{2}) = \frac{1}{8}$

4. One in the left and two in the right:

         (C)–(AB) $(\frac{1}{2})(\frac{1}{2})(\frac{1}{2}) = \frac{1}{8}$
         (B)–(AC) $(\frac{1}{2})(\frac{1}{2})(\frac{1}{2}) = \frac{1}{8}$
         (A)–(BC) $(\frac{1}{2})(\frac{1}{2})(\frac{1}{2}) = \frac{1}{8}$

But while each configuration of the three molecules has the same probability if we distinguish the individual molecules $((\frac{1}{2})(\frac{1}{2})(\frac{1}{2}) = \frac{1}{8})$, if the molecules are *indistinguishable* we have a probability of $3 \times (\frac{1}{8})$ for Cases 3 and 4. This results in the histogram shown in Figure 4.12 for three indistinguishable molecules.

Once again, we note the pattern: (a) There are $n + 1 = 3 + 1 = 4$ possible states (or configurations), and (b) the higher probability cases are those for which there is the most balanced distribution between the two chambers.

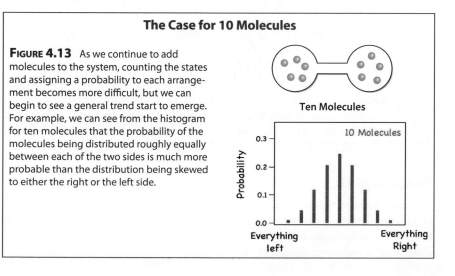

The Case for 10 Molecules

Figure 4.13 As we continue to add molecules to the system, counting the states and assigning a probability to each arrangement becomes more difficult, but we can begin to see a general trend start to emerge. For example, we can see from the histogram for ten molecules that the probability of the molecules being distributed roughly equally between each of the two sides is much more probable than the distribution being skewed to either the right or the left side.

If we now proceed to consider the case for 10 molecules in Figure 4.13, we discover that the counting becomes rapidly more difficult (or tedious!), but the same pattern becomes more pronounced: The histogram peaks in the middle, corresponding to the molecules being distributed equally between the two chambers, the distribution falls off uniformly on both sides of the distribution, and there are $n + 1 = 10 + 1 = 11$ possible states.

So let's extend this to the case for which we have not 10, but 100 molecules, as shown in Figure 4.14. Now the individual counting is best done by computer, but we see that the probability that all molecules reside in the left or in the right chamber is virtually zero; and, moreover, that the probability that the distribution is more than slightly skewed to one chamber or the other is increasingly unlikely as the number of molecules is increased.

The Case for 100 Molecules

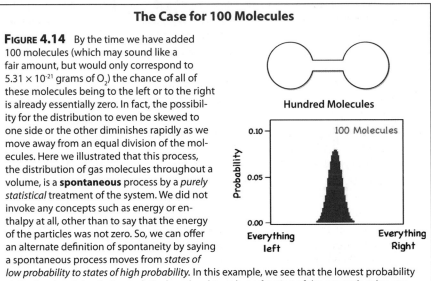

FIGURE 4.14   By the time we have added 100 molecules (which may sound like a fair amount, but would only correspond to 5.31×10^{-21} grams of O_2) the chance of all of these molecules being to the left or to the right is already essentially zero. In fact, the possibility for the distribution to even be skewed to one side or the other diminishes rapidly as we move away from an equal division of the molecules. Here we illustrated that this process, the distribution of gas molecules throughout a volume, is a **spontaneous** process by a *purely statistical* treatment of the system. We did not invoke any concepts such as energy or enthalpy at all, other than to say that the energy of the particles was not zero. So, we can offer an alternate definition of spontaneity by saying a spontaneous process moves from *states of low probability to states of high probability*. In this example, we see that the lowest probability states involve states that are relatively *ordered*, i.e. a large fraction of the gas molecules can be found on either the right or the left of the container; and the states of high probability are configurations where the system is relatively *disordered*.

But the conclusion that we draw from this simple progression is of great practical (and theoretical) importance. It is important because we have illustrated that this process (that of individual molecules seeking a particular distribution between the two chambers) is a spontaneous process based on a purely *statistical* treatment of the system. At no point did we discuss or involve the internal energy, the enthalpy, nor any quantity associated with an energy scale. So we now have another perspective on the concept of spontaneity: A spontaneous process moves from states of lower probability to states of higher probability. We also notice something else of importance: The states of *lowest probability* are characterized by a *more ordered configuration*—for example, if all the molecules are sequestered in one chamber, that is a more ordered configuration; just as is the case of all the clothes in your room being folded and placed in a single drawer! This link between probability, spontaneity, and order will loom large in our understanding of processes in thermodynamics. Imagine now if we move from 100 molecules to a single mole of N_2 in a two chamber system. Now we have 6×10^{23} molecules and the histogram defining the molecule distribution is a *very* narrow spike at the center.

Perspective #2:

We know from experience that when we place a hot object in contact with a cold object, thermal energy will flow spontaneously from the high temperature body to the low temperature body. But why is this? Does it have anything to do with probability? With order and disorder?

To address these questions, we turn again to our abiding strategy that if we have a question—a problem—we just construct first a model that will serve to sharpen the question we wish to answer, and second, a model that will guide us to a coherent answer.

Just as with our previous thought experiment, when we are setting a strategy at the molecular level, we begin with a small number of molecules! In this approach, or model design, we begin with two objects, each comprised of three molecules, to investigate the *direction* of flow of thermal energy (heat). Moreover, we assume that each molecule can occupy a low energy state or a high energy state, and let's designate a molecule in a low energy state as a "blue" molecule and a high energy molecule as a "red" molecule.

If we begin this thought experiment with the initial configuration such that all three molecules of one object are in a high energy state and all three molecules of the other object are in a low energy state; then we can represent the initial configuration as having one "hot" object on the left and one "cold" object on the right.

Now we move the objects such that they are in *physical contact*, allowing energy to flow between the two objects. Since we are not adding to or removing energy from these two objects (each comprised of three molecules) *the total energy of the system does not change*. That is, whatever the distribution of hot molecules and cold molecules is for a given state of the joined systems, the total number of hot (red) molecules must remain at 3, and the total number of cold (blue) molecules must remain at 3. But the number of *possible distributions* of energy among the six molecules after the objects are brought in contact is now not 1, as was the case for the separated objects, but 20, corresponding to the cases for one unit of energy transferred, two units of energy transferred, and three units of energy transferred. The 20 different distributions are shown in the Figure 4.15.

Inspection of the figure reveals several notable facts. *First*, some of the outcomes result from a number of different specific configurations. For example, if just one unit of energy is transferred, we have configurations 2 through 10. If two units are transferred, we have configurations 11 though 19. *Second*, the more ways a configuration can be produced, the greater is the probability that it will occur. *Third*, this leads us to a strategy for quantitatively defining the probability of a given state occurring.

In particular, if we assume all of the 20 possible distributions of energy are equally probable, then the probability of a particular outcome is calculated from the expression

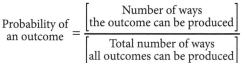

$$\text{Probability of an outcome} = \frac{\left[\begin{array}{c}\text{Number of ways}\\\text{the outcome can be produced}\end{array}\right]}{\left[\begin{array}{c}\text{Total number of ways}\\\text{all outcomes can be produced}\end{array}\right]}$$

We can tabulate the results:

TABLE 4.1

Units of Energy Transferred	Number of Equivalent Ways to Realize the Energy Transfer	Probability of Energy Transfer
0	1	1/20 = 5%
1	9	9/20 = 45%
2	9	9/20 = 45%
3	1	1/20 = 5%

We can now dissect what happened when we moved from (a) our initial condition of two separated objects, one comprised of three high energy (hot) molecules and the other comprised of three low energy (cold) molecules, to (b) our final condition wherein the objects are in physical contact and have equilibrated; this is shown explicitly in Figure 4.15.

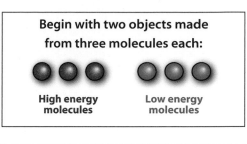

Begin with two objects made from three molecules each:

High energy molecules   Low energy molecules

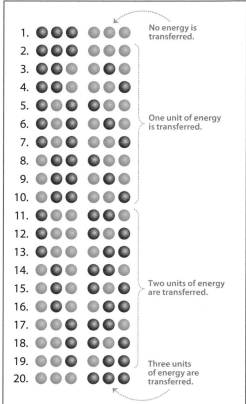

1. No energy is transferred.

One unit of energy is transferred.

Two units of energy are transferred.

Three units of energy are transferred.

FIGURE 4.15 When we bring two bodies together, one hot and one cold, we know from direct experience that thermal energy (heat), will flow from the high temperature body to the low temperature body. We can analyze the probability of energy being transferred when two bodies are brought into contact by modeling the system as two bodies, one of which is comprised of just three molecules. If we bring the hot (red) body in contact with the cold (blue) body, we can transfer zero, 1, 2, or 3 units of energy where each unit corresponds to the energy transferred between a red molecule and a blue molecule. If zero units of energy are transferred, there is just one such configuration; the one representing the case before the bodies were brought into contact. If one unit of energy is transferred, there are nine equivalent configurations available; if two units are transferred there are nine equivalent configurations; and if three units of energy are transferred there is one additional configuration.

We can immediately determine from an inspection of Table 4.1 that there is a 19/20 = 95% probability that some amount of energy will be transferred from the hot object to the cold object. This is for an object with just 3 molecules—that's 2×10^{-23} moles! You can quickly see that if each object has but 10 molecules (6×10^{-22} moles), the probability that energy will flow from the hot object to the cold object greatly exceeds 99%.

Thus, while our model for thermal energy transfer ("heat flow") is very simple, it is powerful indeed because it explicitly demonstrates the role of probability in determining the direction of a spontaneous process. But the conclusion is the same as that for our analysis of molecules distributed between two chambers: spontaneous processes tend to proceed from states of low probability to states of higher probability. The higher probability states are those for which the energy increments (in this case the increment of energy is the amount of energy separating high energy [red] molecules from low energy [blue] molecules) can be stored away in the largest number of different sites or locations. Thus nature drives toward subdividing the total available energy into the smallest possible increments and distributing these increments into the largest possible number of sites or "slots." The higher probability states are those that allow more options, more choices, for hiding a given amount of energy among the available molecules. We can restate the condition for spontaneity: Spontaneous processes proceed in a direction such that energy is dispersed as uniformly as possible.

We also notice another important pattern from our analysis of the two object/three molecules per object model; that is, the higher probability states are those characterized by *maximum disorder*. We can see this quite easily because the initial condition, wherein one object possesses three high energy (hot) molecules and one object possesses three low energy (cold) molecules, was the condition of maximum *order*; a condition rapidly replaced, following our bringing the two objects into physical contact, by a condition of maximum *disorder*.

What we take from this is that in the microscopic world, i.e., at the molecular level, probability and order/disorder play a large part in determining which processes occur and which do not. But as macroscopic systems must reflect what transpires at the molecular level, it follows that *there should be a thermodynamic quantity that relates this microscopic probability to spontaneous processes on the macroscopic scale.*

Entropy

The need to create a coherent, scientifically viable foundation linking the concepts of spontaneous change, inexorable degradation of organized energy to disorganized thermal energy, irreversible change, and finally probability at the molecular level, immediately begs the question: Can such disparate concepts be linked by a single variable, or must we cope with a manifold of variables to quantitatively link these concepts?

Our examples of thermal energy spontaneously flowing through molecular-level (microscopic) interactions and the spontaneous expansion of a gas from low pressure to high pressure, wherein the internal energy of the system remained unchanged, $\Delta U = 0$, suggests that while the direction of spontaneous change left the energy unchanged, the way in which the energy was *distributed* is intimately linked to the direction of spontaneous change. Arguing in analogy with our microscopic case of two objects, the spontaneous flow of thermal energy from volcanic lava as it enters the ocean represents, on the macroscopic scale, a system (lava plus ocean) seeking the greater number of configurations of the microscopic particles among energy levels in a particular state of the system. That is, spontaneous change occurs such that the total available energy is distributed so that the condition of maximum probability is attained. Thus an intrinsic parity emerges involving energy, on the one hand, and how the energy is *distributed* in the system, on the other.

Lava flows into the ocean at Kilauea.   [*Source: USGS, Hawaiian Volcano Observatory*]

The thermodynamic property that quantitatively defines the way in which the energy of a system is distributed among the available microscopic energy levels in a system is called *entropy*. As we will see in the remainder of this chapter, *energy* and *entropy* constitute the pair of state variables that, when taken together, determine whether a process is spontaneous. But more than that, energy and entropy in combination define the relationships between and among spontaneity, reversibility, probability, order/disorder, and the microscopic and macroscopic worlds.

Entropy is designated by the symbol S, and, as in the case of internal energy, U, and enthalpy, H, entropy S is a function of state. It has a unique value for a system with specified temperature, pressure, and composition. The entropy change, ΔS, is the difference in entropy between two states and it has a unique value *independent of path* taken to achieve that state:

$$\Delta S = S_{\text{final}} - S_{\text{initial}}$$

Boltzmann and the Microscopic Formulation of Entropy

One of the remarkable confluences of history occurred between what was learned from studies of the efficiency of the steam engine and what emerged from an analysis of microscopic states of a system. We begin with the latter.

Let's consider again the case of two chambers containing a total of 10 molecules as introduced in Figure 4.13. Except now we consider quantitatively the number of possible *arrangements* beginning with 10 molecules in the left-hand chamber and zero molecules in the right-hand chamber (which we identify as our *initial* state) and progress stepwise through each possible distribution of molecules between the two chambers. This is displayed explicitly in Figure 4.16.

Ludwig Boltzmann was a giant of 19th and early 20th century chemical physics. He was fundamental in developing the foundations for thermodynamics from the microscopic (molecular) perspective and he developed the kinetic theory of gases. Etched in his tombstone, by his own request, was the expression linking entropy, S, to the number of microstates, W.

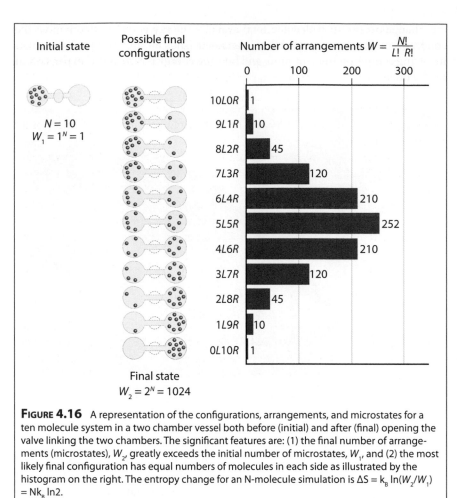

FIGURE 4.16 A representation of the configurations, arrangements, and microstates for a ten molecule system in a two chamber vessel both before (initial) and after (final) opening the valve linking the two chambers. The significant features are: (1) the final number of arrangements (microstates), W_2, greatly exceeds the initial number of microstates, W_1, and (2) the most likely final configuration has equal numbers of molecules in each side as illustrated by the histogram on the right. The entropy change for an N-molecule simulation is $\Delta S = k_B \ln(W_2/W_1)$ $= Nk_B \ln2$.

We designate each possible distribution of the 10 molecules between the two chambers as a specific *configuration*. For example, if there are 7 molecules in the left-hand chamber and 3 in the right-hand chamber we would designate the configuration as 7L, 3R. When we considered our case of 3 molecules distributed between two chambers we came to the conclusion that there were n + 1 possible states, so for n = 3 molecules we had 4 possible configurations. For n = 10 molecules we have n + 1 = 11 possible configurations, as is borne out in Figure 4.16. We also discovered that the *probability* of a given configuration is proportional to the number of possible *arrangements* (ways of choosing how each molecule is placed in each chamber—page 212) that yields a particular configuration. We saw the analogous situation for the transfer of energy between two objects, each of which contained 3 molecules. The probability of a given outcome equaled the number of ways a *given* outcome can be produced divided by the *total* number of ways all outcomes can be produced (page 214).

Thus we seek to determine the number of arrangements of each configuration *relative* to the total number of arrangements available to the system. We begin with the initial configuration, which, as shown in Figure 4.16, we designate as 10L, 0R. The probability that the first molecule is in the left-hand chamber is unity, same for each of the other nine molecules. Thus, because the total number of arrangements possible is equal to the product of the probability for each molecule that makes up that configuration, the total number of arrangements, N_1, for that initial state is

$$W_1 = 1 \cdot 1 \cdot 1 \cdot 1 \cdot 1 \cdot 1 \cdot 1 \cdot 1 \cdot 1 \cdot 1 = 1^{10} = 1$$

If we allow the 10 molecules to expand into the two chambers, each molecule can have two possible arrangements (either the left-hand *or* the right-hand chamber) so the total number of arrangements possible for the case of expansion of the 10 molecules into the two chambers is $W_2 = 2 \cdot 2 \cdot 2 \cdot 2 \cdot 2 \cdot 2 \cdot 2 \cdot 2 \cdot 2 \cdot 2 = 2^{10} = 1024$. This means that the odds of finding 10 molecules in the left chamber (or in the right chamber) is 1 in 1024. Note that if we consider 100 molecules, the odds of finding all 100 in the left chamber (or in the right chamber) are 1 in 2^{100} or approximately 1 in 10^{30}!

What about the number of possible arrangements for the most favored configuration 5R, 5L? If we can calculate this, we know the probability for that specific configuration will be that number of possible arrangements *divided* by the *total* number arrangements for 10 molecules, or 1024. The *number of arrangements* that contribute to 5L, 5R is equal to the number of ways of choosing any 5 of the 10 molecules and placing them on the left. The other 5 must go on the right. It turns out that the *number of arrangements* of the configuration 5L, 5R, which we designate $W(5L, 5R)$, is $N!/(L!R!)$ where $N!$ is "N factorial" and is equal to $N(N-1)(N-2)\ldots(1)$. Similarly, $L! = L(L-1)(L-2)\ldots(1)$ and $R! = R(R-1)(R-2)\ldots(1)$. Thus

$$W(5L, 5R) = 10!/(5!)(5!) = 252$$

This is displayed in Figure 4.16. It is also important to check this against our calculation on page 212 for the case of three molecules in two chambers. In that case, for two molecules on the left and one on the right

$$W(2L, 1R) = 3$$

which is also

$$W(2L, 1R) = \frac{N!}{L!R!} = \frac{3!}{2!1!} = \frac{(3)(2)(1)}{(2)(1)(1)} = 3$$

Now that we have our general expression for W, the number of *arrangements* for a given *configuration*, we can quickly calculate and compare various configurations. For example, $W(6L, 4R) = 10!/6!4! = 210$ and as we consider a greater and greater unbalance, the number of arrangements drops rapidly ($W(8L, 2R) = 45$).

This analysis of specific *configurations*, and the number of possible *arrangements* associated with each configuration corresponds to an important designation in molecular thermodynamics. It is the identification of the states of a system (the configuration) and the corresponding *microstates* of a system (the arrangements) as displayed in Figure 4.16.

It was Ludwig Boltzmann, in one of the profound intellectual leaps in the history of science, who linked probability, disorder, and spontaneous change at the molecular level with spontaneous change and disorder at the macroscopic level. He accomplished this by mathematically linking entropy, S, with the number of microstates, W, possessed by a system of molecules. The relationship he derived turned out to be disarmingly simple:

$$S = k_B \ln W$$

where S is the entropy of the system, k_B is the constant of proportionality between S and the natural logarithm of the number of microstates, W, possessed by the system. Of all his contributions to science, and there were many, it was the union of probability, disorder, and spontaneous change at the molecular level in a starkly simple mathematical expression of which he was most proud. By his request, it was etched on his tombstone.

The number of microstates, W, available to a set of N molecules in a volume of space is displayed in Figure 4.16. We can also consider the number of microstates, W, with respect to the number of ways, *for a given total* energy, that the molecules can be assigned to the available energy states. Let's consider a specific example.

Comparing Entropy to Money Subdivisions

Two ways to count out $2 with paper money.

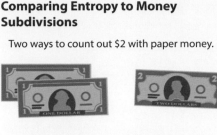

Five ways to count out $2 with coins.

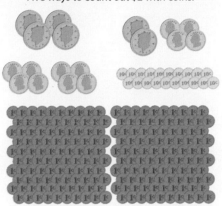

There are a remarkable number of illustrative comparisons between entropy and many common quantities that we are all acquainted with. One example is the way we can break down currency into a multitude of different subdivisions, each with a different degree of disorganization. If we consider, for example, the ways of counting out $2, we can do this with a single $2 bill or two $1 bills. However, if we can use coins, we can represent $2 with: four 50-cent coins, two 50-cent coins in combination with four 25-cent coins, or eight 25-cent coins, or twenty 10-cent coins, or two hundred 1-cent coins. As the number of coins increases, so too does the degree of disorganization and thus the entropy.

Consider the reaction

$$3A \rightarrow 3B$$

where the A molecules are distinguished by the fact that they can take on energies that are multiples of 10 energy units, and B molecules can take on energies that are multiples of 5 energy units, as shown in Figure 4.17.

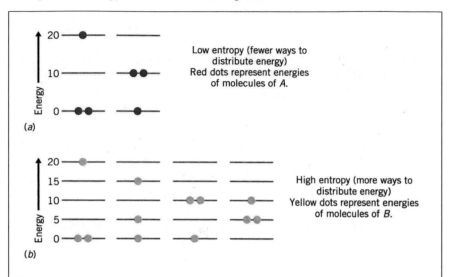

FIGURE 4.17 An important example of the relationship between the entropy of a molecular system, the number of states available for a given amount of energy, and the concept of disorganization is the case of energy distributed within the available energy levels of a molecule. For example, if we have 20 units of energy available to distribute among the energy levels of two different molecules, molecule A with an energy separation of 10 units between states and molecule B with an energy separation of 5 units between states, we can determine the number of ways that 20 units of energy can be distributed within the available energy levels for the two cases. For case A above, there are only two ways to distribute the 20 units of energy. For case B there are four ways to distribute the 20 units of energy. The entropy of case B is higher than that of case A because there are more ways to distribute the same amount of energy in case B. Note also that this broader distribution of energy in case B also represents a more disorganized state. Thus ΔS > 0 when transitioning from case A to case B.

We know that we must count the number of microstates *for a given amount of total energy* because those are the only microstates available to the system. So suppose that the total energy of the reacting mixture is 20 units. Then, with this restriction on total energy, there are two ways to distribute 20 units of energy among the three molecules of A. But there are four ways to distribute 20 units of energy among the three molecules of B, as shown in Figure 4.17.

Thus, the number of microstates of A is two, that of B is four, and thus the entropy, S, of B is higher than the entropy of A simply because there are more ways of distributing the *same amount of energy* in B than in A.

But entropy is a concept tightly connected with the degree of order/disorder in a system. We saw this pattern clearly emerging when we considered the available microstates available to the two objects each with 3 molecules. In the case of our two objects each comprised of three molecules shown in Figure 4.15, the initial state (defined by an object with three high energy [hot] molecules and the other object with three low energy [cold] molecules) was more *ordered* than the final state with energy distributed in 20 different ways. The same was true when we begin with 3, 10, 100, or N_A molecules in one chamber and then allow the system to spontaneously proceed to its final state with molecules equally distributed between the two chambers. It was true for our molecules, one with large energy separation and one with smaller energy separation.

Systems that have a high degree of *order* are in low entropy states and systems with a high degree of *disorder* are high entropy states. This leaves us with a powerful and irrefutable conclusion. If we move from a state of low entropy to a state of high entropy, ΔS is positive *and the change happens spontaneously*:

$$\Delta S = S_{final} - S_{initial} > 0$$

The system proceeds from a low probability, low entropy state spontaneously to a high probability, high entropy condition.

Qualitative Prediction of Entropy Change: Establishing the Sign of ΔS

While entropy is a state variable, and we can and will use it in specific, quantitative calculations, it is of great importance to develop an ability to determine the *sign* of the change in entropy for a host of important cases. It is to this topic that we now turn, because it is the *sign* of ΔS that distinguishes whether entropy change favors a spontaneous process versus a nonspontaneous process.

To begin analyzing the sign of entropy change for various systems, we consider first a simple ordered system of atoms in a lattice. Such a case is displayed in Figure 4.18.

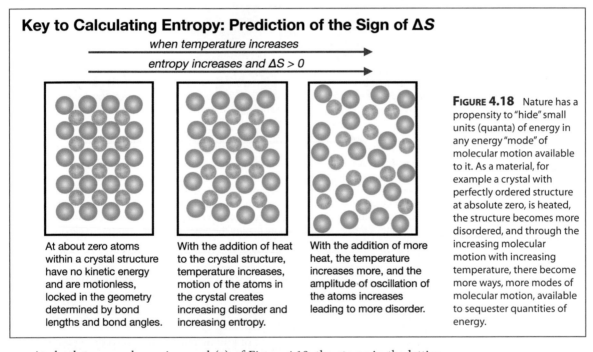

Key to Calculating Entropy: Prediction of the Sign of ΔS

when temperature increases

entropy increases and $\Delta S > 0$

At about zero atoms within a crystal structure have no kinetic energy and are motionless, locked in the geometry determined by bond lengths and bond angles.

With the addition of heat to the crystal structure, temperature increases, motion of the atoms in the crystal creates increasing disorder and increasing entropy.

With the addition of more heat, the temperature increases more, and the amplitude of oscillation of the atoms increases leading to more disorder.

FIGURE 4.18 Nature has a propensity to "hide" small units (quanta) of energy in any energy "mode" of molecular motion available to it. As a material, for example a crystal with perfectly ordered structure at absolute zero, is heated, the structure becomes more disordered, and through the increasing molecular motion with increasing temperature, there become more ways, more modes of molecular motion, available to sequester quantities of energy.

At absolute zero, shown in panel (a) of Figure 4.18, the atoms in the lattice possess no thermal energy and are locked in place. This corresponds to the condition of *maximum order* and minimum energy. As we increase the temperature, the atoms in the lattice gain thermal energy, which, at the molecular level, corresponds to the kinetic energy of vibration and rotation as the atoms in the lattice are displaced from their equilibrium positions. This displacement from equilibrium introduces a random component to the atomic positions and with that a *disordering* of the crystal lattice. At still higher temperatures, the increase in thermal energy manifests in larger amplitude departures from equilibrium positions within the lattice and a requisite increase in disorder. With increasing disorder comes increasing entropy such that as we progress upward in temperature from absolute zero, $\Delta S > 0$ because entropy increases with increasing disorder and disorder increases with increasing temperature. It is important to recognize that as the amplitude of the departure of the atoms from their equilibrium position increases, new "modes" of vibrational and rotational motion become available to the crystal system, and with this increasing number of modes comes an increase in the *number of microstates* within which increments of energy can be sequestered. Thus, while we do not yet have formulas to explicitly calculate the number of these microstates, we do know that with increasing temperature comes an increase in the number of available microstates and, consequently, an increase in entropy.

We turn next to the question of how changes in phase—from solid to liquid to gas—affect the entropy of the system. This progression is shown in Figure 4.19.

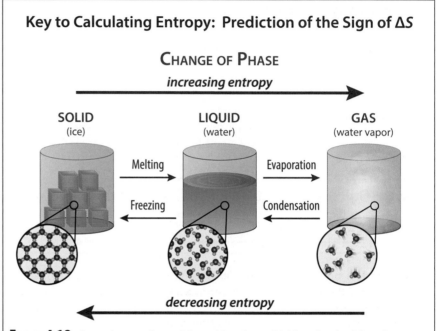

Key to Calculating Entropy: Prediction of the Sign of ΔS

CHANGE OF PHASE

increasing entropy

| SOLID (ice) | LIQUID (water) | GAS (water vapor) |

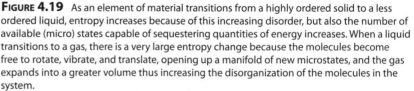

decreasing entropy

FIGURE 4.19 As an element of material transitions from a highly ordered solid to a less ordered liquid, entropy increases because of this increasing disorder, but also the number of available (micro) states capable of sequestering quantities of energy increases. When a liquid transitions to a gas, there is a very large entropy change because the molecules become free to rotate, vibrate, and translate, opening up a manifold of new microstates, and the gas expands into a greater volume thus increasing the disorganization of the molecules in the system.

The solid phase is, of course, characterized by an ordered, strongly bonded, and rigid structure that significantly limits the amplitude of atomic/molecular motion about the equilibrium position set by bond lengths within the solid. As thermal energy is added to the structure of the solid, the atoms bound in the solid take on an increasing amount of kinetic energy and in so doing *increase the disorder of the system*, and entropy *increases* as a result of the increasing number of ways energy can be stored in the available microstates of the system. However, when a phase transition takes place as the system melts to form a liquid, the bonding weakens its control over the motion of the molecules, leading to a dramatically more disordered structure characterized by broken bonds and a far more facile motion of the atoms within the material, and thus a significant increase in entropy in transitioning between a solid and a liquid.

With further heating, the liquid transitions to the gas phase where the atoms and/or molecules that comprise the system are now free to move, and while they are frequently colliding, they are not chemically bonded in any *systematic pattern*. This results in a far more disordered structure and a large increase in entropy. In fact, the unrestricted motion of the molecules allows energy to be sequestered in

Calculating Entropy: Prediction of ΔS
An increase in freedom of molecular motion corresponds to an increase in entropy volume.

The expansion of a gas into a vacuum:
1) A gas in a container separated from a vacuum by a partition
2) The gas at the moment the partition is removed.
3) The gas expands to achieve a more probable (higher entropy) particle distribution.

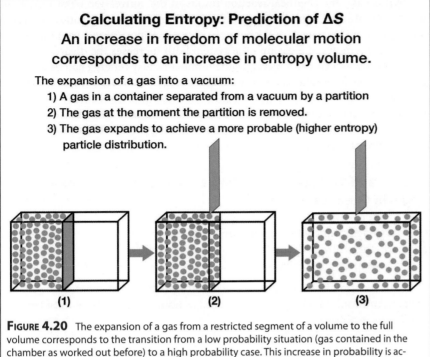

(1)　　　　(2)　　　　(3)

FIGURE 4.20 The expansion of a gas from a restricted segment of a volume to the full volume corresponds to the transition from a low probability situation (gas contained in the chamber as worked out before) to a high probability case. This increase in probability is accompanied by an increase in entropy.

translational motion, rotational motion, and vibrational motion, each of which comprises a manifold of available microstates of the system. This results in the very large increase in entropy in going from a liquid to a gas.

We consider next the case of entropy change resulting from the expansion of a gas from a small volume at high pressure to a larger volume at reduced pressure. We can execute such a process by containing a volume of gas behind a partition that, when withdrawn, allows the gas to expand, as shown in Figure 4.20.

The determination of the change in entropy from such a transition can be gauged by recognizing that, as we deduced from calculating the probability for finding a given number of molecules in one or the other of two chambers (see Figure 4.6), the probability for finding the molecules evenly distributed throughout the volume is much higher than finding the molecules isolated in one-half of the volume. Thus, entropy increases significantly in going from configuration (b) to configuration (c) in Figure 4.20. But we can also predict that $\Delta S > 0$ for this progression from (b) to (c) by recognizing that (c) is a condition of *greater disorder* than (b) and thus *entropy is larger for (c) than for (b).*

What happens to the entropy of a system when a chemical reaction takes place during which reactants are converted to products and the number of moles of product molecules is greater than the number of moles of reactants? Consider the case when bicarbonate of soda (baking soda) is heated—as is the case when bread is baked. The products of the reaction are sodium carbonate, carbon dioxide, and water. This reaction is displayed in Figure 4.21.

Calculating Entropy: Predicting the Sign of ΔS
When a chemical reaction produces or consumes gases, the sign of ΔS is usually easy to predict.

Example: Bicarbonate of soda used in cooking

$$2NaHCO_3(s) \xrightarrow{heat} Na_2CO_3(s) + CO_2(g) + H_2O(g)$$

causes cake to rise　　*adds moisture*

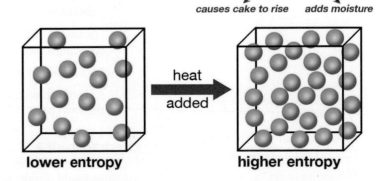

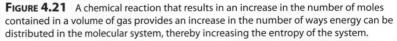

lower entropy　　　　　higher entropy

FIGURE 4.21 A chemical reaction that results in an increase in the number of moles contained in a volume of gas provides an increase in the number of ways energy can be distributed in the molecular system, thereby increasing the entropy of the system.

Solution

1. In this case, wherein a solid reacts to form gas phase species, the highly organized molecular structure of the solid gives way to the highly disorganized combination of independent gas phase molecules. Entropy increases so $\Delta S > 0$.

2. In this case, one gas phase propane molecule reacts with five gas phase oxygen molecules to yield three gas phase CO_2 molecules and four H_2O molecules in the liquid phase. Thus six gas phase molecules are converted to four gas phase molecules and four liquid phase molecules. Thus entropy decreases and $\Delta S < 0$.

3. Conversion of SO_2 to SO_3 involves three gas phase species converting to two gas phase molecules. Entropy decreases so $\Delta S < 0$.

4. This reaction from aqueous cane sugar to solid sucrose represents a phase change from liquid to solid. Entropy decreases so $\Delta S < 0$.

5. In the "water gas shift reaction," two gas phase molecules convert to two other gas phase molecules. Thus, while there will be a shift in entropy because the molecular structure of reactants to products differs somewhat, the entropy change will be small, so $\Delta S \approx 0$.

In this case, the chemical reaction results in the conversion of two moles of bicarbonate of soda (a solid) to one mole of sodium carbonate (a solid), one mole of carbon dioxide (a gas), and one mole of water vapor (also a gas). Thus, the release of two moles of a gas into the system both increases the disorder, by converting a solid to a gas, and increases the number of ways energy can be stored in the translational, vibrational, and rotational energy of the constituent molecules in the system. The result is that $\Delta S > 0$ for such a reaction. Incidentally, as the cake is baked, the CO_2 released forms "air" pockets and the cake will "rise" in the oven, but in addition the water released will be held interstitially, contributing to the moist nature of the product!

Check Yourself 1—Making Qualitative Predictions of Entropy Changes in Chemical Processes

The key to understanding entropy change is predicting the sign of ΔS. The objective here is to determine whether entropy increases, $\Delta S > 0$, for a given chemical reaction; whether entropy decreases, $\Delta S < 0$; or whether it is uncertain, $\Delta S \approx 0$. We consider five important examples:

1. The conversion of ammonium nitrate to nitrogen gas, oxygen gas, and water vapor. Ammonium nitrate, a solid, is used as both an agricultural fertilizer and as an explosive:
$$2NH_4NO_3(s) \rightarrow 2N_2(g) + 2O_2(g) + 4H_2O(g)$$

2. Combustion of propane to produce $CO_2(g)$ and $H_2O(l)$:
$$C_3H_8(g) + 5O_2(g) \rightarrow 3CO_2(g) + 4H_2O(l)$$

3. The reaction of SO_2 with oxygen to produce SO_3:
$$2SO_2(g) + O_2(g) \rightarrow 2SO_3(g)$$

4. Conversion of cane sugar in the liquid phase to sucrose:
$$C_{12}H_{22}O_{11}(aq) \rightarrow C_{12}H_{22}O_{11}(s)$$

5. One of the most important reactions in the conversion of coal to gasoline (C_8H_{18}) involves the "water gas shift reaction":
$$CO(g) + H_2O(g) \rightarrow CO_2(g) + H_2(g)$$

We can summarize by noting that there are four changes, in proceeding from an initial condition to a final condition, that produce an increase in entropy such that $\Delta S = S_f - S_i > 0$:

1. An increase in the temperature of a substance. The increase in temperature means an increased number of accessible energy levels such that energy may be sequestered in the system in a greater variety of ways (vibration, rotation, translation).

2. Formation of a liquid from a solid, resulting in a less ordered structure replacing a more ordered structure.

3. Formation of a gas from a liquid or from a solid.

4. A chemical reaction that produces a larger number of moles of a product than was present in the reacting species.

(a) Conversion of dioxygen to ozone: 3 $O_2(g) \rightarrow 2\ O_3(g)$

(b) Dissociation of N_2O_4 to NO_2: $N_2O_4(g) \rightarrow 2\ NO_2(g)$

(c) Formation of BrCl: $Br_2(l) + Cl_2(g) \rightarrow 2\ BrCl(g)$

Check Yourself 2

Without performing detailed calculations, indicate whether any of the following reactions would occur to a measurable extent at 298 K based upon the change in entropy, ΔS.

Quantitative Treatment of Entropy: Calculating ΔS for a System

We begin our quantitative treatment of entropy by emphasizing again that an understanding of the parity between energy and entropy evolved along two independent lines of reasoning: (1) the molecular-level approach taken by Boltzmann that we have discussed, and (2) studies of the efficiency of the steam engine by James Watt and collaborators. While the details of how efforts leading to the development of more efficient steam engines will enter our discussion of entropy repeatedly, we consider here the issue of how an entropy *change*, ΔS, can be quantitatively measured *in practice* by measuring macroscopic properties of a system. Thus we pursue approach (2) before returning to the microscopic approach developed by Boltzmann.

In Chapter 3, we used the Carnot cycle to calculate the efficiency of a reversible heat engine

$$\varepsilon = \frac{q_h + q_c}{q_h} = \frac{T_h - T_c}{T_h}$$

We calculated this efficiency using reversible isothermal and adiabatic expansion and compression legs on a pV diagram. We concluded in Chapter 3 that *it is of great importance to use this quantitative calculation of efficiency* to establish the fact that

- the efficiency of a Carnot engine depends only on the temperature of the hot and cold reservoirs;
- the most efficient heat engine, which must use irreversible paths on any real pV diagram, will have an efficiency less than that of a Carnot engine; and
- it is impossible to convert heat to work with an efficiency of 100% in a cyclic process.

However, this expression for the efficiency of a Carnot engine in terms of (1) the heat added and subtracted and (2) the temperature of the high and low temperature reservoirs has significant implications for the *quantitative* calculation of the entropy involved in each stage of the cycle. Equation (4.1) can be rewritten such that

$$q_h/q_h + q_c/q_h = T_h/T_h - T_c/T_h$$

so

$$1 + q_c/q_h = 1 - T_c/T_h$$

or

$$q_h/T_h + q_c/T_c = 0$$

We can break up our cycle into a series of incremental steps dq such that dq/T integrated over a *reversible cycle* is

$$\int \frac{dq_h}{T_h} + \int \frac{dq_c}{T_c} = 0$$

This, in turn, implies that there is a quantity, $\int \frac{dq_{rev}}{T}$, that is *conserved* in this complete cycle on the pV diagram.

What is remarkable is that the quantity whose *change* is calculated by $\int \frac{dq_{rev}}{T}$ in a reversible cycle is the entropy, ΔS, such that

$$\Delta S = \int \frac{dq_{rev}}{T} \quad \text{for each segment of the cycle.}$$

Because S is a state variable, the path chosen for the integral must always be the *reversible* path, independent of the path actually taken in a real process.

For a real heat engine, which must also be an irreversible heat engine,

$$\varepsilon_{real} < \varepsilon_{carnot}$$

so

$$q_h/T_h + q_c/T_c < 0$$

for a real, irreversible, spontaneous system.

Because S is a state variable, and thus $(\Delta S)_{cycle}$ must equal zero, we have the important result that for an *arbitrary* cycle:

$$\Delta S \geq \int dq \Big/ T$$

So because real, irreversible engines are the only ones that can actually produce usable work, the inequality above is a requirement all engines must satisfy. Therefore, this relationship is a criterion for the *spontaneous* operation of a real engine.

The most critical result by far from this analysis, however, is that we have deduced the quantitative form of the change in entropy, specifically that

$$\Delta S = \int \frac{dq_{rev}}{T}$$

for a reversible process, and in general, for a *spontaneous* process,

$$\Delta S \geq \int dq \Big/ T.$$

As we have seen in our analysis of the Carnot cycle, isothermal work for an ideal gas depends only on n, T, and volume or pressure ratios. Thus the actual pressures or volumes need not be known, although for simplicity you could easily assume, say, $P_2 = 1$ atm, and proceed to solve the problem. For our case, $V_2 = 3V_1$, the reversible work is

$$w_{rev} = -nRT \ln\left(\frac{V_2}{V_1}\right)$$

$$= -(2 \text{ mol})(1.9872 \text{ cal/K mol})(298 \text{ K})\ln 3$$
$$= -1301 \text{ cal } (-5440 \text{ J})$$

The work ratio *for the same final state* is

$$\frac{w_{rev}}{w_{irrev}} = \frac{-nRT \ln 3}{-nRT\left(1 - \frac{1}{3}\right)} = \frac{3}{2}\ln 3 = 1.65$$

Comparing this to the doubled volume discussed in the text, the reversible advantage is greater; for a volume ratio r, the work ratio is $[r/(r-1)]\ln r$, which grows monotonically as r increases.

Check Yourself 3—Reversible and Irreversible Work

Consider 2 mol of an ideal gas at 298 K expanding isothermally to three times its original volume. Calculate the reversible work in J. Show that the ratio of this work to the irreversible work resulting from an isothermal expansion at 298 K against a constant external pressure equal to the final system pressure is $(3/2) \ln 3$, a number greater than unity. Accompany your calculations with a pV diagram showing the two expansions.

While it is essential to recognize the relationship between the infinitesimal progression along a fully reversible trajectory for the calculation of

$$\Delta S = \int dq_{rev}\Big/ T \ ,$$

the question arises, is there a more tractable expression for ΔS that doesn't require an integral expression? The answer is yes. If the process proceeds along an *isothermal* path, as we did in leg I and leg III of the Carnot cycle, T is a constant and can be withdrawn from the integral

$$\Delta S = \int \frac{dq_{rev}}{T} = \frac{1}{T}\int dq_{rev}$$

But $\int dq_{rev} = q_{rev}$, the total amount of heat added or removed in the reversible process, so for an isothermal trajectory

$$\Delta S = \int \frac{dq_{rev}}{T} = \frac{1}{T}\int dq_{rev} = q_{rev}\Big/ T$$

So, although it took many decades in the 19th century to establish specifically how an entropy change could be measured, it was demonstrated empirically that entropy change, ΔS, can be calculated from two measurable quantities, heat, q, and temperature, T, using the simple expression:

$$\Delta S = q_{rev}/T \qquad\qquad \textbf{(4.1)}$$

where q_{rev} is the heat added to the system in a reversible process and T is the temperature in Kelvin.

At one level, the expression makes intuitive sense, in that as more thermal energy (heat) is added to the system, a larger number of energy levels become available, increasing the entropy of the system as our qualitative treatment demonstrated. But our expression for ΔS also expresses the fact that a given transfer of thermal energy, q, produces *greater disorder for a cold sample* than for a *hot sample*. This also makes intuitive sense because as we progress upward in temperature from absolute zero, a greater fractional change in available states will occur at the lowest temperature. Also, as we consider the loss of heat from, for example, a cup of coffee, we recognize that for a given amount of thermal energy (heat) transferred, q_{trans}, from the high temperature cup to the surroundings, the quantity q_{trans}/T_{hot} at the beginning of the cooling process will be smaller than the quantity q_{trans}/T_{cold} close to the end of the process. Thus,

$$q_{trans}/T_{hot} < q_{trans}/T_{cold}$$

and entropy (of the cup plus surroundings) increases as the cooling process proceeds.

But whereas our empirically determined expression for entropy change stated in terms of observed, macroscopic quantities (Equation. 4.1) appears to be simple, it requires careful consideration. First, we know that S is a state function and, thus, so too is ΔS. But q is not a state function; it depends sensitively on the path by which heat was added to the system. We recognized this explicitly in our discussion of the First Law of Thermodynamics.

A clear example of the path dependence of thermal energy (heat) transfer that follows from the First Law of Thermodynamics is the case of transitioning from an initial state of internal energy, U_i, to a final state of internal energy, U_f, via two different paths: (a) an isochoric process wherein the volume does not change such that $\Delta U = U_f - U_i = q + w = q_V$ because $w = 0$ and $q = q_V$, the heat transferred at constant volume; and (b), a case where the volume does change, $w \neq 0$, and $\Delta U = U_f - U_i = q + w$. These two cases are represented graphically in Figure 4.22.

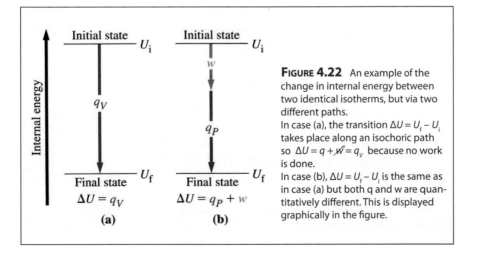

FIGURE 4.22 An example of the change in internal energy between two identical isotherms, but via two different paths.
In case (a), the transition $\Delta U = U_f - U_i$ takes place along an isochoric path so $\Delta U = q + \cancel{w} = q_V$ because no work is done.
In case (b), $\Delta U = U_f - U_i$ is the same as in case (a) but both q and w are quantitatively different. This is displayed graphically in the figure.

Equation 4.1 holds only for a specific path—a path that is reversible; thus the subscript on q_{rev}. A reversible process is one that can be made to reverse its direction when just an infinitesimal change is executed in the opposite direction. Perhaps the most commonplace example of the distinction between a reversible and an irreversible path is the melting of ice. As adjoining Figure 4.23 reminds us,

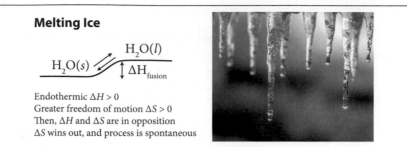

Melting Ice

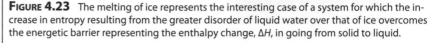

$$H_2O(s) \rightleftarrows \begin{array}{c} H_2O(l) \\ \uparrow \Delta H_{fusion} \end{array}$$

Endothermic $\Delta H > 0$
Greater freedom of motion $\Delta S > 0$
Then, ΔH and ΔS are in opposition
ΔS wins out, and process is spontaneous

FIGURE 4.23 The melting of ice represents the interesting case of a system for which the increase in entropy resulting from the greater disorder of liquid water over that of ice overcomes the energetic barrier representing the enthalpy change, ΔH, in going from solid to liquid.

the melting of ice is spontaneous, but endothermic. If we melt an ice cube at 25°C, the melting process proceeds inexorably to convert all the solid to liquid. That represents an irreversible process, and thus we could not use that experiment to calculate ΔS from $\Delta S = q_{rev}/T$. On the other hand, if we perform the same experiment at 0°C, we can, at any point, stop the melting process and reverse its direction. At 0°C, the melting of ice is thermodynamically reversible and we can use the experiment to determine ΔS for the process. In fact, we can express q_{rev} as $q_{0°C}$ for the case of melting ice and write

$$\Delta S_{melt} = q_{0°C}/T$$

But also notice that if, as is almost always the case, the experiment is carried out at constant pressure, then

$$q_{0°C} = q_p = \Delta H$$

so that

$$\Delta S_{melt} = \Delta H_{fusion}/T$$

where ΔH_{fusion} was examined in some detail in Chapter 3.

Solution
First, we write the conversion from liquid to gas phase as a chemical reaction:

$$H_2O(\ell, 1 \text{ atm}) \rightleftarrows H_2O(g, 1 \text{ atm})$$

$$\Delta H°_{vap} = 40.7 \text{ kJ/mol}$$

Thus

$$\Delta S°_{vap} = \frac{\Delta H°_{vap}}{T} = \frac{40.7 \text{ kJ/mol}}{373 \text{ K}} = 0.109 \text{ kJ/mol} \cdot \text{K}$$

Practice Problem: Chlorofluorocarbons were widely used as both refrigerants and cleaning agents for electronic circuit board fabrication. It was discovered that chlorofluorocarbons, because of their stability as a molecular structure, are not removed in the lower atmosphere but mix into the stratosphere where they are broken apart by ultraviolet radiation. One example is chloroflurocarbon 12, which is CCl_2F_2. The resulting Cl atoms engage in a catalytic reaction cycle that destroys ozone—a topic treated in detail in Chapter 12.

Calculate the standard molar entropy of vaporization, ΔS_{vap}, for CCl_2F_2 if its boiling point is –29.79°C and $\Delta H_{vap} = 20.2$ kJ/mol.

Check Yourself 4—Calculation of Entropy Change for a Phase Change Between Reactants and Products
One of the most common and most important phase changes is the conversion of liquid phase water to gas phase water vapor. This reaction is particularly important in understanding the response of the Earth's climate to increasing use of fossil fuels.
Problem: Calculate the standard molar entropy change when one mole of liquid phase water at 1 atm pressure is converted to one mole of gas phase water at 373K also at 1 atm pressure. The standard molar enthalpy of vaporization is 40.7 kJ/mol.

So if we know the heat required for the process to proceed along the *reversible* path, then we know ΔS for the same process along *any* path, whether it is reversible or irreversible, because ΔS is a state variable. Quite clearly, the converse is not true, i.e., we cannot calculate q from ΔS and T for any path because q is not a state variable and thus is path dependent.

It is worth noting at this point what the behavior of entropy is as a function of temperature. This is displayed graphically in Figure 4.24.

4.26

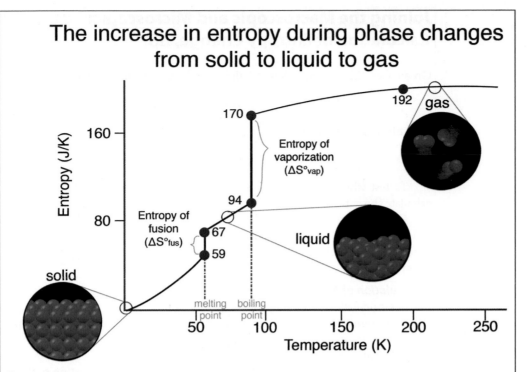

The increase in entropy during phase changes from solid to liquid to gas

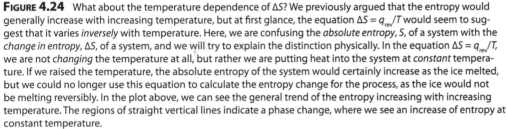

FIGURE 4.24 What about the temperature dependence of ΔS? We previously argued that the entropy would generally increase with increasing temperature, but at first glance, the equation $\Delta S = q_{rev}/T$ would seem to suggest that it varies *inversely* with temperature. Here, we are confusing the *absolute entropy*, S, of a system with the *change in entropy*, ΔS, of a system, and we will try to explain the distinction physically. In the equation $\Delta S = q_{rev}/T$, we are not *changing* the temperature at all, but rather we are putting heat into the system at *constant* temperature. If we raised the temperature, the absolute entropy of the system would certainly increase as the ice melted, but we could no longer use this equation to calculate the entropy change for the process, as the ice would not be melting reversibly. In the plot above, we can see the general trend of the entropy increasing with increasing temperature. The regions of straight vertical lines indicate a phase change, where we see an increase of entropy at constant temperature.

As we have seen before, as the temperature increases, entropy increases smoothly for a given phase (i.e., solid, liquid, or gas) and increases discontinuously at a phase transition.

But Figure 4.24 displays another feature of entropy that is of considerable importance. Notice that, as the temperature approaches absolute zero, so too does the entropy. We noted this qualitatively when we discussed the highly ordered state of a crystal at absolute zero, which represented the state of minimum entropy.

Check Yourself 5

5.00 mol of steam, $H_2O(g)$, cools from 350°C to 100°C at constant pressure without condensing. Calculate ΔS in cal/K and J/K based on ideal gas behavior.

At constant pressure

$$\Delta S = \int_{T_1}^{T_2} \frac{nc_V dT}{T} = nc_V \ln\left(T_2/T_1\right)$$

with c_p replacing c_v, gives

$$\Delta S = nc_p \ln\left(\frac{T_2}{T_1}\right)$$

$$= (5.00 \text{ mol})(3511 \text{ J/K mol})\ln\frac{373K}{623K}$$

$$= -9.03 \text{ kJ/K}$$

The heat capacity of steam increases slowly between 100°C and 350°C, reaching 8.75 cal/K mol; this yields a slightly greater drop in S.

Joining the Macroscopic and Microscopic: Calculation of Entropy Change, ΔS

On the face of it, the expression for the microscopic formation of entropy change

$$\Delta S = S_f - S_i = k_B \ln W_f - k_B \ln W_i = k_B \ln (W_f/W_i)$$

has little to do with the macroscopic formulation

$$\Delta S = q_{rev}/T$$

To test whether we get the same answer, lets select a case wherein we can calculate the entropy change for each formulation and then compare the results.

Consider a system comprised of 1 mol of a gas expanded from 1 L to 2 L at 298 K. We approach the problem first from a statistical approach based on the number of microstates. Then we will repeat the calculation based on the heat absorbed by the system.

1. Calculation of ΔS Based on Microstates

For this calculation we assume an *initial* condition of the 1 mol of gas in a 1L chamber connected by a valve to a second chamber of equal volume. The *final* condition is the expansion of 1 mol of gas into the second chamber such that the 1 mol of gas now occupies 2L of volume.

We have developed all the information we need for the calculation of the entropy change, but let's work through what happens, beginning not with 1 mol of gas (6.02×10^{23} molecules) but with 1 then 2, then 3, then 10 molecules to derive the required formula.

Referring to Figure 4.16, which is a summary of Figures 4.11, 4.12, 4.13, and 4.14, with one molecule, the number of microstates available with just a single chamber available is $W = 1$. When the chamber is opened, the number of microstates available is $W = 2^1 = 2$, or twice as many. With two molecules, initially there is again $W = 1$ microstate available. After opening the valve there are $2^2 = 4$ microstates available. For three molecules there are $2^3 = 8$ microstates available after opening the valve. For ten molecules there are $2^{10} = 1024$ microstates available as displayed explicitly in Figure 4.16.

But we now see a simple pattern emerging. The ratio of microstates available *after* opening the valve to double the volume is

$$\frac{W_{final}}{W_{initial}} = 2^N$$

where N is the number of molecules in the system. For our problem, N happens to be 1 mol or Avogadro's number, N_A, of molecules. Thus

$$\frac{W_{final}}{W_{initial}} = 2^{N_A}$$

The entropy change for the system using the Boltzmann equation is:

$$\Delta S = S_f - S_i = k_B \ln W_{final} - k_B \ln W_{init} = k_B \ln (W_f/W_i)$$

But the Boltzmann constant $k_B = R/N_A$, the ratio of the gas constant to Avogadro's number. We also remember the logarithmic identity that

$$\ln A^X = x \ln A$$

so

$$\Delta S = k_B \ln \left(\frac{W_f}{W_i} \right) = \left(\frac{R}{N_A} \right) \ln 2^{N_A} = \left(\frac{R}{N_A} \right) N_A \ln 2$$

$$= R \ln 2 = \left(8.314 \frac{J}{mol \cdot K} \right) (0.693)$$

$$= 5.76 \frac{J}{mol \cdot K}$$

2. Calculation of ΔS Based on Heat Transfer at Temperature T

The application of the equation

$$\Delta S = q_{rev} / T$$

requires that the process be carried out along a reversible path at constant temperature. We know this formulation well. It is just the first leg of the Carnot cycle; an isothermal, reversible trajectory on a pV plot. If the process is isothermal, then $\Delta U = q + w = 0$ and $q = -w$. If the process is carried out reversibly,

$$q_{rev} = -w_{rev} = nRT \ln (V_f/V_i)$$

Thus for one mol of gas, n = 1, and

$$\Delta S = \frac{q_{rev}}{T} = \frac{nRT}{T} \ln \left(\frac{V_f}{V_i} \right)$$

$$= R \ln \left(\frac{V_f}{V_i} \right)$$

But for our problem here, $V_f = 2$ and $V_i = 1$ so $(V_f/V_i) = 2$ and

$$\Delta S = R \ln (V_f/V_i) = R \ln 2$$

$$= (8.314 \text{ J/mol·K})(0.693) = 5.76 \text{ J/K}$$

Thus we obtain the same answer whether we approach the problem from the microscopic or the macroscopic perspective. This is a very important conclusion because it demonstrates that the quantitative formulation of the entropy is exactly equal whether we employ the Boltzmann equation $\Delta S = k_b \ln(W_f/W_i)$ or the equation $\Delta S = q_{rev}/T$ for a reversible, isothermal process. Remarkable and very useful indeed!

The Second Law of Thermodynamics

We have shown that spontaneous processes are controlled by two compelling tendencies in nature. The first, defined by the *enthalpy* change, ΔH, expresses the tendency to find and follow whatever available pathway leads to a *lower energy state*. An exothermic reaction such as the combustion of octane

$$2C_8H_{18} + 25O_2 \rightarrow 16CO_2 + 18H_2O$$

is such an example. The second, defined by the entropy change, ΔS, expresses the tendency of a system to find and follow whatever available pathway leads to a *higher probability state*. There are cases for which the release of energy, $\Delta H < 0$, and the increase in entropy, $\Delta S > 0$, work in concert, driving a process in the forward direction: Octane combustion both releases energy and forms more molecules, increasing entropy. We have seen that spontaneous processes can also occur where enthalpy increases, $\Delta H > 0$, but as long as the change in entropy is sufficiently large that $-T\Delta S$ exceeds ΔH, the process is spontaneous. An example is the melting of ice. Another example is the combustion of H_2 and O_2 to form H_2O, which is exothermic but entropy decreases because three moles of H_2 and O_2 combine to form two moles of water.

This sets up a head-to-head confrontation: Is the process controlled, as it proceeds forward, by the change in enthalpy, ΔH, or by the change in entropy, ΔS? As we will see after developing the *Second Law of Thermodynamics*, there are actually three factors that determine where a process is spontaneous: ΔH, ΔS, and the temperature, T.

The *Second Law of Thermodynamics* states that whenever a spontaneous event takes place in the universe, the total entropy of the universe must increase.

Gibbs Free Energy

But how do we put this law to practical use?

The first step is to distinguish clearly the difference between the entropy change of the system under consideration, ΔS_{syst}, the entropy change of the surroundings, ΔS_{sur}, and the entropy change of the Universe, ΔS_{total}, where the "universe" constitutes the system and the surroundings. With this delineation, we can write

$$\Delta S_{total} = \Delta S_{syst} + \Delta S_{sur}$$

Let's consider each of these terms, recognizing that our objective here is to relate the entropy change of the *system* to the entropy change of the *universe* by *mathematically eliminating any reference to the surroundings*. The strategy is based on the recognition that while our system can be *specifically defined* and the universe encompasses all entities, the surroundings are not, in general, clearly defined. Thus if we can succeed in eliminating any quantitative reference to the surroundings, we will have eliminated an ill-defined component from the derivation.

Building on the empirically determined fact that the entropy change of the surroundings, ΔS_{sur}, is equal to the heat transferred to the surroundings from the system, q_{sur}, divided by the Kelvin temperature, T, at which the transfer of thermal energy occurred, we can write

$$\Delta S_{sur} = q_{sur}/T$$

But we know from the First Law that

$$q_{sur} = -q_{syst}$$

and, moreover, that at *constant pressure and temperature*,

$$q_{syst} = \Delta H_{syst}$$

for the system. Thus,

$$\Delta S_{sur} = -\Delta H_{syst}/T$$

and we can directly substitute the quantity into our expression for ΔS_{total} such that

$$\Delta S_{total} = \Delta S_{syst} + \Delta S_{sur} = \Delta S_{syst} - \Delta H_{syst}/T$$

Rearranging gives us

$$\Delta S_{total} = (T\Delta S_{syst} - \Delta H_{syst})/T$$

And multiplying both sides by T yields

$$T\Delta S_{total} = T\Delta S_{syst} - \Delta H_{syst}$$

But now we have made considerable progress in that we have, by virtue of the First Law, *eliminated any reference to the entropy change of the surroundings*.

Expressing $T\Delta S_{total}$ using a simple sign change gives us

$$T\Delta S_{total} = -(\Delta H_{syst} - T\Delta S_{syst})$$

But now the statement of the Second Law takes important quantitative form, because if $\Delta S_{total} > 0$ for *any spontaneous event*, then it follows that

$$\Delta H_{syst} - T\Delta S_{syst} < 0$$

for *any spontaneous event*.

But this ostensibly simple relationship carries a very important message because it connects energy, entropy, and temperature to the concept of spontaneous change. In fact, the quantity

$$\Delta H_{syst} - T\Delta S_{syst}$$

is so important to thermodynamics that it defines a new state variable, G, the Gibbs free energy

$$G = H - TS$$

so named after J. Willard Gibbs (1839–1903), who is the father of modern thermodynamics. For changes at *constant temperature and pressure*, this equation becomes

$$\Delta G = \Delta H - T\Delta S$$

and because G is constructed from state variables, it too is a state variable:

$$\Delta G = G_{final} - G_{initial}$$

Now we have a quantitatively powerful restatement of the Second Law:

At constant pressure and temperature, a change can only be spontaneous if it is accompanied by a decrease in the free energy of the system.

Gibbs Free Energy and Spontaneous Change

So with the *Second Law of Thermodynamics*, nature places rigid limits on the way energy and the molecules that comprise matter can be arranged and rearranged. This law sets in place specific terms: For a spontaneous change to occur—change that occurs with no outside intervention—the *combined* entropy of the system and surroundings *cannot* decrease. In fact, for a spontaneous change, the total entropy must increase except for a certain loss of idealized, reversible transformations that never happen in the real world. That is, for all *real* processes, the entropy of the *universe* increases and the Gibbs free energy of the *system* decreases.

Let's return to our equation

$$T\Delta S_{total} = -(\Delta H_{syst} - T\Delta S_{syst})$$

and substitute our expression for the change in Gibbs free energy

$$\Delta G_{syst} = -T\Delta S_{total}$$

linking the change in Gibbs free energy to the system with the entropy change of the universe.

The change in Gibbs free energy for the system, ΔG_{syst}, therefore has with it all the key information about the change in entropy of the universe (for a constant T, P case). As a result, as nature drives a system toward minimum free energy, then the universe is driven toward maximum entropy.

Chemically reacting systems seek the path to the lowest free energy, stopping only when a local energy minimum in free energy is attained and $\Delta G = 0$. Thus a process follows the slope of free energy inexorably downward, moving in whichever direction (toward products from reactants or from products to reactants) leads to a decrease in Gibbs free energy, G. As a result, because the proportion of reactants and products changes during the course of reaction, G and ΔG change continuously until G reaches a minimum and $\Delta G = 0$.

To summarize: When at equilibrium ($\Delta G = 0$) the system of reactants and products is no longer able to lower its free energy by making small changes to the concentrations of those reactants and products. With that, the universe is satisfied; it can acquire no more entropy, at least not from the process at hand. An irreversible change has taken place that leaves the world forever different. The entropy gain can never be taken back.

This drive toward minimizing Gibbs free energy is shown in Figure 4.25.

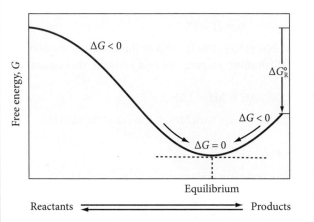

FIGURE 4.25 The chemical system will shift to the right or left as it proceeds inexorably downhill until ΔG = 0 and the system is in equilibrium, having transferred a maximum amount of entropy to the universe in the process.

(a) how much heat is absorbed from the surroundings;

(b) how much work the system does on the surroundings;

(c) the net direction in which the reaction occurs to reach equilibrium;

(d) the proportion of the heat evolved in an exothermic reaction that can be converted to various forms of work.

Check Yourself 6

The free energy change of a reaction can be used to assess:

While the complexity of quantitatively defining the conditions for spontaneous change resulted in a disarmingly simple thermodynamic expression, specifically that

$$\Delta G = G_{\text{final}} - G_{\text{initial}} < 0$$

we recognize immediately that because

$$\Delta G = \Delta H - T\Delta S$$

different *combinations* of ΔH, ΔS, and T will yield different conclusions regarding the spontaneous (or nonspontaneous) nature of a chemical process.

Let's consider the implied control of spontaneity as a function of the sign of ΔH and ΔS, recognizing that if

$$\begin{aligned} \Delta G &< 0 &\quad &\text{spontaneous process} \\ \Delta G &= 0 &\quad &\text{reversible process} \\ \Delta G &> 0 &\quad &\text{nonspontaneous process} \end{aligned}$$

Case 1:

When ΔH is negative (exothermic reaction) and ΔS is positive (disorder increases), an example shown in Figure 4.26, then

$$\Delta G = \Delta H - T\Delta S < 0$$

Because ΔH is negative and ΔS is positive, then $-T\Delta S$ is negative and

$$\Delta G = (-) - (T(+)) < 0$$

at all temperatures. Thus, for this case, the process is spontaneous at all temperatures.

Case 2:

When ΔH is positive (endothermic reaction) and ΔS is negative, so $-T\Delta S$ is positive and

$$\Delta G = (+) - T(-) > 0$$

at all temperatures. Thus, for this case, the process is nonspontaneous at all temperatures.

Case 3:

When ΔH and ΔS have the same sign, then the sign of

$$\Delta G = \Delta H - T\Delta S$$

depends upon the temperature. If ΔH and ΔS are both positive as shown in Figure 4.26, then

$$\Delta G = (+) - (T(+))$$

is the difference between two positive numbers. But with increasing temperature the $-T\Delta S$ term will dominate and the change will be *spontaneous at high temperature* but not at low temperature.

Case 4:

Conversely, if ΔH and ΔS are both negative, then

$$\Delta G = (-) - (T(-))$$

will be negative at low temperatures and thus *spontaneous at low temperatures*. At high temperatures the entropy term becomes dominant, $\Delta G > 0$, and the process is no longer spontaneous. These cases can be summarized in a diagram shown in Figure 4.27.

Referring to Figure 4.27, which of the four cases applies to:

1. $N_2(g) + 3H_2(g) \rightarrow 2NH_3(g)$ $\Delta H° = -92.2$ kJ
2. $2NH_4NO_3(s) \rightarrow 2N_2(g) + 4H_2O(g) + O_2(g)$ $\Delta H = -236.0$ kJ
3. 3. $2N_2O(g) + O_2(g) \rightarrow 4NO(g)$ $\Delta H = 197.1$ kJ
4. $4Fe(s) + 3O_2(g) \rightarrow 2Fe_2O_3(s)$ $\Delta H° = -1651$ kJ
5. $2C(graphite) + 2H_2(g) \rightarrow C_2H_4(g)$ $\Delta H° = 52.26$ kJ

Spontaneous Processes, ΔH and ΔS

We have learned about two thermodynamic variables, ΔH and ΔS, that affect a chemical or physical event. Sometimes they work together; sometimes they work in opposition.

Example 1: Octane combustion

$$2\,C_8H_{18}(l) + 25\,O_2(g) \rightarrow 16\,CO_2(g) + 18\,H_2O(g)$$

$\Delta H < 0$ Exothermic

$\Delta S > 0$ Greater number of molecules generated

Example 2: Melting of ice

$$H_2O(s) \rightarrow H_2O(l) \text{ at } T = 25°C$$

$\Delta H > 0$ Endothermic

$\Delta S > 0$ Greater freedom of motion

FIGURE 4.26 Using our equation for ΔG, we can see how the spontaneity of a reaction depends on the temperature. We see that if $\Delta H > 0$ and $\Delta S < 0$, ΔG is *always* greater than zero, and the reaction in question will never be spontaneous. But if ΔH and ΔS are in opposition, can we drive the reaction toward spontaneity? From our expression, we clearly see that varying the temperature of the process will change ΔG, and we consider cases individually to see how a change in temperature affects ΔG.

Spontaneity and Temperature

$$\Delta G = \Delta H - T\Delta S$$

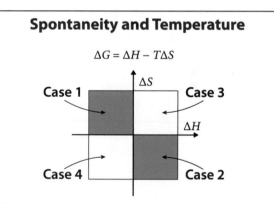

FIGURE 4.27 In the first case, we take $\Delta S > 0$ and $\Delta H > 0$. If $\Delta H > T\Delta S$, the process is not spontaneous. Notice that we always use the Kelvin scale for temperature, so T is always positive. In this scenario, if we increase the temperature sufficiently, we will ultimately reach a point where $\Delta H < T\Delta S$, and accordingly $\Delta G < 0$, and the process will proceed spontaneously at this temperature.

If $\Delta S < 0$ and $\Delta H < 0$, and $\Delta H < T\Delta S$, the process is again not spontaneous. In this case, if we *lower* the temperature, we will reach a temperature at which $\Delta G < 0$, and the reaction will be spontaneous.

We will make one assumption in this discussion, namely that the values of ΔH and ΔS do not change with temperature. This assumption is certainly not true for all cases. However, the approximation holds true in many cases, and in this course we will almost exclusively assume that ΔH and ΔS do not change with temperature.

Problem

1. What is ΔG°_{RX} at 298 K?

2. At what temperature will the reaction become spontaneous?

Solution:

1. The change in Gibbs free energy at 298 K is

$$\Delta G^\circ = \Delta H^\circ - T\Delta S^\circ = +135.6 \text{ kJ} - (298 \text{K})(0.334 \text{ kJ/K})$$
$$= +36.1 \text{ kJ}$$

Thus the reaction is nonspontaneous at 298 K—a fortunate fact since baking powder is sold in stores at STP!

2. To calculate the temperature at which baking soda decomposes to $CO_2(g)$ and $H_2O(g)$, examine first the signs of ΔH° and ΔS°:

$$\Delta H^\circ > 0 \text{ and } \Delta S^\circ > 0$$

Thus as the temperature increases, ΔG° will decrease such that at some elevated temperature, the reaction will become spontaneous when $\Delta G < 0$.

To solve for that threshold temperature, recognize that both ΔH° and ΔS° are independent of temperature to a first approximation, so the temperature at which the reaction becomes spontaneous is, from

$$\Delta G^\circ_T = \Delta H^\circ - T\Delta S^\circ \le 0$$

So for $\Delta G_T = \Delta H^\circ - T\Delta S^\circ = 0$

We have $T = \Delta H^\circ / \Delta S^\circ = 135.6 \text{ kJ} / 0.334 \text{ kJ/K} = 406 \text{ K}$

Practice Problems

When we consider the various cases for the sign of ΔH and of ΔS, we can summarize the determination of whether a given reaction is spontaneous vs. nonspontaneous using the convenient quadrant diagram shown in Figure 4.27. This figure defines the four possible combinations of the sign of ΔH and ΔS.

Check Yourself 7—Using Enthalpy and Entropy Changes to Predict the Direction or Spontaneous Change

The temperature dependence of the determination of whether a reaction is spontaneous or nonspontaneous can be determined from the values of ΔH, ΔS and the temperature. An interesting example is the case of baking soda, $NaHCO_3(s)$, that decomposes into $CO_2(g)$, $H_2O(g)$ and $Na_2CO_3(s)$ in the reaction

$$NaHCO_3(s) \rightarrow Na_2CO_3(s) + CO_2(g) + H_2O(g)$$

$$\Delta H^\circ_{RX} = +135.6 \text{ kJ} \qquad \Delta S^\circ_{RX} = +334 \text{ J/K}$$

Absolute Value for Entropy: The Third Law of Thermodynamics

While entropy may be a more abstract and esoteric concept than energy at this stage in our development of thermodynamics, unlike energy, we can in fact assign a universally acceptable *absolute* value to our scale of entropy!

To establish an absolute value for the entropy of a substance, we must pin the *zero point* entropy of the scale. The entropy of this state is taken to be zero and the change in entropy, ΔS, evaluated as the substance is brought to other states of temperature and pressure and these entropy increments, ΔS, are added to obtain a numerical value of the absolute entropy.

The quantitative existence of an absolute entropy scale is expressed universally as the *Third Law of Thermodynamics*.

The *Third Law*: At absolute zero, the entropy of a perfectly ordered pure crystalline substance is zero: $S = 0$ at $T = 0$ K.

The important point, of course, is that because we know the point at which entropy has the value zero, we can establish, by experimental measurement, the total amount of entropy that a substance has at temperatures above 0 K. This leads to the definition of the *standard molar entropy* of a compound, S°. Standard molar entropy is tabulated for conditions of 298 K and 1 atmosphere (1013 mb, 1013 hPa) pressure. Notice that the reference state for entropy is not at standard conditions (STP is at 273 K)!

It follows from the *Third Law* and the fact that standard molar entropies are tabulated at 298 K that for all elements in their standard states, the absolute entropy will have a positive value.

We can tabulate standard entropies of important substances—this is done in Appendix A. But we can also examine some representative standard entropies to demonstrate the range of values at 298 K, as shown in Table 4.2:

TABLE 4.2

Substance	$S°$ J/mol·K
C(s, graphite)	5.69
Fe(s)	27
S(s)	31.9
CaO(s)	40.0
$H_2O(l)$	69.96
$H_2O(g)$	188.7
$H_2(g)$	130.6
$N_2(g)$	191.5
$O_2(g)$	205.0
$C_2H_4(g)$	219.8
$C_2H_6(g)$	229.5
$C_8H_{18}(l)$	466.9
NaCl(s)	72.38

An examination of Table 4.2 reveals several general patterns in the values of standard entropies. First, the lowest values of standard entropy occur for solids, with the more ordered structures having the lowest values of standard entropy. Second, there is a large increase in standard entropy in going from a liquid to a gas of the same molecule. Third, as the molecule becomes larger (more complex) the standard entropy increases.

Calculation of Entropy Change for a Chemical Reaction

In the same way that we calculated ΔH for a reaction, we can calculate the entropy change for the process by the same procedure. Given the standard molar entropy, $S°$, for a compound, the change in entropy for the reaction, $\Delta S°_{rxn}$ for the reaction

$$aA + bB \rightarrow cC + dD$$

with reactants A, B with stoichiometric coefficients a, b and with products C, D with stoichiometric coefficients c, d, we have

$$\Delta S°_{rxn} = \sum_{p=products} v_p S°_p - \sum_{r=reactants} v_r S°_r$$

where v_p represents the number of moles of each product species and v_r represents the number of moles of each reactant species.

Just as we calculated the enthalpy change, $\Delta H°$, for a chemical reaction, so too, in an exactly analogous way, can we calculate the entropy change for a reaction using tabulated values for the standard entropy of each of the products and of the reactants.

Example:

Urea is an important form of fixed nitrogen that is manufactured from ammonia and carbon dioxide in the process

$$2NH_3(g) + CO_2(g) \rightleftharpoons CO(NH_2)_2(aq) + H_2O(l)$$

When urea is applied to the soil, it produces NH_3 in continuous amounts in a slow release with water in the soil:

$$CO(NH_2)_2(aq) + H_2O(l) \rightarrow CO_2(g) + 2NH_3(g)$$

Problem:

What is the standard entropy change when 1 mole of urea reacts with water?

Step 1:

Recognize that the problem to be solved is the calculation of the standard entropy change $\Delta S°$ given by:

$$\Delta S°_{rxn} = \sum_{products} v_p S°_p - \sum_{reactants} v_r S°_r$$

where v_p is the stoichiometric coefficient for each of the products and v_r is the stoichiometric coefficient for each of the reactants. Thus for the reaction of urea with water

For products: $v_{CO_2} = 1$ $v_{NH_3} = 2$

For reactants: $v_{CO(NH_2)_2} = 1$ $v_{H_2O} = 1$

Step 2:

Look up the standard entropy $S°$ for each of the reactants and products

Substance	$S°$ (J/mol·K)
$CO(NH_2)_2(aq)$	173.8
$H_2O(l)$	69.96
$CO_2(g)$	213.6
$NH_3(g)$	192.5

Step 3:

Execute the calculation

$$\Delta S°_{rxn} = \sum_{products} v_p S°_p - \sum_{reactants} v_r S°_r$$

$$= [213.6 \text{ J/mol·K} + (2)\ 192.5 \text{ J/mol·K}] - [173.8 \text{ J/mol·K} + 69.96 \text{ J/mol·K}]$$

$$= 354.8 \text{ J/mol·K}$$

It is always important to check the sign of ΔS against your qualitative prediction. In this case, the reaction converts two moles of reactant to three moles of product. This implies that the degree of disorganization increases in the course of the reaction such that $\Delta S > 0$. This is in accord with the calculation.

4.36

Check Yourself 8—Calculation of Entropy Change, ΔS, From Standard Molar Entropy Tables

Problem: One of the most important reactions in urban air pollution is the reaction of nitric oxide, NO, with oxygen, O_2, to form nitrogen dioxide, NO_2. The source of NO in the atmosphere is primarily from the combustion of fossil fuels wherein high temperatures within the combustion chambers of the heat engines (gasoline engines, diesel engines, gas turbines, etc.) break the N≡N triple bond, forming N atoms, and breaks the O=O double bond, thereby forming O atoms. The product NO_2 molecule is the yellow haze that is ubiquitous over urban areas. NO_2 is toxic because it forms nitric acid, burning lung tissue. Thus calculate the entropy change for

$$2NO(g) + O_2(g) \rightarrow 2NO_2(g)$$

Solution:
Following the convention that

$$\Delta S^\circ_{rxn} = \sum_{products} \nu_p S^\circ_p - \sum_{reactants} \nu_r S^\circ_r$$

and referring to Appendix A, we can calculate ΔS°_{rxn} given

$$S^\circ_{NO_2(g)} = 240.1 \text{ J/k·mol } S^\circ_{NO(g)}$$
$$= 210.8 \text{ J/k·mol}$$

$$S^\circ_{O_2(g)} = 205.1 \text{ J/k·mol}$$

So we have

ΔS°_{rxn} = 2 × 240.1 J/K·mol – (2 × 210.8 J/K·mol) – 205.1 J/K·mol = –146.5 J/K

It is important to always check the sign of ΔS°_{rxn} to see if it is in agreement with what you predict qualitatively for the reaction.

In the case of our reaction

$$2NO(g) + O_2(g) \rightarrow 2NO_2(g),$$

three moles of gas phase species react to form two moles of products such that entropy decreases as the system becomes more ordered. This implies that $\Delta S^\circ_{rxn} < 0$ in agreement with the calculation.

Practice Problem
Calculate the standard entropy change for the reaction of propane with molecular oxygen to form carbon dioxide and liquid water:

$$C_3H_8 + 5O_2(g) \rightarrow 3CO_2(g) + 4H_2O(l)$$

Calculation of Gibbs Free Energy for a Reaction

Given the ability to calculate the *standard entropy change* for a reaction

$$\Delta S^\circ_{rxn} = \sum_{products} \nu_p S^\circ_p - \sum_{reactants} \nu_r S^\circ_r$$

and the ability to calculate the *enthalpy* change for a reaction

$$\Delta H^\circ_{rxn} = \sum_{products} \nu_p \Delta H^\circ_{f,p} - \sum_{reactants} \nu_r \Delta H^\circ_{f,r}$$

we can now calculate the Gibbs free energy for the reaction

$$\Delta G^\circ = \Delta H^\circ - T\Delta S^\circ.$$

We use the reaction of urea with water to demonstrate the calculation:

$$CO(NH_2)_2 \,(aq) + H_2O(l) \rightarrow CO_2(g) + 2NH_3(g)$$

Step 1:
Calculate ΔS° for the reaction. We have already done this in the problem above.

$$\Delta S^\circ = 354.8 \text{ J/K}$$

Step 2:
Calculate ΔH° for the reaction

$$\Delta H^\circ_{rxn} = \sum_{products} \nu_p \Delta H^\circ_{f,p} - \sum_{reactants} \nu_r \Delta H^\circ_{f,r}$$

using the tables of the heats of formation, ΔH°_f for each of the products and reactants. Thus

ΔH_f°	Molecule
−393.5 kJ/mole	$CO_2(g)$
−46.2 kJ/mole	$NH_3(g)$
−319.2 kJ/mole	$CO(NH_2)_2(aq)$
−285.9 kJ/mole	$H_2O(l)$

$$\Delta H^\circ = [\Delta H^\circ_{f,\,CO_2} + 2\Delta H^\circ_{f,\,NH_3}] - [\Delta H^\circ_{f,\,CO(NH_2)_2} + \Delta H^\circ_{f,\,H_2O}]$$

$$= [-393.5 \text{ kJ/mole} - (2)(46.2 \text{ kJ/mole})] - [-319.2 \text{ kJ/mole} - 285.9 \text{ kJ/mole}]$$

$$=[-485.9 \text{ kJ}] - [-605.1 \text{ kJ}]$$

$$= + 119.2 \text{ kJ}$$

Step 3:

Use $\Delta S°$ and $\Delta H°$ to calculate

$$\Delta G° = \Delta H° - T\Delta S°$$

$$\Delta G° = +119.2 \text{ kJ} - (298 \text{ K})(0.355 \text{ kJ/K})$$

$$= +13.4 \text{ kJ}$$

$$\Delta G° = +13.4 \text{ kJ}$$

Step 4:

Notice what this says about the spontaneous nature of the reaction. At "room temperature," 298.2 K, the reaction is nonspontaneous because $\Delta G > 0$. But, if we elevate the temperature, as is the case in the hot summer sun, then ΔG decreases such that the reaction proceeds, although slowly. Thus there is a timed release of fertilizer, NH_3, as spring moves to summer. However, the excess application of fixed nitrogen to agricultural land is a serious problem in that a large portion is carried away by spring rains and ends up in lakes and aquifers, causing serious over-fertilization of these systems.

---

1. Calculate the change in Gibbs free energy for the reaction $\Delta G°_{rxn}$ at 25°C. Is the reaction spontaneous?

2. If the reaction is not spontaneous, calculate the temperature at which the reaction becomes spontaneous.

3. Referring back to Figure 4.27, to which case does this reaction correspond?

Check Yourself 9—Calculating Gibbs Free Energy Changes and the Prediction of Whether a Reaction is Spontaneous from Calculated Values of ΔH° and ΔS

Problem: The decomposition of gas phase carbon tetrachloride, $CCl_4(g)$, to form solid elemental carbon in graphite form and gas phase molecular chlorine

$$CCl_4(g) \rightarrow C(s, \text{graphite}) + 2 Cl_2(g)$$

has an enthalpy change for the reaction of $\Delta H°_{rxn} = +95.7 \text{ kJ}$ and an entropy change for the reaction of $\Delta S°_{rxn} = +142.2 \text{ J/K}$.

Now that the new thermodynamic variable, G, the Gibbs free energy is defined, and it is the key thermodynamic variable differentiating a spontaneous chemical reaction from a nonspontaneous one, why is ΔG referred to as "free" energy?

To answer this let's examine a specific case. If we consider the reaction of elemental carbon, C(s, graphite), in the form of solid graphite, with molecular hydrogen, $H_2(g)$, we have a reaction that produces methane in the absence of oxygen

$$C(s, \text{graphite}) + 2 H_2(g) \rightarrow CH_4(g)$$

For this reaction

$$\Delta H°_{rxn} = -74.6 \text{ kJ}$$

$$\Delta S°_{rxn} = -80.8 \text{ J/k}$$

$$\Delta G°_{rxn} = -50.5 \text{ kJ}$$

By inspection we see, because $\Delta H°_{rxn} < 0$, the reaction is exothermic and yields 74.6 kJ of thermal energy. But $\Delta G°_{rxn}$ is the quantitative measure of the maximum amount of work that can be extracted from the reaction. Why is $\Delta G°_{rxn}$ significantly *less* than $\Delta H°_{rxn}$?

The answer lies in the fact that $\Delta S°_{rxn}$ is *negative*, so the execution of the reaction results in a *more organized* chemical product relative to the chemical reactants. Carbon in the solid phase reacts with the gas phase hydrogen molecules to produce a *single* methane molecule, CH_4. This means that some of the energy contained in the chemical transformation from carbon and hydrogen to methane was *expended* to *decrease* entropy by the required amount to make the change in entropy of the universe *positive*. The amount of energy available to do work is thus *reduced* by the expenditure required to decrease entropy by the amount $-T(\Delta S^\circ_{rxn})$ where $\Delta S^\circ_{rxn} < 0$ so $-T(\Delta S^\circ_{rxn})$ is positive.

It is clear from all of our work linking $\Delta S = q_{rev}/T \geq q/T$ with the Carnot cycle that the free energy release in a chemical reaction is a theoretical limit to the amount of work that can be extracted from a chemical reaction. In any real (irreversible) process, the amount of energy available to do work is significantly reduced because of the heat lost to the surroundings.

Check Yourself 10—ΔG°_{rxn} for a Stepwise Reaction

Find ΔG°_{rxn} for the following reaction:

$$3\,C(s) + 4\,H_2(g) \rightarrow C_3H_8(g)$$

Use the following reactions with known ΔG's:

$C_3H_8(g) + 5O_2(g) \rightarrow 3CO_2(g) + 4H_2O(g)$	$\Delta G^\circ_{rxn} = -2074$ kJ
$C(s) + O_2(g) \rightarrow CO_2(g)$	$\Delta G^\circ_{rxn} = -394.4$ kJ
$2H_2(g) + O_2(g) \rightarrow 2H_2O(g)$	$\Delta G^\circ_{rxn} = -457.1$ kJ

Solution

To work with this problem, manipulate the reactions with known ΔG°_{rxn}'s in such a way as to get the reactants of interest on the left, the products of interest on the right, and other species to cancel.

Since the first reaction has C_3H_8 as a reactant, and the reaction of interest has C_3H_8 as a product, *reverse* the first reaction and *change the sign* of ΔG°_{rxn}.	$3\,CO_2(g) + 4\,H_2O(g) \rightarrow C_3H_8(g) + 5\,O_2(g)$ $\Delta G^\circ_{rxn} = +2074$ kJ
The second reaction has C as a reactant and CO_2 as a product, just as required in the reaction of interest. However, the coefficient for C is 1, and in the reaction of interest, the coefficient for C is 3. Therefore, *multiply this equation* and its ΔG°_{rxn} by 3.	$3 \times [C(s) + O_2(g) \rightarrow CO_2(g)]$ $\Delta G^\circ_{rxn} = 3 \times (-394.4$ kJ$)$ $= -1183$ kJ
The third reaction has $H_2(g)$ as a reactant, as required. However, the coefficient for H_2 is 2, and in the reaction of interest, the coefficient for H_2 is 4. Therefore, *multiply this reaction* and its ΔG°_{rxn} by 2.	$2 \times [2\,H_2(g) + O_2(g) \rightarrow 2\,H_2O(g)]$ $\Delta G^\circ_{rxn} = 2 \times (-457.1$ kJ$)$ $= -914.2$ kJ
Lastly, rewrite the three reactions after multiplying through by the indicated factors and show how they sum to the reaction of interest. ΔG°_{rxn} for the reaction of interest is then just the sum of ΔG's for the steps.	$3CO_2(g) + 4H_2O(g) \rightarrow C_3H_8(g) + 5O_2(g)$ $\Delta G^\circ_{rxn} = +2074$ kJ $3C(s) + 3O_2(g) \rightarrow 3CO_2(g)$ $\Delta G^\circ_{rxn} = -1183$ kJ $4H_2(g) + 2O_2(g) \rightarrow 4H_2O(g)$ $\Delta G^\circ_{rxn} = -914.2$ kJ $3C(s) + 4H_2(g) \rightarrow C_3H_8(g)$ $\Delta G^\circ_{rxn} = -23$ kJ

Summary Concepts

1. Entropy and Spontaneous Change in Natural Systems

One of the great challenges in the physical sciences is the quantitative investigation of what drives spontaneous change forward in chemical, physical, and biological systems. We are all witness to spontaneous change in the combustion of fuels, in the corrosion of metals, in the melting of ice, and in the flow of heat from high temperature lava to the cool water of the ocean as displayed at right. In some cases those spontaneous changes are accompanied by a release of energy, but in others nature is driving inexorably toward a state of increasing disorganization. An example is when ice melts from a highly organized crystal structure to form a liquid with little internal organization. The thermodynamic variable that captures the degree of disorganization in a system is entropy, S, a state variable of the system.

2. Determination of Probability at the Molecular Level

An understanding of *spontaneous change* at the molecular level begins with an analysis of the *probability* of finding a system in a particular *configuration*. The example at right displays the configurations (a total of 11) available to 10 molecules in a system of two chambers. In the language of molecular thermodynamics, each configuration is referred to as a *state*. The probability of finding the system in a given state is equal to the number of *arrangements* in which a given number of molecules can be found, when distributed randomly, within the available space. The number of arrangements is, in the language of molecular thermodynamics, equal to the number of *microstates*, W. The number of microstates, W, for N molecules distributed with L molecules in the left-hand chamber and R molecules in the right-hand chamber is W = N!/L!R!. The probability that a given state (configuration) will occur is equal to the number of available microstates (arrangements) leading to that configuration divided by the total number of available microstates as displayed explicitly at the right.

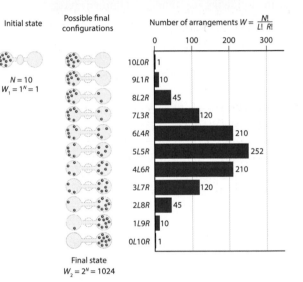

3. Boltzmann and the Microscopic Formulation of Entropy

Thermodynamics is a quantitative subject, and with recognition that entropy change, $\Delta S = S_{final} - S_{initial}$, plays a key role in determining whether a process is spontaneous, it is necessary to develop a quantitative basis for *calculating* entropy. It was Ludwig Boltzmann who developed the remarkably simple expression linking entropy, S, to the number of microstates, W, available to the system. That molecular level formulation of entropy, wherein the Boltzmann constant, k_B, is the proportionality factor between entropy, S, and the natural logarithm of W, provides the basis for coupling probability and spontaneous change:

$$S = k_B \ln W$$

The change in entropy, ΔS, is then

$$\Delta S = k_B \ln(W_f/W_i)$$

The example for heat flow is displayed at right.

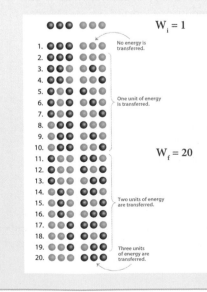

Summary Concepts

4. Predicting the Sign of ΔS

While we develop in this chapter the ability to quantitatively calculate the entropy change, ΔS, for a system, it is important to develop an understanding of whether a given change is associated with an *increase* or a *decrease* in entropy. For this reason we considered four important cases:

- Increasing the temperature of a system leads to an *increase* in entropy.

- The change of state from a solid to a liquid and from a liquid to a gas leads to an *increase* in entropy.

- The expansion of a gas into a vacuum leads to an *increase* in entropy.

- A chemical reaction that converts a small number of molecules to a large number of molecules results in an *increase* in entropy.

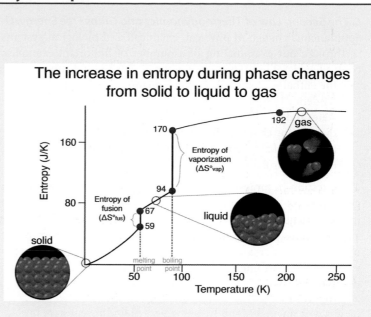

The increase in entropy during phase changes from solid to liquid to gas

5. Calculation of ΔS for a System from the Macroscopic Perspective

Analysis of the Carnot cycle in Chapter 3 results in the discovery that there is a state variable, the change of which is equal to the integral of the heat added *reversibly* divided by the temperature at which the heat is added,

$$\int \frac{dq_{rev}}{T} .$$

That state variable is the entropy such that

$$\Delta S = \int \frac{dq_{rev}}{T}$$

is the change in entropy of the system calculated along a reversible path. For a thermodynamic process that is irreversible, ΔS is equal to or greater than

$$\int \frac{dq}{T}$$

such that in general

$$\Delta S \geq \int \frac{dq}{T}$$

For an isothermal process (T = constant) we can remove the temperature from under the integral

$$\Delta S = \int \frac{dq_{rev}}{T} = \frac{1}{T} \int dq_{rev} = \frac{q_{rev}}{T}$$

While this entropy change, ΔS, was derived from a macroscopic system (the Carnot cycle is based upon a heat engine) this ΔS is the same entropy change as the entropy change deduced at the molecular level by Boltzmann.

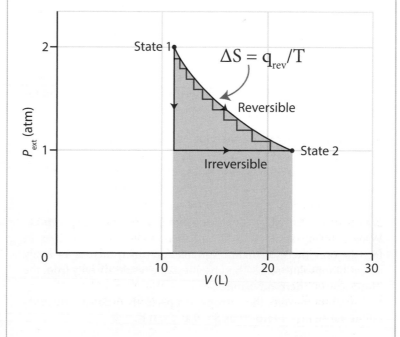

4.41

Summary Concepts

6. The Second Law of Thermodynamics and Gibbs Free Energy, G

Spontaneous change in physical, chemical, and biological systems is, through our examination of a number of important examples, controlled by (at least) two compelling tendencies in nature:

- the enthalpy change, ΔH, which quantitatively expresses the tendency to find and follow the path leading to a lower energy state; and
- the entropy change, ΔS, which quantitatively expresses the tendency of a system to find whatever pathway leads to a state of higher probability, of higher disorder.

The question becomes, how do we quantitatively define whether a given process is spontaneous or nonspontaneous? The answer lies in the statement of the *Second Law of Thermodynamics*: Whenever a spontaneous event occurs in the universe, the total entropy of the universe must increase. This Second Law leads directly to the derivation of the most important thermodynamic variable in chemistry, the state variable

$$G = H - TS,$$

the Gibbs free energy. The Gibbs free energy *change* $\Delta G = \Delta H - T\Delta S$ for conditions of constant pressure and constant temperature.

Key derivation linking the Second Law to Gibbs free energy:

$$\Delta S_{total} = \Delta S_{syst} + \Delta S_{sur}$$

$$\Delta S_{sur} = q_{sur}/T = -q_{syst}/T = -\Delta H_{syst}/T$$

$$\Delta S_{total} = \Delta S_{syst} - \Delta H_{syst}/T$$

$$= (T\Delta S_{syst} - \Delta H_{syst})/T$$

$$T\Delta S_{total} = -(\Delta H_{syst} - T\Delta S_{syst})$$

If $\Delta S_{total} > 0$ for *spontaneous change*,

then $(\Delta H_{syst} - T\Delta S_{syst}) < 0$ for spontaneous change,

and with $\Delta G = \Delta H - T\Delta S$,

$\Delta G < 0$ for spontaneous change.

7. Gibbs Free Energy and Spontaneous Change

Gibbs free energy emerges as the most important thermodynamic variable because it quantitatively defines when a process is spontaneous. Specifically, when

$\Delta G < 0$ the process is spontaneous;

when

$\Delta G > 0$ the process is nonspontaneous;

and when

$\Delta G = 0$ the system is in equilibrium.

This is summarized at right for the case of chemical reactants and products that proceed spontaneously toward the position of equilibrium from either pure reactants or pure products.

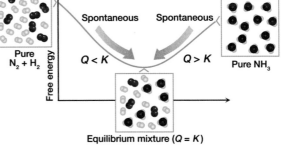

Free Energy Determines the Direction of Spontaneous Change

$$N_2(g) + 3\,H_2(g) \rightleftharpoons 2\,NH_3(g)$$

8. Absolute Value of Entropy: The Third Law of Thermodynamics

While entropy appears to be a rather esoteric concept, it is in fact the only thermodynamic variable for which there is an easily accessible absolute scale. This absolute scale results directly from the Third Law of Thermodynamics:

At absolute zero, the entropy of a perfectly ordered pure crystalline substance is zero. Thus S = 0 at T = 0 K.

Perfect crystal at 0K

W = 1 S = 0

Summary Concepts

9. Entropy Change of a Chemical Reaction

The entropy change for a chemical reaction

$$aA + bB \rightarrow cC + dD$$

is

$$\Delta S^\circ_{rxn} = \sum_{p\,=\,products} v_p S^\circ_p - \sum_{r\,=\,reactants} v_r S^\circ_r$$

where v_p represents the number of moles of each product species and v_r is the number of moles of each reactant species.

Calculating Entropy: Predicting the Sign of ΔS
When a chemical reaction produces or consumes gases, the sign of ΔS is usually easy to predict.

Example: Bicarbonate of soda used in cooking

$$2NaHCO_3(s) \xrightarrow{heat} Na_2CO_3(s) + CO_2(g) + H_2O(g)$$

causes cake to rise   *adds moisture*

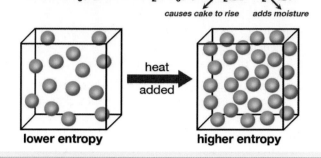

lower entropy   higher entropy

10. Calculation of Gibbs Free Energy for a Reaction

Gibbs free energy change for a chemical reaction, ΔG_{rxn}, can be calculated from thermochemical tables of Gibbs free energy of formation using the same convention for ΔH_{rxn} and ΔS_{rxn}. Therefore

$$\Delta G_{rxn} = \sum_{products} v_p \Delta G^\circ_{f,p} - \sum_{reactants} v_r \Delta G^\circ_{f,r}$$

Alternatively, we could calculate

$$\Delta S^\circ_{rxn} = \sum_{products} v_p S^\circ_p - \sum_{reactants} v_r S^\circ_r$$

and

$$\Delta H^\circ_{rxn} = \sum_{products} v_p \Delta H^\circ_{f,p} - \sum_{reactants} v_r \Delta H^\circ_{f,r}$$

and then calculate

$$\Delta G^\circ_{rxn} = \Delta H^\circ_{rxn} - T\Delta S^\circ_{rxn}$$

CASE STUDY 4.1 Entropy, Free Energy, and the Maximum Amount of Work That Can Be Extracted from a Fuel Cell

A fuel cell is a device wherein the chemical energy contained in a fuel is converted to an electric current. The fuel cell shown in Figure CS4.1a, which we will study in some detail in Chapter 7, reacts H_2 with O_2 to produce an electric current and the product H_2O:

$$2H_2 + O_2 \rightarrow 2H_2O$$

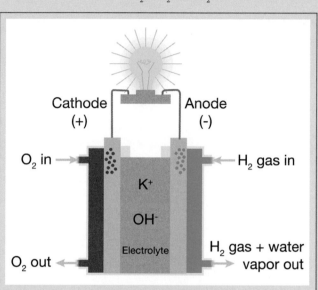

FIGURE CS4.1A Schematic diagram of a fuel cell: The hydrogen fuel cell operates by drawing oxygen from the atmosphere into the cathode and drawing H_2 from a supply tank into the anode. We will study the cathode, anode, and reactions that take place in Chapter 7. However, we can calculate ΔH, ΔS, and ΔG for the reaction and thus calculate the efficiency of the fuel cell from what we have learned in Chapter 4.

While we have determined the maximum amount of work that can be extracted from a heat engine, by virtue of our analysis of the Carnot cycle, how do we calculate the efficiency of a fuel cell? The answer to this question involves the concept of entropy in a very interesting way—specifically entropy limits the amount of work that can be extracted from a chemical reaction; *any* chemical reaction.

We have already calculated that the reaction of molecular hydrogen and molecular oxygen

$$2H_2(g) + O_2(g) \rightarrow 2H_2O(g)$$

releases 482 kJ of energy per mole of O_2, and that energy is released as heat. We can measure this heat release in a calorimeter. But we now know, from our study of entropy, that this reaction results in a *decrease* in entropy because 3 molecules of reactant are converted to 2 molecules of product. There is a further *decrease* in entropy if we execute the reaction at a temperature *below* the boiling point of water (which we can do in a fuel cell) because we convert two gas phase

reactants ($H_2(g)$ and $O_2(g)$) to liquid water, $H_2O(l)$.

The *decrease* in entropy thus limits the amount of energy that can be extracted from the reaction of H_2 and O_2 because $\Delta S_R < 0$ and

$$\Delta G_R = \Delta H_R - T\Delta S_R$$

So to review, the amount of energy available to do *work* is the *free energy* change of the reaction, ΔG_R. If the *heat* release is ΔH_R, then the term $T\Delta S_R$ quantitatively represents the loss of available chemical energy that can be harnessed to do useful work. It is illustrative to recall that $\Delta S_R = Q/T$ so that $Q = T\Delta S_R$ is the "heat value" of the entropy that must be subtracted to calculate the amount of energy available to do useful work.

Thus we can calculate the *maximum efficiency*, η_{max}, for the extraction of useful work with expression

$$\eta_{max} = \Delta G_R / \Delta H_R$$

and because

$$\Delta G_R = \Delta H_R - T\Delta S_R$$

we have

$$\Delta G_R \Big/ \Delta H_R = 1 - T\left(\Delta S_R \Big/ \Delta H_R\right)$$

Notice an important difference here with the expression for the maximum efficiency of the Carnot cycle for a heat engine

$$\eta_{Carnot} = W\Big/Q_{in} = \frac{T_1 - T_2}{T_1}$$

For the Carnot cycle, as T_1 increases for a constant T_2, the efficiency of the Carnot cycle increases. By contrast, as the operating temperature of the fuel cell *increases*, the maximum efficiency of the chemical reaction *decreases*. For the H_2 fuel cell (as we will see, a number of fuels can be used), $\Delta H_R = -476$ kJ at 1000K, while $T\Delta S_R = -84$ kJ, so

$$\Delta G_R = -392 \text{ kJ} \quad \text{per mole of } O_2$$

Thus

$$\eta_{max} = \Delta G_R / \Delta H_R = 0.82$$

The maximum efficiency of the hydrogen fuel cell is thus 82%.

At 300 K, the entropy change is greater because the product H_2O is a liquid and

$$T\Delta S_R = -116 \text{ kJ}$$

However, ΔH also increases to -590 kJ so

$$\Delta G_R = \Delta H_R - T\Delta S_R$$
$$= -590 \text{ kJ} - (-116 \text{ kJ})$$
$$= -474 \text{ kJ}$$

So at 300 K

$$\eta_{max} = \frac{\Delta G_R}{\Delta H_R} = \frac{-474 \text{ kJ}}{-590 \text{ kJ}} = 0.80$$

Thus the maximum efficiency of the hydrogen fuel cell at 300 K is 80%. In both these cases we make the assumption that ΔH and ΔS do not vary appreciably with temperature.

Problem 1

Calculate the maximum efficiency of the hydrogen fuel cell at 300 K based on the thermodynamic data provided in the Appendix of the textbook.

BUILDING QUANTITATIVE REASONING

CASE STUDY 4.2 The Most Probable Distribution Is the Boltzmann Distribution

Adapted From: Ludwig Boltzmann—A Pioneer of Modern Physics 1
Dieter Flamm
Institut für Theoretische Physik der Universität Wien, Boltzmanngasse 5, 1090 Vienna, Austria

In two respects Ludwig Boltzmann was a pioneer of quantum mechanics. First because in his statistical interpretation of the Second Law of Thermodynamics he introduced the theory of probability into a fundamental law of physics and thus broke with the classical prejudice, that fundamental laws have to be strictly deterministic. Even Max Planck had not been ready to accept Boltzmann's statistical methods until 1900. With Boltzmann's pioneering work the probabilistic interpretation of quantum mechanics had already a precedent. In fact in a paper in 1897 Boltzmann had already suggested to Planck to use his statistical methods for the treatment of black body radiation. The second pioneering step towards quantum mechanics was Boltzmann's introduction of discrete energy levels. Boltzmann used this method already in an 1872 paper. One may ask whether Boltzmann considered this procedure only as a mathematical device or whether he attributed physical significance to it. In this connection Ostwald reports that when he and Planck tried to convince Boltzmann of the superiority of purely thermodynamic methods over atomism at the Halle Conference in 1891 Boltzmann suddenly said: "I see no reason why energy shouldn't also be regarded as divided atomically." Before 1900 classical Newtonian mechanics was the prototype of a successful physical theory. As a consequence all physical laws had to be strictly deterministic and universally valid. For most physicists—among them Max Planck—these rules applied also to thermodynamics. In contrast to this view was Boltzmann's statistical interpretation of the second law of thermodynamics which he first formulated in 1877, nearly 50 years before the statistical interpretation of quantum mechanics.

Boltzmann for the first time introduced the theory of probability into a fundamental law of physics and thus broke with the classical prejudice, that fundamental laws of physics have to be strictly deterministic. Because of this prejudice it is no wonder that in the last decades of the 19th century Boltzmann's statistical interpretation of thermodynamics was not accepted by the physics community but met with violent objections coming from physicists and mathematicians.

Ludwig Boltzmann was born in Vienna in 1844 and died in Duino near Trieste in 1906. He studied mathematics and physics at the University of Vienna from 1863 to 1866, where Josef Stefan was his main advisor.

FIGURE CS4.2A While Ludwig Boltzmann emerged as a giant of 19th century science, his ideas were slow to gain acceptance in the physics and chemistry communities. In the upper panel he is shown as a young man and in the lower panel his tombstone is shown engraved with one of his two famous equations.

A central objective in this book is to link the microscopic world of atoms and energy to the macroscopic or global world of observable quantities, whether pressures, temperatures, or constraints on global primary energy generation. A key to understanding physical, chemical, and biological processes brings us back to the issue of understanding what energy is and how molecules and energy interact with each other. A key part of the development of this understanding, it turns out, is the "Boltzmann distribution" and the concepts that surround it.

First we note that Ludwig Boltzmann is famous for the development of two equations. The first is the equation relating the entropy, S, of a system to the number of microstates, W,

$$S = k_B \ln W$$

where k_B is the Boltzmann constant. The importance and application of this equation was treated in detail in the chapter core.

The second famous equation relates the number of molecules, n_i, in energy level i, to the number of molecules n_j in a different (higher) energy level, j, by the expression

$$n_j \big/ n_i = e^{-\left(\varepsilon_j - \varepsilon_i\right)/kT} \qquad \textbf{(Eq 1)}$$

where ε_j is the energy of level j, ε_i is the energy of level i, and $\varepsilon_j - \varepsilon_i > 0$. Thus we seek to understand the relationship between the number of molecules and the energy of the available states *given a certain amount of total energy* available for all the particles in the system to share—known as the "equipartition of energy."

A key characteristic of the Boltzmann distribution is that it provides a straight-forward way to calculate the number of molecules with energy equal to or greater than some energy level j. In particular if the total number of molecules in the system is n, and the number of molecules in the lowest energy level is n_0, the number of molecules in level j or above is $n_{j \text{ or above}}$, and the number of molecules in level j is n_j, then

$$\left(\begin{array}{c}\text{Fraction of the total number of}\\ \text{molecules at or above level } j\end{array}\right) = \frac{n_{j \text{ or above}}}{n}$$

But, for the Boltzmann distribution:

$$n_j \big/ n_i = e^{-\left(\varepsilon_j - \varepsilon_i\right)/kT}$$

$$\frac{n_{j \text{ or above}}}{n} = n_j \big/ n_0 = e^{-\varepsilon_j / k_B T}$$

where ε_j is the energy of level j in joules per molecule. We can equally well express this in joules per mole as

$$n_j \big/ n_0 = e^{-E_A / RT} \qquad \textbf{(Eq 2)}$$

where E_A is the "activation energy" or energy threshold shown in Figure CS4.2b. This provides a remarkably easy way to calculate the fraction of molecules with sufficient energy to pass over the energy barrier on a "reaction coordinate" diagram for a given temperature T, and activation energy E_A.

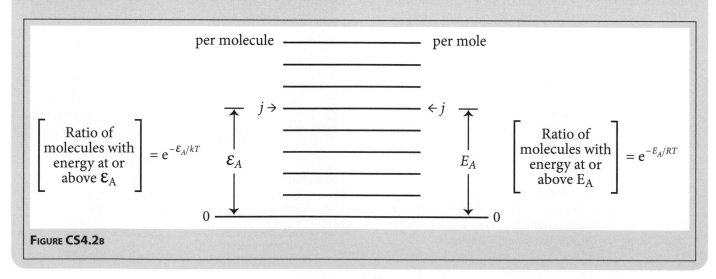

FIGURE CS4.2B

"Boltzmann Distribution" of Velocities

An important result of this Boltzmann distribution

$$n_j/n_0 = e^{-\varepsilon_j/kT} = e^{-E_A/RT}$$

is that the distribution of molecular velocities in a molecular ensemble exhibits the relationship between the fraction of molecules at a given kinetic energy vs. kinetic energy displayed in Figure CS4.2c. As Figure CS4.2c and Eq. 2 above make clear, as the temperature, T, of the molecular ensemble increases, the fraction of molecules with kinetic energy greater than E_A increases exponentially with T.

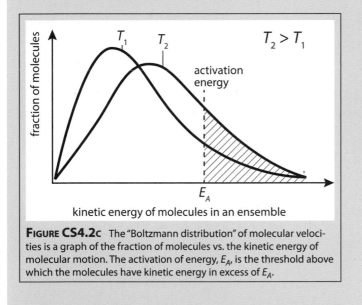

Figure CS4.2c The "Boltzmann distribution" of molecular velocities is a graph of the fraction of molecules vs. the kinetic energy of molecular motion. The activation of energy, E_A, is the threshold above which the molecules have kinetic energy in excess of E_A.

Problem 1

For a certain reaction to occur, the system needs an activation energy of 20 kJ/mole of particles.

a. What fraction of the particles will have enough energy to react at 300 K?

b. At what temperature does the system need to be for 2% of the particles to have enough energy to react?

CASE STUDY 4.3 Methods for Energy Storage

While, as we revealed in Case Study 1.3, the planet has stores of fossil fuels in the form of petroleum, natural gas and coal to supply the energy needs of the human endeavor for centuries, were those hydrocarbons to be combusted as a primary source of energy for even the next few decades, the resulting climate would emulate that of the Eocene, 40 million years ago. During that era sea level was some 100 meters or 300 feet higher than it is today, all glacial systems and the associated water supplies would be eliminated, ocean acidification would increase significantly. In addition, with the previous addition of chlorine and bromine to the atmosphere from CFCs and halons (see Case Study 12.1), ozone loss would increase as would ultraviolet dosage at the ground and there would be little difference in temperature between the equator and the pole (Framework, Chapter 3).

The dominant renewable energy sources that can provide energy demand at the required scale (Framework, Chapter 1) all provide electric power:

- Wind power
- Concentrated solar thermal
- Photovoltaics
- Geothermal
- Hydroelectric

With the exception of geothermal, all these energy sources fluctuate with time. Yet when they are producing power, that power must be used immediately. In fact, one of the great technical challenges of the renewable energy era is that supply and demand must be balanced on the electric power grid with great accuracy under conditions of rapid fluctuations in both supply and demand. The electricity grid in its current form cannot *store* energy.

The solution to this problem at the national level, the regional level, the local level, and the personal level places great importance on how energy can be stored and how that energy storage is coupled to the power grid to balance supply and demand. The technology of energy storage engages virtually all the fundamental principles of chemistry and physics and constitutes a fascinating complement of technological innovation to that of primary energy generation itself.

As we will see, while a device capable of balancing supply and demand on very short time scales must obviously be a *source* of energy when the power from wind or solar thermal diminishes, the combination of energy storage devices must also be able to *extract* energy when supply on the grid exceeds demand. Both problems, it turns out, are serious but highly tractable issues from a technical perspective.

To provide an idea of the fluctuation in demand for electricity during two different months of the year, consider Figure CS4.3a. That figure displays the day-night fluctuations in January and June 2018 in units of kWh/p·day. Note that demand changes on the time scale of a few minutes, so this de-

gree of fluctuation in demand must be balanced by comparable response to supply electric power on the same time scale. Displayed in Figure CS4.3a is a typical graph of demand as a function of time for the national power grid within one time zone with the top panel displaying the day-to-day variation in demand in the winter and the bottom panel showing the same for summer conditions. We do this for a single time zone because as we shift west, the basic power dependence as a function of clock time will shift in one hour segments and, as a result, superpose the rapid spikes that repeat each morning and evening as sunrise and sunset progress across the country.

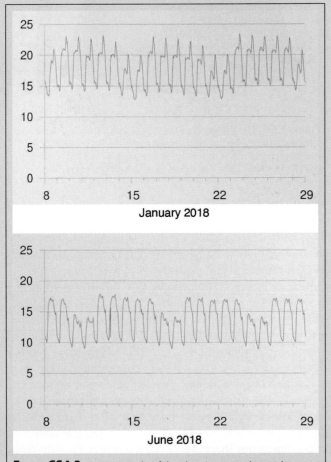

FIGURE CS4.3A An example of the electric power demand over the course of a typical three-week period in January (upper panel) and June (lower panel). While the ordinate is in arbitrary limits, it is important to note the rapid increases in demand each morning and the rapid fall-off in the evening.

The key point here is that the design and operation of the power grid has already evolved to handle these routine but very large fluctuations in demand as a function of time of day.

The issue we must address is: how much do renewable

energy forms fluctuate? To answer this question we engage a recent study that analyzed in detail 5 years of wind velocity information from a large data assimilation system that joins detailed, systematic observations of wind speeds throughout the US (and globally) with a global model of wind velocities with 3 hour time resolution. These models are used for weather forecasting and are thus submitted to careful scrutiny for accuracy. In this study, a portfolio of wind farms were selected throughout the Midwest at the locations displayed in Figure CS4.3b.

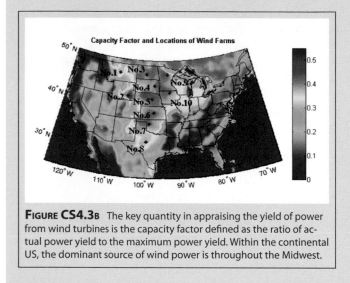

FIGURE CS4.3B The key quantity in appraising the yield of power from wind turbines is the capacity factor defined as the ratio of actual power yield to the maximum power yield. Within the continental US, the dominant source of wind power is throughout the Midwest.

By considering this *combination* of wind farms as a linked ensemble, the large fluctuations at a single point are averaged to provide a reasonable measure of fluctuations in power generation. Figure CS4.3c displays the fluctuations in capacity factor that provides a quantitative characterization of the imbalance in supply that results from actual intermittency in wind power supply. Obviously the introduction of large-scale wind power generation will be done in combination with power generation from a combination of concentrated solar thermal, photovoltaics, etc. This is important because solar power generation, whether concentrated solar thermal or photovoltaic, is *anticorrelated* with wind velocity. The reason is that when a high pressure system dominates the weather for a region, the skies are clear but the wind velocity drops. When a low pressure area dominates, the skies tend to be cloudy but the wind velocity is at a maximum. Thus the combination of wind generated power and concentrated solar thermal/photovoltaic generated power will buffer the delivered power to the grid and thus average over variations on the national level.

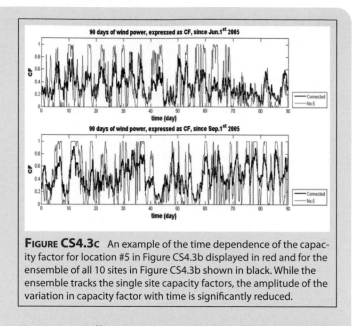

FIGURE CS4.3C An example of the time dependence of the capacity factor for location #5 in Figure CS4.3b displayed in red and for the ensemble of all 10 sites in Figure CS4.3b shown in black. While the ensemble tracks the single site capacity factors, the amplitude of the variation in capacity factor with time is significantly reduced.

However, it is illustrative to consider here the worst case situation by considering delivered wind power alone to judge how much and for how long decreases in power supplied to the grid must be compensated for by stored energy from other sources. Inspection of Figure CS4.3c reveals that the most serious intermittency lasts for approximately 5 days and represents a decrease in power delivered to the grid of a factor of two from the mean of the wind generated electric power delivered to the grid. This provides us with the data needed to calculate the deficit in power that must be made up by alternative methods of energy supply under worst case conditions because solar power would make up a major fraction of the deficit for the reasons noted above. It is also clear from the data in Figure CS4.3c that the periods of a relative lull in wind generating capacity are fairly easy to predict once the frontal high pressure or low pressure system moves over the continental United States. Thus the power grid would be balanced by a *combination* of energy storage contributions, each of which has a different intrinsic time response to power fluctuations on the grid.

Our objective here is to explore the available methods for energy storage. This problem breaks down into two important domains: *long-term* variability and *short-term* fluctuations. For the US case, we can examine the data for the maximum rate of change of supply. This is an important number because it will give us a key metric on how fast we might need to draw power from a stored source to compensate for short-term (minutes to hours) drops in delivered power. This will be a worst case because there will be other sources of renewables that can deliver power to the grid such as concentrated solar thermal, photovoltaics, hydroelectric, etc. that can "buffer" or compensate for variations in wind power depending upon either the *rate* of change, or for long lapses if such a circumstance should occur.

Analyzing the Time Dependence of Power Fluctuations

We can analyze the fluctuations of supply on the power grid by making the following assumptions applied to *wind power* with wind farms distributed across the Midwest sector of the US as shown in Figure CS4.3b. We will assume that in the US wind generates one-third of the total primary energy, which is 1000 GW or 1 TW. In this scenario an additional 1 TW would be supplied by concentrated solar thermal and photovoltaics and 1 TW by a combination of nuclear, hydroelectric, biofuels, fossil fuels with carbon capture and sequestration, geothermal, and tidal.

The first question to answer, then, is what is the maximum rate of change in power supply or power demand that must be contended with? In this analysis it is important to recognize that even under normal conditions on the power grid, very large rates of change of power demand are a fact of life. By focusing on a single time zone, we can clarify the origin of the variation in demand and, in addition, quantify the rate of change in that demand. This will provide a basis for analyzing requirements on the types of stored energy methods that can address fluctuation in supply and demand based upon the daily power fluctuations that the grid must *currently* contend with *every day*. We will use the eastern time zone of the US as an example, because it represents the largest population concentration in the US. The eastern time zone contains about 40% of the US population.

To estimate the rate at which demand changes, consider the rate of increase in demand at the beginning of the day that is displayed in higher time resolution in Figure CS4.3d. As Figure CS4.3d shows, demand in the morning increases by approximately 70% in a period of approximately 4 hours. Therefore, the fractional change in power demand per unit time is 70%/4 hr = 18% per hour. These are important numbers when we analyze the situation on the national power grid.

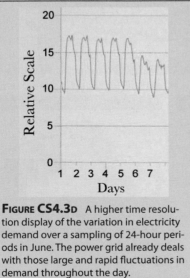

FIGURE CS4.3D A higher time resolution display of the variation in electricity demand over a sampling of 24-hour periods in June. The power grid already deals with those large and rapid fluctuations in demand throughout the day.

If we examine the change in delivered power from the wind generation of the Midwest, the problem we need to deal with is how we supply both short-term fluctuations and longer-term lulls in power delivered to the grid by wind-powered generation of electricity.

Suppose, for example, we have the power generation profile shown in Figure CS4.3c where output, represented by the capacity factor, decreased by 50% for a period of five days. To make up for this deficit, we would have to supply 500 GW for a period of 5 days so we would have to supply, from other sources of energy, an amount of energy equal to:

$$(500 \text{ GW}) \times (5 \text{ days} \times 24 \text{ hr/day}) = 60000 \text{ GWh}$$

There are, in national systems designed to balance demand and supply on the grid, a series of possible energy storage or energy sources that can be turned on or off. These include:

1. Hydroelectric
2. Demand management using electric vehicles
3. Thermal energy storage (TES)
4. Pumped water storage
5. Biofuel and/or waste incineration
6. Control of appliance demand
7. Compressed air energy storage (CAES)
8. Flywheels
9. Natural gas fired power plants with carbon capture

Figure CS4.3e on the following page summarizes the major energy storage reservoirs for the continental US. The figure also delineates the scale of the available energy storage for each category.

Consideration of Figure 4.3e reveals that the dominant energy storage options with regards to capacity for energy storage include thermal energy storage (TES) and compressed air energy storage (CAES). The combination of just these two sources would supply the required 7200 GWh of energy over five days to make up for a worst case deficit in wind generated power. Also of importance are hydroelectric, electric vehicle storage, pumped water storage, and biofuel and/or waste incineration.

What was viewed as a major technology challenge for renewable energy just a very few years ago is now viewed as an infrastructure issue with only minor technology challenges.

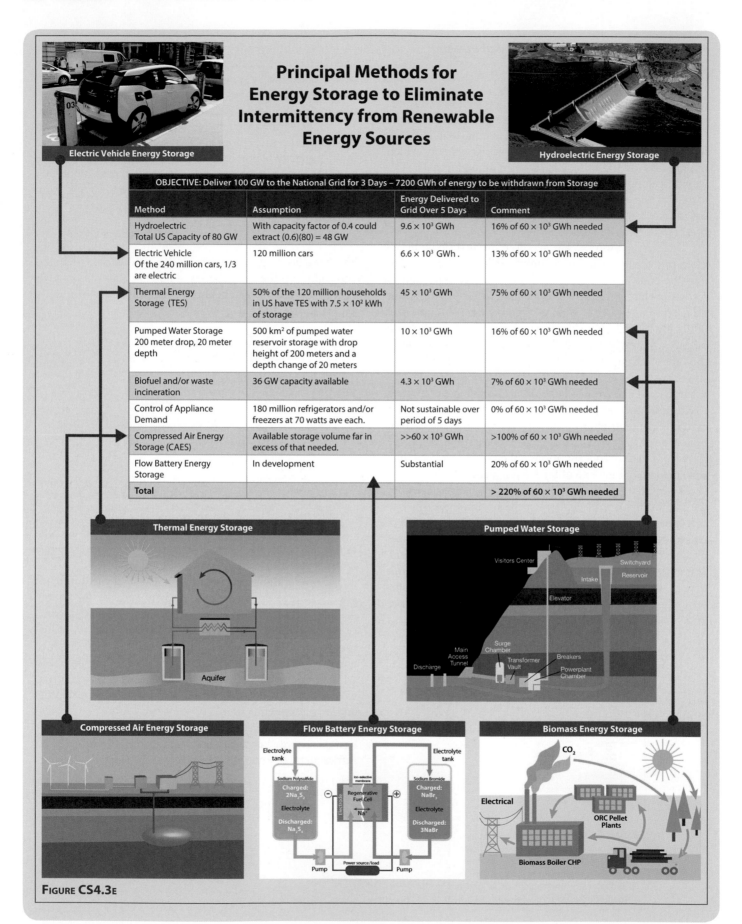

Principal Methods for Energy Storage to Eliminate Intermittency from Renewable Energy Sources

Electric Vehicle Energy Storage

Hydroelectric Energy Storage

OBJECTIVE: Deliver 100 GW to the National Grid for 3 Days – 7200 GWh of energy to be withdrawn from Storage			
Method	Assumption	Energy Delivered to Grid Over 5 Days	Comment
Hydroelectric Total US Capacity of 80 GW	With capacity factor of 0.4 could extract (0.6)(80) = 48 GW	9.6×10^3 GWh	16% of 60×10^3 GWh needed
Electric Vehicle Of the 240 million cars, 1/3 are electric	120 million cars	6.6×10^3 GWh .	13% of 60×10^3 GWh needed
Thermal Energy Storage (TES)	50% of the 120 million households in US have TES with 7.5×10^2 kWh of storage	45×10^3 GWh	75% of 60×10^3 GWh needed
Pumped Water Storage 200 meter drop, 20 meter depth	500 km² of pumped water reservoir storage with drop height of 200 meters and a depth change of 20 meters	10×10^3 GWh	16% of 60×10^3 GWh needed
Biofuel and/or waste incineration	36 GW capacity available	4.3×10^3 GWh	7% of 60×10^3 GWh needed
Control of Appliance Demand	180 million refrigerators and/or freezers at 70 watts ave each.	Not sustainable over period of 5 days	0% of 60×10^3 GWh needed
Compressed Air Energy Storage (CAES)	Available storage volume far in excess of that needed.	$>>60 \times 10^3$ GWh	>100% of 60×10^3 GWh needed
Flow Battery Energy Storage	In development	Substantial	20% of 60×10^3 GWh needed
Total			> 220% of 60×10^3 GWh needed

Thermal Energy Storage

Aquifer

Pumped Water Storage

Visitors Center
Switchyard
Intake
Reservoir
Elevator
Main Access Tunnel
Surge Chamber
Transformer Vault
Breakers
Powerplant Chamber
Discharge

Compressed Air Energy Storage

Flow Battery Energy Storage

Electrolyte tank
Electrolyte tank
Sodium Polysulfide
Ion-selective membrane
Sodium Bromide
Charged: $2Na_2S_2$
Regenerative Fuel Cell
Charged: $NaBr_3$
Electrolyte
Electrolyte
Discharged: Na_2S_4
Na^+
Discharged: $3NaBr$
Pump
Power source/load
Pump

Biomass Energy Storage

CO_2
Electrical
ORC Pellet Plants
Biomass Boiler CHP

FIGURE CS4.3E

EQUILIBRIA AND FREE ENERGY

5

> "If a chemical system at equilibrium experiences a change in concentration, temperature, volume, or total pressure, then the equilibrium shifts to partially counteract the imposed change."
>
> —Henry Louis Le Chatelier

FIGURE 5.1 An analysis of energy sources worldwide demonstrates increasingly that the supply of coal and natural gas is plentiful, but that petroleum is increasingly a supply problem. Either it is increasingly difficult to extract or it is supplied by regions that are politically unstable. Thus there is increasing attention paid to converting either coal or natural gas to liquid fuels using the Fischer–Tropsch process that we will examine in this chapter. The Fischer–Tropsch process, as well as many other industrial processes, depend upon chemical equilibrium for their optimization.

Framework

We can now see a pattern emerging wherein in any spontaneous process—indeed in any natural process—order is replaced by disorder, organized forms of energy become increasingly disorganized forms of energy, and the entropy of the universe inexorably increases. In any natural process, some energy becomes unavailable to do useful work. The logic of this is irrefutable, and one consequence of the degradation of energy is the prediction that, with the passage of time, the universe will approach a state of maximum disorder. As a consequence, matter will become a homogeneous mixture, and thermal energy will have flowed from high temperature regions to low temperature regions until the entire universe is at a single temperature. At that point, no work could be done. All forms of organized energy (chemical, mechanical, nuclear) will have been transformed into thermal energy. There are various names for this final, maximum entropy state, a state that is often referred to as the *heat death* of the universe.

But even in a state of equilibrium there is a dynamic exchange that takes place—only the macroscopic state of the system does not change with the passage of time. A chemical system in equilibrium

$$aA + bB \rightleftarrows cC + dD$$

represents an *unchanging combination of macroscopic properties*: concentration, pressure, temperature, etc. There are no macroscopically apparent changes with time. The equilibrium state thus determines the *extent* to which a reaction takes place. But at the molecular level, the situation is far different. When such a chemical system is in equilibrium, there is a state of dynamic balance, suspended between countervailing forces acting in both the forward and reverse directions to interconnect reactants A and B to products C and D, and *vice versa*.

The relationship is between the expression for equilibrium in terms of the chemical reaction $aA + bB \rightleftarrows cC + dD$ and the

Gibbs free energy surface which, as we have seen in Chapter 4, can be displayed diagramatically as:

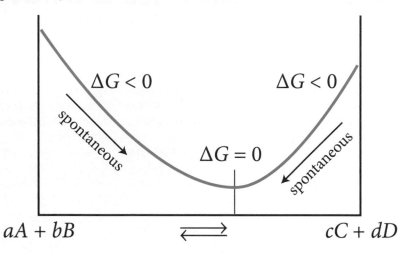

$$\Delta G < 0 \qquad\qquad \Delta G < 0$$

spontaneous $\Delta G = 0$ spontaneous

$$aA + bB \qquad \rightleftharpoons \qquad cC + dD$$

Because the reactions involved in the production of biofuels are inherently equilibrium reactions, the principles associated with equilibrium processes in chemistry play a critical role in the selection of strategic solutions to the global energy problem. Thus command and control of the Gibbs free energy surface that underpins all chemical reactions is linked directly to the challenge of global energy production. This fact constitutes one of the most direct couplings between the molecular level behavior of chemical systems on the one hand, and the macroscopic, indeed global, consequences on the other.

We turn here to the question of how chemical systems behave as they *spontaneously* drive toward a state of lower free energy because *every* chemical reaction obeys the edict

$$\text{Reactants} \rightarrow \text{Products} + \text{free energy}$$

as the reaction seeks to find the equilibrium condition.

This is true until $(\Delta G)_{\text{Reaction}} = 0$ at which point, from a macroscopic perspective, change ceases. Observing a system at equilibrium, we have no discernible way to determine how the system reached equilibrium—all such evidence has been destroyed. At the microscopic, molecular level, however, the situation is far from static.

As we will see, approximately 100 TW of power received from the sun, 6 times the total power consumed by mankind, drives the formation of 100 Gt of organic matter each year through photosynthesis. This provides a critically important component of primary energy generation from biofuels. As we have seen, the transition to electric power for vehicular transportation will decrease the demand for petroleum by 60%. Replacement of oil heat and increased efficiency in the transportation sector will, when taken together, reduce the demand for liquid fuels to 25% of current demand. Thus biofuels become critically important.

Biofuels are the subject of Case Study 5.1 and they comprise a key strategic component of the nation's energy future.

Before moving on from the subject of biofuels, we consider the case of methanol. There has been considerable pressure to switch to compressed natural gas (CNG) in the transportation sector both (1) because the fuel is intrinsically cleaner and yields a greater amount of energy per unit of CO_2 emitted than does petroleum, and (2) because it would reduce US dependence on foreign oil. It also has the potential to reduce fuel costs for transportation because the energy equivalent cost of CNG is approximately $1.55 for a gallon of gasoline.

However, the disadvantage for use in private automobiles is that the CNG

Case Study 5.1 The Technology and the Future of Biofuels

must be stored at high pressure (100 atm) and this implies a considerable increase in weight to store the fuel. For the truck and bus sector this increase in weight becomes far less of an economic issue.

There are, however, critically important options based directly on equilibrium chemistry and controlled by Gibbs free energy.

One of the most remarkable chemical transformations in nature is the ability to interconvert hydrocarbons among both the categories of fuels (e.g. coal, natural gas, gasoline, etc.) and the phases of matter (gas, liquid, solid). These transformations are accomplished by an astute application of the principles of chemical equilibrium, which we address in this chapter, and chemical catalysis, which we will address in subsequent chapters.

We can diagram these remarkable transformations, shown in Figure 5.2, by listing an array of *inputs*, a chemical transformation sequence, and an array of products that are required for supplying the breadth of fuels needed to sustain society.

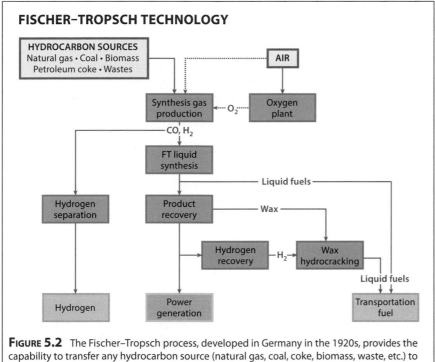

FIGURE 5.2 The Fischer–Tropsch process, developed in Germany in the 1920s, provides the capability to transfer any hydrocarbon source (natural gas, coal, coke, biomass, waste, etc.) to any desired liquid fuel. The steps are displayed here.

The chemical manipulation of both gas phase (e.g. natural gas) and solid phase (e.g. coal) hydrocarbons to produce liquid fuels represents a critically important technique in the setting of energy policy. In particular, addressing the challenges of adjusting and optimizing the blend of fuels to supply transportation demands, where petroleum products are in limited supply or where it is desirable to minimize imports of petroleum, requires the application of the principles of chemical equilibrium.

For example, in the production of gasoline, comprised largely of hydrocarbons with between 5 and 12 carbon atoms, which we will approximate by octane, C_8H_{18}, the primary reaction involves the equilibrium between H_2 and CO as "reactants" and C_8H_{18} and water as "products":

$$17H_2 + 8CO \rightleftharpoons C_8H_{18} + 8H_2O$$

This reaction is executed over a combination of metal catalysts, usually *iron* (Fe) and *cobalt* (Co) and was first developed by Franz Fischer and Hans Tropsch in the 1920s in Germany. Germany is a country with large coal reserves but little petroleum. While Fischer and Tropsch were unaware of the implication at the time of their development, both Germany and Japan used the process extensively in World War II to produce liquid fuels for warfare: tanks, aircraft, submarines, etc. By early 1944 production of gasoline in Germany reached 5 million gallons per day from 25 plants. The remarkable aspect of the Fischer–Tropsch process is that it accepts a wide range of hydrocarbon inputs, converting such compounds as coal, natural gas, garbage, etc. to high quality liquid fuels. We explore here how this is accomplished.

The first step, shown in the upper left of Figure 5.2, in the process is the production of *synthesis gas*, or *syngas,* which is a mixture of H_2 and CO formed from the pyrolysis of coal to form "coke," which is impure carbon denoted by C:

$$C + H_2O \rightleftharpoons CO + H_2 \qquad \Delta H = +131.4 \text{ kJ/mole}$$

The coke in this step is heated to high temperature, approximately 900°C, and the reaction is *endothermic* by 131.4 kJ per mole of coke. As the coke bed cools because heat is required to sustain the endothermic reaction, a temperature is reached at which the reaction forming H_2 and CO no longer proceeds to the right. At that point, O_2 is flowed over the coke bed producing CO_2

$$C + O_2 \rightleftharpoons CO_2 \qquad \Delta H = -393.5 \text{ kJ/mole}$$

followed by reactions between CO_2 and coke

$$CO_2 + C \rightleftharpoons 2CO \qquad \Delta H = +170.5 \text{ kJ/mole}$$

that produce CO. The first of these two reactions is exothermic, the second endothermic, but the *coupled* pair of reactions is exothermic, raising the temperature of the coke bed. When the coke bed reaches 900°C, water is passed across the bed again, producing syngas. The cycle is repeated until the coke is consumed.

In order to control the *ratio* of H_2 to CO, the "water gas shift" reaction is employed

$$CO + H_2O \rightarrow CO_2 + H_2 \qquad \Delta H = -41.0 \text{ kJ/mole}$$

The Fischer–Tropsch reaction is typically executed at temperatures between 150 and 300°C at 3 to 10 atmospheres pressure over iron-cobalt catalysts. The United States Air Force and Navy have a major program in place to produce jet fuel using the Fischer–Tropsch process. In an interesting development, techniques are being developed to produce H_2 by electrolysis of water (which we will study in Chapter 7) using photovoltaics, and then react that H_2 with CO_2 to form hydrocarbon fuels.

Another important example of equilibrium chemistry is the production of methanol as a liquid fuel. This important topic is treated in Case Study 5.2.

An important example of how equilibrium chemistry is used to produce liquid fuels from a combination of hydrocarbon feedstocks, as depicted in Figure 5.2, is the production of methanol (CH_3OH) from natural gas (CH_4). This is an important example because major new supplies of natural gas are being discovered, but it is liquid fuels for transportation that are in high demand. The liquid fuel methanol is stored, transported, and used as a liquid fuel at atmospheric pressure. It has a higher octane rating than gasoline, but because it has an energy density (kWh/ℓ) approximately half that of gasoline, it requires either a more fuel-efficient automobile to achieve the same range from the same volume tank, or it would require a larger tank. We will use methanol production from natural gas (methane) as a specific model for liquid fuel production from fossil fuels.

To this end, we consider first syngas production from natural gas. This process, as was discussed above for the Fischer–Tropsch technique, uses "steam reforming" of methane, wherein methane is reacted in a highly endothermic reaction with steam over a nickel-based catalyst at temperature between 800 and

Case Study 5.2 Ethanol, Net Energy Production, and the Haber–Bosch Process

1000°C at pressures of 20–30 atmospheres, to form CO and H_2. Part of the CO formed reacts with steam in the water gas shift reactions to yield more H_2 and also CO_2. The products of those coupled reactions is a mixture of H_2, CO, and CO_2:

$$CH_4 + H_2O \rightleftarrows CO + 3H_2 \qquad \Delta H = +205.2 \text{ kJ/MJk}$$
$$CO + H_2O \rightleftarrows CO_2 + H_2 \qquad \Delta H = -14.0 \text{ kJ/MJk}$$

Because these reactions are in chemical equilibrium, controlled by Gibbs free energy, the ratio of the chemical species depends on the reaction conditions: temperature, pressure, and the H_2O/CH_4 ratio. The efficiency of syngas generation from methane increases with increasing temperature and decreasing pressure.

Since the overall methane steam reforming process is highly endothermic, heat must be supplied to the system, generally by burning a part of the natural gas used as a feedstock. A key aspect of methanol production from methane is that the process produces excess H_2 over that required by the stoichiometric production of methanol. This has important consequences, because if the production of methanol is combined with a flow of CO_2, for example from carbon extraction from the atmosphere, CO_2 release to the atmosphere can be significantly reduced.

A typical reactor for converting syngas from methane to methanol is shown in Figure 5.3.

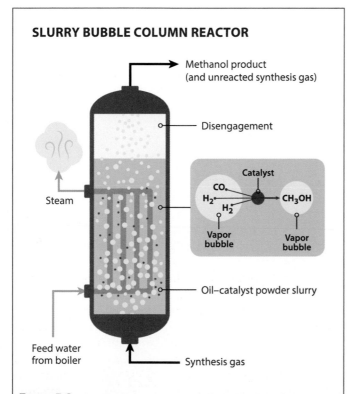

FIGURE 5.3 A modern reactor for methanol production is displayed wherein syngas (CO and H_2) is fed in the bottom of the reactor and methanol (CH_3OH) is extracted from the top of the reactor. The key reaction occurs over a metal catalyst that converts CO and H_2 to CH_3OH.

Chapter Core

Road Map for Chapter 5

We develop the principles of equilibria in chemical reactions by first investigating the concept of an *equilibrium constant*, K_{eq}, that defines the relationship between reactant concentrations and product concentrations under the condition of equilibrium. In developing the concept of the equilibrium constant, we explore how the equilibrium point is reached from a variety of different initial conditions using the concept of a reaction quotient, which is identical in functional form to that of the equilibrium constant, but involves concentrations of reacting species not in equilibrium. We next consider how the *equilibrium constant* expression for each reaction in a sequence of reactions can be used to determine the overall equilibrium constant for a series of coupled reactions. As we will see, this can be done whether the concentration units are given as pressure units or molar units. A key concept in chemical equilibrium studies involves the *response* of a chemical system in equilibrium when that system is *stressed* by changing the conditions of pressure, concentration, temperature, etc. We examine the process of ammonia synthesis as an example of the response of the system to impose external stress. Finally, we examine the relationship between equilibrium constants, spontaneous processes and Gibbs free energy.

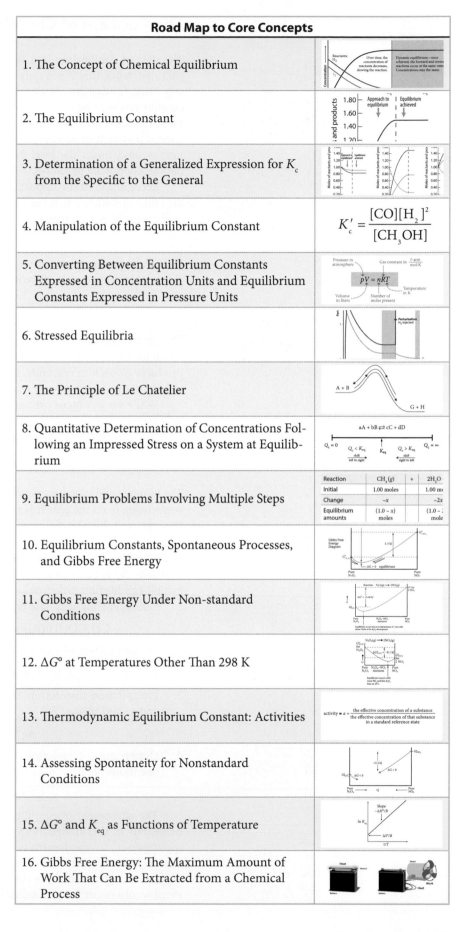

Road Map to Core Concepts

1. The Concept of Chemical Equilibrium

2. The Equilibrium Constant

3. Determination of a Generalized Expression for K_c from the Specific to the General

4. Manipulation of the Equilibrium Constant

$$K'_c = \frac{[CO][H_2]^2}{[CH_3OH]}$$

5. Converting Between Equilibrium Constants Expressed in Concentration Units and Equilibrium Constants Expressed in Pressure Units

6. Stressed Equilibria

7. The Principle of Le Chatelier

8. Quantitative Determination of Concentrations Following an Impressed Stress on a System at Equilibrium

9. Equilibrium Problems Involving Multiple Steps

10. Equilibrium Constants, Spontaneous Processes, and Gibbs Free Energy

11. Gibbs Free Energy Under Non-standard Conditions

12. $\Delta G°$ at Temperatures Other Than 298 K

13. Thermodynamic Equilibrium Constant: Activities

14. Assessing Spontaneity for Nonstandard Conditions

15. $\Delta G°$ and K_{eq} as Functions of Temperature

16. Gibbs Free Energy: The Maximum Amount of Work That Can Be Extracted from a Chemical Process

The Concept of Chemical Equilibrium

The state of equilibrium in a chemical reaction is of such theoretical and practical importance that it not only deserves its own chapter, but also embodies concepts that fundamentally alter how we view chemical transformations. To be in equilibrium is to be in a state of delicate balance, poised between countervailing forward and reverse reactions that appear macroscopically to be at rest. While a chemical system in equilibrium presents an unchanging macroscopic face, at the microscopic, molecular level there is an intense, dynamic competition occurring between opposing processes.

Recognition of the importance of the *dynamic equilibrium* state to which we now turn our attention, is the fact that while we have typically written our generalized chemical equation as

$$aA + bB \rightarrow cC + dD$$

(that distinguishes *reactants* on the left from *products* on the right with the appropriate stoichiometric coefficients), we now express an arbitrary chemical process as

$$aA + bB \rightleftharpoons cC + dD$$

This introduces the bidirectional double arrow where "reactants" (a moles of A and b moles of B) go to "products" (c moles of C and d moles of D) but equally, products go back to reactants. In this dynamic equilibrium, there is nothing special about the designation of reactants vs. products.

While we will focus primarily on chemical reactions that interconnect products and reactants, the principles we develop can be applied equally well to processes such as liquid-gas equilibrium; for example,

$$H_2O(l) \rightleftharpoons H_2O(g)$$

The number of molecules evaporating from the liquid is balanced exactly by those condensing from the gas. This results in a constant vapor pressure over the liquid. We represent this as a *dynamic equilibrium* symbolically by the twin arrows.

The central importance of the concept of equilibrium to modern chemistry and physics can be demonstrated in many ways—and we will explore a number in detail—by specific example. A few examples of chemical transformations that exhibit equilibrium properties at or near room temperature are:

1. Homogeneous acid-base equilibrium in solution:

$$CH_3COOH\ (aq) + H_2O\ (l) \rightleftharpoons H_3O^+(aq) + CH_3COO^-(aq)$$

2. Decomposition of a solid into a solid and a gas:

$$CaCO_3(s) \rightleftharpoons CaO(s) + CO_2(g)$$

3. Precipitation-dissolution, the heterogeneous equilibrium between a solution and a solid phase:

$$Ag^+(aq) + Cl^-(aq) \rightleftharpoons AgCl(s)$$

4. Dimerization-dissolution, which is the formation and subsequent breakup of one molecule from two:

$$ClO(g) + ClO(g) \rightleftharpoons ClOOCl(g)$$

The attainment of equilibrium is a universal imperative driven by the fundamentals of thermodynamics that dictate spontaneous change. An analysis of spontaneous change dominated our discussion in the development of Gibbs free energy, G and ΔG. In the development of our understanding of Gibbs free energy, we recognized that the thermodynamic criteria for determining whether an event would occur is the condition that $\Delta G < 0$. This means that before a chemical reaction begins, the free energy of the reactants, G_R, is greater than that of the products, G_p. As noted in the Framework, if the reaction is to proceed, then the chemical transformation from reactant to product must *release* free energy:

$$\text{Reactants} \rightarrow \text{Products} + \text{Free Energy}$$

The reaction proceeds to the point where $\Delta G = G_p - G_R = 0$ at which point the *equilibrium condition* is satisfied and from the macroscopic perspective, change ceases. All concentrations, pressures, temperature, etc. no longer depend on time. In the natural world there are only two options: (1) the system is in equilibrium and it will remain so unless disturbed by an outside influence, or (2) the system is out of equilibrium, which is the case of all living organisms, and will, in time, achieve equilibrium.

We turn then to address how this equilibrium condition in chemistry is addressed, beginning with an analysis of the equilibrium constant.

The Equilibrium Constant

The quantitative foundation for treating equilibria in chemical systems began with C. M. Guldberg and P. Waage in 1863 with the statement of the *Law of Mass Action* that defined the reaction quotient

$$Q = \frac{[C]^c [D]^d}{[A]^a [B]^b}$$

for our chemical transformation

$$aA + bB \rightleftharpoons cC + dD$$

wherein the lower case letters designate the stoichiometric coefficients and the brackets designate the concentrations. While Q, the reaction quotient, is defined for all conditions, equilibrium or otherwise, the *Law of Mass Action* states that *at equilibrium*, the reaction quotient is equal to a constant

$$Q = K_{eq}(T)$$

and that constant is dependent only on temperature and nothing else. The constant is called the *equilibrium constant*, $K_{eq}(T)$. It is, for a given reaction at a specified temperature, a single number that is established empirically—that is by direct experiment via the determination of reactant/product concentrations *at* equilibrium.

Thus, at equilibrium for a specified temperature

$$K_{eq} = \frac{[C]^c_{eq} [D]^d_{eq}}{[A]^a_{eq} [B]^b_{eq}}$$

where $[A]_{eq}$, $[B]_{eq}$, etc. refer to the concentrations at equilibrium.

The importance of this equilibrium condition, expressed as a single, experimentally determined number, has far-reaching consequences for the diagnosis of chemical systems, be they acid-base neutralization, dissociation, dimerization, precipitation, or another process. *All* of these chemical systems conform to the constraint imposed by the Law of Mass Action at equilibrium. For each explicit, balanced chemical equation, the equilibrium constant unequivocally links the concentrations of reactants/products at equilibrium.

Consider the equilibrium reaction in the formation of methyl alcohol from carbon monoxide and hydrogen:

$$CO(g) + 2H_2(g) \rightleftarrows CH_3OH(g)$$

We can initiate this reaction in a variety of ways by starting from vastly different initial conditions.

Experiment 1:

For example, we can initiate the reaction with 1 mole each of CO and of H_2 with zero moles of CH_3OH in a 10 liter vessel at a given temperature and then wait until the system reaches equilibrium. We know when equilibrium is reached because the macroscopic observables (concentrations of CO, H_2, and CH_3OH; temperature; pressure) become independent of time, as displayed in Figure 5.4a.

But moreover, we can calculate the reaction quotient, Q_c, for the initial conditions, where the subscript "c" refers to the fact that we are carrying out the calculation in *concentration units* because 1 mole of reactant is placed in a 10 liter vessel such that the initial concentrations of CO and H_2 are

$$[CO]_{init} = 0.1 \text{ Molar}$$
$$[H_2]_{init} = 0.1 \text{ Molar}$$

and

$$Q_c = \frac{[CH_3OH]_{init}}{[CO]_{init}[H_2]_{init}^2} = \frac{0}{(0.1)(0.1)^2} = 0$$

As the reaction progresses towards equilibrium, the concentrations of both CO and H_2 *decrease*, while the concentration of CH_3OH *increases* such that Q_c increases from zero to its equilibrium value. From this we can deduce that if $Q_c < K_{eq}$, the reaction will shift from left to right, from reactants to products, until $Q_c = K_{eq}$, at which point the concentrations of reactants and products will cease to change.

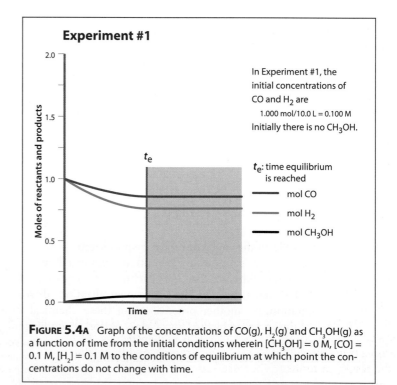

FIGURE 5.4A Graph of the concentrations of CO(g), H_2(g) and CH_3OH(g) as a function of time from the initial conditions wherein $[CH_3OH] = 0$ M, $[CO] = 0.1$ M, $[H_2] = 0.1$ M to the conditions of equilibrium at which point the concentrations do not change with time.

Experiment 2:

Suppose we use an entirely different set of initial conditions. Suppose we use 1 mole of CH_3OH in the 10 liter vessel with no CO or H_2 and allow the reaction to proceed. The evaluation of $[CH_3OH]$, $[H_2]$, and $[CO]$ with time are displayed in Figure 5.4b.

The reaction coefficient at the moment the reaction is initiated is

$$Q_c = \frac{[CH_3OH]_{init}}{[CO]_{init}[H_2]_{init}^2} = \frac{0.10}{(0)(0)} = \infty$$

As the reaction proceeds, the concentrations of CO and of H_2 increase with time as the concentration of CH_3OH decreases. Thus, the value of the reaction quotient will *decrease* until the point is reached where $Q_c = K_{eq}$ and at that point the concentrations of CH_3OH, H_2, and CO will cease to change with time because the condition of equilibrium has been reached. Thus, if, at the initiation of the reaction, $Q_c > K_{eq}$, as the reaction progresses the concentrations of CO and H_2 will increase, the concentration of CH_3OH will decrease, and the reaction will shift from right to left as equilibrium is approached.

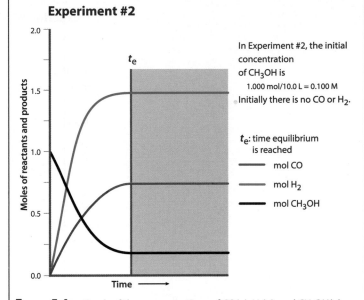

Experiment #2

In Experiment #2, the initial concentration of CH_3OH is
1.000 mol/10.0 L = 0.100 M
Initially there is no CO or H_2.

t_e: time equilibrium is reached

— mol CO
— mol H_2
— mol CH_3OH

FIGURE 5.4B Graph of the concentrations of CO(g), H_2(g), and CH_3OH(g) as a function of time from the initial conditions wherein $[CH_3OH]$ = 0.100 M and $[CO] = [H_2] = 0$ M to the condition of chemical equilibrium where the concentrations do not change with time.

Experiment 3:

Suppose we initiate the reaction with 1 mole each of CO, H_2, and CH_3OH in a 10 liter vessel. This experiment is depicted in Figure 5.4c, showing that CO and H_2 will increase, and CH_3OH will decrease as time progresses.

At the instant the reaction is initiated, the reaction coefficient is

$$Q_c = \frac{[CH_3OH]_{init}}{[CO]_{init}[H_2]_{init}^2} = \frac{0.10}{(0.1)(0.1)^2} = 100$$

As the reaction progresses toward equilibrium, and the concentration of CH_3OH decreases and the concentrations of CO and H_2 increase, Q_c will decrease until $Q_c = K_{eq}$; equilibrium is achieved and the concentrations do not change with time.

What is important to recognize is that while the measured (macroscopic) concentrations of CO, H_2, and CH_3OH do not change, the equilibrium condition represents a *dynamic* state at the molecular (microscopic) level because CO and H_2 are *constantly reacting* to produce methyl alcohol

$$CO + 2H_2 \rightarrow CH_3OH$$

but methyl alcohol is constantly decomposing to form CO and H_2

$$CH_3OH \rightarrow CO + 2H_2$$

and the rates at which these two processes occur exactly balance. We will study the rates of those reactions in Chapter 12, when we study kinetics.

But now we have three experiments and from these results some striking conclusions emerge. Let's first summarize our three experiments graphically, as shown in Figure 5.4d, and then tabulate the results in terms of the observed concentrations at equilibrium.

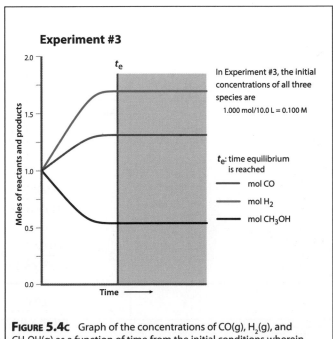

Experiment #3

In Experiment #3, the initial concentrations of all three species are
1.000 mol/10.0 L = 0.100 M

t_e: time equilibrium is reached

— mol CO
— mol H_2
— mol CH_3OH

FIGURE 5.4c Graph of the concentrations of CO(g), H_2(g), and CH_3OH(g) as a function of time from the initial conditions wherein $[CH_3OH]$, $[CO]$, and $[H_2]$ are initially equal to 0.100 M. As in both Figures 5.4a and 5.4b, the concentrations change with time until equilibrium is achieved at which point the concentrations no longer change with time.

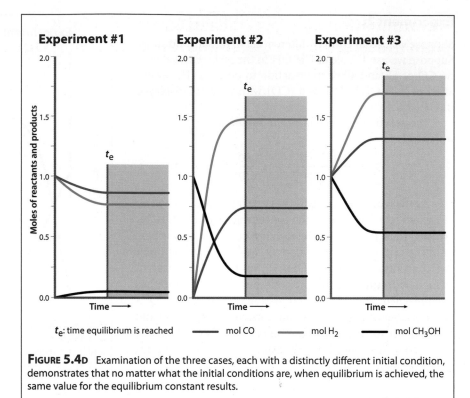

FIGURE 5.4D Examination of the three cases, each with a distinctly different initial condition, demonstrates that no matter what the initial conditions are, when equilibrium is achieved, the same value for the equilibrium constant results.

It is important that we prove experimentally that we have established the correct functional form of the equilibrium constant, K_{eq}. To accomplish this, we analyze the results to find whether (1) there is a single value for the reaction quotient at equilibrium, and (2) what *functional form* of the reactant and product concentrations gives a consistent value for the equilibrium constant. Several attempts to deduce this functional form are given in Table 5.1.

TABLE 5.1

Three attempts to find a constant ratio of equilibrium concentrations in the reaction
$CO + 2H_2 \rightleftarrows CH_3OH$.

Exp	Trial 1: $\dfrac{[CH_3OH]}{[CO][H_2]}$	Trial 2: $\dfrac{[CH_3OH]}{[CO](2 \times [H_2])}$	Trial 3: $\dfrac{[CH_3OH]}{[CO][H_2]^2}$
1	$\dfrac{0.00892}{0.0911 \times 0.0822} = 1.19$	$\dfrac{0.00892}{0.0911 \times (2 \times 0.0822)} = 0.596$	$\dfrac{0.00892}{0.0911 \times (0.0822)^2} = 14.5$
2	$\dfrac{0.0247}{0.0753 \times 0.151} = 2.17$	$\dfrac{0.0247}{0.0753 \times (2 \times 0.151)} = 1.09$	$\dfrac{0.0247}{0.0753 \times (0.151)^2} = 14.4$
3	$\dfrac{0.0620}{0.138 \times 0.176} = 2.55$	$\dfrac{0.0620}{0.138 \times (2 \times 0.176)} = 1.28$	$\dfrac{0.0620}{0.138 \times (0.176)^2} = 14.5$

Determination of a Generalized Expression for K_c from the Specific to the General: Beginning with Our Equilibrium Expression

$$aA + bB \rightleftarrows cC + dD$$

$$CO\ (g) + 2H_2\ (g) \rightleftarrows CH_3OH\ (g)$$

We note first that for Experiment 1 at equilibrium displayed in Figure 5.4d:
$[CO]_{eq} = 0.0911\ M$
$[H_2]_{eq} = 0.0822\ M$
$[CH_3OH]_{eq} = 0.00892\ M$

For Experiment 2 at equilibrium displayed in Figure 5.4b:
$[CO]_{eq} = 0.0753\ M$
$[H_2]_{eq} = 0.151\ M$
$[CH_3OH]_{eq} = 0.0247\ M$

And for Experiment 3 displayed in Figure 5.4d:
$[CO]_{eq} = 0.138\ M$
$[H_2]_{eq} = 0.176\ M$
$[CH_3OH]_{eq} = 0.0620\ M$

What Table 5.1 displays is the calculated values for three different assumptions concerning the functional form for a trial equilibrium constant. Trial 1 assumes that all concentrations appear in the trial equilibrium expression raised to the first power. This assumption yields three different values for each of the experiments—wherein the experiments were defined by the initial concentrations of CO, H_2, and CH_3OH.

Trial 2, calculated in Table 5.1, assumes all concentrations at equilibrium are raised to the first power, but that the *concentration of each species is multiplied by the stoichiometric coefficient* that appears in the balanced chemical reaction. Thus the concentration of CO is multiplied by unity, the concentration of H_2 is multiplied by 2, and the concentration of CH_3OH is multiplied by unity. Values for the three experiments with the assumptions of Trial 2 are tabulated in Table 5.1. For each of the three *different* initial concentrations of CO, H_2, and CH_3OH, three different values for the "equilibrium constant" emerge.

Trial 3 assumes that the concentrations of the chemical species in the equilibrium expression appear *raised to the power of their individual stoichiometric coefficients*. Thus, the CO concentration is raised to the power of one, H_2 concentration to the power of two, and CH_3OH concentration to the power of one. These are tabulated in the third column of the table. But remarkably, the calculated value for the equilibrium constant is the same, within experimental error, for all three experiments—experiments for which the initial concentrations were *very* different, but for which, at equilibrium, the equilibrium constant calculated by raising each of the concentrations to the power of its stoichiometric coefficient, yielded a single value for the equilibrium constant.

This is a remarkable result:

1. Equilibrium is achieved from very different initial conditions but with the same final state expressed by an equilibrium constant expression constructed by assuming that the *concentration* of each of the chemical species involved in the equilibrium is raised to the power of its individual stoichiometric coefficient.

2. By direct observation, the *Law of Mass Action* is demonstrated to hold.

Solution

Write the equilibrium constant expression in terms of activities:

$$K = \left(\frac{a_{CH_3OH}}{a_{CO}\left(a_{H_2}\right)^2} \right)_{eq} = 14.5$$

See page 295 for a discussion of activities. Activities will turn out to be of significant quantitative importance.

Assume that the reaction conditions are such that the activities can be replaced by their concentration values, allowing concentration units to be canceled.

$$K = \left(\frac{[CH_3OH]}{[CO][H_2]^2} \right)_{eq} = 14.5$$

Substitute the known equilibrium concentrations into the equilibrium constant expression:

$$K = \frac{[CH_3OH]}{[CO]\left([H_2]\right)^2} = \frac{1.56}{1.03[H_2]^2} = 14.5$$

Solve for the unknown concentration, $[H_2]$.

$$[H_2]^2 = \frac{1.56}{1.03 \times 14.5} = 0.104$$

$$[H_2] = \sqrt{0.104} = 0.322 \text{ M}$$

Practice Example A: In another experiment, equal concentrations of CH_3OH and CO are found at equilibrium in the reaction $CO(g) + 2H_2(g) \rightleftarrows CH_3OH(g)$. What must be the equilibrium concentration of H_2?

Practice Example B: At a certain temperature, $K = 1.8 \times 10^4$ for the reaction $N_2(g) + 3\,H_2(g) \rightleftarrows 2\,NH_3(g)$. If the equilibrium concentrations of N_2 and NH_3 are 0.015 M and 2.00 M, respectively, what is the equilibrium concentration of H_2?

3. The relationship between the reaction quotient, Q_c, and the equilibrium constant, K_{eq}, is established through the course of a reaction as it approaches equilibrium from very different initial conditions. Specifically, if $Q_c < K_{eq}$, the reaction will progress from left to right until equilibrium is achieved, at which point $Q_c = K_{eq}$. If, on the other hand, $Q_c > K_{eq}$, the reaction will proceed from right to left until $Q_c = K_{eq}$, at which point the reaction has achieved equilibrium.

Check Yourself 1—Relating Equilibrium Concentrations of Reactants and Products

These equilibrium concentrations are measured for the reaction $CO(g) + 2H_2(g) \rightleftarrows CH_3OH[g]$ at 483 K: [CO] = 1.03 M and $[CH_3OH]$ = 1.56 M. What is the equilibrium concentration of H_2?

Manipulation of the Equilibrium Constant

The equilibrium constant, K_{eq}, is so important to the analysis of chemical processes that we digress to consider the relationship between the functional form of the equilibrium constant and the specific form of the chemical reaction to which the equilibrium constant applies.

The first point is that there is a one-to-one relationship between the equilibrium constant and the particular balanced chemical reaction. We have deduced, by direct experiment, that for the reaction

$$CO(g) + 2H_2(g) \rightleftarrows CH_3OH(g) \tag{5.1}$$

the equilibrium constant expressed in concentration units is

$$K_c = \frac{[CH_3OH]}{[CO][H_2]^2} \tag{5.2}$$

But suppose we write the equilibrium reaction as

$$CH_3OH(g) \rightleftarrows CO(g) + 2H_2(g)$$

Following our principle that the equilibrium constant is constructed by taking the *concentrations* of species on the right-hand side of the reaction, raised to the power of their stoichiometric coefficients, *divided by* the product of the concentration of the species on the left-hand side of the equation, raised to the power of their stoichiometric coefficients, we have

$$K_c' = \frac{[CO][H_2]^2}{[CH_3OH]}$$

for reaction 5.2. Thus, we have the result that

1. For each chemical equation defining an equilibrium reaction, there is a unique functional form of the equilibrium constant.
2. When we *reverse* the chemical equation expressing the equilibrium relation between the left-hand and right-hand species, we *invert* the functional form of the equilibrium constant.

Thus, for $\qquad$ $CO(g) + 2H_2(g) \rightleftarrows CH_3OH(g)$ $\qquad$ $K_c = \dfrac{[CH_3OH]}{[CO][H_2]^2}$

and for $\qquad$ $CH_3OH(g) \rightleftarrows CO(g) + 2H_2(g)$ $\qquad$ $K'_c = \dfrac{[CO][H_2]^2}{[CH_3OH]}$

and $\qquad$ $K'_c = \dfrac{1}{K_c}$.

If we multiply the stoichiometric coefficients in a balanced chemical reaction by a common factor (yielding a chemical reaction that is still balanced), then the equilibrium constant, K_c, is raised to the power by which the stoichiometric coefficients were multiplied, so

$$2CO(g) + 4H_2(g) \rightleftarrows 2CH_3OH(g)$$

has a corresponding equilibrium constant

$$K''_c = \frac{[CH_3OH]^2}{[CO]^2[H_2]^4} = (K_c)^2$$

If, on the other hand, we divide the stoichiometric coefficients by 2, such that

$$\tfrac{1}{2}\,CO(g) + H_2(g) \rightleftarrows \tfrac{1}{2}\,CH_3OH(g)$$

then

$$K'''_c = \frac{[CH_3OH]^{1/2}}{[CO]^{1/2}[H_2]}$$

and

$$K'''_c = (K_c)^{1/2}$$

Quite often when we are analyzing a chemical reaction, data may not be available for the specific reaction for which we are interested. In this case, we can assemble a *series of reactions* that, when summed together, yield the overall reaction of interest, but for which we have measured equilibrium constants. For example, suppose we wish to calculate the equilibrium constant for the reaction (which we label reaction 1):

$$NO(g) + O_3(g) \xrightleftharpoons{1} NO_2(g) + O_2(g)$$

We know we can write

$$K_{c_1} = \frac{[NO_2(g)][O_2(g)]}{[NO(g)][O_3(g)]}$$

But suppose we cannot find an equilibrium constant value for K_c for reaction 1, but we are given the equilibrium constants for two other reactions:

$$\tfrac{3}{2}\,O_2(g) \xrightleftharpoons{2} O_3(g) \qquad K_{c_2} = 2.5 \times 10^{-29}$$

$$2NO(g) + O_2(g) \xrightleftharpoons{3} 2NO_2(g) \qquad K_{c_3} = 2.3 \times 10^{12}$$

Our strategy is to rewrite these chemical equations such that when they are added together, they result in the desired net overall reaction.

We note first that our target reaction has O_3 on the left, so we reverse reaction 2:

$$O_3(g) \xrightleftharpoons{2} \tfrac{3}{2}\,O_2(g) \qquad K'_{c_2} = \frac{1}{K_{c_2}} = 4 \times 10^{28}$$

because we must invert the equilibrium constant if we reverse the reaction.

Turning to reaction 3, we note that the desired reaction has NO (g) on the left and NO_2 (g) on the right, so we do not need to reverse the reaction. But we wish to sum reaction 3 with the reverse of reaction 2 to yield reaction 1, and to do this we must multiply reaction 3 by the factor ½ such that reaction 3 becomes

$$NO(g) + \tfrac{1}{2} O_2(g) \rightleftarrows NO_2(g)$$

and this requires that we take the square root of K_{c_3}

$$K'_{c_3} = \sqrt{K_{c_3}} = 1.5 \times 10^6$$

Now we are ready to reassemble our modified forms of reaction 2 and reaction 3 with the modified equilibrium constant for each reaction such that

$$O_3(g) \rightleftarrows \tfrac{3}{2} O_2(g) \qquad K'_{c_2} = \frac{1}{K_{c_2}} = 4 \times 10^{28}$$

$$NO(g) + \tfrac{1}{2} O_2(g) \rightleftarrows NO_2(g) \qquad K'_{c_3} = \sqrt{K_{c_3}} = 1.5 \times 10^6$$

$$\overline{NO(g) + O_3(g) \rightleftarrows NO_2(g) + O_2(g) \qquad K_{c_1} = \left(K'_{c_2}\right)\left(K'_{c_3}\right) = 6 \times 10^{34}}$$

We can quickly prove that the expression for $K_{c_1} = \left(K'_{c_2}\right)\left(K'_{c_3}\right)$ is correct by noting that

$$K_{c_1} = \frac{[NO_2(g)][O_2(g)]}{[NO(g)][O_3(g)]}$$

$$K'_{c_2} = \frac{[O_2(g)]^{3/2}}{[O_3(g)]}$$

$$K'_{c_3} = \frac{[NO_2(g)]}{[NO(g)][O_2(g)]^{1/2}}$$

$$\left(K'_{c_2}\right)\left(K'_{c_3}\right) = \frac{[O_2(g)]^{3/2}}{[O_3(g)]} \frac{[NO_2(g)]}{[NO(g)][O_2(g)]^{1/2}}$$

$$= \frac{[O_2(g)][NO_2(g)]}{[O_3(g)][NO(g)]}$$

Which is indeed equal to K_{c_1}. There are two important conclusions:

1. We can determine the equilibrium constant for a reaction by adding together separate reactions that, when added together, result in the same net overall reaction.
2. When reactions are summed together to form a net reaction, the equilibrium constants are multiplied together to form the resultant equilibrium constant.

Check Yourself 2—Relating K to the Balanced Chemical Equation

The following K value is given at 298 K for the synthesis of $NH_3(g)$ from its elements.

$$N_2(g) + 3 H_2(g) \rightleftarrows 2 NH_3(g) \quad K = 3.6 \times 10^8$$

What is the value of K at 298 K for the following reaction?

$$NH_3(g) \rightleftarrows \tfrac{1}{2} N_2(g) + \tfrac{3}{2} H_2(g) \qquad K = ?$$

Solution

First, reverse the given equation. This puts $NH_3(g)$ on the left side of the equation, where we need it.

$$2 NH_3(g) \rightleftarrows N_2(g) + 3 H_2(g)$$

The equilibrium constant K' becomes

$$K' = 1/(3.6 \times 10^8) = 2.8 \times 10^{-9}$$

Then, to base the equation on 1 mol $NH_3(g)$, divide all coefficients by 2.

$$NH_3(g) \rightleftarrows \tfrac{1}{2} N_2(g) + \tfrac{3}{2} H_2(g)$$

This requires the square root of K'.

$$K = \sqrt{2.8 \times 10^{-9}} = 5.3 \times 10^{-5}$$

Practice Example A: Use data from this example to determine the value of K at 298 K for the reaction

$$\tfrac{1}{3} N_2(g) + H_2(g) \rightleftarrows \tfrac{2}{3} NH_3(g)$$

Practice Example B: For the reaction

$$NO(g) + \tfrac{1}{2}O_2(g) \rightleftarrows NO_2(g)$$

at 184° C, $K = 7.5 \times 10^2$. What is the value of K at 184°C for the reaction $2 NO_2(g) \rightleftarrows 2 NO(g) + O_2(g)$?

Converting between Equilibrium Constants Expressed in *Concentration Units* and Equilibrium Constants Expressed in *Pressure Units*

Suppose we consider the case wherein the equilibrium constant for the oxidation of SO_2 to SO_3 is expressed in concentration units

$$2SO_2(g) + O_2(g) \rightleftarrows 2SO_3(g)$$

Such that

$$K_c = \frac{[SO_3(g)]^2}{[SO_2(g)]^2[O_2(g)]} = 2.8 \times 10^2 \ @ \ 1000 \ K$$

But we wish to express the equilibrium constant in terms of *pressure*, not concentration.

To execute this transformation, we recall that the Perfect Gas Law states that

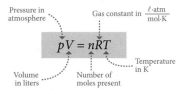

Therefore, we can express concentrations of $SO_2(g)$, $O_2(g)$, and $SO_3(g)$, which we state as moles/liter as

$$[SO_2(g)] = \frac{n_{SO_2}}{V} = \frac{P_{SO_2}}{RT}$$

$$[O_2(g)] = \frac{n_{O_2}}{V} = \frac{P_{O_2}}{RT}$$

$$[SO_3(g)] = \frac{n_{SO_3}}{V} = \frac{P_{SO_3}}{RT}$$

By direct substitution

$$K_c = \frac{[SO_3(g)]^2}{[SO_2(g)]^2[O_2(g)]} = \frac{\left(P_{SO_3}/RT\right)^2}{\left(P_{SO_2}/RT\right)^2\left(P_{O_2}/RT\right)}$$

$$= \frac{P_{SO_3}^2}{P_{SO_2}^2 P_{O_2}} \frac{\left(RT\right)^3}{\left(RT\right)^2}$$

$$= \frac{P_{SO_3}^2}{P_{SO_2}^2 P_{O_2}} RT$$

$$= K_p RT$$

where

$$K_p = \frac{P_{SO_3}^2}{P_{SO_2}^2 P_{O_2}}$$

We can express this interconversion between K_c and K_p in a simple and more general form by considering our general expression for a chemical process in equilibrium

$$aA + bB \rightarrow cC + dD$$

wherein

$$K_c = \frac{[C]^c[D]^d}{[A]^a[B]^b} = \frac{\left(P_C/RT\right)^c\left(P_D/RT\right)^d}{\left(P_A/RT\right)^a\left(P_B/RT\right)^b}$$

$$= \frac{P_C^c P_D^d}{P_A^a P_B^b}\left(\frac{1}{RT}\right)^{c+d-(a+b)}$$

$$= K_p\left(\frac{1}{RT}\right)^{c+d-(a+b)}$$

$$= K_p\left(\frac{1}{RT}\right)^{\Delta n}$$

Alternatively, we can, of course, write

$$K_p = K_c(RT)^{\Delta n}$$

where Δn = change in the number of moles
= (number of moles of product) − (number of moles of reactants)

Another important characterization of the equilibrium constant is that the concentrations of pure liquids and/or the concentration of pure solids are not included in the expression for the equilibrium constant. This is intuitively obvious, because if we take the concentration of water vapor over liquid water in a vessel, the *concentration of water vapor is independent of the amount of liquid water* in the vessel as long as *some* liquid water is present. Thus, for the equilibrium

$$H_2O(l) \rightleftarrows H_2O(g)$$

$$K_{eq} = \frac{[H_2O(g)]}{[H_2O(l)]}$$

We replace the K_{eq} with a measured equilibrium constant that eliminates the concentration of liquid water because the observed equilibrium constant is independent of $[H_2O(l)]$. Thus we write

$$K_{eq} = [H_2O(g)]$$

Solution

Write the equation relating the two equilibrium constants with different reference states.

$$K_p = K_c(RT)^{\Delta n}$$

With $\Delta n = -1$, we have

$$K_p = \frac{K_c}{RT}$$

Substitute the given data and solve.

$$K_p = \frac{2.8\times10^2}{0.08206\times1000} = 3.4$$

Practice Example A: For the reaction $2NH_3(g) \rightleftarrows N_2(g) + 3\,H_2(g)$ at 298 K, K_c = 2.8×10^{-9}. What is the value of K_p for this reaction?

Practice Example B: At 1065°C, for the reaction $2\,H_2S(g) \rightleftarrows 2\,H_2(g) + S_2(g)$, K_p = 1.2×10^{-2}. What is the value of K_c for the reaction

$$H_2(g) + \tfrac{1}{2}\,S_2(g) \rightleftarrows H_2S(g) \text{ at } 1065°C?$$

Check Yourself 3—Illustrating the Dependence of K on the Reference State

Complete the calculation of K_p for the reaction $2SO_2(g) + O_2(g) \rightleftarrows 2SO_3(g)$ knowing that $K_c = 2.8 \times 10^2$ (at 1000 K).

Suppose we wish to consider the important reaction shown in Figure 5.5:

$$CaCO_3(s) \rightleftarrows CaO\ (s) + CO_2(g)$$

Here again the $CO_2(g)$ concentration is independent of the amount of calcium carbonate, $CaCO_3$, and the amount of calcium oxide, CaO, in the vessel as long as both are present. We then write

$$K_c = [CO_2] \quad \text{and} \quad K_p = P_{CO_2}$$
$$K_p = K_c\,(RT)^{\Delta n} = K_c(RT)$$

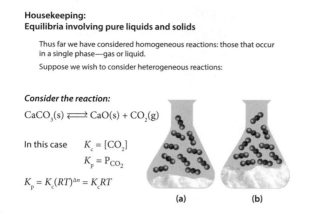

Housekeeping:
Equilibria involving pure liquids and solids

Thus far we have considered homogeneous reactions: those that occur in a single phase—gas or liquid.

Suppose we wish to consider heterogeneous reactions:

Consider the reaction:

$$CaCO_3(s) \rightleftharpoons CaO(s) + CO_2(g)$$

In this case $K_c = [CO_2]$

$K_p = P_{CO_2}$

$K_p = K_c(RT)^{\Delta n} = K_c RT$

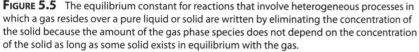

(a)   (b)

FIGURE 5.5 The equilibrium constant for reactions that involve heterogeneous processes in which a gas resides over a pure liquid or solid are written by eliminating the concentration of the solid because the amount of the gas phase species does not depend on the concentration of the solid as long as some solid exists in equilibrium with the gas.

Check Yourself 4—Calculating the Equilibrium Constant for Reactions Involving Pure Liquids and/or Pure Solids

An important factor when we consider how to construct the correct expression for the equilibrium constant for reactions that involve a combination of gas phase, liquid phase, and solid phase reactants and products is to consider what mechanism determines the partial pressure of a given gas phase species in the presence of a pure liquid or a pure solid. Specifically, the partial pressure of a gas over a liquid or solid (see Figure 5.5) does *not* depend upon the amount of the solid over which the gas phase species exists, *as long as there is some of the solid present*. The reason is that molecules in the gas phase (e.g. CO_2) over the solid (e.g. $CaCO_3$) are in a dynamic equilibrium such that gas phase molecules are constantly entering and leaving the solid phase species in the container. Thus, as long as there is some of the solid present, even a very small amount, that dynamic equilibrium will exist. If we double the amount of the solid (e.g. $CaCO_3$) present, there will be no effect on the partial pressure of the gas phase species (CO_2). Thus if the equilibrium between the gas phase species and the solid phase species does not depend on the amount of the solid phase present, the equilibrium constant must be *independent* of the amount of solid present—as long as some of the solid is present. The same principle holds for pure liquids.

Stressed Equilibria

Thus far we have viewed equilibria in chemical systems as a progression that carried us from a set of initial conditions to a final, time independent, state of equilibrium. But in many important examples in real systems, the state of equilibrium will be intruded upon by changes in the conditions impressed on the system in its state of equilibrium. In fact, more frequently than not, we find ourselves in need of analyzing how systems are stressed *after* they reach equilibrium rather than analyzing how they progress from initial conditions to the state of equilibrium.

Consider, for example, one of the most famous and continuously employed equilibria, the Haber process (named for its inventor pictured in Figure 5.6) for "fixing" nitrogen by breaking the N≡N bond to make nitrogen available to biological process.

Problem:
Consider the equilibrium between hydrogen sulfide ($H_2S(g)$) plus $I_2(s)$ as reactants and hydrogen iodide in the gas phase ($HI(g)$) and solid phase sulfur as the products.

$$H_2S(g) + I_2(s) \rightleftharpoons 2\,HI(g) + S(s)$$

Calculate the equilibrium constant in pressure units, K_p, for the reaction if $P_{H_2S} = 9.96 \times 10^{-1}$ atm and $P_{HI} = 3.65 \times 10^{-3}$ atm at 333 K.

Solution:
Inspection of the equilibrium reaction reveals that both I_2 and S are pure solids and thus they do not enter the expression for the calculation of the equilibrium constant.

Thus

$$K_p = \frac{\left(P_{HI}\right)^2}{\left(P_{H_2S}\right)} = \frac{\left(3.65 \times 10^{-3}\right)^2}{9.96 \times 10^{-1}} = 1.34 \times 10^{-5}$$

Practice Problem:
A very important reaction involving both multiphase equilibria as well as human health is the reaction between hydroxyapatite, $Ca_5(PO_4)_3OH$, which is the principal ingredient in mammalian teeth, and an acid solution. Acid is produced in the mouth by bacteria. We can write this reaction as

$$4\,H^+(aq) + Ca_5(PO_4)_3OH \rightleftharpoons 5\,Ca^{2+}(aq) + 3\,HPO_4^{2-}(aq) + H_2O(l)$$

What is the equilibrium constant expression in concentration units?

FIGURE 5.6 Fritz Haber was one of the most famous 20th century chemists, having developed a process for the formation of ammonia, NH_3, from N_2 and H_2 at high pressure and temperature over an iron catalyst. While he developed this method for "fixing" nitrogen (i.e., for breaking the N_2 bond to make nitrogen available for use by plants as a nutrient), the process was used to develop nitrate explosives used extensively in the First World War.

Using iron catalysts and elevated temperatures and pressures, Haber refined the production of ammonia using the equilibrium reaction

$$N_2(g) + 3H_2(g) \rightleftarrows 2NH_3(g)$$

We can immediately write down the reaction quotient, Q, for this reaction in terms of the partial pressures of N_2, H_2, and NH_3:

$$Q_p = \frac{P^2_{NH_3}}{P_{N_2} P^3_{H_2}}$$

If we begin with N_2 and H_2, initially $Q_p = 0$ and, because NH_3 is produced in the reaction, Q_p will increase *until equilibrium is attained* such that

$$Q_p = \frac{P^2_{NH_3}}{P_{N_2} P^3_{H_2}} = K_p$$

at the temperature of the reaction vessel. At that point, the partial pressures of NH_3, N_2, and H_2 are constant with time.

But we know that, at the molecular level, there is a dynamic process occurring wherein the reaction from left to right

$$N_2(g) + 3H_2(g) \rightarrow 2NH_3(g)$$

is rapidly converting N_2 and H_2 to NH_3. At the same time, $NH_3(g)$ is rapidly decomposing to form $N_2(g)$ and $H_2(g)$, such that the *rate* at which N_2 and H_2 react to form $NH_3(g)$ equals the rate of $NH_3(g)$ decomposition. Thus, the time independent equilibrium state is a delicate dynamic balance between reactions carrying the system to the *right* and reactions carrying the system to the *left*:

$$Rate_{\substack{Left \\ to\ Right}} = k_{\substack{Left \\ to\ Right}} [N_2(g)][H_2(g)]^3$$

where $k_{Left\ to\ Right}$ is the *reaction rate constant* (to be fully discussed in Chapter 12) that quantitatively defines the relationship between (1) the flow of N_2 and H_2 molecules through that reaction pathway, referred to as the rate of the reaction, and (2) the reactant concentrations raised to their stoichiometric coefficients.

Suppose now that we employ a piston to compress the reacting mixture by cutting the volume in half. Each of the partial pressures will double in response to the decrease in volume. How will the system respond?

First, we recalculate the reaction quotient immediately following the application of stress applied to the system (i.e. just after the pressure has been doubled by compressing the volume by a factor of two). We know that $(Q_p)_{\substack{Before \\ volume \\ change}} = K_p$ because the system had reached equilibrium.

$$\left(Q_p\right)_{\substack{\text{After} \\ \text{volume} \\ \text{change}}} = \frac{\left(\left(P_{NH_3}\right)_{after} / \left(P_{NH_3}\right)_{before}\right)^2}{\left(\left(P_{N_2}\right)_{after} / \left(P_{N_2}\right)_{before}\right)\left(\left(P_{H_2}\right)_{after} / \left(P_{H_2}\right)_{before}\right)^3}\left(Q_p\right)_{\substack{\text{Before} \\ \text{volume} \\ \text{change}}}$$

$$= \frac{2^2}{2 \cdot 2^3}K_p = \frac{1}{4}K_p$$

This tells us that, in response to the increase in partial pressure of each constituent, $Q < K_p$, such that the *process is no longer in equilibrium*. The system immediately seeks the state of equilibrium, a condition met only when $Q_p = K_p$. Thus, Q must *increase* to achieve this state and this in turn means that P_{NH_3} must increase, a situation that can only be attained by reacting more N_2 and H_2 to form the product, thereby decreasing both the concentration of N_2 and of H_2 in the reactor. The reaction is forced to the right by the spontaneous drive toward equilibrium.

The equilibrium constant, K_p, is *a function only of temperature*, not total pressure, so the system must return to a condition wherein

$$\frac{P_{NH_3}^2}{P_{N_2}P_{H_2}^3} = K_p$$

There is only one equilibrium constant! Out of the myriad possibilities for the concentrations of N_2, H_2, and NH_3 after the equilibrium state is stressed by decreasing the volume and thereby increasing the pressure, the system under stress is forced from one equilibrium state to another with the same equilibrium constant, but with the *position* of the *equilibrium shifted to the right*, forming more $NH_3(g)$ at the expense of $N_2(g)$ and $H_2(g)$.

But notice that we can view what happened when we precipitously stressed our N_2, H_2, NH_3 mixture (by doubling the partial pressures) in two complementary ways. *One way*, an approach we just explored, was to calculate Q after the partial pressures were doubled and then compare that Q to K_p. If Q, just after the stress is applied, is such that $Q < K_p$, we know that the system must shift to the right to regain equilibrium (because at a given temperature there is only one K_p!). A *second way* is to recognize that when a system at equilibrium is subjected to a stress, the system responds by attaining a new equilibrium that partially offsets the impact of the change. Just how did this latter approach play out for the N_2, H_2, NH_3 mixture? Well, as we stressed the system by decreasing the volume and increasing the partial pressures, the reaction "recognized" that *four* molecules (one N_2 and three H_2) react to form just *two* molecules of NH_3 as shown in Figure 5.7. Thus, if more of the reactants on the left-hand side react to form the molecule on the right-hand side, the pressure within the vessel will be reduced, partially compensating for the imposed stress of decreasing the volume and increasing the pressure.

Had we investigated a chemical reaction for which there were fewer molecules involved in the reaction on the left-hand side than the right-hand side, the reaction *would have shifted* to the *left* with *an increase in pressure*.

There are, of course, three possibilities:

1. There are more molecules produced in the reaction than consumed in moving from left to right.

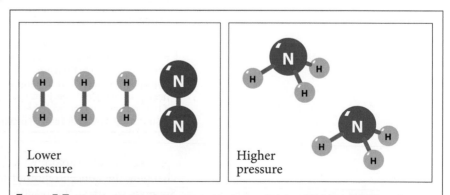

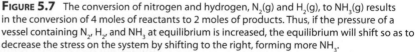

FIGURE 5.7 The conversion of nitrogen and hydrogen, $N_2(g)$ and $H_2(g)$, to $NH_3(g)$ results in the conversion of 4 moles of reactants to 2 moles of products. Thus, if the pressure of a vessel containing N_2, H_2, and NH_3 at equilibrium is increased, the equilibrium will shift so as to decrease the stress on the system by shifting to the right, forming more NH_3.

2. There are an equal number of molecules on the left- and right-hand sides.
3. There are more molecules on the left-hand side than the right-hand side.

The three cases are summarized diagrammatically in Figure 5.8.

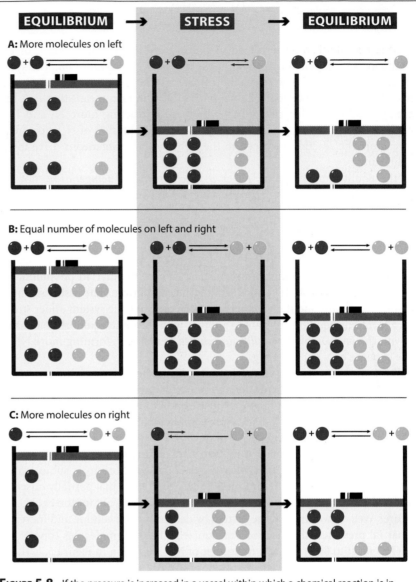

FIGURE 5.8   If the pressure is increased in a vessel within which a chemical reaction is in equilibrium, the equilibrium will shift in a direction so as to reduce the applied stress. Thus, the direction of that shift depends upon whether the reaction increases, decreases, or maintains the number of moles in going from reactants to products; if the reaction decreases the number of moles present from reactants to products (left to right) the reaction will shift toward the right if the pressure in the vessel is increased. If the reaction increases the number of moles present in going from reactants to products (left to right) then the reaction will shift to the left if the pressure is increased. If there is no change in the number of moles, the equilibrium will be unchanged by an increase in pressure.

We can also explore what happens if we stress a system in equilibrium by changing the concentration of just one of the species involved in the reaction. Suppose we stress the reaction $H_2 + I_2 \rightarrow 2HI$ at equilibrium by injecting H_2, as shown in Figure 5.9, into the reactor vessel. For example, suppose we use a high pressure injector to precipitously double the partial pressure of H_2 in the equilibrium mixture. We know that at equilibrium, before injecting the H_2,

$$Q_p = K_p = \frac{P_{HI}^2}{P_{I_2} P_{H_2}}$$

After injection,

$$\left(P_{H_2}\right)_{after} = 2\left(P_{H_2}\right)_{before}$$

such that just after H_2 injection,

$$Q_{p,\,after} = \frac{P_{HI}^2}{P_{I_2}\, 2P_{H_2}} = \frac{1}{2}K_p$$

Thus, since $Q_{p,\,after} < K_p$, the reaction will *shift to the right* to relieve the stress and return to equilibrium (since there is only *one* K_p at a fixed temperature!) But we can also determine the direction of adjustment to attain equilibrium following the stress created by injecting H_2 by recognizing that the system will adjust so as to offset the impact of the applied stress, in this case, by *consuming* H_2. Of course, as a result I_2 will also be consumed and HI produced.

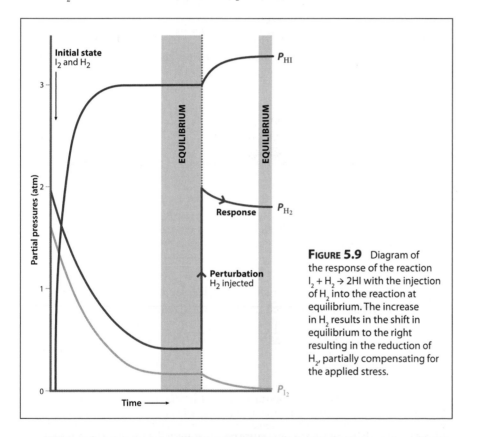

FIGURE 5.9 Diagram of the response of the reaction $I_2 + H_2 \rightarrow 2HI$ with the injection of H_2 into the reaction at equilibrium. The increase in H_2 results in the shift in equilibrium to the right resulting in the reduction of H_2, partially compensating for the applied stress.

We can also stress an equilibrium system by changing the *temperature*. But, as we emphasized in our initial discussion of both the reaction quotient, Q, and the equilibrium constant, $K_{eq}(T)$, the equilibrium constant is a single number that is experimentally determined for a given *temperature*. If the temperature is changed, the value of the equilibrium constant also changes. In fact, as a general rule, equilibrium constants are *very* sensitive to temperature because there is typically a large enthalpy difference, ΔH, as shown in Figure 5.10, between reactants and products. We will see how this affects the equilibrium constant, K_{eq}, quantitatively as we examine the effect of stressing the equilibrium condition by changing temperature.

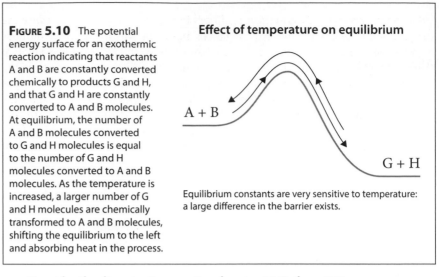

FIGURE 5.10 The potential energy surface for an exothermic reaction indicating that reactants A and B are constantly converted chemically to products G and H, and that G and H are constantly converted to A and B molecules. At equilibrium, the number of A and B molecules converted to G and H molecules is equal to the number of G and H molecules converted to A and B molecules. As the temperature is increased, a larger number of G and H molecules are chemically transformed to A and B molecules, shifting the equilibrium to the left and absorbing heat in the process.

Effect of temperature on equilibrium

A + B

G + H

Equilibrium constants are very sensitive to temperature: a large difference in the barrier exists.

Consider the dimerization reaction forming N_2O_4 from NO_2

$$2NO_2(g) \rightleftarrows N_2O_4(g) + \text{heat}$$

This reaction is 57.2 kJ/mole exothermic, so we can sketch the reaction coordinate for the equilibrium reaction as shown in Figure 5.11.

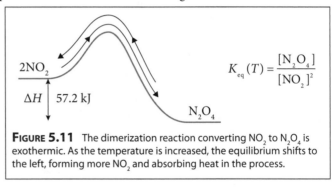

$2NO_2$

ΔH | 57.2 kJ

N_2O_4

$$K_{eq}(T) = \frac{[N_2O_4]}{[NO_2]^2}$$

FIGURE 5.11 The dimerization reaction converting NO_2 to N_2O_4 is exothermic. As the temperature is increased, the equilibrium shifts to the left, forming more NO_2 and absorbing heat in the process.

How will $K_{eq}(T)$ change if the temperature is increased? If we increase the temperature, it means that a greater flow of molecules back up the reaction surface from N_2O_4 to $2NO_2$ will occur such that the reverse reaction

$$N_2O_4(g) + \text{heat} \rightarrow 2NO_2(g)$$

will gain the advantage at a higher temperature. But that means

$$K_{eq}(T) = \frac{[N_2O_4]}{[NO_2]^2}$$

will *decrease* with increasing temperature because, with the increase in temperature, $[N_2O_4]$ decreases and $[NO_2]$ increases. We conclude, correctly, from this example that

1. Raising the temperature of an equilibrium mixture shifts the equilibrium condition in the direction of the endothermic reaction.
2. Raising the temperature of an equilibrium mixture causes the chemical reaction to consume part of that temperature increase by absorbing heat to promote the endothermic process.

This response of the equilibrium state to yet another type of stress fits into a very important emerging pattern. Notice that when we *stressed the equilibrium by changing the pressure*, the equilibrium shifted in a direction such as to *reduce* the magnitude of the change by seeking the condition for which the lower number of moles on either the right-hand or left-hand side of the reaction was favored. Alternatively, when we injected an increase in the concentration of a species on the left-hand side of an equilibrium reaction, the reaction shifted to the right such as to consume part of that additional concentration, thereby reducing the magnitude of stress on the system. When we precipitously increased the temperature of a reaction at equilibrium, the reaction shifted toward the endothermic side of the equilibrium, consuming heat in the process, thereby suppressing the magnitude of the temperature change.

In each case, the system, perturbed from equilibrium, will shift its equilibrium position so as to relieve the applied stress!

The Principle of Le Chatelier

In our development of the analysis of equilibrium processes in chemical reactions, one of the most remarkable and versatile concepts is the Le Chatelier Principle that provides remarkable insight into how a system at equilibrium will respond to an applied external stress such as increasing the concentration of one of the species, changing the temperature, or changing the volume of the container within which the reaction is taking place.

While there are a number of ways of stating the Le Chatelier Principle, two options include:

- For a chemical system at equilibrium, if one or more variables that determine the equilibrium state of the system undergoes a change, a stress, the system will establish a new equilibrium by shifting so as to counter the applied stress.
- If a change in pressure, concentration, or temperature is imposed on a system at equilibrium, the system will establish a new equilibrium condition so as to offset the imposed change.

The power of Le Chatelier's Principle is that it provides a means for determining the *qualitative response*, the *direction* of response, of equilibrium systems to an external stress. Le Chatelier also links into the relationship between the reaction quotient, Q, and the movement of an equilibrium system back toward equilibrium following an imposed stress in the following way. If we write our expression for the equilibrium quotient, Q_c, in terms of the concentrations immediately following the applied stress as

$$Q_c = \frac{[C]^c_{stress} [D]^d_{stress}}{[A]^a_{stress} [B]^b_{stress}}$$

and we note that Q_c can assume values from zero to ∞ for our reaction

$$aA + bB \rightleftarrows cC + dD$$

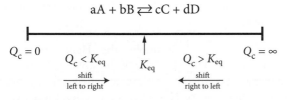

we note that, because K_{eq} lies between the extremes of 0 and ∞, if $Q_c < K_{eq}$, Q_c will seek to *increase* until it satisfies the condition $Q_c = K_{eq}$ and it will do so by virtue of the fact that the reaction shifts, along with Q_c, from *left* to *right* on the diagram

Problem:

Suppose the initial concentrations are

CO: 1.00 M H_2O: 1.00 M
CO_2: 2.00 M H_2: 2.00 M

After equilibrium is achieved, which of the species will increase and which will decrease from their initial conditions?

Solution:

First we sketch the Gibbs free energy surface

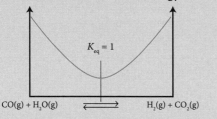

Next we calculate the value of the reaction quotient Q from the functional form

$$Q = \frac{[CO_2(g)][H_2(g)]}{[CO(g)][H_2O(g)]}$$

Since the concentrations are in mol/l we can assume a vessel of volume, V, and write, inserting the given values for the initial number of moles present:

$$Q_{init} = \frac{(2.00/V)(2.00/V)}{(1.00/V)(1.00/V)} = 4.0$$

Next, on our Gibbs free energy surface we note where the value of $Q = 4.0$ resides with respect to the equilibrium value, $K = 1.0$.

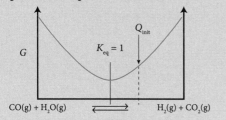

Since $Q_{init} > K_{eq}$, the reaction will shift to the left. Thus at equilibrium the concentrations of CO(g) and H_2O(g) will increase, while the concentrations of H_2(g) and CO_2(g) will decrease.

Practice Problem:

Given that this water gas shift reaction is so important to the production of liquid fuels from other hydrocarbons, suppose equal amounts of H_2O, CO, H_2, and CO_2 are combined in a vessel at 1100 K. After equilibrium is established, in which direction will the reaction shift?

above. On the other hand, if $Q_c > K_{eq}$ as a result of an applied stress, then Q_c must decrease to re-achieve equilibrium and both Q_c *and the reaction* must shift from right to left in the diagram above.

Check Yourself 5—Establishing the Direction of Shift Between Reactants and Products from an Initial State to a Final Equilibrium State

One of the most famous processes that uses multiple-step equilibria is the Fischer–Tropsch (pictured in Figure 5.1) process that converts coal (or any hydrocarbon source) to liquid fuel. A key step in that process is the production of *synthesis gas*, or *syngas*, that is produced by passing water vapor, H_2O(g), at high temperature over elemental carbon:

$$H_2O(g) + C(s) \rightleftarrows CO(g) + H_2(g)$$

A second step, termed the water gas shift reaction, is to further react water vapor, at elevated temperature, with the CO(g) produced in the first reaction to produce H_2(g):

$$CO(g) + H_2O(g) \rightleftarrows H_2(g) + CO_2(g)$$

For this reaction, $K_c = 1$ at a temperature of 1100 K.

Quantitative Determination of Concentrations Following an Impressed Stress on a System at Equilibrium

We are, therefore, moving from the *qualitative* (but *very* important) analysis of the response of equilibrium conditions to stress guided by the Principle of Le Chatelier, to a *quantitative* treatment of such systems when perturbed from their equilibrium state.

If we begin with initial concentrations (or partial pressures) and a balanced equation, in general we wish to develop an approach such that we can systematically calculate concentrations following the application of stress to the equilibrium condition. While such calculations are not complicated, they nevertheless potentially involve a number of variables changing simultaneously.

We employ a three-step process:

1. We use Le Chatelier to establish the direction of change.
2. We set up the reaction quotient, Q.
3. We adjust the concentration or pressures in the reaction quotient to reflect the quantitative extent of the change, and solve the algebraic equation that results.

As is clear from, for example, the Haber process, because the concentrations are raised to the power of the stoichiometric coefficient, these algebraic expressions can become messy when the algebraic expressions have cubic or higher powers. However, (very) often these expressions can be simplified through approximation to yield a tractable calculation.

Check Yourself 6—The Principle of Le Chatelier: Response of System at Equilibrium to an Increase in the Concentration of a Reactant

We consider here the important Haber–Bosch reaction for the production of "fixed nitrogen," in this case $NH_3(s)$, from N_2 and H_2 in the reaction

$$N_2(g) + 3H_2(g) \rightarrow 2NH_3(g)$$

Let's return to our example of the dimerization of NO_2 to form N_2O_4:

$$2NO_2 \rightleftarrows N_2O_4$$

Suppose that at 100°C we have an equilibrium mixture of NO_2 and N_2O_4 for which the partial pressures of the two species are $P_{NO_2} = 3.12$ atm and $P_{N_2O_4} = 1.50$ atm. Then suppose we stress the equilibrium state by a 10-fold *isothermal* (T = constant) expansion. What will be the partial pressures of the NO_2 and N_2O_4 when equilibrium is again established?

We recognize first that

$$K_p = \frac{P_{N_2O_4}}{P_{NO_2}^2} = \frac{1.5}{(3.12)^2} = 0.154$$

If we increase the volume by ×10, then $P_{N_2O_4} = 0.15$ atm and $P_{NO_2} = 0.312$ atm and at that instant the reaction quotient is

$$Q = \frac{0.15}{(0.312)^2} = 1.54$$

and we note that (a) the system is dramatically out of equilibrium; (b) that $Q > K_{eq}$; and, therefore, (c) the reaction will shift to the left to reestablish equilibrium. We can quickly see this either by using our Q, K_{eq}, reaction diagram above or by recognizing that Le Chatelier tells us that as pressure drops the reaction will try to produce more moles to compensate for the increased volume and decreased pressure, thus shifting the reaction to the left toward $2NO_2$. If we denote the *decrease* in $P_{N_2O_4}$ by x, then P_{NO_2} must increase by $2x$ and we can create a Reaction/Initial Condition/Change/Equilibrium (RICE) table in the following way:

Reaction	$2NO_2$	$\rightleftarrows$	N_2O_4
Initial Pressure	0.312		0.150
Change in Pressure	+2x		−x
Equilibrium Pressure	0.312 + 2x		0.150 − x

At the new equilibrium point we thus have

$$K_p = \frac{P_{N_2O_4}}{P_{NO_2}^2} = \frac{0.150 - x}{(0.312 + 2x)^2} = 0.154$$

which yields a quadratic equation

$$0.616x^2 + 1.192x - 0.135 = 0$$

The solution, of course, gives us two roots:

$$x = 0.1073 \quad \text{and} \quad x = -2.043$$

But we know from Le Chatelier that x must be positive from the way we have

Problem:
Suppose we take the system at equilibrium and we impose an increase in $H_2(g)$ on the system. Predict the impact of this applied stress on each of the species present.

Solution:
When we add $H_2(g)$ to the system at equilibrium, the Le Chatelier principle states that the reaction will shift so as to counteract the applied stress.

Thus the reaction will shift to the right in order to reduce the applied stress by converting some of the added H_2 to NH_3. In so doing, N_2 will be consumed in the production of additional NH_3. As a result, N_2 will decrease, NH, will increase, and H_2 will decrease from the condition immediately after the H_2 is added in the establishment of the new equilibrium. However, note that while the reaction shifted to the right to attempt to relieve the applied stress, there will still be more H_2 after equilibrium is reestablished than there was before the H_2 was added as an external stress.

Practice:
Quicklime, $CaO(s)$, is a principal ingredient in the making of concrete, as well as wide use in water purification. $CaO(s)$ is produced from $CaCO_3(s)$ at high temperature in the reaction

$$CaCO_3(s) \rightleftarrows CaO(s) + CO_2(g)$$

This reaction is also an important, and deleterious source of $CO_2(g)$ to the atmosphere.

What is the impact on the equilibrium by:

1. increasing $CaCO_3(s)$
2. increasing $CaO(s)$
3. increasing $CO_2(g)$?

Problem:
Suppose we have an equilibrium mixture of NH_3, N_2, and H_2 in a 1.50 liter flask and the mixture (at equilibrium) is transferred to a flask of volume 5.00 liters. Does the reaction shift to the left or to the right to reestablish equilibrium.

Solution:
If we increase the volume of our equilibrium mixture from 1.5 liters to 5.0 liters, the Principle of Le Chatelier states that the equilibrium will shift so as to reduce the applied stress on the system. Thus, by this principle, the reaction will shift so as to *increase* the pressure to counteract the precipitous *decrease* in pressure from the applied stress. With 4 gas phase molecules on the left-hand side ($3H_2(g) + N_2(g)$) and two gas phase species in the right-hand side ($2NH_3$), the reaction will shift to the left in an attempt to counter the applied stress.

Practice Example:
Looking again at the important *water gas reaction*

$$H_2O(g) + CO(g) \rightleftharpoons H_2(g) + CO_2(g)$$

Explain how the equilibrium responds to

1. changing the total gas phase.
2. changing the volume of the system.

assembled our RICE table. Thus, after equilibrium is reestablished,

$$P_{N_2O_4} = 0.150 - x = 0.150 - 0.1073 = 0.043 \text{ atm}$$
$$P_{NO_2} = 0.312 + 2x = 0.312 + 2(0.1073) = 0.527 \text{ atm}$$

So that

$$Q = \frac{0.043}{(0.527)^2} = 0.154$$

which is indeed K_p!

Check Yourself 7—The Principle of Le Chatelier: Applied to Changing Volume
We again consider the key reaction in the Haber–Bosch process for the production of fixed nitrogen as ammonia

$$3H_2(g) + N_2(g) \rightleftharpoons 2NH_3(g)$$

Equilibrium Problems Involving Multiple Steps

We can also work problems that appear at first blush to be difficult by using what we know about how chemical reactions sum together to yield an "overall" reaction, and how equilibrium constants of each of the reactions that are added together, should be multiplied together to yield an equilibrium constant for the "overall" reaction.

Consider the following: in the manufacture of ammonia (that requires H_2)

$$CH_4(g) + 2H_2O(g) \overset{1}{\rightleftharpoons} CO_2(g) + 4H_2(g)$$

is the chief source of H_2. It is also known that

$$CO(g) + H_2O(g) \overset{2}{\rightleftharpoons} CO_2(g) + H_2(g)$$
$$\Delta H° = -40 \text{ kJ} \qquad K_c = 1.4 \text{ @ } 1000 \text{ K}$$

$$CO(g) + 3H_2(g) \overset{3}{\rightleftharpoons} H_2O(g) + CH_4(g)$$
$$\Delta H° = -230 \text{ kJ} \qquad K_c = 190 \text{ @ } 1000 \text{ K}$$

At 1000 K, 1.0 mole each of CH_4 and H_2O are allowed to reach equilibrium in a 10.0 L vessel.

Calculate the number of moles of H_2 present at equilibrium.

First Issue: We don't have K_c for the overall reaction.
Consequence: We must build it from the component parts and establish K_c.

Adding reaction 2 to the reverse of reaction 3:

$$\left(\Delta H° = -40 \text{ kJ}\right) \quad CO(g) + H_2O(g) \overset{2}{\rightleftharpoons} CO_2(g) + H_2(g) \qquad K_c = 1.4$$

$$\left(\Delta H° = +230 \text{ kJ}\right) \quad H_2O(g) + CH_4(g) \overset{3}{\rightleftharpoons} CO(g) + 3H_2(g) \qquad K'_{c_3} = \frac{1}{K_{c_3}} = 5.3 \times 10^{-3}$$

$$\overline{CH_4(g) + 2H_2O(g) \rightleftharpoons CO_2(g) + 4H_2(g)} \quad K_{c_1} = 7.4 \times 10^{-3}$$

$$\Delta H° = +190 \text{ kJ}$$

Next, set up a table that organizes the *reaction*, *initial* concentration, the *change* in concentration, and the *equilibrium* concentration: RICE.

Reaction	CH_4 (g)	+	$2H_2O$ (g)	$\rightleftarrows$	CO_2 (g)	$4H_2$ (g)
Initial	1.00 moles		1.00 moles		0.0 moles	0.0 moles
Change	$-x$		$-2x$		$+x$	$+4x$
Equilibrium amounts	$(1.0 - x)$ moles		$(1.0 - 2x)$ moles		x moles	$4x$ moles
Equilibrium concentrations	$(1.0 - x)/10$		$(1.0 - 2x)/10$		$x/10$	$4x/10$

But we have

$$K_{c_1} = \frac{[CO_2][H_2]^4}{[CH_4][H_2O]^2} = 7.4 \times 10^{-3}$$

So

$$K_{c_1} = \frac{\left(\dfrac{x}{10}\right)\left(\dfrac{4x}{10}\right)^4}{\left(\dfrac{1.0-x}{10}\right)\left(\dfrac{1.0-2x}{10}\right)^2} = \frac{(x)(4x)^4}{10^2(1.0-x)(1.0-2x)^2}$$

$$= 7.4 \times 10^{-3}$$

$$256x^5 = (7.4 \times 10^{-1})[(1-x)(1-2x)^2]$$

There are a number of ways to solve this equation—for example, a mathematical package from Mathworks on your laptop—but we can use the method of successive approximation to get a back-of-the-envelope solution. We do this using an initial guess, then test to see how the left-hand and right-hand sides equate, adjusting our guess accordingly.

So for

$$256x^5 = (7.4 \times 10^{-1})[(1-x)(1-2x)^2]$$

let's try $x = 0.3$. This gives us

$$(256)(0.3)^5 \stackrel{?}{=} (7.4 \times 10^{-1})[(1-0.3)(1-0.6)^2]$$

$$(256)(2.4 \times 10^{-3}) \stackrel{?}{=} (7.4 \times 10^{-1})[0.112]$$

which reduces to

$$0.614 \stackrel{?}{=} 0.08$$

Thus, our initial guess was somewhat too high, because the left-hand side (LHS) dominates. If we modify our guess to $x = 0.25$, then LHS = 0.25 and RHS = 0.139. A slight decrease to $x = 0.23$ turns out to be a very good approximation.

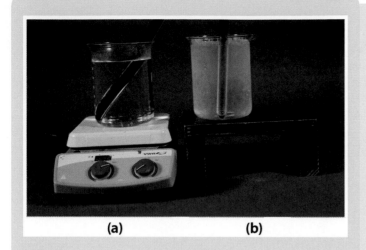

(a)   **(b)**

Solution:

First calculate the number of moles of each species:

$$\text{mol of NO}_2 = (1.56 \text{ g NO}_2)\left(\frac{1 \text{ mol NO}_2}{46.01 \text{ g NO}_2}\right) = 3.391 \times 10^{-2} \text{ mol}$$

$$\text{mol of N}_2\text{O}_4 = (7.64 \text{ g N}_2\text{O}_4)\left(\frac{1 \text{ mol N}_2\text{O}_4}{92.01 \text{ g N}_2\text{O}_4}\right) = 8.303 \times 10^{-2} \text{ mol}$$

Then convert mol to concentration in the 3.00 liter vessel:

$$[\text{NO}_2] = \frac{3.391 \times 10^{-2} \text{ mol}}{3.00 \ \ell} = 1.13 \times 10^{-2} \text{ M}$$

$$[\text{N}_2\text{O}_4] = \frac{8.303 \times 10^{-2} \text{ mol}}{3.00 \ \ell} = 2.77 \times 10^{-2} \text{ M}$$

We can then directly calculate the equilibrium constant in concentration units:

$$K_c = \frac{[\text{NO}_2]^2}{[\text{N}_2\text{O}_4]} = \frac{(1.13 \times 10^{-2} \text{ M})^2}{(2.77 \times 10^{-2} \text{ M})} = 4.61 \times 10^{-3}$$

Practice Example:

Calculate the equilibrium constant, K_c, for the reaction converting hydrogen sulfide to hydrogen gas and gaseous sulfur:

$$2\text{H}_2\text{S} \rightleftarrows \text{S}_2(g) + 2\text{H}_2(g)$$

After attaining equilibrium in a 3.00 liter vessel at 1405 K, it was determined that the number of moles of $\text{H}_2\text{S}(g)$, $\text{H}_2(g)$, and $\text{S}_2(g)$ were:

    $\text{H}_2\text{S}(g)$: 2.78 mol
    $\text{H}_2(g)$: 0.22 mol
    $\text{S}_2(g)$: 0.11 mol

Calculate K_c.

Check Yourself 8—Given the Equilibrium Amounts of Reactants and Products, How is the Value of the Equilibrium Constant, K_c, in Concentration Units Calculated?

An important equilibrium reaction exists between $\text{N}_2\text{O}_4(g)$ and $\text{NO}_2(g)$:

$$\text{N}_2\text{O}_4(g) \rightleftarrows 2\text{NO}_2(g)$$

$\text{N}_2\text{O}_4(g)$ is an important reactant in rocket fuels where it is the oxidizer and hydrazine N_2H_4 is the fuel. NO_2 on the other hand is a critical component of air pollution. It is the common yellow-orange haze so prevalent over urban regions. The equilibrium between $\text{N}_2\text{O}_4(g)$ and $\text{NO}_2(g)$ can be seen directly as shown in the graphic where panel (a) corresponds to the mixture at low temperature where the equilibrium is shifted strongly to the left favoring the (transparent) N_2O_4 whereas at room temperature NO_2 (yellow-orange color) is favored. Suppose we observe the equilibrium reaction at 25°C in a 3.00 liter vessel and determine that there are 1.56 g NO_2 and 7.64 g N_2O_4. Calculate the value of K_c for

$$\text{N}_2\text{O}_4(g) \rightleftarrows 2\text{NO}_2(g)$$

Equilibrium Constants, Spontaneous Processes, and Gibbs Free Energy

Development of the concept of (dynamic) equilibria in chemical systems through the law of mass action and the equilibrium constant, $K_{eq}(T)$, begs the question: If a chemical equilibrium is achieved spontaneously quite apart from the specific concentration (or partial pressures) of reactants on either side of the equilibrium equation

$$a\text{A} + b\text{B} \rightleftarrows c\text{C} + d\text{D}$$

and, in Chapter 3, we recognized from the Second Law of Thermodynamics that the Gibbs free energy

$$\Delta G = \Delta H - T\Delta S$$

determines whether a process is spontaneous, then how are ΔG and K_{eq} related?

$\Delta G < 0$ Spontaneous
$\Delta G > 0$ Nonspontaneous
$\Delta G = 0$ No Motive for Change

To establish the relationship between Gibbs free energy and the equilibrium constant, let's return to the dimerization reaction

$$N_2O_4(g) \rightleftharpoons 2NO_2(g)$$

We know that equilibrium can be reached from either direction. If we begin with only $N_2O_4(g)$, the reaction shifts to the right; if we begin with NO_2 only, the equilibrium reaction shifts to the left. We also know that we can calculate the change in Gibbs free energy for the reaction (see Chapter 4) such that

$$\Delta G_R^\circ = \left(\Delta G_f^\circ\right)_{2NO_2} - \left(\Delta G_f^\circ\right)_{N_2O_4}$$

and we can look up ΔG_f° for N_2O_4 and NO_2 in a table (or calculate $\Delta G_f^\circ = \Delta H_f^\circ - T\Delta S_f^\circ$ from tables of ΔH_f° and ΔS_f°) such that

$$\Delta G_R^\circ = \left(\Delta G_f^\circ\right)_{2NO_2} - \left(\Delta G_f^\circ\right)_{N_2O_4} = (2)(51.8) - 98.3$$

$$= +5.3 \text{ kJ}$$

So for the equilibrium reaction

$$N_2O_4 \rightleftharpoons 2NO_2 \quad \Delta G_R^\circ = +5.3 \text{ kJ}$$

Now we can join our graphical representation of K_{eq} and Q in a very important diagram displayed in Figure 5.12 with our graph of Gibbs free energy, wherein the left-hand side is the Gibbs free energy of formation of N_2O_4 and the right-hand side is the Gibbs free energy of formation of $2NO_2$, shown in Figure 5.13.

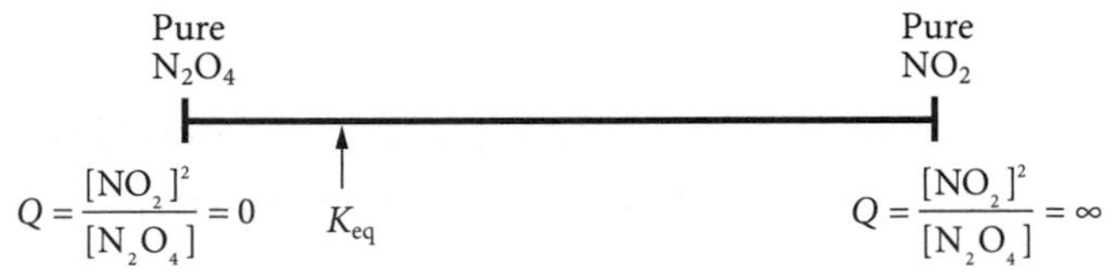

FIGURE 5.12 A very useful diagram that links the extremes of pure reactants (in this case N_2O_4) to pure products (in this case NO_2) with the reaction quotient.

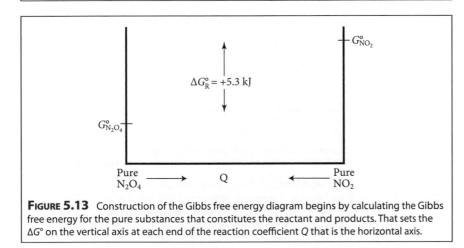

FIGURE 5.13 Construction of the Gibbs free energy diagram begins by calculating the Gibbs free energy for the pure substances that constitutes the reactant and products. That sets the ΔG° on the vertical axis at each end of the reaction coefficient Q that is the horizontal axis.

But what is the form of this Gibbs free energy diagram for intermediate concentrations of N_2O_4 and NO_2 between the case of pure N_2O_4 (left-hand end of diagram) and the case of pure NO_2 (right-hand end of diagram)? Well, we know (from our knowledge that $Q < K_{eq}$) that the spontaneous reaction will proceed to the right from the condition that the mixture is pure N_2O_4 to some intermediate mixture containing both N_2O_4 and NO_2, until $Q = K_{eq}$. We also know, from the fact that the reaction is spontaneous, that $\Delta G = G_{final} - G_{initial} < 0$, and thus Gibbs free energy will decrease in going from pure N_2O_4 to a mixture of N_2O_4 and NO_2, as shown in Figure 5.14.

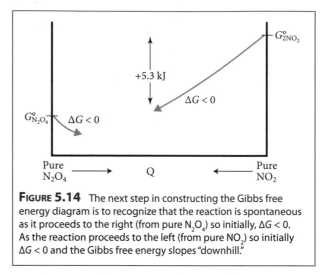

FIGURE 5.14 The next step in constructing the Gibbs free energy diagram is to recognize that the reaction is spontaneous as it proceeds to the right (from pure N_2O_4) so initially, $\Delta G < 0$. As the reaction proceeds to the left (from pure NO_2) so initially $\Delta G < 0$ and the Gibbs free energy slopes "downhill."

By similar logic, if we begin with pure NO_2, the reaction will proceed spontaneously to the left as it seeks the equilibrium condition. We can now construct what is known as a Gibbs free energy diagram by linking the evolution in Gibbs free energy from pure N_2O_4 with the evolution in Gibbs free energy from pure NO_2 as shown in Figure 5.15.

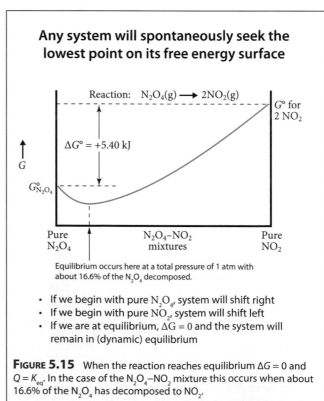

Any system will spontaneously seek the lowest point on its free energy surface

Equilibrium occurs here at a total pressure of 1 atm with about 16.6% of the N_2O_4 decomposed.

- If we begin with pure N_2O_4, system will shift right
- If we begin with pure NO_2, system will shift left
- If we are at equilibrium, $\Delta G = 0$ and the system will remain in (dynamic) equilibrium

FIGURE 5.15 When the reaction reaches equilibrium $\Delta G = 0$ and $Q = K_{eq}$. In the case of the N_2O_4–NO_2 mixture this occurs when about 16.6% of the N_2O_4 has decomposed to NO_2.

But now we are in a position to link our equilibrium reaction to the diagram relating the reaction quotient, Q, to the equilibrium constant, $K_{eq}(T)$, and finally to the Gibbs free energy diagram, as shown in Figure 5.16.

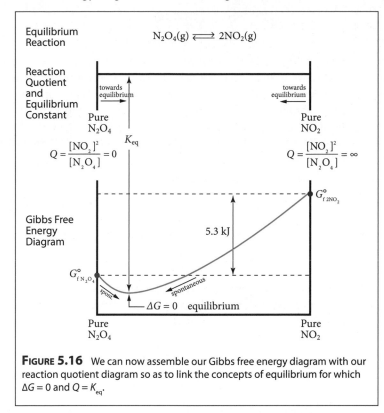

FIGURE 5.16 We can now assemble our Gibbs free energy diagram with our reaction quotient diagram so as to link the concepts of equilibrium for which $\Delta G = 0$ and $Q = K_{eq}$.

Figure 5.16 is a very important diagram because it graphically links $\Delta G°$, the free energy change for a reaction with substances in their standard states (298 K, 1 atm); ΔG, the free energy change under any set of conditions; Q, the reaction quotient under nonstandard conditions; and K_{eq}, the equilibrium constant at the temperature of the reaction. Notice in particular that if $Q < K$, then $\Delta G < 0$ and the reaction will spontaneously shift to the right until $Q = K$ at which point $\Delta G = 0$; if $Q > K$, then again $\Delta G < 0$ and the reaction will spontaneously shift to the left until $Q = K$ at which point $\Delta G = 0$ and equilibrium is achieved.

Gibbs Free Energy Under Nonstandard Conditions

We now have a conceptual (and very important!) understanding of the relationship between Gibbs free energy and the equilibrium constant, but we need a quantitative relationship between ΔG, Q, and K_{eq}. But what is it?

First, we recall from Chapter 3 that the *standard free energy* change, $\Delta G°$, is the free energy change corresponding to reactants and products in their standard states (for a solid or liquid it is the pure element or compound at the pressure of 1 bar (10^5 Pa); for a gas it is the pure gas behaving as an ideal gas at a pressure of 1 bar). Recall that while the temperature is not part of the definition of a standard state, it must be specified in tabulated values of $\Delta H°$, $\Delta G°$, $\Delta S°$, etc., because they depend on temperature. Values are typically tabulated at 298.15 K (25°C) unless otherwise indicated.

Second, we calculate $\Delta G° = \Delta H° - T\Delta S°$, from the tabulated values of the standard free energy of formation, $\Delta G_f°$ (the free energy change for a reaction in which a substance in its standard state is formed from its elements in their reference forms in their standard states). Thus, for a reaction

$$\Delta G_R^\circ = \sum_{products} n_p \Delta G_f^\circ - \sum_{reactants} n_R \Delta G_f^\circ$$

But how do we calculate the relationship of ΔG° to ΔG for *nonstandard* conditions? That is, we wish to describe the relationship between ΔG° (which we can look up in a table) and ΔG for nonstandard conditions. This corresponds to defining ΔG for *any value* of Q. That is for points on the free energy curve that do not correspond to the position of equilibrium where $Q = K_{eq}$. To be specific, let's consider the equilibrium that takes place in the Haber process that "fixes" nitrogen in the equilibrium reaction

$$N_2(g) + 3H_2(g) \rightleftarrows 2NH_3(g)$$

The expressions for ΔG and ΔG° are, of course, $\Delta G = \Delta H - T\Delta S$ and $\Delta G^\circ = \Delta H^\circ - T\Delta S^\circ$. If we consider the case for an ideal gas, we recall that ΔH is only a function of temperature, but is independent of pressure, such that $\Delta H = \Delta H^\circ$. Thus,

$$\Delta G = \Delta H^\circ - T\Delta S$$

But how are ΔS and ΔS° related? Recalling our derivation from Chapter 3 for the isothermal expansion of an ideal gas ($q = -w$ because $\Delta U = 0$) that occurs reversibly, the work of expansion for one mole of an ideal gas is (see Chapter 3)

$$w = -RT \ln (V_f/V_i)$$

Thus with

$$q_{rev} = -w = RT \ln (V_f/V_i)$$

we obtain the entropy change (for the *isothermal* expansion of one mole of an ideal gas)

$$\Delta S = q_{rev}/T = R \ln (V_f/V_i)$$

With this equation, we can calculate the entropy of an ideal gas at any pressure, and, moreover, because $pV = nRT$, $V_f/V_i = P_i/P_f$, so we have

$$\Delta S = S_f - S_i = R \ln (V_f/V_i) = R \ln (P_i/P_f) = -R \ln (P_f/P_i)$$

If we set $P_i = 1$ bar and designate P_i as P° and S_i as S°, then we have, for the entropy at any pressure

$$S = S^\circ - R \ln (P/P^\circ) = S^\circ - R \ln (P/1) = S^\circ - R \ln P$$

Using our Haber process for ammonia production, we can write

$$S_{NH_3} = S_{NH_3}^\circ - R \ln P_{NH_3}$$

$$S_{N_2} = S_{N_2}^\circ - R \ln P_{N_2}$$

$$S_{H_2} = S_{H_2}^\circ - R \ln P_{H_2}$$

Then

$$\Delta S = 2S_{NH_3} - S_{N_2} - 3S_{H_2}$$

becomes, by direct substitution,

$$\Delta S = 2S_{NH_3}^\circ - 2R \ln P_{NH_3} - S_{N_2}^\circ + R \ln P_{N_2} - 3S_{H_2}^\circ + 3R \ln P_{H_2}$$

and by rearrangement

$$\Delta S = 2S_{NH_3}^\circ - S_{N_2}^\circ - 3S_{H_2}^\circ - 2R \ln P_{NH_3} + R \ln P_{NH_3} + 3R \ln P_{H_2}$$

But $\Delta S^\circ = 2S_{NH_3}^\circ - S_{N_2}^\circ - 3S_{H_2}^\circ$, so we have

$$\Delta S = \Delta S^\circ - 2R\ln P_{NH_3} + R\ln P_{N_2} + 3R\ln P_{H_2}$$

$$= \Delta S^\circ - R\ln P_{NH_3}^2 + R\ln P_{N_2} + R\ln P_{H_2}^3$$

$$= \Delta S^\circ + R\ln \frac{P_{N_2}P_{H_2}^3}{P_{NH_3}^2}$$

But now we have a remarkable result, because, from $\Delta G = \Delta H^\circ - T\Delta S$ (see above) we can insert our expression for ΔS such that

$$\Delta G = \Delta H^\circ - T\Delta S = \underbrace{\Delta H^\circ - T\Delta S^\circ} - RT\ln \frac{P_{N_2}P_{H_2}^3}{P_{NH_3}^2}$$

$$= \Delta G^\circ + RT\ln \frac{P_{NH_3}^2}{P_{N_2}P_{H_2}^3}$$

$$= \Delta G^\circ + RT\ln Q$$

This equation, linking ΔG, ΔG°, and the reaction quotient, Q, is one of the most useful and versatile equations in chemical thermodynamics, because Q is evaluated for actual, nonstandard conditions. We can use this equation to determine the spontaneous (or nonspontaneous) *nature of a reaction under any conditions of composition, if the temperature and pressure conditions are constant.*

Moreover, we are now in a position to quantitatively link the equilibrium constant, $K_{eq}(T) = \dfrac{[C]^c[D]^d}{[A]^a[B]^b}$, to ΔG°.

We know that $\Delta G = 0$ at equilibrium *and* we know that $Q = K_{eq}(T)$ *at equilibrium*, so therefore at equilibrium

$$\Delta G = \Delta G^\circ + RT\ln Q$$

becomes

$$0 = \Delta G^\circ + RT\ln K_{eq}(T)$$

or

$$\Delta G^\circ = -RT\ln K_{eq}$$

or

$$K_{eq} = e^{-\Delta G^\circ/RT} = \exp(-\Delta G^\circ/RT)$$

To summarize: There is clearly a connection between the free energy and the equilibrium constant. What is it?

1. Position of equilibrium in a reaction is determined by the *sign* and *magnitude* of ΔG°. Figure 5.17 captures this graphically.

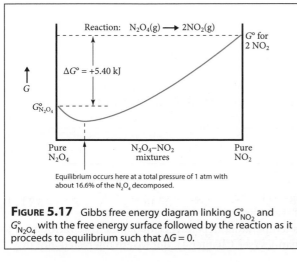

FIGURE 5.17 Gibbs free energy diagram linking $G^\circ_{NO_2}$ and $G^\circ_{N_2O_4}$ with the free energy surface followed by the reaction as it proceeds to equilibrium such that $\Delta G = 0$.

2. The direction in which the reaction proceeds depends on where the system composition stands relative to the minimum on the free energy curve.

3. Quantitatively, the relation between ΔG and $\Delta G°$ is given by

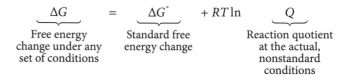

$$\underbrace{\Delta G}_{\substack{\text{Free energy} \\ \text{change under any} \\ \text{set of conditions}}} = \underbrace{\Delta G°}_{\substack{\text{Standard free} \\ \text{energy change}}} + RT \ln \underbrace{Q}_{\substack{\text{Reaction quotient} \\ \text{at the actual,} \\ \text{nonstandard} \\ \text{conditions}}}$$

Thus far we have successfully related ΔG, $\Delta G°$, and Q for nonstandard conditions of concentration, but we have been limited to a fixed temperature, specifically at the conditions for which we could calculate $\Delta G°$ in the equation

$$\Delta G = \Delta G° + RT \ln Q,$$

namely at 298 K and 1 atm pressure. But we know that the equilibrium position will shift if we change the temperature. Le Chatelier showed us how to predict, qualitatively, the shift in equilibrium with increasing temperature: toward the endothermic direction because in so doing we absorb heat by the chemical reaction and thereby compensate for the temperature increase. But how does thermodynamics deal with the temperature change quantitatively?

$\Delta G°$ at Temperatures Other Than 298 K

First, we know that, at 298 K, the position of the equilibrium is determined by the difference between the free energy of pure products and free energy of pure reactants

$$\Delta G°_{298} = \left(\Delta G°_{298}\right)_{\text{products}} - \left(\Delta G°_{298}\right)_{\text{reactants}}$$

We also know that at *any* temperature, T,

$$\Delta G°_T = \left(\Delta G°_T\right)_{\text{products}} - \left(\Delta G°_T\right)_{\text{reactants}}$$

But we don't have tables of $\Delta G°_T$, the change in Gibbs free energy at any temperature, T. This is a potentially serious problem because it means we need a vast array of tables giving us $\Delta G°_T$ for many different temperatures. So how do we compute $\Delta G°_T$? We recognize that

$$\Delta G° = \Delta H° - (298\ \text{K})\ \Delta S° \text{ at standard conditions}$$

and that

$$\Delta G°_T = \Delta H°_T - T\Delta S°_T \text{ at any other temperature.}$$

But ΔH is set by the potential energy surface over which the reaction takes place, which is set by the electronic motion in the orbitals of the atoms that form the bonds in the reactants and products. Since the motion of the electrons that comprise the bonds does not depend on temperature (we will study this in Chapters 8 and 9), ΔH does not, to a rather high degree of accuracy, depend on T! This is a very important result. It means that we can use our tabulated values for $\Delta H°$ at standard conditions for $\Delta H°_T$ at any temperature!

This brings us to the calculation of entropy, $S°$, as a function of temperature. Unlike $H°$, $S°$ is a distinct function of temperature. In fact, we have already examined this temperature dependence of $S°$ when we introduced the Third Law of Thermodynamics (see Chapter 4). But we are interested, not in $S°$, but in $\Delta S°$ as a *function of temperature*. As we can see from Chapter 4, $\Delta S°$ does not change very much as a function of temperature. The conclusion is important and very useful: We can calculate the change in Gibbs free energy at any temperature:

$$\Delta G_T° = \Delta H_{298}° - T\Delta S_{298}°$$

using the listed values for $\Delta H°$ and $\Delta S°$ from our standard tables at 298 K! We need only insert the correct temperature. To demonstrate, we return to our equilibrium reaction

$$N_2O_4(g) \rightleftarrows 2NO_2(g)$$

to work a problem.

Problem:

At 25°C, the value of $\Delta G°$ for the reaction $N_2O_4(g) \rightarrow 2NO_2(g)$ is +5.4 kJ. What is the approximate value of $\Delta G_T°$ for the reaction at 100°C?

Analysis:

At 25°C, we know that $\Delta G° = +5.4$ kJ and the relationship between $\Delta G°$ and the equilibrium position is displayed in Figure 5.18.

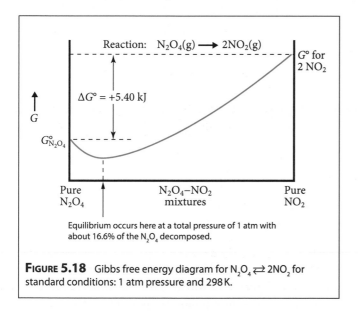

FIGURE 5.18 Gibbs free energy diagram for $N_2O_4 \rightleftarrows 2NO_2$ for standard conditions: 1 atm pressure and 298 K.

To calculate $\Delta G°$ at 100°C, we use the (very good) approximation

$$\Delta G_T° = \Delta H_{298}° - T\Delta S_{298}°$$

So we consult tables to pull out $\Delta H°$ and $\Delta S°$.

For	N_2O_4	NO_2	Units
$\Delta H_f°$	+9.67	+33.8	kJ/mol
$S°$	304	240.5	J/mol

Calculate:

$$\Delta H^\circ = 2\left(\Delta H_f^\circ\right)_{NO_2} - \left(\Delta H_f^\circ\right)_{N_2O_4}$$

$$= 2 \times (33.8) - 9.67 = 57.9 \text{ kJ}$$

$$\Delta S^\circ = 2\left(240.5\right) - \left(304\right) = 177 \text{ J/K}$$

$$\Delta G_{373}^\circ = 57.9 \text{ kJ} - (373 \text{ K})(0.177 \text{ kJ/K})$$

$$= -8.1 \text{ kJ}$$

Conclusion:

At 25°C, $\Delta G^\circ = +5.4$ kJ. At 100°C, $\Delta G^\circ = -8.1$ kJ. This creates a Gibbs free energy diagram that is shown in Figure 5.19. Notice now at 100°C the equilibrium shifts to the right, toward the endothermic direction, just as we would predict from the Principle of Le Chatelier.

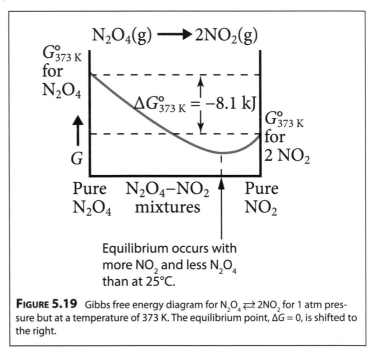

FIGURE 5.19 Gibbs free energy diagram for $N_2O_4 \rightleftarrows 2NO_2$ for 1 atm pressure but at a temperature of 373 K. The equilibrium point, $\Delta G = 0$, is shifted to the right.

An inspection of Figure 5.19 reveals how the temperature dependence of $\Delta G_T^\circ = \Delta H_{298}^\circ - T\Delta S_{298}^\circ$ shifts the equilibrium position of the reaction, shifting it toward the endothermic side, as Figure 5.11 emphasizes.

Check Yourself 9—The Effect of Temperature Change on Equilibrium: A Very Interesting Application of Le Chatelier's Principle

While Le Chatelier's Principle can be directly applied to concentration changes, partial pressure changes, and volume changes in reactants and products, the effect of temperature changes on equilibria are less obvious. As an example, if we consider the reaction of $SO_2(g)$ with $O_2(g)$ to produce $SO_3(g)$

$$2SO_2(g) + O_2(g) \rightleftarrows 2SO_3(g)$$

the enthalpy change for the reaction is $\Delta H^\circ = -197.8$ kJ.

Problem:

If we raise the temperature of the equilibrium mixture will the concentration of $SO_3(g)$ increase or decrease?

Solution:

Le Chatelier's Principle states that the equilibrium will shift so as to diminish the impact of the applied stress. Therefore, if we sketch the reaction coordinate for this exothermic reaction, we have:

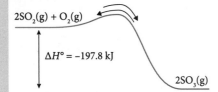

Thus, according to Le Chatelier, if we increase the temperature, the equilibrium will shift so as to remove heat from the system. Thus the equilibrium will shift to the left and the concentration of $SO_3(g)$ will decrease.

Practice:

Given the famous Haber–Bosch reaction for the production of ammonia from H_2 and N_2

$$3H_2(g) + N_2(g) \rightarrow 2NH_3(g)$$

for which $\Delta H^\circ_{RX} = (\Delta H^\circ_f)_{NH3} = -46.2$ kJ/mol, will the yield of $NH_3(g)$ be greater at 100°C or 300°C?

Thermodynamic Equilibrium Constant: Activities

When we derived the equation

$$\Delta G = \Delta G° + RT \ln Q,$$

we used the relationship

$$S = S° - R \ln \frac{P}{P°}$$

$$= S° - R \ln \frac{P}{1}$$

$$= S° - R \ln P$$

That is, we defined a standard state of 1 bar, the reference (pressure) for values of entropy. That means $P/P°$ is dimensionless, as must be the case for any term appearing in a logarithm. We need to use these equations in reference to solutions, solids, etc. This introduces the concept (and the utility) of "activities."

$$\text{activity} \equiv a = \frac{\text{the effective concentration of a substance}}{\substack{\text{the effective concentration of that substance} \\ \text{in a standard reference state}}}$$

Gas phase: $P° = 1$ bar:

$$S = S° - R \ln(P/P°) = S° - R \ln P, \text{ with pressure in bars}$$

Solutions: for 0.1 molar solution:

$$S = S° - R \ln a = S° - R \ln (0.1 \text{ M}/1 \text{ M}) = S° - R \ln 0.1$$

Heterogeneous:

Pure liquids and solids: $a = 1$

For gases, a = gas pressure in bars

For solutes in aqueous solution: a = (molar concentration/1 M)

Key Point: When an equilibrium expression is written in terms of activities, K_{eq} is the *Thermodynamic Equilibrium Constant.*

This is a very important point! In fact, because *activities* must be used whenever K_{eq} is calculated for use in the expression $\Delta G° = -RT \ln K_{eq}$, it is important to practice using the rules given above for the calculation of activities.

Assessing Spontaneity for Nonstandard Conditions

Consider the decomposition of 2-propanol to form acetone and hydrogen:

$$(CH_3)_2CHOH(g) \rightleftarrows (CH_3)_2CO(g) + H_2(g)$$

The equilibrium constant is 0.444 at 452 K. Is the reaction spontaneous under standard conditions? Will the reaction be spontaneous when the partial pressures of 2-propanol, propanone, and hydrogen are 1.0, 0.1, and 0.1 bar, respectively?

Solution:

First we solve for $\Delta G°$, which we need for both parts of the problem:

$$\Delta G° = -RT \ln K_{eq} = -(8.3 \text{ J/mol·K})(452 \text{ K})\ln(0.444) = 3.08 \times 10^3 \text{ J/mol}$$

Because $\Delta G° > 0$, the reaction *will not proceed spontaneously* if all reactants are in their standard states.

Will the reaction be spontaneous when the partial pressures of 2-propanol, propanone, and hydrogen are 1.0, 0.1, and 0.1 bar, respectively?

$$\Delta G = \Delta G° + RT \ln Q$$

$$Q = \frac{P_{(CH_3)_2CO} \, P_{H_2}}{P_{(CH_3)_2CHOH}} = \frac{a_{(CH_3)_2CO} \, a_{H_2}}{a_{(CH_3)_2CHOH}}$$

$$\Downarrow \qquad\qquad \Downarrow$$

$$= \frac{(0.1)(0.1)}{1.0}$$

$$\Delta G = 3.08 \times 10^3 \text{ J/mol} + (8.3 \text{ J/mol·K})(452 \text{ K}) \ln \frac{(0.1)(0.1)}{1.0}$$

$$= 3.08 \times 10^3 \text{ J/mol} + -1.73 \times 10^4 \text{ J/mol}$$

$$= -1.42 \times 10^4 \text{ J/mol}$$

Thus, at 452 K and (1.0, 0.1, 0.1) the reaction will proceed spontaneously.

$\Delta G°$ and K_{eq} as Functions of Temperature

Le Chatelier provided the means for qualitative appraisal of equilibria shifts for changes in temperature. We can now execute this prediction quantitatively:
1.   We know that $\Delta H°$ is (virtually) independent of temperature.
2.   We know that $\Delta S°$ is weakly temperature dependent and we approximate it as temperature independent.
3.   But $T\Delta S°$ is strongly temperature dependent.

The key equation is
$$\Delta G° = \Delta H° - T\Delta S° = -RT \ln K_{eq}$$

Using $\Delta G° = -RT \ln K_{eq}$

and exponentiating
$$K_{eq} = \exp(-\Delta G°/RT)$$

Key Point: Watch carefully the distinction between standard state ($\Delta G°$, $\Delta H°$, $\Delta S°$) and the actual conditions stated in the problem.

Problem 1:

For the reaction
$$2NO(g) + Cl_2(g) \rightleftarrows 2NOCl(g)$$

for which: $\Delta G° = -40.9$ kJ/mol
$\Delta H° = -77.1$ kJ/mol
$\Delta S° = -121.2$ J/mol·K

calculate K_{eq} @ 298 K.

Solution:

$$K_{eq} = \exp\left(\frac{-\Delta G°}{RT}\right) = \exp\left(\frac{40.9 \text{ kJ/mol}}{(8.3 \text{ J/mol·K})298 \text{ K}}\right)$$

$$= 1.5 \times 10^7$$

Problem 2:

For the same reaction, calculate K_{eq} @ 431 K.

Solution:

First calculate

$$\Delta G_{431\,K} = \Delta H° - (431 \text{ K})\Delta S° = -77.1\frac{\text{kJ}}{\text{mol}} - (431 \text{ K})\left(-121.35\frac{\text{J}}{\text{mol·K}}\right)$$

$$= -77.1\frac{\text{kJ}}{\text{mol}} + 52.2\frac{\text{kJ}}{\text{mol}}$$

$$= -24.9\frac{\text{kJ}}{\text{mol}}$$

Then calculate

$$K_{eq} = \exp\left(\frac{-\Delta G_T}{RT}\right) = \exp\left(-\frac{-24.9 \text{ kJ/mol}}{(8.3 \text{ J/mol·K})431\text{K}}\right)$$

$$= 1.05 \times 10^3$$

Overview of Manipulations of the Gibbs Free Energy Equation

As shown in Figure 5.20, we can display the graphical relationship of K_{eq} and temperature from the equation

$$\Delta G° = \Delta H° - T\Delta S° = -RT \ln K_{eq}$$

$K_{eq} = \exp(-\Delta G°/RT)$ at standard conditions.
$K_{eq} = \exp(-\Delta G_T/RT)$ at any temperature,
    but be sure to recalculate $\Delta G_T = \Delta H° - T\Delta S°$.

Calculate K_{eq} at two different temperatures:

Divide original equation by $-RT$.

$$\Delta H° - T\Delta S° = -RT \ln K_{eq}$$

$$\ln K_{eq} = (-\Delta H°/RT) + (\Delta S°/R)$$

$$\ln K_1 = (-\Delta H°/RT_1) + (\Delta S°/R)$$

$$\ln K_2 = (-\Delta H°/RT_2) + (\Delta S°/R)$$

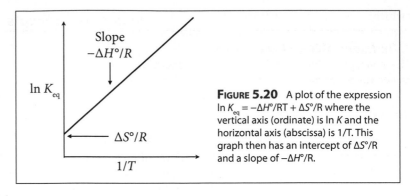

FIGURE 5.20 A plot of the expression ln K_{eq} = −ΔH°/RT + ΔS°/R where the vertical axis (ordinate) is ln K and the horizontal axis (abscissa) is 1/T. This graph then has an intercept of ΔS°/R and a slope of −ΔH°/R.

Subtract:

$$\ln(K_1 / K_2) = \left(\frac{-\Delta H°}{R}\right)\left(\frac{1}{T_1} - \frac{1}{T_2}\right)$$

Gibbs Free Energy: The Maximum Amount of Work That Can Be Extracted from a Chemical Process

Why, incidentally, is ΔG (the maximum amount of work that can be done by a process) called Gibbs *free* energy? A primary use of spontaneous reactions is the production of useful work:

- Steam engines
- Gasoline and/or diesel engines
- Batteries to power cell phones, computers

The key objective is to optimize the amount of work, minimizing the amount of energy lost to heat. The maximum conversion of chemical energy to work occurs when a process is carried out under thermodynamically *reversible* conditions. An important example: We can discharge a battery via two different pathways: sudden or piecemeal. These two cases are displayed in Figure 5.21.

The maximum amount of energy produced by a reaction that can be theoretically harnessed as work is equal to ΔG.

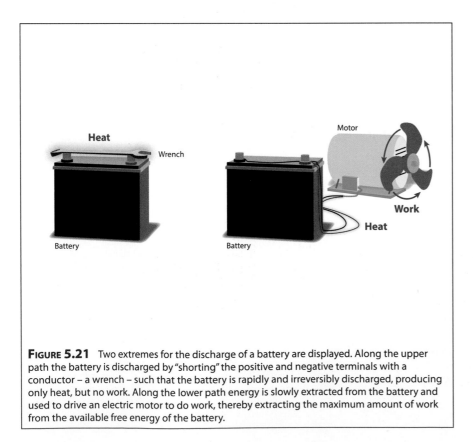

FIGURE 5.21 Two extremes for the discharge of a battery are displayed. Along the upper path the battery is discharged by "shorting" the positive and negative terminals with a conductor – a wrench – such that the battery is rapidly and irreversibly discharged, producing only heat, but no work. Along the lower path energy is slowly extracted from the battery and used to drive an electric motor to do work, thereby extracting the maximum amount of work from the available free energy of the battery.

Summary Concepts

1. The Nature of Equilibrium

The concept of the equilibrium state in chemistry is represented by expressing the generalized form of the chemical reaction as a bidirectional reaction

$$aA + bB \rightleftarrows cC + dD$$

that represents explicitly that there is a dynamic exchange between reactants on the left and products on the right. Just as chemical transformations can proceed from left to right, so too can they proceed from right to left. To be in equilibrium is to be in a state of delicate balance wherein the flow of chemical (or physical) transformations left to right quantitatively equals the flow of transformations from right to left.

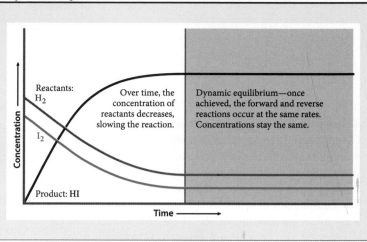

2. The Equilibrium Constant

For a generalized chemical reaction

$$aA + bB \rightleftarrows cC + dD$$

the analysis of the process by which the reaction attains equilibrium begins with the specification of the reaction quotient, Q, where

$$Q = \frac{[C]^c [D]^d}{[A]^a [B]^b}$$

reflecting the stoichiometric coefficients a, b, c, and d in our general chemical reactions. The concentrations [A], [B], [C], and [D] are each raised to the power of the corresponding stoichiometric coefficient. The Law of Mass Action states that *at* equilibrium, the reaction quotient is equal to the equilibrium constant, $K_{eq}(T)$, at the specified temperature:

$$Q = K_{eq}(T)$$

Thus at equilibrium for a specified temperature

$$K_{eq}(T) = \frac{[C]_{eq}^c [D]_{eq}^d}{[A]_{eq}^a [B]_{eq}^b}$$

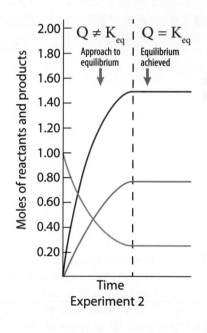

3. Determination of a Generalized Expression for K_c

The functional form of K_c that led to the formulation of the Law of Mass Action depended on the correct determination of the functional form of K_c in terms of the concentrations of reactants and products. The determination of the correct functional form of K_c is established by *experimentally* solving for candidate equilibrium constants under different experimental conditions. The functional form that yields the same equilibrium constant independent of initial condition is the correct formulation.

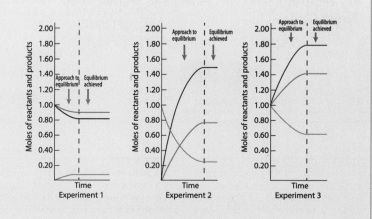

Summary Concepts

4. Manipulation of the Equilibrium Constant

Given that the equilibrium constant is driven by the stoichiometric coefficients and the corresponding concentrations, a given equilibrium constant depends directly on the specific way the chemical reaction is written. The equilibrium constant thus refers specifically to how a chemical reaction is written. If the chemical reaction is reversed, the equilibrium constant is inverted; if the stoichiometric coefficients are divided by 2, the resulting equilibrium constant is the square root of the original one. A listing is given at the right.

1. For each chemical equation defining an equilibrium reaction, there is a unique functional form of the equilibrium constant.

2. When we *reverse* the chemical equation expressing the equilibrium relations between the left-hand and right-hand species, we *invert* the functional form of the equilibrium constant.

3. If we multiply the stoichiometric coefficients in a balanced chemical reaction by a common factor (yielding a chemical reaction that is still balanced), then the equilibrium constant, K_c, is raised to the power by which the stoichiometric coefficients were multiplied.

4. When reactions are *summed* together to form a net reaction, the equilibrium constants are *multiplied* together to form the resultant equilibrium constant.

5. Converting Between the Equilibrium Constant Expressed in Concentration Units vs. Pressure Units

It is clear that reactants and products in the gas phase reach equilibrium, so while we have developed the concept of the equilibrium constant in terms of *concentrations*, there must be a corresponding formulation in terms of the *pressure* of the participating species. The key equation linking concentration to pressure is the Perfect Gas Law:

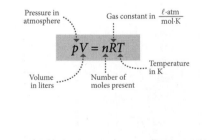

We can express this interconversion between K_c and K_p in a simple and more general form by considering our general expression for a chemical process in equilibrium

$$aA + bB \rightleftarrows cC + dD$$

wherein

$$K_c = \frac{[C]^c[D]^d}{[A]^a[B]^b} = \frac{\left(P_C/RT\right)^c\left(P_D/RT\right)^d}{\left(P_A/RT\right)^a\left(P_B/RT\right)^b}$$

$$= \frac{P_C^c P_D^d}{P_A^a P_B^b}\left(\frac{1}{RT}\right)^{c+d-(a+b)}$$

$$= K_p\left(\frac{1}{RT}\right)^{c+d-(a+b)}$$

$$= K_p\left(\frac{1}{RT}\right)^{\Delta n}$$

Alternatively, we can, of course, write

$$K_p = K_c(RT)^{\Delta n}$$

where Δn = change in the number of moles
= (number of moles of product) − (number of moles of reactants)

Summary Concepts

6. Stressed Equilibria and the Principle of Le Chatelier

This principle was first articulated by the French chemist Henry Le Chatelier (1850–1936) in 1888. The Le Chatelier Principle may be stated in a number of alternative forms:

- When a dynamic equilibrium in a system is upset by a disturbance, the system responds in a direction that tends to *counteract* that disturbance and, if possible, restore equilibrium.
- When an equilibrium system is subjected to a change in temperature, pressure, or concentration of a reacting species, the system responds by attaining a new equilibrium that partially offsets the impact of the change.

The power of Le Chatelier's Principle is that it provides a means for determining the *qualitative response*, the *direction* of response, of equilibrium systems to an external stress. Le Chatelier also links into the relationship between the reaction quotient, Q, and the movement of an equilibrium system back toward equilibrium.

If we write our expression for the reaction quotient, Q_c, in terms of the concentrations immediately following the applied stress as

$$Q_c = \frac{[C]^c_{stress}[D]^d_{stress}}{[A]^a_{stress}[B]^b_{stress}}$$

and we note that Q_c can assume values from zero to ∞ for our reaction

$$aA + bB \rightleftarrows cC + dD$$

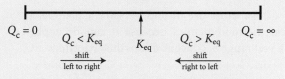

We note that, because K_{eq} lies between the extremes of 0 and ∞, if $Q_c < K_{eq}$, Q_c will seek to *increase* until it satisfies the condition $Q_c = K_{eq}$ and it will do so by virtue of the fact that the reaction shifts, along with Q_c, from left to right on the diagram above. On the other hand, if $Q_c > K_{eq}$ as a result of the applied stress, then Q must decrease to re-achieve equilibrium and both Q and the reaction must shift from right to left in the diagram above.

7. Equilibrium Constraints, Spontaneous Processes, and Gibbs Free Energy

We recognized from the Second Law of Thermodynamics that the Gibbs free energy

$$\Delta G = \Delta H - T\Delta S$$

determines whether a process is spontaneous; thus, how are ΔG and K_{eq} related?

> $\Delta G < 0$ Spontaneous
> $\Delta G > 0$ Nonspontaneous
> $\Delta G = 0$ No Motive for Change

To establish the relationship between Gibbs free energy and the equilibrium constant, let's return to the dimerization reaction

$$N_2O_4(g) \rightleftarrows 2NO_2(g)$$

$$\Delta G^\circ_R = \left(\Delta G^\circ_f\right)_{2NO_2} - \left(\Delta G^\circ_f\right)_{N_2O_4} = (2)(51.8) - 98.3$$

$$= +5.3 \text{ kJ}$$

So for the equilibrium reaction

$$N_2O_4 \rightleftarrows 2NO_2 \quad \Delta G^\circ_R = +5.3 \text{ kJ}$$

But now we are in a position to link our equilibrium reaction to the diagram relating the reaction quotient, Q, to the equilibrium constant, $K_{eq}(T)$, and finally to the Gibbs free energy diagram, as shown in Figure 5.16.

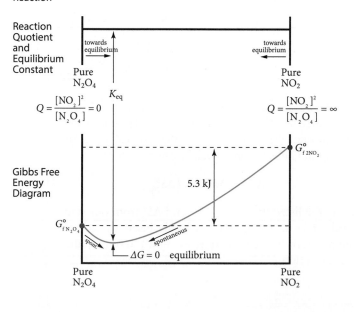

Summary Concepts

8. Gibbs Free Energy Under Nonstandard Conditions

The equation linking ΔG, $\Delta G°$, and the reaction quotient Q:

$$\Delta G = \Delta G° + RT \ln Q$$

is one of the most useful and versatile equations in chemical thermodynamics because it can be applied across a vast range of concentrations that determine ΔG.

We can use this equation to determine the spontaneous (or nonspontaneous) *nature of a reaction under any conditions of composition, if the temperature and pressure conditions are constant.*

Moreover, we are now in a position to quantitatively link the equilibrium constant, $K_{eq}(T) = \dfrac{[C]^c [D]^d}{[A]^a [B]^b}$, to $\Delta G°$.

We know that $\Delta G = 0$ at equilibrium *and* we know that $Q = K_{eq}(T)$ *at equilibrium*, so therefore at equilibrium

$$\Delta G = \Delta G° + RT \ln Q$$

becomes

$$0 = \Delta G° + RT \ln K_{eq}(T)$$

or

$$\Delta G° = -RT \ln K_{eq}$$

or

$$K_{eq} = \exp(-\Delta G°/RT)$$

9. $\Delta G°$ at Temperatures Other than 298 K

It is very important to calculate ΔG at temperatures other than 298 K because most reactions occur at temperatures other than 298 K and both ΔG and K_{eq} are sensitive to temperature.

At temperatures other than 298 K, we write ΔG_T to indicate that we are calculating and using ΔG at nonstandard conditions. Thus

$$\Delta G_T = \Delta H_T - T\Delta S_T$$

ΔH does not, to a rather high degree of accuracy, depend on T! This is a very important result. It means that we can use our tabulated values for $\Delta H°$ at standard conditions for $\Delta H_T°$ at any temperature!

As we can see from Chapter 2, $\Delta S°$ does not change very much as a function of temperature. The conclusion is important and very useful: We can calculate the change in Gibbs free energy at any temperature:

$$\Delta G_T° = \Delta H_{298}° - T\Delta S_{298}°$$

using the listed values for $\Delta H°$ and $\Delta S°$ from our standard tables at 298 K! We need only insert the correct temperature. To demonstrate, we return to our equilibrium reaction.

Reaction: $N_2O_4(g) \longrightarrow 2NO_2(g)$

$\Delta G° = +5.40$ kJ

$G°$ for 2 NO_2

G

$G°_{N_2O_4}$

Pure N_2O_4   N_2O_4 – NO_2 mixtures   Pure NO_2

Shift in temperature from 25°C to 100°C

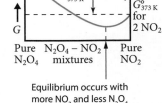

$N_2O_4(g) \longrightarrow 2NO_2(g)$

$G°_{373\,K}$ for N_2O_4

$\Delta G°_{373\,K} = -8.1$ kJ

$G°_{373\,K}$ for 2 NO_2

G

Pure N_2O_4   N_2O_4 – NO_2 mixtures   Pure NO_2

Equilibrium occurs with more NO_2 and less N_2O_4 than at 25°C.

Summary Concepts

10. Activities: The True Thermodynamic Equilibrium Constant

When we derived the equation

$$\Delta G = \Delta G° + RT \ln Q,$$

we used the relationship

$$S = S° - R\ln\frac{P}{P°} = S° - R\ln\frac{P}{1} = S° - R\ln P$$

That is, we defined a standard state of 1 bar, the reference (pressure) for values of entropy. That means $P/P°$ is dimensionless, as must be the case for any term appearing in a logarithm. We need to use these equations in reference to solutions, solids, etc. This introduces the concept (and the utility) of "activities."

$$\text{activity} \equiv a = \frac{\text{the effective concentration of a substance}}{\begin{array}{c}\text{the effective concentration of that substance}\\\text{in a standard reference state}\end{array}}$$

Gas phase: $P° = 1$ bar:

$$S = S° - R\ln(P/P°) = S° - R\ln P, \text{ with pressure in bars}$$

Solutions: 0.1 molar solution:

$$S = S° - R\ln a = S° - R\ln(0.1\,M/1\,M) = S° - R\ln 0.1$$

Heterogeneous:

Pure liquids and solids: $a = 1$

For gases, $a = $ gas pressure in bars

For solutes in aqueous solution: $a = $ (molar concentration/1 M)

Key Point: When an equilibrium expression is written in terms of activities, K_{eq} is the *Thermodynamic Equilibrium Constant.*

This is a very important point! In fact, because *activities* must be used whenever K_{eq} is calculated for use in the expression $\Delta G° = -RT \ln K_{eq}$, it is important to practice using the rules given above for the calculation of activities.

Summary Concepts

Overview of Manipulations of the Gibbs Free Energy Equation

As shown in Figure 5.20, we can display the graphical relationship of K_{eq} and temperature from the equation

$$\Delta G^\circ = \Delta H^\circ - T\Delta S^\circ = -RT \ln K_{eq}$$

$K_{eq} = \exp(-\Delta G^\circ/RT)$ at standard conditions.
$K_{eq} = \exp(-\Delta G_T/RT)$ at any temperature,
    but be sure to recalculate $\Delta G_T = \Delta H^\circ - T\Delta S^\circ$.

Calculate K_{eq} at two different temperatures:

Divide original equation by $-RT$.

$$\Delta H^\circ - T\Delta S^\circ = -RT \ln K_{eq}$$

$$\ln K_{eq} = (-\Delta H^\circ/RT) + (\Delta S^\circ/R)$$

$$\ln K_1 = (-\Delta H^\circ/RT_1) + (\Delta S^\circ/R)$$

$$\ln K_2 = (-\Delta H^\circ/RT_2) + (\Delta S^\circ/R)$$

11. ΔG° and K_{eq} as Functions of Temperature

Le Chatelier provided the means for qualitative appraisal of equilibria shifts for changes in temperature. We can now execute this prediction quantitatively:

1. We know that ΔH° is (virtually) independent of temperature.
2. We know that ΔS° is weakly temperature dependent and we approximate it as temperature independent.
3. But $T\Delta S^\circ$ is strongly temperature dependent.

The key equation is

$$\Delta G^\circ = \Delta H^\circ - T\Delta S^\circ = -RT \ln K_{eq}$$

Using $\Delta G^\circ = -RT \ln K_{eq}$

and exponentiating
$$K_{eq} = \exp(-\Delta G^\circ/RT)$$

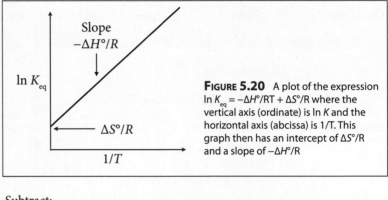

FIGURE 5.20 A plot of the expression $\ln K_{eq} = -\Delta H^\circ/RT + \Delta S^\circ/R$ where the vertical axis (ordinate) is $\ln K$ and the horizontal axis (abcissa) is $1/T$. This graph then has an intercept of $\Delta S^\circ/R$ and a slope of $-\Delta H^\circ/R$

Subtract:

$$\ln(K_1 / K_2) = \left(\frac{-\Delta H^\circ}{R}\right)\left(\frac{1}{T_1} - \frac{1}{T_2}\right)$$

CASE STUDY 5.1   The Technology and the Future of Biofuels

Introduction

The role of biofuels in the current and future mix of primary energy sources has had a tortured excursion from center stage to marginalized and demonized back to the role of a major contributor to future forecasts of the mix of primary energy sources in the US. The cause of those swings of fortune involves both a technological progression and advances in the understanding of sources for biofuel feed stocks, as well as a growing sophistication in judging the end-to-end impact of various primary energy options.

Before addressing the technology that underpins projected growth of biofuels over the next two decades, we briefly summarize (1) the role of biofuels in the current energy mix of the United States and (2) some basic concepts and definitions that constitute the language (old and new) of the biofuel field.

I. Background to Biomass Fuels

1.1 Current Consumption of Biomass Resources

A biofuel is characterized as a fuel whose chemical energy is derived from biological carbon fixation via photosynthesis as depicted graphically in Figure CS5.1a. Biofuels are distinct from fossil fuels in that fossil fuels contain carbon that has been isolated from the active carbon cycle for millennia. Biofuels thus constitute an active partner in the carbon-oxygen cycles of the terrestrial and oceanic biosphere.

A wide range of biomass feed stocks (inputs) are currently employed to generate electricity, to produce transportation (liquid) fuels, and to produce heat by direct combustion. Broadly speaking, there are two categories of biofuels: "first generation" biofuels and "second generation" biofuels.

First generation biofuels, a.k.a. conventional biofuels, are biofuels made from sugar, starch, or vegetable oil. For example biologically produced alcohols such as ethanol are produced by the bacterially mediated fermentation of sugars or starches. Categories of first generation biofuels include, in addition to ethanol, biodiesel, vegetable oil, biogas, or biomethane, syngas, which is a blend of carbon monoxide and hydrogen and other hydrocarbons, and solid biofuels such as wood, sawdust, agricultural waste, grass trimmings, domestic refuse, etc. If this form of biomass is in a suitable form to convert to an alcohol by direct fermentation, to burn directly such as firewood or to be directly extracted (such as vegetable oil by compressional extraction), it is used directly. If the biomass is not in convenient form, it is first "densified" by grinding and compression into pellet size for either direct combustion or chemical transformation.

A key characteristic of first generation biofuels, particularly the use of corn for ethanol production, is that many—though not all—compete with important lines of food production.

Second generation biofuels, such as cellulosic ethanol, algae fuel, biohydrogen, and biomethanol, are produced from cellulosic feedstocks that cannot be used for food because humans cannot digest the cellulosic structure. While the molecular structure of starch and cellulose appear to be virtually identical, the subtle difference is of critical biological importance.

Starch and cellulose differ in the geometry or spatial orientation of the linkage between glucose units that comprise the chain or polymer structure of starch and cellulose. Figure CS5.1b displays the

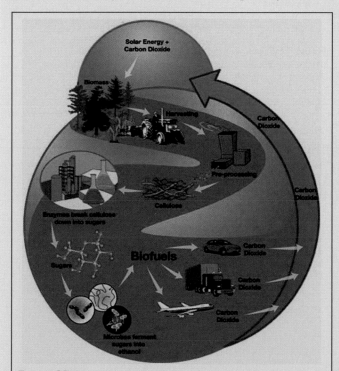

FIGURE CS5.1A   Modern biofuel production represents a cycle linking the input of solar energy to the production of biomass that is then harvested and converted to liquid fuels primarily by microbial action. Historically, biofuels here provided heat by the direct combustion of wood or other bioproducts.

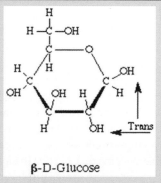

FIGURE CS5.1B   Glucose is a ring hydrocarbon that occurs either as β-glucose shown here with the OH groups in a trans configuration or α-glucose with the OH groups in a cis configuration.

molecular structure of glucose in an aqueous solution. That figure presents both the molecular diagram and the three-dimensional stick model of the glucose structure.

There are two forms of glucose: α-glucose has the OH group diagonally across the ring of the molecule from the HO–CH$_2$–group located on carbon 5, and β-glucose (shown in Figure CS5.1b) that has the hydroxyl groups on the same side of the molecular axis as the HO–CH$_2$–group located on carbon 5.

Starch is formed from a chain of α-glucose molecules that react as shown in Figure CS5.1c, eliminating water and creating a polymer with the bridging oxygen below the plane of the glucose ring.

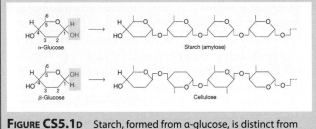

FIGURE CS5.1c   Starch is distinguished by the fact that the bridging oxygen lies below the plane of the hydrocarbon ring while the hydrogens at the binding site lie above the molecular plane.

The same reaction involving β-glucose results in the formation of cellulose, where cellulose is characterized by the bridging oxygen alternating above and below the plane of the glucose ring as shown in Figure CS5.1d.

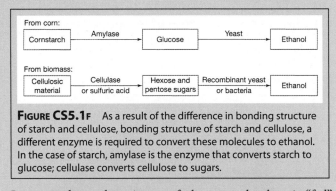

FIGURE CS5.1d   Starch, formed from α-glucose, is distinct from cellulose that is formed from β-glucose resulting in a sequence of bridging oxygens alternating above and below the plane of the hydrocarbon ring for cellulose.

Organisms break down the polymers by catalyzing the reverse reaction using enzymes. However, the catalysts that successfully break down starch have different placements of their catalytic groups at the active site than do catalysts that breakdown cellulose. The enzyme for breaking down starch is amylase; the enzyme for breaking down cellulose is cellulase. What is important and somewhat surprising is that while many organisms (including humans) have amylase, very few have cellulase. Termites and ruminants (e.g., cows) can digest cellulosic structures only because they have digestive systems that harbor the appropriate bacteria.

When ethanol is produced from corn at industrial scale, the starch polymer is degraded to glucose with just amylase that is inexpensive and readily available. Then the glucose is fed to yeast that metabolizes and thereby oxidizes the glucose to ethanol—a.k.a. fermentation.

The breakdown of cellulose, on the other hand, involves two obstacles. The first is that cellulose in plants is protected by a sheath of hemicellulose that is a polymer of C$_5$ sugars and lignin (see Chapter 2). That sheath must first be breached—a process that requires submersion in dilute sulfuric acid at elevated temperatures and pressures. This acid treatment breaks down the hemicellulose polymer to its C$_5$ sugar monomers such as xylose shown in Figure CS5.1e.

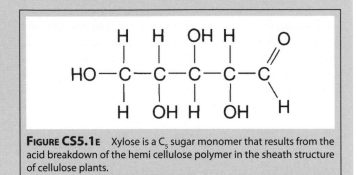

FIGURE CS5.1e   Xylose is a C$_5$ sugar monomer that results from the acid breakdown of the hemi cellulose polymer in the sheath structure of cellulose plants.

We can graphically trace the sequence of processes for the biocatalysis of starch on the one hand and cellulosic material on the other in the formation of ethanol. That is displayed in Figure CS5.1f.

FIGURE CS5.1f   As a result of the difference in bonding structure of starch and cellulose, bonding structure of starch and cellulose, a different enzyme is required to convert these molecules to ethanol. In the case of starch, amylase is the enzyme that converts starch to glucose; cellulase converts cellulose to sugars.

Interestingly, as the mixture of glucose and xylose is "fed" to the yeast that ferments (oxidizes) glucose, the xylose and other C$_5$ sugars cannot be fermented by common yeasts. Work is currently in progress developing strains of bacteria capable of executing the conversion of cellulosic biomass to ethanol in a single step. This would be an important development because 80% of all vegetation worldwide is cellulosic.

II. Biomass Categories for First Generation Biofuels

2.1 Forest-derived Resources

Biomass from forest sources comes primarily from two sources: (1) fuelwood used directly in the commercial and residential sectors and (2) residues resulting from the manufacture of wood products and from forest cleanup. In 2010 the residential, commercial, and electric power sectors combined used approximately 40 million dry tons with the residential sector using 28 million dry tons, the commercial sector using 7 million dry tons, and the electric power sector 5.5 million dry tons. Most fuelwood consumed is in the Northeast and North Central regions and to a lesser extent the Pacific Coast and Southeast.

2.2 Municipal Solid Waste

In the United States alone, 265 million tons of municipal solid waste (MSW) is generated each year. Of this total, approximately one-third is recovered by recycling and compost collection. With respect to renewable energy, about 15% is combusted to produce useable energy. The MSW derived from forest sources include discarded wood, yard trimmings, packaging, paper, and newsprint.

2.3 Agricultural Biomass

With the rapid development of the corn-based ethanol industry and the developing biodiesel industry using oilseed crops, a large fraction of cropland-derived biomass goes to those two sectors. Thus we briefly review those two biofuel production sectors.

2.3.1 Ethanol from Starch

While we will discuss ethanol from corn in more depth in Case Study 5.2, we note that historically United States agriculture has focused on corn production because corn has a high carbohydrate yield when compared with other staple crop categories. This underlies the broad use of corn for feed, ethanol, food, and any array of exports. The very large agricultural region of the Midwest makes this a huge business venture in the US. The highest volume of harvested corn occurred in 2009 and was 13.4 *billion* bushels. Of that total, about 35% was utilized for ethanol production.

Ethanol production in 2010 was 13.2 billion gallons produced from 4.7 billion bushels, which corresponds to approximately 110 million dry tons of corn. A bushel of corn weighs 56 lbs (25.5 kg) at 15% moisture. With current technology, a bushel of corn can produce 2.8 gallons of ethanol. After extraction of the ethanol from the starch in corn, the remaining fiber, protein, vegetable oil, and minerals are used in livestock feed.

A key driver for increasing demand for corn is the 2007 Energy Independence and Security Act that mandates up to 15 billion gallons of corn-based ethanol for the years 2015–2022. As a result of the increased awareness of the intrinsic drawbacks of corn-ethanol, while ethanol corn production increased sevenfold in the last decade (2010–2020) it is only expected to increase to just below 90 million dry tons. About 38% of corn grain produced in 2010 was used in ethanol production. DOE forecasts place ethanol-to-corn proportion to remain stable between 33% and 35% between 2010 and 2020.

The controversy over the adoption of corn based ethanol as a major component of the nation's biofuel supply results from the fact that the net consumption of petroleum based fuels for gasoline and diesel needed to plant, cultivate, and harvest corn and the consumption of natural gas and coal required to produce the fertilizer and supply the water for this crop virtually equals the net energy extracted from the ethanol produced in the process. The now widely accepted reality of this has lead to a major reconsideration of the wisdom of corn based ethanol in the US. This is compounded by the fact that corn-based ethanol competes directly with the demand for food, a significant fraction of which is corn-based in the US and globally.

It is becoming increasingly apparent that the demand placed on corn production by ethanol targets has a number of unintended market, health, and environmental consequences. The demand for corn to produce increasing amounts of ethanol increases the price of corn for the food sector. Corn is a major cereal grain and a primary feed for livestock supplying the meat industry. This has impact both nationally and internationally because the US exports corn.

2.3.2 Biodiesel

Biodiesel is distinct as a fuel from ethanol because of its molecular structure. Figure CS5.1g displays the familiar structure of ethanol in the left-hand panel and an average molecular structure for biodiesel is shown in the right-hand panel.

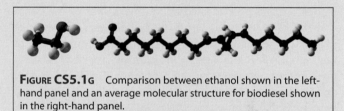

Figure CS5.1g   Comparison between ethanol shown in the left-hand panel and an average molecular structure for biodiesel shown in the right-hand panel.

The sources of biodiesel include vegetable oils, soybeans, and waste fats. Production of biodiesel began in 2005 and increased rapidly through to 2008 as displayed in Figure CS5.1h. Soybean oil has been the primary feedstock for biodiesel, but annual fat and waste oils from restaurants and transportation are increasing in importance. Also displayed in Figure CS5.1h is the fact that the component of biodiesel extracted from soybeans has reached a plateau at approximately 400 million gallons each year—a sizable contribution.

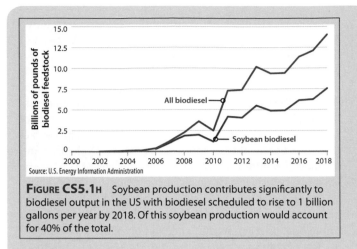

FIGURE CS5.1H Soybean production contributes significantly to biodiesel output in the US with biodiesel scheduled to rise to 1 billion gallons per year by 2018. Of this soybean production would account for 40% of the total.

With this increase in overall biodiesel production, the fractional contribution from vegetable oils that are not soybean based are projected to decrease because of the expense associated with their production. Recycled waste fat and oils is expected to increase in relative importance for biodiesel production.

III. Summary of Categories of Second Generation Energy Crops

Beginning in the late 1970s, a number of perennial grass and woody crops were evaluated in landmark species trials on a wide range of soil types across the United States. A key finding of this research was the development of a manifold of new crop types and the specification of crop management prescriptions for classes of high-potential perennial grasses and specific wood/tree crops. In the past decade, the list of high potential energy crops has been refined and we look here at the options. These energy crops constitute the specific crops that will establish the future potential for biofuel production in the US.

Figure CS5.1i summarizes the important second generation biofuels. As the center panel in Figure CS5.1i shows, second generation biofuels may well become a very important category of primary energy generation in the US by 2030.

IV. Summary

This puts us in position to calculate the importance of biofuel production on both the imports of petroleum into the US for liquid fuels and on the overall spectrum of primary energy generation in the US. We begin with the impact of biofuels on the petroleum imports.

Petroleum imports into the US in 2014 were 9 million barrels a day. Each barrel of crude oil produces, on average, 20 gallons of gasoline. Thus the import of crude oil constitutes a source of 160 million gallons of gasoline each day. As we saw in section II, each dry ton of biomass yields approximately 85 gallons of biofuels. Thus 1 billion dry tons of biomass would yield 85 billion gallons of ethanol per year. With

an energy content of 0.7 relative to gasoline, this corresponds to 60 billion gallons of gasoline equivalent liquid fuel, or 164 million gallons per day.

This could potentially dramatically reduce imported petroleum.

Given that petroleum constitutes 38% of the primary energy generated per year in the US, biofuels would grow to constitute approximately 26% of the total primary energy generated each year in the US. Also of great importance is that biofuels constitute an excellent source of fuel for jet aircraft and can thus eliminate net carbon release from aircraft.

| Problem |

Biofuels are distinct from fossil fuels because the carbon they contain is part of the active terrestrial carbon cycle. Ethanol fuel can be produced from the yeast fermentation of the starches and sugars found in biomass can be modeled by the following:

$$C_6H_{12}O_6(s) \rightleftarrows 2\ C_2H_5OH(l) + 2\ CO_2(g)$$

	$\Delta H°_f$ (kJ/mol)	S° (J/mol K)
$C_6H_{12}O_6(s)$	−1273.3	212.2
$C_2H_5OH(l)$	−277.63	161
$CO_2(g)$	−393.5	213.7

a. Determine the equilibrium constant K_{eq} for the reaction at 298 K.

b. At equilibrium, will the reaction be mostly products or mostly reactants? Is it safe to assume that the reaction goes to completion?

c. One bushel contains 24.5 kg of corn. If 64.4% of this is fermentable starch (glucose), how many gallons of ethanol can you produce? (Density of ethanol: 0.789 g/mL)

SUMMARY OF SECOND-GENERATION BIOFUELS LINKED TO HIGH-YIELD AND LOW-YIELD PRODUCTION FORECASTS

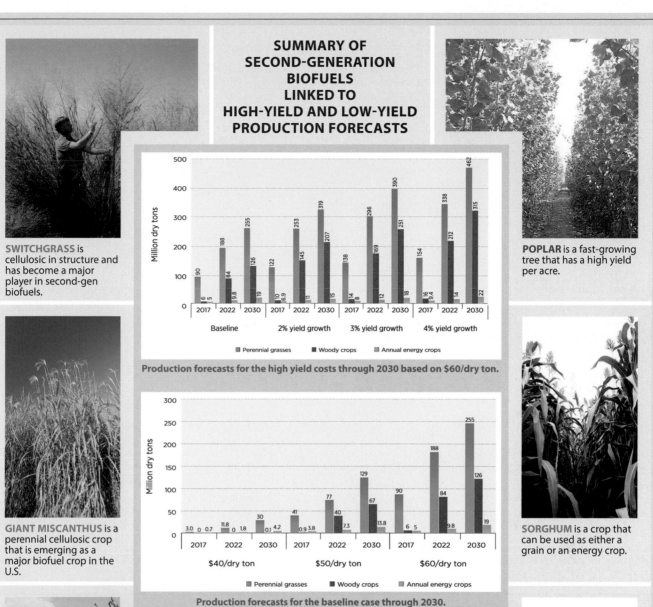

SWITCHGRASS is cellulosic in structure and has become a major player in second-gen biofuels.

GIANT MISCANTHUS is a perennial cellulosic crop that is emerging as a major biofuel crop in the U.S.

Production forecasts for the high yield costs through 2030 based on $60/dry ton.

Production forecasts for the baseline case through 2030.

■ Perennial grasses   ■ Woody crops   ■ Annual energy crops

POPLAR is a fast-growing tree that has a high yield per acre.

SORGHUM is a crop that can be used as either a grain or an energy crop.

SUGARCANE is a large perennial grass with high energy content resulting from the storage of sucrose in the stem.

EUCALYPTUS is fast growing and high yield.

SHRUB WILLOW has become an important biofuel crop in Delaware, Indiana, Illinois, Maryland, Michigan, Minnesota, Missouri, New Jersey, New York, Pennsylvania, South Carolina, Virginia, Vermont, and Wisconsin.

FIGURE CS5.1

CASE STUDY 5.2   Ethanol, Net Energy Production, and the Haber–Bosch Process

Alternative sources of liquid fuels for the transportation sector constitute a very important component in the national effort to transition from fossil-based fuels to a combination of biofuels and liquid fuels derived from natural gas (CH_4). This is particularly true when taken in combination with a new generation of carbon-fiber automobiles that are vastly lower in weight (yet equally safe) and are propelled by electric motors rather than internal engines.

A key example is the production of ethanol—a subject of immense importance to U.S. energy and economic policy. On the face of it, the production of ethanol as a substitute for gasoline appears to be a wise component of any new national energy policy. US citizens consume approximately 150 billion gallons of gasoline a year, which is approximately 500 gallons for each man, woman, and child in this country. If we were to replace gasoline with ethanol, we could potentially achieve multiple objectives: (1) reduce dependence on foreign oil, (2) stabilize (and lower) the cost of liquid fuels for transportation, (3) take advantage of the fact that 40% of the world's corn is grown in the US (and ethanol can be produced from corn), (4) reduce CO_2 emissions because the CO_2 released in ethanol combustion would be "carbon neutral" because that carbon would have been extracted from the atmosphere in the process of photosynthesis to produce the corn.

However, a number of factors, both subtle and not so subtle, conspire to make the ethanol-from-corn story more complicated than it appears on the face of it. We can summarize some of the fundamental issues surrounding ethanol from corn in the figure below.

A key aspect in the corn-ethanol analysis lies in two factors: (1) the energy content per gallon of ethanol is 67% that of gasoline per gallon, and (2) the fossil fuel required to provide the fertilizers, phosphates, transportation, etc., is very nearly equal in energy content to that of the ethanol produced. While the debate rages over whether ethanol represents a net gain over the fossil fuel required to produce it, a number of factors argue against corn-ethanol as a significant component in the liquid fuel mix for the US. Greenhouse gas emissions required in the production of ethanol from corn are virtually as large as if petroleum were burned directly because of the larger emission of more potent greenhouse gases such as nitrous oxide, which is a by-product of fertilizer production and application. In addition, the scale of land use in the US required for significant ethanol production is so large that it places ethanol-for-fuel in direct competition with corn-for-food in the global coupling of energy and food supply.

Rather than using corn as a feedstock for ethanol production, if cellulose based crops such as switchgrass in the US and/or sugar cane in Brazil were used, considerable advantage could be realized both in the net gain in energy per unit of fossil fuel consumed *and* in the reduction of greenhouse gas emissions. However it is turned, corn-based ethanol production will, in the long run, be a small component in the overall energy policy of the US.

A primary reason corn-based ethanol is limited in its potential is that it requires large amounts of fertilizer in the production of corn. An important component in the selection of optimal technology pathways involves the methods by which the crops used to produce biofuels are "fertilized" to maximize yields per acre of farmland. The dominant chemical compound used for crop fertilization is ammonia, NH_3, which supplies usable, or "fixed," nitrogen for agricultural, biological, and manufacturing processes. Thus, the production of ethanol from feedstocks such as corn, sugar cane, or cel-

---

Ethanol Essentials

- The energy content of 1 gallon of gasoline equals the energy content of 1.5 gallons of ethanol.
- The production of ethanol from corn in the US in 2018 was 16 billion gallons.
- The fraction of corn production used for the production of ethanol in the US was 40% in 2018.
- Total gasoline consumption in the US in 2018 was 143 billion gallons.
- To produce 100 gallons of ethanol, one ton of corn kernels are required.
- In most cases, it required a greater amount of fossil fuel energy to produce ethanol than is produced by ethanol.
- Much of the fossil fuel energy expended to produce ethanol is derived from natural gas and coal.
- With respect to the amount of CO_2 released to the atmosphere in the production of ethanol, more CO_2 is added than would be the case if gasoline were used instead of corn ethanol.
- Significant advances in both reducing the energy required to produce ethanol and markedly reducing the CO_2 emitted to the atmosphere would result if cellulose were used rather than corn starch.

lulosic switchgrass depends upon the production of ammonia, and the net amount of ammonia required (and the energy required to produce that ammonia) determines whether ethanol production makes economic and/or strategic sense. Thus, we explore in some detail the equilibria conditions that optimize NH_3 production.

While nature "fixes" nitrogen—the process of breaking the $N\equiv N$ triple bond to produce nitrogen in a form that is usable by plants—by either (1) the elegant enzymatic pathways made available by symbiosis of bacteria in combination with the root structure of certain plants, or (2) the primitive but effective dissociation of N_2 by lightning, humans achieve the objective using the equilibrium reaction

$$3H_2 + N_2 \rightleftharpoons 2NH_3\ (g) \qquad \Delta H_R = -91.8\ kJ$$

The process is simple—it was devised by Fritz Haber in 1913—and worldwide production of ammonia from this mechanism was some 1.4×10^{11} kg in 2019. The Haber process stands as a prime example of the direct application of equilibria principles in pursuit of the highest possible yields. Inspection of the above equilibrium equation reveals three lines of attack:

1. Reduction in $[NH_3]$ that serves to shift the equilibrium to the right. This is accomplished by removing NH_3 from the reactive mixture as it is produced.
2. Increasing the pressure, which, as we will see, shifts equilibrium to the right because 4 moles of reagent are converted to 2 moles of product.
3. Reduction in temperature, which, because the reaction is exothermic, also shifts equilibrium to the right.

In fact, we can, through knowledge of the temperature dependence of the equilibrium constant, generate a graphical representation of the NH_3 yield as a function of temperature—this is displayed in Figure CS5.2a.

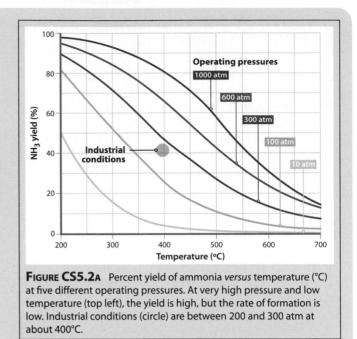

FIGURE CS5.2A   Percent yield of ammonia *versus* temperature (°C) at five different operating pressures. At very high pressure and low temperature (top left), the yield is high, but the rate of formation is low. Industrial conditions (circle) are between 200 and 300 atm at about 400°C.

While this graph strongly suggests an operating regime for the optimization of NH_3 yield in the vicinity of 1000 atm pressure and a temperature of 200°C, it turns out that the forward reaction forming NH_3 is very slow at that temperature. The reason is that there is a barrier in the potential energy surface separating reactants from products—a subject we will treat in some detail in Chapter 12. Because this reaction forming NH_3 from N_2 and H_2 becomes more rapid with increasing temperature, the industrial operation of NH_3 production plants favors temperatures in the vicinity of 400°C. The cost of plant construction, which increases with increasing operating pressure, is such that operating pressures are typically limited to ~200 atm.

The stages in the industrial production of ammonia are designed to comply with the principles of chemical equilibria

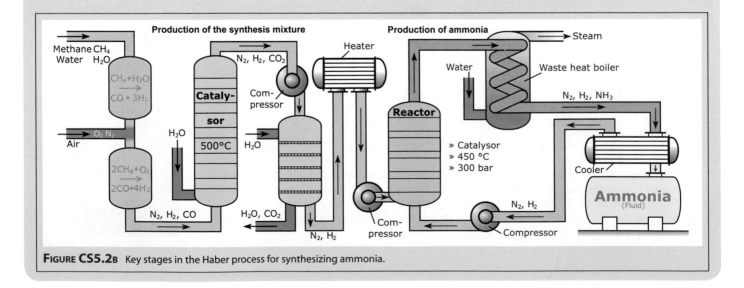

FIGURE CS5.2B   Key stages in the Haber process for synthesizing ammonia.

and with the rates at which reactions take place—the study of chemical kinetics.

Figure CS5.2b captures the primary subsystems used for ammonia production. The first step involves the addition of H_2 and N_2 into a compressor that raises the pressure to 200 atm and the mixture is heated to 400°C. Then the mixture is passed over a bed of iron embedded in a fused mixture of MgO, Al_2O_3, and SiO_2. The emergent equilibrium mixture, which contains approximately 35% NH_3 by volume, is then extracted and cooled by refrigeration coils until the NH_3 condenses and is removed. The unreacted H_2 and N_2 are then fed back into the compressor.

Remarkably, with the immense global production of ammonia reaching 1.4×10^{11} kg in 2019, and with an amount of energy required to produce ammonia of 5.9×10^7 joules/kg, the amount of energy required to produce ammonia each year is $(1.4 \times 10^{11}$ kg/yr$)(5.9 \times 10^7$ joules/kg$) = 7.3 \times 10^{18}$ joules/yr. That is almost 2% of the total global energy demand!

This brings us back to the question of calculating the net amount of energy produced in the generation of ethanol from corn—given that corn requires significant application of ammonia to reach current production levels. So the calculation of energy required to produce a gallon of ethanol *from corn* is as follows. We must account for the energy invested in the production of corn, the transport of corn, and the production of ethanol, which involves fertilizers, irrigation, farm machinery, etc. This amounts to 8.8×10^7 joules/gallon ethanol. However, the energy content of a gallon of ethanol is 8.4×10^7 joules/gallon so that the net energy value (NEV) of ethanol, which is the net energy contained in the final ethanol product *minus* the energy consumed in its production, is

$$\text{NEV} = 8.4 \times 10^7 - 8.8 \times 10^7 = -4 \times 10^6 \text{ J/gallon}$$

The minus sign results because it requires *more* energy (natural gas and oil) to produce ethanol from corn then is contained in the product, and therefore serious doubt has been cast on the wisdom of producing large amounts of ethanol from corn in the U.S. There are many considerations involved in ethanol production from corn including how the products rejected from the ethanol production itself, termed energy "coproducts," are used (animal feed stocks, burnable organic material, etc.); however it remains a fact that ethanol production from corn provides marginal yields with respect to the fossil fuel energy used to produce it. If, on the other hand, ethanol is produced in the U.S. using cellulosic biomass, the figures are quite different. As we noted in the opening of Chapter 2, cellulose is composed of long chains of sugars, and together with lignin, it is responsible for the physical structure of trees, plants, and grass. The issue with cellulosic biomass is the fermentation method used to make ethanol—a process that typically uses either acid decomposition or specifically designed enzymes. A leading candidate for ethanol production is switchgrass, which is a perennial grass native to the Midwest that does not need to be replanted, is pest resistant, has much lower fertilizer demands than corn, and requires only limited irrigation. These factors conspire to significantly reduce the energy invested in raising switchgrass for ethanol production.

Problem

A major contribution to the global increase in population was the use of agricultural fertilizers to increase crop yields globally. The principal source of crop fertilizer is the production of ammonia, NH_3, in the famous Haber–Bosch process.

$$3H_2(g) + N_2(g) \rightleftarrows 2NH_3(g)$$

What the Haber–Bosch process brought to the increased production of $NH_3(g)$ was the ingenious use of metal-catalyzed (iron, SiO_2, Al_2O_3) surfaces at elevated temperature that greatly increased both the yield and the production rate of $NH_3(s)$. The enthalpy for the reaction is

$$(\Delta H°_f)_{NH_3(g)} = \Delta H_{RXN} = -46.2 \text{ kJ/mol } NH_3.$$

a) At 25°C, is this reaction endothermic or exothermic? How does K_p change with temperature?

b) K_c of the Haber–Bosch process is 0.159 at 450°C. Calculate K_p for this reaction at this temperature.

c) For each of the mixtures listed below, indicate whether the mixture is at equilibrium at 450°C. If it is not at equilibrium, indicate the direction in which the mixture must shift to achieve equilibrium.

    i.   105 atm NH_3, 495 atm H_2, 35 atm N_2

    ii.   35 atm NH_3, 595 atm H_2, no N_2

d) To produce the maximum amount of NH_3 at 450°C (you do not need to answer this question in your homework, but think about why we want to increase temperature from 25°C to 450°C), according to Le Chatelier's principle, should the pressure of the system be increased or decreased?

e) At 450°C, if the reaction starts with a mixture of 100 atm N_2 and 300 atm H_2 in the reaction chamber in a factory, what percentage of the N_2 gets converted to the product at equilibrium? (Hint: the equation could be simplified to a quadratic equation by finding a common factor in some terms.)

f) At 450°C, if the reaction starts with a mixture of 200 atm N_2 and 600 atm H_2 in the reaction chamber in a factory, what percentage of the N_2 gets converted to the product at equilibrium?

g) Comparing the numbers you get from e) and f), are your calculations consistent with your prediction in d)?

h) At 450°C, the reaction starts with a mixture of N_2 and H_2 with unknown pressures in the reaction chamber in a factory, and a test of the gas mixture at a time point before the reaction reaches equilibrium finds 120 atm N_2, 360 atm H_2, and 160 atm NH_3 in the chamber. What percentage of the initial N_2 gets converted to the product at equilibrium? (Hint: you have already solved this question if you have finished a–g.)

i) At 450°C, besides changing the pressure of the system, propose another way to increase the yield of NH_3.

5.58

EQUILIBRIA IN SOLUTION
ACID-BASE CONTROL FOR LIFE SYSTEMS

6

FIGURE 6.1 Carbon dioxide exerts strong control over the acid-base balance of the oceans because water reacts with CO_2 to form carbonic acid, H_2CO_3. In addition, living organisms use the carbonate ion CO_3^{2-} in the formation of $CaCO_3$ that constitutes the skeletal structure of most organisms in the food chain. Thus the acid-base balance in the oceans is important to both the inorganic and organic chemistry of the oceans.

Framework

The chemistry of carbon dioxide and water constitutes one of the most important applications of acid-base chemistry in both the physical sciences and the life sciences. Indeed, both the evolution of life processes on Earth and the future of life on the planet as we know it are closely coupled to carbon chemistry in the oceans.

The development of the chemical reactions that couple $CO_2(g)$ in the atmosphere with a series of chemical reactions involving the acidic properties of both fresh water and the carbon chemistry of the oceans begins with the dissolution of atmospheric carbon dioxide in water:

$$CO_2(g) \rightleftharpoons CO_2(aq)$$

This physical dissolution is followed by the chemical reaction forming carbonic acid from $CO_2(aq)$:

$$CO_2(aq) + H_2O(l) \rightleftharpoons H_2CO_3(aq)$$

The net overall reaction is therefore:

$$CO_2(g) + H_2O(l) \rightleftharpoons H_2CO_3(aq)$$

The Gibbs free energy diagram for the overall reaction is displayed in Figure 6.2. As Figure 6.2 displays, the change in Gibbs free energy for the overall reaction is $\Delta G_R^\circ = -8.2$ kJ/mol. Because ΔG_R° is positive, the equilibrium favors the reactants and only a small fraction of the $CO_2(g)$ is converted to $H_2CO_3(aq)$. A small but very important fraction as we will see in this chapter.

end of ice ages, is controlled by the total amount of solar energy received by the northern hemisphere in the summer season, which in turn depends upon the tilt of the Earth's axis (obliquity) with respect to the orbital plane *and* the eccentricity of the orbit itself. This orbital control of climate is referred to as the "Milankovitch Cycle" after Milutin Milankovitch, a Serbian mathematician-engineer. As described in the sidebar, it is the relationship between obliquity and eccentricity of the Earth's orbit (at low CO_2 concentrations), which carries the Earth into and out of glacial periods; that is, between glacial and interglacial periods characterized by significant ice cover followed by conditions of retracted ice, respectfully. With the rapid build-up of CO_2 in the atmosphere following the beginning of the industrial

What has been the cause of the consistent cycling between *glacial periods,* characterized by extensive ice coverage of northern hemisphere continental regions, and *interglacial periods,* characterized by the retraction of large continental glaciers to approximately their current position? The answer involves subtle but important changes in the way the Earth orbits the Sun; specifically changes that repeat in a predictable pattern over tens of thousands of years. In particular, long term variations in the northern hemisphere summer solar intensity and duration are thought to control the degree of glaciation. Three characteristics of the Earth's orbital parameters are involved:

1. The degree of ellipticity of the orbit, also referred to as eccentricity, which is measured by the ratio of the major axis to the minor axis as shown here:

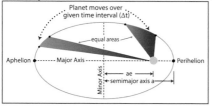

The eccentricity of the Earth's orbit changes on time scales of 100,000 years.

2. The tilt of the Earth's axis of rotation with respect to the plane of the orbit around the Sun. This tilt angle, referred to as the *obliquity*, varies from a minimum of 22.5° to a maximum of 24.5°. The period of changes in obliquity is approximately 40,000 years and the current axis tilt is 23.5°. Of course, it is the tilt of the axis that causes the change of seasons as the Earth orbits the Sun.

3. Changes in the *orientation* of the Earth's axis of rotation with respect to the Earth's orbital plane. This is referred to as *precession* and it affects the apparent location of the North Star, which today is aligned with the Earth's axis of rotation. The period of precession is 26,000 years.

These three effects are displayed graphically in the figure below.

Key points to remember are that (1) continental, northern hemisphere glaciers are sensitive to the amount of sunlight integrated over the duration of the northern hemisphere summer, and (2) as carbon dioxide builds in the atmosphere, the effect of the orbital parameters on climate are suppressed and then virtually eliminated by the trapping of infrared radiation as the CO_2 concentration in the atmosphere increases to 350 ppm and above. Carbon dioxide levels now exceed 390 ppm.

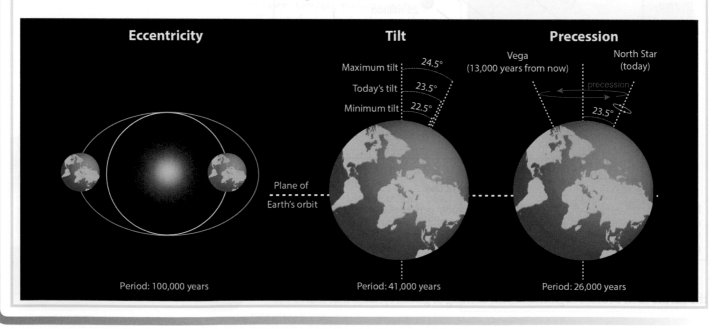

revolution in 1750, the climate system is rapidly transitioning, under fossil-fuel-combustion control (the second leg in Figure 6.5), to the situation characterized by high CO_2 concentrations and significantly warmer temperatures. What is remarkable is that while it required tens of millions of years for the carbonate-silicate weathering reaction to draw the CO_2 concentration down from >1000 ppm to 200 ppm, within this century the CO_2 concentration will reach levels of between 700 and 800 ppm if the international community continues to employ fossil fuel combustion to satisfy its primary energy generation needs. When climate feedbacks are included that involve methane release from clathrates, the greenhouse gas concentrations in the Earth's atmosphere could, in this century, reach levels similar to those not seen for tens of millions of years.

Milankovitch cycles exert dominant control over whether the Earth is in a glacial or interglacial phase at CO_2 concentration *below* 300 ppm. However, amplification of the infrared trapping by increasing water vapor in the atmosphere in response to elevated levels of CO_2 and CH_4 exerts dominant control over Earth's climate at CO_2 concentration levels above ~350 ppm.

It is illustrative to plot the carbon dioxide, nitrous oxide, and methane concentration since the end of the last ice age (approximately 11,000 years ago) as displayed in Figure 6.6. Also shown in Figure 6.6 is the increase in radiative forcing—the increase in infrared trapping resulting from the increase in those greenhouse gas configurations—resulting from those same three molecules. Also of significant importance is the *rate of change* of that forcing resulting from the increase in CO_2, N_2O, and CH_4 since the last ice age. This is displayed in the fourth panel of Figure 6.6.

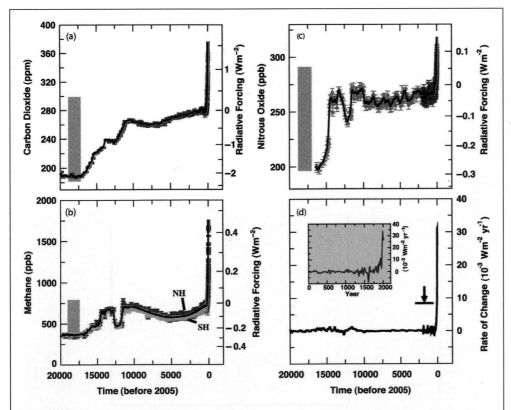

FIGURE 6.6 The tracking of CO_2, N_2O, and CH_4 from the end of the ice age 20,000 years ago until the present. Associated with each of those infrared active gases in the atmosphere is the associated forcing of the climate resulting from the increased trapping of infrared radiation by each molecule. The panel in the lower right displays the *rate of change* of the forcing from those three gases since the last ice age.

Within this context, we can replot the first panel in Figure 6.6 extending the graph to the end of the 21st century, as shown in Figure 6.7, when the CO_2

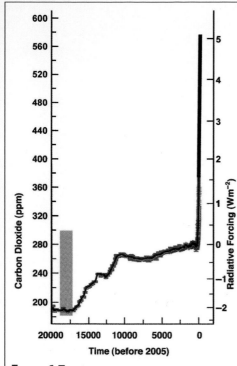

FIGURE 6.7 A plot of carbon dioxide mixing ratio in ppm since the last ice age with the projected CO_2 mixing ratio extended to the end of the century. Also included is the radiative forcing of the climate resulting from that increase in CO_2.

Case Study 6.2
Direct Air Capture of Carbon Dioxide from the Atmosphere

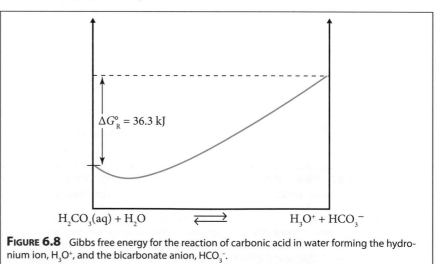

concentrations in the atmosphere will nominally reach 600 ppm with a fossil fuel-based economy, perhaps reaching 800 ppm if there is either accelerated use of fossil fuels or release of methane from clathrates. That increase in forcing will carry the Earth back some 25 million years. Under those conditions (1) the ice structures in both the northern hemisphere (Arctic Ice Cap, Tibetan glacial system, Greenland, Alps, etc.) and the southern hemisphere (Antarctic, Andes, etc.) are unsustainable and (2) the current climate *state* defined by the dynamical structure of the atmosphere and oceans would, with reasonable probability, be fundamentally altered. This altered climate state would be more closely matched to the Eocene of 40 million years ago than to today's climate structure. To review, the climate of the Eocene was characterized by the absence of ice in either hemisphere, sea level 250 feet above present, far warmer and more acidic oceans, and a small temperature difference between the equator and the poles. It is becoming increasingly clear that, in order to return to a stable climate state, we must develop methods to extract currently existing CO_2 from the atmosphere. Case Study 6.2 details one approach.

However, it must also be recognized that, as we described in the Framework to Chapter 3, the methane and carbon dioxide sequestered in the clathrates and permafrost of the high latitude continental regions would, with high probability, be released as the melting of glacial and permafrost continued. This would, in turn, drive both the methane and carbon dioxide concentration to levels unprecedented for 40 million years. It is a fundamental challenge for the physical sciences to decipher just how we would retrace our steps within the next decade or decades to the Eocene 40 million years ago if the world's primary energy generation remains dependent on fossil fuel combustion.

This brings us back to the chemistry of CO_2 and water—specifically the fate of carbonic acid, $H_2CO_3(aq)$, and the acid-base chemistry that proves to be so critical to life systems past, present, and future. Figure 6.2 displays the Gibbs free energy surface for the dissolution of atmospheric carbon dioxide in water. The dissolved CO_2 then reacts with water to form carbonic acid, $H_2CO_3(aq)$. What happens next is the reaction between $H_2CO_3(aq)$ and water, forming the hydronium ion, H_3O^+, and the bicarbonate anion:

$$H_2CO_3(aq) + H_2O \rightleftharpoons H_3O^+ + HCO_3^-$$

This reaction, which is responsible for the *acidic character* of H_2CO_3 by virtue of the formation of the hydronium ion, H_3O^+, has a change in Gibbs free energy of $\Delta G_R = \Delta H_R - T\Delta S_R = 36.3$ kJ/mole at 298 K. The Gibbs free energy surface for this reaction is displayed in Figure 6.8.

FIGURE 6.8 Gibbs free energy for the reaction of carbonic acid in water forming the hydronium ion, H_3O^+, and the bicarbonate anion, HCO_3^-.

While it is common place to write this *acid dissociation reaction* in water, it is important to emphasize the unique role played by water in the extraction, the *theft*, of a hydrogen cation, H^+, from the structure of carbonic acid. In this regard (the ability to effectively extract H^+ from otherwise stable molecular structures in solution), water stands alone from virtually all other solvents. Water is quantitatively different from all other liquids in that:

1. It is small and highly polar because of the large electronegativity difference between oxygen and hydrogen.

2. When interacting with a cation (e.g. H^+ that is electron deficient), water has lone pair electrons to provide a ready concentration of exposed negative charge.

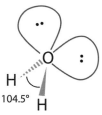

3. Residing at the apex of the V shaped molecular structure, defined by the small hydrogen atoms, is the electronegative center of the oxygen, which is extremely well positioned to closely approach the positive charge of the cation, H^+.

4. For an anion or other *electron rich* entity, water has the partially *positive* hydrogens isolated on the other side of the molecule that create protruding $H^{\delta+}$ "poles" that are very accessible to negative charge centers of other molecules and are predisposed to form noncovalent bonds—termed *hydrogen* bonds.

5. Because of the small size of H_2O and its effective charge separation, water delivers its charge with remarkable efficiency to either cations or anions.

Both because of these unique characteristics as a solvent and because of water's ubiquitous presence in nature, we classify everything else according to its relationship to H_2O:

•   We term those molecules that are drawn to the structure of water as "hydrophilic" (water loving); this includes ionic or polar molecules.

•   We term molecules that are not drawn to water as hydrophobic (water fearing); this includes most nonpolar molecules.

Because of this ability of water to first form carbonic acid in the reaction of CO_2 with H_2O, and then to dissociate the H_2CO_3 to form H_3O^+, the world's oceans have the capacity to extract CO_2 from the atmosphere in such vast amounts that, were the oceans coupled effectively to the atmosphere, CO_2 deposition in the atmosphere from fossil fuel burning could not force changes in the climate through the trapping of infrared radiation, because the CO_2 concentration would not build up in the atmosphere. However, only a tiny fraction of the ocean is in effective contact with the atmosphere—just the upper few hundred meters. As a result, the oceans can *only* take up CO_2 in the quantities released globally by fossil fuel combustion with a time scale of thousands of years. As a result, approximately half the CO_2 released into the atmosphere as a result of fossil fuel combustion remains in the atmosphere for many centuries.

A key technology for shifting permanently from fossil fuel based primary energy generation to non-carbon-based renewable energy is the method of concentrated solar thermal technology. This approach is discussed in Case Study 6.3.

**Case Study 6.3
Concentrated Solar Power**

Chapter Core

Road Map for Chapter 6

We turn now to the development of the central *concepts* of acid-base chemistry by first investigating the acid-base character of pure water and the Gibbs free energy surface associated therewith. It will turn out that the autoionization of water, and the Gibbs free energy for that process, is critically important whenever an acid is added to a base (a.k.a. "neutralization") or a base is added to an acid. We then examine the evolving theory of acid-base reactions before linking acid-base reactivity to the framework of equilibrium chemistry that was the focus of Chapter 6. Then we introduce the terminology of pH, pOH, and pK before examining neutralization reactions. This formulation sets us up to consider important specific cases: the addition of a strong acid to a weak base and of a weak base to a strong acid. We complete the treatment of the core concepts with a discussion of buffer solutions and of titration reactions. The key concepts that we will develop in the Chapter Core are:

Road Map to Core Concepts	
1. Bonding Structure of Water	
2. The Link between the Autoionization of Water and Gibbs Free Energy	
3. Theory of Acid-Base Reactions	
4. Equilibrium and Free Energy in Acidic Solutions	
5. Solutions that are Basic: Manipulation of pOH and pK_a	
6. Reactions of a Strong Acid with a Strong Base	
7. Reaction of a Strong Base with a Weak Acid	
8. Buffer Solutions: The Henderson–Hasselbalch Equation	
9. Titration Reactions	

Introduction

We move our discussion of equilibria to the solution or liquid phase, with a concentration on water as the primary solvent. Nature provides remarkable themes that are laced through many interconnected plots guiding the progression of spontaneous processes that dictate the course of events. These include the inexorable drive toward increasing entropy, the release of free energy in any spontaneous process, the quantum nature of atomic and molecular structure, and the distinction between thermodynamics and kinetics. But the role of proton transfer—the transfer of a hydrogen ion, H^+, from one species to another, often involving water as a solvent—is a theme of broad importance. This is clearly the case for processes central to the cells that sustain life in all known organisms. It is also critical, as we will see, to sustaining living systems in the world's oceans as increasing amounts of CO_2, released by fossil fuel combustion, enter the oceans. The proton, as we will see, organizes the molecular level structure of liquid water, controls life's biochemical pathways, is held in delicate control in the blood of all organisms, constructs and deconstructs polymers, and aids in the synthesis of exquisite calcium containing structures at the intersection of organic and inorganic architectures in living systems as displayed in Figure 6.1.

The Bonding Structure of Water

An investigation of chemical and physical processes that take place in water requires first that we understand the dynamic nature of water itself. This investigation of water and the reactions that are engendered by the unique properties of water are particularly important when it is realized that virtually all processes on Earth take place in this solvent. For a thorough understanding of processes that take place within a solution, we must be careful to distinguish the specific mechanistic role played by the solvent versus the role played by the compound itself.

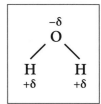

In an aqueous solution, the chemical species added to the mixture is surrounded by water molecules—but those water molecules have a large permanent dipole moment created by the large electronegative character of oxygen relative to that of hydrogen. However, even in pure water, the structure is a dynamic, ever changing combination of monomers, dimers, trimers, etc., because of the hydrogen bonding that constantly attempts to organize the structure of water by constraining the kinetic motion (translation, rotation, vibration) of the individual water molecules. The addition of a proton, H^+, leads to a rearrangement in the structure of water around the proton because the net positive charge of the protons attracts the electronegative (oxygen) end of the surrounding water molecule, as shown in Figure 6.9.

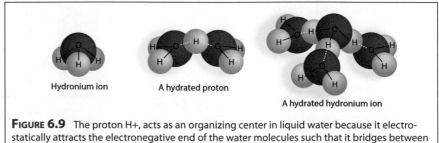

Hydronium ion   A hydrated proton

A hydrated hydronium ion

FIGURE 6.9 The proton H+, acts as an organizing center in liquid water because it electrostatically attracts the electronegative end of the water molecules such that it bridges between the oxygen end of the water molecules forming hydrogen bonded structures.

This rearrangement of the water molecules that determines the local structure around the proton results in the shielding of the protons by the molecules that "hydrate" the proton. The "cage" of water molecules that forms around the proton, which serves as an organizing core, becomes rapidly weaker with radial distance from the H^+ proton.

Another important property of water as a solvent for acid-base chemical reactions is that water can both *donate* and *accept* a proton. In the jargon of acid-base chemistry, this property is referred to as *amphoteric*—the ability to either donate or accept H^+.

This amphoteric character of water leads to a very important result: the self-ionization or *autoionization* of water:

$$H_2O(l) + H_2O(l) \rightleftharpoons H_3O^+(aq) + OH^-(aq)$$

which is an equilibrium reaction with an equilibrium constant, $K_{eq} = 1 \times 10^{-14}$ at 298 K. We know from our study of equilibrium processes that, ignoring the concentration of pure liquids,

$$K_{eq} = [H_3O^+][OH^-] = 10^{-14}$$

So in pure water we have 10^{-7} moles of OH^- and 10^{-7} moles of H_3O^+ in each liter of water. This unit, of moles/liter, is called the *molarity* of the solute. While this equilibrium constant, K_{eq}, is very small and on the face of it, rather unimportant,

---

The link between the autoionization of water and Gibbs free energy

When we analyze the behavior of pure water, an important characteristic that emerges is the self-ionization or autoionization reaction:

$$H_2O + H_2O \rightleftharpoons OH^- + H_3O^+$$

with an equilibrium constant, $K_w = 10^{-14}$ at 298 K. This is a very small equilibrium constant. Expressed as a free energy, we have

$$\Delta G^\circ = -RT \ln K_w = -2.47 \text{ kJ/mole} \ln(10^{-14}) = 79.9 \text{ kJ/mole}$$

This change in Gibbs free energy is rather large, so the free energy diagram for the autoionization reaction has the following form:

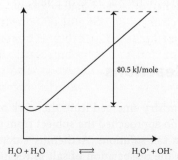

80.5 kJ/mole

$H_2O + H_2O \rightleftharpoons H_3O^+ + OH^-$

The equilibrium is thus very far to the *left* and the reaction proceeds only slightly to the right, with little H_3O^+ + OH^- formed. Of course we could see this immediately because K_w is very small (10^{-14}), so little product (H_3O^+ + OH^-) is formed in the autoionization reaction. However, the magnitude of $K_w (= 10^{-14})$ when viewed from the perspective of autoionization disguises the fact that the reaction in the reverse direction:

$$H_3O^+ + OH^- \rightleftharpoons H_2O + H_2O$$

$$K = 1/K_w = 10^{14}$$

is, in fact, very *large*. The change in Gibbs free energy, DG°, is thus large and negative:

$$\Delta G^\circ = -RT \ln(1/K_w) = -79.9 \text{ kJ/mole}$$

Thus the Gibbs free energy diagram takes the form

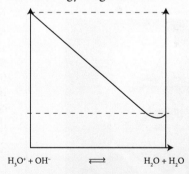

$H_3O^+ + OH^- \rightleftharpoons H_2O + H_2O$

and the reaction proceeds essentially to completion.

But this, we will see, has very significant consequences because any combination of acids or bases added to water first proceeds through "neutralization"—a process that reacts all available H_3O^+ with all available OH^-, forming water as the product, and leaving only the remaining H_3O^+ or OH^-, whichever is in excess.

This is why it is always necessary, when analyzing a reaction of an acid with a base (commonly called a "neutralization reaction"), to account first for this process of OH^-/H_3O^+ "annihilation," resulting in the formation of H_2O.

this is most decidedly not the case. In fact, because we will be investigating the behavior of acids and bases in the "sea" of solvent, this autoionization exerts powerful control over the behavior of acid/base systems. The reason for this is that ΔG for the reverse reaction, $OH^- + H_3O^+ \rightarrow H_2O + H_2O$, is large and negative such that H_3O^+ or OH^- resulting, respectively, from the addition of an acid or a base, results in the formation of water (see sidebar on the previous page).

Because we will be dealing with a very large dynamic range in the concentrations of H_3O^+ and OH^-, we adopt a convention that, rather than constantly writing out the concentrations of H_3O^+ and OH^- in the tedious exponential form, we simplify the nomenclature by introducing logarithmic relations such that

$$pH = -\log[H_3O^+] \qquad\qquad pOH = -\log[OH^-]$$

pH is, therefore, a measure of the relative acidity or basicity of a solution. In fact, in pure water

$$pH = -\log[H_3O^+] = -\log[10^{-7}] = -(-7) = 7$$

If the concentration of H_3O^+ *increases*, for example by adding a substance that dissociates to form a H_3O^+ cation and an anion, the solution becomes more acidic; $[H_3O^+]$ increases and pH decreases. For example, if after the addition of an acid, $[H_3O^+]$ goes from 10^{-7} M to 10^{-6} M, the pH decreases to $pH = -\log[H_3O^+] = -\log[10^{-6}] = 6$. However, because the autoionization of water locks the product

$$K_{eq} = [H_3O^+][OH^-] = 10^{-14}$$

this increase in $[H_3O^+]$ means that $[OH^-]$ must decrease such as to keep the product of $[H_3O^+]$ and $[OH^-]$ constant.

We can then write, because of the equilibrium of water resulting from auto-ionization,

$$
\begin{aligned}
pH + pOH \;\; &= -\log[H_3O^+] + -\log[OH^-] \\
&= -\log[H_3O^+][OH^-] \\
&= pK_{eq} = 14 \\
\text{or} \quad\; pK_w &= 14
\end{aligned}
$$

where henceforth the equilibrium constant for

$$H_2O + H_2O \rightleftharpoons H_3O^+ + OH^-$$

will be designated $K_w = [H_3O^+][OH^-] = 10^{-14}$ at 25°C or 298 K.

Theory of Acid-Base Reactions

The foundations of acid-base chemistry emerged from the work of Svante Arrhenius (1859–1927), Figure 6.10, who approached the subject from the perspective of the conduction of electricity in salt solutions. Arrhenius' studies revealed the nature of ions that resulted from the dissociation of salt into cations (positively charged ions) and anions (negatively charged ions). This foundation of the ionic character of various salts in solution initiated a series of theories regarding acid-base chemistry, summarized in Figure 6.11.

FIGURE 6.10 Svante Arrhenius was one of the most versatile scientists of the modern era. He was the first to recognize that an acid achieves its unique chemical characteristics in water by donating a proton to the solution, and that a base dissociates in water to produce hydroxide ion, OH^-. Arrhenius also deduced the exponential temperature dependence of the rates of chemical reactions and was one of the first scientists to recognize the role of CO_2 in the control of climate.

FIGURE 6.11 The theory of acid-base reactions evolved in steps. The original concept of Arrhenius that acids react in water to produce H⁺ and bases react in water to produce OH⁻ evolved to the Brønsted–Lowry formulation that designated acids as proton donors and bases as proton acceptors. G. N. Lewis generalized the picture of acid-base reactions by designating the acid as an electron pair acceptor and a base as the electron pair donor. All three views of acid-base reactions are used today.

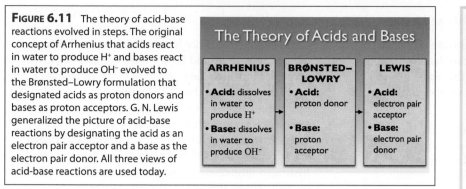

The Theory of Acids and Bases		
ARRHENIUS	**BRØNSTED–LOWRY**	**LEWIS**
• **Acid:** dissolves in water to produce H⁺	• **Acid:** proton donor	• **Acid:** electron pair acceptor
• **Base:** dissolves in water to produce OH⁻	• **Base:** proton acceptor	• **Base:** electron pair donor

According to Arrhenius, as summarized in Figure 6.12, all acids release H⁺ ions in water. For example, when HCl is added to water, HCl dissociates into H⁺ and Cl⁻, and the H⁺ combines with the solvent to form H_3O^+. As we have already seen, H_3O^+ is really a shorthand representation of a structure that involves a number of H_2O molecules organized in a cage around the H⁺ core. Arrhenius identified all bases as those compounds that release hydroxide ions, OH⁻, into water. An example is the addition of ammonia, NH_3, to water. NH_3 reacts with water, extracting a proton to form the ammonium cation, NH_4^+, leaving OH⁻ in solution.

$$NH_3 + H_2O \rightarrow NH_4^+ + OH^-$$

Based on our Gibbs free energy analysis, that excess OH⁻ reacts with the H⁺ until the equilibrium is again established and

$$[H^+] = K_w/[OH^-] = 10^{-14}/[OH^-]$$

All acids release H⁺ ions in water:

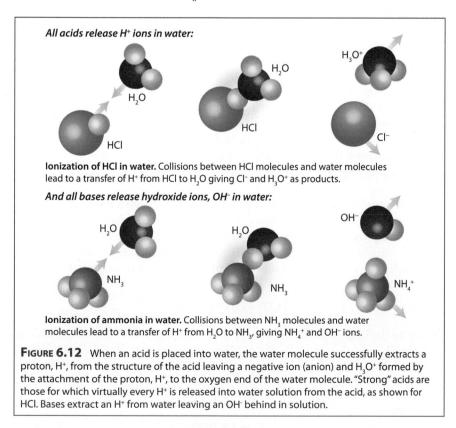

Ionization of HCl in water. Collisions between HCl molecules and water molecules lead to a transfer of H⁺ from HCl to H_2O giving Cl⁻ and H_3O^+ as products.

And all bases release hydroxide ions, OH⁻ in water:

Ionization of ammonia in water. Collisions between NH_3 molecules and water molecules lead to a transfer of H⁺ from H_2O to NH_3, giving NH_4^+ and OH⁻ ions.

FIGURE 6.12 When an acid is placed into water, the water molecule successfully extracts a proton, H⁺, from the structure of the acid leaving a negative ion (anion) and H_3O^+ formed by the attachment of the proton, H⁺, to the oxygen end of the water molecule. "Strong" acids are those for which virtually every H⁺ is released into water solution from the acid, as shown for HCl. Bases extract an H⁺ from water leaving an OH⁻ behind in solution.

Given that Arrhenius approached acid-base chemistry from the broader perspective of electrolytic solutions formed in general from salts, he defined a *salt* as

Representing the autoionization of water in terms of the lone electron pairs in water, hydronium, and hydroxide is illustrative and important to keep in mind. The bonding structure of water (we will treat this in more detail in Chapters 10 and 11) is such that there are two lone pairs with the two H atoms occupying positions separated by 104.5°:

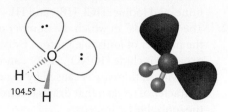

When the autoionization reaction takes place and hydronium, H_3O^+, and hydroxyl, OH⁻, are formed, we express the reaction as

H_3O^+ thus has a single lone pair; OH⁻ has three lone pairs of electrons.

Classifying the relative strengths of acids and bases

While it is necessary to use a table of equilibrium constants when actually working acid/base reaction problems— values that are tabulated in Appendix C of this text—it is instructive to analyze the relative strengths of acids and bases in a systematic way.

Strong Acids. There are two classes of strong acids:
1. The halogen based acids that derive their acidity from the highly electronegative character of chlorine, bromine, and iodine: HCl, HBr, and HI.
2. The "oxoacids" in which the number of O atoms exceeds the number of ionizable protons by two or more. These oxoacids include $HONO_2$, H_2SO_4, and $HClO_4$. The acid behavior in this case is caused by the electronegative character of oxygen that draws electron density from the ionizable hydrogens.

A key point here is that there are *very few* strong acids. We have only listed six here—and that's largely the full inventory! The remaining acids (several hundred) are all weak acids.

Weak Acids. So there are a far greater number of weak acids than strong acids. We identify here four types of acids:
1. The halogen based hydrogen containing diatomic that is not a strong acid: HF. Hydrofluoric acid is not a strong acid because the HF bond is so strong that, even though fluorine is *very* electronegative, it does not easily dissociate in water.
2. Acids in which H is not bonded to O or to a halogen— for example, HCN and H_2S.
3. Oxoacids in which the number of O atoms equals or exceeds by no more than 1 the number of ionizable protons, such as HClO, HNO_2, and H_3PO_4.
4. Carboxylic acids such as acetic acid, CH_3COOH, or benzoic acid, C_6H_5COOH, where the ionizable hydrogen is indicated in red. Note, of course, that it is the hydrogen immediately adjacent to the electronegative oxygen centers that are most easily ionized.

Strong Bases. Water soluble compounds containing the $O^=$ or OH^- ions are the strong bases.
1. M_2O or MOH, where $M \equiv$ Group 1A metals—Li, Na, K, Rb, Cs.
2. MO or $M(OH)_2$, where $M \equiv$ Group 2A metals—Ca, Sr, Ba.

Weak Bases. Many compounds with an electron rich nitrogen atom are weak bases. They are proton acceptors (Brønsted–Lowry bases), not OH^- producers (Arrhenius bases), when dissolved in water. The common structural feature is an N atom with a lone electron pair.
1. Ammonia, NH_3.
2. Amines (RNH_2, R_2NH, R_3N) such as $CH_3CH_2NH_2$, $(CH_3)_2NH$, and $(CH_3H_7)_3N$.

any ionic compound that does not contain either the hydroxide ion, OH^-, or the oxide ion, $O^=$. Ionic compounds that contain OH^- or $O^=$ are bases.

The Brønsted–Lowry theory of acids and bases, introduced in Figure 6.11, generalized the conceptual framework laid down by Arrhenius to consider acid-base reactions as a symmetric reaction of a proton *acceptor* in water and a proton *donor* in water. Figure 6.13 tracks the proton exchange when an acid is added to water.

If we consider the reaction of a proton acceptor in water, the tracking of the proton exchange is shown in Figure 6.14. What links acid-base chemistry inextricably in aqueous solution is that H_3O^+ has a particularly favored reaction partner—specifically the hydroxide ion, OH^-. The two species, H_3O^+ and OH^-, react on virtually every encounter in solution, interlocking the two species, through the reaction

$$H_3O^+ + OH^- \rightarrow 2H_2O \, .$$

Driven by the large negative free energy shown on page 6.11, the equilibrium constant for the reaction is 10^{+14}, which is just the *reciprocal* of the autoionization reaction

$$H_2O + H_2O \rightarrow H_3O^+ + OH^-$$

for which

$$K_w = 1 \times 10^{-14}$$

This is delineated in terms of the Gibbs free energy in the sidebar on page 6.11.

Consider the reaction of a **proton donor** in water:

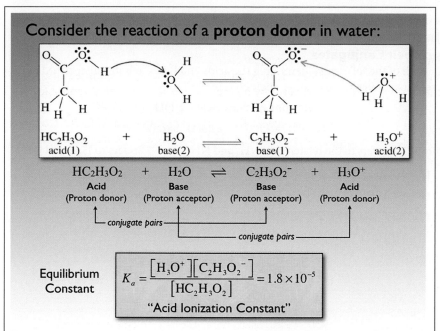

$$HC_2H_3O_2 + H_2O \rightleftharpoons C_2H_3O_2^- + H_3O^+$$

Acid	**Base**	**Base**	**Acid**
(Proton donor)	(Proton acceptor)	(Proton acceptor)	(Proton donor)

conjugate pairs

conjugate pairs

Equilibrium Constant

$$K_a = \frac{[H_3O^+][C_2H_3O_2^-]}{[HC_2H_3O_2]} = 1.8 \times 10^{-5}$$

"Acid Ionization Constant"

FIGURE 6.13 The Brønsted–Lowry theory creates a symmetry between acid behavior and base behavior by identifying the proton as the species exchanged in any acid-base reaction. With this picture the acid (the proton donor) forms the conjugate base following the reaction. The proton acceptor is the base (in this case water) that becomes the conjugate acid after the reaction.

Bronsted - Lowry: Theory of Acids and Bases
Consider the reaction of a **proton acceptor** in water:

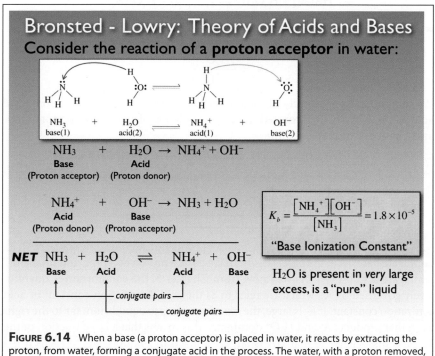

$$NH_3 + H_2O \rightarrow NH_4^+ + OH^-$$

Base	**Acid**
(Proton acceptor)	(Proton donor)

$$NH_4^+ + OH^- \rightarrow NH_3 + H_2O$$

Acid	**Base**
(Proton donor)	(Proton acceptor)

$$K_b = \frac{[NH_4^+][OH^-]}{[NH_3]} = 1.8 \times 10^{-5}$$

"Base Ionization Constant"

NET $NH_3 + H_2O \rightleftharpoons NH_4^+ + OH^-$

Base	**Acid**	**Acid**	**Base**

conjugate pairs

conjugate pairs

H_2O is present in *very large* excess, is a "pure" liquid

FIGURE 6.14 When a base (a proton acceptor) is placed in water, it reacts by extracting the proton, from water, forming a conjugate acid in the process. The water, with a proton removed, becomes the conjugate base after reaction.

Solution

Consider $HClO_2$ in reaction (a). It gives up a proton, H^+, to become ClO_2^-. Therefore, $HClO_2$ is an acid, and ClO_2^- is its conjugate base. Now consider H_2O. It takes the proton from $HClO_2$ and becomes H_3O^+. Thus, H_2O is a base, and H_3O^+ is its conjugate acid. In reaction (b), OCl^- is a base and gains a proton from water. OH^- produced in this reaction is the conjugate base of H_2O. Taken together, reactions (a) and (b) show that H_2O is amphiprotic. Reactions (c) and (d) illustrate the same for $H_2PO_4^-$.

Practice Example A: For each of the following reactions, identify the acids and bases in both the forward and reverse directions.

(a) $HF + H_2O \rightleftarrows F^- + H_3O^+$

(b) $HSO_4^- + NH_3 \rightleftarrows SO_4^{2-} + NH_4^+$

(c) $C_2H_3O_2^- + HCl \rightleftarrows HC_2H_3O_2 + Cl^-$

Practice Example B: Of the following species, one is acidic, one is basic, and one is amphiprotic in their reactions with water: HNO_2, PO_4^{3-}, HCO_3^-. Write the *four* equations needed to represent these facts.

Check Yourself 1—Identifying Brønsted–Lowry Acids and Bases and Their Conjugates

For each of the following, identifying the acids and bases in both the forward and reverse reactions in the manner analogous to

$$NH_3 + H_2O \rightleftarrows NH_4^+ + OH^-$$
$$\text{base(1)}\quad \text{acid(2)} \qquad \text{acid(1)}\quad \text{base(2)}$$

where "1" prefers to the related pair NH_3 and NH_4^+ and "2" refers to H_2O and OH^-

(a) $HClO_2 + H_2O \rightleftarrows ClO_2^- + H_3O^+$

(b) $OCl^- + H_2O \rightleftarrows HOCl + OH^-$

(c) $NH_3 + H_2PO_4^- \rightleftarrows NH_4^+ + HPO_4^{2-}$

(d) $HCl + H_2PO_4^- \rightleftarrows Cl^- + H_3PO_4$

Equilibria and Free Energy in Acidic Solutions

If we represent an acid in a generalized nomenclature as HA, the addition of that acid to water results in the equilibrium reaction

$$HA(aq) + H_2O(l) \rightleftarrows A^-(aq) + H_3O^+(aq)$$

with the equilibrium constant, K_{eq}, given by the "acid dissociation constant"

$$K_a = \frac{[H_3O^+][A^-]}{[HA]}$$

We can immediately identify the acid-conjugate base pair and the base-conjugate acid pair in terms of the Brønsted–Lowry formulation of an acid as a proton donor and a base as a proton acceptor:

$$HA(aq) + H_2O(l) \rightleftarrows A^-(aq) + H_3O^+(aq)$$
$$\text{acid} \qquad \text{base} \qquad \text{conjugate base} \qquad \text{conjugate acid}$$

The equilibrium represents the relative proportion amongst the acid, $[HA]$, its conjugate base, $[A^-]$, the base, $[H_2O]$, that acts as a proton acceptor, and finally the conjugate acid $[H_3O^+]$. As we know from Chapter 5, the equilibrium is quantitatively governed by K_a, which we refer to as the acid dissociation constant or acid ionization constant. If K_a is large, the reaction comes to equilibrium far to the right, such that products A^- and H_3O^+ dominate. This means that a large fraction of the HA has dissociated so that HA is largely ionized in solution, which earns it the designation "strong acid." Notice that there is a potential confusion here because a *strong acid* has a *weak bond* between the anion A^- and the proton H^+ resulting in the successful extraction of H^+ from A^- by the electronegative (oxygen) end of the water solvent.

Conversely, a small K_a indicates a small fraction of the HA acid dissociating to form H_3O^+. This is the case for a weak acid. A schematic representation for this is shown in Figure 6.15.

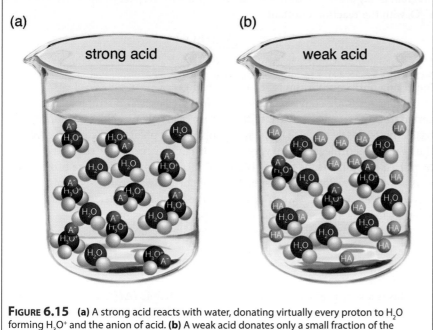

FIGURE 6.15 **(a)** A strong acid reacts with water, donating virtually every proton to H_2O forming H_3O^+ and the anion of acid. **(b)** A weak acid donates only a small fraction of the available protons to water to form H_3O^+.

We generally separate acids into the categories of strong or weak such that if $K_a \geq 1$, the acid is strong; if $K_a \ll 1$, the acid is weak. In practice, however, strong acids (HCl, H_2SO_4, HNO_3, etc.) dissociate to such a degree that the dissociation is virtually complete. As a result the reaction "goes to completion," K_a is very large, we quantitatively ignore the back reaction, and all strong acids are essentially identical.

For weak acids ($K_a \ll 1$), the equilibrium calculation is very important, sometimes complex in multicomponent systems, and, as a reward, both interesting and important to virtually all natural systems. But, as with all systems driven spontaneously, we know that free energy is liberated in going from reactant to product. A strong acid with $K_a \geq 1$, when added to water

$$HA(aq) + H_2O(l) \rightarrow A^-(aq) + H_3O^+(aq)$$

occupies, with its companion reagent $H_2O(l)$, a position of high free energy relative to the products $A^-(aq) + H_3O^+(aq)$. We can sketch the *free energy* surface, as shown in Figure 6.16.

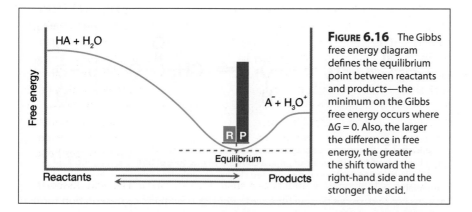

FIGURE 6.16 The Gibbs free energy diagram defines the equilibrium point between reactants and products—the minimum on the Gibbs free energy occurs where $\Delta G = 0$. Also, the larger the difference in free energy, the greater the shift toward the right-hand side and the stronger the acid.

The decrease in free energy drives the reaction (shown in the molecular level diagram in Figure 6.17) "downhill" to the right with respect to the reactants HA + H_2O, with the reaction quotient

$$Q = \frac{[A^-][H_3O^+]}{[HA]}$$

increasing until equilibrium is achieved when $Q = K_{eq} = K_a$.

Strong acids are converted to *weak* conjugate bases and strong bases are converted to weak conjugate acids. In the former case (strong acid → weak conjugate base), this is clearly because in the water solvent, the polar character of water, with the highly electronegative (oxygen) end of H_2O, successfully strips the proton from HA. But this means that A^- is incapable of "winning back" the proton; thus, A^- is a weak base. But, as an inspection of Figure 6.16 reveals, any conjugate species that is referred to as "weak" means, in the language of thermodynamics, that it results in a low free energy relative to its parent species.

Now we are in a position to quantitatively link K_a and $\Delta G°$ via the equation

$$\Delta G° = -RT \ln K_a$$

This is a very useful equation; it links the enthalpic ($\Delta H°$) and entropic ($\Delta S°$) changes with the concentrations [HA], [A⁻], and [H_3O^+] at equilibrium.

We can link this expression for the pH or pK_a of an acid solution by noting that, in general,

$$\ln x = 2.303 \log_{10} x$$

So we can write

$$\Delta G° = -RT \ln K_a = -2.303\, RT \log K_a$$

Thus,

$$pK_a = -\log K_a = \Delta G°/2.303\, RT$$

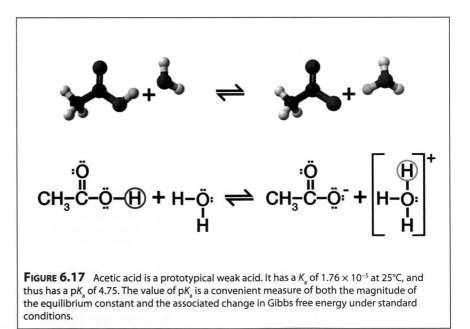

FIGURE 6.17 Acetic acid is a prototypical weak acid. It has a K_a of 1.76×10^{-5} at 25°C, and thus has a pK_a of 4.75. The value of pK_a is a convenient measure of both the magnitude of the equilibrium constant and the associated change in Gibbs free energy under standard conditions.

Solutions That Are Basic: Manipulation of pOH and pK_a

There is an intrinsic bias toward concentrating on the "acidic side," pH < 7, in the development of acid-base chemical processes: first, perhaps because the loss of a proton can be a rather straightforward process and, second, because many applied processes in chemistry are executed in an acidic environment. However, living systems (e.g., human blood) and the global ocean systems are basic in character. So we investigate systems that are characterized by *proton acceptance* (the Brønsted–Lowry (B–L) definition of a base) rather than by proton donation (B–L definition of an acid).

In the aqueous solution of a base

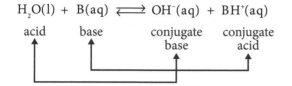

$$H_2O(l) \; + \; B(aq) \; \rightleftharpoons \; OH^-(aq) \; + \; BH^+(aq)$$

acid   base   conjugate base   conjugate acid

H_2O assumes the role of an acid because it donates (loses!) a proton to the base, B(aq), to form the conjugate base OH^-. The equilibrium constant for this process is termed the *base dissociation constant* or *base ionization* constant, K_b, and, as for all equilibrium constants is written

$$K_b = \frac{[BH^+][OH^-]}{[B(aq)]}$$

But the treatment of basic systems follows the same governing protocol as that for acids. Let's review them:

- Water can both donate and accept protons, but in an aqueous solution the *product* of the hydronium ion, H_3O^+, and the hydroxide ion, OH^-, is invariant under any condition imposed by the addition of an acid or a base to the solution

$$[H_3O^+][OH^-] = 1 \times 10^{-14} = K_w$$

which is, as we noted above, referred to as the autoionization constant for water, K_w. Thus the details of how a particular combination of acids or bases is added to a solution is irrelevant to the calculation of the $[H_3O^+][OH^-]$ product: K_w rules, and it is always 1×10^{-14} at 25°C.

We diagram this condition in Figure 6.18 to emphasize the point:

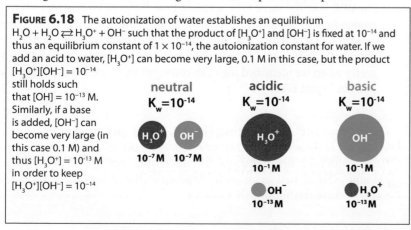

FIGURE 6.18 The autoionization of water establishes an equilibrium $H_2O + H_2O \rightleftharpoons H_3O^+ + OH^-$ such that the product of $[H_3O^+]$ and $[OH^-]$ is fixed at 10^{-14} and thus an equilibrium constant of 1×10^{-14}, the autoionization constant for water. If we add an acid to water, $[H_3O^+]$ can become very large, 0.1 M in this case, but the product $[H_3O^+][OH^-] = 10^{-14}$ still holds such that $[OH^-] = 10^{-13}$ M. Similarly, if a base is added, $[OH^-]$ can become very large (in this case 0.1 M) and thus $[H_3O^+] = 10^{-13}$ M in order to keep $[H_3O^+][OH^-] = 10^{-14}$

neutral $K_w = 10^{-14}$

acidic $K_w = 10^{-14}$

basic $K_w = 10^{-14}$

H_3O^+ OH^-

10^{-7} M 10^{-7} M

H_3O^+

10^{-1} M

OH^-

10^{-1} M

OH^-

10^{-13} M

H_3O^+

10^{-13} M

- A base, dissolved in water, extracts a proton from water, generating a hydroxide ion, OH^-; a *strong base* is so named because it has a more pronounced tendency to accept (steal!) H^+ from water to form BH^+ thereby splitting H_2O into OH^- and BH^+.
- Just as is the case for acids, p(anything) $= -\log_{10}$(anything) and so $-\log_{10}[OH^-] = pOH$ and from

$$K_b = \frac{[OH^-][BH^+]}{[B]} \quad \text{we have} \quad pK_b = -\log_{10}K_b$$

- With increasing strength, the concentration of OH^- increases and, as a result, K_b increases for increasing strength of the base.
- As is clear when conjugate acid-base pairs are considered, a *strong base* has a *weak conjugate acid* and a *weak base* has a *strong conjugate acid*.

Finally, we can tie together the acid dissociation constant, K_a, for the acid dissociation reaction

$$HA + H_2O \rightarrow H_3O^+ + A^- \qquad K_a = \frac{[H_3O^+][A^-]}{[HA]}$$

with the base dissociation constant K_b for the base dissociation reaction to yield an important net reaction:

$$H_2O + A^- \rightarrow OH^- + HA \qquad K_b = \frac{[OH^-][HA]}{[A^-]}$$

$$\overline{\text{Net: } H_2O + H_2O \rightarrow H_3O^+ + OH^- \qquad K_w = [H_3O^+][OH^-]}$$

We have simply used our rule that the equilibrium constant for addition of two reactions to form a net overall reaction is the product of the two equilibrium constants for each of the reactions involved.

$$K_a K_b = \frac{[H_3O^+][A^-]}{[HA]} \frac{[OH^-][HA]}{[A^-]} = [OH^-][H_3O^+] = K_w$$

This, in practice, is a powerful relationship. Just as we always search for constants in dynamic processes in order to simplify complex systems, so too can we quantitatively rationalize the relative strengths of acid-conjugate base, and base-conjugate acid pairs, by recognizing that while K_a and K_b range over many orders of magnitude for various acid/base systems, K_w is invariant. Thus,

$$K_a = K_w/K_b \quad \text{and} \quad K_b = K_w/K_a$$

means that a strong acid (large K_a) is paired with a weak conjugate base (small K_b). In the same vein, a strong base (large K_b) is paired with a weak conjugate acid (small K_a). It is the intrinsic symmetry of proton donation and proton acceptance that simplifies the way we deal with the potentially complex nature of acid-base processes. We have already explored this symmetry when we identified the acid-conjugate base and base-conjugate acid pairs in Figure 6.17.

Check Yourself 2—Relating [H$_3$O$^+$], [OH$^-$], pH, and pOH

When we analyze solutions that are basic, it is essential that we develop a facility with both the mathematical relationships between $[H_3O^+]$, $[OH^-]$, pH and pOH as well as an understanding of the chemical reactions that underlie these relationships.

Problem:

Suppose we carry out a laboratory experiment on two samples: (1) a sample of rainwater and (2) a sample of ammonia. The experimental results:

(1) Rainwater   pH= 4.35
(2) Ammonia   pH = 11.28

Calculate $[H_3O^+]$ in rainwater and $[OH^-]$ in the ammonia.

Solution:

We can calculate $[H_3O^+]$ for rainwater from the pH equation

$$pH = -\log_{10}[H_3O^+] = 4.35$$

Thus
$$\log_{10}[H_3O^+] = -4.35$$
$$[H_3O^+] = 10^{-4.35}$$
$$= 4.5 \times 10^{-5} \text{ M}$$

To calculate $[OH^-]$ in the ammonia solution we employ the equation

$$pOH = -\log_{10}[OH^-]$$

But $pOH = 14.00 - pH = 14.00 - 11.28 = 2.72$

Then from $pOH = -\log_{10}[OH^-]$ we have

$$pOH = -\log_{10}[OH^-] = 2.72 \text{ and}$$
$$\text{so } [OH^-] = 10^{-2.72} = 1.9 \times 10^{-3} \text{ M}$$

Practice:

It is instructive to consider the pH of some of the food and drinks we routinely use. For example:

$(pH)_{\text{Coca-Cola}} = 2.53$ $(pH)_{\text{sugar}} = 5.0$ to 6.2

$(pH)_{\text{coffee}} = 4.5$ $(pH)_{\text{meat}} = 5.2$ to 7.0

$(pH)_{\text{yogurt}} = 4.4$ $(pH)_{\text{egg yolk}} = 6.0$

Calculate the $[H_3O^+]$ and $[OH^-]$ for yogurt.

Neutralization Reactions: The Addition of an Acid to a Base

Three overbearing themes have now emerged in our treatment of acid-base reactions:

1. Acid-base reactions release free energy as they progress from their initial condition to their state of equilibrium.

2. Analysis of the acid-base pairs that emerges naturally from the symmetry of proton donation (acid) matched with proton acceptance (base) in combination with the fact that the product of the acid dissociation constant, K_a, for the *acid* and the base dissociation constant for the *conjugate base* is equal to the autoionization constant for water

$$K_a K_b = K_w$$

means that strong acid → weak conjugate base, strong base → weak conjugate acid, weak acid → strong conjugate base and weak base → strong conjugate acid.

3. When an acid and a base square off in (are added to) an aqueous solution, there is an overwhelming tendency to form water, driven by the release of free energy! This important conclusion emerges directly from the fact that for

$$H_3O^+ + OH^- \rightarrow H_2O + H_2O \qquad K_{eq} = \frac{1}{K_w} = 1 \times 10^{14}$$

This is a *very large equilibrium constant* and because at equilibrium

$$\Delta G° = -RT \ln K_{eq} \approx -80.5 \text{ kJ/mole}$$

it follows that $\Delta G°$ is *large* and *negative* and thus the reaction is driven spontaneously to completion.

The analysis of solutions involving the addition of an acid to a base are, therefore, termed *neutralization reactions* because H_3O^+ and OH^- seek out one another and singlemindedly annihilate each other in the reaction

$$H_3O^+ + OH^- \rightarrow H_2O + H_2O$$

The technique for handling such neutralization reactions is to first analyze the relative strengths of the acids and bases involved. We define, in the process, four categories of such reactions:

1. when a strong acid is added to a strong base;

2. when a strong acid is added to a weak base;

3. when a strong base is added to a weak acid; and, finally

4. when a weak base is added to a weak acid.

Check Yourself 3—Calculating Ion Concentrations in an Aqueous Solution of a Strong Acid

Calculate $[H_3O^+]$, $[Cl^-]$, and $[OH^-]$ in 0.015 M HCl(aq).

Solution:

We can assume that HCl is completely ionized and is the sole source of H_3O^+ in solution. Therefore,

$$[H_3O^+] = 0.015 \text{ M}$$

Furthermore, because one Cl^- ion is produced for every H_3O^+ ion,

$$[Cl^-] = [H_3O^+] = 0.015 \text{ M}$$

To calculate $[OH^-]$, we must use the following fact.

1. All the OH^- is derived from the self-ionization of water.

2. $[OH^-]$ and $[H_3O^+]$ must have values consistent with K_w for water.

$$K_w = \left[H_3O^+\right]\left[OH^-\right] = 1.0 \times 10^{-14}$$

$$(0.015)\left[OH^-\right] = 1.0 \times 10^{-14}$$

$$\left[OH^-\right] = \frac{1.0 \times 10^{-14}}{1.5 \times 10^{-2}} = 0.67 \times 10^{-12} = 6.7 \times 10^{-13} \text{ M}$$

Practice Example A: A 0.0025 M solution of HI(aq) has $[H_3O^+] = 0.0025$ M. Calculate the $[I^-]$, $[OH^-]$, and pH of the solution.

Practice Example B: If 535 mL of *gaseous* HCl, at 26.5°C and 747 mmHg, is dissolved in enough water to prepare 625 mL of solution, what is the pH of this solution?

Strong Acid Reacting with a Strong Base

Given that a strong acid (such as HCl, H_2SO_4, HNO_3) dissociates completely when added to water:

$$HCl + H_2O \rightarrow H_3O^+ + Cl^-$$

and that a strong base (KOH, NaOH) reacts completely to form OH^- when added to water:

$$KOH \xrightarrow{H_2O} OH^- + K^+$$

the neutralization of a strong acid and a strong base is quite simple to analyze. It is a problem of simple counting (stoichiometry). But the way the reaction is envisioned is important for dealing with more complicated systems. Thus we pursue it graphically in some depth.

Suppose we add equal amounts of HCl and KOH, each in its own beaker, to a beaker containing pure water. The beaker of HCl will contain a molar *concentration* of H_3O^+, $[H_3O^+]$, equal to the molar *concentration* of Cl^-, $[Cl^-]$. The KOH beaker will contain a molar *concentration* of OH^-, $[OH^-]$, equal to that of K^+, $[K^+]$.

It doesn't matter if we have added 3 molecules of HCl and 3 molecules of KOH or 3 moles of each.

When these are added together, *but before they react*, we will have a beaker with the following composition:

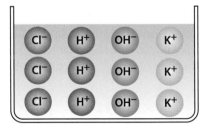

But at the moment the beaker of HCl is mixed with the beaker of KOH the annihilation begins—each H_3O^+ finds an OH^- and the reaction is predestined by the large release of free energy to go to completion:

$$H_3O^+ + OH^- \rightarrow 2H_2O$$

Thus each H_3O^+ annihilates OH^- (or vise versa!) and our head count after reaction is

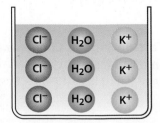

The water so formed disappears into the background solution, leaving no trace of history. The solution is indistinguishable from a solution formed by the addition of KCl salt to pure water.

If we had not added equal amounts of the strong acid (HCl) to the strong base (KOH), but rather, for example, a 0.2 M (mol/liter) solution of HCl to a 0.1 M solution of KOH, the solution to this problem is simple because no equilibrium is involved. The first 0.1 M of HCl goes toward the annihilation of KOH, leaving a 0.1 M HCl solution behind. We can immediately calculate the pH of that mix:

$$[H_3O^+] = 0.1 \text{ M}$$
$$pH = -\log_{10}(0.1 \text{ M}) = -\log(0.1) = -(-1) = 1$$

We can diagram this situation by starting with a 0.1 M KOH solution in a 1 liter beaker:

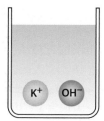

If we add 0.2 moles of HCl to the beaker, we will have, *before* reaction:

But the 0.1 mole of H_3O^+ will find the 0.1 mole of OH^-

$$H_3O^+ + OH^- \rightarrow 2H_2O$$

leaving

We thus have 0.1 M H_3O^+ and the pH = 1. The addition of a strong acid to a strong base does not involve an equilibrium, but rather is a simple process of counting how much H_3O^+ or OH^- is left over *after* neutralization.

Strong Base Reacting with a Weak Acid

The behavior that distinguishes this combination (strong base to weak acid) from the strong base-strong acid combination is that we must recognize that the weak acid component requires an analysis of the *equilibrium behavior* of the weak acid with its conjugate base as depicted in Figure 6.19. Let's consider the weak acid, acetic acid

$$CH_3COOH(aq) + H_2O(l) \rightleftharpoons CH_3COO^-(aq) + H_3O^+ \qquad \textbf{(6.1)}$$
$$K_a = 1.76 \times 10^{-5}$$

Addition of a strong base introduces OH⁻, for example, by adding NaOH because the base splits water to form OH⁻

$$NaOH \xrightarrow{H_2O} Na^+ + OH^-$$

The key point to recognize is that the H_3O^+ from the dissociation of CH_3COOH will react with the OH⁻

$$H_3O^+(aq) + OH^-(aq) \rightleftharpoons 2H_2O(l) \qquad \textbf{(6.2)}$$
$$K = 1/K_w = 1 \times 10^{14}$$

Adding 7.1 and 7.2 yields the net overall reaction:

$$CH_3COOH(aq) + OH^- \rightleftharpoons CH_3COO^- + H_2O$$
$$\text{with } K_{eq} = K_a(1/K_w)$$

which, as Le Chatelier would tell us, shifts the equilibrium to the right, and strongly to the right because $1/K_w = 1 \times 10^{14}$, and $K_a \approx 10^{-5}$ yields $K_{eq} \approx 10^9$.

We can calculate this new equilibrium constant immediately:

$$K = K_a \frac{1}{K_w} = \frac{[CH_3COO^-]}{[CH_3COOH(aq)][OH^-]} = \frac{1.76 \times 10^{-5}}{1 \times 10^{-14}} = 1.8 \times 10^9$$

But this is a remarkable result. It says that, even with a weak acid, and thus a small equilibrium constant, the inexorable drive to form water from H_3O^+ and OH⁻ forces the combined reaction

$$CH_3COOH + OH^- \rightarrow CH_3COO^- + H_2O$$

down the free energy slope to completion. Again we refer to the hand-to-hand combat going on at the molecular level in Figure 6.19.

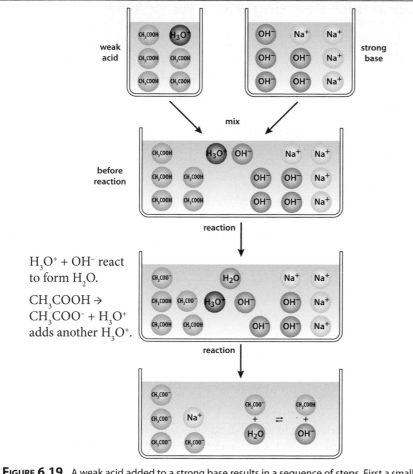

$H_3O^+ + OH^-$ react to form H_2O.

$CH_3COOH \rightarrow CH_3COO^- + H_3O^+$ adds another H_3O^+.

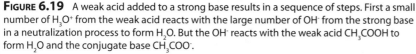

FIGURE 6.19 A weak acid added to a strong base results in a sequence of steps. First a small number of H_3O^+ from the weak acid reacts with the large number of OH⁻ from the strong base in a neutralization process to form H_2O. But the OH⁻ reacts with the weak acid CH_3COOH to form H_2O and the conjugate base CH_3COO^-.

Each H_3O^+ consumed by an OH⁻ to form water is replaced by the dissociation of another CH_3COOH molecule until all that remains is the excess OH⁻, each of the H_3O^+ (and thus CH_3COOH) being eliminated to form H_2O. After each H_3O^+ released by CH_3COOH dissociation has been annihilated by the excess OH⁻ available from the strong base, the net result is to remove all available acetic acid, leaving one OH⁻ removed for each original CH_3COOH in the solution.

But now comes the interesting point. The conjugate base, CH_3COO^- of the *weak* acid, CH_3COOH, is a base. This means that it can compete to extract an H^+ from water to form OH⁻. The CH_3COOH has been eliminated, but the CH_3COO^- that is formed determines the pH of the solution by the equation

$$CH_3COO^- + H_2O \rightarrow OH^- + CH_3COOH$$

We know the resulting solution will be basic, but we need to be quantitative. How do we calculate the pH of such a mixture?

Before we solve *quantitatively* for the pH of this mixture of a base and a weak acid, we consider a unifying theme—the possible circumstances facing a weak acid in aqueous solution:

1. We know that we can just add a weak acid to pure water: This corresponds to HA in equilibrium with a small (and equal) concentration of H_3O^+ and A^-

$$HA + H_2O \rightleftharpoons H_3O^+ + A^-$$

2. We know that we can add a strong base to a weak acid; the OH^- from the strong base will "take out" each H_3O^+ produced from the subsequent dissociation of the HA, leaving the conjugate base A^- with small concentrations of OH^- and HA

$$HA + OH^- \rightleftharpoons A^- + H_2O$$

3. We could add a salt containing the conjugate base, A^-, of the weak acid, leaving a significant concentration of HA (from the weak acid) and A^- from the dissociation of the salt.

The important point is that *all of these cases are fundamentally the same.* They are all equilibrium problems. Let's work them in order.

Case 1:

Consider the calculation of the pH of a weak acid, acetic acid, with 0.1 mole CH_3COOH added to 1 liter of water. We work our standard equilibrium problem with a RICE chart:

Reaction $CH_3COOH + H_2O \rightleftharpoons CH_3COO^- + H_3O^+$ $K_a = 1.76 \times 10^{-5}$

Initial Concentration	0.1 M	0 M	0 M
Change	$-x$	x	x

Equilibrium $K_a = \dfrac{[CH_3COO^-][H_3O^+]}{[CH_3COOH]}$

With $[CH_3COOH]_{initial} = 0.1$ M we have

$$K_a = \frac{x^2}{(0.1 - x)} = 1.76 \times 10^{-5}$$

We can, of course, solve the quadratic equation:

$$x^2 + 1.76 \times 10^{-5}x - 1.76 \times 10^{-6} = 0$$

But we could also try a simplifying assumption that $x << 0.1$ and replace $(0.1 - x)$ with 0.1, yielding a far simpler equation:

$$\frac{x^2}{(0.1 - x)} \approx \frac{x^2}{0.1} = 1.76 \times 10^{-5}$$

or $x^2 = 1.76 \times 10^{-6}$ or $x = [H_3O^+] = 1.33 \times 10^{-3}$

This is indeed much smaller than 0.1, so the approximation is valid within the accuracy of our objective here. We can then calculate the pH

$$pH = -\log_{10}(1.33 \times 10^{-3}) = -(-2.88) = 2.88$$

Thus the pH of this 0.1 M weak acid solution is 2.88, which means that the mix is indeed quite acidic.

Before we relegate this calculation to the bin of excessive simplicity—take note of a couple of points:

- While only 1.3% of the acid is dissociated (we added 0.1 mole CH_3COOH to the liter of water and our equilibrium calculation showed $[H_3O^+] = 1.3 \times 10^{-3}$ M), the balance of $[H_3O^+]$ and $[OH^-]$ has been dramatically shifted. With pure water, $[H_3O^+] = 1 \times 10^{-7}$, so the H_3O^+ *concentration has increased* by a factor of $1.3 \times 10^{-3}/1 \times 10^{-7} = 13,000$!
- Since the autoionization of water requires the product of $[H_3O^+]$ and $[OH^-]$ to be

$$[H_3O^+][OH^-] = 1 \times 10^{-14}$$

it means that

$$[OH^-] = \frac{1 \times 10^{-14}}{1.3 \times 10^{-3}} = 7.6 \times 10^{-12}$$

The 10^{-7} M of OH^- in pure water is swept away by the tsunami of H_3O^+ from the acetic acid, reducing the concentration of OH^- by more than four orders of magnitude!

Case 2:

Now we return to our case of a base added to a weak acid, which leaves us with an excess of OH^- and a considerable concentration of the (strong) conjugate base, CH_3COO^-, of the weak acid, CH_3COOH. But this too is just an equilibrium problem. We write down the *reaction*, state the *initial* condition, define the *change* in concentrations, and write the *equilibrium* expression (a.k.a., RICE diagram):

	Reaction $CH_3COO^- + H_2O \rightleftharpoons$	$OH^- +$	CH_3COOH
Initial Concentration	c	0	0
Change	$-x$	$+x$	$+x$
Equilibrium	$c - x$	x	x

Suppose, as we did for the calculation in Case 1 above, that we have an initial concentration of acetate of 0.1 M. Notice that it doesn't matter how the acetate was formed: We could have added sodium acetate ($NaCH_3COO$) or we could have added a strong base, NaOH—it doesn't matter. But now the equilibrium constant is not K_a, as in Case 1, but

$$K_b = \frac{K_w}{K_a} = \frac{[OH^-][CH_3COOH]}{[CH_3COO^-]} = 5.7 \times 10^{-10}$$

because the K_{eq} equilibrium is $CH_3COO^- + H_2O \rightleftharpoons OH^- + CH_3COOH$, as we derived above for the addition of a strong base to a weak acid.

The first point to note is that, while $K_b = 5.7 \times 10^{-10}$ is a small equilibrium constant, it is *much* greater than $K_w = 1 \times 10^{-14}$; in fact, it is *50,000 times larger*! This increase in the equilibrium constant is a reflection of the strong basic character of the conjugate base (CH_3COO^-) of the weak acid (CH_3COOH). The CH_3COO^- anion successfully splits water to form OH^- far more effectively than H_2O splits itself. Now we can solve this problem just as we have any other equilibrium problem, by inserting equilibrium concentrations from our RICE table

$$K_{eq} = \frac{(x)(x)}{(0.1 - x)} = 5.7 \times 10^{-10}$$

Again, with the approximation $0.1 - x \approx 0.1$, we have

$$x^2 = (5.7 \times 10^{-10})0.1 = 5.7 \times 10^{-11}$$
$$x = [OH^-] = 7.5 \times 10^{-6} \text{ M}$$

Thus, from $[H_3O^+][OH^-] = 1 \times 10^{-14}$ we have

$$[H_3O^+] = \frac{1 \times 10^{-14}}{7.5 \times 10^{-6} \text{ M}} = 1.33 \times 10^{-9} \text{ M}$$
$$pH = -\log_{10}(1.33 \times 10^{-9}) = 8.88$$

This means that the 0.1 M acetate mixture is a weak base with a pH 1.88 units above that of neutral water.

Case 3:

Now we can recognize that we could construct the solution of a weak acid and a base by adding these two ingredients, CH_3COOH and CH_3COO^-, but that the result is identical to simply adding the same molar concentration of the acetate (the conjugate base) by merely adding a salt of that anion, for example, sodium acetate, to pure water. The resulting pH would be the same. It would be impossible to tell, after reaction, how the solution was prepared—by adding a weak base to a strong acid, or by adding a salt of the conjugate base.

Before we examine what happens when we create a "designer solution" by adding both a substantial amount of a weak acid *and* a substantial amount of the conjugate base, CH_3COO^-, by adding the salt of that anion, let's consider how vulnerable Case 1 and Case 2 are to small added amounts of an acid to Case 1 and a base to Case 2. That is, with *only* acid or *only* base, the equilibrium that determines pH is very sensitive to added H_3O^+ or OH^-. This was, of course, just what happened when we added a small amount of acid or base to pure water. The autoionization of water is so weak

$$H_2O + H_2O \rightarrow H_3O^+ + OH^- \quad K_w = 1 \times 10^{-14}$$

that large swings in pH occur with small additions of even a weak acid or base.

To investigate the sensitive nature of a solution containing *only* acid, suppose we take our 0.1 M acetic acid solution, for which we have already determined that $[H_3O^+] = 0.0013$ M and pH = 2.88. Now let's add 0.1 M HCl, a strong acid. What happens? The delicate equilibrium that existed in the acetic acid solution

$$CH_3COOH + H_2O \rightleftharpoons CH_3COO^- + H_3O^+ \quad K_a = 1.7 \times 10^{-5}$$

is obliterated by the 0.1 M $[H_3O^+]$ that results from the complete dissociation of the strong acid

$$HCl + H_2O \rightleftharpoons H_3O^+ + Cl^-$$

The H_3O^+ concentration jumps from 0.0013 to 0.1, a hundredfold increase, and the pH drops from 2.88 to 1.0. The acetic acid is virtually unaffected because only 0.13% of it was dissociated anyway, so while H_3O^+ from HCl dissociation drives the equilibrium of acetic acid to the left (Le Chatelier), the change is inconsequential.

For the case in which we added a base to a weak acid, converting virtually all the acetic acid to its conjugate base (the acetate anion)

$$CH_3COOH(aq) + OH^-(aq) \rightarrow CH_3COO^- + H_2O(l)$$

the equilibrium is pushed almost completely to the right. This was Case 2, above. Now, if we add a base to this mixture, the pH will respond dramatically because there is no acetic acid to neutralize the added base, and a 0.1 M KOH strong base will yield $[OH^-] = 0.1$ M, the pOH will be pOH = $-\log[OH^-]$ = 1, and because $[H_3O^+][OH^-] = 10^{-14}$, we have

$$[H_3O^+] = \frac{10^{-14}}{10^{-1}} = 10^{-13} \quad \text{and} \quad pH = 13$$

But suppose we create a solution comprised of 0.1 M acetic acid and 0.1 M acetate anion, which we can do by adding $NaCH_3COO$ to the acetic acid mixture. Now we have a very different but easily understood situation. If we add acid, H_3O^+, to this mixture, the acetate anion will react with the H_3O^+

$$CH_3COO^- + H_3O^+ \rightarrow CH_3COOH + H_2O$$

If we add a base, the OH^- will react with the acid

$$CH_3COOH + OH^- \rightarrow CH_3COO^- + H_2O$$

From the molecular point of view, we can diagram the battle between the invading H_3O^+ from an added acid in the following way, as shown in Figure 6.20.

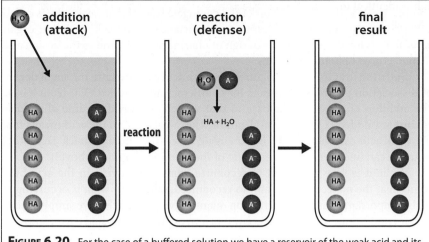

FIGURE 6.20   For the case of a buffered solution we have a reservoir of the weak acid and its conjugate base. When an acid is added, the H_3O^+ reacts with the conjugate base to form water and the weak acid. The pH of the solution is essentially unchanged.

While a weak acid solution has no defense against added H_3O^+, this solution has a readily available means to defend against added H_3O^+, reacting with the conjugate base (the acetate anion)

$$CH_3COO^- + H_3O^+ \rightarrow CH_3COOH + H_2O$$

The hydronium ion is removed, and the "bank" of CH_3OOH is augmented, but while the equilibrium is stressed slightly, the invading H_3O^+ is neutralized.

Suppose now that we attack the solution with a strong base:

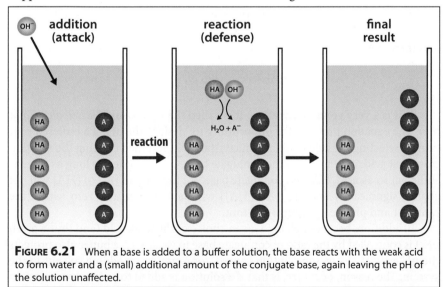

FIGURE 6.21   When a base is added to a buffer solution, the base reacts with the weak acid to form water and a (small) additional amount of the conjugate base, again leaving the pH of the solution unaffected.

In this case, the invading OH^- reacts with the weak acid $CH_3COOH + OH^- \rightarrow CH_3COO^- + H_2O$, removing the hydroxide ion and augmenting the conjugate base. This is described schematically in Figure 6.21. The key point to notice is that the acid (HA) does not react with its conjugate base (A^-) and thus they can coexist in solution.

Buffer Solutions

This technique of building in a defense to limit pH changes by employing a reservoir of a weak acid and a reservoir of the conjugate base of that weak acid, is called a *Buffer Solution* and it is of immense importance to the cells of living organisms, to the global oceans, to the design of pharmaceuticals, and to the design and manufacture of photovoltaic cells.

Buffering uses the "weak" nature of a weak acid as a strength; the small degree of dissociation provides the means for counteracting an added OH$^-$ from a base. The addition of the conjugate base of that weak acid provides a line of defense against any added H$_3$O$^+$ by neutralization. The "weakness" of the acid means that the weak acid can coexist with its conjugate base—they do not attack each other (thus we can line them up on each side of the reaction chambers above!).

While it is very important to recognize what is occurring in the acid-base solution at the molecular level and to recognize qualitatively what is occurring, it is also important to assess the situation quantitatively. To do this, we simply return to our equilibrium treatment. The equilibrium expression for acid dissociation

$$HA + H_2O \rightleftharpoons H_3O^+ + A^-$$

is, of course, simply

$$K_a = \frac{[H_3O^+][A^-]}{[HA]}$$

Taking –log of both sides yields

$$-\log K_a = -\log[H_3O^+] - \log\frac{[A^-]}{[HA]}$$

or

$$pK_a = pH - \log\frac{[A^-]}{[HA]}$$

and therefore

$$pH = pK_a + \log\frac{[A^-]}{[HA]}$$

This is a very versatile equation. It is called the *Henderson–Hasselbalch Equation*. The equation provides the opportunity to "design" the pH of a buffered solution by selecting the acid equilibrium constant, K_a, and then creating the solution from that acid and the salt of the conjugate base. To first order, as long as there is sufficient weak acid, HA, and conjugate base, A$^-$, the ratio of [A$^-$] to [HA] will not change significantly and log ([A$^-$]/[HA]) will be approximately zero within the range ±1 and pH ≈ pK_a within 1 pH unit.

In addition, because pH, for a given choice of the weak acid (and thus a given pK_a) is controlled by the ratio of conjugate base to weak acid, a buffer also protects the pH against changes in dilution caused, for example, by adding solvent to the system. The reason, of course, is that if a solution is added to pure solvent, each of the concentrations, [A$^-$] and [HA], are decreased by the same fraction and pH is unchanged. For example, if the solution is diluted by a factor of 10,

$$\frac{[A^-]/10}{[HA]/10} = \frac{[A^-]}{[HA]}$$

and pH = pK_a + log [A$^-$]/[HA] is unchanged.

We can work *buffer problems* from a number of different angles, but they generally involve the same equation, the Henderson–Hasselbalch Equation, turned various different ways. Consider the most straightforward type of buffer problem—the calculation of pH from a known concentration of a weak acid and the conjugate base. Using our prototype weak acid, CH_3COOH, suppose we create a solution from 1 L of water, 0.05 moles of acetic acid, and 0.05 moles of sodium acetate. What will be the change in pH if we add 0.001 moles of OH^-?

We know that the acid ionization constant, K_a, for the reaction of acetic acid in water is

$$CH_3COOH + H_2O \rightleftharpoons CH_3COO^- + H_3O^+$$

$$K_a = 1.76 \times 10^{-5} \quad pK_a = 4.75$$

Before we add the 0.001 mole of OH^-

$$pH = pK_a + \ln([A^-]/[HA]) = 4.75 + \ln 1 = 4.75 + 0 = 4.75$$

If we add 0.001 mole of OH^-, then OH^- and H_3O^+ will find each other and react to completion such that [HA] *decreases* by 0.001 moles and [A⁻] *increases* by an equal amount such that

$$[HA] = 0.05 - 0.001 = 0.049$$
$$[A^-] = 0.05 + 0.001 = 0.051$$
$$pH = pK_a + \ln([A^-]/[HA]) = 4.75 + \ln(0.051/0.049) = 4.77$$

The conclusion is immediate. The pH changes very little.

The calculation of pH changes for a buffer solution proceeds in two steps. First, we assume that the neutralization reaction proceeds to completion and thereby establishes the new stoichiometric concentrations. Second, these new concentrations are substituted into the Henderson–Hasselbalch Equation. When you are given the concentration of a weak acid and the conjugate base (and you are given or can look up K_a) the Henderson–Hasselbalch Equation provides a very easy method to calculate pH. And, it is very easy to calculate pH for the addition of small amounts of acid or base.

Check Yourself 4—Determining Whether a Solution Is a Buffer Solution

Suppose we create a solution of NH_4Cl–NH_3. Is the solution a buffer solution, and if so, over what pH range would you expect it to behave like a buffer solution?

Calculating pH changes in a Buffer Solution.

Suppose we prepare a solution that is 0.250 M $HC_2H_3O_2$ and 0.560 M $Na_2C_2H_3O_2$ with the volume of 0.300 liters. If we then add (1) 0.0060 mol HCl and (2) 0.0060 mol NaOH to this solution, how will the pH change?

Solution:

The first step is to calculate the pH of the solution before we add the HCl or the NaOH. To do this we use the Henderson–Hasselbalch Equation:

$$pH = pK_a + \log \frac{[C_2H_3O_2^-]}{[HC_2H_3O_2]}$$

$$= 4.74 + \log \frac{0.560}{0.250} = 4.74 + 0.35 = 5.09$$

Solution:
The first step is to determine which species in the solution is capable of neutralizing an acid and which species is capable of neutralizing a base. Given that NH_4Cl will dissociate in water to form NH_4^+ and Cl^-, and that NH_4^+ is an acid, the base neutralization reaction is

$$NH_4^+ + OH^- \rightarrow NH_3 + H_2O$$

and the acid neutralization reaction is

$$NH_3 + H_3O^+ \rightarrow NH_4^+ + H_2O$$

Next, we employ the fact that in all solutions containing NH_4^+ and NH_3 we have

$$NH_3 + H_2O \rightleftharpoons NH_4^+ + OH^-$$

So that

$$K_b = \frac{\left[OH^-\right]\left[NH_4^+\right]}{\left[NH_3\right]}$$

and that $K_b = 1.8 \times 10^{-5}$

If our solution has approximately equal concentrations of NH_4^+ and NH_3, then

$$\frac{\left[NH_4^+\right]\left[OH^-\right]}{\left[NH_3\right]} \approx \left[OH^-\right] = 1.8 \times 10^{-5}$$

$$pOH = -\log_{10}(1.8 \times 10^{-5}) = 4.74$$

and pH = 14.0–pOH = 14.0–4.74 = 9.26

Thus ammonium chloride–ammonia solutions are basic buffer solutions and, using our rule of thumb that buffer solutions operate over a range of two orders of magnitude, the pH range over which this buffer operates is from 8 to 10.

Practice:
Suppose we form a solution by mixing NH_3 and HCl. Can this result in a buffer solution?

Consider first the addition of the 0.006 mol HCl to our buffer solution. The first point to recognize is that we must calculate the number of *moles* of each species involved in the reaction because the addition of the strong acid results in the neutralization (acid/base) reaction going to completion. That is, with the addition of 0.006 moles of H_3O^+, 0.006 moles of CH_3OO^- is converted to C_2H_3OOH in the neutralization reaction

$$C_2H_3OO^- + H_3O^+ \rightarrow C_2H_3OOH + H_2O$$

No equilibrium is involved; the reaction goes to completion. We can tabulate this, as shown in Table 6.1, including recalculating the molar concentrations in moles/liter. Notice that Table 6.1 is not a RICE table; there is no equilibrium involved.

TABLE 6.1

	$C_2H_3O_2^-$	+	H_3O^+	$\rightarrow$	$HC_2H_3O_2$	+	H_2O
Original buffer:	$\underbrace{0.300 \text{ L} \times 0.560 \text{ M}}_{0.168 \text{ mol}}$				$\underbrace{0.300 \text{ L} \times 0.250 \text{ M}}_{0.075 \text{ mol}}$		
Add:			0.0060 mol				
Changes:	−0.0060 mol		−0.0060 mol		+0.0060 mol		
Final buffer amounts:	0.162 mol		(?)		0.0810 mol		
Concentrations:	$\underbrace{0.162 \text{ mol}/0.300 \text{ L}}_{0.540 \text{ M}}$		(?)		$\underbrace{0.0810 \text{ mol}/0.300 \text{ L}}_{0.270 \text{ M}}$		

Thus, our calculation of the pH change for an acid added to a buffer solution involves two steps: (1) the (simple) stoichiometric calculation, and (2) the equilibrium calculation that uses the Henderson–Hasselbalch Equation:

(1) Acid Stoichiometric Calculation:

Let's convert all concentrations to amounts in moles and assume that the neutralization goes to completion. Essentially, this is a limiting reactant calculation, but perhaps simpler than many of those in Chapter 5. In neutralizing the added H_3O^+, 0.0060 mol $C_2H_3O_2^-$ is converted to 0.0060 mol $HC_2H_3O_2$.

(2) Acid Equilibrium Calculation:

We can redetermine the pH using the new equilibrium concentrations:

$$pH = pK_a + \log \frac{[C_2H_3O_2^-]}{[HC_2H_3O_2]}$$

$$= 4.74 + \log \frac{0.540}{0.270} = 4.74 + 0.30 = 5.04$$

This addition of 0.0060 mol HCl lowers the pH from 5.09 to 5.04; this is only a small change in pH.

Calculation of the pH change for the addition of 0.006 moles of the strong base NaOH follows the same two-step sequence:

(1) Base Stoichiometric Calculation:

In neutralizing the added OH^-, 0.0060 mol $HC_2H_3O_2$ is converted to 0.0060 mol $C_2H_3O_2^-$. The calculation of the new stoichiometric concentrations is shown on the last line of Table 6.2.

TABLE 6.2

	$HC_2H_3O_2$	+	OH^-	→	$C_2H_3O_2^-$	+	H_2O
Original buffer:	$\underbrace{0.300\ L \times 0.250\ M}_{0.075\ mol}$				$\underbrace{0.300\ L \times 0.560\ M}_{0.168\ mol}$		
Add:			0.0060 mol				
Changes:	−0.0060 mol		−0.0060 mol		+0.0060 mol		
Final buffer amounts:	0.0690 mol		(?)		0.174 mol		
Concentrations:	$\underbrace{0.0690\ mol/0.300\ L}_{0.230\ M}$		(?)		$\underbrace{0.174\ mol/0.300\ L}_{0.580\ M}$		

(2) Base Equilibrium Calculation:

This is the same type of calculation as for acid, but with slightly different concentrations:

$$pH = 4.74 + \log\frac{0.580}{0.230} = 4.74 + 0.40 = 5.14$$

The addition of 0.0060 mol OH^- *raises* the pH from 5.09 to 5.14—another small change.

Titration Reactions

While the very word conjures up visions of pipettes and beakers, the concept of titration helps to tie together the pH "landscape" shaped by shifting equilibria created by the interplay of H_3O^+, OH^-, weak acids, strong acids, weak bases, and strong bases. What is important for a useful (a.k.a. lasting) understanding of acid-base interplay is to develop an intuition for how acid-base reactions control these pH landscapes. The basic execution of a titration involves the sequential addition of small incremental amounts of an acid to a basic solution, or incremental amounts of a base to an acidic solution. Typically, the pH of the solution is not known, but the pH of the incrementally added solution is—resulting in a determination of the pH of the unknown solution. The "titration curve" is simply a plot of pH on the vertical axis—determined by a pH meter or by pH indicators—against volume of the acid or base added to the mixture on the horizontal axis.

First, we consider the titration of a strong acid with a strong base, and in the process establish the basic form of the titration curve shown in Figure 6.22. We know a great deal from our discussion of neutralization about what will happen in this case. The solution starts out with a pH determined simply by the molar concentration of the strong acid in solution because dissociation is complete. For example, suppose we begin with a 100 mL solution of HCl that is 0.1 M in acid concentration. To this mix we add, in small increments, 0.1 M NaOH and sequentially map out the pH as a function of the volume of NaOH added.

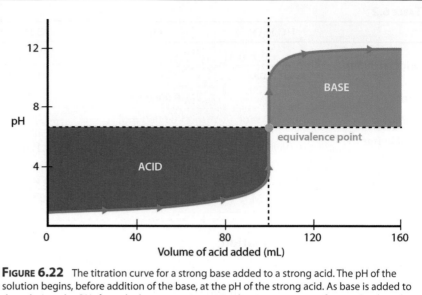

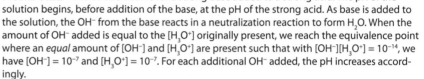

Figure 6.22 The titration curve for a strong base added to a strong acid. The pH of the solution begins, before addition of the base, at the pH of the strong acid. As base is added to the solution, the OH$^-$ from the base reacts in a neutralization reaction to form H$_2$O. When the amount of OH$^-$ added is equal to the [H$_3$O$^+$] originally present, we reach the equivalence point where an *equal* amount of [OH$^-$] and [H$_3$O$^+$] are present such that with [OH$^-$][H$_3$O$^+$] = 10^{-14}, we have [OH$^-$] = 10^{-7} and [H$_3$O$^+$] = 10^{-7}. For each additional OH$^-$ added, the pH increases accordingly.

In this case the titration proceeds from left to right as shown by the arrows, with the neutralization occurring via the added OH$^-$ reacting with the H$_3$O$^+$

$$OH^- + H_3O^+ \rightarrow H_2O + H_2O$$

in a simple lock-step; one-for-one with no subtlety or complications—a simple process of counting, a simple calculation of stoichiometry. The "titration" curve begins prior to addition of the base at pH = –log[H$_3$O$^+$] = –log(0.1 M) = 1 at the left-hand side of the graph and marches through each titration step until the point is approached where the amount of OH$^-$ added begins to approach the amount of H$_3$O$^+$ (0.1 M) originally in the solution. Then the situation changes abruptly at 100 mL of added base because the last H$_3$O$^+$, present in the solution by virtue of the HCl originally placed there, is neutralized by the last OH$^-$ before the solution switches to a condition with OH$^-$ in excess. At the point where the *added* [OH$^-$] = [H$_3$O$^+$] *originally* in place, the equivalence point is reached and the only source of H$_3$O$^+$ and OH$^-$ is the autoionizaton of water

$$H_2O + H_2O \rightarrow H_3O^+ + OH^-$$

As a result, at the equivalence point, pH = –log [H$_3$O$^+$] = 7, just as it is for pure water. The OH$^-$ and H$_3$O$^+$ have sought each other out and annihilated each other in the neutralization reaction. The neutralization is driven by the large, negative free energy in the formation of water from H$_3$O$^+$ + OH$^-$. There is no buffering, no hydrolysis, and no partial dissociation. If we had begun with the KOH 0.1 M solution and proceeded to incrementally add acid instead, as shown in Figure 6.23, the titration curve, which by convention always proceeds from left to right, would have taken the form of Figure 6.23.

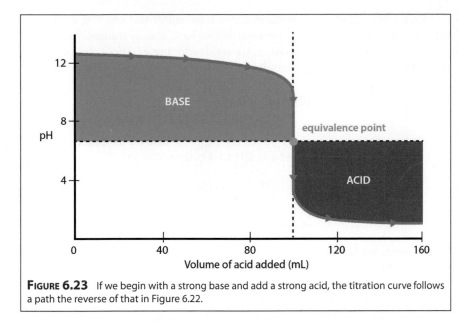

FIGURE 6.23 If we begin with a strong base and add a strong acid, the titration curve follows a path the reverse of that in Figure 6.22.

Titration of a Weak Acid by a Strong Base

Suppose, now, that we examine the titration curve of a weak acid by a strong base. Adopting our canonical acetic acid case, suppose we begin with a 0.1 M solution—we already know the pH of that solution from our previous examples—it is pH = 2.88.

This establishes the initial point, (**a**), on our pH versus volume-of-base-added titration diagram, as shown in Figure 6.24. Now we proceed with the titration, moving incrementally from left to right on the titration plot. Now the "counting" begins, with each OH^- introduced seeking an acetic acid molecule and neutralizing it

$$OH^- + CH_3COOH \rightarrow CH_3COO^- + H_2O$$

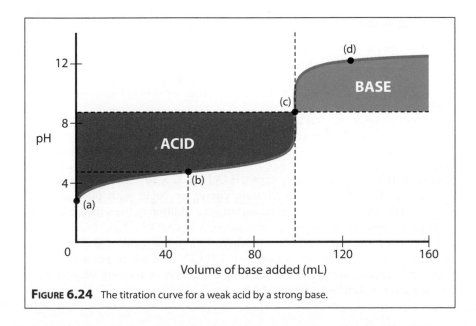

FIGURE 6.24 The titration curve for a weak acid by a strong base.

Each OH⁻ added then produces one acetate anion and one less acetic acid molecule. But this is a pattern, a process, that we have already studied. As we build up the conjugate base concentration, we are developing a buffer solution, with a pH given by the Henderson–Hasselbalch equation:

$$pH = pK_a + \log \frac{[CH_3COO^-]}{[CH_3COOH]}$$

At the point where we have added enough OH⁻ that we have converted half the CH_3COOH to its conjugate base, CH_3COO^-, then

$$[CH_3COO^-] = [CH_3COOH]$$

and

$$pH = pK_a + \log 1 = pK_a$$

at this point, we have reached the center of the "buffer region" of the titration curve, displayed by the vertical dotted line in Figure 6.24, point (b).

As we move incrementally to the right as additional (strong) base is added, the next transition occurs as the point is reached where the added OH⁻ equals the amount of CH_3COOH *originally* in solution before we began the titration. This point is the "equivalence point," point (c) in Figure 6.24, and at that point extremely small increments of added base create very large changes in pH as Figure 6.24 demonstrates. At the equivalence point *all* the CH_3COOH has been converted to CH_3COO^- and any additional OH⁻ will result in hydrolysis, a strong base in water:

$$KOH + H_2O \rightarrow OH^- + H_2O + K^+$$

The pH *at the equivalence point* is determined entirely by the hydrolysis of the conjugate base

$$CH_3COO^-(aq) + H_2O(l) \rightleftharpoons CH_3COOH + OH^-(aq)$$

$$K_b = \frac{[CH_3COOH][OH^-]}{[CH_3COO^-]} = \frac{K_w}{K_a} = \frac{1 \times 10^{-14}}{1.8 \times 10^{-5}} = 5.7 \times 10^{-10}$$

Given that $[CH_3COO^-] = 0.1$ M because all the CH_3COOH has, at the equivalence point, been converted to the conjugate base, and from the hydrolysis reaction we know one OH⁻ is produced for each CH_3COOH present at equilibrium, we have

$$\frac{[OH^-][CH_3COOH]}{[CH_3COO^-]} = \frac{[OH^-]^2}{[CH_3COO^-]} = K_b$$

Then

$$[OH^-] = \sqrt{(K_b)[CH_3COO^-]} = \sqrt{(5.7 \times 10^{-10})\, 0.1\ M}$$

$$= 7.5 \times 10^{-6}$$

Because $[OH^-][H_3O^+] = 10^{-14}$, we know $[H_3O^+] = 1.33 \times 10^{-9}$. Therefore, pH = 8.9.

So, marching through the titration curve, we begin with pure acid (all CH_3COOH); develop the buffer solution with the addition of the strong base until we reach the middle of the buffer zone, where $[CH_3COO^-] = \frac{1}{2}[CH_3COOH]_{initial}$; and then reach the equivalence point, where the original weak acid, CH_3COOH, has been converted to the conjugate base, CH_3COO^-, at which point each additional OH⁻ added corresponds to the simple addition of a strong base to water. This sequence is sketched in Figure 6.25.

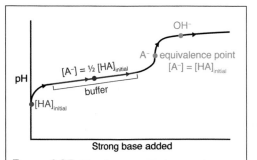

FIGURE 6.25 Titration curve tracing out the progression from a solution of a weak acid [CH_3COOH] with a strong base. The point at which the conjugate base [CH_3COO^-] concentration equals one-half the original acid concentration is shown as well as the equivalence point where [CH_3COO^-] = [CH_3COOH]ₙᵢₜ.

Titration of a Weak Base by a Strong Acid

What about the titration of a weak base with a strong acid? We know, first, that the initial point on our titration curve will be at a pH corresponding to the equilibrium constant of the base at the initial molar concentration—let's assume we have formed a 0.1 M solution of acetate anion, CH_3OO^-. We know

$$K_b = K_w/K_a = \frac{1.0 \times 10^{-14}}{1.76 \times 10^{-5}} = 5.7 \times 10^{-10}.$$

So

$$CH_3COO^- + H_2O \rightleftharpoons OH^- + CH_3COOH$$

$$K_b = \frac{K_w}{K_a} = \frac{[OH^-][CH_3COOH]}{[CH_3COO^-]} = 5.7 \times 10^{-10}$$

$$[OH^-] = 7.5 \times 10^{-6} \text{ M}$$

and $[OH^-][H_3O^+] = 10^{-14}$

so

$$[H_3O^+] = \frac{1 \times 10^{-14}}{[OH^-]} = \frac{1 \times 10^{-14}}{7.5 \times 10^{-6}} = 1.33 \times 10^{-9}$$

So pH $= -\log [H_3O^+] = 8.88$

But with the incremental addition of a strong acid, the buffer zone is created by the conversion of the base to its conjugate acid

$$CH_3COO^- + H_3O^+ \rightarrow CH_3COOH + H_2O$$

We can calculate the pH of the mixture using the same two-step process we developed for calculating the changes in pH of a buffer solution:

1. the stoichiometric conversion of 1 CH_3OO^- to 1 CH_3OOH for each H_3O^+ added;

2. application of the Henderson–Hasselbalch Equation for each new value of $[CH_3OO^-]$ and $[CH_3COOH]$ along the titration curve.

This process occurs until all the CH_3COO^- has been converted to the conjugate acid at the equivalence point. At that point, the pH is governed by the hydrolysis of CH_3COOH

$$CH_3COOH + H_2O \rightleftharpoons CH_3COO^- + H_3O^+$$

and so

Calculating the pH of a Polyprotic Acid

Problem: Compute the concentration of the bicarbonate and carbonate ions, HCO_3^- and CO_3^{2-}, respectively, in a solution where H_2CO_3 is initially 0.05 M.

Inspection of the acid dissociation constants (the equilibrium constants K_{a1} and K_{a2}) reveals that the first dissociation constant is some four orders of magnitude larger than the second. Thus, while K_{a2} will control the concentration of CO_3^{2-}, the pH of the solution will be dictated by the first dissociation.

We can demonstrate this by direct calculation using the same calculation approach we developed for our standard acid equilibrium problems: the RICE chart:

Reaction	H_2CO_3	+	H_2O	$\rightleftharpoons$	HCO_3^-	+	H_3O^+
Initial	0.05 M				0 M		0 M
Change	$-x$				$+x$		$+x$
Equilibrium	$0.05 - x$				x		x
	$\dfrac{[HCO_3^-][H_3O^+]}{[H_2CO_3]} = \dfrac{x^2}{0.05 - x} = K_{a1} = 4.3 \times 10^{-7}$						
	$x = [H_3O^+] = [HCO_3^-] = 0.00015 \text{ M}$						

Considering the next dissociation, we use the same calculation method, but with $[HCO_3^-] = 0.00015$ M.

Reaction	HCO_3^-	+	H_2O	$\rightleftharpoons$	CO_3^{2-}	+	H_3O^+
Initial	1.5×10^{-4} M				0 M		1.5×10^{-4} M
Change	$-x$				$+x$		$+x$
Equilibrium	$1.5 \times 10^{-4} - x$				x		$1.5 \times 10^{-4} + x$
	$\dfrac{[H_3O^+][CO_3^{2-}]}{[HCO_3^-]} = K_{a2} = 5.6 \times 10^{-11}$						

But here we need to recognize that $K_{a2} \ll K_{a1}$, so the $[H_3O^+]$ concentration is not measurably affected by the second dissociation, so $[H_3O^+] = [HCO_3^-]$, so

$$\frac{[H_3O^+][CO_3^{2-}]}{[HCO_3^-]} \approx [CO_3^{2-}] = K_{a2} = 5.6 \times 10^{-11}$$

Thus, at equilibrium: $[H_2CO_3] = 0.05$ M, $[HCO_3^-] = 0.00015$ M, $[CO_3^=] = 5.6 \times 10^{-11}$ M, and the pH is $-\log_{10} [H_3O^+] = -\log_{10} (0.00015)$

Thus   pH = 3.82

Is the change in [CO₃²⁻] resulting from the addition of CO₂ to the atmosphere compensated primarily by change in pH of the ocean?

To test this hypothesis, suppose we take the initial conditions to be pH = 8.23, $[HCO_3^-] = 1.8 \times 10^{-3}$ M, $[CO_3^{2-}] = 2 \times 10^{-4}$ M, at pH = 8.23 and $T = 293$ K, $[H_3O^+] = 5.9 \times 10^{-9}$ M. Suppose now that we add 2×10^{-5} M H_2CO_3 to the ocean. What happens?

Consideration of Case Study 6.1 demonstrates that our key net reaction

$$H_2CO_3(aq) + CO_3^{2-}(aq) \rightarrow 2HCO_3^-(aq)$$

we note that the stoichiometry tells us that

$[CO_3^{2-}]$ *decreases* by 2×10^{-5} M
$[HCO_3^-]$ *increases* by twice that amount, 4×10^{-5} M

But the equilibrium between $[CO_3^{2-}]$ and $[HCO_3^-]$ is controlled by

$$K_{2a} = \frac{[H_3O^+][CO_3^{2-}]}{[HCO_3^-]}$$

so

$$[H_3O^+] = \frac{K_{2a}[HCO_3^-]}{[CO_3^{2-}]}$$

Using $K_{2a} = 6.59 \times 10^{-10}$ M, corresponding to 293 K, we have

$$[H_3O^+] = \frac{(6.59 \times 10^{-10} \text{ M})(1.84 \times 10^{-3} \text{ M})}{1.80 \times 10^{-4} \text{ M}}$$
$$= 6.74 \times 10^{-9} \text{ M}$$

Thus

$$pH = -\log_{10}(6.74 \times 10^{-9} \text{ M})$$
$$= 8.17$$

Thus, indeed the change in $[CO_3^{2-}]$ resulting from adding $[HCO_3^-]$ to the ocean is compensated for in the equilibrium, not by $[HCO_3^-]$, but by $[H_3O^+]$.

Note also that each H_2CO_3 added is consumed such that $\Delta[H_2CO_3] \approx 0$.

$$\frac{[CH_3COO^-][H_3O^+]}{[CH_3COOH]} = \frac{[H_3O^+]^2}{[CH_3COOH]} = 1.8 \times 10^{-5}$$

$$[H_3O^+]^2 = (1.8 \times 10^{-5})(0.1 \text{ M}) = 1.8 \times 10^{-6}$$

$$[H_3O^+] = 1.34 \times 10^{-3}$$

$$pH = -\log_{10}[H_3O^+] = -\log 1.8 + 3 = 2.87$$

After the equivalence point is reached, the titration of a weak base by a strong acid is identical to the addition of a strong acid to water. One H_3O^+ enters the solution for each molecule of acid added. Tracing this progression on the titration curve is the same process, in principle, that we had for the titration of a weak acid by a strong base. Figure 6.26 traces the case for titration of a weak base by a strong acid.

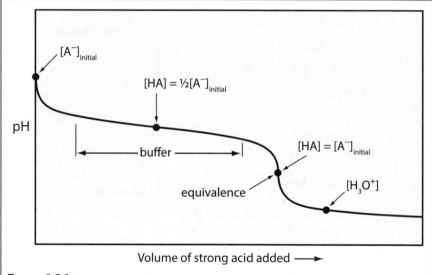

FIGURE 6.26 Titration curve for a strong acid added to a weak base. Shown as before are the initial concentration of the base, the buffer zone, and the equivalence point.

Summary Concepts

1. Bonding Structure of Water

Water molecules have a large permanent dipole moment created by the large electronegative character of oxygen relative to that of hydrogen. However, even in pure water, the structure is a dynamic, ever changing combination of monomers, dimers, trimers, etc., because of the hydrogen bonding that constantly attempts to organize the structure of water by constraining the kinetic motion (translation, rotation, vibration) of the individual water molecules. The addition of a proton, H^+, leads to a rearrangement in the structure of water around the proton because the net positive charge of the protons attracts the electronegative (oxygen) end of the surrounding water molecules, as shown at the right.

Another important property of water as a solvent for acid-base chemical reactions is that water can both *donate* and *accept* a proton. In the jargon of acid-base chemistry, this property is referred to as *amphoteric*—the ability to either donate or accept H^+.

This amphoteric character of water leads to a very important result: the self-ionization or autoionization of water:

$$H_2O(l) + H_2O(l) \rightleftarrows H_3O^+(aq) + OH^-(aq)$$

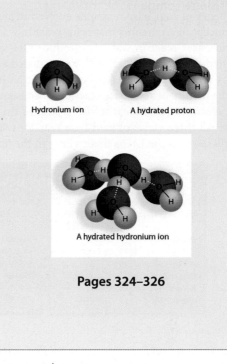

Hydronium ion   A hydrated proton

A hydrated hydronium ion

Pages 324–326

2. The Link between the Autoionization of Water and Gibbs Free Energy

When we analyze the behavior of pure water, an important characteristic that emerges is the self-ionization or autoionization reaction:

$$H_2O + H_2O \rightleftarrows OH^- + H_3O^+$$

with an equilibrium constant, $K_w = 10^{-14}$ at 298 K. This is a very small equilibrium constant. Expressed as a free energy, we have

$$\Delta G° = -RT \ln K_w = -2.47 \text{ kJ/mole} \ln(10^{-14}) = 80.5 \text{ kJ/mole}$$

This change in Gibbs free energy is rather large. The free energy diagram for the autoionization reaction is shown on the right, upper panel.

The equilibrium is thus very far to the *left* and the reaction proceeds only slightly to the right, with little $H_3O^+ + OH^-$ formed. Of course we could see this immediately because K_w is very small (10^{-14}), so little product ($H_3O^+ + OH^-$) is formed in the autoionization reaction. However, the magnitude of $K_w (= 10^{-14})$ when viewed from the perspective of autoionization disguises the fact that the reaction in the reverse direction:

$$H_3O^+ + OH^- \rightleftarrows H_2O + H_2O$$
$$K = 1/K_w = 10^{14}$$

is, in fact, very *large*. The change in Gibbs free energy, $\Delta G°$, is thus large and negative:

$$\Delta G° = -RT \ln(1/K_w) = -80.5 \text{ kJ/mole}$$

Thus the Gibbs free energy diagram takes the form shown at right in the lower panel.

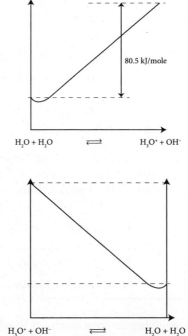

80.5 kJ/mole

$H_2O + H_2O$ $\rightleftarrows$ $H_3O^+ + OH^-$

$H_3O^+ + OH^-$ $\rightleftarrows$ $H_2O + H_2O$

This is why it is always necessary, when analyzing a reaction of an acid with a base (commonly called a "neutralization reaction"), to account first for this process of OH^-/H_3O^+ "annihilation," resulting in the formation of H_2O.

Page 325

Summary Concepts

3. Theory of Acid-Base Reactions

The theory of acid-base reactions evolved in steps. The original concept of Arrhenius that acids react in water to produce H^+ and bases react in water to produce OH^- evolved to the Brønsted–Lowry formulation that designated acids as proton donors and bases as proton acceptors. G. N. Lewis generalized the picture of acid-base reactions by designating the acid as an electron pair acceptor and a base as the electron pair donor. All three views of acid-base reactions are used today. We will develop the Lewis theory later in the text.

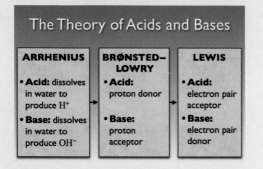

The Brønsted–Lowry theory creates a symmetry between acid behavior and base behavior by identifying the proton as the species exchanged in any acid-base reaction. With this picture the acid (the proton donor) forms the conjugate base following the reaction. The proton acceptor is the base (in this case water) that becomes the conjugate acid after the reaction.

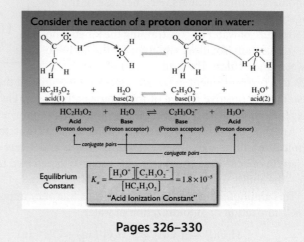

Pages 326–330

4. Equilibrium and Free Energy in Acidic Solutions

If we represent an acid in a generalized nomenclature as HA, the addition of that acid to water results in the equilibrium reaction

$$HA(aq) + H_2O(l) \rightleftharpoons A^-(aq) + H_3O^+(aq)$$

with the equilibrium constant, K_{eq}, given by the "acid dissociation constant"

$$K_a = \frac{[H_3O^+][A^-]}{[HA]}$$

We can immediately identify the acid-conjugate base pair and the base-conjugate acid pair in terms of the Brønsted–Lowry formulation of an acid as a proton donor and a base as a proton acceptor:

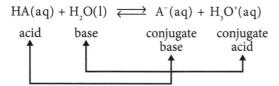

The equilibrium represents the relative proportion amongst the acid, [HA]; its conjugate base, $[A^-]$; the base, $[H_2O]$, which acts as a proton acceptor; and finally the conjugate acid $[H_3O^+]$. As we know from Chapter 5, the equilibrium is quantitatively governed by K_a, which we refer to as the acid dissociation constant or acid ionization constant. If K_a is large, the reaction comes to equilibrium far to the right, such that products A^- and H_3O^+ dominate. This means that a large fraction of the HA has dissociated so that HA is largely ionized in solution, which earns it the designation "strong acid."

The Gibbs free energy diagram defines the equilibrium point between reactants and products—the minimum on the Gibbs free energy occurs where $\Delta G = 0$. Also, the larger the difference in free energy, the greater the shift toward the right-hand side and the stronger the acid.

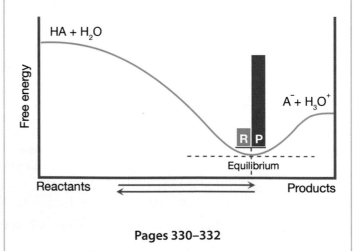

Pages 330–332

Summary Concepts

5. Solutions that are Basic: Manipulation of pOH and pK$_a$

In the aqueous solution of a base

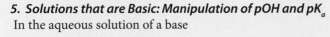
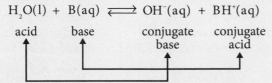

$$H_2O(l) + B(aq) \rightleftharpoons OH^-(aq) + BH^+(aq)$$

acid   base   conjugate   conjugate
base   acid

H_2O assumes the role of an acid because it donates (loses!) a proton to the base, B(aq), to form the conjugate base OH^-. The equilibrium constant for this process is termed the *base dissociation constant* or *base ionization* constant, K_b, and, as for all equilibrium constants is written

$$K_b = \frac{[BH^+][OH^-]}{[B(aq)]}$$

But the treatment of basic systems follows the same governing protocol as that for acids. Let's review them.

Water can both donate and accept protons, but in an aqueous solution the *product* of the hydronium ion, H_3O^+, and the hydroxide ion, OH^-, is invariant under any condition imposed by the addition of an acid or a base to the solution

$$[H_3O^+][OH^-] = 1 \times 10^{-14} = K_w$$

which is, as we noted above, referred to as the autoionization constant for water, K_w. Thus the details of how a particular combination of acids or bases is added to a solution is irrelevant to the calculation of the $[H_3O^+][OH^-]$ product: it rules, and it is always 1×10^{-14} at 25°C.

We diagram this condition in the figure at right to emphasize the point.

The autoionization of water establishes an equilibrium $H_2O + H_2O \rightleftharpoons H_3O^+ + OH^-$ such that the product of $[H_3O^+]$ and $[OH^-]$ is fixed at 10^{-14} and thus an equilibrium constant of 1×10^{-14}, the autoionization constant for water. If we add an acid to water, $[H_3O^+]$ can become very large, 0.1 M in this case, but the product $[H_3O^+][OH^-] = 10^{-14}$ still holds such that $[OH^-] = 10^{-13}$ M. Similarly, if a base is added, $[OH^-]$ can become very large (in this case 0.1 M) and thus $[H_3O^+] = 10^{-13}$ M in order to keep $[H_3O^+][OH^-] = 10^{-14}$.

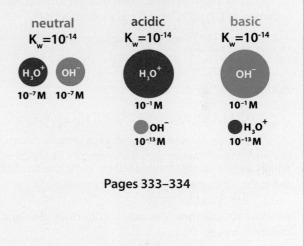

Pages 333–334

6. Reactions of a Strong Acid with a Strong Base

Given that a strong acid (such as HCl, H_2SO_4, HNO_3) dissociates completely when added to water:

$$HCl + H_2O \rightarrow H_3O^+ + Cl^-$$

and that a strong base (KOH, NaOH) reacts completely to form OH^- when added to water:

$$KOH \xrightarrow{H_2O} OH^- + K^+$$

the neutralization of a strong acid and a strong base is remarkably simple. It is a problem of simple counting (stoichiometry). But the way the reaction is envisioned is important for dealing with more complicated systems. Thus we pursue it graphically in some depth.

Suppose we add equal amounts of HCl and KOH, each in its own beaker, to a beaker containing pure water. The beaker of HCl will contain a molar *concentration* of H_3O^+, $[H_3O^+]$, equal to the molar *concentration* of Cl^-, $[Cl^-]$. The KOH beaker will contain a molar *concentration* of OH^-, $[OH^-]$, equal to that of K^+, $[K^+]$.

Addition of Strong Acid to Strong Base

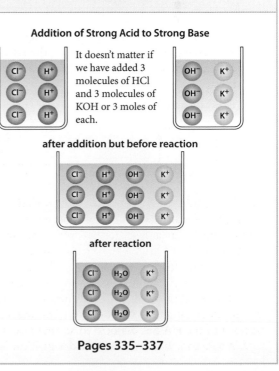

Pages 335–337

Summary Concepts

7. Reaction of a Strong Base with a Weak Acid

The behavior that distinguishes this combination (strong base to weak acid) from the strong base-strong acid combination is that we must recognize that the weak acid component requires an analysis of the equilibrium behavior of the weak acid with its conjugate base as depicted in Figure 6.19. Let's consider the weak acid, acetic acid,

$$CH_3COOH(aq) + H_2O(l) \rightleftarrows CH_3COO^-(aq) + H_3O^+$$

$$K_a = 1.76 \times 10^{-5}$$

Addition of a strong base introduces OH^-, for example, by adding NaOH because the base splits water to form OH^-

$$NaOH \xrightarrow{H_2O} Na^+ + OH^-$$

The key point to recognize is that the H_3O^+ from the dissociation of CH_3COOH will react with the OH^-

$$H_3O^+(aq) + OH^-(aq) \rightleftarrows 2H_2O(l)$$

$$K = 1/K_w = 1 \times 10^{14}$$

Adding yields the net overall reaction:

$$CH_3COOH(aq) + OH^- \rightleftarrows CH_3COO^- + H_2O$$

$$\text{with } K_{eq} = K_a(1/K_w)$$

which, as Le Chatelier would tell us, shifts the equilibrium to the right, and strongly to the right because $1/K_w = 1 \times 10^{14}$, and $K_a \sim 10^{-5}$ yields $K_{eq} \approx 10^9$.

We can calculate this new equilibrium constant immediately:

$$K = K_a \frac{1}{K_w} = \frac{[CH_3COO^-]}{[CH_3COOH(aq)][OH^-]} = \frac{1.76 \times 10^{-5}}{1 \times 10^{-14}} = 1.8 \times 10^9$$

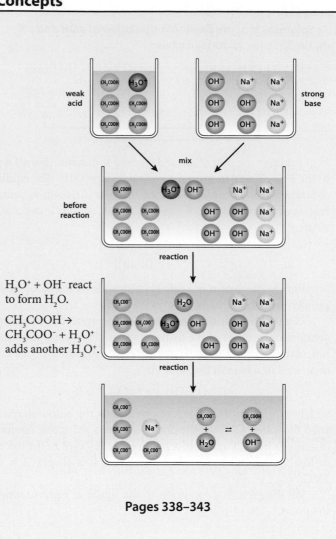

H$_3$O$^+$ + OH$^-$ react to form H$_2$O.

CH$_3$COOH →
CH$_3$COO$^-$ + H$_3$O$^+$
adds another H$_3$O$^+$.

Pages 338–343

6.42

Summary Concepts

8. Buffer Solutions: The Henderson–Hasselbalch Equation

While it is very important to recognize what is occurring in the acid-base solution at the molecular level and to recognize qualitatively what is occurring, it is also important to assess the situation quantitatively. To do this, we simply return to our equilibrium treatment. The equilibrium expression for acid dissociation

$$HA + H_2O \rightleftharpoons H_3O^+ + A^-$$

is, of course, simply

$$K_a = \frac{[H_3O^+][A^-]}{[HA]}$$

Taking –log of both sides yields

$$-\log K_a = -\log[H_3O^+] - \log\frac{[A^-]}{[HA]}$$

or

$$pK_a = pH - \log\frac{[A^-]}{[HA]}$$

$$pH = pK_a + \log\frac{[A^-]}{[HA]}$$

This is a very versatile equation. It is called the *Henderson–Hasselbalch Equation*. The equation provides the opportunity to "design" the pH of a buffered solution by selecting the acid equilibrium constant, K_a, and then creating the solution from that acid and the salt of the conjugate base.

For the case of a buffered solution we have a reservoir of the weak acid and its conjugate base. When an acid is added, the H_3O^+ reacts with the conjugate base to form water and the weak acid. The pH of the solution is essentially unchanged.

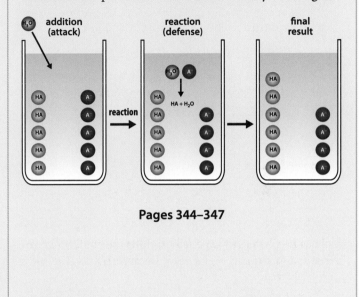

Pages 344–347

9. Titration Reactions

The concept of titration helps to tie together the pH "landscape" shaped by shifting equilibria created by the interplay of H_3O^+, OH^-, weak acids, strong acids, weak bases, and strong bases. What is important for a useful (a.k.a. lasting) understanding of acid-base interplay is to develop an intuition for how acid-base reactions control these pH landscapes. The basic execution of a titration involves the sequential addition of small incremental amounts of an acid to a basic solution, or incremental amounts of a base to an acidic solution. Typically, the pH of the solution is not known, but the pH of the incrementally added solution is—resulting in a determination of the pH of the unknown solution. The "titration curve" is simply a plot of pH on the vertical axis—determined by a pH meter or by pH indicators—against volume of the acid or base added to the mixture on the horizontal axis.

The titration curve for a strong base added to a strong acid. The pH of the solution begins, before addition of the base, at the pH of the strong acid. As base is added to the solution, the OH^- from the base reacts in a neutralization reaction to form H_2O. When an amount of OH^- added is equal to the $[H_3O^+]$ originally present, we reach the equivalence point where an *equal* amount of $[OH^-]$ and $[H_3O^+]$ are present such that with $[OH^-][H_3O^+] = 10^{-14}$, we have $[OH^-] = 10^{-7}$ and $[H_3O^+] = 10^{-7}$. For each additional OH^- added, the pH increases accordingly.

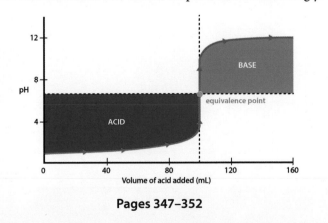

Pages 347–352

CASE STUDY 6.1 Acid Base Reactions, Carbonic Acid Chemistry of the Ocean, and the Control of Climate

Introduction

A watery Sahara? We may not normally hold oceans and deserts in the same light, yet the scarcity of nutrients in the open ocean has inspired numerous comparisons to the bleak inhospitableness of deserts. What infrequent and often ephemeral centers of nourishment exist, such as algal blooms, strongly dictate the distribution of marine creatures. This harsh reality makes the existence of stationary and persistent coral reefs all the more remarkable. Making up a mere 0.1% of the oceans' surface area, coral reefs support nearly a quarter of all known marine species that are closely linked through trophic interdependence.

As of late, much attention has been given to the future of coral reefs in the face of ocean acidifcation, and rightfully so. As we will shortly discuss, yet another side effect of anthropogenically driven carbon dioxide build up in the atmosphere is the increased dissolution of CO_2 into the oceans, and the subsequent formation of carbonic acid. Changing the atmospheric partial pressure of CO_2 places stress on a system that has long been in equilibrium—an equilibrium that has been up until now, conducive to the formation of calcium carbonate.

Through this discussion of carbonate equilibria we will try to gain a better understanding of the threat to marine ecosystems posed by increasing atmospheric CO_2, as well as a whole host of phenomena that are influenced if not controlled by carbonate equilibria including rain and ocean pH, the ability of the ocean to draw down atmospheric CO_2, and climate control through silicate weathering. Let us begin.

A CRS-A   B CRS-B   C CRS-C

375 ppm +1˚C   450-500 ppm +2˚C   > 500 ppm +3˚C

FIGURE CS6.1A Each of these pictures has been taken from different parts of the Great Barrier Reef, and shows the projected progression of ecosystem destruction due to ocean acidification (O. Hoegh, et al., *Science* **318**, no. 1737 [2007]).

In the Lab

If we place a beaker of deionized water on a bench top and let it sit out in the open for an hour, it will equilibrate with the atmosphere. But what does that mean? Let's take $N_2(g)$ as

an example. Gaseous N_2 will begin to dissolve into the water, becoming aqueous N_2

$$N_2(g) \rightarrow N_2(aq)$$

$N_2(g)$ will continue to dissolve into the beaker of water until it has reached its capacity to dissolve $N_2(g)$, in which case $N_2(g)$ will come back out of solution, and return to the air as the equilibrium is established.

$$N_2(g) \leftarrow N_2(aq)$$

Eventually the rate of dissolution will equal the rate of the gas leaving the water, in which case we say the beaker is in equilibrium.

$$N_2(g) \rightleftharpoons N_2(aq)$$

The concentration of $N_2(aq)$ at equilibrium is given to us using Henry's law,

$$[N_2(aq)] = P_{N_2(g)} / KĐ_{H(N_2)}$$

where $P_{N_2(g)}$ is the partial pressure of $N_2(g)$, and $KĐ_{H(N_2)}$ is the Henry's law constant for the process.

We can do the same for CO_2:

$$[CO_2(aq)] = P_{CO_2(g)} / KĐ_{H(CO_2)}$$

However, the solvation of CO_2 is quickly followed by the chemical reaction of CO_2 with H_2O to form carbonic acid.

$$CO_2(aq) + H_2O \rightleftharpoons H_2CO_3(aq)$$

Which is governed by its own equilibrium constant, K_s

$$K_s = \frac{\left[H_2CO_3(aq)\right]}{\left[CO_2(aq)\right]\left[H_2O\right]} = \frac{\left[H_2CO_3(aq)\right]}{\left[CO_2(aq)\right]}$$

where we assume that the solution is dilute enough that the concentration of water is near unity ($[H_2O] = 1$).

As noted in the opening of this chapter, the carbon chemistry of the ocean is initiated by the equilibrium reaction between $CO_2(g)$ in the atmosphere and $CO_2(aq)$ in the ocean. The Gibbs free energy surface for this reaction is displayed in Figure CS6.1b.

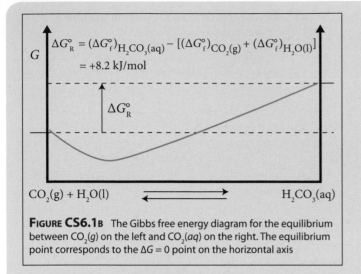

FIGURE CS6.1B The Gibbs free energy diagram for the equilibrium between $CO_2(g)$ on the left and $CO_2(aq)$ on the right. The equilibrium point corresponds to the $\Delta G = 0$ point on the horizontal axis

In many discussions of the carbon chemistry of the oceans, the dissolution of CO_2 and the formation of carbonic acid are combined into a single equilibrium constant, which we will do now.

$$K_H = \frac{K'_{H(CO_2)}}{K_s} = \frac{\dfrac{P_{CO_2(g)}}{[CO_2(aq)]}}{\dfrac{[H_2CO_3]}{[CO_2(aq)]}} = \frac{P_{CO_2(g)}}{[H_2CO_3]}$$

$$K_H = 29.41 \frac{atm}{mol/L}$$

The equilibrium constant, K_H, is known as the Henry's law constant, and represents the observed fact that the concentration of carbonic acid is linearly proportional to the partial pressure of CO_2 over the liquid phase.

Carbonic acid is diprotic, therefore it has two protons that it can donate, and each one has its own equilibrium constant. For the first deprotonation we can write:

$$H_2CO_3 \rightleftharpoons H^+ + HCO_3^-$$

$$K_1 = \frac{[H^+][HCO_3^-]}{[H_2CO_3]} = 4.5 \times 10^{-7} \, mol/L$$

The result is the formation of a hydrogen ion and the bicarbonate ion (HCO_3^-). Likewise for the second deprotonation.

$$HCO_3^- \rightleftharpoons H^+ + CO_3^{2-}$$

$$K_2 = \frac{[H^+][CO_3^{2-}]}{[HCO_3^-]} = 4.7 \times 10^{-11} \, mol/L$$

Which yields another proton and the carbonate ion (CO_3^{2-}).

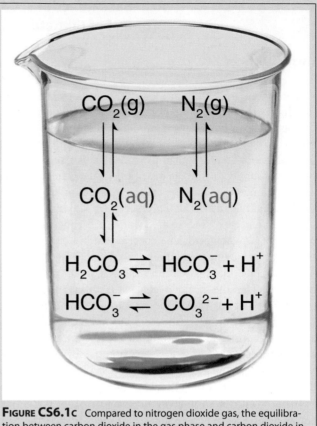

FIGURE CS6.1C Compared to nitrogen dioxide gas, the equilibration between carbon dioxide in the gas phase and carbon dioxide in solution is made more complex by the formation of carbonic acid.

To visualize the relationship between the abundance of carbonic acid, the bicarbonate ion, and the carbonate ion, and pH (Remember: $pH = -\log([H^+])$), it is useful to define a value we will call dissolved inorganic carbon (DIC), which is just the sum of carbonate species:

$$[DIC] = [H_2CO_3] + [HCO_3^-] + [CO_3^{2-}]$$

With some clever substitutions, we can express the ratio of the mole fraction of each of the carbonate species in terms of equilibrium constants and $[H^+]$:

$$\frac{[H_2CO_3]}{[DIC]} = \frac{1}{1 + \dfrac{K_1}{[H^+]} + \dfrac{K_1K_2}{[H^+]^2}}$$

$$\frac{[HCO_3^{-1}]}{[DIC]} = \frac{1}{1 + \dfrac{[H^+]}{K_1} + \dfrac{K_2}{[H^+]}}$$

$$\frac{[CO_3^{2-}]}{[DIC]} = \frac{1}{1 + \dfrac{[H^+]}{K_2} + \dfrac{[H^+]^2}{K_1K_2}}$$

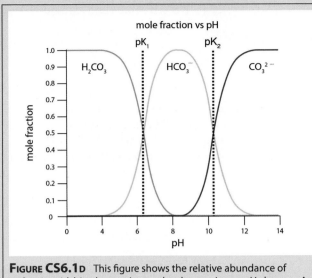

FIGURE CS6.1D This figure shows the relative abundance of carbonic acid, bicarbonate ion, and carbonate ion as pH changes. As we will see later, we can also control pH if we can control the abundance of the different ions. Take note of where the system switches between carbonic acid dominated to bicarbonate dominated, and bicarbonate dominated to carbonate dominated. These transitions occur at $-\log(K_{eq})$.

The plot of mole fraction verses pH is a valuable one to remember. It not only shows the relationship between the abundance of different carbonate species as a function of pH, but also how the pH of a solution can be controlled by controlling the abundance of the different ions. We will return to this when we discuss controls on ocean pH. Before we do that, we should consider the pH of the water that comes from the sky…

Rainwater pH

Since rain precipitates from the atmosphere, it is fair to assume that the water droplets that form are in chemical equilibrium with the atmosphere. Assuming there are no other abundant chemical species that would affect pH, we should be able to calculate the pH of rainwater if we know P_{CO_2}.

We now have 4 equilibrium equations and 5 unknowns if we consider all of the possible dissolved chemical species ($[H_2CO_3]$, $[HCO_3^-]$, $[CO_3^{2-}]$, $[H^+]$, and $[OH^-]$). Since we don't usually know [DIC] a priori, we need one more equation before we can calculate pH of rain water. Since the overall system must be charge neutral, we can write the charge balance equation.

$$[H^+] = [OH^-] + [HCO_3^-] + 2[CO_3^{2-}]$$

If we use the equilibrium expressions in addition to the equilibrium expression for the self ionization of water,

$$K_w = [H^+][OH^-] = 10^{-14}\,\text{mol/L}$$

we can rewrite the charge balance equation solely in terms of P_{CO_2}, $[H^+]$, and equilibrium constants.

$$[H^+] = \frac{K_w}{[H^+]} + \frac{K_1 P_{CO_2}}{K_H [H^+]} + \frac{2K_1 K_2 P_{CO_2}}{K_H [H^+]^2}$$

A simple rearrangement of terms yields a polynomial, the roots of which would be $[H^+]$

$$[H^+]^3 = K_w [H^+] + \frac{K_1 P_{CO_2}}{K_H}[H^+] + \frac{2K_1 K_2 P_{CO_2}}{K_H}$$

If we were feeling lazy, yet enterprising, we would find our favorite polynomial root finder to solve for $[H^+]$. However, in this case we can be clever by using an approximation. Let us start with a couple of justifiable assumptions. Since this is a carbonic acid solution, it is probably reasonable to assume that the concentration of $[OH^-]$ is small enough to be ignored. Likewise, we know from the measured values of K_1 and K_2 that $[HCO_3^-]$ must be much larger than $[CO_3^{2-}]$. This simplifies our charge balance equation nicely, and we are left with

$$[H^+] = \frac{K_1 P_{CO_2}}{K_H [H^+]}$$

which simply is rearranged to yield

$$[H^+] = \sqrt{\frac{K_1 P_{CO_2}}{K_H}}$$

In problem 1 we will calculate the pH of rainwater.

Ocean pH

In the ocean there is an additional equilibrium process that we must concern ourselves with, and that is the equilibrium between calcium carbonate and its ions. Let's write the equilibrium expression, and the equilibrium rate constant called the solubility product.

$$CaCO_3 \rightleftharpoons [Ca^{2+}] + [CO_3^{2-}]$$

$$K = [Ca^{2+}][CO_3^{2-}] = 4.3 \times 10^{-7}\,\text{mol}^2 / \text{L}^2$$

Calcium carbonate is an important constituent in the shells and skeletons of marine organism, like our coral reefs, and the limestone basin walls and floor that these creatures eventually form. This introduces an additional cation which we must take into consideration in the charge balance equation:

$$[H^+] + 2[Ca^{2+}] = [OH^-] + [HCO_3^-] + 2[CO_3^{2-}]$$

In the ocean, the presence of these additional ions will actually shift the carbonic acid equilibrium. Consequently, because dissolved ions in seawater affect carbonate equilibrium, we need to use different equilibrium constants for seawater and bicarbonate, which are

$$K_1 = \frac{[H^+][HCO_3^-]}{[H_2CO_3]} = 9.0 \times 10^{-7}\, mol/L$$

$$K_2 = \frac{[H^+][CO_3^{2-}]}{[HCO_3^-]} = 7.0 \times 10^{-10}\, mol/L$$

Again we can substitute the equilibrium expressions into the charge balance equations to yield an expression for $[H^+]$ in terms of equilibrium constants and P_{CO_2}:

$$[H^+] + \frac{2K_{sp}K_H[H^+]^2}{K_1K_2P_{CO_2}} = \frac{K_w}{[H^+]} + \frac{K_1P_{CO_2}}{K_H[H^+]} + \frac{2K_1K_2P_{CO_2}}{K_H[H^+]^2}$$

We could rearrange this expression to yield a fourth order polynomial in $[H^+]$, or we could be clever again and make our job easier with some educated assumptions. If we make the same assumptions as before about the relative abundances of the different anions, and assume that $[Ca^{2+}]$ is much greater than $[H^+]$, then our charge balance equation simplifies to

$$\frac{2K_{sp}K_H[H^+]^2}{K_1K_2P_{CO_2}} = \frac{K_1P_{CO_2}}{K_H[H^+]}$$

This is handily rearranged to get

$$[H^+] = \sqrt[3]{\frac{K_1^2K_2P_{CO_2}^2}{2K_H^2K_{sp}}}$$

In problem 2 we will calculate the pH of the oceans.

Ocean Acidification

Projections of future carbon emissions put atmospheric CO_2 concentrations of 950 ppm well within possibility by the end of this century. What will this do to the average ocean pH? How much $CaCO_3$ would dissolve in the process? To address these questions, let's begin by calculating the ocean pH with $P_{CO_2} = 950$ ppm using the same expression we did before:

$$[H^+] = \sqrt[3]{\frac{K_1^2K_2P_{CO_2}^2}{2K_H^2K_{sp}}}$$

$$[H^+] = \sqrt[3]{\frac{(9\times10^{-7}\,mol/L)^2(7\times10^{-10}\,mol/L)(950\times10^{-6}\,atm)^2}{2\left(29.41\dfrac{atm}{mol/L}\right)^2(4.3\times10^{-7}\,mol^2/L^2)}}$$

$$[H^+] = 8.8 \times 10^{-9}\, mol/L.$$

This yields a pH of

$$pH = -\log(8.8 \times 10^{-9}) = 8.0$$

This may not seem like much of a change, but remember that pH is on a logarithmic scale. Therefore this change is significant, and would have dramatic consequences for the life in the ocean that has adapted to a less acidic higher pH.

Based upon what we have learned from Le Chatelier, in addition to increased hydrogen ion concentration, we would expect there to be a large amount of calcium carbonate dissolution after increasing P_{CO_2}. To get a feel for how much calcium carbonate (in both the form of limestone and calcium carbonate in living creatures) would dissolve under these circumstances, let's calculate the amount of Ca^{2+} at the present and after we have reached an atmospheric mixing ratio of 950 ppm CO_2 and assume that the difference was the result of calcium carbonate dissolution.

We can obtain the expression for the concentration of calcium ion from our charge balance equation from the ocean.

$$[Ca^{2+}] = \frac{2K_{sp}K_H[H^+]^2}{K_1K_2P_{CO_2}}$$

Using the current P_{CO_2} and the hydrogen ion concentration that we calculated for the modern ocean, we find

$$[Ca^{2+}] = \frac{2\left(4.3\times10^{-7}\,mol^2/L^2\right)\left(29.41\dfrac{atm}{mol/L}\right)\left(4.8\times10^{-9}\,mol/L\right)^2}{(9\times10^{-7}\,mol/L)(7\times10^{-10}\,mol/L)(388\times10^{-6}\,atm)}$$

$$[Ca^{2+}] = 2.4 \times 10^{-3}\, mol/L.$$

And after 950 ppm,

$$[Ca^{2+}] = \frac{2\left(4.3\times10^{-7}\,mol^2/L^2\right)\left(29.41\dfrac{atm}{mol/L}\right)\left(1.0\times10^{-8.0}\,mol/L\right)^2}{(9\times10^{-7}\,mol/L)(7\times10^{-10}\,mol/L)(950\times10^{-6}\,atm)}$$

$$[Ca^{2+}] = 3.3 \times 10^{-3}\, mol/L.$$

The difference is therefore

$$\Delta[Ca^{2+}] = 3.3 \times 10^{-3}\, mol/L - 2.4 \times 10^{-3}\, mol/L$$

$$= 0.9 \times 10^{-3}\, mol/L.$$

FIGURE CS6.1E   This tiny creature is a foraminifer. Foraminifera are just a few of the ocean going creatures with shells made of calcium carbonate.

If for the sake of simplicity we assume that the ocean volume remains constant throughout this process, about 1.3×10^{21} L, we find that the actual mass of calcium carbonate that this would correspond to is quite large,

$$\text{mass}_{CaCO_3} = \frac{(1\,\text{mol CaCO}_3)}{(1\,\text{mol Ca}^{2+})}(9 \times 10^{-4}\,\text{mol/L})(1.3 \times 10^{21}\,\text{L})(100\,\text{g/mol})$$

$$\text{mass}_{CaCO_3} = 1.2 \times 10^{20}\,\text{g CaCO}_3$$

This number is obviously very large. And it does not bode well for the coral reefs we spoke of at the beginning of our discussion, nor the numerous marine animals that secrete calcium carbonate shell. Concern over marine ecosystems therefore is a founded concern for the future, and should not be ignored in our discourse over climate change.

Silicate Weathering

A natural mechanism for the removal of CO_2 from the atmosphere and oceans that is not at the expense of dissolving calcium carbonate does exist. However, it does not act on the time scale needed to enter our discussions of anthropogenic climate change. Over tens of millions of years, the chemical weathering of silicate rock on the land occurs and the process is hastened by increased temperatures, water vapor, and P_{CO_2}

$$CaSiO_3(s) + 2CO_2(aq) + H_2O \rightarrow Ca^{2+}(aq) + 2HCO_2^{-1}(aq) + SiO_2(aq)$$

where $CaSiO_3(s)$ represents a generic silicate rock. The run-off of dissolved minerals helps to replenish the store of Ca^{2+} in the ocean, which increases the rate of calcium carbonate deposition.

$$Ca^{2+}(aq) + 2HCO_2^{-1}(aq) \rightarrow CaCO_3(s) + CO_2(aq) + H_2O$$

The net result is the overall loss of carbon dioxide *from the atmosphere* due to the precipitation of calcium carbonate

$$CaSiO_3(s) + CO_2(aq) \rightarrow CaCO_3(s) + SiO_2(aq)$$

as displayed schematically in Figure CS6.1f.

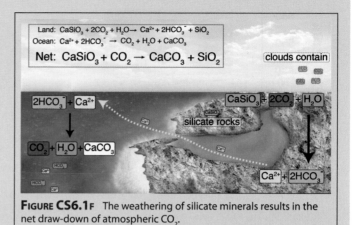

FIGURE CS6.1F   The weathering of silicate minerals results in the net draw-down of atmospheric CO_2.

Even though $CaCO_3$ plays an important role in maintaining the pH of the ocean, it is important to realize that the weathering of carbonate *per se* on land has *no net effect* on the sequestration of CO_2, as shown in Figure CS6.1g.

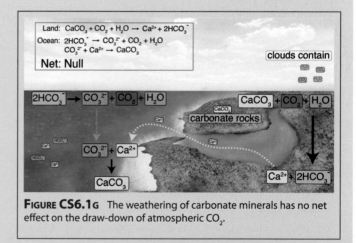

FIGURE CS6.1G   The weathering of carbonate minerals has no net effect on the draw-down of atmospheric CO_2.

Removal of CO_2 from the atmosphere requires the silicate content of crustal rocks on the Earth's surface. Over very long time scales, this is the process that controls atmospheric carbon dioxide removal, and is responsible for the draw down of carbon dioxide following the Eocene (35 to 55 million years ago), as shown in Figure CS6.1h.

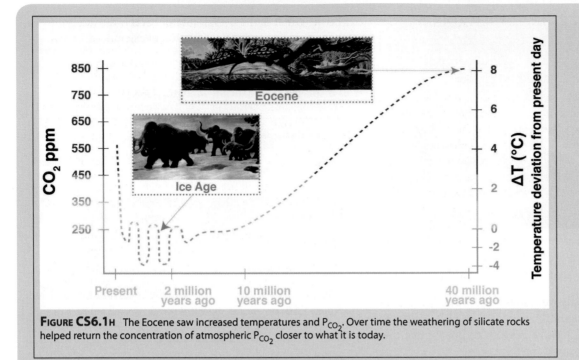

FIGURE CS6.1H   The Eocene saw increased temperatures and P_{CO_2}. Over time the weathering of silicate rocks helped return the concentration of atmospheric P_{CO_2} closer to what it is today.

However, since the rate of weathering is proportional to water content, atmospheric carbon dioxide, and temperature, eventually the weathering process is slowed down. If enough ice forms as a result, the silicate weathering can be shut down, allowing carbon dioxide to build up again by volcanic release, allowing the cycle to start over again.

Problem 1

1. Calculate the pH of rainwater at current CO_2 levels of 420 ppm.

2. Is rainwater acidic or basic?

3. What role does the pH of rainwater play in the weathering of terrestrial rocks?

Problem 2

1. Calculate the pH of the oceans if the atmospheric CO_2 mixing ratio is 270 ppm.

2. Calculate the pH of the oceans if the atmospheric CO_2 mixing ratio is 420 ppm.

3. At the pH level corresponding to 400 ppm, write the *net* reaction for the combination of H_2CO_3, HCO_3^-, CO_3^{2-}, Ca^{2+}, and $CaCO_3$ in sea water.

CASE STUDY 6.2   Direct Air Capture of Carbon Dioxide from the Atmosphere

Introduction

Carbon dioxide added to the atmosphere from the combustion of fossil fuel remains in the atmosphere for time scales of centuries to millennia. Therefore, once the mixing ratio of CO_2 has reached the level wherein irreversible changes to the Earth's climate structure are reached, the only avenue available to return to a stable climate is to extract CO_2 directly from the atmosphere. This process is referred to as Direct Air Capture or DAC. DAC has several important advantages. We review the scale of the issue of DAC for CO_2 removal and a potential method for doing so in this Case Study. The basic approach used in DAC is displayed schematically in Figure CS6.2a. Air high in CO_2 concentration is passed through a system that has a flowing liquid chemical solution that reacts preferentially with CO_2 in the atmosphere reducing its concentration by ~80%. The air reduced in CO_2 reenters the atmosphere on the backside of the air capture system.

Establishing the Need for Direct Air Capture

Capturing CO_2 from the atmosphere is difficult. With two double bonds, CO_2 is an extremely stable molecule. Therefore, it is difficult to break CO_2 bonds and form a new molecule. Further, despite the significance of the 420 ppm atmospheric CO_2 concentration for the forcing of the climate, 400 parts per million means that only 0.04% of a given volume of air is CO_2. Given this degree of dilution, removing any significant amount of CO_2 requires the processing of large volumes of air. Further, a number of thermodynamic barriers exist, which establish relatively steep energy demand for extracting CO_2 from the ambient air, as well as for removing CO_2 from a sorbent (to which atmospheric CO_2 bonds in a DAC reaction). However, despite these challenges, DAC is of key importance for at least two reasons.

First, as noted in the introduction, DAC is the only pos-

sible method to draw down CO_2 levels on a short time scale. Most "solutions" to climate change address reducing future emissions, but there is far less discussion of eliminating CO_2 that is already present in the atmosphere. Therefore, DAC is required to reduce emissions that have accumulated over the past 150 years in order to return to a stable climate state.

Second, what makes DAC important is the fact that this technology is geography independent. Therefore, because atmospheric CO_2 is well mixed, both on local and global scales, DAC technologies can be installed anywhere in the world, and will have an equal effect. The geographic independence of DAC technologies means that CO_2 can be captured directly next to the location where it will be utilized or sequestered. Furthermore, CO_2 extraction can be done at the site of renewable energy sources.

Carbon Dioxide Capture: The Numbers

The first question we must address is the total mass of CO_2 that must be removed from the atmosphere to return to preindustrial levels of CO_2. Currently there are 420 ppm of CO_2 in the atmosphere; in preindustrial times there were 280 ppm.

We first calculate the number of moles of gas (N_2 and O_2) in the atmosphere because 420 ppm means that the mole fraction of CO_2 in air is 420 ppm or 420×10^{-6} or 4×10^{-4}. Thus if we know the number of moles of air molecules there are in the atmosphere we can immediately calculate the number of moles of CO_2.

The mass of the atmosphere is 5.2×10^{18} kg and the atmosphere is 20 percent O_2 and 80 percent N_2. With O_2 at 32 amu and N_2 at 28 amu, the average molecular weight of air is 28.8 amu or $(28.8 \text{ amu})(1.66 \times 10^{-27} \text{ kg/amu}) = 47.8 \times 10^{-27}$ kg/molecule of air $= 4.8 \times 10^{-26}$ kg/molecule. Thus the atmosphere contains

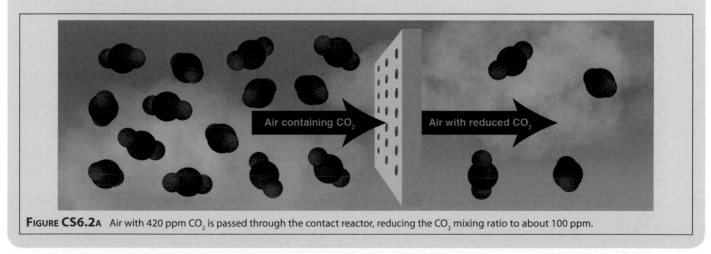

FIGURE CS6.2A Air with 420 ppm CO_2 is passed through the contact reactor, reducing the CO_2 mixing ratio to about 100 ppm.

$$\frac{5.2\times10^{18}\,\text{kg}}{4.8\times10^{-26}\,\text{kg/molecules}}=1.1\times10^{44}\ \text{molecules of air}$$

Converting to moles, we have

$$\frac{1.1\times10^{44}\,\text{molecules}}{6.02\times10^{23}\,\text{molecules/mole}}=1.83\times10^{20}\ \text{moles}$$

If CO_2 *currently* has a mixing ratio of 4×10^{-4}, there are then

$$(4\times10^{-4})(1.83\times10^{20}\ \text{moles})=7.3\times10^{16}\ \text{moles}\ CO_2$$

The number of moles of CO_2 in the preindustrial atmosphere is then, with a mixing ratio of 280 ppm or 2.8×10^{-4}

$$(2.8\times10^{-4})(1.83\times10^{20}\ \text{moles})=5.1\times10^{16}\ \text{moles}\ CO_2$$

The number of moles that must be removed from the atmosphere to return to preindustrial levels is then

$$7.3\times10^{16}\ \text{moles}-5.1\times10^{16}\ \text{moles}=2.2\times10^{16}\ \text{moles}\ CO_2$$

The mass of CO_2 that must be removed can then be calculated. The mass of a single CO_2 molecule is

$$(44\ \text{amu})(1.66\times10^{-27}\ \text{kg})=7.3\times10^{-26}\ \text{kg/molecule}$$

The number of CO_2 molecules that must be removed is

$$(2.2\times10^{16}\ \text{moles}\ CO_2)(6.02\times10^{23}\ \text{molecules/mole})$$
$$=13.2\times10^{39}\ \text{molecules}$$
$$=1.3\times10^{40}\ \text{molecules}$$

Thus the mass of CO_2 that must be removed to return to preindustrial levels is

$$(1.3\times10^{40}\ \text{molecules})(7.3\times10^{-26}\ \text{kg/molecule})$$
$$=9.5\times10^{14}\ \text{kg}$$

It is also of interest to calculate the mass of CO_2 added to the atmosphere in a single year. For example in 2011 the world emitted 32×10^{9} metric tons of CO_2 to the atmosphere. This is $32\times10^{12}\ \text{kg}\ CO_2$ or $3.2\times10^{13}\ \text{kg}\ CO_2$.

Thus the ratio of (1) the mass of CO_2 that must be removed from the atmosphere to reach preindustrial levels to (2) the mass of CO_2 added each year is

$$9.5\times10^{14}\ \text{kg}\ CO_2\Big/3.2\times10^{13}\ \text{kg}\ CO_2=29.7$$

We will approximate this to say that the mass of CO_2 that must be removed to return to preindustrial levels is a factor of 30 more than the amount of CO_2 now added each year. This is an important number to keep in mind.

DAC Basics

The essence of a direct air capture unit is simple. "High carbon" (400 ppm) air must flow through a capture system such that a large fraction of the CO_2 present in the air is extracted, forming a new chemical compound that results in the permanent removal of CO_2 from the atmosphere. While many types of CO_2 trapping are being explored (membranes, physisorption, etc.), the most common method of CO_2 trapping involves passing air over a liquid containing a molecule with which the CO_2 molecule reacts. Once the CO_2 is isolated from the air and bound to a surface or to a different molecule, the CO_2 is either extracted from the substrate or the entire new chemical compound is disposed of, while the "low carbon" (~100 ppm) air is released as exhaust. In practice, almost all companies working on DAC add energy in order to remove the CO_2 from the substrate to form pure CO_2 due to the non-negligible cost of the absorbent. In many cases, the absorbent undergoes a regeneration process, and is returned to the air contactor, where it can absorb more CO_2. Thus, in summary, energy is expended, the concentration of CO_2 in the exhaust air decreases by 75–80%, and a stream of pure (captured) CO_2 remains to be permanently sequestered either under ground or in deposits at the ocean floor.

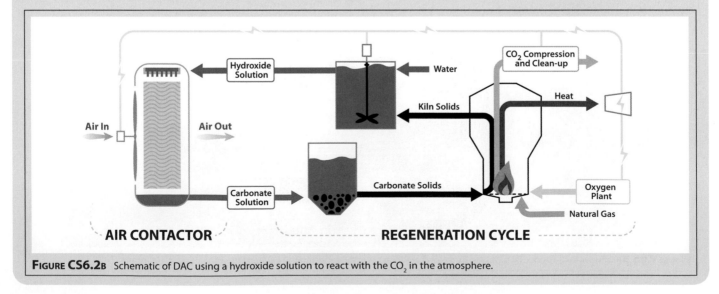

FIGURE CS6.2B Schematic of DAC using a hydroxide solution to react with the CO_2 in the atmosphere.

We will consider here one such example to establish basic parameters for a future world with widespread DAC implementation.

An Example of DAC Technology

The example we will consider here is a "wet-scrubbing" technology that can be used to remove CO_2 from the atmosphere. A diagram of how the process works is shown in Figure CS6.2b.

As seen in Figure CS6.2b, there are two main cycles to this DAC process: CO_2 precipitation in the air contactor, and hydroxide regeneration.

The first step of the CO_2 removal process involves pushing air through the "air-contactor," a high-surface area passage containing an aqueous hydroxide solution that reacts with the CO_2 present in the air. The contactor typically has 10,000 m^2 of surface area allowing for maximum contact of the hydroxide solution with the incoming air. Further, this air contactor surface must be engineered to have optimal air turbulence rates and continued solution refresh rate, in order to obtain constant CO_2 extraction. Moreover, the velocity at which the air travels through the contactor is directly proportional to the amount of CO_2 that can be extracted from the air per second. However, while increased air speed allows more CO_2 to come into contact with the hydroxide solution, the CO_2 must also have time to fully diffuse into the solution before it has passed through the contactor. Therefore, the actual hydroxide composition is important for determining an optimized airflow speed.

The reagent that reacts with CO_2 is NaOH, sodium hydroxide, or KOH, potassium hydroxide. For the purpose of this case study, we will use NaOH as the primary hydroxide used for CO_2 removal.

The overall reaction of a sodium hydroxide with CO_2 is:

$$CO_2(g) + 2\,NaOH(aq) \rightarrow H_2O + Na_2CO_3(aq)$$

The Gibbs free energy of the reaction can be calculated as follows:

$$\Delta G_{rxn} = \Sigma \Delta G°_{f,products} - \Sigma \Delta G°_{f,reactants}$$

In kJ/mole:

$$\Delta G_{rxn} = [-237.14 + (-1051.6)] - [-394.39 + 2(-419.2)]$$
$$= -50.95 \text{ kJ/mole}$$

Therefore, we can see that at STP, the reaction of CO_2 with NaOH(aq) will occur spontaneously.

The second phase of the cycle, after the CO_2 has been captured, involves the release of the CO_2 from the Na_2CO_3 compound and the regeneration of the sodium hydroxide solution. As seen in Figure CS6.2c below, heat is required in order to free the CO_2 from the Na_2CO_3, and both H_2O and Fe_2O_3 must be added in order for the regeneration process to go to completion.

While it is quite difficult to calculate the exact amount of input energy needed to extract 1 mole of CO_2, we will make the plausible estimate here using the fact that experiments have shown that the extraction of 1 ton of CO_2 costs 0.5 tons of CO_2 emission from the natural gas used to fuel the reaction. Using this fact, we can back-calculate the approximate amount of energy needed to remove 1 ton of CO_2.

The MIT Energy Fact Sheet states that 50.3 kg of CO_2 are released per GJ of natural gas used. Therefore:

$$\frac{0.5 \text{ tons } CO_2 \text{ from natural gas}}{1.0 \text{ tons } CO_2 \text{ removed}} \times \frac{1 \text{ GJ natural gas}}{50.3 \text{ kg } CO_2} \times \frac{1000 \text{ kg}}{1 \text{ ton}}$$

$$\approx 10 \text{ GJ/1.0 ton } CO_2 \text{ removed.}$$

This 10 GJ/1 ton of CO_2 is used primarily to power the fans used to ensure a speed of 1.5 m/s airspeed through the air contactor, as well as to heat the Na_2CO_3 in order to release and isolate the CO_2. Later in this case study, we will reuse this parameter to establish the amount of energy necessary to sequester massive amounts of CO_2 from the atmosphere. Moreover, in the future the 10 GJ/t CO_2 will be supplied, not

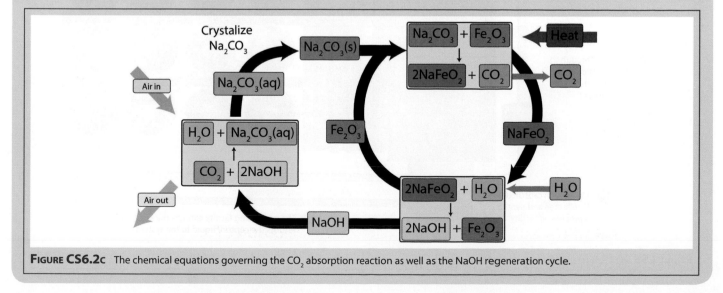

FIGURE CS6.2c The chemical equations governing the CO_2 absorption reaction as well as the NaOH regeneration cycle.

by fossil fuels, but by noncarbon renewables.

In addition to understanding why so much energy is required to remove one ton of CO_2 from the air, it is also important to analyze the volume of air that must be processed in order to extract 1 ton of CO_2. We know that:

- At STP, density of air = 1.275 kg/m³.
- At 400 ppm, CO_2 is 0.06% of air by weight.
- Therefore, in 1 m³ of air, there is 0.000765 kg CO_2.
- Thus, in order to extract 1×10^3 kg (1 metric ton) of CO_2, 1.3×10^6 tons of air must be processed. However, this assumes 100% extraction efficiency. In reality, efficiency is around 75%, so therefore, **1.74×10^6 m³** of air must be passed through the air contactor per ton of CO_2 removed.

Finally, it is also important to perform a spatial analysis to look at how much physical space is required to remove a specified amount of CO_2. A cost effective DAC installation must be able to capture at least one megaton (Mt) of CO_2 per annum. A proposed full-scale, 1 Mt model would be comprised of ten units, each 20 m tall, 8 m deep, and 200 m long. With ten of these units, there would be a combined 38,000 m² of inlet surface area, capable of processing 46,000 m³ of air per second given a 1.5 m/s air speed. Given a 0.75 capture rate, this means that the unit could capture:

$$\frac{46,000 \text{ m}^3 \text{ air}}{1 \text{ sec}} \times \frac{0.000765 \text{ kg CO}_2}{1 \text{ m}^3 \text{ air}} \times 0.75 \text{ eff} = \textbf{26.39 kg/s}$$

Or, given 38,000 m² of inlet space, we can calculate that at 1.5 m/s air speed, each m² of inlet corresponds with 0.69 grams CO_2 per second.

Just to verify these numbers, we can multiply 26.39 kg/s by 3.15×10^7 seconds/year to find that the entire installation would capture 0.8 Mt of CO_2 per year, a reasonable answer given that the installation will likely need maintenance ~15% of the time. In terms of space, this 1 Mt/yr installation would take up a space of 20 m high, 8 m deep, and 2000 m (2 km) long.

In terms of spatial arrangements, it is important to note that the air coming out of the DAC units has a CO_2 concentration of ~100 ppm. Therefore, it is not possible for the DAC units to be arranged one in front of the next, as the low CO_2 exhaust air of one unit would be the intake air of the next unit. Over a distance of 2–3 km, however, the air returns to its well-mixed state of ~400 ppm, so it is possible to arrange thee units consecutively as long as there are a few kilometers between each row. One could imagine these 1 Mt/yr units placed into square of 10 km on each side, with 1 Mt units forming a 10 km "column" that is able to handle 5 Mt/yr, and 5 rows of these columns, with 2.5 km between each row. One of these combined systems would be 10 km × 10 km and would have a capacity of removing 25 Mt CO_2/yr. Finally, it seems logical that these units could be located in regions with high wind velocity so that maximum amounts of air pass through the air contactor with the use of energy-hungry fans.

A schematic of what a single unit would look like is displayed in Figure CS6.2d below.

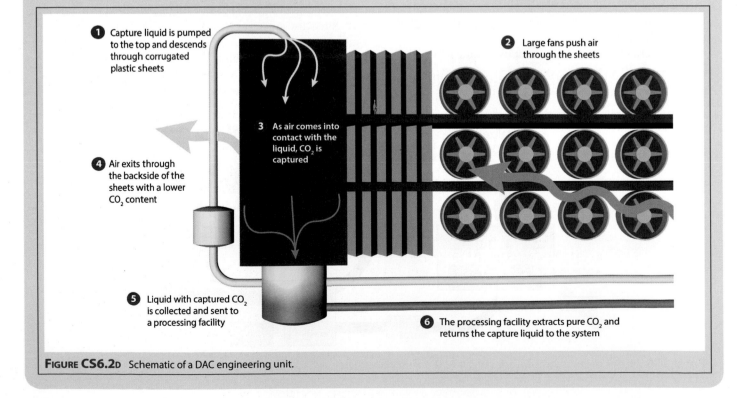

1 Capture liquid is pumped to the top and descends through corrugated plastic sheets

2 Large fans push air through the sheets

3 As air comes into contact with the liquid, CO_2 is captured

4 Air exits through the backside of the sheets with a lower CO_2 content

5 Liquid with captured CO_2 is collected and sent to a processing facility

6 The processing facility extracts pure CO_2 and returns the capture liquid to the system

FIGURE CS6.2D Schematic of a DAC engineering unit.

Finally, one of the most important considerations for any technology, especially one such as DAC, is the cost. Critics of DAC have cited costs as high as $1000/ton CO_2 extracted, while hopeful optimists have cited prices around $10–$15/ton CO_2 extracted. Currently the estimated costs for DAC of CO_2 yields values in the range of $200/ton CO_2.

Now that we have established our basic parameters, we can examine what large-scale implementation of DAC would cost, in terms of energy and financial resources.

Problem 1

1. Using the design and cost estimates from this Case Study, calculate the kWh of energy required to extract the total CO_2 emissions of 2021 from the atmosphere; while the CO_2 emission in 2011 was 32 Gt CO_2/yr, in 2021 it reached 50 Gt CO_2/yr!

2. At $.11/kWh, how much would the energy for that extraction cost?

3. With the global GDP in 2019 of 88×10^{12}, what percent of global GDP would be required to extract the CO_2 emitted in 2019?

Problem 2

1. If we extract CO_2 from its present level of 420 ppm to its preindustrial level of 280 ppm in 20 years, how many Gt CO_2 must be removed each year?

2. Using the design and cost estimates from this Case Study, calculate the kWh of energy required to extract CO_2 from its present mixing ratio of 420 ppm to its preindustrial level of 280 ppm.

3. If this were done in 20 years, how much energy would be required each year?

4. How much would this cost at $.11/kWh?

5. What fraction of current global GDP does this correspond to each year?

CASE STUDY 6.3   Concentrated Solar Power

Introduction

Concentrated solar power or CSP is a class of primary energy generation that uses the visible and near infrared radiation from the Sun in combination with a variety of optical systems to focus the Sun's radiation on collectors that convert those photons to heat that then, typically, produces steam to operate a turbine to produce electricity. Thus a conventional electricity generating plant that employs coal or natural gas is modified by simply replacing the chemically derived (fossil fuel) source of heat with the input of heat from the Sun, employing an optical collection system to concentrate the solar power to achieve the high temperatures required for thermodynamic efficiency in the steam turbine. An array of possible configurations of CSP is displayed in Figure CS6.3a.

CSP is now being widely commercialized on a global basis and the US has installed significant capacity in California, Arizona, and Nevada. By the end of 2013, CSP electricity generating capacity that includes projects either in operation or under construction in 2015 will have the following capacity:

California	1280 MW
Arizona	280 MW
Nevada	110 MW

Thus the US will have some 1.7 GW of installed CSP by early 2016. Our objective in this Case Study is to (1) investigate the technology of various types of CSP and (2) carry out an appraisal of the CSP potential in the US in order to judge the possible role CSP might play in the nation's future energy resources.

FIGURE CS6.3A   There are now an array of techniques for capturing and concentrating solar radiation for the purpose of generating electricity with steam turbine systems. We will review each of the techniques in this Case Study.

Methods of Solar Power Collection and Concentration

There are four principal methods for collecting energy from the Sun and concentrating that radiation to provide the high temperatures (500–1000°C) required for efficient electricity generation using a conventional steam turbine.

The first approach we discuss is the collection of solar radiation by a parabolic trough shown schematically in Figure CS6.3b. A parabolic mirror has the unique characteristic that it takes a parallel rays and focuses all of them at a single "focal point."

Placed at that focal point is a linear tube that carries the circulating high temperature fluid along the axis of the trough and delivers that high temperature fluid either to a steam turbine for the direct generation of electric power or to a heat exchanger that transfers the heat to a second fluid that in turn drives the steam turbine. A typical modern steam turbine system is shown in Figure CS6.3c.

The parabolic troughs are typically aligned in rows as shown in Figure CS6.3d to create a vast array of collection systems covering many acres. In a subsequent section we will consider the optimal location of these solar collection "forms" within the US.

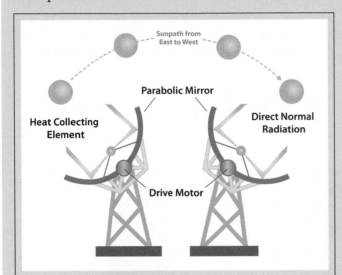

FIGURE CS6.3B The parabolic optical collector takes parallel rays from the Sun and focuses all the radiation at the focal point, concentrating the intensity by a factor of ~500, producing the high temperatures required for a steam turbine.

FIGURE CS6.3D The solar troughs are combined in large arrays covering many square kilometers in "solar farms" capable of providing electricity generation at the GW level.

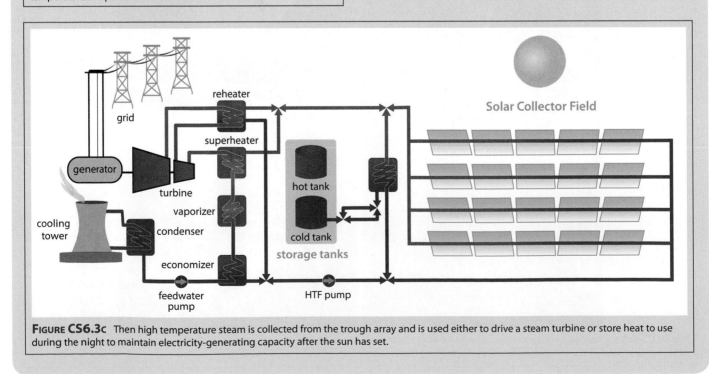

FIGURE CS6.3C Then high temperature steam is collected from the trough array and is used either to drive a steam turbine or store heat to use during the night to maintain electricity-generating capacity after the sun has set.

Compact Linear Fresnel Reflection

A second approach to solar collection and concentration is the Compact Linear Fresnel Reflector (CLFR) that employs long thin segments of mirrors to focus solar photons on either one or multiple absorbers. The reflectors are aligned by a tracking system that optimizes the light intensity at the absorber using the geometry displayed in Figure CS6.3e panel A.

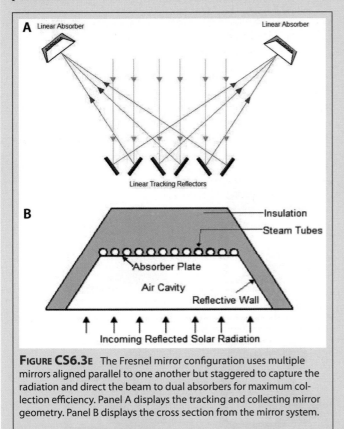

FIGURE CS6.3E The Fresnel mirror configuration uses multiple mirrors aligned parallel to one another but staggered to capture the radiation and direct the beam to dual absorbers for maximum collection efficiency. Panel A displays the tracking and collecting mirror geometry. Panel B displays the cross section from the mirror system.

The long axis of the Fresnel collectors are typically aligned with the north-south axis and rotate about a single axis. The steam tubes that collect the focused solar radiation are installed in the linear absorber as shown in Figure CS6.3e panel B. This use of multiple absorbers shown in Figure CS6.3e panel A has some notable advantages:

- The alternating inclination of the Fresnel collectors minimize the effect of reflectors blocking adjacent reflectors access to sunlight. This significantly improves the collection efficiency over a diurnal cycle.

- Multiple absorbers reduce the amount of ground area required for a given power output.

- The multiple absorbers reduce the length of the steam lines in the absorbers, thereby increasing efficiency by reducing radiation losses from the system.

Solar Power Tower

A third approach that is gaining wide acceptance is the technique of using an array of individually directed mirrors, or heliostats, that direct the Sun's radiation to a central receiver as displayed schematically in Figure CS6.3f.

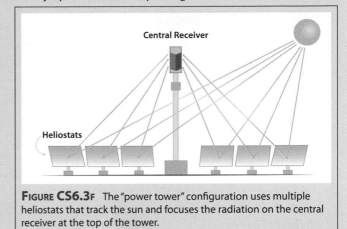

FIGURE CS6.3F The "power tower" configuration uses multiple heliostats that track the sun and focuses the radiation on the central receiver at the top of the tower.

The working fluid in the central receiver, or tower, is heated to between 500°C and 1000°C and then used either directly to power a steam turbine for electric power generation or in conjunction with a heat exchanger from which high pressure steam is extracted for input to a steam turbine. The field of heliostats is typically two square miles in area arranged as shown in Figure CS6.3g.

FIGURE CS6.3G The configuration of the heliostats and the collection tower for modern CSP systems employs as many as 600 heliostats feeding a single tower.

There are two advantages to this approach. One is that the heliostats are rotated about two axes so the solar collection is very efficient over the full extent of the day. Second, the high operating temperature of the system provides the opportunity to store thermal energy (heat) so that the system produces steam for electricity generation through several hours of the night. This is an important feature for reducing the intermittency problem associated with all solar energy generation.

Modern designs using the power tower configuration are operating at a yield of 1 MW per 4 acres of land area (16,000 m²). Spain has been a particularly aggressive developer of solar power using primarily the power tower configuration.

Stirling Engine Dish Design

A fourth configuration, displayed schematically in Figure CS6.3h, uses a parabolic dish design to focus the Sun's radiation at a single focal point. That focused solar radiation is then used to deliver heat to a unique heat engine called a Stirling engine after its inventor (in 1816!) Robert Stirling.

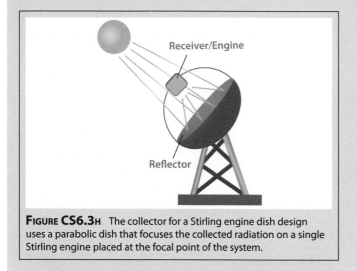

FIGURE CS6.3H The collector for a Stirling engine dish design uses a parabolic dish that focuses the collected radiation on a single Stirling engine placed at the focal point of the system.

A Stirling engine in its original design is a heat engine that uses an *external* heat source such as burned coal, wood, or oil to create a cyclical expansion and compression of air at different temperature levels such that there is a net conversion of heat to mechanical work. This is distinct from *internal combustion* engines that we are familiar with in gasoline and diesel automobile and truck engines. The reason that Stirling engines have experienced a renewal is that the combustion of hydrocarbon fuels can be replaced by concentrated solar radiation or geothermal as an external heat source.

The Stirling cycle of four trajectories on a pV graph is displayed in Figure CS6.3i.

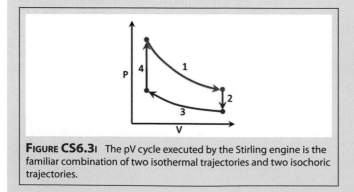

FIGURE CS6.3I The pV cycle executed by the Stirling engine is the familiar combination of two isothermal trajectories and two isochoric trajectories.

Leg 1: The expansion occurs along an isothermal trajectory in response to the external heat supply by the concentrated solar radiation.

Leg 2: The isochoric (constant volume) trajectory during which the gas is passed through the "regenerator" between the hot and cold cylinders in Figure CS6.3j that serves to extract heat for use in the next cycle.

Leg 3: A second isothermal trajectory during which the gas undergoes compression.

Leg 4: A final isochoric trajectory wherein the compressed air flows back through the regenerator and picks up heat as it enters the heated expansion space.

The mechanical device that accomplishes this closed trajectory on the pV diagram is shown in Figure CS6.3j. It is ingenious and represents an important union between theoretical thermodynamics represented in Figure CS6.3i and the practical execution of the "external" heat engine.

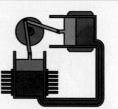
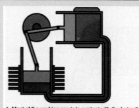
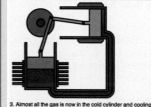

1. Most of the working gas is in contact with the hot cylinder walls, it has been heated and expansion has pushed the hot piston to the bottom of its travel in the cylinder. The expansion continues in the cold cylinder, which is 90° behind the hot piston in its cycle, extracting more work from the hot gas.

2. The gas is now at its maximum volume. The hot cylinder piston begins to move most of the gas into the cold cylinder, where it cools and the pressure drops.

3. Almost all the gas is now in the cold cylinder and cooling continues. The cold piston, powered by flywheel momentum (or other piston pairs on the same shaft) compresses the remaining part of the gas.

4. The gas reaches its minimum volume, and it will now expand in the hot cylinder where it will be heated once more, driving the hot piston in its power stroke.

FIGURE CS6.3J The four primary strokes of the Stirling engine are displayed in order from the top left of the diagram.

This configuration of the Stirling engine (termed the alpha Stirling) is constructed from two power pistons in separate cylinders with the high temperature side heated by concentrated solar thermal radiation and the low temperature side cooled by the atmosphere. This type of engine has a high power-to-volume ratio and is thus well suited to be placed at the focal point of a large solar collection mirror as displayed in Figure CS6.3k.

Available Resources for CSP

We turn now to the analysis of the geographic distribution of available solar radiation in the US. The direct solar radiation in the US in kWh/m²·day is displayed in Figure CS6.3l, and the focus is on the southwestern sector of the US because there is a remarkable amount of available solar radiation available in this region. From the perspective of site selection for CSP, this is the most important consideration because power can be supplied to the rest of the country by a nation-wide power grid. As we will see, CSP has the potential for supplying a very significant fraction of the country's electricity generating capacity.

In order to estimate the available power that can be practically extracted from CSP, it is necessary to remove from consideration all areas with significant population density, all areas that are involved in supplying other resources such as water, etc., all land areas with greater than a 1% slope and all areas that have a contiguous area of less than 1 square kilometer.

FIGURE CS6.3κ As with the parabolic trough collection, the Fresnel collector and the heliostat-tower design, the Stirling dish design uses multiple systems placed together in a solar farm.

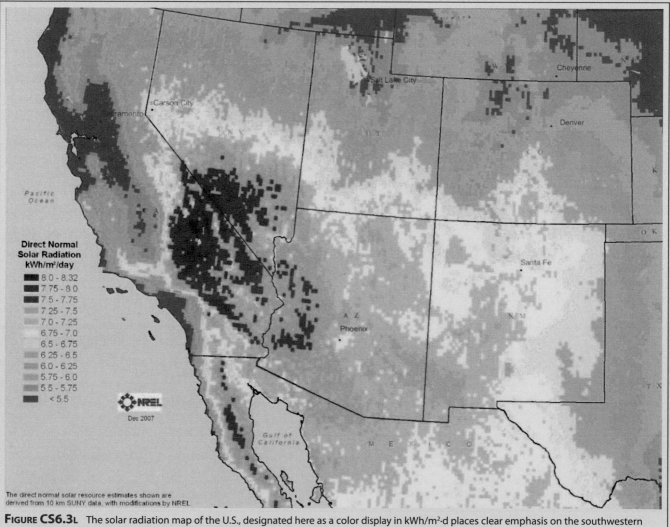

Direct Normal Solar Radiation kWh/m²/day

- 8.0 - 8.32
- 7.75 - 8.0
- 7.5 - 7.75
- 7.25 - 7.5
- 7.0 - 7.25
- 6.75 - 7.0
- 6.5 - 6.75
- 6.25 - 6.5
- 6.0 - 6.25
- 5.75 - 6.0
- 5.5 - 5.75
- < 5.5

NREL
Dec 2007

The direct normal solar resource estimates shown are derived from 10 km SUNY data, with modifications by NREL.

FIGURE CS6.3L The solar radiation map of the U.S., designated here as a color display in kWh/m²·d places clear emphasis on the southwestern part of the country. This map serves to emphasize the very large resources available in the U.S. for CSP.

When this is done, we have left the following table, Table CS6.3a, of available land area, solar capacity, and solar generation capacity for a subset of seven states. We will use this table in problem #2.

TABLE CS6.3A			
State	Land Area (mi²)	Solar Capacity (MW)	Solar Generation Capacity (GWh)
AZ	13,613	1,742,461	4,121,268
CA	6278	803,647	1,900,786
CO	6232	797,758	1,886,858
NV	11,090	1,419,480	3,357,355
NM	20,356	2,605,585	6,162,729
TX	6374	815,880	1,929,719
UT	23,288	2,980,823	7,050,242
Total	87,232	11,165,633	26,408,956

Inspection of Table CS6.3a reveals that while California has the largest region of solar radiation exceeding 7.5 kWh/m²·day, Arizona, New Mexico, and Nevada have the largest solar generating capacity because of the flat terrain, low population density, and available land.

Problem 1

Thermodynamic Efficiency of Concentrated Solar Power

It is an illustrative example of applied thermodynamics to calculate the maximum conversion efficiency of solar-to-work by applying what we know about thermodynamics, radiation collection, and radiative cooling by infrared emission.

If the overall efficiency of a CSP system is η, then that efficiency will be the *product* of the efficiency of the collection device, η_{coll}, and the Stirling efficiency of the heat engine (turbine or Stirling engine) used to generate electricity. Since the efficiency of the electric generator is ~95%, we can ignore it in this calculation.

The Stirling efficiency is just $\eta_{Ster} = 1 - T_c/T_H$. Recall that this is the *maximum* efficiency of a heat engine. In practice the achieved efficiency is approximately ½ the Carnot efficiency. The efficiency of the collection we can express as

$$\eta_{coll} = \frac{Q_{absorbed} - Q_{lost}}{Q_{solar}}$$

where Q_{solar} is the incoming solar flux,
$Q_{absorbed}$ is the heat absorbed by the system, and
Q_{lost} is the heat lost by the system.

Each of those quantities is interesting.

Q_{solar} is equal to the *product* of the solar flux F (in units of w/m²), the concentration ratio, C, expressing the ratio of the area of the collection mirrors divided by the area A of the high temperature collector, and the efficiency of the collection and optics, η_{optics}, for collecting the solar flux incident in the CSP site.

Thus

$$Q_{solar} = \eta_{optics}\, F\, C\, A$$

The energy absorbed by the CSP system, Q_{abs}, is the product of absorptivity of the optical system times Q_{solar}. Thus

$$Q_{absorbed} = \alpha Q_{solar}$$

The loss of heat from the system, Q_{lost}, is primarily from the blackbody radiation of the high temperature system into the surroundings. That is just the Stefan–Boltzmann relationship so

$$Q_{lost} = A\sigma T_H^{\,4}$$

We then have an explicit calculation for the overall efficiency

$$\eta_{overall} = \eta_{coll}\eta_{Ster}$$

1. Write the expression for $\eta_{overall}$ in terms of F, C, T_H, and T_c.

2. If F = 1000 w/m², C = 500, T_H = 1100 K, and T_c = 300 K, calculate η_{coll}, η_{Ster} and $\eta_{overall}$.

Problem 2

1. From Table CS6.3a, determine the power generating capability of the seven listed states in TW (1×10^{12} Watts).

2. What is the approximate electricity generating capacity in the US in TWs? What is the total rate of primary energy generation capacity in the US in TWs?

3. How do the figures in step 2 above compare with step 1?

4. If we wish to produce 1 TW of electric power, what fraction of the available land area would be needed if each state provided an amount of land area equal to the desired electrical energy generation divided by the total available power from all seven states combined?

5. Table CS6.3b provides the total land area of each state. Calculate the fraction of land area needed for each state needed in order to provide the requisite electric energy generation capability for that state.

TABLE CS6.3B

State	Total Land Area of State
AZ	114,006
CA	163,707
CO	104,100
NV	110,567
NM	121,598
TX	268,820
UT	84,904

6. Summarize this by providing:

    • Total land area of the seven states

    _____ mi^2

    • Land area required to generate US electrical demand

    _____ mi^2

    • Fraction of land required to generate US electrical demand

    _____ %

ELECTROCHEMISTRY
THE UNION OF GIBBS FREE ENERGY, ELECTRON FLOW, AND CHEMICAL TRANSFORMATION

7

"When I was in college, I wanted to be involved in things that would change the world. Now I am."

—Elon Musk

FIGURE 7.1 The Tesla Roadster is an example of a new breed of electric cars that are now dominating the design divisions of a number of automobile corporations. These cars are fast, quiet, and extremely efficient. They are a key component in the elimination of dependence on imports of petroleum and, with the introduction of alternative energy, the elimination of carbon dioxide release for the transportation sector.

Framework

The study of the union between chemistry and electricity—the flow of electrons driven spontaneously by free energy release in a chemical reaction—has provided a rich history of remarkable discoveries. These studies constitute the discipline of electrochemistry, a subject that has experienced a dramatic rebirth. That rebirth has been propelled by the emergence of energy production and storage as a dominant problem confronting both science and public policy. A key part of why the rebirth of electrochemistry has been sparked by efforts to advance key technologies associated with global demands for energy is contained in the third diagram in this text—Figure 1.3. That figure traces the flow of electrons, supplied to the power distribution grid by a combination of photovoltaics, wind, and concentrated solar thermal. That primary energy generation is coupled to a com-

bination of batteries, electrolysis cells, fuel cells, and hydrogen storage systems by the power grid. In each of these cases, chemical transformations are coupled to the flow of electrons and visa versa. There is a clear implication that the energy contained in the chemical bond can be harnessed to generate the flow of electrons, and conversely that the flow of electrons, driven by an external voltage source, can enact chemical change.

We are already acutely aware, through our study of thermodynamics, that no chemical change occurs without bringing a requisite increase in disorder to the universe. There is always a cost, a cost to be paid in the currency of free energy, ΔG, such that:

Reactants → Products + Free Energy

With the release of free energy, ΔG, comes the ability to do work. If the flow of electrons is to produce work, we first need a difference in *electrical potential between two points*, just as we need a *difference in gravitational potential energy between two points* that can be converted to kinetic energy. Second, we need a source of electrons; third, a conducting path for those electrons; and fourth, a pump to raise electrons from a region of low electrical potential to a region of high electrical potential. For the case of gravitational potential energy, we can sketch such a cycle, shown in Figure 7.2, displaying a series of mass elements that move in a gravitational field. The "pump" that carries the mass elements to a position of higher gravitational potential energy requires work. When a mass acquires gravitational potential energy, work can be done. This has a direct analogy with the chemical potential shown in Figure 7.3 that drives reactants to products in a chemical reaction, releasing free energy. We can sketch a similar free energy diagram for an electrical potential that drives the flow of electrons through an electric motor, light bulb or electronic device. It was the recognition that the ability to extract useful work from either an *electrical potential* or from a *chemical potential* that lead to the conceptual linking of voltage (that is a measure of electrical potential) to the concept of free energy release, ΔG, of a chemical reaction. This constitutes one of many critically important contributions of electrochemistry to science and society as a whole.

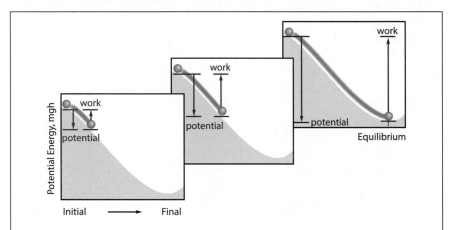

FIGURE 7.2   As we have continued to stress, there is a direct analogy between the release of energy in a chemical reaction and the conversion of potential energy to mechanical work when a mass, *m*, falls through a gravitational field. As time progresses, the internal potential energy, *mgh*, is released when the mass reaches its lowest point on the potential energy surface.

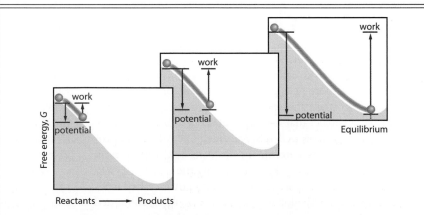

FIGURE 7.3   As electrons flow from the anode (where oxidation occurs liberating those electrons into the external circuit) to the cathode (with the higher standard reduction potential), free energy is released that is capable of doing useful work by virtue of that external circuit.

We can link the concepts of gravitational potential, chemical potential, and electrical potential in a single diagram. Figure 7.4 displays an electrochemical cell, which we will study in this chapter, where the electrochemical cell engages a chemical potential to deliver an electrical potential through an external circuit to an electric motor that in turn performs mechanical work by raising a mass, m, in a gravitational field. As we will see, it was Michael Faraday who used just such an experimental arrangement to *quantitatively* link free energy, the electrical potential, and the gravitational potential to the work done by an electrochemical cell. This provided the conceptual foundation to *quantitatively* link the vertical axes in each of the energy diagrams in Figure 7.4.

Electrochemistry is the study of how electrical potential and Gibbs free energy are quantitatively linked. However, in the spirit of setting the context, we frame the problem by first exploring some of the potent contributions electrochemistry will make to the new global energy structure as we move into the next decade. These examples range from the global scale to the cellular scale to the molecular scale, and each is at the forefront of the global research enterprise in science and technology.

We begin with the automobile, which occupies center stage in the energy debate because it represents a major consumer of petroleum and a remarkable opportunity to infuse technology into the lives of a large fraction of the population. The transition to an electrified transportation system will radically reduce dependence on fossil fuels and thereby improve the strategic and economic position of many nations, including the US, while markedly reducing CO_2 emissions. For these objectives, electrochemistry is essential.

We consider first the energy requirements to propel a gasoline automobile using, as an example, a car that gets 20 miles per gallon. This is close to the US average fuel economy for passenger vehicles. From Case Study 1.1 we know that a liter of gasoline contains 3.5×10^7 joules of energy. If we convert this to the units of kilowatt-hours (kWh), it will allow us to easily calculate the equivalent amount of energy from electricity used to propel an electric automobile with that required to propel a gasoline automobile. We purchase electricity in units of kWh from the electric power grid. A kWh of energy is $(1 \times 10^3$ joules/sec$)(3.6 \times 10^3$ seconds$) = 3.6 \times 10^6$ joules of energy. Thus the energy contained in a liter of gasoline, 3.5×10^7 joules, in units of kWh, is $(3.5 \times 10^7$ joules/liter$)/3.6 \times 10^6$ joules/kWh$) = 9.7$ kWh/liter. We will repeatedly approximate this as 10 kWh/liter. If a vehicle gets 20 miles per gallon, this fuel mileage converts to approximately 8 km/liter because there are approximately 4 liters per gallon and 1.6 km per mile. In order to compare a diverse range of automobiles including passenger cars, SUVs, gasoline-electric hybrids, and all-electric automobiles, we adopt a consistent approach by calculating the *amount of energy* in kWh required to propel a particular vehicle 100 km. We can then quickly calculate, in all cases, the cost to drive each vehicle 100 km, or any other distance. For example, an

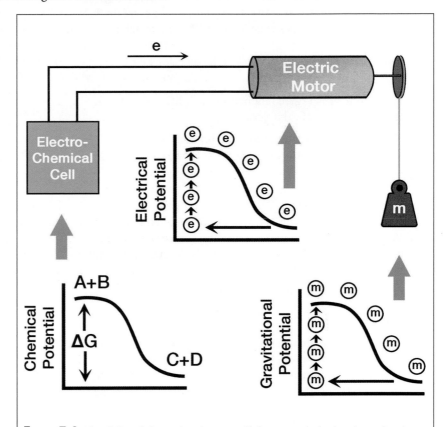

FIGURE 7.4 The ability of electrochemistry to unify key concepts in chemistry, electricity, and mechanical work is one of the major contributions electrochemistry has made to modern science. Because an understanding of electrochemistry lies at the heart of important new *developments* in energy storage and energy transformations, electrochemistry occupies a prominent position in research and technology today. The diagram displayed here links the concept of the chemical potential defined in terms of Gibbs free energy, to the electrical potential defined by the volt, to the gravitational potential of a mass, m, in a gravitational field.

Example 1

For a gasoline powered automobile that gets 20 miles per gallon (the U.S. average for the combined automobile and light truck fleet is 17 mpg), this converts to 8 km/liter because there are approximately 4 liters per gallon and 1.6 km per mile so

$$20 \text{ miles/gallon} = \frac{(20 \text{ miles/gal})}{4 \text{ liters/gal}} = (5 \text{ miles}/\ell)1.6 \text{ km/mile} = 8 \text{ km}/\ell$$

With 10 kWh of energy contained in each liter of fuel, this gives us, for 100 km of driving distance,

$$[\text{energy required to drive 100 km}] =$$
$$(100 \text{ km}/8 \text{ km}/\ell) 10 \text{ kWh}/\ell = 125 \text{ kWh}$$

Example 2

For comparison a large SUV in city driving with high traffic density will achieve a mileage of approximately 10 miles per gallon. This is twice the fuel consumption of our "U.S. average automobile" and thus we need to simply double the energy consumption of Example 1, so

$$[\text{energy required to drive 100 km}] = 2 \times 125 \text{ kWh} = 250 \text{ kWh}$$

Example 3

A hybrid-electric such as the Toyota Prius or Honda Civic hybrid uses an electric motor linked directly to the crankshaft of a gasoline engine, which serves both to assist in powering the automobile but also to recapture energy when decelerating or going downhill by switching the electric motor to a generator and recharging the metal-hydride batteries. The Prius and Civic hybrids get approximately 40 miles per gallon in city driving. They have the important added feature that the gasoline engine switches off when the car is stopped so no fuel is wasted at idle in traffic. Because the hybrid gets 40 miles to the gallon it uses one-half the amount of energy of the "U.S average automobile" of Example 1, so

$$[\text{energy required to drive 100 km}] = \tfrac{1}{2}(125 \text{ kWh}) = 62.5 \text{ kWh}$$

Example 4

Figure 7.5 displays a picture of the Tesla Model S that seats 5 adults and is powered by lithium-ion batteries. It is a large sedan and is all-electric which means that it has no gasoline engine, but it has a range of 300 miles or 480 km. With a battery capacity of 72 kWh of energy, the Model S uses

$$[\text{energy required to drive 100 km}] = 72 \text{ kWh}/480 \text{ km} = 15 \text{ kWh}/100 \text{ km}$$

Example 5

Figure 7.6 shows the Citroen Model C-Zero which is a smaller 4-door family sedan. The C-Zero has a rating per 100 km of

$$[\text{energy required to drive 100 km}] = 10 \text{ kWh}$$

automobile that gets 20 mpg, which, by the above is 8 km/liter, requires an amount of energy to drive 100 km of [(100 km)/(8 km/liter)] (10 kWh/liter) = 125 kWh of energy.

Calculate the amount of energy in kWh required to propel a car 100 km. To begin the comparison, consider five representative cases:

Example 1: The energy required to propel an automobile that gets 20 mpg 100 km.

Example 2: The energy required to propel an SUV in traffic, which gets 10 mpg.

Example 3: The energy required to propel a hybrid electric such as the Toyota Prius, which gets 40 mpg.

Example 4: The energy required to propel a large, all-electric sedan such as the Tesla Model S, shown in Figure 7.5, which requires 15 kWh per 100 km of driving.

Example 5: The energy required to propel a 4-door, all-electric sedan such as the Citroen C-Zero, shown in Figure 7.6, which requires 10 kWh per 100 km of driving.

Inspection of the five examples immediately raises the question: how can there be such a large discrepancy between the energy required to propel a gasoline powered automobile vs. an electric powered automobile? The answer to this question involves three primary considerations. First, a gasoline powered automobile, which is governed by the thermodynamics of a heat engine, is less than 15% efficient in the conversion of chemical energy contained in the gasoline to energy available for moving the automobile. The electric motor, in sharp contrast, is 95% efficient in converting energy stored in the battery to energy available for propelling the car. The second reason that the electric car (as well as the

FIGURE 7.5 The Tesla Model S is a new all-electric automobile built in California that seats five adults and two children and is powered by lithium-ion batteries.

FIGURE 7.6 The Citroen C-Zero represents a new generation of all electric 4-door passenger sedans from Europe that are designed to carry 4–5 adults in comfort with very high efficiency. These cars consume approximately 10 kWh of electrical energy for each 100 km traveled.

hybrid-electric) is more efficient is that it *recaptures* energy during braking and in going downhill. Finally, in traffic, neither the electric nor hybrid consume fuel when stopped. In contrast, the gasoline engine consumes a considerable amount of fuel at idle. Of course, for the all-electric car, the electric power must be supplied to the power socket used to charge the battery. If this is done through hydroelectric power or the use of renewables such as from wind, solar thermal, geothermal, photovoltaics, etc., there is no other energy source involved. If the electricity is generated by a coal or a natural gas fired power plant, then the amount of net energy required depends upon the efficiency of the power plant. In the case of older coal burning power plants this is approximately 35%. For modern gas burning power plants the efficiency approaches 50%.

We can then compare the primary energy extracted from either coal or natural gas by simply multiplying the required amount of energy to go 100km by the inverse of the efficiency: 1/35% ≈ 3 for an old coal burning power plant and 1/50% ≈ 2 for a modern natural gas burning power plant. The result is displayed in Figure 7.7.

Notice in particular that the Tesla Model S, a large all-electric powered sedan, when charged from an old coal burning power plant increases from 15 kWh/100 km to 45 kWh/100 km, which is approximately ¾ of the energy required by the hybrid. With electrical power generated by natural gas, the Model S requires 30 kWh/100 km, or approximately half that required by the hybrid, but one-quarter that of an average US automobile, and one-eighth that required by a large SUV in urban traffic. The Citroen powered by electricity from an aging coal burning power plant requires 30 kWh of (chemical) energy from coal and 20 kWh of energy from natural gas to go 100 km. That corresponds, respectively, to one-quarter the chemical energy consumption of an average U.S. gasoline powered automobile for a coal burning power plant and one-sixth the chemical energy consumption for natural gas generated electricity.

This raises the next question. We know that electricity in the U.S. is generated by a combination of coal and natural gas burning power plants, hydroelectric, geothermal, nuclear and renewables—wind, photovoltaics and concentrated solar thermal. The cost of electric power, as a result, varies significantly across the country. Costs per kWh of electricity for each of the states is displayed in Figure 7.8.

This information allows us to calculate the actual cost of driving the various categories of automobiles. We will take 10 cents per kWh for our average cost of electricity in the U.S. which is close to the national average in 2010, and in addition, this will make it easy to adjust for differences between states. Gasoline is currently about $3.80/gallon, which translates conveniently to $1/liter. With the energy

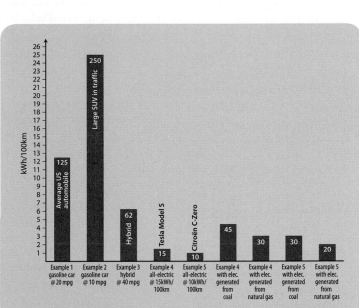

FIGURE 7.7 When comparing the energy consumed by various categories of automobiles what is typically done is to calculate the distance traveled for a given amount of energy contained in the fuel for a gasoline powered automobile. For an electric vehicle, the calculation is usually based on the electrical energy in kWh required to go a given distance. However because in many parts of the country electricity is generated from coal or natural gas, we can relate the electrical energy back to the chemical energy contained in fuel used to generate the electricity. For the case of an old coal burning power plant, the efficiency in converting chemical energy to electricity is about 35%. Thus we multiply the electrical energy a given car uses by a factor of three to determine the corresponding chemical energy. For a modern natural gas fired electrical power plant the efficiency is 50%, so we multiply by a factor of two.

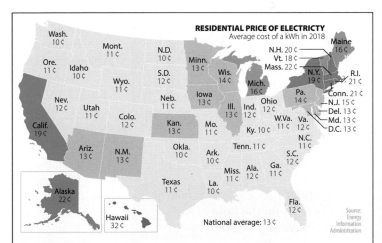

FIGURE 7.8 A remarkable fact in the United States is that the cost of electricity—universally cited units of cents per kWh—varies remarkably from state to state. By region, the highest price for electricity is in the Northeast; the lowest price in the Northwest. The average residential retail price of electricity in the U.S. (2007) was 10.64 cents per kWh. In the US, 48% of the electrical power is generated by the burning coal, 21% by natural gas, 18% by nuclear, 6% by hydroelectric, 3% by renewables (wind, solar, geothermal, etc.), 1% by petroleum.

content of gasoline at 10 kWh/liter, we can directly convert kWh to the cost of driving.

We can take the kWh/100 km from Figure 7.7, and for the case of gasoline powered vehicles calculate cost per 100 km of driving from the expression

$$[\text{cost per 100 km of driving}] = \frac{(\text{kWh}/100\text{ km})}{(\text{kWh}/\text{liter fuel})}(\text{cost fuel}/\text{liter})$$

1. Cost of driving the average US vehicle at 20 mpg 100 km.
2. Cost of driving an SUV 100 km in traffic at 10 mpg.
3. Cost of driving a Prius 100 km at 40 mpg.
4. Cost of driving a Tesla Model S 100 km at 15 kWh/100 km.
5. Cost of driving a Citroën C-Zero 100 km at 10 kWh/100 km.

For the electric automobiles, we simply need to multiply the kWh/100 km by the cost of electricity in kWh. Calculate the cost of purchasing the energy to drive the following five automobiles a distance of 100 km:

1. Gasoline automobile that gets 20 mpg
2. SUV in traffic that gets 10 mpg
3. Prius hybrid that gets 40 mpg
4. Tesla Model S that uses 15 kWh
5. Citroën C-Zero that uses 10 kWh

Assume a gasoline cost of $1.00/liter and a cost of electricity of $.10/kWh.

An inspection of the results of this calculation along with a consideration of the performance of electric cars reveals several key conclusions. First, there is a remarkable savings in driving cost inherent in the electric automobile. As noted before, this results from the high efficiency of the all-electric automobile motor (~95%) and the ability of the electric automobile to recapture energy as well as the low efficiency of the gasoline powered automobile "heat" engine (<15%). So even though electricity in the U.S. is generated primarily by combustion of coal and natural gas, the cost of driving an electric automobile is dramatically lower than that of the gasoline powered automobile. It is important also to note that there is no technical reason why a large SUV could not be powered by an electric motor with no compromise in performance. In a head-to-head comparison of vehicles, the comparison of the calculated costs for the five examples shows that it costs approximately one-eighth as much to drive an electric power vehicle than it does a comparable gasoline powered vehicle.

Battery technology and fuel cells are the focus of Case Study 7.1. Electrochemistry of the cell is treated in Case Study 7.2.

The next question: For an average annual driving distance in the US of 20,000 km (12,500 mile), how much does it cost annually to drive each of our five examples? We can answer this question by referring to our previous calculation of cost and simply multiplying the cost to drive 100 km by 20,000 km/100 km. The results are:

Example 1:	$2500/year	Auto/light truck U.S. average 20 mpg
Example 2:	$5000/year	Large SUV in traffic
Example 3:	$1250/year	Prius hybrid
Example 4:	$300/year	Tesla Model S
Example 5:	$200/year	Citroën C-Zero

The results are worthy of note. Shifting from a large gasoline powered SUV in traffic to a large all-electric sedan in traffic would save approximately $4,700 per year in costs associated with powering the vehicle. Shifting from a sedan with average U.S. fuel economy of 20 mpg to a large all-electric sedan would save $2,200 per year and shifting to a small all-electric sedan would save $2,300 per year. This corresponds to a savings of approximately $200 per month. This constitutes a

**Case Study 7.1
The Electrochemistry of Batteries**

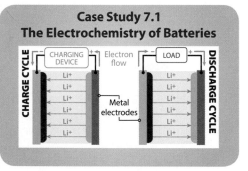

Case Study 7.2   Electrochemistry in the Cell and the Role of ATP in Powering Organisms

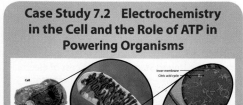

major fraction of an auto loan for an all-electric vehicle as they emerge on the market.

The final calculation we consider is the total annual expenditure in the U.S. for these five examples assuming an annual total driving distance of 3 trillion miles—the total number of miles driven in the U.S. in 2008. Three trillion miles is 4.8×10^{12} km, so for each of our examples:

Example 1:   ($12.50/100 km) 4.8×10^{12} km = $600 billion
Example 2:   ($25.00/100 km) 4.8×10^{12} km = $1.2 trillion
Example 3:   ($6.25/100 km) 4.8×10^{12} km = $300 billion
Example 4:   ($1.50/100 km) 4.8×10^{12} km = $72 billion
Example 5:   ($1.00/100 km) 4.8×10^{12} km = $48 billion

This puts in place a very important question. Suppose the U.S. were to shift from (1) driving the current U.S. national average gasoline powered vehicle that gets 20 mpg, with gasoline costs of $1/liter ($3.80/gallon), to (2) a large all-electric sedan that requires 15 kWh of energy to go 100 km. How much would be saved on a *national* basis? We can determine this by subtracting the amount for Example 4 from the amount for Example 1: $600 \times 10^{12} - 72 \times 10^{12} = $528 billion. That is 528 billion dollars *each year* in national fuel cost expenditures. To emphasize the point, we capture this annual savings graphically in Figure 7.9.

The principles behind both power generation by electromagnetic induction and power delivery by electric motors are the subject of Case Study 7.3.

Shifting from petroleum to electrical energy to power the transportation sector places the energy debate and public policy strategy squarely in the field of electrochemistry for scientific and technical advances. While we have an effective option in lithium-ion batteries discussed in this chapter, how do we advance battery technology to increase the energy storage per unit mass and per unit volume of batteries used in the transportation sector? What are the implications for total electricity consumption if all automobiles and light trucks draw their energy from the electrical power grid? While we can eliminate the need for imported petroleum by switching to electrically powered vehicles, how do we take the next step and eliminate the release of CO_2 added to the atmosphere by fossil fuel combustion? If we respond to the constraints put in place by feedbacks in the climate structure, with its requisite requirements to reduce CO_2 emission, how do we control the balance between supply and demand on the national electricity power grid? An answer to these questions directly engages the study of electrochemistry.

An inspection of Figure CS4.3E emphasizes that when we transition to an energy infrastructure based on sources of primary energy that would both free us of petroleum purchases from other countries and reduce the amount of carbon deposited in the atmosphere, a careful quantitative analysis is required. We must investigate in detail the scientific and technical underpinning of the relationship between the flow of electrons, the chemical storage of energy in batteries, and the generation of electron flow through a "load" by virtue of the energy in chemical bonds. We must also understand the mechanisms by which a flow of electrons can produce a chemical fuel that can store electrical energy reversibly, releasing it on demand. These are the questions that motivate this chapter on electrochemistry.

The next segment in the development of "50 Questions on Global Scale Energy and Power" constitutes Case Study 7.4.

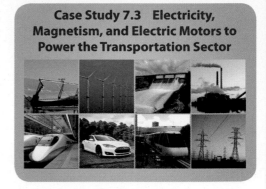

Case Study 7.3   Electricity, Magnetism, and Electric Motors to Power the Transportation Sector

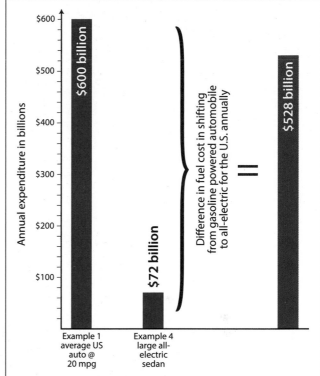

FIGURE 7.9 A number of very high efficiency smaller all-electric cars are coming on the market. The Loremo, built in Europe, consumes 6 kWh of energy per 100 km. The car holds two children. In New York City this translates into a cost of [(6 kWh)/(100 km)] × ($0.171/kWh) = $1.03/100 km. This is to be compared with $23.44/100 km for a large gasoline powered SUV.

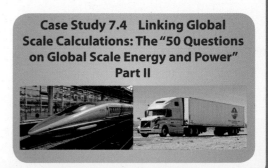

Case Study 7.4   Linking Global Scale Calculations: The "50 Questions on Global Scale Energy and Power" Part II

Chapter Core

Road Map for Chapter 7

In a fundamental sense, the study of electrochemistry is a grand unifier, tying together the concepts of free energy, oxidation-reduction, electronegativity, chemical equilibria, acid-base reactions, thermochemistry, and electron delocalization. Thus in this chapter we begin with a treatment of how electrodes—metals placed in solution—constitute an important example of oxidation-reduction reactions in mixed phases and how *oxidation* at the *anode* is isolated from *reduction* at the *cathode* in an *electrochemical cell* capable of generating an external flow of electrons. This flow of electrons is instigated by the *difference* in the ability of two metals to draw electrons to them. This separation of oxidation and reduction into isolated physical entities connected by an external conductor through which electrons can flow, is called a *galvanic cell* or alternatively a *voltaic cell*. These cells are analyzed in terms of *half-cell reactions* defined by the relative ability of one cell to draw electrons from the other measured by a quantity termed the *standard reduction potential*. This provides the ability to calculate the *voltage* generated by the cell under various conditions. The road map of key concepts is:

Road Map to Core Concepts	
1. The Master Diagram: How Electrochemistry Unifies Thermodynamics	
2. Reviewing Oxidation-Reduction Reactions	
3. Visualization of Oxidation and Reduction at Metal Electrodes	
4. Building a Voltaic Cell from its Components	
5. The Standard Hydrogen Electrode: Setting the Reduction Potential Scale	
6. Linking the Chemical Potential, Gravitational Potential, and Electrical Potential	
7. Cell Potential for Nonstandard Conditions	

Free Energy, Electron Flow, and Electrochemistry

Chemical transformations, which in turn govern changes in all chemical structures, are guided by the movement of electrons either as subtle delocalizations within bonding architectures or by the explicit transfer of electrons, but always in their quest to seek the state of lower free energy. As we have noted, distilled to its fundamentals, any chemical reaction may be expressed as Reactants → Products + Free Energy. Chemical species—in fact all matter in general—at a high chemical potential seek any available path with $\Delta G < 0$ and release free energy in the process. By whatever means matter rearranges the configuration of its bonds, there is a corresponding release of free energy as a result. As developed in Chapter 4, at a constant temperature and pressure—which characterizes the preponderance of processes in nature—that free energy release is Gibbs free energy:

$$\Delta G = \Delta H - T\Delta S$$

Check Yourself 1
As we have seen, Gibbs free energy is one of, if not the, most important thermodyamic variable. It quantitatively links the concepts of spontaneous change, enthalpy, entropy, the equilibrium constant, and the relationship between the change in Gibbs free energy at standard condition as well as under conditions for which reactants and products are not at equilibrium.

As a review from Chapter 4, consider the following

(a) Under standard conditions, what is the relationship between $\Delta G°$, $\Delta H°$, and $\Delta S°$?

(b) At a temperature, T, other than 298 K, can we calculate the change in Gibbs free energy, ΔG_T, from tables of $\Delta H°$ and $\Delta S°$? If so, why? If not, why not?

(c) What is the relationship between $\Delta G°$ and the equilibrium constant K_{eq}?

(d) If the chemical reaction is not at equilibrium, what is the relationship between ΔG and Q?

Electrochemistry is a particularly compelling branch of the physical sciences because it draws together the concepts of oxidation-reduction, electron transfer, free energy release, and electrical potentials in the context of cutting-edge technology for global-scale energy production and storage. For example, were a single technology, lithium-ion batteries, incorporated into plug-in hybrid technology in automobiles and light trucks, the US would reduce its petroleum imports by 60%. While much of electron delocalization and electron transfer in chemical processes is implicit—we cannot observe it directly—electrochemistry makes such processes explicit by *separating* the flow of electrons physically from the reactants and products such that chemical energy can be used to execute work through an external circuit. It is this feature of electrochemistry that effectively links the microscopic (molecular) level battle for electrons to the macroscopic harnessing of creative new ideas to address the challenge of global energy sources for the next century.

Oxidation-Reduction Reactions

We have considered oxidation-reduction reactions in some detail beginning in Chapter 2. The movement of electrons *away* from an atom or element constitutes an oxidation process. It is caused by an atom with *greater electronegativity* successfully extracting electron density from a *less electronegative* species. That "extraction of electron density" can be a subtle delocalization of electron density or an explicit and complete extraction of the electron from the atom or element. No matter the degree of "extraction of electron density," it is all referred to as oxidation. Reduction is, as we have treated in detail for many different cases, the opposite of oxidation. Reduction is the transfer of electron density to or toward an atom or element. As with oxidation, this can be a subtle delocalization of electron density toward another atom or element, or it can be the explicit and complete donation of the electron.

Step 1   *Assign oxidation states to all atoms and identify the substances being oxidized and reduced.*

Step 2   *Separate the overall reaction into two half-reactions: one for oxidation and one for reduction.*

Step 3   *Balance each half-reaction with respect to mass in the following order:*
- *Balance all elements other than H and O.*
- *Balance O by adding H_2O.*
- *Balance H by adding H^+.*

Step 4   *Balance each half-reaction with respect to charge by adding electrons. (The sum of the charges on both side of the equation should be made equal by adding as many electrons as necessary.)*

Step 5   *Make the number of electrons in both half-reactions equal by multiplying one or both half-reactions by a small whole number.*

Step 6   *Add the two half-reactions together, canceling electrons and other species as necessary.*

Check Yourself 2

Oxidation-reduction reactions in the aqueous medium are very important in electrochemistry. To review redox reactions, *balance* the following reaction in an acidic solution:

$$Al(s) + Cu^{2+}(aq) \rightarrow Al^{3+}(aq) + Cu(s)$$

What makes electrochemistry both extremely potent as a source of energy for practical applications *and* conceptually important for understanding oxidation and reduction, is that the electrons involved are explicitly and completely transferred from the site of oxidation to the site of reduction. As we will see, this is so because those electrons flow through an *external circuit* and can be individually counted!

While electrochemistry does not deal exclusively with metals, it is safe to say that because of the ease with which electrons flow in metals *and* the relative ease with which electrons can be removed from metals, metals constitute the foundation for our understanding of electrochemistry.

We can schematically represent the oxidation or reduction process at the interface of a metal immersed in a liquid containing cations (positive ions) of that metal. This is displayed graphically in Figures 7.10 and 7.11. The key interaction occurs at the metal-solution interface. For oxidation:

- a metal atom M(s) on the surface of the electrode may lose n electrons to the electrode, as shown at the atomic level in Figure 7.10, and enter the solution as the cation M^{n+}. The metal atom is *oxidized* in the process.

$$M(s) \rightarrow M^{n+}(aq) + ne^-$$

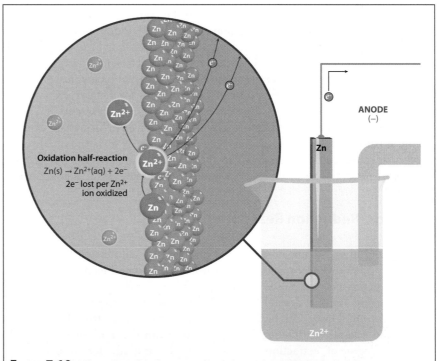

FIGURE 7.10   When viewed at the atomic level, the oxidation that occurs at the anode at the interface between the metal surface and the solution reveals the zinc atom leaving the solid metal structure to enter the solution as Zn^{2+}(aq) with the electron moving into and through the metal electrode. The removal of two electrons from zinc metal, Zn(s), to form the cation, Zn^{2+}(aq), constitutes the oxidation step at the anode.

Or for reduction:

- a metal ion M^{n+} from solution may collide with the electrode, gaining electrons from it, as shown in Figure 7.11, thereby converting the metal cation to a metal atom M(s),

$$M^{n+}(aq) + ne^- \rightarrow M(s)$$

where the metal cation is *reduced* in the process.

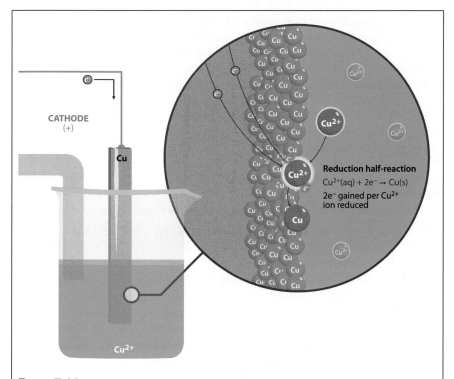

FIGURE 7.11 Reduction, when viewed at the atomic level occurs at the interface between the cathode and the solution containing $Cu^{2+}(aq)$ cations. The union of the copper cations to form copper metal at the electrode surface, $Cu^{2+}(aq) + 2e^- \rightarrow Cu(s)$ constitutes the reduction step at the cathode of the electrochemical half-cell.

Check Yourself 3

Consider Figure 7.10 and 7.11 together. Suppose we connect the beaker in which oxidation occurs, Figure 7.10, to the beaker in which reduction occurs, Figure 7.11, with a conducting wire between the two electrodes.

In which direction do you think the electrons would flow? Explain.

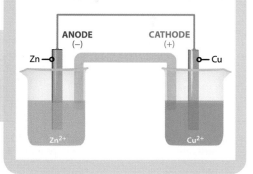

Oxidation-reduction reactions, as displayed in Figures 7.10 and 7.11, lie at the heart of electrochemistry. Oxidation-reduction is a conceptual framework in chemistry wherein one or more electrons are transferred from one species to another. One species is the oxidant, or oxidizing agent, that *takes* electrons from the reductant, or reducing agent. Therefore, *oxidation* denotes a *loss* of electrons; *reduction* denotes a *gain* in electrons.

The oxidizing agent gains electrons causing oxidation to take place such that the oxidizing agent, acting as a recipient for the liberated electrons, undergoes reduction. Consider the case where the metal electrode is a copper rod. The reduction reaction is

Oxidizing agent	Electrons	Reduced species	(reduction)
Cu^{2+}	$+ \quad 2e^-$	$\rightarrow \qquad Cu$	(electron gain)

The oxidation number of the oxidizing agent therefore decreases.

The oxidation reaction involves the reducing agent that loses electrons, so we can write:

Reducing agent		Oxidized species		Electrons	(oxidation)
Zn	$\rightarrow$	Zn^{2+}	$+$	$2e^-$	(electron loss)

The reducing agent's oxidation number increases, going from 0 to +2.

For every electron lost, there is one gained; for every oxidation there is a reduction. It is a two-step process such that oxidation and reduction constitute an inseparable pair—you can't have one without the other, because electrons are neither created nor destroyed in an oxidation-reduction reaction.

We have studied this reciprocity between oxidation and reduction repeatedly before. All oxidation-reduction chemical equations must be balanced with respect to the number of electrons lost in an oxidation step and the number of electrons gained in a reduction step.

In a basic solution.
1. Assign oxidation states.
2. Separate the overall reaction into two half-reactions.
3. Balance each half-reaction with respect to mass.
    - Balance all elements other than H and O
    - Balance O by adding H_2O.
    - Balance H by adding H^+.
    - Neutralize H^+ by adding enough OH^- to neutralize each H^+. Add the same number of OH^- ions to each side of the equation.
4. Make the number of electrons in both half-reactions equal.
5. Add the half-reactions together.
6. Verify that the reaction is balanced.

Check Yourself 4

As a review, balance the following reaction in a basic solution:

$$Cd(s) + 2NiO(OH)(s) + 2H_2O(\ell) \rightarrow Cd(OH)_2(s) + 2Ni(OH)_2(s)$$

Suppose we initiate an oxidation-reduction reaction by placing a coil of clean copper into a solution of silver nitrate. As the reaction proceeds, displayed in Figure 7.12, the clear solution of $Ag^+(aq)$ cations and NO_3^- anions begins to turn a bluish color as copper cations, Cu^{2+}, are released into solution. The *reduction* reaction occurs at the metal-solution interface converting $Ag^+(aq)$ cations in solution to $Ag(s)$ silver atoms on the *surface of the copper coil*.

$$Ag^+(aq) + e^- \xrightarrow{\text{reduction}} Ag(s)$$

The copper atoms at the metal surface, $Cu(s)$, enter solution as Cu^{2+} cations in the oxidation step

$$Cu(s) \xrightarrow{\text{oxidation}} Cu^{2+}(aq) + 2e^-$$

The *net* reaction is the sum of these two steps, balanced according to the number of electrons involved in the overall reaction

$$2\left(Ag^+(aq) + e^- \longrightarrow Ag(s)\right)$$
$$\underline{Cu(s) \longrightarrow Cu^{2+}(aq) + 2e^-}$$
$$2Ag^+(aq) + Cu(s) \longrightarrow 2Ag(s) + Cu^{2+}(aq) \quad \textbf{net reaction}$$

FIGURE 7.12 When clean copper is placed in a solution of silver nitrate ($AgNO_3$), a reaction occurs at the surface of the copper releasing Cu^{2+} ions into solution and forming silver deposits on the copper surface. The result is a messy mixture of silver and copper on the one hand, and a blue solution containing Cu^{2+} cations on the other. While the temperature of the beaker contents increased, no electrical energy could be captured.

It is important to note that

1. The original metal (copper) turns into an amorphous blob of silver and copper.

2. Although the (spontaneous) reaction is exothermic, no usable energy, other than the heat released, can be harnessed from the process resulting from the transfer of electrons in the chemical reaction.

What is unique to electrochemistry is the strategy used to gain control over a process such as the reaction shown in Figure 7.12, that occurs when a metal such as copper is placed in a beaker of silver nitrate, $AgNO_3$. Silver nitrate in solution produces an equilibrium resulting in the release of Ag^+ into solution.

$$AgNO_3 \rightleftharpoons Ag^+ + NO_3^-$$

Insertion of the copper metal into the solution initiates a chemical reaction wherein copper is *oxidized* to form Cu^{2+}

$$Cu(s) \rightarrow Cu^{2+} + 2e^-$$

and Ag^+ is *reduced*

$$Ag^+ + e^- \rightarrow Ag(s)$$

We can also run an experiment where, instead of a solution of $AgNO_3(aq)$, we use zinc nitrate, $Zn(NO_3)_2(aq)$, as contrasted in Figure 7.13. The beaker on the left is a repeat of the experiment depicted in Figure 7.12. A strip of clean copper sheet is inserted into a silver nitrate solution. Ag^+ cations are displaced from the clear silver nitrate solution as silver builds up on the copper surface. In the right-hand beaker of Figure 7.13, a distinctly different occurrence is observed. The same clean copper strip is placed in a beaker of zinc nitrate $Zn(NO_3)_2(aq)$. But *no reaction takes place*. Since the zinc nitrate solution results in the existence of $Zn^{2+}(aq)$ cations in solution, the fact that no chemical reaction occurs means that the reaction between the solid copper, $Cu(s)$, and the zinc cations, Zn^{2+}:

$$Cu(s) + Zn^{2+}(aq) \xrightarrow{\text{reduction}} \text{Products}$$

simply *does not occur*. It is not spontaneous, it does not release free energy.

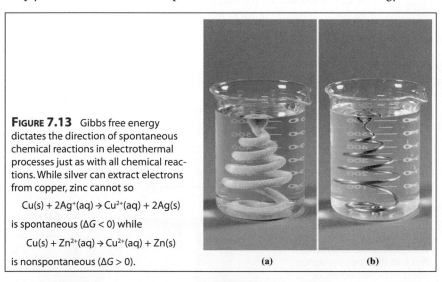

FIGURE 7.13 Gibbs free energy dictates the direction of spontaneous chemical reactions in electrothermal processes just as with all chemical reactions. While silver can extract electrons from copper, zinc cannot so

$Cu(s) + 2Ag^+(aq) \rightarrow Cu^{2+}(aq) + 2Ag(s)$

is spontaneous ($\Delta G < 0$) while

$Cu(s) + Zn^{2+}(aq) \rightarrow Cu^{2+}(aq) + Zn(s)$

is nonspontaneous ($\Delta G > 0$).

(a)   (b)

So while silver cations can extract electrons from copper

$$Cu(s) + 2Ag^+(aq) \rightarrow 2Ag(s) + Cu^{2+}(aq)$$

because $\Delta G < 0$ for this reaction, zinc cations cannot extract electrons from copper, $Cu(s)$, so

$$Cu(s) + Zn^{2+}(aq) \rightarrow Cu^{2+}(aq) + Zn(s)$$

(a) Sketch the Gibbs free energy diagram qualitatively for the reaction in the left-hand beaker:

$$Cu(s) + 2Ag^+(aq) \rightarrow Cu^{2+}(aq) + 2Ag(s)$$

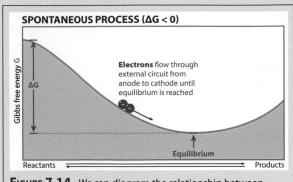

SPONTANEOUS PROCESS (ΔG < 0)

Electrons flow through external circuit from anode to cathode until equilibrium is reached

Gibbs free energy G

ΔG

Equilibrium

Reactants — Products

FIGURE 7.14 We can diagram the relationship between Gibbs free energy and the electrochemical reaction between Cu(s) at the anode and Ag(s) at the cathode. Electrons are transported from the copper anode to the silver cathode, a reaction for which ΔG < 0. Because ΔG < 0, the reaction is spontaneous.

(b) Sketch the Gibbs free energy diagram qualitatively for the reaction on the right:

$$Cu(s) + Zn^{2+}(aq) \rightarrow Cu^{2+}(aq) + Zn(s)$$

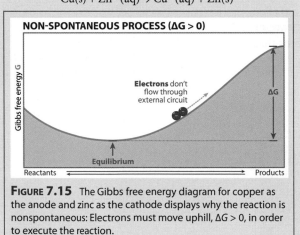

NON-SPONTANEOUS PROCESS (ΔG > 0)

Electrons don't flow through external circuit

Gibbs free energy G

ΔG

Equilibrium

Reactants — Products

FIGURE 7.15 The Gibbs free energy diagram for copper as the anode and zinc as the cathode displays why the reaction is nonspontaneous: Electrons must move uphill, ΔG > 0, in order to execute the reaction.

is *nonspontaneous* and ΔG > 0 for the reaction *as written*, and, therefore, there is no release of free energy in the reaction.

Check Yourself 5

Consider what you see in Figure 7.13. How is this observed contrast captured in a free energy diagram?

But the objective of electrochemistry is to develop strategies wherein electrical energy (free energy) can be *extracted* in an effective manner from such oxidation-reduction reactions. The first step in this strategy is to *isolate* the two oxidation-reduction half reactions, each in a separate cell. It is difficult to overstate the importance of realizing that it was necessary to separate this electrochemical reaction into two separate cells in order to extract electrical energy that can do work from the redox reaction, rather than simply releasing the free energy as heat.

The Galvanic or Voltaic Cell

The basic set up for electrochemistry is the union of Figures 7.10 and 7.11 shown in Figure 7.16: this is called a *galvanic* or *voltaic* cell. It consists of two electrodes. The first electrode, dipped in a solution, has the ability to *release positive ions* from its surface into the solution. A positive ion is an atom stripped of some (or all) of its valence electrons. Positive ions are called *cations*. The second electrode is also dipped in a (generally different) solution and has the ability to *acquire cations* from the solution. The electrode that *releases* cations is called the *anode*; the electrode that acquires cations is called the *cathode*. When a cation is released from the anode, there is an extra negative charge, one or more electrons, that remains behind. Similarly, when a cation is acquired by the cathode, there is an extra positive charge on the cathode. If the anode and the cathode are *connected* by a good conductor, the excess electrons leave the negatively charged anode and travel to the positively charged cathode to restore the charge balance. As long as the release of cations from the anode and the capture of cations by the cathode continue, there is a current

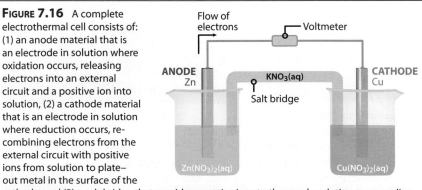

FIGURE 7.16 A complete electrothermal cell consists of: (1) an anode material that is an electrode in solution where oxidation occurs, releasing electrons into an external circuit and a positive ion into solution, (2) a cathode material that is an electrode in solution where reduction occurs, recombining electrons from the external circuit with positive ions from solution to plate-out metal in the surface of the cathode, and (3) a salt bridge that provides negative ions to the anode solution to neutralize the positive ions produced in the oxidation step and (by the salt bridge) positive ions to the cathode solution to neutralize the negative ions produced.

of electrons traveling from the anode to the cathode. If we put a voltmeter into the circuit connecting the anode to the cathode, it will register a voltage. The voltage difference measured in this process is defined to be positive when electrons travel from the anode to the cathode. The architecture of the copper-zinc galvanic cell is shown in Figure 7.16.

Figure 7.16 displays a typical "galvanic cell," the key components of which consists of the following: the anode is made of zinc (Zn), the cathode is made of copper (Cu), and the solutions in each case are $Cu(NO_3)_2$ and $Zn(NO_3)_2$. If we connect these two with a conducting wire and a voltmeter, and if the concentration of the $Cu(NO_3)_2$ and $Zn(NO_3)_2$ are exactly 1 M at temperature 298 K (that is, 25°C, or room temperature), the voltmeter will show 1.10 V. To sustain the current through the external circuit we have to close the loop, which is accomplished by the so called "salt bridge," a tube containing a solution of KNO_3 that dissociates into K^+ ions and NO_3^-. This is connected by two semipermeable membranes to the $Cu(NO_3)_2$ and $Zn(NO_3)_2$ solutions.

Check Yourself 6

Consider Figure 7.16, a galvanic cell with a zinc anode and a copper cathode.

(a) Draw the oxidation step at the atomic scale that is occurring at the interface of the zinc electrode and the $Zn(NO_3)_2$ solution. Include in your drawing the Zn atom, the Zn^{2+} cation, and the relevant electrons.

(b) Repeat a, but for the cathode side. Again include in your drawing the Cu atom, the Cu^{2+} cation, and the relevant electrons.

Now the whole system works as follows: Zn^{2+} cations are released from the anode into the solution containing $Zn(NO_3)_2$, and the extra electrons move from the zinc anode to the external wire; the Zn^{2+} in solution attracts NO_3^- from the salt bridge to balance the charge in the anode solution. To balance the charge on the cathode side, the K^+ cations enter the cathode solution in place of the Cu^{2+} cations removed from solution.

Check Yourself 7

Consider the salt bridge in Figure 7.16.

(a) Sketch a picture of the species in the $KNO_3(aq)$ salt bridge including only the relevant potassium and nitrate species.

(b) Sketch the migration of species from the salt bridge into the anode side of the galvanic cell.

(c) Sketch the migration of species from the salt bridge into the cathode side of the galvanic cell.

At the cathode, the $Cu^{2+}(aq)$ cations in solution combine with the electrons at the interface of the metal and solution that have arrived to that interface from the anode through the external circuit, thereby forming Cu(s) on the surface of the cathode. This makes a complete circuit. As long as there is $Zn(NO_3)_2$ in the anode solution and $Cu(NO_3)_2$ in the cathode solution, the process can keep going, producing current between the anode and cathode. It is that current, that flow of electrons, that can be used for useful purposes. The system is then described by the following reactions:

$$\text{Cathode}: \qquad Cu^{2+}(aq) + 2e^- \rightarrow Cu(s) \qquad \textbf{(7.1)}$$

$$\text{Anode}: \qquad \underline{Zn(s) \rightarrow Zn^{2+}(aq) + 2e^-} \qquad \textbf{(7.2)}$$

$$\text{Net reaction}: \quad Zn(s) + Cu^{2+}(aq) \rightarrow Cu(s) + Zn^{2+}(aq) \qquad \textbf{(7.3)}$$

The functioning of this system depends on the *relative ability* of the copper and zinc electrodes to capture electrons when they are connected through the conducting wire that constitutes the external circuit. In other words, there is a balance in the system which has to do with the intrinsic properties of the metals that comprise the two electrodes. This balance is determined by the so-called "half-reactions." We choose to write these half reactions, for consistency, with the cation always on the left-hand side:

$$Zn^{2+}(aq) + 2e^- \rightarrow Zn(s) \qquad \textbf{(7.4)}$$

$$Cu^{2+}(aq) + 2e^- \rightarrow Cu(s) \qquad \textbf{(7.5)}$$

There is a reason why these half reactions are written with the metal cation as a reactant on the left-hand side of the chemical reaction. The reason is that it is the *ability of that cation to draw electrons to it* in competition with another cation that determines whether an electrochemical reaction pair is spontaneous ($\Delta G < 0$) or is nonspontaneous ($\Delta G > 0$). It is the ability of the cation of one metal to more strongly attract electrons than the cation of another metal that ultimately determines whether an electrochemical cell, as constructed, will in fact produce electricity. This is all about the battle of the cations to gain electrons.

In the galvanic cell described by the net reaction (7.3), the first reaction (7.1) proceeds forwards and the second reaction (7.2) proceeds backwards because $Cu^{2+}(aq)$ cations *attract electrons more strongly* than $Zn^{2+}(aq)$ cations. Recall that the reaction in which a cation is combined with electrons to produce a neutral atom is a *reduction reaction* and the reaction in which a cation and free electrons are produced, is called an *oxidation reaction*. In the present example, the reduction of Cu cations *wins* (proceeds forward) and the reduction of Zn cations *loses* (proceeds backward). To balance the electric charges, for each Zn^{2+} cation that is produced, one Cu^{2+} cation is reduced, because both cations have charge +2.

So let's review the specific processes that bring the electrochemistry (galvanic) cell to life:

- Free energy drives the process downhill wherein copper cations (Cu^{2+}) in solution *successfully extract* electrons from Zn, oxidizing Zn to Zn^{2+} at the anode. This produces two electrons that flow through the external circuit connecting the anode to the cathode. The Gibbs free energy diagram for the process is displayed in Figure 7.17.

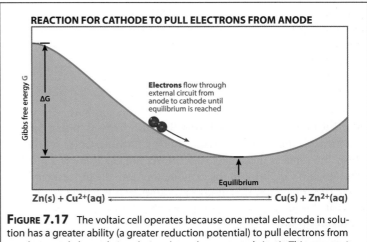

FIGURE 7.17 The voltaic cell operates because one metal electrode in solution has a greater ability (a greater reduction potential) to pull electrons from another metal electrode in solution through an external circuit. This propensity to pull electrons from the anode "down hill" to the cathode means that the electrons move from a position of higher Gibbs free energy to a position of lower Gibbs free energy, as equilibrium is approached, releasing free energy, ΔG.

- Zn atoms release electrons *at the anode* in an oxidation step and those Zn atoms, each stripped of two electrons, enter the solution as Zn^{2+} ions

$$Zn \rightarrow Zn^{2+}(aq) + 2e$$

- Electrons lost by Zn atoms pass through the wire linking the half cells to the cathode where they are acquired by Cu^{2+} cations at the cathode in a reduction step:

$$Cu^{2+}(aq) + 2e \rightarrow Cu(s)$$

We construct our electrochemical cell, designating the anode (Zn) and cathode (Cu), as, respectively, the site of oxidation (anode) and of reduction (cathode) as shown in Figure 7.16.

The Half-Cell Reactions

It is important to sketch out these components to keep the half-cell reactions correctly identified whenever working a problem. (Notice that an important mnemonic is that the order of anode/cathode and oxidation/reduction are in alphabetical order from left to right across the diagram, which is the same direction as the electron motion!)

- As electrons flow from the anode to the cathode, and the cathode collects cations from solution, simultaneously anions (NO_3^-) move from the salt bridge into the Zn half of the cell to electrically neutralize the Zn^{2+} formed at the anode in the oxidation step.
- Cations (K^+) from the KNO_3 salt bridge migrate into the Cu half cell and neutralize the net negative charge resulting from the loss of Cu^{2+} cations.
- As a result, no net charge builds up in either the anode solution nor the cathode solution.

The key point is that electrons seek *whatever* pathway is available to them to find the lowest free energy state that they can, as displayed in Figure 7.17. This *motive force to seek a lower free energy state* generates a potential that serves to extract electrons from the anode and transport them through the external conductor (wire) to the cathode. Just as a mass in a gravitational field has a potential energy equal to *mgh* (where *m* is the mass, *g* is the acceleration of gravity, and *h* is the height above the "ground"), so too will the electron "fall through" the potential created by the free energy release producing useful work as displayed in Figure 7.17. From the perspective of energy and work, the principles are the same; only the names change.

Electrons flow because there is an *electrical potential difference* generated between the two half cells of the electrochemical cell. That *electrical potential difference* is termed the *cell voltage* and it has the units of joule/coulomb, where the coulomb is the unit of charge—a unit we will explore more fully in the following sections.

As noted in Figure 7.16, which depicts our Zn/Cu electrochemical cell, if we place a voltmeter in our external circuit, that voltmeter (if it is accurate) will read 1.103 volts. That voltage is a measure of the *relative ability* of Cu to extract electrons from Zn. This ability to extract electrons by one element in competition with another is expressed as a *Reduction Potential* for it is the measure of an element's ability to steal electrons (*oxidation*) from another element that serves as a *reducing agent*; thus the term *reduction potential*. The measured overall cell potential, determined by a voltmeter connected between the anode and cathode of the electrochemical cell, arises from a *competition*, a tug-of-war, between the two half cells for the electrons wherein Cu *outduels* Zn for the electrons and in the process oxidizes Zn at the anode and reduces Cu^{2+} at the cathode. This competition for electrons, this tug-of-war to extract electrons, is shown in Figure 7.18. Note that it is the more powerful *reducer* that wins! Each element has a *Standard Reduction Potential* when measured under standard conditions (25°C, 1 molar, 1 atm). We will, in the next section, tabulate these reduction potentials. In electrochemistry, we express the *standard reduction potential* as

$$\text{Standard Reduction Potential} = E°.$$

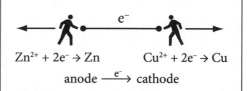

FIGURE 7.18 Whenever two metals are connected by a pathway that allows electrons to flow, a tug-of-war ensues, wherein one of the metals, the metal with the greater reduction potential, successfully pulls electrons away from the metal with the lesser reduction potential. Oxidation occurs at the anode because that electrode loses electrons (in the tug-of-war) to the cathode where reduction occurs.

When two half cells are connected, the *Standard Cell Potential* is the difference between the standard reduction potentials of the two electrodes such that

$$E^\circ_{cell} = \begin{bmatrix} \text{standard reduction potential} \\ \text{of the substance reduced} \end{bmatrix} - \begin{bmatrix} \text{standard reduction potential} \\ \text{of the substance oxidized} \end{bmatrix} \quad \textbf{(7.6)}$$

as summarized in Figure 7.18 and Figure 7.19.

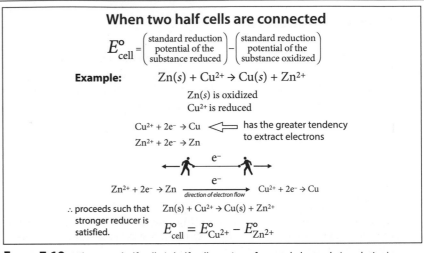

FIGURE 7.19 When two half-cells (a half-cell consists of a metal electrode in solution) are connected with an external circuit (a wire) opening up a pathway for electrons to flow, the electrode with the greater reduction potential pulls electrons from the electrode with the lesser reduction potential in a tug-of-war. The cell potential, E_{cell}, calculated from the individual reduction potentials by subtracting the smaller reduction potential from the larger.

1. Identify the cation with the greater ability to compete for electrons, and sketch a figure indicating the tug-of-war between the respective cation.

2. Identify the element that is reduced. Identify the element that is oxidized.

3. Write the half-cell reactions indicating which reaction occurs at the anode and which reaction occurs at the cathode.

4. Write the equation for the cell voltage, E°_{cell}, as given by Equation 7.6. properly identifying the reduced and oxidized species as is done in the framed equation in Figure 7.19.

Check Yourself 8

Suppose we consider the galvanic cell constructed from an electrode of silver and an electrode of copper. Examine the relationship between Figure 7.19 and Figure 7.14 to answer the following:

The Standard Hydrogen Electrode

There are many possible combinations of anodes and cathodes that could make up a galvanic cell. To characterize the *relative* ability of these systems to produce useful electrical energy, we need a standard way to compare them. This is done by the definition of a *reduction potential* relative to a *standard electrode*. The objective of the standard electrode is to establish a reference value of *zero* for the *reduction potential* of that *standard electrode*. Once the zero of the reduction potential scale is set, the reduction potential for every half-cell configuration is referenced to that standard electrode. The standard electrode is chosen by international decree to be made of hydrogen gas, H_2, and the potential of the following reduction reaction is taken to be *zero*:

$$2H^+(aq) + 2e^- \rightarrow H_2(g) \quad \textbf{(7.7)}$$

Since the free energy release of the reaction is affected by the concentration, the temperature and the pressure, the standard conditions are chosen to be 1 M concentration of H^+, a temperature of 298 K and a pressure of 1 atm of $H_2(g)$. Also, because it is difficult (in fact impossible!) to have an electrode made of pure H_2 at standard conditions, the reaction is done on a Pt catalyst, using H_2 gas at the standard conditions of temperature and pressure. Relative to this potential, the potential of any other electrode is called the *Standard Electrode Potential*. The Standard Hydrogen Electrode is shown in Figure 7.20.

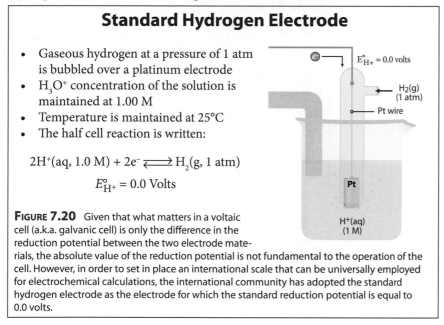

Standard Hydrogen Electrode

- Gaseous hydrogen at a pressure of 1 atm is bubbled over a platinum electrode
- H_3O^+ concentration of the solution is maintained at 1.00 M
- Temperature is maintained at 25°C
- The half cell reaction is written:

$$2H^+(aq, 1.0 \text{ M}) + 2e^- \rightleftharpoons H_2(g, 1 \text{ atm})$$

$$E^\circ_{H^+} = 0.0 \text{ Volts}$$

FIGURE 7.20   Given that what matters in a voltaic cell (a.k.a. galvanic cell) is only the difference in the reduction potential between the two electrode materials, the absolute value of the reduction potential is not fundamental to the operation of the cell. However, in order to set in place an international scale that can be universally employed for electrochemical calculations, the international community has adopted the standard hydrogen electrode as the electrode for which the standard reduction potential is equal to 0.0 volts.

This *standard* hydrogen electrode potential provides the means to set a *quantitative scale* for a vast array of possible anode/cathode combinations and to thereby determine the cell potential and thus the spontaneous direction of electron flow in a wide array of anode and cathode materials to form an electrochemical cell. Table 7.1 lists an array of standard electrode potentials at 25°C in order of *decreasing* reduction potential. Thus, the electrodes at the top of the chart have the greatest ability to attract electrons in a battle with electrodes of lower reduction potential, which appear at the bottom of the table.

However, before we proceed to discuss the design of electrochemical cells, it is important to visualize in detail what occurs at the platinum electrode interface with the solution. This is shown with atomic level resolution in Figure 7.21. Specifically, the platinum electrode plays no chemical role in the reaction. It is chemically inert—it serves only as a stable physical interface through which electrons readily move. As such, it constitutes an important class of *inert* electrodes, a topic we will develop in more detail. However, carefully follow the sequence of events displayed in Figure 7.21. The H^+ cations, which we express as H_3O^+ following Chapter 7, extract electrons from the platinum surface producing $H_2(g)$ that creates a bubble of hydrogen at the surface of two electrodes. The other product, H_2O, enters the solution.

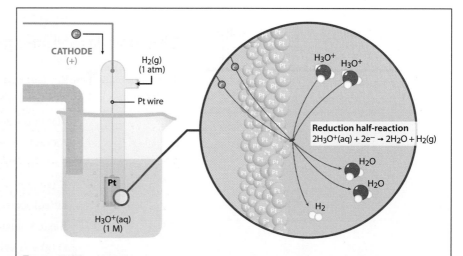

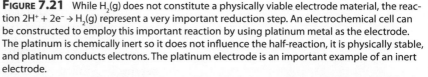

FIGURE 7.21   While $H_2(g)$ does not constitute a physically viable electrode material, the reaction $2H^+ + 2e^- \rightarrow H_2(g)$ represent a very important reduction step. An electrochemical cell can be constructed to employ this important reaction by using platinum metal as the electrode. The platinum is chemically inert so it does not influence the half-reaction, it is physically stable, and platinum conducts electrons. The platinum electrode is an important example of an inert electrode.

TABLE 7.1 Some selected Standard Electrode (Reduction) Potentials at 25°C

Reduction Half-Reaction	$E°$, V
Acidic Solution	
$F_2(g) + 2e^- \rightarrow 2F^-(aq)$	+2.866
$O_3(g) + 2H^+(aq) + 2e^- \rightarrow O_2(g) + H_2O(l)$	+2.075
$S_2O_8^{2-}(aq) + 2e^- \rightarrow 2SO_4^{2-}(aq)$	+2.01
$H_2O_2(aq) + 2H^+(aq) + 2e^- \rightarrow 2H_2O(l)$	+1.763
$MnO_4^-(aq) + 8H^+(aq) + 5e^- \rightarrow Mn^{2+}(aq) + 4H_2O(l)$	+1.51
$PbO_2(s) + 4H^+(aq) + 2e^- \rightarrow Pb^{2+}(aq) + 2H_2O(l)$	+1.455
$Cl_2(g) + 2e^- \rightarrow 2Cl^-(aq)$	+1.358
$Cr_2O_7^{2-}(aq) + 14H^+(aq) + 6e^- \rightarrow 2Cr^{3+}(aq) + 7H_2O(l)$	+1.33
$MnO_2(s) + 4H^+(aq) + 2e^- \rightarrow Mn^{2+}(aq) + 2H_2O(l)$	+1.23
$O_2(g) + 4H^+(aq) + 4e^- \rightarrow 2H_2O(l)$	+1.229
$2IO_3^-(aq) + 12H^+(aq) + 10e^- \rightarrow I_2(s) + 2H_2O(l)$	+1.20
$Br_2(l) + 2e^- \rightarrow 2Br^-(aq)$	+1.065
$NO_3^-(aq) + 4H^+(aq) + 3e^- \rightarrow NO(g) + 2H_2O(l)$	+0.956
$Ag^+(aq) + e^- \rightarrow Ag(s)$	+0.800
$Fe^{3+}(aq) + e^- \rightarrow Fe^{2+}(aq)$	+0.771
$O_2(g) + 2H^+(aq) + 2e^- \rightarrow H_2O_2(aq)$	+0.695
$I_2(s) + 2e^- \rightarrow 2I^-(aq)$	+0.535
$Cu^{2+}(aq) + 2e^- \rightarrow Cu(s)$	+0.340
$SO_4^{2-}(aq) + 4H^+(aq) + 2e^- \rightarrow 2H_2O(l) + SO_2(g)$	+0.17
$Sn^{4+}(aq) + 2e^- \rightarrow Sn^{2+}(aq)$	+0.154
$S(s) + 2H^+(aq) + 2e^- \rightarrow H_2S(g)$	+0.14
$2H^+(aq) + 2e^- \rightarrow H_2(g)$	0
$Pb^{2+}(aq) + 2e^- \rightarrow Pb(s)$	−0.125
$Sn^{2+}(aq) + 2e^- \rightarrow Sn(s)$	−0.137
$Cd^{2+}(aq) + 2e^- \rightarrow Cd(s)$	−0.403
$Fe^{2+}(aq) + 2e^- \rightarrow Fe(s)$	−0.440
$Zn^{2+}(aq) + 2e^- \rightarrow Zn(s)$	−0.763
$Al^{3+}(aq) + 3e^- \rightarrow Al(s)$	−1.676
$Mg^{2+}(aq) + 2e^- \rightarrow Mg(s)$	−2.356
$Na^+(aq) + e^- \rightarrow Na(s)$	−2.713
$Ca^{2+}(aq) + 2e^- \rightarrow Ca(s)$	−2.84
$K^+(aq) + e^- \rightarrow K(s)$	−2.924
$Li^+(aq) + e^- \rightarrow Li(s)$	−3.040
Basic Solution	
$O_3(g) + H_2O(l) + 2e^- \rightarrow O_2(g) + 2OH^-(aq)$	+1.246
$OCl^-(aq) + H_2O(l) + 2e^- \rightarrow Cl^-(aq) + 2OH^-(aq)$	+0.890
$O_2(g) + 2H_2O(l) + 4e^- \rightarrow 4OH^-(aq)$	+0.401
$2H_2O(l) + 2e^- \rightarrow H_2(g) + 2OH^-(aq)$	−0.828

Selected Examples: The Energy Ladder of Reduction Potentials

7.20

The patterns evident in the tabulation of *Standard Electrode Potentials* in Table 7.1 emerge from the pattern of electronegativities across the periodic table reviewed in Figure 7.22. The highly electronegative elements and compounds have the greatest reduction potentials. The halogen compounds, oxygen, and oxides dominate the higher reduction potentials, while the alkali metals and alkaline earth metals with extremely low electronegativity possess the lowest reduction potentials. Just as fluorine, oxygen, chlorine, and nitrogen draw electrons to them in chemical bonds, so too do these species act as *excellent cathodes*, vying very successfully for electrons. In a similar vein, just as lithium, sodium, potassium, etc., donate electrons to ionic bonds, so too do these species readily give up electrons in an electrochemical cell, thereby constituting an excellent anode material where they undergo oxidation. When viewed from the perspective of electronegativity, the periodic table takes the form shown in Figure 7.22.

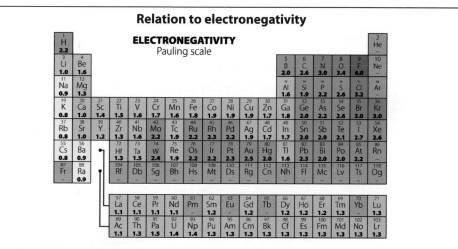

Relation to electronegativity

Electronegativity (tendency of atoms to "grab" electrons) is related to electrochemical potential—relative to H!

FIGURE 7.22   Different elements have varying abilities to draw electrons to them. The relative ability of individual elements in the periodic table to attract electrons is quantified by a scale called the electronegativity scale. The pattern of electronegativity that occurs in the periodic table is shown here. The elements with the greatest electronegativity appear in the upper right of the periodic table (F, O, N, Cl, Br, C, etc.) and the elements with the lowest electronegativity appear in the lower left of the periodic table. Electronegativity of the individual elements is related to the reduction potential of those elements in an electrochemical cell.

Calculation of the Cell Potential

It is important to examine the behavior of a number of electrodes in an electrochemical cell when one of those electrodes is a *standard hydrogen electrode*. To analyze the behavior of such a cell we must first establish which electrode is the anode, the site of oxidation (where electrons are lost), and which electrode is the cathode, the site of reduction (where electrons are gained). We can determine this experimentally using a voltmeter inserted into the external circuit. If we attach the positive lead of the voltmeter to the anode, the meter will read a positive value. If we make a mistake and connect the voltmeter the wrong way, the voltmeter will read a negative voltage. Alternatively, we can consult our table of standard electrode potentials to determine which of the two electrodes has the higher reduction potential. If, for example, we choose Cu as the companion electrode to the standard hydrogen electrode, we see that $E^\circ_{Cu} = +0.340$ volts. Because $E^\circ_{H_2} = 0.0$ volts, we know that the Cu electrode will successfully extract electrons from the hydrogen electrode. This cell is diagramed in Figure 7.23.

Check Yourself 9

It is important to always keep in mind the relative energy (voltage) difference between the half cells involved in an electrochemical cell.

(a) Sketch the hydrogen and copper half cells on an energy diagram indicating which half cell lies at the higher potential.

(b) Connect those two half cells with a conductor and a salt bridge and indicate the direction of electron flow.

Solution

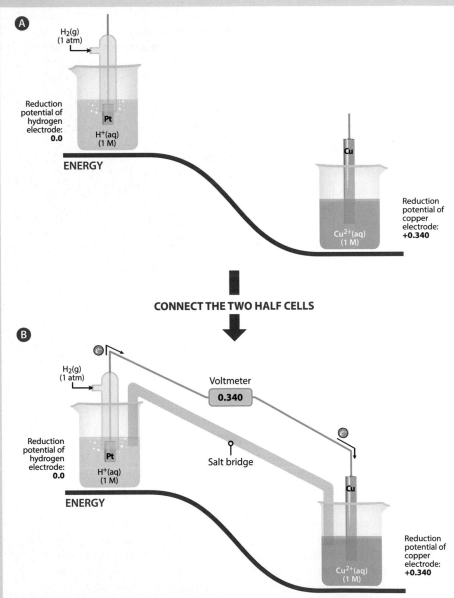

FIGURE 7.23 With the standard reduction potential of the hydrogen electrode set at 0.0 volts, the reduction potential of other electrochemical half-cells are measured relative to the hydrogen electrode. In the case shown here, the copper electrode has a higher standard reduction potential (+0.34 volts) because copper pulls electrons from the hydrogen anode to the copper cathode.

It is instructive to analyze the hydrogen-copper galvanic cell in some detail. Figure 7.23, in the upper panel, represents the two half-cells on an energy diagram *before they are connected*. Because the copper electrode has a higher reduction potential, it has a greater ability to draw electrons to it than does the hydrogen electrode. Thus the copper electrode lies at a *lower* energy than does the hydrogen electrode. If the half-cells are connected by an external circuit consisting of a conducting wire between the hydrogen half-cell and the copper half-cell, and a salt bridge joins the two half-cells, electrons will flow downhill in energy. This is displayed in the lower panel of Figure 7.23.

We can also condense the complete picture presented in Figure 7.23 and place just the half-reactions on an energy diagram

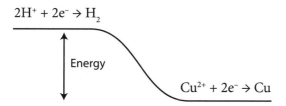

The fact that the $Cu^{2+} + 2e^- \rightarrow Cu$ reaction lies at a lower energy than $2H^+ + 2e^- \rightarrow H_2$ means that in a tug-of-war over electrons, the copper cation will win:

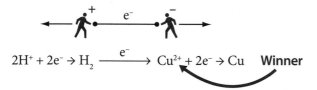

As a result, electrons will flow from the hydrogen half-cell to the copper half-cell. This means that if we are to write the *net reaction* for the electrochemical cell comprised of the two half-cells, then we must *balance* the electrons in the overall reaction. The reason is that electrons cannot be created nor destroyed in the electrochemical cell. Thus, since the copper half-cell reaction *extracts* electrons from the hydrogen half-cell by virtue of the greater reduction potential of Cu^{2+}, we must write the reduction of Cu^{2+}(aq) in the *forward* direction.

$$Cu^{2+}(aq) + 2e^- \rightarrow Cu(s)$$

The hydrogen half-cell, in contrast, must give up electrons to the copper cell, so we write that half-cell reaction backwards because that reaction must *produce* electrons taken up by Cu^{2+}(aq) at the cathode

$$H_2(g) \rightarrow 2H^+(aq) + 2e^-$$

When we add those two reactors to give the net reaction, the electrons cancel.

$$Cu^{2+} + 2e^- \longrightarrow Cu(s)$$
$$H_2(g) \longrightarrow 2H^+(aq) + 2e^-$$

net $\quad \overline{Cu^{2+} + H_2(g) \longrightarrow Cu(s) + 2H^+(aq)}$

To complete the analysis we calculate the *cell potential* by subtracting the *standard reduction potential* of the cell that *loses* electrons from the *standard reduction potential* of cell that gains electrons:

$$E^\circ_{cell} = \begin{bmatrix} \text{standard reduction} \\ \text{potential of the electrode} \\ \text{that } \textit{gains} \text{ electrons} \end{bmatrix} - \begin{bmatrix} \text{standard reduction} \\ \text{potential of the electrode} \\ \text{that } \textit{loses} \text{ electrons} \end{bmatrix}$$

$$= 0.340 \text{ V} - 0.00 \text{ V} = +0.340 \text{ V}$$

Calculate the standard reduction potential for Ag^+ if the standard reduction potential for Cu^{2+} is 0.34 volts.

Solution

This requires that we identify the substance oxidized and the substance reduced. Silver changes from Ag^+ to Ag; its oxidation number decreases from +1 to 0, so Ag^+ is reduced. Similar reasoning tells us that copper is oxidized from Cu to Cu^{2+}. Therefore, according to Equation 7.6,

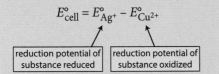

$$E^{\circ}_{cell} = E^{\circ}_{Ag^+} - E^{\circ}_{Cu^{2+}}$$

| reduction potential of substance reduced | reduction potential of substance oxidized |

Substituting values for $E^{\circ}_{Ag^+}$.

$$E^{\circ}_{Ag^+} = 0.46\ V + 0.34\ V$$
$$= 0.80\ V$$

The standard reduction potential of silver ion is therefore +0.80 V.

Does the answer seem reasonable?

We know the cell potential is the difference between the two reduction potentials. The difference between +0.80 V and +0.34 V (subtracting the smaller from the larger) is 0.46 V. Our calculated reduction potential for Ag^+ appears to be correct. We can, of course, cross check our answer against Table 7.1!

Check Yourself 10

The standard cell potential of the silver-copper galvanic cell has a value of 0.46 V. The cell reaction is:

$$2Ag^+(aq) + Cu(s) \rightarrow 2Ag(s) + Cu^{2+}(aq)$$

If, on the other hand, we pair a Zn electrode with our standard hydrogen electrode, the analysis demonstrates that $E^{\circ}_{H_2} = 0.0$ volts, but $E^{\circ}_{Zn^{2+}} = -0.76$ volts. Therefore, we know that, because $E^{\circ}_{H_2} > E^{\circ}_{Zn^{2+}}$, electrons will be extracted from Zn by the H_2 electrode so that the Zn electrode is the anode, the H_2 electrode is the cathode; oxidation occurs at the Zn electrode, and reduction occurs at the Hydrogen electrode. Figure 7.24 sets out the electrochemical cell comprised of Zn as the anode and H_2 as the cathode. The cell voltage is

$$E^{\circ}_{cell} = \left[\begin{array}{c}\text{standard reduction potential} \\ \text{of the substance reduced}\end{array}\right] - \left[\begin{array}{c}\text{standard reduction potential} \\ \text{of the substance oxidized}\end{array}\right]$$

$$= E^{\circ}_{\text{"winner"}} - E^{\circ}_{\text{"loser"}} = E^{\circ}_{reduc} - E^{\circ}_{oxid}$$

$$= E^{\circ}_{cathode} - E^{\circ}_{anode} = +0.76\ \text{volts}$$

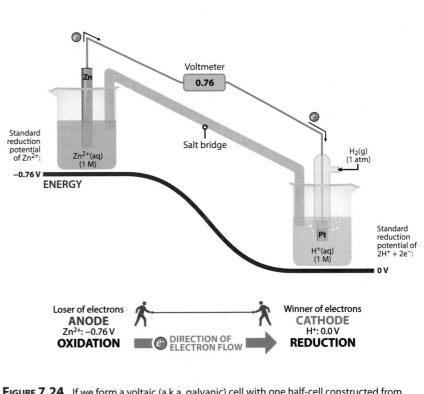

FIGURE 7.24 If we form a voltaic (a.k.a. galvanic) cell with one half-cell constructed from zinc, and the other from hydrogen, we discover that the hydrogen successfully extracts electrons from zinc and thus zinc has a lower standard reduction potential than hydrogen. Thus zinc becomes the anode in the electrochemical cell donating electrons to hydrogen, which is the cathode.

The key points to remember are:
- When the half-cells are connected, this establishes an open path for electrons to flow.
- With the establishment of an open path for electron flow, a tug-of-war is set up between the two half-cells for electrons.

- It is the half-cell with the greater standard reduction potential that has the greater ability to draw electrons to it.
- The half-cell with the *greater* standard reduction potential lies at a *lower* free energy, and electrons move to a state of lower free energy.
- Electrons flow *to* the reducing electrode, which is the cathode.
- Electrons flow *from* the electrode where oxidation occurs which is the anode.
- The cell voltage is the difference between standard reduction potential of the cathode and the standard reduction potential of the anode:

$$E^\circ_{cell} = E^\circ_{winner} - E^\circ_{loser} = E^\circ_{cathode} - E^\circ_{anode}$$

As we discussed in the presentation of Figure 7.17, electrochemical systems are driven in their spontaneous direction toward the release of free energy that in turn is established by the difference in the standard reduction potential between the cathode and the anode with the larger *reduction potential* drawing electrons "downhill" to the cathode. By the same argument, the calculated *cell potential*,

$$E^\circ_{cell} = E^\circ_{cathode} - E^\circ_{anode}$$

can tell us whether a reaction, as written, is spontaneous, i.e. for which $\Delta G < 0$. The key point is that in an electrochemical (galvanic) cell, the calculated cell potential for a spontaneous reaction is always *positive*. Therefore, if the calculated cell potential is negative, the reaction, as written, is spontaneous in the reverse direction.

Check Yourself 11

Determine whether the following reaction is spontaneous *as written*. If it is not spontaneous, write the equation for the reaction that is spontaneous. Consider the reaction

$$Cd^{2+}(aq) + Cu(s) \rightarrow Cd(s) + Cu^{2+}(aq)$$

Discussion of Electrodes

At this point it may appear that the metal from which the electrodes are made must match the cation in solution—namely that in the zinc-copper electrochemical cell the anode must be zinc and the cathode must be copper. However, consider first what the zinc anode and copper cathode look like after an electrochemical cell has operated for a period of time. This is displayed in Figure 7.25. The anode is decidedly eroded away because the reaction

$$Zn(s) \rightarrow Zn^{2+}(aq) + 2e^-$$

has transferred *zinc* metal from the electrode into solution. Thus the anode must indeed be zinc because it is the reactant in the anode (oxidation) half-reaction.

Analysis

First we must determine the *standard reduction potential* for each element or compound involved in the half-cell reaction of the electrochemical cell. We recognize, by the definition of the standard reduction potential that it is the ability of an electrode material to *extract* electrons that determines the reduction potential. Thus we can write, based on an inspection of Table 7.1, that $Cu^{2+}(aq)$ has the *greater* standard reduction potential so we write to half-cell reaction as

$$Cu^{2+}(aq) + 2e^- \rightarrow Cu(s) \qquad E^\circ_{Cu^{2+}} = +0.34 \text{ V}$$

The cadmium half-cell reaction has a standard reduction potential of $E^\circ_{Cd^{2+}} = -0.403$ V. Thus the copper half-cell lies at lower free energy and electrons will flow from the Cd half-cell to the Cu half-cell.

$$Cd^{2+} \xrightarrow{\quad e^- \quad} Cu^{2+}$$

Thus we write the Cd half-cell reaction as

$$Cd(s) \rightarrow Cd^{2+}(aq) + 2e^- \qquad E^\circ_{Cd^{2+}} = -0.403 \text{ V}$$

Combining the two half-reactions we have

$$Cu^{2+}(aq) + 2e^- \longrightarrow Cu(s)$$
$$Cd(s) \longrightarrow Cd^{2+}(aq) + 2e^-$$

net $\quad \overline{Cu^{2+}(aq) + Cd(s) \longrightarrow Cd^{2+}(aq) + Cu(s)}$

Inspection of our original chemical reaction reveals that, as written, the reaction is nonspontaneous. Only in the reverse direction is the reaction spontaneous.

FIGURE 7.25 Examination of the zinc anode and copper cathode reveals that the anode has been eaten away, depositing $Zn^{2+}(aq)$ cations into solution. The buildup of copper metal on the cathode results from the reduction of $Cu^{2+}(aq)$ in the cathode solution. While the anode must be constructed of $Zn(s)$ to supply $Zn^{2+}(aq)$ to the solution, $Cu^{2+}(aq)$ cations will plate out on any cathode that is conductive and sufficiently inert so as not to interface with the reduction of copper cations to form copper metal.

Inspection of the cathode reveals that copper metal, $Cu(s)$, has built up on the surface of the electrode because $Cu^{2+}(aq)$ cations from solution have been reduced by electrons emergent from the cathode:

$$Cu^{2+}(aq) + 2e^- \rightarrow Cu(s)$$

This immediately raises the question: is it necessary that the cathode be composed of the same materials as the metal that plates out on the cathode surface? Must copper metal compose the cathode in order for copper cations to be successfully reduced at the electrode-solution interface? The answer is that the cathode material need not be identical to the metal that plates out on the electrode surface.

The conclusion that the *cathode* material need not be the same as the metal that plates out on it has important implications. What the cathode electrode must provide is (1) a conductive path for electrons coming from the anode through the external circuit, (2) a solid interface at the electrode-solution interface, and (3) the absence of any chemical activity that would impede the union of the cation-electron reduction at the surface. Two cathode materials are particularly good at achieving these objectives. One is platinum and the other carbon. We can depict what happens in a galvanic cell constructed from a zinc anode and a platinum cathode, with the platinum cathode placed in a solution of $Cu(NO_3)_2$ in Figure 7.26.

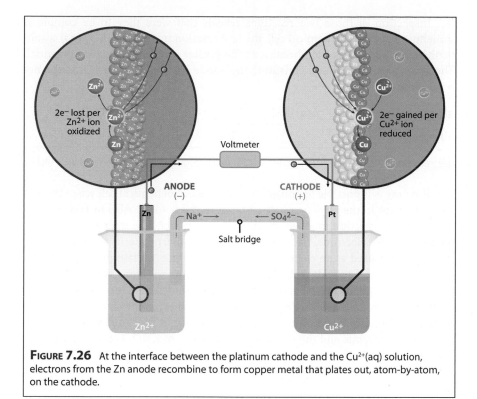

FIGURE 7.26 At the interface between the platinum cathode and the $Cu^{2+}(aq)$ solution, electrons from the Zn anode recombine to form copper metal that plates out, atom-by-atom, on the cathode.

Active vs. Inactive Electrodes

It is important to recognize the role taken by an electrode that is not explicitly involved in an electrochemical half-reaction. An electrode that is not explicitly involved in a half reaction is not oxidized or reduced. It therefore does not appear in the half-reactions that establish oxidation at the anode or reduction at the cathode. *In practice* the role an inert electrode can play in electrochemistry dramatically widens the array of possible electrochemical reactions. The most ubiquitous example of an inert electrode is that of the standard hydrogen electrode. It is obvious that hydrogen, $H_2(g)$, does not constitute a physically viable electrode material. Rather, as we discussed, platinum is used as a mechanically stable, chemically inert interface with the flowing hydrogen gas and the protons, H^+, in solution as depicted in the atomic level schematic in Figure 7.21.

What about other oxidation-reduction reactions that appear in Table 7.1 that do not possess viable electrode materials in their half-reaction? For example the reaction:

$$Mn\text{-}O_4^-(aq) + 8H^+(aq) + 5e^- \rightarrow Mn^{2+}(aq) + 4H_2O(l)$$

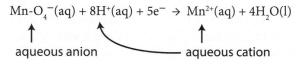

aqueous anion   aqueous cation

This is clearly an aqueous phase oxidation-reduction reaction with a standard reduction potential of +1.51 volts defined in Table 7.1. However, what is the electrode material? There does not appear to be any viable physically stable electrode material in the half-reaction.

Another example from Table 7.1 is the reaction

$$I_2(s) + 2e^- \rightarrow 2I^-(aq)$$

with a standard reduction potential of +0.535 volts.

Inspection of those two reactions reveals that were they to be combined somehow in an electrochemical cell, the first reaction involving $MnO_4^-(aq)$ would be the cathode half-reaction because of the greater standard reduction potential. The second reaction would constitute the anode half-reaction such that the net reaction would be:

Cathode: Reduction half-reaction: $MnO_4^-(aq) + 8H^+(aq) + 5e^- \rightarrow Mn^{2+}(aq) + 4H_2O$

Anode: Oxidation half-reaction: $2I^-(aq) \rightarrow I_2(s) + 2e^-$

Net overall cell reaction: $2MnO_4^- + 16H^+(aq) + 10I^-(aq) \rightarrow 2Mn^{2+}(aq) + 5I_2(s) + 8H_2O(l)$

But how is it possible to construct a viable electrochemical cell when there is no participant in the *reduction* half-reaction that is a solid? Also in this case the *oxidation* half-reaction involves the solid I_2, which lacks the strength to serve as physically stable electrode. In cases such as this, an inert material such as graphite or platinum is used as an electrode. The inert graphite or platinum serves to deliver electrons and provide a stable interface with the solution. In the anode half cell, $I^-(aq)$ is oxidized to $I_2(s)$ on a graphite rod as shown in Figure 7.27. Electrons so released flow to the cathode, which is also an inert graphite electrode. Electrons flow through the graphite reducing $MnO_4^-(aq)$ to $Mn^{2+}(aq)$ at the interface between the graphite and the solution. A salt bridge of KNO_3 is used to maintain charge neutrality with NO_3^- entering the anode half-reaction and K^+ entering the cathode half-reaction. The design of the complete electrochemical cell is displayed in Figure 7.27.

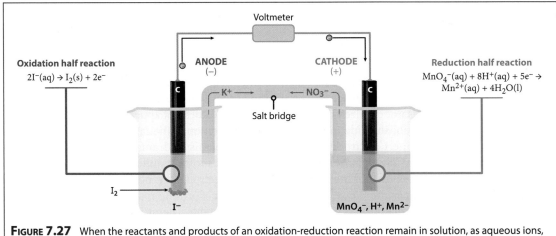

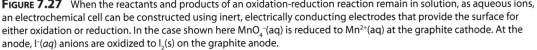

FIGURE 7.27 When the reactants and products of an oxidation-reduction reaction remain in solution, as aqueous ions, an electrochemical cell can be constructed using inert, electrically conducting electrodes that provide the surface for either oxidation or reduction. In the case shown here $MnO_4^-(aq)$ is reduced to $Mn^{2+}(aq)$ at the graphite cathode. At the anode, $I^-(aq)$ anions are oxidized to $I_2(s)$ on the graphite anode.

The idea of using inert electrodes represents a major advance in extending the field of electrochemistry to a far broader compliment of redox reactions. This becomes particularly important in the design of new batteries, as we will see in the Case Studies at the end of the Chapter.

An inspection of Table 7.1 reveals how important the use of inert electrodes is to modern electrochemistry. Nearly half of the reactions listed in Table 7.1 require inert electrodes to function.

Notation for an Electrochemical Cell: A Shorthand Technique

To this point in the chapter, we have taken pains to present for each case a complete drawing of an electrochemical cell showing explicitly the oxidation at the anode, the external circuit, the reduction at the cathode, the direction of electron movement and the salt bridge that maintains charge neutrality. Once the operation of the electrochemical cell is mastered, the focus turns to breaking the voltaic cell down to just represent the species oxidized and the species reduced.

Consider the voltaic cell in Figure 7.28 which uses an inert platinum cathode to mediate the reduction of $MnO_4^-(aq)$ to $Mn^{2+}(aq)$ and an iron anode that is sacrificed in the oxidation reaction $Fe(s) \rightarrow Fe^{2+}(aq) + 2e^-$. The overall redox reaction is

$$5Fe(s) + 2MnO_4^-(aq) + 16H^+(aq) \rightarrow 5Fe^{2+}(aq) + 2Mn^{2+}(aq) + 8H_2O(\ell)$$

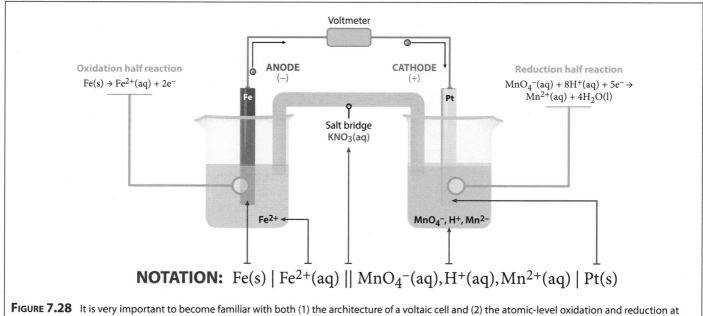

FIGURE 7.28 It is very important to become familiar with both (1) the architecture of a voltaic cell and (2) the atomic-level oxidation and reduction at the anode and cathode respectively. However, once comfortable with the operation of a voltaic cell, the shorthand *notation* for an electrochemical cell represents an important simplification. Here the notation identifies the formation of $Fe^{2+}(aq)$ cations from $Fe(s)$ at the anode, the salt bridge, the reduction of $MnO_4^-(aq)$ to $Mn^{2+}(aq)$ at the cathode, and the inert platinum electrode on the cathode side.

While this is a rather complex reaction that occurs in the two chambers of the voltaic cell in Figure 7.28 the entire structure of the voltaic cell, the anode half-reaction, the cathode half-reaction, the salt bridge, and the inert electrode, can be captured in the electrochemical cell notation displayed below the cell with each segment of the cell *notation* coupled by an arrow to the corresponding segment of the voltaic cell itself. In this condensed notation, the following protocol is used:

1. In keeping with the convention of placing the oxidation half-reaction cell on the left and the reduction half-reaction on the right, so too does the condensed cell *notation* retain this convention. The double vertical line separating the anode and cathode designates the salt bridge.

2. The species oxidized at the anode are separated by a single vertical line to indicate the change in phase. For example in Figure 7.28 the $Fe(s)$ and $Fe^{2+}(aq)$ are separated by a vertical line.

3. For the important example of redox reactions for which species within a given half cell remain in the *same phase*, reactants and products are

Solution

For $Cl_2|Cl^-$ combined with $H^+|H_2$, we find and record the half reactions and their reduction potentials from Table 7.1, and subtract the one with the lower voltage from that with the higher, in accord with Equation 7.6:

$$Cl_2 + 2e^- \rightarrow 2Cl^- \qquad E° = 1.360 \text{ V} \qquad \text{(cathode)}$$
$$H_2 \rightarrow 2H^+ + 2e^- \qquad E° = 0.000 \text{ V} \qquad \text{(anode)}$$
$$\overline{H_2 + Cl_2 \rightarrow 2H^+ + 2Cl^- \qquad 1.360 \text{ V} = E°_{cell}}$$

The line notation for this cell, since $H^+|H_2$ is the anode, is $Pt(s)|H_2|H^+||Cl^-|Cl_2|C(gr)$; in this cell both electrodes are inert. Note that for any cell constructed with a hydrogen electrode, the magnitude of $E°_{cell}$ is given by that of the reduction potential for the other half-reaction, since $E° = 0.000$ V. For $Cl_2|Cl^-$ with $Al^{3+}|Al$, we reuse the $Cl_2|Cl^-$ $E°$, and apply the factors of the least common multiple needed to balance the electrons.

$$Cl_2 + 2e^- \rightarrow 2Cl^- \qquad E° = 1.360 \text{ V} \qquad \text{(cathode)}$$
$$Al \rightarrow Al^{3+} + 3e^- \qquad E° = -1.68 \text{ V} \qquad \text{(anode)}$$
$$\overline{2Al + 3Cl_2 \rightarrow 2Al^{3+} + 6Cl^- \qquad 3.04 \text{ V} = E°_{cell}}$$

For this cell the line notation is $Al|Al^{3+}||Cl^-|Cl_2|C(gr)$.

distinguished from each other with a comma as is the case with $MnO_4^-(aq)$, $H^+(aq)$, and $Mn^{2+}(aq)$ in the cathode side of voltaic cell notation in Figure 7.28.

4. The Pt(s) on the far right of the voltaic cell notation designates the inert electrode material. If an inert electrode material is used on the *anode* side, that electrode appears at the far *left extremity* of the condensed notation separated by a vertical line before listing the reagents in the anode half cell.

Check Yourself 12

A $Cl_2|Cl^-$ half-cell can be built by saturating a NaCl solution with $Cl_2(g)$ and inserting an inert electrode such as Pt(s) or C(gr). What voltage will be generated by a standard $Cl_2|Cl^-$ half-cell combined with a standard hydrogen electrode? With a standard $Al^{3+}|Al$ half-cell? Give the balanced cell reactions and the line notation for these cells.

Maximum Work from a Cell: Gibbs Free Energy

With the ability to calculate the cell voltage for a wide variety of redox reactions, we are in a position to calculate the amount of *work*, the amount of *energy*, that can be extracted from an electrochemical cell. Before we do that, however, recall from Chapter 1 how we expressed the potential energy of a mass, m, in a gravitational field, and how that potential energy is released as the mass falls through the gravitational field, releasing potential energy in the form of kinetic energy. This sequence was depicted in Figures 7.2 and 7.4.

Notice in Figure 7.2 that we could just as easily label the vertical axis "available work" or "kinetic energy released" because, indeed, the *maximum* amount of work, w, that this *mechanical system* can deliver is $w = mgh$. Any friction or dissipation forces intrinsic to the system will *remove* an amount of available work by converting some of that potential energy + kinetic energy into thermal energy (heat) that is *unavailable* to do mechanical work, as we discussed in Chapter 3.

Embodied in this mechanical system is a *gravitational force* that is the genesis of the potential energy and of the system's ability to convert potential energy into available work. Likewise, embodied in the oxidation-reduction reaction that constitutes the motive force in the electrochemical cell is an *electromotive force*. This electromotive force provides the capacity to do purposeful work through the external circuit linking the anode to the cathode: to pump water, move an automobile, heat a building, or light a lecture hall. To understand this release of purposeful energy by an electrochemical system, we seek to formulate how the electromotive force in such a system is related to the force (i.e., mg) in our mechanical system. While there are a number of ways of developing the quantitative parallel between work and energy release in a mechanical system versus work and energy release in an electrochemical system, the most straightforward way is to recognize that we can simply replace the "potential energy" label on the vertical axis of Figure 7.2 with the free energy label, G, for the electrochemical cell, thus creating the free energy diagram that was introduced in Figures 7.3 and 7.4.

We now see a key link between the mechanical (gravitational) system and the electrochemical system because the *maximum* amount of work that can be extracted from the mechanical system is equal to *mgh*, and in the case of the electrochemical system, the maximum amount of work that can be extracted is the free energy, G (under conditions of constant pressure and temperature).

We know, however, that the ability, the *Electromotive Force*, to move electrons is determined by

$$E^\circ_{cell} = E^\circ_{cathode} - E^\circ_{anode}$$

But how do we calculate the number of joules released in an electrochemical system?

So, as has been demonstrated in countless carefully executed experiments based upon the coupled system displayed in Figure 7.4 dating back to the experiments of Michael Faraday, Figure 7.29, the number of joules is equal to the maximum work available resulting from electrons sliding down the free energy surface from the anode to the cathode. But the *charge* of the electron, e, in combination with the number of electrons, n, and the cell voltage, E_{cell}, is what determines the *force* on the electrons, $-n\,e\,E_{cell}$, in the presence of a voltage *difference* between two electrodes. Thus we know that the free energy available to do work will be equal to the product of the charge on the electron, the number of electrons, and the electromotive force between the two electrodes, E_{cell}.

It was Michael Faraday (1791–1867) who made the measurements necessary to relate (1) the free energy (maximum available work) in joules to (2) the cell potential, E°_{cell}. We know that the charge on the electron is measured in coulombs; the charge on a single electron is 1.6×10^{-19} coulombs. The number of electrons in a mole is, of course, 6.022×10^{23} electrons, so the number of coulombs per mole of electrons is

$$F = (1.602 \times 10^{-19} \text{ C/electron})(6.022 \times 10^{23} \text{ electron/mole})$$
$$= 9.6485 \times 10^4 \text{ C/mole}$$

That quantity is the *Faraday Constant*; it is the magnitude of the charge contained in one mole of electrons. Thus, if we have n moles of electrons passing from the anode to the cathode, the total negative charge possessed by those electrons is

$$\text{Total Negative Charge} \equiv -nF \qquad \textbf{(7.7)}$$

The capacity of the external circuit in our electrochemical cell to do work is the product of the total charge passing between the anode and cathode times the voltage between the two electrodes, E_{cell}:

Work = (Cell voltage)(Total charge passing through the circuit)
     = $(E_{cell})(-nF)$

However, Faraday showed that the maximum amount of work that could be extracted from an electrochemical cell was equal to the free energy, ΔG, such that

$$\text{Maximum amount of work} = \Delta G = -nFE_{cell} \qquad \textbf{(7.8)}$$

That provides the quantitative basis for linking a change in *chemical energy*, ΔG, with a matching change in *electrical energy*, $-nFE_{cell}$.

FIGURE 7.29 Michael Faraday (c. 1842) was, and is, regarded as one of the great experimental scientists in human history. Born in 1791 as one of four children, his father, a blacksmith, had moved to London in 1790. With little formal education, Faraday began working with Sir Humphrey Davey of the Royal Institution and Royal Society. He had many discoveries (benzene, chlorine clathrates) and established the laws of electrochemistry developing the terms we use today: anode, cathode, electrode, and ion. His most famous work involved employing electromagnetism to develop generators, motors, and transformers based on the discovery that a changing magnetic field produces an electric field. His contribution to electrochemistry is captured in the "Faraday Constant," F, which is equal to the magnitude of the charge contained in 1 mole of electrons. Specifically, $F = 9.6485 \times 10^4$ coulombs/mole.

We can now tie these concepts together in a single plot, shown in Figure 7.30, of *Gibbs free energy*, ΔG, for an electrochemical cell, in the case for the reaction

$$Zn(s) + Cu^{2+}(aq) \rightleftharpoons Zn^{2+}(aq) + Cu(s)$$

by adding the arrow designating the electrical energy, $-nFE_{cell}$.

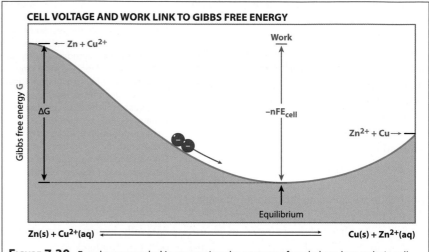

CELL VOLTAGE AND WORK LINK TO GIBBS FREE ENERGY

FIGURE 7.30 Faraday succeeded in measuring the amount of work done by a voltaic cell when a total change of *nF* (where n is the number of moles of electrons and *F* is the number of coulombs of charge per mole of electrons) passes between electrodes with a cell potential of E_{cell}. Recognizing that ΔG for the electrochemical process is the maximum amount of work that can be extracted means that $\Delta G = -nFE_{cell}$, and we can include this on the free energy diagram for the voltaic cell as shown here.

1. Find the work done in pushing electrons through a 100 W light bulb with a potential difference of 120 V for 1 hr. Proceed by finding the current I from the relationship power, $P = I\,E_{cell}$, then find the charge Q = I t in coulombs. By analyzing your calculation, suggest an easier way to do it.

2. How many electrons passed through the light bulb in an hour? How many moles of electrons?

Check Yourself 13
We have been careful to keep the concepts of energy, work, and power clearly defined and frequently applied. Equation 7.8 equates the maximum amount of work, w_{max}, with ΔG and the product of total charge ($-nF = Q$) and the cell voltage, E_{cell}:

$$w_{max} = \Delta G = -nFE_{cell} = QE_{cell}$$

We can also review the logic that links (a) the maximum work that can be extracted from an electrochemical cell to (b) the cell potential, E_{cell}. It is the ability to quantitatively link (1) the *chemical energy*, ΔG, released in a chemical reaction with (2) the electrical energy, $-nFE_{cell}^{\circ}$, contained in an electrochemical cell that succeeds in the coupling of chemical thermodynamics, chemical equilibria, and electrochemical cell voltages. Because of the importance of the link between chemical energy and the cell voltage, we review this union for emphasis.

From a strictly experimental foundation emerged two key results: (1) it was determined that the maximum amount of *work* that can be extracted from a *chemical* reaction is $-\Delta G$, and (2) it was determined that the maximum amount of work that can be extracted from an *electrochemical* cell was nFE_{cell}°.

It was these empirically (experimentally) determined facts that led Faraday to *equate* these two quantities:

$$\Delta G = - nFE_{cell}$$

At *standard conditions* we can write

$$\Delta G^{\circ} = -nFE_{cell}^{\circ}$$

We know from Chapter 4 that the units of ΔG are joules. What about the units of nFE°_{cell}?

$$n = \text{moles of electrons}$$
$$F = 9.65 \times 10^4 \text{ coulombs/mole}$$
$$\text{and } E^\circ_{cell} \text{ is measured in volts}$$
$$\text{where the unit of volts}$$
$$\text{is joules/coulomb.}$$

The *units* of nFE°_{cell} are therefore

$$(\text{moles})(\text{coulomb/mole})(\text{joules/coulomb}) = \text{joules}$$

Thus, as it must be if we are to equate ΔG and $-nFE^\circ_{cell}$, the units of both must be the units of energy; joules.

Link Between K_{eq}, $\Delta G°$, and $E°_{cell}$

This union of $\Delta G°$ and the cell voltage E°_{cell}, is a remarkable result—a remarkable conclusion. It says that at standard conditions, the change in *Gibbs free energy*, $\Delta G°$, is simply equal to the cell potential, $E^\circ_{cell} = E^\circ_{cathode} - E^\circ_{anode}$, times the total number of electrons that pass through the external circuit, $-n$, times a constant F. But this suggests another key relationship because we know, from Chapter 6, that the change in *Gibbs free energy* at *standard conditions* (25°C, 1 atm pressure, 1 M solution), $\Delta G°$, is given by

$$\Delta G° = -RT \ln K_{eq} \qquad (7.9)$$

An obvious question is: is that the *same* Gibbs free energy without any modification? Remarkably, that is so. Combining the above equation with

$$\Delta G° = -nFE^\circ_{cell} \qquad (7.10)$$

we have the key expression that links the cell voltage $E^\circ_{cell} = E^\circ_{cathode} - E^\circ_{anode}$, at standard conditions, to the equilibrium constant, K_{eq}:

$$-nFE^\circ_{cell} = -RT \ln K_{eq} \qquad (7.11)$$

So that

$$E^\circ_{cell} = \frac{RT}{nF} \ln K_{eq} \qquad (7.12)$$

where R = gas constant = 8.3 J/mole·K, F = Faraday constant = 9.65×10^4 coulomb/mole, and n = number of moles of electrons transferred. This expression for E°_{cell} in terms of the equilibrium constant, K_{eq}, immediately implies that from a measured equilibrium constant we can *calculate the cell voltage* or cell reduction potential. But, because E°_{cell} is usually far easier to measure (it only requires a voltmeter), the equilibrium constant can be calculated from the cell voltage, E°_{cell}, at standard conditions. Taking the antilog of each side of Equation 7.12, we have

$$K_{eq} = \exp\left[\frac{nFE^\circ_{cell}}{RT}\right] \qquad (7.13)$$

Solution

For this reaction, 2 electrons are transferred, so

$$n = 2$$
$$F = 9.65 \times 10^4 \text{ C/mole}$$
$$R = 8.3 \text{ J/mol·K}$$
$$T = 298 \text{ K}$$

so

$$K_{eq} = \exp\left[\frac{(2 \text{ moles})(9.65 \times 10^4 \text{ C/mole})(0.32 \text{ J/C})}{(8.3 \text{ J/mol·K})298 \text{ K}}\right]$$

$$= 7.1 \times 10^{10}$$

Check Yourself 14

Calculate K_{eq} for the reaction

$$NiO_2(s) + 2Cl^-(aq) + 4H^+(aq) \rightarrow Cl_2(g) + Ni^{2+}(aq) + 2H_2O$$

The measured cell voltage is

$$E^\circ_{cell} = +0.320 \text{ volts @ } 25°C$$

We now have equations linking K_{eq}, $\Delta G°$, and E°_{cell}. We also have methods for determining each of these by direct observation:

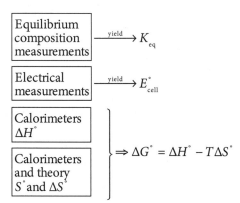

Thus we can summarize these key relationships with a "master diagram" linking $\Delta G°$, E°_{cell}, and K_{eq}, as shown in Figure 7.31.

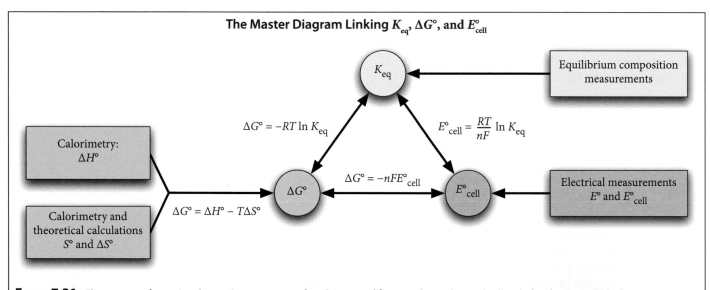

FIGURE 7.31 The concept of equating the maximum amount of work extracted from an electrochemical cell with the change in Gibbs free energy, $-\Delta G$, provided the equality $\Delta G° = -nFE^\circ_{cell}$. But that relationship opens the door for establishing the relationship between E°_{cell} and the equilibrium constant, K_{eq} because $K_{eq} = \exp(-\Delta G°/RT)$ such that $E^\circ_{cell} = (RT/nF) \ln K_{eq}$. This triangle linking $\Delta G°$, E°_{cell}, and K_{eq} is of immense importance to both electrochemistry and to thermodynamics.

Check Yourself 15
Linking between K_{eq}, $\Delta G°$, and $E°_{cell}$

Problem
It is frequently the case, particularly in regions of the western United States, that lead and silver coexist in ore deposits. Because silver is the more valuable of the metals, it is important to employ the fact that lead can displace silver from solution:

$$Pb(s) + 2Ag^+(aq) \rightarrow Pb^{2+}(aq) + 2Ag(s)$$

This provides a chemical method for extracting pure silver. Calculate K and $\Delta G°$, at 298 K (25°C) for this reaction.

Solution
First, recall the relationship between the signs of K_{eq}, $\Delta G°$, and $E°_{cell}$ at standard state

$\Delta G°$	K	$E°_{cell}$	Reaction at standard state condition
			spontaneous
			at equilibrium
			nonspontaneous

Using values of $E°$ from Table 7.1, write the half-reactions:

$$Ag^+(aq) + e^- \rightarrow Ag(s) \qquad E° = 0.80 \text{ V}$$
$$Pb^{2+}(aq) + 2e^- \rightarrow Pb(s) \qquad E° = -0.13 \text{ V}$$

Next calculate $E°_{cell}$:

because $Ag^+(aq)$ has the larger reduction potential, it is the winner, and thus the cathode and is written in the forward direction. To balance the number of electrons, we multiply it by 2 and add it to the reverse of the Pb^{2+} reduction reaction:

$$2Ag^+(aq) + 2e^- \rightarrow 2Ag(s) \qquad\qquad E° = 0.80 \text{ V}$$
$$Pb(s) \rightarrow Pb^{2+}(aq) + 2e^- \qquad\qquad E° = -0.13 \text{ V}$$
$$\overline{2Ag^+(aq) + Pb(s) \rightarrow 2Ag(s) + Pb^{2+}(aq) \quad E°_{cell} = 0.80 \text{ V} - (-0.13 \text{ V}) = 0.93 \text{ V}}$$

From our master diagram in Figure 7.31

$$E°_{cell} = \frac{RT}{nF} \ln K_{eq} = 2.303 \frac{RT}{nF} \log K_{eq}$$

Inspection of the half reaction demonstrates that 2 moles of electrons are transferred for each mole of the reaction so n = 2. Substituting in for R, T, n, and F, we have

$$E°_{cell} = \frac{0.0592 \text{ V}}{2} \log K_{eq} = 0.93 \text{ V}$$

$$\log K_{eq} = \frac{0.93 \text{ V} \times 2}{0.0592 \text{ V}} = 31.42$$

$$\boxed{K = 2.6 \times 10^{31}}$$

Next we calculate $\Delta G°$

$$\Delta G° = -nFE°_{cell} = \left(-\frac{2 \text{ mol } e}{\text{mol r} \times \text{n}}\right)\left(\frac{96.5 \text{ kJ}}{\text{V} \cdot \text{mol } e}\right)(0.93 \text{ V})$$

$$= -1.8 \times 10^2 \text{ kJ}$$

The Death of an Electrochemical Cell: The Nernst Equation

We typically create an electrochemical cell from fresh components. For example, we can assemble a fresh Cu cathode, a fresh Zn anode, fresh 1 M solution of $Zn(NO_3)_2$, fresh 1 M solution of $Cu(NO_3)_2$, and a fresh KNO_3 salt bridge, as shown in Figure 7.16. We read the voltage produced by this new cell at 25°C and it reads 1.10 volts. But we also recognize that, as is the case for all cells (a.k.a. batteries, which we will study in the following sections), these energy sources that power our iPods, iPhones, computers, flashlights, automobile starter motors, cordless power tools, etc., slowly (or rapidly) "go dead" and must be discarded or unless they are "recharged" by some external source of power. How does this process of "going dead" happen?

We know from our discussion of Gibbs free energy that the electrochemical cell operates by virtue of the fact that the difference in the *reduction potentials* of the cathode and anode establish a downhill path from reactants higher in free energy to products lower in free energy. The electrochemical cell's capability to execute work depends upon the fact that the reactants and products are not in equilibrium—just as the cells in your body that sustain you are far from equilibrium. Were your cells in equilibrium, you would not be alive. But an electrochemical cell, in the process of releasing free energy, expends its (high free energy) reactants and moves inexorably toward a state of equilibrium. As the cell approaches equilibrium, its ledge or hill of high chemical potential, initially separating fresh reactants from products, is eroded away by the conversion of reactants to products until the driving force, which depends on the height differential, ΔG, of that potential surface, is eliminated. This is shown schematically in Figure 7.32. The free energy surface becomes flat, equilibrium is attained, and the electrochemical cell is exhausted—the battery is dead, $\Delta G = 0$.

But this is familiar territory for us, because we studied *equilibria* in Chapter 6. We recognized how such reactions progress from their initial conditions to a state of equilibrium. If we consider the redox reaction of our Zn/Cu electrochemical cell, we know the overall reaction is

$$Zn(s) + Cu^{2+}(aq) \rightarrow Zn^{2+}(aq) + Cu(s)$$

with the voltaic cell notation given by $Zn(s) \mid Zn^{2+} \parallel Cu^{2+}(aq) \mid Cu(s)$. But this is a chemical reaction, so by the law of mass action, we can write the reaction quotient, Q, as

$$Q = \frac{[\text{products}]}{[\text{reactants}]} = \frac{[Zn^{2+}(aq)][Cu(s)]}{[Zn(s)][Cu^{2+}(aq)]}$$

We also know that, as long as some of the pure solids $Zn(s)$ and $Cu(s)$ are present (which is to say the pure metals are present in excess), we can drop them from the reaction quotient and write

$$Q = \frac{[Zn^{2+}(aq)]}{[Cu^{2+}(aq)]}$$

With a fresh cell, before any of the reactants $[Cu^{2+}]$ have been converted to products, very little additional $Zn^{2+}(aq)$ will have been formed, and a maximum amount of Cu^{2+} will be available. Thus, the reaction quotient, Q, will be at its minimum value. Each oxidation of Zn to Zn^{2+} erodes away material at the anode and increases the Zn^{2+} concentration in solution. On the other hand, each union of electrons with Cu^{2+} at the cathode (reduction), adds $Cu(s)$ to the cathode, decreasing Cu^{2+} in the $Cu(NO_3)_2$ solution. Steadily the reaction quotient, $Q = [Zn^{2+}]/[Cu^{2+}]$, increases as the reaction progresses toward equilibrium.

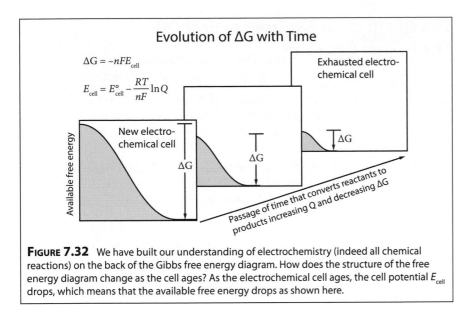

FIGURE 7.32 We have built our understanding of electrochemistry (indeed all chemical reactions) on the back of the Gibbs free energy diagram. How does the structure of the free energy diagram change as the cell ages? As the electrochemical cell ages, the cell potential E_{cell} drops, which means that the available free energy drops as shown here.

But our study of chemical equilibria (see Chapter 6) demonstrated that the relationship between $\Delta G°$, the change in *Gibbs free energy* under *standard conditions*, and ΔG, the change in *Gibbs free energy* under *any* conditions, is given by

$$\Delta G = \Delta G° + RT \ln Q \qquad \textbf{(7.14)}$$

Because $E_{cell} = -\Delta G/nF$, as ΔG goes to zero, so too does the cell voltage, E_{cell}.

However, we are now in a position to rewrite our expression for ΔG under *any conditions* by a direct substitution of $\Delta G = -nFE_{cell}$ and $\Delta G° = -nFE°_{cell}$ to yield the equation of central importance to electrochemistry, the *Nernst equation*:

$$E_{cell} = E°_{cell} - \frac{RT}{nF} \ln Q \qquad \textbf{(7.15)}$$

The Nernst equation relates the cell potential, E_{cell}, to the *progress* of the reaction, expressed by Q.

Before we move on to apply this Nernst equation to specific problems, let's review some important points using our Zn/Cu electrochemical cell shown in Figure 7.26 as an explicit example.

The zinc-copper electrochemical cell, operating at standard conditions (25°C, 1 M, 1 atm), yields a cell voltage, $E°_{cell}$, of 1.103 volts, as shown in the diagram. Because the volt is a joule/coulomb, that voltage difference, generated between the anode and cathode, imparts 1.103 joules of energy to each coulomb of charge transferred from Zn metal to the Cu^{2+} ions (recall that one coulomb is 6×10^{17} electrons!). That means that if 1 coulomb of electrons moves from the anode to the cathode, the maximum work available is 1.10 joules, the change in Gibbs free energy, ΔG. If two coulombs of charge pass from the anode to the cathode, 2.20 joules are available to execute work via the external circuit. But the voltage, $E°_{cell}$, measured in joules per coulomb does *not depend on the size of the cell*. $E°_{cell}$ depends on the concentrations, not on the cell size. $E°_{cell}$ is identical to $\Delta G°$ within a constant, $E° = -\Delta G/nF$, and $E°_{cell}$ is therefore simply related to the equilibrium constant, K:

$$E°_{cell} = \frac{RT}{nF} \ln K_{eq}$$

Just as ΔG goes to zero when Q goes to K_{eq}, so too does E_{cell} go to zero as Q goes to K_{eq}.

WALTHER NERNST was born in West Prussia in 1864. He was responsible for formulating the Third Law of Thermodynamics and he worked extensively with chemical explosives. Nernst worked at the intersection of thermodynamics and electrochemistry and in so doing, developed the *Nernst equation* that relates the cell voltage at standard condition to the cell voltage as concentrations within the cell change.

Check Yourself 16—Application of the Nernst Equation

The Nernst equation is one of the most important in electrochemistry and it results from the straightforward linking of (a) the expression for Gibbs free energy under non-standard conditions

$$\Delta G = \Delta G^\circ + RT \ln Q$$

with (b) the Faraday expression relating cell voltage to ΔG:

$$\Delta G = -nFE_{cell}$$

to yield

$$E_{cell} = E^\circ_{cell} - \frac{RT}{nF} \ln Q$$

Problem:

A galvanic cell that employs electrodes of Ni(s) and Cr(s) with the half reactions

$$Cr^{3+}(aq) + 3e \rightarrow Cr(s)$$
$$Ni^{2+}(aq) + 2e^- \rightarrow Ni(s)$$

is used in a circuit.

Calculate the cell voltage if $[Cr^{3+}(aq)] = 2.0 \times 10^{-3}$ M and $[Ni^{2+}(aq)] = 1.0 \times 10^{-4}$ M.

Solution:

First recognize that because both $[Cr^{3+}(aq)]$ and $[Ni^{2+}(aq)]$ are not 1.0 M, the cell is not at standard conditions. Thus we must use the Nernst equation to calculate the cell voltage.

Next we need to determine the number of electrons involved in the anode-cathode reactions. Finally, we must note that while $Cr^{3+}(aq)$ and $Ni^{2+}(aq)$ enter the calculation of Q in the Nernst equation, the pure solids Cr(s) and Ni(s) do not.

Given that $E^\circ_{Cr3+} = -0.74$ volts and $E^\circ_{Ni2+} = -0.25$ volts, it is Ni^{2+} that has the higher standard reduction potential so the Ni/Ni^{2+} electrode is the cathode and the Cr/Cr^{3+} electrode is the anode. The balanced half-cell reactions are thus

$$2 \times [Cr(s) \rightarrow Cr^{3+}(aq) + 3e] \text{ oxidation}$$
$$3 \times [Ni^{2+}(aq) + 2e \rightarrow Ni(s)] \text{ reduction}$$

$$3\,Ni^{2+}(aq) + 2Cr(s) \rightarrow 3\,Ni(s) + Cr^{3+}(aq)$$

Moreover, six electrons are exchanged in the reaction so n = 6. This provides the information we need to apply the Nernst equation to calculate the cell voltage, E_{cell}, under non-standard conditions:

$$E_{cell} = E^\circ_{cell} - \frac{RT}{nF} \ln Q$$

$$E^\circ_{cell} = -0.25 \text{ V} - [-0.74 \text{ V}] = +0.49 \text{ V}$$

$$n = 6 \qquad R = 8.314 \text{ J/mol} \cdot \text{K} \qquad T = 298 \text{ K}$$
$$F = 9.65 \times 10^4 \text{ C/mol}$$
$$Q = [Cr^{3+}][Ni^{2+}] = 4 \times 10^{-6}$$

So

$$E_{cell} = +0.49 \text{ V} - \frac{(8.314 \text{ J/mol} \cdot \text{K})(298 \text{ K})}{6(9.65 \times 10^4 \text{ C/mol})} \ln (4 \times 10^{-6})$$

$$= 0.49 \text{ V} - 0.0651 \text{ V} = 0.42 \text{ V}$$

It is important to note that E_{cell} is just slightly lower than E°_{cell} even though the cell is far from standard conditions.

The reason is that RT/nF is very small (just 4.28×10^{-3} in this case) and the natural log of Q suppresses changes in Q as the cell ages.

The Master Diagram

We are now in a position to link three major domains of modern chemistry using a disarmingly concise set of equations. Specifically:

1. Measurements of ΔH°, S°, and ΔS° from *thermochemistry* provide the foundation for determining $\Delta G^\circ = \Delta H^\circ - T\Delta S^\circ$, the Gibbs free energy for a vast array of new chemical compounds;

2. Equilibrium *composition* measurements of new materials provide the measurement of K_{eq}; and

3. *Electrical* measurements of E° and E°_{cell} establish the power delivery and storage capabilities for innovative new battery technology, fuel cells, etc.

But we can join these ostensibly separate domains of chemistry by extending our master diagram of Figure 7.31 to include the analysis of how electrochemical cells behave as they "run down" through the process of extracting usable work from the cell.

Thus we assemble our diagram linking K_{eq}, $\Delta G°$, ΔG, $E°_{cell}$, and E_{cell} using our network of equations, as shown in Figure 7.33.

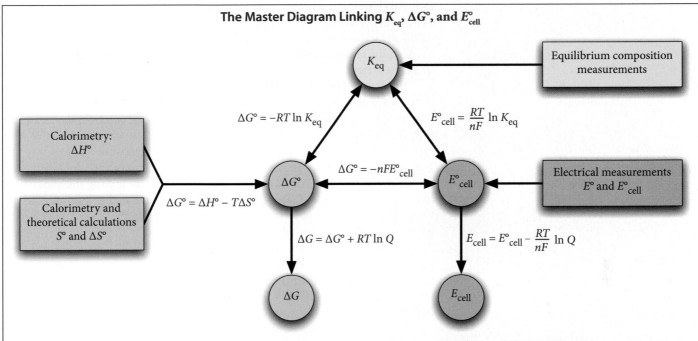

The Master Diagram Linking K_{eq}, $\Delta G°$, and $E°_{cell}$

$$\Delta G° = -RT \ln K_{eq}$$

$$E°_{cell} = \frac{RT}{nF} \ln K_{eq}$$

$$\Delta G° = -nFE°_{cell}$$

$$\Delta G° = \Delta H° - T\Delta S°$$

$$\Delta G = \Delta G° + RT \ln Q$$

$$E_{cell} = E°_{cell} - \frac{RT}{nF} \ln Q$$

Equilibrium composition measurements

Calorimetry: $\Delta H°$

Calorimetry and theoretical calculations $S°$ and $\Delta S°$

Electrical measurements $E°$ and $E°_{cell}$

FIGURE 7.33 We can now extend our Master diagram linking ΔG, $E°_{cell}$, and K_{eq} to nonstandard conditions. In particular the Nernst equation provides the relationship between the cell potential at standard conditions, $E°_{cell}$, and the reaction quotient Q, wherein

$$E_{cell} = E°_{cell} - (RT/nF) \ln Q$$

This greatly extends our ability to dissect different states of both electrochemical systems and thermodynamic systems associated with chemical reactions.

Check Yourself 17

When you use a battery in an electronic device or a flashlight, you know that the device works fine until the battery voltage begins to drop, then the battery goes down very quickly.

The discharge curves for several battery types are displayed here as a plot of cell voltage vs. percent of capacity that has been discharged. As we will see in a Case Study, the chemical reactions in each of the cells are very different, yet the *shapes* of the discharge curves are virtually identical. From the Nernst equation, explain why this is so.

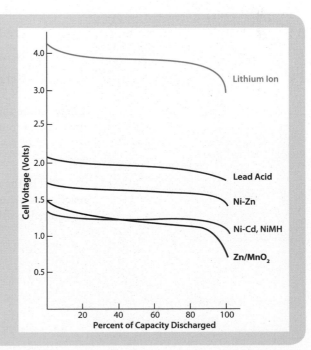

Nonspontaneous Reactions: Driving the Electrochemical Cell Uphill

Electrolysis

We mentioned earlier that nonspontaneous reactions can be driven by applying an external electrical potential. For example, in the first case of a voltaic cell that we described, involving Zn and Cu, with Cu^{2+} being reduced (cathode) and Zn being oxidized (anode), we can imagine running the reaction in the opposite sense by hooking the two electrodes to an external source of electrical potential (voltage), namely a battery.

The source (an external battery) would have to be strong enough (i.e. have a large enough cell potential) to overcome the natural tendency of the system to produce electrical current from the spontaneous chemical reaction. We know that the potential for this reaction is 1.10 V, so the external battery would have to have higher voltage than that. In addition, the battery must be connected in a way that will drive the current in the opposite direction from that of the spontaneous reaction. Recall that in the Zn/Cu voltaic cell, electrons are leaving the anode (Zn) to travel to the cathode (Cu), which implies that Zn acts as the negative pole and Cu acts as the positive pole. In order to drive the reaction backward, the external battery should be connected with its positive pole to Cu and its negative pole to Zn: in this way, the battery will draw electrons from the Cu electrode and send electrons toward the Zn electrode, exactly the opposite of what was happening in the two half reactions involved in the Zn/Cu voltaic cell. The result of this process is that the net reaction (7.3) will be reversed, with Cu dissolving as Cu^{2+} cations into the solution and Zn^{2+} cations from the solution being incorporated into the Zn electrode. All this would take place at the expense of energy provided by the external battery.

There is one case in particular where this kind of reverse reaction is of great importance: the process called *electrolysis*, displayed in Figure 7.34. This is the breakdown of water into oxygen and hydrogen gas using electricity. We can imagine a cell in which the following two reactions take place:

Reduction:	$2[2H_2O(l) + 2e^- \rightarrow H_2(g) + 2OH^-(aq)]$	**(7.22)**
Oxidation:	$2H_2O(l) \rightarrow O_2(g) + 4H^+(aq) + 4e^-$	**(7.23)**
Neutralization:	$4OH^- + 4H^+ \rightarrow 4H_2O$	
Net reaction:	$2H_2O(l) \rightarrow 2H_2(g) + O_2(g)$	**(7.24)**

where, in the third step, four OH^- and four H^+ have combined to form four H_2O molecules. From Table 7.1, we find that the potential for this reaction is:

$$E^\circ_{cell} = E^\circ_{H_2O/H_2} - E^\circ_{O_2/H_2O} = (-0.83 \text{ V}) - (+1.23 \text{ V}) = -2.06 \text{ V}$$

That is, the net reaction is nonspontaneous (the calculated cell voltage for the reaction as written is negative) and an external source of electrical energy (battery) is required to drive it, with a potential *exceeding* +2.06 V.

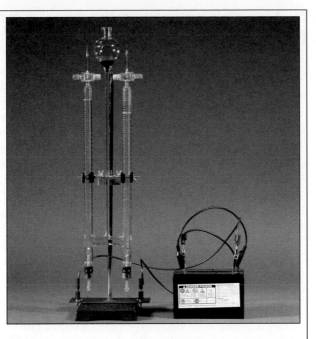

FIGURE 7.34 The electrolysis of water, whereby the passage of current between the anode where the oxidation of water occurs $2H_2O(l) \rightarrow O_2(g) + 4H^+(aq) + 4e^-$ and the cathode where the reduction of water occurs $2H_2O(l) + 2e^- \rightarrow H_2(g) + 2OH^-(aq)$. In the net production of H_2 and O_2 from the electrolysis of water: $2H_2O(l) \rightarrow 2\,H_2(g) + O_2(g)$, we see that two moles of H_2 are formed for each mole of O_2. This we can see from the differing fluid levels in the two test tubes in the figure with the anode on the left (where O_2 is produced) and the cathode on the right (where H_2 is produced).

Pursuing our desire to understand, and thereby simplify, all processes by defining the free energy surface over which they take place, we ask the question: What is the form of the Gibbs free energy surface in the case of electrolysis?

When we allow an electrochemical cell to fully discharge such that it can supply no additional work to an external circuit, it remains "dead," with products and reactants in a specific ratio such that $Q = K_{eq}$ and with *equal free energies* for reactants and products; i.e., $\Delta G = 0$. However, we are not out of options. If we can harness an external energy source to drive the electrochemical cell back up the free energy "hill" to reestablish a difference in free energy between reactants and products, we can thereby bring new life to an old, spent, electrochemical cell.

We can express this coupling of spontaneous and nonspontaneous systems in the free energy diagram of Figure 7.35. This coupled free energy diagram applies equally well to the electrolysis of water to form $H_2(g)$ and $O_2(g)$.

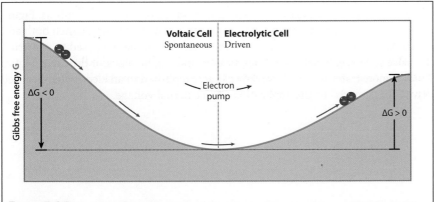

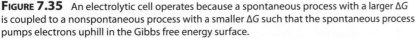

FIGURE 7.35 An electrolytic cell operates because a spontaneous process with a larger ΔG is coupled to a nonspontaneous process with a smaller ΔG such that the spontaneous process pumps electrons uphill in the Gibbs free energy surface.

A key issue in transitioning from a *voltaic cell* to an *electrolytic cell* is to keep clearly in mind what is happening at the anode and cathode of the cell in each case. The key points are these:

(a) The terms anode and cathode are derived directly from the specific half reactions that occur at the interface between the solution and the electrode itself.

(b) The terms anode and cathode are in general not determined by whether the electrode is positive (+) or negative (−).

(c) In particular, the anode is where oxidation occurs in an electrochemical cell.

(d) Electrons are produced at the anode in a half reaction of a *voltaic cell* (such as $Zn(s) \rightarrow Zn^{2+}(s) + 2e$), so the anode of a *voltaic cell* is negative (−1) and electrons leave the anode.

(e) In an electrolytic cell, electrons are extracted from the anode so that the anode in an *electrolytic cell* is positive (+).

(f) For *both* the voltaic cell and the electrolytic cell, electrons enter the cell from the external circuit and the electrode that receives the electrons is the cathode because cations are drawn to that electrode from the solution to accept the incoming electrons.

(g) This provides us with an important fact that is independent of whether we are referring to an electrolytic cell or a voltaic cell:

    1.   Electrons leave the cell at the *anode* to enter the external circuit.

    2.   Electrons enter the cell at the cathode from the external circuit.

We can summarize the designations, contrasting the voltaic cell versus the electrolytic cell.

	Voltaic Cell	**Electrolytic Cell**
Oxidation:	$A \rightarrow A^{+} + e^{-}$ Anode (negative)	$B \rightarrow B^{+} + e^{-}$ Anode (positive)
Reduction:	$B^{+} + e^{-} \rightarrow B$ Cathode (positive)	$A^{+} + e^{-} \rightarrow A$ Cathode (negative)
Overall:	$A + B^{+} \rightarrow A^{+} + B$	$A^{+} + B \rightarrow A + B^{+}$
	$\Delta G < 0$	$\Delta G > 0$
	Spontaneous redox reaction releases energy	Nonspontaneous redox reaction absorbs energy to drive it
	The system (the cell) does work on the surroundings	The surroundings (the source of energy) do work on the system

Electrolysis plays a central role in a vast array of industrial processes. The applications range from the electroplating of chromium onto steel for decoration, to the industrial production of aluminum, to the refining of copper, and to the production of chlorine gas from seawater.

Electroplating is a particularly straightforward example of electrolysis. Figure 7.36a displays the chromium plated components of a motorcycle, which both add cosmetic appeal and serve to suppress the corrosion of the steel used to construct the valve covers, exhaust system, air cleaner, and engine block of the cycle. Figure 7.36b demonstrates how electroplating is accomplished in an electrochemical cell driven "backwards" by the application of an external voltage.

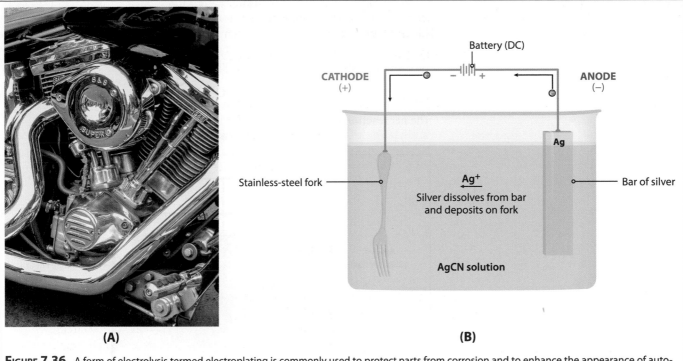

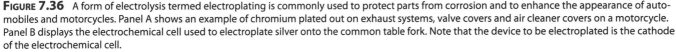

(A)   **(B)**

FIGURE 7.36 A form of electrolysis termed electroplating is commonly used to protect parts from corrosion and to enhance the appearance of automobiles and motorcycles. Panel A shows an example of chromium plated out on exhaust systems, valve covers and air cleaner covers on a motorcycle. Panel B displays the electrochemical cell used to electroplate silver onto the common table fork. Note that the device to be electroplated is the cathode of the electrochemical cell.

Referring to the electroplating cell in panel B of Figure 7.36, the silver cations in solution are reduced at the cathode, leaving a thin coating of silver metal on the surface of the fork. At the anode, electrons are extracted from the silver electrode releasing $Ag^+(aq)$ into solution. Thus silver atoms are transferred from the solid silver anode to the surface of the steel fork, which is the cathode. This results in a fork that *appears* to be solid silver and a fork that does not easily corrode. Both of these attributes raise the value of the fork with a minimal amount of silver—thus constituting a viable industry! For the case of the chromium ("chrome") trim on motorcycles, automobiles, etc., the anode is a bar of chromium and the steel part to be plated is lowered into solution as the cathode.

Production of Aluminum by Electrolysis

Perhaps the most famous example of industrial electrolysis is that of aluminum extraction from the oxide form $Al_2O_3(s)$. Aluminum is the third most abundant element in the Earth's crust after oxygen and silicon. It is a metal of great strength-to-weight ratio and it is a metal that does not continue to corrode because the Al_2O_3 surface oxidation layer is both very hard and very inert chemically. But the inertness of Al_2O_3 combined with the fact that most of the available aluminum occurs in this form made its extraction from aluminum ore very difficult—until a young student at Oberlin College, Charles Hall, discovered how to employ electrolysis! Hall discovered that by heating the mineral cryolite, Na_3AlF_6, to above its melting point, Al_2O_3 would dissolve in the cryolite producing a mixture that conducts electricity, but more importantly, the process could be used to extract pure aluminum. Figure 7.37 diagrams the design of the electrolysis cell. The Al_2O_3, which dissolves in molten cryolite, dissociates to form Al^{3+} and O^{2-}. The cathode reaction is the reduction of aluminum cations to produce pure aluminum, $Al(s)$:

$$Al^{3+} + 3e^- \rightarrow Al(s)$$

which is more dense than the molten mixture of Na_3AlF_6 and Al_2O_3. Thus the aluminum sinks to the bottom of the tank. At the anode, in this case made from carbon, the oxide anion is oxidized to yield free oxygen

$$2O^{2-} \rightarrow O_2(g) + 4e^-$$

The net reaction is

$$4Al^{3+} + 6O^{2-} \rightarrow 4Al(s) + 3O_2(g)$$

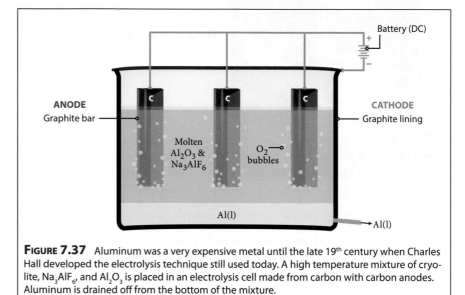

FIGURE 7.37 Aluminum was a very expensive metal until the late 19th century when Charles Hall developed the electrolysis technique still used today. A high temperature mixture of cryolite, Na_3AlF_6, and Al_2O_3 is placed in an electrolysis cell made from carbon with carbon anodes. Aluminum is drained off from the bottom of the mixture.

Corrosion: A Redox Reaction That Causes Problems

As we have emphasized many times in the book, the chemistry of oxygen and the chemistry of water are of central importance to virtually every branch of chemistry. However, we now have a new and important perspective on the role of O_2 and of H_2O in an important class of oxidation-reduction reactions. Consider the reduction of oxygen in the presence of neutral water in Table 7.1

$$O_2(g) + H_2O(l) + 4e^- \rightarrow 4OH^-(aq) \qquad E° = +0.40 \text{ V}$$

Also listed in Table 7.1 is the reduction of oxygen in the presence of acidic water

$$O_2(g) + 4H^+(aq) + 4e^- \rightarrow 2H_2O(l) \qquad E° = +1.23 \text{ V}$$

which has a markedly higher standard reduction potential, +1.23 V vs. + 0.40 V. Thus the reduction of oxygen both in neutral water and in aqueous acid are spontaneous and for the acidic case, strongly so. With $\Delta G° = -nF\, E°_{cell}$, the acidic free energy release for the reduction of water is   .

$$\Delta G° = -(4)\,(9.65 \times 10^4)(1.23 \text{ V}) = -4.7 \times 10^5 \text{ J} = -4.7 \times 10^2 \text{ kJ}$$

This is a *large* free energy release. Even the reduction of water in the absence of an acid has a free energy release of $\Delta G° = -150$ kJ for each mole of O_2.

The next point to notice is that the reduction potentials of all the metals in Table 7.1 are less than that of oxygen reduction in neutral water except for silver,

$Ag^+(aq) + e^- \rightarrow Ag(s)$ (+0.800 V) and iron, $Fe^{3+}(aq) + e^- \rightarrow Fe^{2+}(aq)$ (+0.771 V). For the case of O_2 reduction in an *acidic* solution, all reduction potentials of metals lie below that of O_2 reduction in water. As a result the oxidation of all metals listed in Table 7.1 will be spontaneous when paired with the reduction of O_2 in water that is in contact with air because the CO_2 in the atmosphere reacts in water to form carbonic acid. The process of oxidation of metals is termed corrosion, and it is a serious problem for objects constructed from common metals except aluminum that reside outside in the presence of water and oxygen: bridges, cars, light fixtures, buildings, etc.

Some metals, such as aluminum, form impermeable oxides that protect the metal against further corrosion. Aluminum in the presence of oxygen, acid, and water forms a surface layer of Al_2O_3 that is chemically inert and a durable surface coating. The oxides of iron, however, are not mechanically stable, forming flakes of $Fe_2O_3 \cdot nH_2O$ that is commonly called "rust," the hydrated form of Fe_2O_3.

Rusting is a redox reaction that is initiated by the simple oxidation of iron:

$$Fe(s) \rightarrow Fe^{2+}(aq) + 2e^- \qquad E° = -0.44 \text{ V}$$

But this is where the corrosion of iron becomes an intriguing story. The oxidation step of solid iron occurs in small defects of the metal, as shown in Figure 7.38. Because it is these sites that initiate the oxidation step, these defects are termed the anodic domain of what will become a coupled electrochemical cell.

Because the metal and water conduct electricity, the electrons produced at the anode move through the metal to the cathodic domain where they react with H^+ present in the water in the presence of O_2 in the air to reduce O_2 to $H_2O(l)$:

$$O_2(g) + 4H^+(aq) + 4e^- \rightarrow 2H_2O(l)$$

Again the acidic character of water results primarily from the carbonic acid present in the water resulting from CO_2 in the atmosphere.

Combining these two half-reactions results in the net conversion of Fe(s) to $Fe^{2+}(aq)$

$$2Fe(s) \rightarrow 2Fe^{2+}(aq) + 4e^- \qquad E°_{cell} = -0.45V$$
$$\underline{O_2(g) + 4H^+ + 4e^- \rightarrow 2H_2O(l) \qquad E°_{cell} = +1.23V}$$
$$2Fe(s) + O_2(g) + 4H^+ \rightarrow 2Fe^{2+}(aq) + 2H_2O(l) \quad E_{cell} = E°_{cathode} - E°_{anode}$$

What happens next is that from the anodic domain, sketched schematically below in Figure 7.38, the $Fe^{2+}(aq)$ cations diffuse through the water film on the metal surface to the cathodic domain where they undergo the next stage in oxidation from Fe^{2+} to Fe^{3+} in the form of $Fe_2O_3 \cdot n H_2O$ via reaction with oxygen

$$4Fe^{2+}(aq) + O_2(g) + (4+2n) H_2O(l) \rightarrow 2 Fe_2O_3 \cdot nH_2O + 8H^+(aq)$$

The "rust" is the structurally weak $2Fe_2O_3 \cdot nH_2O$ material whose molecular form is a *hydrate* with the number of bound water molecules dependent on the amount of moisture present.

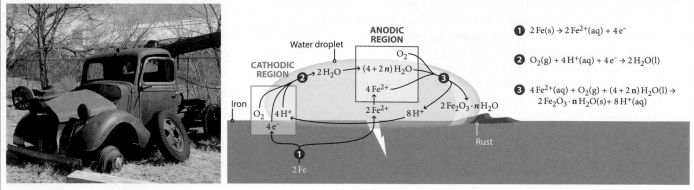

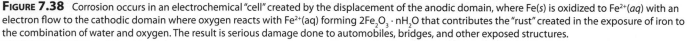

FIGURE 7.38 Corrosion occurs in an electrochemical "cell" created by the displacement of the anodic domain, where Fe(s) is oxidized to $Fe^{2+}(aq)$ with an electron flow to the cathodic domain where oxygen reacts with $Fe^{2+}(aq)$ forming $2Fe_2O_3 \cdot nH_2O$ that contributes the "rust" created in the exposure of iron to the combination of water and oxygen. The result is serious damage done to automobiles, bridges, and other exposed structures.

The right-hand panel in Figure 7.38 deserves careful consideration because it involves important categories of chemical reactions. In particular, notice the combined roles that oxygen and water play in both the chemical changes *and* the physical construct of the corrosion "cell." While oxygen is the dominant oxidizing agent, water (with H^+ added from carbonic acid formed from reaction of CO_2 with water), is necessary to provide a pathway for the flow of charge from the anode to the cathode. Water is also important as a reagent in the second stage of reaction that oxidizes $Fe^{2+}(aq)$ to Fe_2O_3. Because pure water is not a good conductor of electricity, the addition of either a salt or an acid speeds the rate of corrosion. Thus salt added to roadways in the winter accelerates the corrosion of metals in the structure of automobiles and acidity in rain accelerates the weakness of bridge structures.

Given what we know about electrochemical cells, in many instances corrosion of valuable structural parts can be slowed by the use of sacrificed metals attached to the desired part of the structure or by plating a layer of a metal such as zinc over the metal to be protected.

Problem:

1. Given that the electrified current would be supplied by generators at the surface, how many coulombs of electrical charge would supply the 7×10^8 mol of H_2?

2. Given that the total pressure at the depth of the *Titanic* is 300 atmospheres and the temperature is 4°C, what voltage is required to drive the electrolysis of water to produce H_2 and O_2?

3. Calculate then the minimum electrical energy to raise the *Titanic* from the sea floor to the surface.

A view of the bow and railing of the RMS *Titanic*. *Image copyright Emory Kristof/National Geographic.* http://oceanexplorer.noaa.gov/

Check Yourself 18

The discovery of the *Titanic* by Robert Ballard in 1985 ignited the idea of floating the shipwreck to the surface using electrochemistry. The idea was to place pontoons inside the hull of the ship and to then employ electrolysis to inflate these pontoons with H_2 by placing the cathode inside the pontoon and using electricity supplied from the surface. Calculations demonstrated that approximately 7×10^8 mol of hydrogen gas would be required to lift the ship to the surface.

Summary Concepts

1. Electrochemistry is the subject that links the concepts of stoichiometry, thermodynamics, oxidation-reduction reactions, chemical potentials, Gibbs free energy, acid-base behavior, and photochemistry within a single framework. In addition, electrochemistry connects the principles of chemical reactivity to the forefront of modern technology in an era strongly influenced by demand for greater production of primary energy, energy storage, and efficient energy transformations displayed in of Figure CS4.3E to support modern society. The quantitative relationship between ΔG, $\Delta G°$, K_{eq}, E_{cell}, and $E°_{cell}$ can be summarized in a single coupled diagram.

The Master Diagram Linking K_{eq}, $\Delta G°$, and $E°_{cell}$

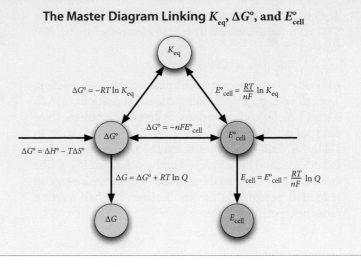

$$\Delta G° = -RT \ln K_{eq}$$

$$E°_{cell} = \frac{RT}{nF} \ln K_{eq}$$

$$\Delta G° = -nFE°_{cell}$$

$$\Delta G° = \Delta H° - T\Delta S°$$

$$\Delta G = \Delta G° + RT \ln Q$$

$$E_{cell} = E°_{cell} - \frac{RT}{nF} \ln Q$$

An understanding of Electrochemistry is built upon a foundation of the following concepts:

2. Oxidation-reduction reactions, which can be balanced by the half-reaction approach wherein the oxidation half-reaction is analyzed separately from the reduction half-reaction and then the half-reactions are added, balancing the number of electrons such that the net reaction neither produces nor destroys electrons. Redox reactions in acidic or basic media can also be balanced using the sequence given here.

1. Assign oxidation states

2. Separate the overall redox reaction into two half-reactions, one of oxidation the other for reduction

3. Balance each half-reaction: balance all elements except H and O, balance O with H_2O, balance H by adding H^+ (for a basic solution neutralize H^+ by adding enough OH^- to neutralize each H^+)

4. Balance each half-reaction with respect to charge

5. Balance number of electrons

6. Add the half-reactions

3. Analysis of the oxidation and reduction reactions at the atomic level at the anode and cathode respectively of an electrochemical cell. This atomic level visualization underpins the concepts defining how a spontaneous overall reaction comprised of the sum of an oxidizing half-reaction and a reducing half-reaction can release electrical energy through an external circuit.

Oxidation half-reaction
$Zn(s) \rightarrow Zn^{2+}(aq) + 2e^-$ ANODE Zn(s)   Zn(NO₃)₂(aq)

CATHODE Cu(s)   Cu(NO₃)₂(aq) $Cu^{2+}(aq) + 2e^- \rightarrow Cu(s)$
Reduction half-reaction

4. The design of a voltaic cell that employs a separated anode and cathode joined by a salt bridge to maintain charge neutrality and an external circuit to deliver the flow of electrons to a specific load to produce light or accomplish work. The flow of electrons is measured in amperes where one ampere is one coulomb of electrons per second. The voltage generated by the cell is measured in units of the volt, which is equal to one joule per coulomb.

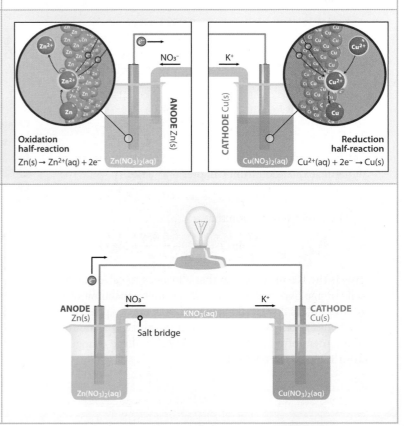

Summary Concepts

5. The quantitative voltage scale that is set by definition to be 0.00 volts for a standard hydrogen electrode for the reaction

$$2H^+(aq) + 2e \rightarrow H_2(g)$$

All cell voltages are measured against this half-reaction. The cell voltage, $E°_{cell}$, under standard conditions (298 K, 1 atm, 1 molar concentrations) is determined by the standard reduction potential listed for each half-reaction in Table 7.1. This establishes the cathode reduction potential, $E°_{cathode}$, and the anode reduction potential, $E°_{anode}$. The cell voltage is calculated by subtracting the half-reaction reduction potential at the anode from the half-reaction reduction potential at the cathode:

$$E°_{cell} = E°_{cathode} - E°_{anode}$$

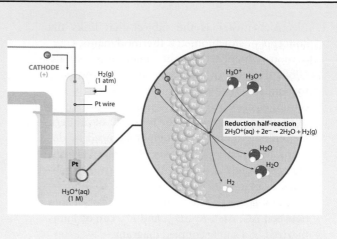

6. The key link between the free energy change of a spontaneous chemical reaction under standard conditions, $\Delta G°$, and the maximum work that can be extracted from an electrochemical cell, $-nFE°_{cell}$, establishes the equality

$$\Delta G° = -nFE°_{cell}$$

But because, from our studies of thermodynamics in Chapter 4, we know

$$\Delta G° = -RT \ln K_{eq}$$

so we can link the cell voltage directly to the equilibrium constant K_{eq}

$$E°_{cell} = \frac{RT}{nF} \ln K_{eq} = \frac{0.0592}{n} \log K$$

using the values for R, T, and F and the identity $\ln x = 2.303 \log x$.

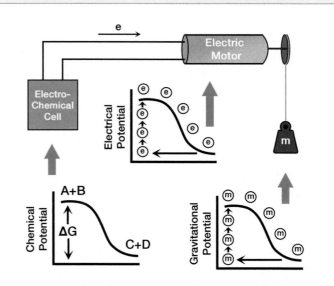

7. For nonstandard conditions, we can employ the thermodynamic expression from Chapter 6 to write

$$\Delta G = \Delta G° + RT \ln Q$$

and by direct substitution that

$$E_{cell} = E°_{cell} - \frac{RT}{nF} \ln Q = E°_{cell} - \frac{0.0592}{n} \log Q$$

This is the Nernst equation that allows us to calculate the cell voltage over a wide range of reactant concentrations. An example is displayed for the case of a zinc-copper electrochemical cell with the reaction quotient Q spanning nine orders of magnitude with the cell voltage changing from 1.26 volts to 0.98 volts.

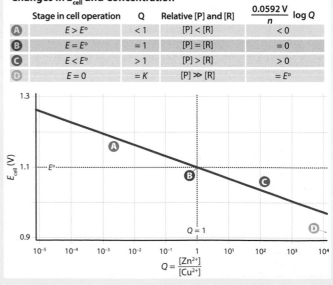

CASE STUDY 7.1   The Electrochemistry of Batteries

Battery design constitutes a top priority in chemical research today. One of the most challenging aspects of modern energy distribution systems such as that displayed in of Figure CS4.3E is the storage of energy. Energy storage that optimizes energy content per unit mass and per unit volume is extremely important for electrification of the transportation sector–particularly for automobiles and light trucks. Energy storage that optimizes energy content per unit of cost is important for balancing production and demand on the power grid.

Fuel cells are important for converting chemical fuels—for example hydrogen—directly to electricity. Fuel cell technology provides, thereby, an important option for storing energy by using excess electrical power to produce hydrogen by electrolysis, storing the hydrogen, and then producing electricity from that stored hydrogen with a fuel cell to supply the power grid during periods of peak demand.

We will explore the frontiers of battery technology in a subsequent chapter, but we must first explore current battery technology.

Batteries, that date back to the 19th century as shown in Figure CS7.1a, consist of electrochemical cells that involve a spontaneous net reaction. There are various type of batteries, many employed in various common devices and applications. They are classified in two categories: *primary cells* in which the reaction is *not reversible*, and when the reactants have been completely transformed into products, the cell is dead; and *secondary cells*, in which the chemical reaction *can be reversed* by passing electrons through the cell in the opposite direction, thus "recharging" the battery.

Lead-Acid Storage Battery

As a first example, we look at the so-called lead-acid storage battery displayed in Figure CS7.1b, in which the cathode is lead oxide, $PbO_2(s)$, and the anode is lead, $Pb(s)$, with a sulfuric acid, H_2SO_4 solution (about 35% by mass sulfuric acid in water). The acidic nature of the electrolyte produces HSO_4^- (aq), which is involved in *both* the reduction and oxidation half reactions:

Reduction: $PbO_2(s) + 3H^+(aq) + HSO_4^-(aq) + 2e^- \rightarrow PbSO_4(s) + 2H_2O(l)$

Oxidation: $Pb(s) + HSO_4^-(aq) \rightarrow PbSO_4(s) + H^+(aq) + 2e^-$

Net reaction: $PbO_2(s) + Pb(s) + 2H^+(aq) + 2HSO_4^- \rightarrow 2PbSO_4(s) + 2H_2O(l)$

which has a cell potential

$$E^\circ_{cell} = E^\circ_{PbO_2/PbSO_4} - E^\circ_{PbSO_4/Pb} = (+1.70\ V) - (-0.31\ V) = +2.01\ V$$

The advantage of this cell is that the reaction can be run backwards when connected to an external source of electrical potential, which recharges the battery. This type of battery is used extensively in conventional cars, typically with six cells in series, to generate a total voltage of 12.1 V. It stores the necessary energy to start the car engine, and is then recharged by the alternator when the engine is running. Its disadvantage is that it is bulky and heavy, and uses potentially harmful substances (lead, sulfuric acid).

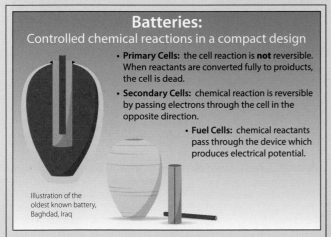

Batteries:
Controlled chemical reactions in a compact design

- **Primary Cells:** the cell reaction is **not** reversible. When reactants are converted fully to proiducts, the cell is dead.
- **Secondary Cells:** chemical reaction is reversible by passing electrons through the cell in the opposite direction.
- **Fuel Cells:** chemical reactants pass through the device which produces electrical potential.

Illustration of the oldest known battery, Baghdad, Iraq

FIGURE CS7.1A   Batteries are devices that store electrochemical energy. They are critical to developments in modern technology where the usefulness of innovations in electronics, automobiles, aircraft, etc. are dependent upon the amount of energy stored per kg of battery mass. The oldest known battery was discovered in Baghdad, Iraq. We, categorize modern batteries according to three designations: primary cells that are not reversible (nonrechargeable), secondary cells that are reversible (rechargeable), and fuel cells wherein chemical reactants pass through the cell, converting their chemical potential to electric current.

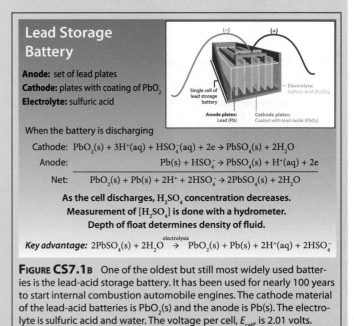

Lead Storage Battery

Anode: set of lead plates
Cathode: plates with coating of PbO_2
Electrolyte: sulfuric acid

Single cell of lead storage battery

Electrolyte: Sulfuric acid (H_2SO_4)

Anode plates: Lead (Pb)   Cathode plates: Coated with lead oxide (PbO_2)

When the battery is discharging

Cathode: $PbO_2(s) + 3H^+(aq) + HSO_4^-(aq) + 2e \rightarrow PbSO_4(s) + 2H_2O$

Anode: $Pb(s) + HSO_4^- \rightarrow PbSO_4(s) + H^+(aq) + 2e$

Net: $PbO_2(s) + Pb(s) + 2H^+ + 2HSO_4^- \rightarrow 2PbSO_4(s) + 2H_2O$

As the cell discharges, H_2SO_4 concentration decreases. Measurement of $[H_2SO_4]$ is done with a hydrometer. Depth of float determines density of fluid.

Key advantage: $2PbSO_4(s) + 2H_2O \xrightarrow{electrolysis} PbO_2(s) + Pb(s) + 2H^+(aq) + 2HSO_4^-$

FIGURE CS7.1B   One of the oldest but still most widely used batteries is the lead-acid storage battery. It has been used for nearly 100 years to start internal combustion automobile engines. The cathode material of the lead-acid batteries is $PbO_2(s)$ and the anode is $Pb(s)$. The electrolyte is sulfuric acid and water. The voltage per cell, E_{cell}, is 2.01 volts.

Alkaline Battery

Alkaline batteries, shown in Figure CS7.1c, dominate the primary cell market. They are sold in AAA, AA, C, and D cell sizes for everything from flashlights to radios. In alkaline batteries, the following reactions take place:

Reduction: $2MnO_2(s) + H_2O + 2e^- \rightarrow Mn_2O_3(s) + 2OH^-(aq)$

Oxidation: $Zn(s) + 2OH^-(aq) \rightarrow ZnO(s) + H_2O + 2e^-$

Net reaction: $MnO_2(s) + Zn(s) \rightarrow ZnO(s) + Mn_2O_3(s)$

with KOH as the electrolyte. This produces a voltage of +1.54 V. The anode is zinc metal and the cathode is actually an inert rod of graphite in contact with the manganese oxide, which is being reduced (Mn is in a +4 oxidation state in MnO_2 and in a +3 oxidation state in Mn_2O_3). The advantage of this cell is that zinc dissolves slowly in the basic medium, which gives the cell a long useful life when it is in operation. This is the basic design of many common batteries used in various devices (note that most of these batteries have a rating of 1.5 V).

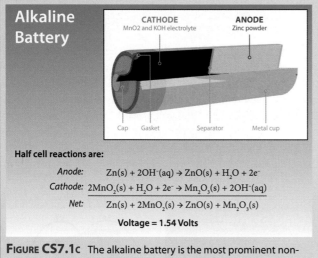

Alkaline Battery

CATHODE
MnO2 and KOH electrolyte

ANODE
Zinc powder

Cap   Gasket   Separator   Metal cup

Half cell reactions are:

Anode: $Zn(s) + 2OH^-(aq) \rightarrow ZnO(s) + H_2O + 2e^-$

Cathode: $2MnO_2(s) + H_2O + 2e^- \rightarrow Mn_2O_3(s) + 2OH^-(aq)$

Net: $Zn(s) + 2MnO_2(s) \rightarrow ZnO(s) + Mn_2O_3(s)$

Voltage = 1.54 Volts

FIGURE CS7.1c The alkaline battery is the most prominent non-rechargeable battery used today—primarily for portable electronics, flashlights, etc. The anode material is $Zn(s)$ and the cathode material is MnO_2 (s). The electrolyte is KOH. The cell voltage, E_{cell}, is 1.5 volts.

Nickel-Cadmium Battery

A typical secondary cell—that is, rechargeable battery—involves a nickel compound as the cathode and cadmium as the anode:

Reduction: $2NiO(OH)(s) + 2H_2O + 2e^- \rightarrow 2Ni(OH)_2(s) + 2OH^-(aq)$

Oxidation: $Cd(s) + 2OH^-(aq) \rightarrow Cd(OH)_2(s) + 2e^-$

Net reaction: $Cd(s) + 2NiO(s) + 2H_2O(l) \rightarrow 2Ni(OH)_2(s) + Cd(OH)_2(s)$

This cell produces a voltage of 1.4 V. Ni is reduced at the cathode because it has an oxidation state of +3 in NiO(OH) and +2 in $Ni(OH)_2$. The advantage of this design is that it can be recharged many times, because the reactants and products are solids that easily adsorb on the electrode surfaces.

Nickel-Metal Hydride Battery

While lead-acid, alkaline, and Ni-Cd batteries have dominated the battery market for decades, battery technology is now moving forward rapidly. One class of modern high performance batteries is called the metal hydride and involves the following reactions:

Reduction: $NiO(OH)(s) + H_2O + 2e^- \rightarrow Ni(OH)_2(s) + OH^-(aq)$

Oxidation: $MH(s) + OH^-(aq) \rightarrow M(s) + H_2O + 2e^-$

Net reaction: $MH(s) + NiO(OH)(s) \rightarrow Ni(OH)_2(s) + M(s)$

where MH(s) is a solid hydride of some metal, M. The electrolyte is a solution of KOH. As in the nickel-cadmium battery, Ni is reduced at the cathode from an oxidation state of +3 to one of +2. At the anode, the reactant is hydrogen, which is removed from the hydride MH. This is a secondary cell used in many devices such as cell phones and computers. The great advantage of these batteries is that they can store 50% more energy per unit volume than other cells, such as the nickel-cadmium battery. This important class of battery is diagrammed and described in Figure CS7.1d.

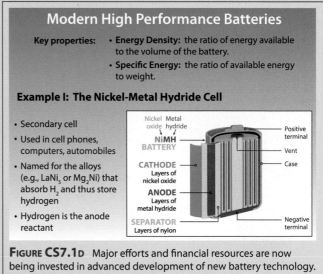

Modern High Performance Batteries

Key properties:
- **Energy Density:** the ratio of energy available to the volume of the battery.
- **Specific Energy:** the ratio of available energy to weight.

Example I: The Nickel-Metal Hydride Cell

- Secondary cell
- Used in cell phones, computers, automobiles
- Named for the alloys (e.g., $LaNi_5$ or Mg_2Ni) that absorb H_2 and thus store hydrogen
- Hydrogen is the anode reactant

Nickel   Metal
oxide   hydride

NiMH BATTERY

Positive terminal

Vent

Case

CATHODE
Layers of nickel oxide

ANODE
Layers of metal hydride

SEPARATOR
Layers of nylon

Negative terminal

FIGURE CS7.1d Major efforts and financial resources are now being invested in advanced development of new battery technology. The nickel-metal hydride cell has been refined for use in rechargeable electronics, cordless tools, and the current line of hybrid automobiles.

The nickel metal hydride cell is the energy storage technology used in the Toyota Prius and Honda Civic hybrid automobiles, as described in Figure CS7.1e.

Toyota Prius   **Honda Civic**

FIGURE CS7.1E During the 1980s and 1990s the Japanese automakers Toyota and Honda began the development of hybrid technology that couples an electric motor/generator to the crankshaft of an internal combustion engine. A large battery (nickel-metal hydride) drives the electric motor to assist the internal combustion engine, but also receives electric charge from the same motor in generator mode when the car is going downhill or is braking. This recaptures the kinetic energy of the vehicle motion and stores the energy as electrochemical energy. When the car is not moving in traffic or at a stop light, the internal combustion engine shuts off to conserve fuel. The two vehicles, the Prius hybrid on the left and Honda Civic hybrid, get 50 mpg on the highway and 40 mpg in town.

Lithium-Ion Battery

Perhaps the most important category of battery technology sitting on the cusp of full scale acceptance is the lithium-ion battery. We have seen before that Li is the element with the lowest value of standard electrode potential (–3.05 V). In this sense, it is an ideal candidate for an anode. It is also rather light in weight, which always is a desirable feature in batteries. Its main drawback is that Li reacts with water strongly, so it cannot be used in the usual type of battery cell that involves a water electrolyte solution.

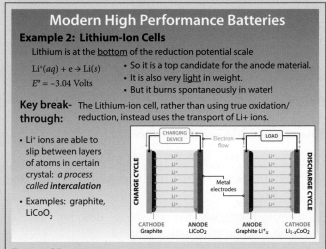

Modern High Performance Batteries

Example 2: Lithium-Ion Cells

Lithium is at the <u>bottom</u> of the reduction potential scale

$Li^+(aq) + e \rightarrow Li(s)$
$E° = -3.04$ Volts

- So it is a top candidate for the anode material.
- It is also very light in weight.
- But it burns spontaneously in water!

Key breakthrough: The Lithium-ion cell, rather than using true oxidation/reduction, instead uses the transport of Li+ ions.

- Li⁺ ions are able to slip between layers of atoms in certain crystal: *a process called intercalation*
- Examples: graphite, LiCoO₂

FIGURE CS7.1F The lithium-ion battery is the focus of major efforts to refine the technology for an array of applications extending from computers to plug-in hybrid automobiles and light trucks. The large improvement in energy storage per kg of battery mass over the nickel-metal hydride battery opens the possibility of propelling an automobile for up to 60 miles without recharging—the so-called plug in hybrid. This technology alone offers the opportunity to eliminate up to 60% of US imports of petroleum.

In lithium batteries, as shown in cross section in Figure CS7.2f, instead of the standard design, there is transport of Li ions through a porous medium. A typical porous medium is graphite, which consists of layers of atoms strongly bonded among themselves within each layer, but weakly interacting across layers. This makes it possible for the Li ions to be stored between the layers, a process called *intercalation*. Another material in which Li ions can intercalate is $LiCoO_2$. In a lithium battery during charging, Li^+ ions leave $LiCoO_2$ and travel through the electrolyte (liquid $LiPF_6 = Li^+ + PF_6^-$) to the graphite electrode. In the discharging phase, the reverse takes place. The charging-discharging process simply moves Li^+ ions between the two electrodes. The two reactions can be described as follows:

$$\text{Charging: } LiCoO_2 + C_6 \rightarrow Li_{1-x}CoO_2 + Li_xC_6$$
$$\text{Discharging: } Li_{1-x}CoO_2 + Li_xC_6 \rightarrow Li_{1-x+y}CoO_2 + Li_{x-y}C_6$$

This is an efficient process for generating electrical current but cannot go on forever, because at each discharging step a small fraction of the available Li^+ ions is lost ($x \neq y$ in the second reaction above). These features of the Li battery are summarized in Figure CS7.1g.

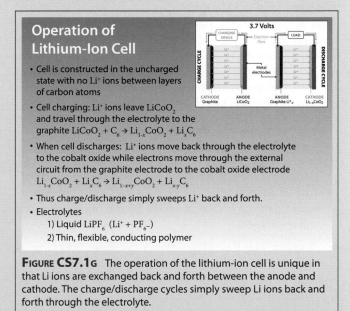

Operation of Lithium-Ion Cell

- Cell is constructed in the uncharged state with no Li⁺ ions between layers of carbon atoms
- Cell charging: Li⁺ ions leave $LiCoO_2$ and travel through the electrolyte to the graphite $LiCoO_2 + C_6 \rightarrow Li_{1-x}CoO_2 + Li_xC_6$
- When cell discharges: Li⁺ ions move back through the electrolyte to the cobalt oxide while electrons move through the external circuit from the graphite electrode to the cobalt oxide electrode $Li_{1-x}CoO_2 + Li_xC_6 \rightarrow Li_{1-x+y}CoO_2 + Li_{x-y}C_6$
- Thus charge/discharge simply sweeps Li⁺ back and forth.
- Electrolytes
  1) Liquid $LiPF_6$ ($Li^+ + PF_6^-$)
  2) Thin, flexible, conducting polymer

FIGURE CS7.1G The operation of the lithium-ion cell is unique in that Li ions are exchanged back and forth between the anode and cathode. The charge/discharge cycles simply sweep Li ions back and forth through the electrolyte.

Lithium-ion batteries are a critical part of the emerging technology for the coming era of "plug-in hybrids," which are capable of operating for some 40 to 60 miles unassisted by recharging from an internal combustion engine. An example of a plug-in hybrid is the Chevy Volt—a car released for production in 2012, and shown in Figure CS7.1h.

FIGURE CS7.1H General Motors has made a major investment in plug-in hybrid technology with the introduction of the Chevy Volt. What is interesting about this venture is that first of all, the car was designed to combine style and performance. Second, the rate at which the Volt will be brought to market depends not on the refinement of the car itself, but on the Li ion batteries required to propel the car between charges—currently expected to be 40 miles.

Problem 1

Analyze a lead-acid battery

A car battery is based on an electrochemical reaction between lead and sulfuric acid. It is a stable and robust energy storage system and can be regenerated many times. The properties of a lead-acid battery can be determined from the analysis of the electrochemical reactions that occur in the battery. The overall cell equation for a lead-acid battery is:

$$PbO_2(s) + Pb(s) + 2H^+(aq) + 2HSO_4^-(aq) \rightarrow$$
$$2PbSO_4(s) + 2H_2O(l)$$

a. Using the table of standard reduction potentials, calculate $E°$ for a lead-acid battery at 25°C. If a car battery contains 6 such cells in series, what is its approximate total voltage?

b. Explain why the voltage slowly decreases over time as a car battery is discharged, and then at a point near the end of the discharge, the voltage drops very quickly. *Hint*: consider the Nernst equation.

c. If the starter motor draws a current of 300 amperes, and it takes 2.0 seconds to start the car, calculate the *mass* of $PbSO_4$ that is produced in the battery each time the car is started.

Problem 2

Diagnose a battery problem

It's a beautiful winter morning in Boston (T = −5°F), and you are on your way to work. You get in your car and attempt to start the engine, but the starter motor will not turn over.

a. You remember that the car battery is a lead-acid storage battery, in which the following oxidation-reduction reaction takes place:

$$PbO_2(s) + Pb(s) + 2H^+(aq) + 2HSO_4^-(aq) \rightarrow$$
$$2PbSO_4(s) + 2H_2O(l)$$

You find $\Delta H° = -374.4$ kJ/mol and $\Delta S° = 40.01$ J/mol K for the above reaction. You then remember how important free energy is for any process of this type. Calculate $\Delta G°$ for the lead-acid battery reaction at the time when you tried to start the car.

b. You look in your car's manual and find that the ignition spark, which starts the car, cannot occur if the battery has a potential below 12.00 V. You want to find out if the car is not starting for this reason or because of some other problems. You inspect the battery and find that it is made of six individual lead acid storage cells in series. Calculate the standard cell reduction potential $E°_{cell}$ for the battery at the time when you tried to start the car, and determine if this is the cause behind why the car will not start.

c. You are now very anxious to get to work. What can you do to fix the problem with the car?

Problem 3

Design an alkaline battery

An **alkaline** battery with two unknown **electrodes** X_2O and Y has the following overall equation.

$$Y(s) + X_2O(s) \rightarrow YO(s) + 2X(s)$$

a. Write the balanced half-reactions for the cathode and the anode.

b. Explain briefly why the cell voltage does not decrease significantly as the reaction proceeds.

c. When this battery was used for 1 hour at a current of 0.300 A, it was found that the weight of Y decreased by 0.366 g, while the weight of X increased by 1.207 g. Identify X and Y.

Problem 4

Design a pacemaker

Artificial pacemakers are used to regulate the beating of the heart. Such devices, once implanted, should be able to function continuously for long periods of time to minimize the need for repeated surgeries. They are usually powered with a lithium iodide battery. The overall reaction in a typical lithium iodide battery and some relevant thermodynamic data are:

$$2Li(s) + I_2(s) \rightarrow 2LiI(s)$$

Substance	$S°$(J/K mol)	$\Delta H°_f$ (kJ/mol)
Li (s)	29.1	−238.6
I$_2$ (s)	116.1	−110.5
LiI (s)	85.65	−270.0

a. Lithium iodide batteries are not to be confused with lithium-ion batteries. Explain the difference between the two.

b. Calculate $E°_{cell}$ of a pacemaker at a normal body temperature (37°C).

c. Assume that the pacemaker must generate a 10 mA pulse lasting 2 ms to trigger each heartbeat, and that over the lifetime of the battery the average heart rate is 80 beats/minute. The electrodes are manufactured with 1.5 g of lithium and 30 g of iodine, respectively. How long will the battery last before it needs to be replaced? Assume that E_{cell} remains approximately the same over the lifetime of the battery.

CASE STUDY 7.2 Electrochemistry in the Cell and the Role of ATP in Powering Organisms

Electrochemistry is critical to the ability of the cell to create membrane structures that can separate charge, generate proton concentration gradients, induce oxidation on one side of the membrane and reduction on the other. The structure of cells turns out to be designed to apply the principles of electrochemistry in a variety of ways, but one of the most important is the ability to couple the chemical bond energy in food to generate an electrochemical potential, $E°$.

We analyze this step in terms of the cell as an electrochemical model.

Coupling Bond Energy to Electrochemical Potential. We know from our "Master Diagram," Figure 7.33, that the Gibbs free energy release, ΔG, is related to the cell potential via the equation

$$\Delta G°' = -nFE°'$$

where the superscript on $E°'$ is used to denote standard conditions within *biological systems*, specifically at pH = 7. The cell employs a controlled energy release by passing electrons along the electron-transport chain (ETC) that is situated along the inner membrane of the mitochondrion—a subunit of the cell that produces the cell's energy. The sequence linking the cell structure to the mitochondrion to the membrane containing the ETC sequence is shown in Figure CS7.2a.

The reaction that provides the chemical energy (free energy) that sustains this ETC sequence is the simple oxidation of hydrogen:

$$H_2 + \tfrac{1}{2} O_2 \rightarrow H_2O$$

But, as we saw, H_2 does not exist in the cell, but rather nicotinamide adenine dinucleotide hydride, or NADH. The reaction sequence is as follows: the oxidizing agent, NAD^+, requires two protons and two electrons in the process of oxidizing molecules in food in the half-reaction

$$NAD^+(aq) + 2H^+(aq) + 2e^- \rightarrow NADH(aq) + H^+(aq)$$

At the mitochondrial inner membrane (Figure CS7.2a), the NADH(aq) and H^+(aq) transfer the two electrons to the first redox couple of the ETC and release two protons.

$$NADH(aq) + H^+(aq) \rightarrow NAD^+(aq) + 2H^+(aq) + 2e^-$$
$$E°' = -0.315 \text{ V}$$

As the electrons are guided down the chain of redox couples they finally reduce O_2 to H_2O:

$$\tfrac{1}{2} O_2(aq) + 2e^- + 2H^+(aq) \rightarrow H_2O(l)$$

$$E^b = 0.815 \text{ V}$$

Net: $NADH(aq) + H^+(aq) + \tfrac{1}{2} O_2(aq) \rightarrow NAD^+(aq) + H_2O(l)$

$$E°'_{overall} = 0.815 \text{ V} - (-0.315 \text{ V}) = 1.13 \text{ V}$$

The result is that for every *mole* of NADH that enters the redox chain, the free energy equivalent of 1.13 V is available, so

$$\Delta G°' = -nFE°' = -(2 \text{ mol e}^-/\text{mol NADH})$$
$$\times (96.5 \text{ kJ/V} \cdot \text{mol} \cdot \text{e}^-)(1.130 \text{ V})$$
$$= -218 \text{ kJ/mol NADH}$$

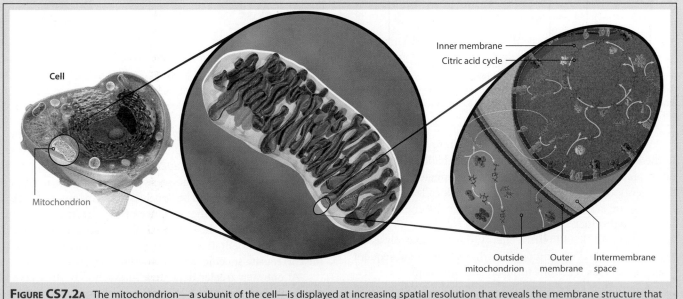

FIGURE CS7.2A The mitochondrion—a subunit of the cell—is displayed at increasing spatial resolution that reveals the membrane structure that serves to separate the oxidizing reaction of NADH(aq) with H^+(aq) from the reduction reaction of O_2(aq) with H^+(aq) forming H_2O.

Turning to the role of ATP in powering living organisms, the variety of living forms that have evolved over Earth's history number in the millions. But virtually all of these organisms use the same reaction to provide the Gibbs free energy to drive a manifold of nonspontaneous reactions to sustain life. The reaction of central importance is the hydrolysis of the molecule adenosine triphosphate (ATP) to form adenosine diphosphate (ADP):

$$ATP^{4-} + H_2O \rightleftharpoons ADP^{3-} + HPO_4^{2-} + H^+$$

$$\Delta G^{\circ\prime} = -30.5 \text{ kJ}$$

While this release of free energy is not large on the scale of all chemical reactions (the release of free energy in the combustion of octane is $\Delta G^\circ = -10,000$ kJ), within the manifold of chemical reactions that take place within living organisms, −30 kJ of free energy release is indeed large. So what aspects of the ATP bond structure relative to ADP yield this large free energy difference—thus making ATP a uniquely potent supplier of free energy? The answer lies in the phosphate portions of ATP, ADP, and HPO_4^{2-}.

Figure CS7.2b diagrams the bonding structure of ATP, ADP, and the phosphate anion. The first reason for the large energy release is that at pH ~7, the triphosphate portion of ATP has four negative charges grouped closely together. Because of this, ATP has a large charge repulsion intrinsic to its structure. This charge repulsion is reduced in ADP, as shown in Figure CS7.2b in the upper panel.

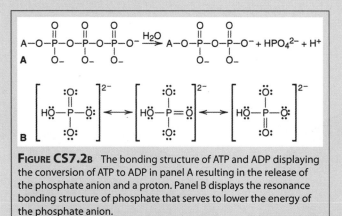

FIGURE CS7.2b The bonding structure of ATP and ADP displaying the conversion of ATP to ADP in panel A resulting in the release of the phosphate anion and a proton. Panel B displays the resonance bonding structure of phosphate that serves to lower the energy of the phosphate anion.

The second reason for the large free energy release is that in the HPO_4^{2-} product, there is an electron delocalization that takes place within the bonds of the phosphate that lowers the energy of HPO_4^{2-} in solution. This delocalization forms the resonant structure that serves to lower the energy of the phosphate product.

Given this 30.5 kJ of free energy, the ATP hydrolysis reaction to form ADP can couple with a series of subsequent nonspontaneous reactions to sustain an organism.

A common second step is the metabolic breakdown of glucose with the addition of a phosphate group to glucose in the dehydration reaction

$$Glucose + HPO_4^{2-} \rightleftharpoons [glucose\text{-}phosphate]^{2-} + H_2O$$

$$\Delta G^{\circ\prime} = +13.8 \text{ kJ}$$

If the hydrolysis of ATP is coupled to this glucose-phosphate reaction, the net reaction, after cancelling HPO_4^{2-} and H_2O, yields the net reaction

$ATP^{4-} + H_2O \rightleftharpoons ADP^{3-} + HPO_4^{2-} + H^+$	$\Delta G^{\circ\prime} = -30.5$ kJ
$Glucose + HPO_4^{2-} \rightleftharpoons [glucose\text{-}phosphate]^{2-} + H_2O$	$\Delta G^{\circ\prime} = +13.8$ kJ
$Glucose + ATP^{4-} \rightleftharpoons [glucose\text{-}phosphate]^{2-} + ADP^{3-} + H^+$	$\Delta G^{\circ\prime} = -16.7$ kJ

But in the case of the ATP → ADP net reaction, there is more involved in guiding the course of the reaction than simply occupying common physical proximity. The reaction converting ATP to ADP is so slow under those conditions that it is not of biochemical interest.

What must be added to the process is a way of "docking" the glucose and the ATP in a geometrically optimal configuration to instigate reaction. This process of enzyme catalyzed reactions can be viewed at a number of levels of molecular complexity. On the strictly diagrammatic level, shown in Figure 7.2c, the catalyst (in this case the enzyme hexose kinase) provides the required docking site.

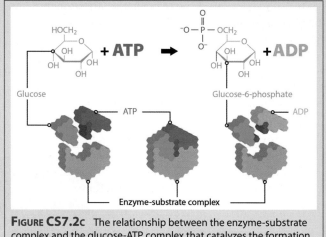

FIGURE CS7.2c The relationship between the enzyme-substrate complex and the glucose-ATP complex that catalyzes the formation of glucose-phosphate and ADP.

When ATP is properly positioned, the glucose is retained in position such that the [glucose-phosphate]⁻ anion forms as the ADP product forms, and both products decouple from the enzyme. If the enzyme were not present, the reactants (ATP and glucose) would be largely isolated in their respective solvent "cages." With the presence of the enzyme active site (the "docking" site), the reaction is fostered by a combination of proximity, orientation, bond distance, and, in many cases, the correct site specific acid or base characteristic.

At one step greater detail, the interaction of the enzyme active site and the ATP-ADP-[glucose-phosphate]²⁻ interaction is shown in Figure CS7.2d. In this more detailed view,

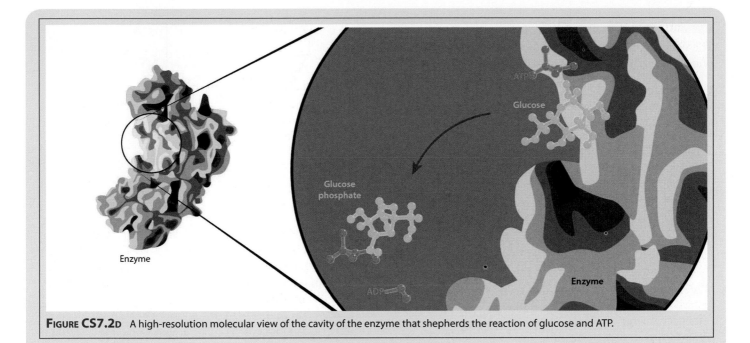

FIGURE CS7.2D A high-resolution molecular view of the cavity of the enzyme that shepherds the reaction of glucose and ATP.

the ATP molecule (shown here as ADP-O-PO$_3$H) is (weakly) bonded to the enzyme in such a way that the cavity of the enzyme active site contains the glucose. The terminal hydrogen on the glucose transfers to the terminal oxygen on the ADP as the phosphate group replaces the terminal hydrogen on the glucose, forming the [glucose-phosphate]$^{2-}$ anion.

Problem 1

The body would rapidly run out of ATP if there were not some process by which it could be regenerated from ADP. The hydrolysis of 1,3-diphosphoglycerate^{4-} (1,3-DPG^{4-}) to 3-phosphoglycerate^{3-} (3-PG^{3-}) drives the conversion of ADP^{3-} into ATP^{4-}. The hydrolysis reaction has a large negative $\Delta G°$ value. Calculate $\Delta G°$ for the reaction

ADP^{3-} + 1,3-diphosphoglycerate^{4-} →
$$3\text{-phosphoglycerate}^{3-} + \text{ATP}^{4-}$$

from the following changes in standard free energy:

(1) 1,3-DPG^{4-}(aq) + H$_2$O(l) →
  3-PG^{3-}(aq) + HPO$_4{}^{2-}$(aq) + H$^+$(aq) $\Delta G° = -49.0$ kJ

(2) ADP^{3-}(aq) + HPO$_4{}^{2-}$(aq) + H$^+$(aq) →
  ATP^{4-}(aq) + H$_2$O(l) $\Delta G° = 30.5$ kJ

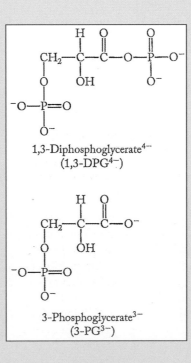

1,3-Diphosphoglycerate^{4-}
(1,3-DPG^{4-})

3-Phosphoglycerate^{3-}
(3-PG^{3-})

Problem 2

The conversion of glucose into lactic acid drives the phosphorylation of 2 moles of ADP to ATP:

$$C_6H_{12}O_6(aq) + 2HPO_4^{2-}(aq) + 2ADP^{3-}(aq) + 2H^+(aq) \rightarrow$$
$$2CH_3CH(OH)COOH(aq) + 2ATP^{4-}(aq) + 2H_2O(l)$$
$$\Delta G° = -135 \text{ kJ/mol}$$

What is $\Delta G°$ for the conversion of glucose into lactic acid?

$$C_6H_{12}O_6(aq) \rightarrow 2CH_3CH(OH)COOH(aq)$$

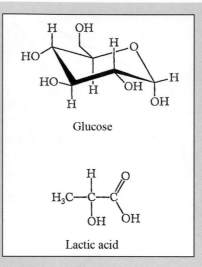

Glucose

Lactic acid

CASE STUDY 7.3 Electricity, Magnetism, and Electric Motors to Power the Transportation Sector

Introduction

As the implications of continued use of fossil fuels to meet primary energy generation demands becomes more widely understood, there will be a rapidly increasing demand for the generation of electric power. This increased demand for electric power means that the sources of primary energy generation will shift to the incorporation of wind power generation, concentrated solar thermal power generation, and geothermal power generation into the nation's power grid. All these primary energy sources involve the conversion of mechanical energy or thermal energy to electric power through the use of generators that are very similar to the large generators currently used in coal and natural gas burning power plants as well as in nuclear and hydroelectric power plants. The ubiquitous application of electric generators begs the question: how do they work? What are the principles of electricity and magnetism upon which their performance depends? Electric motors are required to convert an electrical potential into mechanical work to propel automobiles, high speed rail systems, and other modes of transport. How do electric motors work? How are they different from generators? How are they similar?

This combination of energy *generation* and energy *end use* is summarized in Figure CS7.3a, with the top row of fig-ure panels focused on energy *production* and the bottom row focused on energy *uses*.

The interaction of a magnetic field with a moving charge or the creation of a magnetic field by a moving charge establishes an intimate linkage between electric fields, magnetic fields, and associated forces that challenge any effort at clearly separating cause and effect. Thus, we choose to begin by examining the force exerted on a moving charge by a time independent magnetic field.

We will assume, therefore, that we have a magnetic field with both a magnitude and direction that is invariant with time, such as that established between the pole pieces of a permanent magnet as shown in Figure CS7.3b. By direct experimentation we can determine the net force on a charge q passing through a magnetic field B. The force exerted by the magnetic field B on the charge q turns out to depend not only on the charge q, but also on both the direction of motion of the charge with respect to the magnetic field and the velocity of the charge. As displayed in Figure CS7.3b, a positive charge moving across magnetic field lines, B, experiences a force perpendicular to both the velocity of the particle, v, and the direction of the magnetic field. The direction of the force on a positive charge, +q, is determined by the *right-hand rule*: if the fingers of the right hand are extended in the direction

FIGURE CS7.3A As the consequences of burning fossil fuels are more widely understood, there will be a shift in both the production of electricity and the use of electricity. The upper row of figures represents four important *sources* of electric power generation: concentrated solar thermal, wind, and hydroelectric do not release carbon dioxide; in the last panel is a coal burning power plant that does release carbon dioxide. In all four cases, mechanical energy is converted to electrical energy by an electric *generator*. In the lower panels the use and distribution of electric power is represented: high-speed rail, electric cars, electric monorail systems, and finally power lines used to distribute electrical energy through a "power grid."

of the particle's velocity and the fingers are rotated into the direction of the magnetic field, the positively charged particle will experience a force, f, in the direction in which the thumb is extended. For a negative charge, the force is in the opposite direction as is also shown in Figure CS7.3b.

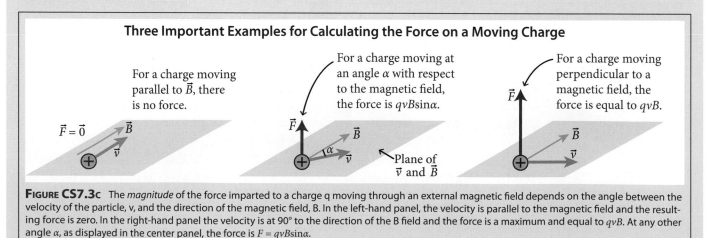

FIGURE CS7.3ʙ The application of the right-hand rule to determine the *direction* of force on a charge +q moving with velocity v through a fixed magnetic field B. If the fingers of the right hand are pointed in the direction of the positively charged particle's direction and the fingers are rotated into the direction of the B field, the thumb points in the direction of the force on the positively charged particle. If the particle is negatively charged, the force is in the opposite direction.

Figure CS7.3c displays the result of three experiments. In the left panel of Figure CS7.3c the charge is moving parallel to the magnetic field and that charge experiences *no net force*. When the charge moves perpendicular to the magnetic field, as shown in the right panel of Figure CS7.3c, the force is a *maximum* and is equal in magnitude to the product q v B and in a direction perpendicular to the direction of the B field—the case reviewed in Figure CS7.3b. In the center panel of Figure CS7.3c, the direction of the particle is movement is at an angle α with respect to the B field. In this case the magnitude of the force on the charge q is qvB sin α and the direction is perpendicular to the plane containing the B field direction and the direction of motion of the charged particle.

The magnetic field B exerts a force on a moving charge q, and the magnitude of that force has the following properties:

1. When a charge is at rest in a magnetic field, no force is delivered to the charge.
2. If the charge is moving parallel to or antiparallel to the direction of the magnetic field, no force is applied to that charge by the B field.
3. When a charged particle is moving through a magnetic field, the force applied to that charge by the magnetic field is perpendicular to both v and B.
4. If the charge q is negative, the direction of the force applied to the charge by the magnetic field is opposite of the case for a positive charge.
5. The most convenient way of defining the relationship between the velocity v, the magnetic field B, and the force applied to the charged particle is to use the "right-hand rule." As displayed in Figure CS7.3b, if you extend the fingers of the right hand in the direction of the motion of the charge, q, and you rotate the fingers into the direction of the magnetic field, B, then the force is aligned with the direction of the thumb.

Magnetic Force on a Current-Carrying Wire

We can easily determine the force exerted on a current carrying wire by an externally applied magnetic field by a straightforward application of the principle outlined above for the forces on a single charge. To begin, we calculate the force that a uniform magnetic field exerts on a long straight wire that carries a current I through a magnetic field B. As we can deduce from our experiments with single charges, if the wire is parallel to the magnetic field, no force will be exerted on the wire. If the wire is perpendicular to the magnetic field, as shown in Figure CS7.3d, with the magnetic field pointing *into* the plane of the figure (as designated by the Xs in the

FIGURE CS7.3ᴄ The *magnitude* of the force imparted to a charge q moving through an external magnetic field depends on the angle between the velocity of the particle, v, and the direction of the magnetic field, B. In the left-hand panel, the velocity is parallel to the magnetic field and the resulting force is zero. In the right-hand panel the velocity is at 90° to the direction of the B field and the force is a maximum and equal to *qvB*. At any other angle *α*, as displayed in the center panel, the force is F = qvBsinα.

figure), each charge has a force qvB on it as a result of the motion of the charges through the field B. With no current flowing, as shown in the left panel, there is no force exerted on the wire. With a current *I* flowing in the wire, there is a force, F, exerted in the plane of the figure in the direction defined by the right-hand rule—the force is to the left as displayed in the right panel of Figure CS7.3d.

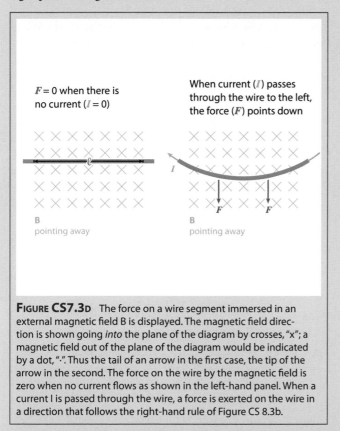

FIGURE CS7.3D The force on a wire segment immersed in an external magnetic field B is displayed. The magnetic field direction is shown going *into* the plane of the diagram by crosses, "x"; a magnetic field out of the plane of the diagram would be indicated by a dot, "·". Thus the tail of an arrow in the first case, the tip of the arrow in the second. The force on the wire by the magnetic field is zero when no current flows as shown in the left-hand panel. When a current I is passed through the wire, a force is exerted on the wire in a direction that follows the right-hand rule of Figure CS 8.3b.

The magnitude of the force can be directly calculated by recognizing that a section of wire of length ℓ contains charge carriers with total charge q moving with speed, v. The current I, by definition, is simply the charge q divided by the time it takes for the charge to flow out of that section of the wire: specifically I = q/Δt. The time required is $\Delta t = \ell/v$ where ℓ is the length of the wire and v is the velocity of charges moving through the wire. This gives us an expression for I as a function of q, v, and ℓ:

$$I = \frac{q}{\Delta t} = \frac{q}{\ell/v} = \frac{qv}{\ell}$$

Thus qv = I L and from F = qvB, we have

$$F = q v B = I \ell B$$

for the force on a wire of length ℓ carrying a current I perpendicular to a magnetic field of strength B. This disarmingly simple expression for the force on a wire of length ℓ carrying a current I perpendicular to the magnetic field B provides an opportunity to examine the units involved in the expression.

We know that force in SI units is measured in units of Newtons, N. We also know that current I is measured in Amperes, A, and length ℓ is measured in meters, m. Thus the SI units of the B field must have units of N/A·m. The SI unit of magnetic field strength is the Tesla, where 1 T = 1 N/A·m.

We can now summarize what we know about the force exerted by a magnetic field on a wire carrying current I as follows. From direct experiment we know:

- The force exerted on the wire is proportional to the current in the wire
- The force is proportional to the length of wire that is within the magnetic field, B
- The force is proportional to the strength of the magnetic field
- The force depends on the angle between the B field and the current direction

As a result, we can write the proportionality: F α I ℓ B sinΘ where Θ is defined in Figure CS7.3e.

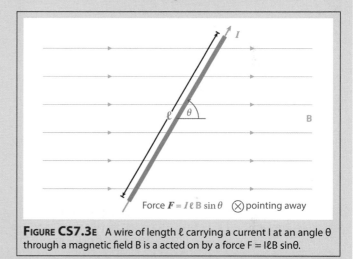

FIGURE CS7.3E A wire of length ℓ carrying a current I at an angle θ through a magnetic field B is a acted on by a force F = IℓB sinθ.

Because the field strength, B, is *defined* in SI units by the force exerted on a moving charge, the proportionality factor between the force, F, and I ℓ B sinΘ in equation (1) above is unity, so we can directly equate the force F with the quantity I ℓ B sinΘ. Thus we have the key result that: **F = I ℓ B sinΘ**.

Forces and Torques on Current Loops

A key objective of this Case Study is to understand the principles underpinning the operation of an electric motor, so we now explore how this calculated *force* on a wire carrying a current I through a magnetic field can be used to develop the *torque* required to propel a car, drive a water pump or otherwise provide the mechanical means for doing work.

The key idea that transforms the simple force on a segment of wire that carries a current through a magnetic field, to a *torque* is that of a current loop, as displayed in Figure CS7.3f.

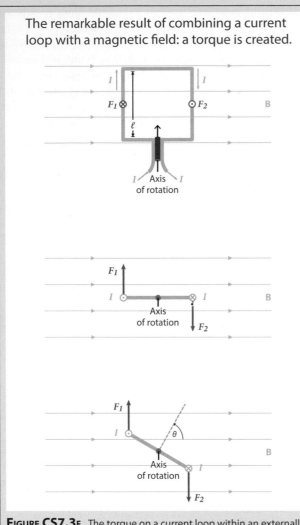

The remarkable result of combining a current loop with a magnetic field: a torque is created.

FIGURE CS7.3F The torque on a current loop within an externally imposed magnetic field, B, results from the fact that the *direction* of the force is in the opposite direction on the two sides of the loop carrying current across the direction of the magnetic field. In the upper panel of Figure CS 8.3f, the left-hand side of the current loop has force F = IℓB directed *into* the page (indicated by the ⊗) and the right-hand side has force F = IℓB directed *out of* the page (indicated by the ⊙). As shown in the center panel, this results in forces F1 and F2 operating in opposite directions resulting in a torque that depends on the angle of the loop, θ, with respect to the magnetic field direction as shown in the lower panel.

When a current flows in a closed loop of wire immersed in an externally imposed magnetic field, the magnetic force on the moving charge carried by the wire loop can produce a torque as shown in panel A of Figure CS7.3f. With the B field in the plane of the diagram pointing to the right, the current moving through the wire, also in the plane of the diagram, will create a force F = IℓB = IaB where a is the length of the current loop perpendicular to the B field in panel A. The direction of the force will be into the plane of the diagram, as indicated in panel A. The current segment of length b that is parallel to the B field exerts no force because the current carried by the wire is parallel to the B field. The segment of the loop of length a carrying current from top to bottom in panel A will have a force of magnitude IaB, but the direction of that force will be out of the plane of the diagram, in the *opposite* direction to that of the wire segment on the left-hand side carrying current from the bottom to the top of panel A.

Because the force created by the current flowing through the loop in a uniform magnetic field is equal but opposite in direction on each side of the loop, the oppositely directed forces apply a *net torque* on the current loop about the axis of rotation shown in panel A. Panel B defines the force vectors operating on the current loop as viewed along the axis of rotation and perpendicular to the B field. If the current loop is allowed to rotate by an angle θ in the B field, the direction of the force vectors and their magnitude is still perpendicular to the B field but the torque is reduced. The reason is that torque is defined by the *product* of the length of the moment arm, r, from the center of rotation to the point at which the force is applied perpendicular to the vector connecting that point to the center of rotation as displayed in Figure CS7.3f.

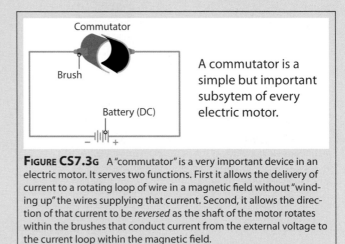

A commutator is a simple but important subsytem of every electric motor.

FIGURE CS7.3G A "commutator" is a very important device in an electric motor. It serves two functions. First it allows the delivery of current to a rotating loop of wire in a magnetic field without "winding up" the wires supplying that current. Second, it allows the direction of that current to be *reversed* as the shaft of the motor rotates within the brushes that conduct current from the external voltage to the current loop within the magnetic field.

We can calculate the torque on the current loop in Figure CS7.3f by calculating the torque for each segment of wire (of length a) perpendicular to the magnetic field multiplied by the moment arm (length b/2) from the point of force to center of rotation. Thus the force on each segment is

$$F = IaB$$

and the torque from each segment is

$$\tau = \left(\frac{b}{2}\right)IaB$$

Adding the torque from each of the two segments gives a total torque, τ_T, of

$$\tau_T = \left(\frac{b}{2}\right)IaB + \left(\frac{b}{2}\right)IaB = IabB$$

But the area of the current loop is A = ab, so we have the simple expression for the torque developed by the current loop in terms of the current carried by the wire, I, the area of the current loop, A, and the applied magnetic field, B.

$$\tau_T = IabB = IAB$$

If there are N turns in the current loop, then the torque is increased in proportion to the number of turns, N, such that:

$$\tau_T = NIAB$$

because for each turn, the current is increased by I.

As panel C of Figure CS7.3f makes clear, as the current loop rotates about the axis of rotation in the fixed magnetic field, the moment arm decreases as the angle θ decreases so the moment arm b/2 must be replaced by (b/2) sin θ, and thus $\tau_\theta = \tau_T \sin \theta = NIAB \sin \theta$. As a result, when the current loop rotates to the point where the plane of the current loop aligns with the magnetic field, B, the torque on the loop τ_θ = NIAB sin θ = 0 because θ = 0° and thus sin θ = 0.

Therefore, while we have solved the problem of how to generate a torque using a loop of wire immersed in a magnetic field, we have a device that only supplies torque for ½ a rotation. We need to modify the design such that the shaft of the device rotates continuously. This requires a device known as a "commutator" which is displayed in CS7.3g.

The commutator provides two crucial functions. First, it supplies the required voltage to the current loop(s) as the current loop rotates in the magnetic field so the wires do not "wind-up" as the current loop continues to rotate. This is accomplished by having the ends of the current loop slide, with minimal friction, inside the curved metal electrodes that are split so as to apply the positive side of the applied voltage to one electrode and the negative side of the applied voltage to the other electrode as shown in Figure CS7.3h.

Second, and equally important, the commutator serves

to systematically *reverse* the polarity of the applied voltage just as the current loop reaches the point where the plane of the current loop is perpendicular to the direction of the magnetic field. It is at that point where the force imparted but the magnetic field on the current loop is perpendicular to the axis of rotation (as displayed in Figure CS7.3h) and thus it is the point of zero torque. If the current is reversed through the current loop at that point, then the force vectors on each side of the wire loop are reversed and a torque is then applied in the same direction as the rotation. That torque then increases as θ increases from 0° toward 90° such that τ = NIAB sin θ. The polarity of the voltage and thus the direction of the current is then fixed for the next 180° of rotation to keep the direction of the torque the same, although the magnitude of the torque varies as sin θ of the angle between the plane of the current loop and the direction of the magnetic field.

The configuration of the commutator and the "brushes" between which the commutator rotates are displayed in panel A of Figure CS7.3i along with the torque, τ, as a function of time as the wire loop rotates in the magnetic field.

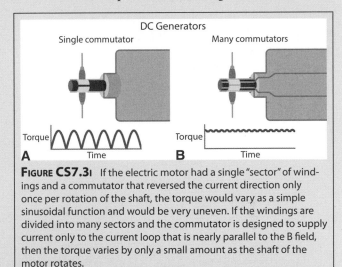

FIGURE CS7.3i   If the electric motor had a single "sector" of windings and a commutator that reversed the current direction only once per rotation of the shaft, the torque would vary as a simple sinusoidal function and would be very uneven. If the windings are divided into many sectors and the commutator is designed to supply current only to the current loop that is nearly parallel to the B field, then the torque varies by only a small amount as the shaft of the motor rotates.

As we observed above, the torque delivered by the motor with a single plane of wire loop is given by

$$\tau = NIAB \sin \theta$$

A motor with a single plane of "windings" on the armature would provide a very uneven torque as a function of shaft angle as shown in panel (a) and would thus deliver very uneven power to the motor shaft. This problem is dramatically reduced by using a design that uses many "sectors" of windings, as displayed in panel (b) of Figure CS7.3i. Each sector is, in this design, supplied with the full current to its individual winding when the plane of that winding is nearly perpendicular to the magnetic field such that the torque is very near the maximum $\tau_T = NIAB \sin \theta \approx NIAB$ because θ ≈ 90° and sin θ ≈ 1. This results in a torque delivered by the motor shaft that is far smoother as shown in panel (b) of Figure CS7.3i.

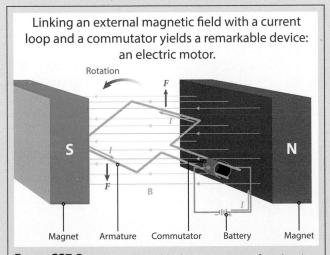

FIGURE CS7.3h   We can assemble the components of an electric motor into a single diagram by showing the magnetic pole pieces that impose an external magnetic field, B, the current loop that carries the current, I, through the magnetic field that creates a torque on the current loop, and finally, the commutator that connects the battery to the current loop through the sliding brushes that reverse the current direction as the plane of the loop becomes perpendicular to the magnetic field direction.

Problem 1

We observed the video of the drag race between a BMW M5 and the Tesla Model S. At the end of that race, the moderator noted that the Model S produced 420 hp.

Calculate the amperage delivered to the electric motor of the Model S.

Problem 2

The Tesla Model S develops 450 lb·ft of torque. If the Model S motor has 100 windings and the current loop has an area of 0.2 m², what is the magnetic field strength of the motor?

CASE STUDY 7.4   Linking Global Scale Calculations: The "50 Questions on Global Scale Energy and Power" Part II

We continue from Case Study 3.4 the development of the quantitative relationships involving global energy and power that are essential for wise stewardship of national security, international relations and informed public policy decisions in a modern democracy. The condensed summary of these key quantitative relationships at the global scale is contained within the set of 50 questions—questions that any informed member of modern society must know in order to proactively engage in both national and international discourse. In the development of these "50 Questions" segments, we will repeatedly revisit previous sectors of the text and previous Case Studies.

FIGURE CS7.4B   There is a remarkable difference in the energy expenditure per km between freight transport by train and by truck. This difference constitutes an important consideration in national energy policy.

Problem 1

Energy required to propel gas vs. electric automobiles

How much chemical energy in kWh is required to propel an average US sedan 100 km? How much energy is required to propel an electric car 100 km? What is the energy consumed per person per day driving an SUV in traffic vs. driving an electric car in traffic? If you drive 20,000 km a year, how much does it cost to drive a gasoline powered car at 8 km/liter (20 mpg) and $1/liter, for a year. How much per year does it cost to drive an SUV at 4 km/liter? How much does it cost to drive an electric car 20,000 km a year? The total number of km driven per year in the US is 3.8×10^{12} km. How much does it cost if everyone drives an 8km/liter gasoline car? If everyone switches to an electric car, how much would the US spend per year?

FIGURE CS7.4A   There are now designs for SUVs in both all electric and gasoline. Shown in the left panel is the Chevrolet Tahoe and in the right panel a Tesla Model X.

Problem 2

Power consumption by different modes of transportation

How much power is required to sustain a medium/large jet liner in flight? What is the power output of the first stage of the Saturn V rocket that launched men to the moon? What is the ratio of energy consumption in kWh/t-km for shipping by truck vs. shipping by rail? For shipping by air vs. shipping by rail?

Problem 3

Comparison of energy release

What is the ratio of the energy released in a medium sized rain storm to the energy release in the first atomic bomb? What is the ratio of the energy released in a hurricane to the energy release in the first atomic bomb? What is the ratio of the energy contained in 100 tons of coal to the energy contained in the energy generated by the Grand Coulee Dam in a year? What is the ratio of the energy contained in the fuel of a fully fueled jet liner to the kinetic energy of the airliner in flight? What is the ratio of the energy contained in the gas tank of an automobile to the kinetic energy of the car at 100 km/hr? What is the power output of the first stage of the Saturn V rocket that launched men to the moon?

FIGURE CS7.4C   Development of an understanding of energy scales and power scales involves important considerations between different sources of primary energy. For example, the energy release per kg between nuclear fission, shown in the upper panel, and coal, shown in the lower panel.

Length Scales

What is the ratio of the diameter of an atom to the diameter of the nucleus? What is the ratio of the diameter of the Sun to the diameter of the Earth? How does the height of Mt. Everest compare with the depth of the deepest part of the ocean? What is the ratio of distance to the next galaxy compared to the diameter of our galaxy? What is the ratio of the diameter of a virus to the length of a bacterium? What is the ratio of the length of a carbon–carbon bond to the thickness of the cell wall? What is the ratio of the length of a C–C bond to the width of a carbon nanotube?

Setting the Context: Spatial Scales in Meters

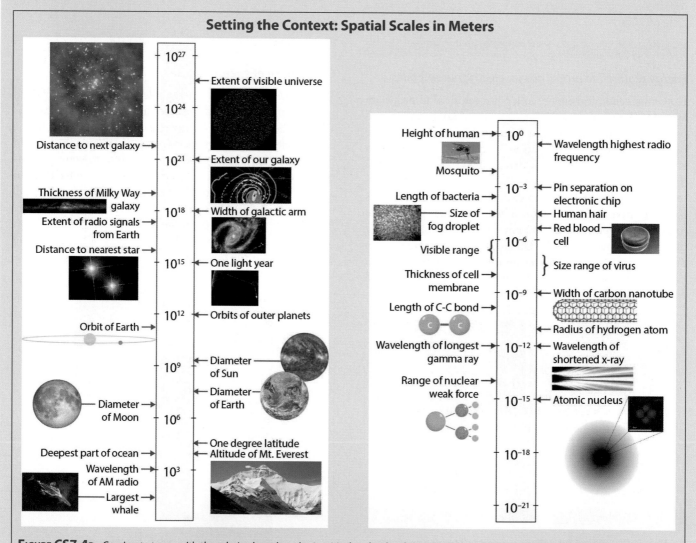

FIGURE CS7.4D Coming to terms with the relative length scales is critical to the development of how natural systems are constructed and of how they exchange energy. Natural systems are architected by building up sequentially from the scale of the nucleus to the scale of the universe and it is important to relate the size of the prominent structures within those boundaries. The reason an understanding of length scales is important for understanding energy transformations in nature is that electromagnetic radiation is a primary mechanism for converting energy of one form into energy of another form and electromagnetic radiation is defined by its frequency, ν, and its wavelength, λ, such that the product $\lambda\nu = c$, where c is the speed of light. As we will see, this places great importance on knowing the wavelength, λ, of electromagnetic energy relative to all spatial scales in nature.

Problem 5

Carbon release

How many tons of carbon does each US citizen add to the atmosphere each year? How many tons of carbon does a 20 mpg gasoline auto emit each year? How many tons of carbon does the US add to the atmosphere each year? How many tons of carbon does China add to the atmosphere each year? When did the carbon release from China equal that of the US? When will China add twice the carbon to the atmosphere that the US adds? How many tons of carbon does the global economy add to the atmosphere each year?

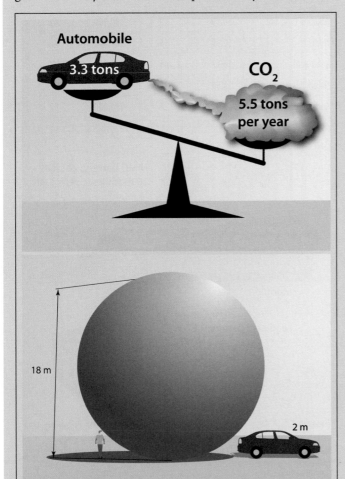

FIGURE CS7.4E We can compare the mass and the volume of CO_2 emitted by a number of different devices and for the global economy as a whole. In the first panel the mass of CO_2 emitted by an automobile is displayed showing that a mass of CO_2 emitted from an automobile driven (the national average) 20,000 km per year by the automobile that nearly twice the mass of the automobile. The corresponding *volume* of CO_2 emitted by the automobile at STP is displayed in the second panel—a sphere 18 m in diameter containing pure CO_2.

Problem 6

Comparison of energy content

How much energy in kWh is contained in a liter of gasoline? In a liter of ethanol? What is the ratio of the energy content per kg in TNT vs. gasoline? Ratio of energy content per kg in a lead-acid battery to that of gasoline?

TABLE CS7.4A Energy per kg for various objects and substances.			
	Kilowatt-hours	Compared to TNT	kJ
Bullet (1000 ft/s)	0.005	0.015	1.8×10^1
Auto battery	0.016	0.046	5.8×10^1
Computer battery	0.053	0.15	1.9×10^2
Alkaline battery	0.079	0.23	2.8×10^2
TNT	0.343	1	1.2×10^3
High explosive (PETN)	0.528	1.5	1.9×10^3
Chocolate chip cookies	2.6	7.7	9.4×10^3
Coal	3.2	9.2	1.2×10^4
Butter	3.7	11	1.3×10^4
Ethanol	3.2	9	1.2×10^4
Gasoline	5.3	15	1.9×10^4
Natural gas (methane)	6.9	20	2.5×10^4
Hydrogen	14	40	5.0×10^4
Asteroid (30 km/s)	57	165	2.1×10^5
^{235}U	11 million	32 million	4.0×10^{10}

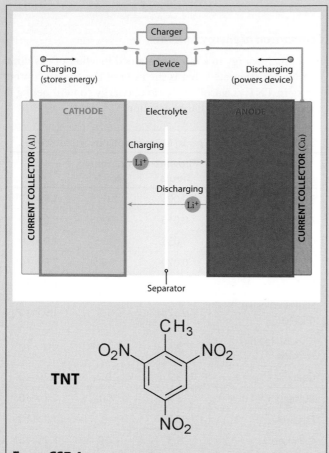

FIGURE CS7.4F It is important to know the ratios of the energy content per kg of the major energy sources from automobile batteries to nuclear fuel. Shown in the top panel is the schematic of a lithium-ion battery and in the bottom panel the molecular structure of TNT.

Comparison of energy content in different battery types

How many watt-hours (Wh) of energy per kg are contained in a full charged lead-acid battery? In a NiMH battery? In a state-of-the-art lithium-ion battery? What is the range of a Tesla Model S with current production Li⁺ batteries? By what fraction is that predicted to increase over the next few years?

TABLE CS7.4B

Battery type	Cost $ per Wh	Wh / kg	Joules / kg	Wh / liter
Lithium-ion	$0.47	180	460,000	230
Alkaline long-life	$0.19	110	400,000	320
NiMH	$0.99	95	340,000	300
Lead-acid	$0.17	41	146,000	100
NiCad	$1.50	39	140,000	140
Carbon-zinc	$0.31	36	130,000	92

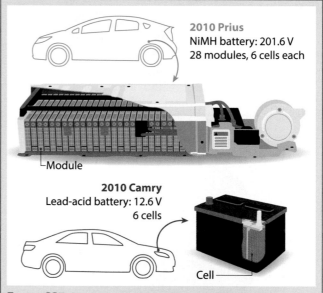

2010 Prius
NiMH battery: 201.6 V
28 modules, 6 cells each

Module

2010 Camry
Lead-acid battery: 12.6 V
6 cells

Cell

FIGURE CS7.4G There is now extreme pressure in the research and technology sectors to advance the energy content per unit mass of batteries. This is the primary determinant that establishes the distance an electric car can go between charges. Shown in the upper panel is the arrangement of metal-hybrid batteries used in a gasoline-electric hybrid automobile. In the lower panel is the lead-acid battery used to start the engine in a gasoline automobile.

Problem 8

Comparison of power consumption per capita in different countries

What is the average power consumption of every man, woman and child in the U.S.? In Western Europe? In China? In India? In Sub-Saharan Africa? The population of Africa is predicted by the UN to increase from 1.2 billion to 3.7 billion; if per capita power consumption of Africans increases from the present 200 watts to the current western European average, by what fraction would current power consumption in Africa increase? By what fraction would world power consumption increase?

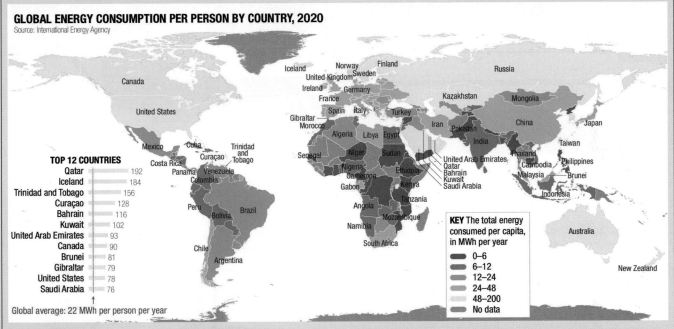

GLOBAL ENERGY CONSUMPTION PER PERSON BY COUNTRY, 2020
Source: International Energy Agency

TOP 12 COUNTRIES

Country	MWh
Qatar	192
Iceland	184
Trinidad and Tobago	156
Curaçao	128
Bahrain	116
Kuwait	102
United Arab Emirates	93
Canada	90
Brunei	81
Gibraltar	79
United States	78
Saudi Arabia	76

Global average: 22 MWh per person per year

KEY The total energy consumed per capita, in MWh per year

- 0–6
- 6–12
- 12–24
- 24–48
- 48–200
- No data

FIGURE CS7.4H   While it is important to determine specifically the average power consumption per capita in example nations such as the US, China, Western Europe, India, Brazil, Kenya, Bangladesh, etc., it is also important to consider the *pattern* of energy consumption per year (the power consumption) across the globe.

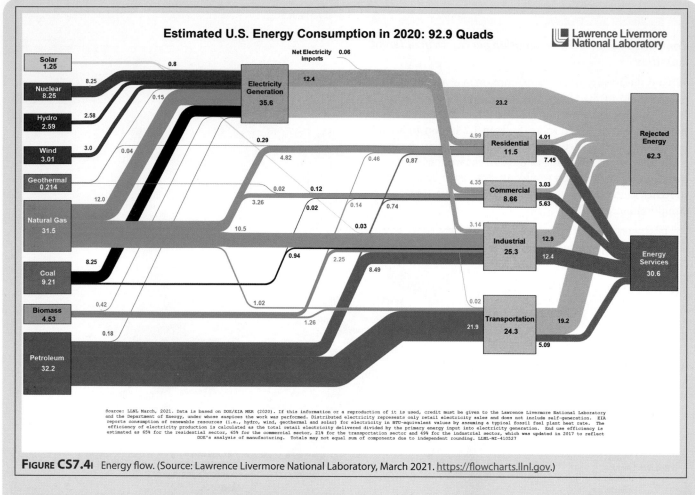

FIGURE CS7.4I   Energy flow. (Source: Lawrence Livermore National Laboratory, March 2021. https://flowcharts.llnl.gov.)

Problem 9

What are the principal (top 5) primary energy sources in the US? What fraction of primary energy generation goes to electricity production? What fraction of primary energy generation is wasted? Why is this fraction so high? What fraction of primary energy generation goes to transportation? What fraction of primary energy generation that goes to transportation is wasted?

Problem 10

When 100 units of chemical energy is expended in a typical power generation plant, how many units of electrical energy is generated? How many units of the original 100 units is typically lost in transmission? Of the remaining power delivered to the home, if that electrical energy is used to power an incandescent light, how many of the original 100 units of chemical energy is released as visible light?

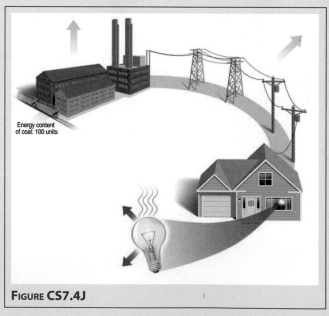

FIGURE CS7.4J

CASE STUDY 7.5   Linking Acid-Base Reaction, Oxidation-Reduction, and Electrochemistry

Remarkably, the most common battery in use today is the lead-acid storage battery that is universally used with a "starter-motor" to start automobile, truck, and bus engines —both gasoline and diesel. As noted in Case Study 7.1, the lead-acid battery half-reactions are:

Cathode: $PbO_2(s) + 3H^+(aq) + HSO_4^-(aq) + 2e^- \rightarrow PbSO_4(s) + 2H_2O(l)$

Anode: $Pb(s) + HSO_4^-(aq) \rightarrow PbSO_4(s) + H^+(aq) + 2e^-$

NET: $PbO_2(s) + Pb(s) + 2H^+(aq) + 2HSO_4^- \rightarrow 2PbSO_4(s) + 2H_2O(l)$

Inspection of the half-reactions of the lead-acid battery reveals the following:

1. The solid cathode is $PbO_2(s)$.
2. The anode is pure lead, $Pb(s)$.
3. The solution is sulfuric acid, $H_2SO_4(aq)$.
4. Reduction occurs at the cathode.
5. Oxidation occurs at the anode.
6. The reaction is spontaneous at standard conditions (otherwise the lead-acid battery would not be in general use) and as also noted in Case Study 7.1, the cell potential under standard conditions is
$E^\circ_{cell} = E^\circ_{PbO_2/PbSO_4} - E^\circ_{PbSO_4/Pb} = +2.01$ V
7. The bisulfate anion, HSO_4^-, is involved as a reactant in both the oxidation and reduction reactions.

Before we analyze the behavior of the lead-acid battery under nonstandard conditions, we consider first the oxidation-reduction reactions that take place, respectively, at the anode and cathode. Consider the graphical representation of what occurs within the lead-acid battery in Figure CS7.5a.

We first analyze the change in oxidation state of Pb in PbO_2 when it reacts at the cathode to form $PbSO_4$:

$$PbO_2 + 3H^+(aq) + 2HSO_4^-(aq) + 2e^- \rightarrow PbSO_4 + 2H_2O(l)$$

In PbO_2, the oxidation state of Pb is determined by applying our rules outlined in Case Study 2.4 (shown here as Table CS7.5a).

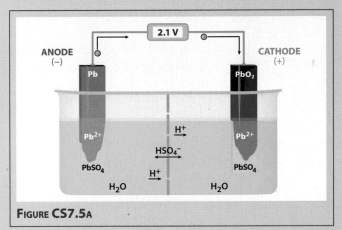

FIGURE CS7.5A

TABLE CS7.5A (From Case Study 2.4, Table CS2.4a)

Rules of oxidation states

1. As noted, the oxidation state of an atom in a free element is zero

2. The oxidation state of a monoatomic ion is equal to its charge

3. Because electrons are neither produced nor destroyed in a redox reaction, the *sum of the oxidation states* of all atoms in:
   - A neutral molecule or molecular formula is 0
   - An ion is equal to the charge of the ion

4. Metals, when they form compounds, have positive oxidation states
   - Group 1A metals always have an oxidation state of +1
   - Group 2A metals always have an oxidation state of +2

5. When *non-metals* form compounds, those non-metals are assigned values according to the table displayed here. Within this table, there is an ordering of importance with the most important rule beginning at the top

non-metals →

Oxidation States of Nonmetals		
Nonmetal	**Oxidation State**	**Example**
Fluorine	− 1	MgF_2 − 1 ox state
Hydrogen	+ 1	H_2O + 1 ox state
Oxygen	− 2	CO_2 − 2 ox state
Group 7A	− 1	CCl_4 − 1 ox state
Group 6A	− 2	H_2S − 2 ox state
Group 5A	− 3	NH_3 − 3 ox state

Given that there is neither fluorine nor hydrogen involved in PbO_2, we assign an oxidation state of -2 to each oxygen and so, recognizing that PbO_2 is neutral so the sum of oxidation states must be zero, the oxidation state of Pb in PbO_2 must be $+4$.

On the product side, we similarly analyze the oxidation state of Pb in $PbSO_4$. The sulfate anion has an oxidation state of -2, so the oxidation state of Pb in $PbSO_4$ is $+2$. Therefore, Pb in PbO_2 is reduced from an oxidation state of $+4$ to an oxidation state of $+2$ at the cathode. Analysis of the anode reaction

$$Pb(s) + HSO_4^-(aq) \rightarrow PbSO_4 + H^+ + 2e$$

reveals that $Pb(s)$ goes from an oxidation state of 0 to an oxidation state of $+2$. Thus Pb is oxidized at the anode.

However, in order to analyze the behavior of the lead-acid battery under nonstandard conditions, we need to calculate not only $\Delta G°$ from which we can determine $E°$ from $-nFE° = \Delta G°$, but we need both $\Delta H°$ and $\Delta S°$ so we can calculate $\Delta G_T°$ at temperatures other that 295 K and thus we can calculate

$$\Delta G = \Delta G° + RT \ln Q$$

for nonstandard conditions.

For the overall reaction of the lead-acid battery

$$PbO_2(s) + Pb(s) + 2H^+(aq) + 2HSO_4^-(aq) \rightarrow$$
$$2PbSO_4(s) + 2H_2O(l)$$
$$\Delta H_R° = -374.4 \text{ kJ/mol}$$
$$\Delta S_R° = +40.01 \text{ J/mol·K}$$

Inspection of the overall reaction reveals that

$$Q = \frac{1}{\left[H^+(aq)\right]^2 \left[HSO_4^-(aq)\right]^2}$$

Given that we wish to investigate the behavior of the lead-acid battery under different conditions such as the degree of discharge and the behavior under decreasing and increasing temperature, we employ the Faraday equation

$$\Delta G = -nFE_{cell}$$

to calculate E_{cell} under nonstandard conditions. Specifically

$$\Delta G = \Delta G° + RT \ln Q$$

so, substitution for $\Delta G = -nFE_{cell}$ we have the Nernst equation $-nFE_{cell} = -nFE°_{cell} + RT \ln Q$ or

$$E_{cell} = E°_{cell} - \frac{RT}{nF} \ln Q$$

In order to analyze the behavior of the lead-acid battery under a range of conditions, we first ask: "At what temperature is the lead-acid battery no longer spontaneous reactants and products are at standard state?" To address this part of

the analysis, we recognize first that if reactants and products are at standard state, then $\Delta G_T° = \Delta H° - T\Delta S°$ where $\Delta G_T°$ is at standard conditions except for the temperature. And thus we can solve for the temperature at which $\Delta G_T° = 0$ with $\Delta H° = -374.4$ kJ/mol and $\Delta S° = +40.01$ J/mol·K.

Then from

$$\Delta G_T° = \Delta H° - T\Delta S°$$

it is clear that the right-hand side of the equation is negative for all possible temperatures because T is measured in Kelvin, and the minimum temperature is T = 0K so $\Delta G_T°$ is spontaneous at all temperatures.

To demonstrate this explicitly we can solve for T at $\Delta G_T° = 0$ at which point the reaction is no longer spontaneous. Inserting our values for $\Delta H°$ and $\Delta S°$, we have

$$\Delta G_T° = \Delta H° - T\Delta S°$$
$$= -374.4 \text{ kJ/mol} - T(.04001 \text{ kJ/mol·K}) = 0$$

So the temperature at which $\Delta G_T° = 0$ and the reaction is no longer spontaneous is T = -9360 K.

Since this is negative and physically impossible, we see explicitly that the lead-acid battery reaction is spontaneous at all (possible) temperatures.

Next we analyze under what condition the net overall reaction would become more spontaneous, i.e. what factors would make $\Delta G = \Delta G° + RT \ln Q$ more negative. We know from the above analysis that

$$Q = \frac{1}{\left[H^+(aq)\right]^2 \left[HSO_4^-(aq)\right]^2}$$

Thus if we increase $[H^+(aq)]$ we decrease $RT \ln Q$ and $\Delta G = \Delta G° + RT \ln Q$ becomes more negative.

What about changing the temperature? This has significant practical importance because as nearly everyone has experienced, when conditions are very cold, your car will not start because the battery lacks sufficient voltage to successfully operate the starter motor. Analysis of the temperature dependence requires that we consider two factors.

First we have

$$\Delta G = \Delta G° + RT \ln Q$$

and

$$\Delta G° = -RT \ln Q$$

Combining these two expressions gives us

$$\Delta G = -RT \ln K_{eq} + RT\ln Q$$
$$= RT(\ln Q - \ln K_{eq})$$
$$= RT \ln(Q/K_{eq})$$

But now it is instructive to sketch the free energy surface for the reaction. For a (standard) temperature of 298 K

$$\Delta G = -374.4 \text{ kJ/mol} - (298)(0.040 \text{ kJ/mol·K})$$
$$= -386.4 \text{ kJ/mol}$$

Thus the equilibrium point lies very close to the complete conversion of reactants to products and our Gibbs free energy surface takes the following form:

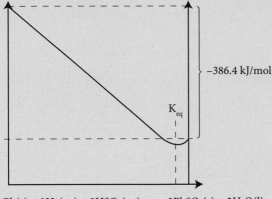

$PbO_2(s) + Pb(s) + 2H^+(aq) + 2HSO_4^-(aq)$ $2PbSO_4(s) + 2H_2O(l)$

When the battery is fully charged, Q lies well to the left of K_{eq} and $Q/K_{eq} \ll 1$ and thus

$$\Delta G = RT \ln Q/K_{eq} < 0$$

For example if $Q/K_{eq} = .01$ then $\ln(Q/K_{eq}) = -4.61$.

However, as the battery discharges, Q/K_{eq} increases, so when we calculate ΔG at, for example, $Q/K_{eq} = 0.5$, we have $\Delta G = RT \ln (0.5) = -0.69$.

Now $\Delta G = -0.69RT$

Thus as we decrease T, ΔG becomes less negative and thus

$$E_{cell} = -nF\Delta G \quad \text{decreases}$$

This is why, as a lead-acid battery becomes increasingly discharged, it is more susceptible to failure at low temperature.

QUANTUM MECHANICS, WAVE-PARTICLE DUALITY, AND THE SINGLE ELECTRON ATOM

8

"The task is not so much to see what no one has yet seen; but to think what nobody has yet thought, about that which everybody sees."

—Erwin Schrödinger

FIGURE 8.1 The observed position of individual iron atoms on flat copper arranged in a "corral" and detected with a scanning tunneling microscope (STM) has opened the atomic scale to the direct observation of electron density and atomic and molecular geometry.

Framework

Figure 8.1 displays a pattern of individual iron atoms on a copper surface by imaging the electron density with a sharpened tip of tungsten used to map the topography of the electron density surrounding individual atoms. If Figure 8.1 shows the position of individual atoms made of electrons, protons, and neutrons, all of which are particles, where does the apparent wave pattern come from? The image looks like a pond into which a stone has been thrown. How can this be?

Atomic structure and the relationship between the spatial distribution of electrons in separated atoms vs. those same atoms in a molecular structure has always been a central challenge for chemistry. While we will develop an understanding of atomic structure, and the remarkable interplay between the wave properties of electrons and that of electromagnetic radia-

tion, we move first to the forefront of research that is rapidly revealing revolutionary new developments in visualizing, imaging, and the physical manipulation of individual atoms.

Within the last few years, techniques have been developed that are capable of detailing the position of individual atoms as well as the distribution of electron density that constitutes the bonding geometry that establishes the *structure* of exotic new materials related to energy generation, nanoscale computing, memory storage devices, and drug delivery. Figure 8.1 is an image of individual iron atoms arranged on the surface of flat copper. The specific position of the "corral" of iron atoms was arranged by moving individual iron atoms with the tip of an extremely sharp needle.

Figure 8.2 displays the sequence by which the corral of iron

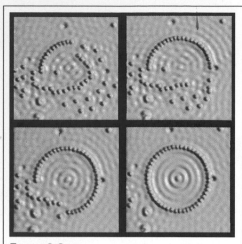

FIGURE 8.2 The sequential placement of iron atoms in a ring using STM to create specific geometrical arrangements at the atomic scale.

atoms was assembled one by one. The technique is called scanning tunneling microscopy (STM). While the iron atoms form the circle or corral, the wave pattern of electron density confined within the corral represents direct evidence of the wave nature of electrons. Specifically, while electrons are viewed as identifiable particles, they possess a wave-particle duality that constitutes the foundation of *quantum mechanics*—the scientific foundation that underpins our understanding of atomic structure and of molecular bonding.

While it is both remarkable and counterintuitive, a major component in the solution of global scale energy demand for what will be 10 billion inhabitants of the planet lies within the atomic scale spatial domain dominated by quantum mechanics. There are many reasons for this: (1) the interaction of photons of light with molecular structures engages directly the wave properties of the electron and the particle properties of light, (2) the development of new materials at the scale of 10 to 100 atoms provides remarkable control over electrical properties for high speed computing and high density information storage, and (3) chemical processes at the 10 to 100 atom scale can be controlled by molecular *self-assembly* where the architecture of the system is controlled by molecule-molecule interactions opening up the new field of nanochemistry and the ability to design large structures from the "bottom up"—to build from the molecular level to the macroscopic.

When we come to understand and to utilize the principles that dictate the behavior of electrons and atoms in molecular bonding—the domain of quantum

FIGURE 8.3 Image of a microscopic array of the top side of a silicon wafer based photovoltaic cell with a low reflectivity surface. The surface appears blue-violet due to the silicon nitride antireflective coating. The photons of light from the sun generate electrons that are collected by the <40 micron silver conducting "wires" that appear gold-colored in the image.

mechanics—we open new technologies that can very effectively and inexpensively generate electrical power from sunlight, and can create new materials capable of significantly reducing energy demand through far more efficient automobiles, high speed train systems and buildings. Technologies based on the principles of quantum mechanics can bring hundreds to thousands of people into meetings linked by the movement of electrons rather than the transport of materials or of people.

Thus we extend our exploration to the realm of the individual atom. From the principles that govern the structure of the individual atom, we will gain remarkable insight into the world of "nanomaterials" that constitute structures on the scale of 10^{-9} meters where the principles of quantum mechanics play a central role. Figure 8.3 displays the image of a device developed to convert solar photons to electrons and to then collect those electrons with "wires" only $\sim 10^{-6}$ m in width.

In Chapter 2 we reviewed the development of the atomic view of matter by John Dalton. But humans could not actually "see" individual atoms until the early 1980s. Then with the development of the STM—individual atoms of iron manipulated on a surface of copper opened a new era of visualization at the atomic level.

An inspection of Figure 8.1, taken by an STM, immediately raises three questions: (1) What are the waves that appear between the central point within the corral of iron atoms, (2) Why does the name of the device that made the observation contain the word "tunneling," and (3) Why does the apparent wave pattern within the corral of iron atoms virtually mimic the waves created by throwing a stone into a pond, resulting in concentric rings of waves in water?

With the discovery of the electron by J. J. Thomson in 1897, electrons were viewed as *particles*. With the development of electromagnetic theory in the 19th century and the observed behavior of light, it was taken as a fundamental tenet of physics that electromagnetic radiation was a wave phenomenon. Electrons were particles, light was a wave. Any student of the physical world knew that to be true!

But as we will see, the behavior of an electron on the scale of an atom has far more in common with a violin string than it does with a particle such as the stone thrown into a pond that creates waves. But, as we will also see, the wave nature of the electron allows it to penetrate into places forbidden by what we know in our macroscopic world—such as how when an object strikes a wall of concrete it will stop, unable to penetrate the barrier.

Electrons, in sharp contrast, can penetrate into barriers without damaging themselves or the barrier they penetrate. This is called "tunneling"—an electron passing through a barrier and appearing on the other side unscathed. On the macroscopic scale it is as if we were riding on a roller coaster, as shown in Figure 8.4, and instead of moving from point A, with initial potential energy of mgh and kinetic energy of zero that would carry us (in the absence of friction) to point C, we actually "tunneled" through from point C to point E on the other side of the (energy) barrier! This phenomenon is treated in Case Study 8.1.

In order to take an active and effective role in the big questions confronting modern society, we must be in command of the world of electrons and how they

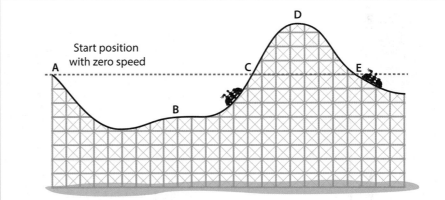

FIGURE 8.4 Conservation of energy dictates that *macroscopic* devices, such as a roller coaster, are constrained by conservation of energy and the physical barrier of the track to remain between points A and C, if the roller coaster begins at A with zero kinetic energy. In the world of quantum mechanics controlled by the *wave properties* of electrons, electrons can "tunnel" through barriers and emerge on the other side of the barrier. It is as though the roller coaster can reach point E without having to go over the barrier through point D.

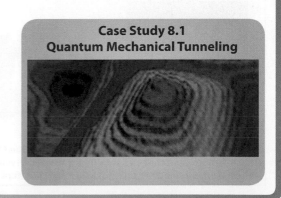

Case Study 8.1
Quantum Mechanical Tunneling

FIGURE 8.5 A computer image of the tungsten tip of an STM approaching within atomic dimensions of atoms on a surface. The electron flow (electric current) from the tungsten tip to the atoms in the surface is a sensitive function of the distance from the probe tip to the electron cloud around each atom as well as the voltage between the tip of the probe and the surface.

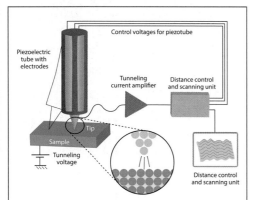

FIGURE 8.6 The operation of the STM employs not only the sharpened tungsten tip shown in Figure 8.5, but also the use of piezoelectric crystals that either expand or contract depending on the voltage applied to the crystal. This provides the ability to very precisely control the vertical and horizontal position of the tip by simply changing the voltage across the piezoelectric crystals controlling the horizontal and vertical position of the tungsten tip with respect to the surface.

behave on the scale of single atoms to tens of atoms to thousands of atoms. The behavior of electrons at the nanoscale defines developments in energy generation, capture and storage, information technologies, computer design, medical imaging, battery technology, drug delivery, catalytic processes, molecular motors, and genetic mapping.

In 1986 Gerd Binnig and Heinrich Rohrer were awarded the Nobel prize in physics for recognizing a simple fact: If they could create a metal tip with a terminus comparable to the dimension of an atom, they could use this tunneling effect of electrons to "map" the presence of individual atoms on a surface by bringing the metal tip within approximately an atomic radius of an atom on the surface. They could measure the position of that atom by observing the current flow (the number of electrons passing per unit time) between that tip and the atom on the surface by holding the metal tip at a fixed electrical potential (voltage) difference with respect to the surface.

Although it took Binnig and Rhorer three years of intensive work to develop the extremely sharp metal tips required and to develop extremely precise positional control (both vertically and horizontally), their new *scanning* (movement of the tip) *tunneling* (electron tunneling) *microscope* (ability to see small objects) or STM proved to be exquisitely sensitive to features at the atomic scale. A new era was born. Not only could science now resolve spatial features at the atomic level, science could dream of new objects to construct at the atomic scale and to then view the handiwork directly.

Figure 8.5 shows a computer reconstruction of an STM needle point fabricated to achieve atomic level resolution. The operation of the STM is depicted in Figure 8.6, wherein the tip of the probe approaches the array of atoms on the surface and, by virtue of the voltage difference between the tip and the surface, electrons flow from the tip to the surface. The tunneling of the electrons is extremely sensitive to the potential barrier, which is controlled at a fixed voltage difference by the distance between the probe tip and the electron cloud of the atom on the surface. Thus, as the STM tip is moved across the surface, the flow of the electrons from the tip to the surface rises sharply when passing over the top of an atom. As the tip moves to a position between two atoms on the surface, the current decreases sharply because of the increased distance between the probe and any surface atom. By systematically observing the current through the STM tip as a function of horizontal position, the surface topography is determined.

The STM, and variations of the STM, have become both the "eyes" and the "hands" of cutting edge research because the high spatial resolution images carry information heretofore inaccessible to science and the ability to manipulate the position of individual atoms on a surface has opened a new field of atomic level architecture. We consider first the development of high density data storage and the remarkable potential increase in the density of stored information using atomic

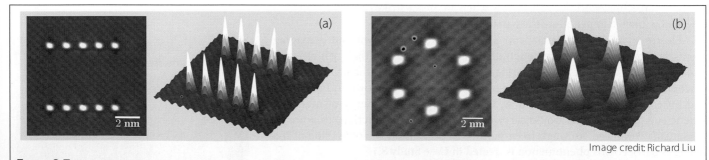

Image credit: Richard Liu

FIGURE 8.7 STM image of atom manipulation with an STM tip. In panel (a) Bi adatoms are moved to form the "0" and "1" patterns on the surface of CeBi. Those patterns can be used to store information at very high density on the surface. Panel (b) displays another configuration of Bi adatoms on the surface of CeBi demonstrating the ability of STM to execute atom manipulation to construct multiple functional structures on the surface.

level nanostructures. Figure 8.7 displays an STM image of atom manipulation with STM tip. Bi adatoms are moved to form the "0" and "1" patterns on the surface of CeBi displayed in panel (a). The term "adatom" is commonly used in surface science to refer to an atom that lies on a crystal surface. Those patterns of adatoms in panel (a) can be used to store information on the surface at very high density serving the important objective of markedly increasing the amount of information storage per unit area. The distance between individual Bi atom centers is approximately 1 nm in the image, defining the remarkable spatial resolution of the STM. Panel (b) demonstrates the ability of STM to execute atom manipulation to construct a variety of functional structures on the surface for a range of applications in this rapidly developing field of research.

Another domain that atomic level visualization using STM technology has opened up is in the area of carbon nanotubes, which can be formed with a range of diameters and lengths, creating the ability to "wire up" circuits on the scale of the atomic dimension. Figure 8.8 displays an STM image of these nanotubes of different diameters on a gold (Au) surface. The ability to both synthesize carbon nanotubes and to manipulate them on metal as well as electrically insulating surfaces has lead to the ability to tailor these carbon nanotubes into molecular scale nanowires with increasingly effective control over nanowire growth and geometry. Figure 8.9 displays nanowires with triangular joints linked to straight sections of controllable length. This provides the technological foundation for the fabrication of molecular scale nanocircuits that will revolutionize electronic circuit design.

With an increasing level of sophistication in the fabrication of nanowires, it is now possible to create radially layered tubes for a new generation of photovoltaic devices for converting sunlight to electrical power. These devices, shown in Figure 8.10, are capable of higher collection efficiency and they have a potentially lower cost of fabrication. This opens up the possibility of developing low cost techniques for converting sunlight to electrical power. The fundamentals of energy generation by solar radiation is treated in Case Study 8.2.

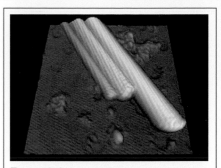

FIGURE 8.8 The advent of carbon nanotubes fabricated at selected diameters and in controlled lengths opens the possibility of far smaller electric circuits.

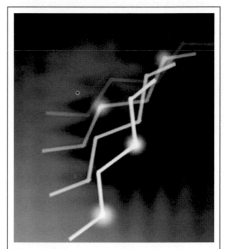

FIGURE 8.9 It is now possible to create nanotubes with articulated geometries to facilitate the wiring of nanocircuits critical to the continuing miniaturization of electronics and computers.

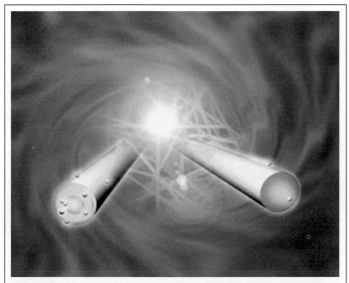

FIGURE 8.10 By fabricating nanowires from radially layered tubes that are far more efficient at gathering photons, the efficiency for converting solar photons to electric power can be increased considerably over planar arrays.

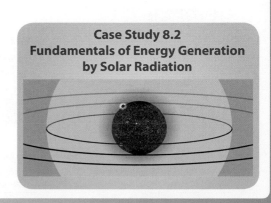

Case Study 8.2
Fundamentals of Energy Generation by Solar Radiation

Chapter Core

Road Map for Chapter 8

Development of the quantum theory of atomic structure begins with a treatment of the distinction between properties associated with waves and the properties associated with particles. At the macroscopic level of systems that we both observe on a day-to-day basis and manipulate with our hands, there is no ambiguity between waves and particles. As the behavior of electromagnetic radiation—light—was investigated with increasingly sophisticated experimental systems, it became clear that certain phenomena could only be explained by assigning particle behavior to light; these particles of light were termed "photons." The fact that electromagnetic waves could possess particle behavior proved difficult for science to accept—but the evidence was irrefutable.

Along initially separate lines of investigation, it became increasingly clear that electrons, always assumed to be particles, exhibited wave properties when confined to spatial domains on the atomic scale. In order to reconcile the observed behavior of light emitted by atoms at discrete wavelengths, the structure of the atom was proposed—the Bohr atom—with electrons occupying discrete orbits about the nucleus. With the hypothesis put forward by the French physicist Louis de Broglie that the electron was in fact a "matter wave" with wavelength $\lambda = h/p$ (where h is Planck's constant and p is the momentum of the electron) came the advent of quantum mechanics and the birth of modern physics. Quantum mechanics recognized that at the atomic scale, electrons were controlled by their wave properties, and those wave properties were in turn defined mathematically by the wavefunction, ψ, of the electron. The wavefunction in turn defined the distribution of electron density about the nucleus. The electrons thus occupy discrete geometric domains called "orbitals" that can be uniquely defined by "quantum numbers" that define the energy, shape, and orientation of the orbitals.

In the core of this chapter we thus explore the concepts displayed here.

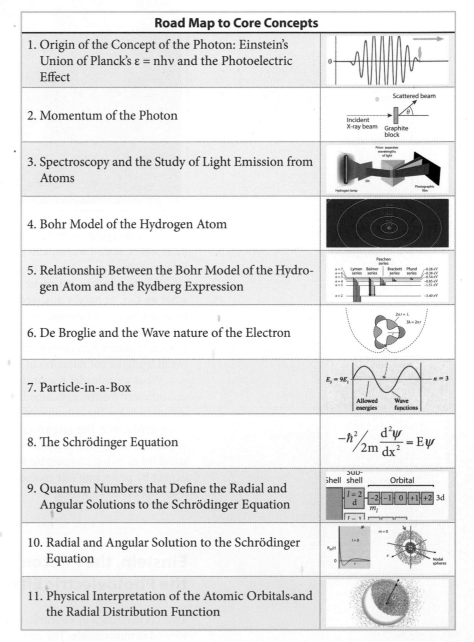

Road Map to Core Concepts
1. Origin of the Concept of the Photon: Einstein's Union of Planck's $\varepsilon = nh\nu$ and the Photoelectric Effect
2. Momentum of the Photon
3. Spectroscopy and the Study of Light Emission from Atoms
4. Bohr Model of the Hydrogen Atom
5. Relationship Between the Bohr Model of the Hydrogen Atom and the Rydberg Expression
6. De Broglie and the Wave nature of the Electron
7. Particle-in-a-Box
8. The Schrödinger Equation
9. Quantum Numbers that Define the Radial and Angular Solutions to the Schrödinger Equation
10. Radial and Angular Solution to the Schrödinger Equation
11. Physical Interpretation of the Atomic Orbitals and the Radial Distribution Function

Waves and Particles: From Separation to Union

Separation of the Concepts of Waves and Particles

As observing creatures we are guided by what we see and what we experience. We first experiment by throwing stones into a lake, observing that the impact of the stone leaves a circular pattern of waves emanating from the point where the stone struck the water as shown in Figure 8.11. Obviously, while the impact of the stone resulted in the genesis of the wave pattern, they—the particle and the wave—are clearly independent and fully distinguishable at the spatial scale of our direct observation, the macroscopic world.

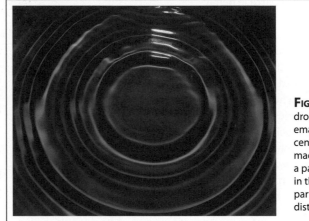

FIGURE 8.11   When a stone is dropped into water, waves emanate radially from the center outward. At the macroscopic level, the stone is a particle and the disturbance in the water is a wave. The particle and the wave are distinct.

We also come to recognize that waves occur in many forms: ocean waves, light waves, sound waves; and that waves transmit energy, momentum, heat, and information from a source to a destination.

What we believe to be true, even what we imagine, is based almost entirely on what we can, or have, visualized. Electromagnetic radiation, light, is real and sensible, yet lacking in substance it appears to carry with it no momentum, no physical impulse. In stark contrast, matter can be physically grasped, it can be given a trajectory, and when matter collides with matter, considerable consequences can result from the combination of mass and velocity.

What we consider reasonable, what we term common sense, is built upon the accumulated evidence obtained in the macroscopic world around us. Our intuitive ability to judge the outcome of future events is constructed from observations of the macroscopic. But just as our intuition leaves us unprepared to define the consequences of a planet like ours with a population of 10 billion people, each with a standard of living comparable to developed countries today, so too does what we expect by projecting our macroscopic intuition on the microscopic—the atomic level—prove to be wrong.

Einstein, the Photon, and the Union of Planck and the Photoelectric Effect

Two cornerstones of scientific thinking were, at the opening of the 20th century, viewed as unassailable. The first, having stood the test of 250 years of experimental verification, were Newton's laws of motion and the mechanics associated therewith. The second were the laws governing electromagnetic radiation developed by Maxwell and published in 1862. Newtonian mechanics explained the behavior of particles. Maxwell's equations explained the wave properties of electromagnetic radiation. Particles were particles and waves were waves. The distinction was clear-cut and unequivocal. Particles were categorized as entities that possessed well-defined masses, trajectories, velocities, energies, and momenta. Electromagnetic waves, in sharp contrast, possessed no apparent mass, could not be specifically located, and were best characterized in terms of their *characteristic* frequency and wavelength. Waves also possessed the characteristics that they could interact, such as to produce constructive (additive) and destructive (subtractive) interference because the amplitude of two or more waves summed algebraically at any point in space.

As discussed in Chapter 1, the work of Max Planck in the study of radiation from blackbody cavities and the resulting "ultraviolet catastrophe" marked the beginning of the end for the clear distinction between the wave nature and the particle nature of natural phenomena.

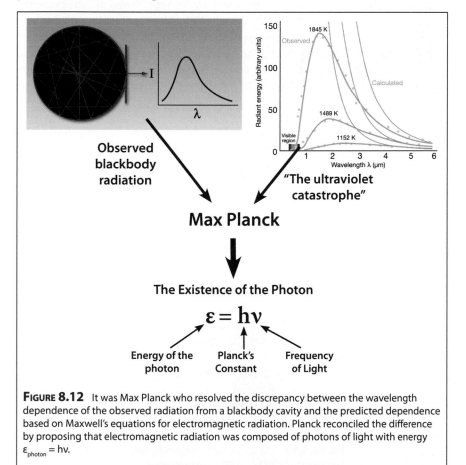

FIGURE 8.12 It was Max Planck who resolved the discrepancy between the wavelength dependence of the observed radiation from a blackbody cavity and the predicted dependence based on Maxwell's equations for electromagnetic radiation. Planck reconciled the difference by proposing that electromagnetic radiation was composed of photons of light with energy $\varepsilon_{photon} = h\nu$.

Recall from Chapter 1 that the only way Planck could reconcile the observed wavelength dependence of the intensity of radiation from a blackbody cavity was if light was comprised of individual energy bundles with energy $\varepsilon = nh\nu$ where ε was the energy of the individual photons, n was an integer, ν was the frequency of the radiation from the blackbody, and h was a constant.

Solution

We can first convert the wavelength of the light from nanometers to meters and then apply the equation linking wavelength, frequency, and the speed of light: $\lambda v = c$.

$$\lambda = 589 \text{ nm} \times \frac{1 \times 10^{-9} \text{ m}}{1 \text{ nm}} = 5.89 \times 10^{-7} \text{ m}$$

$$c = 2.998 \times 10^{8} \text{ m s}^{-1}$$

$$v = ?$$

Rearrange $\lambda v = c$ to the form $v = c/\lambda$, and solve for v.

$$v = \frac{c}{\lambda} = \frac{2.998 \times 10^{8} \text{ m s}^{-1}}{5.89 \times 10^{-7} \text{ m}} = 5.09 \times 10^{14} \text{ s}^{-1}$$

$$= 5.09 \times 10^{14} \text{ Hz}$$

Practice Example A: The light from red LEDs (light-emitting diodes) is commonly seen in many electronic devices. A typical LED produces 690 nm light. What is the frequency of this light?

Practice Example B: An FM radio station broadcasts on a frequency of 91.5 megahertz (MHz). What is the wavelength of these radio waves in meters?

Check Yourself 1

Relating Frequency and Wavelength of Electromagnetic Radiation. Most of the light from a sodium vapor lamp has a wavelength of 589 nm. What is the frequency of this radiation?

Prior to and during the intensifying study of blackbody radiation leading to Planck's highly controversial conclusion, Heinrich Hertz (for whom the unit of frequency is named) was carrying out a series of experiments using light of different wavelengths incident on metal surfaces with the apparatus shown in Figure 8.13. These observations investigated the emission of electrons from the metal surfaces as Figure 8.13 shows. These emitted electrons were called "photoelectrons" because they were electrons liberated from the metal surface by light. The concept of light as a wave would imply that, with the impinging radiation, energy would be imparted to the metal surface until sufficient energy was "stored up" to release the photoelectron from the surface of the metal. With this wave picture of light there would be a delay as the energy was stored in the metal. That delay would decrease as the intensity of light increased, and radiation with longer wavelengths could release a photoelectron simply by increasing the delay between when the radiation first impinged on the surface and when the electron was released.

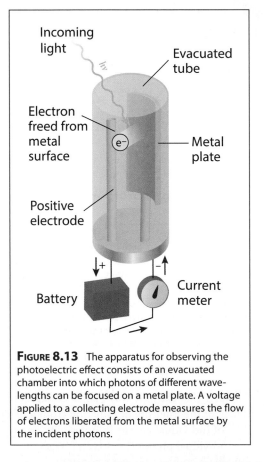

FIGURE 8.13 The apparatus for observing the photoelectric effect consists of an evacuated chamber into which photons of different wavelengths can be focused on a metal plate. A voltage applied to a collecting electrode measures the flow of electrons liberated from the metal surface by the incident photons.

Yet this is not what was observed. Rather than a threshold for photoelectron release determined by the total amount of energy added to the metal surface by the impinging electromagnetic wave, the *threshold* for photoelectron release was determined by the *frequency* of the radiation. Below a threshold frequency, v_0, *no* photoelectrons were emitted, no matter how long the light impinged on the metal surface. Each metal possessed a different threshold frequency, ν_0. In addition, when that threshold frequency, ν_0, was reached, there was no delay before photoelectrons were ejected from the surface. It was as though a single "particle" of light knocked a single photoelectron from the surface of the metal. When the photoelectron's kinetic energy was plotted against the frequency of the impinging light, as shown in Figure 8.14, there was an intercept, ν_0, and a slope that represented a linear increase in the relationship between maximum kinetic energy of the photoelectron and the *frequency* of the incident radiation. Furthermore, if the intensity of the radiation was increased but the frequency was kept constant, a *greater number* of photoelectrons was emitted from the surface, *but the maximum kinetic energy of the photoelectrons was proportional to the frequency of that radiation.*

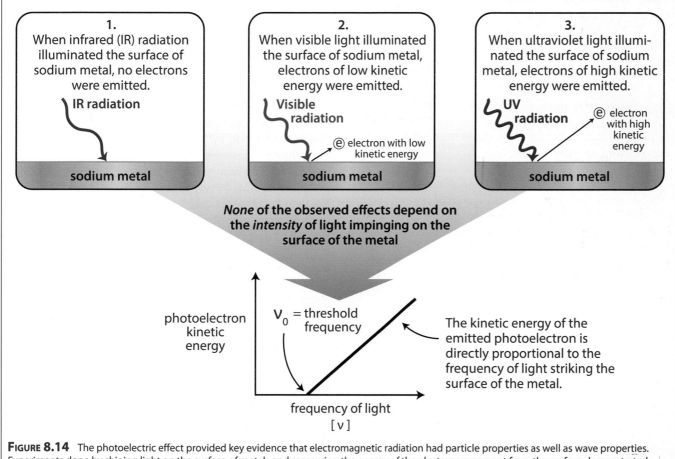

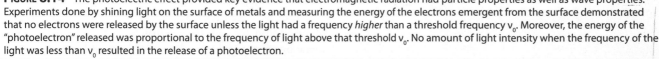

FIGURE 8.14 The photoelectric effect provided key evidence that electromagnetic radiation had particle properties as well as wave properties. Experiments done by shining light on the surface of metals and measuring the energy of the electrons emergent from the surface demonstrated that no electrons were released by the surface unless the light had a frequency *higher* than a threshold frequency ν_0. Moreover, the energy of the "photoelectron" released was proportional to the frequency of light above that threshold ν_0. No amount of light intensity when the frequency of the light was less than ν_0 resulted in the release of a photoelectron.

The observation that the emission of the electrons from the surface of the metal was *independent* of the intensity of radiation falling on the surface of the metal was in direct conflict with Maxwell's theory of light, which was built upon the wave nature of electromagnetic radiation. That wave-based theory of radiation stipulated that the energy delivered by electromagnetic radiation was *proportioned to the intensity of the light.* Thus, given this wave formulation of light, how could both the *threshold* for photoelectron emission *and the kinetic energy of the emitted electrons* be independent of the intensity of light impinging on the metal surface?

The suggestion for a solution to this dilemma emerged from Planck's hypothesis that the radiant energy of the light striking the surface of the metal arrives in

Solution

(a) To use Planck's equation, we need the frequency of the radiation. This we can get from the equation $\lambda v = c$ after first expressing the wavelength in meters.

$$v = \frac{c}{\lambda} = \frac{2.998 \times 10^8 \, \text{m s}^{-1}}{242.4 \times 10^{-9} \, \text{m}} = 1.237 \times 10^{15} \, \text{s}^{-1}$$

Planck's equation is written for one photon of light. We emphasize this by including the unit, photon^{-1}, in the value of h.

$$E = hv = 6.626 \times 10^{-34} \frac{\text{J s}}{\text{photon}} \times 1.237 \times 10^{15} \, \text{s}^{-1}$$

$$= 8.196 \times 10^{-19} \, \text{J/photon}$$

(b) Once we have the energy per photon, we can multiply it by the Avogadro's number to convert to a per-mole basis.

$$E = 8.196 \times 10^{-19} \, \text{J/photon} \times 6.022 \times 10^{23} \, \text{photons/mol}$$

$$= 4.936 \times 10^5 \, \text{J/mol}$$

(This quantity of energy is sufficient to raise the temperature of 10.0 L of water by 11.8 °C.)

Practice Example A: The protective action of ozone in the atmosphere comes through ozone's absorption of UV radiation in the 230 to 290 nm wavelength range. What is the energy, in kilojoules per mole, associated with radiation in this wavelength range?

Practice Example B: Chlorophyll absorbs light at energies of 3.056×10^{-19} J/photon and 4.414×10^{-19} J/photon. To what color and frequency do these absorptions correspond?

discrete packets such that the *energy* of each packet of light delivered was given by the Planck formula

$$E = hv$$

Check Yourself 2

Using Planck's Equation to Calculate the Energy of Photons of Light. For radiation of wavelength 242.4 nm, the longest wavelength that will bring about the photodissociation of O_2, what is the energy of **(a)** one photon, and **(b)** a mole of photons of this light?

We can sketch this confluence of ideas by asking another question: *What did the combination of the photoelectric effect and blackbody radiation tell us?* This linkage is diagrammed in Figure 8.15. It was Einstein in 1905 who put these lines of evidence together and postulated that these quanta of energy, E = hv, implied that the energy of light was delivered to the metal surface as *particles* of light energy, which he termed *photons*. This interpretation by Einstein created a direct link between the collision of a *single* photon with energy E = hv, and a single electron within the metal, releasing that electron at any energy *above* the threshold energy $E_0 = hv_0$. Any photon with energy less than $E_0 = hv_0$ was unable to release the electron. Any photon with energy greater than $E_0 = hv_0$ could release an electron from the surface of the metal and if an electron was released, it would leave the surface with kinetic energy equal to the *difference* between the original energy of the incident photon, E = hv, and the threshold energy $E_0 = hv_0$ required to free the electron. Thus, with KE = kinetic energy of the liberated electron, we can write

$$KE = hv - hv_0 = h(v - v_0) \qquad \text{for } v > v_0$$

and

$$KE = 0 \qquad \text{for } v \le v_0$$

This quantity $E_0 = hv_0$, the minimum energy required to release an electron from the metal surface, is *different* for each metal and is termed the *work function* of the metal.

Thus we have the energy level diagram for the interaction of the photon, hv, with the electron bound to the surface of the metal by an energy hv_0 as shown in Figure 8.16.

In summary, the electron is bound in a potential energy well of depth hv_0. If the photon has energy $hv > hv_0$, the electron can escape the potential energy well and leave the surface of the metal with kinetic energy, KE, equal to the *difference* in energy between the incident photon, hv, and the binding energy, hv_0.

To Einstein the evidence was overwhelming. Specifically that light must exhibit particle properties—that light must be comprised of what he termed "photons." Yet the very idea that light could exhibit both wave and particle properties was, for the majority of the scientific community, unimaginable.

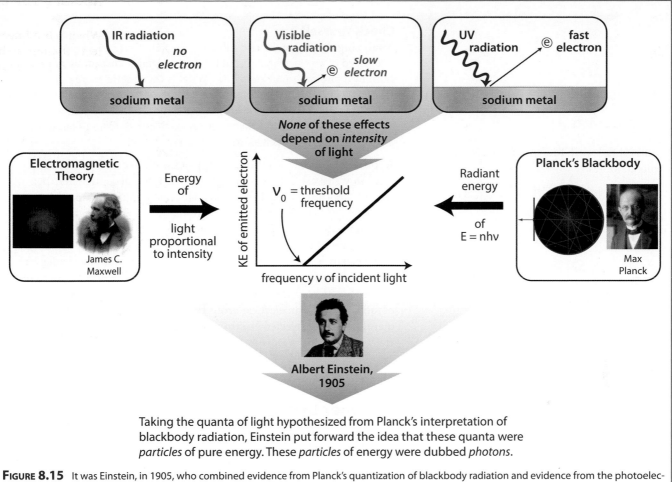

Taking the quanta of light hypothesized from Planck's interpretation of blackbody radiation, Einstein put forward the idea that these quanta were *particles* of pure energy. These *particles* of energy were dubbed *photons*.

FIGURE 8.15 It was Einstein, in 1905, who combined evidence from Planck's quantization of blackbody radiation and evidence from the photoelectric effect, to hypothesize that light consisted of particles, or *photons*.

This study of the interaction of light with the surface of metals became a significant field of research. The linkage between photons of frequency ν and wavelength $\lambda = c/\nu$ with electron ejection from metal surfaces introduced yet another unit of energy—the electron volt or eV. One electron volt is the energy acquired by an electron when it falls through a potential of 1 volt. As we will see, it is very convenient unit of energy for many applications.

The charge on the electron, as we discussed in Chapter 2, is 1.602×10^{-19} coulomb or 1.602×10^{-19} C. We also know that one volt is equal to a joule/coulomb or $1\ \text{V} = 1\ \text{J/C}$. Thus if an electron volt is the energy of one electron falling through a potential of 1 volt

$$1\ \text{eV} = (1.602 \times 10^{-19}\ \text{C})1\ \text{V} = 1.602 \times 10^{-19}\ \text{J}$$

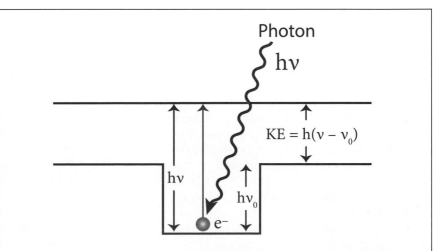

FIGURE 8.16 The electron is held within a potential energy well of depth $h\nu_0$ by the metal surface. When a photon of energy $h\nu > h\nu_0$ is absorbed, the electron is released from the metal surface with kinetic energy $KE = h\nu - h\nu_0 = h(\nu - \nu_0)$.

Solve:

(a) From $E_k = \frac{1}{2}mv^2$,

$$E_k = \frac{1}{2} \times (9.109 \times 10^{-31}\,\text{kg}) \times (6.68 \times 10^5\,\text{m} \cdot \text{s}^{-1})^2$$

$$= 2.03 \times 10^{-19}\,\text{J}$$

(b) Convert the work function from electron-volts to joules.

$$2.29\,\text{eV} \times \frac{1.602 \times 10^{-19}\,\text{J}}{1\,\text{eV}} = 3.67 \times 10^{-19}\,\text{J}$$

From

$$\frac{1}{2}m_e v^2 = h\nu - h\nu_0, \quad h\nu = h\nu_0 + \frac{1}{2}m_e v^2 = h\nu_0 + E_k,$$

$$h\nu = 3.67 \times 10^{-19}\,\text{J} + 2.03 \times 10^{-19}\,\text{J} = 5.70 \times 10^{-19}\,\text{J}$$

so

$$\nu = \frac{5.70 \times 10^{-19}\,\text{J}}{h} = \frac{5.70 \times 10^{-19}\,\text{J}}{6.63 \times 10^{-34}\,\text{J} \cdot \text{s}} = 8.60 \times 10^{14}/\text{s}$$

Now use $\lambda = c/\nu$:

$$\lambda = \frac{(3.00 \times 10^8\,\text{m} \cdot \text{s}^{-1}) \times (6.626 \times 10^{-34}\,\text{J} \cdot \text{s})}{5.70 \times 10^{-19}\,\text{J}}$$

$$= 3.49 \times 10^{-7}\,\text{m} \quad \text{or} \quad 349\,\text{nm}$$

(c) To find the longer wavelength of radiation able to eject an electron, set $E_k = 0$ in Eq. 5, so $h\nu = h\nu_0$, and therefore $\lambda = ch/h\nu_0$.

$$\lambda = \frac{(3.00 \times 10^8\,\text{m} \cdot \text{s}^{-1}) \times (6.626 \times 10^{-34}\,\text{J} \cdot \text{s})}{3.67 \times 10^{-19}\,\text{J}}$$

$$= 5.42 \times 10^{-7}\,\text{m} \quad \text{or} \quad 542\,\text{nm}$$

Practice Example 1: The work function of zinc is 3.63 eV. What is the longest wavelength of electromagnetic radiation that could eject electrons from zinc?

[*Answer:* 342 nm]

Practice Example 2: The speed of an electron that is emitted from the surface of a sample of zinc by a photon is 785 km·s⁻¹. (a) What is the kinetic energy of the ejected electron? (b) The work function of zinc is 3.63 eV. What is the wavelength of the radiation that caused photoejection of the electron?

Check Yourself 3

Analyzing the Photoelectric Effect.

The speed of an electron emitted from the surface of a sample of potassium by a photon is 668 km·s⁻¹. (a) What is the kinetic energy of the ejected electron? (b) What is the wavelength of the radiation that caused photoejection of the electron? (c) What is the longest wavelength of the electromagnetic radiation that could eject electrons from potassium? The work function of potassium is 2.29 eV.

Approach (a) Find the kinetic energy of the ejected electron from $E_k = 1/2\,m_e v^2$. To use SI base units, first convert the speed to meters per second. (b) The energy of the ejected electron is equal to the difference in energy between the incident radiation and the work function. The photon needs to provide enough energy to eject the electron from the surface (the work function) as well as to impart a velocity of 668 km·s⁻¹ to the electron. Convert that value of the work function into joules and use $h\nu = KE + h\nu_0$ to determine the value of $h\nu$ for the photon. Then use $\lambda\nu = c$ to convert that energy to wavelength. (c) The longest wavelength of radiation that can eject electrons from a substance is the wavelength that results in the ejected electron having zero kinetic energy.

Momentum of the Photon

The interaction of photons with electrons demonstrates that the photon possesses energy and it can transfer that energy to an electron. But if the photon has energy and it can be treated as a particle, does it possess momentum? Does it possess mass?

To answer these questions, we can obtain the energy of the photon from Einstein's equation $E = mc^2$ where the mass refers to the inertial mass of the photon that is a result of the energy possessed by the photon. We know that the energy of the photon is $E = h\nu$, so equating those two expressions for the energy of the photon yields

$$E = mc^2 = h\nu = (h\,c/\lambda)$$

$$\text{since } \lambda\nu = c$$

The *momentum* of the photon is mc so

$$p_{phot} = mc = h/\lambda_{phot}$$

So while the *rest mass* of the photon is zero, this calculation of the momentum of the photon suggests that the momentum of the photon is not only nonzero, but should be a measurable quantity.

The momentum of the photon was measured in 1923 by Compton, using a beam of x-rays (photons) incident on a sample of graphite as shown in Figure 8.17a. When photons were observed after they passed through the graphite block, there were some photons with an identical wavelength as those incidents on the graphite block. Those photons had passed through the graphite block without colliding with any material. However, there was another group of photons emergent from the back side of the graphite block that were characterized by a longer wavelength. Those photons, Compton hypothesized, had a longer wavelength because they had lost energy by collision with electrons in the graphite. Compton designated those photons (those with lower energy) as scattered photons with wavelength λ_s as shown in Figure 8.17b.

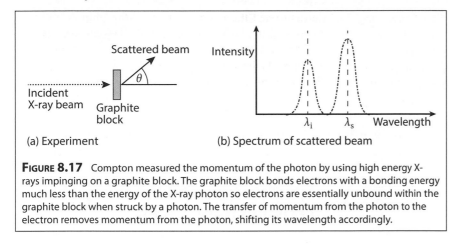

(a) Experiment

(b) Spectrum of scattered beam

FIGURE 8.17 Compton measured the momentum of the photon by using high energy X-rays impinging on a graphite block. The graphite block bonds electrons with a bonding energy much less than the energy of the X-ray photon so electrons are essentially unbound within the graphite block when struck by a photon. The transfer of momentum from the photon to the electron removes momentum from the photon, shifting its wavelength accordingly.

The key is that the energy binding outer electrons in graphite is *very small* compared to the energy of the x-ray photon. Therefore the electron behaves as a free electron suspended in the block of graphite.

If we assume that we have a simple system of a photon (with energy hν and momentum $p_{phot} = h/\lambda_{phot}$) and an electron (with mass m_e), then conservation of energy and momentum in this collision yields a wavelength shift of

$$\lambda_s - \lambda_i = \Delta\lambda = h/m_e c \,(1 - \cos\theta)$$

where the geometry of the collision is displayed in Figure 8.17a. In our *expression for Δλ*, the wavelength shift between the incident and scattered photon is $\Delta\lambda$, m_e is the mass of the electron, θ is the angle measured from the incident direction, h is Planck's constant, and c is the speed of light. But the key point is this: the *observed* wavelength shift $\Delta\lambda = \lambda_s - \lambda_i$ was in excellent agreement with the *calculated* wavelength shift when the momentum of the x-ray photon was taken to be h/λ. This was compelling evidence that the photon indeed carries a momentum of p = h/λ.

The fact that a photon of zero rest mass has a defined momentum that is directly observable held great importance for the development of the idea of wave-particle duality of photons.

It was Einstein, in 1905, who explained these results by proposing that *light* behaves as a *particle*—a particle he termed a *photon* of light with energy E = hν, where ν is the frequency of radiation and h is a constant—Planck's constant, in fact, the same quantity that Max Planck invoked to explain blackbody radiation.

$$p = h/\lambda$$

An important consequence of recognizing that the photon has momentum p = h/λ is that, just as the energy of a photon is quantized, so too is its momentum. Momentum is imparted by photons to their target in discrete parcels of h / λ in the same way energy is delivered in units of hν. What ties these ideas together is Planck's constant h, which was developing into a quantity of great fundamental importance. Planck's constant thus represented an indivisible amount of energy associated with a quantum of light but linked as well to the structure of matter.

The immutable barrier between particle behavior and wave behavior, built from observations of the macroscopic world, was beginning to crumble as science pressed toward the microscopic. To possess mass, momentum, and physical impact is to be a particle. To have a wavelength, to interfere, and to diffract is to be a wave. To be a *photon*, however, is to posses both the attributes of a particle and of a wave. The photon can collide like a projectile yet interfere like a wave.

Answer: The momentum of the *photon* is given by the equation:

$$p = \frac{h}{\lambda} = \frac{6.626 \times 10^{-34} \text{ J s}}{0.17 \times 10^{-9} \text{ m}} = 3.9 \times 10^{-24} \text{ kg m s}^{-1}$$

In this problem the velocity of the electron is much less than the velocity of light and relativistic corrections to the mass need not be made. Under these circumstances the momentum of the electron, *p*, can be obtained from the classical equation relating energy and momentum:

$$E = \frac{1}{2}mv^2 = \frac{p^2}{2m}$$

where v is the velocity and *m* is the mass of the electron. Thus:

$$p = \sqrt{2mE}$$

The energy is equal to the electronic charge multiplied by the voltage. Thus:

$$E = (1.602 \times 10^{-19} \text{ C}) \times (100 \text{ V}) = 1.602 \times 10^{-17} \text{ J}$$

And thus the momentum of the *electron* is

$$p = \sqrt{2mE} = \sqrt{2 \times \left(9.109 \times 10^{-31} \text{ kg}\right) \times \left(1.602 \times 10^{-17} \text{ J}\right)}$$
$$= 5.40 \times 10^{-24} \text{ kg m s}^{-1}$$

It can be seen that the momenta of the *electron* and the X-ray *photon* have similar magnitudes.

Check Yourself 4

Calculate the momentum of an X-ray photon with a wavelength of 0.17 nm. How does this compare with the momentum of an electron that has been accelerated through a potential difference of 100 volts?

Spectroscopy and the Study of Light Emission from Atoms

While human sentiment gave grudgingly to the idea that light could somehow exhibit the characteristics of a particle, no one believed in the early part of the 20th century that particles could exhibit wave character. But information was pouring in from yet another domain, that of spectroscopy—the study of light emitted by atoms and molecules and separated into specific wavelengths by devices such as a prism as shown in Figure 8.18.

This spectroscopic evidence was extracted from a variety of systems—some using high temperature flames to vaporize metals and some using discharge tubes that applied a high voltage between two electrodes within a sealed tube containing a specific gas such as hydrogen, helium, or neon. This discharge tube was similar to the one J. J. Thomson used in the discovery of the electron, and it was also used increasingly by placing small amounts of metal, such as mercury or strontium, within the sealed tube. In all cases, spectacular *individual emission lines* were revealed within the spectra obtained by the prism apparatus used to separate or "disperse" the light emitted from the discharge tube. For each element contained within the discharge tube, a unique set of lines were observed by the "spectrometer" as shown in Figure 8.18. It was the series of *discrete*, separated emission lines of limited number from a given element that began to suggest that the *pattern* of the emission lines could be quantitatively deciphered and analytically or mathematically represented.

The spectrum of atomic hydrogen was the first to yield a simple relationship between the wavelength of emitted light from the atom and a mathematical expression linking all the observed lines in the spectrum to a single simple equation. These lines in the spectrum of atomic hydrogen extend from the infrared to the visible region to the "vacuum ultraviolet" so named because, as we will see, molecular oxygen absorbs radiation between 100 and 200 nm. Thus O_2 must be pumped out of any system observing radiation in this wavelength interval.

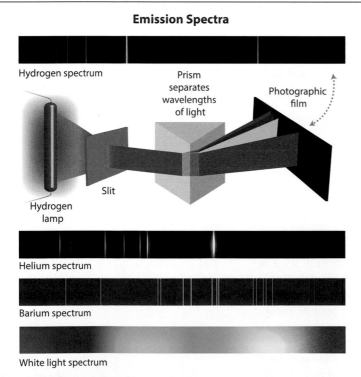

Emission Spectra

Hydrogen spectrum

Prism separates wavelengths of light

Photographic film

Slit

Hydrogen lamp

Helium spectrum

Barium spectrum

White light spectrum

FIGURE 8.18 The remarkable, sharply defined emission lines that emanate from electrical discharge tubes that are separated in wavelength by a prism proved to be critically important to developing an understanding of atomic structure.

For that reason a vacuum is used to study radiation in the region between 100 and 200 nm, thereby creating the term *vacuum ultraviolet*.

The full spectrum of atomic hydrogen is displayed in Figure 8.19 showing the "ultraviolet series" in the vicinity of 100nm, the "visible series" from 400 to 750 nm and the "infrared series" from 800 to 2000 nm.

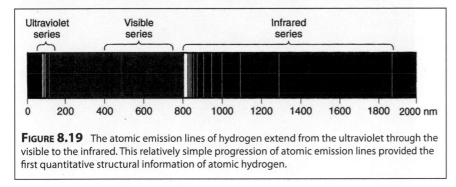

FIGURE 8.19 The atomic emission lines of hydrogen extend from the ultraviolet through the visible to the infrared. This relatively simple progression of atomic emission lines provided the first quantitative structural information of atomic hydrogen.

What was discovered was that within a given series, the wavelength of the emitted light could be calculated from the disarmingly simple expression

$$\frac{1}{\lambda} = R_H \left(\frac{1}{n_f^2} - \frac{1}{n_i^2} \right)$$

where λ was the specific wavelength of emitted radiation, R_H was a constant, called the *Rydberg* constant, n_f was the integer common to each *series* of spectral lines (i.e. infrared series, visible series, etc.), and n_i was a set of integers corresponding to each of the several lines in the series. The Rydberg constant for hydrogen always had the same value, independent of the series; its value was determined to be $R = 1.097 \times 10^7$ m^{-1}. The "visible series" had a value of $n_f = 2$ such that all lines in the series could be calculated with $n_i = 3, 4, 5, 6$, etc. The "ultraviolet series" corresponded to a value of $n_f = 1$ such that all lines in the series could be calculated with $n_i = 2, 3, 4, 5$, etc. The Rydberg equation $1/\lambda = R_H(1/n_f^2 - 1/n_i^2)$ brought both a *quantitative foundation* to studies of molecular structure and the clear recognition that science could not explain *why* the equation worked! There was no model for atomic structure and no coherent theory to link the observed spectra of atoms to a picture of the atom's architecture.

Bohr Model of the Hydrogen Atom

A young Danish physicist, Niels Bohr, went to England immediately after receiving his PhD to work with Rutherford who had just proposed his nuclear model of the atom. Bohr was interested in building a new model of the atom using three lines of evidence: (1) the Rutherford picture of negatively charged electrons attracted to a far more massive positively charged nucleus, (2) the spectroscopic information defining the sharp atomic emission lines from the hydrogen atom, and (3) the concept of the photon of light with energy $E = h\nu$ where ν is the frequency of the radiation.

From those distinct lines of evidence Bohr put forward a radically new model of the atom with the following assertions:

1. The hydrogen atom consists of a negatively charged electron that can occupy one of several *circular orbits* about the central, positively charged nucleus, as shown in Figure 8.20.

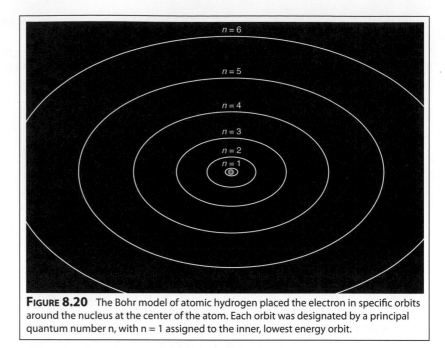

FIGURE 8.20 The Bohr model of atomic hydrogen placed the electron in specific orbits around the nucleus at the center of the atom. Each orbit was designated by a principal quantum number n, with n = 1 assigned to the inner, lowest energy orbit.

2.   Those atoms have only certain allowable energy levels called "stationary states" with each of these stationary states corresponding to a fixed circular orbit of the electron around the nucleus. These distinct states, these distinct orbits, are designated by a *quantum number* n, which can take integer values n = 1, 2, 3, …

3.   When the atom is in one of these stationary states, it does not emit electromagnetic radiation unless it jumps in a discrete step to a stationary state of lower energy, or it absorbs a photon and jumps to a stationary state of higher energy.

4.   Each stationary state has a discrete, well-defined energy, E_n. These atomic energy levels are quantized with the lowest energy level, E_1, being the energy of *ground state,* and all other stationary states lying at a higher energy such that $E_1 < E_2 < E_3 < E_4 …$

5.   When the atom transitions from one stationary state to another, the electron jumps to a new orbit only by *absorbing* or *emitting* a photon of energy equal to the *difference* between the energy of two stationary states. The energy difference between the initial and final state is

$$\Delta E = E_f - E_i$$

However, because ΔE is negative for emission, in order to calculate the frequency of the emitted photon, we take the absolute value of ΔE so that

$$|E_f - E_i| = |\Delta E| = h\nu$$

where ν is the frequency of the absorbed or emitted photon as displayed in Figure 8.21.

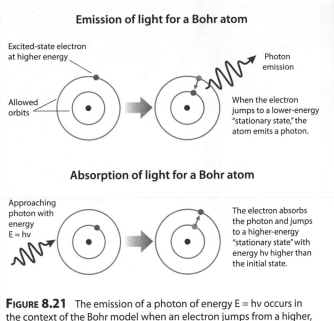

Emission of light for a Bohr atom

Excited-state electron at higher energy

Photon emission

Allowed orbits

When the electron jumps to a lower-energy "stationary state," the atom emits a photon.

Absorption of light for a Bohr atom

Approaching photon with energy E = hv

The electron absorbs the photon and jumps to a higher-energy "stationary state" with energy hv higher than the initial state.

Figure 8.21 The emission of a photon of energy E = hv occurs in the context of the Bohr model when an electron jumps from a higher, excited state orbit to a lower orbit. Absorption of a photon occurs when a photon promotes the electron from a lower energy orbital to a higher energy orbital.

6. Atoms will seek the lowest energy state, emitting a photon in each *transition* from a higher energy stationary state to a lower energy stationary state until the lowest, or ground state, energy level is reached.

While Bohr's model was based fundamentally on Rutherford's nuclear model of the atom, the Bohr model added two new and important ideas. The first is that only certain electron orbits (the "stationary states") can exist. The second linked Einstein's photon with the energy of the jump of the electron from one orbit to another. Moreover, these emitted (or absorbed) photons of energy hv corresponded to the frequency of light emitted from the atoms that appear in the Rydberg relationship between the wavelength of the light and the integers designating the series in the hydrogen atom's emission. This critically important link between the electron orbits, the spectroscopic data, the Rydberg equation, and the concept of the photon set a new course for scientific understanding. This union is displayed in Figure 8.22.

Solution
The specific data for the Rydberg equation are $n_i = 5$ and $n_f = 2$.

$$\frac{1}{\lambda} = R_H \left(\frac{1}{n_f^2} - \frac{1}{n_i^2} \right)$$

$$= 1.097 \times 10^7 \, \text{m}^{-1} \left(\frac{1}{2^2} - \frac{1}{5^2} \right)$$

$$= 2.30 \times 10^6 \, \text{m}^{-1}$$

$$\lambda = 4.34 \times 10^{-7} \, \text{m}$$

Practice Example A: Determine the wavelength of light absorbed in an electron transition from $n = 2$ to $n = 4$ in a hydrogen atom.

Practice Example B: Consider Figure 8.22 and determine which transition produces the longest wavelength line in the Lyman series of the hydrogen spectrum. What is the wavelength of this line in nanometers and in angstroms?

Check Yourself 5
Calculating the Wavelength of a Line in the Hydrogen Spectrum.
Determine the wavelength of the line in the Balmer series of hydrogen corresponding to the transition from $n = 5$ to $n = 2$.

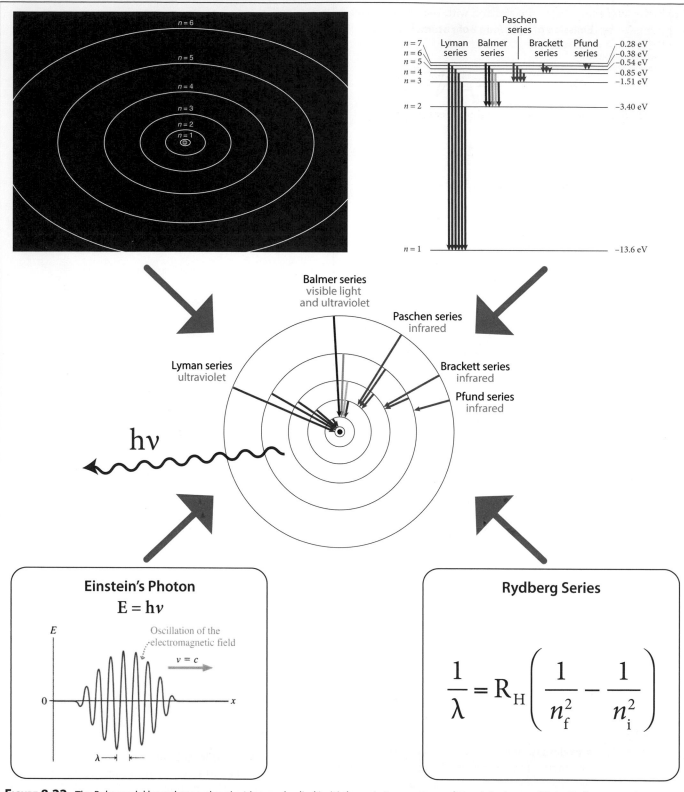

FIGURE 8.22 The Bohr model brought together the ideas embodied in (1) the emission spectrum of atomic hydrogen, (2) the Rydberg equation quantitatively defining the wavelength of the photons emitted from the hydrogen atom, and (3) the circular orbits that represented the stationary states of the Bohr hydrogen atom.

What Bohr's model of the hydrogen atom did was provide a basis for applying the laws of physics to the motion of the electron in these circular orbits. Specifically the *total* energy, E, of the electron with velocity v moving in the potential energy created by the Coulomb attraction of the nucleus is: E = Kinetic Energy + Potential Energy = KE + PE

$$E = KE + PE = \frac{1}{2}mv^2 - \frac{e^2}{4\pi\varepsilon_0 r}$$

In addition, the circular trajectory of radius r required that the Coulomb force of attraction,

$$F = \frac{e^2}{4\pi\varepsilon_0 r^2}$$

balance the centripetal force, mv^2/r, such that

$$e^2/4\pi\varepsilon_0 r^2 = mv^2/r$$

as displayed in Figure 8.23.

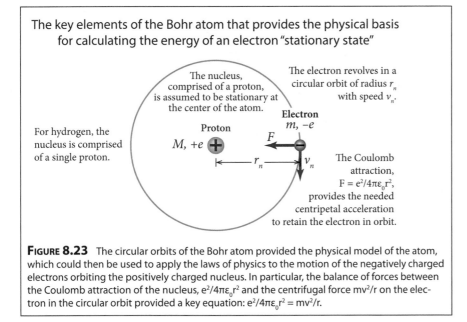

The key elements of the Bohr atom that provides the physical basis for calculating the energy of an electron "stationary state"

For hydrogen, the nucleus is comprised of a single proton.

The nucleus, comprised of a proton, is assumed to be stationary at the center of the atom.

The electron revolves in a circular orbit of radius r_n with speed v_n.

Electron
$m, -e$

Proton
$M, +e$

r_n

v_n

The Coulomb attraction, $F = e^2/4\pi\varepsilon_0 r^2$, provides the needed centripetal acceleration to retain the electron in orbit.

FIGURE 8.23   The circular orbits of the Bohr atom provided the physical model of the atom, which could then be used to apply the laws of physics to the motion of the negatively charged electrons orbiting the positively charged nucleus. In particular, the balance of forces between the Coulomb attraction of the nucleus, $e^2/4\pi\varepsilon_0 r^2$ and the centrifugal force mv^2/r on the electron in the circular orbit provided a key equation: $e^2/4\pi\varepsilon_0 r^2 = mv^2/r$.

Bohr knew that if his model of the atom was correct, it must quantitatively predict the wavelength of emitted radiation with the same functional form as that of the Rydberg equation

$$\frac{1}{\lambda} = R_H\left(\frac{1}{n_f^2} - \frac{1}{n_i^2}\right)$$

but moreover, it must provide a way of *calculating* the Rydberg constant, R_H, from first principles.

Bohr hypothesized that the angular momentum of the electron, L, which is given classically by the expression

$$L = mvr$$

must be quantized in analogy with the photon such that

$$L_n = mv_n r_n = n\,h/2\pi = n\hbar$$

where the allowed velocities are v_n and the allowed orbital radii are r_n. The quantity $\hbar = h/2\pi$ appears so frequently that we introduce it here for future use.

The balance of Coulomb and centrifuged forces (Figure 8.23) provided the expression for the quantized velocities for a given quantum number, n:

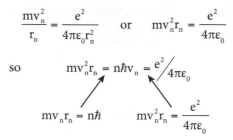

$$so \qquad mv_n^2r_n = n\hbar v_n = {e^2}\!\big/{4\pi\varepsilon_0}$$

$$mv_nr_n = n\hbar \qquad mv_n^2r_n = \frac{e^2}{4\pi\varepsilon_0}$$

Thus, solving for v_n yields

$$v_n = \frac{1}{n}\frac{e^2}{4\pi\varepsilon_0 \hbar}$$

But, from above

$$r_n = \frac{e^2}{4\pi\varepsilon_0 mv_n^2}$$

so, using expression for v_n,

$$r_n = \frac{4\pi\varepsilon_0 \hbar^2}{me^2}n^2$$

The allowed values of the radius, r_n, and of the velocity, v_n, provided the key expressions needed to check the predictions of the Bohr model against the *observed* Rydberg expression. When the values of v_n and r_n were substituted into the expression of the *total energy* of the electron in orbit, we have

$$KE = \tfrac{1}{2}mv_n^2 = \tfrac{m}{2}\frac{1}{n^2}\frac{e^4}{\left(4\pi\varepsilon_0\right)^2 \hbar^2}$$

and

$$PE = \frac{-e^2}{4\pi\varepsilon_0 r_n} = \left(\frac{-e^2}{\left(4\pi\varepsilon_0\right)}\right)\frac{me^2}{4\pi\varepsilon_0 \hbar^2}\frac{1}{n^2}$$

so the total energy

$$E_n = KE + PE = -\frac{me^4}{2\left(4\pi\varepsilon_0\right)^2 \hbar^2}\frac{1}{n^2} = -\frac{1}{\varepsilon_0^2}\frac{me^4}{8n^2h^2}$$

Thus the energy *difference* between two electron orbits with initial quantum number n_i and final quantum number n_f would be

$$\Delta E = E_f - E_i = \frac{-me^4}{8\varepsilon_0^2 h^2}\left[\frac{1}{n_f^2} - \frac{1}{n_i^2}\right]$$

When emitting a photon

$$\Delta E < 0 \quad so$$

$$|\Delta E| = -(E_f - E_i)$$

$$= \frac{me^4}{8\varepsilon_0^2 h^2}\left[\frac{1}{n_f^2} - \frac{1}{n_i^2}\right]$$

The question is, how does this compare *quantitatively* with the *observed* Rydberg relationship from the atomic emission of hydrogen:

$$\frac{1}{\lambda} = R_H\left[\frac{1}{n_f^2} - \frac{1}{n_i^2}\right]$$

Observations of atomic hydrogen emission demonstrated that

$$R_H = 1.097 \times 10^7 \text{ m}^{-1}$$

Since $\Delta E = h\nu$ and $\lambda\nu = c$, the emitted energy of a photon from a quantum jump of the Bohr atom would be

$$\left(\Delta E\right)_{BOHR} = h\nu = \frac{me^4}{8\varepsilon_0^2 h^2}\left[\frac{1}{n_f^2} - \frac{1}{n_i^2}\right]$$

and for the Rydberg equation

$$\left(\Delta E\right)_{RYDBERG} = h\left(\frac{c}{\lambda}\right) = hc\left(\frac{1}{\lambda}\right) = hc\,R_H\left[\frac{1}{n_f^2} - \frac{1}{n_i^2}\right]$$

Thus the comparison between *theory*, the Bohr factor

$$\frac{me^4}{8\varepsilon_0^2 h^2}$$

and the *observation*, the Rydberg constant R_H times hc, hcR_H became the key test of the validity of the Bohr hypothesis. Inserting the known quantities for the electron mass, the electron charge, the permeability of space, ε_0, and the Planck constant h, yields

$$\frac{me^4}{8\varepsilon_0^2 h^2} = \frac{\left(9.109\times10^{-31}\text{ kg}\right)\left(1.602\times10^{-19}\text{C}\right)^4}{(8)\left(8.854\times10^{-12}\,C^2\middle/N\cdot m^2\right)^2\left(6.626\times10^{-34}\text{J}\cdot\text{sec}\right)}$$

$$= 2.179\times10^{-18}\text{J} = 13.60 \text{ ev}$$

The corresponding Rydberg expression is

$$hcR_H = (6.626 \times 10^{-34}\text{ J}\cdot\text{sec})(2.998 \times 10^8\text{ m/sec})(1.097 \times 10^7\text{ m}^{-1})$$

$$= 2.179 \times 10^{-18}\text{ J} = 13.60 \text{ eV}$$

The equality of the expression for $\left(\Delta E\right)_{BOHR}$ and $\left(\Delta E\right)_{RYDBERG}$ represented a triumph for the Bohr theory. Not only did the Bohr model correctly express the energy difference with respect to the quantum numbers, n, but the independently derived (from first principles) factor $me^4/8\varepsilon_0^2 h^2$ *quantitatively* matched the Rydberg constant times hc.

Check Yourself 6
Using the Bohr Model. Determine the kinetic energy of the electron ionized from a Li^{2+} ion in its ground state, using a photon of frequency 5.000×10^{16} s^{-1}.

The Bohr model represented an important stability point, or niche, in the march toward an understanding of atomic structure and thus of chemical bonding. It was an important model of the atom because it represented a comprehensible geometry for the atom, it provided quantitatively accurate *predictions* for the specific energy levels of the hydrogen atom, and it linked those concepts to the observed emission spectrum of atomic hydrogen.

The conventions associated with the assignment of specific energies to the Bohr orbits of atomic hydrogen are worth careful consideration. Bohr wrote down the expression for calculating the energy level of the hydrogen atom as

Solution
The energy of the electron in the Li^{2+} ion is calculated using the Rydberg equation.

$$E_1 = \frac{-3^2 \times 2.179\times10^{-18}\text{J}}{1^2} = -1.961\times10^{-17}\text{J}$$

The energy of a photon of frequency 5.000×10^{16} s^{-1} is

$$E = h\nu = 6.626\times10^{-34}\times\frac{\text{J s}}{\text{photon}}\times5.000\times10^{16}\text{ s}^{-1}$$

$$= 3.313 \times 10^{-17}\text{ J photon}^{-1}$$

Thus 1.96×10^{-17} J is required to ionize Li^{2+}, so in order to calculate the kinetic energy of the electron, that amount of energy must be subtracted from the energy of the incident photon:

$$KE_e = 3.31 \times 10^{-17}\text{ J} - 1.96 \times 10^{-17}\text{ J} = 1.35 \times 10^{-17}\text{ J}$$

Practice Example A: Determine the wavelength of light emitted in an electron transition from $n = 5$ to $n = 3$ in a Be^{3+} ion.

Practice Example B: The frequency of the $n = 3$ to $n = 2$ transition for an unknown hydrogen-like ion occurs at a frequency 16 times that of the hydrogen atom. What is the identity of the ion?

$$E = -2.18 \times 10^{-18} \, J\left(\frac{z^2}{n^2}\right)$$

where Z is the charge of the nucleus, n is the quantum number of the orbit, and the constant 2.18×10^{-18} J is calculated directly from the Rydberg equation

$$\frac{1}{\lambda} = R_H\left(\frac{1}{n_f^2} - \frac{1}{n_i^2}\right)$$

with $R_H = 1.097 \times 10^7 m^{-1}$ for the wavelength of emission lines from atomic hydrogen and with $E_{photon} = h\nu = hc/\lambda$.

For hydrogen Z = 1 and, for the energy of the ground state, n = 1, and we have

$$E_{grd} = -2.18 \times 10^{-18} \, J(Z^2/n^2) = -2.18 \times 10^{-18} \, J$$

But notice the *sign convention* of the ground state energy: it is *negative*. This is an important convention in defining the energy of electrons in an atomic structure *or* in a chemical bond. An electron free from the attractive Coulomb potential of the proton, with no kinetic energy, is taken to have a total energy, E, of zero. As the electron is drawn into the *attractive* electrostatic field of the nucleus, the energy *decreases*. It is as though we defined the potential energy of a mass m to be zero at ground level, which we typically do, and then lowered that mass into a well of depth h and assigned the energy of the mass lowered into the well to be at a potential energy of –mgh, as shown in the left panel Figure 8.24. The mass at a "well depth" of h meters is thus at a negative potential energy and only when the mass is raised to the surface is its energy equal to zero. In direct analogy, any electron with energy *less than zero* is captured within the "potential well" by the Coulomb attraction between the nucleus and the electron as shown in the right-hand panel of Figure 8.24.

Notice that for the Bohr model of the hydrogen atom, as the electron jumps from a *stationary state* with a larger quantum number n, to a stationary state with smaller quantum number n, the energy, E, of that electron becomes a *larger negative number* and as a result, the atom becomes *more stable*.

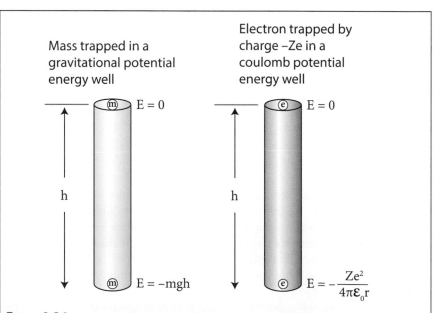

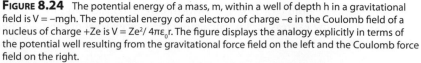

FIGURE 8.24   The potential energy of a mass, m, within a well of depth h in a gravitational field is V = –mgh. The potential energy of an electron of charge –e in the Coulomb field of a nucleus of charge +Ze is V = Ze²/ 4πε₀r. The figure displays the analogy explicitly in terms of the potential well resulting from the gravitational force field on the left and the Coulomb force field on the right.

Problem with the Bohr Model

While the quantitative prediction of the Bohr model for the absolute value of the energy differences between quantum states of the hydrogen atom represented a triumph of science at the time, the very idea of a *stationary state* raised a major problem for scientific thought because the electron, in its orbit about the nucleus, was accelerating (moving in a curved orbit, as shown in Figure 8.23), so, by the laws of electromagnetism, that electron must continually emit electromagnetic radiation. If the electron emits electromagnetic radiation, then by the law of conservation of energy it must spiral into the nucleus as it loses energy, leading to the collapse of the atom. The *ad hoc* and highly arbitrary *postulate* of Bohr that such stationary states exist was unacceptable within the laws of electromagnetic radiation. It was not a satisfying state of affairs.

But here human intuition, built upon our experience with objects approximately our own size, proves inadequate to step into the world of the atomic scale. In order to cross over to the "other side," to the scale of the atom, it was necessary to be presented with a hypothesis that violated the fundamentals of our intuition.

Returning to Figure CS1.4c, we recall that the analysis of blackbody radiation and then the photoelectric effect forced Planck and Einstein to conclude that electromagnetic waves were actually comprised of photons of light—that there was an intrinsic duality of wave and particle behavior associated with light. This was a preamble for the reassessment of the behavior of electrons confined on the spatial scale of the atom.

The de Broglie Wavelength of the Electron

In the early 1920s there was considerable unrest in the segment of the scientific community that was trying to reconcile this Bohr picture of the atom.

It was a young French nobleman by the name of Louis de Broglie who, in 1923, joined three lines of evidence to present a fundamentally new view of the electron when that electron was *confined* to spatial scales on the order of the atomic size. *First,* de Broglie was curious about the fact that the wavelengths of the spectral lines of hydrogen were described quantitatively by *integers,* as they are in the Rydberg equation and in the Bohr formula for $\Delta E = E_f - E_i$ of the emitted or absorbed photon. The only other place where integers appeared in physics was when wave motion was *confined* within boundaries, such as a violin string that is a standing wave. *Second,* he was aware that the wave nature of electromagnetic radiation had undergone a fundamental reformulation in terms of the inherent *particle* characteristics of light; of Einstein's photon. *Third*, de Broglie considered the possibility that if waves (electromagnetic radiation) could be particles, could particles (electrons) also be waves?

How could the wavelength of the electron be expressed? De Broglie knew that the Planck expression for the photon was

$$E = h\nu = h(c/\lambda)$$

and that Einstein's theory of relativity relating energy and mass was

$$E = mc^2$$

Louis de Broglie

Setting the equations for *energy* equal to each other, yielded a simple expression for the wavelength of a photon:

$$\lambda_{photon} = h/mc = h/p_{photon}$$

where the momentum of the photon is $p_{photon} = mc$. What would be the corresponding wavelength of the electron? By analogy with the photon, the wavelength for a material particle (the electron) would be

$$\lambda_{el} = h/mv = h/p_{el}$$

De Broglie proposed in his PhD thesis in 1924 that "matter waves," with $\lambda_{el} = h/p_{el}$, described the electron. While experts close to the problem had tried unsuccessfully to reconcile Newton, Maxwell, and Einstein based on scientifically held views of the time, de Broglie was young and not so constrained.

His hypothesis, even before it was substantiated experimentally three years later, had an immediate and, it turned out, irreversible impact. De Broglie's standing "electron wave" linked to the orbits of Bohr with the simplest of analogies. Each "stationary state" of Bohr corresponds to a standing wave of the electron, confined to the circumference of the orbit. Only certain orbits fulfill the condition of the standing wave, which in turn demands that an integer number of wavelengths fit into the circumference of the orbit such that $n\lambda_e = 2\pi r$ with n = 1, 2, 3. This union of the Bohr model of the atom with the de Broglie hypothesis of the wave property of the electron is displayed in Figure 8.25.

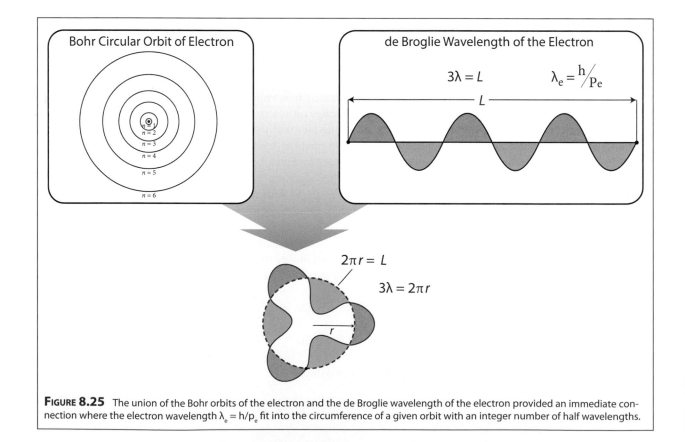

FIGURE 8.25 The union of the Bohr orbits of the electron and the de Broglie wavelength of the electron provided an immediate connection where the electron wavelength $\lambda_e = h/p_e$ fit into the circumference of a given orbit with an integer number of half wavelengths.

Solution
The electron mass, expressed in kilograms, is 9.109×10^{-31} kg. The electron velocity is $v = 0.100 \times c = 0.100 \times 3.00 \times 10^8$ m s^{-1} = 3.00×10^7 m s^{-1}. Planck's constant $h = 6.626 \times 10^{-34}$ J s = 6.626×10^{-34} kg m^2 s^{-2} s = 6.626×10^{-34} kg m^2 s^{-1}. Substituting these data into the equation $\lambda_{el} = h/p_{el}$

$$\lambda_{el} = \frac{6.626 \times 10^{-34} \text{ kg m}^2 \text{ s}^{-1}}{\left(9.109 \times 10^{-31} \text{ kg}\right)\left(3.00 \times 10^7 \text{ m s}^{-1}\right)}$$

$$= 2.42 \times 10^{-11} \text{ m} = 24.2 \text{ pm}$$

Check Yourself 7
Calculating the Wavelength Associated with a Beam of Particles. What is the wavelength associated with electrons traveling at one-tenth the speed of light?

Practice Example A: Assuming Superman has a mass of 91 kg, what is the wavelength associated with him if he is traveling at one-fifth the speed of light?

Practice Example B: To what velocity (speed) must a beam of protons be accelerated to display a de Broglie wavelength of 10.0 pm?

Nature of Waves and the Wave Equation

Two key aspects emerged from de Broglie's hypothesis that the electron possessed an intrinsic wavelength, $\lambda_{el} = h/p_{el}$, and that this wavelength was confined to the spatial domain surrounding the nucleus, as shown in Figure 8.25. This combination of wave properties and confinement, while never applied in the context of a particle, was a mathematics problem treated rigorously by the developing mathematics of the 19th century, decades before de Broglie put forward his hypothesis. What happened following de Broglie's contention that the electron had an integer wavelength was an explosion of work that engaged previously developed mathematics with the application associated with the properties of the electron in the context of atomic structure.

Before we treat the hydrogen atom explicitly with this new "wave mechanics," it is well worth examining the character of waves confined by physical boundaries that establish the characteristics of the wave properties of the electron.

We all know a wave when we see it; ocean waves, waves in a jump rope, waves in a field of grass, waves in an oscillating string on a concert bass. We also know the shape of a sine function in mathematics. It, in fact, reflects rather accurately what we know a wave to be: a repetitive oscillation with an ordered sequence of peaks and valleys. The wave has an amplitude, A, and a wavelength, λ, as shown in Figure 8.26.

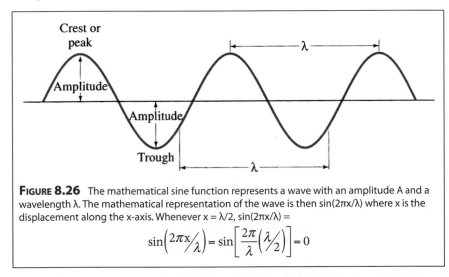

FIGURE 8.26 The mathematical sine function represents a wave with an amplitude A and a wavelength λ. The mathematical representation of the wave is then sin($2\pi x/\lambda$) where x is the displacement along the x-axis. Whenever $x = \lambda/2$, sin($2\pi x/\lambda$) =

$$\sin\left(2\pi x/\lambda\right) = \sin\left[\frac{2\pi}{\lambda}\left(\lambda/2\right)\right] = 0$$

We can represent that wave by a mathematical function, which we call a *wavefunction*, that connects what we know a wave looks like with the mathematical function that quantitatively represents the amplitude, A, the wavelength, λ, and the displacement, x, along the direction of propagation as displayed in Figure 8.27.

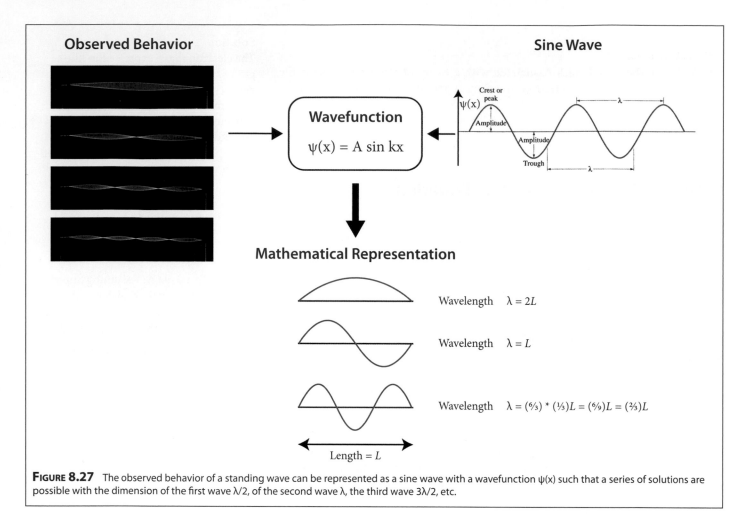

Observed Behavior

Sine Wave

Wavefunction

$$\psi(x) = A \sin kx$$

Mathematical Representation

Wavelength $\lambda = 2L$

Wavelength $\lambda = L$

Wavelength $\lambda = (6/3) * (1/3)L = (6/9)L = (2/3)L$

Length = L

FIGURE 8.27   The observed behavior of a standing wave can be represented as a sine wave with a wavefunction $\psi(x)$ such that a series of solutions are possible with the dimension of the first wave $\lambda/2$, of the second wave λ, the third wave $3\lambda/2$, etc.

The constant k simply allows us to represent the wavelength, λ, in units of the displacement x such that we can have any number of wavelengths in a given unit of distance within the domain occupied by the wave.

The question is, if $\psi(x)$ = A sin kx is the *wavefunction*, is there an equation for which the wavefunction, $\psi(x)$, is a solution? If so, then this equation would constitute a partnership: namely the *wave equation* for which the *wavefunction* is a solution.

We can, of course, try guessing at an equation for which wavefunction $\psi(x)$ = A sin kx is a solution, but we consider first what the functional form of that wave equation *must be* to mathematically *generate* the oscillating behavior in the wavefunction.

If we begin at x = x_0 = 0 in Figure 8.28, we see that the *gradient* in $\psi(x)$, d $\psi(x)$/dx, is positive, but that as we progress along x to x = x_1, the gradient decreases such that

$$\left(\frac{d\psi(x)}{dx}\right)_{x=x_0} > \left(\frac{d\psi(x)}{dx}\right)_{x=x_1}$$

If the wave is to form, then $\psi(x)$ must return to the x-axis by bending around until it reaches the x-axis; then it must begin to bend in the *opposite* direction when $\psi(x)$ becomes negative so as to return again to the x-axis, repeating this sequence in each wavelength.

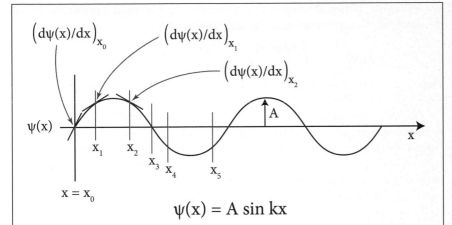

$$\psi(x) = A \sin kx$$

FIGURE 8.28   The wave equation provides the mathematical relationship for which the *wavefunction* is a solution. The wave equation thus represents the oscillatory, repetitive behavior of the wave. If

$$\left(\frac{d\psi(x)}{dx}\right)_{x_i}$$

is the slope of the wave at any point then the curvature of the wave is

$$\frac{d}{dx}\left(\frac{d\psi(x)}{dx}\right) \quad \text{or} \quad \frac{\Delta\left(\frac{d\psi(x)}{dx}\right)}{\Delta x}.$$

As the figure shows, the *curvature*,

$$\frac{\Delta\left(\frac{d\psi(x)}{dx}\right)}{\Delta x}$$

is negative when $\psi(x)$ is positive and the wave bends around toward the x axis. As soon as the wave passes through the x-axis, it becomes negative and the curvature becomes positive. This bends the wave back toward the x-axis.

The bending, or *curvature*, in $\psi(x)$ is the *change* in the slope (the gradient) as a function of x. Thus

$$\text{CURVATURE} = \frac{d}{dx}\left(\frac{d\psi(x)}{dx}\right) = \frac{\left(\left(\frac{d\psi(x)}{dx}\right)_{x_1} - \left(\frac{d\psi(x)}{dx}\right)_{x_0}\right)}{\Delta x}$$

However, since

$$\left(\frac{d\psi(x)}{dx}\right)_{x_0} > \left(\frac{d\psi(x)}{dx}\right)_{x_1}$$

then

$$\left(\frac{d\psi(x)}{dx}\right)_{x_1} - \left(\frac{d\psi(x)}{dx}\right)_{x_0} < 0$$

and since

$$\frac{d}{dx}\left(\frac{d\psi(x)}{dx}\right) = \frac{d^2\psi(x)}{dx^2}$$

we have the result that the *curvature* is negative as long as $\psi(x)$ is positive.

$$d^2\,\psi(x)\,/\,dx^2 < 0 \text{ as long as } \psi(x) > 0$$

We can check this by noting that

$$\left(\frac{d\psi(x)}{dx}\right)_{x_0} > \left(\frac{d\psi(x)}{dx}\right)_{x_1} > \left(\frac{d\psi(x)}{dx}\right)_{x_2} > \left(\frac{d\psi(x)}{dx}\right)_{x_3}$$

in Figure 8.28.

When $\psi_{(x)}$ passes through zero and becomes negative, $d\,\psi(x)/dx$ begins to *increase* from its maximum negative value and

$$\frac{d}{dx}\left(\frac{d\psi(x)}{dx}\right) = \frac{d^2\psi(x)}{dx^2} > 0$$

Thus when $\psi(x)$ is positive, $d^2\psi(x)/dx^2$ is negative and when $\psi(x)$ is negative, $d^2\psi(x)/dx^2$ is positive.

The requirement for *oscillatory motion* of $\psi(x)$ thus establishes an important relationship between $d^2\psi(x)/dx^2$ and $\psi(x)$, namely that they are of opposite sign but also that the *curvature*, $d^2\psi(x)/dx^2$, determines how quickly $\psi(x)$ returns to the x-axis. That is, the curvature, $d^2\psi(x)/dx^2$, determines how many wavelengths fit within a given displacement x along the x-axis.

Since the oscillatory behavior of our wave bears considerable resemblance to a sine wave, if we take the second derivative of $\psi(x) = A \sin kx$, we have

$$\frac{d}{dx}\psi(x) = \frac{d}{dx}(A \sin kx) = Ak \cos kx$$

$$\frac{d^2\psi(x)}{dx^2} = \frac{d}{dx}(Ak \cos kx) = -Ak^2 \sin kx$$

But then $\psi(x)$ is a solution to the differential equation

$$\frac{d^2\psi(x)}{dx^2} = -k^2\psi(x)$$

This is indeed the wave equation for which $\psi(x)$ is the solution. This $\psi(x) = A \sin kx$ is referred to as the *wavefunction* of the *wave equation*.

What we will see is that this *wave equation* and this wavefunction form the basis for the mathematical representation of an array of different phenomena extending from ocean waves to electromagnetic radiation to stringed musical instruments to the behavior of electrons in an atom or molecule. We simply need to adapt the equation to the particular objectives, select the most appropriate coordinate system, and physically identify k for our particular application.

We can interpret the physical meaning of k by noting that the repetitive nature of $\psi(x) = A \sin kx$ means that

$$A \sin kx = A \sin k(x + \lambda) = A \sin (kx + k\lambda)$$

We also know that $\sin \theta = \sin (\theta + 2\pi)$ so $k\lambda$ must equal 2π and therefore

$$k\lambda = 2\pi \text{ and } k = 2\pi/\lambda$$

Thus our *wavefunction* is

$$\psi(x) = A \sin\left(\frac{2\pi x}{\lambda}\right)$$

and our *wave equation* is

$$\frac{d^2\psi(x)}{dx^2} = -\left(\frac{2\pi}{\lambda}\right)^2 \psi(x)$$

What is remarkable is the ease with which this general treatment of the mathematical representation of a wave adapts to the wave properties of the electron. As we deduced above, $k = 2\pi/\lambda$. However, the de Broglie's hypothesis that the electron possess a wavelength given by

$$\lambda_{el} = h/p_{el}$$

means that for our electron with $\lambda_{el} = h/p_{el}$ our wave equation

$$\frac{d^2\psi(x)}{dx^2} = -\left(\frac{2\pi}{\lambda}\right)^2 \psi(x)$$

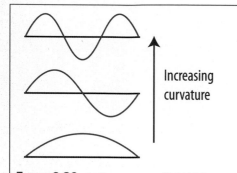

FIGURE 8.29 As the curvature $d^2\psi(x)/dx^2$ increases in magnitude, more and more half-wavelengths fit between the boundaries at x = 0 and x = L. As the curvature increases, the energy in the wave increases.

becomes

$$\frac{d^2\psi(x)}{dx^2} = -\left(\frac{2\pi p}{h}\right)^2 \psi(x)$$

But the *kinetic energy,* E_k, of a particle (the electron) is

$$E_k = \tfrac{1}{2}\,mv^2 = p^2/2m$$

So our wave equation becomes

$$\frac{d^2\psi}{dx^2} = -\frac{2m}{\hbar^2}E_k\psi(x)$$

Where again $\hbar = h/2\pi$. We can multiply our new wave equation through by $-\hbar^2/2m$ to yield

$$-\hbar^2\!\Big/\!2m\,\frac{d^2\psi}{dx^2} = E_{el}\psi(x)$$

This wave equation expresses mathematically a very important physical relationship between the *curvature* of the wavefunction, $d^2\psi(x)/dx^2$, and the kinetic energy of the electron, E_k; specifically that as the curvature of the wavefunction increases, the kinetic energy of the electron increases. Increasing curvature also means, of course, that more wavelengths fit into the same spatial dimensions. More curvature, more energy, and more wavelengths per unit distance as shown graphically in Figure 8.29.

We have developed this wave equation for a particle (an electron) with kinetic energy, $E_k = \tfrac{1}{2}\,mv_2 = p^2/2m$. However, when an electron is attracted to the positive charge of the nucleus (to the proton in a hydrogen atom) it is clearly in a potential energy field. How, then, is the potential energy considered in this wave equation such that we can use the wave equation to describe the electron in the hydrogen atom?

Before we answer this question, we consider a very important example for which the potential energy can be taken as zero.

Particle-in-a-Box: An Important Example

To consider this problem of an electron confined to physical boundaries within which the potential energy, V, is zero, and the resulting wave properties associated therewith, we consider an electron captured in a potential energy well with walls infinitely high in potential such that the electron cannot escape from the confinement. We further assume that the potential energy *between* the walls of the potential well (one wall located at x = 0 and the other at x = L) is zero. This potential is sketched in Figure 8.30.

If the electron lies within the boundaries of this potential well, the wave defining the behavior of the electron is confined and is thus a standing wave within a one dimensional box—just as a violin string is confined between the bridge and the neck of the violin. The resonance that allows a violin string to sustain a musical note results from the matching of the wavelengths of the oscillating string of the instrument with the dimensions of the confinement. As we saw above, because the electron has a wavelength $\lambda_{el} = h/p_{el}$, the mathematical function that describes the electron confined within a dimension L is termed the *wavefunction* of the electron, $\psi(x)$.

The wavefunction of either the violin string, or the electron in a one-dimension box, is

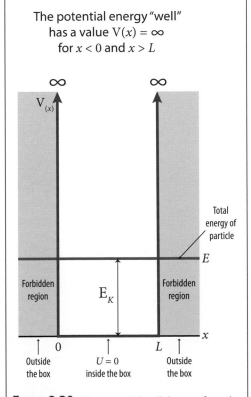

The potential energy "well" has a value $V(x) = \infty$ for $x < 0$ and $x > L$

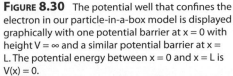

FIGURE 8.30 The potential well that confines the electron in our particle-in-a-box model is displayed graphically with one potential barrier at x = 0 with height V = ∞ and a similar potential barrier at x = L. The potential energy between x = 0 and x = L is V(x) = 0.

$$\psi(x) = A \sin kx = A \sin\left(\frac{n\pi x}{L}\right) \quad \text{because } k = \frac{n\pi}{L}$$

where n is an integer n = 1, 2, 3, ..., L is the dimension of the box, and x is the position between x = 0 at one wall and x = L at the other. The quantity A is termed the *normalization factor* and is, as we will see, determined such that the *probability* of finding the electron in the box is equal to unity.

Before we move to determine the allowed energies of this electron in a one-dimensional box, we consider how to calculate the normalization factor, N, that quantitatively establishes the probability for finding the electron somewhere in the box. De Broglie's hypothesis was that the electron had a wavelength $\lambda_{el} = h/p_{el}$ and that, when confined to the dimension of an atom, it was controlled primarily by its wave properties. However, the physical interpretation of the electron wavefunction was a subject of intense debate. It was Max Born in 1928, three years after de Broglie's hypothesis, who put forward the supposition that the square of the wavefunction, $\psi(x)^2$, represented the *probability* that the electron would be found at position x in space. Applied to an electron confined to a potential well with infinite walls from which it cannot escape, the probability of finding the electron somewhere within the box must be unity. Thus,

$$\int_{x=0}^{x=L} \psi^2(x)dx = 1$$

with our wavefunction

$$\psi(x) = A \sin(n\pi x/L)$$

this gives us

$$\int_{x=0}^{x=L} A^2 \sin^2\left(\frac{n\pi x}{L}\right)dx = 1$$

We can integrate this expression using the trigonometric identity

$$\sin^2\theta = \tfrac{1}{2}(1 - \cos 2\theta)$$

Thus, pulling the *normalization constant* A from under the integral and using our trigonometric identity, yields

$$\int_{x=0}^{x=L} \psi^2(x)dx = A^2 \int_{x=0}^{x=L} \sin^2\left(\frac{n\pi x}{L}\right)dx = \frac{A^2}{2}\int_{x=0}^{L}\left[1 - \cos\left(\frac{2n\pi x}{L}\right)\right]dx$$

$$= \tfrac{1}{2}A^2\left[L - \left(\frac{L}{2n\pi}\right)\left\{\sin(2n\pi) - \sin 0\right\}\right] = 1$$

Given that $\sin(2n\pi) = \sin 0 = 0$, we can easily solve directly for the *normalization* constant because our equation collapses to

$$\tfrac{1}{2} A^2 L = 1$$

and thus the normalization term that ensures that the electron exists somewhere within the box is:

$$A = \sqrt{\frac{2}{L}}$$

Thus our properly "normalized" wavefunction for the electron in a box is

$$\psi(x) = \sqrt{\frac{2}{L}} \sin\left(\frac{n\pi x}{L}\right) \quad \text{with } n = 1, 2, 3, ...$$

This raises the question: what are the allowed energies of the particle (the electron) in the square well potential? The total energy of the electron within the square well is the sum of the kinetic and potential energy

$$E_{el} = E_k + V(x) = \tfrac{1}{2}\,mv^2 + V(x)$$

However, because $V(x) = 0$ between $x = 0$ and $x = L$, the total energy of the electron is

$$E_{el} = E_k = \tfrac{1}{2}\,mv^2 = p^2{}_{el}/2m$$

The confined electron wave of wavelength

$$\lambda_{el} = h/p_{el}$$

must obey the (boundary) condition that an integer (n) number of half-waves, $n\,\lambda_{el}/2$, must fit within the dimensions of the box, L, so

$$n\,\lambda_{el}/2 = L$$

And so

$$E_{el} = \frac{P_{el}^2}{2m} = \frac{1}{2m}\left(\frac{h}{\lambda_{el}}\right)^2 = \frac{n^2 h^2}{8mL^2}$$

Before we continue with this important case of an electron trapped in a box with $V(x) = 0$ between two walls of infinite height at $x = 0$ and $x = L$, it is important to see whether the quantized values of the allowed energies agree quantitatively with our wave equation for the electron

$$-\frac{\hbar^2}{2m}\frac{d^2\psi(x)}{dx^2} = E_{el}\psi(x)$$

with

$$\psi(x) = \sqrt{\frac{2}{L}}\sin\left(\frac{n\pi x}{L}\right)$$

we have

$$\frac{d^2\psi(x)}{dx^2} = -\left(\frac{n\pi}{L}\right)^2\psi(x) = \left(\frac{2\pi}{\lambda_{el}}\right)^2\psi(x) = \left(\frac{2\pi p_{el}}{h}\right)^2\psi(x)$$

$$L = \frac{n\lambda_{el}}{2} \qquad \lambda_{el} = \frac{h}{p_{el}}$$

But

$$\left(\frac{2\pi p_{el}}{h}\right)^2 = \left(\frac{P_{el}}{\hbar}\right)^2 = \left(\frac{P_{el}^2}{2m}\right)\frac{2m}{\hbar^2} = E_{el}\left(\frac{2m}{\hbar^2}\right)$$

so

$$\frac{d^2\psi(x)}{dx^2} = -\left(\frac{2\pi p_{el}}{h}\right)^2\psi(x)$$

becomes

$$\frac{d^2\psi(x)}{dx^2} = -\frac{2m}{\hbar^2}E_{el}\psi(x)$$

or

$$-\frac{\hbar^2}{2m}\frac{d^2\psi(x)}{dx^2} = E_{el}\psi(x)$$

Thus, for our particle-in-a-box, the wavefunction, wave equation, and quantized energies are correct.

We can display these wavefunctions

$$\psi_n(x) = \sqrt{\frac{2}{L}}\sin\left(\frac{n\pi x}{L}\right)$$

and the associated quantized energies

$$E_n = \frac{n^2 h^2}{8mL^2}$$

in the square well potential as shown in Figure 8.31.

It is important to note that the case for n = 0 is not allowed. The reason is that were n = 0, the wavefunction would be equal to zero everywhere in the box

$$\psi_{n=0}(x) = \sqrt{\frac{2}{L}}\, \sin\!\left(\frac{n\pi 0}{L}\right)_2 = \sqrt{\frac{2}{L}}\, \sin 0 = 0$$

This would mean, because $\psi_n^2(x)$ is the probability for finding the particle somewhere in the box, that $\psi_n^2(x) = 0$ so there would be no particle in the potential well of the system. Therefore, the minimum energy corresponds to n = 1 and the energy of this "ground state" is

$$E_1 = \frac{n^2 h^2}{8mL^2} = \frac{h^2}{8mL^2}$$

We can, as displayed in Figure 8.32, show both the wavefunctions $\psi_n(x)$ and the probability $\psi_n^2(x)$ for finding the particle (the electron) at any position x within the potential well of the box.

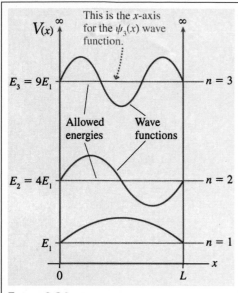

FIGURE 8.31 The solutions to the wave equation are the wavefunctions ψn(x) for which the energy En = n²h²/8mL² corresponding to each quantum number, n, is given for the corresponding wavefunction.

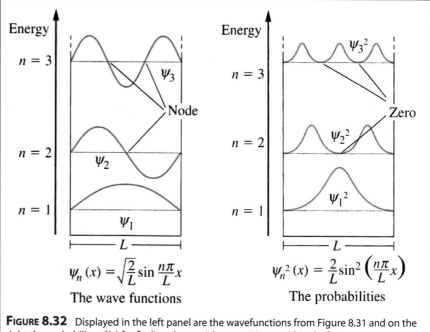

$$\psi_n(x) = \sqrt{\frac{2}{L}}\sin\frac{n\pi}{L}x$$

The wave functions

$$\psi_n^2(x) = \frac{2}{L}\sin^2\!\left(\frac{n\pi}{L}x\right)$$

The probabilities

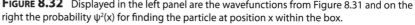

FIGURE 8.32 Displayed in the left panel are the wavefunctions from Figure 8.31 and on the right the probability ψ²(x) for finding the particle at position x within the box.

Recognizing that the *probability* of finding the electron at a position x is given by $\psi_n^2(x)$, the n = 1 quantized level is characterized by a maximum probability of finding the electron at the center of the well at x = L/2. For the next higher quantized level, n = 2, the probability of finding the electron at the center of the well is zero because $\psi_2(x)$ is zero and thus so too is $\psi_2^2(x)$ for x = L/2. For the next higher wavefunction, $\psi_3(x)$, there are two nodes and thus two points with zero probability of finding the electron along the x-axis.

The dependence of the energy level on the principal quantum number n is also important. For n = 1, the energy is nonzero; it is equal to

$$E_{n=1} = \frac{h^2 n^2}{8L^2 m_e} = \frac{h^2}{8L^2 m_e}$$

The energy of the next quantized level, n = 2, is

$$E_{n=2} = \frac{h^2 n^2}{8L^2 m_e} = \frac{4h^2}{8L^2 m_e}$$

and the next quantized level has energy

$$E_{n=3} = \frac{h^2 n^2}{8L^2 m_e} = \frac{9h^2}{8L^2 m_e}$$

The energy *separation* between energy levels thus *increases* with increasing principal quantum number, n. In fact we can write the expression for the separation between energy levels with the general expression

$$\Delta E_{n \to n+1} = E_{n+1} - E_n = \frac{h^2}{8L^2 m_e}\left[(n+1)^2 - n^2\right] = \frac{h^2}{8L^2 m_e}(2n+1)$$

Also of great importance is the dependence of the energy of the quantized levels on the *width* of the well, L. If we enlarge the well width to 2L from L, the energies become

$$E_n = \frac{h^2 n^2}{8(2L)^2 m_e}$$

Thus the energy of each quantized level has *decreased* by a factor of 4 by increasing the well width by a factor of 2 as shown in Figure 8.33. This, as it turns out, is related to why a molecular bond forms when two atoms combine to form a molecule; the wavefunction of the electrons "spreads out" and the energy of the ensemble of electrons and protons decreases such that the molecule is more stable than the individual atoms.

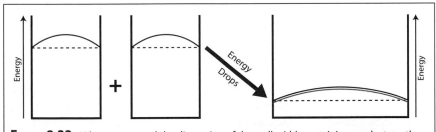

FIGURE 8.33 When we expand the dimension of the well *width* containing an electron, the energy of that electron wave decreases as the square of the well width. Thus if we double the width of the well, the lowest energy level of the electron decreases by a factor of four. This represents the simplest model of the formation of the molecular bond from two atoms.

Approach: The lowest energy level has $n = 1$, and so we can use the equation

$$\Delta E = E_{n+1} - E_n = \frac{h^2}{8m_e L^2}\left[(n+1)^2 - n^2\right]$$

with $n = 1$ and m = the mass of the electron. The energy difference is carried away as a photon of radiation; we set the energy difference equal to hv and express v in terms of the corresponding wavelength by using $\lambda = c/v$.

Solution: From our equation above with $n = 1$, $2n + 1 = 3$

$$E_2 - E_1 = \frac{3h^2}{8m_e L^2}$$

From $E_2 - E_1 = hv$,

$$hv = \frac{3h^2}{8m_e L^2}, \quad \text{so} \quad v = \frac{3h}{8m_e L^2}$$

From $\lambda = c/v$

$$\lambda = \frac{c}{\left(3h/8m_e L^2\right)} = \frac{8m_e c L^2}{3h}$$

Now substitute the values of each quantity:

$$\lambda = \frac{8 \times \overbrace{\left(9.109\,39 \times 10^{-31}\,\text{kg}\right)}^{m_e} \times \overbrace{\left(2.998 \times 10^8\,\text{m}\cdot\text{s}^{-1}\right)}^{c} \times \overbrace{\left(1.50 \times 10^{-10}\,\text{m}\right)^2}^{L=150\,\text{pm}}}{3 \times \underbrace{\left(6.626 \times 10^{-34}\,\text{J}\cdot\text{s}\right)}_{h}}$$

$$= \frac{8 \times 9.109\,39 \times 10^{-31} \times 2.998 \times 10^8 \times \left(1.50 \times 10^{-10}\right)^2}{3 \times 6.626 \times 10^{-34}} \underbrace{\frac{\text{kg}\cdot\text{m}\cdot\text{s}^{-1}\cdot\text{m}^2}{\text{kg}\cdot\text{m}^2\cdot\text{s}^{-2}\cdot\text{s}}}_{\text{J}}$$

$$= 2.47 \times 10^{-8}\,\text{m}$$

Analysis: This wavelength corresponds to 24.7 nm. The experimental value for the actual transition in a hydrogen atom is 122 nm. Although there is a big discrepancy, an atom does not have the hard boundaries that confine a particle in a box, and is three-dimensional. The fact that the predicted wavelength has nearly the same order of magnitude as the actual value suggests that a quantum theory of the atom, based on a more realistic three-dimensional model, should give good agreement.

Practice Example: Use the same model for helium but suppose that the box is 100 pm in width, because the atom is smaller. Estimate the wavelength of the same transition.

Check Yourself 8
Calculating the Energies of a Particle in a Box

Very simple expressions can often be used to estimate the order of magnitude of an important property without doing a detailed calculation. For example, treat a hydrogen atom as a one-dimensional box of length 150 pm (the approximate diameter of the atom) containing an electron. Predict the wavelength of the radiation emitted when the electron falls to the lowest energy level from the next higher energy level.

Uncertainty in the Position of the Electron in the Square Well Potential

In the formulation of classical mechanics, an object, a macroscopic particle, has a trajectory specified as a function of time to any desired degree of accuracy fully determined by the initial conditions of that trajectory and the force field within which it moves. By virtue of the fully specified position and velocity, both the momentum and the location of the particle in space can be calculated simultaneously at any time.

With the introduction of the wave nature of electrons, there emerges fundamental constraints on the accuracy with which the position and the momentum of the electron can be simultaneously determined. We will examine the ramifications of this fundamental constraint for a number of cases, but consider first the case of the electron in a box, our particle-in-a-box, wherein the electron is confined within a potential well as described above. The available energy levels are given by

$$E_n = \frac{n^2 h^2}{8mL^2}$$

While we can calculate the energy exactly, we do not know if, at any given instant, the particle is moving from left to right or right to left.

With the momentum of the particle given by

$$p^2_{el} = 2m\,E_{el}$$

we have

$$p = \pm\sqrt{2mE_{el}} = \pm\frac{nh}{2L}$$

reflecting the fact that we don't know the direction of particle motion.

There is also uncertainty in the position of the

particle because we cannot accurately locate the particle within the envelope of its wavefunction. We can adopt a number of assumptions regarding the particle's position, say within half the width of the box, or L/2, but this is somewhat arbitrary. The fundamental point is that the wave nature of the particle prevents locating it exactly along the x-axis of the potential well in Figure 8.32. If the uncertainty in momentum along the x-axis is Δp_x, then for the lowest energy state $\Delta p_x = (h/2L)$ the uncertainty in the product of momentum and position is given by

$$\Delta p_x \Delta x \approx \left(\frac{h}{2L}\right)\left(\frac{L}{2}\right) = \frac{h}{4}$$

Notice that if we attempt to increase the accuracy of determining the position of the particle by shrinking the dimension of the box, the uncertainty in the momentum of the particle increases proportionately.

There are other implications of this uncertainty in the product of position and momentum. For example, the energy of the particle cannot equal zero because then $p_x = 0$ and that would force Δx to infinity.

Werner Heisenberg developed the idea that measuring the *position* of an electron disturbs its *momentum* and measuring the momentum of an electron can result in altering its position. This leads to a detailed quantitative analysis of the *coupled* uncertainty of measurements of momentum and position that lead to the *Heisenberg Uncertainty Principle*.

The principle states that for position x and momentum along x, p_x, the product of the errors in the measurement of x, Δx, and in the measurement of p_x, Δp_x, can never be smaller than h/4π such that

$$\Delta x \, \Delta p_x \geq h/4\pi.$$

This is a purely quantum mechanical effect. It means that the more precisely we measure the position of the electron, the less precisely we can know the momentum, and vice versa.

Check Yourself 9

Suppose that by some means an electron in a hydrogen atom with a total energy of –3.40 eV is determined to be between 2.0 and 2.2 Å away from its proton. By using $\Delta p \, \Delta x \geq h/4\pi$, estimate the resulting uncertainty in radial momentum of the electron, and compare this to the root-mean-square average total momentum obtained from $K = p^2/2m = -E$.

Solution:

$\Delta p_r = h/(4\pi r) = 6.63 \times 10^{-34}$ J s/[(4π)(0.2 × 10⁻¹⁰ m)]

$\quad = 2.64 \times 10^{-24}$ kg m s⁻¹

$p = [2mK]^{½} = [(2)(9.11 \times 10^{-31}$ kg)(3.40 eV)(1.60 × 10⁻¹⁹ J/eV)]^{½}

$\quad = 9.96 \times 10^{-25}$ kg m s⁻¹

So Δp is more than twice as large as p itself. This is another way of explaining the impossibility of the Bohr orbits, with their fixed total momentum, and of understanding the breadth of the radial distribution functions.

Solution:

$\Delta p_r = h/(4\pi\Delta x) = 6.63 \times 10^{-34} \text{ J s}/[(4\pi)(1.0 \times 10^{-3} \text{ m})]$

$\qquad = 5.28 \times 10^{-32} \text{ kg m s}^{-1}$

$\qquad p = (0.140 \text{ kg})(92 \text{ mph})(1609 \text{ m/mi})(1 \text{ hr}/3600 \text{ s})$

$\qquad = 5.76 \text{ kg m s}^{-1}$

The uncertainty in momentum in this case influences only the 32nd decimal place of the total! The completely unmeasurable error means that ballplayers and you in your everyday life don't have to cope with intrinsic uncertainties that are so critically important at the atomic scale.

Check Yourself 10
The Uncertainty Principle Applied to Macroscopic Objects

To hit a baseball squarely, a batter must locate the 140 g ball (by eye) within about 1 mm. Use the Heisenberg Uncertainty Principle to find the uncertainty in momentum of the ball, and compare this to the total momentum of a 92 mph fastball.

To summarize, we emphasize several key points

1.   As the width of the box, L, increases, the energy decreases for a given n and mass, m. This important fact follows from the relationship between the *curvature*, $d^2\psi(x)/dx^2$, and the kinetic energy of the particle, E_k, in the wave equation

$$-\hbar^2/2m \frac{d^2\psi(x)}{dx^2} = E_k\psi(x)$$

2.   Because n can take on only integer values, the energy of the particle (the electron) is *quantized*. That is, the energy of the particle is restricted to a series of discrete values called energy levels (Figure 8.32).

3.   The shape of the wavefunction of a confined particle contains important information. Consider the wavefunction for the first two allowed energy levels, n = 1 and n = 2.

$$\text{For n} = 1 \qquad E_1 = h^2/8mL^2$$

And $\psi_1^2(x)$, the probability of finding the particle at x, peaks in the center of the box as shown by the shaded area in Figure 8.32.

$$\text{For n} = 2 \quad E_2 = 4h^2/8mL^2 = \frac{h^2}{2mL^2}$$

And $\psi_2^2(x)$ reaches a maximum on either side of the center of the box with zero probability of finding the particle in the center of the box. The point where $\psi^2(x) = 0$ is called a *node* of the wavefunction.

4.   A particle in a contained space cannot have zero energy, thus a particle can never be motionless. It must have energy if it is to exist.

5.   The fact that the electron confined in a box is described by a wavefunction, $\psi_n(x)$, and that the probability of finding the electron at a point x within the box, $\psi_n^2(x)$, is *distributed* throughout the dimension of the box, places a very important constraint on the accuracy of simultaneously determining the *position* of the electron and its *momentum*.

The Schrödinger Equation

To this point we have combined two lines of reasoning to arrive at the point where we can mathematically describe the behavior of an electron in a confined one-dimensional box. Those two lines of reasoning are (1) the de Broglie hypothesis that the electron has a wavelength $\lambda_{el} = h/p_{el}$ and (2) the application of the mathematics developed to formulate both a wave equation and the associated wavefunctions that are a solution to that wave equation. The allowed, quantized, energies of the electron result from the *boundary conditions* imposed on the solution by the square-well potential (V(x) = 0 between x = 0 and x = L, V(x) = ∞ elsewhere).

Development of the new "wave mechanics" to describe the behavior of these new "matter waves" occurred remarkably quickly following the hypothesis put forward by de Broglie in 1924. In the winter of 1925, Erwin Schrödinger, a mathematical physicist, pictured in Figure 8.34, gave a lecture discussing some of the implications of de Broglie's hypothesis. At that meeting Peter Debye, a well-known physical chemist, rose at the end of Schrödinger's talk and remarked, "What is this foolishness? If there are waves they must obey a wave equation!" The reaction by Schrödinger was swift and decisive. He recognized the truth in Debye's words and turned full time to working out the *wave equation* for electrons confined in the Coulomb potential of the proton.

FIGURE 8.34 Erwin Schrödinger, by his work on the wave equation for the electron, became a giant of 20th century science.

The first problem Schrödinger had to contend with was how to treat the *combination* of the kinetic energy, E_k, and potential energy, V, of the electron in the wave equation. When the potential energy is zero (V = 0), the total energy $E = E_k + V$ reduces to $E = E_k$. Under that condition, E_k is constant and the wave equation is just our wave equation for the electron-in-a-box

$$-\hbar^2\!\!\left/2m\right. \frac{d^2\psi(x)}{dx^2} = E_k\psi(x)$$

This suggests that if the mathematical "operation"

$$-\hbar^2\!\!\left/2m\right. \frac{d^2}{dx^2}$$

is applied to the wavefunction, $\psi(x)$, such that

$$-\hbar^2\!\!\left/2m\right. \frac{d^2\psi(x)}{dx^2}$$

is executed mathematically, the result is the determination of the allowed values of the *kinetic energy*, E_k, where the potential energy is zero.

This concept of an "operator," in this case

$$-\hbar^2\!\!\left/2m\right. \frac{d^2}{dx^2},$$

was very familiar to mathematical physics at the time. It had been applied by Hamilton to solve problems in classical (Newtonian) mechanics decades earlier. In this formulation, the "Hamiltonian" operator, Ĥ (where the "hat" on H designates it as an operator), for the general case of a particle in motion, represented mathematically the sum of the kinetic and potential energy.

Thus, arguing by analogy for the case where V(x) = 0, the Hamiltonian represents just the kinetic energy of the particle and

$$-\frac{\hbar^2}{2m}\frac{d^2\psi(x)}{dx^2} = \hat{H}\psi(x)$$

where again $\hat{H}$ is the Hamiltonian operator

$$\hat{H} = -\frac{\hbar^2}{2m}\frac{d^2}{dx^2}$$

Schrödinger's objective, however, was to formulate a wave equation that could mathematically describe the electron in an atom under the influence (the confinement) of the Coulomb potential of the nucleus. This meant that Schrödinger needed to solve the problem for which the potential energy was no longer zero, but varied in space as a function of position.

When $V \neq 0$, then the total energy E is given by $E = E_k + V$ and while E_k is no longer constant, the total energy E is. After considerable effort, Schrödinger expressed the wave equation of the electron in a potential V as

$$-\frac{\hbar^2}{2m}\frac{d^2\psi(x)}{dx^2} + V(x)\psi(x) = E\psi(x)$$

for one dimension, x, by analogy with the Hamiltonian operator formulation. What is commonly done is to express the left-hand side of the equation in terms of the "Hamiltonian operator"

$$\hat{H} = -\frac{\hbar^2}{2m}\frac{d^2}{dx^2} + V(x)$$

such that the one-dimensional Schrödinger equation is then given by

$$\hat{H}\psi(x) = E\psi(x)$$

The concept of an "operator" in mathematics is quite familiar in calculus because, for example, when we take the derivative of some function, f(x), we "operate" on f(x) with the derivative such that

$$\left(\frac{d}{dx}\right)f(x) = \frac{df(x)}{dx}$$

and (d/dx) is the "operator."

What Schrödinger was after, however, was not a wave equation in one dimension, but a wave equation in three dimensions appropriate to the electron in the hydrogen atom moving in the Coulomb field of the proton. As it turns out, the three-dimensional wave equation can be deduced from the one-dimensional case by a straightforward modification to the Hamiltonian, extending it to three dimensions:

$$\hat{H} = -\frac{\hbar^2}{2m}\left(\frac{\partial^2}{\partial x^2} + \frac{\partial^2}{\partial y^2} + \frac{\partial^2}{\partial z^2}\right) + V(x, y, z)$$

The formulation expresses the fact that the derivatives, $\partial^2/\partial x^2$, etc., are *partial derivatives*, which means that the second derivative is taken with respect to x while holding y and z constant.

Thus, while we will not solve the Schrödinger equation,

$$-\frac{\hbar^2}{2m}\left(\frac{\partial^2\psi}{\partial x^2} + \frac{\partial^2\psi}{\partial y^2} + \frac{\partial^2\psi}{\partial z^2}\right) + V(x, y, z)\psi(x, y, z) = E\psi(x, y, z)$$

we will examine solutions to the equation that establish the *spatial distribution* of the electron around the proton. We limit ourselves, therefore, to the important case of electrons in the Coulomb potential of the nucleus.

The Hydrogen Atom

Given what we developed for the wave equation, wavefunctions, boundary condition, and quantized energies appropriate to the particle-in-a-box, when we consider the case of an electron confined by the Coulomb attraction of the nucleus, we can expect some important analogies. First, the combination of wave behavior and confinement will result in only those energies for which the wave fits within the spatial dimension of the potential. Second, the potential will have a spatial domain defined by the mathematical representation of that potential energy.

In fact, from our solution of the Bohr atom, we can write the potential energy of an electron in the Coulomb field of a single proton (which is the case for a hydrogen atom) as

$$V(r) = \frac{(-e)(+e)}{4\pi\varepsilon_0 r} = \frac{-e^2}{4\pi\varepsilon_0 r}$$

where r is the distance from the electron to the nucleus and ε_0 is the permeability of space. We can diagram this potential in three dimensions as displayed in Figure 8.35.

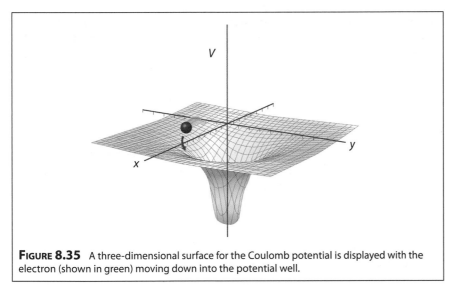

FIGURE 8.35 A three-dimensional surface for the Coulomb potential is displayed with the electron (shown in green) moving down into the potential well.

Just as is the case for a gravitational field, the Coulomb potential is a *central* or *radial* force field and is a function only of r.

We can, with this mathematical expression for the potential energy, insert V(r) into the Schrödinger wave equation, find the wavefunctions that are solutions, and determine the allowed energies.

While we could use our Schrödinger equation in Cartesian coordinates, by expressing the potential energy in Cartesian coordinates:

$$V(r) = V(x, y, z) = \frac{-e^2}{4\pi\varepsilon_0 \left(x^2 + y^2 + z^2\right)^{1/2}}$$

it turns out to be more mathematically convenient (in the long run!) to write the potential energy in terms of r, $V(r) = -e^2/4\pi\varepsilon_0 r$, and to convert the wave equation from Cartesian coordinates to polar coordinates. Figure 8.36 shows explicitly the coordinate conversion: $x = r \sin\theta \cos\phi$ and $z = r \cos\theta$.

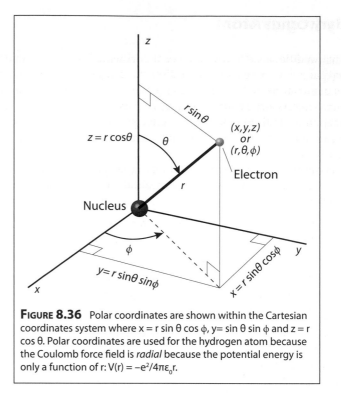

FIGURE 8.36 Polar coordinates are shown within the Cartesian coordinates system where x = r sin θ cos φ, y= sin θ sin φ and z = r cos θ. Polar coordinates are used for the hydrogen atom because the Coulomb force field is *radial* because the potential energy is only a function of r: V(r) = −e²/4πε₀r.

While the Schrödinger equation in three dimensions in polar coordinates appears to be more complex than the same equation in Cartesian coordinates, the solutions (the wavefunctions) are simpler when expressed in polar coordinates.

As we saw, our Hamiltonian operator in Cartesian coordinates is

$$\hat{H} = -\frac{\hbar^2}{2m}\left(\frac{\partial^2}{\partial x^2} + \frac{\partial^2}{\partial y^2} + \frac{\partial^2}{\partial z^2}\right) - \frac{e^2}{4\pi\varepsilon_0\left(x^2 + y^2 + z^2\right)^{1/2}}$$

in polar coordinates it is

$$\hat{H} = -\frac{\hbar^2}{2m}\left(\frac{1}{r^2}\frac{\partial}{\partial r}\left(r^2\frac{\partial}{\partial r}\right) + \frac{1}{r^2\sin\theta}\frac{\partial}{\partial\theta}\left(\sin\theta\frac{\partial}{\partial\theta}\right) + \frac{1}{r^2\sin^2\theta}\frac{\partial^2}{\partial\phi^2}\right) - \frac{e^2}{4\pi\varepsilon_0 r}$$

And Schrödinger's equation in polar coordinates is written

$$\hat{H}\psi\left(r,\theta,\phi\right) = E\psi\left(r,\theta,\phi\right)$$

What is remarkable is that this equation had already been solved, over 100 years earlier, by the French mathematicians Laguerre and Legendre. At the time they had no idea that their solution had anything to do with the wave properties of the electron and that their work would revolutionize the fields of chemistry and physics, but so it is.

Schrödinger began developing solutions to the wave equation by assuming that the wave equation could be separated into radial and angular functions such that

$$\psi(r,\theta,\phi) = R(r)Y(\theta,\phi)$$

where R(r) is the radial wave equation that is a function only of r, and Y(θ, φ) is the angular wavefunction that is a function only of θ and φ. Following Laguerre and Legendre, Schrödinger demonstrated that the wave equation ψ(r, θ, φ) indeed separated into the product of two wavefunctions, one a function only of r and the other a function only of θ and φ.

Boundary Conditions

Boundary conditions are what transform mathematical abstraction, a general mathematical solution, to physical reality: the bridge and neck of a violin, the acoustics of a concert hall, the quantized energy level of a particle-in-a-box, and the quantized levels of the hydrogen atom. Because we seek a wavefunction, $\psi(r, \theta, \phi)$, that captures the three-dimensional wave properties of the electron confined by the Coulomb attraction of the nucleus, $\psi(r, \theta, \phi)$ must exhibit certain characteristics.

1. The wavefunction must possess just one value for a given position in space. The reason is that $\psi^2(r, \theta, \phi)$ is the probability for finding the electron at a given position and there can be only one value for that probability.

2. The wavefunction must vary smoothly in space such that there are no discontinuities in ψ or its first derivative. The reason is that ψ as well as $\partial\psi/\partial r$, $\partial\psi/\partial\theta$, and $\partial\psi/d\phi$ must be defined at every point in space.

3. The integral of ψ^2 over all of space must be equal to unity such that the electron exists within the volume considered (the atom). Thus the wavefunction must remain finite.

These requirements on $\psi(r, \theta, \phi)$ (single valued, smooth, and finite everywhere) constitute the boundary conditions for acceptable solutions to the Schrödinger equation for the hydrogen atom.

It was the application of those boundary conditions by Schrödinger to the array of solutions to the general equation discovered by Laguerre and Legendre that selected the physically acceptable solutions for an electron in a hydrogen atom. Therefore, the difficult part of the task for Schrödinger was first to account for the potential energy that varied in space, V(r), and then to employ the boundary conditions on the physical problem to identify the proper formulation of the wavefunction for the electron in the hydrogen atom from an array of possibilities.

Once that correct wavefunction was identified, operating on that wavefunction with the Hamiltonian resulted directly in the determination of the quantized energy levels of the electron in the hydrogen atom.

Energy Levels of the Hydrogen Atom

As we established in the previous section, Bohr had cast the structure of the hydrogen atom in terms of a planetary model with the electron occupying one of several circular orbits around the nucleus. Combined with the formula for the energy *difference*, $\Delta E = h\nu$, between two states of the hydrogen atom, which was equal to the energy of the photon emitted in jumping from an upper energy level to a lower one, Bohr could calculate the Rydberg constant for hydrogen, R_H, in the Rydberg formula

$$\frac{1}{\lambda} = R_H \left[\frac{1}{n_1^2} - \frac{1}{n_2^2} \right] \quad n_1 = 1, 2, 3$$

$$n_2 = n_1 + 1, n_1 + 2, \text{ etc.}$$

with

$$R_H = \frac{m_e e^4}{8h^3 \varepsilon_0^2}$$

The question for Schrödinger was: how did the predictions of his wave equation and associated wavefunctions and quantized energies square with the Bohr and Rydberg results?

Radial and Angular Solutions to the Schrödinger Equation

When Schrödinger applied the boundary conditions that in turn selected the correct solutions to the wave equation for the specific case of the hydrogen atom, the solutions were found to be specified by integers such that $\psi(r, \theta, \phi)$ was given by

$$\psi_{n,\ell,m}(r,\theta,\phi) = R_{n,\ell}(r)Y_{\ell,m_\ell}(\theta,\phi)$$

where the integers n, ℓ, m_ℓ specified or imposed the *standing wave* boundary conditions on motions in three dimensions. The three dimensions defined by the polar coordinates r, θ, and ϕ in Figure 8.36 for the Coulomb potential displayed in Figure 9.35.

As it turns out, these integers n, ℓ, m_ℓ, which specify the explicit functional form of the wavefunctions that are solutions to the Schrödinger equation for the hydrogen atom, each have an important physical interpretation.

The quantum number n is called the *principal quantum number*. When Schrödinger solved for the allowed energies of the hydrogen atom, the energies were

$$E_n = -\frac{m_e e^4}{8h^2\varepsilon_0^2}\frac{1}{n^2}$$

These allowed energies, and the energy *difference* between energy levels predicted by the Schrödinger equation for the hydrogen atom, was therefore:

$$\Delta E = E_f - E_i = \frac{-m_e e^4}{8h^2\varepsilon_0^2}\left(\frac{1}{n_f^2} - \frac{1}{n_i^2}\right)$$

What was immediately apparent was that this expression for the energy of the photon emitted from the hydrogen atom matched the energy given by the Bohr expression.

In addition, because, as we have seen, Bohr's model predicted the correct value for the Rydberg constant, there was now an entirely new model of the hydrogen atom based on the wave characteristics of the electron that agreed with the spectroscopic data. The energy levels of the hydrogen atom were thus uniquely determined by the single integer, n.

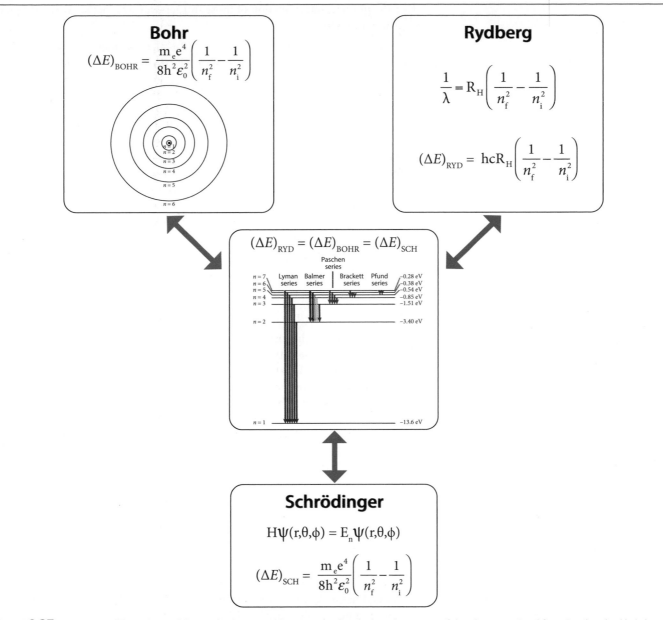

FIGURE 8.37 A key test of the Bohr model was whether it could quantitatively calculate the energy of the photon emitted from (or absorbed by) the hydrogen atom, $(\Delta E)_{\text{BOHR}} = (\Delta E)_{\text{RYD}}$. With the development of the Schrödinger equation, the question became: can the equation calculate the energy levels of the hydrogen atom such that $(\Delta E)_{\text{SCH}} = (\Delta E)_{\text{BOHR}} = (\Delta E)_{\text{RYD}}$ such that the theoretical calculation would quantitatively match the observed energy separation in the spectrum of atomic hydrogen?

Quantum Numbers That Define the Radial and Angular Solutions to the Schrödinger Equation

With the application of the boundary conditions appropriate to the electron confined by the Coulomb potential of the atomic nucleus, and the recognition that the wavefunction could be separated into radial and angular functions such that

$$\psi_{n,\ell,m_\ell}(r,\theta,\phi) = R_{n,\ell}(r)\,\psi_{\ell,m_\ell}(\theta,\phi)$$

the issue is to determine

1. The physical interpretation of those integers n, ℓ, m_ℓ that we will call "quantum numbers";

2. The spatial *shapes* of $R_{n, \ell}(r)$ and $\psi_{\ell,m_\ell}(\theta,\phi)$ because these spatial functions determine the position in space of the electron about the nucleus; and

3. The explicit mathematical function for $R_{n, \ell}(r)$ and $\psi_{\ell,m_\ell}(\theta,\phi)$ for specific values of n, ℓ, m_ℓ.

While we can treat these issues in any order, and all three are important, we will take them in the order presented above. So we begin with the physical interpretation of the quantum number n.

As noted above, n is termed the *principal quantum number* because it is the primary determinant of the energy corresponding to a given solution to the Schrödinger equation. It is also the principal quantum number because, as we will see, it determines the allowed values of the other two quantum numbers ℓ and m_ℓ.

As noted above, the energy E that Schrödinger determined from his wave equation for the hydrogen atom depended only on this principal quantum number, n, and E_n was calculated by the formula

$$E_n = -\frac{me^4}{8\varepsilon_0 h^2}\left(\frac{1}{n^2}\right)$$

which is identical to the formula determined by Bohr. This is made explicit in Figure 8.38. As is the case with Bohr's expression, n can take the values n = 1, 2, 3, ...

Of course, since Bohr's expression agreed with the Rydberg expression, which is based on direct experimental evidence, the Schrödinger expression for E_n *must necessarily* yield the same functional form.

As we will see, the three quantum numbers n, ℓ, and m_ℓ form a "nested set." That is, they are linked according to rules set by the mathematical relationship between $R_{n, \ell}(r)$ and $Y_{\ell,m}(\theta,\phi)$. The quantum number ℓ is termed the *azimuthal quantum number* or the *angular momentum quantum number*. The allowed values of ℓ are limited by the value of n such that ℓ = 0, 1, 2, ... n − 1.

The third quantum number, m_ℓ, is called the *magnetic quantum number* because it defines the behavior of the hydrogen atom in a magnetic field. The value of m_ℓ is limited by the value of the angular momentum quantum number, ℓ, such that m = −ℓ, −ℓ + 1, −ℓ +2, ... −1, 0, 1, ...ℓ − 1, ℓ. Thus m_ℓ, for a given value of ℓ, spans from −ℓ to + ℓ in integer steps through zero.

Before discussing the spatial shapes of $R_{n, \ell}(r)$, and $Y_{\ell,m_\ell}(\theta,\phi)$ we summarize the nested nature of these three quantum numbers in Figure 8.38. The key point is that for a given value of the principal quantum number, n, the allowed values of ℓ and m_ℓ are determined because ℓ is limited to values of 0, 1, 2, ...n −1. Thus if n = 3 then ℓ is limited to ℓ = 0, 1, 2, etc. as shown in Figure 8.38. Once the value of ℓ is set, the *range* of m_ℓ is also set. This is the reason m_ℓ is given the subscript ℓ, such that m_ℓ ranges from −ℓ through zero to +ℓ as shown in the Figure 8.38.

For example, with n = 3, ℓ can take the values ℓ = 0, 1, 2. For each value of ℓ, m_ℓ can assume a range of values from −ℓ through zero to +ℓ. For example, as Figure 8.39 delineates, for n = 3, ℓ = 2, m_ℓ ranges from −2 to +2 in integer increments.

Moreover, in the language of atomic structure, the *shell* is designated by n, the *subshell* is designated by the value of ℓ with ℓ = 0 referred to as the "s subshell," ℓ = 1 referred to as the "p subshell" and ℓ = 2 referred to as the "d subshell." For each value of ℓ, the value of m_ℓ designates the particular *orbital* within which the electron resides. We can summarize the nested relationship between n, ℓ and m_ℓ as displayed in Figure 8.38, and we can tabulate the relationship between these quantum numbers as shown in the table in Figure 8.39.

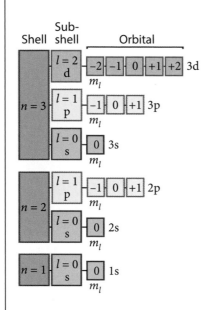

FIGURE 8.38 The quantum numbers, n, ℓ, m_ℓ form a nested set such that the principal quantum number n places bounds on the azimuthal quantum number ℓ restricting ℓ to ℓ = 0, 1, 2...n − 1. For a given value of ℓ, the magnetic quantum number m_ℓ can only assume the values m_ℓ = −ℓ, −ℓ + 1...0...ℓ − 1, ℓ.

Quantum Numbers for Electrons in Atoms				
Name	**Symbol**	**Values**	**Specifics**	**Indicates**
principal	n	1, 2, . . .	shell	size
orbital angular momentum*	l	$0, 1, \ldots, n-1$	subshell: $l = 0, 1, 2, 3, 4, \ldots$ s, p, d, f, g, . . .	shape
magnetic	m_l	$l, l-1, \ldots, -l$	orbitals of subshell	orientation
spin magnetic	m_s	$+\frac{1}{2}, -\frac{1}{2}$	spin state	spin direction
*Also called the azimuthal quantum number.				

FIGURE 8.39 We can summarize the quantum numbers n, ℓ, and m_l with respect to name, symbol, allowed range of values, shell/subshell/orbital designation, and the geometric or physical characterization indicated for each quantum number.

In summary, the principal quantum number designates the shell size; the angular momentum quantum number, ℓ, designates the subshell shape; and the magnetic quantum number, m_l, designates the individual orbital's orientation.

What is most important for understanding the behavior of the atomic orbitals that contain the electrons in a given atomic structure, and for gaining an understanding of how atoms form bonds with other atoms to create molecules, is the *spatial geometry* of the individual orbitals corresponding to a specific set of quantum number n, ℓ, and m_l. While each of the orbitals has a corresponding mathematical representation for the radial part, $R_{n,\,l}(r)$ and the angular part, $Y_{l,m}(\theta,\phi)$, it is the *geometry* of each of the corresponding orbitals that is most important to understand at this point. Thus we investigate the geometry of the orbitals corresponding to a given set of quantum numbers n, ℓ, and m_l. Becoming familiar with the relationship between the quantum number designation and the geometry of the corresponding radial and angular orbitals provides the critical insight into the relationship between the structure of those orbitals and their interaction with photons and with bonding structures in molecules.

Check Yourself 11

List all possible values for the unspecified quantum number or numbers for the cases: $n = 2$; $n = 3$; $n = 4$, $\ell = 2$.

Geometry and Spatial Characteristics of the Three-Dimensional Waves of the Hydrogen Atom

We build up our knowledge of the geometry of the atomic orbitals by beginning with the lowest energy atomic orbital, for which the corresponding quantum numbers are n = 1, ℓ = 0, and $m_l = 0$. We can define the geometry of this lowest energy orbital by systematically presenting the dependence of the radial and angular parts of the wavefunction $\psi_{n,\ell,m_l}(r,\theta,\phi) = R_{n,\ell}(r)\psi_{\ell,m_l}(\theta,\phi)$ in three dimensions. For the case n = 1, ℓ = 0, and $m_l = 0$, the wavefunction is thus $\psi_{n,\ell,m_l}(r,\theta,\phi) = \psi_{100}(r,\theta,\phi) = R_{1,0}(r)Y_{0,0}(\theta,\phi)$. We can represent ψ_{100} as a function of r, the distance from the nucleus as shown in Figure 8.40a. If this function ψ_{100} is rotated about the vertical axis it forms a cone as shown in Figure 8.40b, and the projection of this cone when viewed from the top is displayed in Figure 8.40c. The three-dimensional representation is displayed in Figure 8.40d as a cloud of points with the density of points representing the amplitude of the wavefunction ψ_{100}. This cloud of points

Solution:
Using the preceding ℓ-rule, for $n = 2$, ℓ can be 0 or 1; using the m_ℓ-rule for each ℓ, for ℓ = 0, m_ℓ can only be 0, while for ℓ = 1, m_ℓ can be –1, 0, or 1. Writing the possible combinations as a string of three digits, $n\ell m_\ell$ = 200, 211, 210, and 21–1. For $n = 3$, ℓ may be 0, 1, or 2, and we have $n\ell m_\ell$ = 300, 311, 310, 31–1, 322, 321, 320, 32–1, and 32–2. When n and ℓ are specified, then only m_ℓ may vary, so for $n = 4$, ℓ = 2 we have $n\ell m_\ell$ = 422, 421, 420, 42–1, and 42–2. Note that for $n = 2$, four combinations are allowed, while for $n = 3$, nine are allowed. In general there are n^2 combinations for a given n. For a given n and ℓ, m_ℓ can take on $2\ell + 1$ possible values, and the number of combinations is therefore $2\ell + 1$. Right now this seems like just "playing with numbers," but this numerology will prove to be pivotal in understanding the structure of atoms and the periodic table.

is represented in Figure 8.40e by the boundary to ψ_{100} within which 90% of the density of points resides.

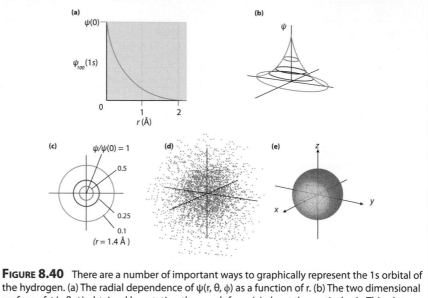

FIGURE 8.40 There are a number of important ways to graphically represent the 1s orbital of the hydrogen. (a) The radial dependence of $\psi(r, \theta, \phi)$ as a function of r. (b) The two dimensional surface of $\psi(r, \theta, \phi)$ obtained by rotating the graph from (a) about the vertical axis. This plot represents the orbital amplitude in the x-y plane. (c) The contour map of the 2-D surface in (b) as viewed from the top. (d) The three-dimensional cloud with density of points proportional to the amplitude of $\psi(r, \theta, \phi)$ at each position in space about the nucleus at the origin of the coordinate system. (e) The 3-D boundary within which 90% of the cloud in (d) resides.

As we will see when we begin to use these orbitals to understand how atoms interact with photons and how these atomic orbitals combine to form molecular bonds, the spatial geometry of each orbital that corresponds to a unique set of quantum numbers n, ℓ, m_ℓ, is of critical importance. At the same time, it is important to realize that the solution to the Schrödinger equation provides the explicit mathematical function $\psi_{100}(r, \theta, \phi)$ describing the radial and angular dependence of the orbital for each set of quantum numbers. In particular, for the orbital in Figure 8.40 for which n = 1, ℓ = 0, and m_ℓ = 0, the solution to the Schrödinger equation is

$$\psi_{100}(r, \theta,\phi) = R_{1,0}(r)Y_{0,0}(\theta,\phi) = \left[\frac{2e^{-r/a_0}}{a_0^{2/3}}\right]\left[\frac{1}{2\pi^{1/2}}\right]$$

where the first bracket is

$$R_{1,0}(r) = \frac{2e^{-r/a_0}}{a_0^{2/3}} \quad \text{with} \quad a_0 = \frac{4\pi\varepsilon_0\hbar^2}{m_e e^2}$$

which is the radius of the lowest energy orbit of the Bohr hydrogen atom. The second bracket is the solution

$$Y_{0,0}(\theta,\phi) = \frac{1}{2\pi^{1/2}}$$

that for the specific case of ℓ = 0 and m_ℓ = 0 is a constant, independent of θ and ϕ. Since ψ_{100} is independent of θ and ϕ, it is *spherically symmetric*. This spherical symmetry reflects the fact that at any particular distance r from the nucleus, $\psi(r, \theta, \phi)$ does not depend on θ or ϕ. In the (history based) jargon of orbital designations, this ψ_{100} orbital is termed the 1s orbital corresponding to n = 1 and the s representing ℓ = 0 as indicated in Figure 8.39.

Before moving on to sequentially higher values of n, with the corresponding options for ℓ and m_ℓ, it is important to note that the 1s orbital has, in addition to spherical symmetry, the characteristic that (1) it is a decaying exponential with respect to the distance r from the nucleus, and (2) the wavefunction ψ_{100} has no *nodes* (where a node represents the passage of the wavefunction ψ through zero). Recall that the same was true of the lowest energy level of the one-dimensional particle-in-a-box, Figure 8.31.

As we progress to increasing values of n, and therefore to a widening manifold of corresponding ℓ and m_ℓ values, the allowed array of orbital geometries, with each orbital corresponding to a specific set of n, ℓ, and m_ℓ values, expands accordingly. When we consider the allowed geometries for n = 2, we have allowed values of $\ell = 0$ and $\ell = 1$. For $\ell = 0$ we have only one allowed value of m_ℓ, that of $m_\ell = 0$. This corresponds to the wavefunction $\psi_{200}(r,\theta,\phi) = R_{2,0}(r)Y_{0,0}(\theta,\phi)$ shown in Figure 8.41 on the top row. This graphical representation demonstrates that $R_{2,0}(r)$ has a node where $R_{2,0}(r)$ equals zero at a distance $r = 2a_0$ from the nucleus. The angular dependence, $Y_{0,0}(\theta,\phi)$ is spherically symmetric and thus $Y_{0,0}$ is a constant, independent of angle around the nucleus.

With n = 2 and $\ell = 1$, we open the options of $m_\ell = -1, 0,$ and $+1$, which correspond to three orbitals that share the same *shape* (determined by $\ell = 1$) but have three different *orientations* (determined by $m_\ell = 0, \pm 1$). The three resulting orbitals are displayed in Figure 8.41 in the second, third, and fourth rows.

Check Yourself 12

Calculating the Probability of Finding an Electron at a Certain Location:
Suppose the electron is in a 1s orbital of a hydrogen atom. What is the probability of finding the electron in a small region a distance a_0 from the nucleus relative to the probability of finding it in the same small region located right at the nucleus?

Anticipate: We should expect a lower probability because the wavefunction decays exponentially with distance from the nucleus.

Approach: The probability density is independent of angle when $\ell = 0$. We need to compare the probability densities at the two locations. To do that, we take the ratio of the squares of the wavefunction at the two locations.

Solve:
The *ratio* of the probability that the electron is found at the nucleus vs. at $r = a_0$ is:

$$\frac{\text{Probability density at } r = a_0}{\text{Probability density at } r = 0} = \frac{\psi^2(a_0)}{\psi^2(0)}$$

From $\psi^2(r,\theta,\phi) = (1/\pi a_0^3)e^{-2r/a_0}$

$$\frac{\psi^2(a_0)}{\psi^2(0)} = \frac{(1/\pi a_0^3)\overbrace{e^{-2a_0/a_0}}^{e^{-2}}}{(1/\pi a_0^3)\underbrace{e^0}_{1}} = e^{-2} = 0.14$$

Analyze: As expected, the probability of finding the electron in a small region at a distance a_0 from the nucleus is lower than at the nucleus itself: the probability is only 14% of that of finding the electron in a region of the same volume located at the nucleus.

Practice Example A: Calculate the same ratio but for the more distant point at $r = 2a_0$, twice as far from the nucleus.
[**Answer:** 0.018]

Practice Example B: Calculate the same ratio but for a point at $3a_0$ from the nucleus.

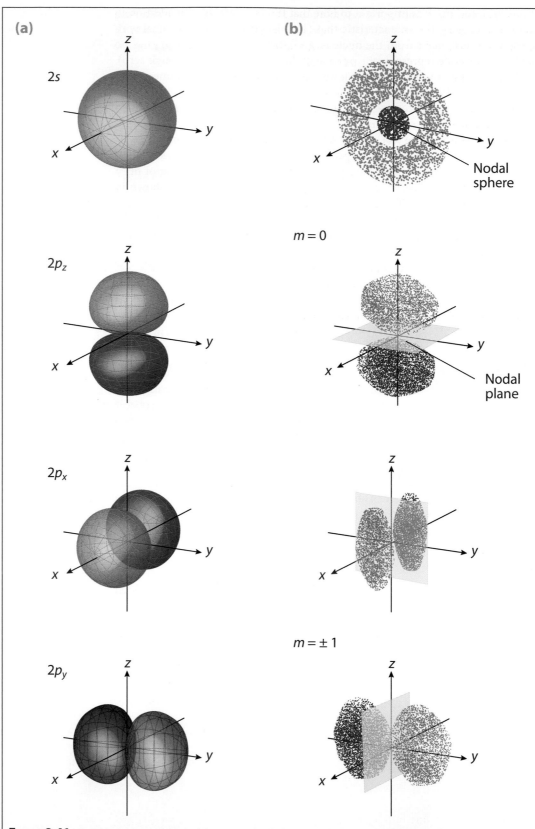

(a) 2s

2p$_z$

2p$_x$

2p$_y$

(b) Nodal sphere

$m = 0$

Nodal plane

$m = \pm 1$

FIGURE 8.41 Graphical representation of the n = 2 orbitals for atomic hydrogen. (a) The three-dimensional boundary within which resides 90% of the wave amplitude; the color of the lobe represents the phase of the wave. (b) Two-dimensional cuts through the 3-D clouds of density proportional to orbital amplitude. Of particular importance is that the phase of the wave is indicated by the color of the spots that represent the amplitude of the wavefunction ψ(r, θ, φ). Also indicated are the nodal planes that separate the positive and negative phases of the wave.

8.50

Just as was the case for the wavefunction ψ_{100}, for each specific set of quantum numbers n, ℓ, m_ℓ, there is a unique atomic orbital and a unique wavefunction. Therefore for each n, ℓ, m_ℓ, there is a specific mathematical function defining the radial and angular dependence of that orbital.

While our intuition for how electrons are distributed in space around the atom's nucleus is built primarily on the *shape* (defined by ℓ) and orientation (defined by m_ℓ), we can write down the mathematical function that corresponds to each of the orbitals shown in Figure 8.41.

For the quantum numbers n = 2, ℓ = 0, and m_ℓ = 0 the wavefunction is

$$\psi_{210}\left(r, \theta, \phi\right) = \underbrace{\left[\frac{1}{2\sqrt{6}}\left(\frac{1}{a_0}\right)^{3/2}\left(\frac{r}{a_0}\right)e^{-r/2a_0}\right]}_{R_{21}(r)}\underbrace{\left[\left(\frac{3}{4\pi}\right)^{1/2}\cos\theta\right]}_{Y_{10}(\theta,\phi)}$$

This mathematical function, the wavefunction, corresponds to the p_z *orbital* in the second row of Figure 8.41. It has lobes in θ and ϕ oriented along the z-axis. However, note that the wavefunction is independent of ϕ, the azimuthal angle, and is, as a consequence, *cylindrically symmetric*. This wavefunction is designated the $2p_z$ orbital.

This leaves us with two more wavefunctions, namely the two orbitals corresponding to n = 2, ℓ = 1, and m_ℓ = ±1. The geometry of these two wavefunctions are displayed in the next-to-last and last row of Figure 8.41. The $2p_x$ orbital has lobes aligned along the x axis, and the $2p_y$ orbital has lobes aligned along the y axis. The corresponding wavefunctions are, for p_x

$$\psi_{2,1,+1}\left(r, \theta, \phi\right) = \underbrace{\left[\frac{1}{2\sqrt{6}}\frac{1}{a_0^{5/2}}re^{-r/2a_0}\right]}_{R(r)}\underbrace{\left[\left(\frac{3}{4\pi}\right)^{1/2}\sin\theta\cos\phi\right]}_{Y(\theta,\phi)}$$

$$= \frac{1}{4}\left(\frac{1}{2\pi a_0^5}\right)^{1/2}re^{-r/2a_0}\sin\theta\cos\phi$$

and for p_y

$$\psi_{2,1,-1}\left(r, \theta, \phi\right) = \frac{1}{4}\left(\frac{1}{2\pi a_0^5}\right)^{1/2}re^{-r/2a_0}\sin\theta\sin\phi$$

Note that, as the graphical representation in Figure 8.41 suggests, p_x and p_y differ only in their orientation expressed in the angular factor.

While it may seem peculiar to revert to Cartesian coordinates for both the graphical visualization of the various orbitals and the orbital designations (p_x, p_y, p_z), when the *geometry* of the orbitals is visualized, this turns out to be the most useful form, and it is virtually universal in both chemistry and physics. The reason the p_x orbital is so designated comes from the fact that x = r sin θ cos ϕ in the conversion between Cartesian and polar coordinates and the angular dependence of $Y_{\ell,m_\ell}(\theta,\phi)$ for the p_x orbital has the same functional form. The same analogy holds for the p_y and p_z orbitals.

Check Yourself 13

Sketch radial wavefunctions $R_{nl}(r)$ on a single set of labeled axes for 1s, 2s, and 3s orbitals, explicitly calculating and indicating the values of r (in units of a_0 and in angstroms) at the nodal positions. How are these nodes manifested in the 3-D orbitals?

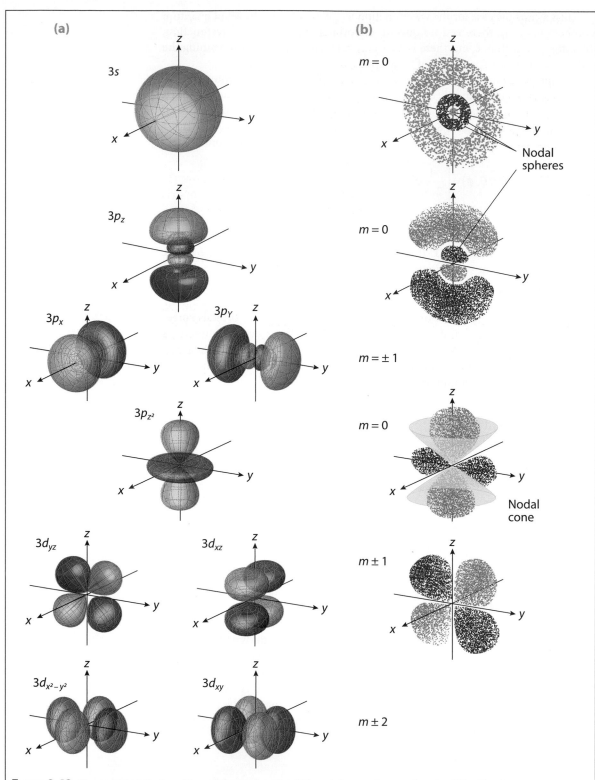

FIGURE 8.42   The graphical display of the radial wavefunction, R(r), and the angular wavefunction Y(θ, φ) for the principal quantum number n = 3. The figures follow from the designation in Figure 8.41. (a) Displays the surfaces that include 90% of the wave amplitude for the 3s, 3p, and 3d orbitals. The phase of the wave is designated by the color. (b) Displays the two-dimensional cuts through the three-dimensional amplitude clouds.

When we consider the orbitals that arise for n = 3, we again refer back to Figure 8.38. With n = 3, we have three subshells: ℓ = 0 or the s subshell, ℓ = 1 or the p subshell, and ℓ = 2 or the d subshell. For each of these values of ℓ, there are the corresponding values of m_ℓ. Specifically for $\ell = 0$, there is only one option, $m_\ell = 0$. For $\ell = 1$, there are 3 possibilities: $m_\ell = 0, \pm 1$ corresponding to the $3p_x$, $3p_y$, and $3p_z$ orbitals. For $\ell = 2$, there are 5 possibilities: $m_\ell = 0, \pm 1, \pm 2$ corresponding to the $3d_z^2$, $3d_{xz}$, $3d_{yz}$, $3d_{xy}$, and $3d_{x^2 - y^2}$. All of these are displayed in Figure 8.42.

While we will not write down the mathematical functions for the $n = 3$ wavefunctions (they become increasingly complex) it is very important to identify the emerging *patterns* in the geometry of those s, p, and d orbitals. Consider first the $n = 3$ orbitals with respect to the $n = 1$ and $n = 2$ cases. The 3s orbital has two radial nodes, compared to zero radial nodes for the 1s orbital and one radial node for the 2s orbital. This is a pattern: an ns orbital has $n - 1$ radial nodes. As ℓ increases from 0 to $n - 1$, the number of nodes $(n - 1)$ remains the same, but in the progression from $\ell = 0$ to $\ell = n - 1$, ℓ of the radial nodes are exchanged for angular nodes. Thus, for example, the 3p orbitals have 2 total nodes with one radial node and one angular node. The 3d orbitals have 2 total nodes with zero radial nodes and 2 angular nodes.

A second and equally important pattern emerges when we compare the *spatial extent* of the orbitals. This is done in Figure 8.43, which displays, with the same spatial scale, the relative sizes of the 1s; 2s and 2p; and 3s, 3p, and 3d orbitals. The first conclusion from inspecting this figure is that the principal quantum number, n, controls, to first order, the *size* of the orbitals. This reflects what the simple Bohr orbits represented: that the orbital radii were proportional to n^2. Second, as noted earlier, ℓ controls the *angular shape* of the orbitals and m_ℓ controls the *orientation* of the orbitals.

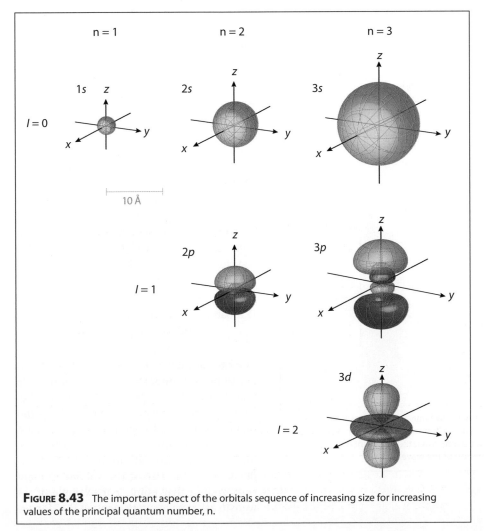

FIGURE 8.43 The important aspect of the orbitals sequence of increasing size for increasing values of the principal quantum number, n.

Check Yourself 14

Make a plot on a common set of labeled axes of the (unnormalized) angular functions $Y_{10}(\theta) = \cos\theta$ and $Y_{10}(\theta) = \frac{3}{2}\cos^2\theta - \frac{1}{2}$ versus θ for $0 < \theta < \pi$. Explicitly find the nodes in these functions, and indicate them on your plot. How are these nodes manifested in the corresponding 3-D orbitals $2p_z$ and $3d_{z^2}$?

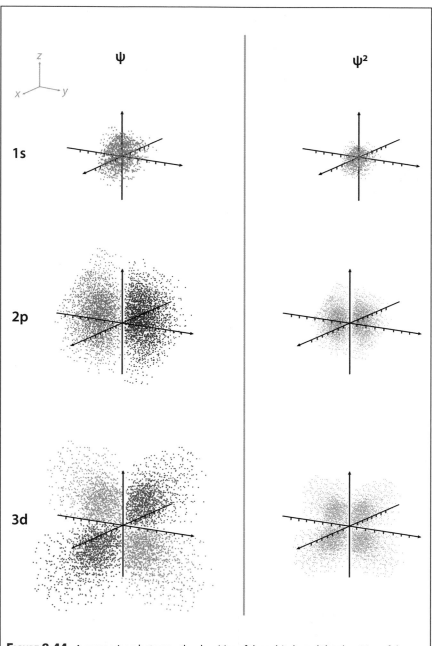

FIGURE 8.44 A comparison between the densities of the orbitals and the densities of the *square* of the wavefunctions that determine the probability of finding the electron in a region of space around the nucleus. Notice a very important point: When the wavefunction, $\psi(r, \theta, \phi)$ is squared to determine $\psi^2(r, \theta, \phi)$, the phase information of the wave is lost.

Physical Interpretation of the Schrödinger Wavefunction $\psi_{n,\ell,m_\ell}(r,\theta,\phi)$

Schrödinger was a mathematical physicist. The remarkable advance in our quantitative understanding of atomic structure that followed the introduction of the Schrödinger equation led to his concentration on extending that equation to more complicated, multi-electron systems rather than the development of a deeper physical interpretation of the wavefunction $\psi_{n,\ell,m_\ell}(r,\theta,\phi)$. Niels Bohr, however, was intent on developing a more physical picture of what ψ_{n,ℓ,m_ℓ} represented and how that wavefunction related to experimentally observable quantities. Bohr encouraged Max Born to pursue this question of what the wavefunction represented physically and it was indeed Born who developed the insight that related the mathematics of Schrödinger to the physical reality of the wavefunction. Born used the wave theory of light, in which the intensity of light is proportional to the square of the wave amplitude, to build his analogy to the wavefunction of the electron. Born put forward the hypothesis that the probability of finding an electron at a given point in space is equal to the square of the wavefunction as we noted in our discussion of the particle-in-a-box. Thus

[Probability of finding the electron at r,θ,ϕ] = $P(r,\theta,\phi) = \psi^2_{n,\ell,m_\ell}(r,\theta,\phi)$

The careful interpretation of $P(r, \theta, \phi)$ turns out to be very important because of the role that electron density plays in the chemical behavior of the atoms of different elements and in the structure of chemical bonding. First it must be recognized that the probability of finding the electron at any point r, θ, ϕ depends upon the *size* of the volume element surrounding that point. For example, if we were to ask how many students at the University were exactly 6 feet tall, it would be immediately apparent that there is no correct answer to the question because the more precise the requirement for "exact" becomes, the smaller would be the number given as an answer. If we were to provide a range of, say 6.0 ± 0.1 feet, then we would have a sensible question to answer.

To emphasize the point, $P(r,\theta,\phi) = \psi^2_{n,\ell,m_\ell}(r,\theta,\phi)$ is called the *probability density* of the electron in space. The probability of finding the electron in a volume element, dv, is then equal to the *product* of the probability density times the size of the volume element.

Thus

[Probability of finding the electron in volume element dv] =

$$P(r,\theta,\phi)dv = \psi^2_{n,\ell,m_\ell}(r,\theta,\phi)dv$$

In Cartesian coordinates $dv = dxdydz$. In polar coordinates $dv = r^2 \sin^2\theta dr d\theta d\phi$.

We can represent $\psi^2(r, \theta, \phi)$ graphically for a single electron in a given orbital (that is for a given value of n, ℓ, m_ℓ) as shown in Figure 8.44. That figure represents both $\psi(r, \theta, \phi)$ and $\psi^2(r, \theta, \phi)$ as probability clouds where the density of points represents the magnitude of ψ on the left-hand column and ψ^2 on the right-hand column. Three cases are shown, the 1s, 2p, and 3d orbitals. Note that while the nodes in $\psi(r, \theta, \phi)$ are preserved in $\psi^2(r, \theta, \phi)$, the *phase* of the wavefunction is lost when the probability density $\psi^2(r, \theta, \phi)$ is calculated. This will turn out to be important when we add atomic orbitals together.

When we investigate the chemical behavior of various multielectron atoms in the next chapter, it will turn out that the probability that an electron (in a given orbital) is found a certain distance from the nucleus is very important. This probability is called the *radial distribution function* and it is calculated by multiplying the probability density (the square of the wavefunction, $\psi^2(r, \theta, \phi)$) by the volume of the spherical shell, $\psi\pi r^2 dr$, and then averaging over θ and ϕ. This is displayed graphically in Figure 8.45.

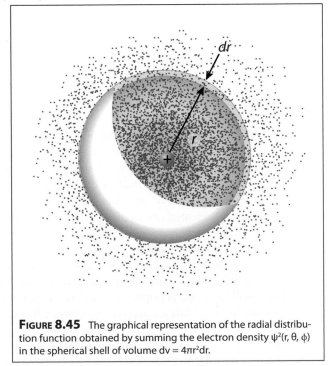

FIGURE 8.45 The graphical representation of the radial distribution function obtained by summing the electron density $\psi^2(r, \theta, \phi)$ in the spherical shell of volume $dv = 4\pi r^2 dr$.

The averaging over θ and ϕ cancels the 4π from the volume element $4\pi r^2 dr$ and the radial distribution function is then given by

$$Prob(r) = r^2 R^2_{n,\ell}(r)$$

Examples of the radial distribution function for n = 1, n = 2, n = 3, and n = 4 are displayed in Figure 8.46. There are several important things to recognize in this figure.

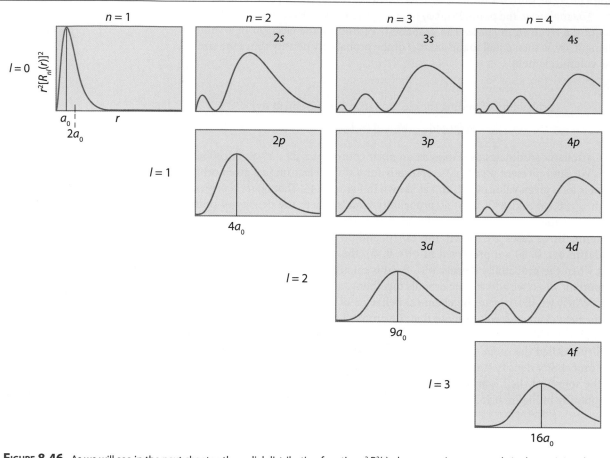

FIGURE 8.46 As we will see in the next chapter, the radial distribution function $r^2 R^2(r)$ plays a very important role in determining the chemical behavior of a given atom—a given element in the Periodic Table. Shown here are the radial distribution functions for the four lowest energy levels of the hydrogen atom. Notice that the distance scales for n = 3 and 4 have been compressed. Each point where $r^2 R^2(r)$ touches the r-axis corresponds to a node. The n = 3 and 4 curves decay smoothly to zero at large r.

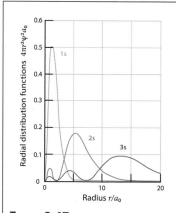

FIGURE 8.47 When the radial distribution functions $r^2 R^2(r)$ are superimposed on the same radial scale, measure in units of r divided by the Bohr radius, a0, it becomes quantitatively clear how the penetration of the 1s orbital dominates the 2s and how the 2s dominates the 3s.

First, note that $r^2 R^2_{1,0}(r)$ peaks at $r = a_0$, $r^2 R^2_{2,0}(r)^2$ peaks at $r = 4a_0$, and $r^2 R^2_{3,0}(r)^2$ peaks at $r = 9a_0$. This was true for the Bohr model of circular orbits for the electron trajectory in the hydrogen atom, specifically that $r_n = n^2 a_0$, where a_0 is the Bohr radius. However, in the Schrödinger model of the atom, rather that a specific orbital trajectory for the electron as a particle, the distribution of the electron is smeared out over a considerable range in radius as a result of the wave nature of the electron. Second, because the magnitude of the radial distribution function defines the amount of time the electron spends at a given radius, the electron spends most of its time in the *outer lobe* of the distribution. Third, it is the s orbital that penetrates most closely into the nucleus. This, it will turn out, is important for determining the effective (positive) charge of the nucleus that an electron "sees" when it is one of many electrons in a multielectron atom. Finally, notice how the most probable radius increases rapidly (by n^2) with increasing n. This is clearly revealed when we superimpose the radial distribution function for the 1s, 2s, and 3s electrons as shown in Figure 8.47.

Summary Concepts

1. Origin of the Concept of the Photon: Einstein's Union of Planck's ε = nhν and the Photoelectric Effect

With the hypothesis by Planck (based on the wavelength dependence of blackbody radiation) that electromagnetic radiation is quantized *and* the studies of the photoelectric effect demonstrating that electrons are liberated from metal surfaces only by radiation above a threshold frequency, Einstein postulated that light is composed of *photons*. Photons are individual bundles or particles of radiation with energy ε = hν where ν is the frequency of the radiation.

2. Momentum of the Photon

If the photon is a particle of light, and has energy ε = hν, does this imply that the photon has momentum?

The fact that the photon has momentum was demonstrated conclusively by Arthur Compton using high-energy photons, x-rays, impinging on graphite. The observed wavelength shift of the scattered photons represented the momentum transfer from the photons to the electrons in the graphite.

The momentum of the photon is given by $p_{phot} = h/\lambda_{phot}$.

3. Spectroscopy and the Study of Light Emission from Atoms

Light emitted from atoms in both high temperature flames and high voltage electrical discharges provided a powerful way of studying the structure of atoms. In particular, the emission spectra from atomic hydrogen revealed three distinct series of lines in the ultraviolet, visible, and infrared spectral ranges. Analysis of these hydrogen emission lines by Balmer, Paschen, and Lyman lead to a systematic formula for calculating the emitted wavelength, λ, in terms of a single constant and integers, n:

$$\frac{1}{\lambda} = R_H \left(\frac{1}{n_f^2} - \frac{1}{n_i^2} \right)$$

where R_H is the Rydberg constant for hydrogen, n_f is the integer defining the final state and n_i is the integer defining the initial state.

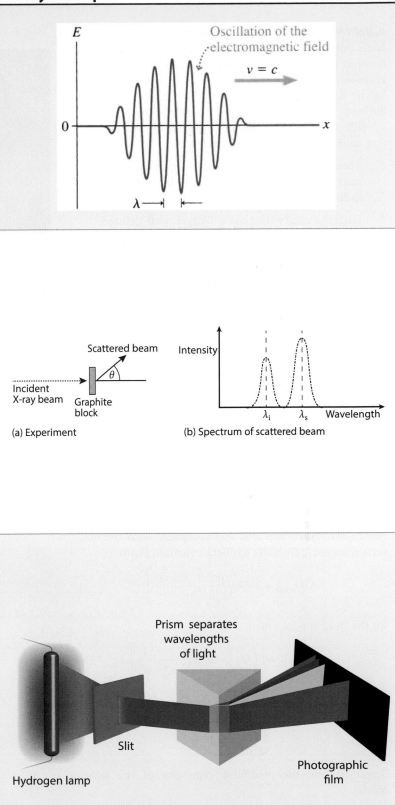

(a) Experiment

(b) Spectrum of scattered beam

Summary Concepts

4. Bohr Model of the Hydrogen Atom

Niels Bohr combined the Rutherford model of the nuclear atom with information from the pattern in the sharp emission lines of atomic hydrogen which were defined by the simple Rydberg expression for the wavelength of emitted radiation. Bohr's model postulated that:

1. Electrons were confined in the atom by *circular* orbits around the central nucleus.

2. These orbits correspond to only certain allowable *stationary states* within which the electron will not radiate.

3. Each stationary state has a discrete, well-defined energy, E_n.

4. When the atom transitions from one stationary state to another, it either absorbs or emits a photon of energy $\Delta E = E_f - E_i = h\nu$ where ν is the frequency of the emitted or absorbed photon.

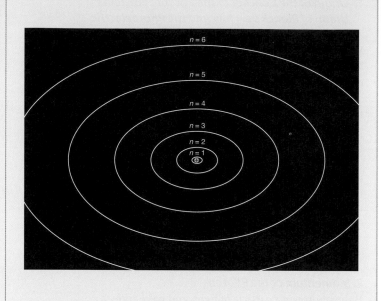

5. Relationship Between the Bohr Model of the Hydrogen Atom and the Rydberg Expression

The Bohr model of the hydrogen atom with the electron moving in well-defined circular orbits about the central nucleus (as planets move about the Sun) gave explicit structure to the atom. Moreover, the Bohr model provided the equation for the total energy of the electron and the equation for the balance of forces on the electron in a circular orbit. This allowed the energy difference between two orbits to be calculated and compared with the observed wavelength of the Rydberg equation. From

$$\left(\Delta E\right)_{\text{BOHR}} = h\nu = \frac{me^4}{8\varepsilon_0^2 h^2}\left[\frac{1}{n_f^2} - \frac{1}{n_i^2}\right]$$

for the Bohr model, and

$$\left(\Delta E\right)_{\text{RYDBERG}} = h\left(\frac{c}{\lambda}\right) = hc\,R_H\left[\frac{1}{n_f^2} - \frac{1}{n_i^2}\right]$$

for the Rydberg equation based on observations of the emission spectrum of atomic hydrogen, it turned out that

$$\frac{me^4}{8\varepsilon_0^2 h^2} = hcR_H$$

demonstrating the predictive capability of the Bohr model.

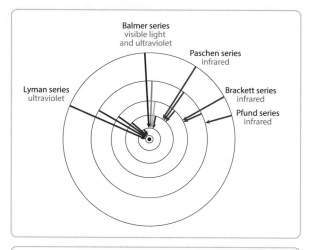

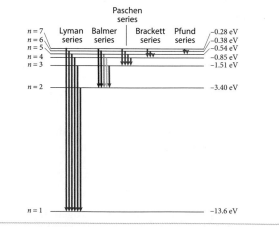

Summary Concepts

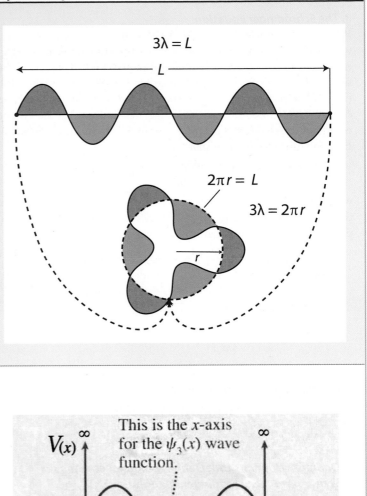

6. De Broglie and the Wave Nature of the Electron

Louis de Broglie, in his PhD thesis, postulated that if light could exhibit particle properties, where the photon is a particle of light with momentum $p_{phot} = h/\lambda_{phot}$, then the electron could possess wave properties where $\lambda_{el} = h/p_{el}$.

7. Particle-in-a-Box

The wave nature of the electron confined within a potential well is described by the combination of (1) the wave equation and (2) the boundary conditions, namely the walls of infinite potential at $x = 0$ and $x = L$ with the potential energy $V(x) = 0$ between the walls of the potential. The boundary conditions that the wave amplitude must be zero at $x = 0$ and $x = L$ and thus that an integral number of half wavelengths, $n \lambda/2$, fit between the potential walls requires that $kx = n\pi x/L$. The requirement that the electron exists within the square well potential means that the probability of finding the electron at position x, $\psi^2(x)$, integrated over the available range of x be equal to 1:

$$\int \psi^2(x)dx = 1 .$$

This establishes the normalization constant, $N = \sqrt{2/L}$. Thus the normalized wavefunction for an electron in the infinite potential well is

$$\psi(x) = \sqrt{2/L} \sin\left(\frac{n\pi x}{L}\right)$$

When this *wavefunction* is substituted into the wave equation, the quantized energy levels are given by

$$E_n = \frac{n^2 h^2}{8mL^2}$$

Summary Concepts

8. The Schrödinger Equation

With the hypothesis by de Broglie that the electron possessed a wavelength $\lambda_{el} = h/p_{el}$, Schrödinger addressed the problem of determining the *wave equation* that would mathematically describe the wave behavior of the negatively charged electron in the Coulomb field of the positively charged nucleus described by the potential energy $V(r) = -Ze^2/4\pi\varepsilon_0 r$. Schrödinger recognized that the kinetic energy of the electron is proportional to the *curvature* of the wavefunction, $d^2\psi(x)/dx^2$, such that

$$-\hbar^2\!\!\Big/\!2m \; \frac{d^2\psi(x)}{dx^2} = E_k\psi(x)$$

where E_k is the kinetic energy of the electron. With the inclusion of the potential energy term, $V(n) = -Ze^2/4\pi\varepsilon_0 r$, and the conversion to polar coordinates, the equation becomes

$$-\hbar^2\!\!\Big/\!2m \; \frac{d^2\psi(x)}{dx^2} + V(x)\psi(x) = E\psi(x)$$

$$-\hbar^2\!\!\Big/\!2m\left[\frac{1}{r^2}\frac{\partial}{\partial r}\left(r^2\frac{\partial}{\partial r}\right) + \frac{1}{r^2\sin^2\theta}\frac{\partial^2}{\partial\phi^2}\left(\sin\theta\frac{\partial}{\partial\theta}\right) + \frac{1}{r^2\sin^2\theta}\frac{\partial^2}{\partial\phi^2}\right]\psi(r,\theta,\phi)$$

$$+ \frac{-Ze^2}{4\pi\varepsilon_0 r}\psi(r,\theta,\phi) = E\psi(r,\theta,\phi)$$

This Schrödinger equation can be written in a more convenient form in terms of the Hamiltonian $\hat{H}$ as

$$\hat{H}\psi(r,\theta,\phi) = E\psi(r,\theta,\phi)$$

where the quantized energies are given by E.

9. Quantum Numbers that Define the Radial and Angular Solutions to the Schrödinger Equation

When the Schrödinger equation is submitted to the boundary conditions imposed by the Coulomb field of the proton in the nucleus of the hydrogen atom, three quantum numbers result from the confinement of the wave properties of the electron in three dimensions. Those quantum numbers constitute a nested array of integers:

1. The principal quantum number n designates the "shell" occupied by the electron. The principal quantum number is the primary determinant of the energy of the electron with the quantized energy levels for the hydrogen atom given by

$$E_n = \frac{-m_e e^4}{8\varepsilon_0^2 h^2}\left(\frac{1}{n^2}\right)$$

2. The *angular momentum* or *azimuthal* quantum number ℓ designates the subshell occupied by the electron that in turn designates the shape of the atomic orbital. The azimuthal quantum number ℓ can assume the values $\ell = 0, 1, 2\ldots n-1$. Thus the values of ℓ are constrained by the principal quantum number n.

3. The quantum number m_ℓ, termed magnetic quantum number designates the shape of the individual orbital occupied by an electron. Allowed values of m_ℓ are, for a given value of ℓ: $-\ell$, $-\ell+1$, $-\ell+2$, $\ldots 0, 1, 2, +\ell$.

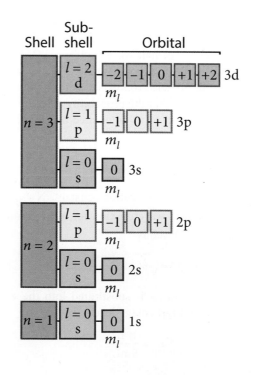

Summary Concepts

10. Radial and Angular Solution to the Schrödinger Equation

The wavefunction, $\Psi_{n,\ell,m_\ell}(r,\theta,\phi)$, that is a solution to the Schrödinger equation can be factored into the *product* of two wavefunctions. The radial solution, $R_{n,\ell}(r)$, is a function only of the radial distance of the electron from the nucleus. The angular solution to the Schrödinger equation, $Y_{\ell,m_\ell}(\theta,\phi)$ is a function only of the polar angle θ and the azimuthal angle ϕ. The total wavefunction, $\Psi_{n,\ell,m_\ell}(r,\theta,\phi)$, is then given by the product

$$\Psi_{n,\ell,m_\ell}(r,\theta,\phi) = R_{n,\ell}(r)Y_{\ell,m_\ell}(\theta,\phi)$$

With the specification of the quantum numbers n, ℓ, and m_ℓ, the size (determined by n), the shape (determined by ℓ), and the orientation (determined by m_ℓ) are established. An example for the n = 3, ℓ = 0 case is shown to the right.

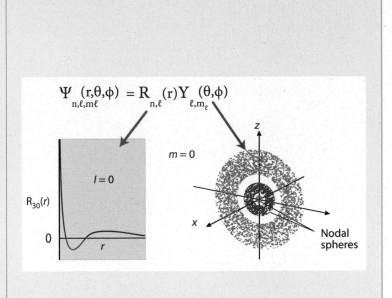

$$\Psi_{n,\ell,m\ell}(r,\theta,\phi) = R_{n,\ell}(r)Y_{\ell,m_\ell}(\theta,\phi)$$

11. Physical Interpretation of the Atomic Orbitals and the Radial Distribution Function

While the wavefunction, $\Psi_{n,\ell,m_\ell}(r,\theta,\phi)$, as a solution to the Schrödinger equation provides the mathematical foundation for the wave behavior of the electron confined by the Coulomb potential of the proton, the physical interpretation of the wavefunction was developed by Max Born. Born linked the wavefunction of the electron to the *probability* of finding the electron at position r, θ, ϕ through the square of the wavefunction, $\psi^2(r, \theta, \phi)$. In particular, the probability density, P(r, θ, ϕ), at position r, θ, ϕ is given by the square of the wavefunction

$$P(r,\theta,\phi) = \psi^2_{n,\ell,m_\ell}(r,\theta,\phi)$$

and the probability the electron exists in a volume element $dv = dxdydz = r^2 \sin^2\theta dr d\theta d\phi$ is

$$P(r,\theta,\phi)dv = \psi^2_{n,\ell,m_\ell}(r,\theta,\phi)r^2\sin^2\theta dr d\theta d\phi$$

The total radial probability is obtained by integrating (adding up) the probability in each shell of thickness dr. When this is plotted as a function of r, it defines the total radial probability, $P(r) = r^2 R^{(r)}_{n,\ell}$.

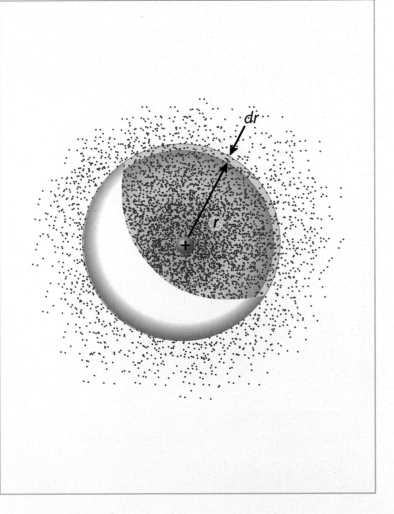

CASE STUDY 8.1   Quantum Mechanical Tunneling

Because atomic behavior is so unlike ordinary experience, it is very difficult to get used to, and it appears peculiar and mysterious to everyone—both to the novice and to the experienced physicist. Even the experts do not understand it the way they would like to, and it is perfectly reasonable that they should not, because all of direct, human experience and of human intuition applies to large objects. We know how large objects will act, but things on a small scale just do not act that way, so we have to learn about them in a sort of abstract or imaginative fashion and not by connection with our experience.

—Richard Feynman, *The Feynman Lectures on Physics*

Perhaps nowhere is this both (1) a more appropriate admonition yet (2) of immense practical importance to modern science and technology than in the case of "tunneling" in quantum mechanical systems. One of the most startling consequences of the de Broglie wave hypothesis for the electron when incorporated with the mathematics of the Schrödinger equation, was the realization that electrons can tunnel through a potential energy barrier that is forbidden to classical motion of macroscopic objects. Some graphical examples of what we will explore in this Case Study are displayed in Figure CS8.1a.

We received our first glimpse of this in the somewhat

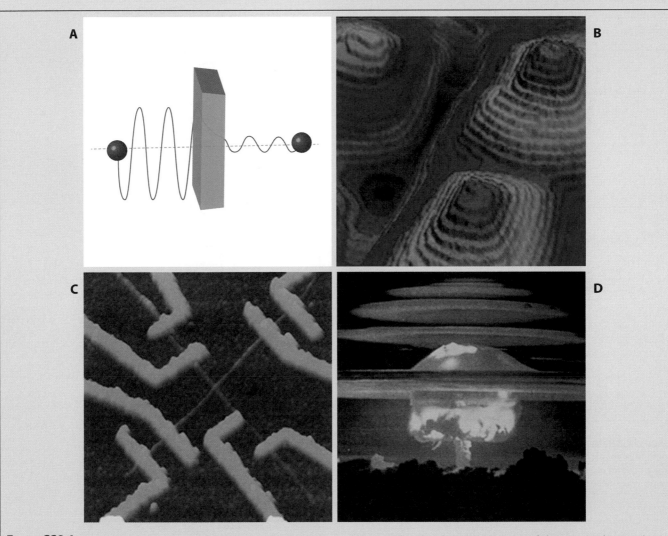

FIGURE CS8.1A   A vast array of new scientific and technical advances are based on the fact that the wave properties of electrons (and protons) allow them to penetrate, or tunnel through, potential energy barriers that would prevent microscopic objects from penetrating. Panel A shows the wave property of an electron impinging from the left on a potential energy barrier, and tunneling through to the other side. Panel B is a scanning tunneling microscope (STM) image of spiral gallium antimonide structures grown on a gallium arsenide substrate. Panel C displays an image of crossed carbon nanotubes arranged on a silicon surface. Panel D is a picture of an atomic bomb explosion. All examples depend upon quantum mechanical tunneling.

whimsical display in Figure 8.4 wherein if we released our roller coaster at point A, and ignored small frictional losses, the car would pass over the barrier at B and ascend the next grade (potential barrier) to point C. At that point, the conservation of mechanical energy dictates that the car would reverse direction, having insufficient energy to pass over the barrier. Because of this barrier, the region from C through D to E is "classically forbidden."

However, at the molecular scale, because of the intrinsic wave property of electrons, there is a finite probability that the electron will tunnel through a repulsive barrier. We can contrast the classical picture by considering the case of an electron colliding with a barrier in which the kinetic energy of the electron is insufficient to overcome the potential energy barrier created by an electric field as shown in the upper panel of Figure CS8.1b. In the classical picture, the barrier region and the region to the right-hand side of the barrier is forbidden by the laws of energy conservation.

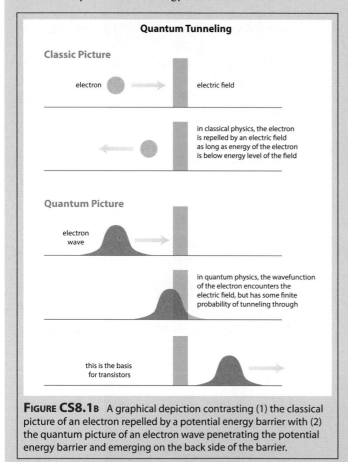

Quantum Tunneling

Classic Picture

electron   electric field

in classical physics, the electron is repelled by an electric field as long as energy of the electron is below energy level of the field

Quantum Picture

electron wave

in quantum physics, the wavefunction of the electron encounters the electric field, but has some finite probability of tunneling through

this is the basis for transistors

FIGURE CS8.1B   A graphical depiction contrasting (1) the classical picture of an electron repelled by a potential energy barrier with (2) the quantum picture of an electron wave penetrating the potential energy barrier and emerging on the back side of the barrier.

In the quantum picture, in sharp contrast, the wave property of the electron means that there is a nonzero probability that the wave will penetrate the potential energy barrier and emerge on the backside of the barrier. Our objectives here are first to develop the quantitative analysis of how such remarkable behavior emerges naturally from the wave properties of the electron, and second to explore how critical this

behavior is to the operation of transistors, integrated circuits, scanning tunneling microscopes, and an array of new electronic devices.

An Electron In-A-Box: Transitioning from a Box with Infinite Potential Walls to Finite Potential Walls

We have already solved for the quantized energy levels for an electron trapped in a potential energy well with walls of infinitely high energy in the chapter core. Recall that well was characterized as shown in Figure CS8.1c with the potential V = 0 between x = 0 and x = L and the potential V = ∞ outside the "walls" of the box.

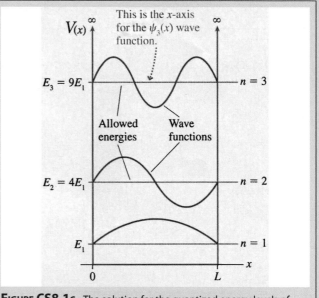

FIGURE CS8.1C   The solution for the quantized energy levels of an electron confined within a potential well with an infinitely high potential energy barrier at x = 0 and x = L.

The wavefunction for the electron, normalized to ensure the probability for finding the electron is equal to 1, is

$$\Psi_n(x) = \sqrt{\frac{2}{L}} \sin\left(\frac{n\pi x}{L}\right)$$

and the energy levels are

$$E_n = \frac{n^2 h^2}{8mL^2}$$

We consider next what happens when the potential, V(x), is again zero between x = 0 and x = L, but that *outside* the potential well, the potential energy is *finite*. This configuration is shown in Figure CS8.1d.

In classical mechanics, the particle would be strictly trapped within the boundaries of the potential well *if the kinetic energy* of the particle was less than the *potential energy* of the barrier. Thus a classical particle would never be observed for x < 0 or for x > L—the classically forbidden zone.

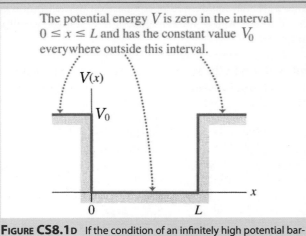

The potential energy V is zero in the interval $0 \leq x \leq L$ and has the constant value V_0 everywhere outside this interval.

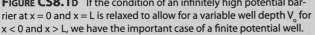

FIGURE CS8.1D If the condition of an infinitely high potential barrier at $x = 0$ and $x = L$ is relaxed to allow for a variable well depth V_0 for $x < 0$ and $x > L$, we have the important case of a finite potential well.

To solve this problem for the quantum mechanical case, where solutions must obey the Schrödinger equation

$$-\hbar^2\!\!\Big/\!2m \, \frac{d^2\Psi(x)}{dx^2} + V(x)\Psi(x) = E\Psi(x)$$

we break the problem into domains; the region between $x = 0$ and $x = L$, for which $V = 0$, and everywhere else, where $V = V_0$.

In the region between $x = 0$ and $x = L$ where $V = 0$, the Schrödinger equation is simply

$$-\hbar^2\!\!\Big/\!2m \, \frac{d^2\Psi(x)}{dx^2} = E\Psi(x)$$

and the energy of the free electron (the electron free of any influence by the potential energy) is just

$$E = \frac{\hbar^2 k^2}{2m} \quad \text{and so} \quad k = \sqrt{2mE}\Big/\hbar$$

where k is the wave number $k = 2\pi/\lambda$. Inside the well where $V = 0$ our general solution is

$$\Psi(x) = A \sin k\,x + B \cos k\,x$$

$$= A \cos \sqrt{2mE}\Big/\hbar \, x + B \cos \sqrt{2mE}\Big/\hbar \, x$$

where A and B are constants, and represent the wave amplitude.

But what about the region outside the potential energy barrier, where $V = V_0$? In this region, the Schrödinger equation with nonzero potential energy must be used:

$$-\hbar^2\!\!\Big/\!2m \, \frac{d^2\Psi(x)}{dx^2} + V_0\Psi(x) = E\Psi(x)$$

We can rearrange the equation:

$$\frac{d^2\Psi(x)}{dx^2} = \frac{2m(V_0 - E)}{\hbar^2}\Psi(x)$$

In this domain the quantity $(V_0 - E)$ is positive, so the gen-

eral solution to the equation is $\psi(x) = Ce^{kx} + De^{-kx}$ where $k = [2m(V_0 - E)]^{1/2}/\hbar$ and C and D are constants but with *different values* in the $x < 0$ and $x > L$ regions.

This is a very important result. While the wavefunction is sinusoidal *inside* the well where $V = 0$, the wavefunction is *exponential* outside the well! But one criterion that the wavefunction must meet is that it remains *finite* at all points in space, thus D must equal *zero* for $x < 0$ and C must equal zero for $x > L$. If this were not the case, $\psi(x)$ would diverge to infinity.

The next consideration we apply in our solution of joining the domains inside and outside the potential well is a very interesting one. First $\psi(x)$ and $d\,\psi(x)/dx$ must be *continuous* at $x = 0$ and at $x = L$. If either $\psi(x)$ or $d\,\psi(x)/dx$ were to change discontinuously at the edge of the potential well, $d^2\,\psi(x)/dx^2$ would be *infinite* at that point. But we know the energy of the electron is proportional to $d^2\,\psi(x)/dx^2$, so the energy of the electron would be infinite at that point in space—a physically unacceptable situation. Thus both $\psi(x)$ and $d\,\psi(x)/dx$ must be smoothly varying and continuous at the potential energy wall.

But the matching of both ψ and $d\,\psi(x)/dx$ at the boundaries between the domain where $V = 0$ and $V = V_0$ is possible only for certain values of the total energy E. This is a somewhat more complicated manifestation of the simple quantization we saw for the infinite square well (Figure CS8.1c) where we had only to match the wavelength of the electron, λ, to the well width, L. This gave us the relationship

$$L = \frac{n\lambda}{2}$$

and from the de Broglie wavelength $\lambda_e = h/p_e$ the relationship

$$p_e = h/\lambda_e = \frac{nh}{2L} \quad \text{and} \quad E = p_e^2\Big/2m = \frac{n^2h^2}{8mL^2}.$$

However, while the wavefunction of the electron in a finite potential well looks very much like the electron wavefunction for the infinite well, there is no simple formula for the energy levels corresponding to the simple formula for the infinite square well.

However, we can graphically display the solutions of the electron wave in the finite potential well. Thus is displayed in Figure CS8.1e in panel a. The center panel of Figure CS8.1e displays the energies of the quantized levels for the case where V_0 is six times the value of the lowest energy level for the infinite potential well case. The probability distribution, $\psi^2(x)$, is displayed in panel c on the far right of Figure CS8.1e. An inspection of Figure CS8.1e reveals the following: first, qualitatively, the *shape* of the wavefunction within the well, where $V = 0$, is very similar to the wavefunction for the infinite potential well case. The wavefunction for the lowest energy, E_1, has a single maximum at the center of the well and dies off in magnitude toward the wall of the potential. The wavefunction corresponding to the next higher energy level has a node in the middle of the well with two maxima of opposite phase on either side of the center node and the amplitude dies off toward the well.

Second, in sharp contrast to the case of the infinite potential case, the wavefunction for the finite potential well has a nonzero amplitude at the wall of the finite potential, and the wave amplitude dies off exponentially with increasing penetration depth into the wall. Importantly, because $\psi^2(x)$ represents the probability of finding the electron at any point in space, this means that there is a *nonzero* probability of finding the electron *outside* the potential well at positions x < 0 and x > L. This is a purely quantum mechanical effect—it means that in a system controlled by quantum mechanical behavior, electrons can penetrate into classically forbidden territory! This characteristic of barrier penetration by electrons is the foundation upon which much of modern electronics is based.

Before developing these ideas further, let's compare the finite and infinite square well solutions in a little more detail.

Contrasting Infinite and Finite Square Well Solutions

The recognition that the wavefunction for the finite potential square well is not zero at x = 0 and x = L means that the wavelength of the sinusoidal part of each wavefunction is *longer* than is the case for the infinite square well potential. This in turn means that the *curvature* of the wavefunction is reduced for the finite square well when compared with the comparable energy level for the infinite potential square well. If the curvature is reduced, so too is the energy reduced for a given quantum number. This is immediately evident on the inspection of Figure CS8.1e, panel b. A similar conclusion can be drawn immediately from the de Broglie relation p_e = h/ λ_e. As λ_e increases, the momentum of the of the electron decreases, and so too does the energy. The result is that each energy level, including the ground state, is *lower* for the

finite potential square well in comparison to the infinite potential square well of the same width L for a given quantum number n.

A second important distinction between the finite potential case and the infinite potential case is that the finite potential case has only a *finite* number of bound states and thus a finite number of energy levels. The number of bound states depends on V_0 and thus on the depth of the well. A very useful way of quantifying the energy levels for a finite well of depth V_0, without doing the complicated mathematics associated with the finite potential case, is to scale the comparison to the ground state (lowest energy level) for the infinite potential case

$$E_\infty = \frac{n^2 h^2}{8 m_e L^2} = \frac{\pi^2 \hbar^2}{2mL^2}$$

The key comparisons are these:
- When V_0 is significantly larger than E_∞, which corresponds to a very deep well, there are a large number of bound states and the energies of the lowest quantized energy levels are very similar in energy to the infinite square well case.
- When V_0 is only a few times E_∞ there are a small number of bound states and those states are significantly lower in energy than are the corresponding energy levels for the infinite potential case.
- No matter how shallow the well, how small V_0, there is always one bound state within the well. There is no state with E = 0.
- As the well becomes shallower, the electron wave penetrates more deeply into the wall of the potential.

Referring again to Figure CS8.1e, the case for which V_0 = 6 E_∞ is a case in point. There are three bound states, all displayed in panel a. For each quantum number n, the energy of each bound state is expressed in terms of V_0 and in terms of E_∞ as shown in panel b.

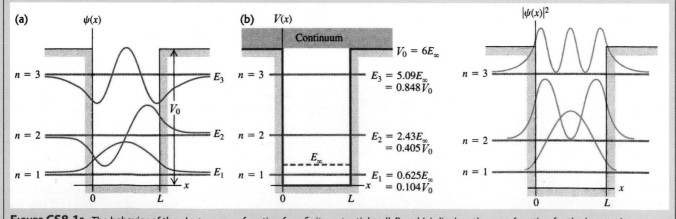

FIGURE CS8.1E The behavior of the electron wavefunction for a finite potential well. Panel (a) displays the wavefunction for the lowest three energy levels, panel (b) provides the energy of those first three energy levels in comparison with the lowest energy level of the infinite square well potential and panel (c) displays the probability, ψ^2, of finding the electron at any position x.

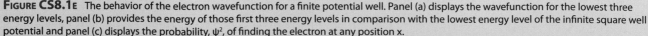

For the finite potential well case, there are states for which E is *greater* than V_0. In this domain the electron is not bound but is *free* to move through all values of x. These free particle states form a *continuum* of states rather than a discrete set of states with definite energy levels. In practice, those are important states, and they have important characteristics displayed in Figure CS8.1f.

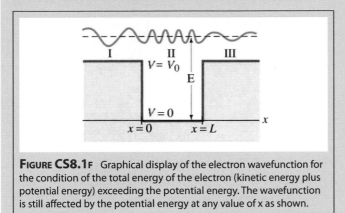

FIGURE CS8.1F Graphical display of the electron wavefunction for the condition of the total energy of the electron (kinetic energy plus potential energy) exceeding the potential energy. The wavefunction is still affected by the potential energy at any value of x as shown.

The wavefunction for the free electrons where $E > V_0$ is sensitive to the shape of the potential that it "sees" below it as is clear from Figure CS8.1f. In particular, in domain I and III in Figure CS8.1f, the wavelength of the electron is *greater* than in domain II because the kinetic energy of the electron, equal to $E - V_0$, is greater in domain II and thus the *curvature* of the wavelength in domain II is greater, decreasing the wavelength of the electron.

Application of Tunneling: The Scanning Tunneling Microscope

We have already seen the remarkable ability of the scanning tunneling microscope (STM) to observe the position of individual atoms—such a display is shown in Figure 8.1. The resolution of the STM is approximately 0.1 nm. The insight afforded by the STM replaced standard optical imaging techniques using visible light that could only resolve objects of dimension equal to or larger than the wavelength of light—about 500 nm. With a dimension for a typical atom of about .5 nm, optical techniques in the visible have a resolution 1000 times less than that required to image an individual atom. Electron microscopes are limited by the de Broglie wavelength of a free electron—about 10 times the dimension of the atom.

With the invention of the STM in 1981, the situation changed dramatically. For the first time the positioning of single atoms could by determined in remarkable detail. An example of the geometry of the STM is displayed in Figure CS8.1g. The microscope tip, made of a drawn point of tungsten, is positioned approximately 0.5 nm above the position of an atom on a metal surface. The position of the STM probe tip is moved both vertically and horizontally using a material

that lengthens or contracts depending on the voltage applied. A voltage is applied between the tip and an individual atom on the surface and the current flow from the tip to the surface that is proportioned to the *probability* of electrons tunneling through the air gap between the tip and the atoms on the surface.

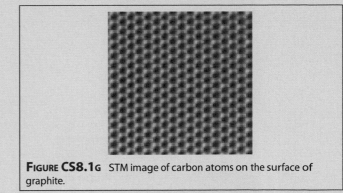

FIGURE CS8.1G STM image of carbon atoms on the surface of graphite.

The image in Figure CS8.1G displays the STM image of carbon atoms residing on a surface of graphite. Many different combinations of atoms and base materials can be imaged because the STM uses the tunneling current between the metal tip and the electron orbitals of the atoms on the surface—and all atoms have electrons in orbitals. Figure CS8.1H shows an STM image of bismuth strontium calcium copper oxide (BSCCO) imaged with atomic resolution. BSCCO is a cuprate superconductor, an important category of high-temperature superconductors sharing a two-dimensional layered structure. The structure of BSCCO provides the basis for high temperature superconductivity up to 108K, making these structures important for advances in the application of superconductors.

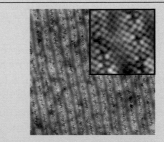

FIGURE CS8.1H STM image of BSCCO, which is a high temperature superconductor with Tc = 91 K. The bright spot corresponds to Bi atoms. Electrons in BSCCO form copper-pairs that can carry current without any loss, thus creating the capability for superconductivity.

So in the context of our analysis of quantum mechanical barrier tunneling, how does the STM work?

As shown in Figure CS8.1i, panel a, a conducting needle with a (very) sharp tip that is just a few atoms across is brought within approximately 0.5 nm of the surface to be imaged. A great deal of development and refinement has gone into preparing the needle tips and controlling the dimensional adjustment of the needle tip over the surface to be imaged. However, these practical problems have been overcome to the point where images such as those in Figures 8.1, CS8.1a, and CS8.1g have become marquee displays of modern scientific prowess. So we turn to the issue of how the STM works.

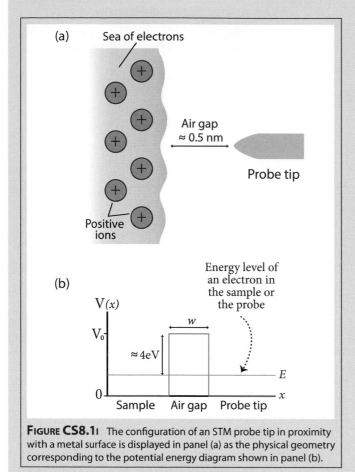

FIGURE CS8.1I The configuration of an STM probe tip in proximity with a metal surface is displayed in panel (a) as the physical geometry corresponding to the potential energy diagram shown in panel (b).

To analyze the principles behind the STM, recall first our analysis of the photoelectric effect (Figure 8.16) wherein an electron, bound to the surface of a metal by an amount of energy, E_o, the work function, is liberated from the metal surface by a photon. Typically, the work function is between 4 and 5 eV. The energy of the electron in the metal is thus an amount E_o *less* than what its energy would be just *outside* the metal surface. We can capture this in the context of our probe (or needle) tips brought to the vicinity of metal surface by drawing a potential energy diagram that includes the metal surface, the air gap, and the probe tip as shown in panel b of Figure CS8.1i. The key point is that the *air gap* represents a potential energy barrier—the energy of the electron on the metal surface is 4 eV less than the energy of the electron on the air gap. We could liberate the electron by using a photon with energy hν equal to or greater than 4 eV, to lift the electron *over* the barrier. *Or*, we could use the quantum me-

chanical wave property of the electron to tunnel *through* the barrier (the air gap), provided that the dimension of the air gap is sufficiently small to yield a measurable current flow.

In operation, the probe tip is biased slightly positive in voltage, shown in Figure CS8.1j, and the *current* is recorded as the probe tip is scanned across the surface.

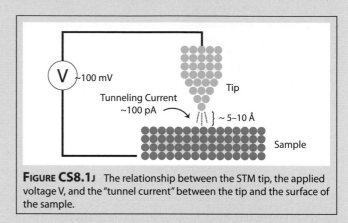

FIGURE CS8.1J The relationship between the STM tip, the applied voltage V, and the "tunnel current" between the tip and the surface of the sample.

The tunneling probability is exponentially dependent on the *gap* between the STM probe tip and the nearest atom. As the tip scans over the proximity of the closest atom on the surface, the gap decreases by ~0.1 nm and the current *increases*. When the tip moves into a position between two atoms, the gap increases and the current decreases. Modern STMs can detect changes in the gap of as little as 0.001 nm, which is approximately 1% of an atomic diameter. The STM images are generated using a computer to systematically scan the surface.

The STM has revolutionized modern scientific research and has revolutionized the technology of materials so important for innovation in nanotechnology, micro- and nanoelectronics, drug delivery, and laser systems just to note a few.

Problem 1

Electron in a Square Well

An electron is trapped in a square well with a width of 0.50 nm (comparable to a few atomic diameters). (a) Find the ground-level energy if the well is infinitely deep. (b) If the actual well depth is six times the ground-level energy found in part (a), find the energy levels. (c) If the atom makes a transition from a state with energy E_2 to a state with energy E_1 by emitting a photon, find the wavelength of the photon. In what region of the electromagnetic spectrum does the photon lie? (d) If the electron is initially in its ground level and absorbs a photon, what is the minimum energy the photon must have to liberate the electron from the well? In what region of the spectrum does the photon lie?

BUILDING A GLOBAL ENERGY BACKBONE

CASE STUDY 8.2 Fundamentals of Energy Generation by Solar Radiation

Energy from the Sun, as we have seen, exceeds by nearly four orders of magnitude the total energy per unit time—the power—consumed by humans on Earth. But how does this large ratio of power received by the Earth from the Sun to power consumed by humanity translate in practical terms to providing power to the Earth's population distribution? Could solar power provide a practical solution at scale to sustain the human endeavor, replacing fossil fuels?

To answer this question we *first* note the various means by which energy received from the sun as visible radiation can be converted to primary energy generation. The principal techniques for generating power from sunlight are:

1. Solar thermal that involves the heating of water using roof panels on buildings or homes.
2. *Concentrated* solar thermal that uses reflectors or lenses to focus the light to achieve high temperatures to provide steam to drive a turbine to generate electricity that is then linked to the power grid.
3. Solar photovoltaics that convert photons from the sun directly to electrical power that can be used either locally or connected to the power grid.

Second, we must track the solar energy received by the Earth to calculate the actual power received per unit area in various regions of the world to accurately calculate power delivered to various population centers. We begin with the quantitative analysis of how to calculate the power delivered.

As we saw in Chapter 1 (see pages 17–20), the Sun produces some 1.2×10^{37} joules of energy each year, and an overwhelming fraction of that energy is radiated into the blackness of space. If we wish to calculate the fraction of this energy that enters the Earth's climate system each year we need to calculate (1) the amount of energy per year incident on the Earth by virtue of its proximity to the Sun and (2) the

fraction of that solar energy reflected from the Earth back to space before it enters the climate system. We treat these issues in order.

(1) The amount of solar energy incident on the Earth.

We know from experience that if we were to place a card with a hole in it **1 meter** from a light bulb, as shown in Figure CS8.2a, and then we *increased* the distance from the card to the light bulb, the fraction of light emitted by the bulb that passed through the hole would *decrease*.

Suppose we were to move the card with the hole in it to a distance of **10 meters** from the bulb—by how much would the *fraction* of energy per unit time emitted by the bulb that passes through the hole *decrease*? We can deduce this from a very simple "thought experiment" that goes like this: Suppose we begin with our light bulb modeled as a glowing sphere with an output of P_0 joules/sec (P_0 is the power in watts!) and suppose we construct a series of concentric spheres with different radii about that glowing sphere, as shown in Figure CS8.2b.

The first thing to realize is that because no energy is absorbed, the total amount of energy per unit time (joules/sec) is the *same* at each distance from the glowing sphere at the center. But the *area* of each sequential sphere ($4\pi r_0^2$, $4\pi r_1^2$, $4\pi r_2^2$) is increasing as the *square* of the distance from the center of each sphere. Thus, $P_0 = P_1 = P_2$—the energy per unit time (the power) passing through each sphere is the same. But if we need to know how much energy per unit time

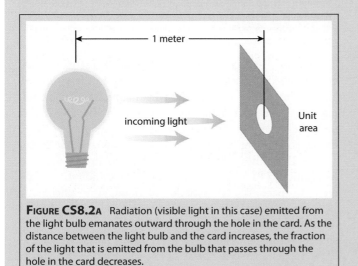

FIGURE CS8.2A Radiation (visible light in this case) emitted from the light bulb emanates outward through the hole in the card. As the distance between the light bulb and the card increases, the fraction of the light that is emitted from the bulb that passes through the hole in the card decreases.

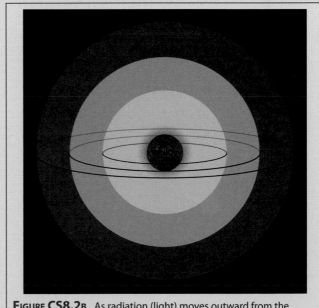

FIGURE CS8.2B As radiation (light) moves outward from the source, the same amount of energy per unit time passes through each of the concentric spheres.

(joules/sec) is passing **through the hole** in our card as we move it to greater distances from the light bulb, then we need to calculate the energy per unit time passing through a *unit area on the surface of each sphere*. If S is the energy *per unit time per unit area*, then the energy per unit time per unit area at each distance r_0, r_1, r_2 is just

$$S_0 = P_0/\text{area of sphere} = P_0/4\pi r_0^2$$
$$S_1 = P_0/4\pi r_1^2$$
$$S_2 = P_0/4\pi r_2^2$$

Or, because P_0 is the same for each sphere and setting each quantity equal to P_0, we have

$$P_0 = 4\pi r_0^2 S_0$$
$$P_0 = 4\pi r_1^2 S_1$$
$$P_0 = 4\pi r_2^2 S_2$$

and

$$4\pi r_0^2 S_0 = 4\pi r_1^2 S_1 = 4\pi r_2^2 S_2$$

so $\quad S_1 = S_0(r_0^2/r_1^2) \qquad S_2 = S_0(r_0^2/r_2^2)$

We can easily generalize this for *any* distance, r, so that

$$S = S_0(r_0/r)^2$$

This "inverse square law" dependence provides a very simple way to calculate energy (joules) per unit area (m^2) at any distance from a source. We can also calculate *power* (joules/sec) per unit area (m^2) at any distance from a source using the identical approach. The energy per unit time per unit area (joules/sec·m^2 or watts/m^2) is often referred to as the *flux* of energy.

From Chapter 1 we know that the Sun generates 3.9×10^{26} watts and we know from above that this same amount of energy per unit time passes through the surface of concentric spheres with the Sun at the center, no matter what the radius of the sphere. Thus, the *fraction* of that total number of watts intercepted by the Earth is equal to the area intercepted by the Earth, as shown in Figure CS8.2c, divided by the area of the sphere with radius equal to this distance from the Sun to the Earth (1.5×10^{11} m).

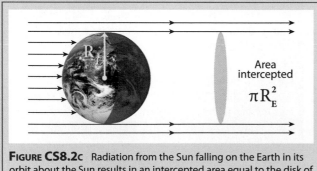

FIGURE CS8.2c Radiation from the Sun falling on the Earth in its orbit about the Sun results in an intercepted area equal to the disk of the Earth with area πR_E^2.

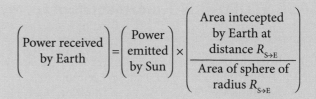

$$\begin{pmatrix} \text{Power received} \\ \text{by Earth} \end{pmatrix} = \begin{pmatrix} \text{Power} \\ \text{emitted} \\ \text{by Sun} \end{pmatrix} \times \begin{pmatrix} \text{Area intecepted} \\ \text{by Earth at} \\ \text{distance } R_{S\rightarrow E} \\ \hline \text{Area of sphere of} \\ \text{radius } R_{S\rightarrow E} \end{pmatrix}$$

where $R_{S\rightarrow E}$ is the radius from the Sun to the Earth. Thus the power received by the Earth from the Sun each year is

$$\begin{pmatrix} \text{Power received} \\ \text{by Earth} \end{pmatrix} = \left(3.9 \times 10^{26} \text{ watts}\right)\left(\frac{\pi R_E^2}{4\pi R_{S\rightarrow E}^2}\right)$$

$$= \left(3.9 \times 10^{26} \text{ watts}\right)\left(4.6 \times 10^{-10}\right)$$

$$= 1.79 \times 10^{17} \text{ watts} = 179 \text{ petawatts}$$

(2) The fraction of the solar energy reflected from the Earth back to space.

When Earth is viewed from space, what we see is a spectacular image of oceans, continents, ice caps, clouds, and a stunning blue luminescence that blend to create one of the most remarkable photogenic sights in our collective experience, as shown in Figure CS8.2d.

FIGURE CS8.2D The Earth appears from Space as a bright orb of blue, white, green, and brown because the Earth reflects back to Space approximately 30% of the incident solar radiation falling on it.

That image is created by a fraction of the incoming solar energy reflected back from Earth to space in the visible region of the spectrum. Consider what happens to the incoming shortwave solar radiation in the visible. The 179 PW of incident radiation received from the Sun is first broken down into a component reflected back by clouds and the atmosphere. This removes 50 PW that is returned directly to space. In addition, 7 PW is reflected by the surface and returned directly to space. The combination of cloud, atmosphere, and surface reflection is the image you see when looking at Earth from space. After reflection of 50 PW from the cloud/atmosphere component and 7 PW from the surface, we have 179 PW − (50 + 7) PW = 122 PW of energy entering the climate system, as summarized in Figure CS8.2e. Of this 122 PW, 80 PW is absorbed by the surface comprised of the land and ocean, and 42 PW is absorbed by the clouds and atmosphere. That is, within this model defining the input of heat to each of the component reservoirs, all of the incoming energy from the Sun is accounted for.

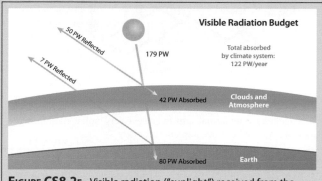

FIGURE CS8.2E Visible radiation ("sunlight") received from the Sun amounts to 179×10^{15} joules/yr, or 179 pettawatts; of that, 50 PW is reflected back to space by the atmosphere and clouds, and 7 PW more are reflected from the Earth's surface. These components of the visible radiation budget are what you "see" in photographs of the earth from space. 42 PW from the Sun are absorbed by the clouds and atmosphere directly wherein the electromagnetic radiation from the Sun is converted into thermal energy (heat) that increases the internal energy of the atmosphere. The remaining 80 PW are absorbed by the Earth's surface, which includes the ocean, continents, etc.

When we account for the amount of power reflected back to space by the Earth—the albedo—the Earth reflects approximately 33% of the solar radiation incident in it in the visible back to space so

$$\begin{pmatrix} \text{Power entering} \\ \text{Earth/atmos-} \\ \text{phere system} \end{pmatrix} = \begin{pmatrix} \text{Power received} \\ \text{by the} \\ \text{Earth} \end{pmatrix} (1 - 0.33)$$

$$= \left(1.79 \times 10^{17} \text{ watts}\right)\left(0.67\right)$$

$$= 1.24 \times 10^{17} \text{ watts} = 122 \text{ PW}$$

But what does this mean in terms of the power delivered per unit area (W/m^2) if we live in New York City or Seattle or Minneapolis or Denver or Phoenix?

First, to get a global view, we note that this total power delivered must be divided by the area of the Earth to calculate the *average* amount of energy received per unit area. The area of the Earth is $5.1 \times 10^8 \text{ km}^2 = 5.1 \times 10^{14} \text{ m}^2$. Thus the average power delivered by the sun per unit area is 1.22×10^{17} watts$/5.1 \times 10^{14} \text{ m}^2 = 239 \text{ W/m}^2$. This is an important number because it gives us an idea of the *average* amount of power received from the sun per unit area on the Earth's surface. However, in order to convert this to actual power received in specific locations by actual solar receptors we must account for such factors as cloud cover, the angle of the sun, the season, etc. It is also important to recognize

that direct sunshine at midday on a cloudless day is approximately 1000 W/m^2. We can immediately calculate this to reasonable accuracy by recognizing that with ~1350 watts/m^2 falling in the disc of Figure CS8.2c, and an albedo of 33%, this gives us $(1350)(1 - 0.33) = 950 \text{ W/m}^2$.

If we are considering the power per unit area delivered to a flat panel – for example a solar thermal panel on a roof to heat water or a photovoltaic (PV) panel we must calculate the amount of power lost by the fact that the angle between the direction to the Sun and the plane of the panel is not 90° but some smaller angle. Figure CS8.2g gives some examples. The latitude of New York City is N 40°47′; the latitude of Denver, CO is N 39°45′; the latitude of Seattle is N 47°37′; the latitude of Phoenix, AZ is N 33°29′; and the latitude of Los Angeles is N 34°3′. The latitude of Nairobi, Kenya is N 1°16′; the latitude of Mogadishu, Somalia is N 2°2′; the latitude of Rio de Janeiro, Brazil is S 22°57′; the latitude of London is N 51°32′; and the latitude of Shanghai, China is N 31°10′. We can display graphically the impact of latitude on power received per unit area by a flat solar panel parallel to the ground in CS8.2f, choosing the examples of Seattle and Nairobi.

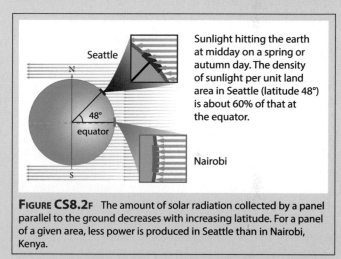

FIGURE CS8.2F The amount of solar radiation collected by a panel parallel to the ground decreases with increasing latitude. For a panel of a given area, less power is produced in Seattle than in Nairobi, Kenya.

For example the angle between the incoming solar radiation and the land surface *reduces* the power delivered per unit area by 33% in Seattle, 18% in Phoenix, and 1% in Nairobi.

Next, and perhaps most obvious, we must account for the fact that it is not noon all day! This reduces the average power delivered per day to about 36% of the midday maximum for regions in the middle of the continental United States. All these factors, including cloud cover, have lead to the direct measure of the average power delivered by the sun per square meter for a number of locations in the U.S. and globally. This is verified experimentally at each of these sites.

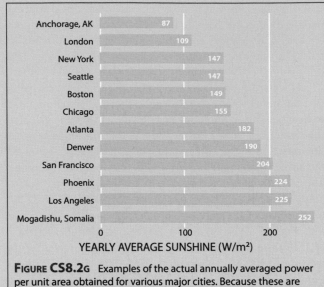

YEARLY AVERAGE SUNSHINE (W/m²)

FIGURE CS8.2G Examples of the actual annually averaged power per unit area obtained for various major cities. Because these are experimentally determined numbers, they include the effects of latitude, cloud cover, etc.

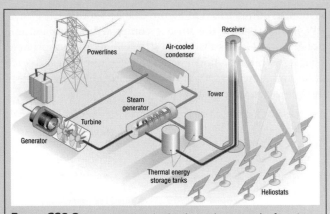

FIGURE CS8.2H Concentrated solar thermal operates by focusing solar radiation onto a small area to achieve very high temperatures, generating steam to drive a steam turbine.

Armed with the amount of power per unit area actually delivered by the Sun for a variety of locations, we can proceed to estimate the actual yield from the three options identified at the outset.

Solar Thermal

As noted, in the language of solar energy technology, "solar thermal" is taken to mean the heating of water using panels installed on the roofs of homes or buildings. This heated water is used for boosting the temperature of water for bathing or for preheating water for heating the house or building. It is the least sophisticated form of energy production from the sun, but it is also one of the most efficient because it can deliver approximately 60% of the power available from the sun as heated water. See problem 1 at the end of this Case Study.

Concentrated Solar Thermal

The term "concentrated solar thermal" in the jargon of global energy technology refers to the use of large reflectors that concentrate solar radiation to produce high temperature steam to drive turbines for electrical power generation. A diagram of a typical system is displayed in Figure CS8.2h. There are a number of options for converting the concentrated solar power to electricity, notably including pressurized water driving a steam turbine, shown in Figure CS8.2h. In general, concentrated solar thermal is a technology that operates at large scale as shown in the two panels of Figure CS8.2i.

FIGURE CS8.2I There are a number of different designs available for concentrated solar thermal power generating stations. Panel A is a parabolic reflector with power collection at a single point: the tower extending up to the center of the mirror. Panel B uses a parabolic reflector to heat a tube of flowing fluid at the focus.

A reasonably accurate figure for delivered concentrated solar thermal power per unit area for desert based systems is 60 W/m². These concentrated solar facilities are typically constructed to produce on the order of 100 MW per installation. In the United States, those facilities are destined for construction in Arizona, Southern California, and New Mexico. Because these concentrated solar thermal systems provide large amounts of power, they will see their full potential only when linked to a national grid. But we seek here to ask the question: How realistic is the idea of generating a significant fraction of the power demands of the U.S. using concentrated solar thermal? See problem 2 at the end of this Case Study.

Solar Photovoltaics (PVs)

We will develop a thorough understanding of photovoltaics as the text progresses, but at this juncture we seek to determine the power generating potential for each category of solar energy. Photovoltaic (PV) panels convert solar photons to electrons, where the electrons are at a higher electrical potential such that each PV cell generates a voltage and when connected to a "load" it supplies a flow of electrons just as does a coal fired or concentrated solar thermal electrical power generating plant. The PV cells are arranged in an array within a solar panel and the solar panel can be placed on the roof of houses or buildings or situated in "solar farms" where hundreds of panels are arranged to maximize the capture of solar energy—usually in places with high delivered solar power per unit area, such as in desert locations. These solar panels are also very important for supplying power to remote locations off the power grid. See problem 3 at the end of this Case Study.

Problem 1

The calculation of the average power for low temperature solar thermal for two examples: Chicago and Los Angeles. We will assume the solar thermal panels are mounted on the south facing roof and that the *angle* of that roof gives us approximately the delivered power per unit area given in Figure CS8.2g. We take the efficiency of conversion of solar power to heated water of 60%, and that we have solar panels with an area of 10 m² per person in the house. In *Chicago* how much power would we net as a yearly average? In Los Angeles?

We next need to convert this to our preferred units—the number of kWh per person per day delivered to the home. Since a watt is one joule/sec and there are 3.1×10^7 seconds/year and there are 365 days/yr, for *Chicago* how many kWh/p·d would we generate? For Los Angeles?

One important point to emphasize here is that while this energy per person per day is a significant contribution to the personal energy budget, it is low-grade energy in that it can be used to heat water, but it can not be used to generate electricity. It is energy that cannot be stored for extended periods and it is not necessarily delivered in a timely manner—it peaks in the summer when less heating is called for. However, it is one of the lowest cost contributions to the energy budget of a home or office and thus constitutes a universally important component of a sustainable energy structure. When you travel through Europe, you will notice the large number of homes with low temperature solar thermal panels on the roof.

Problem 2

We turn next to the question of determining whether concentrated solar thermal is capable of generating a major fraction of the primary energy consumed by the United States.

To answer this question, let's assume that we wish to generate a third of the total primary energy of the U.S. by this method. What fraction of the area of the state of Arizona would be required? One-third of the total power consumption of the US corresponds to 1 TW, 1×10^{12} watts, of power.

a. If we realize 15 W/m² from concentrated solar thermal, what area of concentrated solar thermal collectors are required?

b. The area of the state of Arizona is 2.95×10^5 km² so what percentage of the state does this correspond to?

c. Of course it is our objective to keep track of what this means in terms of kWh/p·d for citizens of the U.S. With a population of 3.1×10^8 people, 365 days in a year, and 24 hours in a day, this 1×10^{12} watts converts to how many kWh/p·d?

Problem 3

Typical solar panels have efficiencies of between 10% and 30% depending on the technology and cost. Calculate using Figure CS8.2g the average power delivered by a 20% efficiency south-facing solar panel on a residential home, assuming 100 m^2 of solar panels for the following cities:

Anchorage:

New York:

Phoenix:

Miami:

QUANTUM MECHANICS OF MULTIELECTRON SYSTEMS AND THE LINK BETWEEN ORBITAL STRUCTURE AND CHEMICAL REACTIVITY

9

"If quantum mechanics hasn't profoundly shocked you, you haven't understood it yet."

—Niels Bohr

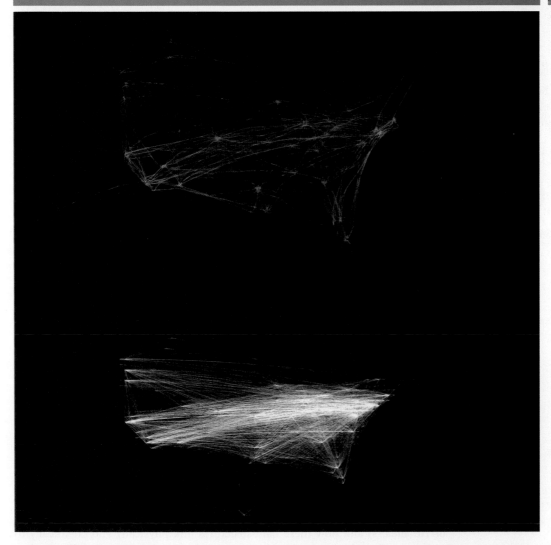

FIGURE 9.1 Displayed are images that represent large data sets with the upper margins representing a visualization of flight patterns of commercial aircraft over the continental United States and the lower image displaying bit flows on the internet developed at Carnegie Mellon University.

Framework

The internet represents a development of immense importance to modern society—that much is obvious. What is less obvious, but of great importance, is that the worldwide web opens up the conceptual revolution of transporting electrons around the planet instead of chemicals, conference attendees, and materials. Figure 9.1 contrasts the flight paths of commercial aircraft in the United States with the pattern of electron movement on the internet. The energy required to execute the aircraft flights in the upper panel of Figure 9.1 represents the combustion of 6×10^9 barrels of petroleum per year, representing an energy expenditure of 3500 billion kWh of energy with the corresponding addition of 3×10^8 tons of CO_2 to the atmosphere each year. The energy required to operate the internet within the US each year, in contrast, is 350 billion kWh, the energy for which can easily be generated by renewable sources. While aircraft will be used to transport individuals for vacations, for certain business matters, and to universities, the development of teleconferencing and the development of high speed rail systems for both passengers and freight holds the promise of significantly reducing dependence on fossil fuels at the same time communication pathways are dramatically opened up. International exchanges between peoples of the world hold the possibility of establishing an entirely new venue of societal exchange among countries. And it is the *movement of electrons* that will foster these developments.

Remarkably, it is quantum mechanics, operating as it does at the atomic scale, that brought and continues to bring a revolution in our understanding that has in turn opened a gateway to new innovation in the command and control of new materials with properties that were unimagined just a few years ago. Quantum mechanics introduced three key attributes of the behavior of electrons at the atomic scale: (1) the wave properties of the electron when confined to the spatial scale of atoms and molecules, (2) the electron spin that dictates the number of electrons allowed in any atomic (or molecular) orbital, and (3) the placement of electrons in, and the transfer of electrons between, both occupied energy levels and unoccupied energy levels. The relationship between electronic structure and chemical reactivity is treated in Case Study 9.1.

As we progress from the single electron hydrogen atom to the case of multielectron systems, the laws of quantum mechanics become more complicated to apply, but far more powerful for the creative process of developing new materials and new architectures, opening the pathway to integrated designs of new systems. The power of developing a hierarchy of scale from the atomic level to the molecular level to the cellular level is the hallmark of biological systems—systems that evolved for a very specific, and often sophisticated, objective, yet systems designed to self-assemble! In fact, those systems capable of self assembly are those systems with the greatest complexity and interconnectedness. As we have noted, the photosynthetic chloroplast in a green plant is in fact a solar cell constructed as a hierarchy of atomic and molecular level subsystems. The ATP produced by the mitochondria in a cell is a chemical energy converter of remarkable adaptability, yet it is self-assembled.

It is tempting to jump immediately from the properties of individual multielectron atoms to theories of the formation of the molecular bond. But revolutionary advances in technology have refocused attention and a new generation of innovative research on the reanalysis of how characteristics of individual *multielectron* atoms map into larger dimensions to study nanostructures (dimensions of 10^{-9} meters) and how these nanostructures link physically and chemically to architectures on the microscale (dimensions of 10^{-6} meters). As we will see, when we begin to seriously examine the design of integrated systems that require a hierarchy of connected subsystems linking scales from the atomic to the macroscopic,

Case Study 9.1
The Union of Electronic Structure and Chemical Reactivity in Multielectron Atoms

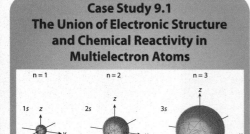

FIGURE 9.2 Shown are the first transistors with the left panel a replica of the point-contact transistor invented by Bardeen and Brattain. The Shockley junction transistor is shown in the right panel and while it is less elegant architecturally, it was easier to fabricate.

FIGURE 9.3 The three inventors of the transistor at the time they announced the development. Shockley is on the left, Brattain in the center, and Bardeen on the right.

we must view the problem in terms of an array of bonding forces that operate over a large *range* of length scales and over a large *range* of bonding energies.

As we will see, quantum mechanics and associated restrictions on how electrons are allowed to enter the energy levels associated with atomic orbitals sets the stage for understanding how the associated hierarchal assembly constitutes the foundation for two critically important lines of development. The first category is the pathway to *integrated chemical systems*, which include (1) the study of catalysts that provide control over the rate at which chemical reactions proceed, (2) the development of photochemical cells for converting sunlight to electrical power, (3) the development of new materials for dramatically improving energy storage capability of batteries, and (4) new membranes capable of improving the performance of fuel cells that can convert hydrogen and oxygen to electrical power. The second gateway opened by studies of the properties of multielectron systems is that of *integrated physical systems*, which include (1) the assembly of atoms into specifically tailored conductors, semiconductors, and insulators that constitute the key components for (2) the development of transistors, diodes, and integrated circuits for (3) the development of computers, memory storage, and high resolution graphical displays that (4) allow the development of the integrated system we now call the "internet." The first transistor, shown in Figure 9.2 was developed using a *combination* of multielectron elements by Shockley, Brattain, and Bardeen shown in Figure 9.3.

As the understanding and ability to control combinations of multielectron elements developed, it became possible to create increasingly complex integrated circuits, microprocessors, and computers—the first microprocessor is shown in Figure 9.4.

What is remarkable to watch is how the the quantum mechanical behavior at the atomic level "maps" onto the global reach of the internet and to think carefully about how

FIGURE 9.4 The first microprocessor was constructed by Ted Hoff at Intel. It combines all the components of a programmable computer on a single chip. The chip shown here has the dimensions of approximately 3 mm by 4 mm and contains approximately 2000 transistors.

the unfolding revolution in nanotechnology and nanochemistry bridges the union between the microscopic, atomic scale behavior of materials and the macroscopic manifestation of integrated systems capable of advancing societal objectives: human health, human communication, and an improving global standard of living through sustainable energy sources, distribution, and storage.

While we must come to understand the scientific foundation that underpins multielectron atoms before we can understand the direct impact of quantum mechanics on modern society, we can set in place the chain of interlocking concepts that, when taken together, establishes the union of quantum mechanics with modern society. This sequence of connected elements is shown in Figure 9.5.

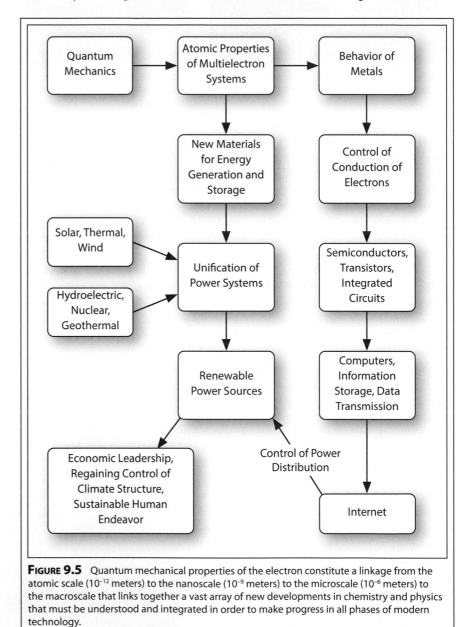

FIGURE 9.5 Quantum mechanical properties of the electron constitute a linkage from the atomic scale (10^{-12} meters) to the nanoscale (10^{-9} meters) to the microscale (10^{-6} meters) to the macroscale that links together a vast array of new developments in chemistry and physics that must be understood and integrated in order to make progress in all phases of modern technology.

This mapping of quantum mechanics onto the current explosive developments in nanotechnology has its roots in the understanding of multielectron systems. As we have seen, it was Dimitri Mendeleev, seeking a coherent structure for teaching the behavior of chemical elements to students of inorganic chemistry, who created the first Periodic Table that assigned repetitive patterns to chemical behavior. Mendeleev succeeded in grouping all the known elements (63 at the time) into distinct families and in so doing he not only radically simplified the study of chemistry, he also identified gaps in his periodic table. Mendeleev recognized that these gaps were as yet undiscovered elements. Because he could use the patterns of chemical behavior to predict the chemical characteristics of the missing elements, they were far more easily searched for and identified. Indeed, gallium, scandium, and germanium—some of the most important elements for the development of innovative electronic devices today—were discovered in his lifetime as a direct result of his ability to predict their chemical and physical behavior.

But why did the elements exhibit this periodic behavior? Why was there this unbroken sequence of elements going from the lightest (hydrogen) to the heaviest? It was a full half century before the physicist Wolfgang Pauli, shown in Figure 9.6, articulated his "exclusion principle" stating that no two electrons could share the same (i.e. identical) set of quantum numbers. This brought order to the considerable confusion regarding how the sequence of multielectron elements was "built up" by adding an additional electron as the atomic number of the elements progressed to larger and larger atoms.

The answer to how the full complement of quantum numbers, including electron spin, defined the sequence of building up the multielectron atom's designation of orbital occupancy is a central theme in this chapter. It is also of considerable interest to recall how the intrinsic spin of the electron came to be understood. It was well known from studies of electromagnetism in the 19th century that a loop of wire carrying an electric current produces a magnetic field. That is, a current moving in a wire loop acts just like a small magnet—it has a "north pole" and a "south pole" and this separation between the poles of the magnet constitutes a *magnetic moment* that establishes the intrinsic axis of the magnet in space. As we will see in this chapter, the famous experiments by Otto Stern and Walther Gerlach demonstrated that the magnetic moment of the electron is quantized in space. That is, the value of the magnetic moment of the electron cannot assume any value such that there is a continuum of possible values of the magnetic moment, but rather the magnetic moment of the electron can take only specific, discrete values. This spatial quantization of the magnetic moment of the electron is a strictly quantum mechanical phenomenon—it has no analog in classical physics. But what wasn't realized at the time Stern and Gerlach announced their discovery in 1922 was that the magnetic moment of the electron was *not a result of its motion in orbit about the nucleus, but rather a result of the electron spin about its own axis!*

The hypothesis that the electron had a spin, angular momentum, analogous to the spin of the Earth on its axis, was put forward by two young physicists, George Uhlenbeck and Sam Goudsmit. Uhlenbeck and Goudsmit proposed the idea of intrinsic electron spin in 1925, a full year before Schrödinger published his wave equation that provided the mathematical solution defining the electron orbitals as discussed in Chapter 8. Uhlenbeck and Goudsmit realized that their suggestion of an intrinsic electron spin was, to put it gently, controversial. Thus they sought advice from their professor Paul Ehrenfest. Ehrenfest thought over their hypothesis, but decided it was best that they consult the regarded expert on the subject, Hendrik Lorentz. Lorentz agreed to consider their proposition, and after mulling it over, expressed the strong view to Uhlenbeck and Goudsmit that there were too many reasons why such an idea as an intrinsic electron spin was incompatible with any known behavior in physics. This lead Uhlenbeck and Goudsmit to reconsider publishing the hypothesis of electron spin so they asked Ehrenfest to withdraw their paper. However, too late! Ehrenfest had already submitted the paper for publication on their behalf and it was already in print. That lead Ehrenfest to say: "You are both young enough to allow yourselves such foolishness." The impact of electron spin on the magnetic properties of the elements is featured in Case Study 9.2 on rare earth elements.

FIGURE 9.6 Wolfgang Pauli (1900–1958) proposed the now famous "Pauli Exclusion Principle" in 1925, which stated that no two electrons in an orbital (atomic or molecular) can share the same set of quantum numbers n, ℓ, m_ℓ, and m_s. Pauli received the Nobel Prize in 1945..

Case Study 9.2
Rare Earth Elements: Applications, Chemistry, Resources & Production

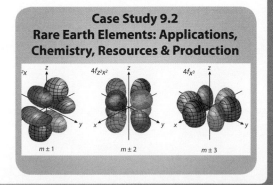

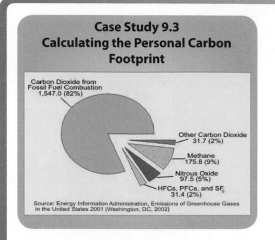

**Case Study 9.3
Calculating the Personal Carbon Footprint**

Carbon Dioxide from
Fossil Fuel Combustion
1,547.0 (82%)

Other Carbon Dioxide
31.7 (2%)

Methane
175.8 (9%)

Nitrous Oxide
97.5 (5%)

HFCs, PFCs, and SF$_6$
31.4 (2%)

Source: Energy Information Administration, Emissions of Greenhouse Gases
in the United States 2001 (Washington, DC, 2002)

But there was more to be learned. At the same time Uhlenbeck and Goudsmit put forward their hypothesis of intrinsic electron spin, Ralph Kronig, also a young physicist, had put forward the same idea—except that when he approached Wolfgang Pauli with his idea, Pauli convinced him that such an idea could not be correct. Kronig acquiesced and did not publish. The remarkable irony, of course, is that it is on the shoulders of *electron spin* that the famous Pauli Exclusion Principle is built!

With this, we turn our attention to the construction and characterization of the multielectron atom. Multielectron atoms hold the key to reducing our carbon footprint, a subject featured in Case Study 9.3.

Chapter Core

Road Map for Chapter 9

Quantum mechanics was developed first to quantitatively describe the structure of the hydrogen atom—one electron and one proton—an accomplishment that set the world of chemistry and physics on a gateway to dramatic change. The impact of the changes that quantum mechanics has brought, and continues to bring, to both the core of scientific thought and to the emerging objectives of society as a whole are striking indeed. That era is marked by the innovation, exploration, and insight brought to bear on the atomic scale, nanoscale, and macroscale leading to new methods of controlling the interaction of light with the molecular structure, profound advances in miniaturization of electronics and data storage, revolutionary new imaging techniques, vectored drug delivery, and unprecedented new techniques for energy generation and storage. A critically important bridge from the atomic to the molecular to the nanoscale depends upon an understanding of the multielectron atom. Thus we proceed to develop an understanding of the quantum mechanics of increasingly complex atomic structure with the following topics.

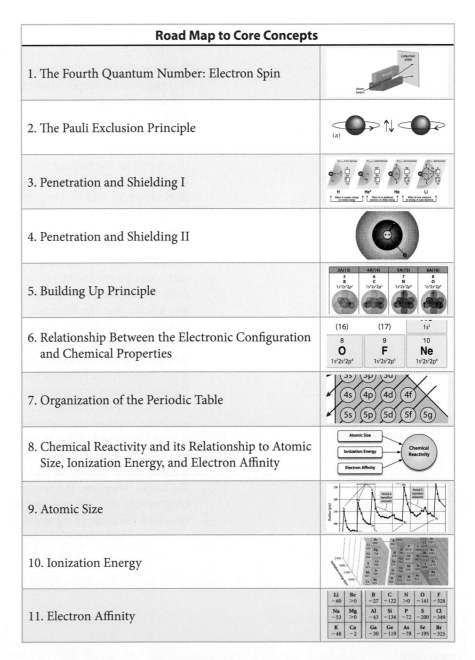

Road Map to Core Concepts
1. The Fourth Quantum Number: Electron Spin
2. The Pauli Exclusion Principle
3. Penetration and Shielding I
4. Penetration and Shielding II
5. Building Up Principle
6. Relationship Between the Electronic Configuration and Chemical Properties
7. Organization of the Periodic Table
8. Chemical Reactivity and its Relationship to Atomic Size, Ionization Energy, and Electron Affinity
9. Atomic Size
10. Ionization Energy
11. Electron Affinity

9.7

Multielectron Atoms

Development of the theory underpinning our understanding of atomic structure was built on the back of the one electron system: atomic hydrogen. The reasons for this were many. But two important ones include the simplified emission spectrum of H atoms, which allowed a simple formulation of the wavelength of the emitted lines in terms of integers, and the quantitative success of the Bohr model using the quantized photon energy, $\Delta E = h\nu$, from Planck and Einstein. A third reason for the importance of atomic hydrogen to the development of theories of atomic structure was the fact that the Schrödinger wave equation predicted the radial position of the maximum probability for finding the 1s electron at the *same radius* as the first Bohr radius in hydrogen deduced from the emission lines of atomic hydrogen. As we have seen, *confinement* of the electron matter wave leads to *quantization* that leads to *eigenfunctions* that in turn determine the *energies* of the orbitals in three dimensions that require three quantum numbers to uniquely define the energy, shape, and orientation—the principal quantum number n, the angular momentum quantum number ℓ, and the magnetic quantum number m_ℓ respectively.

But questions remained unanswered concerning atoms that contained more than one electron. Why didn't subsequent electrons added to the single electron hydrogen atoms simply "stack up" in the lowest orbital? Why could the Schrödinger equation calculate the exact energies of the hydrogen atom, but not of the multielectron atoms?

Answers to these questions were emerging nearly simultaneously with the hypothesis that the electron had wave properties and that the distribution of electron density about the nucleus could be calculated with the Schrödinger equation. The key experiment that set in place the foundation for assigning quantum numbers to electrons in multielectron atoms was done in 1922 by Otto Stern and Walther Gerlach at the University of Frankfurt *before* de Broglie hypothesized the wave property of electrons. That experiment used a beam of silver atoms passed through a magnetic field as shown in Figure 9.7. The beam of silver atoms, emergent into a vacuum from a high temperature oven was captured on a target, a glass plate that could be used to image the pattern of the spatial distribution of silver atoms deflected by the magnetic field. At the time of the Stern–Gerlach experiment, it was known that electrons possessed angular momentum and it was thought that angular momentum would lead to a *magnetic moment* intrinsic to the individual atoms. By passing individual atoms through a magnetic field that increased in strength as a function of the distance from the magnetic pole piece shown in Figure 9.7, a force would be imparted to the silver atom according to the *orientation* of the angular momentum of the electrons in orbit about the nucleus and thus to the *orientation* of the magnetic moment of the electrons. Since the orientation of the magnetic moment of the electrons would, from a classical perspective, be random, it would be expected that the distribution of silver atoms on the glass target would be random—that is with a continuously varying probability that would produce a continuous "smear" of silver atoms deflected by the magnetic field striking the target. However, the results of the Stern–Gerlach experiment demonstrated conclusively that just *two spatially separated spots* appeared on the glass target struck by the beam of Ag atoms as shown in Figure 9.7!

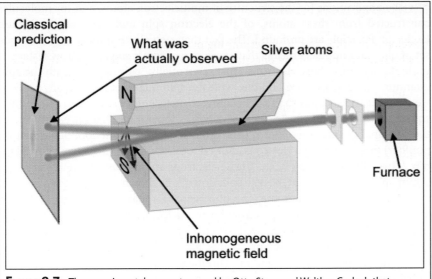

FIGURE 9.7   The experimental apparatus used by Otto Stern and Walther Gerlach that demonstrated for the first time that angular momentum is quantized in the world of quantum mechanics. This was done by firing a beam of silver atoms through a magnetic field that was stronger close to one pole piece than at the other. This "inhomogeneous" magnetic field applies a net force perpendicular to the beam direction on a magnetic moment. It was the separation of the beam of silver atoms into just two spots on the target that was the first evidence for the spatial quantization of angular momentum in quantum mechanics.

The silver atoms exhibited a *spatial quantization* of angular momentum, a spatial quantization of the magnetic moment. The experiment was direct justification for the hypothesis that angular momentum is quantized in a system that obeys the laws of quantum mechanics. It is important to note that this experiment was performed just *before* quantum mechanics was born. It was not until 1925 that Samuel Goudsmit and George Uhlenbeck put forward the hypothesis that the electron possessed a magnetic moment resulting from its *intrinsic spin*, rather than a magnetic moment resulting from its motion in orbit about the nucleus and that the electron possessed a spin quantum number, m_s, which could assume only one of two values, ± ½. It was Wolfgang Pauli in 1926 who first articulated the implications for atomic structure of the remarkable discovery that the magnetic moment of the electron is quantized (Stern–Gerlach) and that the magnetic moment results from the intrinsic spin of the electron (Goudsmit & Uhlenbeck): the *Pauli Exclusion Principle*, which states that *no two electrons can share the same set of quantum numbers*. We summarize the quantum numbers of electrons in multielectron atoms in Table 9.1—they are the three quantum numbers of the electron orbitals, n, ℓ, and m_ℓ that we developed in Chapter 8, with the addition of the spin quantum number m_s = +½ or –½ defining the *direction* of the electron spin.

TABLE 9.1   Summary: Quantum Numbers of Electrons in Atoms

Name	Symbol	Permitted Values	Property
principal	*n*	positive integers (1, 2, 3, ...)	orbital energy (size)
angular momentum shape of wavefunction	ℓ	integers from 0 to *n*–1	orbital shape (The ℓ values 0, 1, 2, and 3 correspond to *s*, *p*, *d*, and *f* orbitals, respectively.)
magnetic	m_l	integers from –ℓ to 0 to +ℓ	orbital orientation
spin	m_s	+1/2 or –1/2	direction of e⁻ spin

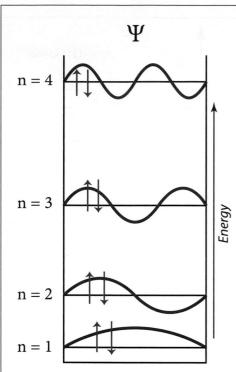

FIGURE 9.8 When electrons are inserted into the eigenstates, the wavefunctions, of a square well potential, the Pauli Exclusion Principle states that no two electrons can share the same set of quantum numbers. Thus for each value of the principal quantum number n, one electron must enter the wavefunction with spin up, $m_s = +\frac{1}{2}$, and one electron must enter the wavefunction with spin down, or $m_s = -\frac{1}{2}$.

The implications for the structure of multielectron atoms, and for molecules constructed from these atoms, of the electron spin and the associated Pauli Exclusion Principle are profound. The fact that each electron must have a unique set of quantum numbers means that as we move from a single electron system to multiple electrons, those electrons must sequentially "pair up," one with the spin quantum number $m_s = +\frac{1}{2}$ and one with the spin quantum number $m_s = -\frac{1}{2}$. If we consider the solution to the Schrödinger equation for *electrons* in our infinite square well potential—the "particle in a box"—this means that for each value of the principal quantum number n (one dimensional wavefunction, one quantum number) there will be two electrons allowed for n = 1, there will be two allowed for n = 2, two allowed for n = 3, etc. Those electrons are inserted, one with $m_s = +\frac{1}{2}$ ("spin up") and one with $m_s = -\frac{1}{2}$ ("spin down"), as shown in Figure 9.8.

When we move from the one-dimensional square well potential to the three-dimensional Coulomb potential for multielectron configurations, the Pauli Exclusion Principle applies as well—each electron must have a unique set of quantum numbers n, ℓ, m_ℓ, and m_s. That is, no two electrons can have the same set of quantum numbers n, ℓ, m_ℓ, and m_s. This means that as we move from hydrogen, to helium, to lithium, to beryllium, etc. through the periodic table, electrons will be inserted sequentially into shells and subshells in order of increasing energy with a maximum of two electrons, one with spin up ($m_s = +\frac{1}{2}$) and one with spin down ($m_s = -\frac{1}{2}$) *up to the total number of electrons possessed by that element in the periodic table.*

While for the single electron case, the energy of each subshell was determined uniquely by the principal quantum number n (i.e. the 2s and 2p subshells were of identical energy; the 3s, 3p, and 3d were of identical energy), in multielectron atoms, orbitals with the same n but different ℓ have *different* energies. The reason for this is that the electron-nuclear attraction is more important for low ℓ (low angular momentum), whereas the electron-electron repulsion is more significant for high ℓ (high angular momentum). Thus the 2p orbitals are higher in energy than the 2s orbital.

We can diagram the electron-nucleus and electron-electron interaction in Figure 9.9 for the first three elements H, He, and Li. Also displayed in Figure 9.9 are the sequentially assigned orbitals for each electron, a pattern we will repeat for the assignment of electrons to orbitals that constitutes the *electron configuration* for each element. The binding energy of the outermost electron is indicated in each case.

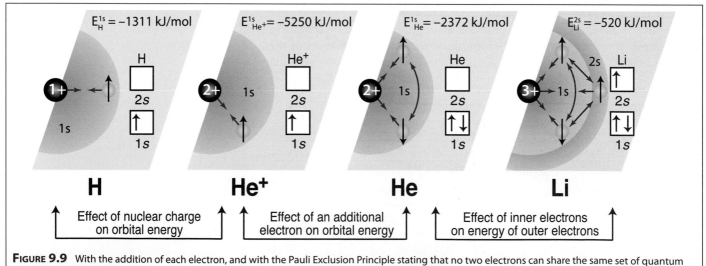

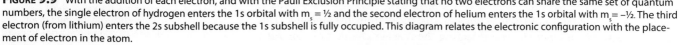

FIGURE 9.9 With the addition of each electron, and with the Pauli Exclusion Principle stating that no two electrons can share the same set of quantum numbers, the single electron of hydrogen enters the 1s orbital with $m_s = \frac{1}{2}$ and the second electron of helium enters the 1s orbital with $m_s = -\frac{1}{2}$. The third electron (from lithium) enters the 2s subshell because the 1s subshell is fully occupied. This diagram relates the electronic configuration with the placement of electron in the atom.

Notice in particular in Figure 9.9 that the energy of the electron (E_H^{1s}, $E_{He^+}^{1s}$, E_{He}^{2s}, E_{Li}^{2s}) is markedly different for each of the four cases. These differences turn out to be of great importance for understanding patterns of chemical reactivity across the periodic table. Thus we seek to understand those differences before building up the periodic table.

Penetration, Shielding, and Effective Nuclear Charge, Z_{eff}

With the introduction of multielectron configurations for all elements other then hydrogen in the periodic table, it becomes essential that we develop an intuition for how the electron-electron (e-e) *repulsion* is balanced against the electron-proton (e-p) *attraction* that binds the electrons to the nucleus. We need to understand this competitive balance between e-p attraction and e-e repulsion because it is this balance that determines the energy of the individual orbitals in the structure of any atom and it is this balance that determines the size of the atom, the ability of a given atom to retain its own electrons and the ability of a given atom to draw other electrons to it. These are the characteristics that ultimately determine the *chemical behavior* of atoms and the chemical behavior of the molecules formed from them.

We can break this problem down by considering two ideas: penetration and shielding. For our purposes here, *penetration* is taken to mean a measure of how close an electron comes to the nucleus—it is important because the Coulomb force between a single electron and the proton or protons in the nucleus is given, in the absence of other electrons, by

$$F_{e-p} = \frac{Ze^2}{4\pi\varepsilon_0 r^2}$$

where Z is the atomic number of the atom in question, e is the charge in the electron, and r is the distance between the electron and the protons in the nucleus. The *potential energy* of the electron a distance r from the Z protons is then given by

$$V = \frac{-Ze^2}{4\pi\varepsilon_0 r} .$$

Thus the potential energy becomes more negative with increasing Z and decreasing r. We can emphasize the dramatic effect of increasing Z on the energy of the 1s orbital by extending the pattern set forth in Figure 9.9 to include Z = 1 to Z = 3 for a single electron. As Figure 9.10 makes clear, the energy of the 1s orbital decreases (becomes more negative and thus more stable) from –1311 kJ/mol for H(1s) to –11,815 kJ/mol for Li²⁺(1s). This is an increase of nearly an order of magnitude. As Z increases, the force between the electron and the nucleus increases, for a given distance r, and the *energy* of the electron becomes more negative. But the increasing attractive force between the nucleus with an increasing number of protons and the electron draws the electron closer to the nucleus, decreasing r and in turn making its potential energy more negative. The result is that the 1s electron of Li²⁺ is far more stable than the 1s orbital of H and in turn the electron is far more difficult to remove from Li²⁺ than from H, as Figure 9.10 shows.

The second consideration in analyzing the energy of orbitals in a multi-electron system is the issue of *shielding*. Shielding is a measure of how electrons that occupy a spatial domain closer to the nucleus serve to reduce the attraction between the nucleus and electrons at greater distances from the nucleus. This distinction between an *inner electron* shielding the nuclear charge from an *outer electron* is shown in Figure 9.11. In this case, with Z = 3, the inner electrons with

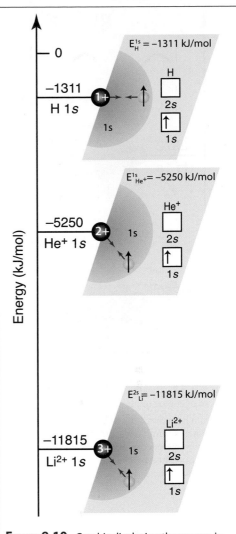

FIGURE 9.10 Graphic displaying the energy, in kJ/mol, required to remove an electron from H (one proton), from He⁺ (two protons), and from Li²⁺ (three protons).

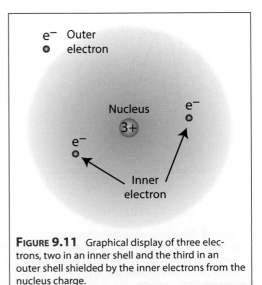

FIGURE 9.11 Graphical display of three electrons, two in an inner shell and the third in an outer shell shielded by the inner electrons from the nucleus charge.

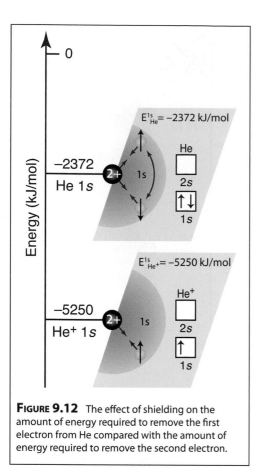

FIGURE 9.12 The effect of shielding on the amount of energy required to remove the first electron from He compared with the amount of energy required to remove the second electron.

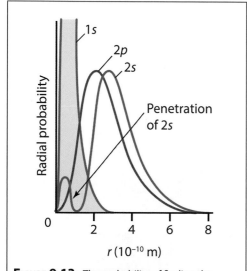

FIGURE 9.13 The probability of finding the electron at a distance r from the nucleus for the 1s, 2s, and 2p atomic orbitals. The radial probability defines the degree of penetration of the electron in the respective orbitals.

charge –e serve to shield (or "screen") the charge on the nucleus from the outer electron. For example, if the inner electrons are in the 1s orbital of Li and the outer electron is in the 2s orbital of Li, the inner electrons will experience the attraction of three protons, reducing their orbital radius below that of the 1s electron in H. The outer electron, on the other hand, will see a reduced attraction to the nucleus because of the shielding by the inner electrons and will move to a *larger* distance from the nucleus as result.

Penetration implies reaching closer to the nucleus, achieving greater stability by virtue of a greater negative energy binding the electron to the nucleus. An outer electron that is *shielded* by another electron residing at a smaller distance from the nucleus is, as a result, less stable, and more easily removed as a result of a less negative binding energy. While we have placed our two separated orbitals, the 1s and the 2s, in Figure 9.10 to make a clear distinction between *penetration* and *shielding*, we can also compare quantitatively the binding energy of a single electron in He⁺ and an electron in the filled $1s^2$ orbital of He as shown in Figure 9.12. The presence of the second 1s electron in He decreases the energy required to remove one electron from the system from 5250 kJ/mole to 2372 kJ/mole as a result of both shielding *and* penetration—a very large difference.

The concepts of penetration and shielding come together in a single variable called the effective nuclear charge, Z_{eff}. This quantity turns out to be very important for understanding the *chemical behavior* of isolated atoms and for the molecules comprised of those atoms. Z_{eff} can range from a maximum of the atomic number Z, corresponding to an inner electron that "sees" only the bare nucleus, to a minimum of unity. We will return often to the concepts of penetration and shielding and the manifestation of both in the form of Z_{eff}.

Drawing from Chapter 8, we can use the radial density of the probability of finding an electron at a given radius to draw important conclusions about the balance between penetration and shielding. Most strongly penetrating are the s orbitals, and it is instructive to examine the superposition of s and p orbitals by graphing the radial probability as a function of distance r from the nucleus as shown in Figure 9.13. The 1s orbital will always be of lowest energy because it dominates the most deeply penetrating domain close to the nucleus. But the competition between the 2s and 2p orbitals is less straightforward. While the 2p radial probability peaks at a *smaller* radius from the nucleus, the 2s orbital is of greater negative energy (greater stability, lower energy) than the 2p orbital because of the deeply penetrating lobe of the 2s orbital as shown in Figure 9.13. Electrons in the p, d, and f orbitals possess higher angular momentum, which results in greater radial distances from the nucleus (for the same principal quantum number n) resulting in a smaller and smaller effective nuclear charge Z_{eff} when we progress from p to d to f.

We can compare the radial probability as a function of the distance r from the nucleus for the n = 3 orbitals as shown in Figure 9.14. Figure 9.14 displays the 3p radial probability and the 3s radial probability emphasizing the fact that the lower angular momentum (3s) orbital is characterized by an electron distribution closer to the nucleus. So while the s orbital *penetrates* more effectively than the p orbital, the p is better *shielded* than the s. Thus we are left with the general conclusion

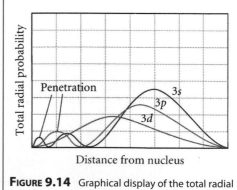

FIGURE 9.14 Graphical display of the total radial probability as a function of the distance from the nucleus for the 3s, 3p, and 3d orbitals.

that for a given principal quantum number n, the orbitals will be ordered in energy from lower to higher angular momentum such that for the sequence from s to p to d to f we have the energy ordering $E_s < E_p < E_d < E_f$.

This results in an important relationship between (1) the ordering of the energies of orbitals with the same n value but different ℓ values and (2) the penetration of the electron in a given subshell. In particular, for a given principal quantum number, n,

- s electrons penetrate closer to the nucleus than p electrons
- p electrons penetrate closer to the nucleus than do d electrons
- d electrons penetrate closer to the nucleus than do f electrons

Thus the ns orbitals are lower in energy than the (n-1)d orbitals and so the 4s orbital fills before the 3d, the 5s before the 4d, etc. This provides the basis for the simple (but powerful) diagram that determines energy ordering displayed in Figure 9.15.

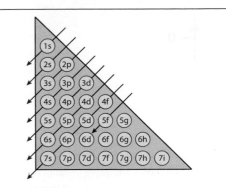

FIGURE 9.15 While the ordering of orbitals does not follow a simple sequence in the principle quantum number n, there is a simple diagrammatic way of keeping track of the subshell ordering as shown in the diagram.

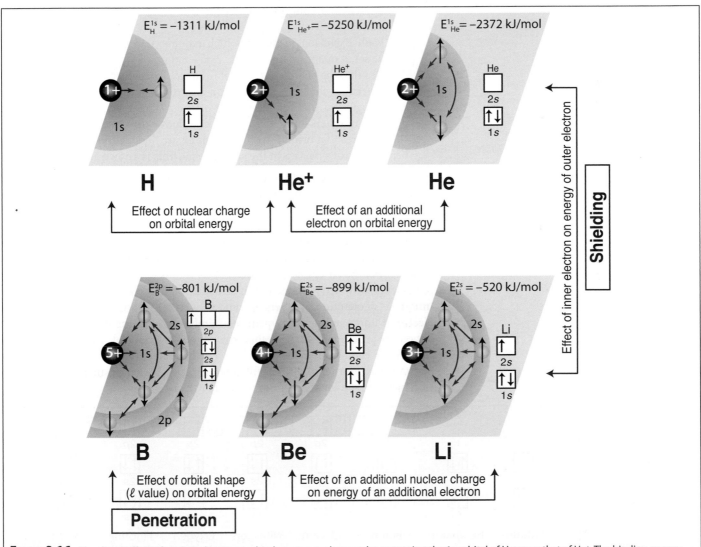

FIGURE 9.16 The direct effect of nuclear charge on orbital energy can be seen by comparing the 1s orbital of H versus that of He⁺. The binding energy of a 1s electron in He vs. that in H reflects the minimal shielding by one 1s electron on the other 1s electrons in the same orbital. Comparing Li to He demonstrates the strong shielding of the nuclear charge by the 1s electrons on the 2s electron of Li. With the addition of the second 2s electron in Be, the screening of the 2s electrons by He 1s electrons does not fully compensate for the additional nuclear charge. With the addition of the first 2p electron in boron, the shielding of the 1s² and 2s² inner electrons significantly reduce the binding energy of the 2p electron.

We can summarize the close connection between shielding, penetration and effective charge, Z_{eff}, by considering Figure 9.16. That figure traces:
1. The effect of nuclear charge on orbital energy;
2. The effect of an additional electron (and proton) on orbital energy;
3. The effect of inner electron on the energy of an outer electron (shielding);
4. The effect of an additional nuclear charge on the energy of an additional electron; and
5. The effect of orbital shape (ℓ value) on orbital energy (penetration).

The concepts of penetration and shielding taken together then determine the effective nuclear charge, Z_{eff}, experienced by the valence electron.

Building Up the Periodic Table

With this background we are now in a position to work our way through the periodic table inserting the electrons into the correct orbitals with increasing atomic number. The conceptual basis for this progression uses three primary considerations:
1. The Pauli Exclusion Principle, which requires that each electron possess a unique set of quantum numbers n, ℓ, m_ℓ, and m_s.
2. The idea that the energy *ordering* of orbitals is controlled by the combination of penetration, shielding and electron-electron repulsion.
3. The recognition that for degenerate orbitals—those with the same n and ℓ quantum numbers but with different m_ℓ values—the electron-electron repulsion requires that orbitals with different m_ℓ are filled with unpaired electrons before electrons pair up in the same orbital. This is known as Hund's rule.

With this we are ready to determine the ground state *electron configuration* for each element in the periodic table. This is a process of *building up* (termed the Aufbau principle) from the elements with the smallest number of electrons to those with the largest. Each time we add a proton to the nucleus, we add an electron to the orbital structure and the energy ordering of those orbitals is such as to maximize the attractions and minimize the repulsions, thereby minimizing the energy of the ensemble of electrons and protons.

We can now proceed, using this structure of identifying each orbital with its unique set of quantum numbers n, ℓ, and m_ℓ with the insertion of each subsequent electron with its unique spin quantum number m_s. Picking up where Figure 9.9 left off, we proceed along the second row of the periodic table in Figure 9.17 with the *electron configuration* associated with each element (e.g. Boron $1s^2 2s^2 2p$, etc.) and the energy-ordered orbitals filled sequentially in energy such that each electron has a unique set of quantum numbers n, ℓ, m_ℓ, and m_s.

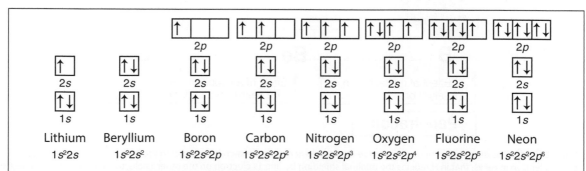

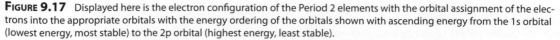

FIGURE 9.17 Displayed here is the electron configuration of the Period 2 elements with the orbital assignment of the electrons into the appropriate orbitals with the energy ordering of the orbitals shown with ascending energy from the 1s orbital (lowest energy, most stable) to the 2p orbital (highest energy, least stable).

There are some important things to note in Figure 9.17. First, in the isolated atom, the energy of each of the orbitals with the same value of ℓ is the same. Thus given a designated principal quantum number n, and angular momentum quantum number ℓ, all orbitals with that n and ℓ have exactly the same energy—these orbitals are referred to as "degenerate" because they are all of equal energy. Second, in filling the orbitals with $\ell \geq 1$, electrons enter orbitals with *different* values of m_ℓ *before* the electrons pair up to fill the remaining orbitals. The reason for this is both simple and very important! Because electron-electron repulsion dominates the interaction energy of the system within a given set of quantum numbers n, ℓ, and m_ℓ, the configuration that *maximizes* the *distance between electrons* is the configuration of lowest energy. This is clear if we revisit the geometry of, for example, the p orbitals in Figure 8.41. The double lobed geometry of the p_x, p_y, and p_z orbitals aligned along the orthogonal axes of the Cartesian coordinate system provides a greater distance between two electrons, wherein each electron is located within its own p orbital, than if two electrons are located within a *single* p orbital. Third, the number of electrons occupying each of the orbitals is designated by the superscript on the orbital sequence in the configuration designation. Thus, for example oxygen has 4 electrons in the 2p orbital, so we designate that electron configuration $2p^4$; boron has 2 electrons in the 2s orbital, $2s^2$; etc.

We can represent this progressive addition of electrons in a variety of graphical approaches, and we will do so with greater detail for the first two periods of the periodic table and then seek to simplify the notation for Period 3 through 8. Figure 9.18 captures the sequence of building up from hydrogen through neon—the first 10 elements in the periodic table, demonstrating the energy ordering of the orbitals and the shape and orientation of the orbitals graphically. The filling of orbitals 1s, 2s, and 2p is shown indicating the number of electrons in each orbital and the

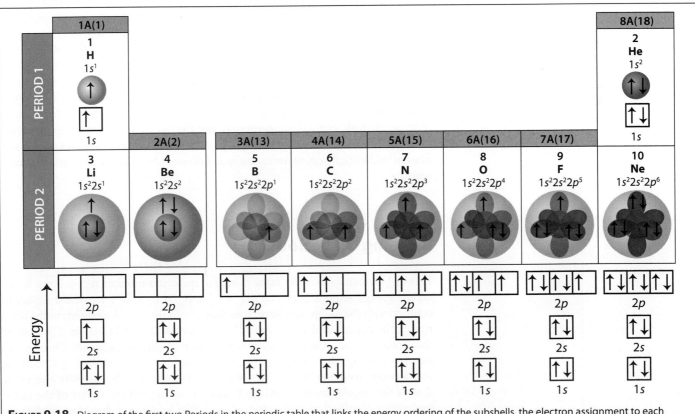

FIGURE 9.18 Diagram of the first two Periods in the periodic table that links the energy ordering of the subshells, the electron assignment to each subshells for each atom, and the geometrical distribution of the s and p subshells. The occupancy of the orbitals is indicated by the degree of shading of the specific orbitals.

energy ordering for each element with the *filling* of the orbitals indicated by the *degree of shading* with doubly occupied orbitals shaded darker. Notice that we order the subshells *vertically* in order of *increasing* energy in the designation of the electron configuration. We will do this in all cases to keep track of both the electron occupancy *and* the energy ordering. The reason for this will become evident when we begin to combine atomic orbitals to form molecules in subsequent chapters.

With the 1s orbital filled with two electrons, $m_s = +\frac{1}{2}$ and $-\frac{1}{2}$, the first Period is complete with He, the smallest noble gas. Lithium, the crucial element in modern battery development, places its third electron in the 2s orbital to establish the $1s^2 2s^1$ configuration with beryllium filling the 2s orbital to establish the $1s^2 2s^2$ configuration. The next available orbital is the 2p, which can accept a total of six electrons into the degenerate p_x, p_y, and p_z orbitals as shown in Figure 9.18. So boron inserts its 5th electron into the $2p_x$ orbital, carbon inserts its 6th electron into the $2p_y$ orbital to minimize the electron-electron repulsion that would occur if that electron was paired into the same p_x orbital. Nitrogen inserts its 7th electron into the $2p_z$ orbital for the same reason. Now each 2p orbital has a single, unpaired electron and with oxygen, the 8th electron has a choice. Either it can be "promoted" to the 3s orbital, or it can be paired with the electron already in the p_x orbital. Because the empty 3s orbital is higher in energy than the half filled $2p_x$ orbital, the 8th electron of oxygen chooses to pay the "pairing-up" cost of entering the half occupied $2p_x$ and so oxygen has the $2s^2 2p^4$ electronic configuration, as shown in Figure 9.18. Fluorine with 9 electrons inserts the next electron in the sequence into the $2p_y$ orbital and neon completes the filling of the 2p orbitals with its 10th electron into the $2p_z$ orbital as shown in Figure 9.18.

It should be emphasized that we designate the p_x, p_y, and p_z orbitals with subscripts simply to keep track of the electron pairing in the electron configuration. The designation is arbitrary because the 2p orbitals are all of equal energy — they are degenerate. It is also important to remember that the three 2p orbitals; p_x, p_y, and p_z correspond to the magnetic quantum numbers $m_\ell = -1, 0, +1$, but that assignment correllating the axes definition (x, y, z) to the value of m_ℓ is also arbitrary. The important point is that all 2p orbitals possess the same angular momentum for they all have n = 2, ℓ = 1. It's just that, in the world of quantum mechanics, all angular momenta are quantized along a spatial axes! The Stern–Gerlach experiment demonstrated that principle for electron spin, which results from an intrinsic angular momentum of the electron; but the spatial quantization of angular momentum is a general property of quantum mechanical systems and one that is distinctly counterintuitive based on our observation of the macroscopic world!

Building Up Period 3

Period 3 elements share the same principal quantum number n = 3. Thus, based on the energy sequencing so far ($E_{1s} < E_{2s} < E_{2p} < E_{3s} < E_{3p}$) we would expect to contend with the 3s, 3p, and 3d orbitals that would involve 18 elements in the third row of the periodic table ($3s^2\, 3p^6\, 3d^{10}$ corresponding to 18 electrons). We stated that energy of the orbitals increases with principal quantum number n because an electron's average distance from the nucleus increases with increasing n resulting in a weaker attraction and higher (less negative) energy. The case was also made that as the angular momentum (ℓ) increased, so too did the radius of the radial probability distribution of the electron such that $E_s < E_p < E_d$ based upon the progressive decrease in electron *penetration* and a corresponding increase in *shielding*. We could already see, however, from Figure 9.14, that the battle between deeper

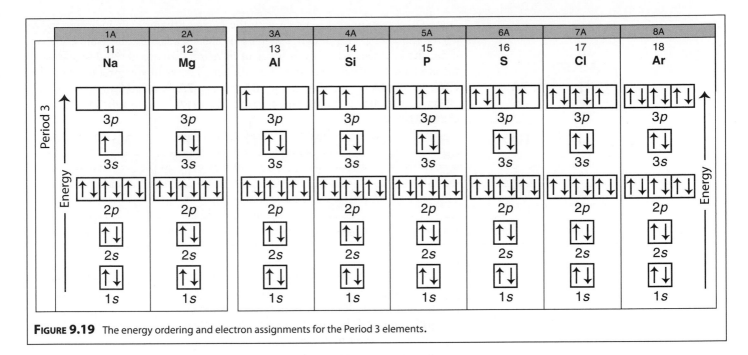

FIGURE 9.19 The energy ordering and electron assignments for the Period 3 elements.

penetration inherent in the s orbital and the effect of modest increases in the radial probability distribution with increasing angular momentum (ℓ), implies that the possibility exists for a reversal in energy ordering of an ns orbital with an (n–1) d or (n–1)f orbital. Such is the case with the 4s and 3d orbital. The 4s electron is better able to penetrate than the more strongly shielded 3d orbitals such that Z_{eff} is greater for the 4s than the 3d rendering the 4s *lower* in energy (more stable) than the 3d. Thus Period 3 does not include the 3d orbitals and as a consequence has 8 members rather than 18. Figure 9.19 delineates the full electron configuration of the 8 members of Period 3: sodium through argon indicating the energy ordering of the orbitals. *We note again and will adopt the convention that whenever diagramming the electron configuration, the orbitals will be ordered vertically with increasing energy.* Notice that for Period 3, the insertion of electrons into the energy ordered orbitals follows exactly the same principles that held for the Period 2 elements wherein the 3p orbitals are filled first with the electrons unpaired (Hund's rule) until the $3p_x$, $3p_y$, and $3p_z$ orbitals each have an unpaired electron, then the electrons pair up to complete the Period.

With the addition of sodium, magnesium, aluminum, silicon, phosphorus, sulfur, chlorine, and argon in Period 3, we have now surveyed the electronic configuration of 18 elements in the periodic table and this affords us the opportunity to examine the relationship between our electron configuration and the *chemical behavior* of those elements—a central objective of chemistry. The vertical columns in the periodic table comprise the *Groups*, and we can condense and summarize the electronic configuration periodic table, as shown in Figure 9.20, and note that within each group, the *configuration* of the outer electrons is the same with the exception of the principal quantum number, n. Examination of Figure 9.20 reveals the following *correlations* between the *electronic configuration* in each Group and the *chemical properties* of that Group.

1. In Group 1A, lithium has a three proton nucleus with the inner $1s^2$ electrons shielding the proton charge from the third electron in the outer 2s orbital. Calculations demonstrate that the attraction of the $1s^2$ electrons by the 3 protons in the nucleus draws them in, penetrating to small radii and shielding the 2s electron such that $Z_{eff} = 1.3$ for the *third* electron in lithium. This leaves the 2s electron in lithium at a high energy (i.e. a relatively small negative potential energy), making its removal relatively easy. The next element down the Group

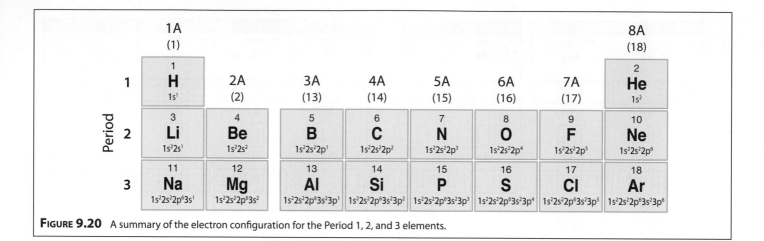

FIGURE 9.20   A summary of the electron configuration for the Period 1, 2, and 3 elements.

1 column is sodium with 10 electrons in its $1s^22s^22p^6$ core and the 3s electron significantly shielded ($Z_{eff} = 2.5$) by the penetration of the core electrons. Both lithium and sodium are highly reactive metals that form ionic compounds with nonmetals because they easily donate their electrons, and all react vigorously with water to displace H_2.

2. At the other end of the periodic table, the Group 7A elements, the halogens fluorine and chlorine, have the same np^5 configuration and both are highly reactive *nonmetals* that form ionic compounds with metals (KCl, MgF_2), homonuclear covalently bonded diatomic molecules with each other (F_2, Cl_2), and covalent compounds with hydrogen (HF, HCl). But the p^5 electrons *penetrate* effectively to relatively small distances from the nucleus and as a result are less shielded by the inner electrons. As a result Z_{eff} is 5.5 for fluorine and roughly as large for chlorine. Thus the Group 7A elements react by *removing* electrons from other elements—they are strong oxidizing agents.

3. The Group 8A elements in Figure 9.20, helium, neon, and argon, share fully filled electron configurations, and are virtually inert. They have no propensity to donate electrons and no propensity to extract electrons. They are the noble gases—unreactive monatomic gases.

Building Up Period 4

With the insertion of the 4s orbital at lower energy (more stable) than the 3d orbital, we now have a Period with 18 elements that fill out the $4s^23d^{10}4p^6$ configuration spanning from potassium $[1s^22s^22p^63s^23p^6]$ $4s^1$ to krypton $[1s^22s^22p^63s^23p^6]$ $4s^23d^{10}4p^6$. The first thing we typically do, for the sake of brevity, is to condense the argon core $[1s^22s^22p^63s^23p^6]$ into the abreviation [Ar] such that the electron configuration for potassium becomes [Ar] $4s^1$ and the configuration for krypton becomes [Ar] $4s^23d^{10}4p^6$.

But now the close competition between the energy ordering of the 4s and 3d orbitals manifests itself as we build up the Period 4 elements. As shown in Figure 9.21, potassium [Ar] $4s^1$ and calcium [Ar] $4s^2$ electron configurations reflect the fact that the 4s orbital lies lower in the energy than the 3d—when the 3d is unpopulated. But the next element, scandium, places the next electron in the 3d orbital such that we have the configuration Sc: [Ar] $4s^2$ $3d^1$ but with the addition of the third electron, the 3d orbitals displace the 4s in energy ordering as shown in Figure 9.21. Experiments have conclusively demonstrated that the 4s electron in scandium is higher in energy than the 3d electron. This observed fact highlights

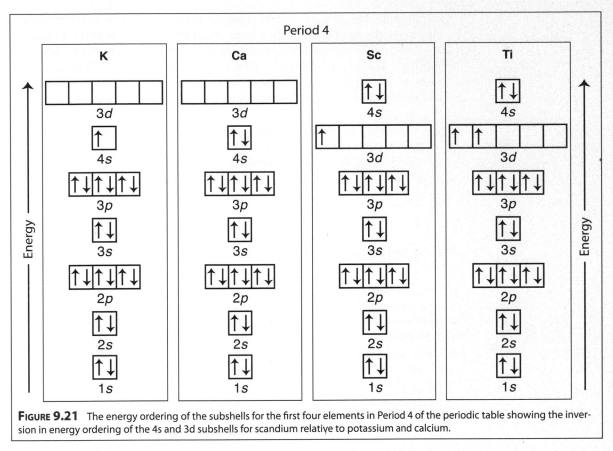

FIGURE 9.21 The energy ordering of the subshells for the first four elements in Period 4 of the periodic table showing the inversion in energy ordering of the 4s and 3d subshells for scandium relative to potassium and calcium.

two important points:

1. The presence of an electron in an orbital affects the energy spacing of all outer electrons because of the interplay between *penetration* and *shielding* that in turn control the effective nuclear charge, Z_{eff}, that a given electron "sees."

2. An atom's total energy is not simply equal to the sum of its orbital energies. The electron-electron repulsion forces each electron to account for the existence of all other electrons.

Thus scandium arranges its electrons so as to minimize the energy of the ensemble of all electrons and protons such that the two electrons remain in the 4s orbital and the third electron enters the 3d that is *lower* in energy than the 4s.

We can now complete the electron configuration through the filling of the d orbital in Period 4 as displayed in Figure 9.22. Note that because of the close

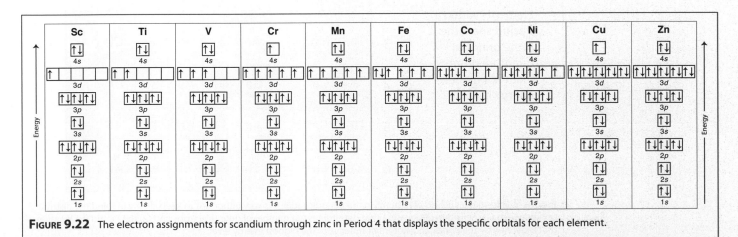

FIGURE 9.22 The electron assignments for scandium through zinc in Period 4 that displays the specific orbitals for each element.

9.19

energy separation of the 4s and 3d subshells, chromium adds to the next (the 6th) electron to the 3d orbital but also moves one of the 4s electrons to the 3d yielding a $3d^5$ electron configuration. Copper adds the tenth electron to the 3d orbital by moving a 4s electron to the 3d orbital as shown in Figure 9.22. This rearrangement, which creates a half filled subshell ($3d^5$) for Cr and a filled 3d subshell for Cu($3d^{10}$), decreases the total energy of the electron-proton ensemble by reducing the electron-electron repulsions.

Zinc, with its $3d^{10}$ configuration closes the n = 3 shell. The next orbitals to be filled in the valence shell of Period 4 are the 4p orbitals, starting with gallium (z = 31, [Ar] $4s^23d^{10}4p^1$) and ending with the noble gas krypton (z = 36, [Ar] $4s^23d^{10}4p^6$).

Organization of the Periodic Table

Our objective is to use the intrinsic organization of the periodic table to develop an understanding and an intuition for *chemical reactivity* based upon *similarities* and *patterns* revealed by systematic domains within the periodic table.

Before we employ the insight gained by the radial and angular geometry of the *s, p, d,* and *f* orbitals that set the energy ordering of the subshells, it is important to recall the distinction between the Periods and the Groups that constitute the organizing structure of the periodic table. Figure 9.23 summarizes those key distinctions, and it is well worth noting them as they are a recurring theme.

In particular, note that elements within a given Period, the horizontal rows of the table, share the same principal quantum number, *n,* of their highest energy level. Elements within a Group, the vertical columns of the table, have the same

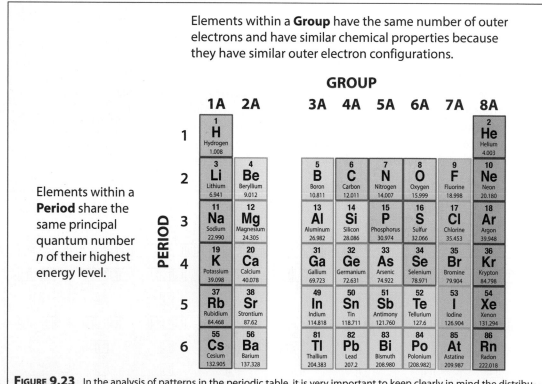

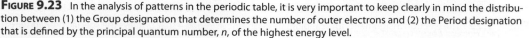

FIGURE 9.23 In the analysis of patterns in the periodic table, it is very important to keep clearly in mind the distribution between (1) the Group designation that determines the number of outer electrons and (2) the Period designation that is defined by the principal quantum number, *n,* of the highest energy level.

number of outer electrons and have similar chemical properties because they have similar outer electron configurations.

Next we note the correlation between the s, p, and d block elements within the periodic table and the corresponding s, p, and d atomic orbital shapes that distinguish the specific geometry of the outer electrons in each case.

In developing the pattern recognition that guides our systematic understanding of chemical reactivity, we can divide the whole of the periodic table into blocks according to the orbital angular momentum quantum number ℓ, as displayed in Figure 9.24. That is, we partition the periodic table into the s, p, d, and f blocks and we identify the systematic behavior *within each block*. This links chemical behavior to orbital shape and size that are in turn controlling factors in electron penetration and shielding that in turn control Z_{eff}, recalling that Z_{eff} controls the propensity for specific elements to donate or extract electrons or electron density from other elements.

Note from Figure 9.24 that the s and p blocks constitute the "main group" elements. These main group elements capture, to a remarkable degree, the full range of chemical reactivity across the periodic table.

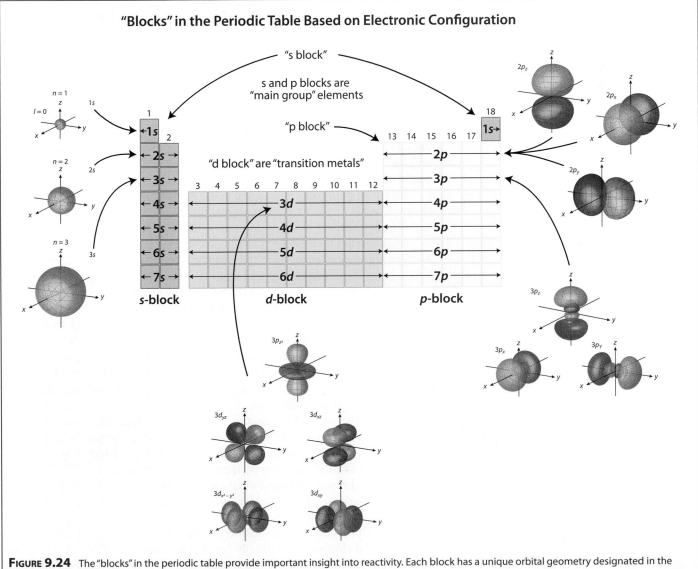

FIGURE 9.24 The "blocks" in the periodic table provide important insight into reactivity. Each block has a unique orbital geometry designated in the figure. Both the radial and angular geometry of the s, p, and d orbitals that in turn define the block designations in the periodic table play an important role in determining chemical properties of the elements.

Referring to Figure 9.24, the s block takes up the first two columns. The chemical reactivity of those elements is determined by one or two electrons respectively in the valence *ns* orbital. Lithium is grouped with cesium extending from 1s to 6s with beryllium grouped with barium extending from $2s^2$ and $6s^2$ valence orbitals. Because of the shielding of the core noble gases that precede them in the periodic table, Z_{eff} for each of those 2s–6s electrons is very small, and thus these alkali metals easily donate the electron. As a result, the alkali metals are very easily oxidized and thus acquire oxidation states of +1. The next column in the s block includes the metals beryllium and magnesium and the alkaline earth metals calcium and strontium, which also easily surrender their s electrons to chemical encounter. Thus they typically acquire oxidation states of +2 in their respective compounds.

The p block of elements, with the six columns corresponding to the np^1 through np^6 electron configurations, represent a remarkable range in chemical behavior. In fact a majority of chemical research has been expended on the understanding of the remarkable variability in chemical behavior of the p block elements alone.

Joining Periodic Behavior to Chemical Reactivity

Our central objective is to develop an understanding of how the chemical properties of the elements, that evolve across the periodic table as patterns, can be traced to the insight in atomic structure that evolves from the quantum mechanical structure of orbitals.

To this end we address the linkage between the organization of the periodic table and the s, p, d orbitals as they inform the linkage between electron shielding and penetration; atomic size; effective nuclear charge, Z_{eff}; ionization energy (IE) and electron affinity (EA); and chemical reactivity. This coupling is displayed graphically in Figure 9.25.

Before we go into any more detail in our pursuit of those factors that influence chemical behavior, we review four points that are intrinsic to the periodic table:

1. Elements within a Group (a column) in the periodic table have similar chemical properties because they have similar outer electron configurations.
2. The sequence of elements in the periodic table is built up by filling orbitals in order of increasing energy, and while that ordering of energy is not simply a sequence in principal quantum number n, the energy ordering can be coherently represented and is displayed in Figure 9.15.
3. The chemical behavior of the elements in the periodic table is based upon three categories of electrons
   a. The *core* electrons that constitute the inner, lower energy levels of an atom. While the core electrons do not engage directly in a chemical reaction, they do influence the penetration and shielding of all electrons in the atom.
   b. The *outer* electrons that occupy the highest (least stable) energy level. Those are the electrons that must directly "see" or are seen by, electrons in other atoms as the internuclear distance between two potentially reactive partners (atoms) decreases.
   c. The *valence* electrons that are involved directly in forming molecules comprised of individual atoms. In the case of the main group elements, the valence electrons and the outer electrons are one-and-the-same. When considering the transition elements, while all the d electrons are formally counted as valence electrons, it is often the case that some of those d electrons are not involved in the actual formation of bonding structures in certain classes of compounds.

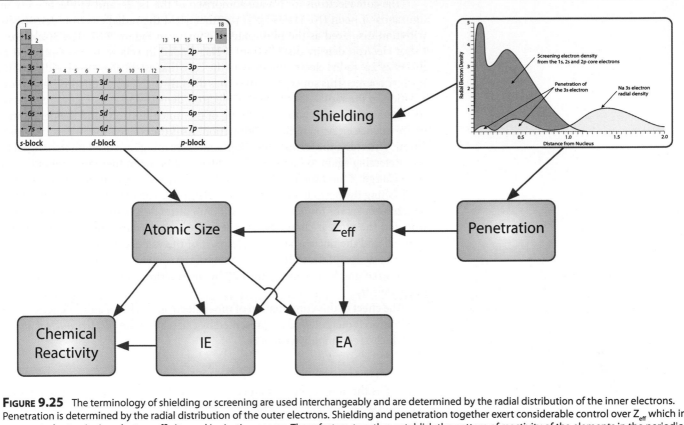

FIGURE 9.25 The terminology of shielding or screening are used interchangeably and are determined by the radial distribution of the inner electrons. Penetration is determined by the radial distribution of the outer electrons. Shielding and penetration together exert considerable control over Z_{eff} which in turn controls atomic size, electron affinity, and ionization energy. These factors, together, establish the pattern of reactivity of the elements in the periodic table.

4. The distinction between the Group number (which for main group elements is the *number* of *outer electrons*) and the Period number (the principle quantum number n of the highest energy level) is the central organizing concept in the periodic table. The Group number determines the column and the Period number determines the row for a given element. Notice that the princip quantum number squared (n^2) provides the total number of orbitals in that energy level.

Electron Shielding and Penetration

We turn now to the question of which factors determine the periodic properties of atoms and how those periodic properties control chemical reactivity.

Treatment of this issue depends to a remarkable degree on how lower-energy inner electrons shield higher-energy outer electrons from experiencing the full Coulomb charge of the nucleus. The general picture for a multielectron system then is displayed in Figure 9.26.

In particular Figure 9.26 graphically displays the fact that the Coulomb interaction between the electron of interest and the positively charged nucleus is controlled by the inner electrons in the atom that reside *between* the nucleus and the electron of interest. The degree to which an electron experiences the Coulomb attraction of the nucleus depends upon two key factors: (a) the *shielding* or *screening* provided by the inner electrons, and (b) the radial penetration of the orbital within which the electron of interest resides. This relationship is displayed graphically in Figure 9.27 for the case of the element sodium, Na.

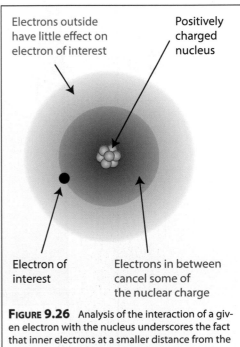

FIGURE 9.26 Analysis of the interaction of a given electron with the nucleus underscores the fact that inner electrons at a smaller distance from the nucleus than the electron of interest have a very significant impact on the energy of the electron of interest. In sharp contrast, outer electrons at a larger distance from the nucleus have a marginal effect on the energy of the electron of interest.

The core electrons for Na are comprised of the 1s, 2s, and 2p electronic configuratios of neon (Ne [$1s^2 2s^2 2p^6$]) together yield the radial electron density distribution displayed as the pink shaded domain in Figure 9.27. The 3s electron's radial electron density distribution of Na, shaded in yellow, is superimposed on the $1s^2 2s^2 2p^6$ radial electron density of the core that establishes the shielding of the nuclear charge. This superposition relates directly to the quantum mechanical calculation of the radial distribution of electron density and thus Figure 9.27 serves to emphasize the interplay of shielding of the inner electrons and the penetration of the valence electron radial wavefunction. As is the convention in descriptions of atomic structure, the terms "shielding" and "screening" are used interchangeably.

Referring again to Figure 9.27, this balance between the screening of the nuclear charge, Z, and the penetration of the valence electron is very important for establishing the periodic behavior of the elements. The reason is that the degree of penetration serves to increase the Coulomb attraction between the valence electron and the nuclear charge because it reduces the distance between the valence electron and the nucleus and it places part of the valence electron's radial density inside a fraction of the radial distribution of the screening electron density. This is emphasized graphically in Figure 9.27 by the multiple lobes of the Na 3s electron radial density.

The effect of this combination of inner electron shielding and valence electron penetration is quantitatively represented by calculating the *effective nuclear charge*, Z_{eff}, which is commonly written as

$$Z_{eff} = Z - S$$

where Z is the atomic number equal to the number of protons in the atom and S is the *shielding constant*. Because of the importance of Z_{eff} for an understanding of

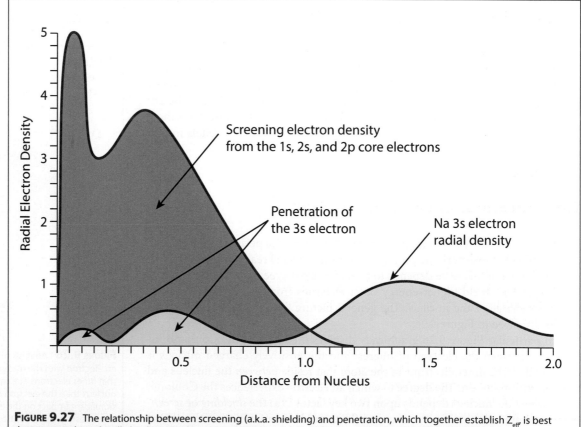

FIGURE 9.27 The relationship between screening (a.k.a. shielding) and penetration, which together establish Z_{eff} is best demonstrated graphically by distinguishing between the radial distributions of the inner electrons shown in pink and the degree of penetration of the valence electron shown in yellow for the case of sodium.

atomic size, ionization energy, electron affinity, and chemical reactivity, considerable effort has been expended in carrying out the quantum mechanical calculations of S in order to establish Z_{eff} for each element in the periodic table.

Figure 9.28 presents calculated values of Z_{eff} for the s and p block "main group" elements, and we will repeatedly refer to this figure as we explore key patterns in the periodic table.

Calculated Values of Z_{eff} for the s and p Block "Main Group" Elements

H 1.0							**He** 1.8
Li 1.27	**Be** 1.92	**B** 2.42	**C** 3.14	**N** 3.83	**O** 4.45	**F** 5.10	**Ne** 5.76
Na 2.51	**Mg** 3.31	**Al** 4.06	**Si** 4.28	**P** 4.89	**S** 5.48	**Cl** 6.12	**Ar** 6.76
K 3.50	**Ca** 4.40	**Ga** 6.22	**Ge** 6.78	**As** 7.45	**Se** 8.29	**Br** 9.03	**Kr** 9.34
Rb 4.99	**Sr** 6.07	**In** 8.47	**Sn** 9.10	**Sb** 10.0	**Te** 10.8	**I** 11.6	**Xe** 12.4
Cs 6.36	**Ba** 7.58	**Tl** 10.8	**Pb** 12.4	**Bi** 13.3	**Po** 14.2	**At** 15.2	**Rn** 16.1

FIGURE 9.28 A considerable amount of calculational effort is required to establish specific values of Z_{eff} for the main group elements shown here for each element. Notice in particular the trends in Z_{eff} across a Period and down a Group.

Periodic Trends in Atomic Size

Continuing the development of the coupling of shielding, penetration, Z_{eff}, and atomic size as they inform an understanding of chemical reactivity, we turn next to the important question of atomic size. As we will see, the size of an atom in important ways affects an element's chemical activity as well as exerting influence on the geometry of chemical bonding.

While there are various definitions of atomic size, and it is important to recognize these definitions, determination of the size of the atom varies little depending on the chosen definition. In Figure 9.29 are displayed four examples of how atomic radii are defined. In panel A, the case of a noble gas in a frozen matrix is shown corresponding to the fact that a chemical bond is not formed, but rather the frozen crystal structure is maintained by weak Van der Waals forces. In this case the atomic radius is simply one-half the length between nuclei of the element.

In panel B the case of a covalent bond between two identical atoms is displayed—in this case two Cl atoms. The atomic radius is taken as one-half the internuclear distance of the Cl_2 molecule. In panel C the case of a molecular bond comprised of two different atoms, Cl and C, is displayed. In this case the radius

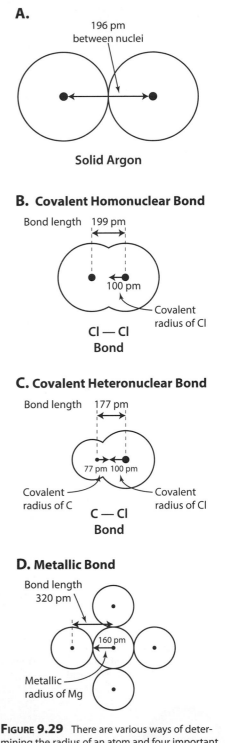

FIGURE 9.29 There are various ways of determining the radius of an atom and four important cases are shown here. For solid argon, the interatomic distance in the crystal structure is taken as twice the atomic radius. In a homonuclear molecule that atomic radius is half the distance from nucleus to nucleus. For a covalent heteronuclear molecule the internuclear distance is subtracted from the covalent radius of Cl to determine the radius of carbon. In a metallic bond it is just half the distance between the internuclear distance in the metallic crystal.

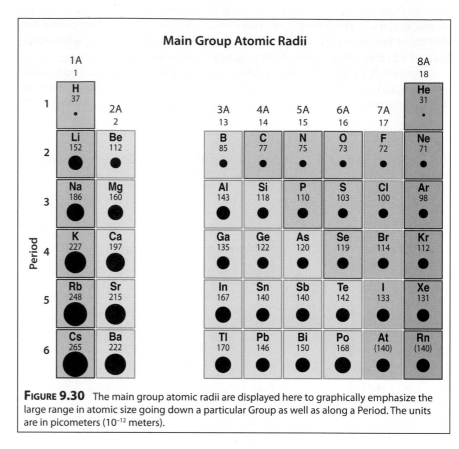

FIGURE 9.30 The main group atomic radii are displayed here to graphically emphasize the large range in atomic size going down a particular Group as well as along a Period. The units are in picometers (10^{-12} meters).

of the C atom is taken as the bond length between nuclei minus the radius of the covalent Cl atom: 177 pm – 100 pm = 77 pm. Finally, panel D presents the case of metallic bonds—in this case for magnesium, in which the atomic radius is simply one-half the metallic bond length.

Atomic size is controlled by (1) the number of electrons and protons, (2) the radial and angular geometry of the outer electrons, and (3) the effective nuclear charge, Z_{eff}. As a comparison between Figure 9.28 and Figure 9.30 reveals for a given radial and angular geometry of the atomic orbitals along a given period (horizontal row) of the periodic table, as the number of electrons increases, Z_{eff} increases and the size of the atom decreases. As we progress down a given Group (vertical column) in the periodic table, the principal quantum number, n, increases with the addition of a new shell and the size of the atom increases.

Thus as we progress down a given Period in the main group elements both Z_{eff} and the atomic radius increase.

As a result of careful measurements we can summarize the main group atomic radii as displayed in Figure 9.30 in units of picometers (pm) or 10^{-12} meters.

Notice, in particular, in Figure 9.30, the remarkably systematic trend in atomic radii both with respect to increasing size as you progress down a given Group (column) and with respect to decreasing size as you progress from left to right along a given Period.

It is also important to emphasize that among members of a Group, the differences in the size of the atom plays a key role in determining the chemical behavior of the element. Taking Group 4A as an example, there is a stark difference in the chemical properties of carbon vs. lead as summarized in Figure 9.31.

We can summarize the periodic pattern of atomic radii across the periodic table in Figure 9.32. The precipitous increase in atomic radius occurs for each of the alkali metals as we progress to the next integer value of the principal quantum number, n. Following that precipitous rise in atomic radius for each new Period,

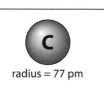

Carbon (C)
Nonmetal
Gaseous soluble oxides
Inert to acid

radius = 77 pm

Lead (Pb)
Metal
Solid insoluble oxides
Reacts with nitric acid

radius = 146 pm

FIGURE 9.31 A clear example of the role of atomic size in determining the differences in the chemical behavior of members of the same group. Displayed here are two members of Group 4A with carbon in Period 2 and lead in Period 6.

(for each new row in the periodic table) the atomic radius decreases as Z_{eff} rapidly increases across a given Period.

Therefore, the observed pattern in atomic size results from two opposing influences. As the principal quantum number, n, increases, the outer electrons are farther from the nucleus, decreasing the Coulomb force between the nucleus and the added electron

$$F_{nuc-e} = \frac{Z_{eff}e^2}{4\pi\varepsilon_0 r^2}$$

As we move across a given Period, Z_{eff} increases such that the outer electrons are pulled closer to the nucleus thereby decreasing the atomic radius.

It is the combination of atomic size (i.e. radius) and effective nuclear charge, Z_{eff}, that determines, in large measure, (a) the energy required to remove an electron from the atom, the "ionization energy," IE, and (b) the ability of a given atom to attract an electron, the "electron affinity," EA. As we will repeatedly see, it is the IE and EA of an atom that in turn controls its chemical reactivity. We consider, therefore, the trends in IE and EA across the periodic table.

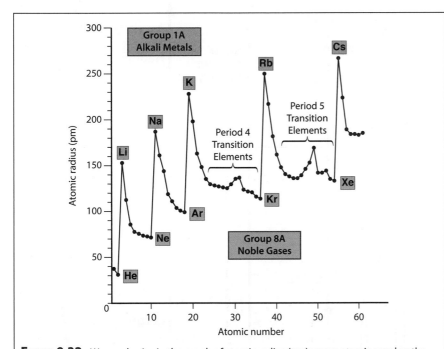

FIGURE 9.32 We emphasize in the graph of atomic radius (pm) versus atomic number the sharp transition between the noble gas radius and the following alkali metal in the periodic table. Also note the sharp drop in atomic radius in going from the alkali metal to the adjoining alkaline earth.

Periodic Trends in Ionization Energy

Although not obvious, more than any other atomic property, the ionization energy, IE, ties together the orbital model and periodicity and exerts surprising control

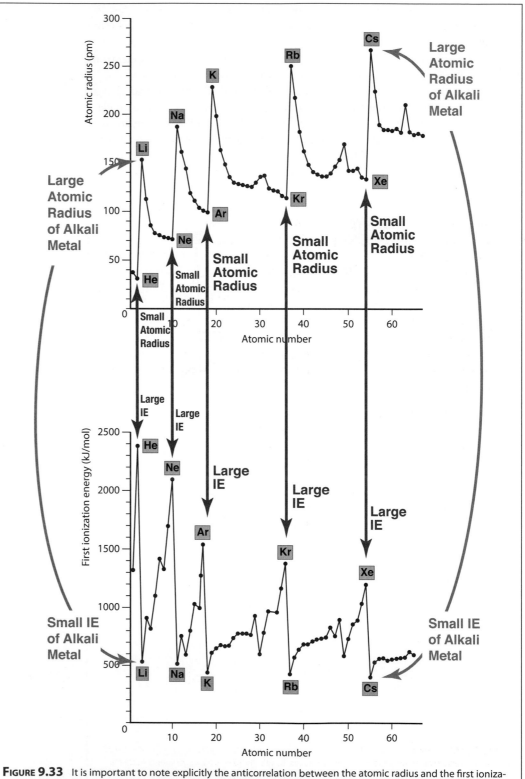

FIGURE 9.33 It is important to note explicitly the anticorrelation between the atomic radius and the first ionization energy—a relationship that holds throughout the periodic table.

over chemical reactivity. Thus we explore patterns in ionization energy across the periodic table in some detail.

Extracting an electron from an atom requires energy because the electrons in a stable atom reside in bound states. This is made explicit by examining Figures 9.9, 9.11, 9.12, and 9.16. The first ionization energy, IE_1, is defined to be the energy needed to remove the most weakly bound electron in an atom, which is the electron in the highest occupied atomic orbital. Thus the first ionization energy, IE_1, is defined as the energy, $\Delta E = IE_1$, required to remove an electron from an atom wherein

$$\text{Atom} \rightarrow A^+ + e \qquad \Delta E = IE_1 > 0$$

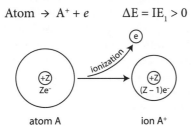

atom A   ion A⁺

To remove a second electron

$$A^+ \rightarrow A^{++} + e$$

requires an amount of energy $\Delta E = IE_2 > 0$.

There is a roughly inverse relationship between IE_1 and atomic size. As we proceed down a given Group in the periodic table, the distance from the nucleus to the outer electron increases. Thus because the Coulomb interaction decreases as $1/r^2$, the attraction between the outer electron and the nucleus decreases, thus reducing IE_1. This important covariance between atomic size and the first ionization energy, IE_1, is displayed in Figure 9.33.

We summarize the main group trends in IE_1 graphically in Figure 9.34. Note first the general increase in IE_1 as we progress from left to right along a Period in the periodic table and the general decrease in IE_1 as we progress down a Group in the periodic table.

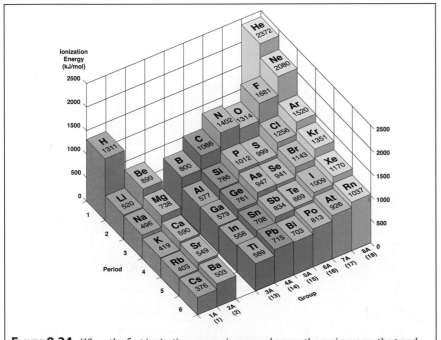

FIGURE 9.34 When the first ionization energy is mapped across the main group, the trend is clear. IE_1 decreases as we proceed down a Group, but IE_1 increases as we proceed across a Period. Important counterexamples include nitrogen to oxygen and beryllium to boron. See text for a discussion of those cases.

As we progress across a given Period, the effective charge, Z_{eff}, increases so the atomic radii become smaller such that the valence electron becomes more difficult to remove, increasing IE_1 across a given Period. Important exceptions to this pattern include, in Period 2, oxygen and boron, and in Period 3, aluminum and sulfur. The decrease in IE_1 for oxygen results from the fact that the addition of the $2p^4$ electron into a shared p orbital increases the electron-electron repulsion, *raising* the energy of that electron and thereby *decreasing* IE_1 for that valence electron. The decrease in IE_1 for boron relative to beryllium results from the fact that the 2p orbital of B is of higher energy than the $2s^2$ orbital of Be. The argument for the drop in IE_1 for aluminum (3p) versus magnesium ($3s^2$) is the same as for B relative to Be. The drop from $P(3p^3)$ to $S(3p^4)$ is caused by the pairing of the fourth electron in the 3p subshell with another electron in the same subshell increasing Coulomb repulsion and decreasing IE_1 for S relative to P.

We can gain insight into the relationship between the energy of an $n\,\ell$ subshell, the effective nuclear charge Z_{eff} and the principal quantum number, n, by noting that the energy of the Bohr orbit (see Chapter 8) is

$$E_n = -\frac{Z_{eff}^2 e^2}{2a_0 n^2}$$

where a_0 is the Bohr radius

$$a_0 = \frac{h^2}{4\pi^2 m_e e^2} \; .$$

Each passage of a chemical Period corresponds to a jump from n to $n+1$. As the Period is traversed, n is held fixed while Z_{eff} increases, thereby lowering the orbital energy and increasing IE_1. This sharp differentiation in IE_1 as n jumps to n + 1 is displayed quantitatively in Figure 9.35.

Element	Number of valence electrons	IE_1 [kJ/mol]	IE_2 [kJ/mol]	IE_3 [kJ/mol]
Li	1	520	7300	11810
Be	2	899	1760	14850
B	3	801	2430	3660
C	4	1086	2350	4620
N	5	1402	2860	4580
O	6	1314	3390	5300
F	7	1681	3370	6050
Ne	8	2081	3950	6120
Na	1	469	4560	6910
Mg	2	738	1450	7730
Al	3	578	1820	2750
Si	4	786	1580	3230
P	5	1012	1900	2910
S	6	1000	2250	3360
Cl	7	1251	2300	3820
Ar	8	1521	2670	3930

FIGURE 9.35 Trends in the first (IE_1), second (IE_2), and third (IE_3) ionization energies are tabulated to show trends across the periodic table as well as across the first-to-third ionization energy.

9.30

The second ionization energy, IE_2, corresponds to removing the highest energy (most easily removed) electron from A^+; IE_3 the most easily removed electron from A^{++}, etc. Because Z_{eff} increases with the removal of each subsequent electron, the sequence of ionization energies increase rapidly in the progression from IE_1 to IE_2 to IE_3, etc.

We can make this explicit by tabulating the subsequent ionization energies for lithium vs. sodium in Figure 9.35.

As illustrated below, the increase in ionization energy from left to right in the periodic table and the increase from top to bottom imply a general pattern in IE_1 increasing from lower left to upper right. Thus any element lying to the right or above a given element in the periodic table will, with minor exceptions, have a higher IE as displayed in Figure 9.36.

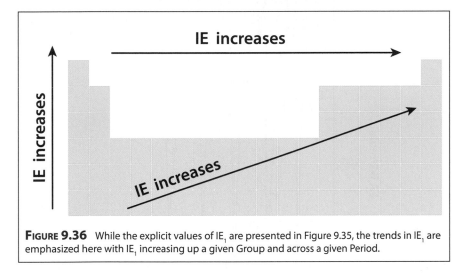

FIGURE 9.36 While the explicit values of IE_1 are presented in Figure 9.35, the trends in IE_1 are emphasized here with IE_1 increasing up a given Group and across a given Period.

As indicated in the figure above, as we descend a given Group (column) in the periodic table, IE falls slowly. The Bohr equation for the orbital energy

$$E_n = -\frac{Z_{eff}^2 e^2}{2a_0 n^2}$$

does not make this trend obvious because as we proceed down a Group, both Z_{eff} and n increase, tending to offset (cancel) the tendency in each. However, the screening (see Figure 9.27) allows only a modest increase in Z_{eff} allowing the n dependence of the orbital energies to dominate.

Periodic Trends in Electron Affinity

We discuss EA after ionization energy because while it plays a critical role in chemical reactivity along with atomic size and ionization energy, the pattern of EA across the periodic table is not as straightforward as the others.

The *electron affinity* (EA) for a singly charged *negative ion* is defined as the energy required to remove the highest energy (least tightly bound) electron, leaving a neutral atom.

Thus for the reaction

$$A^- \rightarrow A + e^- \quad \Delta E = E(A) - E(A^-) = EA$$

A key point to note is that EAs are invariably smaller than IEs because the extra electron that creates the anion from the neutral atom enters an atom for which the number of protons and electrons is equal such that the Coulomb attraction is just the residual effective charge of the atom with all electrons in place. Alternatively stated, the potential energy that creates the atom-electron pairing in the anion, A^-, is much weaker than for an $e\text{-}A^+$ pairing that forms a stable atom.

So, in general across the periodic table, the added electron to form the anion provides increased electron-electron repulsion and therefore $EA << -E_{HOAO}$ where

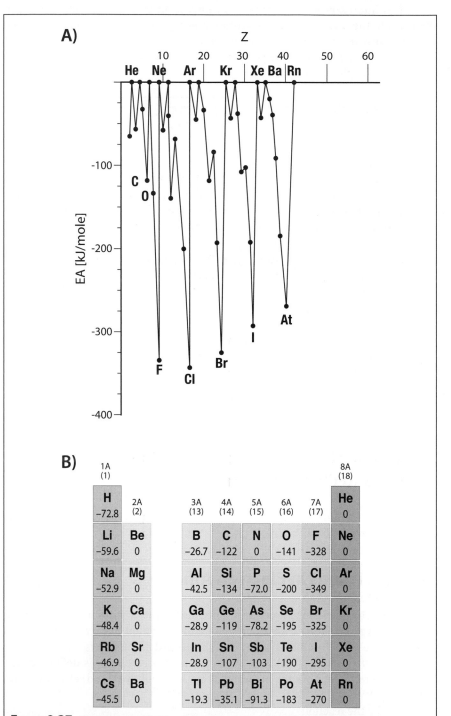

FIGURE 9.37 The electron affinities of the main group elements exhibit a similar pattern to that of the ionization energy with the halogens displaying the largest (negative) values of EA because of the correspondingly large value of Z_{eff}. However, the EAs are significantly smaller in magnitude when compared with the corresponding IEs so subtleties in other effects related to atomic structure cloud a simple pattern in EAs.

E_{HOAO} is the energy of the highest occupied atomic orbital of the neutral atom.

The most common atomic negative ions are halogen anions, the halide ions F^-, Cl^-, Br^-, and I^-, wherein the extra electron enters a "hole" in the p shell aided by (stabilized by) the very large Z_{eff} of the Group 7A elements.

We can graph the measured EAs of the elements in two equivalent ways as displayed in Figure 9.37 where panel A plots EA versus atomic number, Z, and panel B assigns the EA value of each element in the main group of the periodic table. It is important to note that those atoms (elements) for which EA = 0 corresponds to the cases where it actually costs energy to add the electron to a filled shell or subshell.

Linking Periodic Trends in IE and EA

When we compare trends in IE, and EA across the periodic table, three important patterns emerge:

1. Elements in Group 6A (oxygen through polonium) and 7A (the halogens) exhibit large ionization energies and highly negative electron affinities. These elements in particular both strongly retain their electrons and strongly attract electrons in neutral atoms. Thus they exhibit *very large* values of IE – EA, and they readily form negative ions.

2. Elements in Group 1A and 2A exhibit low ionization energies and either small or zero electron affinities. Thus Group 1A and 2A lose electrons from their neutral atoms easily and have little or no ability to attract electrons to their neutral atoms. Thus they exhibit *very small* values of IE – EA and they readily form positive ions.

3. The Group 8A (noble gas) elements have very large ionization energies and electron affinities of zero such that noble gases neither lose nor gain electrons to their closed shells and are thus chemically inert, thereby earning the designation "noble gases."

Electronegativity: Unify the Concepts of Ionization Energy and Electron Affinity

In 1932 Linus Pauling introduced the idea of *electronegativity* that quantitatively defines the ability of an element in the periodic table to draw electrons to it when engaged in a chemical bond. While we will turn to chemical bonding in the next chapter, it is instructive to point out here that the electronegativity (EN) of a given atom is one of the most important concepts in chemistry. In 1934, the American physicist Robert S. Mulliken developed an approach to electronegativity based entirely on the two key properties of an atom in the periodic table, the ionization energy, IE, and the electron affinity, EA, such that

$$EN = \frac{IE - EA}{2}$$

While we will develop the concept of electron affinity when we examine chemical bonding, it is important to note here that:

1. Patterns in IE and EA across the periodic table translate directly into the character of chemical bonding because the resulting electronegativity, EN, establishes the propensity for a given atom (element) to either draw electron density from a neighboring atom or donate electron density to a neighboring atom in a chemical bond.

1. As we move from left to right across a Period in the periodic table, the effective charge, Z_{eff}, seen by valence electron *increases*.
2. As we move from top to bottom down a given Group in the periodic table, the effective charge, Z_{eff}, seen by valence electron *increases*.
3. As we move from left to right across a Period in the periodic table, the first ionization energy, EI_1, *increases* significantly.
4. As we move from top to bottom down a given Group in the periodic table, the first ionization energy, EI_1, *decreases*.
5. As we move from left to right across a Period in the periodic table, metallic character *decreases*.
6. As we move down a Group in the periodic table metallic behavior *increases*.

2. Remarkably, the quantitative property of electronegativity in an isolated atom, determined by IE and EA of the isolated atom, is to a large degree maintained even when the atom is taking part in a chemical bond, making electronegativity a very valuable concept as the EN for an element can be assigned (to a close approximation) without specific reference to the molecular bond in question.

Trends in the Chemical Behavior of Metals

Metals are the subject of increasingly intensive study with an increasingly effective blend of experimental sophistication and theoretical capability. The chemistry of metals hold the promise for the development of new materials, new methods for generating energy, and new catalytic pathways for selective chemistry in organo-metallic research linking the organic and inorganic worlds. As Figure 9.38 makes clear, virtually the entire left-hand three-quarters of the periodic table is comprised of metals. Metals as a category tend to reflect light, are excellent conductors of electricity, and are excellent conductors of heat. Metals can be drawn into long, thin wires (ductile), can be pounded into thin sheets (malleable), and tend to donate electrons to nonmetals. Their chemical behavior pervades nearly all modern research into the properties of innovative new materials.

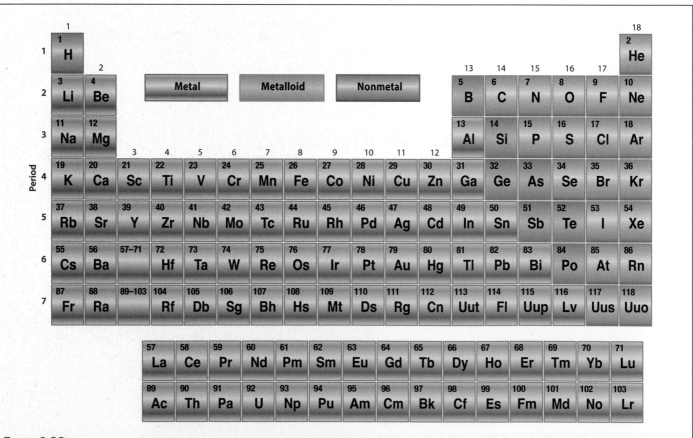

FIGURE 9.38 A key point to emphasize when considering patterns in the periodic table is that a large fraction of the elements fall in the category of metals with a far smaller fraction of nonmetals and a very limited number of metalloids.

Summary Concepts

1. The Fourth Quantum Number: Electron Spin

The hypothesis by Goudsmit and Uhlenbeck that the electron possesses an intrinsic spin that produces a magnetic moment gives rise to a fourth quantum number, m_s, called the *spin quantum number*. The spin quantum number can assume only one of two possible values $m_s = \pm\frac{1}{2}$. The electron spin was detected by Stern and Gerlach using a magnetic field to split a beam of metal atoms.

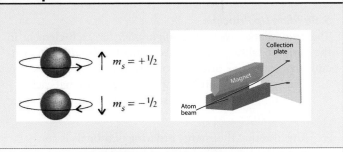

2. The Pauli Exclusion Principle

Wolfgang Pauli put forward the hypothesis that within a given orbital (defined as we have seen by a specific set of quantum numbers n, ℓ, and m_ℓ) no two electrons can share the same set of quantum numbers n, ℓ, m_ℓ, and m_s. Thus each orbital can contain a maximum of two electrons, one with spin quantum number $m_s = +\frac{1}{2}$ and one with $m_s = -\frac{1}{2}$.

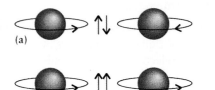

3. Penetration and Shielding I

When we consider multielectron atoms and their chemical properties it is essential to develop an intuition for how electron-electron (e-e) repulsion is balanced against electron-proton (e-p) attraction. It is this balance of forces that determines the energy ordering of individual orbitals in the structure of multielectron atoms. It is also this balance that determines the size of the atom, the ability of a given atom to retain its own electrons as well as the ability of a given atom to draw electrons to it from another atom.

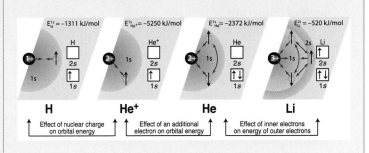

4. Penetration and Shielding II

The size of the atom and the ability of a given atom to retain electrons or draw other electrons to it ultimately determines the chemical behavior of atoms of a given element. Two ideas are particularly important for determining chemical behavior: penetration and shielding.

1. Penetration is taken to mean a measure of how close an electron comes to the nucleus. This is important because the *force* of proton-electron attraction is $F_{p\text{-}e} = -Ze^2/4\pi\varepsilon_0 r^2$ and this increases as the square of distance between the electron and the nucleus decreases.

2. Shielding is a measure of how electrons that occupy a spatial domain closer to the nucleus serve to reduce the attraction between the nucleus and the electrons at greater distances from the nucleus.

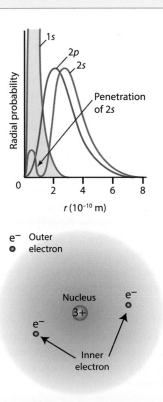

Summary Concepts

5. Building Up Principle

The ground state *electron configuration* for each element in the periodic table is determined using the *building up* principle that uses three primary considerations:

1. The Pauli Exclusion Principle that requires that each electron possess a unique set of quantum numbers n, ℓ, m_ℓ, and m_s.

2. The idea that the energy ordering of orbitals is controlled by the combination of penetration, shielding and electron-electron repulsion.

3. The recognition that for degenerate orbitals (those with the same n and ℓ quantum numbers but with different m_ℓ values) the electron-electron repulsion requires that orbitals with different m_ℓ values are filled with *unpaired* electrons before electrons pair up in the same orbital.

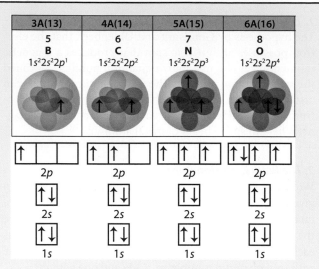

6. Relationship Between the Electronic Configuration and Chemical Properties

These are important correlations between the *electronic configuration* in each Group and the *chemical properties* of that Group.

1. In Group 1A, lithium has a three proton nucleus with the inner $1s^2$ electrons shielding the proton charge from the third electron in the outer 2s orbital. This leaves the 2s electron in lithium at a high energy (i.e. relatively small negative potential energy), making its removal relatively easy. The next element down the Group 1 column is sodium with 10 electrons in its $1s^2 2s^2 2p^6$ core and the 3s electron significantly shielded ($Z_{eff} = 2.5$) by the penetration of the core electrons. Both lithium and sodium are highly reactive metals that form ionic compounds with nonmetals because they easily donate their electrons, and all react vigorously with water to displace H_2.

2. At the other end of the periodic table, the Group 7A elements, the halogens fluorine and chlorine, have the same np^5 configuration and both are highly reactive *nonmetals* that form ionic compounds with metals (KCl, MgF_2), homonuclear covalent compounds with hydrogen (HF, HCl). But the p^5 electrons *penetrate* effectively to relatively small distances from the nucleus and as a result are less shielded by the inner electrons. As a result Z_{eff} is 5.5 for fluorine and roughly as large for chlorine. Thus the Group 7A elements react by *removing* electrons from other elements—they are strong oxidizing agents.

3. The Group 8A elements in Figure 9.18, helium, neon, and argon, share fully filled electron configurations, and are virtually inert. They have no propensity to donate electrons and no propensity to extract electrons. They are the noble gases—unreactive monatomic gases.

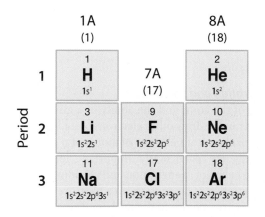

Summary Concepts

7. Organization of the Periodic Table

While the ordering of orbitals does not follow a simple sequence in the principal quantum number n, there is a simple diagrammatic way of keeping track of the subshell ordering as shown in the diagram at the right. This diagram is used by following the sequence of subshells along the axis of the arrow.

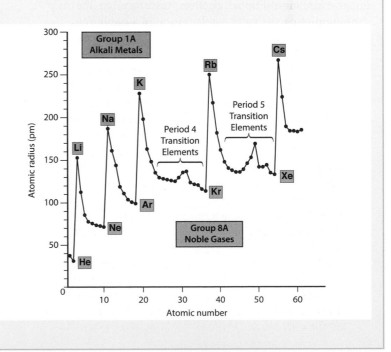

8. Chemical Reactivity and its Relationship to Atomic Size, Ionization Energy, and Electron Affinity

Because the patterns of reactivity are strongly influenced by the willingness of an element to give up one or more electrons (the ionization energy), the ability of an element to draw electrons to it (the electron affinity), and the size of the atom that reflects how closely bound orbitals are held to the nucleus, it is important to link those characteristics to the pattern of chemical reactivity across the periodic table.

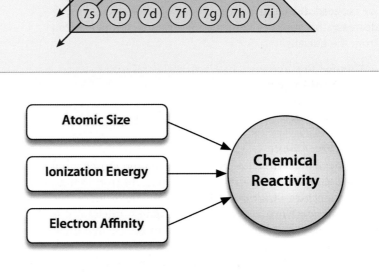

9. Atomic Size

There are two broad patterns determining atomic size that are evident across the periodic table. First, when we progress along a Period from left to right, the effective nuclear charge *increases*. Each added proton increases, Z, but the electrons in the p orbital are unable to shield additional electrons that are inserted as we progress across the Period so Z_{eff} continues to increase. As Z_{eff} increases, the electrons in the outer shells are bound more tightly and are drawn closer to the core electrons, making the outer electrons more difficult to remove. Thus, across the Period, the size of the atoms decreases, they become more difficult to ionize and more able to draw electron density from other atoms toward them. Second, as we progress down a Group, the size of the atom increases simply because we are adding more electrons, the principal quantum number is increasing, and the outer electrons are more easily removed because they are further from the nucleus.

Summary Concepts

10. Ionization Energy

The energy required to move an electron from a specific orbital and move it to a position such that the bonding energy drops to zero is termed the ionization energy of that electron from a specific orbital. The first ionization energy, IE_1, is the minimum energy required to remove from the atom the most weakly bound electron. We are primarily interested in IE_1, the first ionization energy, because it is a key factor determining the chemical reactivity of an atom. The pattern in IE_1 vs. atomic number Z is easy to understand because it follows directly from the trend in Z_{eff}—as Z_{eff} increases so does the first ionization energy. As the size of the atom increases as we proceed down a given Group, the first ionization energy decreases because of the increasing distance of the electron from the nucleus.

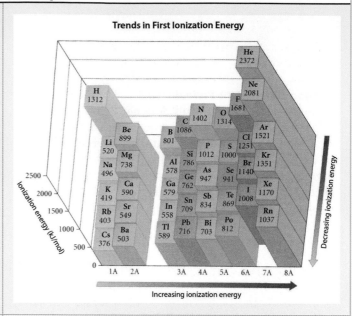

Trends in First Ionization Energy

11. Electron Affinity

By virtue of the fact that Z_{eff} is greater than zero, most neutral atoms are capable of adding an additional electron to their orbital structure—with the exception of the noble gases. The energy release when a neutral atom acquires an electron is called the electron affinity, EA. Patterns in EA are important because they affect patterns in chemical reactivity. Moving from left to right in the periodic table, electron affinity *increases* (becomes more negative). Halogens are particularly hungry for an additional electron, transforming the ns^2p^5 configuration to the ns^2p^6 configuration of a noble gas. Electron affinity tends to decrease as the atomic size increases.

Electron Affinities (kJ/mol)

1A	2A		3A	4A	5A	6A	7A
H -73							
Li -60	**Be** >0		**B** -27	**C** -122	**N** >0	**O** -141	**F** -328
Na -53	**Mg** >0		**Al** -43	**Si** -134	**P** -72	**S** -200	**Cl** -349
K -48	**Ca** -2		**Ga** -30	**Ge** -119	**As** -78	**Se** -195	**Br** -325
Rb -47	**Sr** -5		**In** -30	**Sn** -107	**Sb** -103	**Te** -190	**I** -295

CASE STUDY 9.1   The Union of Electronic Structure and Chemical Reactivity in Multielectron Atoms

Before proceeding to develop quantitative reasoning around concepts related to multielectron atoms, we consider the question:

What did the hydrogen atom tell us about atomic structure that applies to the structure of all elements in the periodic table?

As we will see when we move to consider multielectron atoms, science owes a great debt to the simple hydrogen atom. One proton, one electron; but the key gateway to the structure of all atoms because from this single proton and single electron there emerges a stable atomic structure—it does not collapse and it demonstrates that the home for the electron is the orbital that is drawn from the wavefunction that is a solution to the Schrödinger equation.

While there are many facets involved in the answer to this question, one of the most important is that the hydrogen atom orbitals, the solution to the Schrödinger wave equation, provided a remarkable *array* of wavefunctions depending on the particular set of quantum numbers. These quantum numbers include the principal quantum number n, the angular momentum quantum number ℓ and the magnetic quantum number m. Just to review, a selection of these angular hydrogen wavefunctions are recalled in Figure CS9.1a.

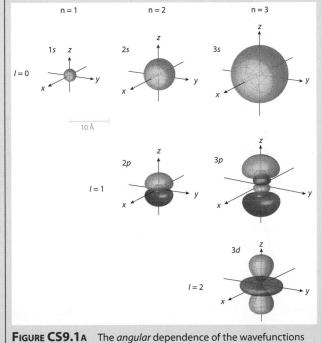

Figure CS9.1a   The *angular* dependence of the wavefunctions for n = 1, 2, and 3 with candidate ℓ = 1 and ℓ = 2 angular wavefunctions that displays both the geometry of the angular wavefunctions and the number of angular nodes.

But recall that there are limitations on ℓ determined by the value of n. To explore this we pose the following question:

Which of the following are impossible combinations of n and ℓ: 1p, 2s, 3f, 3g, 4f, 5g?

To address this question, recall that the orbitals are designated as nℓ, with the spdf letter code used for ℓ:

Orbital	n	ℓ
1p	1	1
2s	2	0
3f	3	3
3g	3	4
4f	4	3
5g	5	4

Since ℓ is restricted to the values 0, 1, 2, ... , $n - 1$, the 1p ($\ell = n$), 3f ($\ell = n$), and 3g ($\ell > n$) orbitals do not exist.

For each set of allowed quantum numbers, as Figure CS9.1a reminds us, there is a unique geometry and orientation defining the 1s, 2s, 2p, 3s, 3p, and 3d orbitals. We also are reminded that while the ground (lowest energy) state of the hydrogen atom has but one electron in the 1s orbital, that electron can be excited to higher energy orbitals such as the 2p or 3d orbital. Each of these orbitals has both a *radial* dependence and an angular/orientational dependence. Figure CS9.1b summarizes the important radial part of the wavefunction.

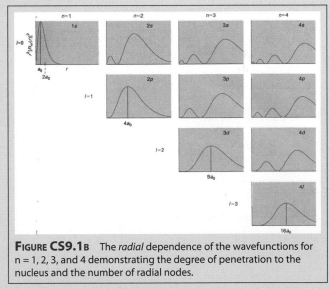

Figure CS9.1b   The *radial* dependence of the wavefunctions for n = 1, 2, 3, and 4 demonstrating the degree of penetration to the nucleus and the number of radial nodes.

We can focus briefly on the distinction between these quantum numbers n, ℓ, and m by drawing on an analogy with the *classical motion* of the electron in orbit about the nucleus. To do this we first distinguish between the energy contained in the "in-and-out" or vibrational motion of the electron vs. the energy contained in the "round-and-round" rotational motion of the electron. This is depicted in Figure CS9.1c for the case of s and p orbitals. In general, the amount of energy that the electron has is a mixture of this in-and-out motion and the round-and-round motion *depending on the values of n and ℓ.* The amount of energy contained in the angular mode is given by the *angular momentum quantum number,* ℓ, and the number of nodes in the angular wavefunction specifies this. For example, for n = 2, ℓ can equal zero (2s) or ℓ can equal 1 (2p). The angular wavefunction has zero nodes for the 2s and one node for the 2p. Thus the electron in the 2p orbital has *higher angular momentum* than the 1s. However, the number of *radial nodes* is a measure of the energy contained in the in-and-out mode, and the number of those radial nodes is given by n − ℓ − 1. So for n = 3 and ℓ = 0, the 3s, the number of radial nodes is 3 − 0 − 1 = 2. Thus *all* the energy is tied up in the in-and-out mode, none in the angular momentum mode. For the case n = 3, ℓ = 2, the number of radial nodes is n − ℓ − 1 = 3 − 2 − 1 = 0 and *no energy* is tied up in the in-and-out mode, all the energy is in the round-and-round (angular momentum) mode.

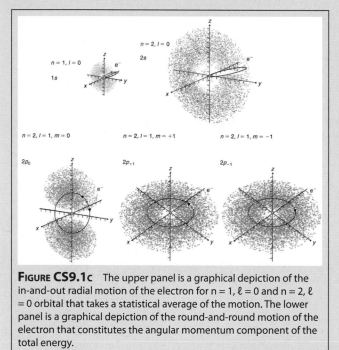

Figure CS9.1c The upper panel is a graphical depiction of the in-and-out radial motion of the electron for n = 1, ℓ = 0 and n = 2, ℓ = 0 orbital that takes a statistical average of the motion. The lower panel is a graphical depiction of the round-and-round motion of the electron that constitutes the angular momentum component of the total energy.

Consider the following question:

How many radial nodes are contained in the 1s, 2s, 3s, 4s, and 5s orbitals?

To answer this question, we have just a simple repetition of the exercise above. But there is a point worth making: s orbitals are spherically symmetric. They have no angular nodes.

Since each s orbital is free of angular nodes (ℓ = 0), the full complement of n − 1 nodes is exclusively radial:

Orbital	n	ℓ	Angular Nodes	Radial Nodes	Total
1s	1	0	0	0	0
2s	2	0	0	1	1
3s	3	0	0	2	2
4s	4	0	0	3	3
5s	5	0	0	4	4

Notice how the number of nodes increases with orbital energy.

The third quantum number, m, the "magnetic quantum number," is really a "projection quantum number," because it defines the *orientation* of the angular component of the wavefunction. That is, it specifies the projection of the angular momentum on a specific axis. For a given ℓ, m = 0 requires that the plane of rotation contain the z-axis, as shown for the options available to the 2p orbitals in Figure CS9.1c. When m = ℓ, the plane of rotation lies in the xy plane. The *sign* of m determines the direction of rotation. For m = +1 the rotation of the electron in orbit is counterclockwise, for m = −1 the rotation of the electron in orbit is clockwise. Thus the allowed values of m_ℓ depend upon the value of ℓ that in turn depends upon the value of n.

This raises the question:

How many orbitals are contained within the fifth shell? Name them and specify quantum numbers for each.

To answer this, we just follow the rules. With $n = 5$, the azimuthal quantum number ($ℓ = 0, 1, …, n − 1$) ranges from 0 to 4. In each subshell there are then $2ℓ + 1$ orbitals with magnetic quantum numbers $m_\ell = 0, ±1, ±2, …, ±ℓ$.

The s subshell, as always, supports only one magnetic orientation. The p subshell, as always, supports three. The d subshell, five. The f subshell, seven. etc. Quantum numbers are summarized in the table below:

Orbital	n	$ℓ$	m_ℓ
5s	5	0	0
5p	5	1	−1 0 1
5d	5	2	−2 −1 0 1 2
5f	5	3	−3 −2 −1 0 1 2 3
5g	5	4	−4 −3 −2 −1 0 1 2 3 4

Count them up: $1 + 3 + 5 + 7 + 9 = 25 = 5^2$. There are always n subshells and n^2 orbitals in a shell.

From an analysis of the orbitals of the hydrogen atom, we are equipped with a remarkable array of possible orbitals with a manifold of available radial and angular geometries and associated energies and quantum numbers. But in the hydrogen atom, we have only one electron! However, there are 116 elements in the periodic table. If we are to embrace the chemistry of the elements, where do all those electrons go?

So hydrogen with one electron has this large array of available orbitals. Lowest in energy is, of course, the *ground state* 1s orbital and then the second shell (n = 2) with the 2s, $2p_x$, $2p_y$, and $2p_z$ all with the same energy in hydrogen because there is *no preferred direction in space with a single electron*. To the electron in hydrogen, all orbitals with a specific value of n and ℓ have the same energy—all directions look the same to the electron in hydrogen. Sitting next in energy are the nine orbitals for n = 3: 3s, $3p_x$, $3p_y$, $3p_z$, $3d_z{}^2$, $3d_{x^2-y^2}$, $3d_{xy}$, $3d_{xz}$, and $3d_{yz}$. But, once we move past hydrogen to *any* other element, where do the electrons go?

Before we proceed to add electrons, we should answer the question:

Consider the one-electron species H, He⁺, and Li²⁺ in their ground states. (a) Which atom has the smallest radius? (b) Compute the ionization energy for each.

Turning to the first question: start by noting that: (1) H has a nuclear charge of +1 (Z = 1), He⁺ has a nuclear charge of +2, and Li⁺² a nuclear charge of +3. (2) The ground state for each species is the 1s orbital. (3) one electron orbital energies

$$E_n = -R_\infty \frac{Z^2}{n^2}$$

depend strictly on Z and n. With that, we have all the information we need.

(a) *Radius.* In H, one proton attracts one electron. In He⁺, two protons attract one electron. In Li²⁺, three protons attract one electron. The progressively stronger attraction pulls the 1s orbital in more tightly, leaving Li²⁺ as the smallest of the three. The radii increase in the order Li²⁺, He⁺, H.

(b) *Ionization energy.* To ionize the atom, we must supply sufficient energy to promote the electron from n = 1 to n = ∞. The ionization energy is therefore

$$E_\infty - E_1 = -R_\infty Z^2 \left(\frac{1}{\infty^2} - \frac{1}{1^2} \right) = +R_\infty Z^2$$

where $R_\infty = 2.18 \times 10^{-18}$ J/atom (or 1.31×10^3 kJ/mol). The sign is positive. We put energy *into* the atom to tear off its electron.

For H, the requisite energy is R_∞. For He⁺, where Z = 2, the amount is four times greater (2²); for Li²⁺, nine times greater (3²). The numbers scale directly with Z^2:

	ΔE	Per Atom	Per Mole
H	R_∞	2.18×10^{-18} J	1.31×10^3 kJ
He⁺	$4R_\infty$	8.72×10^{-18} J	5.24×10^3 kJ
Li²⁺	$9R_\infty$	1.96×10^{-17} J	1.18×10^4 kJ

It is also important to review the distinction between kinetic energy and potential energy. A free electron moves with velocity v = 1.00×10^6 m s⁻¹ before being captured by a hydrogen nucleus. Eventually it lands in the 1s orbital. What is the change in energy?

To answer this question, we begin by noting that the free electron's energy, entirely kinetic, is initially ½mv². The bound electron's energy is $E_n = -R_\infty (Z^2/n^2)$. Trapped by the nucleus, then, the electron suffers a *loss* equal to

$$E = E_{bound} - E_{unbound} = -R_\infty \frac{Z^2}{n^2} - \frac{1}{2} mv^2$$

Substituting the quantities Z = 1, n = 1, m = 9.11×10^{-31} kg, v = 1.00×10^6 m s⁻¹, and $R_\infty = 2.18 \times 10^{-18}$ J, we have,

$$E = -R_\infty - \frac{1}{2} mv^2$$

$$= -2.18 \times 10^{-18} J - \frac{1}{2} \left(9.11 \times 10^{-31} \, kg \right) \times \left(1.00 \times 10^6 \, ms^{-1} \right)^2$$

$$= -2.18 \times 10^{-18} J - 4.56 \times 10^{-19} \, J = -2.64 \times 10^{-18} \, J$$

Where does the energy go?

The last step before we address the multielectron atom involves one of the great surprises in the development of quantum mechanics: the spin of the electron.

This raises the question: **What did the Stern–Gerlach experiment tell us?**

Prior to the formulation of quantum mechanics in 1925 with the introduction of the de Broglie wavelength λ_e = h/p for the electron, and the Schrödinger wave equation that defined the orbitals we have just discussed, Stern and Gerlach in 1922 had demonstrated, for a series of elements with a single unpaired electron (such as sodium or silver), that a beam of that atom split into *two distinct beams* when passed through an inhomogeneous magnetic field as shown in Figure 9.7. As an inspection of Figure 9.17 reveals, sodium has an unpaired 3s electron. But if the only unpaired electron in sodium (or silver) is in an "s" orbital with ℓ = 0, then the only unpaired electron possesses *no orbital rotational motion*. It has no angular momentum. The fundamental cause of the splitting of the beam of sodium atoms into two beams when passed through an inhomogeneous magnetic field was not clear in 1922. What was clear was that there was undeniably quantization, separation, along a preferred direction in space and that the preferred direction was established by the direction of the magnetic field.

It was known from classical physics that a circulating charge, a charge in a circular orbit, will generate a magnetic field. Based on the Bohr model of the atom, with its fixed planetary orbits, it was natural to ascribe this behavior of sodium or silver atoms in a magnetic field as simply the result of the interaction between the externally applied inhomogeneous magnetic field and the *magnetic field generated by the electron in orbit* in one of two directions:

clockwise or counterclockwise as shown in Figure CS9.1d. But, as we have just seen, quantum mechanics makes clear that an "s" orbital has *no angular momentum* associated with the electron. Thus following the Bohr theory of the atom with defined "planetary" orbits for the electron, it was reasonable to ascribe the splitting of the beam of sodium atoms in the Stern–Gerlach experiment to the magnetic fields generated by the motion of the electron in orbit. What *was* mysterious was the clear *spatial quantization* along the axis defined by the magnetic field. This brings us to the next question:

Why was the union of (1) the Stern–Gerlach experiment with (2) the hypothesis by Goudsmit and Uhlenbeck that the electron possessed intrinsic angular momentum, or spin, so important for an understanding of atomic structure in general and multielectron atoms in particular?

Exercise:

What mysteries did the union of Goudsmit–Uhlenbeck hypothesis, shown schematically in Figure CS9.1e, that the electron possesses an intrinsic spin, clear up?
Answer:

1. The physical basis for the Stern–Gerlach experiment
2. The structure of multielectron atoms
3. The periodic properties of the elements
4. The Aufbau principle

The key result was that a new quantum number, a *fourth* quantum number must be added to the set of three: n, ℓ, and m. That fourth quantum number was the *spin quantum number* m_s that can have only one of two values, $+\frac{1}{2}$ or $-\frac{1}{2}$, depending on the *projection* of the intrinsic electron spin on any axis. This is displayed graphically in Figure CS9.1e.

What did the union of the Stern–Gerlach experiment, the Goudsmit and Uhlenbeck hypothesis of intrinsic electron spin and the Pauli Exclusion Principle tell us?

By any reasonable analysis of the classical laws of electrostatic interactions, as we add electrons they would all seek to minimize their distance from the nucleus in search of the position of lowest potential energy and they would balance that tendency against the electrostatic repulsion by all other electrons seeking the same minimum in potential energy. But it was Pauli who articulated the principle that brought order to this chaotic picture when he stated: *No two electrons in the same atom may occupy the same quantum state, lest they be identical in both space and spin.* This brought coherence and order to the assignment of electrons to available orbitals in multielectron atoms. If the spins are the same, both up ($m_s = +\frac{1}{2}$) or both down ($m_s = -\frac{1}{2}$), then those two electrons must occupy different orbitals (i.e. they must not be in an orbital with the same n, l, m). If they are in the same orbital (same n, ℓ, m_ℓ) then one must be spin up ($m_s = +\frac{1}{2}$) and one must be spin down ($m_s = -\frac{1}{2}$).

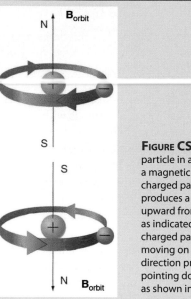

FIGURE CS9.1D   A charged particle in a circular orbit generates a magnetic field. By convention, a charged particle moving clockwise produces a magnetic field pointing upward from its plane of motion as indicated in the upper panel. A charged particle in a circular orbit moving on a counterclockwise direction produces a magnetic field pointing downward from that plane as shown in the lower panel.

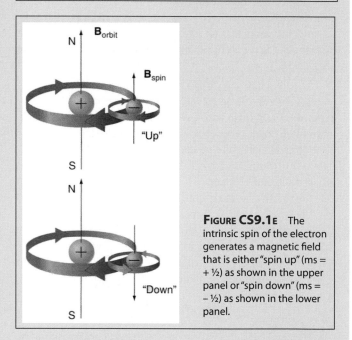

FIGURE CS9.1E   The intrinsic spin of the electron generates a magnetic field that is either "spin up" (ms = + ½) as shown in the upper panel or "spin down" (ms = – ½) as shown in the lower panel.

With this principle, all electrons must compete to get as close as possible to the nucleus, they all repel each other and they cannot coexist in an orbital *unless* they are of opposite spin. They both *repel* each other and they *exclude* each other. The analysis of this is Problem 1 at the end of this Case Study.

We now have our model, not only of the simple electron hydrogen atom, but also of the multielectron atom. The spatial domain is set by the orbital geometry, the occupancy is set by the Pauli Exclusion Principle that allows only electrons of opposite spin to occupy each room—each orbital. From this perspective we have constructed a "condominium" with the partners in each domain defined by their spin.

So emerge the orbitals of our multielectron atom. But with the addition of electrons, even the second electron in helium, come complications. On the face of it, to proceed from the hydrogen atom to He, should be straightforward:

1. Add one proton to the nucleus and one more electron to the 1s orbital with opposite spin
2. Sort out the electron-electron repulsion and the electron-proton attraction
3. Write the Schrödinger equation
4. Solve the Schrödinger equation

However, while the *strategy* is easy to lay out, the *execution* proves problematic—even for helium. As we pile on more electrons and protons, the situation gets even more difficult. This fact led one of the great mathematical chemical physicists of the 20th century, Paul A. M. Dirac, to exclaim:

> "The underlying physical laws necessary for...the whole of Chemistry are thus completely known, and the difficulty is only that the exact application of these laws leads to equations much to complicated to be soluble."

With the advent of modern high speed computers, we can now make considerable progress, but we cannot solve the Schrödinger equation exactly for multielectron atoms to this day.

The fact that the Schrödinger equation for multielectron atoms cannot be solved exactly does not mean we cannot develop an insightful and highly effective understanding of multielectron atoms and their chemical behavior. To the contrary, we gain an appreciation for chemical behavior by dissecting those aspects of electron behavior that control a manifold of chemical interplay.

For multielectron atoms, no longer does the energy of the orbital depend *only* on the principal quantum number n—the spherical symmetry of the one electron system is broken by the presence of the second electron. In helium an electron in the 1s orbital lies *closer* to the nucleus than an electron in the 2p orbital (see Figure CS9.1b) and thus it *penetrates* more effectively. In addition, an electron in the 2p orbital has more angular momentum and is thus more distant from the nuclear charge. So two important considerations emerge. First, the 1s electron screens or shields the 2p electron from the positive charge of the nucleus by inserting negative charge between the nucleus and the 2p electron. Second, the higher angular momentum moves the 2p electron further from the nucleus. As a result the 2p electron sees a diminished nuclear charge such that the full nuclear charge Z, is replaced by a smaller *effective nuclear charge* Z_{eff}. We can write this reduction in nuclear charge as

$$Z_{eff} = Z - S .$$

The result is that for each shell (same n, different ℓ) the orbital energies increase in the order

$$E_{ns} < E_{np} < E_{nd} < E_{nf} .$$

Electrons with higher angular momentum penetrate less and are screened more, thus their energies increase.

We therefore consider how we can calculate Z_{eff} by using what we know about the hydrogen atom and thinking about reasonable methods of approximation.

Suppose the He atom was a hydrogen atom. Then, using the formula for the energy of an electron, we know that the orbitals have energies

$$E_n = -R_\infty \frac{Z^2}{n^2}$$

where $R_\infty = -1312$ kJ mol^{-1}. The energy needed to ionize the one electron from hydrogen's 1s orbital is therefore R_∞:

$$I_1 = E_\infty - E_1 = 0 - (-R_\infty) = R_\infty$$

Now consider helium and its *two* 1s electrons, attracted to a nucleus containing two protons (Z = 2). Competing for the same nuclear charge, the electrons screen one another. They repel. Exactly how much we cannot say, but we certainly can establish upper and lower limits for ionization energy.

Lower limit: Pretend that each 1s electron screens the other completely, interposing a full charge of –1 between its orbital mate and the nucleus. If so, the screened electron sees roughly an effective nuclear charge of +1 (Z_{eff} = 2 – 1); and, if it does, we have a pseudo-hydrogen atom. The lowest possible ionization energy is R_∞, the value expected for a single electron bound to a nucleus with Z_{eff} = 1.

Upper limit: At the other extreme, pretend that there is no screening at all. Assume that each electron interacts with the *full* nuclear charge so as to make Z_{eff} = 2. The ionization energy, proportional to

$$Z_{eff}^2$$

then quadruples to $4R_\infty$.

To complete this analysis, see Problem 2 at the end of this Case Study.

As it turns out, these concepts of penetration, screening or shielding, and effective nuclear charge play directly into the chemical behavior of multielectron atoms so we pursue some important examples of how those factors affect atomic radii, ionization energy, and electron affinity.

ATOMIC RADII

For elements in the s and p blocks, we rely on two broad guidelines. (1) Atomic size generally decreases across a row, consistent with a progressive *deshielding* of the valence electrons. Left to right, Z_{eff} grows larger as the ns and np subshells are sequentially filled. The atom contracts. (2) Top to bottom, the principal quantum number increases. Radii expand as the new valence shell is layered upon a larger and larger core. See Problem 3.

IONIZATION ENERGY

The same trends apply. Across: Shielding decreases. Z_{eff} increases. Atoms become smaller and more difficult to ionize. Down: Atoms grow larger. The valence electrons, more distant now, are easier to remove. See Problem 4.

ELECTRON AFFINITY

Trends for electron affinity are not as sharply drawn as those for radius and ionization energy, but still we ask the same kinds of questions. Which atom has the higher effective nuclear charge? Which system can better accommodate the added electron? Must the electron be added to a singly occupied orbital, so as to increase the electron-electron repulsion? Will an extra electron *open* a subshell (say from ns^2 to ns^2np^1) or *close* a subshell (as from ns^2np^5 to ns^2np^6)? How does the attached electron alter the balance between attractions and repulsions in the system? To illuminate these issues, see Problem 5.

Problem 1

Which of the following p^2 configurations should realize the lowest energy? Which should realize the highest? Which are forbidden?

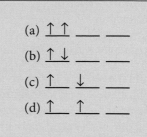

Problem 2

Without attempting a detailed calculation, estimate the first ionization energy for a ground-state helium atom. Determine the lowest and highest values possible.

Problem 3

Which atom in each pair has the smaller radius? (a) Be, N. (b) B, Br. (c) C, Si. (d) Al, Ca.

Problem 4

Which atom or ion in each pair has the larger ionization energy? (a) C, N. (b) N, O. (c) O, S. (d) Na, Na⁺.

Problem 5

Which atom in each pair has the stronger electron affinity? (a) S, Cl. (b) Be, B.

BUILDING A TECHNOLOGY BACKBONE

CASE STUDY 9.2   Rare Earth Elements: Applications, Chemistry, Resources, and Production

Introduction

The rare earth elements have emerged from their initial discovery in 1787 in a mine in the village of Ytterby, Sweden and over a century of obscurity and marginal involvement in commerce and industry, to a position of extreme importance to modern technology with associated economic impact. As Figure CS9.2a graphically illustrates, the rare earths play a central role in jet engine design, laser systems, light emitting diodes, high speed rail systems, X-ray imaging, magnetic resonance imaging, wind turbine design and production, electric cars, and flat screen TVs.

The rare earth elements (hereafter REEs) occupy the position in the periodic table of the *inner transition f block elements* as displayed in Figure CS9.2b. It is customary to designate the Period 6 rare earths as the "lanthanides," which include lathanum, cerium, praseodymium, neobium, promethium, samarium, europium, gadolinium, terbium, dysprosium, holmium, erbium, thulium, ytterbium, and lutetium. Convention adds scandium (^{21}Sc) and yttrium (^{39}Y) to the list of rare earths because of their very similar chemical properties.

Before moving on to discuss the characteristics of these critically important elements, note that these inner transition elements in Period 6 begin the filling of the f orbitals for which l = 3, so m_l = −3, −2, −1, 0, +1, +2, +3, so there are seven f orbitals and thus 14 rare earth elements in addition to lathanum, scandium (Period 4, Group 3B), and yittrium

FIGURE CS9.2A   With the advent of high technology sectors in jet engine design, high-speed rail, diodes, high performance electric automobiles, and high performance wind turbines the demand for rare earth elements has expanded dramatically.

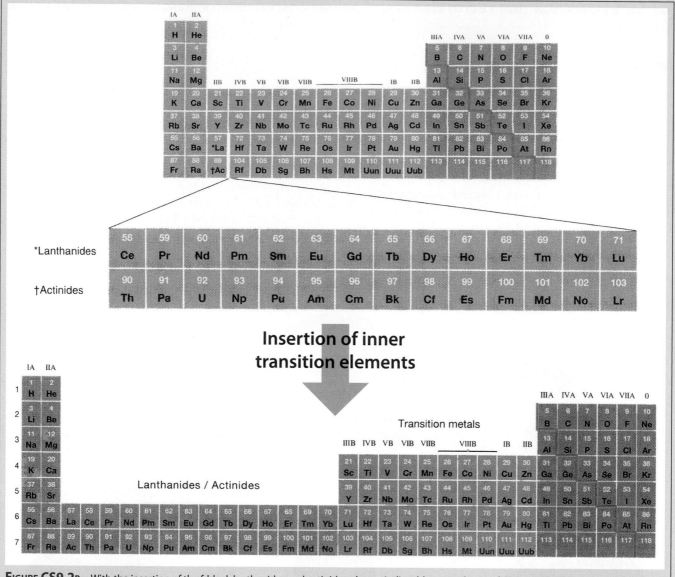

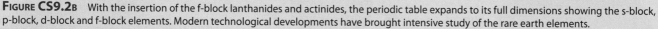

Insertion of inner transition elements

FIGURE CS9.2B With the insertion of the f-block lanthanides and actinides, the periodic table expands to its full dimensions showing the s-block, p-block, d-block and f-block elements. Modern technological developments have brought intensive study of the rare earth elements.

(Period 5, Group 3B). The rare earths all have similar atomic radii and tend to form ions with a 3+ charge, so they can easily substitute for one another in crystal lattices. This results in their being found interlaced within the same ore deposits making their extraction as purified metals more complex.

It is important to examine the electron configuration of the rare earths because they serve as an important review of the causes for orbital energy ordering and the associated chemical behavior of all elements in the periodic table. The first point is that beginning with Period 3, n = 3, the energetic effects of screening become comparable to the energy differences between shells. The result is that high-ℓ orbitals of a given n may lie higher in energy than a low-ℓ orbital of the next higher n. Thus the principal quantum number, n, *no longer rules the energy ordering* as it did in hydrogen

(where the ℓ levels are all degenerate and the energy depends *only* on n) and as it does in n = 1 and n = 2 in multi electron systems. As we progress to n = 4, 5, and 6, the splitting and interweaving of orbital energies becomes pronounced, but also the energy *separation* between orbitals becomes increasingly smaller. The effect of this is displayed in Figure CS9.2c.

Let's review quickly the structure of the periodic table displayed in Figure CS9.2b with the energy of the orbitals in Figure 9.2c that determines the electron configuration. Considering the Period 3 elements, Na to Ar in Figure CS9.2b, there are only 8 elements instead of 18 that would logically occur because n = 3, so ℓ = 0, 1, 2, and there are then s, p and d orbitals with 2, 6, and 10 electrons each. However, as Figure CS9.2c shows, the *4s orbital lies lower in energy* than the 3d, so the 3d subshell does not begin to fill until Period

4, and the 4s orbital fills before the 3d. We saw this explicitly in the discussion of Period 3 and Period 4 in the chapter core. The 3d subshell would have filled before the 4s except the screening by the n = 1 and n = 2 shells and the deeper penetration of the 4s determines the energy ordering. Next in energy following the 3d subshells, as Figure CS9.2c shows, the 4p subshell so the periodic table completes Period 4 with Ga through Kr. The insertion of the 5s orbital at a lower energy than either the 4d or the 5p means that Period 5 has the same sequence of orbitals and elements as Period 4 except that both n and ℓ advance by an integer. Therefore, for both Period 4 and Period 5, the ns, (n −1)d and np orbitals fill sequentially giving 18 elements.

This brings us to the rare earth or lanthanide series. From the orbital energy ordering of Figure CS9.2c we see that the 6s orbital lies lower in energy than the 4f and 5d orbitals. Therefore Cs and Ba initiate Period 6 and fill the 6s orbital. Based on the energy ordering in Figure CS9.2c we would expect the next electron, for lanthanum (La) would go in the 4f orbital but it doesn't, it occupies the 5d orbital. It is the relationship between the screening by the inner electrons and the differing geometry of the d an f orbitals that creates this "anomalous behavior," inverting the orbital ordering between the 5d and the 4f. We saw a similar anomalous behavior in both Cr and Cu in Period 4 and it also occurs in Nb, Mo, Ru, Rh, Pd, and Ag in Period 5.

Following La, in the f block of Period 6, Ce also departs from the "standard" energy/orbital ordering of Figure CS9.2c. The electron configuration of Ce is [Xe] $6s^2\ 4f^1\ 5d^1$. Following Ce, the 4f orbitals fill in sequence through Eu [Xe] $6s^2\ 4f^7$ but Gd is anomalous because the 5d orbital fills with the next electron so the configuration of Gd is [Xe] $6s^2\ 4f^7\ 5d^1$. Just to remind ourselves what the full electron configuration is for the Period elements, including the rare earths, the first few examples are displayed in Figure CS9.2d.

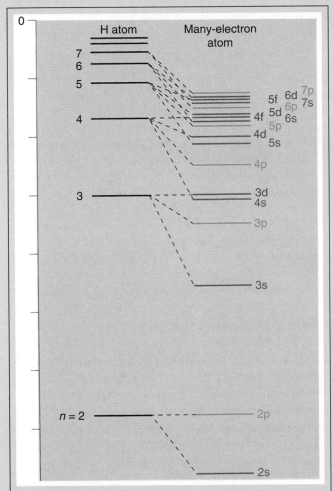

FIGURE CS9.2c One of the most useful diagrams in chemistry summarizes the ordering of the subshells of elements in the periodic table. As both the principal quantum number n and angular momentum quantum number ℓ increase, the energy separation of the subshells becomes smaller and smaller leading to "annotators behavior" in the order of electron configurations.

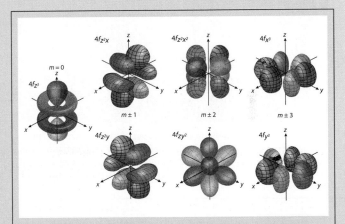

FIGURE CS9.2e The wavefunctions for the seven f orbitals are displayed with their coordinate designators. As the angular momentum increases from p to d to f, the electrons in the higher angular momentum orbitals are less and less effective in shielding the outer electrons form the nuclear charge.

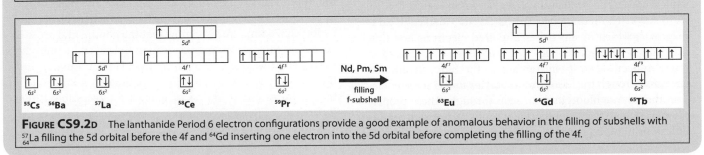

FIGURE CS9.2d The lanthanide Period 6 electron configurations provide a good example of anomalous behavior in the filling of subshells with $^{57}_{57}$La filling the 5d orbital before the 4f and $^{64}_{64}$Gd inserting one electron into the 5d orbital before completing the filling of the 4f.

The f orbitals, displayed in Figure CS9.2e, are characterized by higher angular momentum, less penetration into the vicinity of the nucleus and a diminished ability to screen the charge of the nucleus. As a result, the filling of the 4f subshell *prior* to the 5d subshell of causes an increase in Z_{eff} and thus the valence orbitals shrink in size as we progress from left to right along Period 6. This is termed the "lanthanide contraction."

Uses of the Rare Earths

As noted in the introduction, the importance of REEs spans fields as diverse as medicine, information technology, energy generation, high speed transportation, electric automobiles, super conductors, lasers, and camera lens manufacturing just to name a few. For example, consider the following MRI images of the defective blood-brain barrier in a stroke victim's head shown in Figure CS9.2f.

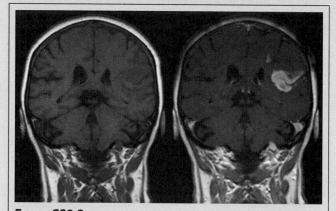

FIGURE CS9.2F   Because of the high spin of members of the lanthanide series leading to paramagnetic behavior in combination with attraction to water molecules preventing passage of trace elements across the blood-brain barrier, these elements are powerful tracers of internal bleeding.

The image on the right is clearly more helpful than that on the left in highlighting the problematic area. The difference is that, before taking the image on the right, the patient's bloodstream was injected with a gadolinium contrast agent. Gadolinium (^{64}Gd) is a rare earth element with 64 protons; its position in the lanthanide series shown in Figure CS9.2b. Its electron configuration ([Xe] $6s\ 5d^1\ 4f^7$) makes it paramagnetic, which means it has unpaired electrons and thus responds to a magnetic field. Gd is also greatly attracted to water molecules, so it will remain in the bloodstream and not pass through the blood-brain barrier *unless* there is a lesion that allows blood itself to seep through. These properties make it a perfect contrast agent for locating brain tumors or lesions in blood vessels via magnetic resonance imaging.

REEs are vital in other technologies, too. Europium is the only known red-colored phosphor, so it is the only material currently used for manufacturing the colored cathode-ray tubes found in computer screens and televisions. Hard drives and speakers incorporate neodymium magnets—a type of permanent magnet composed of an alloy of iron, boron, and the rare earth neodymium. The combination of lightness, strength and high spin resulting from the large number of unpaired spins in the 4f subshell in neodymium magnets also makes these magnets extremely well suited for high torque, low volume, low mass electric motors, like those being used in the new generation of electric cars discussed in Chapter 7.

In many cases, electric cars also depend upon supplies of lanthanum, another rare earth metal, because it is a typical component of rechargeable batteries. To get an idea of the importance of rare earths in modern automobile technology, consider this: each hybrid Toyota Prius requires 1 kilogram of neodymium for its motor and 10 kilograms of Lanthanum for its nickel-metalhydride battery.

Wind-turbines use neodymium magnet–based motors; compact fluorescent light bulbs use europium; lasers depend upon neodymium, samarium, europium, gadolinium, dysprosium, and erbium; flat-screen televisions depend upon rare earth phosphors. The modern list of REE uses is lengthy, and as technology advances, it will only grow longer.

Abundance and Supply

It would be useful, then, to estimate the world's REE supply to find out how "rare" these elements actually are. The result is surprising given this category of element's name! Earth's crust actually proves to be home to a vast mass of REEs. The upper crustal abundance of the most common rare earth, cerium, is higher than that of copper (64 ppm, compared to 60 ppm), for example, and neodymium is over 8000 times as abundant as gold (26 ppm/0.0031 ppm). Table 1 summarizes the list of rare earths, their atomic numbers, and their upper crust abundance.

Estimating Earth's total supply of REEs:

Given the abundance in the chart above, we could estimate the upper-crust supply of REEs if we knew the mass of the upper crust. Let us assume that the Earth is spherical, of radius $R_E = 6378$ km. We also know that the continental crust is 30–50 km thick. However, the deepest hole ever dug into the Earth's crust was only 12.3 km (at Kola Hole in Russia), and that the deepest mines reach to approximately 4 km. So let's take 3 km as our thickness of interest for accessible REEs. Referring to Figure CS9.2g, we can estimate the mass of the accessible crust as follows:

TABLE CS9.2A

REEs, atomic numbers, and abundances

Element	Symbol	Atomic Number	Upper Crust Abundance, ppm[*]	Chondrite Abundance, ppm[†]
Yttrium	Y	39	22	na[‡]
Lanthanum	La	57	30	0.34
Cerium	Ce	58	64	0.91
Praseodymium	Pr	59	7.1	0.121
Neodymium	Nd	60	26	0.64
Promethium	Pm	61	na	na
Samarium	Sm	62	4.5	0.195
Europium	Eu	63	0.88	0.073
Gadolinium	Gd	64	3.8	0.26
Terbium	Tb	65	0.64	0.047
Dysprosium	Dy	66	3.5	0.30
Holmium	Ho	67	0.80	0.078
Erbium	Er	68	2.3	0.20
Thulium	Tm	69	0.33	0.032
Ytterbium	Yb	70	2.2	0.22
Lutetium	Lu	71	0.32	0.034

* *Source:* Taylor and McClennan 1985

† *Source:* Wakita, Rey, and Schmitt 1971.

‡ na = not available.

Mass of 3 km crust = (density of crust) × (volume of crust)

$$= (2.8 \text{ g/cm}^3)[4/3 \ \pi \ (R_E^3 - R_{intern}^3)]$$
$$= (2.8 \text{ g/cm}^3)(\text{kg}/1000 \text{ g})(100^3 \text{cm}^3/\text{m}^3)(1000^3 \text{m}^3/\text{km}^3) \times$$
$$[4/3 \ \pi \ (6378^3 - 6375^3) \text{ km}^3] = 4.3 \times 10^{21} \text{ kg}$$

However, what really interests us is the mass of REEs accessible for technical and industrial use. That does not include any of the REEs found in oceanic crust, or even most of the continental crust. So before we continue, let's restrict our calculation specifically to identified US reserves and research some more helpful figures:

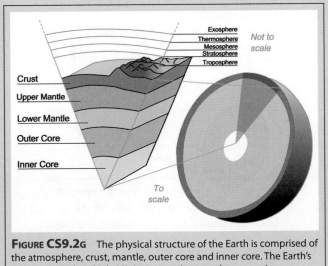

FIGURE CS9.2G The physical structure of the Earth is comprised of the atmosphere, crust, mantle, outer core and inner core. The Earth's crust is 30–50 km thick. Mining operations reach to a maximum depth of approximately 4 km.

If we add the values in the fourth column of Table CS9.2a above to get the total rare earth oxide content in the US, we get 11.7×10^9 kg REEs. For comparison, the total mass of rare earth oxides estimated to be found within Chinese borders is about 36 million metric tons, or 33×10^9 kg.

Though Chinese reserves are almost triple those of the United States, if we compare US reserves to the global demand for REEs in 2010, the US reserves of 11.7 million tons is significant: in 2010, global consumption was about 130,000 tons. By that measure, the US could supply the world for the next century. From the perspective of supplying the US, where demand is at about 10,200 tons, we could meet our current demand for approximately a thousand years.

Production

If America has such a significant mass of rare earth elements, then why are we dependent entirely upon imports? The reason is that mining and purification of the rare earths are challenging endeavors from both an entrepreneurial and an environmental perspective.

The first step in establishing any type of a mining operation is exploration. However, REE exploration has been minimal in the United States—the Mountain Pass reserves, for example, were actually discovered accidentally during a USGS search for uranium. Once reserves are identified, a company must be willing to invest up to $500 million for testing, construction, and other operation costs. This is before the company has sold any product to make revenue. That up-front investment, as a result, creates a significant barrier to entry.

Once ore has been extracted, it must be processed extensively. Techniques for separation include gravitational, magnetic, and electrostatic techniques, in combination with strong acid/strong base leaching and selective precipitation. The need for large quantities of HCl, NaOH, H_2SO_4, etc. makes waste disposal an expensive and demanding process, as well as an environmental hazard. Equally problematic is the fact that most REE ores contain a good deal of thorium, which is radioactive, and whose decay products include other radioactive elements like radon. Monazite, for example, is a yellow-brown, resinous mineral that was once the principal source of rare earth oxides. However, because it contains 4–10% thorium, extensive environmental regulations drove monazite purification off the market in the 1980s. In 1992, Mitsubishi Chemical had to quietly close its operations in Malaysia and begin the long, expensive clean-up of radioactive wastes. What exactly went wrong is unclear, but Mitsubishi has had to invest about $100 million in clean-up operations. In order to get rid of the material that is rumored to be causing leukemia cases in the local community, engineers have cut the top off a hill three miles away in a forest reserve, buried the material inside the hill's core, and then entombed it under more than 20 feet of clay and granite. And the major REE production site in the United States—Mountain Pass Mine in the Mojave Desert—faced regulatory issues in 1998. A spill accident involved waste with a high radioactive element content, so the California EPA suspended the mine's actions. It was closed in 2002.

Unfortunately, then, REE processing is an expensive and sophisticated operation.

Current global supply

Necessary environmental concerns, combined with cheap labor and abundant supply in China, have allowed that country to take over the world's supply of REEs. As a result, China is now responsible for some 97% of the current, global supply of rare earth elements, while the United States argues about whether or not to reopen Mountain Pass with stronger environmental regulations. The evolution of rare earth production as a function of time is displayed in Figure CS9.2h. Until 1948, Brazil and India were the primary global source for rare earths. In 1950, extensive deposits of monazite were discovered in South Africa and that country became the dominant exporter of of rare earths. From the mid 1960s to the mid 1980s, the Mountain Pass rare earth mine in California lead world production. Beginning in 1985, China rapidly grew in importance, until it now commands 95% of the world supply market for rare earths.

Despite the challenging financial and environmental concerns, the world needs REEs for uses as diverse as MRIs

and batteries—as explored earlier—so the sharp decline in non-Chinese production of REEs is now an international concern. In 2002, the USGS Fact Sheet on Rare Earth Elements included the warning, "Availability of Chinese REE to US markets depends on continued stability in China's internal politics and economy, and its relations to other countries." Changes in the level of export from China in the last two years has demonstrated the importance of its grip on global REE production. China has cracked down on environmental hazards associated with the mining of rare earth elements. Though the environmental hazards are obviously quite significant, some say "environmental concerns" are a pretext, and that China is actually using its control over REE exports to shift its economy from low-value goods to high-value goods. In other words, by monopolizing the supply of raw materials needed for producing high-tech merchandise, China will be able to attract producers of such merchandise into its borders. Whatever the reason, the country has significantly restricted supply: The Ministry of Industry and Information Technology has cut the country's target output from rare earth mines by 8.1 percent in 2010 and is forcing mergers of mining companies in a bid to improve technical standards, according to the government-controlled China Mining Association, a government-led trade group. In late 2010, the nation continued to limit supplies, causing prices to skyrocket as displayed in Figure CS9.2i.

China even imposed an embargo against Japan, forcing the island nation to develop recycling techniques to try to bolster its REE supply. In 2015 the US began to rapidly increase the extraction of rare oxides.

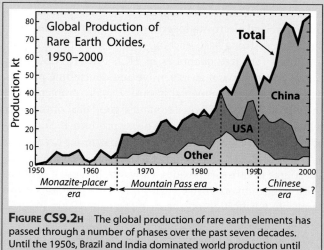

FIGURE CS9.2H   The global production of rare earth elements has passed through a number of phases over the past seven decades. Until the 1950s, Brazil and India dominated world production until South Africa took the lead until the mid-1960s when the Mountain Pass mine on the California-Nevada border grew in prominence. Since 1990 China has taken the lead such that today 95% of rare earth elements are mined in China.

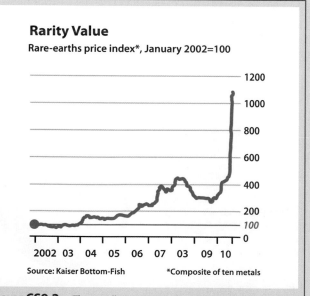

FIGURE CS9.2I   The rapidly increasing demand for rare earth elements in the global high technology industry combined with China's environmental and export policies have driven up the rare earth price index dramatically.

Conclusions

Wind power, electric cars, batteries, CFCs, etc. are all in need of rare earth elements, so the world must develop a broadly cast method for meeting this demand—a method that does not rely entirely upon China, nor entirely upon an environmentally destructive refining processes.

What then are the practical realities of this objective? Regarding non-Chinese resources, the issue is not supply but rather complications arising from collateral damage in extraction. Though China possesses the largest fraction of estimated global REE reserves, Australia, Canada, and the United States have significant enough fractions to avoid a total Chinese monopoly, if these other nations could tap their respective supplies. In an attempt to do so, the Australian Lynas corporation is currently working to open a processing plant in Malaysia, and Molycorp is planning to reopen the Mountain Pass Mine in California.

But just as important as expanding the world supply of REEs is doing so in a responsible manner. Though the wind turbines and hybrid vehicles will help us reduce our dependence on fossil fuels, advanced chemical technology must be employed to limit toxic damage. In the meantime, perhaps instead of relying entirely upon new ores and additional processing, more effective recycling techniques could be investigated to make use of the REEs collecting currently in discarded electronics.

The take-home point is this: the issue is not the global supply of rare earth elements—we have plenty. The issue is to find a way to extract them cleanly and economically and that is the reason the chemistry of rare earth elements is so critical to the future.

Problem 1

Tabulated below are the categories of principal uses of rare earth elements (REEs) globally. For each category, list the REEs associated with each application.

Hybrid and Electric Vehicles

REEs: _____

A typical electric vehicle contains approximately 28 kg of REEs. Many forecasts for electric vehicles show rapid growth rates, up to 60% per annum.

Catalytic Converters

REEs: _____

Automotive catalytic converters transform the primary pollutants in engine exhaust gases into nontoxic compounds. Rare earth element coatings on the catalyst surface play a critical role in the chemical reactions within a catalytic converter whereby CO and ozone are oxidized to CO_2 and O_2. REEs also enable reactions to be run at high temperatures where efficiency is gained, at a significantly lower cost than Pt metal group alternatives.

Wind Power Generators

REEs: _____

Wind power generation provides a clean alternative to hydrocarbon fuels, and assists in reducing the output of greenhouse gas emissions. New wind power generator designs are larger (up to 7 megawatts) by the inclusion of REE magnets enabling gearless generators for better reliability and online performance. Large wind turbine generators require 2 tonnes of high strength magnets, which contain approximately 30% REEs.

Designs exist for wind turbines to be built with batteries installed to allow stable power output. Such designs greatly increase REE consumption in this important green technology.

Hard Disk Drives

REEs: _____

Rare earth element permanent magnets are installed in all computer hard drives, CD-ROM drives, AND DVDs. The voice coil motor controls the arm that reads and writes information onto the disk and a rare earth magnet gives better control allowing thinner tracks and more information storage. More stable spinning is achieved with a rare earth magnet driving the spindle of the disk drive.

Flat Panel Display Screens

REEs: _____

Cathode ray tube (CRT) and plasma televisions as well as computer monitors are coated with phosphors, which when subject to low pressure UV excitation generate the primary colors red, blue, and green. Red colors require phosphors of the rare earth element europium, for which there is no alternative. The introduction of plasma televisions saw the development of a new blue europium phosphor that retains brightness ten times longer than previous blue phosphors. Phosphors generating the color green are doped with the rare earth element terbium. A combination of these primary color phosphors are used to create the white 'backlight' used in LCD screens and the energy efficient 'tri-phosphor' light bulbs.

Flat screen and plasma televisions provide a new and growing field of consumption of REEs, the traditional supply of which is extremely restricted and concentrated within China.

In addition to the consumption of REEs in phosphors, cerium oxide, through its unique physical and chemical properties, has been the building block upon which glass polishing has been based for over 40 years. It is a very efficient polishing compound as it removes glass both by its chemical dissolution and by mechanical abrasion.

Polishing powders are used to polish cathode ray tubes (CRT), plasma (PDP), and liquid crystal display (LCD) glass, along with many other circumstances where highly polished glass is required.

Medical Applications

REEs: _____

Rare earth elements play a key role in a wide range of health and medical applications, including drug treatments, diagnostic techniques and equipment. REEs act as catalysts in biomedical and chemical research, are used as tracing agents during imaging, and in laser and radioisotopic treatment for cancer.

The largest single use of rare earth elements in the medical industry is the incorporation of REE permanent magnets to generate high strength magnetic fields for MRI imaging. 700 kg of magnets are consumed in each MRI machine, which are playing an increased and growing role in routine diagnostics. REE permanent magnets are replacing expensive traditional systems of wire coils in liquid helium, where the helium super cooled the wire coil to reduce the electrical resistance.

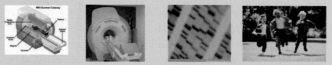

Portable Electronics and Small Motors

REEs: _____

Neodymium and samarium magnets are much more powerful than traditional ferric magnet alternatives. These magnets providing much better performance under a wider range of operating conditions, thereby allowing effective miniaturization and production of compact, lightweight and powerful motors.

In portable electronics, including Apple's popular iPod music player, tiny neodymium magnets are used to drive the speakers within earphones resulting in smaller, lighter speakers with faster base response and lower overall distortion. Such technology would not be possible without the role played by rare earth elements.

CASE STUDY 9.3   Calculating the Personal Carbon Footprint

Analysis of an individual's "carbon footprint" has become a widely used means for measuring the amount of CO_2 emitted to the atmosphere by an individual *each year*. As such, it has a great deal in common with the personal energy budget in kWh/p·d that we developed in Case Study 2.1. But these are important differences. It should be recognized first that the objective of calculating a "carbon footprint" is actually to determine the amount of *climate forcing* resulting from the greenhouse gas trapping of infrared radiation. Thus what we seek is to determine the net impact on greenhouse gas trapping resulting from our various activities: purchasing and driving a car, heating our homes, purchasing food, purchasing goods and services, etc. Thus it is important to recognize that while carbon dioxide is the dominant greenhouse gas added to the atmosphere, it is not the only one. Figure CS9.3a captures this quantitatively by breaking out the fractional contributions to the global greenhouse gas emission that includes, in addition to CO_2, methane, nitrous oxide, sulfur, hexafluoride, CFCs, HFCs, and other miscellaneous contributions.

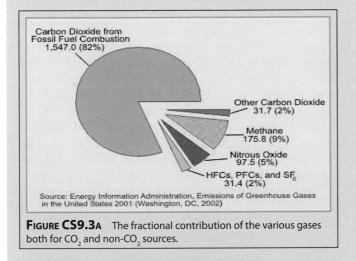

FIGURE CS9.3A   The fractional contribution of the various gases both for CO_2 and non-CO_2 sources.

The inclusion of the other greenhouse gases is typically done by calculating the carbon dioxide *equivalent*, CO_2 e, for each activity that constitutes an individual's contribution to climate forcing.

We will, therefore, use the unit of tons of CO_2 equivalent per year, t CO_2 e/y, to calculate the spectrum of personal usage globally. We will take the world view first, examining the contribution to per capita t CO_2 e/yr for an array of countries in the categories of food, shelter, construction, clothing, manufactured products, transportation, services and trade. After we establish the baseline for a selected set of representative countries, we will examine, quantitatively, specific choices for different lifestyles in the U.S. as we did in Case Study 2.1 for personal energy usage.

The Global Picture

We examine first the range of annual per capita carbon release in t CO_2 e/p for a representative range of countries in Figure CS9.3b. The first point to note from an examination of Figure CS9.3b is the large range in annual per capita carbon release that ranges 28.6 t CO_2 e/p in the US to 1.1 t CO_2 e/p in Bangladesh and Uganda. Note that Western European countries, whose standard of living is approximately equivalent to that of the US, have a per capita carbon release about one-half that of the US. China has a per capita carbon release about 4.5 times less than the US and India 10 times less.

The next step in the analysis of the carbon footprint is to consider the various categories of goods and services that comprise the demand for carbon. For this purpose we break the total contribution to carbon emission into eight categories:

- Construction
- Shelter
- Food
- Clothing
- Manufactured Products
- Mobility
- Service
- Trade

Those categories support three primary sectors
1. Households
2. Government
3. Investment

The household segment, which includes heat and light, and where appropriate, in the place of work, is fairly straightforward; goods and services that sustain the standard of living for all individuals in their place of residence or work whether that is a single family home, an apartment, or a dormitory room. The government sector includes administration, defense, nondefense equipment purchases, etc. The investment sector includes roads, building construction, electronic infrastructure, transportation associated with design and construction, furniture, lighting, etc. associated with any new commercial construction or infrastructure.

Figure CS9.3c summarizes the annual CO_2 release, the nonCO_2 release, and the total greenhouse gas release in units of gigatons carbon equivalent, Gt CO_2 e, for each sector—household, government, and investment—indicating the contribution for each category as a global annual average.

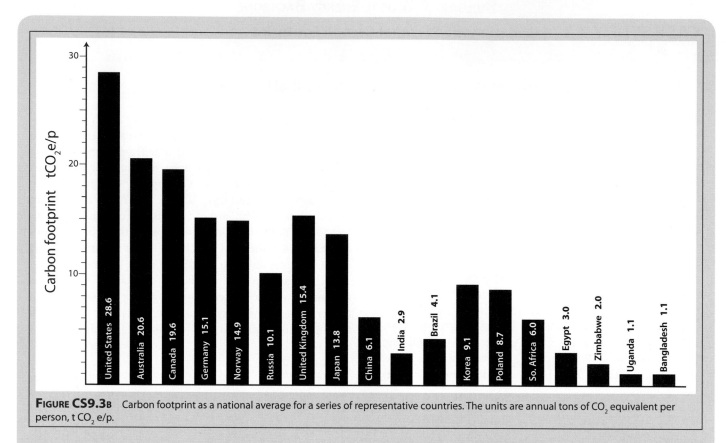

FIGURE CS9.3B   Carbon footprint as a national average for a series of representative countries. The units are annual tons of CO_2 equivalent per person, t CO_2 e/p.

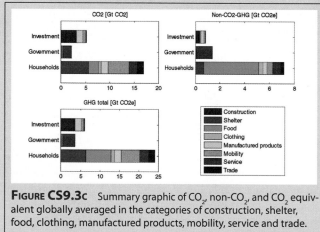

FIGURE CS9.3C   Summary graphic of CO_2, non-CO_2, and CO_2 equivalent globally averaged in the categories of construction, shelter, food, clothing, manufactured products, mobility, service and trade.

The most important point to note is that, at the global level, 72% of greenhouse gas emissions are in support of household consumption, 10% to government consumption, and 18% to investments in building and other infrastructure. As a global average, food accounts for 20% of GHG emissions; the heating, cooling, operation, and maintenance of offices and residences 19%; and transportation some 17%.

But just as there is a very large dynamic range in greenhouse gas release per capita in different countries (see Figure CS9.3b) so too is there a large dynamic range in per capita income in different countries. A sampling of per capita income for the same countries summarized for GHG release is presented in Figure CS9.3d. Methods for determining per capita income figures vary significantly. We use here the information from the World Bank, which uses a variant on the Purchasing Power Parity (PPP) method that attempts to use the U.S. as the base for comparing the price of a basket of goods in each country, then establishing the approximate per capita income from that comparison. This results in the distribution for the year 2003 as displayed in Figure CS9.3d.

A global analysis of how the GHG emission in t CO_2 e per capita increases as a function of per capita income results in some consistent trends and considerable coherence. The global analysis of each of the eight categories as a function of per capita income is shown in Figure CS9.3e.

An inspection of Figure CS9.3e reveals that the category with the least rapid increase with increasing per capita income is food. While food costs dominate in countries with low per capita income, food ranks near the bottom in countries *with high per capita income. On the other hand*, transportation costs and manufactured products dominate in countries with high per capita income. This was clearly reflected in our analysis of personal energy budgets in Case Study 2.2.

Given this global view as a context, we next consider the calculation of carbon footprints for a number of cases in the United States. We consider five cases:

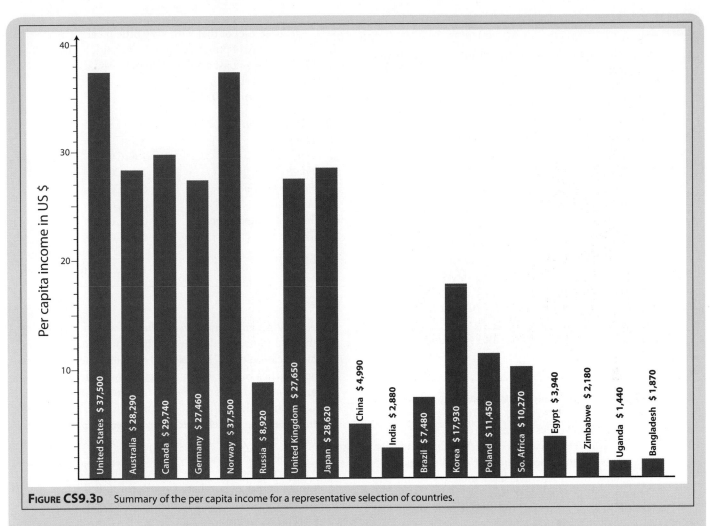

FIGURE CS9.3D Summary of the per capita income for a representative selection of countries.

Case 1:

1. A person living in *California* within a family of three in a three bedroom house.
2. One car in the family driven 12,000 miles per year.
3. "Standard" US diet with a mix of chicken, fish, and beef.
4. Typical array of purchased goods for middle class.
5. One coast-to-coast airplane trip per year.
6. Nominal recycling.
7. Train travel of 1000 miles per year.

The annual carbon footprint for Case 1:

1.	*House:*	4.7 tCO_2e/p
2.	*Car:*	5.6 tCO_2e/p
3.	*Food:*	4.1 tCO_2e/p
4.	*Purchased Goods:*	6.7 tCO_2e/p
5.	*Air Travel:*	0.8 tCO_2e/p
6.	*Recycling and Waste:*	1.2 tCO_2e/p
7.	*Train:*	0.1 tCO_2e/p
	Total	23.2 tCO_2e/p

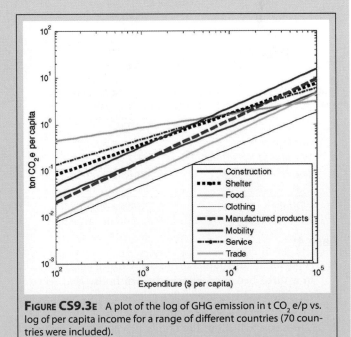

FIGURE CS9.3E A plot of the log of GHG emission in t CO_2 e/p vs. log of per capita income for a range of different countries (70 countries were included).

Case 2:

1. A person living in *Massachusetts* within a family of three in a three bedroom house.
2. One car in the family driven 12,000 miles per year.
3. "Standard" US diet with a mix of chicken, fish, and beef.
4. Typical array of purchased goods for middle class.
5. One coast-to-coast airplane trip per year.
6. Nominal recycling.
7. Train travel of 1000 miles per year.

The annual carbon footprint for Case 2:

1.	*House:*	11.0 tCO_2e/p
2.	*Car:*	5.6 tCO_2e/p
3.	*Food:*	4.1 tCO_2e/p
4.	*Purchased Goods:*	6.7 tCO_2e/p
5.	*Air Travel:*	0.8 tCO_2e/p
6.	*Waste and Recycling:*	1.2 tCO_2e/p
7.	*Train:*	0.1 tCO_2e/p
	Total	29.5 tCO_2e/p

Case 3:

1. A person living in *Massachusetts* within a family of three in a large house
2. SUV driven 15,000 miles per year.
3. "Standard" US diet with a mix of chicken, fish, and beef.
4. Typical array of purchased goods for middle class.
5. Fairly intense business travel that includes round trip to Beijing, China from NY, two round trip flights to western Europe, four round trip flights from NY to San Francisco, four round trip flights from NY to Chicago, ten round trip flights from NY to Wash., DC.
6. Nominal recycling.
7. Train travel of 1000 miles per year.

The annual carbon footprint for Case 3:

1.	*House:*	22.0 tCO_2e/p
2.	*Car:*	13.0 tCO_2e/p
3.	*Food:*	4.1 tCO_2e/p
4.	*Purchased Goods:*	8.0 tCO_2e/p
5.	*Air Travel:*	17.0 tCO_2e/p
6.	*Waste and Recycling:*	1.2 tCO_2e/p
7.	*Train:*	0.1 tCO_2e/p
	Total	65.4 tCO_2e/p

Case 4:

1. A person living in *Massachusetts* within a family of three in a three bedroom attached town house that recycles carefully and has used all available energy savings approaches on the house structure, windows, heating, etc.
2. One car, a hybrid, in the family driven 12,000 miles per year.

3. Vegetarian diet or diet with very little meat.
4. Typical array of purchased goods for middle class.
5. One coast-to-coast airplane trip per year.
6. Careful recycling.
7. Train travel of 1000 miles per year.

The annual carbon footprint for Case 4:

1.	*House:*	4.3 tCO_2e/p
2.	*Car:*	1.6 tCO_2e/p
3.	*Food:*	1.1 tCO_2e/p
4.	*Purchased Goods:*	6.7 tCO_2e/p
5.	*Air Travel:*	0.8 tCO_2e/p
6.	*Waste and Recycling:*	0.5 tCO_2e/p
7.	*Train:*	0.1 tCO_2e/p
	Total	15.1 tCO_2e/p

Case 5:

1. A person living *anywhere in the US* within a family of three in a three bedroom attached town house that recycles carefully and has used all available energy savings approaches on the house structure, windows, heating, and *heats and cools the house with a heat pump.*
2. One car, a plug in/hybrid, in the family driven 12,000 miles per year and charged from the power grid with renewable energy sources (wind, concentrated solar thermal, nuclear, hydro, etc.).
3. Vegetarian diet.
4. Limited array of purchased goods for middle class.
5. All business via teleconferencing with no air travel.
6. Careful recycling.
7. High speed rail travel of 5000 miles per year.

The annual carbon footprint for Case 5:

1.	*House:*	1.0 tCO_2e/p
2.	*Car:*	0.2 tCO_2e/p
3.	*Food:*	1.1 tCO_2e/p
4.	*Purchased Goods:*	3.0 tCO_2e/p
5.	*Air Travel:*	0.0 tCO_2e/p
6.	*Waste and Recycling:*	0.5 tCO_2e/p
7.	*Train:*	0.5 tCO_2e/p
	Total	6.3 tCO_2e/p

9.56

Problem 1

It is very important to study and to contrast the 5 cases treated in this Case Study because the analysis clearly identifies both the principal contributors to the carbon foot print and the impact that life choices make in the total carbon footprint for an individual.

Therefore, construct a bar graph with the vertical axis in units of tCO_2e/person and the horizontal axis separated into the 5 cases delineated in the Case Study.

- The (vertical) bar graph for each case is to be broken down into the categories: house, car, food, purchased goods, air travel, recycling and waste, and train travel.

- Indicate each of the categories by color, by arrows, or whatever way you wish to make it clear what contribution each category makes to each case.

- For clarity it is suggested that the vertical axis use the full range of the long dimension of a sheet of 8½ x 11 inch paper or graph paper.

Theories of Molecular Bonding I
Valence Electron Configuration, Electron Sharing, and Prediction of Molecular Shape

"There are ancient cathedrals which, apart from their consecrated purpose, inspire solemnity and awe. Even the curious visitor speaks of serious things, with hushed voice, and as each whisper reverberates through the vaulted nave, the returning echo seems to bear a message of mystery. The labor of generations of architects and artisans has been forgotten, the scaffolding erected for the toil has long since been removed, their mistakes have been erased, or have become hidden by the dust of centuries. Seeing only the perfection of the completed whole, we are impressed as by some superhuman agency. But sometimes we enter such an edifice that is still partly under construction; then the sound of hammers, the reek of tobacco, the trivial jests bandied from workman to workman, enable us to realize that these great structures are but the result of giving to ordinary human effort a direction and a purpose.

"Science has its cathedrals, built by the efforts of a few architects and many workers...."

—Gilbert Newton Lewis, from the preface to
Thermodynamics and the Free Energy of Chemical Substances

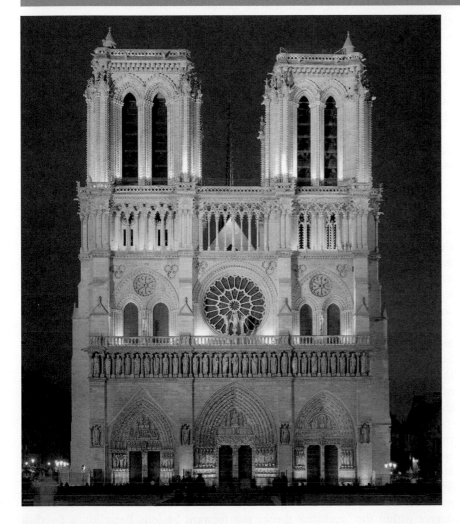

Figure 10.1 Notre Dame Cathedral, Paris, France.

Framework

The evolution of theories to understand and to explain the nature of the chemical bond represents a case study in scientific development entangled with the personalities associated therewith. At the center of developments attempting to bring rationality and order to the apparent complexity of molecular bonding in the early part of the 20th century was Gilbert Newton Lewis. G. N. Lewis, a physical chemist, was well known initially for the reformulation of thermodynamics in a form quantitatively applicable to problems at the intersection of chemistry and physics. Lewis was born in Weymouth, Massachusetts in 1875 and the family moved to Nebraska in 1884 when he was 9 years old. Lewis, home schooled until departing for college, began his studies at the University of Nebraska but transferred to Harvard in 1892 when his father, a lawyer, brought the family back to Boston. Lewis graduated with a degree in chemistry in 1895. From his college years, Lewis showed little interest in the descriptive branches of chemistry—he received a D in advanced organic chemistry as a senior—but gravitated strongly toward a blend of chemistry, physics, and mathematics. Following graduation and a year teaching at Phillips Andover Academy, Lewis returned to Harvard as a Ph.D. candidate studying under T. W. Richards. The focus of Lewis's graduate work was on chemical thermodynamics, where his grasp of physics and mathematics was considerably more sophisticated than that of his mentor. Richards was, however, a superb experimentalist and was the first American chemist to receive a Nobel Prize for his remarkably accurate determination of atomic weights. But increasingly Lewis realized that while the laws of thermodynamics controlled the *direction* of spontaneous change through the release of free energy, thermodynamics had little to say about processes inside molecules during a chemical reaction. Within the domain of chemistry there were $> 10^6$ molecular structures, but no coherent way of understanding or predicting what governed the architecture of those structures.

Now, if the story of the evolution of thinking about, and reaching an understanding of, the nature of the chemical bond that Lewis led did not lend valuable insight into an intuitive and predictive ability to determine bonding structures, we would not frame this problem. What is remarkable about the Lewis theory of bonding, which we visited briefly in Chapter 2 and will develop in some detail in this chapter, is that while the Lewis theory of bonding came after the discovery of the electron (J. J. Thomson 1897), the Lewis theory preceded the development of quantum mechanics. This is remarkable because quantum mechanics was so revolutionary that it rendered obsolete much of the scientific literature addressing the formation of the chemical bond written prior to 1925. But it did not eliminate the Lewis electron dot structure, nor the fundamental idea of shared electron pair bonding in molecules. So here is what happened.

First, we note what the state of thinking was in 1915 regarding the structure and underlying physical principles determining the formation of molecular bonding when Lewis published his paper on electron pair bonding. As early as the 1860s, chemists were drawing chemical formulas in a way recognizable to any of us today—the nomenclature of atoms linked together by lines drawn between them as displayed in Figure 10.2.

At the time, chemical reactions were recognized to be rearrangements in the placement of atoms into different molecular structures, with the number of specific atoms occurring in the reactants being equal to the number of specific atoms present in the product molecules—atoms are neither created nor destroyed in a chemical reaction. As far as what held the reactant molecules, the product molecules, or any molecule together, the abiding explanations were to a large degree within the "dualistic" theory put forward by Jöns Berzelius, a Swedish chemist. The dualistic theory held that atoms were *electrically charged* and that

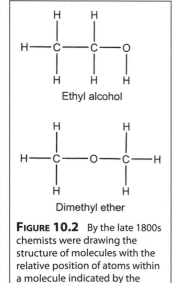

FIGURE 10.2 By the late 1800s chemists were drawing the structure of molecules with the relative position of atoms within a molecule indicated by the placement of those atoms linked by straight lines.

molecules were held together by the Coulomb attraction of those atoms. How the atoms somehow acquired a charge when they went from a neutral, isolated atom to a participant in a molecular bond was a problem that went largely unaddressed and clearly not understood. When J. J. Thomson discovered the electron in 1897, the prevailing thought was that this discovery lent substantial weight to the Berzelius dualistic theory in that this new negatively charged particle was the key to bonding because it was the entity *transferred* from one atom to the other in the formation of the chemical bond. This transfer of the electron resulted in the Coulomb force between the resulting positive ion and negative ion in the molecular bond.

This dualistic theory would explain the formation of sodium chloride: the sodium atom would transfer an electron to a chlorine atom creating an Na^+ cation and a Cl^- anion. The typical descriptive notation at the time diagrammed the chemical bond created by this electron transfer as an arrow from the donating atom to the receiving atom

$$Na \rightarrow Cl$$

Which meant simply that the electron was transferred:

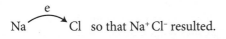 so that $Na^+ Cl^-$ resulted.

What was particularly telling is that at the time, 1900–1915, it was strongly believed that for atoms with comparable abilities to attract electrons, the electrons could be transferred in *either* direction to form the bond. For example, for the case of NCl_3, two possible "electromeres" (named in analogy with isomers, which are molecules with the same number and identity of atoms but arranged in different geometries) were possible depending on the direction of electron transfer. One electromere could be created by transferring an electron from each chlorine to the nitrogen; the other electromere could be formed by transferring three electrons from the nitrogen to each of the chlorine atoms as sketched in Figure 10.3.

The key point within this picture of molecular bonding formation is that in order to generate the strength in the bond so formed, the electron must be transferred to create the Coulomb attraction. But what force would act not just to move one or more electrons from one atom to another, but to move an electron in the opposite direction within the same molecular structure? As a measure of the degree to which this concept of the electron transfer as the prerequisite for the formation of the molecular bond and the associated idea of the electromere, were central to the chemical thinking of the day, two leading chemists in the US, William Noyes and Julius Stieglitz—chairs of the departments of chemistry at the University of Illinois and the University of Chicago, invested nearly two decades in the search for electromeres! No rational explanation within the domain of physics existed for such an idea. It was as though some mysterious force required the electron to decide which nucleus to attach itself to, thereby generating the Coulomb attraction needed to form the molecular bond. In a sense, this idea that "something different happens" at the atomic and molecular level that violated the fundamental tenets of physics as it was known in the first two decades of the 20th century turned out to be correct. But it had little to do with the dualistic theory, electromeres, etc.

Remarkably, as early as 1902, in course material prepared for lectures he gave at Harvard with T. W. Richards, Lewis was developing a new type of diagram wherein he kept explicit track of the "new" electrons, discovered by Thomson seven years before, in the electronic structure of the atoms in the periodic table. This diagram by Lewis placed the electrons at the corners of cubes as shown in Figure 10.4.

Cl		Cl
↓		↑
N ← Cl and		N → Cl
↑		↓
Cl		Cl

FIGURE 10.3 Theories of bonding prior to the work of G. N. Lewis and prior to the development of quantum mechanics involved hypothesized "electromeres" that were formed by the transfer of charge, electrons, in both directions (indicated by the arrows) so as to form a chemical bond from the Coulomb attraction between the atoms that resulted from the attraction of positively and negatively charged atoms within the molecule.

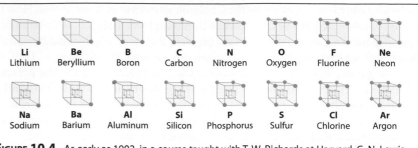

FIGURE 10.4 As early as 1902, in a course taught with T. W. Richards at Harvard, G. N. Lewis was developing a cubic model for atoms that placed an electron at the corner of the cube, increasing by one electron in steps across the second and third Periods of the periodic table. With the filling of the 8 electrons in neon of the second Period, that cube moved to the *inside* of the cube representing the sites for electron placement in building up the third Period of the periodic table.

Just as is done in the periodic table, as we progress from hydrogen to neon in the first row, an electron is added, but in Lewis's scheme, to the corner of each new cube corresponding to the next heavier element. But, as we progress to the second row of the periodic table, the cube with eight electrons, Ne, moves to the interior of the next set of cubes that progressively represent the elements from sodium to argon as shown in Figure 10.4. At the time (1902) Lewis was teaching the class with T. W. Richards, Richards referred to these ideas of Lewis's as "twaddle," which may well have been why Lewis didn't publish this cubic model of electron designation in atomic structure. The tenor of this may also have been related to why Lewis left Harvard, with a brief interlude in the Philippines where he continued to develop both his new formulation of thermodynamics and his cubic model of the atom, for MIT where he remained from 1906 to 1912. There he worked closely and productively with Arthur Noyes before being offered the position of chemistry department chairman and dean of the College of Chemistry at the University of California, Berkeley in 1912.

It was in 1916, four years after arriving at Berkeley, that Lewis published his paper "The Atom and the Molecule" where he both rejected the dualistic framework that required an explicit electron transfer to create a molecular bond via the formation of Coulomb interaction between ion pairs, and introduced the idea that the molecular bond results from the sharing of electrons. But in particular, Lewis took the clear position in that paper that it was the *sharing of electron pairs* that created the stable structure of the molecular bond. What lead Lewis to this formulation of the electron pair bond was the union of his cubic picture of the atom to form molecules using this explicit electron counting intrinsic to his cubic model of the atom. What particularly concerned Lewis was that first, while it might be possible to rationalize such species as NaCl in terms of an electron transfer, a vast proportion of molecular structures occurred with an even number of electrons. Molecular structures with an unpaired electron were very rare. Second, how about molecules such as H_2, O_2, and Cl_2? Why would an O atom arbitrarily give up an electron to the other O atom, leaving one with 5 valence electrons and the other with 7 valence electrons?

The key transition in Lewis's thinking on the matter occurred in 1915 when he realized that if he joined his cubic model of the atom to form molecules, he could explain this *sharing* of electron pairs by linking his cubic structures in different ways. In the case of the single bond in Cl_2, he linked his cubic model for the Cl atom along an edge as shown on the right in Figure 10.5. For the double bond in O_2, he linked his cubic model for the O atom along a common face as shown on the left in Figure 10.5.

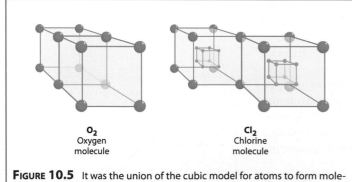

FIGURE 10.5 It was the union of the cubic model for atoms to form molecules by sharing electron pairs that led G. N. Lewis to propose his theory of the sharing of a pair, or pairs, of electrons to form a stable molecular bond.

In each of the molecules shown in Figure 10.5, both atoms have eight electrons in their outer cubes, and of primary importance, they have achieved this through the *sharing* of electron pairs. In the course of developing this theory of the shared electron pair bond, Lewis evolved a shorthand diagram that eliminated the need to explicitly draw the cubic structure around each atom with the electrons located at the corners of each cube. Each bond was represented as a shared pair of dots between two atoms, and the structure that was most stable resulted when each atom was surrounded by eight electrons—except hydrogen, which required but two electrons to form a stable bond. Thus, the original cubic structure evolved into the electron dot structure as shown in Figure 10.6.

Thus, 1916 was the turning point for Lewis and the developing understanding of the structure of the molecular bond. In "The Atom and the Molecule" he laid out the picture of the shared electron pair bond—the covalent bond model—of molecular structure.

This concept of the sharing of an electron pair, that electron pairs would remain immovably suspended between atoms in a molecular structure, was so antithetical to the principles of physics at the time that the notion was dismissed out of hand by the physics community. Why would two negatively charged *particles* that repel each other mysteriously attract each other in a molecular bond? Perhaps more remarkable in retrospect, that seminal 1916 paper on the shared electron bond was largely ignored by the *chemistry* community. The reason was that Lewis had solved a fundamental problem that most chemists of the time didn't even know existed. That is, the nature of the molecular bond, how electrons pair to form molecular structures, was simply not regarded as central to chemical research because the preponderance of research in chemistry was focused on the determination of *structure*, rather than on *why* those structures existed. But as Case Study 10.1 makes clear, the Lewis shared electron pair concept clarifies more than just why molecular structures form. It also clarifies many aspects of why molecules *react* the way they do.

At the time, Lewis was excused for such extraneous meanderings because the backbone of his scientific contributions continued to be in the field of chemical thermodynamics where a chain of continuing developments ensured his scientific stature, and he was rapidly building one of the most potent departments of chemistry in the world at Berkeley. Soon after publishing "The Atom and the Molecule" in 1916, Lewis became involved directly in the war effort, where he was commissioned as a major and shipped out to France as director of the Chemical Warfare Service (CWS) laboratory in France to counter the German's use of toxic gas on the battle fields of Europe. A major architect of the German chemical warfare developments was another chemist we have met—Fritz Haber, the developer of catalytic techniques for the production of fixed nitrogen for fertilizer.

In this period between 1912 and the early 1920s, there was a clear need to join the separate worlds of physics and chemistry in order to reconcile (1) the emerging recognition of the nuclear structure of the atom—that of negatively charged electrons in orbit about a positively charged nucleus, and (2) the clear ability of the Lewis approach of static electron pairs for predicting stable molecular structures. In large measure, these communities—physics on the one hand, chemistry on the other—passed one another in the night, scarcely recognizing each other. However, *both* the evolving Bohr model of the atom and Lewis's model of the molecular bond violated fundamental laws of classical electromagnetic theory. An electron in a Bohr orbit undergoes centripetal acceleration—the laws of electromagnetism

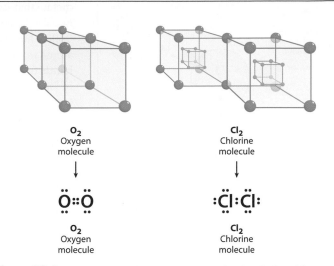

O₂
Oxygen
molecule

↓

Ö::Ö

O₂
Oxygen
molecule

Cl₂
Chlorine
molecule

↓

:C̈l:C̈l:

Cl₂
Chlorine
molecule

FIGURE 10.6 Simplification of the cubic model of molecular bond formation led Lewis to develop his electron dot formation of bonding structure. It was the simplified electron dot structure that became widely adopted.

Case Study 10.1
Lewis Structures and Insight into Chemical Reactivity

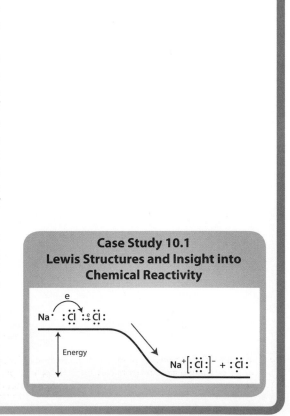

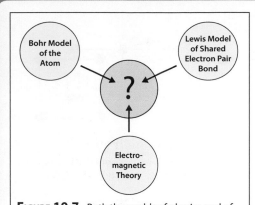

FIGURE 10.7 Both the worlds of physics and of chemistry were faced with a serious conundrum prior to the development of quantum mechanics. The Bohr atom, emerging from the physics community, involved electrons in circular orbits about the positively charged nucleus—electrons that were accelerating by virtue of the centripetal force between the electron and proton. The accelerating charge would radiate electromagnetic energy to space and collapse into the nucleus according to electromagnetic theory. The Lewis model of static electrons forming a chemical bond would suffer the same fate: they would be drawn into the nucleus and the molecule would collapse.

dictate that *any* accelerating charge must emit electromagnetic radiation, thereby radiating energy to its surroundings, and by conservation of energy, losing energy itself. Thus, electrons in orbit about a nucleus would spiral into the nucleus and atoms would not exist as described by the Bohr model. The physics world had its hands full! As we saw in Chapter 8, Bohr was forced to hypothesize the existence of "stationary state" orbits for electrons in an atom, which was a decidedly *ad hoc* conjecture.

On the other hand, the static electron pair or pairs in the Lewis model would simply collapse into the positively charged nucleus, and the entire molecular structure would suffer the same fate as the Bohr atom.

Yet, unquestionably, Bohr's model quantitatively described the hydrogen atom, and Lewis's model explained the patterns of stable molecular structures. We can summarize the profoundly untenable state of affairs during this period (1912–1924) with a diagram tracing the various schools of thought that converged into irreconcilable confusion centered on the formation of the molecular bond in Figure 10.7. This was a world without quantum mechanics, and, to engage in a profound understatement, it was not a happy situation!

With the departure of G. N. Lewis for the war effort, his work on the electron structure of molecular bonding was suspended and was largely ignored by both the physics and chemistry communities. Except for one important case. The extremely well known physical chemist Irving Langmuir saw very early that the shared electron bond that Lewis described in his 1916 paper was an important conceptual advance, and the two had discussed Lewis's shared electron pair concept of the molecular bond at the New York symposium on the structure of matter in the fall of 1916. As early as 1917, Langmuir cited the concept in a paper, and in 1918 he began to independently incorporate Lewis's line of thinking into his own work on the formation of chemical structures on surfaces. In 1919 and 1920, Langmuir "rediscovered" Lewis's ideas with what Langmuir called his "octet theory." The idea that molecules form physically stable octets or groups of eight electrons. Langmuir, the inventor of the gas filled electric light bulb, the hydrogen welding torch, and the mercury condensation vacuum pump, while working at the General Electric laboratories in Schenectady, NY, was a compelling speaker, whereas G. N. Lewis was not. In the spring of 1919, Langmuir presented a series of eloquent lectures at the American Chemical Society meeting in Buffalo, NY, and the National Academy of Sciences meeting in Washington, DC, that launched this "new" shared electron, octet rule, covalent bond theory straight into the limelight. It was then referred to as the Lewis–Langmuir theory—a designation that Lewis found rather irritating.

In 1923, Lewis published the seminal paper "Valence and the Structure of Atoms and Molecules." In that paper Lewis articulated his ideas of the shared electron bond in considerable detail, and he made it clear that it was he who had both understood and then made clear in the scientific literature the importance of the shared electron pair bond. While Lewis had focused on developing an intuition about molecular bond structure through the use of his electron dot structures, Langmuir had opted for a more formal mathematical equation to represent the octet structures in stable molecular structures. By 1923, it was both Lewis's dot structure diagrams *and* the emerging realization that it was in fact Lewis who was the true architect of the shared electron pair chemical bond, that cemented the designation of the term Lewis theory in perpetuity. It should be emphasized, however, that Langmuir made major contributions to the Lewis theory that went well beyond its resurrection from obscurity. Langmuir coined the terms "covalent bond" and "octet rule" and the idea that if more than one structure could be drawn that satisfied the octet rule, the preferred (lowest energy) structure was that for which adjoining atoms had the least difference in "formal charge"—an idea we will develop in the following chapter.

Just two years after Lewis published his "Valence and the Structure of Atoms and Molecules," the de Broglie view of the wave properties of the electron exploded onto the scene followed by the Schrödinger equation that mathematically defined the orbitals within which electrons reside and the Pauli Exclusion Principle that dictated electron pairs with opposite spin within orbitals. The reasons for the "stationary state" Bohr orbits *and* for the "static electrons" occurring in shared pairs yielding the lowest potential energy configuration all became clear.

In 1927, Walter Heitler and Fritz London performed the first quantum mechanical calculation of the chemical bond, applying the Schrödinger equation to the combination of two protons and two electrons to calculate the bond strength of the simplest molecule, H_2. At the time, these calculations were done by dozens of individuals, coordinated in the same room, using mechanical hand calculators—but the worlds of physics and chemistry had formed a union in the form of theoretical chemical physics that is today standing at the forefront of efforts to develop revolutionary pathways for the generation of energy at the scale required for modern society. Case Study 10.2 sets in place a key emerging component of renewable energy: wind power. As we will see, extracting power from the wind will be of increasing importance, particularly in the US and China.

What must be gathered from this development of the Lewis structure of the shared electron pair molecular bond is that while modern computers of immense calculational power are routinely used to execute increasingly complicated electronic structure calculations in the gas, liquid, and solid states, we do not develop an intuition about molecular bonding structure without developing an understanding of the Lewis structure for molecular bonding. It is the way we can deduce the most stable molecular structure, the structure of lowest potential energy, on the back of an envelope. It has survived both (1) the advent of quantum mechanics—which made irrelevant virtually all the chemical literature on bonding written prior to 1925, and (2) the advent of the high powered digital computer that has heralded a new union between theoretical chemistry and such areas as material science, nanochemistry, and photoelectrochemistry.

An important example of the role played by different bonding structures in nature is examined in Case Study 10.3 which spotlights how important a type of bond—the hydrogen bond—is to the human body's defense against diseases such as cancer.

It has been said that all truth passes through three stages. First it is ridiculed or ignored or both. Second, it is violently opposed by the established structure of the day. Third, it is accepted as being self-evident. While Lewis evolved his thinking regarding the central role played by the shared electron covalent chemical bond over many years, based purely on chemical rationale rather than physical principles, his theory not only passed from the first to the final stage in the progression above in less than ten years, it is a perspective, a framework of thinking, that is as valuable today as it was when it emerged.

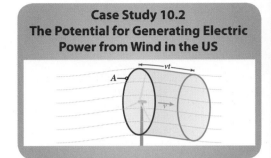

Case Study 10.2
The Potential for Generating Electric Power from Wind in the US

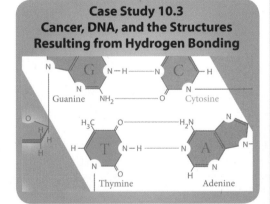

Case Study 10.3
Cancer, DNA, and the Structures Resulting from Hydrogen Bonding

Chapter Core

Road Map for Chapter 10

In this chapter we develop the theoretical basis for the molecular bond that provides a means for understanding both the structure of the molecular bond and the electron assignments in the Lewis structure of bonding theory. We develop the following concepts.

Road Map to Core Concepts	
1. Electronegativity	
2. Types of Chemical Bonds	
3. Representation of Valence Electrons with the Corresponding Lewis Electron Dot Diagram	
4. Lewis Structure for Ionic Bonds	
5. Lewis Structures for Covalent Bonds	
6. Lewis Structures and Resonance Structures	
7. Method of Formal Charge	
8. Limitations to the Lewis Theory	
9. Valence Shell Electron Pair Repulsion: VSEPR Theory	

The Structure of the Molecular Bond

Given the existence of electrons, protons, and neutrons, it is not difficult to imagine the existence of atoms simply because electrons and protons attract, capturing each other to form quantized wave structures. As we have seen, an isolated atom represents the remarkable balance between the quantum mechanical principles of electron matter waves with associated quantized energies and angular momenta on the one hand, and classical Coulomb attraction between opposite charges on the other. However, while an atom can exist alone, it usually chooses not to. It chooses, rather, to form *bonds* with other atoms. We can, to first order, appreciate why this may be so from a simple argument based upon our electron-in-a-box model that demonstrated a sequence of eigenvalue energies corresponding to the sequence of wavefunctions ordered upward from n=1 at the lowest energy (most stable) with the energy of each n level given by

$$E_n = \frac{h^2 n^2}{8L^2 m_e^2}$$

as displayed in Figure 8.31.

But we also recognize that because the energies E_n depend *inversely* on the width of the potential well that confines the electron, as L *increases*, the energy E_n *decreases*. This was displayed explicitly in Figure 8.33. While we will develop increasingly potent theories for the details of chemical bonding, it remains fundamentally true that in the microscopic domain of quantum mechanics, molecules form from atoms because the formation of a molecule provides a means for the wavefunctions of atoms to spread out, to delocalize, thereby reducing the energy of the *combined* wavefunction as shown in Figure 10.8. This figure, in panel a, represents the simplest model of the chemical bond wherein two atoms, each with an approximate square well potential, combine to form a molecule wherein the electron wavefunction spreads out across the larger dimension of the molecule. As we saw, a wavefunction that spreads out, decreases in energy. This *decrease* in energy corresponds to a more stable configuration of electrons and protons and thus a stable union results. Two isolated atoms choose to form a molecule because that molecular geometry allows the wavefunction of each electron to spread to larger spatial domain resulting in decreased energy—the atoms capture each other to reduce their combined energy.

The lower panel in Figure 10.8 (panel b) highlights the importance of electrostatics—Coulomb attraction—in determining the *distribution* of the wavefunction when two atoms of different energy of different ability to attract electrons combine to form a molecule. The atom with the greater ability to attract its valence electron (or electrons), is capable of extracting electron density from the atom that less strongly binds its valence electrons resulting in a shift in the *combined* wavefunction, a delocalization of electron density toward the atom with greater electronegativity.

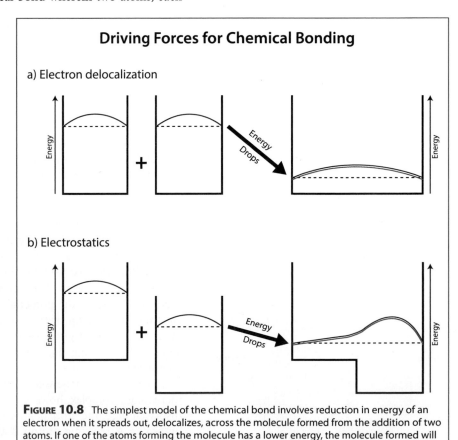

FIGURE 10.8 The simplest model of the chemical bond involves reduction in energy of an electron when it spreads out, delocalizes, across the molecule formed from the addition of two atoms. If one of the atoms forming the molecule has a lower energy, the molecule formed will draw electron density preferentially toward the atom that more strongly attracts an electron.

Thus atoms combine when given the chance to create larger structures. To first order, nature does not distinguish between atoms or molecules. The protons, neutrons, and electrons are assembled by their electrical forces of attraction balanced against the electron-electron repulsion and the proton-proton repulsion within the context of the wave properties of the electrons themselves.

As we will see, whether in atoms or in molecules, electrons obey the Schrödinger equation. But while the principles of quantization of angular momentum remain invariant with the formation of molecules, important changes take place as a result of the union of atomic wavefunctions to form molecular structures.

First, as we will see, the *spherical symmetry* inherent in the atom, which leads to the degeneracy of the orbitals with the same angular momentum quantum number, ℓ, is lost with the formation of the chemical bond. Considering the simplest case of a diatomic molecule, the spatial structure of the molecule has an axis of symmetry—the internuclear axis between the atoms in the molecular bond.

Second, the existence of a defined spatial axis and an internuclear distance means that the diatomic molecule cannot only move through space (translation), a molecule can also rotate and vibrate as depicted here respectively in panels a, b, and c.

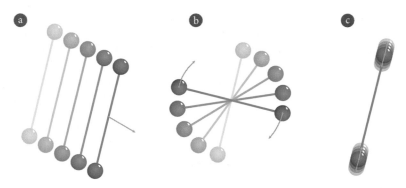

These rotations and vibrations of the molecule are *degrees* of *freedom* unavailable to the isolated atom. In order to construct the molecular bond such that we can answer key questions related to bond strengths and bond geometry, let's consider what happens as two atoms approach one another, seeking the possibility of a more stable configuration by forming a bond with the requisite release of energy. Just as we began our study of atomic structure with the study of atomic hydrogen, we begin our investigation of the chemical bond with the formation of H_2 from two separate hydrogen atoms that we will designate as atom A and atom B. As we bring the separated H atoms together, the electron associated with proton A begins to sense the pull of proton B, albeit through the shielding of the electron associated with atom B so that the electron on A is attracted to atom B with a charge Z_{eff} at that internuclear distance. Similarly the electron on B senses the Z_{eff} of proton A and there is a net attraction. With decreasing internuclear distance, the electron associated with proton A begins to *penetrate* the shielding of the electron associated with proton B and the net attraction between the nuclei increases, lowering the potential energy of the system of two protons and two electrons. This union of the two electrons and two protons from the separated atom A and atom B provides the new wavefunction for each of the electrons to spread out, lowering the energy of the ensemble of electrons and protons that comprise the molecule.

The interaction between the two electrons and the two protons becomes increasingly complex, however, as the distance between the points decreases. As displayed here, the *attraction* between each electron and the two nuclei is offset by proton A repelling proton B and each electron repels the other. But the *net* result, the small difference between strong attraction (electron-proton) and strong

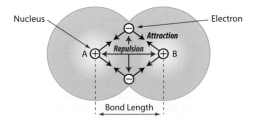

10.10

repulsion (proton-proton and electron-electron), is attraction, the release of energy as the potential energy of the system of electrons and protons *decreases* with decreasing internuclear distance. We can sketch the potential energy of this system of electrons and protons as a function of internuclear distance as displayed in Figure 10.9.

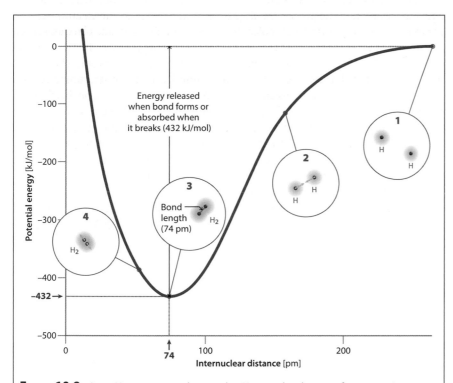

FIGURE 10.9   As an H atom approaches another H atom, the electron of one atom is attracted to the nucleus of the other and visa versa. This attraction begins to *reduce* the potential energy of the ensemble of electrons and protons that constitute the H_2 molecule relative to that of the separated atoms. The attraction strengthens until the decreasing distance between the nuclei begins to rapidly increase the proton-proton repulsion and the potential energy curve begins to increase with *decreasing* internuclear distance. As the electrons are compressed into a smaller and smaller volume with decreasing internuclear distance, they too will increasingly contribute to the increase in the potential energy curve at small internuclear distance.

The *decrease* in potential energy with internuclear distance doesn't continue unchecked. While the potential energy of the system decreases as the electrons delocalize and spread out to minimize the system energy across the totality of the forming H_2 molecule, the proton-proton *repulsion* begins to increase with decreasing internuclear distance. As the internuclear distance drops, the electrons are forced into a smaller and smaller spatial extent, and the electron-electron repulsion *increases*. The result, as displayed in Figure 10.9 is that the potential energy passes through a minimum and then, with further decrease in internuclear distance, begins to rise rapidly.

The net result is that the H atoms are pulled together at larger internuclear distances, pushed apart at small internuclear distance and in between resides the "equilibrium internuclear distance" at point 3 in Figure 10.9.

It is important to emphasize here that in the formation of the chemical bond, energy is neither created nor destroyed, but energy is released in the forming of the bond

$$H + H \rightarrow H_2 + \text{energy}$$

The energy release, by virtue of the bond formation itself, is equal to the depth of the well shown in Figure 10.9 at the equilibrium internuclear distance. While we developed this potential energy diagram for H$_2$, the basic shape of the potential energy "surface" of any chemical bond between two atoms or between an atom and a larger molecule will have the same basic shape. The next question is: What is the relationship between the electronic wavefunctions shown in Figure 10.8 and the potential energy curve shown in Figure 10.9? The answer is that when we have two identical nuclei, as is the case for H$_2$, the ability of the electron-proton combinations to draw electron density to them is equal, so the electrons are shared equally between the nuclei and we have the situation depicted in the upper panel of Figure 10.8.

This electron sharing is termed a *covalent* bond, and the relationship between the simplified bond model in Figure 10.8 and the picture of the electron distribution about the two nuclei of the homonuclear ("same nuclei") molecule is displayed in Figure 10.10a. Thus, while the electron distribution about each atom delocalizes within the structure of the molecule, thereby lowering the potential energy of the ensemble of electrons and protons, the probability density of the electrons within the new molecular bond is equally distributed between the two nuclei.

If a molecule is formed from two atoms, one of which has its valence orbitals more closely bound (thus lowering the energy levels of the valence in one of the atoms), the delocalized wavefunction of the resulting molecule will have the molecular wavefunction shifted preferentially toward the atom with the lower energy orbitals, as shown in Figure 10.10b. This leads to the formation of a "polar covalent" bond that has greater electron density centered on the atom with more tightly bound (lower energy) electrons in the separated atoms. In the extreme case where there is a *large* energy difference between the orbitals of the separated atoms, the molecule formed will have delocalization of the combined wavefunction toward the atom with lower energy orbitals. This case is displayed in Figure 10.10c, and is called an "ionic" bond because the bond is created in large measure by the

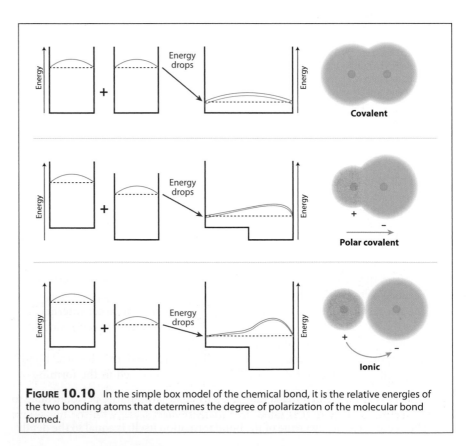

FIGURE 10.10 In the simple box model of the chemical bond, it is the relative energies of the two bonding atoms that determines the degree of polarization of the molecular bond formed.

simple Coulomb attraction between the cation that has donated electrons into the anion that has extracted electrons.

All three bond types—covalent, polar covalent, and ionic—however, share the same qualitative shape for their potential energy surface representing the relationship between potential energy and internuclear distance as shown in Figure 10.9.

A fascinating aspect of chemistry is that the chemical behavior of molecules depends, in large measure, on the spatial distribution of electron density in the chemical bond. We also developed a sense for the linkage of electron *penetration* and *shielding* in individual atoms that lead to the concept of the effective charge, Z_{eff}, that a valence electron "sees" in an atom. It is the variation of Z_{eff} across the periodic table that provides a systematic pattern for atomic size, for the first ionization energy, IE, for electron affinity, EA, and for trends in metallic behavior as summarized in Figure 10.11.

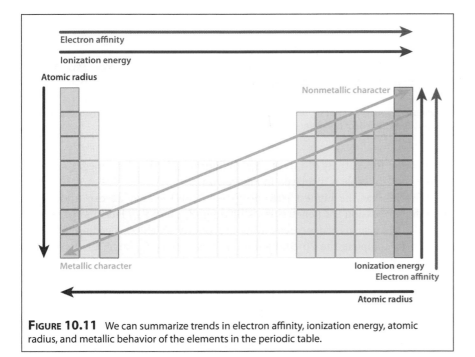

FIGURE 10.11 We can summarize trends in electron affinity, ionization energy, atomic radius, and metallic behavior of the elements in the periodic table.

How do these concepts of effective nuclear charge, first ionization energy, electron affinity, and atomic size translate over to help us understand the chemical behavior of molecules formed from those atoms? A significant part of the answer to this important question is that the propensity for an atom to draw electron density to it in a chemical bond exhibits a coherent pattern or tendency that is, to first order, independent of the particular pairing of atoms involved in the bond. That is, we can find a quantity that can be assigned to an atom in the periodic table, that represents a quantitative measure of that atom's ability to draw electron density to it *independent of the identity of the other atom.*

The ability of a given atom to draw electrons to itself *in a chemical bond* is called *electronegativity.* The concept of electronegativity provides a useful way to estimate the degree to which electron density is delocalized in a chemical bond toward a particular atom in that bond structure. Linus Pauling, an American chemist, succeeded in assembling a coherent scale defining the electronegativity for each element in the periodic table. The Pauling electronegativity scale is an empirical system by which the ability of a particular atom in the periodic table to draw electron density to it is rated on a *dimensionless* scale, with the maximum

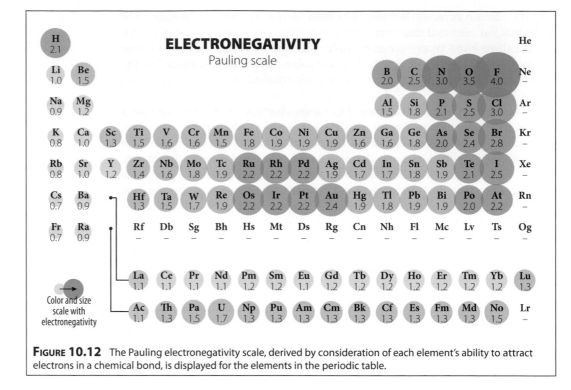

FIGURE 10.12 The Pauling electronegativity scale, derived by consideration of each element's ability to attract electrons in a chemical bond, is displayed for the elements in the periodic table.

value of electronegativity of 4.0 assigned to fluorine (F). The electronegativity scale for the periodic table is displayed in Figure 10.12.

Fluorine has the highest electronegativity (EN), which we can understand from our discussion of penetration, shielding, Z_{eff}, and EI_1. Fluorine has a very *large* Z_{eff} because its valence electrons are poorly shielded, needing only one electron to complete the 2p configuration. Not only is Z_{eff} large for the valence electrons of fluorine, the fluorine atom is small so it attracts available negative charge more strongly than any other element. That large EN for fluorine is in stark contrast with, for example, rubidium (Rb) or cesium (Cs), which are each large with strong shielding of the valence electrons. Rb and Cs have ENs of 0.8 and 0.7 respectively. Hydrogen, which is able to both *extract* electron density in a chemical bond from alkali (Group 1A) and alkaline earth (Group 2A) elements and *donate* electron density to nonmetals in the upper right of the periodic table, has an EN of 2.1.

1	H 2.1	Increasing electronegativity →																
2	Li 1.0	Be 1.5											B 2.0	C 2.5	N 3.0	O 3.5	F 4.0
3	Na 0.9	Mg 1.2											Al 1.5	Si 1.8	P 2.1	S 2.5	Cl 3.0
4	K 0.8	Ca 1.0	Sc 1.3	Ti 1.5	V 1.6	Cr 1.6	Mn 1.5	Fe 1.8	Co 1.9	Ni 1.9	Cu 1.9	Zn 1.6	Ga 1.6	Ge 1.8	As 2.0	Se 2.4	Br 2.8
5	Rb 0.8	Sr 1.0	Y 1.2	Zr 1.4	Nb 1.6	Mo 1.8	Tc 1.9	Ru 2.2	Rh 2.2	Pd 2.2	Ag 1.9	Cd 1.7	In 1.7	Sn 1.8	Sb 1.9	Te 2.1	I 2.5
6	Cs 0.7	Ba 0.9	La 1.0	Hf 1.3	Ta 1.5	W 1.7	Re 1.9	Os 2.2	Ir 2.2	Pt 2.2	Au 2.4	Hg 1.9	Tl 1.8	Pb 1.9	Bi 1.9	Po 2.0	At 2.2
	I	II	[←			Transition metals						→]	III	IV	V	VI	VII

Increasing electronegativity

FIGURE 10.13 A summary of the trends in the Pauling electronegativity scale for the first six periods of the periodic table.

The electronegativity (EN) scale is both extremely useful in practice and remarkable in the fact that, for all the possible combinations of atom-atom pairs that exist in the myriad of possible molecules, a single value of EN can be assigned to each element in the periodic table. Figure 10.13 summarizes the trend in EN across and down the periodic table. The electronegativity scale uses a combination of the first ionization energy, IE_1, and electric affinity, to establish the electronegativity ranking displayed in Figure 10.12 and 10.13. For example, a species A with small IE_1, and small EA will easily surrender an electron but will not compete to extract an electron. Confronted with a species B with higher IE_1, and high EA, A will transfer electron density to B in the formation of a chemical bond. Atoms characterized by small IE_1 and EA are the metals that have, as a result, low values of electronegativity. Atoms in the upper right of the periodic table, the non-metals, have high EN values.

While values of EN are unitless, what matters in the formation of a chemical bond is the *difference* in EN between two atoms involved in a chemical bond. The larger the difference in EN between two species, the more polar will be the bond formed from those species. For example, in the formation of O_2, while the EN for oxygen is very large (3.5) the *difference* in electronegativity in the bond is $\Delta EN_{bond} = EN_O - EN_O = 3.5 - 3.5 = 0$. The bond so formed is covalent with electron density balanced evenly between the two O atoms in the O_2 bond so formed. For the case of NaF, the EN of Na is 0.9 and the EN of F is 4.0, so for the NaF bond $\Delta EN_{NaF} = EN_F - EN_{Na} = 4.0 - 0.9 = 3.1$. This is a very large *difference* in EN and thus the bond will be highly *polar*.

We can relate the range of *differences* in EN between two species in a bond to the range in bond character from nonpolar covalent, to polar covalent, to ionic as shown in Figure 10.14.

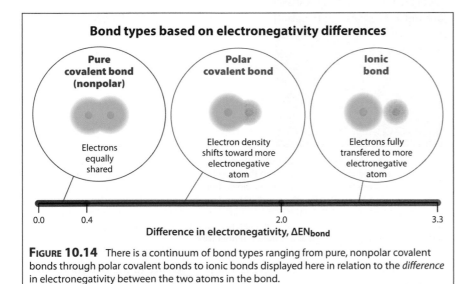

FIGURE 10.14 There is a continuum of bond types ranging from pure, nonpolar covalent bonds through polar covalent bonds to ionic bonds displayed here in relation to the *difference* in electronegativity between the two atoms in the bond.

Types of Chemical Bonds

Our identification and analysis of covalent, polar covalent, and ionic bond structure, which emerges directly from the concept of electronegativity, establishes the foundation for the three types of chemical bonds found in nature. Figure 10.15 displays examples of the three types: ionic bonding, covalent bonding, and metallic bonding.

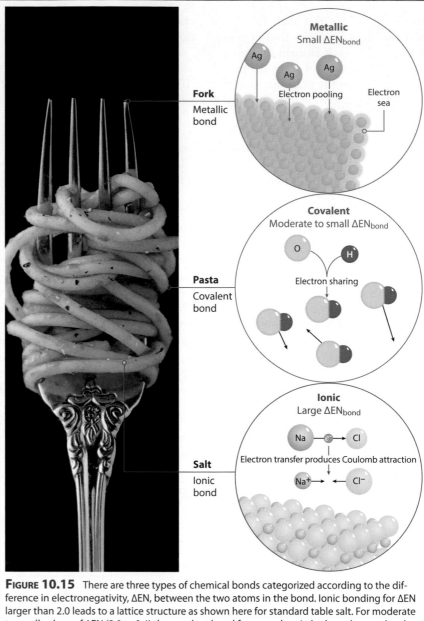

FIGURE 10.15 There are three types of chemical bonds categorized according to the difference in electronegativity, ΔEN, between the two atoms in the bond. Ionic bonding for ΔEN larger than 2.0 leads to a lattice structure as shown here for standard table salt. For moderate to small values of ΔEN (2.0 to 0.4) the covalent bond forms such as in hydrocarbon molecules found in pasta. For small ΔEN for the bond *and* small values of EN for the atoms, metallic bonding occurs wherein the outer electrons are loosely bound to the positive cation core. This type of bond is found in metals such as a standard dinner fork.

Ionic bonding results when there is a large *difference* in the electronegativity of the two atoms that comprise the chemical bond. When metals on the left-hand side of the periodic table (with small EN) bond to nonmetals on the upper right of the periodic table (with large EN), electron density is extracted from the metal and drawn to the nonmetal forming a cation (the metal ion) and an anion (the nonmetal ion). The cation and anion are thus attracted to each atom by the Coulomb force.

When a nonmetal forms a chemical bond with another nonmetal, both partners have large, or fairly large, ENs and thus the *difference* in EN is fairly modest. Thus the valence electrons are *shared* between the atoms participating in the bond. The "shared electron" bond is, as we have seen, termed the *covalent bond*. However, as Figure 10.14 demonstrates, there is a continuum of the degree of electron delocalization from strongly ionic (e.g. NaCl) to polar covalent (e.g.

HF) to covalent (e.g. NO) depending on the *difference* in electronegativity between the two atoms that form the molecular bond. The third type of chemical bond is the *metallic bond* that, as the name suggests, occurs when metal atoms bond to each other in an ordered lattice as depicted in Figure 10.15. Metals have a low electronegativity and a low ionization energy and thus lose their electrons easily. The simplest model for metallic bonding is termed the "electron sea" model because the valence electrons of metals in a lattice delocalize throughout the lattice structure such that the metal cations are surrounded by a continuous "sea" of delocalized electrons. This freedom of movement of electrons throughout the metal lattice is the reason metals reflect light, conduct electricity, and conduct heat. In each case, it is the freedom of movement of electrons in the metal lattice that is responsible.

Metals are also capable of being pounded into thin sheets (a characteristic termed malleability) and metals can be drawn into long wires (termed ductility). The reason for this emerges from the "electron sea" model because while the metal cation in the sea of electrons can be displaced laterally with respect to a neighbor, chemical bonds are not explicitly broken, but rather seamlessly shifted as displayed in Figure 10.16.

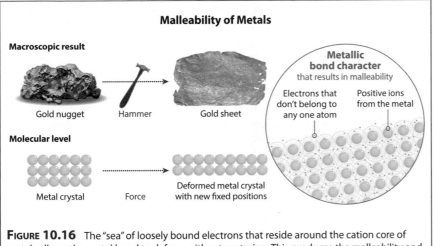

FIGURE 10.16 The "sea" of loosely bound electrons that reside around the cation core of metals allows the metal bond to deform without rupturing. This produces the malleability and ductility of metals. It also explains the excellent conductance of both electricity and heat in metals.

We can summarize the relationship between the *types of atoms* that comprise a chemical bond and the *type of chemical bond* that results from the union of the two atoms:

Type of Atom	Type of Bond	Electron Movement
Metal and nonmetal	Ionic	Electron transferred
Nonmetal and nonmetal	Covalent	Electron shared
Metal and metal	Metallic	Electron delocalized through lattice

While the delocalization of electron density throughout the lattice structure of a metal is typically treated theoretically using the "electron sea" model at various levels of sophistication, metal-metal bonds tend to exhibit far less variety in chemical behavior than chemical bonds that explicitly bind electrons within a molecular structure. We consider here models used to understand chemical bonds that involve electrons bound within a specific chemical bond. Thus we consider a series of increasingly detailed models applied to covalent and ionic bonds that respectively involve the sharing of valence electrons or the transfer of an electron from one atom to its partner.

Representation of Valence Electrons in a Chemical Bond

A key part of the development of quantitative reasoning in science is built upon a foundation of physical models that are as simple as possible, but still capture the essence of the factors that control the behavior of a given system.

The development of theories regarding molecular bonding have been developed to span from the simple to the complex, but they are all *models* that employ simplifications. The key judgment is to know when to apply a particular model to a particular problem, depending upon the scientific question to be answered or the scientific hypothesis to be tested. This evolution in complexity spans the range from simply keeping track of the valence electrons in a chemical bond (Lewis electron dot symbols) to solutions to the Schrödinger equation for many electron systems that demand (and often exceed) the capacity of the world's largest supercomputers. If the evolution from the simple models through increasingly complex models did not represent an effective progression that contributed significantly to our chemical intuition, we would not take the trouble to present that sequence of increasingly sophisticated models here because computing power is now available to easily solve the Schrödinger equation for important systems on a desktop computer. But there is an important difference between solving an equation with a computer on the one hand, and understanding how the formation of molecular bonds is tied to chemical behavior on the other. An understanding of the sequence of increasingly complex models of the chemical bond will provide very useful insight into the different character of chemical bonds that comprise molecular structure, and as a result, a fundamental grasp of chemical behavior that opens doors to an array of exciting opportunities.

As we have seen, the inner "core" electrons are bound so tightly (have such large *negative* potential energies) that they do not engage in the exchange of electron density with other atoms in a molecular bond. We thus focus attention on the outer *valence* electrons that are far less tightly held, and are thus the ones that are engaged in the formation of the chemical bond.

As we noted in Chapter 2 and in the Framework section of this chapter, it was G. N. Lewis who developed a systematic way of both (1) representing the valence electrons in an atom and (2) representing those same electrons in the chemical bond formed from those atoms.

We already know how to determine the *electron configuration* for each atom in the periodic table. What the Lewis electron dot symbol represents is a simplified representation of the detailed electron configuration in a form that keeps track of the valence electrons in both the atom and in the bonding structure of the molecules formed from those atoms. We simply need to link the electron configuration to the Lewis dot structure. In order to do this, we need to first isolate the core electrons from the valence electrons. We can do this by linking the elements in a given *Period* with the corresponding electron configuration. The separation between the *core* electrons and the *valence* electrons occurs with the progression from a principal quantum number n, to the next higher principal quantum number n+1.

This is shown for the Period 2 (i.e. n=2) elements in Figure 10.17. The core electrons in this case are the $1s^2$ electrons, which are the most tightly bound electrons in each of the Period 2 elements. It is the n=2 electrons that constitute the *valence* electrons from lithium (Li) to the noble gas neon (Ne). The *reason* for the distinct demarcation in energy between inner *core* electrons and outer *valence* electrons is shown in the left-hand panel of Figure 10.17. Specifically, with the progression from the n=1 to the n=2 principal quantum number, there exists a dramatic *drop* in the first ionization energy, IE_1, of the next electron, in this case lithium (Li), resulting from the shielding of the closed shell He $1s^2$ electrons and the increase in radius of the 2s electron that reduces the potential energy of

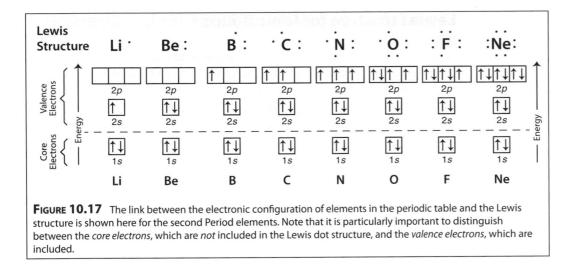

FIGURE 10.17 The link between the electronic configuration of elements in the periodic table and the Lewis structure is shown here for the second Period elements. Note that it is particularly important to distinguish between the *core electrons*, which are *not* included in the Lewis dot structure, and the *valence electrons*, which are included.

the Li 2s electron thereby reducing how tightly bound that electron is to the Li nucleus. Displayed at the top of each of the elements electron configuration is the corresponding Lewis structure. Notice that the Lewis structure involves *only* the valence electrons and those electrons are organized according to whether the orbitals are singly occupied or doubly occupied. It is very important when constructing a Lewis diagram to first clearly identify the valence electrons.

Moving to the next Period in the periodic table, Period 3, Figure 10.18 steps sequentially through the elements corresponding to n=3 from sodium, Na, to argon, Ar. The first ionization energy, IE_1, for Ne ($1s^2 2s^2 2p^6$) is 2100 kJ/mole while IE_1 for Na ($1s^2 2s^2 2p^6 3s^1$) is 500 kJ/mole. The *core* electrons of Na ($1s^2 2s^2 2p^6$) are tightly held and do not engage in molecular bond formation, while the *valence* electrons are dramatically less tightly bound, and those valence electrons are displayed in Figure 10.18 for each element in Period 3.

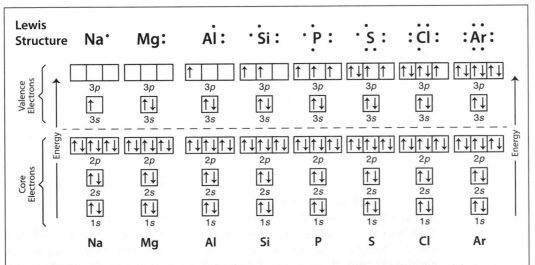
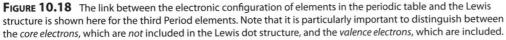

FIGURE 10.18 The link between the electronic configuration of elements in the periodic table and the Lewis structure is shown here for the third Period elements. Note that it is particularly important to distinguish between the *core electrons*, which are *not* included in the Lewis dot structure, and the *valence electrons*, which are included.

Lewis Structure for Ionic Bonds

While we will focus the application of Lewis structure models on the vast array of covalent bonds in chemistry, the Lewis structure can be used for ionic bonding as well. In fact we begin with ionic bonding because it is simple, yet provides an introductory framework for building Lewis structures for a diverse range of molecules.

Consider, as an example, the bonding structure for potassium chloride, KCl. We can write the Lewis symbols for potassium and for chlorine as shown in Figure 10.19.

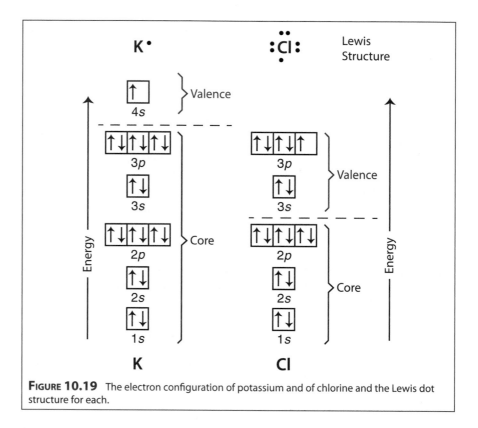

FIGURE 10.19 The electron configuration of potassium and of chlorine and the Lewis dot structure for each.

When potassium and chlorine bond, the electronegativity of Cl (EN=3.0) is so much greater than that of K(EN=0.8) that potassium effectively transfers its electron to Cl

$$K\cdot + :\ddot{C}l: \longrightarrow K^+ \Big[:\ddot{C}l:\Big]^-$$

This transfer of an electron from potassium gives chlorine an octet of electrons and as a result a closed shell electron configuration. The transfer of the electron from potassium leaves it without a valence electron, but with a closed n=3 shell electron configuration. However, because K has donated an electron, it becomes a cation, K^+, and Cl, receiving an electron, becomes an anion, Cl^-. The Coulomb attraction of the cation to the anion *decreases* the potential energy of K^+Cl^- below that of the separated atoms, resulting in the formation of an *ionic* bond, as displayed in Figure 10.20. It is common practice to write the cation without brackets and the anion with brackets.

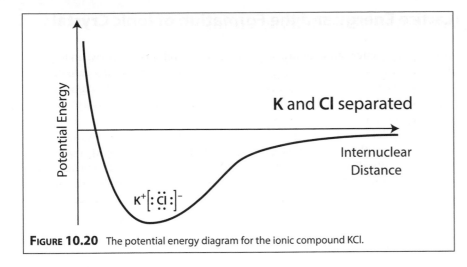

FIGURE 10.20 The potential energy diagram for the ionic compound KCl.

The Lewis model of ionic bonding has, it turns out, considerable predictive power. For example, if we use Lewis structures to determine the bonding between magnesium, Mg, and chlorine, Cl, we note that, from Figure 10.18, magnesium has the Lewis structure **Mg:** and chlorine **:Cl:**. However, a chlorine atom can only accept a single electron, so how is it possible to leave both Mg and Cl with a closed shell octet? The answer is that a single magnesium atom bonds to two chlorine atoms

$$\text{Mg:} + \begin{matrix} \ddot{\ddot{Cl}}: \\ \ddot{\ddot{Cl}}: \end{matrix} \longrightarrow Mg^{2+}\left[:\ddot{Cl}:\right]_{2}^{-} \longrightarrow MgCl_2$$

to form the ionic bond joining the magnesium cation with two chlorine anions to form $MgCl_2$. The key observation is that it *is* indeed $MgCl_2$ that is the observed structure in nature. Notice that with an EN of 1.2 for magnesium (and EN for Cl of 3.0) the large *difference* in electronegativity results in the transfer of the electrons from magnesium to the two chlorine atoms.

What does Lewis theory predict for the bond structure of aluminum (Figure 10.18, **Al:**) with oxygen (Figure 10.17, **·O:**)? First, the difference in electronegativity for those two elements is $\Delta EN = EN_O - EN_{Al} = 3.5 - 1.5 = 2$ so aluminum will transfer an electron to oxygen. But each oxygen can receive just two electrons to fill its octet, while each aluminum has three electrons to donate in order to attain a closed shell octet of its own. The solution, by the prediction of Lewis's theory, is that *two* aluminum atoms bond to *three* oxygen atoms with the following electron transfer

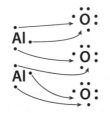

The oxide of aluminum, Al_2O_3, is ubiquitous in nature, in the industrial production of aluminum metal, and in abrasives because of its intrinsic hardness, and it is a primary reason why the durability of aircraft, boats, and housing structures makes aluminum so valuable.

Energy Required to Transfer Electron

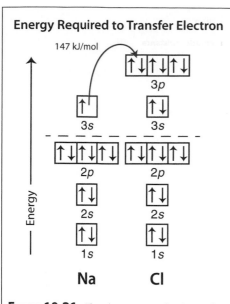

FIGURE 10.21 The electron transfer shown for the electron configuration of Na and of Cl in the formation of the ionic bond in NaCl.

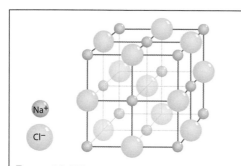

FIGURE 10.22 The crystal structure of NaCl showing the alternating position of the Na⁺ and Cl⁻ cation and anion.

Lattice Energy and the Formation of Ionic Crystals

There is a very interesting feature of the *energy* trade-off that occurs when ionically bonded molecules coalesce to form crystal structures. Consider the case of common table salt, sodium chloride, that is formed from elemental sodium (a solid metal) and gas phase chlorine atoms Na(s) + Cl(g) → NaCl(s) $\Delta H°_f$ = −411 kJ/mole. At first glance, it would seem reasonable to assume that the *exothermicity* of the reaction results from the energy release of the electron transfer to form the ionic bond. However, the first ionization energy (EI₁) of sodium is +496 kJ/mol and the electron affinity (EA) of Cl is −349 kJ/mole. Thus the transfer of the electron, as shown in Figure 10.21, is +147 kJ/mole. It is *endothermic*! It requires the input of energy to form the NaCl bond. Why, then, is the reaction of elemental sodium and chlorine exothermic?

The answer lies in the Coulomb attraction between the cation Na⁺ and anion Cl⁻ when the crystal structure is formed with alternating Na⁺ and Cl⁻ ions as shown in Figure 10.22. Thus it is the reduction in potential energy, resulting from the Coulomb attraction as energy is *expended* to transfer the electrons from the sodium to the chlorine while at the same time energy is *released* as the cation and anion are attracted toward each other to form the crystal structure. We can summarize the sequence of energy *expended* to form the cation and anion from Na(s) and Cl(g) followed by the *release* of energy resulting from the Coulomb attraction of the cation-anion pairs that form the lattice resulting in the formation of the NaCl crystal structure, shown as a two-step process in Figure 10.23.

The lattice energy, ΔH_{lat}, corresponding to the wide variety of metal-nonmetal ionic bonds range over large values for two principal reasons:

1. The potential energy *release* in the crystal lattice formation is proportional to the product of the charge on the cation times the charge on the anion and inversely proportional to the distance between the ions. As a result of the range in bond lengths there is a corresponding range in lattice energy.

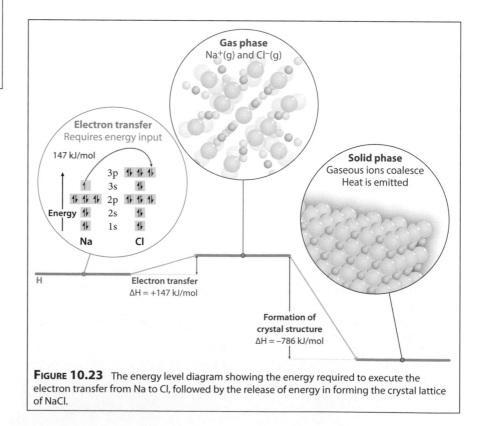

FIGURE 10.23 The energy level diagram showing the energy required to execute the electron transfer from Na to Cl, followed by the release of energy in forming the crystal lattice of NaCl.

2. The *product* of the ionic charge, which appears in the numerator of the potential energy expression, exerts powerful control over the lattice energy. For example Na^+F^- has a bond length of 231 pm and $Ca^{2+}O^{2-}_2$ has a bond length of 239 pm, nearly identical. But the lattice energy for NaF is $(\Delta H_{lat})_{NaF} = -910$ kJ/mol while that for CaO is $(\Delta H_{lat})_{CaO} = -3414$ kJ/mol reflecting the factor of 4 larger product of the ionic charge for CaO.

3. The trend in lattice energy as a function of the ionic bond length in a series of Group 1A cations with Cl^- demonstrates clearly (see sidebar) that as the internuclear distance increases as a result of increasing size of the cation, the lattice energy decreases markedly.

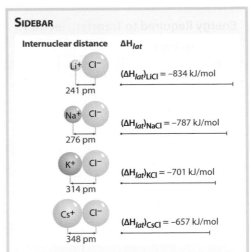

SIDEBAR

Internuclear distance $\quad \Delta H_{lat}$

Li^+ Cl^- $(\Delta H_{lat})_{LiCl} = -834$ kJ/mol
241 pm

Na^+ Cl^- $(\Delta H_{lat})_{NaCl} = -787$ kJ/mol
276 pm

K^+ Cl^- $(\Delta H_{lat})_{KCl} = -701$ kJ/mol
314 pm

Cs^+ Cl^- $(\Delta H_{lat})_{CsCl} = -657$ kJ/mol
348 pm

The lattice energy (energy release when the gas phase cation-anion pair form the ionic crystal) as a function of internuclear distance for a range of Group 1A cations with Cl^-. As the nuclear distance increases, the lattice energy decreases markedly.

Lewis Structures and Covalent Bonding

While *ionic bonding* of main group elements represents a clear and straightforward application of Lewis bonding theory, it is the treatment of *covalent* bonding using Lewis theory that constitutes a simple but powerful foundation for a series of increasingly complex approaches to bonding theory.

As an introduction to covalent bonding using Lewis theory, we briefly summarize the central elements of that theory:

1. Valence electrons, because they are as a group less tightly bound to the nucleus, play a fundamental role in chemical bonding. Thus it is important to carefully distinguish the inner core electrons from the valence electrons. The Lewis structure is a systematic way of representing only the valence electrons of an atom, and that separation between the core electrons and the valence electrons is displayed in Figures 10.17 and 10.18.

2. Electrons are either *transferred* or they are *shared* in such a way that each atom in the bond acquires a stable electron configuration. This stable electron configuration is usually a filled shell. For main group elements this filled shell is an octet (ns^2np^6).

3. Ionic bonds result when there is a *transfer* of one more electrons from one atom (of low electronegativity) to another atom (of high electronegativity) in a chemical bond.

4. When electrons are *shared* in a chemical bond, a *covalent* bond results—typically between atoms with small to moderate *differences* in electronegativity.

5. Lewis theory is very effective in the analysis of main group elements—those for which the filled valence shell is octet. For d and f orbital systems, other bonding models are more appropriate. But that is important—simple but powerful models always have their limitations, and so it is with Lewis theory.

Lewis Structures for Covalent Bonds

Chemical behavior is controlled by the distribution of electrons about the nuclear centers in the bonding structure of a molecule. Indeed, the position and ordering of the nuclei in a bonding structure is controlled by the electron distribution that leads to the lowest possible potential energy of the ensemble of electrons and nuclei that comprise a molecular bonding structure. When we develop the Lewis electron dot theory for molecules, it is important to remember that the Lewis theory is a model that helps select the molecular structure of lowest potential energy. Thus, while the procedures for writing a Lewis structure appear to be simply a list of rules, those rules represent a technique for predicting a structure of lowest potential energy. As is typically the case, we begin with the simple application of Lewis theory for diatomic molecules and then build toward polyatomic structures; always keeping in mind that the Lewis model brings important chemical insight, but it is not without its limitations.

Lewis Structures for Single Covalent Bonds: Diatomics

If we extract the diagram linking the *electron configuration* of a chlorine atom, Cl, with its Lewis dot structure, Figure 10.24, we observe first that only the valence electrons are displayed in a Lewis diagram. Second, that there are seven electrons in the valence shell. As we will see, setting the correct bonding structure in a Lewis diagram always begins and ends with the counting of electrons! Third we recognize that the difference in electronegativity between the bonding atoms is zero because Cl_2 is a homonuclear molecule—the bonding partners are the same. Thus the bond has no intrinsic polarity and the electrons are shared equally between the Cl atoms. Fourth, Lewis theory postulates that for elements in the second and third row of the periodic table, the outer shell of electrons must be an *octet* of electrons.

For the case of Cl_2, we have a total of $2 \times 7 = 14$ valence electrons so if we are to form a bonding structure for Cl_2 wherein each Cl atom in the molecule is surrounded by 8 electrons, then the Cl atoms must each share an electron from the other Cl atom. Thus the Lewis electron dot diagram for Cl_2 is

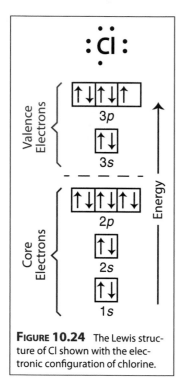

FIGURE 10.24 The Lewis structure of Cl shown with the electronic configuration of chlorine.

where the two electrons forming the bond between the two atoms are shared.

If we were to write the Lewis structure for HCl, we would first count valence electrons, 7 for Cl and 1 for H. While completing the closed shell configuration for Cl requires an octet (8), completing the closed shell configuration for H requires just two electrons, a duet. Thus by sharing a pair of bonding electrons, both H and Cl complete their respective closed shell configuration:

Because the electronegativity of Cl exceeds that of H, electron density will delocalize from the H atom to the Cl atom, but the Lewis diagram is *not* written to reflect this delocalization unless the electron is *transferred* as is the case for an ionic bond previously discussed.

There is an important distinction to be made in each of the Lewis diagrams we have written: the distinction between the *bonding* pair of electrons and the electron not involved in the shared electron bond. This latter category of electrons is termed the "lone pair" electrons. The distinction is highlighted to emphasize the point for Cl_2 and for HCl.

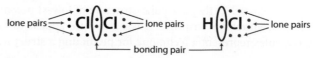

The distinction between the bonding electron pairs and the lone pairs will become increasingly important as we examine more complex bonding structures and the relationship between molecular *shape* and bonding structure.

Lewis Structures for Single Covalent Bonds: Polyatomic Molecules

Let's next consider two of the most important atoms that sustain life on the planet: oxygen and hydrogen. To determine the Lewis structure of bonds involving those two elements we first count valence electrons: six electrons for oxygen and one electron for hydrogen, a total of seven.

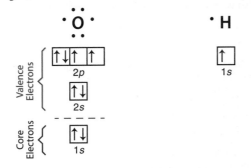

We can write the Lewis dot structure for diatomic OH:

However, while H has the requisite two electrons to form a closed shell, oxygen is left with seven electrons in the outer shell.

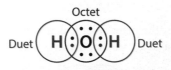

Thus, Lewis theory deems this structure unacceptable because it violates the octet rule, the outer electron shell of oxygen has seven, not eight electrons. Thus Lewis theory predicts that there is a structure of *lower potential energy* than OH—what is it?

We recognize that if we were to add another hydrogen atom to the structure, that hydrogen atom would contribute one shared bonding electron to O such that the octet rule would be satisfied for oxygen and the duet rule for hydrogen.

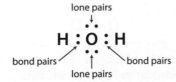

Thus Lewis theory predicts that the lowest potential energy (i.e. most stable) structure involving hydrogen atoms bonded to oxygen atoms is H_2O. Indeed it is.

It is also important to distinguish between the bonding pairs of electrons and the lone pair electrons in water.

It is the convention that *electron pair bonds* in Lewis structures are represented by a straight line to emphasize the distinction between electron pairs that are bonding vs. lone pairs, so we typically use this convention because it contains more information *and* it is quicker to write!

$$H - \overset{..}{\underset{..}{O}} - H$$

We should stop to reflect upon the fact that, while Lewis theory is very simple, it has already exhibited significant predictive power — it has given us the first order structure for water and it has distinguished between bonding electron pairs and lone pairs not directly involved in the bonding structure.

This emerging ability of Lewis structures to predict why specific combination of atoms form stable molecules while other do not, begs the question: what about H_3O?

We can immediately write a candidate Lewis structure.

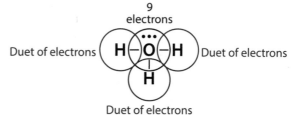

Because the 9 electrons in the shell of oxygen violates the octet rule, Lewis theory predicts that H_3O is not stable. *However*, if we remove one electron to form the *cation*

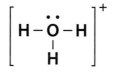

we are left with the hydronium ion H_3O^+. We know the hydronium ion is stable; it is the central species in acid-base chemistry!

Lewis theory also provides insight into other possible combinations of hydrogen-oxygen bonding. For example if we create the bonding structure of O_2, we combine two oxygen atoms with 6 electrons each, for a total of 12 electrons:

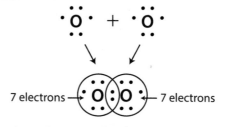

But this structure violates the octet rule! It is not an acceptable Lewis structure. If, alternatively, we draw a structure that involves *four* shared bonding electrons

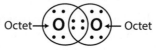

then we can satisfy the Lewis octet rule with a double bond.

$$\overset{..}{\underset{..}{O}} = \overset{..}{\underset{..}{O}}$$

As we develop more sophisticated models of bonding based upon quantum mechanics we will reveal inadequacies in this Lewis structure for O_2, but for simple models of bonding, this is an acceptable Lewis structure. But this structure of O_2 immediately suggests alternative possibilities for bonding structures between oxygen and hydrogen. Specifically if each oxygen contributes six electrons and each hydrogen contributes one electron, the double bond in O_2 can be replaced by a single bond and the *seven* outer shell electrons in our first guess at the Lewis structure

$$\cdot \ddot{O} : \ddot{O} \cdot$$

can share the 7th electron with hydrogen to form the structure

$$H : \ddot{O} : \ddot{O} : H$$

that satisfies the octet rule for oxygen and the duet rule for hydrogen:

$$H - \ddot{O} - \ddot{O} - H$$

This is the structure of a very important category of compounds, the *peroxides*. In this case hydrogen peroxide H_2O_2 is a widely used disinfectant as well as a powerful catalytic agent in biological systems. The hydroxide structure of chlorine

$$:\ddot{Cl} : \ddot{O} : \ddot{O} : \ddot{Cl} :$$

or

$$:\ddot{Cl} - \ddot{O} - \ddot{O} - \ddot{Cl} :$$

is the species singularly responsible for the Antarctic ozone hole as we will see in Chapter 12.

Lewis Structures and Bonding Character

This idea, intrinsic to the Lewis theory, that the attraction between two covalently bonded atoms is due to the sharing of one or more *electron pairs* is very important to the development of chemical intuition. The shared electron pair, or pairs, idea implies that each bond links a *specific pair* of atoms. As a consequence, the fundamental units within covalently bonded compounds are atom pairs leading to a bonding structure dominated by forces *within* molecular structures. This is in stark contrast to ionic bonding, which creates structures that are not *directional* between atoms in a specific molecule; rather, ionic bonding creates a *lattice* of positive and negative charges among an entire *array* of ions.

Thus for covalently bonded molecules, the bonding interactions *within* the molecules (intramolecular forces) are far stronger than the interactions *between* molecules (intermolecular forces). As a consequence, when a molecular compound undergoes a *phase transition*, such as when water boils or ice melts, the H_2O molecular structure remains intact, but only the weaker forces between H_2O molecules play a role in the phase transition.

Constructing Lewis Structures For Polyatomic Molecular Compounds

We now develop a strategy for writing the Lewis structure for a molecular compound. As we will see, this process depends primarily on practice, and as the molecules become more complicated, it is helpful to establish a consistent sequence of steps to sort out the details.

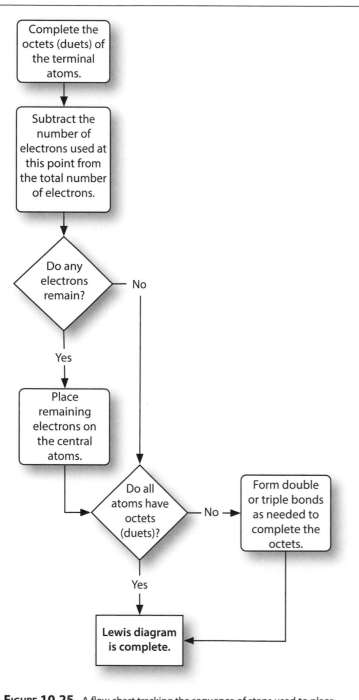

Step 1: Draw the Correct Skeletal Structure for the Molecule

For simple molecules, this step in the sequence is easy. As the molecular structure becomes increasingly complicated, this step becomes increasingly difficult. It is also important to recognize that the only way to be certain that the correct skeletal structure has been identified is to verify the structure by experiment. However, there are two important guidelines. The first is that hydrogen atoms are almost invariably *terminal* atoms because hydrogen has a single electron and thus cannot form multiple bonds that is required of a central atom. The second important guideline is that the *more electronegative* atoms reside in the *terminal* positions, the less electronegative in the central positions.

Step 2:

Bookkeeping; we need to count the total number of valence electrons contributed by each atom to the molecular structure. Given that Lewis structures are almost exclusively for elements in Period 1 through 3, the number of valence electrons for any main group element is equal to its group number in the periodic table. In practice it is essentially necessary to memorize Figures 10.17 and 10.18. A small price in practice!

Step 3:

Place electrons within the skeletal structure giving octets to the atoms other than hydrogen, and duets to the hydrogen atoms. Begin this process by placing two electrons between every pair of atoms—those are the minimal number of bonding electron pairs. Next distribute the remaining electrons as lone pairs beginning with the *terminal* atoms and then moving to the central atoms.

Step 4:

Count the electrons around each atom and the total number of available electrons. If the available electrons have been used up and any atoms lack an octet, form double or triple bonds as necessary to satisfy the octet (duet) rule.

In Figure 10.25, we can summarize this final step in a logic diagram that captures the sequence in completing the Lewis diagram.

Let's work out the Lewis structure for three of the most important polyatomic molecules to life using this strategy: H_2O, CO_2, and O_3 in Table 10.1.

Is it correct to have two different Lewis structures for the same molecule? Is it physically plausible that a mol-

FIGURE 10.25 A flow chart tracking the sequence of steps used to place electrons in their correct positions within the bonding structure of molecules in the first three periods of the periodic table.

TABLE 10.1

	Steps in the Procedure	Lewis Structure for H_2O	Lewis Structure for CO_2	Lewis Structure for Ozone
1	Correct skeletal structure for the molecule.	Hydrogen atoms are terminal atoms. Thus O is the central atom H O H	Oxygen is more electronegative than carbon; carbon is the central atom O C O	All atoms are the same O O O
2	Determine total number of electrons for the Lewis structure by adding the valence electrons contributed by each atom. See Figures 10.17 and 10.18.	Each hydrogen contributes one electron: 2 electrons from hydrogen, oxygen is Group 6 so the total is $6 + 2 = 8$ electrons	Carbon has 4 valence electrons, oxygen 6. Total number of electrons: $6 + 6 + 4 = 16$ electrons	Total number of electrons: 6 electrons from each oxygen atom 3×6 = 18 electrons
3	Distribute electrons beginning with bonding electrons, then assigning lone pairs to terminal atoms, then to lone pairs on the central atom. Check to see whether each atom has an octet (duet for hydrogen atoms.)	Bonding electrons first H : O : H 4 electrons left; lone pairs on terminal atoms next H : O : H None needed; lone pairs on central atom next H :O: H Zero electrons left	Bonding electrons first O : C : O 12 electrons left; lone pairs on terminal atoms next :O : C : O: Zero electrons left	Bonding electrons first O : O : O 4 electrons used, 14 remain; lone pairs on terminal atoms next :O : O : O: 16 electrons used, two remain :O : O : O:
4	If an atom lacks an octet, form double or triple bonds as necessary to give them octets.	All atoms have an octet or duet. No electrons left; Lewis diagram complete	Carbon lacks an octet, so move terminal lone pairs to central atom to form double bonds :O: C :O: Octet Octet (O :: C :: O) Octet :O = C = O: Correct Lewis Structure	Central atom does not have an octet, create double (or triple) bonds on central atom Octet Octet (:O : O :: O:) Octet This would imply that two different structures exist for ozone :O — O̤ = O: and :O = O̤ — O:

ecule could have an asymmetric structure wherein one of the O–O bonds in ozone is a double bond and one a single bond? Experiments have shown conclusively that the bonds in ozone are identical. How can such an observation be reconciled with the predictions of the Lewis model? The answer is that, as we will see in Chapter 10, electrons are capable of delocalizing within the bond structure of a molecule such that the two bond structures represented by the two Lewis structures of O_3 are of equal energy.

$$:\ddot{O} - \ddot{O} = \ddot{O}: \quad \overset{\text{Resonance}}{\longleftrightarrow} \quad \ddot{O} = \ddot{O} - \ddot{O}:$$

and are thus "resonance structures." That is, neither the structure on the left nor the structure on the right actually exist, but rather a *hybrid* of the two structures exists with the extra electron pair delocalized along the spine of the O_3 molecules. Ozone is actually a bent molecule and we draw the bonding structure of ozone with the delocalized electron pair spread along the backbone of the molecule.

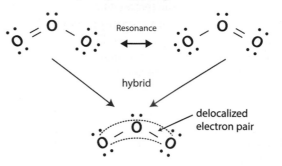

As we will see, this resonance structure, which leads to the concept of a hybrid structure involving the delocalization of electron density over an extended domain in the bonding structure of a molecule, is quite common in chemistry. It is a manifestation of the fact that electrons distribute themselves within the structure of a molecule such as to minimize the potential energy of the ensemble of all electrons and protons in the structure. If the electrons "need" to delocalize to reduce the potential energy of the molecule system, they will do just that.

One of the most important applications of modern chemistry is the development of the techniques to enhance the productivity of crops worldwide. An important compound for fertilizing agricultural crops is ammonium nitrate, NH_4NO_3, which is produced from ammonia, NH_3, and various forms of nitrogen oxides. As noted in Chapter 5, the annual ammonia production is 150 billion kg produced in the Haber process that reacts N_2 with H_2 over an iron catalyst at elevated temperature and pressure:

$$N_2 + 3H_2 \rightarrow 2NH_3$$

The ammonia is oxidized to ammonium nitrate fertilizer and applied to crops worldwide. The use of ammonium nitrate fertilizers has become so widespread that the majority of fixed nitrogen entering the Earth's biosphere is from synthetic fertilizers. A key problem in chemistry today is the development of more efficient ways of producing fixed nitrogen and in controlling the amount of fixed nitrogen used. Let's examine the Lewis structures for NH_3, NH^+_4, and NO^-_3 using our stepwise procedure in Table 10.2..

Inspection of the derived Lewis diagram for NO_3^- reveals, as was the case for ozone, that the nitrate anion has a single bonding pair of electrons between two of the oxygen atoms and the central nitrogen, and a double bonding pair of electrons between one of the oxygens and the central nitrogen. However, experiments reveal that all three oxygen-nitrogen bonds in NO^-_3 are equivalent, so the actual structure of the nitrate anion is a *hybrid* of the three structures, with the electron pair that forms the double bond *delocalized* across the structure of the anion:

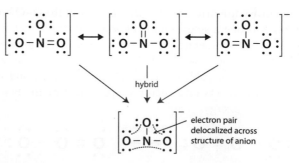

While the determination of the correct Lewis structure generally results in either a single structure or a hybrid of equivalent resonance structures, all with the same positioning of the atoms within the structure (the only differences being the location of single, double and triple bonds within the same structure), there are important cases where there are structures with alternative *positions* of the atoms within the inferred molecular structure. How do we decide which structure is the correct one? Which one has the lowest potential energy?

Consider the case of nitrosyl chloride, NOCl, which is an important constituent in the powerful acid *aqua regia* that is a mixture of concentrated nitric acid and

TABLE 10.2

	Steps in the Procedure	Lewis Structure for NH_3	Lewis Structure for NH_4^+	Lewis Structure for NO_3^-
1	Correct skeletal structure for the molecule.	Hydrogen is always terminal so nitrogen is the central atom H N H H	Hydrogen's terminal so nitrogen is the central atom H H N H H	Nitrogen is least electronegative so is central atom O O N O
2	Determine total number of electrons for the Lewis structure by adding the valence electrons contributed by each atom. See Figures 10.17 and 10.18.	Nitrogen has 5 valence electrons and hydrogen has 1. $5 + 3 \times 1 = 8$ electrons	Nitrogen has 5 valence electrons, hydrogen has 1, and because NH_4^+ is a +1 Cation, subtract 1 electron $5 + 4 \times 1 - 1 = 8$ electrons	Nitrogen has 5 valence electrons, oxygen has 6 valence electrons, and 1 negative charge adds 1 electron $5 + 3 \times 6 + 1 = 24$
3	Distribute electrons beginning with bonding electrons and then assigning lone pairs to terminal atoms then to lone pairs on the central atom. Check to see whether each atom has an octet (duet for hydrogen atoms.)	Bonding electrons added first H : N : H H This uses 6 of 8 electrons. Last 2 electrons added to central N *Octet / Duet — H N H — Duet / H Duet*	Bonding electrons H H : N : H H Eight electrons available, eight electrons used *Duet / Octet — H N H — Duet / H Duet* Octet and duets satisfied	Bonding electrons first O O : N : O Distribute remaining electrons first to terminal atom : O : : O : N : O : 24 electrons used but central N does not have octet move a lone pair to form double bonds : O : : O : N :: O :
4	If an atom lacks an octet, form double or triple bonds as necessary to give them octets.	The central nitrogen has an octet, the hydrogen a duet. All electrons accounted for. Correct Lewis structure	As a convention enclose the Lewis structure in brackets with charge of ion in upper right $\left[\begin{array}{c} H \\ H-N-H \\ H \end{array} \right]^+$	Enclose in brackets with charge designated $\left[\begin{array}{c} :\ddot{O}: \\ :\ddot{O}-N=\ddot{O}: \end{array} \right]^-$

hydrochloric acid. NOCl reacts with water to form HCl and it photodissociates to form the nitric oxide radical, NO, and the atomic chlorine radical, Cl:

$$NOCl + h\nu \rightarrow NO + Cl$$

Let's submit nitrosyl chloride to our sequence for determining the Lewis structure of the molecule.

At this point we are left with four structures, each of which satisfy the Lewis structure octet rule, yet only one of them is correct. We are thus left with a choice. Either we can do a full quantum mechanical structure calculation on a computer to determine which structure is of lowest potential energy, or we can employ the methods of *formal charge* to gain insight into which structure is the correct one, an approach developed by Irving Langmuir.

TABLE 10.3

	Steps in the Procedure	Lewis Structure for NOCl
1	Correct skeletal structure for the molecule	We recognize immediately that the common chemical formula for nitrosyl chloride is not the correct skeletal placement because it places the most electronegative atom in the central position. Nitrogen has an electronegativity of 3.0, but so too does Cl. Thus we have two possible skeletal structures: Choice A: O Cl N Choice B: O N Cl
2	Number of electrons	Nitrogen has 5 valence electrons, oxygen has 6, and chlorine 7, so we have a total of $5 + 6 + 7 = 18$ Electrons
3	Distribute the electrons	We assign the bonding electrons first Choice A: O : Cl : N Choice B: O : N : Cl Next place the electrons on the terminal atoms Choice A: :O : Cl : N: Choice B: :O : N : Cl: This uses 16 of the available 18 electrons. Place the last 2 electrons on the central atom
4	If an atom lacks an octet, form double or triple bonds	Both structures violate the octet rule, so we move an electron pair from either the terminal oxygen or the terminal nitrogen to form a double bond for choice A. For choice B either an electron pair from the terminal oxygen or from the terminal chlorine. All four of these structures satisfy the octet rule. Can they all be correct? No, only one structure of nitrosyl chloride exists, only one is the lowest potential energy.

Method of Formal Charge

Just as the Lewis dot structure is a bookkeeping technique for electrons in a molecular structure, so too is the *formal charge* method. The method assumes, regardless of the electronegativity of a given atom, that the electrons in a bond are shared equally—that the bonds are fully covalent with no delocalization of electron density toward a more electronegative member of a bonding pair. The technique is to assign half the bonding electrons to one atom in the bonding pair, and half the bonding electrons to the other atom in the bonding pair. After assigning the bonding electrons, the lone pair electrons on each atom are assigned fully to that atom. We therefore calculate the *formal* charge of *each atom* in a molecular structure with the formula

$$\begin{array}{c}\text{Formal}\\\text{charge}\end{array} = \begin{array}{c}\text{Number}\\\text{of valence}\\\text{electrons}\end{array} - \left(\begin{array}{c}\text{Number}\\\text{of lone pair}\\\text{electrons}\end{array} + \begin{array}{c}\text{½ number}\\\text{of bonding}\\\text{electrons}\end{array}\right)$$

We can do this for each of our candidate structures for nitrosyl chloride, indicating the formal charge within a circle associated with each atom:

TABLE 10.4

	Structure A$_1$			Structure A$_2$		
Number of valence e⁻	6	7	5	6	7	5
– number of lone pair e⁻	4	2	6	6	2	4
– ½ (number of bonding e⁻)	2	3	1	1	3	2
Formal charge	0	+2	–2	–1	+2	–1

	Structure B$_1$			Structure B$_2$		
Number of valence e⁻	6	5	7	6	5	7
– number of lone pair e⁻	4	2	6	6	2	4
– ½ (number of bonding e⁻)	2	3	1	1	3	2
Formal charge	0	0	0	–1	0	+1

The next question is, how do we use the *formal charge* to distinguish between the various competing structures? To determine the molecular structure of lowest potential energy (most stable), we apply the following rules:

1. The sum of the formal charges in a Lewis structure must equal zero for a neutral molecule and must equal the net charge on a cation or anion.
2. A smaller formal charge on an individual atom is favored over larger formal charges. Molecular structures with the smallest formal charge represent the structures of lowest potential energy and are thus the most stable.
3. Negative formal charges should appear on the most electronegative atoms.
4. Structures having formal charges of the same sign on adjoining atoms are not favored.

We are now in a position to use the *formal charge* analysis to select the favored (lowest potential energy, most stable) structure for nitrosyl chloride from the four candidate possibilities (each of which satisfy the Lewis octet rule). First, we note that rule 1 is satisfied for all four structures; each structure has a sum of formal charges equal to zero. The requirement (#2) that a smaller formal charge on each atom is favored over a larger formal charge eliminates both structure A_1 and structure A_2. The third requirement that negative formal charges should appear on the most electronegative atom is also violated by both structure A_1 and structure A_2. Structure B_1 is the structure with the smallest formal charge, which is zero for each atom. Thus the structure

$$\ddot{O} = \ddot{N} - \ddot{Cl} :$$

is the favored structure. It is indeed the structure verified by experiment.

Limitation to the Lewis Theory

The major limitation to the Lewis structure analysis of chemical bonding is that it is of use primarily for Period 1 through 3 elements of the periodic table. However, since a vast range of chemical behavior occurs within these first three periods, a great deal of critically important chemical intuition is captured by Lewis theory. However, there are, within the first three periods of the periodic table, some important exceptions to the octet rule.

Exceptions to the Octet Rule: Free Radical Structures

There exists a class of molecular structures called *free radicals* that have an odd number of valence electrons—specifically an unpaired electron in the valence shell of the structure. One of the most important of those is the *hydroxyl radical*, OH, which has an electron dot structure

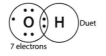

The hydroxyl radical, often written **OH·** to indicate the existence of an unpaired electron, is *physically stable* but chemically highly reactive because it seeks to form an electron pair bond with any of a number of species. The hydroxyl radical is a major oxidizing agent in the attack on DNA and is thus implicated in both the aging of organisms and in the initiation of cancer. Hydroxyl is also the primary oxidizing agent in the atmosphere and is centrally responsible for the catalytic destruction of ozone in the Earth's stratosphere as will be discussed in Chapter 12.

Another important radical is nitric oxide

$$\cdot \ddot{N} = \ddot{O} :$$

which is both a key signaling species in biological systems and a key catalytic species in the *production* of ozone in the lower atmosphere (troposphere) and a key catalytic species in the *destruction* of ozone in the stratosphere.

Exception to the Octet Rule: Expanded Valence Shells

While we have built our analysis of Lewis structures on the concept of the octet rule (duet for H), there are some important examples wherein the octet rule is broken by having 10 or 12 electrons around a central atom, rather than 8. While those structures are not common, they are very important because they involve both phosphoric acid and sulfuric acid and are thus of considerable importance to modern chemistry. Consider first the phosphoric acid molecule, H_3PO_4. We can quickly write down a Lewis structure. We have 4 oxygen atoms, 3 hydrogen atoms and a phosphorus atom. Phosphorus has a lower electronegativity than oxygen so the skeleton is

The total number of electrons is, with 6 for oxygen, 5 for phosphoros, and 1 for hydrogen

$$4 \times 6 + 1 \times 5 + 3 \times 1 = 32 \text{ electrons}$$

Inserting the bonding electrons

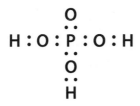

uses 14 of the 32 electrons and the addition of lone pair electrons to the oxygen atoms

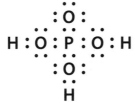

uses up 18 more for a total of 14+18 = 32 electrons. All electrons are accounted for, all octets are satisfied, as are all duets.

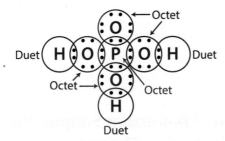

However when we calculate the *formal charge* we find that the central P atom has a formal charge of +1, the oxygen that is not bonded to a hydrogen has a formal charge of −1 and the remaining oxygens have a formal charge of zero. However, if we move one of the lone pairs on the oxygen that is not bound to a hydrogen, we have the structure

When we calculate the formal charge on the phosphorus and oxygen atoms, they are identically zero. Thus the *formal charge* analysis suggests that the second structure is preferable, but it violates the octet rule! We thus have a conflict between the octet rule of Lewis and the formal charge analysis. The dilemma can only be settled unequivocally by observation. Experimental analysis of the structure demonstrates that the bond length of the oxygen–phosphorus bond with the oxygen not bound to the hydrogen is *less* than that of the oxygen bounded to the hydrogens. Thus the structure with 10 electrons around the central phosphorus is the structure with lower potential energy and thus the more stable structure. The method of formal charge trumps the octet rule in this case.

Sulfuric acid, H_2SO_4, is another important case in point of "expanded valence shell" structures. If we construct a Lewis structure using our normal protocol, we would deduce the structure

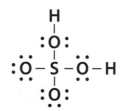

An assignment of formal charge gives a value of +2 to the central sulfur and −1 to the oxygens without a bonded hydrogen. We can eliminate the formal charge by modifying the structure by moving the lone pair electrons from the oxygens that are not bonded to the hydrogen to form double bonds between the sulfur and the oxygens.

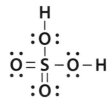

This structure results in a zero formal charge on all oxygens and the central sulfur. This results in a *expanded valence shell* for sulfuric acid—the central sulfur has 12 electrons in its valence shell. Experimental evidence suggests that the double bonded, *expanded valence shell* structure is correct based on an analysis of bond lengths: the single bond O-S bond length is 154 pm and the double bond O-S bond length is 143 pm.

Determination of Molecular Shapes: Valence Shell Electron Pair Repulsion Theory

Chemical behavior, chemical reactivity, and physical behavior are very sensitive to molecular shape, molecular size, the size of the atoms that constitute the molecular structure, and the charge distribution within the molecular structure itself. Does it matter that H_2O is bent rather than linear? Does it matter that methane is tetrahedral rather than planar? As we have seen, and will increasingly see, it matters indeed.

When we consider the chemical behavior of a molecular structure, we become concerned immediately with the distribution of electron density about the nuclei that depends to first order upon the electronegativity of the constituent atoms, on the existence or absence of lone pair electrons, and on the *shape* of the skeletal structure of the molecule.

10.36

A prime example is the water molecule. Water is not only ubiquitous in nature, it is unique with respect to all other liquids. Water is small and it is highly polar. It is a particularly strong solvent capable of interacting with both positive and negative solutes. For an anion or in fact any electron-rich molecular structure, water offers its partially positive hydrogens—the $H^{\delta+}$ end members that are separated by the bond angle in water that make them easily accessible as diagramed here.

**Water: the small but highly polar molecule
that is ubiquitous in nature**

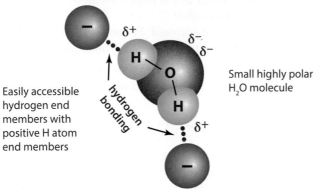

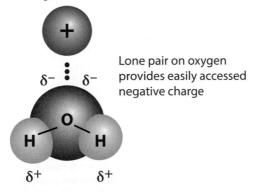

Easily accessible
hydrogen end
members with
positive H atom
end members

Small highly polar
H_2O molecule

For a cation, or an electron-poor atom, the lone pair on the oxygen provides an easily available concentration of exposed negative charge. The highly electronegative oxygen is, by virtue of the bent configuration of water, ideally positioned to engage the positive charge of the cation:

Lone pair on oxygen
provides easily accessed
negative charge

Thus we must develop an ability to determine molecular *shapes*, and then combine knowledge of the fundamental architecture of the molecule with our knowledge of electronegativity *and* size of the constituent atoms in the bonding structure of the molecule to link that structure with the chemical and physical properties of the molecule in question. We seek, therefore, a method, a model, with which to link electronic structure, molecular geometry and chemical properties.

Atoms form molecules because the spatial options provided by the molecular structure offers opportunities to distribute the electron wavefunctions over less confined spatial domains, thereby lowering the (potential) energy of the ensemble of electrons and protons relative to that of the separated atoms. But within that larger context that drives atoms to bond to form molecules, electrons will use every option available to minimize the (potential) energy of the system. They will seek to balance their role in maximizing the shielding of proton-proton repulsion while minimizing the electron–electron repulsion by maximizing the distance between neighboring electrons. We saw this property for electrons to maximize the distance between themselves in the determination of the electronic configuration of atoms wherein, with the addition of electrons to orbitals using the building up (aufbau) principle, electrons entered unfilled subshells (those with different magnetic quantum numbers) rather than pairing up within the same subshell.

The reason was simply that electrons in different subshells were farther apart than those in the same subshells, thus reducing Coulomb repulsion.

The model that provides a method for establishing the shape of molecules is commonly termed the Valence Shell Electron Pair Repulsion model or VSEPR model. This model is based on the simple concept that *electron groups* that surround an atom repel each other through Coulombic forces and that repulsion of electron groups controls the geometry—the shape—of a molecule. We define an "electron group" as a lone pair, a single bond, a double bond, a triple bond, or a single electron in the case of a radical. For molecules having just one central atom, the molecular geometry depends upon

1. the number of electron groups around the central atom
2. how many of those groups are bond pairs and how many are lone pairs

Let's consider what VSEPR theory says about the geometry of the H_2O molecule. First, while the Lewis structure tells us nothing about the shape of the molecule

$$H : \overset{\cdot\cdot}{\underset{\cdot\cdot}{O}} : H$$

it does tell us how many *electron groups* surround the central atom. There are four electron groups surrounding the oxygen atom, two bonding groups and two lone pairs. The geometry that maximizes the *distance* between these four electron groups is the tetrahedral structure shown in Figure 10.26:

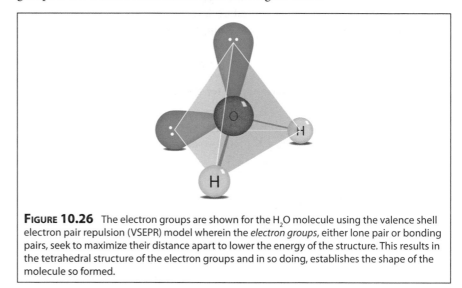

FIGURE 10.26 The electron groups are shown for the H_2O molecule using the valence shell electron pair repulsion (VSEPR) model wherein the *electron groups*, either lone pair or bonding pairs, seek to maximize their distance apart to lower the energy of the structure. This results in the tetrahedral structure of the electron groups and in so doing, establishes the shape of the molecule so formed.

The tetrahedral structure has four triangular faces, and each of the four electron pairs in the H_2O points to one of the four vertices of the tetrahedral. The two bonding pairs occupy the others.

Several features of the geometrical figure of the bonding structure of water are important for setting up the VSEPR theory in the practice:

1. It is the number of *electron groups* that determines the first order geometry of the structure. This is termed the *electron-group geometry*.
2. The number of *electron groups* is determined from the Lewis structure of the molecule.
3. Once the basic architecture of the molecule is set by the *electron group geometry*, the *molecular* geometry is set by the relationship between the number of electron groups and the number of *lone pairs* in the structure. Thus for water, there are two lone pairs and four electron groups so the *molecular geometry* is bent with an angle between the central atom and the vertices of the tetrahedral.

4. The lone pair electrons occupy a greater volume of space than do the bonding electrons and thus the bond angles in the actual molecule deviate from that of a perfect tetrahedral because the lone pair-lone pair repulsion exceeds the lone pair-bond pair repulsion, which in turn exceeds the bond pair-bond pair repulsion. Thus the lone pair-lone pair angle exceeds 109.5°. In fact, the bond angle in H_2O is measured to be 104.5 degrees.

The backbone of VSEPR theory is thus the linkage between (1) the Lewis structure that establishes the *number of electron groups,* and (2) the *electron group geometry* that is a direct consequence of the *number of electron groups.* Before we move to lay out the full strategy of the VSEPR model, let's consider two more examples, carbon dioxide CO_2 and the formaldehyde molecule CH_2O.

On the face of it, CO_2 appears to be similar to H_2O. Carbon dioxide is a triatomic molecule that is comprised of nonmetals. However, we now know that the first two steps in using the VSEPR model to determine the *shape* of a molecule is to (1) generate the Lewis structure and (2) determine the *electron group geometry* by counting the *number of electron groups.* Thus for CO_2, the skeletal structure places the least electronegative species in the central position

$$\textbf{O} \quad \textbf{C} \quad \textbf{O}$$

and there are $2 \times 6 + 1 \times 4 = 16$ total electrons. Inserting the bonding electrons first

$$\textbf{O} : \textbf{C} : \textbf{O}$$

and terminal electrons second initiates the Lewis structure determination:

$$: \overset{..}{\underset{..}{\textbf{O}}} : \textbf{C} : \overset{..}{\underset{..}{\textbf{O}}} :$$

All electrons are used up but the octet rule for the central carbon is violated

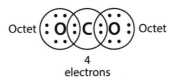

Now we move an electron pair from each of the terminal oxygens to the central carbon

and we have the correct Lewis structure

$$\overset{..}{\underset{..}{\textbf{O}}} = \textbf{C} = \overset{..}{\underset{..}{\textbf{O}}}$$

To determine the *electron group geometry* we must count the *number of electron groups* on the central atom. Remember that an *electron group* is a lone pair, one bonding pair, a double bonding pair, a triple bonding pair or a single electron. For the central carbon in CO_2, there are two electron groups, *each* electron group is a double bond in this case. The maximum angle between two electron groups (that *minimizes* the electron-electron repulsion and therefore *lowers* the potential energy of the structure) is a *linear* geometry with a bond angle of 180°.

Thus a molecule with a central atom with two electron groups has an *electron group geometry* that is *linear.*

If we consider the molecular shape of formaldehyde, CH_2O, we first generate the correct Lewis structure, placing the hydrogens in the terminal position and the less electronegative atom in the center

There are $1 \times 6 + 1 \times 4 + 2 \times 1 = 12$ electrons. Placing the bonding electrons first

then adding the terminal electrons

which leaves the central carbon with just six electrons

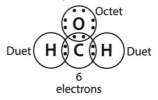

We can satisfy the octet rule by moving a lone pair of electrons into the bonding position between the oxygen and carbon creating correct Lewis structure.

To determine the *electron group geometry*, we count the number of electron groups around the central carbon. There is a double bond (a double bond counts as one electron group) and two single bonds. Thus a total of *three* electron groups. The electron group geometry that maximizes the distance between the electron groups is the trigonal planar geometry, which has a bond angle of 120° between the electron groups.

Spectroscopic analysis of the formaldehyde molecule shows that the bond angles deviate somewhat from 120°. The HCO bond angles are 121.9° and the HCH bond angle is 116.2°. The reason is that the double bond contains more electron density than the single bond and thus the repulsion is greater between a double bond and a single bond than between two single bonds. This expands the HCO bond angle and "squeezes" the HCH bond angle.

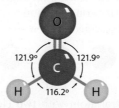

So these examples establish the relationship between (1) the *number* of elec-

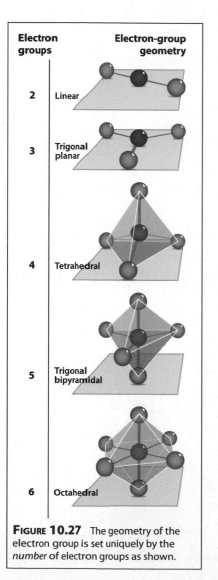

Electron groups	Electron-group geometry	
2	Linear	
3	Trigonal planar	
4	Tetrahedral	
5	Trigonal bipyramidal	
6	Octahedral	

FIGURE 10.27 The geometry of the electron group is set uniquely by the *number* of electron groups as shown.

tron groups and (2) the *electron group geometry* that maximizes the distance between electron groups and thereby minimizes the potential energy of the ensemble of electrons and protons in the structure. We have thus established the *electron group geometry* for 2, 3, and 4 electron group molecules. We have also put in place an approach that can determine the architecture of the molecule independent of the specific type of electron group—lone pair, single bond, multiple bond, or single electron. What about molecular structure with 5 or 6 electron groups around the central atom? The *electron group geometry* for those cases is determined simply by answering the question: What geometry maximizes the distance between 5 electron groups around the central atom? For the case of 6 electron groups we must answer the same question: the answer is pure geometry. For 5 electron groups the *trigonal bipyramid* structure maximizes the distance between electron groups; for 6 electron groups the octahedral structure maximizes the distance between electron groups. Thus we have a one-to-one relationship between the *number of electron groups* and the *electron group geometry*! The relationship is shown in Figure 10.27.

Once we have established the *electron group geometry*, the next question is: What is the *molecular* geometry? What is the shape of the molecule itself? What are the bond angles between the actual atoms that comprise the molecular structure? The answer to these questions involves taking each of the *electron group geometries* and then placing the atoms, the lone pairs or the single electron (in the case of a radical) in the *electron group* position.

Consider first the case of 3 electron groups. We already know that the *electron group geometry* is trigonal planar. What about the molecule geometry? If we have zero lone pairs, we have three atoms bonded into the central atom. This is the case for formaldehyde. The *molecular geometry* is *trigonal planar*. If we have one lone pair, the *molecular geometry* is *bent*. These cases are shown in Figure 10.28.

If we have 4 electron groups, we have three possibilities: zero lone pairs, 1 lone pair, or 2 lone pairs. This case is displayed in Figure 10.29.

A number of patterns can now be recognized. First, we need simply determine the

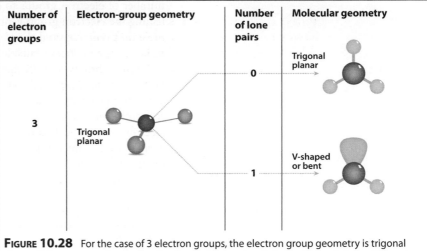

FIGURE 10.28 For the case of 3 electron groups, the electron group geometry is trigonal planar. If there are zero lone pairs, the shape of the molecule is also trigonal planar. If there is a lone pair electron group, the molecule is V-shaped or bent.

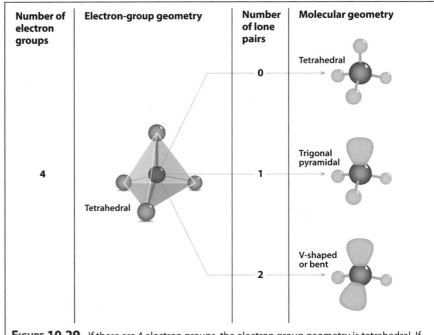
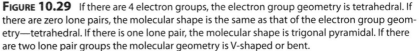

FIGURE 10.29 If there are 4 electron groups, the electron group geometry is tetrahedral. If there are zero lone pairs, the molecular shape is the same as that of the electron group geometry—tetrahedral. If there is one lone pair, the molecular shape is trigonal pyramidal. If there are two lone pair groups the molecular geometry is V-shaped or bent.

number of electron groups to establish the *electron group geometry*. After establishing the electron group geometry, if the number of lone pairs is zero, then the *molecular geometry* and the *electron group geometry* are identical. The number of molecular geometries available is equal to the number of electron groups minus 1.

Proceeding to VSEPR theory for 5 electron group systems, we link the number of *electron groups* to the *electron group geometry* just as we did for the 2, 3, and 4 electron group systems. But for the 5 electron group systems, we must explicitly recognize that the *lone pair* electron group *spreads out* more than the bond pair electron group and this fact aids in the selection of the option for the *molecular geometries* that are available for a given *electron group geometry*. Figure 10.30 demonstrates the role played by lone pair repulsion in limiting the available molecular geometry.

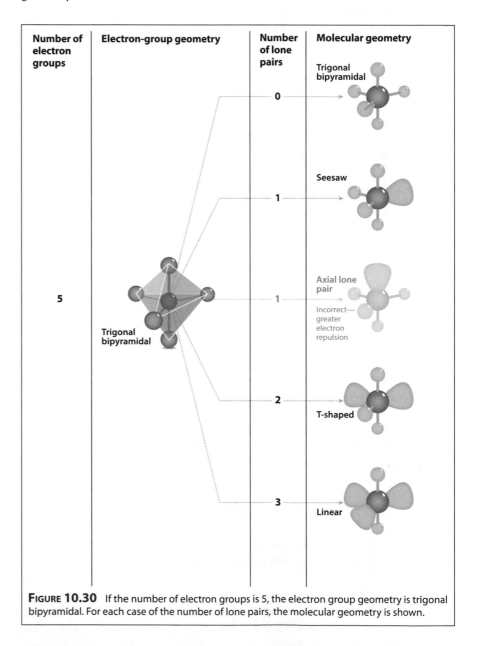

FIGURE 10.30   If the number of electron groups is 5, the electron group geometry is trigonal bipyramidal. For each case of the number of lone pairs, the molecular geometry is shown.

The key point in choosing the correct structure for the 1 lone pair case is to recognize that the electron group repulsion is greater (and thus the potential energy is higher) for the axial lone pair placement because (1) the angle between the lone pair and the bond pair electron groups is 90° whereas (2) for the equatorial lone pair the angle is 120° between the lone pair and the equatorial groups *and* (3) that there are two not three bond pair electron groups at 90° to the lone pair. For the case of both the two lone pair and the three lone pair structures, it is the larger 120° angle between electron groups that maximizes the distance between electron groups and thus minimizes the potential energy of the structure with the lone pairs in the equatorial position.

For the six electron group case, the electron group geometry is set—it is the octahedral and the important cases are for 0, 1, and 2 lone pairs. This results in *molecular geometries* that are octahedral, square pyramidal and square *planar* respectively as shown in Figure 10.31.

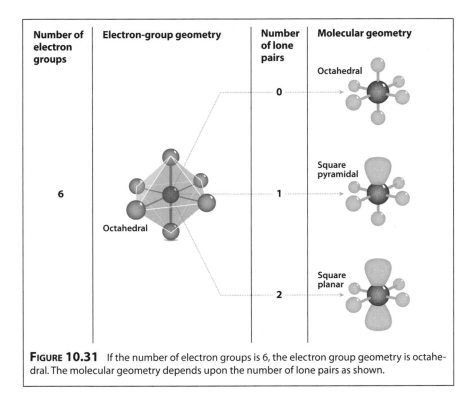

Number of electron groups	Electron-group geometry	Number of lone pairs	Molecular geometry
6	Octahedral	0	Octahedral
		1	Square pyramidal
		2	Square planar

FIGURE 10.31 If the number of electron groups is 6, the electron group geometry is octahedral. The molecular geometry depends upon the number of lone pairs as shown.

Shapes of Molecules: Bond Lengths and Bond Energies

We are now in a position to determine the bond angles in broad classes of molecules, but in the determination of *molecular shape*, we must also consider the *length* of the chemical bonds. The length of a given chemical bond is dependent on many factors, some easily recognized from the structure such as whether the bond is single, double, or triple. Other factors that control bond length involve the size of the atoms, the electronegativity of the species involved, etc. In general there is no simple way of predicting bond length—they must be measured or calculated using advanced computational methods. Of course bond lengths depend upon the details of the molecular structure within which the bond resides, but there are average bond lengths that emerge when a given bond is analyzed across a broad range of molecules. A table of some of the most important average bond energies and lengths is presented in Table 10.5. These bond lengths are given as the distance between the nuclei in the bond in units of picometers (pm), which is 10^{-12} meters.

Interestingly, bond lengths range from 74 pm for H_2 to 266 pm for I_2. Examination of Table 10.5 reveals several important features.

1. As the number of bonds increases the bond length decreases. If C–C bond has a bond length of 154 pm, the C=C bond a length of 134 pm, and the C≡C bond a length of 120 pm. The C–O bond has a length of 143 pm and the C=O bond 120 pm.

2. The C–C bond, N–N bond, and O–O bond are all approximately 150 pm, and the H–C, H–N, and H–O bonds decrease in bond length with increasing electronegativity of the species bonded to hydrogen.

One of the key reasons for highlighting bond lengths, in addition to the important role bond lengths play in the *shape* of molecules, is the relation between bond *length* and bond *energy*. As we have worked out, the bond energy is the energy required to remove two atoms (or molecular entities) bound in a potential energy well, and to separate them to a distance such that they are no longer interacting.

For example, if we break the bond in N_2, one of the strongest bonds in nature, we must invest 946 kJ of energy to dissociate a mole of N_2:

$$N \equiv N(g) \rightarrow N(g) + N(g) \qquad \Delta H = 946 \text{ kJ/mole}$$

To remove a hydrogen atom from methane we must invest 438 kJ/mole of energy

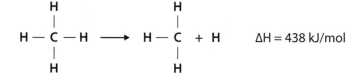

However, as we noted, the enthalpy required to break bonds does depend on the molecular structure. For example, if we wish to remove a hydrogen atom from a selection of halocarbon species, we find that for

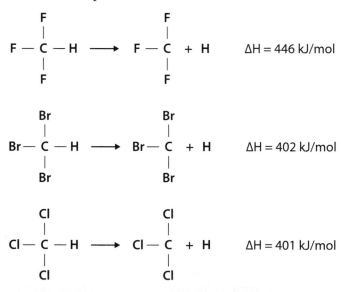

Thus, when we list bond energies, those bond energies are *average* bond energies, bond energies taken over a large number of molecules containing that particular bond. Nevertheless, while we give up some accuracy when we use average bond energies, we gain considerable utility in deciphering patterns in chemical behavior by condensing a vast number of molecules into a small number of characteristic bond strengths and bond lengths. Note that the bond energies depend on the type of bond (single, double, triple), the kind of atom invoked in the bond, and the size of the atom.

When we compare the bond *energy* with the bond *length*, it becomes clear that when we progress from single to double to triple bonds, the bond length *decreases* and the bond energy *increases*. Also in general, as the bond length increases, the bond strength decreases.

One practical application of the bond energies (enthalpies) is that we can quickly estimate, with considerable accuracy, the enthalpy change for a reaction, ΔH_R, using the table of bond strengths using our familiar equation

$$\Delta H_R = \Sigma \Delta H \text{ bonds broken} + \Sigma \Delta H \text{ bonds formed.}$$

TABLE 10.5

Average Bond Energies

Bond	Bond Energy (kJ/mol)	Bond	Bond Energy (kJ/mol)	Bond	Bond Energy (kJ/mol)
H—H	436	N—N	163	Br—F	237
H—C	414	N=N	418	Br—Cl	218
H—N	389	N≡N	946	Br—Br	193
H—O	464	N—O	222	I—Cl	208
H—S	368	N=O	590	I—Br	175
H—F	565	N—F	272	I—I	151
H—Cl	431	N—Cl	200	Si—H	323
H—Br	364	N—Br	243	Si—Si	226
H—I	297	N—I	159	Si—C	301
C—C	347	O—O	142	Si=O	368
C=C	611	O=O	498	Si=Cl	464
C≡C	837	O—F	190	S—O	265
C—N	305	O—Cl	203	S=O	523
C=N	615	O—I	234	S=S	418
C≡N	891	F—F	159	S—F	327
C—O	360	Cl—F	253	S—Cl	253
C=O	736*	Cl—Cl	243	S—Br	218
C≡O	1072			S—S	266
C—Cl	339				

*799 in CO_2

Summary Concepts

1. Electronegativity

Electronegativity describes, and attempts to quantitatively define, the ability of one atom of an element in the periodic table to compete for electron density in a chemical bond with another atom of another element in the periodic table. The remarkable aspect of the electronegativity scale is that it can be constructed so as to be generally applicable independent of the pairing of the two atoms as they compete with each other for electron density. The electronegativity scale devised by Linus Pauling is displayed at right. The electronegativity of an atom, A, EN_A, is proportional to the difference between the ionization energy, IE_A, of element A and the electron affinity, EA_A, of that element such that $EN_A \propto IE_A - EA_A$.

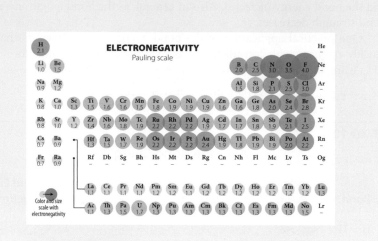

2. Types of Chemical Bonds

There are three major categories of chemical bonding, the fundamentals of which result from the large range of electronegativities across the periodic table: (1) ionic bonding that results from the formation of a chemical bond between a metal with very low electronegativity and a nonmetal with high electronegativity; (2) covalent bonding between two nonmetals with comparable electronegativity; and (3) metallic bonding between atoms, each with low electronegativities, where electrons are delocalized through the lattice of metal cations.

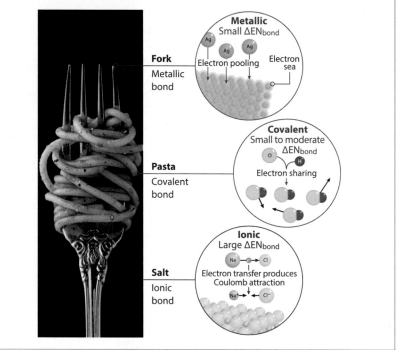

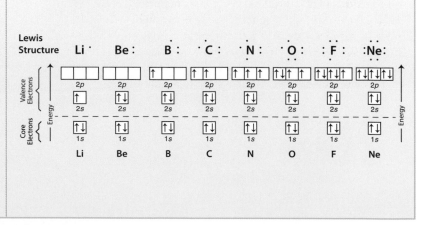

3. Representation of Valence Electrons with the Corresponding Lewis Electron Dot Diagram

A key concept in chemical bonding is that core electrons in an atom are not directly involved in the formation of chemical bonds, but rather the outer or valence electrons are the important participants in bonding. It is, therefore of great importance to be able to identify the valence electrons in an electronic configuration. Core electrons are those for which a shell of electrons is complete. Valence electrons are those for which a subshell within a shell remains unfilled. An example is shown at right with the corresponding Lewis electron dot diagram.

Summary Concepts

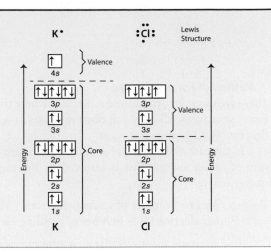

4. Lewis Structure for Ionic Bonds

While Lewis electron dot structures are used primarily for determining the molecular structure for the vast array of covalent bonds, Lewis structures are used widely for ionic bond structures as well.

On the right we show the core and valence electrons for potassium and chlorine and the resulting Lewis structure for KCl.

5. Lewis Structures for Covalent Bonds

To determine the Lewis structure for bonds involving two elements we first count the *total* number of valence electrons. For example for O and H we count valence electrons as shown at right. Next we write the Lewis dot structure for OH as shown at right. While H has the requisite two electrons to form a duet, oxygen is left with seven electrons in its outer shell. Lewis theory deems this structure unacceptable because it violates the octet rule—the outer electron shell of oxygen has seven not eight electrons. If we were to add another hydrogen atom, the octet rule for oxygen would be satisfied. Lewis theory predicts that the lowest potential energy (most stable) structure involving hydrogen atoms bonded to oxygen is H_2O, which indeed it is.

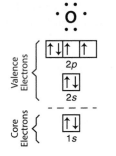

6. Lewis Structures and Resonance Structures

When the Lewis structure is worked out for ozone, the octet rule results in two different structures for O_3, one with a double bond on one side of the central oxygen and a single bond in the other. As shown at right, this would mean that two different structures exist for ozone. Experimental evidence, however, demonstrates that the two bonds in O_3 are in all ways equivalent. This dilemma is reconciled by forming a hybrid structure wherein an electron pair is delocalized across the O_3 structure as shown at right.

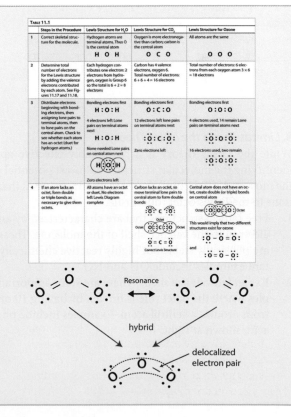

10.47

Summary Concepts

7. Method of Formal Charge

There are cases for polyatomic molecules where there are multiple Lewis structures, all of which obey the octet rule. An example is that for NOCl as shown at right.

In the case, the question becomes how to determine the structure of *lowest energy*. This is resolved by the method of formal charge where

Formal charge = Number of valence electrons – (Number of lone pair electrons + ½ number of bonding electrons)

To determine the molecular structure of lowest potential energy, we apply the following rules after determining the formal charge for each atom:

1. The sum of the formal charges in a Lewis structure must equal zero for a neutral molecule and must equal the net charge on a cation or anion.
2. A smaller formal charge on an individual atom is favored over larger formal charges. Molecular structures with the smallest formal charge represent the structures of lowest potential energy and are thus the most stable.
3. Negative formal charges should appear on the most electronegative atoms.
4. Structures having formal charges of the same sign on adjoining atoms are not favored.

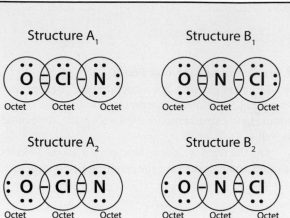

Structure A₁   Structure B₁

Structure A₂   Structure B₂

TABLE 11.4

	Structure A₁ $\ddot{O} = \ddot{Cl} - \ddot{N}:$			Structure A₂ $:\ddot{O} - \ddot{Cl} = \ddot{N}$		
	⓪	⊕2	⊖2	⊖1	⊕2	⊖1
Number of valence e⁻	6	7	5	6	7	5
– number of lone pair e⁻	4	2	6	6	2	4
– ½ (number of bonding e⁻)	2	3	1	1	3	2
Formal charge	0	+2	–2	–1	+2	–1

	Structure B₁ $\ddot{O} = \ddot{N} - \ddot{Cl}:$			Structure B₂ $:\ddot{O} - \ddot{N} = \ddot{Cl}$		
	⓪	⓪	⓪	⊖1	⓪	⊕1
Number of valence e⁻	6	5	7	6	5	7
– number of lone pair e⁻	4	2	6	6	2	4
– ½ (number of bonding e⁻)	2	3	1	1	3	2
Formal charge	0	0	0	–1	0	+1

8. Limitations to the Lewis Theory

While the Lewis theory is a remarkably effective approach for identifying the molecular structure of lowest potential energy (i.e. the most stable structure) there are important cases that are exceptions to the Lewis octet (duet) rule.

1. Free radical structures that are characterized by an unpaired electron in the valence shell of the molecule. These radicals are physically stable but highly reactive chemically. Important examples include OH and NO.
2. Expanded valence shells. There are some important examples where the octet rule is broken by having 10 or 12 electrons around a central atom. Examples include phosphoric acid, shown at right.

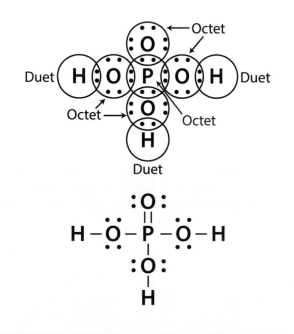

Summary Concepts

9. Valence Shell Electron Pair Repulsion: VSEPR Theory

Chemical behavior, chemical reactivity and the physical behavior of molecules are very sensitive to molecular shape, molecular size and the charge distribution within the molecular structure. The VSEPR model is based on the simple concept that *electron* groups that surround an atom repel each other through Coulombic forces and that repulsion of electron groups controls the geometry of the molecule. We define an "electron group" as a lone pair, a single bond, a double bond, a triple bond, or a single electron in the case of a radical.

For example, if the number of electron groups is equal to 4, the electron group geometry is tetrahedral and the *molecular geometry* is determined by the number of lone pairs as shown.

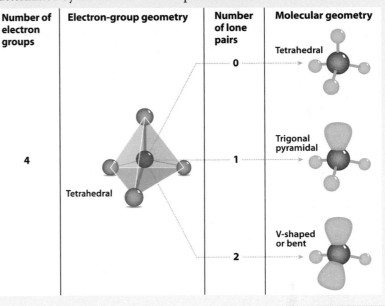

Number of electron groups	Electron-group geometry	Number of lone pairs	Molecular geometry
4	Tetrahedral	0	Tetrahedral
		1	Trigonal pyramidal
		2	V-shaped or bent

Quantum mechanics in combination with today's high power computers can calculate, to a reasonable degree of accuracy, the bond lengths, bond angles and, most importantly, the distribution of electron charge in a molecular structure. So why do we spend time developing the Lewis dot formulation of chemical bonding? Of what use is the concept of "formal charge"? Why VSEPR theory? VSEPR theory is, after all, only an approximation. The answer is that electrons, small as they are, determine chemical behavior. An intuition about *how* chemical reactions take place is built upon a grasp of what instigates electrons to shift in any chemical transformation leading from reactants to products. Thus, in the process of bridging from the revolutionary developments of quantum mechanics for multielectron atomic systems to the quantum mechanics of molecular bonding, we pause to examine the nature of the chemical bond from the perspective of the electrons and the implications the assignment of electrons between bond and lone pairs has on the deduced shape of the resulting molecule. As we will see in Chapter 11, the *shape* of a molecule is important for strategic decisions regarding how quantum mechanics is best implemented to optimize the accuracy in the calculated properties of complex molecules.

So while we will revisit many of the chemical reaction types discussed here in subsequent chapters, we begin by viewing the world of chemical reactions through the eyes of the Lewis structure for chemical bonds.

A Chemical Reaction from the Perspective of the Lewis Structure

We consider first the reaction between sodium and chlorine to form NaCl

$$Na(g) + Cl_2(g) \rightarrow NaCl(g) + Cl(g)$$

In terms of the Lewis structures, we write the reaction as

$$\text{Na}^{\cdot} + \ddot{\underset{\cdot\cdot}{\text{Cl}}}-\ddot{\underset{\cdot\cdot}{\text{Cl}}}: \longrightarrow \text{Na}^{+}\left[:\ddot{\underset{\cdot\cdot}{\text{Cl}}}:\right]^{-} + :\ddot{\underset{\cdot\cdot}{\text{Cl}}}:$$

If we compare this reaction between gas phase sodium atoms and Cl_2 with that of the reaction between two atoms, one of sodium the other of atomic chlorine,

$$\text{Na}^{\cdot} + \cdot\ddot{\underset{\cdot\cdot}{\text{Cl}}}: \longrightarrow \text{Na}^{+} + \left[:\ddot{\underset{\cdot\cdot}{\text{Cl}}}:\right]^{-}$$

it is more difficult to see how the products

$$\text{Na}^{+}\left[:\ddot{\underset{\cdot\cdot}{\text{Cl}}}:\right]^{-} \quad \text{and} \quad :\ddot{\underset{\cdot\cdot}{\text{Cl}}}:$$

would be formed in the molecular reaction because the chlo-

rine atoms in

$$:\ddot{\underset{\cdot\cdot}{\text{Cl}}}:\ddot{\underset{\cdot\cdot}{\text{Cl}}}:$$

are bonded such that each chlorine has an octet of electrons. It fully satisfies the Lewis octet rule.

However, Lewis structures are all about how we can determine the resulting molecular structure with the *lowest potential energy*—we always seek ways to figure out the structure with the lowest potential energy without having to use a large computer to solve the multielectron Schrödinger equation.

So how does this work using the Lewis structure approach? A key insight is to recognize that the drive of Cl atoms is to minimize their energy by achieving a noble gas electron configuration—that is what dictates the octet rule in Lewis structures. But a chlorine anion,

$$\left[:\ddot{\underset{\cdot\cdot}{\text{Cl}}}:\right]^{-}$$

is a more complete octet than each of the Cl atoms in a

$$:\ddot{\underset{\cdot\cdot}{\text{Cl}}}:\ddot{\underset{\cdot\cdot}{\text{Cl}}}:$$

molecule wherein they must share an electron in order to complete the octet. So in order to lower the potential energy of the *ensemble* of atoms, one Cl atom relinquishes one of the shared electrons to the other Cl atom *in exchange* for the complete electron transfer shown in Figure CS10.1a.

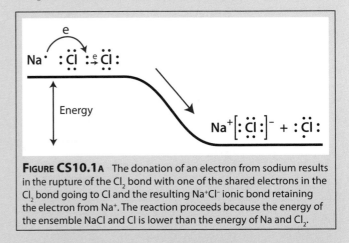

FIGURE CS10.1A The donation of an electron from sodium results in the rupture of the Cl_2 bond with one of the shared electrons in the Cl_2 bond going to Cl and the resulting $Na^{+}Cl^{-}$ ionic bond retaining the electron from Na^{+}. The reaction proceeds because the energy of the ensemble NaCl and Cl is lower than the energy of Na and Cl_2.

This is an example of the general rule that the acceptor molecule in an electron transfer reaction suffers a bond rupture if that molecule began as a species satisfying the Lewis (noble gas) octet rule.

Before we work through a number of important categories of reactions that Lewis structures help elucidate, we highlight a number of *general* rules that emerge from the Lewis formalism:

1. Most Group 1 and Group 2 metals give away electrons in

redox reactions with water. The reason is that the polar structure of water draws the weakly bound electron(s) from those metals, releasing significant energy in the formation of products.

2. Metals and nonmetals form ionic bonds with each other because of the much larger electron affinity of the nonmetals.

3. Nonmetals *share* electrons with each other because of the roughly equal electron affinity among nonmetals.

4. Metals react with water to yield basic solutions and H_2 in the gas phase.

5. Nonmetals react with water to give acidic solutions and finally, from our example above.

6. If a metal reacts with an acceptor *molecule* that itself satisfies the Lewis octet rule, that molecule will suffer a bond rupture.

Let's see how the Lewis structures provide insight across an array of chemical reaction categories.

Lewis Picture for Acid-Base Reactions

The Lewis formulation extends this picture of acid-base chemistry by generalizing acid-base behavior as a class of electron transfer reactions. In particular the Lewis acid-base formulation stipulates that the important factor in an acid-base reaction is the attainment of a *new shared pair of electrons* in a *new* polar covalent bond as displayed in Figure CS10.1b.

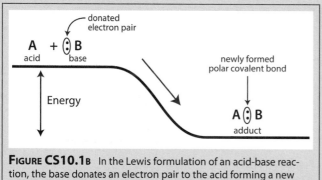

FIGURE CS10.1B In the Lewis formulation of an acid-base reaction, the base donates an electron pair to the acid forming a new shared electron bond in the adduct.

Thus for Lewis,

- A *base* is *any* species that *donates* an electron pair.
- An *acid* is *any* species that *accepts* an electron pair and the creation results in the formation of a new species, the *adduct*.

The species **A** and **: B** can be neutral or charged.

For example, consider the neutralization reaction between a proton and the hydroxide ion in the formation of the adduct H_2O as shown in Figure CS10.1c.

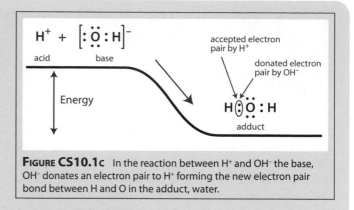

FIGURE CS10.1C In the reaction between H^+ and OH^- the base, OH^- donates an electron pair to H^+ forming the new electron pair bond between H and O in the adduct, water.

In this case H^+ is just a particular example of an electron pair acceptor.

Lewis expressed strongly (as he often did) his objection to hanging the definition of acid-base reactions on the presence of the proton as the defining feature of an acid: "*To restrict the group of acids to those substances which contain hydrogen interferes as seriously with the systematic understanding of chemistry as would the restriction of the term oxidizing agent to those substances containing oxygen.*"

A key attribute of the Lewis picture of acid-base reactions is that it vastly expands the class of acids. One such example is the reaction of quicklime (CaO) with SO_2 (shown in Figure CS10.1d) wherein the latter is emitted in the combustion of coal:

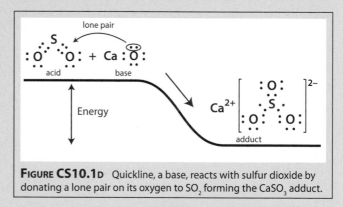

FIGURE CS10.1D Quickline, a base, reacts with sulfur dioxide by donating a lone pair on its oxygen to SO_2 forming the $CaSO_3$ adduct.

In this case no hydrogen ion is involved, but the base is CaO, which donates an electron pair to the acid, SO_2, forming $CaSO_3$ as the adduct.

When quicklime is added to water, it reacts to form calcium hydroxide, a base as displayed in Figure CS10.1e.

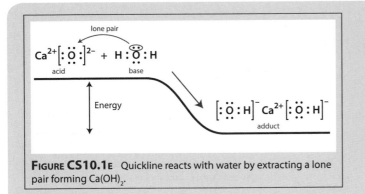

FIGURE CS10.1E Quickline reacts with water by extracting a lone pair forming $Ca(OH)_2$.

Another example is the case of SO_2, when added to water, that produces weak acid H_2SO_3 shown in Figure CS10.1f:

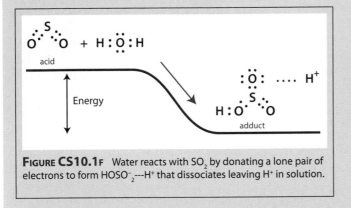

FIGURE CS10.1F Water reacts with SO_2 by donating a lone pair of electrons to form $HOSO_2^- \text{---}H^+$ that dissociates leaving H^+ in solution.

Reactions of Metal and Nonmetals with H_2O

Just as we stressed in Case Study 2.2, the chemistry of water and of oxygen lies at the heart of a large fraction of chemistry—both in natural systems and in laboratory systems. We note again in Figure CS10.1g the structure of water displaying oxygen and hydrogen with bent geometry, featuring a strongly electronegative end and a strongly electropositive end that has, as we will see, major consequences:

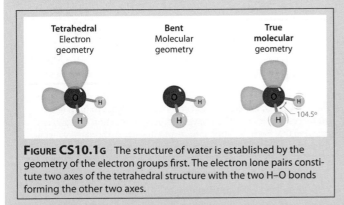

FIGURE CS10.1G The structure of water is established by the geometry of the electron groups first. The electron lone pairs constitute two axes of the tetrahedral structure with the two H–O bonds forming the other two axes.

The separation of charge results from the electronegativity *difference* between O ($\chi = 3.5$) and H ($\chi = 2.2$). It is this

dual capability to attract cations to the electronegative end of H_2O and attract anions to the electropositive end of H_2O that underpins the versatility of water as a solvent. When salts (which invariably contain a metal donor and nonmetal acceptor) are placed in water, cations and anions result. A key part of the chemical vocabulary is the naming of these cations and anions. We can make a short list of the anions that repeatedly appear, and are summarized in Table CS10.1a. Given the importance of vocabulary to any discussion or any thought process, it may be wise to simply conquer this list once and for all with a set of flashcards!

TABLE CS10.1A Commonly encountered anions.

Formula	Name	Protonated Form(s)	Names(s)
H^-	Hydride	H_2	Hydrogen(g)
F^-	Fluoride	HF	Hydrogen fluoride (g); hydrofluoric acid (aq)
Cl^-	Chloride	HCl	Hydrogen chloride (g); hydrochloric acid (aq)
O^{2-}	Oxide	OH^-	Hydroxide
S^{2-}	Sulfide	HS^-	Hydrogen sulfide (bisulfide)
N^{3-}	Nitride	NH_2^-	Amide
OH^-	Hydroxide	H_2O	Water
CO_3^{2-}	Carbonate	HCO_3^-	Hydrogen carbonate (bicarbonate)
		H_2CO_3	Carbonic acid [also CO_2(aq)]
$C_2H_3O_2^-$	Acetate	HC_2HO_2	Acetic acid
SiO_3^{2-}	Silicate	H_4SiO_4	Silicic acid ($H_2SiO_3 \cdot H_2O$)
NO_3^-	Nitrate	HNO_3	Nitric Acid
PO_4^{3-}	Phosphate	HPO_4^{2-}	Monohydrogen phosphate
O_2^{2-}	Peroxide	H_2O_2	Hydrogen Peroxide
SO_4^{2-}	Sulfate	HSO_4	Hydrogen sulfate (bisulfate)
SO_3^{2-}	Sulfite	HSO_3^-	Hydrogen sulfite (bisulfite)
		H_2SO_3	Sulfurous acid [also SO_2(aq)]
ClO_4^-	Perchlorate	$HClO_4$	Perchloric acid

A key part of understanding chemical *reactions*, as distinct from chemical *structure*, is to develop an intuition concerning how electrons *move* in a chemical reaction. Thus, in what follows, we will use a sequence of arrows to indicate the movement of electrons between the Lewis structures of reactants that result in the formation of the new covalent bonding structure of the adduct.

In the reaction of sodium metal with water,

$$2Na(s) + 2H_2O(l) \rightarrow 2NaOH(aq) + H_2(g)$$

we break the reaction down to a sequence of electron donater-acceptor steps displayed in Figure 10.1h.

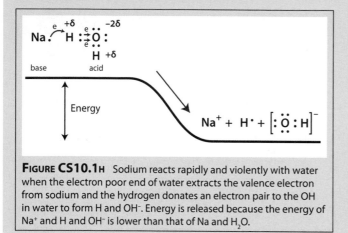

FIGURE CS10.1H Sodium reacts rapidly and violently with water when the electron poor end of water extracts the valence electron from sodium and the hydrogen donates an electron pair to the OH in water to form H and OH⁻. Energy is released because the energy of Na⁺ and H and OH⁻ is lower than that of Na and H₂O.

The H atom produced, **H·**, reacts with another **H·** to produce gas phase H₂, which burns in air to create an explosion. Thus reaction is a flagship example of general rules 1, 4, and 6 above, because while H₂O satisfies the Lewis octet rule, it undergoes bond breakage in accepting the electron from sodium.

The prototype reaction above between sodium, an alkali metal, and water is exactly what happens between any of the alkali atoms (Na, Li, K, Rb, or Cs) and water. This chemical similarity results from (1) their common ns electron configuration and (2) their low electron affinity that makes them excellent electron donors.

Group 2A metals, the alkaline earths, are characterized by their ns² valence configurations and their somewhat higher electron affinities.

Thus the Period 1 member, calcium, is the electron donor in the reaction

$$Ca(s) + 2H_2O(l) \rightarrow Ca(OH)_2(aq) + H_2$$

which, when we track the movement of electrons in a Lewis diagram, yields a two step process initiated by the reaction shown in Figure CS10.1i where the electron moves from Ca to H₂O with the electron pair going to form OH⁻.

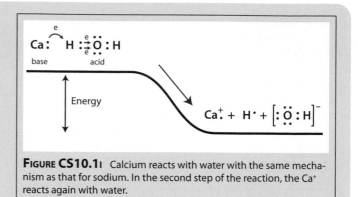

FIGURE CS10.1I Calcium reacts with water with the same mechanism as that for sodium. In the second step of the reaction, the Ca⁺ reacts again with water.

The reaction of alkaline earths in Group 2A with water, in contrast with the alkali atoms, is not true for every element in the Group. Only the heavier atoms, Ca through Ra, react spontaneously with water. Be does not react at all with water, and Mg only slowly. A key understanding as to why this is true resides in the ionization energy of the metal and tracks back to our discussion of shielding and penetration in multielectron systems. Recalling that the ionization energy is a measure of the *ease* of electron removal, both Be and Mg have a higher IE than Ca through Ra.

What about the reaction of water with nonmetals? Let's examine a reaction that is of central importance to the energy-climate link, the reaction of water with carbon dioxide to form carbonic acid:

$$H_2O + CO_2 \rightleftharpoons H_2CO_3$$

When we diagram the electron shift(s) associated with this reaction in a Lewis formulation, we require two steps. The first step is an electron donation from the lone pair of water (the base) to the carbon of CO₂ as shown in Figure CS10.1j.

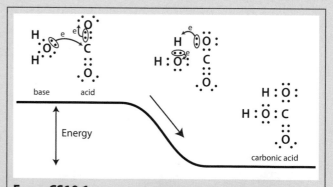

FIGURE CS10.1J Water and carbon dioxide react to form the adduct carbonic acid. This is a very important reaction in seawater, drawing CO₂ from the atmosphere and making carbonate available for inclusion into biological structures in the ocean.

Following the donation from the lone pair of water, the next steps are by the transfer of an electron pair to oxygen in CO₂ and the donation of an electron pair from the oxygen in CO₂ to form a new hydrogen-oxygen bond as diagramed in Figure CS10.1j.

It is this remarkable inorganic reaction between a non-metal and water that makes carbon available to the organic/biological process in the world's oceans that creates the skeletal structures for all living things in the ocean. It is also the removal process that extracts massive amounts of CO_2 from the atmosphere as we add CO_2 to the atmosphere by the combustion of fossil fuels.

So, keeping track of reactivity to this point we note that alkali metals and the heavier alkaline earths react with water as summarized in Figure CS10.2k.

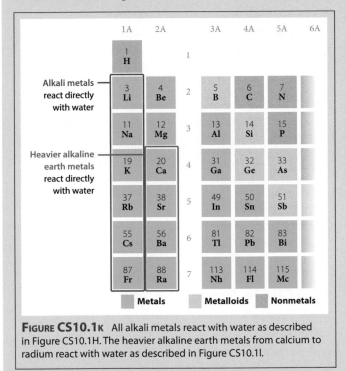

FIGURE CS10.1K All alkali metals react with water as described in Figure CS10.1H. The heavier alkaline earth metals from calcium to radium react with water as described in Figure CS10.1l.

Heavier alkaline earth metals react directly with water to form M^+ cations, H_2 in the gas phase and OH^- in the aqueous phase. Alkali metals react directly with water in a Lewis acid-base reaction forming an M^+ metal cation and H_2, leaving OH^- in aqueous solution, thus forming a base.

Reaction of Metals with Acids

As we witnessed with Be and Mg, when the IE of a metal reaches approximately 600 kJ/mol, that element no longer reacts directly with water. Thus the clear demarcation between magnesium and calcium in the reactivity of alkaline earths with water. The first ionization energies of key elements in the periodic table are displayed in Figure CS10.1l.

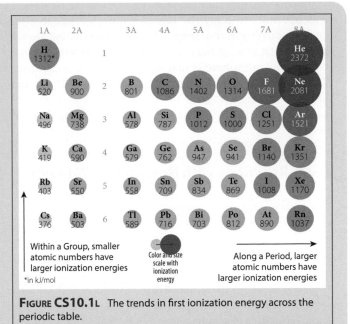

FIGURE CS10.1L The trends in first ionization energy across the periodic table.

However, metals in Group 3A and 4A along with beryllium and magnesium react with protons, H^+, in solution, i.e. react with *acidic* solutions. For example, magnesium reacts to produce the cation of the metal and H_2 in the gas phase:

$$Mg(s) + 2H^+(aq) \rightarrow Mg^{2+}(aq) + H_2(g)$$

The reason these metals with higher ionization energies react with an acidic solution is that the proton, $H^+(aq)$, has a *far greater ability to extract electrons* than does neutral water. This opens up a large segment of the periodic table to acidic reaction with metals in Groups 2A, 3A, and 4A as shown in Figure CS10.1m.

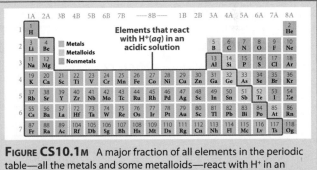

FIGURE CS10.1M A major fraction of all elements in the periodic table—all the metals and some metalloids—react with H^+ in an acidic solution.

It is important to recall where the proton in the acidic solution came from. When an acid is placed in water, it dissociates

$$HCl \xrightarrow{H_2O} H^+ + Cl^-$$

Thus we can write the complete reaction as

$$Mg(s) + 2H^+(aq) + 2Cl^-(aq) \rightarrow Mg^{2+}(aq) + 2Cl^-(aq) + H_2(g)$$

It is important to write out this full equation to keep track of all species, including the *spectator* ions, Cl⁻(aq), that do not enter directly in the chemical transformation, but are for keeping track of what has actually occurred in the solution.

The key point here is that by adding H⁺ to an aqueous solution through the addition of an acid, the reactivity of that solution then encompasses a very large segment of the periodic table because H⁺ can extract electrons from all those metals.

Water Reactions with Nonmetallic Elements

As we have seen, what instigates the reaction between water and the alkaline metals and the heavier alkaline earths is water's ability to extract electrons from those elements because of their very low ionization energy. If we spike that aqueous solution with protons by adding an acid, the greatly enhanced ability to extract electrons extends reactivity of an aqueous solution to encompass virtually all metals in the periodic table. This encompasses a very large fraction of all *elements* in the periodic table.

But what about the reaction of water with *nonmetals*? The first point to note is that water is a *nonmetal*. Water has the ability to react with a nonmetal only if it can extract electrons from the nonmetal or if the nonmetal can extract electrons from water. The result is that water does not react with nonmetals except for the case of fluorine gas. Chlorine gas reacts very weakly when bubbled through water to form HCl and hypochlorous acid, HClO:

$$H_2O(l) + Cl_2(g) \rightarrow HClO(aq) + HCl(aq)$$

The reaction products are acidic, and we can track the electron movement in the Lewis structure formulation as displayed in Figure CS10.1n.

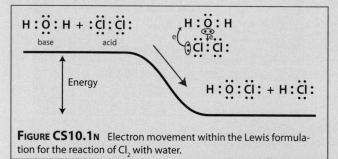

FIGURE CS10.1N Electron movement within the Lewis formulation for the reaction of Cl_2 with water.

Thus, except for the extremely electronegative elements fluorine and chlorine, water, a nonmetal, does not react with other nonmetals. But recall that the reaction between water and metals was dramatically enhanced by the addition of H⁺—the reason is that the reactions of water with metals depends upon the more electronegative species (the oxygen end of water) *extracting* electrons from the metal. But when water reacts with another nonmetal, we only have a reaction if water *donates* an electron to a more electronegative species (fluorine or chlorine).

This raises the question: What happens if we spike our aqueous solution, *not* with an electron extractor (H⁺), but with an electron donor? Will that strategy induce reactions between nonmetals and water?

Since F_2 reacts with water, and Cl_2 reacts very weakly with water, but neither Br_2 nor I_2 react with water, if we add a base (an electron donor) to water will a reaction occur? The answer is, yes indeed. Molecular bromine and molecular iodine both react with water in the presence of a base:

$$Br_2(aq) + 2\ OH^- \rightarrow BrO^- + Br^- + H_2O$$

Interestingly this reaction is followed by the self-reaction of BrO^- anions in solution to form Br^- and BrO_3^-

$$3\ BrO^- \rightarrow 2\ Br^- + BrO_3^-$$

This *self-reaction* of a species is termed *disproportionation*.

The reaction of the halogen with water or OH⁻ results from the highly electronegative halogen attacking the *lone pair* on water or the hydroxide anion as diagrammed in Figure CS10.1o.

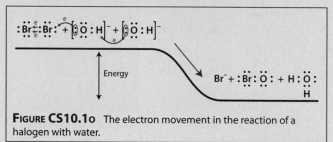

FIGURE CS10.1O The electron movement in the reaction of a halogen with water.

This gives us some chemical perspective. When metals were involved, it was the positive domain of the electron deficient hydrogen in water that was the center of interest. It was the electron deficient hydrogen that attacked the metal, inducing the donation of the electron from the metal, releasing energy in the process of supplying electron density to the electron deficient hydrogen as diagrammed in Figure CS10.1p.

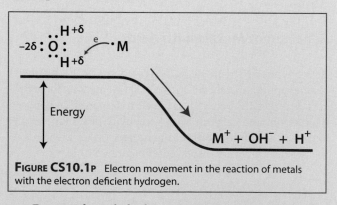

FIGURE CS10.1P Electron movement in the reaction of metals with the electron deficient hydrogen.

For metals with higher ionization potentials, the *incomplete* extraction of electron density from the hydrogen end of water (because the electronegativity of oxygen (χ =

3.5) cannot fully overcome the electronegativity of hydrogen (χ = 2.2)) limits the electron extraction capability of the electron deficient end of water. However, spiking the aqueous solution with an acidic solution provides protons that have a far greater ability to steal electrons, and thus essentially *all* metals are attacked by H$^+$ in that specie's quest for electrons.

In principle, the attack of halogens (nonmetals) on water is all about electron hungry (high electronegativity) species attempting to steal electrons, and in so doing, instigating a chemical reaction with its requisite release of energy. So it is the *lone pair* electrons on water that provide the site of attack, for they are the most vulnerable source of electron density in the structure of the water molecule. Thus, since electrons are the prize that the highly electronegative halogens (nonmetals) seek, if a base is added, providing the electron rich OH$^-$ to the solution, halogens less electronegative than fluorine or chlorine are presented with an easy target—the extra electron on OH$^-$.

In the jargon of chemistry, species that extract electron density are called *electrophilic*, or "electron loving." They might more aptly be called "electron stealing." In contrast, metals, with their loosely bound electrons, seek domains that are electron poor, or "nucleus rich" and are thus termed nuclophilic or "nucleus loving." In the end, however, these "philicities" are euphemisms for the theft of electrons—no more and no less.

We have, to this point, taken a very large part of the periodic table and investigated the reaction of those elements with water, and with water spiked with H$^+$ to enhance reactivity with metals (electron donors) and with OH$^-$ to enhance reactivity with a far smaller number, but important, nonmetals (electron extractors). But in this analysis, including tracking electron shifts with Lewis structures, we can see a *pattern* emerging wherein a vast array of chemical reactivity revolves around *oxygen* and *hydrogen*, albeit centered thus far on their remarkably versatile behavior within the *structure* of water with its hydrogen (electron deficient) end and its oxygen (electron rich) end with lone pair electrons as an additional dimension in the competitive market place of electron exchange.

It is a logical step, then, to extend this discussion in two related and important directions:

1. An analysis of *oxides* that occur along Periods in the periodic table and
2. Analysis of *hydrides* that occur along Periods in the periodic table.

The Chemistry of Binary Oxides

We have swept the full extent of the periodic table with respect to its reactivity *patterns* with water and its counterparts with H$^+$ and OH$^-$ spiked into the aqueous solution. Given that a vast fraction of all reactions occur in water, this is a big step forward.

Underscoring the fact that an intuitive understanding of chemical reactions is all about *patterns* in chemical reactivity, we next pursue the patterns in chemical reactivity by cutting horizontally across the periodic table examining the oxygen containing *binary* ("containing two elements") compounds, selecting the third Period as an example highlighted in Figure CS10.1q.

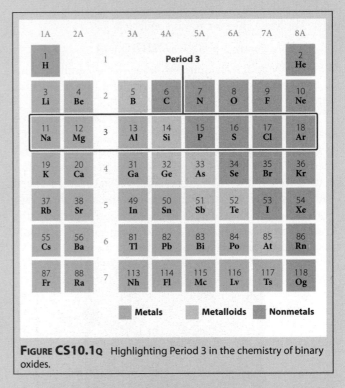

FIGURE CS10.1Q Highlighting Period 3 in the chemistry of binary oxides.

From the third Period, we have the following binary compounds with oxygen:

$$Na_2O, MgO, Al_2O_3, SiO_2, P_2O_5, SO_3, \text{ and } Cl_2O_7.$$

The *nonmetal* Period Three elements form other oxides; those listed here are the most fully oxidized. We first think about electrons and electronegativity. Oxygen is more electronegative than all elements in Period Three. If these oxygen containing binary compounds were ionic, each of the third Period elements would give up all their valence electrons to oxygen. The first two binary oxides are in fact ionic. Na$_2$O and MgO are ionic. But the remainder are polar covalent becoming less polar, as expected from the trend in electronegativity as we proceed to the right across Period Three. How do these binary oxides react with water? As we can guess by analogy with the reaction of sodium with water, Na$_2$O is an ionic compound

$$Na_2O \rightarrow 2Na^+ + O^{2-}$$

$$Na\!:\!\ddot{O}\!:\!Na \longrightarrow 2Na^+ + \left[:\!\ddot{O}\!:\right]^{2-}$$

and so we write the balanced chemical equation Na$_2$O + H$_2$O → 2NaOH.

The Na$^+$ is a *spectator* cation in water, but the oxide anion is a strong *base*, donating an electron pair to water, a Lewis acid:

$$[:\ddot{O}:]^{2-} \xrightarrow{e} H\ddot{O}: \longrightarrow [:\ddot{O}:H]^- + [:\ddot{O}:H]^-$$

The sodium hydroxide is soluble in water forming a strong base. For MgO, the reaction is similar:

$$MgO + H_2O \rightarrow Mg(OH)_2(aq)$$

However, Mg(OH)$_2$ is quite insoluble so the resulting solution is weakly basic. Mg(OH)$_2$ is a commonly used stomach antacid—*milk of magnesium*.

But what a difference *one* step across Period Three makes! Aluminum forms the familiar binary oxide Al$_2$O$_3$ that is extremely insoluble in water *and* in acid. Al$_2$O$_3$ is remarkably durable, is used as an abrasive in sandpaper, as a cutting tool and in toothpaste. Large crystals can be doped with metal impurities to form gemstones such as ruby or sapphire. Aluminum is the most abundant element in the Earth's crust when measured by mass and more surprisingly, only oxygen and silicon are more abundant in the crust.

Ruby

The second most abundant element in the Earth's crust is silicon that is the metalloid transition from metals to nonmetals in the periodic table. Its binary oxide, SiO$_2$ is also extremely inert—in fact it is the glass that chemical beakers are made from and the primary substance in glass. The reason Al$_2$O$_3$ and SiO$_2$ are inert is that the increasing electronegativity of aluminum and of silicon draws the lone pairs on oxygen to more tightly hold them—fending off attack from any electrophilic intruder. The other possibility, attack by the oxygen end of water on the more electropositive Al or Si, is thwarted by the highly stable network of bonds in those oxides. For example, just as carbon forms –C–C– chains that dominate organic chemistry, Si forms –Si–O– chains and groupings that repeat in a wide variety of silicates, particularly in minerals in the Earth's crust. Three examples include the minerals zircon, hemimorphite, and beryl that are constructed as shown in Figure CS10.1r.

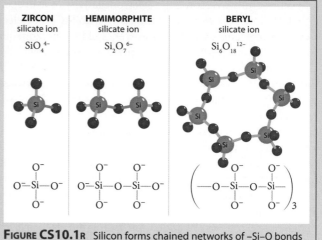

FIGURE CS10.1R Silicon forms chained networks of –Si–O bonds that create a wealth of different geometries.

The next binary oxides in Period Three are the phosphorus containing oxides that react with water to form phosphoric acid

$$P_4O_{10}(s) + 6H_2O(l) \rightarrow 4H_3PO_4(aq)$$

This tetraphosphorus decaoxide, P$_4$O$_{10}$, is formed by burning solid phosphorus, P$_4$(s), in excess oxygen

$$P_4(s) + 5O_2(g) \rightarrow P_4O_{10}(s)$$

and has the structure shown in Figure CS10.1s.

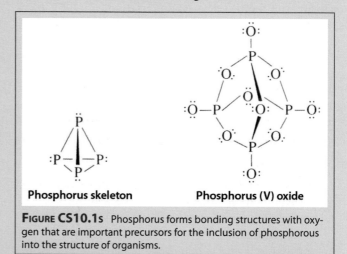

Phosphorus skeleton   Phosphorus (V) oxide

FIGURE CS10.1S Phosphorus forms bonding structures with oxygen that are important precursors for the inclusion of phosphorous into the structure of organisms.

Unlike the silicates, these binary oxygen compounds with phosphorus do not form bonded networks. So the reaction of P$_4$O$_{10}$(s) with water producing phosphoric acid has the following structure displayed in Figure CS10.1t:

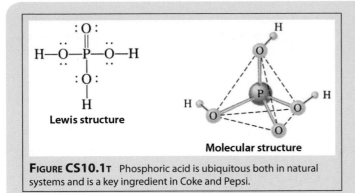

FIGURE CS10.1T Phosphoric acid is ubiquitous both in natural systems and is a key ingredient in Coke and Pepsi.

This structure results from the simple protonation of the phosphate ion

Phosphorus, phosphoric acid, and phosphate have critically important roles in modern chemistry and chemical biology that we will explore in subsequent chapters. The point here is to look at *patterns* in chemical reactivity. When water attacks the electropositive P atom, it does so with its lone pair as diagrammed in Figure CS10.1u.

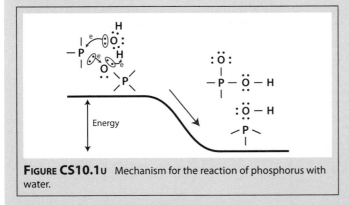

FIGURE CS10.1U Mechanism for the reaction of phosphorus with water.

The next Period Three element, sulfur, shares a great deal in common with phosphorus. It reacts with water to produce sulfuric acid

$$SO_3 + H_2O \rightarrow H_2SO_4(aq)$$

Analyzed from the perspective of Lewis structures, SO_3 is a resonance structure

In the reaction with water, it is the lone pair on the oxygen of water that goes after the vulnerable sulfur

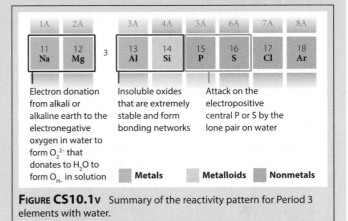

Thus, rather than breaking the S-O bond, the double bond is converted to a single bond with the "acidic" hydrogen attached. Thus we have the clear groupings of Period Three elements with water as displayed in Figure CS10.1v.

FIGURE CS10.1V Summary of the reactivity pattern for Period 3 elements with water.

The final binary oxide in Period Three is $Cl_2O_7(l)$, which is a viscous, explosive yellow oil that reacts with water to form perchloric acid

$$Cl_2O_7(l) + H_2O(l) \rightarrow 2HClO_4(aq)$$

Perchloric acid is a strong, inorganic (non-carbon-containing) acid. When added to water it readily gives up its hydrogen. Then relative acidic strengths of $HClO_4$, H_2SO_4, and H_3PO_4 are $HClO_4 > H_2SO_6 > H_3PO_4$. Because we are interested in *reaction patterns* and their causes, we can link this strength of acidity (willingness to donate an H^+ when added to water) to the degree of electronegativity of the nonmetal central atom in the structure. The most electronegative is Cl, so in the polar covalent bond between H and Cl, electron density has been *extracted* from the H, giving it a highly electropositive character leaving it *exposed* to extraction by the electronegative end of H_2O. As the electronegativity of the central atom decreases, so does the electropositive character of the acidic hydrogen, leaving it less vulnerable to attack by the oxygen end of H_2O. We turn finally to the hydrides.

Reactivity Patterns of the Period Two Hydrides

We are in pursuit of reactivity trends that we can link back to coherent patterns in electronegativity across the periodic table and for which we can use Lewis structures to make sense of those reactivity patterns. When we add the hydrides into the mix we know first of all that some of these hydrogen containing compounds will donate one or more hydrogens into aqueous solution forming an acid.

But why do some hydrides act as acids while others do not? Why is H_2S acidic but CH_4 is not? Why is HCl a strong acid, but its partner HF is not?

In order to make this systematic, let's run across the Period Two elements, examining patterns in chemical behavior of the binary hydrides. The second row elements for which we examine the hydrides are recalled in Figure CS10.1w.

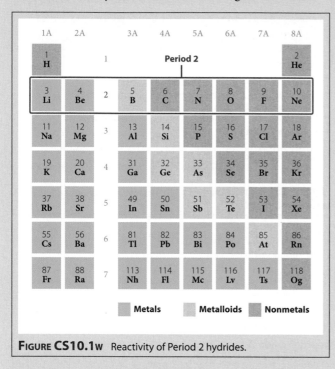

FIGURE **CS10.1w**   Reactivity of Period 2 hydrides.

What makes hydrogen unique is that it is a Period all to itself. If we remove an electron, we have an ion with no electrons, similar to that of the alkali metals. If we add an electron, it becomes noble-gas-like as do halogens with an additional electron. The electronegativity of hydrogen, 2.2, lies in the vicinity of the metal/nonmetal boundary in the periodic table. Thus, when it forms compounds with other elements in Period Two or Period Three, it may be expected to extract electrons from as many elements on the left-hand side of the periodic table as it donates to on the right-hand side. We can list the binary hydrides of Period 2 from Group 1A to Group 7A with the nonhydrogen element labeled first:

LiH(s) Lithium Hydride	BeH$_2$(s) Beryllium Hydride	B$_2$H$_6$(g) Diborane	CH$_4$(g) Methane	NH$_3$(g) Ammonia	OH$_2$(l) Water	FH Hydrogen Fluoride

In the lexicon of chemistry, only the first two and last one are named logically. The others are so central to the history of chemistry that they have names coined from the historical beginnings of chemistry.

Because hydrogen is not highly electronegative, *none* of

the compounds are fully ionic. LiH is strongly polar covalent with hydrogen extracting electron density from lithium, $Li^{+\delta}H^{-\delta}$. Hydrogen fluoride is strongly polar covalent with hydrogen donating electron density to fluoride, $F^{-\delta}H^{+\delta}$.

When lithium hydride reacts with water, the hydrogen in LiH donates an electron pair to the electron deficient end of H_2O as that hydrogen donates an electron pair to the newly formed OH^-, producing $H_2(g)$ as shown in Figure CS10.1x.

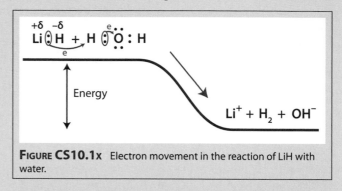

FIGURE **CS10.1x**   Electron movement in the reaction of LiH with water.

Moving to beryllium hydride and diborane, both react slowly with water, but a slightly acidic aqueous solution increases the rate of the reaction and produces H_2. This exhausts the metals and metalloids in Period Two. The nonmetal hydrides have very different reactivity. Given that the electronegativity of carbon is only 0.3 units greater than that of hydrogen, CH_4 is an example of a balanced (nonpolar) covalent bond with no lone pairs. As a result, it is completely nonreactive with water. Ammonia, when placed in water, is an example of a Lewis base because it donates its lone electron pair to water as shown in Figure CS10.1y.

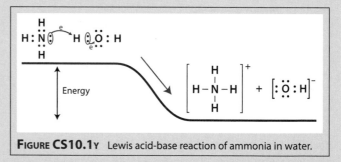

FIGURE **CS10.1y**   Lewis acid-base reaction of ammonia in water.

While we have used water as our reference reactant, it would be logical to jump over it to HF. However, water reacts with itself to a small degree in a Lewis acid-base reaction diagrammed in Figure CS10.1z.

This self-reaction of water is of immense importance to acid-base reactions because it forces the product of the hydronium ion concentration $[H_3O^+]$ and the hydroxide concentrate $[OH^-]$ to be a constant

$$[H_3O^+][OH^-] = K$$

The final hydride, HF, is the only hydride polar enough (resulting from the very high electronegativity of fluorine) to donate its hydrogen to water.

$$HF + H_2O \rightarrow F^- + H_3O^+$$

This is accomplished by using the lone pair on the oxygen of water.

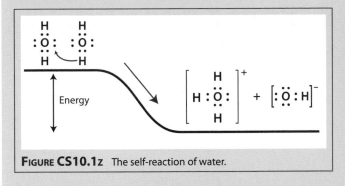

FIGURE CS10.1z The self-reaction of water.

"To restrict the group of acids to those substances which contain hydrogen interferes as seriously with the systematic understanding of chemistry as would the restriction of the term oxidizing agent to those substances containing oxygen."

The most flexible and widely applicable definition of an acid-base reaction is that of the Lewis formulation:

The Lewis acid-base formulation states that the important factor in an acid-base reaction is the attainment of a new shared pair of electrons in a new polar covalent bond.

Therefore:

- A base is any species that _____
- An acid is any species that _____
- The product of a Lewis acid-base reaction is called an _____.

One of the most important reactions involved in the critically important reduction of SO_2 from coal-fired power plants is the reaction between lime ($CaO(s)$) and $SO_2(g)$. This reaction is an important example of:

- A Lewis acid-base reaction
- The fact that a Lewis acid-base reaction can occur between two different phases.

For the reaction of $CaO(s)$ with $SO_2(g)$, sketch the Free Energy surface for the reaction and identify the acid and the base as well as the donated electron pair and the adduct formed.

Consider the reaction of potassium, K, with water.
1. Write the balanced chemical reaction.
2. Sketch the reactants and products on a free energy diagram and note the transfer of electrons that occurs in the reaction.

Why is IE of Be higher than IE of Ra? Explain in terms of the multielectron analysis of Chapter 4.

As we saw so clearly in Chapter 6, the chemistry of $CO_2(g)$ with water, both freshwater and seawater, constitutes one of the most important reactions in global climate change.
1. Write the balanced chemical equation for $CO_2(g)$ reacting with water.
2. Sketch the free energy diagram for $CO_2(g)$ reacting with water and note carefully the electron exchange sequence that takes place to form H_2CO_3.

Why do the heavier alkaline earth metals react with water, where as the lighter alkaline earths (Be and Mg) do not react with water?

a. Write out the complete reaction for magnesium, Mg(s), reacting with nitric acid in water. Be sure to include both the metal and the acid.
b. Approximately what fraction of the periodic table reacts with acids?
c. Why does the addition of an acid to water so significantly increase the number of elements that will react with the acidic solution?

a. Draw the electron exchange that occurs when F_2 is bubbled through water.
b. Is the product solution acidic or basic?
c. What other nonmetals does water react with?

Problem 8

Draw the Lewis structure for Al_2O_3.

Problem 9

Why doesn't the reaction $NH_3 + H_2O \rightarrow NH_2^- + H_3O^+$ occur instead?

BUILDING A GLOBAL ENERGY BACKBONE

CASE STUDY 10.2 The Potential for Generating Electric Power from Wind in the US

Introduction

A central axiom of any discussion of energy options is that the *scale* of potential energy generation, the order of magnitude of the available power, must be considered first. In all the discussion of renewable power generation from wind, is it feasible to produce a major fraction of current demand in the U.S. from this source? How would we accurately calculate the available amount of wind power? How much would it cost? Could it be reasonably integrated into the energy structure of the U.S.? Given that, in 2010, wind generated power constituted less than 2% of the total primary energy generation in the U.S., specifically what are the relevant numbers?

Defining the Objective

We begin by setting our objective of 1 terrawatt (1 TW) or 1 × 10¹² watts of power generated from land based wind turbines within the continental United States. This constitutes one third of the total power demand in the U.S.—a country that consumes 3 TW of power which is approximately 10.5 kW for every man, woman and child in the U.S. Recall that with a global power consumption of 15 TW and a global population of 6.5 billion, the average per capita power consumption world wide is 2.3 kW—about 20% of the US per capita power consumption.

The Fundamentals of Wind Power Generation

Before we move to establish how much power we can generate in the U.S. from wind, we need to consider the physics behind how the kinetic energy contained in wind is actually converted to electrical power. On the face of it, we know that the kinetic energy of a mass in air moving at velocity v is $KE = \frac{1}{2} mv^2$. But what is m? How do we calculate it?

To establish how we calculate the coupling between the motion of the atmosphere, a fluid moving with velocity v, and the blades of a wind turbine, we consider a cylinder of air passing through a hoop with an area equal to the circle swept out by the rotor blades. The wind turbine we have selected is the GE 2.5 MW, which has a blade length of 50 meters. Thus the area swept out is $A = \pi r^2$, with r = 50 m. We can represent graphically the volume of air moving past the rotor blades in time t by starting at t = 0 with the cylinder of air coincident in space with our hoop of area $A = \pi r^2$ as shown in the upper panel of Figure CS10.2a.

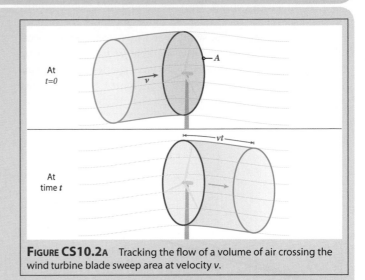

FIGURE CS10.2A Tracking the flow of a volume of air crossing the wind turbine blade sweep area at velocity v.

At time t later, that cylinder of air will have moved a distance equal to vt as displayed in the bottom panel of Figure CS10.2a.

The kinetic energy of this volume of air is then

$$KE = \tfrac{1}{2}\, mv^2 = \tfrac{1}{2}\, \rho\, [vt]\, Av^2$$

where s is the density of the air (kg/m³) and [vt]A is the volume of air in m³. That is, the mass passing through the hoop, which is the same as the mass passing the plane of the turbine blades, is m = ρ [vt] A. Thus the kinetic energy of the air passing through the sweep area of the blades is

$$KE = \tfrac{1}{2}\, mv^2 = \tfrac{1}{2}\, \rho\, Av^3 t$$

The *power* delivered is the energy per unit time or

$$P = \text{Energy}\big/\text{time} = KE\big/t = \tfrac{1}{2}\frac{mv^2}{t} = \tfrac{1}{2}\frac{\rho Av^3 t}{t} = \tfrac{1}{2}\rho Av^3$$

The density of air at sea level is 1.25 kg/m³. At an altitude of 4000 ft above sea level, the atypical altitude of the high plains of the central United States, the density of air is close to 1 kg/m³. This is a very handy number to remember.

Therefore, the power delivered by the wind *per square* meter of the blade aerial sweep for a wind speed of 10 m/sec is

$$\tfrac{1}{2}\rho v^3 = \tfrac{1}{2}\left(1.0 \text{ kg/m}^3\right)(10 \text{ m/sec})^3$$
$$= 500 \text{ W/m}^2$$

For a wind speed of 5 m/sec, the power delivered is

$$\tfrac{1}{2}\, \rho\, v^3 = \tfrac{1}{2}\, (1.0 \text{ kg/m}^3)(5 \text{ m/sec})^3 = 63 \text{ W/m}^2$$

As we will see, not all of this kinetic energy is successfully captured by the turbine blades.

Wind speed counts! Thus we should look, albeit briefly, at the dependence of wind speed on height above the ground. The primary reason that wind speed increases rapidly above the ground is that there is considerable *friction* generated by the roughness of the grasses, trees, etc. experienced by the air moving over land. The upper panel of Figure CS10.2b presents a plot of wind speed vs. the $\log_{10}$ of the height above the ground.

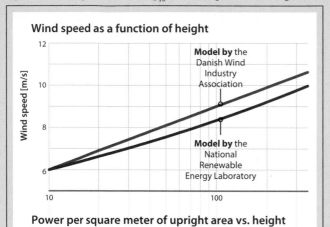

Wind speed as a function of height

Model by the Danish Wind Industry Association

Model by the National Renewable Energy Laboratory

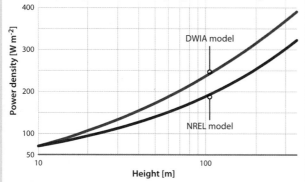

Power per square meter of upright area vs. height

DWIA model

NREL model

Height [m]

FIGURE CS10.2B   The dependence of wind speed on height above the ground and the dependence of the power density of wind vs. height. The upper panel displays the wind speed; the lower panel displays the power density.

As a rule of thumb, increasing the height above the ground by a factor of two increases wind speed by 10%. But, as the lower panel of Figure CS10.2b shows, the power delivered by the wind increases by 30%.

Selection of a Wind Turbine

It is now clear that if we are to carry out a quantitative analysis of the wind power potential for the U.S. we will need to select a wind turbine, because the length of the turbine blades *and* the height of the turbine hub above the ground are key quantities in the calculation. For our calculations here, we choose the new generation of wind turbines built by GE—in particular the 2.5 MW turbine with three blades of 50 meter length and a hub height of 100 meters shown in Figure CS10.2c.

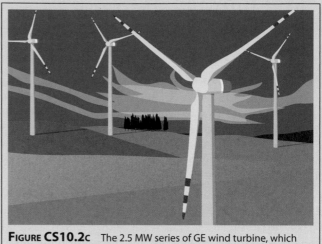

FIGURE CS10.2C   The 2.5 MW series of GE wind turbine, which constitutes the selected turbine design for the calculations here.

The rating of a wind turbine is typically measured by the maximum or "peak power" that the system will deliver. Wind turbines are designed to begin producing power at wind speeds of approximately 3 m/sec, to increase in power output rapidly with increasing wind speed (recall the basic formula goes as v^3) and to plateau at the rated peak power. At wind speeds greater than 25 m/sec, the blades are "feathered" to stop rotation to protect the turbine under high wind conditions. The output of the GE 2.5 MW turbine as a function of wind speed is shown in Figure CS10.2d.

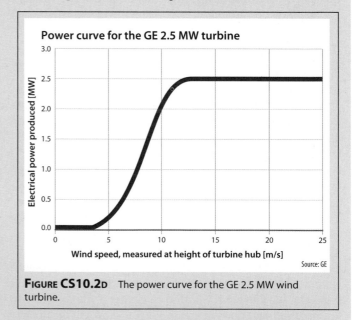

Power curve for the GE 2.5 MW turbine

Wind speed, measured at height of turbine hub [m/s]

Source: GE

FIGURE CS10.2D   The power curve for the GE 2.5 MW wind turbine.

Wind Power Calculations for the Continental U.S.

The next and most difficult part of the calculation is to determine the wind speed as a function of geographic position and height above the ground for everywhere in the U.S. with adequate temporal and spatial resolution to accurately represent the wind speed and air density passing through the blade sweep area of a GE 2.5 MW turbine placed at any point in the U.S. For this calculation we follow the recent work by Xi Lu and colleagues (Lu, Xi, M.B. McElroy and J. Kiviluoma, PNAS 106, 10933, 2009). They used a simulation of global wind fields from the fifth generation Goddard Earth Observatory System Data Assimilation System (GEOS-5 DAS). This approach uses a state-of-the-art weather climate model, but includes thousands of observations from ground-based, aircraft, balloon and satellite platforms. The derived wind fields are tested against the entire array of observations systematically coupled to the atmospheric model. The wind fields are determined every six hours within grid boxes 66 km × 50 km which corresponds to a resolution of 2/3° in longitude by ½° in latitude. The lowest layers of the model are at the ground, 71 meters, 201 meters, and 332 meters. The wind field is interpolated between these points and superimposed on the sweep area of the GE 2.5 MW turbine.

The power generated by the 2.5 MW turbine is given by

$$P = \tfrac{1}{2}\, \rho\, \pi r^2 f_p\, v^3$$

where the efficiency factor, f_p, is taken into account in the turbine power curve displayed in Figure CS10.2d. Assembly of the wind turbine system is shown in Figure CS10.2e.

The next issue to settle is: how close together can these large wind turbines be located with respect to each other? If the turbines are placed too close together, they will interfere with each other. They will interfere both because each turbine extracts kinetic energy from the moving atmosphere (reducing the wind speed downwind of the turbine) *and* because the presence of turbines creates turbulence that significantly reduces the effective velocity of the air motion past the turbine rotors. On the other hand, the amount of power extracted per unit area of land is important to the owners of the property upon which the wind turbines are placed.

Both calculations of atmospheric flow in the presence of this class of wind turbine and direct field trials have demonstrated that to reduce power loss to less than 20%, a wind turbine must be placed 7 rotor diameters or more downwind of its partner and greater than 4 rotor diameters across wind. In practice this means that each 2.5 MW wind turbine requires 0.28 km² of land.

A final consideration is given to those regions that are not suitable for installation of wind turbines—most notably forested regions, water and/or permanent snow-bound areas.

The calculated wind generating potential for the continental U.S. using the 2.5 MW GE wind turbine and the wind

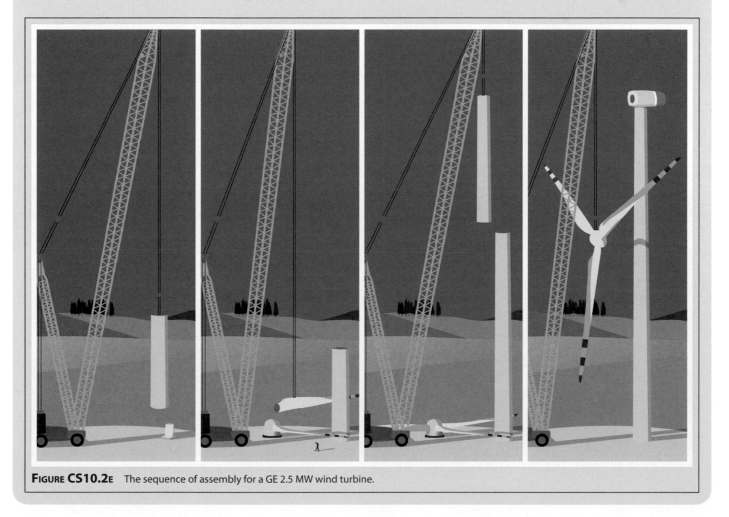

FIGURE CS10.2E   The sequence of assembly for a GE 2.5 MW wind turbine.

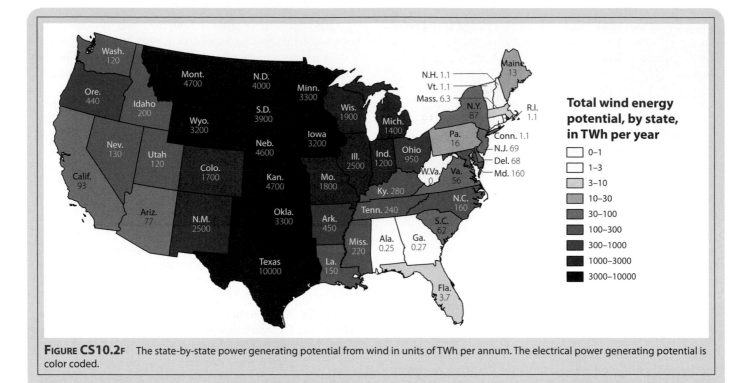

Total wind energy potential, by state, in TWh per year

- ☐ 0–1
- ☐ 1–3
- ☐ 3–10
- ☐ 10–30
- ☐ 30–100
- ☐ 100–300
- ☐ 300–1000
- ☐ 1000–3000
- ☐ 3000–10000

FIGURE CS10.2F   The state-by-state power generating potential from wind in units of TWh per annum. The electrical power generating potential is color coded.

fields as calculated above is displayed as a state-by-state map in Figure CS10.2f in units of terawatt hour, TWh, per annum.

A key issue, given this very large wind generated potential for electric power, is the balance of production and demand as well as the need for power distribution nationally. This issue will be explored repeatedly in subsequent Case Studies. It will require a smart grid for national distribution as first displayed in Figure CS4.3E of the text, and other forms of renewable energy generation and storage, also not-

ed in Figure CS4.3E, such as concentrated solar thermal as well as energy storage that can be drawn from during periods of peak demand.

While we have focused here on the big picture of wind generated power in large regions of the Midwest, it is important to recognize that very important sources of wind power also exist along selected ridges such as shown here in the hills above San Francisco Bay in Figure CS10.2g.

FIGURE CS10.2G   Wind turbines located in the hills above San Francisco Bay demonstrating the importance of localized high yield sites to the national strategy.

Problem 1

It was our stated objective to develop wind power in the US so as to provide 1 TW of power to the electrical grid. Figure CS10.2f provides a state-by-state breakdown of the number of TWh per annum generated from each state.

a. Calculate the number of TWh per annum realized from 1 TW of power production from wind.
b. Compare the number of TWh/yr provided by wind power in Texas alone to 1 TW. What about from North and South Dakota alone?
c. Given the wind power production from all states indicated by the red and orange states in Figure CS10.2f, calculate the fraction of that power generation from wind that is required to produce 1 TW of power to the grid.

CASE STUDY 10.3 Cancer, DNA, and the Structures Resulting from Hydrogen Bonding

We have already observed the importance of hydrogen bonding in the structure of water. The bonding between the electronegative oxygen end of the water molecule and the electron deficient hydrogen end of the water molecule serves to organize the molecules of water in the liquid phase. In the solid phase of water (ice), it is the hydrogen bonds that establish the crystal structure of water as displayed in Figure CS10.3a.

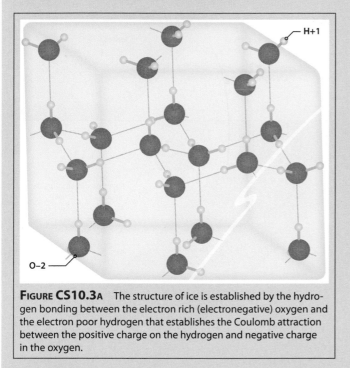

FIGURE CS10.3A The structure of ice is established by the hydrogen bonding between the electron rich (electronegative) oxygen and the electron poor hydrogen that establishes the Coulomb attraction between the positive charge on the hydrogen and negative charge in the oxygen.

A single hydrogen bond is weak in comparison with a fully developed covalent or ionic bond. Hydrogen bonds fall in the range of 10 kJ/mole whereas covalent bonds range from 100 to 500 kJ/mole. However, a key attribute of hydrogen bonding is that a *large number* can occur within a single molecular structure. Many large molecules can have tens to hundreds of hydrogen bonds within their structure resulting in remarkable architectures. Thus, collectively multiple hydrogen bonds can establish a strong bonding structure that controls a potentially complex three-dimensional structure. A key example is the DNA molecule, as displayed in Figure CS10.3b.

As Figure CS10.3b shows, DNA is constructed from two long, helical strands containing many thousands of atoms and molecules. The double helix is linked across its structure by hydrogen bonds between base pairs, the sequence of which encodes information defining the genetic signature of an organism.

A key consideration in the structure of DNA, complementary to its ability to encode genetic information, is the

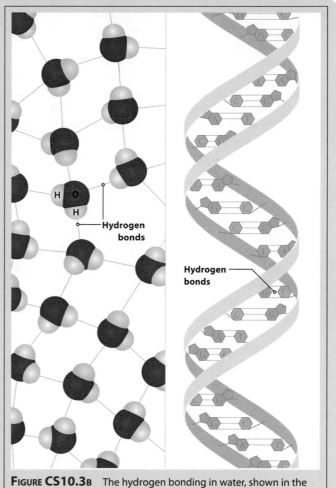

FIGURE CS10.3B The hydrogen bonding in water, shown in the upper panel, is the same mechanism that bonds the base pairs across the double helix of DNA.

balance between the cumulative strength of many hydrogen bonds with the weakness of *individual* hydrogen bonds. If an interaction between two molecules involves a limited number of hydrogen bonds, the structure can be altered quite easily. This aspect of hydrogen bonding plays an important role in enzyme-mediated reactions in biological systems—a subject we will develop sequentially through this text. This balance between the strength of many hydrogen bonds and the vulnerable nature of a small number of hydrogen bonds serves as a preamble to our discussion of the causes of cancer as well as an organism's ability to defend itself against cancer.

Mutations: The Link between the Structure of DNA and Cancer

Genetic information is encoded in DNA for the synthesis of proteins. When a gene is expressed, DNA is unchanged in the process. However, on occasion, in approximately 1 case in 10^8 transcriptions, a *mutation* may occur where a mutation is defined as a heritable change in genetic material. As a result of the mutation, the structure of DNA is changed by an unintended alteration in the sequence of base pairings along the spine of DNA. This alteration is passed from the parent DNA to daughter cells during cell division.

However, mutations cut both ways. On the positive side, mutations supply the variation that enables species to evolve and become more proficient at coping with their adversaries and their environment. Mutations drive evolutionary change. The genes within modern species evolved over billions of years. Once a species has acquired the attributes that provide it with the ability to compete successfully within its surroundings, random mutations are more likely to disrupt genes than to enhance the species ability to survive.

Because mutations can be quite harmful, all species are armed with mechanisms to repair damaged DNA. Such DNA repair strategies are among the most remarkable processes found in the natural world. These DNA repair mechanisms are capable of reversing DNA damage before a permanent mutation can occur. Were DNA repair mechanisms not as effective as they are, mutations would be so prevalent that few, if any, species would survive.

We will develop the mechanisms for DNA repair in chapters following our full development of molecular bonding. We turn here to the context of cancer as a major disease affecting humans.

Cancer

Cancer is a disease inherent in all multicellular organisms and is characterized by uncontrolled cell division. Over 1 million individuals in the United States alone are diagnosed with cancer each year. About half that number, or 500,000, die each year from the disease. In about 10% of cancers, a higher predisposition to develop the disease is associated with inherited traits. However, research has shown that nearly 90% of cancers do not involve genetic changes that are passed from parents to offspring. Instead, cancer is an acquired condition that occurs later in life as a result of exposure to carcinogens. Carcinogens are chemical or photochemical agents that induce mutations that in turn lead to the development of cancer. The most common form of cancer is skin cancer. The mechanism for the onset of skin cancer is the severing of the hydrogen bonded base pairs across the spine of the double helix by the absorption of a UV photon as shown in Figure CS10.3c. The breakage of the base pair linking the base pairs results in the formation of the thymine dimer as shown in Figure CS10.3c.

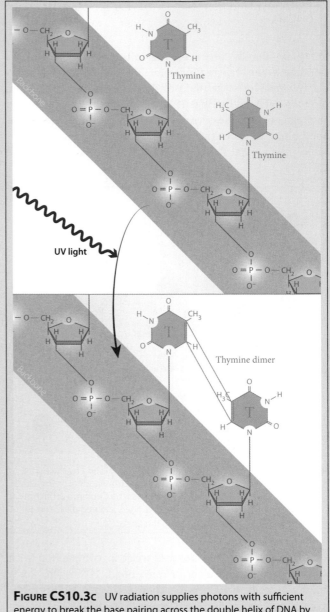

Figure CS10.3c UV radiation supplies photons with sufficient energy to break the base pairing across the double helix of DNA by forming a thymine dimer, which, as a result, alters the genetic code of the DNA.

Mechanisms for Cancer

As noted above, cancers occur when normal mechanisms that limit cell growth and division are disrupted. It has been demonstrated that several mutations are required to transpose a normal cell into a cancerous one. This requires that multiple mutations are required and explains why there is a long latency period, typically decades, between exposure to a substance capable of inducing a mutation and the actual detection of the cancer. Another important consideration is that because of the probabilistic nature of mutations, the risk of cancer increases with age. As a result, while children and young adults do develop cancer, the disease is primarily one of old age. In addition, with increasing life expectancy, the probability for increasing cancer cases increases significantly.

To review briefly, deoxyribonucleic acid (DNA) gains its acidic character because it releases hydrogen ions, H^+, into solution and thus has a net negative charge at neutral pH. We can consider the structural features of DNA at different levels of complexity. With reference to Figure CS10.3d:

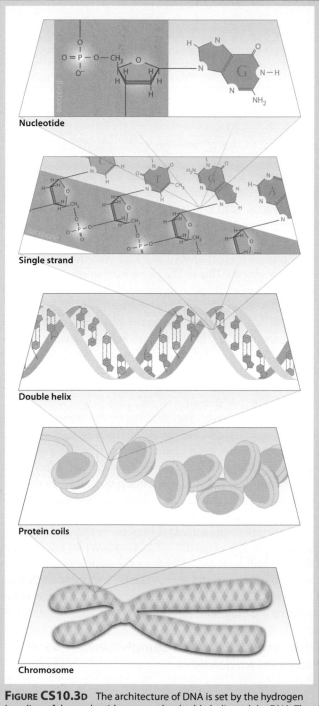

FIGURE CS10.3D The architecture of DNA is set by the hydrogen bonding of the nucleotides across the double helix and the DNA. The DNA strand associates with selected proteins to form a chromosome as shown in the lower sector of the diagram.

1. Nucleotides, shown in Figure CS10.3e, are the building blocks of DNA and are constructed from the covalent linkage of these nucleotides in a linear chain, as shown in Figure CS10.3f.

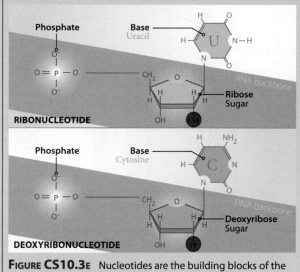

FIGURE CS10.3E Nucleotides are the building blocks of the DNA structure and are themselves constructed of a phosphate group, a sugar, and a base. The nucleotide is distinguished by both the structure of the sugar and the structure of the base.

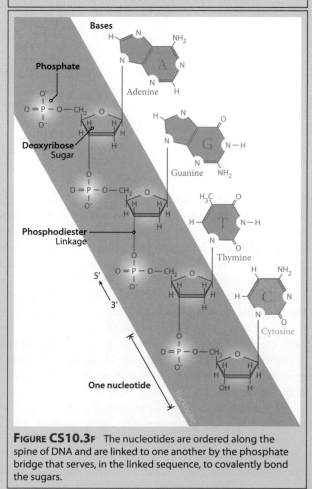

FIGURE CS10.3F The nucleotides are ordered along the spine of DNA and are linked to one another by the phosphate bridge that serves, in the linked sequence, to covalently bond the sugars.

2. Two strands of DNA can be hydrogen bonded to each other to form the double helix with two structural categories of bases: the double ring purines and the single ringed pyrimidines, as shown in Figure CS10.3g.

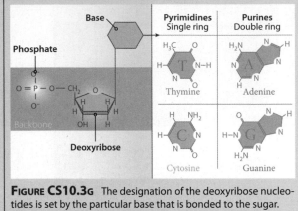

FIGURE CS10.3G The designation of the deoxyribose nucleotides is set by the particular base that is bonded to the sugar. There are four unique bases—two double ring purines and two single ring pyrimidines as shown in the figure.

3. There are two purines that constitute the bases that cross-link the nucleotides single strand of the DNA backbone: adenine (A) and guanine (G), shown in Figure CS10.3g. There are also two pyrimidines that are bases capable of cross-linking the nucleotide strand of DNA: thymine (T) and cytosine (C), also displayed in Figure CS10.3g.

4. In living cells, DNA is associated with an array of different proteins to form chromosomes. This association of proteins with DNA organizes the extended DNA strands into compact architectures that are incorporated into the cell nucleus.

5. Finally, a genome is the complete complement of an organism's genetic encoding.

So the nucleotide is composed of three components: a phosphate, a five member sugar (pentose), and a nitrogen containing base that repeats along the nucleotide strand as shown in Figure CS10.3h. A numbering system has been established to identify the attachment sites of the base and the phosphate to the deoxyribose sugar as shown in Figure CS10.3h.

This numbering system is important for understanding the sequencing along the nucleotide strand, so we quickly summarize. With reference to Figure CS10.3h, in the sugar ring, carbon atoms are numbered in a clockwise direction beginning at the bonding site of the base to the deoxyribose, which is also the carbon atom to the immediate right of the oxygen atom in the sugar ring. The second, third and fourth carbon atoms are in the ring; the fifth carbon atom lies above the ring. All carbon atom numerical designations are primed: 1′, 2′, 3′, 4′, and 5′ to distinguish the number of carbons associated with the sugar. Carbon atoms in the ring structure of the bases are not primed. Thus a base

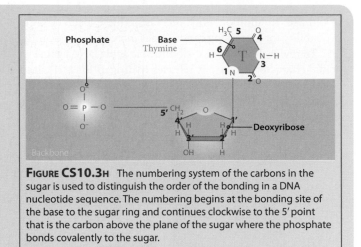

FIGURE CS10.3H The numbering system of the carbons in the sugar is used to distinguish the order of the bonding in a DNA nucleotide sequence. The numbering begins at the bonding site of the base to the sugar ring and continues clockwise to the 5′ point that is the carbon above the plane of the sugar where the phosphate bonds covalently to the sugar.

is attached to the 1′ position and a phosphate to the 5′ position.

Consider again Figure CS10.3f, which displays a short strand of DNA with four nucleotides. A key characteristic of this linkage is to notice that a phosphate group connects two sugar molecules together such that the phosphate and sugar molecules form the backbone of the DNA strand. The bases project inward from the backbone; the backbone is negatively charged due to the phosphate anion. A key structural feature of a nucleic acid strand is the orientation of the nucleotides. Each phosphate is covalently bonded to the 5′ carbon in one nucleotide and to the 3′ carbon on the other (see Figure CS10.3f). In a strand, all sugar molecules are oriented in the same direction, which means that all 5′ carbons in each sugar ring lies above the 3′ carbons. This establishes a directionality based on the orientation of the sugar molecules within that strand.

Referring again to Figure CS10.3f, the direction of the strand is designated 5′ to 3′ in going from top to bottom; the 5′ end of the DNA strand has a phosphate group and the 3′ end has a hydroxyl (–OH) group. The genetic information contained in DNA is the information contained in the *specific sequence* of bases. Referring again to Figure CS10.3f, the sequence of bases is thymine-adenine-cytosine-guanine or TACG. To orient the sequence of bases to the directionality, the strand is abbreviated 5′–TACG–3′. While the bases in DNA are hydrogen bonded across the double helix, the nucleotides *within* a strand are covalently bonded such that the sequence of bases cannot be altered, cannot be rearranged except by a mutation.

DNA Structure and the Complementary Base Pairing of Nucleotides

Watson and Crick, shown in Figure CS10.3i, proposed the structure of DNA as the double helix with the sugar-phosphate backbone on the outside with specific complementary base pairing in the inside. The key feature of the base pairing (which Watson and Crick worked out after a number of false starts) is the specificity of the hydrogen bonding between the bases.

An adenine base in one strand links across the double helix with two hydrogen bonds to a thymine base on the opposite strand as displayed in the uppermost sector of Figure CS10.3j. A second specific hydrogen bonded pair is the three-bond guanine link to cytosine displayed in Figure CS10.3j. The double hydrogen bonded AT pair and the triple bonded GC pair constitute the AT/GC rule of DNA bare pair bonding. According to this AT/GC rule, purines always bond with pyrimidines, keeping the width of the double helix approximately constant along the length of the double helix.

As a result of the AT/GC rule, the base sequences of two DNA strands are complementary to one another such that if you know the sequence along one strand, you can predict the sequence in the opposite strand. So, if one strand has the sequence 5′–GCGGATTT–3′, the opposite strand must be 3′–CGCCTAAA–5′. This also dictates the fact that two strands of a DNA double helix must be antiparallel. Inspection again of Figure CS10.3j demonstrates that if one strand runs in the 5′ to 3′ direction from top to bottom, the other strand must run from 3′ to 5′ bottom to top.

Figure CS10.3i James Watson and Francis Crick, in 1953, were the first to determine the detailed bonding structure of DNA. Their work opened a profoundly new arena of research that has served to bring together the fields of chemistry, biology, and physics, which has fundamentally reshaped the scientific endeavor.

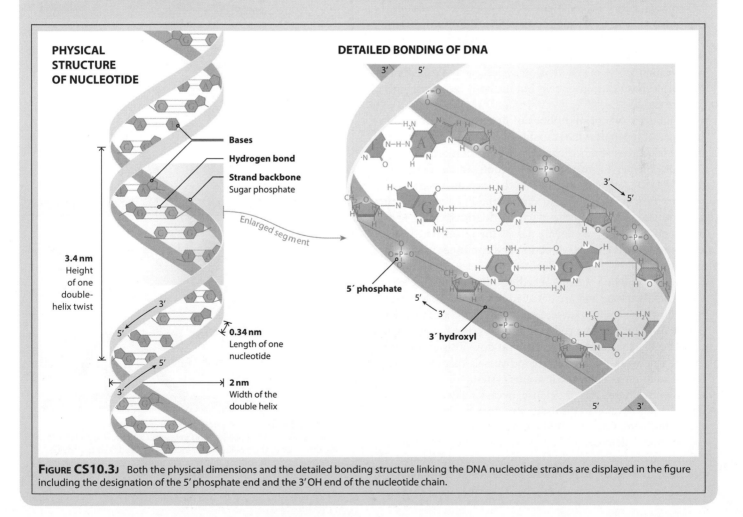

Figure CS10.3j Both the physical dimensions and the detailed bonding structure linking the DNA nucleotide strands are displayed in the figure including the designation of the 5′ phosphate end and the 3′OH end of the nucleotide chain.

Mechanisms for Cancer Initiation

Having established the structure of DNA and the critical importance of the sequence of base pairs that contain the genetic code for cell reproduction, we return to the question of how cancers can be initiated within a living organism. The key initial step in cancer development is that mutations occur in the cell's DNA at positions that specify the synthesis of key regulating proteins. An important consideration is that *several* such mutations are required to transform a normal cell into a cancerous cell.

A mutation occurs when DNA is incorrectly transcribed during cell division. Of course, maintaining the correct genetic encoding requires the correct pairing of bases when a new DNA strand is copied from the original. If an incorrect base is incorporated in the sequence along the spine of the double helix, then, unless repaired, the error will be passed to subsequent generations of the cell. If that incorrect base is an integral part of a gene, the error is then introduced into the protein for which the gene codes and this can lead to a misfunction of the protein.

What is important to realize is that while mutations are more-or-less random, occurring in 1 in 10^8 transcriptions, there is some probability that they will occur at key sites coding for regulatory proteins and even a smaller probability that this will lead to the alteration of the cell resulting from the accumulation of enough critical mutations. However, because our bodies contain billions of cells, and because we live through scores of cell division cycles, each of us harbors precancerous or cancerous cells but the body has remarkable lines of defense.

There are two primary lines of defense. The first is the presence of repair enzymes capable of detecting and correcting incorrect base pairs. Enzymes, as we will more fully develop in Chapter 14, are capable of "proofreading" the newly formed DNA. The enzyme detects the improper base pairing, reverses its direction and digests the linkages between nucleotides at the end of the newly made strand in the 3′ to 5′ direction. Once the enzyme passes the mismatched base and removes it, it changes direction and continues to synthesize the new, now corrected, DNA in the 5′ to 3′ direction.

The second line of defense is the body's immune system. Cancer cells can be and are detected by the immune system by virtue of the ability to detect variations in the surface molecules of the cancerous cells. In addition, the fast-growing cancer cells require a large local blood flow. This in turn requires a network of "foreign" blood vessels that can be detected and destroyed by the immune system.

Against these lines of defense, occasionally a cancerous domain may develop in the tissue structure in a sequence depicted in Figure CS10.3k. While the probability of overcoming the body's defense mechanism is small, the probability is a function of several factors. One important consideration is genetics. Individuals may inherit a defect that compromises

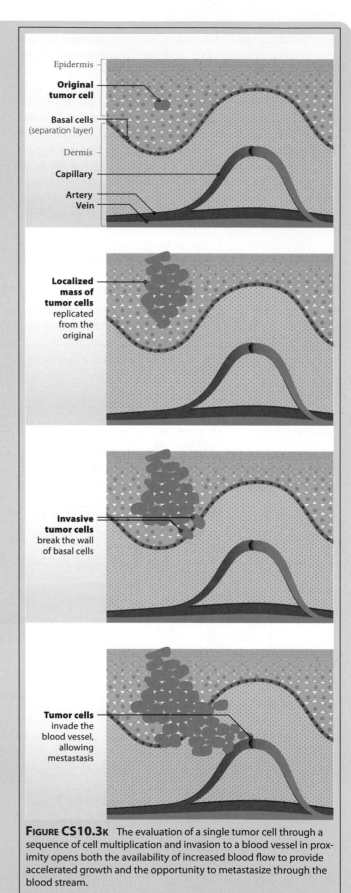

FIGURE CS10.3K   The evaluation of a single tumor cell through a sequence of cell multiplication and invasion to a blood vessel in proximity opens both the availability of increased blood flow to provide accelerated growth and the opportunity to metastasize through the blood stream.

the normal lines of defense outlined above. Examples include a faulty repair enzyme that allows mutations to pass undetected or uncorrected. Another possibility is that there may be an inherited, preexisting mutation in a gene for one of the regulatory proteins, opening the possibility of linking with other accidental mutations. This genetic mapping is now uncovering an array of genes in which mutations increase the probability of developing specific cancers as well as the propensity for other diseases involving the cardiovascular and circulating systems.

Exposure to Carcinogens: Chemically Induced Cancers

As noted in the beginning of the Case Study, the vast proportion of cancers are caused by exposure to chemical or photochemical factors in our environment. Research is becoming increasingly detailed in the identification of the specific mechanisms responsible at the molecular level. As a general rule carcinogens can operate in two distinct ways. *First*, they can induce mutations by attacking the hydrogen bonding linking the DNA bases. *Second*, they can increase cancer probability indirectly, as for example, alcohol consumption, which is linked to liver cancer because it induces cell proliferation in the liver. The liver is the organ wherein alcohol is metabolized.

The chemistry involved in cancer initiation is interesting in its own right, in addition to being important for human health. There are, broadly speaking, two requirements for mutagens.

1. They must react with the DNA base pairs in ways that alter their hydrogen bonding structure with the complementary base. Because bases are inherently *electron rich*, the mutagens are usually electrophiles.

2. The carcinogens (the mutagens) must gain access to the nucleus, because that is where the DNA resides. Interestingly, many electrophiles are not carcinogens simply because they react with other molecules and are chemically transformed (deactivated) before they can reach the nucleus. A key consequence of this fact is that carcinogens are not reactive themselves, but rather are converted to reactive intermediates by the body's own biochemistry.

A key ability of the body is the capability of removing foreign chemicals either directly through excretion or by chemically transforming these "xenobiotics." An important example is the hydroxylation of lipophilic (affinity for fats) organic compounds. Consider the example of benzanthracene ($C_{18}H_{14}$), a molecule that has four-linked benzene rings as shown in Figure CS10.3l and is carcinogenic.

The hydroxylation of benzanthracene is accomplished by enzyme activation that forms an intermediate epoxide that then pursues one of two paths. Either the epoxide reacts

with DNA or it is detoxified. It is detoxified by hydroxylation that replaces the epoxide oxygen bridge with a hydroxyl group that in turn serves to increase water solubility. In addition, the presence of the reactive –OH group serves as a reaction center attaching other hydrophilic groups that further increase water solubility, promoting excretion by the kidneys. Hydroxylation involves the insertion of one of the O atoms from O_2 into the C-H bond of a hydrocarbon with the remaining O atom being reduced to H_2O by supplying two electrons from a biological reductant:

$$O_2 + \text{–C–H} + 2e^- + 2H^+ \rightarrow \text{–C–O–H} + H_2O$$

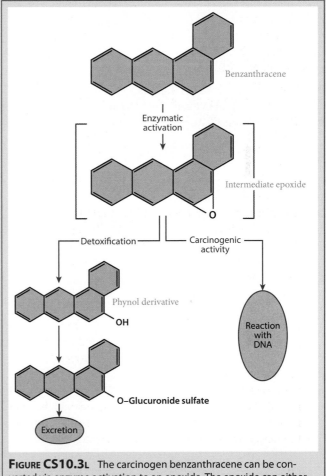

FIGURE CS10.3L   The carcinogen benzanthracene can be converted via enzyme activation to an epoxide. The epoxide can either (1) react with DNA, leading to the possibility of genetic alteration or (2) pass through a sequence of chemical activation steps leading to excretion from the body and the elimination of the molecule from the body.

What is interesting about this important reaction is that the chemical transformation must simultaneously form the bonding structure between the carbon and the OH radical as it forms the H_2O product. If a free O atom is formed in isolation, the high reactivity intrinsic to the O atom (a free radical) can damage surrounding tissue.

The hydroxylation is carried out, as shown in Figure CS10.3m, by an important class of enzymes, cytochrome P450, that contains a heme group to bond the O_2, just as it does to transport O_2 in the blood stream. However, the P450 not only bonds the O_2, it binds the xenobiotic invader at an adjacent site leading to hydroxylation. The issue is that the epoxide precursor is a potent electrophile and because the precursor is generated *inside the cell*, the possibility exists that the epoxide, which is a potent electrophile, will reach the nucleus by diffusion thereby potentially reacting with DNA before it is converted to the hydroxylated intermediate. It is by virtue of this explicit mechanism that polycyclic aromatic hydrocarbons (PAH) are carcinogenic.

Problem 1

The mutagenic effect of UV light is:
a. the alteration of cytosine bases to adenine bases;
b. the formation of purine dimers that interfere with genetic expression;
c. the breaking of the sugar-phosphate backbone of the DNA molecule;
d. the formation of pyrimidine dimers that disrupt DNA replication;
e. the deletion of thymine bases along the DNA molecule.

Problem 2

Discuss three ways that alterations in DNA structure can be repaired.

THEORIES OF MOLECULAR BONDING II

QUANTUM MECHANICAL BASED THEORIES OF COVALENT BONDING

11

"If you want to have good ideas, you must have many ideas. Most of them will be wrong, and what you have to learn is which ones to throw away."

—Linus Pauling

FIGURE 11.1 In green plants, the manganese-containing oxygen-evolving complex (OEC) in Photosystem II (blue ribbons) plucks electrons from water molecules. The electrons hop to a tyrosine (Y_z, yellow), a chlorophyll (P680, green), a pheophytin (Pheo, orange), and two quinones (Q_A and Q_B, magenta), not to mention an iron atom (Fe), and eventually cross the chloroplast's membrane (gray) to a second light-absorbing system, Photosystem I, which reunites them with their protons.

Framework

The *water splitting reaction*, termed by Professor Harry Gray of California Institute of Technology as the most important reaction in nature and for the future of the planet:

$$2H_2O \rightarrow 4H^+ + 4e^- + O_2$$

appears to be a reasonably simple reaction. Yet to effectively execute this reaction to both initiate the release of oxygen to support life on the planet and to provide the chemical energy needed to sustain the supply of organic material to create the biological world as we know it, nature had to develop *photosynthesis*. To understand photosynthesis, we must develop a more sophisticated theory of molecular bonding based on quantum mechanics that explicitly links *molecular orbitals* with the specific electrons that occupy those orbitals.

Photosynthesis is, by almost any measure, one of the chemical miracles of life. It is responsible for the conversion of photons of light from the sun that supply energy to convert two

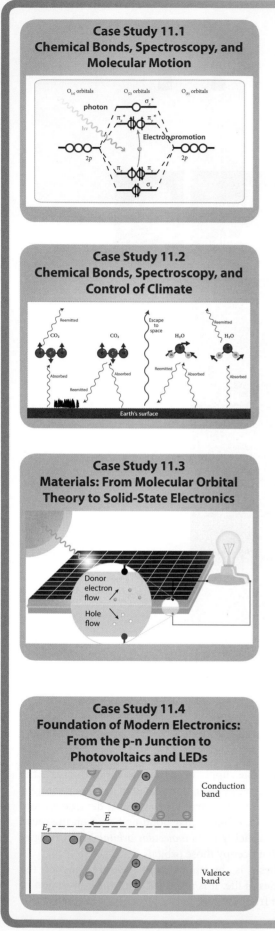

Case Study 11.1
Chemical Bonds, Spectroscopy, and Molecular Motion

Case Study 11.2
Chemical Bonds, Spectroscopy, and Control of Climate

Case Study 11.3
Materials: From Molecular Orbital Theory to Solid-State Electronics

Case Study 11.4
Foundation of Modern Electronics: From the p-n Junction to Photovoltaics and LEDs

inorganic substances, H_2O and CO_2, into virtually all the organic material we see in the world around us. Yet as we shall see, early research in photosynthesis provided remarkable insight into chemical processes in a far more general sense with very simple experiments. A full understanding of photosynthesis is a major research area today motivated by the potential of photosynthesis to supply a significant component to sustainable energy. Remarkably, we still don't understand fundamental aspects of photosynthesis, but the use of femtosecond lasers and the most advanced quantum mechanical calculations hold the promise of solving one of the mysteries of nature. When we come to understand the photosynthesis system, one of nature's engineering marvels, we will gain considerable insight into how quantum mechanics links to the macroscopic world. Quantum mechanics also determines how CO_2 and H_2O in the atmosphere control climate as described in Case Study 11.1.

The water splitting reaction is not only the most important single reaction in nature; it is also the explicit link between molecular level quantum mechanics and global energy production, storage and distribution. Figuring out a way to execute the water splitting reaction at the molecular level is one of the great challenges for chemistry and physics in the coming decades. We thus frame the problem of photosynthesis.

Early experiments in photosynthesis were some of the most important in the history of chemical understanding. As early as the mid 1600s, the Flemish physician Baptista van Helmont carried out experiments wherein he studied the mass acquired by a young willow tree as it grew, the mass of the soil within which it grew, and the mass of water added to the soil over five years. After 5 years the mass of the willow tree had increased by 75 *kilograms*, but the soil had lost only 56 *milligrams*. Van Helmont concluded that the tree did not constitute its organic structure from what it extracted from the soil. He did, however, incorrectly conclude that the bark, wood, roots, and leaves come from the water alone, which he had added over those five years. We now know, of course that it was the CO_2 extracted directly from the atmosphere that combined with H_2O to form the hydrocarbon molecular structure of the plant.

In the late 1700s, Joseph Priestley carried out a series of famous experiments in which he first placed a burning candle in a closed chamber and it extinguished quickly. He repeated the experiment but included a mint plant in the chamber that was placed in the sunlight. Again the candle extinguished. But after a few days, he could *relight* the candle in the chamber with the mint plant, but not in the chamber without the mint plant. At the time Priestley concluded that plants restore to the air whatever the burning candle removed. These experiments not only demonstrated that photosynthesis releases O_2, the experiments demonstrated this *before* O_2 was identified.

In the late 18th century and early 19th century a series of observations were made. Jan Ingenhousz, a Dutch physician, immersed green plants in water and discovered that when illuminated with light, they released bubbles of oxygen. Next, Jean Senebier, a Swiss botanist, discovered that CO_2 is a necessary ingredient for plant growth. Nicolas-Théodore de Saussure, a Swiss chemist, demonstrated that the other required ingredient was indeed water.

With the developments in the latter part of the 19th century, particularly the understanding of Gibbs free energy, it became well understood that photosynthesis was the reaction of CO_2 and H_2O, in the presence of sunlight, to form hydrocarbon compounds, represented by the critically important molecule glucose:

$$6CO_2 + 6H_2O + \text{light energy} \rightarrow C_6H_{12}O_6 + 6O_2$$

The change in Gibbs free energy for this reaction is (a very large):

$$\Delta G = +2870 \text{ kJ}$$

What is remarkable about the natural photosynthesis mechanism is that the solar photon is captured and its energy is carefully marshaled to achieve a very sophisticated objective, that of chemical synthesis. That is, the energy of the solar photon is harnessed to build very complicated organic molecules from the *inorganic* molecules CO_2 and O_2. Just to emphasize this point we can diagram photosynthesis from the perspectives of the energy transformations that are possible. We know three possibilities for the fate of the energy contained in a solar photon after being absorbed by chlorophyll: (1) that photon can be reemitted as a photon of the same or lower energy; (2) that photon can be absorbed by the plant and converted to an increase in the internal energy of the plant material—the conversion of electromagnetic energy to heat; or (3) the desired objective, and by for the most complicated, is to capture the energy of that photon and couple it into the mechanism of molecular synthesis. Those possible pathways are displayed in Figure 11.2.

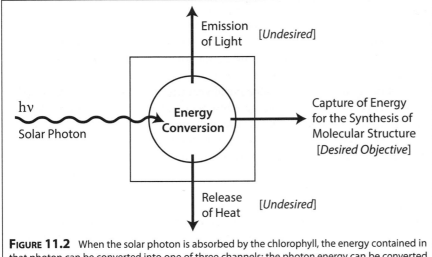

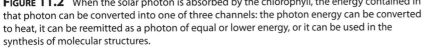

FIGURE 11.2   When the solar photon is absorbed by the chlorophyll, the energy contained in that photon can be converted into one of three channels: the photon energy can be converted to heat, it can be reemitted as a photon of equal or lower energy, or it can be used in the synthesis of molecular structures.

The mechanism of photosynthesis can be divided into two stages. The first stage involves the absorption of photons from the sun to produce the O_2 and the energy containing molecules such as ATP and NADPH (nicotinamide adenine dinucleotide phosphate). The second stage, the Calvin cycle, involves the coupling of energy from ATP and NADPH to synthesize an array of organic structures that sustain the plant. We focus here on the reactions that involve photons from the sun to create ATP.

The conversion of solar photons into usable chemical energy begins with the light gathering ability of the chlorophyll molecule displayed in Figure 11.3. An understanding of the molecular bonding structure of chlorophyll requires an understanding of the quantum mechanical bonding theories presented in this chapter. The absorption of a photon of light by chlorophyll initiates a series of energy conversions beginning with the removal of an excited (high energy) electron produced within the chlorophyll molecule by the absorption of the photon from the sun, and the transfer of that high energy electron to another molecule where the electron is more stable. When this transfer of the high-energy electron to another pigment molecule occurs, the energy contained in the electron is successfully captured because the electron does not readily drop to a lower energy level releasing that energy as heat or light as diagrammed in Figure 11.2.

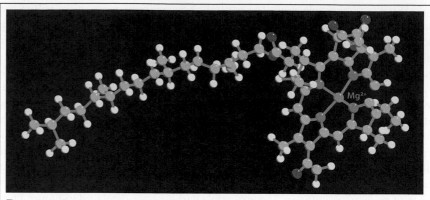

FIGURE 11.3 The structure of the chlorophyll molecule. The segment of the molecule containing the Mg²⁺ center that absorbs light occupies the right-hand segment of the molecule as shown.

The chlorophyll molecule shown in Figure 11.3 contains a structure called a porphyrin ring that forms a molecular cage around the Mg^{2+} magnesium cation that occupies the right-hand sector of Figure 11.3. A detail of the Mg^{2+} porphyrin is shown in Figure 11.4. Within the porphyrin ring an electron can follow a pathway wherein it can *delocalize* among a number of atoms in the structure, providing a temporarily stable holding site for the high-energy electron. How do molecular bonding structures allow an electron to be captured? An understanding of this requires the development of molecular orbital theory in this chapter.

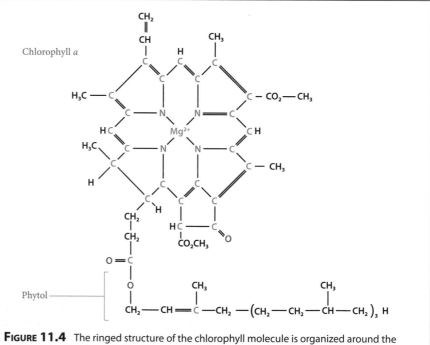

FIGURE 11.4 The ringed structure of the chlorophyll molecule is organized around the Mg2+ center. The structure is categorized by the five central 5-atom rings.

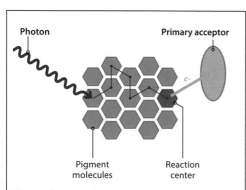

FIGURE 11.5 The energy contained in the excited electron created by the absorption of a solar photon is passed sequentially along a chain of pigment molecules by resonance energy transfer (RET) to the reaction center.

The energy contained in the excited electron is transferred among pigment molecules by a process called resonance energy transfer (RET) until it reaches a particular chlorophyll called P680, which leaves that pigment molecule in an excited state designated P680*, which is at the site of the reaction center shown in Figure 11.5. The high energy electron in P680* is transferred to the primary electron acceptor, as shown schematically in Figure 11.5, where it is stable. This step oxidizes P680* so that it becomes P680⁺, the cation. A keypoint is that the transfer of an electron from P680* to the primary electron acceptor is *remarkably fast*. It occurs in less than a few picoseconds (10^{-12} seconds). Because this transfer of the electron occurs so quickly, the excited electron does not have sufficient time to release its energy in the form of heat or light.

Now a very important step in photosynthesis occurs. A low energy electron from the structure of H_2O is transferred to P680$^+$ to reduce it to P680 as shown in Figure 11.6 resulting in the "splitting" of water

$$2H_2O \rightarrow 4H^+ + 4e^- + O_2$$

that releases 4H$^+$ into the thylakoid interior of the cytoplast and O_2 into the atmosphere. The 4 electrons are balanced by the reaction

$$4P680^+ + 4e^- \rightarrow 4P680$$

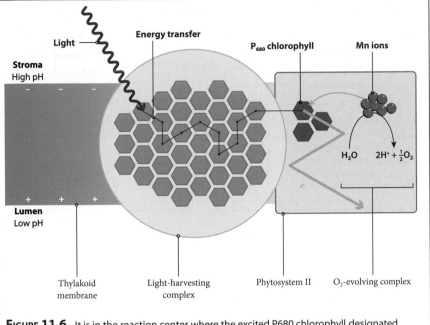

FIGURE 11.6   It is in the reaction center where the excited P680 chlorophyll designated P680* transfers an electron to the electron acceptor forming P680$^+$. It is the P680$^+$ cation that oxidizes water, splitting H_2O into $2H^+ + 2e^-$ and O in the water splitting reaction $2H_2O \rightarrow 4H^+ + 4e^- + O_2$.

The release of the electron from P680* to the primary electron acceptor *and* the splitting of water to return P680$^+$ to P680 occurs in the "reaction center" of Photosystem II, which is a protein complex that is the initial step in converting *photon energy* from the sun into *chemical energy* in the plant. This protein complex is the only known protein complex that can oxidize water, resulting in the release of O_2 to the atmosphere.

While we will return to photosynthesis as a key contextual structure for understanding the most important principles in chemistry, we complete this Framework by noting that the chemical energy captured from the separation of H$^+$ and e$^-$ in the water splitting reaction is used to synthesize the molecule adenosine triphosphate (ATP) in the thylakoid membrane. The structure of this molecule is shown in Figure 11.7.

Cells have the capacity to extract energy from molecules (primarily sugars) through a coupled series of reactions that release energy in a controlled sequence that can either release heat directly or control that energy to be used for muscular control, nerve function, etc. In almost all organisms this transfer of energy along chains of coupled reactions is accomplished by ATP. The energy content of ATP that is readily transferred is in the phosphate subunit of the molecule, highlighted in red in Figure 11.7a. The conversion of ATP to adenosine diphosphate, or ADP, shown in Figure 11.7b, occurs in the reaction with water

$$ATP \xrightarrow{\text{H}_2\text{O}} ADP + energy$$

and energy is released in going from two phosphate-phosphate bonds to one phosphate-phosphate bond.

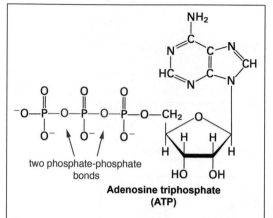

FIGURE 11.7A One of the most important molecules used in nature to control the flow and the coupling of energy within living organisms is adenosine triphosphate, or ATP. The containment of available energy for biological systems is in the double bond of the phosphate groups linked to the adenosine structure.

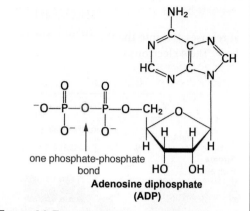

FIGURE 11.7B With the elimination of one of the high energy phosphate bonds, ATP is converted to adenosine diphosphate, or ADP which, as a consequence, contains less chemical energy than does ATP.

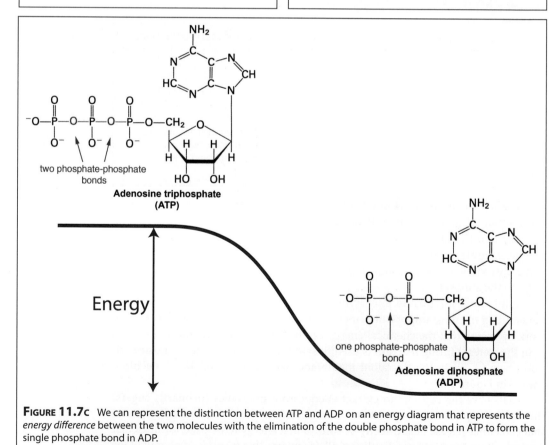

FIGURE 11.7C We can represent the distinction between ATP and ADP on an energy diagram that represents the *energy difference* between the two molecules with the elimination of the double phosphate bond in ATP to form the single phosphate bond in ADP.

Inspection of Figure 11.7a reveals that an ATP molecule has two phosphate-phosphate bonds, which are often written in short hand form as Ap~p~p where the red highlighted bonds correspond to those in Figure 11.7a. Inspection of Figure 11.7b reveals that an ADP molecule has one phosphate-phosphate bond.

We can represent the relationship between ATP and ADP on an energy diagram, a "reaction coordinate," just as we did in Figure 1.11 for the reaction $2H_2 + O_2 \to 2H_2O$; but retaining the molecular structure.

If the phosphate group in ATP were to break in this reaction with water in a single step, which we can write in short hand by representing the high energy phosphate bonds as p~p, then the reaction may be written

$$Ap{\sim}p{\sim}p + H_2O \to Ap{\sim}p + P_i + H^+ + \text{energy}$$

where P_i is an inorganic phosphate and H^+ is a proton. In this case, the energy difference displayed in Figure 11.7c would appear as heat—an increase in the disorganized energy of motion contained in the product molecules ADP, phosphate, etc. This disorganized energy is contained in the kinetic energy of motion of the products, ADP, P_i, etc. and is rapidly transferred to the surrounding solution and appears as an increase in temperature reflecting the release of energy contained in the bond structure of ATP.

However, nature does not typically use the energy difference between ATP and ADP in this simple, uncontrolled way. Rather, within the cellular structure of organisms, nature uses the energy difference between ATP and ADP to drive controlled reactions that convert CO_2 and H_2O to sugars—most notably glucose $C_6H_{12}O_6$.

We can represent this coupled set of reactions (the details of which will emerge in the chapters that follow) in terms of the energy *released* from the conversion of ATP to ADP and the energy *required* to form glucose from carbon dioxide and water in a single diagram, displayed in Figure 11.8. Specifically how chemical reactions couple together to use the available energy from one reaction to "drive" reactions that form a desired product is a key part of the study of thermochemistry.

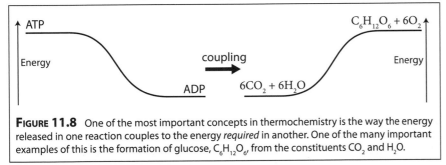

FIGURE 11.8 One of the most important concepts in thermochemistry is the way the energy released in one reaction couples to the energy *required* in another. One of the many important examples of this is the formation of glucose, $C_6H_{12}O_6$, from the constituents CO_2 and H_2O.

We can also write the coupling of the reaction displayed in Figure 11.8 in shorthand form as

$$6CO_2 + 6H_2O \xrightarrow{\quad ATP \quad ADP \quad} C_6H_{12}O_6 + 6O_2$$

which indicates that the *energy* to synthesize the bonds in glucose, $C_6H_{12}O_6$, is supplied by the energy difference between ATP and ADP.

A critically important product of photosynthesis is ATP, formed from ADP using the energy contained in a solar photon. Plants then use that energy difference between ATP and ADP to construct hydrocarbon molecules for the *structure* of the plant (such as lignin or cellulose) and for the feeding of the plant (such as glucose).

Animals consume the products of plant photosynthesis and "burn" them for energy in the process of *respiration*; using glucose as our prime example

$$C_6H_{12}O_6 + 6O_2 \to 6CO_2 + 6H_2O + \text{energy}$$

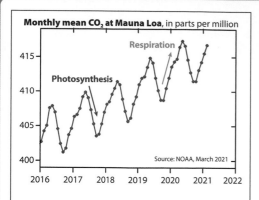

FIGURE 11.9 We can examine the observations of CO_2 from Mauna Loa, Hawaii, which reveal the remarkable signature of photosynthesis and respiration as a function of season in the northern hemisphere. In the spring, CO_2 decreases as it is drawn from the atmosphere to supply carbon to the foliage developing with the season. In the fall, CO_2 is returned to the atmosphere when that foliage dies and is oxidized back to CO_2 and H_2O in the process of respiration.

But the energy in that reaction can either be released as heat, to maintain our body temperature at 38°C, or can be coupled to synthesize ATP from ADP within the cells of animals in the generalized reaction

$$C_6H_{12}O_6 + 6O_2 + 38P_i + 38ADP \rightarrow 6CO_2 + 6H_2O + 38ATP$$

This reaction, driven by the energy contained in the bonding structure of glucose, is found in all higher plant and animal cells. Thus the overall result of glucose *respiration*

$$C_6H_{12}O_6 + 6O_2 \rightarrow 6CO_2 + 6H_2O$$

is the exact reverse of the *photosynthesis* reaction in which glucose is formed:

$$6CO_2 + 6H_2O \rightarrow C_6H_{12}O_6 + 6O_2.$$

Respiration and photosynthesis are the two major processes constituting the *carbon cycle* in nature: sugars and oxygen produced by plants are the raw materials for respiration and the generation of ATP by plant and animal cells alike; the end products of respiration, CO_2 and H_2O, are the raw materials for the photosynthetic production of sugars and oxygen. The only net source of energy in the cycle is sunlight.

The scale of energy trapped per unit time by photosynthesis worldwide is remarkable—approximately 100×10^{12} watts or 100 TW. Remember that the power required to sustain the human endeavor worldwide is 15 TW, so photosynthesis requires about six times the power consumed by the global economic system. But in addition to the energy captured by photosynthesis, the *mass* of organic material created *each year* from, CO_2 and H_2O in the atmosphere by photosynthesis is 100×10^9 tonnes (100 Gt).

We can directly observe the scale of the impact of photosynthesis by looking at the seasonal record of CO_2 in the atmosphere obtained from Mauna Loa, Hawaii, displayed in Figure 11.9. The seasonal *decrease* in CO_2 in the spring is a direct result of carbon dioxide extraction from the atmosphere by plants as their foliage and growth progress into summer. The seasonal *increase* in CO_2 in the fall reflects the oxidation of plant material after it dies with the onset of cold temperatures.

A key strategy in the development of alternative, renewable energy sources is the development of biofuels that use the organic materials formed in photosynthesis to produce steam for heating or driving turbines to generate electricity. A major debate resides around the competition for organic material as a food to supply humans and farm animals worldwide versus the use of biomass as a fuel.

We can summarize the cycle linking ATP and ADP in Figure 11.10 which emphasizes that the sole primary source of energy creating ATP from ADP is the energy contained in solar photons. However, that energy difference between ATP and ADP is used by nature to execute broad categories of objectives critical to life on the planet.

The developing technology of photovoltaic cells that convert photons from the sun to electric power is described in Case Study 11.2.

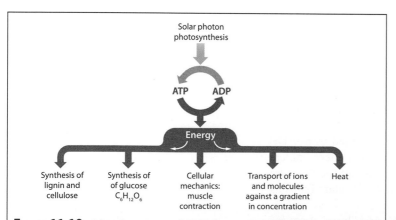

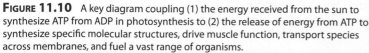

FIGURE 11.10 A key diagram coupling (1) the energy received from the sun to synthesize ATP from ADP in photosynthesis to (2) the release of energy from ATP to synthesize specific molecular structures, drive muscle function, transport species across membranes, and fuel a vast range of organisms.

Chapter Core

Road Map for Chapter 11

The quantum mechanical formulation of covalent bonding constitutes a cornerstone of modern chemistry. In this chapter we develop molecular bonding in the context of quantum mechanics using two complementary approaches. First we develop the valence bond (VB) formulation of molecular bonding that combines atomic orbitals (and the electrons associated with two atomic orbitals) to form bonding structures with the electrons assigned to those atomic orbitals. In order to achieve agreement with observed molecular structures, the atomic orbitals used to create the molecular bonds are often "hybridized" by a judicious combination of atomic orbitals *before* creating the overlap of those atomic orbitals to form the molecular bond. Second, we develop the molecular orbital (MO) theory of chemical bonding that takes linear combinations of atomic orbitals to form new molecular orbitals that are distributed such as to reduce the potential energy of the molecular structure. Electrons are inserted into those new MOs in order of their ascending energy according to the Pauli Exclusion Principle.

Road Map to Core Concepts	
1. Formation of a Molecular Bond from Atomic Orbitals	
2. Mixing of Atomic Orbitals to Form Hybrid Orbitals	
3. Different Combinations of Atomic Orbitals Yield Different Geometries of the Hybridized Orbitals: the sp^2 and sp Examples	
4. Hybridization of More Than Two Atomic Orbital Types	
5. Molecular Orbital (MO) Theory	
6. Combining Atomic Orbitals To Form Bonding and Antibonding Molecular Orbitals	
7. Molecular Orbital Structure of Molecular Oxygen	
8. Molecular Orbital Structures of Heteronuclear Molecules	
9. Molecular Orbital Bonding Structure of Benzene: Delocalization	

Valence Bond Theory: Orbital Overlap and the Name of the Chemical Bond

We have now come to a significant juncture in molecular bonding theory—the point at which we must join the quantum mechanical picture of atomic orbitals with empirically based approaches that guide our intuition concerning the molecular structure of lowest potential energy: Lewis structures and VSEPR theory. The thread running through the Lewis and VSEPR models of representing the bonding structure of molecules is the abiding drive of electrons in any assemblage of atoms to "find" the configurations of lowest potential energy.

We have developed in Chapter 8 a facility with quantum mechanics to the extent that we know that electrons in isolated atoms are described by s, p, d, and f orbitals, which are *wavefunctions;* solutions to the Schrödinger wave equation. For each of those wavefunctions there is a corresponding energy representing the energy of that electron in the specific orbital. But now we wish to make the leap from atomic structure to molecular structure:

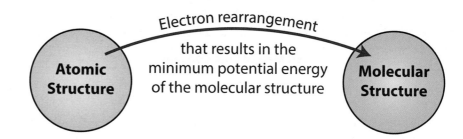

But when we turn to how *molecules* are constructed from the combination of those atomic orbitals, we are confronted with the question: How does the ensemble of electrons and protons that constitutes a molecular structure minimize the potential energy of the ensemble? Or more pointedly, how can we determine the bonding structure that represents the minimum energy configuration?

While Lewis structures appear on the surface to be just a set of rules for the placements of electrons in a molecular architecture, what in fact Lewis developed was a way to find the molecular structure of lowest potential energy. Of course, since all molecules seek to find their potential energy surface of minimum energy, what Lewis created was an electron pair bonding arrangement guided by countless examples of molecules with structures resulting directly from minimizing their potential energy. The technique of *formal charges* applies to the cases where there is ambiguity in the Lewis diagram leading to the determination of the true minimum energy. The method of formal charges attempts to resolve the ambiguity over which molecular structure is of minimum energy. But Lewis structures are, by nature, two-dimensional so insight into the three-dimensional structure is advanced by the use of the valence shell electron pair repulsion (VSEPR) approach detailed in Chapter 10. As we will see, this overview of the Lewis structure to determine the electron configuration of minimum energy (aided in some cases by analysis of formal charge) in combination with the VSEPR picture of the *geometry* of the molecular bonding structure, plays a surprisingly important role in the application of quantum mechanics to molecular bonding.

But we need a theory of bonding that recognizes the central role of quantum mechanics, of the wave properties of electrons, in order to explain *why* the structures deduced from Lewis theory and VSEPR theory are indeed the ones of lowest potential energy. It is important to realize that while a number of our most important molecules have tetrahedral or trigonal bipyramid structures, *none* of the s, p, or d orbitals are oriented such that they could rationalize such geometries.

This is a serious problem if we are to use just atomic orbitals that are a solution to the Schrödinger equation for isolated atoms to describe molecular bonding. In addition to molecular bonding geometry, for modern chemical applications we must also develop an understanding of the spectroscopic behavior of molecules and the magnetic properties of molecules. None of those critically important aspects of molecular bonding emerge from non-quantum mechanical theories.

First, we examine the critical role that Linus Pauling played in the union of Lewis structures with the emerging development of quantum mechanics as it was applied to *molecular* structures. Pauling, it turns out, was the unifier of two major, yet unreconciled, communities within the domain of chemical thought when quantum mechanics burst on the scene in 1925. A major component of chemical reasoning had just realigned with the Lewis structure formulation in the mid-1920s. The chemistry community was using it to great advantage in bringing coherence to an otherwise disjointed myriad of chemical facts that had characterized the field for decades. When the Schrödinger equation emerged as a profoundly successful solution to the hydrogen atom, and was at least a reasonable approximation to atoms with a few electrons, the question immediately arose: If the Schrödinger equation works for atoms, does it also work for molecules?

The next year, 1927, Heitler and London published their work on the quantum mechanics of molecular bonding using H_2 as the prototype molecule. Heitler and London built their model of the H_2 molecule arguing by simple analogy. As we have seen, the Schrödinger wave equation is written as

$$\hat{H}\Psi_n = E_n \Psi_n$$

where Ψ_n is the wavefunction for the electron, $\hat{H}$ is a mathematical operator that represents the sum of the kinetic and potential energies of the electron, and the energy of the atomic levels are given by the values of E_n.

The *wavefunctions*, Ψ_n, for the square well potential from Chapter 8 are

$$\Psi_n(x) = \sqrt{\frac{2}{L}} \sin\left(\frac{n\pi x}{L}\right)$$

and the wave equation for which these wavefunctions are a solution is

$$\frac{d^2\Psi}{dx^2} = -\left(\frac{2\pi}{\lambda}\right)^2 \Psi$$

To obtain the Schrödinger equation, we substitute for the de Broglie wavelength of the electron

$$\lambda_{el} = h/p_{el}$$

and we have

$$\frac{d^2\Psi}{dx^2} = -\left(\frac{2\pi p}{h}\right)^2 \Psi$$

But

$$E_{kinetic} = p^2/2m$$

So

$$\frac{d^2\Psi}{dx^2} = -\frac{4\pi^2 p^2}{h^2}\Psi = -\frac{4\pi^2 2mE_k \Psi}{h^2}$$

Or

$$-\frac{\hbar^2}{2m}\frac{d^2\Psi}{dx^2} = E_k\Psi$$

This is the Schrödinger equation for electrons in a potential for which the potential energy $V(x)$ is zero for $0 < x < L$ and is infinite everywhere else.

But if the electron is moving in a potential that is not zero, we can generalize the Schrödinger equation to include the potential energy term by recognizing that

LINUS PAULING

in the Schrödinger equation

$$H\Psi_n = E_n \Psi_n$$

the Hamiltonian H is the sum of the kinetic and potential energy and that the E_n are the eigenvalues for the *total* energy of the electron. Thus, we can rewrite our equation for the electron in the square well potential by adding the potential energy terms to the kinetic energy term:

$$-\frac{\hbar^2}{2m}\frac{d^2\Psi}{dx^2} + V(x)\Psi(x) = E_{total}\Psi(x)$$

However, if we are considering an electron moving in a 3-dimensional potential, then we need to extend the one dimensional Schrödinger equation to three dimensions:

$$-\frac{\hbar^2}{2m}\left[\frac{\partial^2\Psi}{\partial x^2} + \frac{\partial^2\Psi}{\partial y^2} + \frac{\partial^2\Psi}{\partial z^2}\right] + V(x,y,z)\Psi(x,y,z) = E_{total}\Psi(x,y,z)$$

The Coulomb potential is the potential between an electron and a nucleus with charge Z as a function of distance, r, between the two charged bodies:

$$V(r) = -Z\,e^2/r$$

The dependence on r suggests that we convert from Cartesian coordinates to polar coordinates. Substitution of the values for x, y, and z in polar coordinates transforms the Schrödinger equation to

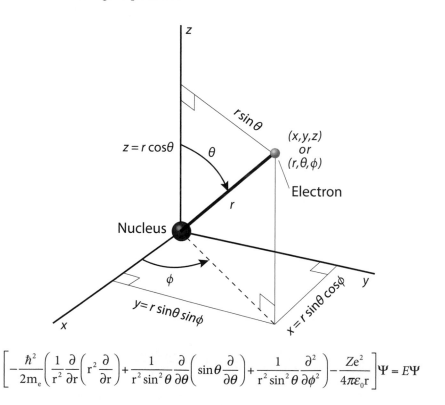

$$\left[-\frac{\hbar^2}{2m_e}\left(\frac{1}{r^2}\frac{\partial}{\partial r}\left(r^2\frac{\partial}{\partial r}\right) + \frac{1}{r^2\sin^2\theta}\frac{\partial}{\partial\theta}\left(\sin\theta\frac{\partial}{\partial\theta}\right) + \frac{1}{r^2\sin^2\theta}\frac{\partial^2}{\partial\phi^2}\right) - \frac{Ze^2}{4\pi\varepsilon_0 r}\right]\Psi = E\Psi$$

What Heitler and London did was to advance this treatment to the molecular case by rewriting the potential energy term to include the interaction of the electrons with themselves *and* the two electrons with the two hydrogen nuclei. This can be done for the case of two H atoms with the two electrons centered on their individual nuclei as shown in Figure 11.11.

11.12

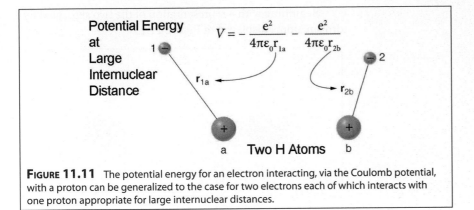

FIGURE 11.11 The potential energy for an electron interacting, via the Coulomb potential, with a proton can be generalized to the case for two electrons each of which interacts with one proton appropriate for large internuclear distances.

In this case the potential energy of the systems involves just two terms, electron 1 moving in the Coulomb potential of nucleus a and electron 2 moving in the Coulomb potential of nucleus b.

However, as we move the two nuclei close together, we introduce a new combination of interactions that includes the interaction of electron 1 with both nuclei a and b, the interaction of electron 2 with both nuclei a and b, and the interaction of the two nuclei and the two electrons. This yields a potential energy term that contains six terms as Figure 11.12 shows.

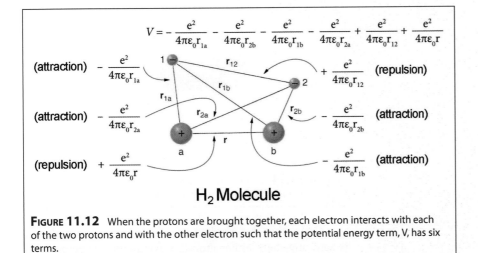

FIGURE 11.12 When the protons are brought together, each electron interacts with each of the two protons and with the other electron such that the potential energy term, V, has six terms.

Writing down a potential energy *function* is rather straight forward. The more challenging step is to write down a wavefunction that when substituted into the Schrödinger equation yields a realistic potential energy surface for the H_2 molecule.

Here, Heitler and London made a first approximation that the wavefunction for the *molecule* would have some mathematical resemblance to the wavefunction for the atom. As an initial guess, they constructed the H_2 wavefunction as the *product* of (1) the wavefunction for electron 1 in the Coulomb field of nucleus a, $\psi_{1s_a}(r_1)$ and (2) the wavefunction for electron 2 in the Coulomb field of nucleus b, $\Psi_{1s_b}(r_2)$

$$\Psi_{mol}(r_1, r_2) = \Psi_{1s_a}(r_1) \cdot \Psi_{1s_b}(r_2)$$

We can picture what this wavefunction would be like by superimposing (multiplying together) two 1s hydrogen atomic orbitals.

Beginning with a representation of the two separated hydrogen atoms displayed in Figure 11.13

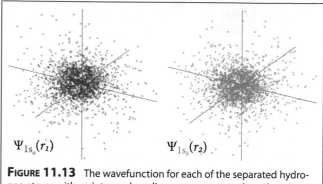

$\Psi_{1s_a}(r_1)$ $\Psi_{1s_b}(r_2)$

FIGURE 11.13 The wavefunction for each of the separated hydrogen atoms with an internuclear distance great enough so there is no overlap of the individual wavefunctions.

that represents the wavefunction for the separated atoms, we can decrease the internuclear distance to form the molecular wavefunction from the *product* of the two atomic wavefunctions as shown in Figure 11.14.

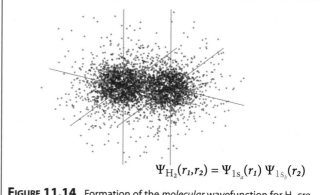

$\Psi_{H_2}(r_1,r_2) = \Psi_{1s_a}(r_1)\,\Psi_{1s_b}(r_2)$

FIGURE 11.14 Formation of the *molecular* wavefunction for H_2 created from the hydrogen atom wavefunctions by taking the product.

Notice that this product wavefunction for the molecule is zero anywhere either atomic wavefunction is zero and it tends to a maximum along the internuclear axis in the region between the nuclei. This would tend to reduce the nuclear proton-proton repulsion and enhance the electron-proton attraction such that there is the possibility of forming a stable bond—a minimum in the potential energy curve for the molecule.

When this guess for the mathematical form for the wavefunction of molecular H_2, $\Psi_{mol}(r_1, r_2)$, was inserted into the Schrödinger wave equation, the calculated energy as a function of internuclear distance did indeed have a minimum, but it was a minimum far shallower than the experimentally determined depth of the potential energy surface for H_2 obtained by spectroscopic evidence. This comparison is displayed in Figure 11.15.

Heitler and London then recast the mathematical structure of the molecular wavefunction by hypothesizing that if the Lewis electron pair bond represented a true *sharing* of the

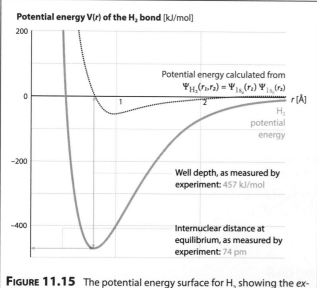

Potential energy V(r) of the H₂ bond [kJ/mol]

Potential energy calculated from
$\Psi_{H_2}(r_1,r_2) = \Psi_{1s_a}(r_1)\,\Psi_{1s_b}(r_2)$

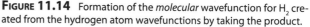

H₂
potential
energy

Well depth, as measured by
experiment: 457 kJ/mol

Internuclear distance at
equilibrium, as measured by
experiment: 74 pm

FIGURE 11.15 The potential energy surface for H_2 showing the *experimentally observed* potential energy as a function of internuclear distance (solid line) and the *calculated* potential energy surface using the H_2 *molecular* wavefunction from Figure 11.14.

two electrons, then the appropriate molecular wavefunction should reflect the *indistinguishable* character of the shared electrons in that bond. This could be represented mathematically by a molecular wavefunction that exchanged the electrons with respect to their originally identified nuclei. If $\Psi_{1s_a}(r_1)$ represented electron 1 within the Coulomb influence of nucleus a, then $\Psi_{1s_a}(r_2)$ would represent electron 2 within the Coulomb influence of nucleus a. Thus the molecular wavefunction for the case of both electrons shared by both nuclei would be

$$\Psi_{mol}(r_1, r_2) = \frac{1}{\sqrt{2}} \left[\Psi_{1s_a}(r_1)\,\Psi_{1s_b}(r_2) + \Psi_{1s_b}(r_1)\,\Psi_{1s_a}(r_2) \right]$$

The $1/\sqrt{2}$ in front of the brackets is just to ensure that the probability of finding an electron anywhere in space is 1. This is called the "normalization" term.

When this wavefunction was substituted into the Schrödinger equation, the potential energy curve for H_2 exhibited a far greater well depth as shown in Figure 11.16.

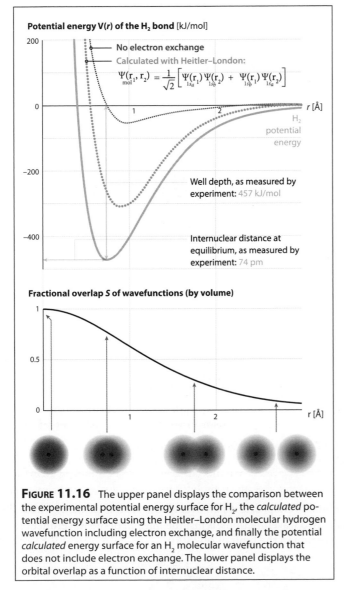

Potential energy V(r) of the H₂ bond [kJ/mol]

— No electron exchange
— Calculated with Heitler–London:
$$\Psi_{mol}(r_1, r_2) = \frac{1}{\sqrt{2}} \left[\Psi_{1s_a}(r_1)\,\Psi_{1s_b}(r_2) + \Psi_{1s_b}(r_1)\,\Psi_{1s_a}(r_2) \right]$$

H₂ potential energy

Well depth, as measured by experiment: 457 kJ/mol

Internuclear distance at equilibrium, as measured by experiment: 74 pm

Fractional overlap S of wavefunctions (by volume)

FIGURE 11.16 The upper panel displays the comparison between the experimental potential energy surface for H₂, the *calculated* potential energy surface using the Heitler–London molecular hydrogen wavefunction including electron exchange, and finally the potential *calculated* energy surface for an H₂ molecular wavefunction that does not include electron exchange. The lower panel displays the orbital overlap as a function of internuclear distance.

This was a major breakthrough in the development of quantum mechanics in pursuit of a quantitative understanding of the chemical bond. While the Heitler–London bond strength of approximately 3.2 eV was considerably less than the observed 4.75 eV, it was also clear that this very approximate molecular wavefunction that represented the exchange or sharing of electrons between the nuclei contained many of the essential aspects of a shared electron bond.

Of particular importance was the fact that this rewriting of the molecular wavefunction that included the *exchange* of electrons between the two nuclei was an entirely quantum mechanical phenomenon. It has no classical analog because it could not be described by an electrostatic interaction among electrons or protons. Recall that the potential energy term in the Schrödinger equation was the same for both Figure 11.15 and Figure 11.16—only the wavefunction was changed. This formulation of the wavefunction that allowed the electron to spread, to delocalize, across the molecule by allowing the exchange or sharing of the electron pair in the bond lowers the energy of the ensemble of electrons and protons.

A key aspect of the formulation of the wavefunction for the molecular bond from the individual atomic wavefunctions was the emphasis placed on the importance of the *overlap* of the atomic orbitals intrinsic to the molecular bond. Electrons can only be exchanged between atoms and shared in pairs, if their orbitals overlap in a region of space, as shown in Figure 11.16. The cumulative overlap of two waves of two wavefunctions, is referred to as the *orbital overlap*, s, and it is quantitatively represented as the sum of all overlap contributions, volume element by volume element as shown in Figure 11.17.

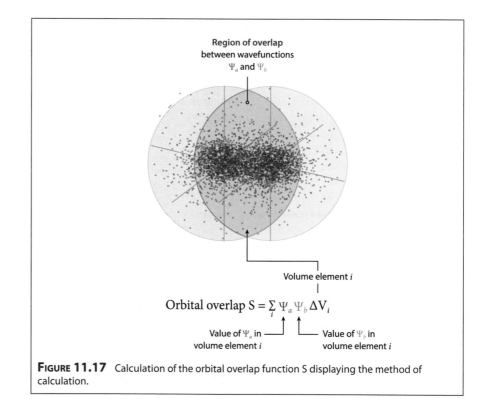

FIGURE 11.17   Calculation of the orbital overlap function S displaying the method of calculation.

The orbital overlap, S, as a *sum* over all the *volume elements* within the overlap region of the two wavefunctions, is just the *integral* of the wavefunction taken over the overlap region in the limit of small Δv:

$$S = \int_{overlap} \Psi_a \Psi_b \, dv$$

The value of this orbital overlap, S, is shown as a function of internuclear distance for the H_2 molecule in the lower panel of Figure 11.16. Figure 11.16 emphasizes that as the *orbital overlap* between two atomic orbitals *increases*, the exchange contribution increases the depth of the potential energy surface (and the strength of the chemical bond) until the repulsive force of the two nuclei becomes dominant at small internuclear distance and the potential energy surface increases.

Pauling, while still a very young man, recognized that:

1. Molecules exhibited similar discrete emission line spectra as atoms, although molecular spectra were considerably more complicated than atomic spectra.

2. The clear quantum mechanical implications of the Heitler–London electron exchange force for molecular wavefunctions included the sharing of electrons in a chemical bond.

3. The Pauli Exclusion Principle applied to electrons in the molecular orbitals as well as the atomic orbitals.

4. The *orbital overlap* was of key importance in understanding the structure and strength of the chemical bond.

These lines of evidence convinced Pauling that quantum mechanics was to play the central role in understanding the nature of the chemical bond, and he set out to formulate the union between the new quantum mechanics and the older insight brought by the Lewis formation of the shared electron chemical bond. Pauling proceeded to develop the valence bond (VB) model of the molecular bond based on the formulation of quantum mechanics using the orbital overlap of wavefunctions drawn directly from the atomic orbitals of the atoms involved in the molecular bond. In practice, Pauling proceeded to superimpose the orbitals drawn from quantum mechanics on the bonding architecture that emerged from the Lewis diagrams. When applied to a broad class of molecular systems, this revolutionized the understanding of molecular bonding.

The fundamental premise in what is termed the valence-bond (VB) theory of molecular bonding is that a covalent bond forms when the wavefunctions of two atoms of opposite spin *constructively* add in the region between two nuclei. Because the electron wavefunction has both an amplitude and a phase, the region where two atomic wavefunctions constructively add is commonly termed the "orbital overlap" region. From the analysis of the relationship between (1) the orbital overlap of two hydrogen atoms and (2) the potential energy surface defining the *decrease* in potential energy with *increasing* orbital overlap in Figure 11.16, we can deduce a very important fact: that with *increasing* overlap of two orbitals, the strength of the bond formed (the decrease in potential energy) *increases* until the proton-proton repulsion at very small internuclear distance begins to dominate the attraction of the two atoms that form the molecular bond.

Thus valence bond (VB) theory is based on the development of combinations of atomic orbitals that combine to form orbital overlap between two atoms such that a chemical bond is formed containing two electrons. We can extend the case of the H_2 bond formed from the overlap of two H atoms by considering the case of a 1s H atom orbital combining with the $3s^2 3p^5$ electron configuration of a Cl atom. The VB picture of this bond, shown in the top panel of Figure 11.18, would overlap the $1s^1$ orbital of H with the 3p orbital of Cl.

The valence bond picture of two Cl atoms bonding to form Cl_2 is shown in the bottom panel of Figure 11.18. In each case the half filled orbitals overlap to form an electron pair bond. For the case of HCl, the 1s electron of hydrogen combines with the half filled 3p orbital of Cl. For Cl_2, the half filled 3p orbital of one Cl atom combines with the half filled 3p orbital of the other Cl atom to form the overlap region containing an electron pair that constitutes the Cl-Cl bond. A key point to note is that the *constructive addition* of two electron wavefunctions overlap in space. Thus, because the 3s orbital of the Cl atom does not extend out from the nucleus far enough to overlap either the 1s orbital of H or the 3p orbital of Cl, the bond is formed in each case by the overlap with the 3p orbital of Cl that extends a considerable radial distance from the Cl nucleus.

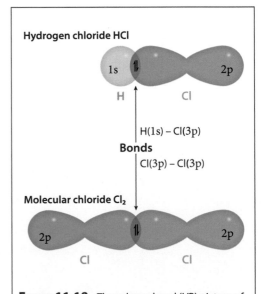

FIGURE 11.18 The valence bond (VB) picture of the molecular bond in HCl where the electron pair bond between H and Cl is shown in the top panel with the 1s orbital of H overlapping the 3p orbital of Cl. The VB picture of the Cl_2 bond showing the shared electron pair from the 3p orbital of each Cl atom shown in the bottom panel.

Based on the development of VB theory thus far, we have a remarkably simple quantum mechanical picture of molecular bonding: in VB theory, a covalent chemical bond is the overlap of protruding half-filled s, p, d, or f atomic orbitals that form electron pair bonds. Yet, what is fundamentally new is that we are treating electrons as waves—with the atomic orbitals explicitly forming the atom-atom bonds that constitute a molecule. Building on the idea that bonds are formed in molecules by the constructive overlap between orbitals of adjacent atoms, we can, in the simplest possible way, build molecular structures by combining atomic orbitals just as those orbitals are configured in the separated atoms! It is not at all obvious that when electron waves combine, or "mix," to form a molecular structure that they would "remember" the geometry of the wavefunction in the separated atoms, but in a large number of cases they do.

Molecular Shape and the Concept of Bond Hybridization

Remarkably, while many molecular structures appear to obey this simple picture of molecular bonds formed from the overlap of atomic orbitals just as they exist in the separated atoms, we can quickly identify some very important cases for which this approach fails.

Case 1: Methane, CH$_4$, is a molecule with tetrahedral geometry:

In methane, each of the C-H bonds is of equal length and the same strength. Each bond extends from the central atom (carbon) to the vertices of the tetrahedral and each H-C-H bond angle is 109.5°.

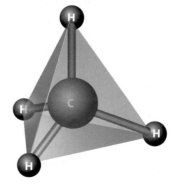

Case 2: Ethylene (C$_2$H$_4$) is a planar molecule:

Ethylene has four equivalent carbon-hydrogen bonds of bond length 109 pm and a carbon-carbon double bond of length 134 pm. Each HCH bond angle is 120°.

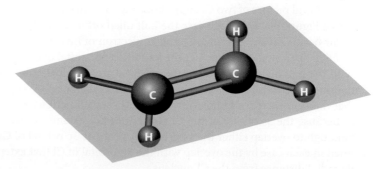

Case 3: Acetylene (C_2H_2) is a linear molecule with a triple bond between the carbon atoms:

The two carbon-hydrogen bonds are equivalent and have a bond length of 106 pm; the carbon-carbon triple bond has a bond length of 120 pm.

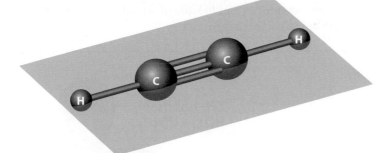

On the face of it, these examples present a significant problem for VB theory using the atomic orbitals of carbon, which have p orbitals occupying the three orthogonal axes p_x, p_y, and p_z. How can such a geometry be reconciled with a tetrahedral structure? With a planar geometry involving a carbon-carbon double bond? What is also mystifying is that we have three molecules, three different molecular geometries, yet one kind of atom, carbon, forming bonds with one other kind of atom, hydrogen, to produce three distinct structures. We know that if we deconstructed the methane, ethylene, and acetylene molecules into atoms we would have our $1s^2 2s^2 2p^2$ carbon atoms with orthogonal p orbitals and our 1s hydrogen atoms back again. But how can a carbon atom, which possesses a 2s and three 2p valence orbitals, make 4 indistinguishable molecular bonds in methane? Is the VB model, which is built on the assumption that we need simply overlap the atomic orbitals of isolated atoms to form molecules, inadequate to explain the simplest hydrocarbon molecule—methane?

To first order we know why the methane molecule rearranged its configuration—it was simply seeking the molecular geometry and electron distribution that would result in the lowest potential energy configuration. How would it achieve this configuration given only 2s and 2p orbitals to work with? Is this a violation of our VB model that must be built from atomic orbitals of the constituent atoms?

The key point is that the wavefunctions, the 2s and 2p orbitals of carbon, can *mix* in any manner that will reduce the potential energy of the *ensemble* of electrons and nuclei of the methane molecule. The electrons are not locked into the orbitals of the separated atoms. The electron wavefunctions can add together and reshape into any configuration that lowers the potential energy of the resulting molecular structure.

sp³ Hybridization and the Structure of Methane

If we combine one s orbital and three p orbitals of the central atom we create a new geometrical structure. We can represent this beginning with the electron configuration of carbon that corresponds to s and p *orbitals* of the *isolated carbon atom* and mix them, resulting in four *hybridized* orbitals of *carbon* in the *molecular structure* of *methane*, as shown in Figure 11.19.

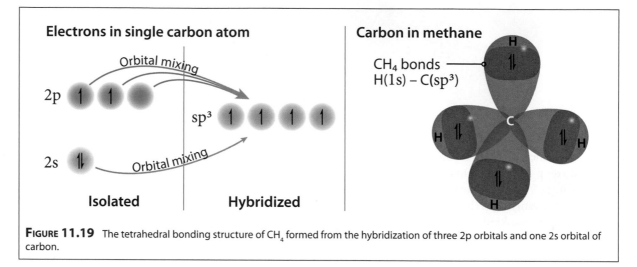

Figure 11.19   The tetrahedral bonding structure of CH_4 formed from the hybridization of three 2p orbitals and one 2s orbital of carbon.

Those four resulting hybridized orbitals of carbon are all of *equal energy*; they are each 25%:75% composites of s and p character with a geometry pointing to the 4 vertices of a tetrahedron. Each of these degenerate (equal energy) orbitals is designated as an sp³ hybrid orbital and, as the electron configuration in Figure 11.19 demonstrates, each contains a single unpaired electron. Thus the bonding structure of the central carbon atom in methane accepts the single 1s electron from H resulting in four electron pair bonds to form CH_4. This *mixing* of atomic orbitals of the separated *atoms* to form hybrid orbitals in the *molecule formed* from those atoms, explicitly demonstrates the wave nature of the electron that results in its ability to combine spatially in whatever manner is required to lower the potential energy of the assembled molecule.

This revelation that atomic orbitals can mix to create new geometrical arrangements in a molecular construction is the key concept linking (1) the development of the wave properties of the electron in Chapters 8 and 9 with (2) the deduced shape of molecules that emerges from the union of Lewis theory and VSEPR theory in Chapter 10. There are a number of ways of diagrammatically representing the hybridization of atomic orbitals to create the bonding structure of the assembled molecule. Figure 11.19, which links the energy ordering of the electron configuration of the separated atoms to the four equivalent sp³ hybridized orbitals and displays the resulting molecular geometry of CH_4, is one approach. This approach is simple, concise, and geometrically correct. It links the electronic configuration of the separated atoms to the hybridized orbitals of the central bonding carbon and to the structure of methane including the C-H bonds, wavefunctions and electron occupation.

We can also represent the mixing of atomic orbitals in a step wise mathematical addition of the atomic orbitals with different *coefficients* of orbital addition that represents the different *phases* of adding the electron waves of the p_x, p_y, and p_z orbitals. This systematic addition of different wave phases results in the four hybrid sp³ orbitals pointing to the four vertices of the tetrahedron as shown in Figure 11.20.

The Combination of One Unhybridized s Orbital and Three Unhybridized p Orbitals

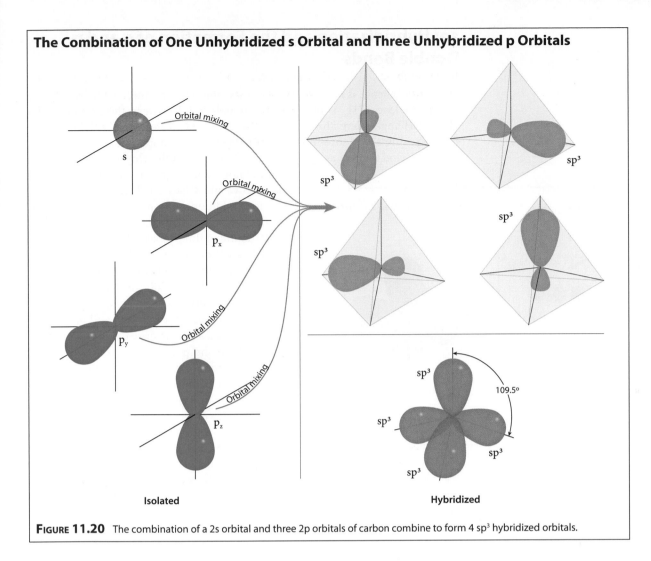

Isolated   **Hybridized**

FIGURE 11.20 The combination of a 2s orbital and three 2p orbitals of carbon combine to form 4 sp³ hybridized orbitals.

Why do the p_x, p_y, and p_z orbitals "choose" this combination of phases to create the hybrid orbitals from the s and p atomic orbitals? Because the coefficients of the wave mixing of the p_x, p_y, and p_z orbitals is "chosen" by the electrons in order to minimize the potential energy of the ensemble of hybrid orbitals *in the assembled molecule.*

In Figure 11.20, once we have established the geometry of the four resulting sp³ orbitals, we can assemble those orbitals around the central carbon atom to construct the correct orientation of the tetrahedral structure. For clarity, it is common practice to ignore the small negative lobe in each of the sp³ orbitals when configuring the final ensemble of orbitals in the molecule as shown on the right-hand side of Figure 11.20.

It is worth noting that electrons are not mathematicians—they do not solve the Schrödinger wave equation to establish the proper orbital within which they reside. Electrons are simply masters of energy minimization and they are remarkably quick-witted at what they do. As we will see in Chapter 12, it takes an ensemble of electrons only ~10^{-15} seconds to transition from one geometrical distribution to another—for example from an unhybridized orbital to a hybridized orbital. This is orders of magnitude faster than the calculations can be done by the world's fastest computer. It is the far slower motion of the nuclei that determines the rate at which atomic orbitals transition to molecular orbitals or the time required for molecular structures to be interconverted to other molecular structures in a chemical reaction.

sp² Hybridization and the Formation of σ and π Double Bonds

Our next challenge is to try to figure out the bonding structure in ethylene. As we noted, ethylene, C_2H_4, is planar with a double bond between the carbons and a 120° HCH angle. Here, the revelation that we could create a tetrahedral structure from an s orbital and three p orbitals provides both the hope that almost anything is possible and the hint as to how to do it. If we are to solve this bonding structure we know, from Figure 11.20 for the sp³ case, that we must find coefficients for the judicious addition of p_x, p_y, and p_z orbitals with the 1s orbital that will give us a bonding structure of three lobes, all in the same plane, that are 120° apart. We immediately know, then, that the combination of p orbitals that constitutes the hybridization *cannot* include a p orbital perpendicular to that plane. We must work with the s orbital and just two of the p orbitals—we choose the p_x and p_y orbitals. From the point of view of the electron configuration, this implies that the 2s orbital of carbon will mix with two of the 2p orbitals as shown in Figure 11.21 that displays the electronic configuration of the s-p mixing.

We can track the formation of these sp² orbitals from the separated atomic orbitals explicitly using the analogous diagram to Figure 11.20 introduced for the sp³ hybridization. In the case of sp² hybridization, the mixing of the p orbitals such as to minimize the potential energy of the assembled molecule is displayed in Figure 11.22. The mixing of the p orbitals is represented by the *coefficients* that define both the amplitude and phase of the electron waves that are condensed to form the hybrid orbitals.

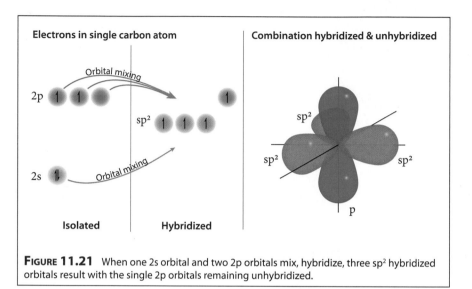

FIGURE 11.21 When one 2s orbital and two 2p orbitals mix, hybridize, three sp² hybridized orbitals result with the single 2p orbitals remaining unhybridized.

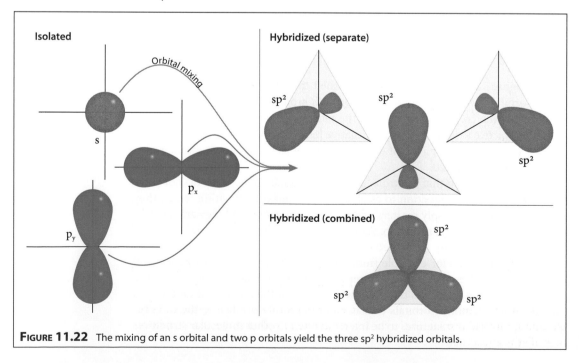

FIGURE 11.22 The mixing of an s orbital and two p orbitals yield the three sp² hybridized orbitals.

We are now positioned to define the bonding structure of ethylene using the three degenerate sp² orbitals that all lie in a plane with an angle of 120° between each of the orbitals. First we form a carbon-carbon bond with the overlap of an sp² orbital from each of the carbon atoms as shown in Figure 11.23.

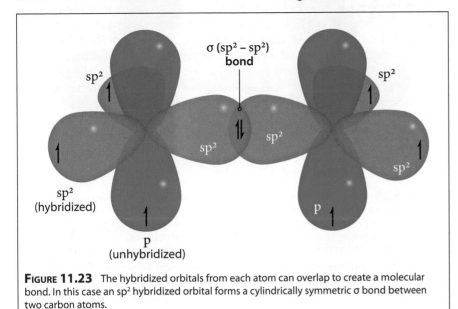

FIGURE 11.23   The hybridized orbitals from each atom can overlap to create a molecular bond. In this case an sp² hybridized orbital forms a cylindrically symmetric σ bond between two carbon atoms.

These two sp² hybridized carbon atoms establish one sp²–sp² σ bond in a head-on arrangement creating one electron pair bond. The plane of the molecular structure is determined by the plane of the sp² hybridized orbitals with the unhybridized p orbital perpendicular to that plane.

Having established the first carbon-carbon bond with this electron pair σ bond, the unhybridized p orbitals form a π bond that locks the remaining sp² orbitals into a common plane. This electron pair π bond in combination with the σ bond establishes the carbon-carbon double bond, as shown in Figure 11.24.

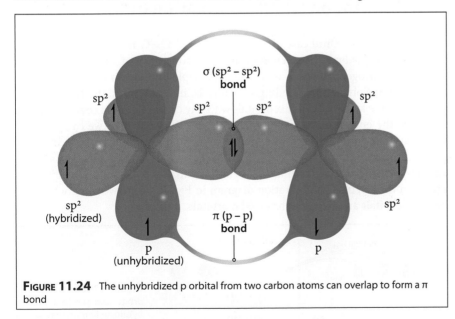

FIGURE 11.24   The unhybridized p orbital from two carbon atoms can overlap to form a π bond

The four hydrogen atoms then form electron pair σ bonds with the four protruding sp² hybrid orbitals in the molecular plane, completing the bonding structure of ethylene as displayed in Figure 11.25.

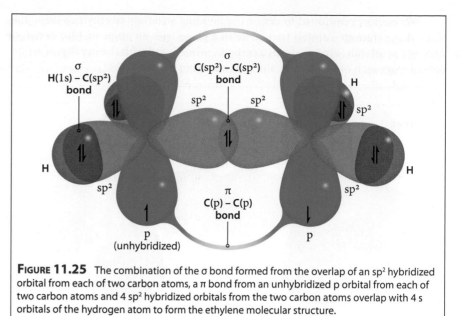

Figure 11.25 The combination of the σ bond formed from the overlap of an sp² hybridized orbital from each of two carbon atoms, a π bond from an unhybridized p orbital from each of two carbon atoms and 4 sp² hybridized orbitals from the two carbon atoms overlap with 4 s orbitals of the hydrogen atom to form the ethylene molecular structure.

A planar structure of ethylene is thus constructed with a backbone consisting of the C=C double bond. The stronger of the two carbon-carbon bonds is that formed from the sp²–sp² σ overlap, primarily because of the extension of the sp² hybrid orbitals along the axis connecting the carbon nuclei that provides greater *overlap* for the σ electron pair bond. In addition the electrons reside *between* the nuclei providing shielding of the proton-proton repulsion. The second carbon-carbon bond, the weaker of the two, is a π bond formed from the unhybridized p-p orbital overlap. This bond is not only weaker, it is also more exposed to attack by electron hungry species such as free radicals or Lewis acids.

sp Hybridization and the Formation of Triple Bonds

We turn next to our third case, that of acetylene, H-C≡C-H, that is a linear molecule with a triple bond between the central carbons, each of which is bonded to a terminal hydrogen. How can we construct a triple bond from the s and p orbitals of carbon? Solving this bonding structure is, in many ways, the easiest of the three cases. We know we can mix orbitals with any combination of coefficients that minimize the energy of the molecular structure. We know that we can combine both σ and π bonding to form the molecular structure, and we have no inherent geometric problems linking the orthogonal p_x, p_y, and p_z orbitals to create a linear structure. If we mix the s orbital of beryllium with *one* p orbital of beryllium, as shown in the electron configuration diagram in Figure 11.26, we have two sp hybridized orbitals and two unhybridized p orbitals.

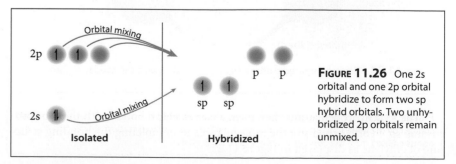

Figure 11.26 One 2s orbital and one 2p orbital hybridize to form two sp hybrid orbitals. Two unhybridized 2p orbitals remain unmixed.

We can track the creation of the resulting *two* sp hybrid orbitals from the one s orbital and the one p orbital (remember we always get the same number of hybrid orbitals as atomic orbits were combined to form them) as shown in Figure 11.27. When the two sp orbitals are combined we have a linear orbital extending 180° from each other with the carbon atom at the center.

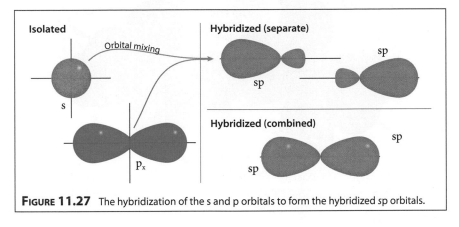

FIGURE 11.27   The hybridization of the s and p orbitals to form the hybridized sp orbitals.

When we combine the two lobe sp orbital with the unhybridized p orbitals we have the geometry displayed in Figure 11.28. This orientation of orbitals (note that the *phase* of the hybrid orbitals are both *positive* because the small negative phase orbital is suppressed as can be seen by inspection of Figure 11.28) provides for carbon-carbon σ bonding along the internuclear axis, and carbon-hydrogen bonding at the terminal positions.

The carbon-carbon σ bonds are shown in Figure 11.29 along with the two π bonds formed from the unhybridized p orbitals of the carbon.

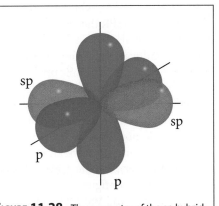

FIGURE 11.28   The geometry of the sp hybridized and unhybridized p orbitals of carbon.

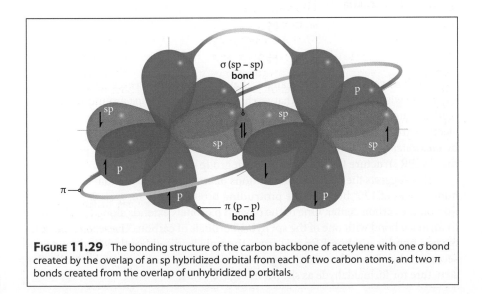

FIGURE 11.29   The bonding structure of the carbon backbone of acetylene with one σ bond created by the overlap of an sp hybridized orbital from each of two carbon atoms, and two π bonds created from the overlap of unhybridized p orbitals.

With the addition of the terminal hydrogen, the VB model of the bonding structure in acetylene is complete as shown in Figure 11.30.

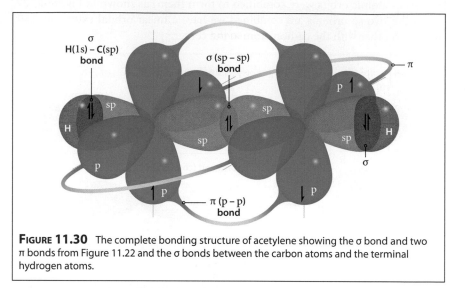

FIGURE 11.30   The complete bonding structure of acetylene showing the σ bond and two π bonds from Figure 11.22 and the σ bonds between the carbon atoms and the terminal hydrogen atoms.

Let's review what we have done by considering another molecular structure, formaldehyde, CH_2O, which we used as a case study in our discussion of Lewis structure and VSEPR theory in Chapter 10. First, quantum mechanics tells us the molecular orbitals that we have to work with—specifically s and p orbitals from carbon, s and p orbitals from oxygen, and s orbitals from hydrogen. At this point, we have our tool kit of atomic orbitals, but we don't know how to minimize the potential energy of the ensemble of electrons and nuclei in the completed molecules. We do know that the electrons will arrange their wave properties so as to find the minimum potential energy structure. While the electrons know the answer, we, at this point, do not. Thus we have two choices: Option 1 is to employ a reasonably powerful computer to solve an electron structure calculation of the Schrödinger equation to determine the potential energy surface of the most stable (lowest energy) molecular configuration. Option 2 is to use what we know from our study of Lewis structures and VSEPR theory. Reviewing quickly (see Chapter 10) the Lewis structure places hydrogen at the terminal positions and the least electronegative atom (that's carbon) at the center of the molecule and inserts electrons to satisfy the octet/duet criteria

$$\begin{array}{c} H \diagdown \\ \quad\quad C = \overset{\displaystyle ..}{\underset{\displaystyle ..}{O}} \\ H \diagup \end{array}$$

We don't yet know the bond angles that constitute the lowest energy structure, but the Lewis diagram tells us that we have 3 electron groups around the central (carbon) atom and we have two lone pairs on the oxygen. VSEPR theory sets the electron group geometry to *minimize* the electron group-electron group repulsion by *maximizing* the distance between those entities: the trigonal planar geometry is the VSEPR structure for the three electron group case.

This suggests the planar configurations of bonding orbitals about carbon with bond angles of 120° between the protruding orbitals—this suggest sp^2 hybridization on the carbon center. One of the oxygen p orbitals extends along the C-O axis to form a σ bond with one of the sp^2 hybrid orbitals of carbon. The second carbon-oxygen bond is a π bond formed from the unhybridized bond on carbon and a second unhybridized bond on oxygen. This forms the backbone of the molecular structure for formaldehyde as shown in Figure 11.31.

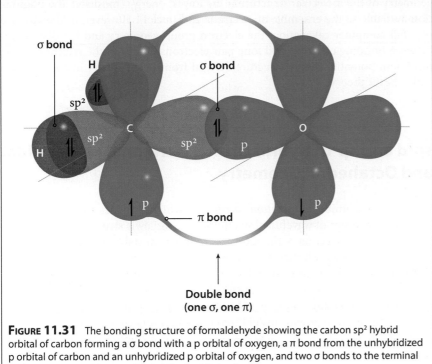

FIGURE 11.31 The bonding structure of formaldehyde showing the carbon sp² hybrid orbital of carbon forming a σ bond with a p orbital of oxygen, a π bond from the unhybridized p orbital of carbon and an unhybridized p orbital of oxygen, and two σ bonds to the terminal hydrogen atoms with two sp² hybridized orbitals of carbon.

The complete molecule structure is achieved by adding the two hydrogen 1s orbitals to the unpaired sp² hybrid orbitals of carbon and labeling each of the orbital overlaps as shown in Figure 11.32.

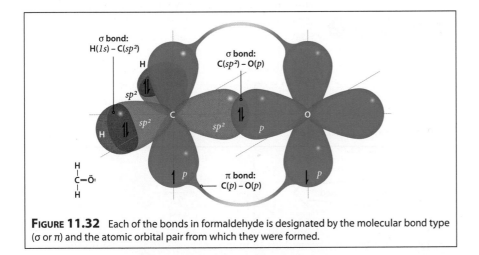

FIGURE 11.32 Each of the bonds in formaldehyde is designated by the molecular bond type (σ or π) and the atomic orbital pair from which they were formed.

From one perspective, Lewis theory and VB theory are analogous: the structure of formaldehyde is set by the central carbon atom, which serves to organize four bonds: two single bonds to the terminal hydrogen atoms and a double bond to oxygen. VB theory provides extremely important additional information. VB theory distinguishes between the structure of the σ bond and that of the π bond. This distinction becomes increasingly important when we consider the chemical characteristics of molecular structures *and* the analysis of how photons interact with the molecular structure.

Notice again that quantum mechanics provides the tool box of atomic orbitals of the separated atoms, and provides the concept of the constructive and deconstructive addition of electronic waves within a molecular structure such that the geometry of the molecular structure is the lowest energy (most stable configuration available to the ensemble of electrons and nuclei). However, in the absence of a full computer calculation, the electron group geometry and the distinction between bonding electrons and lone pair electrons guides in the selection of the minimum potential energy structure derived from Lewis theory in combination with VSEPR theory.

sp³d and sp³d² Hybrid Orbitals: Trigonal Bipyramidal and Octahedral Geometry

Thus from the union of *quantum mechanics*, which provides the atomic orbitals and the concept of wavefunction addition, with Lewis structures and VSEPR, which guide the selection of the geometry of the most stable molecular structure, we have formulated VB theory with remarkable success. To this point we have developed an understanding of linear, trigonal planar, and tetrahedral structures. But we know from observation that trigonal bipyramidal and octahedral electron geometries are common in many molecular structures. Can this concept of atomic orbital hybridization provide insight into those bonding configurations?

The way we can answer these questions is to think about what hybridization

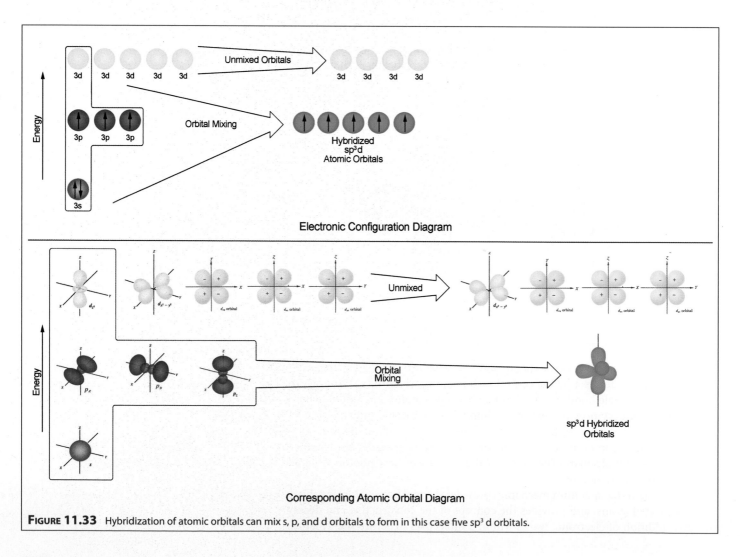

FIGURE 11.33 Hybridization of atomic orbitals can mix s, p, and d orbitals to form in this case five sp³ d orbitals.

of atomic orbitals has taught us about the creation of quite different geometries for bonding within molecules.

We would first ask, can we manipulate the *coefficients* of s and p orbitals so as to create a trigonal bipyramidal backbone structure by hybridization of just these two wavefunction geometries? After trying a number of different possibilities we would conclude that we cannot generate the *five* orbitals required. In fact we know that we can only generate n hybrid orbitals from n atomic orbitals. Because we only have four atomic orbitals given one s orbital and three p orbitals, we know immediately that we can't achieve the five required orbitals of the trigonal bipyramidal configuration. Do we have any other options? Examination of the periodic table, with an eye on the atomic orbitals associated therewith provides a hint. Period 4 elements require d orbitals to define their electron configurations. What would happen if we hybridized s, p, *and* d atomic orbitals to create an orbital structure in a molecular bonding geometry? On the face of it, this is not an easy question to answer, but we do know how we would express this hybridization with our energy ordered electron configuration diagram. Testing the hybridization of the 3s, 3p, and one 3d atomic orbital, we would diagram the problem as shown in Figure 11.33.

This is a remarkable result. By borrowing one d atomic orbital (which we might have guessed since we needed another atomic orbital to throw into the mix and the d orbitals were the next in energy) and mixing this d atomic orbital with three p atomic orbitals and one s atomic orbitals, we create 5 sp^{3}d equivalent

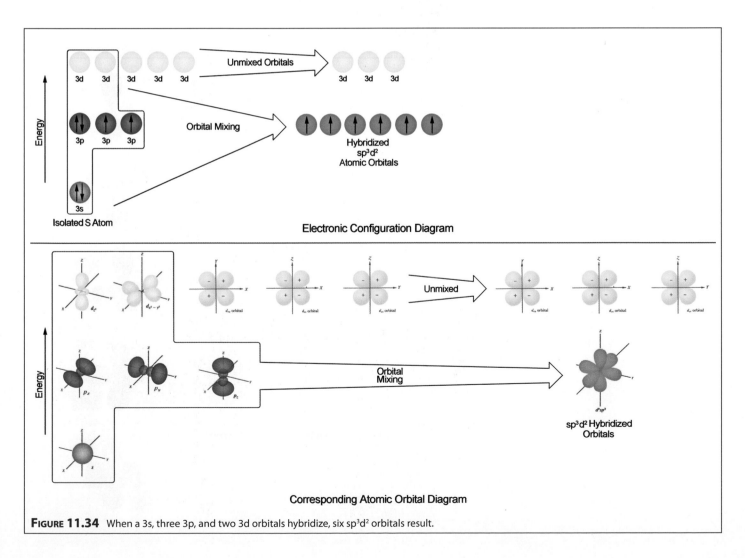

FIGURE 11.34 When a 3s, three 3p, and two 3d orbitals hybridize, six sp^{3}d^2 orbitals result.

hybrid orbitals with electron density protruding from the central atom in a trigonal bipyramidal configuration! We simply have to get the proper coefficients of each atomic wavefunction to yield the correct geometry of the five resulting sp^3d hybrid orbitals. This underscores again the fact that we get the same numbers of hybrid orbitals as atomic orbitals we mix to form those hybrids.

This suggests that if we are to create a set of six hybrid orbitals that are required to establish an octahedral structure, we need to hybridize six atomic orbitals. Given our success of creating the 5 hybrid orbital trigonal bipyramidal structure by engaging a d orbital, can we create a 6-hybrid bond octahedral structure by engaging a second d orbital to form a sp^3d^2 hybrid? The electron configuration diagram is straight forward to draw as is shown in Figure 11.34.

We now have command of the VB hybrid structures that cover a vast array of molecular geometries. In addition, we have brought the insight of quantum mechanics to the bonding *structure* of those molecules. It is the shape of a molecule, the electron distribution within the molecules, the character of the chemical bonds involved and the size of the species involved in the molecule that determines the chemical behavior of a molecule. We have made great progress on the back of a few simple principles! We summarize in Table 11.1 the relationship between the number of electron groups, the electron group geometry from VSEPR theory, and the hybridization of atomic orbitals that create the required geometry for the *observed* molecular structure.

TABLE 11.1 Hybridization Scheme from Electron Geometry

Number of Electron Groups	Electron Geometry (from VSEPR Theory)		Hybridization Scheme
2	Linear	sp	
3	Trigonal planar	sp^2	120°
4	Tetrahedral	sp^3	109.5°
5	Trigonal bipyramidal	sp^3d	90° 120°
6	Octahedral	sp^3d^2	90° 90° 90°

Molecular Orbital Theory and Electron Delocalization

Modern research in chemistry and physics and modern devices that emerge from this research are increasingly dependent on a more detailed picture of molecular bonding. In particular, breakthroughs in chemistry and physics require an understanding of how photons interact with electrons in molecular structures, how magnetic properties of materials can be modified to suit particular objectives, and how electron conduction in both organic and inorganic structures can be manipulated. All of this requires an increasing degree of sophistication in models of chemical bonding and thus of molecular structure.

A key conceptual foundation that underpins the development of more sophisticated models of chemical bonding is that electrons, when given the opportunity, will employ the full spatial dimension of a molecular structure to spread out over, or to delocalize within, in an effort to minimize the potential energy of the ensemble of electrons and nuclei. It is this tendency of electrons to *delocalize* from the confinement of *atomic orbitals* to the geometrical flexibility of extended *molecular orbitals* that sets the stage for the molecular orbital (MO) class of bonding theories. This is in stark contrast to VB theory, which we have seen is successful, but confines electrons to atomic orbitals, only allowing them the ability to mix specific orbitals to form hybrid orbitals.

We saw that the solution to the Schrödinger equation for *atomic structure* yielded the complete wavefunction for multielectron system with a *single nucleus*. There is no limitation intrinsic to the Schrödinger equation preventing a full solution for multielectron/multinucleus systems. Thus, when the Schrödinger equation is solved for a molecule, a *molecular* orbital results, with the electrons delocalized in whatever configuration yields the structure with the lowest potential energy given the constraint that no two electrons share the same set of quantum numbers.

Before we develop this new *molecular orbital* approach to bonding, we review three characteristics of electron waves:

1. Electrons exhibit wave properties that under confinement of a nucleus yield standing wave solutions to the Schrödinger equation and electrons seek to minimize the energy of their configuration. Therefore, the waves associated with multiple electrons will add the amplitude and phase of their wavefunctions within their spatial proximity, while obeying the Pauli Exclusion Principle such as to minimize the energy of the combined ensemble of electrons and nuclei.

2. Adding the wavefunctions of the individual electrons together *constructively* yields a region of higher electron density between the nuclei in a molecular structure. This can create a *bonding* molecular orbital distributed across two or more nuclei.

3. Subtracting wavefunctions of individual electrons results in the *destructive* interference of two wavefunctions that can reduce the electron density between two nuclei in a molecular structure resulting in a *nonbonding* molecular orbital distributed across two or more nuclei.

The specifics of how wavefunctions attributed to individual *atoms* combine to form wavefunctions of a *molecule* bears some careful consideration. We must add two waves (or wavefunctions) by correctly combining *both* the amplitude and phase of those individual waves to find the resulting multielectron *wavefunction of the molecule*. The electron density, proportional to the *square* of the resulting wavefunction, is what controls the forces and thus the potential energy surface that defines the strength and character of a chemical bond.

To this end we consider carefully the addition of two atomic wavefunctions to create a molecular orbital by exploring ways that we can add the wavefunctions of two hydrogen atoms to form the H_2 molecule. While this represents a simple case, it engages a large fraction of the key concepts involved in *molecular orbital* (MO) development.

We begin with a hydrogen atomic wavefunction ψ_A for atom A and a wavefunction ψ_B for atom B, which, at large internuclear distance do not interact. We can sketch this condition as shown in Figure 11.35. The diagram sketches the *radial* wavefunction for the 1s orbital of atomic hydrogen for each atom.

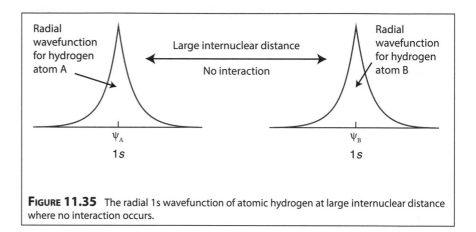

FIGURE 11.35 The radial 1s wavefunction of atomic hydrogen at large internuclear distance where no interaction occurs.

As we move these two hydrogen atom atomic orbitals to smaller and smaller internuclear distance, the wavefunction of atom A begins to interact with the wavefunction of atom B. In quantum mechanical language, the wavefunctions of the atoms begin to "mix." If the waves add constructively we create a domain of partial overlap of the wavefunctions of the two hydrogen atoms and the potential energy of the combined orbitals begins to decrease because the electron in each atom "sees" the Z_{eff} of the other atom which begins to draw the atoms together as shown in Figure 11.36.

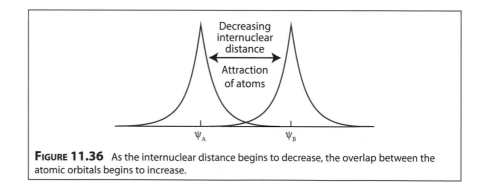

FIGURE 11.36 As the internuclear distance begins to decrease, the overlap between the atomic orbitals begins to increase.

As the internuclear distance between the hydrogen atom atomic orbitals decreases, the individual wavefunctions begin to overlap creating, by virtue of the constructive interference between the wavefunctions, an increase in electron density between the two nuclei as shown in Figure 11.37.

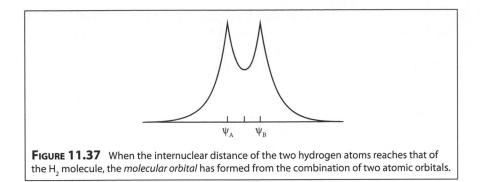

FIGURE 11.37 When the internuclear distance of the two hydrogen atoms reaches that of the H_2 molecule, the *molecular orbital* has formed from the combination of two atomic orbitals.

We can assemble Figures 11.35, 11.36, and 11.37 to form a composite figure as a function of potential energy, demonstrating the decrease in potential energy as a function of decreasing internuclear distance. This is displayed in Figure 11.38.

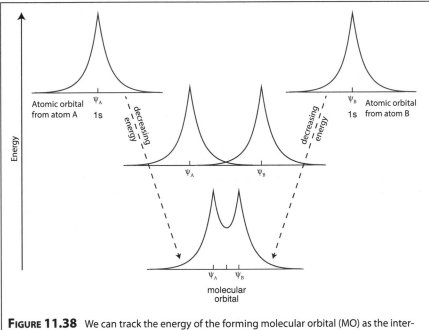

FIGURE 11.38 We can track the energy of the forming molecular orbital (MO) as the internuclear distance decreases carrying the molecular formation from the separated atoms to the fully formed molecular orbital.

While in this case the new molecular orbital $\psi_{H_2} = \psi_A + \psi_B$ is the solution of the Schrödinger equation, the electron density distribution that determines the bond strength is given by the *square* of the wavefunction such that

$$\begin{pmatrix} \text{Probability} \\ \text{of finding} \\ \text{the electron} \\ \text{at any point} \end{pmatrix} = \psi_{H_2}^2 = \left(\psi_A + \psi_B \right)^2 = \psi_A^2 + \psi_B^2 + 2\psi_A\psi_B$$

We can represent the electron distribution graphically and display the buildup of electron density between the nuclei as shown in Figure 11.39. Thus the square of the molecular orbital wavefunction is the contribution from the wavefunction of atom A plus the contribution from the wavefunction of atom B *plus* the term $2\psi_A\psi_B$, which contributes electron density in the domain between the two nuclei.

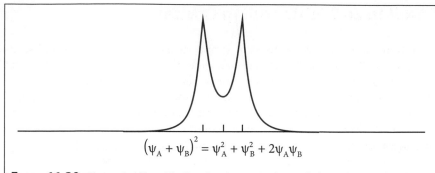

$$\left(\psi_A + \psi_B\right)^2 = \psi_A^2 + \psi_B^2 + 2\psi_A\psi_B$$

FIGURE 11.39 The probability of finding the electron in the newly formed molecular orbital is calculated from the square of the sum of the wavefunctions for the two atoms $(\psi_A + \psi_B)^2 = \psi_A^2 + \psi_B^2 + 2\psi_A\psi_B$.

Thus we have the *molecular orbital* (MO) for molecular hydrogen and this extended domain available to both electrons leads to two effects. First, the electrons become indistinguishable from each other. Second, the fact that the electrons are free to delocalize throughout the full extent of the newly formed molecule serves to *lower* the potential energy of the ensemble of 2 electrons and 2 protons relative to the two isolated hydrogen atoms. Thus a stable bond is formed, and a new *molecular orbital* is created. A key point is that this new *molecular orbital* assumes the same role for the molecule that the *atomic orbital* played for the atom. Namely it is a solution, in its own right, to the Schrödinger equation that provides (1) a mathematical description of the spatial domain occupied by the electrons, (2) the energy of the electrons in the individual molecular orbitals, and (3) the angular momentum and spin of the electrons in each orbital.

The use of atomic orbitals of the separated atoms to form the molecular orbitals of the molecule formed therefrom is called the "linear combination of atomic orbitals" or LCAO model. The typical designation of the union of molecular orbital (MO) theory with the linear combination of atomic orbitals is the LCAO-MO model. While it is important to recognize exactly how we have combined the two atomic wavefunctions to generate a lower energy molecular wavefunction by the *constructive* addition of the electron waves, we can simplify our graphical representation of this union by using the diagram in Figure 11.40 that captures the union of the two 1s orbitals of atomic hydrogen to form a molecular orbital of H_2 that is cylindrically symmetric about the internuclear axis and add the probability distribution of the 1s orbitals of H to form the bonding MO of H_2. As was the case for the VB orbitals, the MO formed is cylindrically symmetric about the internuclear axis; therefore, the bonding MO of H_2 is a σ bond. In addition, because the σ bond MO is formed from two 1s atomic orbitals, the MO is designated a σ_{1s} with the subscript designating the atomic orbitals from which the MO was formed. We can diagram this union of 1s atomic orbitals of H to form the σ_{1s} MO of H_2 as shown in Figure 11.40.

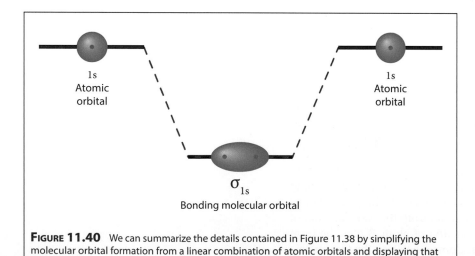

1s
Atomic
orbital

1s
Atomic
orbital

σ_{1s}

Bonding molecular orbital

FIGURE 11.40 We can summarize the details contained in Figure 11.38 by simplifying the molecular orbital formation from a linear combination of atomic orbitals and displaying that simplification as a molecular orbital diagram as shown here.

11.34

Bonding and Antibonding Orbitals

What is of great importance, however, is that the atomic orbitals (AO$_s$) of hydrogen, because they are waves, can add *destructively* as well as constructively. That is, we can form an MO by subtracting the wavefunctions of one H atom (atom B) from the wavefunction of the other H atom (atom A). With respect to the radial wavefunction this *destructive* union of the two atomic orbitals as a function of internuclear distance and energy, the analogous diagram to Figure 11.38 for constructive addition is shown for *destructive interference* in Figure 11.41.

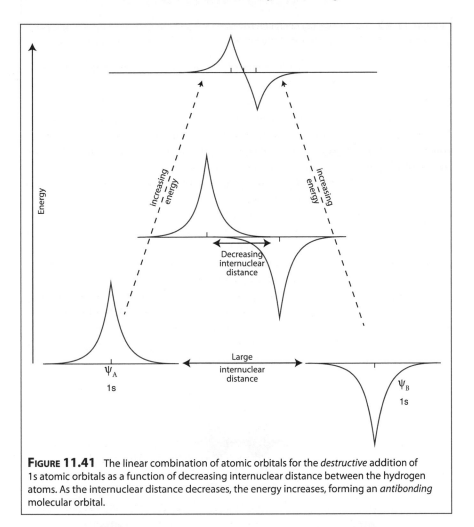

FIGURE 11.41 The linear combination of atomic orbitals for the *destructive* addition of 1s atomic orbitals as a function of decreasing internuclear distance between the hydrogen atoms. As the internuclear distance decreases, the energy increases, forming an *antibonding* molecular orbital.

This figure displays the fact that if we combine, if we *mix*, two electron wavefunctions *destructively*, the resultant MO will *increase* in potential energy with decreasing internuclear distance. This increase in potential energy results from the fact that the electron density, which goes as $(\psi_A - \psi_B)^2 = \psi_A^2 + \psi_B^2 - 2\psi_A\psi_B$, reduces to zero between the nuclei, thus exposing the Coulomb *repulsion* of the two nuclei. We can express the probability of finding the electron in the new *molecular orbital*, $\psi_{H2}^2 = (\psi_A - \psi_B)^2 = \psi_A^2 + \psi_B^2 - 2\psi_A\psi_B$, as a function of internuclear distance as shown in Figure 11.42.

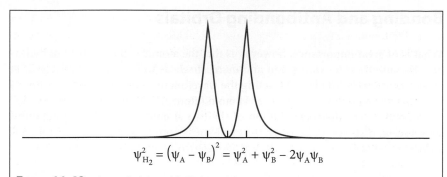

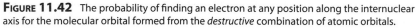

$$\psi^2_{H_2} = \left(\psi_A - \psi_B\right)^2 = \psi^2_A + \psi^2_B - 2\psi_A\psi_B$$

FIGURE 11.42 The probability of finding an electron at any position along the internuclear axis for the molecular orbital formed from the *destructive* combination of atomic orbitals.

Because the energy of this H₂ *molecular orbital* is *higher* than that of the two separated hydrogen *atomic* orbitals, we define this new molecular orbital as an *antibonding* orbital and designate the orbital as a σ_{1s}^* orbital. As always the σ designates an orbital that is cylindrically symmetric about the internuclear axis, the 1s subscript designates the atomic orbital from which the molecular orbital was constructed, and the asterisk superscript designates the *anti* bonding character of the molecular orbital.

We can summarize the two ways to combine or mix wavefunctions of two separated hydrogen atoms: either (1) *constructive* interference leading to an *increase* in electron density between the two atoms that comprise the molecule and a *decrease* in potential energy created by the net attraction between the atoms or (2) *destructive* interference leading to a *decrease* in electron density between the two atoms and an *increase* in potential energy created by the net repulsion between the atoms. *Constructive* interference of the two 1s hydrogen atomic orbitals leads to a bonding σ_{1s} *molecular orbital*. *Destructive* interference of the two 1s atomic orbitals leads to an *antibonding* σ_{1s}^* molecular orbital. Using our representation of atomic and molecular orbitals that depict the three dimensional configuration of the orbitals as we did in Figure 11.40, but in this case for both the constructive addition and destructive addition of the atomic orbitals, we capture the energy ordering of the union of atomic orbitals to form molecular orbitals in Figure 11.43.

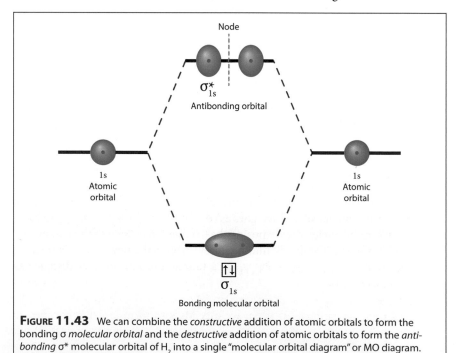

FIGURE 11.43 We can combine the *constructive* addition of atomic orbitals to form the bonding σ *molecular orbital* and the *destructive* addition of atomic orbitals to form the *antibonding* σ* molecular orbital of H₂ into a single "molecular orbital diagram" or MO diagram.

Two additional points highlighted in Figure 11.43 deserve careful consideration. First, note that the σ_{1s}^* antibonding orbital has a plane of zero electron density between the two nuclei. This is referred to as a "nodal plane" and, as we will see, the number of nodal planes in a wavefunction, all else being equal, determines the *energy ordering* of molecular orbitals. The greater the number of nodal planes the higher the potential energy. Second, the *molecular orbital* is a wave and thus it has both an amplitude and a *phase*. It is critically important to keep track of the phase of any electron wave because when any orbital combines with or *mixes* with another, the resulting wavefunction is created by the *interference* of those waves.

Once we have fully defined how this combination of bonding and antibonding molecular orbitals are created from the constructive and destructive addition of atomic orbitals respectively, we can now summarize the creation of the molecular orbital of H_2, in a shorthand diagram representing the energy *evolution* and energy ordering of the atomic and molecular orbitals in what is generally termed an "MO diagram." This MO diagram for H_2 is displayed in Figure 11.44.

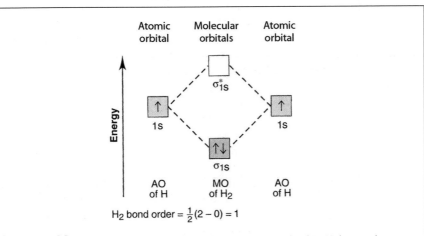

FIGURE 11.44 It is common practice to create the molecular orbital (MO) diagram by creating the molecular orbitals simply as energy ordered boxes with the electron occupation of those boxes explicitly represented as shown here. The *shape* of the molecular orbital is understood from the bonding, σ, nomenclature and the antibonding, σ*, nomenclature and the atomic orbitals from which the molecular orbital was formed appear as a subscript, σ_{1s} and σ_{1s}^*.

This raises the question: how are electrons inserted into the newly formed molecular orbitals formed from the *linear combination of atomic orbitals* (LCAO)? The answer is that electrons are inserted into MOs just as they were into AOs:

1. MOs are filled in their *order* of increasing energy.

2. A MO can be occupied by a maximum of two electrons, each with opposite spin—a manifestation of the Pauli Exclusion Principle.

3. When electrons are inserted into MOs of equal energy, the electrons first enter unoccupied MOs before they pair up into the same MO. This is a manifestation of the fact that electrons seek to maximize the distance between themselves in order to minimize the potential energy of the ensemble.

There is one other aspect of Figure 11.44 that is important. Notice that the σ_{1s}^* antibonding orbital of H_2 *exists* even though it is *unoccupied*. This simple fact becomes very important when we (1) add or subtract electrons from an MO forming anions or cations, (2) when we mix MOs to form larger molecules, or (3) when we "promote" electrons to a higher orbital with a photon of light. While the very existence of "antibonding" molecular orbitals may seem counterintuitive, as we will see, these antibonding orbitals are actually lower in energy than the bonding orbitals formed from atomic orbitals of the next higher principal quantum number.

This sets the ground work for developing the relationship between three important concepts: (1) the geometry of the formation of the molecular bond from the separated atomic orbitals by considering the probability distribution of electrons about the nuclei as a function of internuclear distance, (2) the evolving molecular orbital diagram that defines the energy splitting between the bonding and antibonding molecular orbitals as a function of internuclear distance, and (3) the potential energy surface for the bonding and antibonding molecular orbitals. Figure 11.45 tracks the evolution of these three facets of molecular bonding at three different internuclear distances.

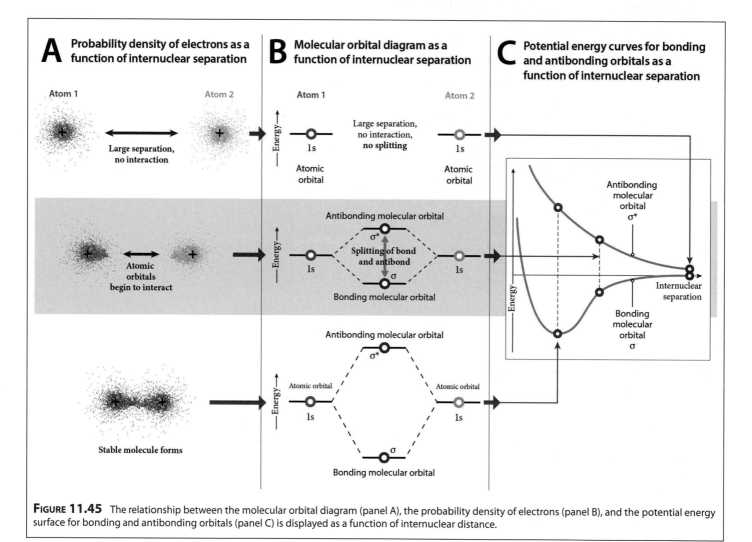

FIGURE 11.45 The relationship between the molecular orbital diagram (panel A), the probability density of electrons (panel B), and the potential energy surface for bonding and antibonding orbitals (panel C) is displayed as a function of internuclear distance.

Panel A of Figure 11.45 displays the relationship between the probability distribution of electrons as the internuclear distance decreases from the condition of large separation for which there is no overlap through the condition of marginal overlap leading to the formation of the chemical bond to the final condition of maximum bond strength represented by the equilibrium internuclear distance. Panel B of Figure 11.45 delineates the evolution of the molecular orbital diagram from *large* internuclear separation (for which there is no interaction of the valence electrons) through *intermediate* internuclear distance wherein the overlap is developing between the 1s atomic orbitals leading to a separation in energy between the bonding, σ_s, and antibonding, σ_s^*, orbitals. Finally at the equilibrium internuclear distance, the σ_s and σ_s^* orbitals achieve their maximum separation and the potential energy reaches its deepest point. Panel C tracks the bonding and antibonding potential energy surfaces from the condition of large separation where the bonding and antibonding potential energy surfaces are of equal energy (i.e. degenerate) because there is no overlap of the atomic orbitals through to the condition of minimum energy of the bonding molecular orbital at the equilibrium internuclear distance.

As more electrons are added with increasing atomic number, the molecular orbital diagram becomes more complicated but the basic relationships defined in Figure 11.45 remain. We turn now to the case of molecular oxygen to develop the LCAO-MO theory for the combination of s and p orbitals.

Molecular Orbital Structure of Molecular Oxygen

Development of the bonding structure of oxygen from the molecular orbital perspective builds on and uses the concepts developed for H_2. We have developed a perspective defining how the bonding and antibonding orbitals of a molecule are formed from the atomic orbitals of the separated atoms, and how electrons are inserted into the lowest lying molecular orbitals with opposite spin. This insertion of electrons continues until all available electrons are accounted for. We also recognize that the potential energy surface of the bonding orbital has a corresponding bound state (an *"electronic state"* because it is defined by the orbitals of the *electron*) and the antibonding orbital has a corresponding repulsive potential energy surface. Armed with this perspective we are now in a position to address the orbital structure of oxygen—the most important molecule in the evolution of life forms on Earth.

First, we recall that p *atomic orbitals* can be combined in two distinct ways to form *molecular* orbitals. Namely, the p orbitals can be combined "head-on" along an axis that passes through the nuclei of the two atoms such that the p orbitals combine either *constructively* or *destructively*. This is shown explicitly in Figure 11.46. This results in a molecular orbital that is symmetric under rotation (that is it does not *change* when rotated about the axis linking the nuclei) for either *bonding* when the p orbitals (the electron waves) are added *constructively* or antibonding if the p orbitals are added *destructively*. These molecular orbitals formed from the addition of two 2p orbitals are designated σ_{2p} when they are bonding, and σ_{2p}^* when they are antibonding because they are formed from 2p orbitals of atomic oxygen that are aligned along the x-axis joining the nuclei. While the designation of the x-axis as the internuclear axis is arbitrary, we will use this convention throughout our discussion of MO theory.

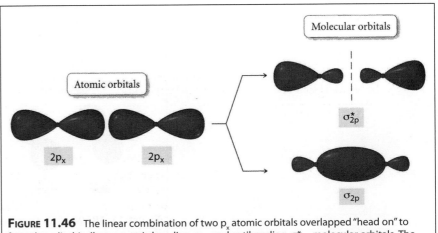

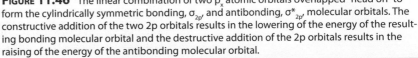

FIGURE 11.46 The linear combination of two p_x atomic orbitals overlapped "head on" to form the cylindrically symmetric bonding, σ_{2p}, and antibonding, σ_{2p}^*, molecular orbitals. The constructive addition of the two 2p orbitals results in the lowering of the energy of the resulting bonding molecular orbital and the destructive addition of the 2p orbitals results in the raising of the energy of the antibonding molecular orbital.

But we also know that the p atomic orbitals can be combined to form a molecular orbital by bringing the p orbitals together *perpendicular* to the axis of the p orbital lobes to form both a bonding and an antibonding molecular orbital as displayed in Figure 11.47. The side-by-side orientation of the p_y orbitals in Figure 11.47 is in sharp contrast to the head on orientation of the p_x atomic orbitals forming the σ bonding in Figure 11.46. As a consequence of adding the p_y orbitals sideways to form the overlap of atomic orbitals that creates the molecular bond, the bonding pattern occurs above and below the internuclear axis.

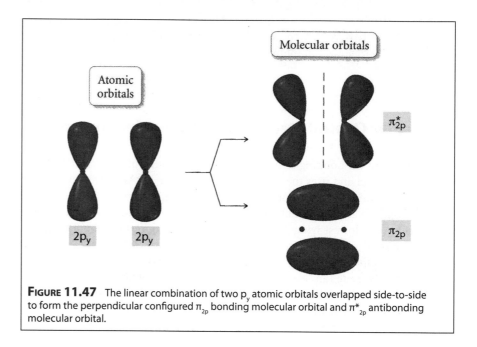

FIGURE 11.47   The linear combination of two p_y atomic orbitals overlapped side-to-side to form the perpendicular configured π_{2p} bonding molecular orbital and π^*_{2p} antibonding molecular orbital.

The bonding geometry is termed a π_{py} orbital. The Greek π is used to denote a bond structure formed from atomic orbitals *perpendicular* to the internuclear axis. The antibonding molecular orbital is designated the π^*_{py} molecular orbital. We are now in a position to form molecular orbitals from both s atomic orbitals and p atomic orbitals. However, there is another pair of p orbitals of oxygen that align along the z-axis as shown in Figure 11.48. These are the bonding π_{pz} and antibonding π^*_{pz} molecular orbitals respectively.

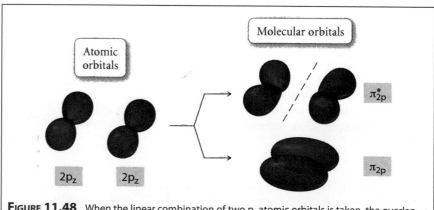

FIGURE 11.48   When the linear combination of two p_z atomic orbitals is taken, the overlap is side-to-side as was the case for the p_y atomic orbitals. The bonding π_{pz} and antibonding π^*_{pz} molecular orbitals are analogous to the π_{py} and π^*_{py} bonding and antibonding orbitals except they are rotated by 90° about the internuclear axis.

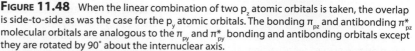

Molecular Orbital Structure and the Potential Energy Structure

This provides the information needed to connect three important aspects of MO theory together in the same diagram: (1) the geometry of the atomic orbitals as they form the corresponding molecular orbitals with decreasing internuclear distance between the atoms, (2) the molecular orbital diagram that defines the energy ordering and energy spacing of the molecular orbitals as a function of decreasing internuclear distance as the molecular bonds form, and (3) the potential energy surface of bound O_2 that defines the binding energy of the molecule as a function of internuclear distance.

Three "snapshots" of the sequence linking these three aspects of molecular orbital theory for three different internuclear distances are shown in Figures 11.49, 11.50, and 11.51. Figure 11.49 displays the molecular orbital geometry, molecular orbital diagram, and potential energy surface corresponding to the case of *large* separation between the oxygen atoms such that there is zero overlap between the orbitals, no splitting of the molecular orbital energies, and no decrease in the potential energy surface defining the bonding between the two separated oxygen atoms.

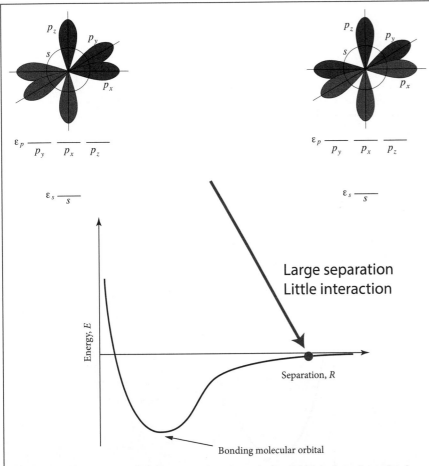

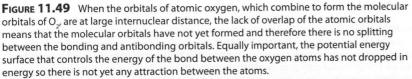

FIGURE 11.49 When the orbitals of atomic oxygen, which combine to form the molecular orbitals of O_2, are at large internuclear distance, the lack of overlap of the atomic orbitals means that the molecular orbitals have not yet formed and therefore there is no splitting between the bonding and antibonding orbitals. Equally important, the potential energy surface that controls the energy of the bond between the oxygen atoms has not dropped in energy so there is not yet any attraction between the atoms.

Figure 11.50 displays the case for decreasing internuclear distance as the atomic orbitals begin to interact. This condition of overlap between the atomic orbitals begins to split the molecular orbitals so formed by decreasing the energy of the bonding orbitals and increasing the energy of the antibonding orbitals. This "splits" the energy levels of the forming molecular bond with the σ_{2p} bonding orbital being lowest in energy because of the *greater overlap* between the p_x atomic orbitals aligned along the internuclear axis. The $\sigma_{2p}*$ antibonding overlap is higher in energy because of the large repulsion that results from the large overlap of the p_x orbitals combined with the absence of electron density between the two nuclei.

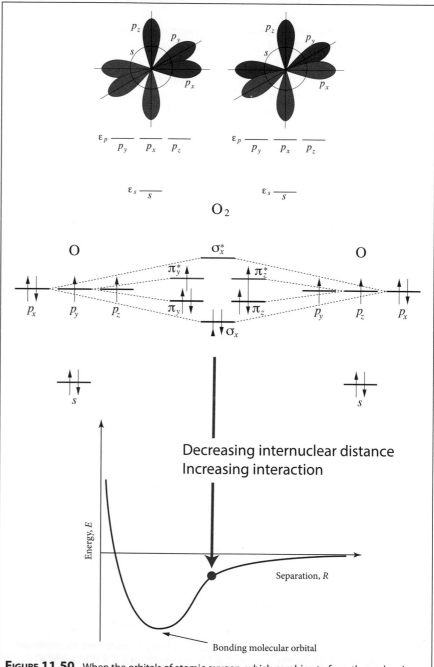

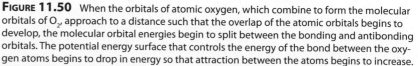

FIGURE 11.50 When the orbitals of atomic oxygen, which combine to form the molecular orbitals of O_2, approach to a distance such that the overlap of the atomic orbitals begins to develop, the molecular orbital energies begin to split between the bonding and antibonding orbitals. The potential energy surface that controls the energy of the bond between the oxygen atoms begins to drop in energy so that attraction between the atoms begins to increase.

11.42

The evolving potential energy surface of O_2 corresponds to the point on the surface representing the decrease in energy relative to the interactive energy at large internuclear distance. This decrease in energy of the potential energy surface represents the binding energy of the O_2 bond at that internuclear distance.

Figure 11.51 displays the molecular orbital geometric configuration when the internuclear distance has decreased to the position of maximum splitting between the bonding orbitals with respect to their ("unsplit") energies at large internuclear distance. At this point, the bond has reached its maximum strength represented by the maximum depth of the potential energy well of the molecule's potential energy

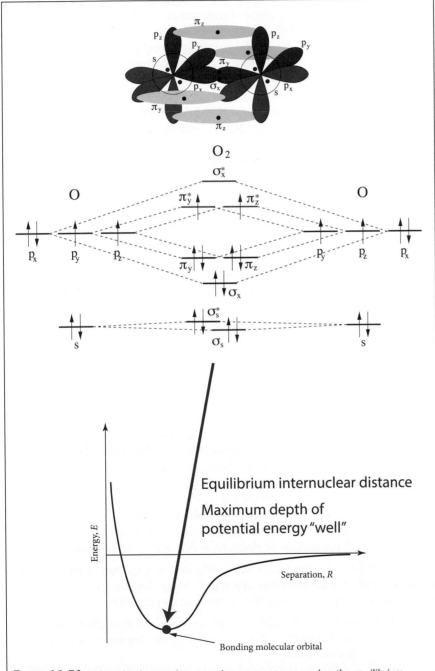

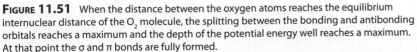

FIGURE 11.51 When the distance between the oxygen atoms reaches the equilibrium internuclear distance of the O_2 molecule, the splitting between the bonding and antibonding orbitals reaches a maximum and the depth of the potential energy well reaches a maximum. At that point the σ and π bonds are fully formed.

surface. Note that the splitting of the σ_s and σ_s^* orbitals formed from the mixing of the 2s atomic orbitals of the separated oxygen atoms is small because the greater extension of the p_x orbitals prevents any significant overlap of the 2s orbitals.

Development of MO theory for H_2 and for O_2 establishes the essential concepts in MO theory wherein we have formed those molecular orbitals by a simple adding together or "mixing" of atomic orbitals; the so called linear combination of atomic orbitals-molecular orbital (LCAO-MO) theory. It is important to recognize that we need not be constrained or restricted to using those explicit mathematical functions, the wavefunctions for the isolated atoms. We can use any mathematical function we choose that can be optimized in the molecular structure to minimize the potential energy of the ensemble of electrons and nuclei in the molecular structure. It just happens that to a greater or lesser degree, depending on the molecule, the atomic wavefunctions are a good (sometimes very good) first approximation that can be successfully mixed, added together constructively or destructively, to yield the resulting molecular orbitals.

Given this discussion of the molecular orbitals of H_2 and of O_2, we are in a position to quickly work our way through the Period 2 diatomic molecules formed from the same two atoms in each case: the homonuclear diatomic Li_2, Be_2, C_2, N_2, O_2, F_2, and Ne_2.

As we move through the Period 2 homonuclear cases, we will employ the basic tenets of the LCAO-MO theory:

- Molecular orbitals are constructed by adding together, by mixing, combinations of atomic orbitals. The *number* of MOs generated from a selected set of AO will always equal the number of AOs in the set.
- The atomic orbitals must be added constructively or destructively to establish the full array of molecular orbitals available to the molecule. The MOs formed from constructively added AOs will form bonding MOs. The MOs formed from the destructive addition of AOs will form *antibonding* MOs.
- Although the complete set of bonding MOs and antibonding MOs will in most cases provide more orbitals than there are available electrons to fill them when electrons are assigned to the MOs the electrons are inserted into the lowest energy MOs first with a maximum of two electrons, paired with opposite spins, assigned for each MO.
- Electrons entering MOs that are of equal energy will enter empty orbitals before pairing up with electrons of opposite spin in the same MO. This follows Hund's rule, which reflects the fact that electrons in separate MOs are at greater distance than electrons confined to the same MO.
- The only electrons considered in the creation of a molecular orbital from atomic orbitals are the valence shell electrons of the atoms. Inner shell electrons are ignored as they are too tightly bound to engage in bonding.
- Finally, the bond order in a diatomic molecule is calculated for an MO configuration by simply *subtracting* the number of electrons in *antibonding* orbitals from the number of electrons in *bonding* orbitals divided by 2.

Notice before we assemble the MO configuration of the Period 2 homonuclear molecules, that MO theory brings new insight to bonding because:

1. We are keeping track of specific electrons in specific MOs.

2. Each electron that enters a bonding molecule orbital strengthens the molecular bond; each electron that enters an antibonding weakens or destabilizes the molecular bond.

3. For each electron configuration in the available MOs, there is a corresponding potential energy surface representing the *ensemble* of electrons and nuclei in the molecular structure.

11.44

Molecular Orbital Structure of Homonuclear Diatomics

We proceed to construct the MOs of the diatomic molecules formed from the elements in the second Period of the periodic table. In our development of the MO bonding structure for O_2, we have introduced the σ_{2s}, σ_{2s}^*, π_{2p}, and π_{2p}^* MO structures constructed respectively from the atomic orbitals with principal quantum number n=2 and for angular momentum subshells s and p. We have also developed arguments for the *energy ordering* of those MOs for molecular hydrogen and for molecular oxygen.

In particular, for H_2 we demonstrated that the MO ordering was σ_{1s} as the most stable, lowest energy orbital and that σ_{1s}^* was the higher energy orbital such that the two available electrons entered the σ_{1s} MO and the σ_{1s}^* molecular orbital was empty. The bond order for H_2 is thus ½(2 – 0) = 1.

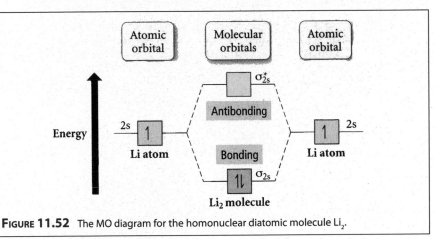

FIGURE 11.52 The MO diagram for the homonuclear diatomic molecule Li_2.

The first two homonuclear diatomics in Period 2 are Li_2 and Be_2. We consider only the valence electrons of the atoms when constructing the MOs so we construct the appropriate MOs from the two 2s atomic orbitals in each case to create the σ_{2s} bonding orbital and σ_{2s}^* antibonding orbital in exact analogy with H_2. For Li_2 we have a 2s electron from one Li atom and a 2s electron from the other Li atom so the two electrons enter the bonding σ_{2s} orbital as shown in Figure 11.52. Thus Li_2 has a bond order of ½ (2–0) = 1 and is a stable molecular configuration.

For Be_2, we have *exactly the same* MO diagram, but in this case we have a total of 4 electrons two 2s electrons from each Be atom. The Be_2 molecule has a bond order of ½ (2 – 2) = 0. Thus MO theory predicts that Be_2 is unstable. Indeed it is not observed in nature. Figure 11.53 displays the MO diagram for Be_2.

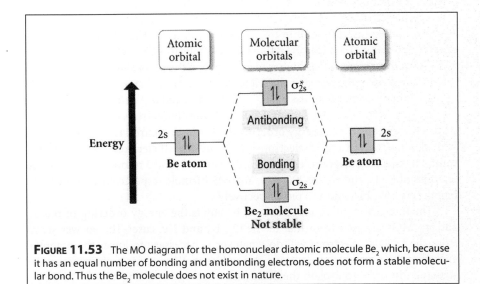

FIGURE 11.53 The MO diagram for the homonuclear diatomic molecule Be_2, which, because it has an equal number of bonding and antibonding electrons, does not form a stable molecular bond. Thus the Be_2 molecule does not exist in nature.

For oxygen, the energy ordering of the MOs, developed in the previous section, is reviewed in Figure 11.54. As that figure indicates, the energy ordering of MOs for F_2 and Ne_2 is the same as for O_2. Once we have constructed the MO

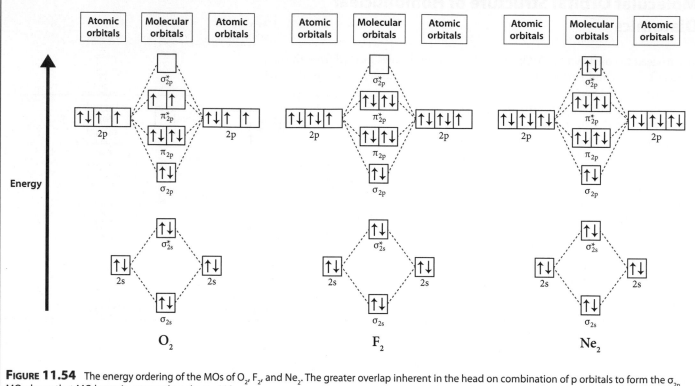

FIGURE 11.54 The energy ordering of the MOs of O_2, F_2, and Ne_2. The greater overlap inherent in the head on combination of p orbitals to form the σ_{2p} MO places that MO lower in energy than the π_{2p} MO.

diagram we simply insert the total number of electrons in the MOs until we have accounted for all valence electrons. Molecular oxygen has $2 \times 6 = 12$ valence electrons and a bond order of $\frac{1}{2}(8 - 4) = 2$. Molecular fluorine has $2 \times 7 = 14$ valence electrons and a bond order of $\frac{1}{2}(8 - 6) = 1$. Ne_2 has $2 \times 8 = 16$ valence electrons and a bond order of $\frac{1}{2}(8 - 8) = 0$. Thus Ne_2 is unstable and no bond forms.

Turning to the final three homonuclear diatomics in Period 2, we have B_2, C_2, and N_2. It would seem to be a straight forward matter. We would simply count the total number of valence electrons in B_2, C_2, and N_2 (6, 8, and 10 electrons respectively) and then insert them into the MO *energy ordering* that we used for O_2, F_2 and Ne_2 in Figure 11.54. However, there is a complication. Experimental evidence has shown that the σ_{2p} MO is higher in energy than the π_{2p} orbital in B_2, C_2, and N_2 and thus the *energy ordering* of the MOs for these three diatomics is σ_{2s}, σ_{2s}^*, π_{2p}, σ_{2p}, π_{2p}^*, and σ_{2p}^* rather than the σ_{2s}, σ_{2s}^*, σ_{2p}, π_{2p}, π_{2p}^*, and σ_{2p}^* ordering appropriate for O_2, F_2, and Ne . Thus we sequentially insert the 16 electrons of B_2 into the MO diagram with the proper energy ordering as shown in Figure 11.55, then proceed with the insertion of the 18 electrons of C_2 and the 10 electrons of N_2 into that same MO ordering diagram. The bond orders are 1, 2, and 3 respectively. The bond energies of B_2, C_2 and N_2 are 290, 620, and 649 kJ/mole respectively and the bond lengths are 159, 131, and 110 pm respectively.

The interesting question, of course, is why is the energy ordering of the σ_{2p} and π_{2p} MOs inverted from that of the O_2, F_2, and Ne_2 case? The answer stems from the fact that electrons will do whatever is necessary to lower the potential energy of the ensemble of electrons and nuclei that constitute a given molecular structure. In order to explain the difference in energy ordering of MOs in B_2, C_2, and N_2, we need to reexamine how we created the MOs from the AOs. In particular a fundamental assumption of the LCAO-MO theory is that, to create an MO, we simply add pairwise the s and p orbitals of the 2nd period elements to form the homonuclear diatomics. That is specifically what Figures 11.52, 11.53, and 11.54

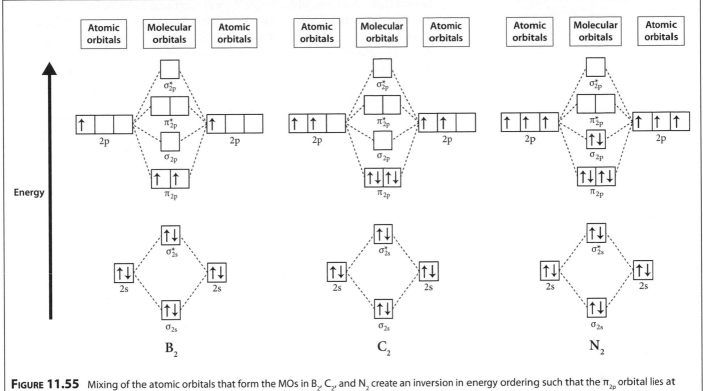

Figure 11.55 Mixing of the atomic orbitals that form the MOs in B_2, C_2, and N_2 create an inversion in energy ordering such that the π_{2p} orbital lies at lower energy than the σ_{2p}.

demonstrate. However, as we have seen from our treatment of hybrid orbitals, if atomic orbitals lie close to one another in energy *and* they have favorable overlap geometry, they will mix together and, in so doing, the resulting wavefunction leads to a lowering of the potential energy of that structure. In the case of B_2, C_2, and N_2, 2s and 2p orbitals mix to create the MOs of the combined molecule. If we had at our disposal a full computer calculation we would see the mixed character of 2s and $2p_x$ atomic orbitals along the internuclear axis that forms the σ bond. What differentiates LCAO-MO theory from the more general MO theory is the simple pair-wise addition of atomic orbitals, a simplification that is easily eliminated in computer based electron structure calculations that allow s-p mixing of the atomic orbitals that constitute the basis for the computer based molecular orbital calculations. For our purposes here, it is important to realize that s-p mixing can affect the energy ordering of the σ and π MOs of the molecule, that s-p mixing is significant in B_2, C_2, and N_2, and that s-p mixing is not significant in O_2, F_2, and Ne_2 as shown in Figure 11.55.

Molecular Orbital Structure of Heteronuclear Molecules

When we combine the atomic orbitals of two identical atoms, the energies of the valence atomic orbitals are identical; for example in the Period 2 homonuclear examples the 2s orbitals of one of the atoms is equal in energy to that of the other 2s orbital as are the 2p orbitals of equal energy. However, when we combine the atoms of different elements, we must contend with the fact that atoms with different effective nuclear charge, Z_{eff}, possess differing abilities to draw electronic charge toward them. An atom with a larger Z_{eff} will exert a greater Coulomb attraction on any electronic charge distribution in its vicinity and thus the corresponding valence atomic orbital will be of lower potential energy than an atom with a smaller Z_{eff}. If we consider the Period 2 elements, this serves to *shift* the 2s and 2p atomic orbitals of atoms with a higher Z_{eff} *downward* in a MO diagram relative to an atom with a small Z_{eff}. Considering again our Period 2 elements, if we plot Z_{eff} vs. atomic number we recall the pattern as reviewed here in Figure 11.56.

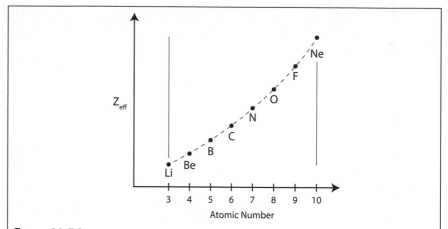

FIGURE 11.56 The effective nuclear charge, Z_{eff} increases with atomic number from lithium to neon.

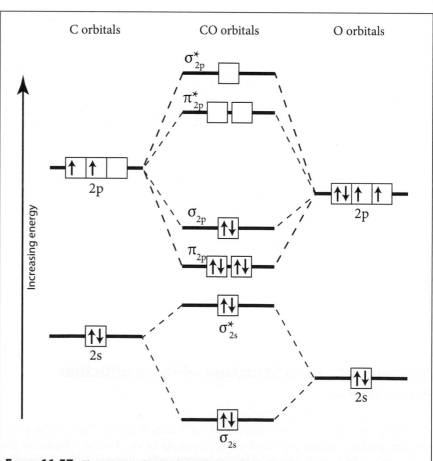

FIGURE 11.57 The atomic orbitals of oxygen lie at somewhat lower energy than those of carbon so, when the MOs are formed from the atomic orbitals of carbon and oxygen, electron density is drawn to the lower energy (more electronegative) atom. The energy ordering of the bonding MOs is the same as that for C_2.

If we consider carbon monoxide, CO, the Z_{eff} for O atoms is larger than Z_{eff} for C atoms so the atomic orbitals of oxygen are at a lower energy than the atomic orbitals of carbon. In addition, the s-p mixing in carbon is such that the energy ordering of the MOs is the same as that for diatomic carbon. Thus we have an MO diagram for CO shown in Figure 11.57.

Nitric oxide is one of the most important molecules in nature. It plays a significant role in biological systems as a "signaling molecule"—it is released by one system in the human body to trigger a reaction in another system of the human body. For example the inner lining of blood vessels use nitric oxide to signal the surrounding smooth muscle to relax, resulting in increased blood flow. As we will see in the MO diagram of NO, nitric oxide has an unpaired electron in its valence shell and is thus very reactive as it seeks to form an electron pair bond with any molecule with which it collides. NO has a lifetime in human systems of a few seconds, but because it is small, it diffuses easily across membranes.

Nitric oxide is also critically important to air pollution in urban centers where it is central to the photochemical *production* of ozone, and in the *destruction* of ozone in the Earth's stratosphere. At ground level, NO is produced in high temperature combustion processes such as in gasoline and diesel engines. At high temperature (> 1300 K), N_2 and O_2 are chemically broken apart to form N and O atoms that react to form NO. In the stratosphere, NO reacts with ozone, O_3, to form NO_2 and O_2. The NO_2 then reacts with atomic oxygen to reform NO. The net result is to convert ozone to molecular oxygen, which can, in the stratosphere, lead to net ozone destruction through the reaction sequence:

$$NO + O_3 \longrightarrow NO_2 + O_2$$
$$NO_2 + O \longrightarrow NO + O_2$$
$$\text{Net:} \quad \overline{O_3 + O \longrightarrow O_2 + O_2}$$

Atomic nitrogen has a Z_{eff} slightly less than that for atomic oxygen so the energy of the atomic orbitals of N are somewhat *higher* in energy than the corresponding atomic orbitals of O. The energy ordering of the molecular orbitals of O have the same ordering as that for CO and are shown in Figure 11.58.

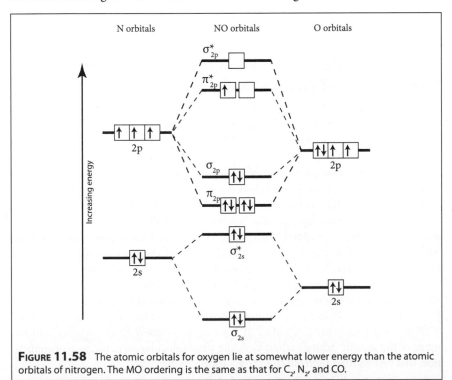

FIGURE 11.58 The atomic orbitals for oxygen lie at somewhat lower energy than the atomic orbitals of nitrogen. The MO ordering is the same as that for C_2, N_2, and CO.

Another important example involves the MO structure of two atoms with *very different* atomic orbital energies. For example, the fluorine atom has a far greater Z_{eff} than does atomic hydrogen such that the 2p atomic orbitals of fluorine lie at much lower energy than the 1s orbital of hydrogen. This large energy difference between the 1s and 2p atomic orbitals, shown in the MO diagram of HF in Figure 11.59, results in the lack of mixing between the p_x and p_y orbitals of fluorine and the formation of only the σ bonding MO between the 1s orbital of H and the 2p orbital of F. This underscores the fact that in order for two orbitals to mix, they must both overlap spatially *and* be of similar energy.

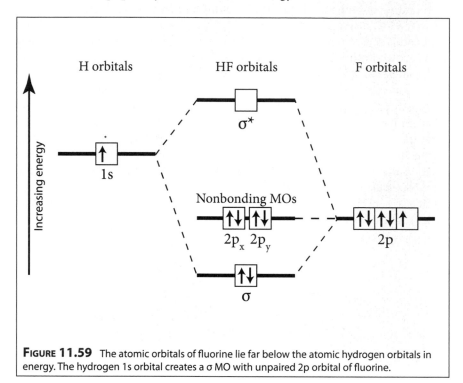

FIGURE 11.59 The atomic orbitals of fluorine lie far below the atomic hydrogen orbitals in energy. The hydrogen 1s orbital creates a σ MO with unpaired 2p orbital of fluorine.

Molecular Orbital Theory Applied to Benzene: The Central Role of Delocalization

As we have witnessed, the LCAO-MO theory opened important doors into the workings of molecular bonding. With MO theory we can now explicitly identify which electrons go into which orbitals in the assembled molecules, we can understand the magnetic properties of molecules, we can identify the geometry of specific molecular orbital types such as σ and π bonds and we can link the bonding and antibonding MOs to the potential energy surface of the molecule itself. We also recognize that the lingering attachment to the explicit geometrical form of the atomic orbitals that are added to form the MOs restrict the ability of the resulting MOs to fully exploit the degree of delocalization of electron density throughout the full architectural form of the molecule. We have built the molecular orbitals from bricks shaped like atomic orbitals, but electrons have a remarkable propensity to flow into whatever geometric form serves to lower the potential energy of the ensemble. This does not represent a shortcoming of the Schrödinger equation, it simply represents the limitation of sticking too closely to the spatial confinement of the atomic orbitals when building the molecular structure—a simplification that results directly from the LCAO model.

If we drop the LCAO restriction and treat a molecule as an ensemble of electrons and nuclei and solve the Schrödinger equation without bias toward specific atomic orbital geometries, the result is a true *molecular orbital* theory. From the Schrödinger equation will emerge molecular orbitals for the entire structure and we then fill those MOs according to the Pauli exclusion principle and Hund's rules. A state-of-the-art calculation of the molecular orbital structure of benzene is a case in point, as shown in Figure 11.60.

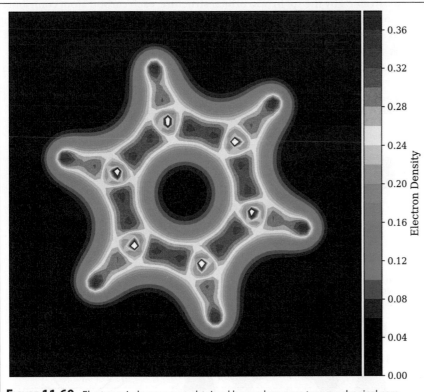

FIGURE 11.60 Electrons in benzene as obtained by modern quantum mechanical computational methods: distribution of the valence electrons on the plane of the molecule; the colors indicate the concentration of electrons (the vertical bar on the right gives the values of electron density for each color).

Benzene is a molecule of great importance to modern chemistry and it is a molecule in possession of remarkable stability due to its unique bonding structure. For our purposes here, however, it serves as a unifying entity that links together all of our molecular bonding models.

This one-bond-at-a-time treatment of multiatom electronic structures works remarkably well. Far better in fact than we might well expect. There are, however, clear examples demonstrating the limitations to this electron pair bond-by-bond picture. There are examples of molecules for which electron density is uniformly distributed over the entire molecule—such an example is benzene. Is there any vestige of our VB or MO picture in such a molecule?

Experiments have demonstrated that benzene, C_6H_6, is a planar molecule with a remarkable hexagonal geometry set in place by a backbone of carbon atoms. *All* carbon-carbon bonds are equal length (140 pm) and the C-C-C bond angle in each case is 120°. Valence bond theory guides an important initial assessment of the structure of benzene. The framework of the carbon backbone is comprised of a σ bonded structure that is a network of sp^2–sp^2 hybrid orbitals as shown in Figure 11.61A. But the unhybridized p orbitals of carbon protrude perpendicular to the plane of the σ bonding structure as displayed in Figure 11.61B.

Constrained by the sp^2 hybrid orbital structure, each carbon atom contributes 3 electrons to the formation of the σ bonds—one electron each to the adjoining carbons and the third electron to the hydrogen atom bonded into the carbon backbone as shown in the upper panel of Figure 11.61.

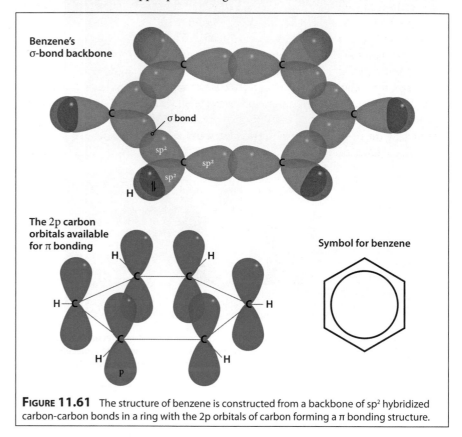

FIGURE 11.61   The structure of benzene is constructed from a backbone of sp^2 hybridized carbon-carbon bonds in a ring with the 2p orbitals of carbon forming a π bonding structure.

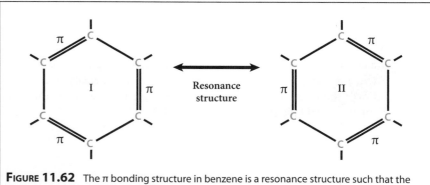

FIGURE 11.62   The π bonding structure in benzene is a resonance structure such that the electrons in the π bonding structure delocalize in a ring forming the cyclic structure of benzene.

11.52

Now things become interesting when we consider the bonding structure resulting from the unhybridized p orbitals that lie perpendicular to the plane of the molecule. The unhybridized p orbitals are clearly situated for π molecular bonding, and when we link them we immediately see that there are two equivalent structures that constitute a resonance structure, which we saw in Chapter 10 for ozone, nitrates, etc. This is displayed in Figure 11.62.

So how do we couple this idea of a resonance structure with the *molecular orbital* picture of molecular bonding? The first point to emphasize is that the picture of having to choose between either a *double bond* or a *single bond* is a manifestation of the concept of a *localized* bond: a one-bond-at-a-time treatment of molecular structure. The second point is that we have six p orbitals, and if we combine them all, we will have six π molecular orbitals. Each of the p orbitals has one electron, so we insert these six electrons into the lowest energy orbitals first, inserting them sequentially into the energy-ordered MOs that result. The MOs for benzene as viewed from the top are displayed in Figure 11.63. The lowest energy MO is the delocalized π system that rings the molecule. Next higher in energy lie two π bonding orbitals shown in Figure 11.63 that are of equal energy. Each of these three π bonding orbitals contains two electrons such that all bonding orbitals are filled. Thus one full σ bond plus approximately half the normal π overlap holds together each pair of carbons as the π electrons delocalize in a continuous ring encompassing the carbon atoms in the backbone of the benzene molecular structure.

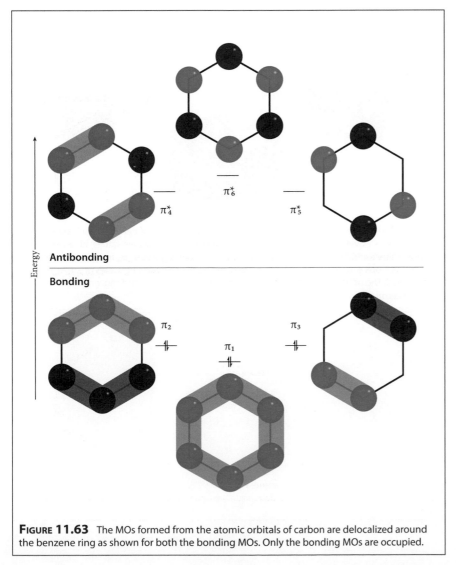

FIGURE 11.63 The MOs formed from the atomic orbitals of carbon are delocalized around the benzene ring as shown for both the bonding MOs. Only the bonding MOs are occupied.

The antibonding π orbitals exist and are shown in the upper panel of Figure 11.63. Note that the antibonding orbitals have alternating phases around the benzene carbon ring and thus there are nodes between the carbon atoms thereby *decreasing* electron density and *increasing* the potential energy of the structure. None of the antibonding orbitals are filled.

Once we have diagrammed the MO of benzene from the top, we can make sense of the full MO computer calculations of the actual shape and phase of the π bonding structure that is shown in Figure 11.64. It is typically this computer generated MO picture of molecules that is displayed in the scientific literature.

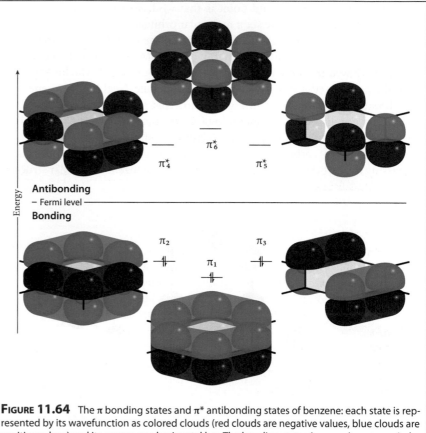

FIGURE 11.64 The π bonding states and π* antibonding states of benzene: each state is represented by its wavefunction as colored clouds (red clouds are negative values, blue clouds are positive values) and its energy as a horizontal bar. The bonding states (π_1, π_2, π_3) are occupied by two electrons each (with up and down spins), while the antibonding states (π_4, π_5, π_6) are unoccupied. The occupied and unoccupied states are separated by the long, grey horizontal line, the Fermi level.

Summary Concepts

1. Formation of a Molecular Bond from Atomic Orbitals

When we address quantitatively how molecules are constructed from the combination of atomic orbitals we must develop a strategy for how to mathematically represent the ensemble of electrons and protons that constitute a molecular structure of minimum potential energy. Heitler and London developed an approach to molecular bonding that explicitly used the new *wavefunctions* of the hydrogen atom from the Schrödinger equation to create a molecular wavefunction, ψ_{MOL}

$$\psi_{MOL}\left(r_1, r_2\right) = \psi_{1s_a}\left(r_1\right)\psi_{1s_b}\left(r_2\right)$$

where $\psi_{1s_a}(r)$ is the 1s hydrogen atomic orbital of atom a and $\psi_{1s_b}(r_2)$ is the hydrogen atomic orbital of atom b. They found that they could only reproduce the *observed* bond strength of H_2 if they used a molecular wavefunction representing the *sharing* of the electron between the two nuclei

$$\psi_{MOL}\left(r_1, r_2\right) = \frac{1}{\sqrt{2}}\left[\psi_{1s_a}\left(r_1\right)\psi_{1s_b}\left(r_2\right) + \psi_{1s_b}\left(r_1\right)\psi_{1s_a}\left(r_2\right)\right]$$

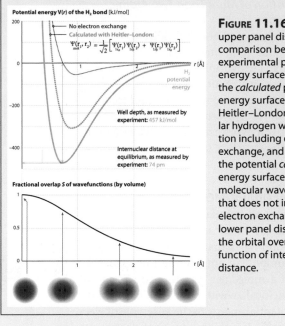

FIGURE 11.16 The upper panel displays the comparison between the experimental potential energy surface for H_2, the *calculated* potential energy surface using the Heitler–London molecular hydrogen wavefunction including electron exchange, and finally the potential *calculated* energy surface for an H_2 molecular wavefunction that does not include electron exchange. The lower panel displays the orbital overlap as a function of internuclear distance.

2. Mixing of Atomic Orbitals to Form Hybrid Orbitals

Very important molecular structures such as the tetrahedral methane, CH_4, the planar ethylene, C_2H_4, and the linear acetylene, C_2H_2, molecules cannot be formed from the simple addition of atomic orbitals to form molecular bonds. Pauling developed the idea that atomic orbitals can *mix* in any manner that will reduce the potential energy of the *ensemble* of electrons and nuclei. For example, if we combine one s orbital with three p orbitals we create a new geometrical structure resulting in four *hybridized* orbitals of carbon with electron density extending to the vertices of a tetrahedron, as shown at right.

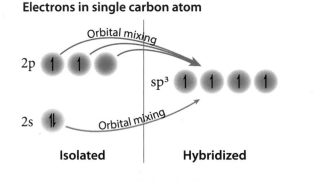

3. Different Combinations of Atomic Orbitals Yield Different Geometries of the Hybridized Orbitals: the sp^2 and sp Examples

Atomic orbitals can mix so as to minimize the energy of molecular structures that result from the addition of the hybridized orbitals. When, for example, one 2s orbital and two 2p orbitals mix, a hybridized, three lobe, sp^2 structure results with the single 2p orbital remaining unhybridized. That sp^2 hybridization establishes the geometry for the planar bonding in ethylene, C_2H_4, and the double bond between the carbon atoms in ethylene. When one 2s orbital and one 2p orbital hybridize to form two sp hybrid orbitals, two unhybridized 2p orbitals remain unmixed. This establishes the linear, triple bonding in acetylene, C_2H_2.

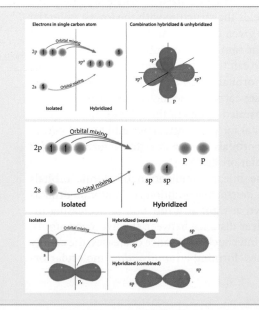

Summary Concepts

4. Hybridization of More Than Two Atomic Orbital Types

While the most common hybridization occurs between s and p orbitals, a molecular structure will mix (i.e. hybridize) any combination of atomic orbitals in an effort to create the molecular structure of minimum energy. Thus, for example, one s orbital, three p orbitals, and a d orbital of phosphorus will hybridize to form five sp^3d orbitals that set the structure of PCl_5.

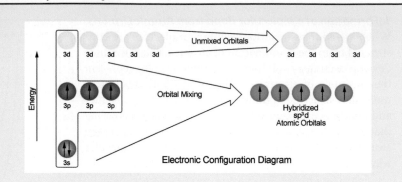

5. Molecular Orbital (MO) Theory

A key conceptual foundation that underpins the development of more sophisticated models of chemical bonding is that electrons, when given the opportunity, will employ the full spatial dimension of a molecular structure to spread out over, or to delocalize within, in an effort to minimize the potential energy of the ensemble of electrons and nuclei. It is this tendency of electrons to *delocalize* from the confinement of *atomic orbitals* to the geometrical flexibility of extended *molecular* orbitals that sets the stage for the molecular orbital (MO) class of bonding theories.

We saw that the solution to the Schrödinger equation for *atomic structure* yielded the complete wavefunction for multielectron system with a *single nucleus*. There is no limitation intrinsic to the Schrödinger equation preventing a full solution for multielectron/multinucleus systems. Thus, when the Schrödinger equation is solved for a molecule, a *molecular* orbital results, with the electrons delocalized in whatever configuration yields the structure with the lowest potential energy given the constraint that no two electrons share the same set of quantum numbers.

We can track the energy of the forming molecular orbital (MO) as the internuclear distance decreases carrying the molecular formation from the separated atoms to the fully formed molecular orbital for the constructive addition of two 1s atomic orbitals, shown at right, to form a *bonded* molecular orbital.

The linear combination of atomic orbitals for the *destructive* addition of 1s atomic orbitals is shown at right as a function of decreasing internuclear distance between the hydrogen atoms. As the internuclear distance decreases, the energy increases, forming an *antibonding* molecular orbital.

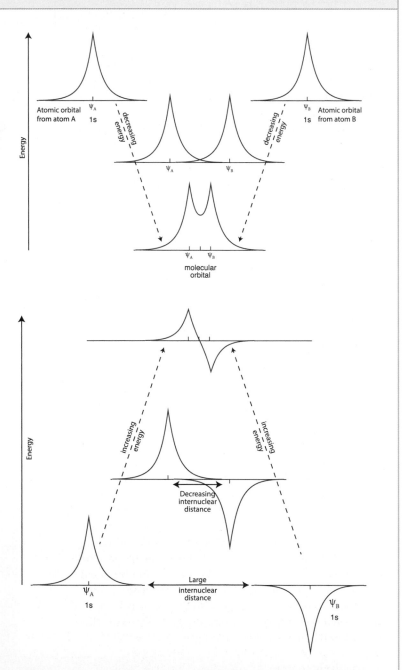

Summary Concepts

6. Combining Atomic Orbitals To Form Bonding and Antibonding Molecular Orbitals

Constructive interference of the two 1s hydrogen atomic orbitals leads to a bonding σ_{1s} *molecular orbital*. *Destructive* interference of the two 1s atomic orbitals leads to an *antibonding* σ_{1s}^* molecular orbital.

Panel A displays the relationship between the probability distribution of electrons as the internuclear distance decreases from the condition of large separation, for which there is no overlap, through the condition of marginal overlap leading to the formation of the chemical bond with maximum bond strength represented by the equilibrium internuclear distance. Panel B delineates the evolution of the molecular orbital diagram from *large* internuclear separation (for which there is no interaction of the valence electrons) through an *intermediate* internuclear distance wherein the overlap is developing between the 1s atomic orbitals leading to a separation in energy between the bonding, σ_s, and antibonding, σ_s^*, orbitals. Finally at the equilibrium internuclear distance, the σ_s and σ_s^* orbitals achieve their maximum separation and the potential energy reaches its deepest point.

7. Molecular Orbital Structure of Molecular Oxygen

The figure at right displays the case for decreasing internuclear distance as the atomic orbitals begin to interact. This condition of overlap between the atomic orbitals begins to split the molecular orbitals so formed by decreasing the energy of the bonding orbitals and increasing the energy of the antibonding orbitals. This "splits" the energy levels of the forming molecular bond with the σ_{2p} bonding orbital being lowest in energy because of the *greater overlap* between the p_x atomic orbitals aligned along the internuclear axis. The σ_{2p}^* antibonding overlap is higher in energy because of the large repulsion that results from the large overlap of the p_x orbitals combined with the absence of electron density between the two nuclei. The evolving potential energy surface of O_2 corresponds to the point on the surface representing the decrease in energy relative to the interactive energy at large internuclear distance. This decrease in energy of the potential energy surface represents the binding energy of the O_2 bond at that internuclear distance.

Development of MO theory for H_2 and for O_2 establishes the essential concepts in MO theory wherein we have formed those molecular orbitals by a simple adding together, or "mixing" of atomic orbitals; the so called linear combination of atomic orbitals-molecular orbital (LCAO-MO) theory. It is important to recognize that we need not be constrained or restricted to using those explicit mathematical functions, the wavefunctions for the isolated atoms. We can use any mathematical function we choose that can be optimized in the molecular structure to minimize the potential energy of the ensemble of electrons and nuclei in the molecular structure. It just happens that to a greater or lesser degree, depending on the molecule, the atomic wavefunctions are a good (sometimes very good) first approximation that can be successfully mixed, added together constructively or destructively, to yield the resulting molecular orbitals.

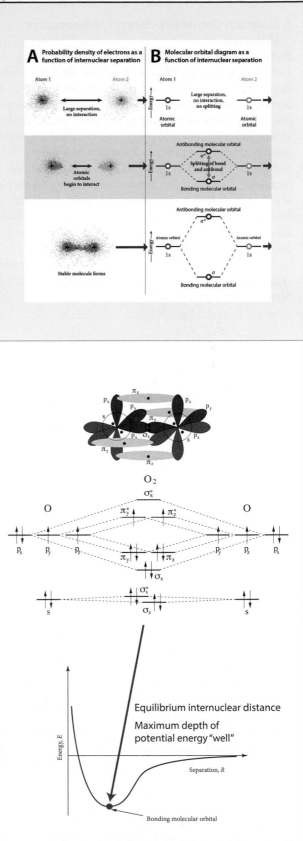

Summary Concepts

8. Molecular Orbital Structures of Heteronuclear Molecules

Nitric oxide is one of the most important molecules in nature. It plays a significant role in biological systems as a "signaling molecule"—it is released by one system in the human body to trigger a reaction in another system of the human body. For example, the inner lining of blood vessels use nitric oxide to signal the surrounding smooth muscle to relax, resulting in increased blood flow. As we will see in the MO diagram of NO, nitric oxide has an unpaired electron in its valence shell and is thus very reactive as it seeks to form an electron pair bond with any molecule with which it collides. NO has a lifetime in human systems of a few seconds, but because it is small, it diffuses easily across membranes.

The atomic orbitals for oxygen lie at somewhat lower energy than the atomic orbitals of nitrogen. The MO ordering is the same as that for C_2, N_2, and CO.

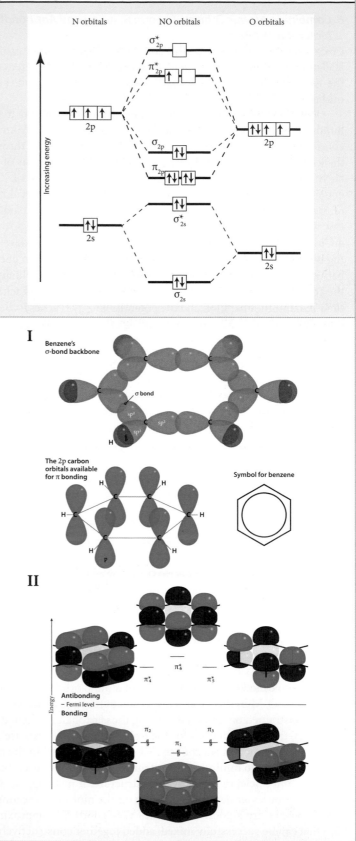

9. Molecular Orbital Bonding Structure of Benzene: Delocalization

I. Experiments have demonstrated that benzene, C_6H_6, is a planar molecule with a remarkable hexagonal geometry set in place by a backbone of carbon atoms. *All* carbon-carbon bonds are equal length (140pm) and the C-C-C bond angle in each case is 120°. Valence bond theory guides an important initial assessment of the structure of benzene. The framework of the carbon backbone is comprised of a σ bonded structure that is a network of sp²–sp² hybrid orbitals as shown in the top panel. But the unhybridized p orbitals of carbon protrude perpendicular to the plane of the σ bonding structure as displayed in the bottom panel.

II. The **π** bonding states and **π*** antibonding states of benzene: Each state is represented by its wavefunction as colored clouds (red clouds are negative values, blue clouds are positive values) and its energy as a horizontal bar. The bonding states (**π₁**, **π₂**, **π₃**) are occupied by two electrons each (with up and down spins), while the antibonding states (**π₄**, **π₅**, **π₆**) are unoccupied. The occupied and unoccupied states are separated by the long, grey horizontal line, known as the Fermi level.

CASE STUDY 11.1 Chemical Bonds, Spectroscopy, and Molecular Motion

With the development of quantum mechanics came the ability to accurately calculate bond strengths, bond angles, and the *distribution* of electronic charge within the bonding structure. This provided the quantitative foundation to calculate specifically how a given molecular structure would interact with electromagnetic radiation.

As we have seen, there is an explicit relationship between the period of motion of the charge distribution within a molecule and the wavelength of light, of the photon, that can be absorbed or emitted by a molecule. In particular, because light moves with a specific velocity, c, and because the relationship between the wavelength, λ, and the frequency, ν, of light is set by the relationship $\lambda\nu = c$, the motion of charge within the molecule must match the frequency of that emitted radiation. A molecule creates, or releases, a photon in the ultraviolet by an electron jumping from one energy level to another. If those two energy levels are separated by an energy difference, ΔE, that photon, produced by an electron *jump* between levels, will have energy

$$\Delta E = h\nu_{photon}$$

But in order to create that photon, that electron can only generate the electromagnetic wave by moving in space, oscillating, with a period of electronic motion $\tau_{el} = 1/\nu_{photon}$.

Thus the frequency of the emitted photon, $\nu_{ph} = c/\lambda_{photon}$, is *locked* to the frequency of the electronic motion. It follows that the period of motion of the electron, τ_{el}, responsible for the emission of the photon with wavelength λ_{photon} must satisfy the relationship:

$$\Delta E/h = \nu_{photon} = c/\lambda_{photon} = 1/\tau_{el}$$

We therefore have a close relationship between the structure of the chemical bond, the potential energy surface controlling the motion of charge, and the period of motion of the charge. For electronic motion, we can express this as a union between (1) the molecular orbital (MO) structure with the promotion of the electron by the absorption of a photon and (2) the transition of the molecule from the ground electronic state to an electronic state higher in energy. This is displayed in Figure CS11.1a.

Note that this locking together of the period of electronic motion, τ_{el}, and the wavelength of light emitted, λ_{photon}, when the electron jumps to another electronic energy level, has two key consequences:

1. The match between the wavelength of the photon and the period of motion of the electron holds whether the photon is emitted or absorbed.

2. The electromagnetic radiation, the photon, can interact

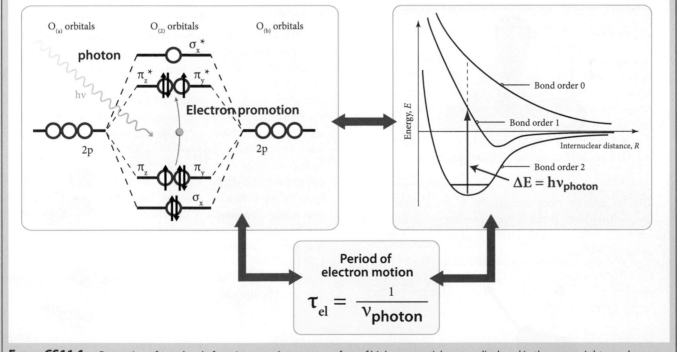

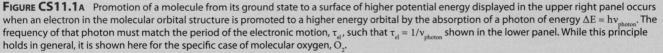

FIGURE CS11.1A Promotion of a molecule from its ground state to a surface of higher potential energy displayed in the upper right panel occurs when an electron in the molecular orbital structure is promoted to a higher energy orbital by the absorption of a photon of energy $\Delta E = h\nu_{photon}$. The frequency of that photon must match the period of the electronic motion, τ_{el}, such that $\tau_{el} = 1/\nu_{photon}$ shown in the lower panel. While this principle holds in general, it is shown here for the specific case of molecular oxygen, O_2.

with the molecule if and only if there is a spatial *change* in the distribution of charge as a result of that interaction. If the electric field of the photon cannot invoke a change of charge distribution within the molecule, the photon cannot be absorbed. If the distribution of charge within a molecule does not change, a photon cannot be emitted.

However, quantum mechanics requires that a molecule can only change its energy state in discrete steps—in quantum steps. Moreover, these quantum steps can be taken either as a result of movement of an *electron*, as in the case above, *or* by the movement of the molecule's *nuclei*, which alters the distribution of charge within the molecule.

This immediately raises the question: How can the motion of nuclei lead to the motion of charge? We already have two examples to draw from. One is the change in the electronic state of a hydrogen atom or the change in the electronic state of a molecule described above. The other is the emission of radiation from a blackbody—the thermal energy of the molecules that comprise solid matter. Solid matter that contains a combination of electrons and protons (negative and positive charges) that emits a continuum of electromagnetic radiation depending on temperature.

So how does the motion of *nuclei* lead to the absorption or emission of radiation? Let's take the water molecule as an example. First, we know the structure of the water molecule is built upon a tetrahedral structure with two lone pairs and two of the four sp³ hybrid orbitals bonded to the 1s atomic hydrogens.

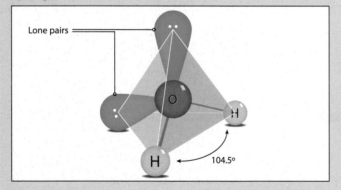

The far greater electronegativity of the oxygen draws electron density toward the oxygen, leaving the hydrogen atoms electron deficient. This creates a separation of negative and positive charge:

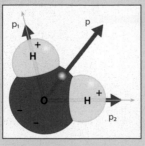

This separation of charge created by the greater electronegativity of oxygen leads to a charge separation that can be represented by a vector extending from the center of the two partial positive charges on the hydrogen to the partial negative charge in the oxygen. This displacement of charge is called the "dipole moment" of the molecule and it can be simplified to a charge separation through the axis of the molecule.

When this charge separation "sees" an electric field, the charges are forced apart in one phase of the electromagnetic wave and forced together in the next phase as shown in Figure CS11.1b.

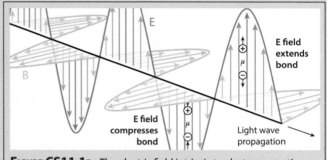

FIGURE CS11.1B The electric field intrinsic to electromagnetic radiation applies a force to a dipole that transfers energy from the photon to the vibrational motion of the molecule. Energy is transferred only if the frequency of oscillation of the electromagnetic field matches the frequency of the vibrational motion of the molecule.

Because the structure of water can accept *only* quantized radiation, radiation whose frequency matches the energy difference $\Delta E = h\nu_{photon}$ between two quantized energy levels, *only* photons with wavelength $\lambda_{photon} = c/\nu_{photon}$ will be absorbed by the *nuclear* motion of the molecule. What then are the frequencies of the nuclear motions of the water molecule, and what is the character of the molecular motion associated with each of those frequencies? Of the many different molecular motions available to the two hydrogen atoms and the oxygen atom in the structure of water, there are three distinct "modes" of vibrational motion. Those are referred to as "normal modes" because any combination of vibrational motion can be uniquely represented as a combination of those normal modes. The three normal modes are displayed in Figure CS11.1c. They are designated top to bottom in that figure as ν_1, the symmetric stretch; ν_2, the bending mode; and ν_3, the asymmetric stretch.

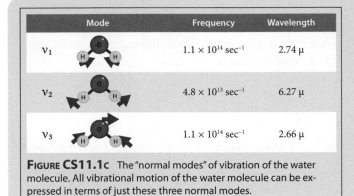

Mode		Frequency	Wavelength
ν_1		1.1×10^{14} sec^{-1}	2.74 µ
ν_2		4.8×10^{13} sec^{-1}	6.27 µ
ν_3		1.1×10^{14} sec^{-1}	2.66 µ

FIGURE CS11.1c The "normal modes" of vibration of the water molecule. All vibrational motion of the water molecule can be expressed in terms of just these three normal modes.

Listed in Figure CS11.1c are the frequencies of each of the normal modes and the corresponding wavelengths of photons that are either absorbed or emitted by the mode. Once we have the frequency of the mode, we can calculate the *period*, τ_{vib}, associated with the nuclear motion and the *wavelength* of photons that are either absorbed or emitted by the mode. Those frequencies, vibrational periods and wavelengths are summarized in Table CS11.1a.

TABLE CS11.1A

Normal Mode	Frequency	Vibrational Period	Wavelength
Symmetric stretch	1.09×10^{14} sec^{-1}	9.17×10^{-15} sec	2.74 µ
Bending	4.80×10^{13} sec^{-1}	2.08×10^{-14} sec	6.27 µ
Asymmetric stretch	1.13×10^{14} sec^{-1}	8.85×10^{-15} sec	2.66 µ

Before we consider the spectral interval over which the water molecule absorbs radiation, we consider what it is that determines the frequencies of that motion. First, we note that in general a molecular bond is represented by a potential energy surface with the characteristic shape displayed in Figure CS11.1d.

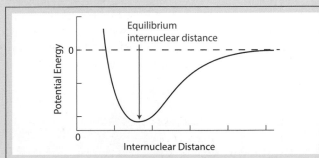

FIGURE CS11.1d Any chemical bond between two atoms has, qualitatively, the same general shape in terms of the potential energy as a function of internuclear distance. At large internuclear distance the potential energy is zero, but, as the bond forms, the potential energy drops until it reaches a minimum at the equilibrium internuclear distance. At smaller internuclear distances, the potential energy increases rapidly forming a repulsive "wall."

As we have seen, the characteristic shape of a chemical bond is a potential energy that decreases with decreasing internuclear distance until it reaches a minimum. This minimum in the potential energy surface, referred to as the equilibrium internuclear distance, is characterized by the condition that the change in potential energy divided by the change in internuclear distance, $\Delta V/\Delta r$, is equal to zero. Thus, at that equilibrium internuclear distance, no force is exerted on the nuclei. As the internuclear distance *decreases* from that equilibrium internuclear distance, the potential energy surface increases rapidly. The force felt by the nuclei, a force that seeks to push them apart, can be represented by a simple expression:

$$F = -k \times \left(\frac{\text{displacement from equilibrium}}{\text{internuclear distance}}\right) \quad \textbf{Eq.CS11.1.1}$$

As the displacement from the equilibrium internuclear distance increases, so too does the force that tries to return the nuclei to a separation equal to the equilibrium internuclear distance. This is true whether the bond length is greater than or less than the equilibrium internuclear distance. The restoring force operating on the nuclei is proportioned to the *slope* of the potential energy surface, $\Delta V/\Delta r$. If the "walls" of the PES are steep, the restoring force is large and the constant k in equation CS11.1.1 is proportionally larger. This is a characteristic true of all nuclear motion in a molecular structure, and thus of the normal modes of the water molecule. It is the *shape* of the potential energy surface that forces the nuclei back to their equilibrium position and that determines the force constant of the restoring force.

But how is the force constant, which depends upon the slope of the potential energy surface, related to the frequency of the vibrational motion of the nuclei? Using the analogy of the diatomic molecule again, the frequency of the vibrational motion depends upon the spring constant k, and the mass of the atoms in the molecular structure. Specifically, the frequency of vibration, ν_{VIB}, is given by the simple expression

$$\nu_{VIB} = \left(\frac{1}{2\pi}\right)\sqrt{\frac{k}{\mu}}$$

where k is the spring constant and µ is the "reduced mass" of the two atoms in the bond:

$$\mu = \frac{m_1 m_2}{m_1 + m_2}$$

The key points are:
1. The steeper the potential energy surface, the greater the restoring force attempting to return the nuclei in the molecule to their equilibrium position.
2. The spring constant k is proportionally larger for steeper potential energy surfaces.
3. The frequency of a vibrational mode increases with

increasing spring constant and decreases for increasing mass of the atoms in a chemical bond. Light atoms connected by stiff bonds have high frequency vibrations. Heavy atoms linked by weak bonds have low frequency vibrations.

We are now ready to connect the vibrational motion of the water molecule to the electromagnetic field in order to understand how the nuclear motion of molecules interacts with light.

An inspection of Table CS11.1a reveals that with a frequency of the three normal modes $v_1 = 1 \times 10^{14}$ sec^{-1}, $v_2 = 4.8 \times 10^{13}$ sec^{-1}, and $v_3 = 1.1 \times 10^{13}$ sec^{-1}, the *period* of molecular vibration is in the range 2.1×10^{-14} sec to 9.1×10^{-13} sec. This is typical of polyatomic molecules. The wavelength of light absorbed and/or emitted by the water molecules in its vibrational modes are $\lambda_1 = 2.74$ μ, $\lambda_2 = 6.27$ μ, and $\lambda_3 = 2.66$ μ.

We can create a table of normal modes, frequencies, vibrational periods, and wavelengths for another triatomic (3 atom) molecule, CO_2, as displayed in Table CS11.1b.

TABLE CS11.1B

Normal Mode	Frequency	Vibrational Period	Wavelength
v_1	4.2×10^{13} sec^{-1}	2.3×10^{-14} sec^{-1}	7.2 μ
v_2	2×10^{13} sec^{-1}	5×10^{-14} sec^{-1}	15 μ
v_3	7.0×10^{13} sec^{-1}	1.4×10^{-14} sec^{-1}	4.3 μ

Notice that the frequency, vibrational periods, and wavelengths fall in roughly the same range as those of water. It should be noted that CO_2 has 4 normal modes, while H_2O has three. This difference results from the fact that CO_2 is a *linear molecule* and it can bend in two distinct directions that are perpendicular to each other. Water, a molecule that is bent, does not have two equivalent directions of bending that are independent but equal in energy. This is an example of a consistent characteristic of nuclear motion in polyatomic molecules. Linear polyatomic molecules with n atoms have 3n – 5 vibrational normal modes. Nonlinear polyatomic molecules have 3n – 6 normal modes. Thus CO_2, which is linear, has 3n – 5 = (3)3 – 5 = 4 normal modes. Water, which is bent, has 3n – 6 = (3)(3) – 6 = 3 normal modes of vibration.

An inspection of Table CS11.1a and CS11.1b reveals that H_2O and CO_2 absorb radiation in their normal modes throughout the wavelength region 2.7 μ to 15 μ. If we

examine the blackbody radiation curve for the Earth's surface, which has an average temperature of 288 K, we see that this blackbody emission peaks in the region of the CO_2 and H_2O v_2 bending modes.

Problem 1

Electronic transition in most molecules occurs in the "near UV" and "vacuum UV." The near UV region is typically between 200 and 350 nm. The vacuum UV is so designated because of the very strong O_2 absorption between 100 and 200 nm—requiring that an instrument operating below 200 nm must be "pumped out," evacuated of O_2, in order for photons to pass through the instrument.

Select a representative UV wavelength, 200 nm, and calculate the frequency of electromagnetic radiation, the energy of the corresponding photon, and the period of electronic motion corresponding to that electronic transition.

CASE STUDY 11.2 Chemical Bonds, Spectroscopy, and Control of Climate

The Link Between Molecular Spectroscopy and Climate

We can summarize the absorption of infrared radiation by H_2O and CO_2 by graphically displaying the percent of absorption on the vertical axis against wavelength on the horizontal axis as shown in Figure CS11.2a. The vibrational and rotational bands of H_2O and CO_2 are displayed along with the absorption band of the bending mode of ozone at 10 μ. The 100% absorption regions means that photons emitted by the Earth's surface are absorbed by the infrared active gases before leaving the atmosphere.

Those absorbed photons then deliver their energy to those molecules, inducing vibrational and rotational energy within the molecular structure of, for example, water. Some of that vibrational and rotational energy is transferred to the translational kinetic energy of the surrounding atmosphere. But H_2O and CO_2 are both strong absorbers of IR radiation *and* strong emitters of IR radiation. In fact, in regions of the atmosphere those molecules act to cool the atmosphere by emitting more IR radiation than they absorb from lower levels of the atmosphere. The result is that the emission of IR radiation downward to the surface acts to warm the surface by absorbing upwelling IR and emitting it downward, while at the same time, molecules that are in the upper troposphere and lower stratosphere cool these regions of the atmosphere by emitting IR radiation to space. So this sets up an absorp-

tion-reemission sequence that returns a significant fraction of the IR radiation downward toward the ground. This establishes a pathway for the IR photons shown in Figure CS11.2b.

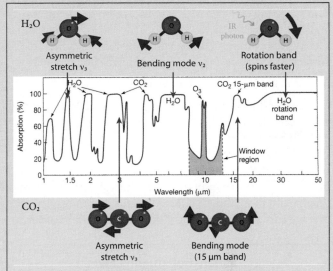

FIGURE CS11.2A The absorption of infrared radiation upwelling from the Earth's surface is caused by H_2O and CO_2 absorbing infrared radiation in their vibrational modes of symmetric stretch, bending and asymmetric stretch. Water also absorbs in its rotational mode. This diagram displays the percent absorption as a function of wavelength for a photon leaving the surface in an upward direction. Ozone also contributes to the absorption of infrared radiation at about 10 μ.

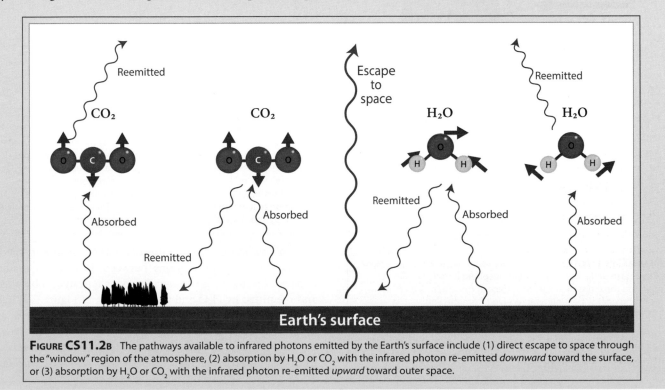

FIGURE CS11.2B The pathways available to infrared photons emitted by the Earth's surface include (1) direct escape to space through the "window" region of the atmosphere, (2) absorption by H_2O or CO_2 with the infrared photon re-emitted *downward* toward the surface, or (3) absorption by H_2O or CO_2 with the infrared photon re-emitted *upward* toward outer space.

Depending on their wavelength, photons can escape directly to space through the "window region" between 8 and 14 μ, they can be absorbed by H_2O or CO_2 and reemitted downward toward the surface, or they can be absorbed by H_2O or CO_2 and emitted upward toward space.

But the key question is, at what wavelength does the Earth's surface emit IR radiation, and how does that IR emission spectrum coincide with the absorption spectrum of H_2O and CO_2 (as well as O_3)?

We know from Chapter 1 that the flux of energy from a surface, any surface, is given by the Stefan–Boltzmann law

$$F = A\sigma T^4$$

The Earth, over time, must balance the energy per unit time received from the Sun and the energy per unit time emitted by the Earth in the IR. In the absence of an atmosphere, this is a straightforward calculation. The Earth receives 179 PW (179×10^{15} watts) from the sun, so it must radiate that same amount to space. This calculation is shown graphically in Figure CS11.2c. Without an atmosphere, the average surface temperature of the Earth would be 255 K.

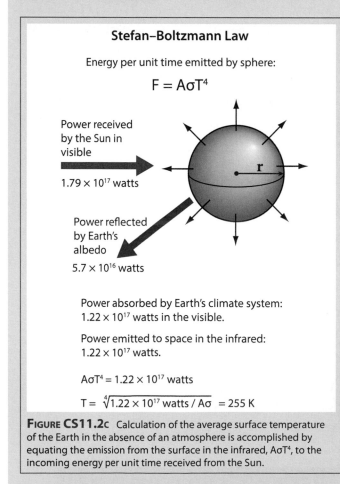

Stefan–Boltzmann Law

Energy per unit time emitted by sphere:

$$F = A\sigma T^4$$

Power received by the Sun in visible

1.79×10^{17} watts

Power reflected by Earth's albedo

5.7×10^{16} watts

Power absorbed by Earth's climate system:
1.22×10^{17} watts in the visible.

Power emitted to space in the infrared:
1.22×10^{17} watts.

$A\sigma T^4 = 1.22 \times 10^{17}$ watts

$T = \sqrt[4]{1.22 \times 10^{17} \text{ watts} / A\sigma} = 255$ K

FIGURE CS11.2c Calculation of the average surface temperature of the Earth in the absence of an atmosphere is accomplished by equating the emission from the surface in the infrared, $A\sigma T^4$, to the incoming energy per unit time received from the Sun.

The question we must answer is: what happens when we place CO_2 and H_2O in the atmosphere? The first step in answering this is to align the wavelengths of the *absorption*

of CO_2 and H_2O with the blackbody *emission* curve of the Earth's surface. The Earth's surface emits with a blackbody distribution of intensity appropriate to its temperature as shown in the bottom panel of Figure CS11.2d.

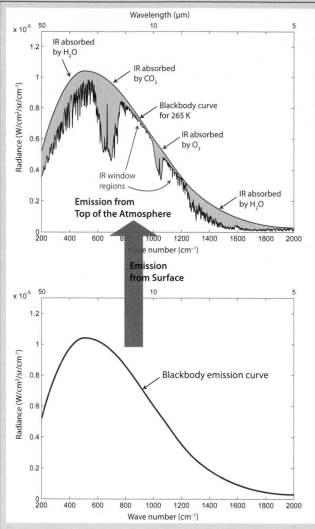

FIGURE CS11.2d The infrared radiation emitted as blackbody radiation from the Earth's surface displayed in the lower panel, must pass through the H_2O, CO_2, and O_3 in the atmosphere in order to balance the incoming energy from the Sun. Absorption in the infrared by these infrared active molecules reduces the amount of infrared radiation that escapes to space in the spectral regions where these molecules absorb. Therefore, the spectral distribution of infrared radiation that escapes to space is clearly evident in the absorption by H_2O, CO_2, and O_3 as displayed in the upper panel of the figure.

This radiation, emitted upward, either escapes to space through the window regions between 10 and 20 μ and between 7.5 and 9.5 μ, or the radiation is repeatedly absorbed and reemitted by H_2O and CO_2 as shown in Figure CS11.2b until it escapes in the higher atmosphere to space. This results in a distribution of radiation from the *top of the atmosphere* that follows this general shape of the blackbody emission from the surface but with IR radiation removed through absorption by H_2O and CO_2 as displayed in the upper panel of Figure CS11.2d. The blue areas denote the absorption of IR radiation by H_2O and CO_2, with a small contribution from O_3 at about 10 μ. This is the absorbed IR radiation that returns to the Earth's surface to increase the surface temperature.

We can create a simple model of how this works using the diagram in Figure CS11.2e. This simple model takes *incoming* power per unit area that the Earth receives from the Sun, $F_S(1 - A)/4$, where F_S is the flux of solar radiation received by the Earth in watts/m^2 and A is the Earth's albedo. This incoming solar flux is then equated with the flux of radiation emitted to space by the Earth in the IR. The flux of radiation emitted by the Earth per unit area, σT_0^4, is the flux of radiation emitted per unit area *at the surface* of the Earth. The absorption by H_2O and CO_2 is approximated by a single layer, well above the Earth's surface, that resides at some lower temperature T_1. The fraction of IR radiation absorbed by the layer of H_2O and CO_2 is f, such that the IR radiation transmitted through the layer is $\sigma T_0^4 (1 - f)$. But strong absorbers of radiation, of photons in the infrared, are also strong emitters in the IR.

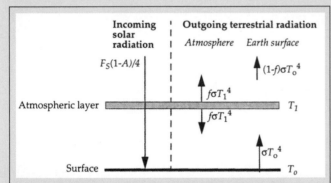

FIGURE CS11.2E We can establish a "climate model" by representing the absorption of infrared radiation by H_2O, CO_2, and O_3 as taking place in a single layer of the atmosphere at temperature T_1, a single layer above the surface that absorbs a fraction, f, of the outgoing infrared radiation emitted at temperature T_0 from the surface. That absorbing layer reemits radiation both downward and upward. By equating the incoming radiation from the Sun in the visible spectral region to the outgoing radiation from the Earth in the infrared, we can calculate the surface temperature in the presence of absorption in the atmosphere by H_2O, CO_2, and O_3.

The ability of H_2O and CO_2 to absorb determines f, the amount of radiation removed from the outward flow of IR from the surface. The IR absorbed by the H_2O and CO_2 in the atmosphere must be reemitted in the IR. IR radiation is emitted from the absorbing layer, both upward *and* downward according to the temperature of the layer, T_1, and the strength of the emission, f. Thus an amount of power per unit area of $f\sigma T_1^4$ is emitted both upward and downward. Thus the *outgoing* infrared radiation from the Earth is given by:

$$\sigma T_0^4 (1 - f) + f\sigma T_1^4$$

But that outgoing IR radiation must balance the incoming radiation from the Sun in the visible from Figure CS11.2d

$$F_S (1 - A)/4 = \sigma T_0^4(1 - f) + f \sigma T_1^4$$

However, there is one more equation available to us. The energy emitted by the absorbing layer must equal the energy absorbed, so

$$2 f\sigma T_1^4 = f\sigma T_0^4$$

So we have, solving for T_0, the expression

$$T_0 = \sqrt[4]{2}\, T_1$$

Inserting this expression for T_1 into our energy balance expression gives

$$F_S (1 - A)/4 = \sigma T_0^4 (1 - f) + f\sigma T_0^4/2$$

Solving for the Earth's surface temperature we have

$$T_0 = \left[\frac{F_S (1 - A)}{4\sigma \left(1 - \frac{f}{2}\right)} \right]^{1/4}$$

We thus have a direct link between the bonding structure of H_2O and CO_2, the spectroscopy of H_2O and CO_2, and a climate model of the Earth.

Problem 1

Given the simple (but powerful) model of the Earth's climate presented in this Case Study, calculate the Earth's surface temperature if the fraction of IR radiation absorbed by H_2O and CO_2 is f, and f = 0.77.

CASE STUDY 11.3   Materials: From Molecular Orbital Theory to Solid-State Electronics

I. Introduction

Materials play a very important role in human society. This is reflected in the fact that the phases of human civilization are usually characterized by the type of materials used to make the basic tools and even the not quite basic but still important ornaments and other objects of everyday life. Examples are displayed in Figure CS11.3a. We refer to the Stone Age as the period when humans had mastered the formation of tools from stone, extending up to 5000 years BC, which was followed by the Bronze Age (roughly from 5000 to 1000 years BC) and the Iron Age (from 1000 years BC through the 20th century AD). In the middle of the 20th century, humans began to extensively use materials that are not found in nature as such, or in pure enough forms to be used for the purposes of modern technology.

The characteristic material of the current era, on which most electronic devices are based, is silicon. Silicon is a semiconductor used for the manufacturing of microchips for computers but also in other applications like solar cells in photovoltaic devices, two examples of which are shown in Figures CS11.3b and CS11.3c. Materials similar to silicon, like gallium arsenide, are used for electronic devices in CD and DVD players, cell phones, etc. The operation of these devices is based on what are called "doped semiconductors," a class of materials with very specific electronic properties. The applications of these materials have affected human civilization in literally all of its aspects. The era of human civilization from the late 20th century to date has been dubbed the "Silicon Age." Perhaps a more appropriate term would be the "Quantum Artificial Materials Age," since silicon is only one example of a broad class of materials that are artificial (in the sense they are not found in nature but must be manufactured artificially, often with great difficulty and at great cost), and whose operation is based on purely quantum mechanical effects.

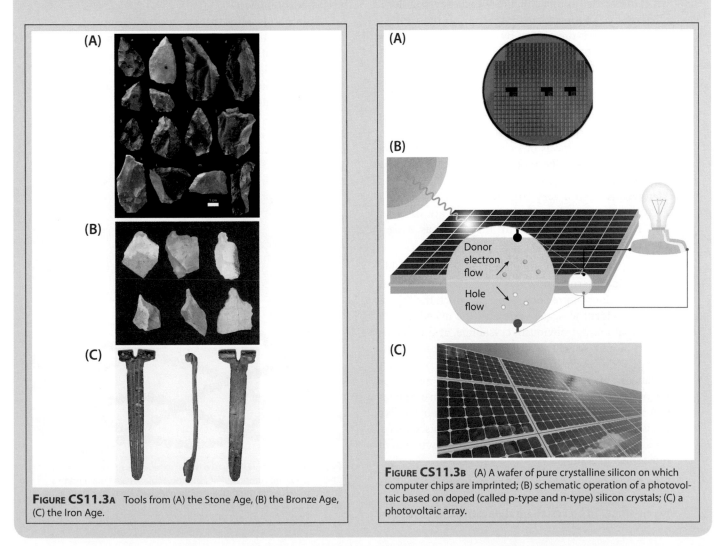

FIGURE CS11.3A   Tools from (A) the Stone Age, (B) the Bronze Age, (C) the Iron Age.

FIGURE CS11.3B   (A) A wafer of pure crystalline silicon on which computer chips are imprinted; (B) schematic operation of a photovoltaic based on doped (called p-type and n-type) silicon crystals; (C) a photovoltaic array.

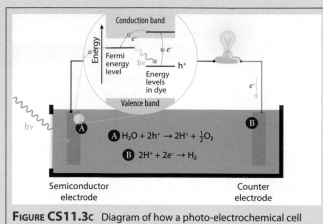
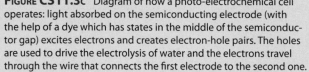

FIGURE CS11.3c Diagram of how a photo-electrochemical cell operates: light absorbed on the semiconducting electrode (with the help of a dye which has states in the middle of the semiconductor gap) excites electrons and creates electron-hole pairs. The holes are used to drive the electrolysis of water and the electrons travel through the wire that connects the first electrode to the second one.

It is difficult to imagine an aspect of modern life that does not depend in a crucial manner on advanced materials—just think of the number of times each day you use a device like your cell phone, your computer, your iPod, even a car, which nowadays is mostly run by computers (around the year 2000, the computer components in cars started exceeding in value that of the steel content, and this trend is accelerating). And this does not include the countless other objects that you use every day (from ceramic plates, to sun glasses, to the airplane you board to travel, to your tennis racket) that involve artificial materials of various levels of sophistication.

There are several *aspects* of materials that are different from simpler structures, like molecules consisting of a few atoms, that we have discussed so far.

a. Materials contain a very large number of atoms. In a *cubic centimeter* of a typical solid there are of order 10^{22}–10^{24} atoms (same order of magnitude as Avogadro's number).

b. The atoms in materials are usually arranged in highly symmetric patterns, known as crystal structures. This is most obvious in precious stones, in which the crystal order is evident at scales visible to the naked eye. In solids that are not obviously crystalline, at the microscopic, atomic scale, the atoms are more or less regularly placed with crystalline order extending over small distances (micrometers, 1 μm = 10^{-6} m), and these microcrystals are joined at their surfaces in irregular patterns. This is the case for many common objects like utensils, chocolate bars, candles, etc. As a consequence, the structure of crystals is the basis for the discussion of most solid structures.

c. The presence of crystalline order produces another important effect: the electrons that form the link between atoms in a crystal are **delocalized**, that is, they spread throughout the entire solid. This applies to each and every individual electron in a crystal. It is a consequence of the quantum mechanical nature of electrons, and it has important implications for the behavior of solids.

d. Solids exhibit an enormous range of physical properties. For example, their electrical conductivity can range from very good conductors to very good insulators, with a difference of the two extremes of behavior spanning a range of 30 orders of magnitude! Solids also have a wide range of mechanical properties (from very brittle to very ductile), as well as a number of other properties like magnetic behavior, thermal behavior, superconducting behavior, etc. The huge range of physical properties is what makes them useful in applications.

II. Categories of Solids

We consider first the various categories of solids, which may be classed as follows:

1. Amorphous solids that are "glasses" comprised of long molecules that become entangled to form the solid. The structure of amorphous solids lack the long-range order found in highly ordered solids. A familiar example of an amorphous solid is common glass used to make bottles, windows, etc. The material is brittle and it breaks into incoherent shapes when sufficient force is applied, as displayed in Figure CS11.3d.

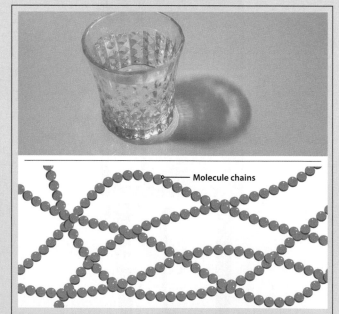

FIGURE CS11.3d Of the two major categories of solids, the amorphous solids are comprised of long molecular chains that create a strong structure by the process of tangling those chains. There is no long-range systematic order in these "glasses."

2. A second general category of solids is represented by an array of crystalline structures that are all characterized by an *ordered* arrangement of atoms that comprise the solid. Within this crystalline category there are four prominent examples, as displayed in Figure CS11.3e.

    a. **Ionic crystals** have cations and anions at specific lattice sites. Ionic crystals, such as table salt, are relatively hard; they have high melting points and are brittle. These properties reflect opposite charge as well as repulsion that occurs when ions of like charge are in close proximity. Ionic compounds do not conduct electricity in their solid states, but when melted or dissolved they conduct electricity well.

    b. **Molecular crystals** are solids in which the lattice sites are occupied either by atoms (as in solid argon or krypton) or by molecules (as in solid CO_2, SO_2, or H_2O). If the molecules of such solids are relatively small, the crystals tend to be soft and have low melting points because the particles in the solid experience relatively weak intermolecular attractions. In crystals of argon, for example, the attractive forces are exclusively London forces. In SO_2, which is composed of polar molecules, there are dipole-dipole attractions as well as London forces. And in water crystals (ice) the molecules are held in place primarily by strong hydrogen bonds.

    c. **Covalent crystals** are solids in which lattice positions are occupied by atoms that are covalently bonded to other atoms at neighboring lattice sites. The result is a crystal that is essentially one gigantic molecule. These solids are sometimes called **network solids** because of the interlocking network of covalent bonds extending throughout the crystal in all directions. A typical example is diamond. Covalent crystals tend to be very hard and to have very high melting points because of the strong attractions between covalently bonded atoms. Other examples of covalent crystals are quartz (SiO_2, found in some types of sand) and silicon carbide (SiC) a common abrasive used in sandpaper.

    d. **Metallic crystals** have cations at lattice sites surrounded by mobile electrons. Metallic crystals have properties that are quite different from those of the other three types. Metallic crystals conduct heat and electricity well, and they have the luster characteristically associated with metals.

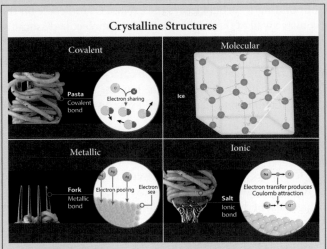

FIGURE CS11.3E The second category of solids is the crystalline structure. There are four primary examples: ionic crystals, molecular crystals, covalent crystals, and metals.

III. Building Solids from Atoms

Molecular orbital theory provides the quantum mechanical basis for a striking array of phenomena associated with the interaction of photons with molecular structure. Figure CS11.3f demonstrates the coupled concepts treating two important examples. As we progress from the upper left corner of Figure CS11.3f across the top of the figure, we recall the case for molecular oxygen, O_2, which has a molecular orbital structure constructed from the p orbitals of atomic oxygen forming the bonding sigma orbitals, σ^b, and the bonding π^b orbitals as discussed in the Chapter core. In addition, the antibonding π^* orbitals and the antibonding σ^* orbitals are formed from the destructive interference of the p orbitals of opposite phase. Given the eight electrons in the p atomic orbitals of the two oxygen atoms, we insert those electrons in pairs into the lowest orbitals filling the bonding orbitals that account for six of the eight electrons. The last two electrons enter the antibonding π^* orbital, one electron in one of the available π^* orbitals, the other in the other π^* orbital such that the O_2 molecule has two unpaired electrons. This sets the foundation for understanding the interaction of a photon with the molecular structure of O_2. In a diatomic molecule such as molecular oxygen, an electron is promoted from a bonding orbital to an antibonding orbital. The range of wavelengths of the photon that can be absorbed is determined by the relationship between the ground state potential energy surface and the excited state potential energy: specifically the *energy difference* between those two states and the *shape* of the upper potential energy surface with respect to the wavefunction of the ground vibrational state as displayed graphically in the upper right panel of Figure CS11.3f. Absorption of a photon of sufficient energy to promote the electron to an antibonding π^* orbital leaves the excited O_2 molecule on the repulsive wall of the upper state and the molecule dissociates to two oxygen atoms: $O_2 + h\nu \to O + O$.

When photons interact with a solid, we can understand the process by again using the molecular orbital formulation, but with a different model for the binding structure. It is the development of the "band model" for solids to which we now turn.

Band Model for Bonding in Solids

First note that when we developed the MO theory of bonding for molecules, those bonds were, in general, comprised of highly directional σ and π bonds. When we first discussed categories of bonding types (ionic, covalent, metallic) in Chapter 10, it was pointed out that metals are characterized by cations of the metal embedded in a sea of electrons. The bonding theory that has emerged to describe the structure

of metals and other solids is the *band theory of solids*. We turn now to a development of this band theory by building up from individual atomic orbitals.

To build this "band model" we begin with two lithium atoms as the first example. When we add the 2s atomic wave-functions for the lithium atoms we form two molecular orbitals as shown in the upper panel of Figure CS11.3g. This is virtually identical to the formation of H_2 from two hydrogen atoms. The only difference at this point is that the bonding σ_{1s} and antibonding σ^*_{1s} orbitals of H_2 are formed from the linear combination of two 1s atomic orbitals. In the case of Li_2, the bonding and antibonding orbitals of Li_2 are formed from two 2s atomic orbitals.

When we add the third Li atom, we have three atomic orbitals, each of which can be occupied by up to two electrons, one spin up and one spin down. The molecular orbital lowest in energy is the orbital formed from the constructive

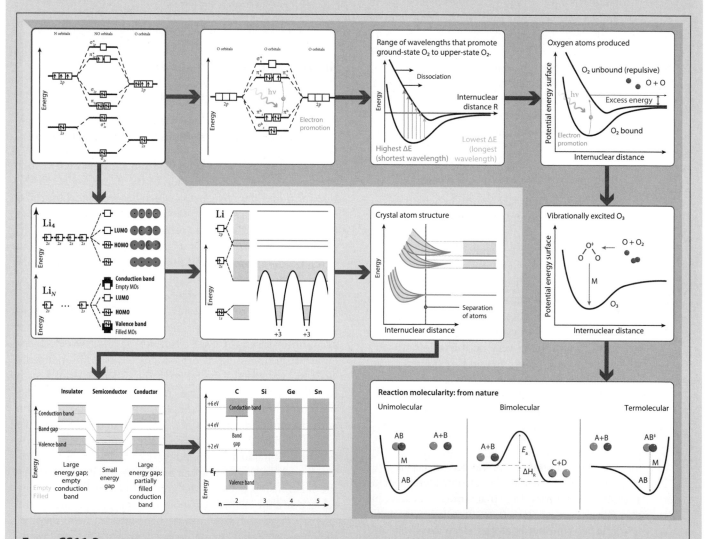

FIGURE CS11.3F Molecular orbital (MO) theory or the linear combination of atomic orbitals (LCAO) theory constitutes the basis for an understanding of two objectives: (1) the absorption of photons by a molecule and (2) the band theory of solids. Objective (1) is traced from the upper left across the top of the diagram to the lower right of the diagram. Objective (2) is traced from the upper left down the left-hand side across to the center of the diagram down to the band model in the lower left.

interference of the three 2s atomic orbitals as shown in the second panel down in Figure CS11.3f that is displayed at higher resolution in Figure CS11.3g.

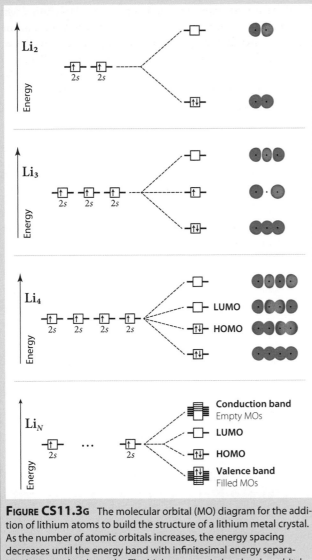

FIGURE CS11.3G The molecular orbital (MO) diagram for the addition of lithium atoms to build the structure of a lithium metal crystal. As the number of atomic orbitals increases, the energy spacing decreases until the energy band with infinitesimal energy separation between levels results. The highest occupied molecular orbitals (HOMO) are filled and the lowest unoccupied molecular orbitals (LUMO) are empty. They meet at the Fermi level, E_f.

The molecular orbital lowest in energy in the second panel down in Figure CS11.3g contains two electrons. The next orbital higher in energy has a node between the two outer Li atoms and it contains one electron. The next higher molecular orbital has two nodes and is unoccupied. These three molecular orbitals of Li_3 follow the same model structure and phase relation as the simple wavefunctions for a particle-in-a-box as displayed in Figure CS11.3h. A key point is that, with Li_3, the electrons are exhibiting delocalization across the dimension of the molecular structure of Li_3.

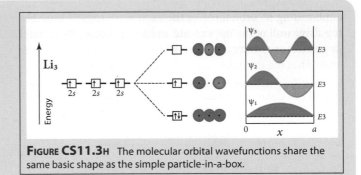

FIGURE CS11.3H The molecular orbital wavefunctions share the same basic shape as the simple particle-in-a-box.

With the addition of the fourth Li atom there are now 4 molecular orbitals formed with electrons occupying the lower two molecular orbitals and the upper two molecular orbitals are vacant, as shown in Figure CS11.3g. As we add additional Li atoms, three characteristics of the bonding structure emerge

1. The energy *spacing* between the energy levels of the molecular orbitals decreases as the number of atomic orbitals (and thus electrons) is increased.

2. The electrons are free to delocalize across the dimension of the molecular orbital, which in turn corresponds to the dimension of the metal crystal.

3. There is a clear transition between the occupied molecular orbitals and the unoccupied molecular orbitals.

4. The clearly defined transition point between the filled molecular orbitals and the empty molecular orbitals is termed the Fermi level, E_f.

5. Below the Fermi level, the continuum of filled molecular orbitals (MO_s) constitutes a band of closely spaced energy levels called the *valence band*.

6. Above the Fermi level, the continuum of empty MOs is called the *conduction band*.

This constitutes the *band theory of solids*, and this model makes two important predictions. *First*, all the occupied orbitals are strongly bonding, leading to substantial physical stability of the resulting solid. Second, the energy spacing between the highest occupied MO (HOMO) and the lowest unoccupied MO (LUMO) is very small, such that only a very small increment of energy is required to promote an electron from the valence band to the conduction band. This explains the high conductance of metals—the electrons are free to move through the crystal lattice via the conduction band such that when a voltage is applied to the metal, electrons flow freely from the negative pole of the battery to the positive pole of the battery.

To complete our picture of the band structure in a lithium crystal, we consider the relationship between the energy bands in a lithium crystal and the coulomb potential that the electrons "see" from the +3 charge of the nucleus in Figure CS11.3i. As developed above, the atomic orbitals give rise to a band of energies: The 1s band is low lying and captured in the Coulomb well. The 1s band is narrow because the outer 2s electron prevents any significant overlap of the 1s atomic orbitals in the crystal.

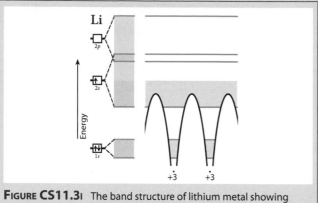

FIGURE CS11.3ı The band structure of lithium metal showing the bands developed from the 1s, 2s, and 2p atomic orbitals are displayed with the Coulomb potential of the individual lithium atoms in the metal crystal. The valence band is half full and the electrons are free to move or delocalize across the extent of the metal crystal.

The band structure resulting from the combination of 2s orbitals has the valence band that fills half the available MOs—up to the Fermi level, E_f. But that provides a "sea" of electrons that are able to delocalize across the full dimension of the crystal structure. The empty 2p band overlaps in energy with the unoccupied segment of the 2s band. This overlap occurs when the metal-metal bond energy is comparable to the valence orbital energy splitting. If we move to beryllium, with the additional electron in each atom, the 2s band is filled, but because the 2s and 2p bands overlap, the unoccupied 2p band is available as a conduction band and thus Be is a good conductor.

Nonmetallic Solids

While we developed our band model using a metal as the atomic species that combines to form the molecular orbitals that in turn evolve into a continuum of states that comprise a band, it is important to investigate the available elements more broadly to extend this band model to nonmetallic species. This is most effectively done by investigating the band structure of a coherent series of cases.

To do this we explore the bonding in solids by examin-

FIGURE CS11.3ᴊ We are concerned with the elements at the interface of the metals and nonmetals in the periodic table. Metals are conductors; nonmetals are insulators. Semiconductors lie in-between and are often referred to as metalloids in the periodic table.

ing the Group 4A elements as displayed in Figure CS11.3j. This allows us to investigate the metal-nonmetal divide because the elements become progressively more metallic in character as we move down the Group.

In Group 4A, carbon, silicon, and germanium (C, Si, and Ge) all form solids that are three-dimensional networks in which all the atoms have their normal valence, each making four bonds with neighboring atoms. Perhaps the most familiar example is the structure of diamond in which each C is tetrahedrally (sp^3) bonded to four nearest neighbor carbons. This geometry is displayed in Figure CS11.3k and is identical to the bonding structure of methane with the hydrogens replaced by carbon atoms.

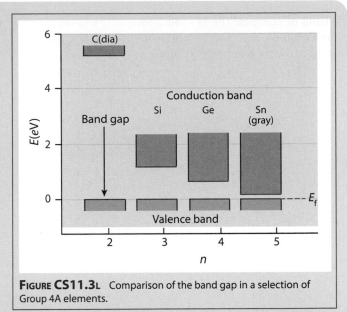

FIGURE CS11.3L Comparison of the band gap in a selection of Group 4A elements.

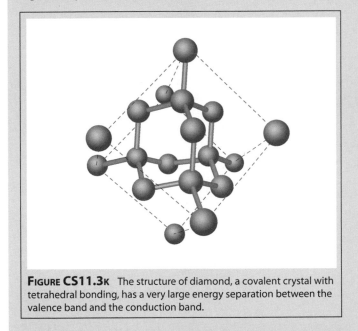

FIGURE CS11.3K The structure of diamond, a covalent crystal with tetrahedral bonding, has a very large energy separation between the valence band and the conduction band.

While the structure of diamond is considerably more complex than Li metal, the common feature is that the entire diamond crystal is a single giant molecule. When we combine carbon atomic orbitals to form the MOs of the diamond crystal, a sizable energy gap or *band gap* occurs between the highest occupied molecular orbital (HOMO) and lowest unoccupied molecular orbital (LUMO). This translates in the solid to a large energy gap between the (occupied) valence band and the (unoccupied) conduction band. In fact the *band gap* in diamond is a remarkably large 5.5 eV separating the valence band and conduction band. It is for this reason that diamond is such a good electrical insulator.

Stepping down Group 4A, the metallic character of the elements is increasing, with the band gap decreasing to 1.1 ev for Si, 0.8 eV for Ge, and 0.15 eV for Sn. The band gaps for C(dia), Si, Ge, and Sn are summarized graphically in Figure CS11.3l. The small band gap in silicon lies at the heart of why Si has found tremendous usefulness in the electronics industry—an issue we will explore here.

Problem 1

a. Beginning with molecular orbital theory of molecular bonding, explain the origin of the concept of the "band gap."

b. Why does Si have a larger band gap than Ge?

Problem 2

a. What aspect of molecular bonding structure distinguishes crystalline structures?

b. Briefly discuss what distinguishes ionic crystals from molecular crystals from covalent crystals from metallic crystals.

CASE STUDY 11.4 Foundation of Modern Electronics: From the p-n Junction to Photovoltaics and LEDs

IV. The p-n Junction

The reason silicon has become so important in electronic circuit design is that small amounts of judiciously selected impurities implanted in the Si crystal can be used to adjust the band gap of the resulting "doped" crystal structure.

To investigate this technique of doping semiconductors, consider the case of germanium, the next element below Si in Group 4A. In order to understand how the doping of a crystal lattice can alter the band gap, consider the case of a few parts-per-million of gallium (Ga from Period 3A) implanted in a crystal of germanium (Ge from Period 4A). Those seeded Ga atoms will randomly replace Ge atoms in the lattice as shown in Figure CS11.4a.

Because gallium is from Group 3A, it has one less electron than germanium, and this leaves vacancies or "holes" that form a narrow *empty band* just at the top of the valence band. This is shown on an energy level diagram in Figure CS11.4a. Rather than a band gap of 1 eV, electrons in the valence band "see" an energy gap of 0.01 eV, so electrons, albeit in limited numbers because of the part-per-million mixing ratio of gallium in germanium, are more easily elevated or promoted into those holes. The net result is that this doped material has far greater electrical conductance than does pure germanium, and the degree of electrical conductance can be "tuned" by adjusting the amount of gallium doped into the germanium lattice. In the language of semiconductor physics, germanium with a Group 3A impurity is referred to as a "p-type" semiconductor where p designates the positive holes created by the doping impurity—in this case gallium. The addition of a Group 3A impurity to a Group 4A crystal lattice is not the only way to tune the conduction properties of a material.

If a Group 5A impurity is doped into a Group 4A crystal lattice, electrons are *added* to the crystal structure as displayed in Figure CS11.4b for the case of arsenic (As) doped into germanium. These few excess electrons add a valence electron to a narrow filled band at the lower edge of the conduction band, again significantly increasing the conductance, decreasing the resistance of the material. This doping method results in an n-type semiconductor with an energy ordering shown in Figure CS11.4b. Electrons need only jump 0.01 eV in energy to enter the conductor band, rather than the full band gap of 1 eV.

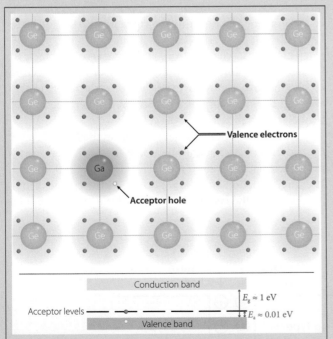

FIGURE CS11.4A When a trace of (Group 3A) gallium is doped into a crystal lattice of (Group 4A) germanium, the gallium has one fewer valence electron than germanium, so gallium provides an acceptor "hole" into which an electron can move. When the electron moves to the left, the hole moves to the right. This means that a limited number of acceptor levels exist just above the valence band and thus the conductance of the material and the Fermi level can be "tuned" by the addition of the impurity.

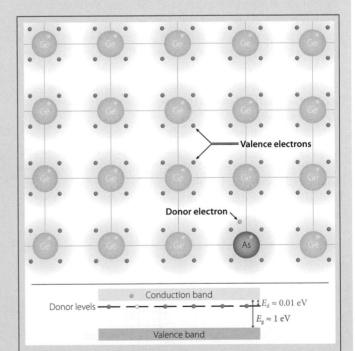

FIGURE CS11.4B With the addition of trace amounts of (Group 5A) arsenic to (Group 4A) germanium, an extra electron is introduced into the crystal structure, raising the Fermi level to the lower edge of the conduction band and thereby placing a limited number of electrons in the conduction band where they are free to move through the metal crystal.

On the face of it, this control over the resistance of a solid-state material by doping impurities may seem inconsequential, but this is far from the case. The reason is that when the p-type semiconductor is used in *combination* with an n-type semiconductor, a unique and powerful device results. If an n-type semiconductor is brought in direct contact with a p-type semiconductor, a p-n junction is created. The p-n junction has the unique attribute that it conducts electrons in only *one* direction, from the n side to the p side. The p-n junction is the heart of a vast range of devices that are central to modern life:

1. transistors for electronics

2. rectifier diodes for power systems

3. light emitting diodes (LEDs) for lighting

4. solar cells for the production of electricty from sunlight

Something interesting happens when a p-doped and an n-doped semiconductor are placed in contact. Consider the interface between the two sides. The extra electrons on the n-type side will "see" the hole states on the p-type side and will cross the interface to fill those states. When an electron leaves the n-type side, it leaves a hole behind. As a consequence, the motion of electrons from the n-type to the p-type side is equivalent to the motion of holes in the reverse direction. Once a certain number of electrons have transferred from the n-type to the p-type side (and the same number of holes have traveled in the opposite direction), a layer of negative charges is set up just inside the p-type side and a layer of positive charges is set up just inside the n-type side. These layers strongly repel more charges of the same sign to make the transition from one to the other side. The region over which the charges have made this transition is known as the **depletion region**.

Consider diagrammatically what happens when the p-type material is brought in contact with the n-type material. Before the materials are brought into electrical contact we can sketch both the physical position of the two materials and the energy diagram referenced to the Fermi level as shown in Figure CS11.4c.

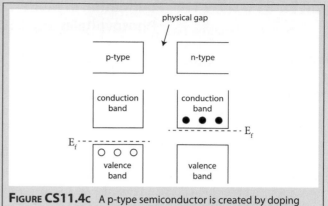

FIGURE CS11.4c A p-type semiconductor is created by doping a trace amount of an element with acceptor "holes" in the crystal structure so the valence band has an excess of positive holes. Thus the designation "p-type." An n-type semiconductor is created by doping a limited number of atoms with an excess electron such that the conduction band is occupied by a limited number of electrons.

When the p-type material is brought into electrical contact with the n-type material, the excess electrons in the n-type material move toward the p-type material and the holes in the p-type material move toward the n-type material across the newly formed "junction" between the two materials as shown in the left-hand panel of Figure CS11.4d.

The result is an interface or "junction" between the p-type and n-type with excess electrons on the p side of the newly formed p-n junction and excess holes on the n side of the junction as shown in the right-hand panel of Figure CS11.4d.

This shift in charge at the junction serves to block any further charge shift because the surface of excess electrons repels any further movement of electrons from the n to the p side and the excess of holes blocks any further migration of the holes from the p side to the n side.

The next question is: what happens to the Fermi levels of the two materials when the junction is formed? There are two principles that guide us to the answer to this question:

1. The formation of the junction affects the charge distribution in the *immediate vicinity* of the junction, but has

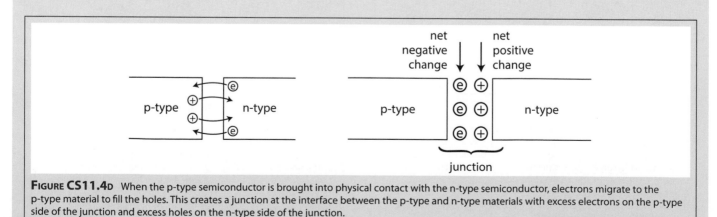

FIGURE CS11.4d When the p-type semiconductor is brought into physical contact with the n-type semiconductor, electrons migrate to the p-type material to fill the holes. This creates a junction at the interface between the p-type and n-type materials with excess electrons on the p-type side of the junction and excess holes on the n-type side of the junction.

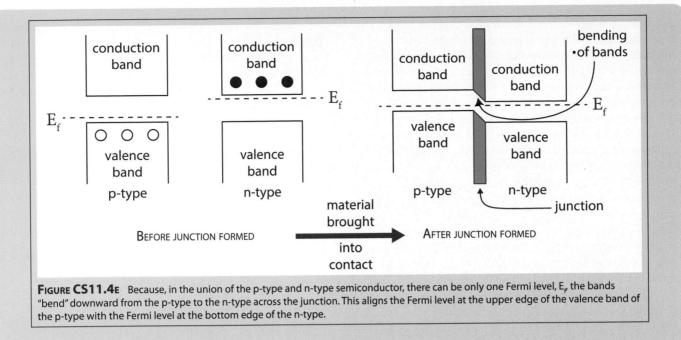

FIGURE CS11.4E Because, in the union of the p-type and n-type semiconductor, there can be only one Fermi level, E_f, the bands "bend" downward from the p-type to the n-type across the junction. This aligns the Fermi level at the upper edge of the valence band of the p-type with the Fermi level at the bottom edge of the n-type.

no effect on either the p- or n-type material within the bulk of those materials.

2. There can be only *one* Fermi level across the junction because there cannot be a discontinuity in energy—if there were, the charges rearrange to eliminate the discontinuity in energy. Thus, by bringing the two materials, one p-type and the other n-type, into contact, the Fermi level, E_f, is established across the junction.

However, we know from Figures CS11.4a and CS11.4b that the Fermi level in a p-type material coincides with the *upper* edge of the valence band but the Fermi level in an n-type material coincides with the *lower* edge of the conduction band. This situation must be reconciled with the fact that the Fermi level must be single valued and continuous across the junction. As we bring the p-n junction together from the separated state we can envision a sequence taking place as the electrical contact is made as shown graphically in Figure CS11.4e.

Because there must be a single Fermi level separating the valence band from the conduction band there occurs an effect called *band bending* that is a direct result of the electric field set up by the layer of electrons on the p-side of the junction and the layer of holes on the n-side of the junction. Thus the electric field is a result of the movement of charge when the electrical contact was made in the formation of the junction by bringing the materials into physical contact as shown in Figure CS11.4f.

So we can now summarize in a single diagram displayed in Figure CS11.4f what has occurred in the formation of our p-n junction. On the left is the bulk p-type material with a slight (part-per-million) excess in holes—a.k.a. the p side. On the right is the bulk n-type material with a slight (part-

per-million) excess in electrons—the n side of the p-n junction. The junction has a slight excess of electrons on the edge of the junction on the p-side and a slight excess of holes on the n-side that generates an electric field E across the junction with field direction toward the p side as shown in Figure CS11.4f.

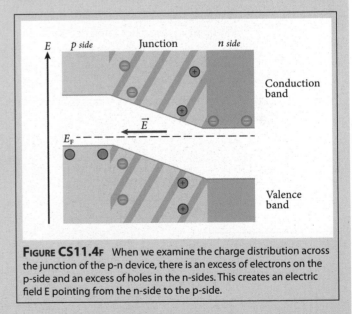

FIGURE CS11.4F When we examine the charge distribution across the junction of the p-n device, there is an excess of electrons on the p-side and an excess of holes in the n-sides. This creates an electric field E pointing from the n-side to the p-side.

The Fermi level, E_f, is invariant (i.e. at the same energy) across the junction and this results in the "bending" of the bands on the n side down with respect to the p side—the phenomena of band bending is common in the discussion of p-n junctions.

At this point, it may not be immediately apparent why the p-n junction constitutes the basis for all modern electronics from cell phones to solar cells to computers. To demonstrate why this is, consider what happens when a battery is attached to the p-n junction. In contrast to the case of the metal wire, now it makes a huge difference on how we attach the battery to the system. If the positive pole on the battery is attached to the p-type side, then it will attract the electrons that have passed to the p side from the n side. This will free up the hole states where these electrons had been residing, which will allow more electrons from the n-type side to cross the interface and replenish the negative charges in the depletion region, and this process will keep going as long as the battery is connected to the system. Equivalently, the holes will be attracted to the negative pole of the battery, liberating some states on the n-type side where more holes can jump from the p-type side. This way of attaching the battery to the p-n junction is called **forward bias** and corresponds to current flowing through the device (electrons to the left and holes to the right).

In contrast to this, the opposite polarity has a very different effect: if the positive pole of the battery were attached to the n-type side it would repel the holes on this side of the interface, and the negative pole on the p-type side would repel the electrons on this side. The net effect would be that no current could flow through the device. This is called **reverse bias**.

The device in the forward bias mode, with current flowing through it, represents the state 1 in a digital circuit, whereas in the reverse bias mode, with no current flowing through it, represents the 0 state. This is all that is needed to build up a representation of any number in binary code. This is the basis of operation of all digital electronic devices.

The condition of a forward bias applied to the p-n junction is displayed in Figure CS11.4g.

When the voltage is reversed the current is very small and in the opposite direction as shown in Figure CS11.4h. But the "gate" created by the simple "on state" resulting from the forward bias condition that corresponds to a 1 and the "off state" corresponding to a 0 in binary code that constitutes the basis for all modern computers, is only the beginning of the power that the p-n junction brings to modern electronics.

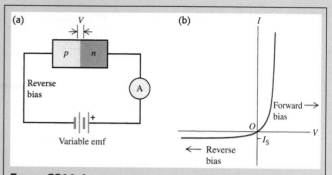

FIGURE CS11.4H When the positive pole of the battery is connected to the n-side of the p-n junction, the electric field across the junction is increased and electrons are blocked from crossing into the p-side and the p-n junction is reverse biased.

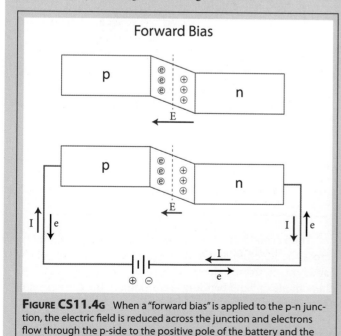

FIGURE CS11.4G When a "forward bias" is applied to the p-n junction, the electric field is reduced across the junction and electrons flow through the p-side to the positive pole of the battery and the holes flow through the n-side to the negative poles of the battery.

11.78

V. Semiconductor Devices and Light

We are already familiar with the relationship between the *absorption* of photons of light and the elevation of an electron from a lower energy state to a higher energy state—the Bohr atom was our first example. We also saw that when an electron in an atom jumps to a lower energy level a photon is *emitted*. Recall our diagram from Chapter 8 that briefly summarizes these two processes, as shown in Figure CS11.4i.

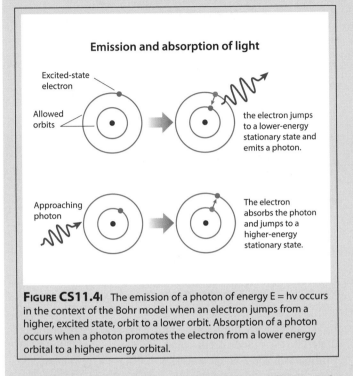

Emission and absorption of light

Excited-state electron

Allowed orbits

the electron jumps to a lower-energy stationary state and emits a photon.

Approaching photon

The electron absorbs the photon and jumps to a higher-energy stationary state.

Figure CS11.4i   The emission of a photon of energy E = hν occurs in the context of the Bohr model when an electron jumps from a higher, excited state, orbit to a lower orbit. Absorption of a photon occurs when a photon promotes the electron from a lower energy orbital to a higher energy orbital.

We also recognize that the photoelectric effect, a study of which was a critical part of the recognition that light possessed particle properties (the photon) as well as wave properties, was the absorption of a photon followed by the ejection of an electron from the surface of a metal. From an *energy perspective*, we can capture the relationship between the photon absorption and the ejection of the electron in Figure CS11.4j that shows the potential energy well that retains the electron at the surface of the metal. Whatever excess energy is available after liberating the electron from the metal surface goes into the kinetic energy of the ejected electron.

But we are interested here in developing an understanding of the interaction of photons with the p-n junction of a semiconductor device. To this end we can express the absorption of a photon by an electron-hole pair in terms of the potential energy well and the annihilation (absorption) of an incident photon leading to the ejection of the electron as shown in Figure CS11.4k panel A. We can express this process as $\boxed{h \mid e} + h\nu \rightarrow h^+ + e^-$ where the absorption of the photon by the electron-hole pair results in the physical sepa-

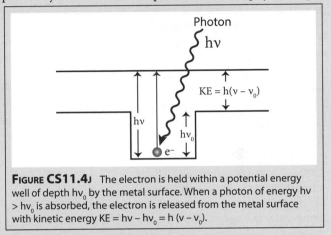

Photon

$h\nu$

$KE = h(\nu - \nu_0)$

$h\nu$

$h\nu_0$

e^-

Figure CS11.4j   The electron is held within a potential energy well of depth $h\nu_0$ by the metal surface. When a photon of energy hν > $h\nu_0$ is absorbed, the electron is released from the metal surface with kinetic energy KE = hν − $h\nu_0$ = h (ν − ν_0).

ration of the hole and the electron. The reverse process is the recombination of a hole and an electron that results in the emission of a photon of light:

$$h^+ + e^- \rightarrow \boxed{h \mid e} + h\nu$$

Expressed as an energy diagram, this process is displayed in Figure CS11.4k panel B. The electron recombines with the hole that constitutes the potential well, releasing the photon. The question here is: how does this relate to the process that occurs when a p-n junction emits (a light emitting diode or LED) or absorbs (a photovoltaic cell) a photon. We consider first the LED.

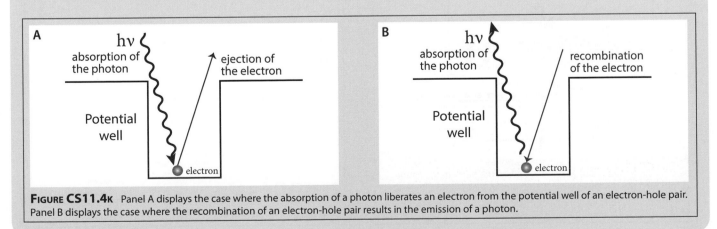

A $h\nu$ absorption of the photon   ejection of the electron

Potential well

electron

B $h\nu$ absorption of the photon   recombination of the electron

Potential well

electron

Figure CS11.4k   Panel A displays the case where the absorption of a photon liberates an electron from the potential well of an electron-hole pair. Panel B displays the case where the recombination of an electron-hole pair results in the emission of a photon.

Semiconductor Devices and Light: The LED

When a p-n junction is forward biased, holes, h^+, are pushed from their p region to the junction and electrons, e^-, are pushed from their n region to the junction. Within the junction region the electrons recombine with the holes thereby emitting light with a wavelength corresponding to the band gap. A combined energy level diagram with the configuration of the p-n junction is shown in Figure CS11.4l. The reason the emitted photon is equal in energy to the band gap is made apparent in Figure CS11.4l: the electron falls through an energy separation equal to the band gap within the junction and energy is conserved in the recombination process. Thus the emitted photon is (approximately) equal in energy to the band gap.

The technology of LEDs has progressed remarkably since the 1960s when they were first investigated intensively. This progress is summarized graphically in Figure CS11.4m which displays the *logarithm* of the emitted intensity per device-unit as a function of calendar time from 1968 to 2001. LEDs, because of their inherent efficiency, vastly exceed the light output per unit of power compared to incandescent light bulbs, and significantly exceed the efficiency of compact fluorescent devices (CFDs). For that reason, they are destined to take over the lighting market in the coming decades.

The relationship between the band gap of the p-n junction and the wavelength of the emitted photon from the LED immediately suggests the "tuning" of the semiconductor material to the desired color of the emitted light. A summary of the current wavelength ranges with the specific semiconductor materials is presented in Table CS11.4a.

TABLE CS11.4A   Available wavelengths emitted from the modern class of LEDs now range from the ultraviolet to the infrared.

Color	Wavelength [nm]	Voltage drop [ΔV]	Semiconductor material
Infrared	$\lambda > 760$	$\Delta V < 1.9$	Gallium arsenide (GaAs) Aluminium gallium arsenide (AlGaAs)
Red	$610 < \lambda < 760$	$1.63 < \Delta V < 2.03$	Aluminium gallium arsenide (AlGaAs) Gallium arsenide phosphide (GaAsP) Aluminium gallium indium phosphide (AlGaInP) Gallium(III) phosphide (GaP)
Orange	$590 < \lambda < 610$	$2.03 < \Delta V < 2.10$	Gallium arsenide phosphide (GaAsP) Aluminium gallium indium phosphide (AlGaInP) Gallium(III) phosphide (GaP)
Yellow	$570 < \lambda < 590$	$2.10 < \Delta V < 2.18$	Gallium arsenide phosphide (GaAsP) Aluminium gallium indium phosphide (AlGaInP) Gallium(III) phosphide (GaP)
Green	$500 < \lambda < 570$	$1.9^{[47]} < \Delta V < 4.0$	Indium gallium nitride (InGaN) / Gallium(III) nitride (GaN) Gallium(III) phosphide (GaP) Aluminium gallium indium phosphide (AlGaInP) Aluminium gallium phosphide (AlGaP)
Blue	$450 < \lambda < 500$	$2.48 < \Delta V < 3.7$	Zinc selenide (ZnSe) Indium gallium nitride (InGaN) Silicon carbide (SiC) as substrate Silicon (Si) as substrate – (under development)
Violet	$400 < \lambda < 450$	$2.76 < \Delta V < 4.0$	Indium gallium nitride (InGaN)
Purple	multiple types	$2.48 < \Delta V < 3.7$	Dual blue/red LEDs, blue with red phosphor, or white with purple plastic
Ultraviolet	$\lambda < 400$	$3.1 < \Delta V < 4.4$	Diamond (235 nm)[48] Boron nitride (215 nm)[49][50] Aluminium nitride (AlN) (210 nm)[51] Aluminium gallium nitride (AlGaN) Aluminium gallium indium nitride (AlGaInN) – (down to 210 nm)[52]
White	Broad spectrum	$\Delta V = 3.5$	Blue/UV diode with yellow phosphor

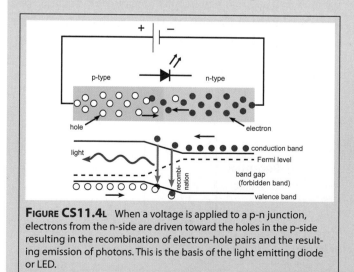

FIGURE CS11.4L   When a voltage is applied to a p-n junction, electrons from the n-side are driven toward the holes in the p-side resulting in the recombination of electron-hole pairs and the resulting emission of photons. This is the basis of the light emitting diode or LED.

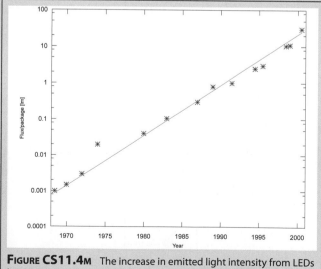
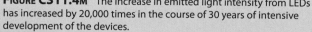

FIGURE CS11.4M   The increase in emitted light intensity from LEDs has increased by 20,000 times in the course of 30 years of intensive development of the devices.

Semiconductor Devices and Light: The Photovoltaic Cell

The reverse process from the LED is the photovoltaic effect. At its most fundamental level, the photovoltaic cell uses the process shown in Figure CS11.4k panel A where a photon is absorbed, separating an electron-hole pair into an isolated electron and an isolated hole that then drives the flow of current in an eternal circuit, converting incident photons (primarily from the Sun) into electric power. But how, specifically, does this work at the semiconductor materials level?

When a photon impinges on a piece of silicon, the photon can pass through the silicon, the photon can reflect off the surface, or the photon can be absorbed by the silicon. If the photon energy exceeds the silicon band gap, that photon can generate an electron-hole pair when the energy of the photon is transferred to an electron in the lattice. While the electron is normally bound in the valence band and thus cannot migrate in position, when it is promoted to the conduction band it is free to move within the semiconductor. In a similar manner, with the promotion of the electron to the conducting band, a hole, h^+, is created in the valence band of the silicon. The existence of a hole in the silicon allows electrons in the valence band to move into the hole, which creates an adjoining hole. Thus the hole can migrate through the material *thereby creating* mobile electron-hole pairs within the semiconductor. We can represent this process of electron promotion from the valence band to the conduction band resulting from the absorption of a photon in Figure CS11.4n.

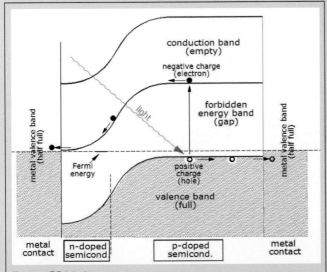

Figure CS11.4n The energy level diagram for the photovoltaic formed from a silicon p-n junction is displayed with the promotion of the electron from the valence band of the p-doped side to the conduction band. The promoted electrons flow to the electrode on the n-side.

The solar cell is constructed from a large p-type panel of silicon, shown in Figure CS11.4o, that has been doped with a trace amount of boron (B). That p-type silicon panel is in direct contact with an n-type silicon panel doped with arsenic (As).

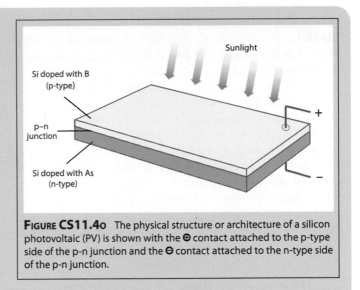

Figure CS11.4o The physical structure or architecture of a silicon photovoltaic (PV) is shown with the ⊕ contact attached to the p-type side of the p-n junction and the ⊖ contact attached to the n-type side of the p-n junction.

As discussed in our introduction to the p-n junction above, when that p-n junction is formed by bringing the p-type semiconductor in direct contact with the n-type semiconductor, diffusion of electrons from the n-type to the p-type occurs that generates the band bending as shown in Figure CS11.4n for our solar cell.

When the metal contacts are connected to the junction as shown in Figure CS11.4n and Figure CS11.4o, the electrons promoted to the conduction band by the absorption of photons create a negative charge on the n-type side of the solar cell and a positive charge on the p-side. When connected to an external circuit, electrons on the n-side pass through the external load until they reach the p-type semiconductor-metal contact. They then recombine with the hole that has passed through the p-type semiconductor. The voltage generated is equal to the *difference* in energy of the promoted electron in the conduction band and the hole in the valence band as shown in Figure CS11.4n.

Transistors

While the invention of the transistor by Bardeen, Brattain, and Shockley in 1948 was primarily responsible for the dawn of the modern technology age, we consider it last in our treatment of electronic devices based on the p-n semiconductor junction. A bipolar junction transistor includes two p-n junctions that form a sandwich configuration as shown in Figure CS11.4p. This three-layer sandwich configuration is comprised of the emitter (E), the base (B), and the collector (C).

The "load" in this circuit is displayed in Figure CS11.4p as a simple resistor although it could be a light source, an electric motor or any other device that uses electric power. When the emitter voltage, V_e, is zero, the current from the collector to the load is very small because the n-p junction between the base and the collector is reverse biased (see Figure CS11.4h). However, when V_c is increased to some threshold voltage above zero (usually ~0.6 volts), the p-n junction between the emitter and the base is forward biased. Importantly, most of the holes moving from the emitter to the base travel *through* the base to the second (n-p) junction between the base and the collector. At the B-C junction the holes "see" the electric field caused by the collector voltage V_c in Figure CS11.4p. The majority of the holes thus pass, not out of the base into the left-hand side of the circuit, but through the collector to the load, R. The key point is that the voltage applied to the emitter, V_e, exerts sensitive control over the flow of holes to the collector and thus to the load. Because, as is typically the case, the collector voltage, V_c, is considerably larger than the emitter voltage, V_e, the power dissipated in the load will be much larger than the power supplied to the emitter circuit by V_e. Thus the transistor serves as a *power amplifier*.

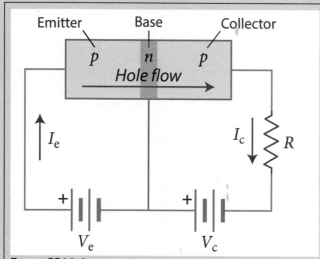

FIGURE CS11.4P When $V_e = 0$, the current is very small. When a potential V_e is applied between emitter and base, holes travel from the emitter to the base. When V_e is sufficiently large, most of the holes continue into the collector.

Problem 1

What is a p-n junction and how does it work?
a. Sketch the p-n junction including the distribution of holes and of electrons. Indicate which is the conduction band and which is the valence band.
b. Indicate the direction of the electric field within the junction.
c. Sketch in the Fermi level.

Problem 2

From what you know about a p-n junction, how does an LED work? Why is it so much more efficient than an incandescent light bulb, and how much more efficient is it?

Problem 3

How do an LED and a photovoltaic device differ? How are they similar?

KINETICS

THE PRINCIPLES THAT GOVERN THE RATE AT WHICH CHEMICAL REACTIONS OCCUR

12

"Climate change and ozone depletion are two global issues that are different but have many connections. In the ozone depletion case, we managed to work with decision makers effectively so that an international agreement called the Montreal Protocol was achieved."

—Mario J. Molina

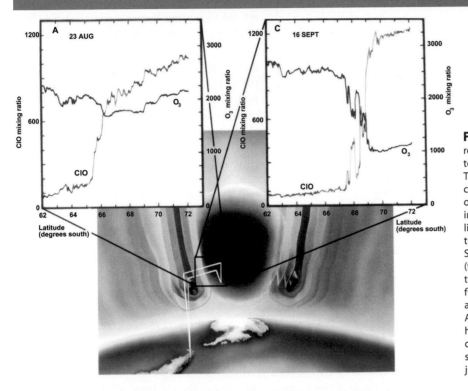

FIGURE 12.1 Chemical reactions require different amounts of time to transform reactants to products. The amount of time required for a chemical process to occur depends on the concentration of the reacting species, the probability per collision that a reaction will take place, the temperature, and other factors. Shown here is a case where ClO (formed from the photodissociation of chlorine containing chlorofluorocarbon (CFC) molecules) reacts with ozone in the region over Antarctica, which causes the "ozone hole" within which large amounts of ozone are destroyed by very small amounts of ClO in a period of just three weeks.

Framework

The fields of molecular bonding (molecular orbital theory), photochemistry (the interaction of photons with molecular structure), and kinetics are closely related. Chemical kinetics is a study of the **rate** at which chemical transformations take place, and it is a subject central to modern chemistry in a number of key areas. In particular kinetics:

- defines key pathways in organic synthesis, polymers, materials and drug development,
- constitutes the key to determining metabolic rates in living organisms,
- controls pollutants from energy sources and is thus central to human health,
- is key to developments of nanoelectronics and computer memory fabrication,

- establishes the foundation of forensic medicine,
- is central to patents, legislation, and international negotiations,
- is a critical diagnostic for restructuring world energy strategy.

Kinetics is also central to the rapidly developing connections between science and public policy. An example of this is the potent catalytic process linking the global scale destruction of ozone to the release of chlorofluorocarbons (CFCs) that are a product of chemical synthesis on an industrial scale.

The bonding structure of molecular oxygen, its photochemistry, and its kinetics provide a unifying theme that will be our framework for kinetics.

Bonding Structure and Photodissociation of Molecular Oxygen

To understand the mechanism by which a photon of light interacts with the orbital structure of molecules—the critically important topic of photochemistry—we consider specifically the case of molecular oxygen.

The first step in breaking the molecular bond in O_2 by the absorption of light involves the "promotion" of one of the valence electrons of O_2 to a molecular orbital of higher energy—specifically a nonbonding orbital.

To understand this, we return to the molecular orbital structure of O_2. This molecular orbital architecture is summarized in Figure 12.2. To briefly review, we have the configuration of MOs into which the twelve valence electrons for molecular oxygen must be inserted. Thus the first four, with opposite spins, go into the σ_s^b and σ_s^* orbitals. This leaves eight electrons. The next two with opposite spins go into the σ_x^b MO. The next four go into the degenerate π_z^b and π_y^b MOs. This leaves two remaining electrons. The next open MOs are the π_z^* and π_y^*, each of which could hold two electrons of opposite spin. Thus we could either insert one electron in each of the π_z^* and π_y^* MOs or both electrons in one or the other of the π_z^* or π_y^* antibonding orbitals.

Our intuition tells us that, given the strong repulsion between electrons in a chemical bond, the lowest energy configuration would place the last two electrons in *different* MOs because that would maximize the distance between them thereby minimizing their Coulomb repulsion. Thus we would place each of the last two electrons in the π_z^* and π_y^* antibonding MOs.

Also, there is direct experimental verification that this is the correct MO assignment because with one each of the last two electrons in the antibonding π_y^* and π_z^* orbitals, this leaves the O_2 molecule with two unpaired electrons, each of which have a magnetic moment. An unpaired electron, by virtue of the magnetic moment of the electron, will be attracted by a magnetic field, a phenomenon known as paramagnetism. Thus if O_2 has two such unpaired electrons it would be strongly *attracted* by a magnetic field, and this is indeed the case, as Figure 12.3 shows. Had the final two electrons been placed, with opposite spins, into one or the other of the π_x^* and π_y^* antibonding orbitals, the spin pair would not interact strongly with the external magnetic field as the O_2 molecule would not be attracted to the magnetic field.

We turn again to the question of the absorption of a photon by O_2, recognizing that the electric field that constitutes the photon acts on the molecular orbital structure of O_2 by promoting an electron to a higher energy orbital. But in the case of O_2, we have a number of orbitals to choose from. Experiments and theoretical calculations have demonstrated that the O_2 bond is broken to form oxygen atoms by the promotion of one of the electrons in the π_z^b bonding orbital (or the π_y^b bonding orbital) to the π_z^* antibonding orbital (or the π_y^* antibonding orbital) leaving an electron in the π_z^* antibonding orbital and one unpaired electron in the π_z^b bonding orbital. Thus, promotion of this single electron, resulting from the absorption of the photon, $h\nu$, transitions the O_2 molecule from a bound state to a state with sufficient energy to dissociate. In terms of the potential energy surface

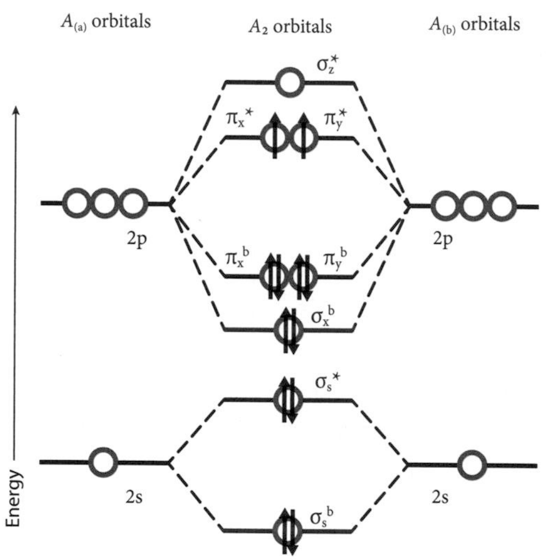

FIGURE 12.2 Diagram of the energy ordering of the *molecular orbitals* formed from the combination of atomic orbitals of two identical atoms. The *atomic orbital* energy ordering of the separated atoms (2s and 2p) are shown on the left- and right-hand sides; the molecular orbital energy levels are shown in the center. The 2s atomic orbitals combine to form the σ_s^b (bonding) and σ_s^* (antibonding) molecular orbitals. The 2p atomic orbitals combine to form the σ_z^b (*bonding* from 2p orbitals) molecular orbital and σ_z^* (*antibonding* from 2p orbitals, but added *destructively*) molecular orbital. The 2p atomic orbitals also combine to form the π^b (*bonding* from 2p) molecular orbital and the π^* (*antibonding* molecular orbitals from 2p, but added destructively) molecular orbital.

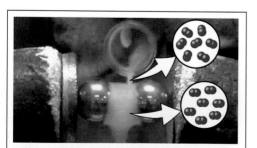

FIGURE 12.3 Unpaired electron spins are held by (attracted to) a magnetic field, creating a characteristic called *paramagnetism*. This simple experiment demonstrates that the assignment of unpaired spins in the π^* molecular orbital of molecular oxygen, as shown in Figure 12.2, is, in fact, the correct assignment.

(PES), the O_2 molecule moves from a **bound** PES to a repulsive PES resulting in the formation of two free oxygen atoms. The photon absorption is represented in shorthand by the photochemical reaction

$$O_2 + hv \rightarrow O + O$$

with a corresponding molecular orbital picture shown in Figure 12.4.

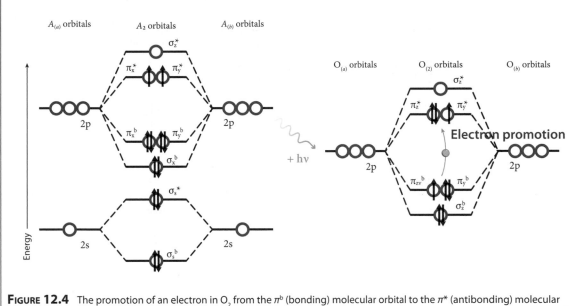

FIGURE 12.4 The promotion of an electron in O_2 from the π^b (bonding) molecular orbital to the π^* (antibonding) molecular orbital results in the photodissociation of O_2 as a result of that photon absorption event. The electron promotion is shown by the vertical arrow linking the π^b orbital with the π^* orbital.

So the promotion of that electron by the absorption of a photon leaves the O_2 molecule in a repulsive state.

However, inspection of the electron occupancy of the molecular orbitals of O_2 following the promotion of an electron induced by the absorption of a photon reveals that the bond order of the O_2 bond is equal to 1 even after absorption of the photon. This implies that the O_2 molecule should be bound (i.e. stable) even after promotion of an electron from a bonding to an antibonding orbital. Yet the O_2 molecule dissociates when the electron is promoted. How is that possible? To understand this, and many other aspects of photochemistry, we consider the orbital period of the electron.

Orbital Period of the Electron

The first question we ask is, how fast do processes occur at the molecular level? To address this question, we seek to determine how long it takes for the electron to orbit the nucleus in the valence orbital of a typical molecule. We can estimate the period of the electron in orbit by a number of methods, but perhaps the most straightforward is to recall that when light is absorbed or emitted from an atom or molecule, the **change** in energy is given by $\Delta E = hv$, as first described by Bohr for the hydrogen atom, where h is Planck's constant and v is the frequency of the light in $\sec^{-1}$ emitted or absorbed when the electron jumps between two quantized orbits. This idea was developed in Chapter 8. As with all electromagnetic radiation, the **product** of the wavelength of the light, λ, and the frequency of the light, v, is equal to the speed of the light, c:

$$\lambda v = c$$

with λ measured in [meters], ν measured in [sec^{-1}], and c measured in [m/sec]. Returning to the case of the photodissociation of molecular oxygen, we note that the absorption of a photon in the vicinity of 200 nm, which corresponds to the ultraviolet region of the electromagnetic spectrum, allows us to calculate the frequency of that radiation ν using the previous equation. We also know from Bohr that $E = h\nu$ and we know from de Broglie that the wavelength of the electron is:

$$\lambda = h/p$$

where p is the momentum of the electron such that $p = m\nu$ where m is the electron mass (at rest) and ν is the velocity of the electron in orbit. But if we represent the kinetic energy of the electron as $E = \frac{1}{2}m\nu^2$, and we use the wavelength as an approximation for the dimension (circumference) of the orbit, we have:

$$E = \frac{1}{2}m\nu^2 = \frac{1}{2}(m\nu)\nu = \frac{1}{2}p\nu = \frac{1}{2}\left(\frac{h}{\lambda}\right)\frac{\lambda}{\tau} = \frac{1}{2}\frac{h}{\tau}$$

In deriving this expression, the momentum, p, is replaced by (h/λ) from de Broglie and ν is determined by dividing the circumference of the orbit, approximated by the wavelength of the electron, by the period of the orbit, τ. The result is the approximation, $E \approx \frac{1}{2}h/\tau$, which is also equal to $h\nu$, from Bohr, such that $E \approx \frac{1}{2}h/\tau \approx h\nu$. Now we can calculate the energy, E, of the ultraviolet photon that dissociates O_2 at ~200 nm by using the Bohr equation

$$E = h\nu = h\left(\frac{c}{\lambda}\right) = 6.6\times10^{-34}\,\text{J}\cdot\text{sec}\left(\frac{3\times10^8\,\text{m/sec}}{2\times10^{-7}\,\text{m}}\right) \approx 1\times10^{-18}\,\text{joules}$$

Then from our kinetic energy expression

$$E = \frac{1}{2}\frac{h}{\tau} \approx 1\times10^{-18}\,\text{joules}$$

We can solve directly for τ and we have:

$$\tau = \frac{h}{(2)(1.0\times10^{-18}\,\text{joules})} = \frac{6.6\times10^{-34}\,\text{J}\cdot\text{sec}}{2\times10^{-18}\,\text{J}} = 3\times10^{-16}\,\text{sec}$$

That is a **very** short period of time, but it is the unit of time against which all atomic and molecular processes are measured. Why is this important? Returning to the potential energy surface defining the promotion of the electron in O_2 from a bonding orbital to an antibonding orbital, we draw the arrow designating promotion to the excited, repulsive state as a **vertical** line. This means that the internuclear distance of O_2 *does not change* as the electron jumps, promoting O_2 to the repulsive state. However, there is another important implication of the very rapid motion of the electron that we will discuss shortly: It is that we can draw these potential energy surfaces as a single surface representing a single relationship between the *potential energy* and the *internuclear distance*. This effectively separates the potential energy from all forms of kinetic energy that take place on the surface. It is this fact that makes the potential energy surface so important and so versatile.

The fact that the O_2 molecule transitions from its lowest energy bound state to the higher energy potential energy surface along a vertical path for which there is no change in internuclear distance means that as a result of the electron repulsion, the O_2 molecule is positioned on the repulsive wall of the upper surface. But at that energy, the O_2 molecule will dissociate into two oxygen atoms, completing

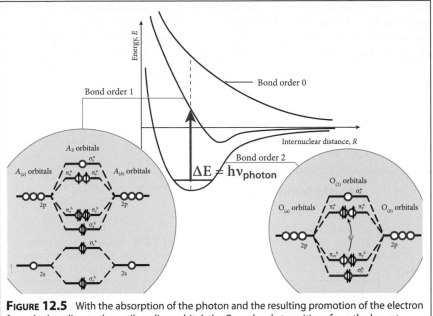

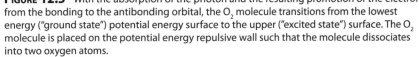

FIGURE 12.5 With the absorption of the photon and the resulting promotion of the electron from the bonding to the antibonding orbital, the O_2 molecule transitions from the lowest energy ("ground state") potential energy surface to the upper ("excited state") surface. The O_2 molecule is placed on the potential energy repulsive wall such that the molecule dissociates into two oxygen atoms.

the dissociation process. The relationship between the molecular orbital diagram and the potential energy surfaces is shown in Figure 12.5.

Figure 12.5 summarizes graphically a number of very important concepts. First, in the upper left corner the formation of molecular orbitals from the component atomic orbitals is summarized with the electrons properly inserted into the orbitals of O_2. Second, the absorption of a photon by the electronic structure of O_2 is represented by the incoming photon and the promoted electron from the bonding $\pi_z^{\,b}$ oribital to the antibonding $\pi_z^{\,*}$ orbital in the lower left panel. Third, the correlation between the configuration of electrons inserted into the molecular orbitals of O_2 and the corresponding potential energy surface is shown in the right-hand panel. This is done for both the ground state and promoted state potential energy surface. Finally, the fact that the period of the electron is much shorter than the period of nuclear motion means that the promotion of the electron results in the molecule moving vertically (i.e. with no change in internuclear distance) on the potential energy diagram from the ground state to the promoted state. From the repulsive wall of the upper (promoted) potential energy surface, the O_2 molecule is unbound and dissociates into two oxygen atoms. The dissociation of O_2 from the excited state occurs because the total energy of the O_2 molecule exceeds the depth of the well of the upper potential energy surface.

Vibrational, Rotational, and Translational Motion of a Molecule: Motion on a Potential Energy Surface

Before completing our treatment of O_2 photodissociation, we investigate more fully the categories of *kinetic energy* defining the types of motion a molecule undergoes *on a potential energy surface*. Molecules can move through space, which is represented by **translational** motion. Molecules can rotate in space, which is represented by **rotational** motion, but molecules can also oscillate along the internuclear axis, which is represented by **vibrational** motion.

Vibrational energy exchange occurs in discrete steps because the vibrational motion on the molecular scale, like electronic motion, also has a wave associated with it so that allowed *vibrational energy levels* only occur in discrete steps within the lowest potential energy surface of the molecular "electronic state" represented by the specific molecular orbitals that are occupied, as Figure 12.6 explicitly shows.

We can represent the vibrational motion of a diatomic molecule in general on the potential energy surface because vibrational energy is simply the extension and contraction of the chemical bond along the internuclear axis.

But the wave nature of the vibrational states means that only discrete vibrational energy levels are available for the molecule to occupy, as shown in Figure 12.6.

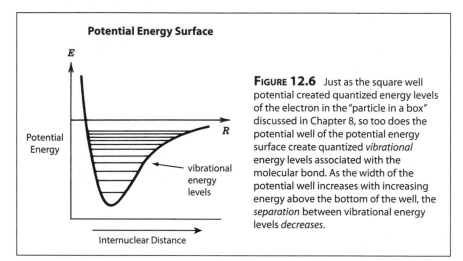

FIGURE 12.6 Just as the square well potential created quantized energy levels of the electron in the "particle in a box" discussed in Chapter 8, so too does the potential well of the potential energy surface create quantized *vibrational* energy levels associated with the molecular bond. As the width of the potential well increases with increasing energy above the bottom of the well, the *separation* between vibrational energy levels *decreases*.

Just as the vibrational energy levels are segmented or "quantized," so too are the *rotational energy levels* of the molecule. These rotational transitions are of much lower energy than either the vibrational energy levels or the electronic energy levels. But more than that, for each vibrational energy level, the molecule can possess a number of rotational energy levels such that these rotational energy levels "stack" on top of each of the vibrational levels. Thus if we look with greater detail at the potential energy surface, we can see how these energy levels are assembled, as displayed in Figure 12.7.

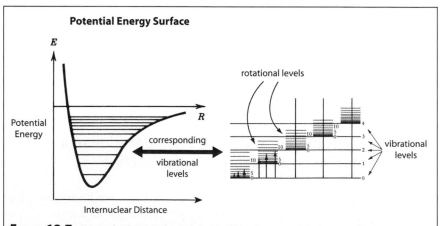

FIGURE 12.7 For *each vibrational energy level* within the potential energy well of a molecular bond, there are a manifold of rotational energy levels that are quantized. As the molecule rotates more rapidly, each quantized rotational level is separated from the lower quantized rotational level by a greater and greater amount.

Because $\Delta E = h\nu$ and the energy spacing between vibrational energy levels is much less than between electronic states, the vibrations of the molecule take place on a time scale considerably longer than the orbital period of the electron. The vibrational period of a typical molecule is about 10^{-14} to 10^{-13} seconds, about 100 times longer than the orbital period of the electron. This large difference between the orbital period of an electron in orbit about the nucleus and the vibrational period of a typical molecule begs the question of what the characteristic time scale is for other forms of molecular kinetic energy such as the **rotational** period of a molecule. It turns out that the rotational period of a typical molecule is approximately 100 times longer than the vibrational period of the molecule and thus the rotational period is some 10,000 times longer than the orbital period of the electron. While transitions between vibrational levels correspond to photon wavelengths in the infrared region of the spectrum ($\sim$10 μ), transitions between rotational levels correspond to photon wavelengths in the far infrared ($\sim$100 μ).

To summarize, the period of the electron in orbit is $\tau_{el} = 10^{-16}$–10^{-15} sec, the period of a single molecular vibration is $\tau_{vib} = 10^{-14}$–10^{-13} sec, and the period of a single molecular rotation is $\tau_{rot} = 10^{-12}$–10^{-11} sec.

However, the fact that **electronic** motion is so much faster than any other motion associated with the molecule has very important consequences. It means, for example, that we can draw a **fixed** potential energy surface that remains invariant when molecules "dance" over the top of that surface, either by rotation or by vibration. The reason is that for any given internuclear distance, the electrons move so fast that the energy of the ensemble of electrons and nuclei is minimized in a period short compared with any other change in the system. It also means, as a corollary, that we can **separate** the energy of vibration, rotation, translation, from that of the motion of the electron. That is why it is called a **potential energy surface**: It represents the invariant surface over which all molecular motion moves, whether it is vibrational, rotational, or translational.

Our diagram of O_2 photodissociation must therefore be refined to include quantized vibrational energy levels as shown in Figure 12.8.

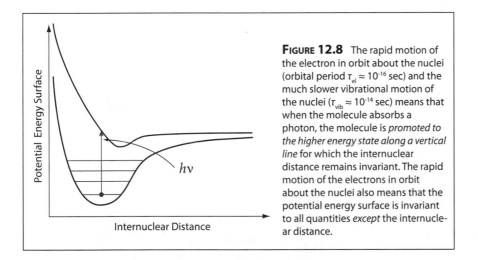

FIGURE 12.8 The rapid motion of the electron in orbit about the nuclei (orbital period $\tau_{el} \approx 10^{-16}$ sec) and the much slower vibrational motion of the nuclei ($\tau_{vib} \approx 10^{-14}$ sec) means that when the molecule absorbs a photon, the molecule is *promoted to the higher energy state along a vertical line* for which the internuclear distance remains invariant. The rapid motion of the electrons in orbit about the nuclei also means that the potential energy surface is invariant to all quantities *except* the internuclear distance.

Therefore, we return to the reaction $O_2 + h\nu \rightarrow O + O$ and note that this process is captured by the transition from one vibrational energy level on the "ground state" (or lowest) potential energy surface to a higher energy PES, as shown in Figure 12.8. So after O_2 is promoted by the absorption of a photon of light to the upper repulsive potential energy surface, the molecule begins to move "outward" to a larger internuclear distance on the time scale of a vibration, and as that internuclear distance increases, it represents a larger and larger *separation* between the two oxygen atoms that comprise the O_2 molecules such that the product of the "reaction" of a photon with an O_2 molecule is the formation of two oxygen atoms $O_2 + h\nu \rightarrow O + O$ that separate on the unbound potential energy surface. They are then "free" oxygen atoms. The excess energy, E_{excess}, shown in Figure 12.9, appears as the kinetic energy of translation (i.e., $\frac{1}{2}mv^2$) of the separating oxygen atoms. This sequence is shown in Figure 12.9.

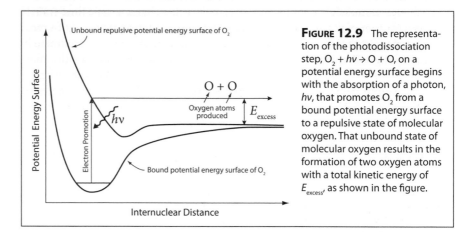

FIGURE 12.9 The representation of the photodissociation step, $O_2 + h\nu \rightarrow O + O$, on a potential energy surface begins with the absorption of a photon, $h\nu$, that promotes O_2 from a bound potential energy surface to a repulsive state of molecular oxygen. That unbound state of molecular oxygen results in the formation of two oxygen atoms with a total kinetic energy of E_{excess}, as shown in the figure.

This is an example of a very important category of reaction, the **unimolecular** reaction, because the reaction involves just *one* molecule, in this case, O_2. Our next objective is to build on our knowledge of the photodissociation of O_2 in order to understand how ozone is formed because it is essential to the evolution of life as we know it.

The Formation of Ozone

Ozone is, as seen in Chapter 10, a molecule containing three oxygen atoms, and its Lewis dot structure gives us (we have $3 \times 6 = 18$ electrons):

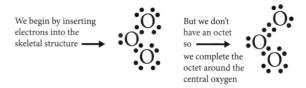

However, with this Lewis structure, we have *unequivalent* oxygen bonds because this implies a double bond on one oxygen, but not the other. Thus we employ the resonance structure between two structures of *equal* energy (i.e., two degenerate structures).

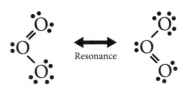

We can understand the electron *delocalization* in the bonding structure that creates the resonance because the electronic motion is so much *faster* than the vibrational motion that the electrons choose their position within the molecule to *minimize* the energy of the ensemble of electrons and nuclei on the time scale of 10^{-16} seconds! So how do we form ozone from molecular oxygen and sunlight?

First we photodissociate O_2 to form oxygen atoms: $O_2 + h\nu \rightarrow O + O$. The atomic oxygen then reacts with the abundant O_2 to form O_3 in the reaction $O + O_2 \rightarrow O_3$.

The question then is: what potential energy surface does this occur on? We can represent the bound O_3 molecule as a bound chemical species such that, as the O atom and oxygen molecule approach, they enter the bound region represented by the formation of the potential energy well of the ozone molecule. But if left to move freely, the newly formed ozone molecule will strike the repulsive wall of the potential energy surface and *rebound*, reforming a separated oxygen atom and oxygen molecule, as shown in Figures 12.10 and 12.11. We write this as the *reversible* reaction $O + O_2 \rightleftharpoons O_3$.

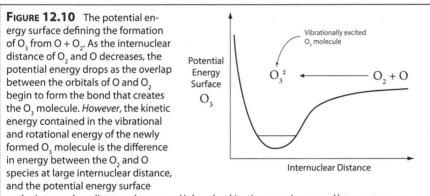

FIGURE 12.10   The potential energy surface defining the formation of O_3 from $O + O_2$. As the internuclear distance of O_2 and O decreases, the potential energy drops as the overlap between the orbitals of O and O_2 begin to form the bond that creates the O_3 molecule. *However*, the kinetic energy contained in the vibrational and rotational energy of the newly formed O_3 molecule is the difference in energy between the O_2 and O species at large internuclear distance, and the potential energy surface as the internuclear distance decreases. *Unless that kinetic energy is removed* by one or more collisions with another molecule, the energy will localize in an $O–O_2$ bond of O_3, and the newly formed O_3 molecule will "fall apart" into $O + O_2$, resulting in no net reaction.

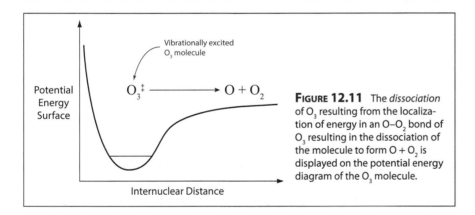

FIGURE 12.11   The *dissociation* of O_3 resulting from the localization of energy in an $O–O_2$ bond of O_3 resulting in the dissociation of the molecule to form $O + O_2$ is displayed on the potential energy diagram of the O_3 molecule.

No O_3 product is formed in this "reaction"; it simply represents an elastic collision between an atom and a molecule. We need some help to form the ozone that protects all DNA on the Earth's surface from solar ultraviolet radiation. That help comes from the collision of that $O–O_2$ complex with another molecule we designate M, capable of removing vibrational energy from that O_3 molecule and "dropping" it into the bound "well" of the potential energy surface. This collisional partner, M, does not interact chemically with the new O_3 molecule; it simply removes *vibrational energy* from the bond structure of O_3 such that the ozone molecule is trapped in the well of the potential energy surface. Thus we write the reaction for the *formation* of ozone that includes the "chaperone" collisional partner, M:

$$O + O_2 + M \rightarrow O_3 + M.$$

And we represent the process on the potential energy surface as shown in Figure 12.12.

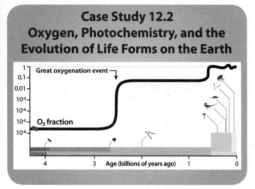
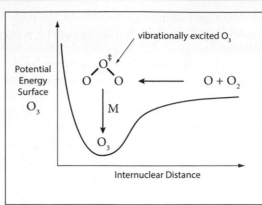

FIGURE 12.12 The two-step process forming O_3 from the reaction of $O + O_2$ is shown on a potential energy surface. The first step is the formation of O_3 resulting from the collision of O and O_2. However, that newly formed O_3 has a large amount of kinetic energy present as vibrational and rotational energy in the bond structure of O_3. If a third body, *M*, collides with that vibrationally and rotationally excited O_3 molecule, removing energy, a stable O_3 molecule will be formed.

The Molecularity of a Chemical Reaction

This is an important result. We have explicitly identified the potential energy surface over which the formation of ozone has occurred and we have identified exactly what molecule, the "chaperone" *M*, collided with what molecular "complex," $O_3^‡$ (where the superscript ‡ indicates vibrational excitation), to form the stable product molecule O_3. Whereas we designate the photodissociation of O_2 as a *unimolecular* process because (a) we know exactly what potential energy surface is involved and (b) there is only one molecule involved, we designate the ozone formation step

$$O + O_2 + M \rightarrow O_3 + M$$

as a *termolecular* reaction because (a) we know exactly what potential energy surface is involved, and (b) the reaction involves *three molecules*. Notice, however, that the collision of the "chaperone" molecule, M, need not be *simultaneous* with the collision of the oxygen atom with O_2. The molecule will vibrate many times (sometimes many thousands of times) *before all the energy is localized* in a *single bond leading to its rupture*. In general, many collisions between the $O_3^‡$ vibrationally excited molecule and the chaperone are required to initiate the stabilization of the ozone molecule, but in many cases vibrational energy is removed in a period short compared with energy localization to break a chemical bond. Thus, these termolecular reactions are very common in nature and are essential for life on the planet.

In the presence of ultraviolet light, ozone will photodissociate, just as O_2 did, so we must consider the step $O_3 + h\nu \rightarrow O + O_2$, which is also a *unimolecular* reaction since only *one molecule* is involved in the "reaction." However, the production of both O atoms and O_3 by virtue of sunlight breaking the O_2 bond results in another very important reaction that *reforms* the O_2 bond:

$$O + O_3 \rightarrow O_2 + O_2.$$

This is a reaction of the type with which we are very familiar. It is a *bimolecular* reaction because it involves two reactant molecules, atomic oxygen and ozone, and multiple products ($O_2 + O_2$ in this case) that are capable of removing the excess energy in the form of translational kinetic energy. The bimolecular reaction occurs on a potential energy surface that we are familiar with, and one that is shown in Figure 12.13.

The potential energy surface of a *bimolecular* reaction has some very important features. First, the surface is formed by bringing two reacting molecules in from large internuclear separation where there is no interaction between the va-

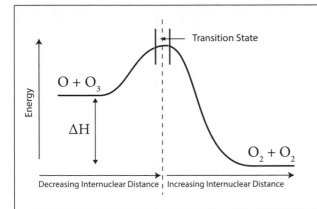

FIGURE 12.13 We can also represent the reaction of atomic oxygen, O, with ozone, O_3, on a potential energy surface. In this case, as O and O_3 approach on a collision course as the internuclear distance decreases, a repulsive potential develops because of the Coulomb repulsion of the valence electrons in each species, creating an energy barrier that must be surmounted if the reaction is to take place. At the peak of the barrier, the transition state defines the highly restricted zone at minimum internuclear distance where the bond rearrangement occurs. As the molecules move through the transition state where the new bonds are formed, the products ($O_2 + O_2$) move down a repulsive potential surface such that the product molecules gain kinetic energy.

lence electrons of the individual species. As the reactants begin to approach in close proximity, the electron-electron repulsion of the valence electrons begins to generate a repulsive "barrier" to the reaction, but the electrons are moving so rapidly relative to the period of the collision that the system moves along the path of minimum energy. As the reacting species reach the *minimum* internuclear distance (providing the kinetic energy of collision is adequate to reach the top of the barrier) which corresponds to the *maximum* energy of the potential energy surface, shown in Figure 12.13, they will either bounce off of one another or the bonds will rearrange (on the time scale of electronic motion) to form the new bonds of the *product molecules*. The product molecules will move down the repulsive potential energy surface, releasing the potential energy in the form of vibrational, rotational, and translational energy.

Thus we have identified three distinct, but very important types of chemical reactions and identified a very important concept: the *molecularity* of a chemical reaction. These are summarized in Figure 12.14.

Reactions that occur on a specific potential energy surface are termed *elementary reactions*. Elementary reactions and the associated concept of molecularity occupy a position of great importance in the study of chemical kinetics.

Molecularity: Determined by Mother Nature

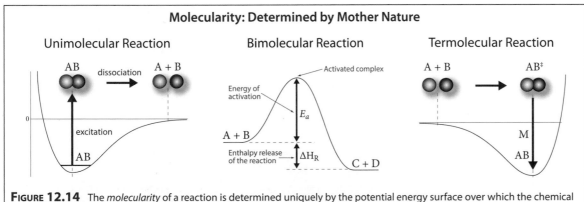

FIGURE 12.14 The *molecularity* of a reaction is determined uniquely by the potential energy surface over which the chemical transformation occurs. The dissociation of a *single* molecule into products is called a *unimolecular* reaction. The unimolecular process is shown in the upper panel as a two-step process: excitation followed by dissociation. The reaction of *two* molecules, A + B, to form multiple products is called a bimolecular reaction. The potential energy diagram for a bimolecular reaction is shown in the middle panel. A termolecular reaction involves three molecules: A and B that collide to form AB‡, and a third "chaperone" molecule, M, that removes the excess kinetic energy to form the stable product.

Chapter Core

Road Map for Chapter 12

We address the subject of chemical kinetics by first defining what is meant by the *rate, R, of a chemical reaction*. This requires us to define the distinction between the rate of an *overall reaction* vs. the rate of an *elementary reaction*. We then turn to the question of how the *rate* of a reaction is actually determined in the laboratory and how that leads to the concept of an observed *reaction rate constant, k_{obs}*. The determination of a reaction rate constant leads to the development of the concept of a *rate law* that allows us to systematically and quantitatively represent how fast one species reacts with another. The concept of a rate law then leads to the issue of the functional form of the species concentrations as they appear in the statement of the rate law that in turn defines the *order of the reaction*. As we will emphasize, there is a crucial difference between the *order of a reaction* and the *molecularity of a reaction*. One of the most useful and powerful approximations in chemical kinetics, the *steady state approximation*, is discussed and applied to some important cases. The temperature dependence of reaction rate constants is then presented in the framework of the Arrhenius expression that provides a systematic way to calculate reaction rate constants over a wide range of temperatures.

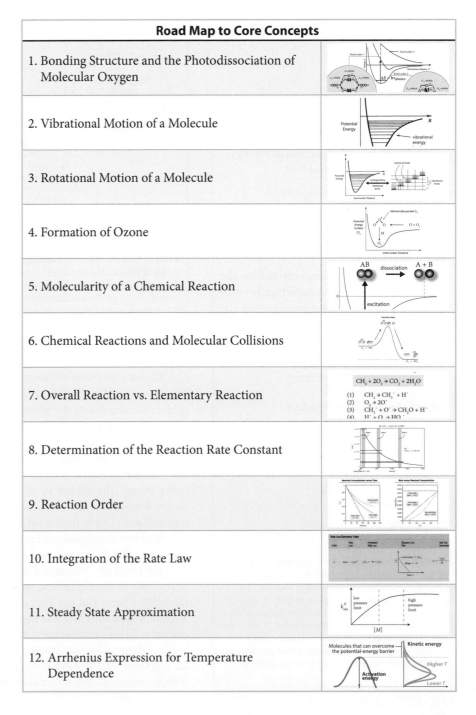

Road Map to Core Concepts	
1. Bonding Structure and the Photodissociation of Molecular Oxygen	
2. Vibrational Motion of a Molecule	
3. Rotational Motion of a Molecule	
4. Formation of Ozone	
5. Molecularity of a Chemical Reaction	
6. Chemical Reactions and Molecular Collisions	
7. Overall Reaction vs. Elementary Reaction	
8. Determination of the Reaction Rate Constant	
9. Reaction Order	
10. Integration of the Rate Law	
11. Steady State Approximation	
12. Arrhenius Expression for Temperature Dependence	

Kinetics

As mentioned, kinetics constitutes the foundation for an immense array of topics in modern chemistry as displayed in Figure 12.15.

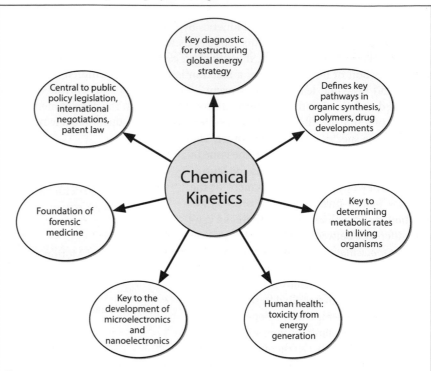

FIGURE 12.15 Chemical kinetics is a branch of chemistry that is important for the understanding of many topics in modern chemistry. The rate at which a chemical reaction occurs significantly affects pathways in organic synthesis that in turn establishes the strategy for developments in pharmaceuticals. Living organisms are sustained by chemical reactions, the rates of which are temperature dependent. New materials are synthesized using a careful blend of reactions that take place in the gas phase and on surfaces. Solutions to and an understanding of the global structure of energy generation are built upon a foundation of chemical kinetics.

Kinetics is therefore about *rates* in chemical and physical (and biological) processes; it is also about *time scales* in these processes and about *timing*, so we turn to the examination of the timing in these transitions on the molecular scale. Let's return to the potential energy surfaces and their associated bonding and antibonding molecular orbitals involved in the photodissociation of molecular oxygen.

We have developed a picture of the wave properties of electrons that forms the foundation for the atomic orbitals that constitute the "home" for electrons in those atoms. The geometry of those atomic orbitals establishes the structure of molecules because the molecular orbitals are built from the constructive (bonding) and destructive (antibonding) combination of atomic orbitals.

We have developed the basic idea concerning molecular orbitals and we have explored what happens when a molecule absorbs a photon of light by analyzing how electrons are "promoted" to a higher orbital, often resulting in the dissociation of the molecule into atoms or fragments of molecules. We have calculated the orbital period of the electron and discovered that it is short compared with any other type of molecular motion—vibration, rotation, or translation. We have discovered three important types of chemical reactions that are uniquely defined by the potential energy surface over which the reaction takes place: unimolecular reactions that involve a single reacting molecule; bimolecular reactions that involve two reacting molecules; and termolecular reactions that involve three

---

Key Reaction in Extraction of Energy from Natural Gas

$$CH_4 + OH \rightarrow CH_3 + H_2O$$

Rate of the Reaction: *Reactants*

$$R = -\frac{\Delta[OH]}{\Delta t} = -\frac{\Delta[CH_4]}{\Delta t}$$

The rate of the reaction *counts* the number of times the reaction is completed per unit time by counting the *change* in the number of molecules per cubic centimeter, per unit time.

$$\frac{\Delta(molecules/cm^3)}{\Delta t}$$

Rate of the Reaction: *Products*

$$R = +\frac{\Delta[CH_3]}{\Delta t} = \frac{\Delta[H_2O]}{\Delta t}$$

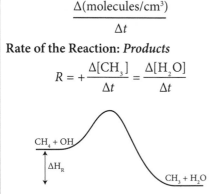

The rate of a reaction, *R*, is a single quantity that establishes the number of times per second that a reaction is *completed*. The rate of the reaction, *R*, thus is a quantity that applies to both reactants *and* products. Because reactants are *disappearing*, their rate is taken as the negative value of the time rate of change of the reactants. Because products are *appearing*, their rate is taken as the positive value of the time rate of change of the products.

molecules. We have applied these principles to the photochemistry of oxygen and to the formation of ozone from molecular oxygen in Earth's atmosphere. We turn now to the question of how the rate of a chemical reaction is defined and how it is determined.

Chemical Reactions and Molecular Collisions

The development of molecular orbital theory, as we have just seen, opens the pathway for understanding how photons interact with molecular structures by promoting one (or more) electron(s) to a higher energy orbital. This promotion of an electron leads to the idea of photodissociation of a single molecule on a specific potential energy surface. Thus the idea of a unimolecular reaction and, more generally, of the *molecularity* of a chemical reaction—unimolecular, bimolecular, and termolecular reactions, each occurring on a specific potential energy surface.

While unimolecular reactions and termolecular reactions are very common in nature, indeed our survival depends on them, the bimolecular reaction is often the centerpiece of chemical reaction studies. The reason is that they are very common, and they involve the *collision* of two molecules (or an atom and a molecule or just two atoms) that are the *reactants* leading to the formation of two products as displayed in Figure 12.12 and Figure 12.13.

A key point therefore, is that a chemical reaction cannot occur unless a *collision* occurs between two reactants that may or may not engage in a bond rearrangement leading to products of the reaction.

So we can immediately deduce the fact that because a reaction requires a collision, a chemical reaction cannot occur at a rate faster than the rate at which reactant molecules collide. If we could shrink to the size of a molecule as it collides with others, it would be a very revealing ride. It would also be a wild ride. First, as we know from Chapter 1, we would be moving at the speed of a high velocity bullet: 500 m/sec or approximately 1000 miles per hour. Suppose for example we could ride around on an ozone molecule (a convenient shape!) as shown in Figure 12.16 in a mixture of O_3 and NO.

Electrons would be moving within the structure of ozone measured on a time scale of 10^{-16} seconds—the period of electron motion. The O_3 molecule would execute a full vibration every 10^{-14} seconds and our "ride" would be pitching end over end, rotating every 10^{-12} seconds. The immediate question for us is: if we were in this mixture of O_3 and NO at one atmosphere pressure, how many times would we collide with another molecule of ozone or NO?

The *rate* at which we would collide with other molecules in this mixture, which we will call Z_{molec}, is just the number of collisions per second:

Z_{molec} = rate of collisions taking place = number of collisions/sec.

FIGURE 12.16 If we could ride around on the back of an ozone molecule, it is important to imagine what we would experience.

Fortunately, we have already addressed this problem. In Chapter 1 we calculated the number of collisions between molecules in a gas and the wall of the vessel containing those molecules. We determined that the *approximate* collision rate between molecules in a gas and the *wall*, which we will call, Z_{wall}, is

$$Z_{wall} \approx N/{\Delta t} = {Nv_x}/{L} = Nv_x A/V$$

where N is the number of gas molecules in the volume, V; Δt is the period between collisions of a single molecule moving with velocity v_x in the x direction; and A is the area of the container wall. If [X] is the concentration of molecules in the gas (equal to N/V), and $\bar{v}$ is the average velocity of the molecules in the gas, which we will approximate as $\bar{v} \approx v_x$, then $Z_{wall} = [X]\bar{v}A$. We can use this expression to estimate the collision rate between molecules. The first step is to replace the area of the wall, A, by the collision cross section between two molecules each of radius r that defines a mutual collision area of $\pi(2r)^2 = \pi d^2$. This is shown in Figure 12.17. Thus for molecular collisions we have as a reasonable *approximation* for the number of collisions per second

$$Z_{molec} = [X]\bar{v}\pi d^2$$

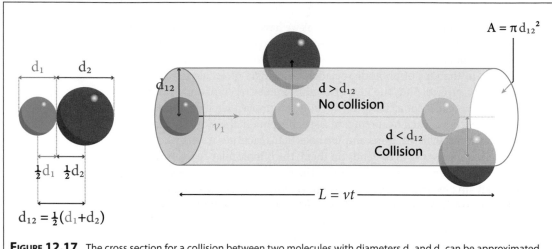

FIGURE 12.17 The cross section for a collision between two molecules with diameters d_1 and d_2 can be approximated as ½ the sum of the two diameters. Then the area of the cross section for the collision between the two molecules is πd^2_{12} where $d_{12} = ½ (d_1 + d_2)$.

At one atmosphere pressure, [X] is given by 3×10^{19} molecules/cm³, the average velocity is 5×10^4 cm/sec and a typical molecular diameter is $d = 3 \times 10^{-8}$ so $\pi d^2 \approx 30 \times 10^{-16}$ cm². Thus

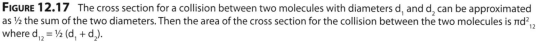

$$Z_{molec} = [X] \bar{v} \pi d^2 = (3 \times 10^{19} \text{ molec/cm}^3)(5 \times 10^4 \text{ cm/sec})(30 \times 10^{-16} \text{ cm}^2)$$

$$= 5 \times 10^9 \text{ collisions/sec}$$

Thus in this ride on the back of an ozone molecule you would experience 5 billion collisions every second.

Check Yourself 1

Before moving on to consider what it takes to actually realize a *chemical reaction* from one of these collisions, we look at one of these collisions; we look at one other question.

From the vantage point of the ozone molecule with a dimension of ~3 $\times 10^{-8}$ cm $= 3 \times 10^{-10}$ m $= 0.3$ nm $= 3Å$, how far apart are the other molecules in the mixture? This question is easily answered because (1) we know how fast we are moving (5×10^4 cm/sec) and (2) we know the period of time between collisions

$$\tau_{coll} = 1/Z_{molec} = \frac{1}{5 \times 10^9 \text{ collisions / sec}}$$

$$= 2 \times 10^{-10} \text{ sec}$$

Thus, the distance between molecules in the mixture at STP, which we will call S_{mol} for molecular separation is just

$$S_{mol} = v\tau_{coll} = (5 \times 10^4 \text{ cm/sec})2 \times 10^{-10} \text{ sec}$$

$$= 1 \times 10^{-5} \text{ cm} = 1 \times 10^{-7} \text{ m}$$

$$= 100 \text{ nm}$$

$$= 1000 \text{ Å}$$

That is roughly 300 molecular diameters.

Now that we have established the *rate* of molecular collisions, we can now examine what is involved in the chemical reaction between an ozone molecule and a nitric oxide molecule. In order to execute a chemical reaction, the colliding O_3 and NO molecules must approach close enough that a bond transformation takes place. To accomplish this, the colliding molecules must overcome the repulsive barrier created by the electron-electron repulsion of the valence electrons in O_3 and NO. This repulsion establishes the height of the barrier separating reactants (O_3 and NO) and products (NO_2 and O_2) as shown in Figure 12.18.

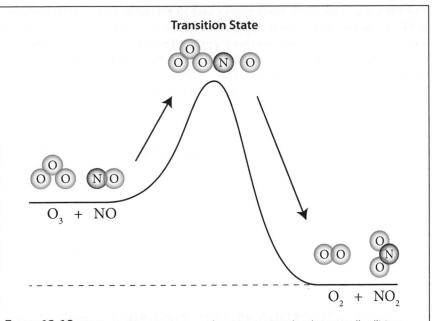

FIGURE 12.18 While a collision is necessary for a reaction to take place, not all collisions lead to a reaction. First the kinetic energy of the colliding molecules must be sufficiently high to surmount the repulsive barrier created by the electron repulsion of the valence electrons. This is necessary such that the nuclei of the transition state structure are in close enough proximity for electron rearrangement leading to products. Second, the alignment of the molecules must be favorable to yield the correct geometry of the products.

However, even if the colliding NO/O_3 pairs do possess sufficient kinetic energy to position the nuclei close enough for a bond rearrangement to take place, the geometric *configuration* of the positions of the oxygen and the nitrogen nuclei may make the required bond rearrangement difficult. As a result, no reaction takes place. On the potential energy surface, this *nonreactive* encounter would take the form displayed in Figure 12.19.

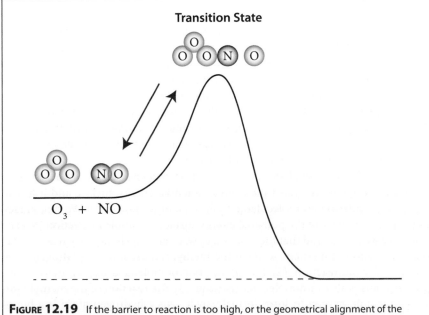

FIGURE 12.19 If the barrier to reaction is too high, or the geometrical alignment of the molecules at the transition state is not conducive to product formation, the reactants will rebound from one another, resulting in no net chemical reaction.

Thus, if we examine a mixture of O_3 and NO at the molecular level in a vessel at STP we will see collisions taking place at the rate of 5 billion a second, but only a very few will result in a reaction taking place. If we could measure the rate of the chemical reaction, we could immediately deduce the *probability per collision* that a reaction would take place because if R_{obs} is the observed rate of the reaction and Z_{mole} is the rate of molecular collisions, then the probability of a reaction taking place per collision, P_{react}, is just

$$P_{react} = R_{obs}/Z_{molec}$$

The probability per collision of a reaction taking place, P_{react}, ranges from 1 to 1 in 10^{12}, even for important reactions. That is a span in reaction probability of *twelve orders of magnitude*. So it is clear, if we are to understand a system of competing reactions, then we must understand the probability that the collisions between individual molecules will result in the formation of a given set of product molecules. Thus we must find a way to measure, to observe directly, the rate of a chemical reaction, R_{obs}.

The Overall Reaction vs. the "Elementary Reaction"

We use the reaction of methane with the hydroxyl radical as our prototype reaction to establish how the chemical kinetics of reactions in general are analyzed. We note first that the reaction is bimolecular. This means that it has a *specific potential energy surface* over which the reactants move in the course of the chemical reaction to form the products, $CH_3 + H_2O$. Because the reaction proceeds over a single, *well defined* potential energy surface, the reaction is termed an "elementary reaction"; meaning that it is not a combination or a sequence of reactions that, when taken together, yield an "overall" chemical reaction, as is often the case. Rather, an "elementary reaction" is associated with a *specific potential energy surface*. In practice, it is not always easy to determine if a reaction as written is indeed an elementary reaction. Sometimes it takes hundreds of hours of experimental and theoretical work to prove a reaction is, in fact, an elementary reaction. But once established, the statement that a reaction is an elementary reaction means that the only entities (molecules or atoms) involved in the reaction are the ones cited as reactants and products.

To drive home the important distinction between an *overall reaction* and an *elementary reaction*, consider what happens in the combustion of natural gas to form the final oxidized products CO_2 and H_2O, which we generally write as $CH_4 + 2O_2 \rightarrow CO_2 + 2H_2O$. Inspection of the mechanism in the sidebar that breaks the *overall reaction*, $CH_4 + 2O_2 \rightarrow CO_2 + 2H_2O$, into 11 "*elementary reactions*" each of which occurs on a single potential energy surface, reveals the large number of individual steps involved in the overall reaction.

Each of the individual steps in the reaction sequence displayed at left is an *elementary reaction*—it occurs on a single potential energy surface and it has an explicit molecularity: unimolecular, bimolecular, or termolecular. We can, in fact, identify the character of the potential energy surface by simple inspection for each step, once we are assured that the reaction as written is an elementary reaction: **(1)** The dissociation of methane to form the methyl radical and the hydrogen atom is a unimolecular reaction. **(2)** The dissociation of molecular oxygen to form two oxygen atoms is also a unimolecular reaction. **(3)** The reaction of the methyl radical with atomic oxygen to form formaldehyde plus a hydrogen atom is a bimolecular reaction. **(4)** The reaction of a hydrogen atom with O_2 to form the perhydroxyl radical is a termolecular reaction. **(5)** The reaction of a hydrogen atom with the perhydroxyl radical to form hydroxyl radicals is a bimolecular reaction. **(6)**

What actually happens in the combustion of natural gas?

$$CH_4 + 2O_2 \rightarrow CO_2 + 2H_2O$$

(1)	$CH_4 \rightarrow CH_3^{\cdot} + H^{\cdot}$
(2)	$O_2 \rightarrow 2O^{\cdot}$
(3)	$CH_3^{\cdot} + O^{\cdot} \rightarrow CH_2O + H^{\cdot}$
(4)	$H^{\cdot} + O_2 \rightarrow HO_2^{\cdot}$
(5)	$H^{\cdot} + HO_2^{\cdot} \rightarrow 2OH^{\cdot}$
(6)	$CH_2O + OH^{\cdot} \rightarrow CHO^{\cdot} + H_2O$
(7)	$CHO^{\cdot} \rightarrow CO + H^{\cdot}$
(8)	$CO + OH^{\cdot} \rightarrow CO_2 + H^{\cdot}$
(9)	$H^{\cdot} + H^{\cdot} \rightarrow H_2$
(10)	$O^{\cdot} + H_2 \rightarrow OH^{\cdot} + H^{\cdot}$
(11)	$OH^{\cdot} + H^{\cdot} \rightarrow H_2O$

NET: $CH_4 + 2O_2 \rightarrow CO_2 + 2H_2O$

It is routine to write the combustion of a hydrocarbon, such as methane or octane or any other combustible material, in the form hydrocarbon $+ O_2 \rightarrow H_2O + CO_2$. But even for the simplest case, $CH_4 + 2O_2 \rightarrow CO_2 + 2H_2O$, this is a net, overall reaction. It does not occur on a specific potential energy surface. Rather, the reaction of CH_4 with O_2 is comprised of a *sequence* of reactions, each of which takes place on a specific potential energy surface. The sum of those "elementary reactions"—the specific unimolecular, bimolecular, and termolecular reactions—then constitutes the net overall reaction.

The reaction of formaldehyde with the hydroxyl radical to form the CHO radical and water is a bimolecular reaction. (7) The dissociation of CHO to form carbon monoxide and the hydrogen atom is a unimolecular reaction. (8) The reaction of a carbon monoxide with the OH radical to form carbon monoxide and the hydrogen atom is a bimolecular reaction. (9) The reaction of hydrogen atoms to form molecular hydrogen is a termolecular reaction. (10) The reaction of oxygen atoms with molecular hydrogen to form hydroxyl radical and hydrogen atoms is a bimolecular reaction. Finally, (11) the reaction of hydroxyl radicals with atomic hydrogen to form water is a termolecular reaction. Notice a very important point, *cancellation* occurs such that the *sum* of the eleven elementary reactions equals the single net *overall reaction* $CH_4 + O_2 \rightarrow CO_2 + H_2O$.

Determination of the Rate of a Chemical Reaction

We return to the dissection of the kinetics of methane and the hydroxyl radical. How do we observe the *rate* of disappearance of one species in the presence of another? There are many ways, but one important possibility, shown schematically in Figure 12.20, would be to design a system whereby a laser pulse is used to form an initial OH concentration, $[OH]_{init}$, in a mixture of an inert gas, for example argon. The other reactant, CH_4, would have been mixed with the argon before adding the mixture to the reaction vessel. Then, following the formation of OH by the laser pulse, for example by the photodissociation of hydrogen peroxide,

$$H_2O_2 + h\nu \rightarrow OH + OH$$

the OH thus formed would react with CH_4 in the bimolecular reaction

$$CH_4 + OH \rightarrow CH_3 + H_2O.$$

The decay of [OH] with time, by reaction with CH_4, is displayed in Figure 12.21.

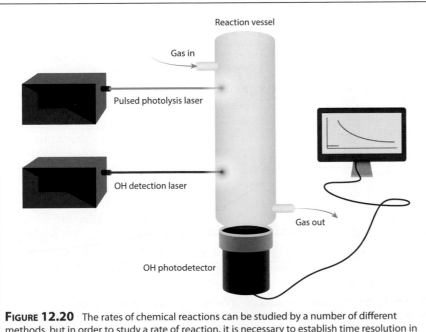

FIGURE 12.20 The rates of chemical reactions can be studied by a number of different methods, but in order to study a rate of reaction, it is necessary to establish time resolution in the measurement of the disappearance of one chemical species in the presence of another. One way to achieve this time resolution is to use a laser pulse to form one of the reactants and then monitor the rate of decay of that molecule with a second laser. This technique is called "flash photolysis."

With the ability to resolve in time the decay of OH in the presence of CH_4, we are in a position to actually determine the rate of the reaction, R_{obs}. We make a strategic decision first, by recognizing that if the concentration of methane, $[CH_4]$ (where the square brackets indicate the molecular concentration in molecules/cm^3), is adjusted such that $[CH_4] \gg [OH]_{init}$, then $[CH_4]$ will change *very little* through the course of the reaction $CH_4 + OH \rightarrow CH_3 + H_2O$. So we now observe, with a laser, the disappearance of OH in the presence of CH_4, and we plot those data as a function of time. The time axis is determined by knowing the time over which the decay of OH has occurred. The results from an actual experiment are shown in Figure 12.21.

In order to analyze the kinetics of OH loss in the presence of CH_4, we first break the decay curve of OH into zones to calculate the *rate* of the reaction

$$R_{obs} = -\Delta[OH]/\Delta t$$

that measures the change in OH concentration, $\Delta[OH]$, over a time increment, Δt. Inspection of Figure 12.21 reveals that $-\Delta[OH]/\Delta t$ in each of the three zones is

Zone 1: $-\Delta[OH]/\Delta t = \dfrac{8 \times 10^9 \text{ molecules/cm}^3}{2 \times 10^{-4} \text{ sec}} = 4 \times 10^{13} \text{ molecules/cm}^3 \text{ sec}$

Zone 2: $-\Delta[OH]/\Delta t = \dfrac{3.8 \times 10^9 \text{ molecules/cm}^3}{2 \times 10^{-4} \text{ sec}} = 1.9 \times 10^{13} \text{ molecules/cm}^3 \text{ sec}$

Zone3: $-\Delta[OH]/\Delta t = \dfrac{1 \times 10^9 \text{ molecules/cm}^3}{2 \times 10^{-4} \text{ sec}} = 5 \times 10^{12} \text{ molecules/cm}^3 \text{ sec}$

Two conclusions emerge. First, the slope of the [OH] decays with time; $\Delta[OH]/\Delta t$ decreases with time as the reaction proceeds. Second, nothing is constant in the analysis of the reaction of OH with methane: the concentrations are changing, the slope of the concentration with time is changing, and time is changing. Yet we seek to determine a coherent expression defining the probability that an OH radical will react with a CH_4 molecule at a given temperature and pressure. Is there any quantity that *is* constant in the course of the reaction?

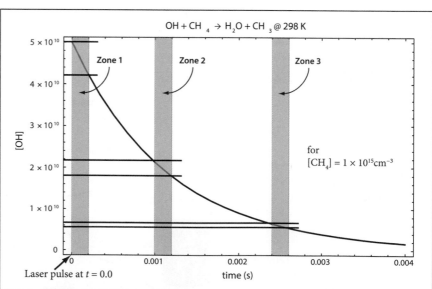

FIGURE 12.21 Following the formation of OH by flash photolysis, for example by using a laser pulse to break apart hydrogen peroxide ($H_2O_2 + h\nu \rightarrow OH + OH$), OH decays with time in the presence of CH_4 in the reaction $CH_4 + OH \rightarrow CH_3 + H_2O$. The concentration of hydroxyl, [OH], drops very rapidly at first, as shown in zone 1. The rate of the reaction $R = -\Delta[OH]/\Delta t$ is the slope of the [OH] decrease with time, with the minus sign designating that OH is removed in the reaction. The rate of the reaction, R, decreases as time progresses. Thus, $R = -\Delta[OH]/\Delta t$ is less in zone 2 than zone 1, etc.

12.20

Determination of the Reaction Rate Constant

After inspection of Figure 12.21, we notice first that the rate of the reaction, $R = -\Delta[OH]/\Delta t$, is *changing continuously* throughout the course of the reaction. Initially, OH is disappearing rapidly; but as the reaction progresses, the rate decreases until at times greater than about 4×10^{-3} seconds, the rate has slowed to something approaching zero.

As a first attempt, we try dividing the rate of the reaction, $-\Delta[OH]/\Delta t$, by the OH concentration within each time zone.

As the OH concentration decays out to longer times, the measurement of $\Delta[OH]$ becomes more and more difficult and thus less and less accurate. But if we divide the rate of OH decay, $\Delta[OH]/\Delta t$, by the average OH concentration within each corresponding time increment, the rate $\Delta[OH]/\Delta t$ divided by [OH] remains **unchanged** throughout the course of the reaction!

Zone 1: $\quad \dfrac{-\Delta[OH]/\Delta t}{[\overline{OH}]} = \dfrac{4 \times 10^{13}\ \mathrm{mol}/\mathrm{cm}^3 - \sec}{4.6 \times 10^{10}\ \mathrm{mol}/\mathrm{cm}^3} = 8.7 \times 10^{2}\ \sec^{-1}$

Zone 2: $\quad \dfrac{-\Delta[OH]/\Delta t}{[\overline{OH}]} = \dfrac{1.9 \times 10^{13}\ \mathrm{mol}/\mathrm{cm}^3 - \sec}{2.0 \times 10^{10}\ \mathrm{mol}/\mathrm{cm}^3} = 9.5 \times 10^{2}\ \sec^{-1}$

Zone3: $\quad \dfrac{-\Delta[OH]/\Delta t}{[\overline{OH}]} = \dfrac{5 \times 10^{12}\ \mathrm{mol}/\mathrm{cm}^3 - \sec}{6 \times 10^{9}\ \mathrm{mol}/\mathrm{cm}^3} = 8.3 \times 10^{2}\ \sec^{-1}$

But that means that we have *identified a constant*, which we shall call k_{obs}, the observed *reaction rate constant*:

$$\frac{-\Delta[OH]}{\Delta t} \Big/ [OH] = k_{obs}$$

We are careful to label k_{obs} with the subscript "obs" to indicate that we have obtained this information through *direct observation*. We will soon see why the fact that we have *observed* this reaction rate constant is so important.

We can now write that the rate of the reaction, R, is just

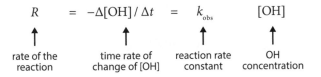

$$R \quad = \quad -\Delta[OH]/\Delta t \quad = \quad k_{obs} \quad [OH]$$

| rate of the reaction | time rate of change of [OH] | reaction rate constant | OH concentration |

But if k_{obs} is a constant, let's calculate what it is throughout the course of the reaction. To do this, we break the time increments into systematic steps, as displayed in Figure 12.22, and carry out the calculation at each step. This is displayed in Figure 12.22 and tabulated in Table 12.1.

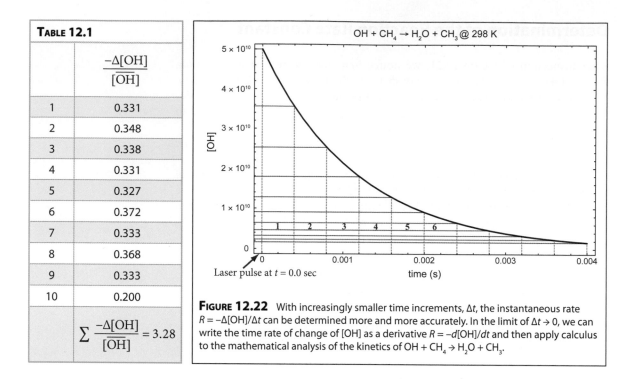

TABLE 12.1	
	$\dfrac{-\Delta[OH]}{[\overline{OH}]}$
1	0.331
2	0.348
3	0.338
4	0.331
5	0.327
6	0.372
7	0.333
8	0.368
9	0.333
10	0.200
$\sum \dfrac{-\Delta[OH]}{[\overline{OH}]} = 3.28$	

Figure 12.22 With increasingly smaller time increments, Δt, the instantaneous rate $R = -\Delta[OH]/\Delta t$ can be determined more and more accurately. In the limit of $\Delta t \to 0$, we can write the time rate of change of [OH] as a derivative $R = -d[OH]/dt$ and then apply calculus to the mathematical analysis of the kinetics of $OH + CH_4 \to H_2O + CH_3$.

We then calculate the rate over the course of the reaction, noting that we can rearrange the rate equation such that

$$\frac{-\Delta[\overline{OH}]}{\Delta t} = k_{obs}[OH] \quad \text{is rewritten} \quad \frac{-\Delta[OH]}{[\overline{OH}]} = k_{obs}\Delta t.$$

We check to see that indeed, within experimental error, $\Delta[OH]/[OH]$ is constant and then we can calculate k_{obs} by summing $\Delta[OH]/[OH]$ for each increment and dividing by the time over which the decay of OH took place. So

$$\sum_{i=1}^{10} \left(\frac{-\Delta[OH]}{[\overline{OH}]} \right)_i = \sum_{i=1}^{10} k_{obs}\Delta t_i = k_{obs}\sum_{i=1}^{10} \Delta t_i$$

So

$$k_{obs} = \sum_{i=1}^{10} \frac{-\Delta[OH]}{[\overline{OH}]} \bigg/ \sum \Delta t_i$$

But calculating each of the increments of $\Delta[OH]/[\overline{OH}]$ is quite tedious! So we examine the distinction between these *average* rates and the *instantaneous* rate as shown in Figure 12.23. As the OH concentration decays, the slope of $\Delta[OH]/\Delta t$ decreases such that whenever an average rate is calculated, that average rate will depart from the instantaneous rate at the beginning and end of the time increment Δt over which the average rate is taken. In the limit of very small time increments we can define the *derivative* of the OH concentration with time, as the time increment Δt decreases toward zero. This allows us to reformulate kinetics within the mathematical structure of *calculus*, providing powerful tools and considerable simplification. We thus represent in the limit as $\Delta t \to 0$, the average rate $-\Delta[OH]/\Delta t$ approaching the instantaneous rate given by the *derivative* of the OH concentration with respect to time as

$$\lim_{\Delta t \to 0} \left(\frac{-\Delta[OH]}{\Delta t} \right) = -d[OH] \Big/ dt$$

It follows from our experimentally deduced expression that

$$R_{obs} = \frac{\lim}{\Delta t \to 0} \left(-\frac{\Delta[OH]}{\Delta t} \right) = -\frac{d[OH]}{dt} = k_{obs}[OH]$$

The expression Rate of Reaction = R_{obs} = –d[OH]/dt = k_{obs}[OH] has an important place in the study of chemical kinetics. It is termed the *Rate Law* that defines the relationship between the rate of reaction and the concentration of the reactant, in this case the OH radical. As the rate law expresses, the proportionality constant between the observed rate of reaction, R_{obs} and the reactant concentration, [OH], is the *observed reaction rate constant*, k_{obs}.

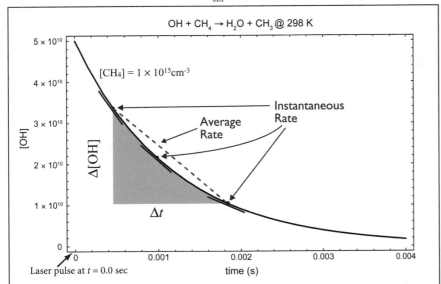

FIGURE 12.23 The decay of [OH] in the presence of CH_4 can be expressed as an average rate, Δ[OH]/Δt, over an arbitrarily long period. But because the rate R = –Δ[OH]/Δt is decreasing throughout the course of the reaction, as the time interval Δt becomes smaller, the determination of the rate at a particular instant of time becomes more accurate. As Δt approaches the limit of Δt → 0, the average rate approaches the instantaneous rate.

The Reaction Rate Order: Determination of the Effect of Concentration on Reaction Rate

Because the concentration of OH, [OH], appears in the experimentally determined rate law raised to the first power, [OH]¹, the rate law is referred to as *first order* with respect to [OH].

But now what happens if we double the concentration of methane in our laboratory system such that [CH_4] = 2×10^{15} molecules/cm³ rather than 1×10^{15} molecules/cm³, as was the case in the first experiment. When we rerun the experiment at twice the concentration of CH_4, the decay plot shows a rather significant difference, as displayed in Figure 12.24.

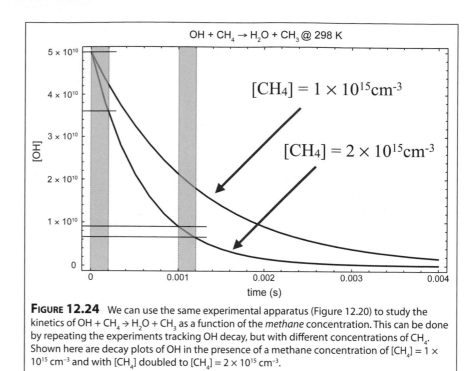

FIGURE 12.24 We can use the same experimental apparatus (Figure 12.20) to study the kinetics of OH + CH$_4$ → H$_2$O + CH$_3$ as a function of the *methane* concentration. This can be done by repeating the experiments tracking OH decay, but with different concentrations of CH$_4$. Shown here are decay plots of OH in the presence of a methane concentration of [CH$_4$] = 1 × 10^{15} cm^{-3} and with [CH$_4$] doubled to [CH$_4$] = 2 × 10^{15} cm^{-3}.

We know how to analyze the decay of OH in the presence of the reactant CH$_4$ in specific "zones" of time.

Zone 1

$$\frac{-\Delta[\text{OH}]/\Delta t}{[\overline{\text{OH}}]} = \frac{7 \times 10^{13}\ \text{mol}/\text{cm}^3 - \text{sec}}{4.2 \times 10^{10}\ \text{mol}/\text{cm}^3} = 1.8 \times 10^3\ \text{sec}^{-1}$$

Zone 2

$$\frac{-\Delta[\text{OH}]/\Delta t}{[\overline{\text{OH}}]} = \frac{1.5 \times 10^{13}\ \text{mol}/\text{cm}^3 - \text{sec}}{8 \times 10^9\ \text{mol}/\text{cm}^3} = 1.8 \times 10^3\ \text{sec}^{-1}$$

This demonstrates in comparison with our first experiment that, since

$$\frac{-\Delta[\text{OH}]/\Delta t}{[\overline{\text{OH}}]} = k_{\text{obs}},$$

the reaction rate constant has *doubled* for a *doubling* of [CH$_4$] because

$$\frac{-\Delta[\text{OH}]/\Delta t}{[\overline{\text{OH}}]}\ \text{has doubled:}$$

The value of the reaction rate constant, k_{obs}, has increased from 9 × 10^2 sec^{-1} to 1.8 × 10^3 sec^{-1}. That is, when [CH$_4$] is doubled, k_{obs} is also doubled. This is an important observation because, from previous experiments, we demonstrated that

$$\frac{-d[\text{OH}]}{dt} = k_{\text{obs}}[\text{OH}].$$

But we now see that $k_{\text{obs}} \propto$ [CH$_4$]; that is the *first-order* reaction rate constant, which we now designate as $k^{\text{I}}_{\text{obs}}$, is in fact equal to the *product* of another constant multiplied by the concentration of methane in the reaction zone, [CH$_4$]. Thus we can write

$$k^{\text{I}}_{\text{obs}} = k^{\text{II}}_{\text{obs}}[\text{CH}_4]$$

where $k^{\text{II}}_{\text{obs}}$ is the *second-order* reaction rate constant. So we can write the *rate law* for the reaction as

$$\frac{-d[OH]}{dt} = k^{II}_{obs}[CH_4][OH]$$

Note that the units of the second-order reaction rate constant k^{II}_{obs} are,

$$\left[\frac{cm^3}{molecule \cdot sec}\right]$$

from a unit analysis of our second-order rate law.

In chemical kinetics, it is very important to keep an eye on the *units* of the reaction rate constants because the units of the reaction rate constant tell you the *order* of the reaction.

We can diagnose the *overall* order of a reaction by a simple test. If we plot the *natural logarithm* of the OH concentration, ln[OH], as a function of time in our laser experiment, we notice a key characteristic of the reaction. *First*, the plot of ln[OH] *versus* reaction time is *linear* over the course of the reaction. *Second*, the *slope* of ln[OH] *versus* t is proportional to $[CH_4]$, but still linear. This is shown in Figure 12.25.

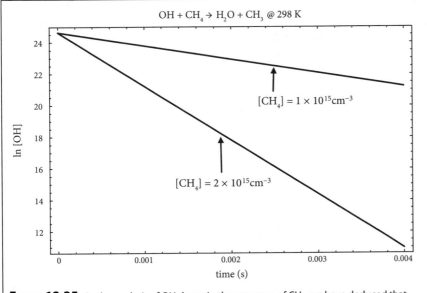

FIGURE 12.25   In the analysis of OH decay in the presence of CH_4, we have deduced that $-(d[OH]/dt)/[OH]$ is a *constant*, k_{obs}. Integration of this expression results in the expression $\ln[OH]_f - \ln[OH]_i = -k_{obs}t$. Graphing this result emphasizes that a plot of ln[OH] versus t is *linear* with a slope equal to $-k_{obs}$. In addition, the slope is proportional to the methane concentration, $[CH_4]$.

This raises the question: can the rate law for a reaction $A + B \rightarrow C + D$ be written in a general form

$$Rate = R = k[A]^m[B]^n$$

where m and n are the *order* of the reaction with respect to the concentration of A and B respectively? The answer is, in general, yes and it is the convention to define:

m = reaction order with respect to A
n = reaction order with respect to B

and the *overall order* of the reaction = sum of the individual orders = $m + n$.

This is where it is important to distinguish between the experimentally determined reaction rate, R_{obs}, and the general form of the rate law $R = k[A]^m[B]^n$ that has not yet been tested experimentally to establish the *order* of the reaction with respect to [A] and the order of the reaction with respect to [B].

Thus far we have built our understanding of the rate law upon experimental observation. We have arranged the experimental conductors such that the concentration of one species, OH, was much less than its reactive partner, CH_4. This forced the reaction to be first order with respect to OH:

$$R_{obs} = -d[OH]/dt = k^I_{obs}[OH]$$

We then reran the experiment at different concentrations of CH_4, demonstrating through direct observations that the rate of the reaction was proportional to $[CH_4]$ such that

$$R_{obs} = -d[OH]/dt = k^I_{obs}[OH] = k^{II}_{obs}[CH_4][OH].$$

This begs the question: if there are first-order reactions and second-order reactions, are there zero-order reactions? Are there third-order reactions? The answer is yes to both questions. So let's look at each of these in order.

---

1. The *rate* counts the number of times the reaction is completed per unit time:

$$R = \frac{-d[OH]}{dt} = \frac{-d[CH_4]}{dt}$$
$$= \frac{d[H_2O]}{dt} = \frac{d[CH_3]}{dt}$$

The *rate law* for a chemical reaction establishes the *functional form* of the reactant concentration with respect to the time rate of change of reactant concentration. The general form of the rate law is related to the *rate* of the reaction, R, by:

$$R = \frac{-d[OH]}{dt} = k[CH_4]^m[OH]^n$$

2. NO! The *order* of the reaction can only be established *by experiment*:

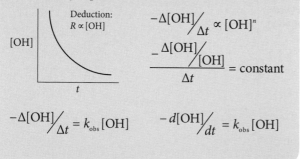

$$-\frac{\Delta[OH]}{\Delta t} = k_{obs}[OH] \qquad -\frac{d[OH]}{dt} = k_{obs}[OH]$$

Check Yourself 2

1. What is the difference between the *rate of a chemical reaction* and the *rate law for a chemical reaction*?

$$CH_4 + OH \rightarrow H_2O + CH_3$$

2. Can the general form of the *rate law* be used to establish the order of the reaction?

The Behavior of Zero-Order, First-Order, Second-Order, and Third-Order Kinetics

Zero-Order Reactions

A very important category of chemical reactions follows the rate law

$$R = -d[A]/dt = k[A]^0 = k$$

wherein the rate of the reaction does not depend upon the concentration of the reacting species, yet the reactant A is decaying with time. Examples include the kinetics of the catalytic converter on a car or many examples of biological catalysts termed enzymes. If we take the example of the catalytic converter on an automobile, displayed in Figure 12.26, toxic species such as CO, NO, and unburned gasoline (C_xH_y) are converted to chemically inert substances such as CO_2, H_2O, and N_2.

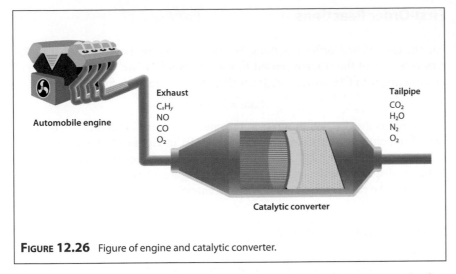

FIGURE 12.26   Figure of engine and catalytic converter.

When we examine what is occurring inside the catalytic converter, the first thing we notice is a meshed network that contains a large surface area of finely divided metals and metal oxides (Pt, Pd, V_2O_5, Cr_2O_3, and CuO). On the surface of those metal substrates, the incoming species from the automobile exhaust dissociates into carbon atoms, oxygen atoms, and nitrogen atoms, which are bound to the metal surface as shown for the case of NO in Figure 12.27.

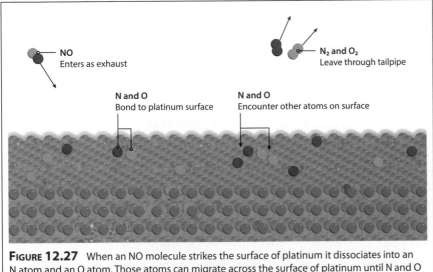
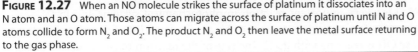

FIGURE 12.27   When an NO molecule strikes the surface of platinum it dissociates into an N atom and an O atom. Those atoms can migrate across the surface of platinum until N and O atoms collide to form N_2 and O_2. The product N_2 and O_2 then leave the metal surface returning to the gas phase.

The dissociation of the toxic species is then followed on the metal surface by the recombination of the separated atoms into their most thermodynamically stable form: CO_2, H_2O, and N_2. But from the perspective of the *kinetics*, each of the NO molecules simply disappears onto the surface of the metal, which has a virtually limitless number of sites available to dissociate NO into N atoms and O atoms. The removal rate of NO is thus independent of the concentration of NO. The concentration of NO decreases, linearly, with time as shown in the left-hand panel of Figure 12.28.

$$-d[NO]/dt = k[NO]^0 = k$$

Another important example is the catalytic destructions of ozone within the Antarctic ozone hole—a subject treated in some detail in Case Study 12.1.

First-Order Reactions

For the case of first-order reactions, the rate of reaction is proportional to the concentration of the reactant raised to the first power as we saw for our case of OH reacting with CH_4 when $[CH_4] \gg [OH]$. So in general for a first-order reaction

$$Rate = k[A]^1$$

As a result, as the reaction progresses, the rate slows as the concentration of A decays. This behavior is reflected specifically for OH in Figure 12.21 and in general for species A in Figure 12.28. The rate of the reaction, R, versus reactant concentration is displayed in the right-hand panel of Figure 12.28, reflecting that *linear* increase in the rate as a function of the concentration of A.

Second-Order Reactions

Following the logic of zero-order and first-order kinetics, second-order reactions are characterized by a rate of reaction that is proportional to the square of the reactant concentration:

$$Rate = k[A]^2$$

As a result the rate of the reaction decreases more rapidly with time than is the case for first-order reactions. This is reflected graphically in the left-hand panel of Figure 12.28. In the right-hand panel of Figure 12.28 the rate of the reaction as a function of the concentration of A is seen to increase quadratically with increasing [A]. There is an important distinction to be made with respect to second-order reactions, a distinction that involves the difference between a reaction that is second order overall (such as the reaction of OH with CH_4) and a reaction that is second order with respect to a specific reactant. We know how to express the order of the reaction for OH with CH_4:

$$R_{obs} = -\frac{d[OH]}{dt} = k_{obs}^{II}[OH][CH_4]$$

The reaction is first order with respect to OH, first order with respect to CH_4, second order *overall*. We successfully determined the order of the reaction and the kinetics of the reaction by judicious control of the *relative concentrations* of the reagents.

However, the reaction of OH with itself

$$OH + OH \rightarrow H_2O + O$$

which is a very important reaction in photochemical systems and in fabrication techniques for microcircuits, cannot be manipulated into first-order kinetics because OH is reacting with itself. The rate of the reaction must be second order.

$$R = -\frac{1}{2}\frac{d[OH]}{dt} = k[OH]^2$$

This rate law is second order with respect to [OH] *and* second order overall. Notice also that the factor ½ appears because two OH radicals are removed each time the reaction is completed. Stated another way, if we are observing the disappearance of OH, the reaction is completed at one-half the rate at which OH is disappearing. This is an important point in kinetics and we consider it carefully after treating third-order kinetics.

We can summarize the disappearance of the reacting species with time for zero, first, and second order as shown schematically in Figure 12.28 as well as the rate of the reaction versus reactant concentration in the right-hand panel of Figure 12.28.

12.28

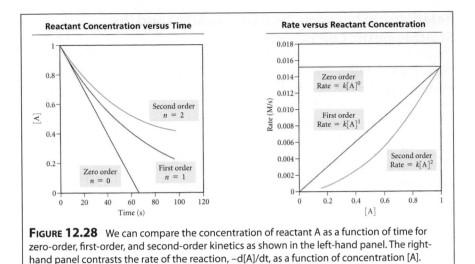

FIGURE 12.28 We can compare the concentration of reactant A as a function of time for zero-order, first-order, and second-order kinetics as shown in the left-hand panel. The right-hand panel contrasts the rate of the reaction, –d[A]/dt, as a function of concentration [A].

Third-Order Reactions

We know that third-order reactions must exist because we have examined the formation of ozone on the potential energy surface of Figure 12.12

$$O + O_2 + M \rightarrow O_3 + M$$

The rate law for the reaction is, if we are following the disappearance of O atoms in the presence of ozone

$$Rate = -\frac{d[O]}{dt} = k_{obs}^{III}[M][O_2][O]$$

We will examine this very important class of reactions in this chapter, but it is clear that we can control the reagent concentrations such that $[O] << [O_2], [M]$. This results in *observed* first-order dependence with respect to [O]:

$$R_{obs} = -\frac{d[O]}{dt} = k_{obs}^{I}[O]$$

Then we can rerun the experiment at different O_2 concentrations and determine the second-order reaction rate constant

$$R_{obs} = -\frac{d[O]}{dt} = k_{obs}^{II}[O_2][O]$$

Finally we can vary the concentration of that chaperone molecule, M, that serves as a chemically inert participant to remove vibrational energy from the vibrationally excited $O_3^\ddagger$ as discussed in the Framework, to determine the third-order reaction rate constant

$$R_{obs} = -\frac{d[O]}{dt} = k_{obs}^{III}[M][O_2][O]$$

These examples drive home the point that (1) the *rate law* defines the relationship between the *rate of reaction* and the reactant concentration, (2) the order of the reaction depends upon experimental conditions, and (3) the order of a reaction can only be determined by experiment.

We turn next to the relationship between the stoichiometry and the statement of the rate law.

Relationship between the Rate Equation and the Stoichiometric Coefficients

There is a key relationship between the stoichiometric coefficients and the rate equation that is made clear by this second-order reaction of OH reacting with itself to produce $H_2O + O$. In particular, when we write a chemical reaction, the rate of the reaction, R, equals the number of times the chemical reaction, as written, is completed each second. However, if one of the reactants is removed at *twice* the rate the reaction is completed, then when that rate equation is written in terms of the time rate of change of that species, the rate of disappearance of that species must be multiplied by a factor of ½. For the reaction

$$OH + OH \rightarrow H_2O + O$$

OH will be disappearing at twice the rate that the reaction is completed. Thus, when the rate is written in terms of d[OH]/dt, the rate is

$$R = -\frac{1}{2}\frac{d[OH]}{dt}$$

When the rate is written in terms of $d[H_2O]/dt$ or d[O]/dt, only one of each of those products is created each time the reaction is completed so

$$R = \frac{d[H_2O]}{dt} = \frac{d[O]}{dt}$$

Because there is only one rate, R, at which a reaction is completed, all those expressions for the rate must be equal, and thus

$$R = -\frac{1}{2}\frac{d[OH]}{dt} = \frac{d[H_2O]}{dt} = \frac{d[O]}{dt}$$

Another (and important) way of looking at this is that if we were *observing* the rate of disappearance of OH in the reaction

$$OH + OH \rightarrow H_2O + O$$

the OH radical would be disappearing at twice the rate at which the reaction is completed and therefore at twice the rate that H_2O or O are forward.

We can generalize this important fact in terms of the stoichiometric coefficients of the reaction

$$aA + bB \rightarrow cC + dD$$

The rate equation for this reaction would be

$$R = -\frac{1}{a}\frac{d[A]}{dt} = -\frac{1}{b}\frac{d[B]}{dt} = \frac{1}{c}\frac{d[C]}{dt} = \frac{1}{d}\frac{d[D]}{dt}$$

Integration of the Rate Law: Defining the Concentration as a Function of Time

When we look forward in time, we would like to understand, to predict, how the concentration of a given reactant will change with time. This ability to predict engages all time scales:

- The dissociation of a molecule on a metal surface that occurs on the order of seconds on a catalytic surface.

- The removal of alcohol from the bloodstream that occurs on the order of hours.
- The removal of heavy metals from a coal burning power plant that occurs on the order of weeks.
- The removal of chlorofluorocarbon from the stratosphere that occurs on the order of decades.
- The removal of CO_2 added to the atmosphere by fossil fuel combustion that occurs on the time scale of centuries.

The relationship between concentration and time is greatly aided by our formulation of the rate law in terms of the *derivative* of the concentration because we can then use calculus to *integrate* that rate law to determine the concentration as a function of time. We consider here in order, zero-order, first-order, and second-order integrated rate laws.

Zero-Order Integrated Rate Law

We know the functional form of the zero-order rate law

$$R = -\frac{d[A]}{dt} = k[A]^0 = k$$

To integrate this rate expression, we rearrange

$$d[A] = -kdt$$

and integrate both sides

$$\int_{[A]_o}^{[A]_t} d[A] = -\int_{t=0}^{t} kdt$$

to yield the relationship between $[A]_t$ at time t, the initial concentration $[A]_o$, and the rate constant:

$$[A]_t - [A]_o = -kt$$

or

$$[A]_t = -kt + [A]_o$$

When $[A]_t$ is graphed against time, t, we have, as shown in Figure 12.29, the decay of [A] as a function of time.

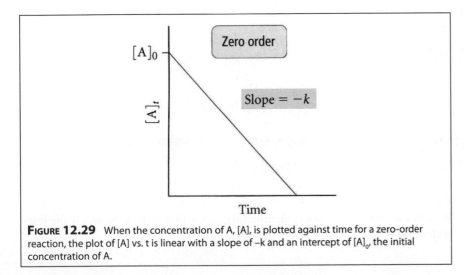

FIGURE 12.29 When the concentration of A, [A], is plotted against time for a zero-order reaction, the plot of [A] vs. t is linear with a slope of –k and an intercept of $[A]_o$, the initial concentration of A.

The integrated form of the zero-order rate law is thus a linear relationship between the concentration of A, [A], and time with a slope of –k and an intercept of [A]$_o$. The integral form of the zero-order rate law reveals some very important characteristics of these reactions. First, note that unlike the first-order reaction, there is a clear *endpoint* to the reaction. For if

$$[A]_t = -kt + [A]_0$$

then when

$$[A]_t = 0 \qquad t_{end} = [A]_0/k$$

and literally *all* the reactant is removed. It is for this reason that catalytic converters remove virtually all the NO and CO from the exhaust of an automobile. It is also why, as we will see in Case Study 12.1, chlorine and bromine compounds can destroy all the ozone in the stratosphere of the Earth.

First-Order Integrated Rate Law

Consider our experimentally determined rate law for OH reacting with CH$_4$:

$$R_{obs} = \frac{-d[OH]}{dt} = k^I_{obs}[OH]$$

If we wish to calculate the concentration of OH as a function of time throughout the course of the reaction, we must employ calculus to integrate the equation. First we rearrange the equation by dividing both sides by [OH] and multiplying both sides by dt:

$$-\frac{d[OH]}{[OH]} = k^I_{obs}dt$$

Then we integrate the left-hand side from the initial OH concentration at t = 0, [OH]$_o$, to the final OH concentration at time t, [OH]$_t$. The right-hand side is integrated from t = 0 to t:

$$-\int_{[OH]_o}^{[OH]_t} \frac{d[OH]}{[OH]} = \int_{t=0}^{t} k^I_{obs}\, dt$$

The left-hand side is just a specific form of the general integral

$$\int_{x_1}^{x_2} dx/x = \ln x_2 - \ln x_1$$

But ln x$_2$ – ln x$_1$ = ln(x$_2$/x$_1$) = –ln(x$_1$/x$_2$). Thus we can integrate both sides.

$$-\int_{[OH]_o}^{[OH]_t} d[OH]/[OH] = \int_{t=0}^{t} k^I_{obs}dt$$

$$\downarrow \text{integrate} \qquad\qquad \downarrow \text{integrate}$$

$$-\ln\left([OH]_t/[OH]_o\right) = k^I_{obs}t$$

We can use the mathematical identity ln(x$_2$/x$_1$) = ln x$_2$ – ln x$_1$ to recast the integrated rate law

$$-\ln\left([OH]_t \Big/ [OH]_o\right) = k^I_{obs} t$$

to give $\ln[OH]_t = -k^I_{obs} t + \ln[OH]_o$. This equation is just the slope-intercept equation for y as a function of x, y = m x + b, where m is the slope of y vs. x and b is the intercept. Thus the relationship between $[OH]_t$ and t can be represented graphically by plotting $\ln[OH]_t$ vs. t as shown in Figure 12.30.

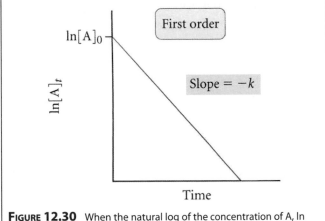

FIGURE 12.30 When the natural log of the concentration of A, ln [A], is plotted against time for a reaction, if the observed kinetics is first order, that plot of ln [A] vs. time will be linear. The slope of the line is equal to –k, and the intercept is ln [A]$_0$ where [A]$_0$ is the initial concentration.

Next lets consider the integral form of the first-order rate law in the form

$$\ln([OH]_t/[OH]_o) = -k^I_{obs} t$$

If we exponentiate this expression we have

$$[OH]_t = [OH]_o\, e^{-k^I_{obs} t}$$

Expressed this way, it raises the question: how can we succinctly express the time required to remove a certain fraction of the original concentration? If we could do this, it would provide a very straightforward way to compare how fast a given reaction or process is occurring relative to other examples.

Suppose we ask: how long would it take for half the original [OH] to disappear? We can use the above equation to solve for this time, which we will refer to as the *half-life* of the reaction, $t_{1/2}$. As it will turn out, this half-life is important in kinetics as well as in other fields such as radioactive decay.

Check Yourself 3
Expressing the Percent of Reactant Consumed in a First-Order Reaction:
Use a value of k = 7.30×10^{-4} s^{-1} for the first-order decomposition of H_2O_2(aq) to determine the percent H_2O_2 that has decomposed in the first 500.0 seconds after the reaction begins.

Solution:
The ratio $[H_2O_2]_t/[H_2O_2]_0$ represents the fractional part of the initial amount of H_2O_2 that remains *unreacted* at time t. Our problem is to evaluate this ratio at t = 500.0 s.

$$\ln\frac{[H_2O_2]_t}{[H_2O_2]_0} = -kt = -7.30\times10^{-4}\,\text{s}^{-1} \times 500.0\,\text{s} = -0.365$$

$$\frac{[H_2O_2]_t}{[H_2O_2]_0} = e^{-0.365} = 0.694 \quad \text{and} \quad [H_2O_2]_t = 0.694[H_2O_2]_0$$

The fractional part of the H_2O_2 remaining is 0.694, or 69.4%. The percent of H_2O_2 that has decomposed is 100.0% – 69.4% = 30.6%.

Practice Example A: Consider the first-order reaction A → products, with k = 2.95×10^{-3} s^{-1}. What percent of A remains after 150 s?

Practice Example B: At what time after the start of the reaction is a sample of H_2O_2(aq) two-thirds decomposed? K = 7.30×10^{-4} s^{-1}?

If t=$t_{1/2}$ is the time required to remove ½ the original OH concentration then

$$[OH]_t = \tfrac{1}{2} [OH]_o$$
$$\text{and} \quad t = t_{1/2}$$

so

$$[OH]_t = \frac{1}{2} [OH]_o = [OH]_o e^{-k_{obs}^I t_{1/2}}$$

Solving for $t_{1/2}$ we have

$$\frac{1}{2}[OH]_o = [OH]_o e^{-k_{obs}^I t_{1/2}}$$

so

$$\frac{1}{2} = e^{-k_{obs}^I t_{1/2}}$$

and therefore

$$t_{1/2} = \frac{-\ln\left(\frac{1}{2}\right)}{k_{obs}^I} = \frac{0.693}{k_{obs}^I}$$

This is an important result. It says that the period of time to remove ½ the concentration of the reactant is *independent* of the initial concentration. This means that in every time increment $0.693/k_{obs}^I$, the concentration of the reactant decreases by one-half. In our experiment observing the decay of OH in the presence of a methane concentration of $[CH_4] = 1 \times 10^{15}$ cm^{-3}, [OH] dropped to half the original concentration in

$$t_{1/2} = \frac{0.693}{k_{obs}^I} = \frac{0.693}{9 \times 10^2 \text{ sec}^{-1}} = 7.7 \times 10^{-4} \text{ sec}$$

But in each subsequent 7.7×10^{-4} seconds, the concentration dropped by a factor of two again.

Since we focused on the half-life of the first-order reaction, what is the half-life of a zero-order reaction? If $[A]_t = \tfrac{1}{2} [A]_o$ then we have

$$[A]_t = \tfrac{1}{2} [A]_o = -kt_{1/2} + [A]_o$$

$$\text{and} \quad t_{1/2} = [A]_o / 2k$$

In contrast to the behavior of first-order kinetics where $t_{1/2}$ is independent of the initial concentration, in the case of zero-order kinetics, the half-life is proportional to the initial concentration so if we begin with twice as much reactant, it will take twice as long to remove half of it.

12.34

Check Yourself 4

Using the Integrated Rate Law for a First-Order Reaction and the use of M = moles/liters as a Concentration Unit: $H_2O_2(aq)$, initially at a concentration of 2.32 M, is allowed to decompose. What will $[H_2O_2]$ be at $t = 1200$ s? Use $k = 7.30 \times 10^{-4}$ s^{-1} for this first-order decomposition.

Solution:

We have values for three of the four terms in equation $\ln[A]_t = -kt + \ln[A]_0$

$k = 7.30 \times 10^{-4}$ s^{-1} $\quad t = 1200$ s

$[H_2O_2]_0 = 2.32$ M $\quad [H_2O_2]_t = ?$

The demonstration of first-order behavior depends on the observed dependence of concentration vs. time. In general, the order of a reaction can only be determined by direct experiment. An example is shown in the table.

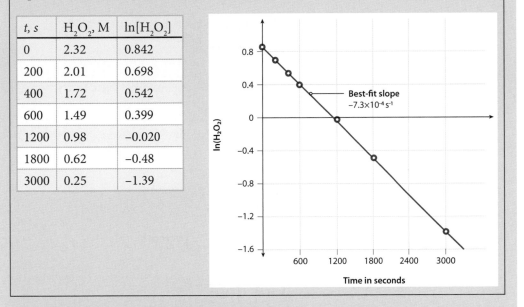

t, s	H_2O_2, M	$\ln[H_2O_2]$
0	2.32	0.842
200	2.01	0.698
400	1.72	0.542
600	1.49	0.399
1200	0.98	−0.020
1800	0.62	−0.48
3000	0.25	−1.39

which we substitute into the expression

$\ln[H_2O_2]_t = -kt + \ln[H_2O_2]_0$

$\qquad = -(7.30 \times 10^{-4}$ s$^{-1} \times 1200$ s$) + \ln 2.32$

$\qquad = \qquad\qquad -0.876 \qquad\qquad + 0.842 = -0.034$

$[H_2O_2]_t = e^{-0.034} = 0.967$ M

This calculated value agrees well with the experimentally determined value of 0.98 M.

Practice Example A: The reaction A → 2 B + C is first order. If the initial $[A] = 2.80$ M and $k = 3.02 \times 10^{-3}$ s^{-1}, what is the value of $[A]$ after 325 s?

Practice Example B: Use data tabulated in the above figure together with the equation $\ln[A]_t = -kt + \ln[A]_0$ to show that the decomposition of H_2O_2 is a first-order reaction.

[*Hint*: Use a pair of data points for $[H_2O_2]_0$ and $[H_2O_2]_t$ and their corresponding times to solve for k. Repeat this calculation using other sets of data. How should the results compare?]

Second-Order Integrated Rate Law

In this case, for example the reaction of OH with itself

$$OH + OH \rightarrow H_2O + O_2$$

we have the rate law

$$-d[OH]/dt = k_{obs}^{II}[OH]^2$$

When we rearrange

$$-d[OH]/[OH]^2 = k_{obs}^{II}\,dt$$

and integrate from $[OH]_o$ to $[OH]_t$

$$\int_{[OH]_o}^{[OH]_t} \frac{d[OH]}{[OH]^2} = \int_{t=0}^{t} k_{obs}^{II}\,dt$$

we use the general integral from

$$\int_{x_1}^{x_2} dx/x^2 = \frac{1}{x_1} - \frac{1}{x_2}$$

so

$$-\int_{[OH]_o}^{[OH]_t} \frac{d[OH]}{[OH]^2} = \int_{t=0}^{t} k_{obs}^{II}\,dt$$

$\downarrow$ integrate $\downarrow$ integrate

$$\frac{1}{[OH]_t} - \frac{1}{[OH]_o} = k_{obs}^{II}t$$

or

$$\frac{1}{[OH]_t} = k_{obs}^{II}t + \frac{1}{[OH]_o}$$

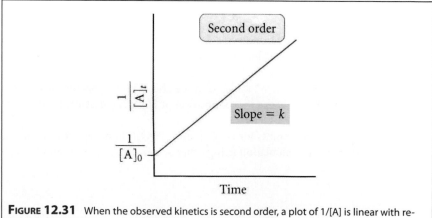

FIGURE 12.31 When the observed kinetics is second order, a plot of 1/[A] is linear with respect to time with an intercept 1/[A]$_0$ where [A]o is the initial concentration and a slope equal to the observed reaction rate constant, k.

12.36

If we graph the concentration of [OH] versus time, the most convenient form is to plot $1/[OH]$ vs. t so the slope of the line is equal to k^{II}_{obs} and the intercept is $1/[OH]_0$, as shown in Figure 12.31.

As we did with the zero-order and first-order cases, we can calculate the half-life of a reactant in a second-order reaction by setting $[OH]_t = \frac{1}{2}[OH]_o$ and $t = t_{\frac{1}{2}}$ so the above equation gives us

$$\frac{1}{[OH]_t} = \frac{1}{\frac{1}{2}[OH]_o} = k^{II}_{obs}t_{\frac{1}{2}} + \frac{1}{[OH]_o}$$

Solving for $t_{1/2}$

$$k^{II}_{obs}t_{\frac{1}{2}} = \frac{2}{[OH]_o} - \frac{1}{[OH]_o} = \frac{1}{[OH]_o}$$

and

$$t_{\frac{1}{2}} = \frac{1}{k^{II}_{obs}[OH]_o}$$

Thus the half-life of a second-order reaction depends inversely on the initial reactant concentration.

We can summarize the relationship between the order of a reaction, the rate law, the units of the reaction rate constant, the integral form of the rate law, the graphical relationship between concentration and time, and the expression for the half-life of the concentration, all displayed in Figure 12.32.

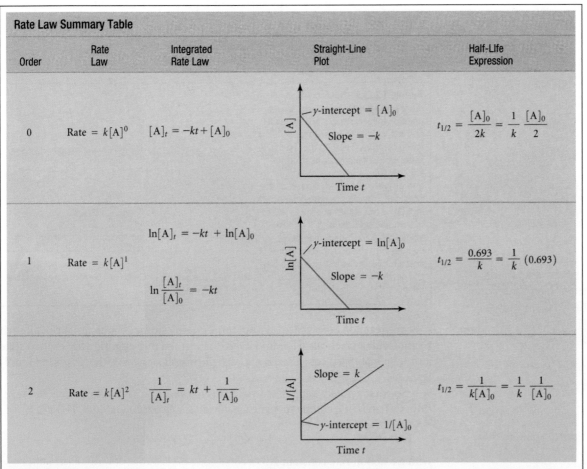

Rate Law Summary Table

Order	Rate Law	Integrated Rate Law	Straight-Line Plot	Half-Life Expression
0	Rate $= k[A]^0$	$[A]_t = -kt + [A]_0$	*y*-intercept $= [A]_0$, Slope $= -k$ ([A] vs Time *t*)	$t_{1/2} = \dfrac{[A]_0}{2k} = \dfrac{1}{k}\dfrac{[A]_0}{2}$
1	Rate $= k[A]^1$	$\ln[A]_t = -kt + \ln[A]_0$ $\ln\dfrac{[A]_t}{[A]_0} = -kt$	*y*-intercept $= \ln[A]_0$, Slope $= -k$ ($\ln[A]$ vs Time *t*)	$t_{1/2} = \dfrac{0.693}{k} = \dfrac{1}{k}(0.693)$
2	Rate $= k[A]^2$	$\dfrac{1}{[A]_t} = kt + \dfrac{1}{[A]_0}$	Slope $= k$, *y*-intercept $= 1/[A]_0$ ($1/[A]$ vs Time *t*)	$t_{1/2} = \dfrac{1}{k[A]_0} = \dfrac{1}{k}\dfrac{1}{[A]_0}$

FIGURE 12.32 We can summarize, for observed zero-order, first-order, and second-order kinetics, the rate law, the integrated rate law, the plot of concentration vs. time, and the half-life expression for each case.

12.37

Answer:

- Plot ln [A] vs. time:

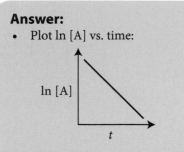

- If ln [A] is linear, the reaction is first order with respect to species A and the *overall* order of the reaction is first order.

Question:

Can the rate law for a chemical reaction be constructed from the stoichiometric coefficient?

$$aA + bB \rightarrow cC + dD$$

Answer:

- No. The rate law depends on *experimental conditions*:

$$\frac{-d[OH]}{dt} = k_{obs}[CH_4][OH]$$

First order with respect to [OH]
First order with respect to [CH4]
Overall order of reaction
$1 + 1 = 2 \qquad k_{obs} = k^{II}_{obs}$

- If we arrange the experiment such that $[CH_4] \gg [OH]$:
$$k^{II}_{obs}[CH_4] \approx constant = k^{I}_{obs}$$

$$\frac{-d[OH]}{dt} = k^{II}_{obs}[CH_4][OH] = k^{I}[OH]$$

Overall reaction becomes first order.

- Stoichiometry is invariant; it is set by bonding structure. Therefore, we cannot in general determine the rate law from the stoichiometric coefficients.

Check Yourself 5
Question:
Given the observed concentration of a reacting species as a function of time throughout the course of a reaction, what is the most direct test to determine whether the reaction is first order?

Steady State Approximation

We have now explored the kinetics of chemical reactions from (1) the molecular perspective in our analysis of the *molecularity* of chemical reactions, and (2) the experimental perspective in terms of the observed *order* of a chemical reaction; we turn to how the union of those two approaches are used in practice.

One of the most useful and powerful approximations in chemistry is the **Steady State Approximation**, which takes explicit advantage of the fact that within many chemical mechanisms reside elementary reactions that rapidly produce and remove reactive intermediates for which the **loss rate** of the species is limited by its **production rate**.

Consider the important case of the hydroxyl radical, OH, reacting with NO_2 shown on its potential energy surface in Figure 12.33. The only thermodynamically allowed pathway for this reaction is the termolecular reaction

$$OH + NO_2 + M \rightarrow HONO_2 + M,$$

forming nitric acid.

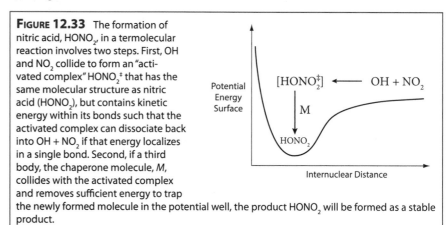

FIGURE 12.33 The formation of nitric acid, $HONO_2$, in a termolecular reaction involves two steps. First, OH and NO_2 collide to form an "activated complex" $HONO_2^{\ddagger}$ that has the same molecular structure as nitric acid ($HONO_2$), but contains kinetic energy within its bonds such that the activated complex can dissociate back into OH + NO_2 if that energy localizes in a single bond. Second, if a third body, the chaperone molecule, *M*, collides with the activated complex and removes sufficient energy to trap the newly formed molecule in the potential well, the product $HONO_2$ will be formed as a stable product.

The reaction takes place on a potential energy surface characteristic of all such **termolecular** reactions wherein $HONO_2$ is formed in a vibrationally excited state and then stabilized into the potential well by collision with the chaperone molecule, *M*.

So what actually happens to yield this net reaction? First, to initiate the reaction, OH and NO_2 collide to form the vibrationally excited adduct, $HONO_2^{\ddagger}$:

$$OH + NO_2 \xrightarrow{k_1} HONO_2^{\ddagger}$$

with a reaction rate constant k_1.

But as we have seen, if the vibrational energy becomes localized in the (weakest) bond that links OH to NO_2, the adduct will break apart into the original species OH and NO_2:

$$HONO_2{}^\ddagger \xrightarrow{k_{-1}} OH + NO_2$$

with a reaction rate constant k_{-1}.

But it is also possible to stabilize the $HONO_2{}^\ddagger$ adduct by collision with the chaperone molecule, M, to yield product nitric acid molecule, $HONO_2$:

$$HONO_2{}^\ddagger + M \xrightarrow{k_2} HONO_2$$

with reaction rate constant, k_2.

We can assemble these three steps to yield the net overall reaction

$$OH + NO_2 \underset{k_{-1}}{\overset{k_1}{\rightleftarrows}} HONO_2^\ddagger$$

$$\underline{HONO_2^\ddagger + M \xrightarrow{k_2} HONO_2 + M}$$

$$\text{Net:} \quad OH + NO_2 + M \to HONO_2 + M$$

We seek an expression that is correct for all pressures, that is for all concentrations of the chaperone species, M.

Step 1:

We write the **rate** of the reaction in terms of the elementary step that produces the product $HONO_2$:

$$R = \frac{d[HONO_2]}{dt} = k_2[M][HONO_2^\ddagger] = \frac{-d[OH]}{dt}$$

Step 2:

But we must *calculate the concentration* of the adduct, $HONO_2{}^\ddagger$, in order to make this a useful analysis. So we write the rate equation for the production of $HONO_2{}^\ddagger$

$$\text{Production Rate} = P_{HONO_2^\ddagger} = \frac{d[HONO_2^\ddagger]}{dt} = k_1[OH][NO_2]$$

Step 3:

Then we write the rate equation for the **removal** of $HONO_2{}^\ddagger$

$$\text{Loss Rate} = L_{HONO_2^\ddagger} = \frac{d[HONO_2^\ddagger]}{dt} = k_{-1}[HONO_2^\ddagger] + k_2[M][HONO_2^\ddagger]$$

Step 4:

The *net* time-rate-of-change of the adduct concentration, $[HONO_2{}^\ddagger]$, is then given by the production rate, $P_{HONO_2{}^\ddagger}$, minus the loss rate, $L_{HONO_2{}^\ddagger}$, such that

$$\frac{d[HONO_2^\ddagger]}{dt} = P_{HONO_2^\ddagger} - L_{HONO_2^\ddagger} = k_1[OH][NO_2] - k_{-1}[HONO_2^\ddagger] - $$
$$k_2[M][HONO_2^\ddagger]$$

Step 5:

But now the key "steady state" approximation:

1. $HONO_2{}^\ddagger$ is both produced and destroyed very rapidly on the time scale of the overall reaction converting OH and NO_2 to $HONO_2$.

2. If the *difference* between $P_{HONO_2{}^\ddagger}$ and $L_{HONO_2{}^\ddagger}$ is small compared with the magnitude of $P_{HONO_2{}^\ddagger}$ or of $L_{HONO_2{}^\ddagger}$, then

$$\frac{d[\text{HONO}_2^{\ddagger}]}{dt} = P_{\text{HONO}_2^{\ddagger}} - L_{\text{HONO}_2^{\ddagger}} \approx 0$$

That is, the *net* time-rate-of-change of $[\text{HONO}_2^{\ddagger}]$ is approximately equal to zero.

Then, if this is true, we can solve for the concentration of the adduct, $\text{HONO}_2^{\ddagger}$, by setting the time rate of change of $[\text{HONO}_2^{\ddagger}]$ equal to zero!

$$\frac{d[\text{HONO}_2^{\ddagger}]}{dt} = P_{\text{HONO}_2^{\ddagger}} - L_{\text{HONO}_2^{\ddagger}} = 0 \quad \text{so}$$

$$\frac{d[\text{HONO}_2^{\ddagger}]}{dt} = k_1[\text{OH}][\text{NO}_2] - k_{-1}[\text{HONO}_2^{\ddagger}] - k_2[M][\text{HONO}_2^{\ddagger}] = 0$$

Step 6:

We can then use the fact that $k_1[\text{OH}][\text{NO}_2] - k_{-1}[\text{HONO}_2^{\ddagger}] - k_2[M][\text{HONO}_2^{\ddagger}] = 0$ to solve for $[\text{HONO}_2^{\ddagger}]$:

$$[\text{HONO}_2^{\ddagger}] = \frac{k_1[\text{OH}][\text{NO}_2]}{k_{-1} + k_2[M]}$$

Then, inserting the expression for $[\text{HONO}_2^{\ddagger}]$ back into the rate expression,

$$R = \frac{-d[\text{OH}]}{dt} = \frac{d[\text{HONO}_2]}{dt} = k_2[M][\text{HONO}_2^{\ddagger}] = \frac{k_2[M]k_1[\text{OH}][\text{NO}_2]}{k_{-1} + k_2[M]}$$

Rearranging, we have the rate expression for the disappearance of [OH] expressed in terms of a second-order reaction rate constant, $k_{\text{obs}}^{\text{II}}$

$$R = \frac{-d[\text{OH}]}{dt} = k_{\text{obs}}^{\text{II}}[\text{NO}_2][\text{OH}] \quad \text{where} \quad k_{\text{obs}}^{\text{II}} = \frac{k_1 k_2[M]}{k_{-1} + k_2[M]}$$

This is a very important result because it defines, quantitatively, the functional dependence of the second-order reaction rate constant, $k_{\text{obs}}^{\text{II}}$, on the concentration of $[M]$ and of the individual reaction rate constants k_1, k_{-1} and k_2. If we graph $k_{\text{obs}}^{\text{II}}$ vs. $[M]$, we can identify two limits. At the high pressure (large $[M]$) limit, $k_2[M] \gg k_{-1}$ so that we can ignore k_{-1} in the denominator. At this limit,

$$k_{\text{obs}}^{\text{II}} = \frac{k_1 k_2[M]}{k_{-1} + k_2[M]} \cong \frac{k_1 k_2[M]}{k_2[M]} = k_1.$$

The other limit, the low pressure (small $[M]$) limit for which $k_2[M] \ll k_{-1}$ such that $k_2[M]$ can be ignored in the denominator of $k_{\text{obs}}^{\text{II}}$, gives us

$$k_{\text{obs}}^{\text{II}} = \frac{k_1 k_2[M]}{k_{-1} + k_2[M]} = \frac{k_1 k_2}{k_{-1}}[M].$$

Thus when we graph $k_{\text{obs}}^{\text{II}}$ vs. $[M]$, we have a range of $[M]$ for which the slope of $k_{\text{obs}}^{\text{II}}$ vs. $[M]$ increases linearly for small $[M]$. At large $[M]$, $k_{\text{obs}}^{\text{II}}$ approaches a limit that is independent of $[M]$. This is sketched in Figure 12.34.

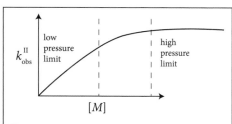

Figure 12.34 When the second-order reaction rate constant, $k_{\text{obs}}^{\text{II}}$, is plotted against the concentration, [M], the pressure dependence of the reaction rate separates into three domains. At low pressure, where $k_2[M] \gg k_{-1}$, the observed reaction rate constant $k_{\text{obs}}^{\text{II}}$ increases proportionality with [M]. At high pressures, where $k_2[M] \gg k_{-1}$, the observed reaction rate constant, $k_{\text{obs}}^{\text{II}}$, is independent of [M]. Between the high- and low-pressure regimes the observed reaction rate constant, $k_{\text{obs}}^{\text{II}}$, transitions smoothly between the two limits.

Concept of the Rate Limiting Step

Within this treatment of the steady state approximation is a very important concept in chemistry—the Rate Limiting Step. Notice that at high pressures, for which

12.40

$k_2[M] \gg k_{-1}$, the collisional removal by the chaperone, M, occurs at a faster rate then the dissociation step, k_{-1}. Under these conditions the rate at which the product $HONO_2$ is formed is "rate limited" by the rate of foundation of the adduct, $HONO_2^{\ddagger}$. That is why the observed rate law takes the simple form

$$-\frac{d[OH]}{dt} = k_{obs}^{II}[NO_2][OH] = k_1[NO_2][OH]$$

where k_1 is the initial formation step in our steady state mechanism.

This is an important example of how, in a chemical mechanism that is comprised of a *series* of elementary steps, it is the slowest step in the sequence that determines the overall rate of the reaction.

We examine this with another example of the application of the steady state approximation:

Example: Using the steady state approximation for the testing of proposed mechanisms against the observation of a rate law.

Consider the reaction of NO(g) and O_2(g)

$$2NO(g) + O_2(g) \rightarrow 2NO_2(g)$$

Laboratory experimentation demonstrates that the *observed* rate law is

$$R_{obs} = -\frac{d[O_2]}{dt} = k_{obs}[NO]^2[O_2]$$

Without making an assumption about the relative rates of each step in the mechanism, we propose the three-step sequence:

$$NO + NO \underset{k_2}{\overset{k_1}{\rightleftarrows}} N_2O_2$$

$$N_2O_2 + O_2 \xrightarrow{k_3} 2NO_2$$

$$\text{Net: } 2NO + O_2 \longrightarrow 2NO_2$$

noting that reaction (2) is just the *reverse* of reaction (1). We note that the *observed* rate law contains $[O_2]$, so we propose that reaction (3) is rate limiting:

$$R_{proposed} = k_3[N_2O_2][O_2]$$

Then we use the steady state approximation to calculate the concentration of the N_2O_2 intermediate. Therefore,

$$\frac{d[N_2O_2]}{dt} = P_{N_2O_2} - L_{N_2O_2} = 0$$

But

$$P_{N_2O_2} = k_1[NO][NO]$$

$$L_{N_2O_2} = k_2[N_2O_2] + k_3[N_2O_2][O_2]$$

So

$$\frac{d[N_2O_2]}{dt} = k_1[NO][NO] - k_2[N_2O_2] - k_3[N_2O_2][O_2] = 0$$

Solving for $[N_2O_2]$

$$k_1[NO][NO] = (k_2 + k_3[O_2])[N_2O_2]$$

$$[N_2O_2] = \frac{k_1[NO][NO]}{k_2 + k_3[O_2]}$$

But now we have, by virtue of the steady state approximation, an expression for the *proposed* rate law:

$$R_{proposed} = k_3[O_2][N_2O_2] = k_3[O_2]\frac{k_1[NO][NO]}{k_2 + k_3[O_2]}$$

If reaction (3) is "rate limiting," then

$$k_2[N_2O_2] \gg k_3[N_2O_2][O_2]$$

so

$$k_2 \gg k_3[O_2]$$

and

$$R_{proposed} = k_3[O_2]\left[\frac{k_1[NO][NO]}{k_2 + k_3[O_2]}\right] = k_3[O_2]\left[\frac{k_1}{k_2}[NO]^2\right]$$

But the experimentally observed rate is

$$R_{obs} = k_{obs}[O_2][NO]^2,$$

so the proposed mechanism is consistent with the observed kinetics and the observed reaction rate constant

$$k_{obs} = \frac{k_1 k_3}{k_2}.$$

Arrhenius Expression for Temperature Dependence

We all have a great deal of experience with the temperature dependence of chemical reactions. Nearly all foods that are open to the atmosphere will spoil much more rapidly as temperatures increase. That is, of course, why we have refrigeration in our homes and virtually everywhere else. We also notice that adhesives such as epoxy cure more rapidly at elevated temperature and materials age more rapidly at high temperature. It was Arrhenius who first systematically studied this behavior, and as he investigated more and more chemical systems with respect to the rate at which reactions proceeded as a function of temperature, he began to see a consistent pattern.

What Arrhenius noticed was that the reaction rate constant, k, changed very little over a fairly broad range of temperatures and then as the rate began to increase it became rapidly faster with increasing temperature, as shown in Figure 12.35.

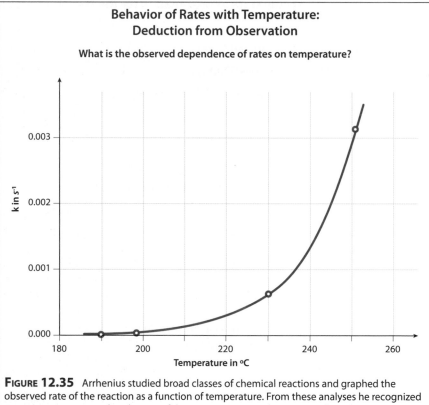

FIGURE 12.35 Arrhenius studied broad classes of chemical reactions and graphed the observed rate of the reaction as a function of temperature. From these analyses he recognized a consistent pattern—that the rate of the reaction increased exponentially with increasing temperature.

Arrhenius recognized, after investigating a large number of different reactions, that the reaction rate constant, k, seemed to increase exponentially with temperature. That is, it exhibited such a strong dependence on temperature that a simple polynomial such as T^2 or T^3 could not capture the dramatic increase with temperature. So Arrhenius began to play around with different functions in an attempt to mathematically "model" this temperature dependence of k. He recognized that if he plotted the **logarithm** of the observed rate against the **inverse** of temperature, the graph had a simple slope-intercept form—the slope was largely invariant with increasing $(1/T)$!

This meant that the **intercept** in this "Arrhenius plot" corresponded to the reaction rate constant at high temperature, and the slope was a constant such that it was possible to write

$$\ln k = -\frac{C}{T} + \ln A$$

where C had to have units of temperature (Kelvin) and $\ln A$ was the intercept such that, by taking the antilog, he could write

$$k_{obs} = A\exp\left[-\frac{C}{T}\right]$$

But from Boltzmann's work Arrhenius also knew that the gas constant, R, had units of [Joule/mole·K] so that if the constant $C = E_a/R$, where E_a was the *activation energy*, then the observed reaction rate constant was simply

$$k_{obs} = A\exp\left[-\frac{E_a}{RT}\right]$$

The slope of the "Arrhenius Plot" was $-E_a/R$ such that we can capture this all in simple graphical form, as shown in Figure 12.36.

This was a very important deduction with far reaching consequences. **First**, it was remarkably simple and remarkably general—it was successfully applied across a vast range of chemical processes. **Second**, it clearly implied that there was an energy barrier to these chemical reactions that must be surmounted in order to execute the chemical transformation. It is also implied that the velocity or energy distribution of the molecules within the reacting medium had an exponential dependence on temperature.

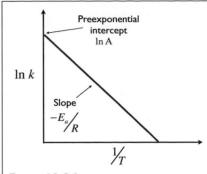

FIGURE 12.36   As a result of studying the rates of chemical reactions as a function of temperature, and deducing the exponential dependence of k on temperature, Arrhenius expressed the relationship between the reaction rate constant, k, and the temperature, T, as $k = Ae^{-E_a/RT}$, where A was the temperature independent "preexponential" and E_a is the "barrier height" energy. Exponentiation results in $\ln k = \ln A - E_a/RT$ such that when $\ln k$ is graphed against $1/T$, the intercept is $\ln A$ and the slope is $-E_a/R$, where R is the gas constant.

Relating Molecular Motion to the Arrhenius Expression

At the same time Arrhenius was developing his expression for the temperature dependence of chemical reactions, Boltzmann was working on the problem of molecular motion and the relationship between the kinetic energy of molecules, the temperature of a molecular ensemble, and the pressure exerted by those molecules. Consider first a vessel that contains N molecules in a volume V. In addition, suppose all of those molecules are moving at the same speed, v (units of m/sec).

Now let's focus in on a section of the vessel wall and consider a volume extending from the wall a distance v, which is how far a molecule will move in *one second* if it is moving directly toward the wall.

All molecules moving toward the wall at a distance equal to or less than v meters will hit the wall in one second. That number, however, is equal to the number of molecules per unit volume times the volume of vA, where A is the cross sectional area. Thus, we can write

$$\left(\begin{array}{c}\text{number of}\\\text{molecules in}\\\text{the volume}\end{array}\right) = \left(\frac{N}{V}\right)vA$$

Now we make an approximation (an approximation that is quite accurate) that 1/6 of all N molecules in the vessel are moving toward the wall. But in order to determine the pressure exerted by those molecules striking the wall, we must determine the force imparted by each of those collisions. To do this, we use the fact that the force imparted by each collision is equal to the **change** in **momentum** divided by the period of the collision for each collision.

However, each collision with the wall corresponds to a momentum change of mv to stop the molecule as it hits the wall, and *another* mv to accelerate the molecule back up to speed v as it leaves the wall in the opposite direction. Thus, the collision event delivers a momentum change of $2mv$ that is the force transmitted to the wall by the molecule.

So, when we consider all the molecules in the volume, we can write

$$\text{Force} = \begin{pmatrix} \text{number of molecules} \\ \text{striking the wall} \\ \text{per sec} \end{pmatrix} \begin{pmatrix} \text{momentum} \\ \text{imparted by} \\ \text{each molecule} \end{pmatrix} = \frac{1}{6}\left(\frac{N}{V}\, vA\right)(2mv)$$

But pressure is just force per unit area, so

$$P = \left(\frac{\text{Force}}{A}\right) = \frac{1}{3}\left(\frac{N}{V}\right) mv^2$$

As Boltzmann recognized, all molecules do not move at the same speed and we must take the *average* of the *square* of the molecular velocities. We write

$$P = \frac{1}{3}\left(\frac{N}{V}\right) m\overline{v^2}$$

This is quite a discovery: By examining individual collision events between a molecule and the vessel wall, we have deduced an expression for the pressure exerted by the ensemble of molecules contained in the vessel.

There is another crucial expression that gives us the pressure exerted by a gas: It is the "Perfect Gas Law."

$$PV = nRT$$

where, as always, P is the pressure, V is the volume, n is the number of moles, R is the gas constant, and T is the temperature in K.

If $PV = nRT$, then we can solve for

$$P = \left(\frac{n}{V}\right) RT$$

and equate it to our molecular level deduction

$$P = \frac{1}{3}\left(\frac{N}{V}\right) m\overline{v^2}$$

such that

$$P = \left(\frac{n}{V}\right) RT = \frac{1}{3}\left(\frac{N}{V}\right) m\overline{v^2} \quad \text{or} \quad nRT = \frac{1}{3} Nm\overline{v^2}$$

If we assume we have 1 mole ($n = 1$), which is Avogadro's number, N_A, then

$$\frac{1}{3} N_A\, m\overline{v^2} = RT$$

However, $N_A m = M$, the **molar mass**, so

$$\overline{v^2} = \left(\frac{3RT}{M}\right)$$

And writing $v_{rms} = \sqrt{\overline{v^2}}$ for the root-mean-square velocity of the molecules gives us

$$v_{rms} = \sqrt{\overline{v^2}} = \left(\frac{3RT}{M}\right)^{1/2}$$

This suggests the union of the molecular world, as Boltzmann saw it, and the macroscopic world, as Arrhenius viewed chemical reactions.

12.44

If we plot the velocity distribution of molecules there is a very characteristic shape to such a function, as displayed in Figure 12.37. But we can also represent that velocity in terms of the kinetic energy of the molecular motion $E = \frac{1}{2} mv^2$ and then plot the number of molecules as a function of kinetic energy. Not surprisingly, that molecular distribution has the same basic shape. It rises from zero at zero energy, passes through a maximum, and then "tails off" at high energy.

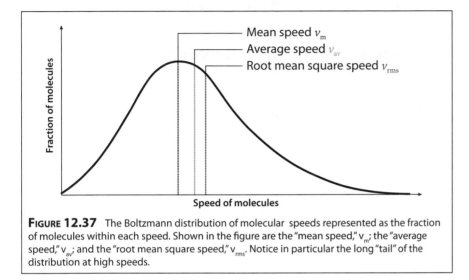

FIGURE 12.37   The Boltzmann distribution of molecular speeds represented as the fraction of molecules within each speed. Shown in the figure are the "mean speed," v_m; the "average speed," v_{av}; and the "root mean square speed," v_{rms}. Notice in particular the long "tail" of the distribution at high speeds.

This allows us to link three concepts, shown in Figure 12.38, into a single picture.

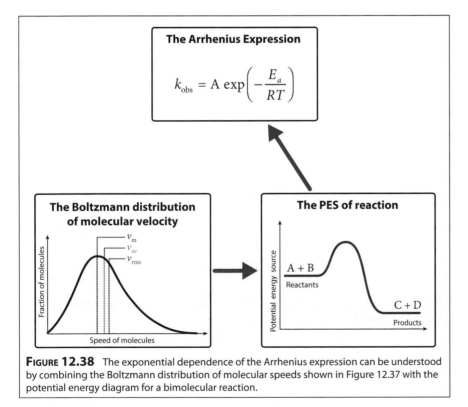

FIGURE 12.38   The exponential dependence of the Arrhenius expression can be understood by combining the Boltzmann distribution of molecular speeds shown in Figure 12.37 with the potential energy diagram for a bimolecular reaction.

So now if the vertical axis is energy, as it is in the potential energy diagram, and we **rotate** the Boltzmann distribution so as to align its energy axis with that of the PES, as shown in Figure 12.39, we identify the energy of activation, E_a, as it appears in the Arrhenius expression. We see that only those molecules in the high energy tail of the Boltzmann distribution have sufficient energy to surmount the barrier.

Notice, however, that this is the case for *bimolecular reactions*; unimolecular and termolecular reactions do not, in general, follow the same temperature dependence.

Manipulation of the Arrhenius Expression

Answer:

Determine the reaction rate constant at two different temperatures and then use the Arrhenius expression:

Measurement #1 of k at temperature T_1

$$\ln k_1 = \frac{-E_a}{RT_1} + \ln A$$

Measurement #2 of k at temperature T_2

$$\ln k_2 = \frac{-E_a}{RT_2} + \ln A$$

Subtract $\ln k_2$ from $\ln k_1$

$$\ln k_1 - \ln k_2 = \left(\frac{-E_a}{RT_1} + \ln A \right) - \left(\frac{-E_a}{RT_2} + \ln A \right)$$

or

$$\ln\left(\frac{k_1}{k_2} \right) = \frac{E_a}{R}\left(\frac{1}{T_2} - \frac{1}{T_1} \right)$$

And solve for E_a.

Check Yourself 6
Question:
How can we determine the barrier height of a reaction from observation of the reaction rate constant at two different temperatures?

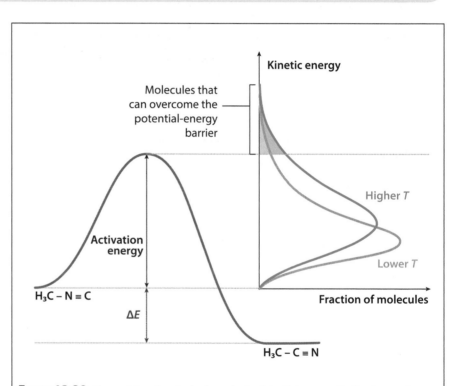

FIGURE 12.39 Recognizing that the *horizontal* axis of the Boltzmann distribution can be represented as the kinetic energy of the molecules in a gas and that the *vertical* axis of the potential energy diagram is energy, we can join the two diagrams to represent the number of molecules with kinetic energy large enough to get over the potential energy barrier of the bimolecular reaction.

12.46

Summary Concepts

1. Bonding Structure and the Photodissociation of Molecular Oxygen

The figure at right summarizes graphically a number of very important concepts. First, in the upper left corner the formation of molecular orbitals from the component atomic orbitals is summarized with the electrons properly inserted into the orbitals of O_2. Second, the absorption of a photon by the electronic structure of O_2 is represented by the incoming photon and the promoted electron from the bonding π_{zb} oribital to the antibonding π_z^* orbital in the lower left panel. Third, the correlation between the configuration of electrons inserted into the molecular orbitals of O_2 and the corresponding potential energy surface is shown in the right-hand panel. This is done for both the ground state and promoted state potential energy surface. Finally, the fact that the period of the electron is much shorter than the period of nuclear motion means that the promotion of the electron results in the molecule moving vertically (i.e. with no change in internuclear distance) on the potential energy diagram from the ground state to the promoted state. From the repulsive wall of the upper (promoted) potential energy surface, the O_2 molecule is unbound and dissociates into two oxygen atoms.

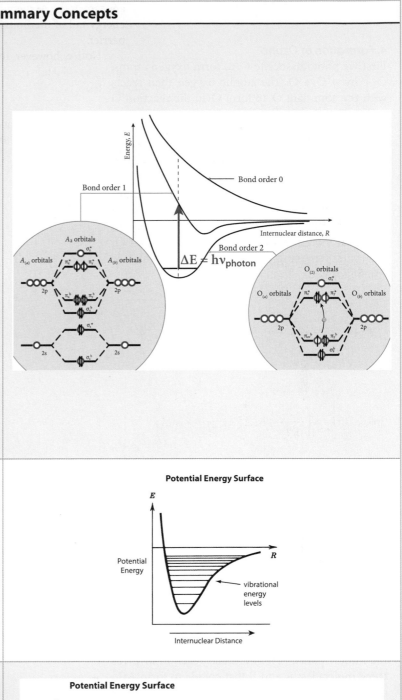

2. Vibrational Motion of a Molecule

Vibrational energy exchange occurs in discrete steps because the vibrational motion on the molecular scale, as is the case for electronic motion, has a wave associated with it. The allowed *vibrational energy levels* only occur in discrete steps within the lowest potential energy surface of the molecular "electronic state" represented by the specific molecular orbitals that are occupied.

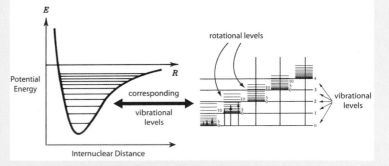

3. Rotational Motion of a Molecule

Just as the vibrational energy levels are segmented or "quantized," so too are the *rotational energy levels* of the molecule. These rotational transitions are of much lower energy than either the vibrational energy levels or the electronic energy levels. Furthermore, for each vibrational energy level, the molecule can possess a number of rotational energy levels such that these rotational energy levels "stack" on top of each of the vibrational levels. Thus if we look with greater detail at the potential energy surface, we can see how these energy levels are assembled.

12.47

Summary Concepts

4. Formation of Ozone

First we photodissociate O_2 to form oxygen atoms: $O_2 + h\nu \rightarrow O + O$. The atomic oxygen then reacts with the abundant O_2 to form O_3 in the reaction $O + O_2 \rightarrow O_3$.

The two-step process forming O_3 from the reaction of $O + O_2$ is shown on a potential energy surface. The first step is the formation of O_3 resulting from the collision of O and O_2. However, that newly formed O_3 has a large amount of kinetic energy present as vibrational and rotational energy in the bond structure of O_3. If a third body, M, collides with that vibrationally and rotationally excited O_3 molecule, removing energy, a stable O_3 molecule will be formed.

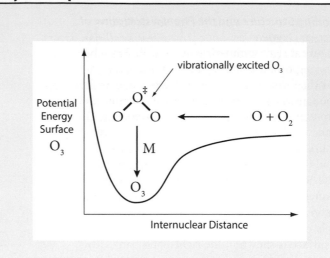

5. Molecularity of a Chemical Reaction

The *molecularity* of a reaction is determined uniquely by the potential energy surface over which the chemical transformation occurs. The dissociation of a *single* molecule into products is called a *unimolecular* reaction. The unimolecular process is shown in the upper panel as a two-step process: excitation followed by dissociation. The reaction of *two* molecules, A + B, to form multiple products is called a bimolecular reaction. The potential energy diagram for a bimolecular reaction is shown in the middle panel. A termolecular reaction involves three molecules: A and B that collide to form AB‡, and a third "chaperone" molecule, M, that removes the excess kinetic energy to form the stable product.

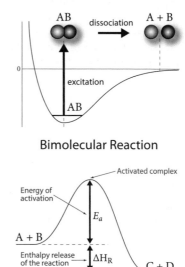

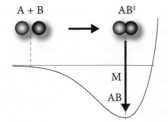

Summary Concepts

6. Chemical Reactions and Molecular Collisions

A chemical reaction cannot occur unless a *collision* occurs between two reactants that may or may not engage in a bond rearrangement leading to products of the reaction.

We can immediately deduce the fact that because a reaction requires a collision, a chemical reaction cannot occur at a rate faster than the rate at which reactant molecules collide.

However, even if the colliding molecules do possess sufficient kinetic energy to position the nuclei close enough for a bond rearrangement to take place, the geometric *configuration* of the positions of the oxygen and the nitrogen nuclei may make the required bond rearrangement difficult. As a result, no reaction takes place.

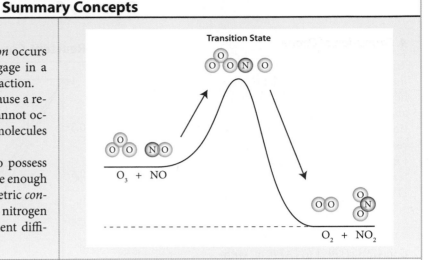

7. Overall Reaction vs. Elementary Reaction

It is routine to write the combustion of a hydrocarbon, such as methane or octane or any other combustible material, in the form hydrocarbon $+ O_2 \rightarrow H_2O + CO_2$. But even for the simplest case, $CH_4 + 2O_2 \rightarrow CO_2 + 2H_2O$, this is a net, overall reaction. It does not occur on a specific potential energy surface. Rather, the reaction of CH_4 with O_2 is comprised of a *sequence* of reactions, each of which takes place on a specific potential energy surface. The sum of those "elementary reactions"—the specific unimolecular, bimolecular, and termolecular reactions—then constitutes the net overall reaction.

$$CH_4 + 2O_2 \rightarrow CO_2 + 2H_2O$$

$$
\begin{array}{rl}
(1) & CH_4 \rightarrow CH_3{}^{\cdot} + H^{\cdot} \\
(2) & O_2 \rightarrow 2O^{\cdot} \\
(3) & CH_3{}^{\cdot} + O^{\cdot} \rightarrow CH_2O + H^{\cdot} \\
(4) & H^{\cdot} + O_2 \rightarrow HO_2{}^{\cdot} \\
(5) & H^{\cdot} + HO_2{}^{\cdot} \rightarrow 2OH^{\cdot} \\
(6) & CH_2O + OH^{\cdot} \rightarrow CHO^{\cdot} + H_2O \\
(7) & CHO^{\cdot} \rightarrow CO + H^{\cdot} \\
(8) & CO + OH^{\cdot} \rightarrow CO_2 + H^{\cdot} \\
(9) & H^{\cdot} + H^{\cdot} \rightarrow H_2 \\
(10) & O^{\cdot} + H_2 \rightarrow OH^{\cdot} + H^{\cdot} \\
(11) & OH^{\cdot} + H^{\cdot} \rightarrow H_2O \\
\hline
\end{array}
$$

NET: $CH_4 + 2O_2 \rightarrow CO_2 + 2H_2O$

8. Determination of the Reaction Rate Constant

In order to analyze the kinetics of OH loss in the presence of CH_4, we first break the decay curve of OH into zones to calculate the *rate* of the reaction

$$R_{obs} = -\Delta[OH]/\Delta t$$

that measures the change in OH concentration, $\Delta[OH]$, over a time increment, Δt.

However, if we divide the rate of OH decay, $-\Delta[OH]/\Delta t$, by the average OH concentration within each corresponding time increment, a remarkable fact emerges: The rate $-\Delta[OH]/\Delta t$ divided by $\overline{[OH]}$ remains *unchanged* throughout the course of the reaction! But that means that we have *identified a constant*, which we shall call k_{obs}, the observed *reaction rate constant*:

$$\frac{-\Delta[OH]}{\Delta t}\Big/\overline{[OH]} = k_{obs}$$

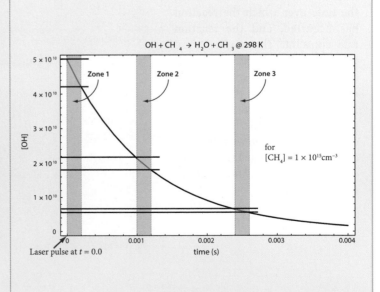

12.49

Summary Concepts

9. Reaction Order

The reaction order is an *experimentally determined* relationship between the rate of the reaction and the constituent concentrations upon which the rate of the reaction depends. The generalized form of the reaction rate is written, for the reaction A + B → C + D as

$$R_{obs} = k\,[A]^m\,[B]^n$$

Where the order of the reaction with respect to [A] is m, with respect to [B] is n, and m + n with respect to the overall order of the reaction. Examples of the dependence of concentration on time is shown at left for zero-order, first-order, and second-order reactions.

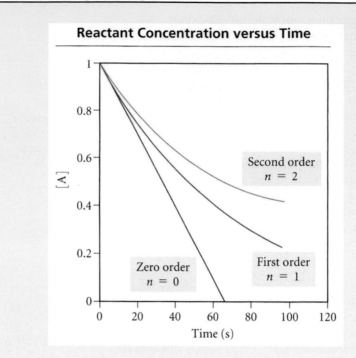

10. Integration of the Rate Law

The integration of the rate law provides the mathematical relationship between the concentrations, the time over which the reaction has occurred, and the reaction rate constant. The mathematical form of the integrated rate law depends upon the experimentally determined *order* of the reaction. Displayed at right are the integrated rate laws, the dependence of the concentration as a function of time, and the half-life for zero-order, first-order, and second-order reactions.

Rate Law Summary Table

Order	Rate Law	Integrated Rate Law	Straight-Line Plot	Half-Life Expression
0	Rate = $k[A]^0$	$[A]_t = -kt + [A]_0$	*y*-intercept = $[A]_0$; Slope = $-k$	$t_{1/2} = \dfrac{[A]_0}{2k} = \dfrac{1}{k}\dfrac{[A]_0}{2}$
1	Rate = $k[A]^1$	$\ln[A]_t = -kt + \ln[A]_0$ $\ln\dfrac{[A]_t}{[A]_0} = -kt$	*y*-intercept = $\ln[A]_0$; Slope = $-k$	$t_{1/2} = \dfrac{0.693}{k} = \dfrac{1}{k}(0.693)$
2	Rate = $k[A]^2$	$\dfrac{1}{[A]_t} = kt + \dfrac{1}{[A]_0}$	Slope = k; *y*-intercept = $1/[A]_0$	$t_{1/2} = \dfrac{1}{k[A]_0} = \dfrac{1}{k}\dfrac{1}{[A]_0}$

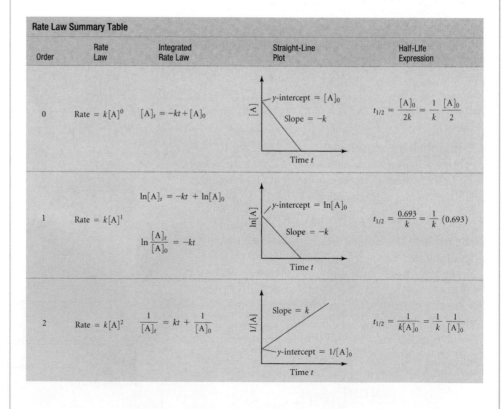

12.50

Summary Concepts

11. Steady State Approximation

One of the most useful and powerful approximations in chemistry is the **steady state approximation** that takes explicit advantage of the fact that within many chemical mechanisms reside elementary reactions that rapidly produce and remove reactive intermediates for which the **loss rate** of the species is limited by its **production rate**.

The steady-state approximation is used when there is an intermediate in a sequence of chemical reactions that is rapidly produced and removed such that the adducts production rate, P, and loss rate, L, are much greater in magnitude than the *difference* between P and L. Then, if the concentration of the adduct is [A], we can use the approximation

$$d[A]/dt = P - L = 0$$

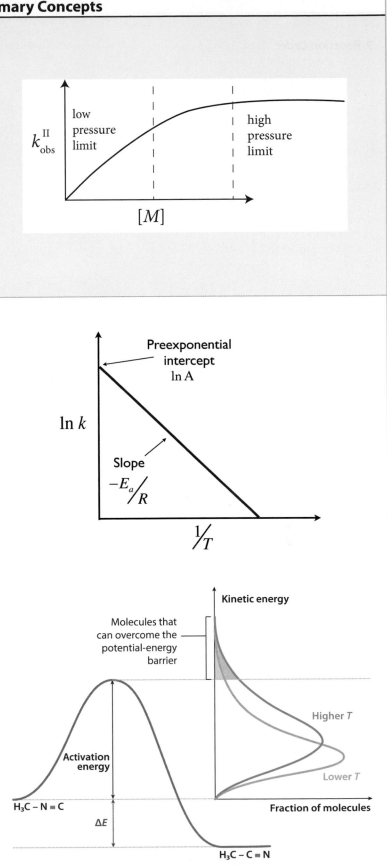

12. Arrhenius Expression for Temperature Dependence

The Arrhenius expression for the temperature dependence of a chemical reaction is written in the form

$$k = Ae^{-E_a/RT}$$

where k is the reaction rate constant, A is the temperature *independent* preexponential, E_a is the energy barrier (activation energy) for the reaction, R is the gas law constant and T is the temperature.

The Arrhenius expression can be most easily graphed by plotting the natural log of the reaction rate constant, ln k, versus 1/T such that the intercept (at infinite temperature, 1/T = 0) is the pre-exponential A and the slope of the line is $-E_a/R$. This graphical representation is shown at right.

The reason for the exponential dependence can be seen by plotting the distribution of molecular energy on the vertical axis of the reaction coordinate for the reaction.

CASE STUDY 12.1   Kinetics, Catalysis, Free Radicals, and the Antarctic Ozone Hole

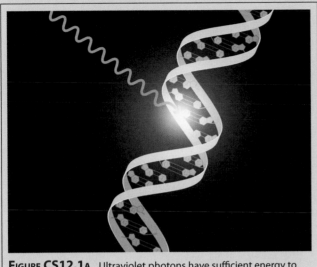

FIGURE CS12.1A   Ultraviolet photons have sufficient energy to break the base pairing across the spine of the DNA double helix. This results in the formation of dimers along each strand such that the genetic code cannot be copied, resulting in the increased probability of cancer.

Figure CS12.1a captures schematically the collision of a UV photon with the structure of DNA. A UV photon has sufficient energy to break the base pair across the double helix of DNA. The only reason organisms can survive on the Earth's surface is that these UV photons are absorbed in the "near UV" between 200 and 300 nm by ozone that exists in the stratosphere. The existence of ozone in the stratosphere depends in a very sensitive way on the kinetics of reactions that both produce O_3 and that destroy O_3. In this case study, we first analyze the reactions that control the ozone distribution and then consider the cause of the antarctic ozone hole.

This discussion involves kinetics, photochemistry, and catalytic reactions as a coupled network of reactions. A catalyst is a substance (gas, liquid, or solid) that accelerates the rate of a chemical transformation without being consumed in the process. The most familiar example of this is the "catalytic converter" on your car that converts literally hundreds of kilograms of CO and NO to CO_2 and N_2 without altering its own chemical structure. Perhaps the most important example of catalysis in the gas phase involves the control of ozone in the Earth's stratosphere. To understand how catalysis and photochemistry work together, we begin by reviewing the photochemistry of molecular oxygen that we developed in the Framework of Chapter 12:

- Molecular orbitals and electron promotion from bonding to antibonding orbitals is instigated by the absorption of a photon.

- *Unimolecular* decomposition of O_2 by the absorption of an ultraviolet photon leads to the formation of atomic oxygen.

$$O_2 + h\nu \xrightarrow{k_1} O + O \qquad \textbf{(Reaction 1)}$$
Reaction rate constant k_1

- *Termolecular* formation of O_3 occurs when the atomic oxygen reacts with O_2.

$$O + O_2 + M \xrightarrow{k_2} O_3 + M \qquad \textbf{(Reaction 2)}$$
Reaction rate constant k_2

- *Unimolecular* decomposition of O_3 by photodissociation leads to the reformation of atomic oxygen and molecular oxygen.

$$O_3 + h\nu \xrightarrow{k_3} O + O_2 \qquad \textbf{(Reaction 3)}$$
Reaction rate constant k_3 (photons in the near UV that also break the base pairing across the spine of DNA)

- *Bimolecular* removal of O_3 and O to reform the O_2 bond completes the cycle.

$$O_3 + O \xrightarrow{k_4} O_2 + O_2 \qquad \textbf{(Reaction 4)}$$
Reaction rate constant k_4

Notice that ozone exists in a narrow layer in Earth's atmosphere because there are a limited number of photons with enough energy to break the bond in molecular oxygen

$$O_2 + h\nu \xrightarrow{k_1} O + O \qquad \textbf{(Reaction 1)}$$

So when those photons are removed, the source of atomic oxygen, and thus of ozone, through the reaction

$$O + O_2 + M \xrightarrow{k_2} O_3 + M \qquad \textbf{(Reaction 2)}$$

goes to zero and no more ozone can be produced.

Thus, as we come **down** through the atmosphere from above, as is the case with UV photons from the sun, the **rate** of reaction 1, R_1, where

$$R_1 = \frac{-d[O_2]}{dt} = \frac{1}{2}\frac{d[O]}{dt} = k_1[O_2]$$

(photodissociation reactions are written in terms of a first-order reaction rate constant in units of [sec^{-1}]), increases as $[O_2]$ increases until all the UV photons are used up and, as a result, R_1 goes to zero. This is why ozone increases as we come down through the atmosphere, until it reaches a peak, and then decreases as we go lower in the atmosphere, thereby forming a layer (the "ozone layer") in the stratosphere about 25 kilometers above the ground. This sequence is summarized in Figure CS12.1b.

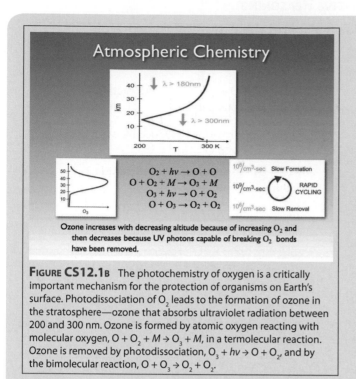

FIGURE CS12.1B The photochemistry of oxygen is a critically important mechanism for the protection of organisms on Earth's surface. Photodissociation of O_2 leads to the formation of ozone in the stratosphere—ozone that absorbs ultraviolet radiation between 200 and 300 nm. Ozone is formed by atomic oxygen reacting with molecular oxygen, $O + O_2 + M \rightarrow O_3 + M$, in a termolecular reaction. Ozone is removed by photodissociation, $O_3 + h\nu \rightarrow O + O_2$, and by the bimolecular reaction, $O + O_3 \rightarrow O_2 + O_2$.

But a careful comparison between the rates of the individual reactions in oxygen photochemistry (Reactions 1–4 above) and the amount of ozone in the stratosphere demonstrated that there was only about one-third the amount of ozone present in the stratosphere when compared with what would be predicted by this "pure oxygen" reaction scheme. Thus, something else was removing ozone, because we know (1) how much O_2 there is in the atmosphere, (2) how many UV photons are received from the sun, and (3) laboratory measurements of the reaction rate constants k_1, k_2, k_3, and k_4.

This led to a very important realization. With the evolution of life came methane, CH_4, produced by the anaerobic decomposition of organic matter in swamps:

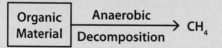

But atomic oxygen, O, in any electronically excited state (that is, in any state for which an electron is promoted to a higher orbital which we will designate O*) reacts very rapidly with methane to produce one of the most important radicals in chemistry and biochemistry: the hydroxyl radical OH:

$$CH_4 + O^* \rightarrow OH + CH_3$$

But one of the reasons OH is so important within natural systems is not only that it is so reactive (as a result of the unpaired electron in its valence shell), but also because it engages in catalytic reaction cycles, such as

$$O_3 + OH \xrightarrow{k_5} HO_2 + O_2 \qquad \textbf{(Reaction 5)}$$

$$HO_2 + O_3 \xrightarrow{k_6} OH + O_2 + O_2 \qquad \textbf{(Reaction 6)}$$

Net: $O_3 + O_3 \rightarrow O_2 + O_2 + O_2$

This pair of reactions is *catalytic* because the two reactions, when taken together,

a) convert O_3 to O_2
b) leave the concentration of OH and HO_2 **unaffected**.

This means that a very small amount of OH can potentially destroy a **very large** amount of ozone. But how **fast** will that pair of reactions remove ozone? Well, because there are (many) other reactions interconverting OH and HO_2 within the system and because the reaction rate constants k_5 and k_6 are very different, the **rate** of one of the reactions (reaction 5 or 6) will be fast **compared** to the other.

Normally in chemistry we are interested in fast reactions, but in a serial process such as a catalytic cycle

$$OH + O_3 \xrightarrow{k_5} HO_2 + O_2 \qquad \textbf{fast step}$$

$$HO_2 + O_3 \xrightarrow{k_6} OH + O_2 + O_2 \qquad \textbf{slow step}$$

Net: $O_3 + O_3 \rightarrow O_2 + O_2 + O_2$

the rate of the overall process is equal to the **slow** step in the sequence. This is the **rate limiting step** and it is of immense importance. It defines (a.k.a. restricts) the rate at which the catalytic process occurs. By knowing the concentrations of OH, HO_2, k_5, and k_6, it can be easily determined that the slower of the two steps

$$R_5 = \frac{-d[OH]}{dt} = k_5[O_3][OH]$$

$$R_6 = \frac{-d[HO_2]}{dt} = k_6[O_3][HO_2]$$

is reaction 6, so the rate of the *net reaction* $O_3 + O_3 \xrightarrow{k_{net}} O_2 + O_2 + O_2$ in the catalytic cycle is simply

$$R_{net} = 2\,k_6[HO_2][O_3],$$

where the coefficient 2 results from the fact that two molecules of O_3 are removed each time the catalytic pair is complete.

But this tells us something very important: A very small amount of HO_2 can accelerate the removal of ozone if (and only if)

$$2\,k_6[HO_2][O_3] \geq 2\,k_4[O][O_3]$$

This is, in fact, the case in Earth's stratosphere. In fact, throughout the evolution of life at Earth's surface and in its oceans, these hydrogen radicals, produced from methane, have modulated, but not completely destroyed, ozone in the stratosphere.

But this also tells us something else: namely, that if a substance is released into the atmosphere that in turn produces a radical that instigates another catalytic cycle whose rate limiting step is on the order of $2k_6[HO_2][O_3]$, that catalytic cycle will alter the concentration of ozone in the stratosphere. This brings us to the subject of ozone destruction by compounds, produced by human industry and by chemical synthesis, that are released at the surface but find their way to the stratosphere.

In 1928 a chemist by the name of Thomas Midgley, Jr., working for General Motors Corporation, developed a number of carbon-halogen compounds, which found broad use as cleaning agents, refrigerants, propellants in aerosol cans, agents for "blown foam," as well as a number of medical applications based on the nontoxic attributes of these "chlorofluorocarbons" or "CFCs." But the remarkable chemical stability of the carbon-chlorine and carbon-fluorine bonds, which was the reason these compounds were inert and nontoxic, had unpredictable consequences. Through the 1930s, 1940s, 1950s, 1960s, and 1970s, use worldwide of these compounds increased exponentially.

In the middle of the twentieth century, a global study was initiated—a study entitled the *International Geophysical Year* or IGY. This study, begun in 1958, was a two year effort to build a database on key scientific observations from the far reaches of the globe, including Antarctica and the Arctic. As part of those global observations, measurement stations were established on the Antarctic continent, one of them by Cambridge University (Cambridge, England) at Halley Bay. That measurement station began to systematically gather data on sea level, temperature, precipitation, and ozone concentrations in the stratosphere. All of those observations were obtained from the ground; stratospheric ozone amounts were obtained by observing the absorption of solar radiation in the same spectral region within which ozone absorbs UV and thereby protects inhabitants of the planet—200 to 300 nm.

Those observations, displayed in Figure CS12.1c, were continued each year, beginning in 1958, and they proved to be of critical importance for society. As the 1950s gave way to the 1960s and then the 1970s, researchers from the British Antarctic Survey wintered over at Halley Bay, keeping careful records of many geophysical quantities, but there was little change, save the natural variability of these quantities. Then in the mid 1970s, the British Antarctic Survey began to record significant drops in the amount of stratospheric ozone in the month following the return of sunlight to Antarctica. That month was October, and the October monthly mean ozone concentration continued to decline significantly each year through the remainder of the 1970s and into the mid 1980s, as shown in Figure CS12.1c.

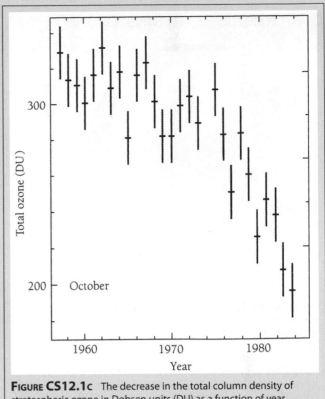

FIGURE CS12.1c The decrease in the total column density of stratospheric ozone in Dobsen units (DU) as a function of year between 1958 and 1984 constitutes one of the most important observations in the history of Earth observations. The measurements were obtained by the British Antarctic Survey at Halley Bay on the Antarctic continent and constitute the discovery of what was to become known as the "Antarctic Ozone Hole."

Then one of the most shocking discoveries in the annals of science was announced in 1985 by the British journal *Nature*, wherein the British Antarctic Survey revealed the data record for ozone over Antarctica from 1958 until 1984 showing a reduction in the October monthly mean of more than 30 percent! This was an amount of ozone loss that, had it occurred over populated regions of the globe, would have induced an epidemic in skin cancer cases, as the incidence of such cases is extremely dependent on UV dosage.

This discovery set off both a firestorm of concern *and* a wave of scientific hypotheses addressing potential mechanisms responsible for such massive loss of ozone. It also triggered intense scrutiny of satellite data; observations that should have detected the loss of ozone years earlier when the first instruments capable of observing the total ozone content of the stratosphere were placed in Earth orbit in the 1970s. The first published satellite measurements revealed a massive void in stratospheric ozone concentration over a region larger than the Antarctic continent. Those images, an example of which is shown in Figure CS12.1d, led to the phenomenon being dubbed the "Antarctic ozone hole," and the name stuck.

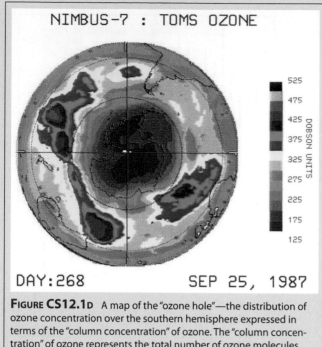

FIGURE CS12.1D A map of the "ozone hole"—the distribution of ozone concentration over the southern hemisphere expressed in terms of the "column concentration" of ozone. The "column concentration" of ozone represents the total number of ozone molecules contained in a column extending from Earth's surface to the outer edge of the atmosphere. The units are "Dobson units" named after a pioneer in global ozone measurements. A column concentration of 500 Dobson units corresponds to a column 0.5 cm high at standard temperature and pressure.

FIGURE CS12.1E The NASA ER-2 aircraft, which is a very high altitude research aircraft used to study processes in the stratosphere of Earth. The aircraft was originally designed as a military surveillance aircraft—a spy plane—used to gather information from very high altitude prior to the development of satellites. It was originally called the U-2.

The debate grew over what was causing such a precipitous reduction in ozone, whether it would spread to the larger global context and whether it was natural or human induced. The three leading hypotheses were that:

1. As sunlight returned to the Antarctic, the heating of the atmosphere by the sun lead to the rising of air in the stratosphere, sweeping ozone up and out of the stratosphere to midlatitudes.
2. The magnetic field of the Earth focused high energy protons and helium ions into the stratosphere over the Antarctic and triggered chemical reactions that removed ozone; in particular the nitrogen species NO and NO_2 were implicated because they can catalytically remove ozone.
3. Compounds released at the surface by human activity penetrated the stratosphere and lead to the destruction of ozone—in particular chlorine and bromine from chlorofluorocarbons and halon compounds were somehow involved.

The United States responded by mounting exploratory missions on both the ground and by aircraft. The aircraft mission employed the famous U-2 spy plane, which could fly high enough to reach the stratosphere. It was outfitted with instruments that could test the various hypotheses outlined above. The aircraft flights revealed a rather dramatic series of events, beginning with the first flight, timed for just after the sun appeared for the first time on the horizon over Antarctica. The U-2, renamed the ER-2 by NASA, was equipped with instruments to study the ozone problem over Antarctica and is displayed in Figure CS12.1e.

As the aircraft, shown in Figure CS12.1e, passed through what is called the winter "polar jet" shown in Figure 12.1 that physically isolates the Antarctic stratosphere from the outside world, the ozone concentration changed very little (as indicated in the August 23rd plot shown in the left panel of Figure 12.1), but the ClO concentration increased dramatically, reaching levels ten times that present outside this "vortex" region. Just three weeks later, on September 16th (shown in the upper right panel of Figure 12.1), the ozone concentration had decreased by 60% inside the vortex in the region of highly amplified ClO. In fact, on smaller scales, every region high in ClO was found to be low in ozone and every region low in ClO was found to be high in ozone.

So how does society deal with such a situation? The anticorrelation of ClO and O_3 suggests (but does not prove) that ClO is somehow affecting O_3. But how exactly *is* ClO affecting O_3? There are some 1.2 parts per billion (ppb) of ClO, but there are 3000 ppb of ozone, so how could ClO possibly remove ozone? Also, what determines the rate at which ozone is lost? Where did the ClO molecule come from? Before society acts, these questions must be answered—particularly because many people regarded the ozone loss issue as a hoax. However, whatever a person's views were, scientific evidence was necessary for informed public policy decisions because such policy has major economic consequences.

To answer these questions, we turn to our analysis of chemical kinetics—the study of how **fast** chemical reactions take place. To develop an understanding of how chemical

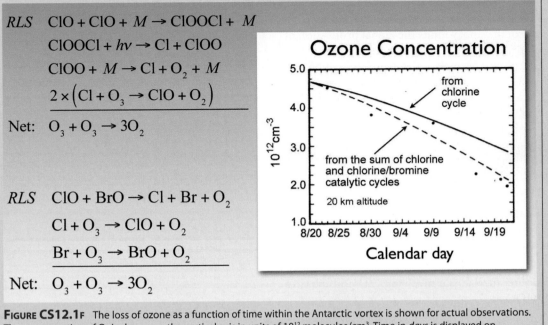

$$RLS \quad ClO + ClO + M \rightarrow ClOOCl + M$$
$$ClOOCl + h\nu \rightarrow Cl + ClOO$$
$$ClOO + M \rightarrow Cl + O_2 + M$$
$$2 \times \left(Cl + O_3 \rightarrow ClO + O_2 \right)$$
$$Net: \quad O_3 + O_3 \rightarrow 3O_2$$

$$RLS \quad ClO + BrO \rightarrow Cl + Br + O_2$$
$$Cl + O_3 \rightarrow ClO + O_2$$
$$Br + O_3 \rightarrow BrO + O_2$$
$$Net: \quad O_3 + O_3 \rightarrow 3O_2$$

FIGURE CS12.1F The loss of ozone as a function of time within the Antarctic vortex is shown for actual observations. The concentration of O_3 is shown on the vertical axis in units of 10^{12} molecules/cm³. Time in *days* is displayed on the horizontal axis. The rate of ozone loss, $d[O_3]/dt$, is plotted based on the rate limiting step of the chlorine dimer mechanism. When the rate of removal of the bromine-chlorine catalytic step is *added* to the dimer catalytic step, the *rate* of ozone loss, shown by the dashed curve, is equal to the observed rate of ozone loss.

kinetics applies to this problem, we return to the reactions of oxygen and ozone in the evolution of life on Earth to remind ourselves that chemical reactions, at their most fundamental level, can be classified according to their **molecularity**: unimolecular, bimolecular, and termolecular; and each of those reaction categories has associated with it a *specific type of potential energy surface*.

We recognize that an important objective of studies in the physical sciences is to guide key societal objectives. For example, how do we maintain a UV dosage level at Earth's surface so as to allow the complex organisms that inhabit the surface to remain alive and well? It is important to recognize that there is a contentious battle between those who are focused on stewardship of the planet, not necessarily with regard for a specific economic intent, and those who have a serious economic stake in any federal legislation to limit the production of a given chemical. In fact, then, we are focused on how to dissect the scientific issues linking what happens at the molecular level with what happens in a Senate hearing.

Following the discovery of the dramatic loss of ozone over Antarctica by the British Antarctic Survey, Mario Molina and his wife, Luisa, then at the Jet Propulsion Laboratory in Pasadena, California, proposed a catalytic mechanism implicating chlorine radicals:

$$ClO + ClO + M \xrightarrow{k_7} ClOOCl + M \quad \text{slow (rate limiting step)}$$
$$ClOOCl + h\nu \xrightarrow{k_8} Cl + ClOO \quad \text{slow (but equal to rate of reaction 7)}$$
$$ClOO + M \xrightarrow{k_9} Cl + O_2 + M \quad \text{fast}$$
$$2 \times \left(Cl + O_3 \xrightarrow{k_{10}} ClO + O_2 \right) \quad \text{fast}$$
$$\overline{O_3 + O_3 \rightarrow O_2 + O_2 + O_2}$$

And Mike McElroy and Stephen Wofsy of Harvard proposed a catalytic cycle implicating chlorine and bromine:

$$ClO + BrO \xrightarrow{k_{11}} Cl + Br + O_2 \quad \textbf{slow (rate limiting step)}$$
$$Cl + O_3 \xrightarrow{k_{12}} ClO + O_2 \quad \textbf{fast}$$
$$Br + O_3 \xrightarrow{k_{13}} BrO + O_2 \quad \textbf{fast}$$
$$\overline{O_3 + O_3 \rightarrow O_2 + O_2 + O_2}$$

The observed concentrations of ClO and BrO and the rate of disappearance of ozone measured by the U-2 aircraft within the Antarctic vortex established quantitatively that the rate limiting steps, occurring on explicitly defined potential energy surfaces, were responsible for the observed rate of ozone loss. This is summarized in Figure CS12.1f.

Thus, while the observed anticorrelation of O_3 and ClO (shown in Figure 12.1) was a dramatic visual, what made the case compelling in a court of public policy *and* in a court of science was the unequivocal proof through an understanding of catalytic reactions. In particular, specific catalytic cycles were identified that control chemical transformations at a rate dictated by the **rate limiting step** in each of those catalytic cycles. This was the scientific foundation underlying the Montreal Protocol that today limits the global production of CFCs and halons. It has now been demonstrated that had immediate controls not been placed on CFC production and distribution in the 1990s, large increases in UV dosage at the Earth's surface would now have occurred with serious human health implications.

Problem 1

We consider here the catalytic cycles that destroy ozone in the Antarctic vortex.

a. Write the rate expression for ozone loss

$$\left(-\frac{d[O_3]}{dt} \right)_{ClO/ClO} = \ ?$$

for the Molina mechanism involving ClO and the ClO dimer ClOOCl.

b. Write the rate expression for ozone loss

$$\left(-\frac{d[O_3]}{dt} \right)_{ClO/BrO} = \ ?$$

for the McElroy–Wofsy mechanism involving ClO and BrO.

c. Given your answer in (a) and (b), what is the total rate of ozone loss for the combination of (a) and (b)?

CASE STUDY 12.2 Oxygen, Photochemistry, and the Evolution of Life Forms on the Earth

The history of life, as it evolved from very primitive beginnings and grew to exert powerful control over the chemical and physical structure of the Earth system, is a remarkable yet unsolved problem. It began like this. After a delay of some one billion years following the formation of the proto-Earth, the first life forms began to emerge. Microfossils resembling modern cyanobacteria have been found in 3.5 billion year old rocks. The circumstances related to the first life forms have been, and continue to be, the subject of a great debate. There have emerged a number of plausible paths leading to the inexorable march toward more complex and intricate biological architectures, exhibited by the vast array of species that now inhabit the planet.

One line of reasoning begins with the observed fact that a stunning number of organic structures exist in the interstellar medium—most of these molecular structures have been observed in the microwave region using ground-based receiving antennas, as shown in Figure CS12.2a.

These precursors to life, or indeed life forms themselves, may have been delivered to the Earth's surface by comets, which are composed largely of ice and provide a potential harbor for primitive life forms. Another school of thought is organized around the concept that the molecules that constitute the building blocks of more complex structures (carbon, nitrogen, oxygen, etc.) were present in the early atmosphere, and the energy release from photochemical reactions initiated by ultraviolet radiation or from lightning triggered the synthesis of the amino acids that polymerized to produce proteins—the building blocks of all living structures. Yet another line of reasoning cites the existence of deep-sea vents as the most probable domain for fostering early life forms. These vents are located along regions of the sea floor which are spreading, driven by the motion of tectonic plates. These vent regions provide a remarkable array of temperatures, chemical environments and protection from destructive UV radiation. Studies of these regions have intensified over the past two decades revealing intricate ecological infrastructure that fostered bacteria capable of using sulfur contained in the vent water to develop an ecosystem that supports a dense population of shellfish and worms. There is growing evidence that these life forms, in the vicinity of deep-sea vents, shown in Figure CS12.2b, constitute the early life forms some 3 billion years ago.

A key component of the evolution of the Earth system to what we regard today

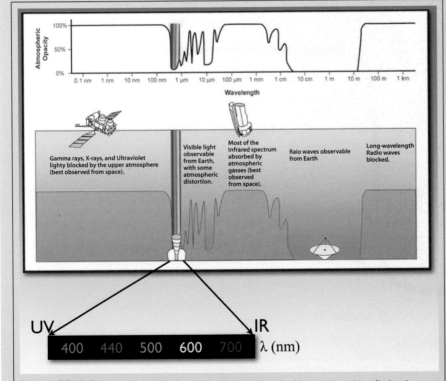

FIGURE CS12.2A The ability of the atmosphere to absorb electromagnetic radiation is a critical part of sustaining life at the Earth's surface. The opacity of the atmosphere, expressed here as complete absorption of radiation for 100% opacity and no absorption of radiation for 0% opacity, is displayed in the upper panel for wavelengths from 0.1 nm to 1 km. This extends from the wavelengths of gamma rays and x-rays (0.1 nm) at the short wavelength extreme to radio waves at the long wavelength extreme. The lower panel indicates the visible region, where the human eye is sensitive; the middle infrared, where CO_2, H_2O, etc., have banded absorption; and the microwave region, where the atmosphere is transparent to radiation.

FIGURE CS12.2B Deep sea vents, where hot water rich in mineral nutrients emerges from the ocean floor, constitute a fertile chemical environment for the origin of molecular structures critical to the evolution of early life forms.

as a planet possessing immense beauty and diversity, was the buildup of free oxygen in the atmosphere. This growth of oxygen in the atmosphere, from a minor species some 3.5 billion years ago, when its mole-fraction in the atmosphere was approximately 10×10^{-6} (that is ten-parts-per-million), to its role over the past 600 million years as a major species that grew to comprise some 20 percent of the atmosphere 350 million years ago, is displayed in Figure CS12.2c.

Molecular oxygen arose in concert with the evolution of early life forms. The primitive Earth possessed an atmosphere influenced primarily by gases injected directly by volcanic activity. Once the surface of the Earth had cooled sufficiently to condense water, the primary molecules in the atmosphere were CO_2 and N_2. Our two nearest neighbors in the Solar System, Venus and Mars, have, respectively, 96 and 95 percent CO_2 atmospheres, and, respectively, 4 percent and 3 percent nitrogen. This early atmosphere of the Earth is usually referred to as the Archean atmosphere as it is this simple outgassed atmosphere that is associated with the old-

est known rocks; those of the Precambrian period that are primarily igneous in composition.

The very first organism must have been **heterotrophic**, that is an organism that can assimilate organic compounds from their environment, because these organisms could not, and today cannot, synthesize their own food and thus are dependent on complex organic substances for nutrition. Prior to 3 billion years ago, these early organisms must have obtained their energy from reactions other than respiration, as there was no available oxygen.

In particular the reaction pathways may well have mimicked the sequences that occur today in anaerobic systems; those biochemical systems that operate in the absence of molecular oxygen.

In varied and dramatic ways, life on Earth is fueled by oxidation-reduction (redox) reactions exhibiting striking variety and creativity. As developed in Chapter 2, redox reactions are chemical reactions in which electrons are transferred between atomic or molecular "sites," and the organism

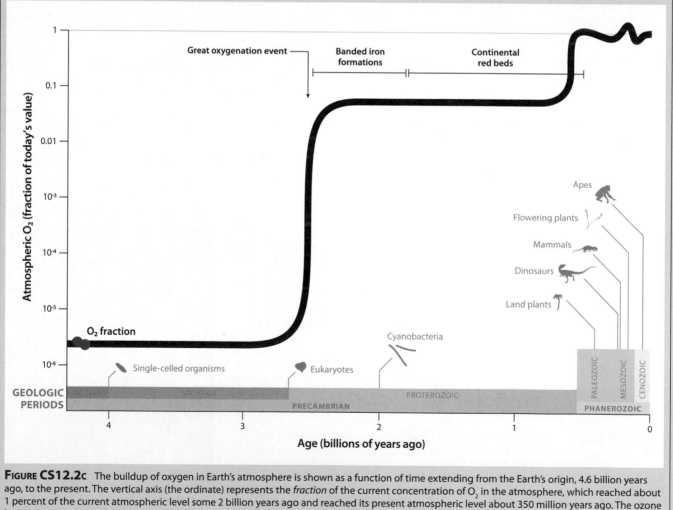

FIGURE CS12.2c The buildup of oxygen in Earth's atmosphere is shown as a function of time extending from the Earth's origin, 4.6 billion years ago, to the present. The vertical axis (the ordinate) represents the *fraction* of the current concentration of O_2 in the atmosphere, which reached about 1 percent of the current atmospheric level some 2 billion years ago and reached its present atmospheric level about 350 million years ago. The ozone build up is represented in terms of the fraction of the total number of ozone molecules currently present in a "column" extending from Earth's surface to the outer extremity of the atmosphere.

captures the energy, released in the exchange of one or more electrons. In the course of life's evolution, organisms have "learned" through the development of specific proteins and associated membranes to channel this energy release into biochemical pathways that sustain life.

There is a coherent pattern to these redox processes, and they are worthy of careful consideration. In an oxygen rich environment, termed aerobic, the dominant biological redox process is respiration

$$(CH_2O) + O_2 \rightarrow CO_2 + H_2O$$

where the generalized organic molecule is represented schematically by (CH_2O). The key point to focus on is that the carbohydrate molecules, (CH_2O), **provide electrons** for the reduction of molecular oxygen. All advanced life forms employ the oxidation of organic material for their primary energy source.

When free oxygen is not available, as was largely the case for the first 2.5 billion years of Earth's existence, and is the case in many aquatic systems today, many other redox reactions can be utilized by lower life forms such as bacteria. As we have seen, early bacterial forms appeared some 3 billion years ago, sustained entirely on the back of a series of redox reactions that cascade to sequentially lower redox potentials producing CO_2, N_2, H_2S, CH_4, and other molecular species of great importance to the development of the early atmosphere.

There is strong evidence that photosynthesis evolved quite early, allowing the critical transition to organisms capable of synthesizing their own organic molecules from CO_2—the **autotrophs**. In addition to fossil evidence, carbon isotope measurements on fossil organic carbon show photosynthesis to be at least 3.5 billion years old.

The signature isotopic marker for organic material formed in photosynthetic processes is the depleted stable ^{13}C isotope relative to ^{12}C resulting from the slower rate of capture of the heavier isotope by the CO_2 fixing enzyme ribulose bisphosphate carboxylase.

But the key result of the emergence of autotrophic organisms was the *release* of O_2. While the large class of anaerobic bacteria survived under difficult conditions in the initial stages of evolution, their species were very sensitive to O_2—they could not survive in its presence. What saved these early life forms from destruction by the O_2 produced by autotrophic organisms were the many oxidizable minerals in the Earth's crust, most notably sulfur and iron. These minerals extracted the oxygen formed by the early autotrophs and chemically bound the oxygen, allowing these early species to prosper.

The silicate minerals of the mantle yielded elevated levels of Fe^{2+} in the early ocean. And Fe^{2+} is quite soluble, providing an oxidation pathway to Fe^{3+}, primarily as precipitates of $Fe(OH)_3$. Fe_2O_3, ferric oxide, has been discovered in sedimentary rocks as old as 3.5 billion years and constitutes the first evidence of *oxygen production* in Banded Iron Formations in which siliceous sediment is sandwiched in with Fe_2O_3 as shown in Figure CS12.2d.

As autotrophs grew in abundance, the O_2 they produced gradually overwhelmed the Fe^{2+} supply of the oceans and the increased O_2 that resulted began to oxidize exposed minerals on land, most importantly iron pyrite, FeS_2 producing $Fe(OH)_3$ and H_2SO_4. The discovery of Red Beds, deposits of Fe_2O_3 shown in Figure CS12.2e, emerged about 2 billion years ago and marked the end of Banded Iron Formations.

FIGURE CS12.2D Ferric oxide, Fe_2O_3, in Banded Iron Formations shown here, in which silicate minerals containing Fe^{2+} are sandwiched between bands of ferric oxide, are as old as 3.5 billion years. These discoveries indicate the existence of the first oxygen producing organisms.

FIGURE CS12.2E Red Beds formed from deposits of Fe_2O_3 extend back some 2 billion years, when O_2 production from autotrophs had increased to the point where the supply of reduced iron (Fe^{2+}) was overwhelmed in favor of the oxidized form (Fe^{3+}).

The earliest and simplest bacteria and blue green algae were the prokaryotes shown in Figure CS12.2f, characterized by the *absence of a nuclear membrane* and by DNA that was not organized into chromosomes. These simpler cell architectures were replaced in the first billion years of Earth history by eukaryotes, distinguished by a distinct membrane-bound nucleus displayed in Figure CS12.2g. It was a development of profound importance as the evolution of more complex multicellular organisms resulted; organisms that could survive in the presence of free oxygen. This took place about 3 billion years along the evolutionary path from the Earth's origin—about 4.6 billion years ago. Eukaryotes could survive on O_2 at or above 1 percent of the present atmospheric level (PAL), a threshold that was passed approximately 2 billion years ago. At this juncture, free oxygen production accelerated with the evolution of chloroplasts in the widening array of eukaryotes—chloroplasts are organelles that are capable of organic synthesis driven by sunlight, the "dawn" of photosynthesis! But of great importance for the evolution of life at the surface was the formation of the oxygen allotrope ozone (O_3) in the atmosphere. This was a critically important development because O_3 absorbs sunlight in the ultraviolet region between 200 and 300 nm, permitting the migration of living forms from the ocean (water strongly absorbs ultraviolet) to the continents. Sunlight in the 200 to 300 nm range breaks the base-pairing of nucleic acids across the spine of the DNA double helix, thereby scrambling the genetic code of the affected cell as displayed in Figure CS12.1a. Fossils of the evolving manifold of multicellular organisms have been found in sedimentary rocks that are 680 million years old but the rise of "modern" species of green plants that drove oxygen levels upward to present levels date from some 400 million years ago.

Development of oxygen within the Earth system is traced out in Figure CS12.2c. As limited oxygen levels completed the oxidation of Fe^{2+} to Fe^{3+} in the oceans, the Banded Iron Formations ended approximately 2 billion years ago, followed by the occurrence of continental Red Beds extending from 2 billion years ago until 500 million years ago. Atmospheric O_2 began to build following the formation of the Banded Iron Formations and reached 21% some 350 million years ago. An important point is that the present atmospheric reservoir is approximately 2 percent of the cumulative production of O_2, with the balance tied up in the oxidation of minerals and the burial of organic material.

What is also of considerable importance is that the basic photosynthetic step

$$CO_2 + H_2O \xrightarrow{\text{photosynthesis}} (CH_2O) + O_2$$

that results in a net production of O_2 and its release as free oxygen into the atmosphere, is the result of the burial of the (reduced) organic product. The buried carbon is reflected in the "fossil fuel" reservoirs in the Earth's crust.

Oxygen dominates the unique chemistry of life's history

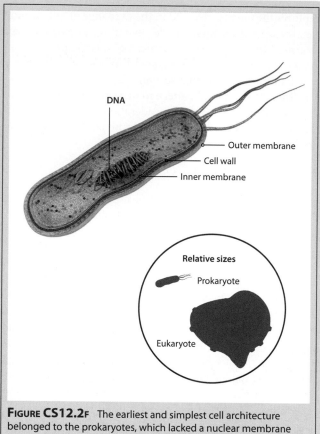

FIGURE CS12.2F The earliest and simplest cell architecture belonged to the prokaryotes, which lacked a nuclear membrane within the cell. In addition, the DNA of the cell was not organized into chromosomes.

FIGURE CS12.2G The cell structure of the eukaryotes is distinguished by the existence of a membrane-bound nucleus. This provided much greater control within the cell, as the channels between the outer membrane of the cell and the nucleus provided much more sophisticated control of cell function.

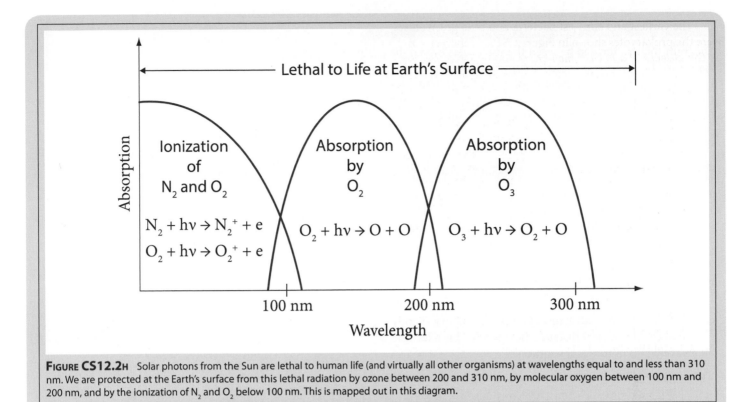

FIGURE CS12.2H Solar photons from the Sun are lethal to human life (and virtually all other organisms) at wavelengths equal to and less than 310 nm. We are protected at the Earth's surface from this lethal radiation by ozone between 200 and 310 nm, by molecular oxygen between 100 nm and 200 nm, and by the ionization of N_2 and O_2 below 100 nm. This is mapped out in this diagram.

on the planet. Elemental oxygen is avid for electrons—it has the highest electronegativity of any element except fluorine. Therefore, oxygen combines with most other elements to form stable oxides.

The emergence of molecular oxygen as a dominant component of the Earth's atmosphere had two profound effects on the development of species. The first was that organisms had a plentiful supply of an oxidizer that could be combined with a range of hydrocarbon "fuels" to sustain the organism. This provided the opportunity to develop large organisms that could employ a circulatory system to carry both the fuel in the form of sugars, and oxygen attached to hemoglobin to supply oxygen within the cellular structure of the organism. This led to the development of large ocean creatures that brought complicated ecosystems that included sharks, dolphins, whales, etc. embedded within the world's oceans. Second, and more remarkably, the ability of oxygen to absorb solar ultraviolet photons in the spectral range between 200 and 300 nm is the sole reason both animals and plants could venture out of the protection of the ocean onto land and survive. We have already developed an understanding of molecular orbital structure, specifically for oxygen. When a photon is absorbed by a molecule, an electron within that molecular orbital structure is "promoted" to a higher energy state—to an upper potential energy surface. Quite often, that upper potential energy surface leads to the dissociation of the molecule into molecular fragments. This mechanism of electron promotion by an ultraviolet photon leads to the dissociation of O_2. In the case of O_2, the molecule has only one

choice when it is "photodissociated"—the formation of two atoms of atomic oxygen. What is truly remarkable about molecular oxygen is that the position of the upper potential energy surface is located at *just the right position* in energy and internuclear distance with respect to the ground state so as to remove all solar ultraviolet photons between 100 and 200 nm before they reach the ground. The implications of this are truly remarkable. This relationship between shifts in the upper potential energy surface and the corresponding position and range of the absorption of ultraviolet photons is critically important to the existence of life at the surface of the Earth.

But the remarkable chain of events linking the molecular structure of O_2 to the existence of life on the planet continues. The atomic oxygen produced in the photodissociation of O_2 reacts with the O_2 in the atmosphere to produce ozone, O_3. But ozone is the *only species* in the atmosphere that absorbs solar ultraviolet photons between 200 and 300 nm, photons that are themselves lethal to organisms at the surface because, as depicted in Figure CS12.1a, they break the base pairing across the double helix of DNA. Figure CS12.2h schematically captures the absorption region of ozone that extends protection of the Earth's surface to wavelengths slightly higher than 300 nm—a wavelength marginally safe for organisms to survive at the surface. As we can see, a study of specifically how photons interact with the structure of molecules is critically important to both life as it has evolved on Earth and to present and future developments linking scientific and technical developments to societal objectives.

Problem 1

Diagram the relationship between the molecular orbital diagram for O_2 and the potential energy surfaces for O_2 that indicate the promotion of the electron from the ground electronic state of O_2 to the excited state of O_2 that is responsible for absorption of solar radiation between 100 and 200 nm.

Problem 2

a. Using the potential energy surfaces defining the absorption of UV radiation by O_2 between 100 and 200 nm, sketch what would occur if the upper potential energy surface was shifted to a slightly larger internuclear distance.

b. How would this shift affect the wavelength position of the O_2 absorption?

c. If the upper (excited) state of O_2 was shifted to a slightly smaller internuclear distance, how would this affect the wavelength of the O_2 absorption?

12.64

NUCLEAR CHEMISTRY
ENERGY, REACTORS, IMAGING, AND RADIOCARBON DATING

13

"It is a profound and necessary truth that the deep things in science are not found because they are useful; they were found because it was possible to find them."

—J. Robert Oppenheimer

FIGURE 13.1 The United States advanced test reactor at the Idaho National Laboratory. The core design of this light-water reactor employs square cross section fuel rods with round control rods inserted into the center of the fuel rods. The blue glow is the Cherenkov radiation that is emitted when a charged particle penetrates into a material (in this case electrons into water) at a velocity greater than the speed of light in that medium. The high velocity electrons are produced in the nuclear reaction that occurs in the fuel rods that contain the ^{235}U isotope.

Framework

A recurring theme in the effort to understand the global structure of primary energy generation and then define a path forward engaging a new blend of energy sources at the required scale, is the necessity to reanalyze all options. It is for this reason that we have critiqued, thus far, the potential for renewable energy from wind, concentrated solar thermal, solar photovoltaics, hydroelectric, tidal, and geothermal. Nuclear energy currently contributes about 6% to the global total generation of primary power generation. However, the fractional contribution of nuclear power in each nation varies remarkably. Virtually without exception, nuclear power is used to generate electricity and we can judge the relative contribution of nuclear power in each country by comparing the fraction of electrical power generated by nuclear for a number of important examples. These cases are summarized in Table 13.1 with the notable examples of France with 75% of electrical power generated by nuclear, the U.S. with 20%, Sweden with 47%, Canada with 13%, Russia with 14%, Japan with 36%, and China with <1%.

Nuclear power is the most highly developed alternative to hydrocarbon combustion as a primary energy source. It is also, along with tidal and geothermal, the only form of alternative primary energy generation that does not depend on the sun.

The key consideration with respect to the potential role for nuclear power in the generation of primary power in the U.S. is to reanalyze in a rational framework how nuclear stands up to alternatives that include fossil fuels, wind, concentrated solar thermal, photovoltaics, hydro, biofuels, etc.

First, we note that the extraction of energy through *nuclear reactions*, that must, by their definition involve *changes in the structure of the nucleus*, falls into two distinct categories. Nuclear *fusion* involves the union of two smaller nuclei to form a larger nucleus of a different element. It is the reaction of hydrogen nuclei to form helium nuclei at very high pressure and temperature at the core of the sun that supplies the Earth with 1.5×10^{10} kWh/year of energy. Second, nuclear *fission* involves the splitting of heavy nuclei into smaller nuclei that releases energy which we are familiar with in nuclear power plants and in most nuclear weapons.

We can distinguish these two categories of nuclear power generation by referring to a graph of the binding energy per nucleon vs. atomic number of each of the elements in the periodic table that is displayed in Figure 13.2. We plot this with a vertical axis that displays the binding energy as *increasingly negative* as the binding energy per nucleon increases. This is identical to the convention we have used for the potential energy of electron "capture" by a positively charged nucleus.

Nuclear fusion thus occupies the left-hand side of the energy diagram in Figure 13.2. Nuclear fission occupies the right-hand side of Figure 13.2. Iron, ^{56}Fe, sits at the minimum of the binding energy per nucleon as a function of atomic mass. Both fusion and fission possess a common characteristic, which is that the *energy provided* per nuclear reaction is approximately one million times the energy provided per molecule in a chemical reaction involving hydrocarbon fuels. This has important implications for a planet that must sustain 10 billion people with a constantly increasing standard of living with its requisite increase in demand for energy. Specifically, that the amount of fuel—the volume of fuel—and of waste that must be dealt with are both one million times less for nuclear energy than for fossil fuel combustion.

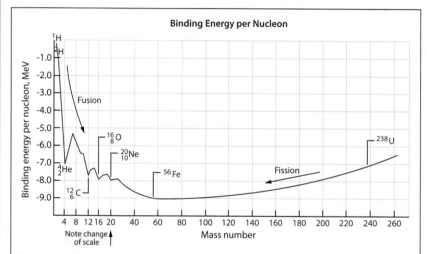

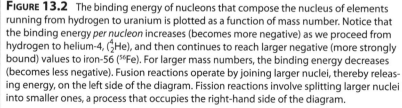

FIGURE 13.2 The binding energy of nucleons that compose the nucleus of elements running from hydrogen to uranium is plotted as a function of mass number. Notice that the binding energy *per nucleon* increases (becomes more negative) as we proceed from hydrogen to helium-4, ($^{4}_{2}$He), and then continues to reach larger negative (more strongly bound) values to iron-56 (^{56}Fe). For larger mass numbers, the binding energy decreases (becomes less negative). Fusion reactions operate by joining larger nuclei, thereby releasing energy, on the left side of the diagram. Fission reactions involve splitting larger nuclei into smaller ones, a process that occupies the right-hand side of the diagram.

To place this in perspective, *each person* in the U.S. consumes approximately 32 kg of fossil fuel *per day* with 8 kg as coal, 8 kg of oil, and 16 kg of natural gas. This translates into an amount equal in weight to 7 gallons of milk that is extracted from the Earth, transported, processed, and distributed to the consumer, and then burned. This combustion produces approximately 60 kg of CO_2/day for each person as waste that is dumped into the atmosphere. That is 60 kg of carbon dioxide *per person, per day*!

The amount of naturally occurring uranium consumed in a typical fission reaction required to provide the same amount of energy as 32 kg of fossil fuel is 4 grams, and the weight of the waste is one-half of one gram. This does not mean that the disposal of nuclear waste does not involve problems, it is simply a statement of the relative magnitude of the mass into and out of fossil fuel combustion plants vs. nuclear power plants. We will explore the issue of waste subsequently.

Sustainable Power

The question we consider first is: does nuclear fission represent a sustainable form of energy? Will it reasonably provide 50 years, 100 years, or 1000 years of primary power, and by what means will it do so?

While there are a number of ways of calculating the answer to those questions, and in fact a number of ways of extracting nuclear energy from various sources, we will examine some of the most important. First, we focus on nuclear fission—the splitting of large radioactive nuclei into smaller nuclei—because sustained nuclear *fusion* is still under development. Uranium is the primary fuel for fission so we will analyze supplies of the isotope ^{235}U, which can be used as a nuclear fuel in a reactor, although there are a number of alternatives.

Interestingly, the majority of extractable uranium resides in the oceans—seawater contains 3.3 mg of uranium per cubic meter, which translates into 4.5 billion tons of available uranium in the world's ocean. Because the turnover time of the oceans is approximately 1000 years, much of this is not accessible on the time scale of a century. But of greater importance, while Japan has experimented with extraction methods to remove uranium from water, an economically feasible method has not been worked out on the scale of the needed amounts of uranium to supply the required number of nuclear reactors. Thus we will focus on the uranium ore that can be mined from the ground. There is, by current estimates, approximately 5 million tons of uranium in the ground and another 22 million tons in phosphate deposits for a total of 27 million tons of mineable uranium available globally.

Uranium ore can be used in two different types of nuclear reactors. The most common type of reactor is the "once-through" design that extracts energy primarily from the ^{235}U isotope that makes up but 0.7% of the extracted uranium; the other isotope, ^{238}U, makes up the other 99.3%. The other reactor type is the fast breeder reactor (see Sidebar 13.2) that converts ^{238}U to fissionable plutonium, ^{239}Pu, and thereby produces approximately 60 times as much energy from the same mass of uranium ore. We will examine the sustainability of both types of reactors.

We consider first the "once-through" reactor with a power generating capacity of 1 GW which would use 160 tons of uranium per year. Given that the U.S. has 3.4×10^5 tons of known uranium reserves (there may well be considerably more that is yet undiscovered), we can consider the balance between the sustainable period of time nuclear power generation could be supported by known uranium deposits vs. a given fraction of the total power demand of the U.S that is supplied by nuclear power. We have developed the case that the U.S. will decrease its primary energy intensity (PEI) as rapidly as the *product* of population and per capita income increase in the U.S., so the total power demand from the U.S. remains at approximately 3 TW through 2050. If we wish to generate 200 GW of power from nuclear energy, we would require

$$\left[\frac{\left(2 \times 10^{11} \text{ watts}\right)}{\left(1 \times 10^9 \text{ watts/power plant}\right)}\right] = 2 \times 10^2 \text{ power plants}$$

That is approximately twice the number of reactors currently in operation in the U.S. If each of the 200 power plants used 160 tons of uranium a year, we would consume

$$(200 \text{ power plants}) (160 \text{ tons ore/year}) = 3.2 \times 10^4 \text{ tons/year}$$

At this rate of consumption, our uranium ore supply would last for

$$\frac{3.4 \times 10^5 \text{ tons}}{3.2 \times 10^4 \text{ tons/yr}} = 10 \text{ years}$$

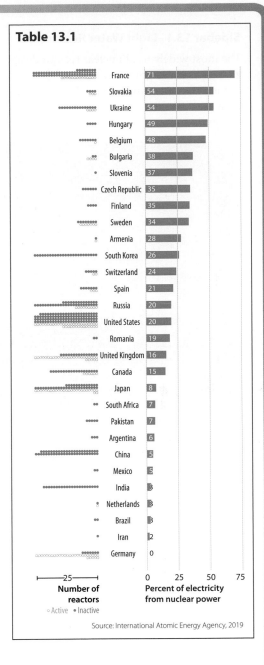

Table 13.1

	Number of reactors	Percent of electricity from nuclear power
France		71
Slovakia		54
Ukraine		54
Hungary		49
Belgium		48
Bulgaria		38
Slovenia		37
Czech Republic		35
Finland		35
Sweden		34
Armenia		28
South Korea		26
Switzerland		24
Spain		21
Russia		20
United States		20
Romania		19
United Kingdom		16
Canada		15
Japan		8
South Africa		7
Pakistan		7
Argentina		6
China		5
Mexico		5
India		3
Netherlands		3
Brazil		3
Iran		2
Germany		0

○ Active   • Inactive

Source: International Atomic Energy Agency, 2019

Sidebar 13.1   Light Water Reactor

The most widely used nuclear reactor design in the US and globally is the "light water reactor" that uses the neutron induced fission of uranium-235, ^{235}U, which produces barium, ^{141}Ba, ^{92}Kr, three neutrons, and 3.2×10^{-11} joules of energy as summarized in the figure below:

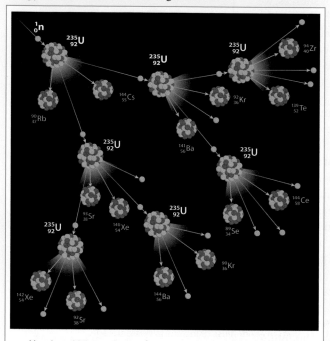

Uranium-235 constitutes about one percent of total uranium.

$$^{235}_{92}U + ^{1}_{0}n \rightarrow ^{236}_{92}U \rightarrow \text{Fragments} + \text{Neutrons} + 3.2 \times 10^{-11} \text{ J}$$

This 3.2×10^{-11} Joules is from a *single nucleus.*

What about from 1 gram of $^{235}_{92}U$?

$$\left(1g^{235}U\right)\left(\frac{1 \text{mol}^{235}U}{235g^{235}U}\right)\left(\frac{6.022 \times 10^{23} \text{atoms}^{235}U}{1 \text{mol}^{235}U}\right)\left(\frac{3.2 \times 10^{-11} \text{J}}{\text{atom}^{235}U}\right)$$

$$= 8.2 \times 10^{10} \text{ J} = 8.2 \times 10^{7} \text{ kJ}$$

This is equivalent to 3 tons of coal!

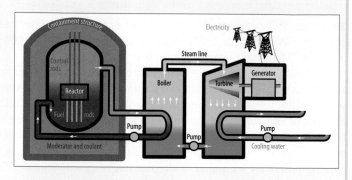

What is interesting about this reactor design is that it is based on the peculiar properties that define the relationship between the energy of the neutrons and the nuclei of the elements involved in the reactor core. The intrigue begins with the ^{235}U that is just 0.7% of the naturally occurring uranium; the remaining 99.3% is ^{238}U. **First** of all, the probability that a ^{235}U will capture a neutron to initiate the nuclear fission reaction depends sensitively on the *velocity* of that neutron. A slow moving or "thermal" neutron is captured by ^{235}U with high probability; a high velocity neutron has a very low probability of initiating a fission reaction with ^{235}U. **Second**, ^{238}U does not initiate a fission reaction no matter what the neutron velocity is, but ^{238}U absorbs high velocity neutrons but not slow ones. To sustain the nuclear reaction, neutrons are required in considerable number, so the fuel rods are quite thin to allow the fast neutrons from the fission of ^{235}U to escape without being absorbed by the dominant ^{238}U. **Third**, as noted above, the neutrons need to be slowed to be captured by the ^{235}U to initiate the fission reaction. In the terminology of nuclear physics, a "moderator" is needed to slow the neutrons and that role is played by the water within which the fuel rods and control rods are submerged. It is the hydrogen atoms in the water that slow the neutrons without absorbing them, such that the slowed neutrons re-enter the fuel rods where they are not absorbed by the ^{238}U because they are moving so slowly, but they are effectively captured by the ^{235}U, initiating another fission reaction. **Fourth**, this reactor design is termed a "light" water reactor because the hydrogen is sufficiently effective at slowing neutrons, causing some neutrons to be lost in the process, that the ratio of ^{235}U to ^{238}U must be increased from the naturally occurring 0.7% to about 3% to sustain the reaction. If "heavy" water, D_2O, is used, the deuterium is less effective at slowing the neutrons, and fewer are lost, so that uranium with naturally occurring 0.7% ^{235}U is adequate to sustain nuclear fission. The problem with heavy water reactors is that D_2O is difficult to produce in the quantities needed. It is easier to "enrich" the ^{235}U to ^{238}U ratio in the uranium fuel rods. Thus the name "pressurized light water nuclear reactor."

On the face of it, 3.2×10^{-11} joules does not seem like much energy; however, when we calculate the amount of energy in one gram of ^{235}U, it works out to be 8.2×10^{10} joules, which is the chemical energy content of 3 tons of coal!

The reactor design centers on the reactor core that contains the "fuel rods" that contain the uranium and the "control rods" that contain cadmium or boron and absorb neutrons that emerge from the fuel rods, thereby controlling the *rate* of the nuclear reaction. But the fuel rods and control rods are literally submerged in water which is directly heated by the thermal energy released in the fission of ^{235}U to a temperature of ~300°C and passed through a heat exchanger that creates the steam that drives a turbine to generate electricity. The schematic of the pressurized light water reactor is shown here.

While further exploration would certainly reveal more uranium deposits, a recurring theme is that uranium used in "once through" reactors is not in abundant supply. If the same uranium ore were used in a fast breeder reactor, which consumes both the ^{235}U *and* the ^{238}U isotope in the ore that is converted to fissionable ^{239}Pu, sixty times as much energy is produced from the same mass of uranium ore reserves such that the U.S. could produce 200 GW of power (one-half of our current electricity demand) for a period of 600 years.

It is important to keep in mind the sequence of events in the cycle of nuclear fuel, which is summarized in Figure 13.3. The uranium ore is extracted in a conventional mining process and that ore is then converted to UF_6—a fuel that consists of 99.3% ^{238}U. In the U.S., which employs "once-through" light water (H_2O) reactors—see Sidebar 13.1 defining light water vs. heavy water (HDO) reactors—the fuel is then enriched such that ^{235}U is 3% of the uranium fuel and ^{238}U is 97%. That enriched fuel is then used in the reactor to produce steam that drives the turbines to produce electricity. The used fuel is then either stored as waste or chemically processed to separate the plutonium and the recoverable uranium that is reused. We will discuss the issue of nuclear water storage shortly, as we are currently considering the issue of sustainability of nuclear fuel. We

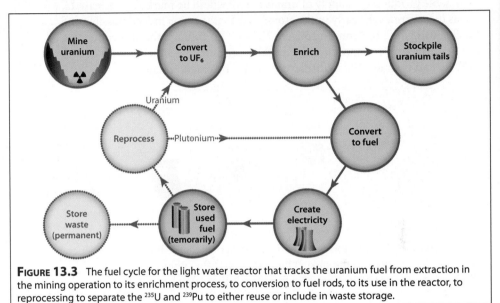

FIGURE 13.3   The fuel cycle for the light water reactor that tracks the uranium fuel from extraction in the mining operation to its enrichment process, to conversion to fuel rods, to its use in the reactor, to reprocessing to separate the ^{235}U and ^{239}Pu to either reuse or include in waste storage.

have now calculated that with an increase in nuclear power generation by a factor of two, such that one-half of all electrical power generation in the U.S. comes from nuclear plants, our current supply of uranium ore would give us a 10 year supply in a "once-through" light water reaction and 600 years in a fast breeder reactor.

But the question of sustainable nuclear energy is incomplete without a discussion of thorium as a nuclear fuel. Thorium-232, ^{232}Th, is very similar to uranium except that there is almost three times as much minable thorium and, of particular importance, thorium can be burned completely in a nuclear power plant. Recall that in contrast, only 0.7 percent of uranium can be used in a nuclear reactor.

Thorium reactors deliver 3.6 billion kWh of heat per ton of thorium. So a 1 GW nuclear reactor requires 6 tons of thorium per year at an efficiency of 40%. U.S. thorium resources are approximately 640 thousand tons so if we use thorium as the nuclear fuel to produce 50% of our current electricity needs, we would consume

$$\left[\frac{0.2 \text{ TW}}{1.0 \text{ GW/reactor}}\right] 6 \text{ tons/reactor-yr} = 1200 \text{ tons/yr}$$

Thus our thorium resources would sustain this demand for

$$\frac{\left(6.4 \times 10^5 \text{ tons } ^{232}Th\right)}{\left(1.2 \times 10^3 \text{ tons}^{232} \text{ Th/yr}\right)} = 400 \text{ years}$$

It is highly probable that the U.S. will move to use thorium as a nuclear fuel if public policy chooses to adopt nuclear energy as a key component of the national primary energy generation mix.

Sidebar 13.2 Fast Breeder Reactors

If nuclear power is to be sustainable, it must present an opportunity for significant power generation over many decades, if not a century or more. The problem with ^{235}U fueled reactors is that, unless decidedly more uranium-235 is discovered, or unless an efficient and economically viable method is developed to extract ^{235}U from seawater, the supply will be exhausted in a few decades at the scale required to supply a significant fraction of global energy demands. However, if the dominant uranium isotope, ^{238}U, could be converted to a fissionable product, the supply of uranium fuel would be markedly increased. It was discovered very early in the atomic age that this was indeed possible and feasible using fast neutrons to convert ^{238}U to ^{239}Pu through the absorption of a fast neutron by ^{238}U to produce the unstable product ^{239}U which in turn decays to ^{239}Np and then to ^{239}Pu as displayed here in the reaction sequence.

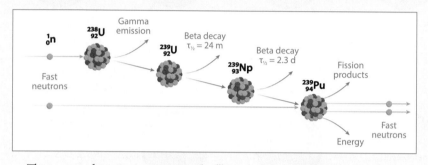

The reason the reaction is termed a "breeder" is that ^{239}Pu fission produces 3 neutrons. One goes to the next ^{239}Pu fission, and the other two go to the production of more ^{239}Pu from ^{238}U fuel. The primary design difference distinguishing the light water reactor from the breeder reactor is that the water that "moderates" the neutrons in the former is replaced by liquid sodium. While liquid sodium is an excellent fluid for heat transfer from the reactor core, it is also much less effective at slowing the neutrons from the ^{239}Pu production reactions so those neutrons are available to "breed" more ^{239}Pu from blankets of uranium ore placed within the core of the reactor as shown here in the schematic design of the reactor.

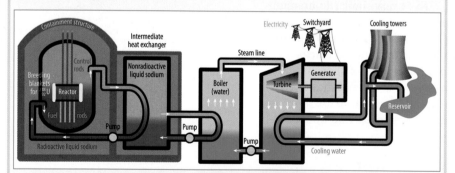

The potential problem with the breeder reactor is that ^{239}Pu is a fissionable product that can be used not only to fuel civilian nuclear power plants, but also to build bombs. If the number of ^{239}Pu reactors grows from the present one hundred or so to numbers in the thousands, it becomes increasingly difficult to keep track of the ^{239}Pu, and thus the opportunity to use it for destructive purposes becomes more difficult to control. It is also true that the technology required to build a viable ^{239}Pu bomb is considerable so the technological infrastructure to accomplish the feat is limited to nations with considerable sophistication in their mastery of nuclear engineering as discussed in the text.

The Issue of Safety

Nuclear energy generation for peaceful purposes emerged virtually simultaneously with the use of nuclear weapons as bombs and nuclear weapons used as strategic elements in international relations—most notably the Cold War. Thus, distinguishing between (1) nuclear reactors used for civilian electric power generation and (2) nuclear weapons, in the public's mind, began in a cloud of confusion and distrust. However, as the court of public opinion judges the relative merits of various forms of primary energy generation, the facts will become increasingly important as global demand for energy dramatically expands and the impact of fossil fuel combustion on human health and on the forcing of the climate structure becomes more clearly defined.

Compounding the close association between nuclear power generation and nuclear weapons was the advent of two nuclear accidents that occurred in the late 1970s and 1980s. The first was the Three Mile Island accident that occurred on March 28, 1979 near Harrisburg, Pennsylvania. This was a case of a rather minor mechanical failure compounded by multiple human errors. Under normal operation, cooling water is used to extract heat from the reactor core (recall that the efficiency of the reactor for electricity generation is ~40%). The primary cooling pump suffered a mechanical failure. However, the backup pump was left with its main supply valve closed, so, by operator error, it was inoperable. With the termination of cooling water flow, control rods were immediately inserted into the reactor core, stopping the nuclear reaction. However, the radioactive fission fragments in the fuel rods that are a by-product of the primary nuclear reaction continued, in the absence of circulating cooling water, to heat the core. At that point, a technician turned *off* an emergency core cooling system because he misread the information in the central control room thinking that the reactor was full of water. This was a second operator error. As the uranium fuel rods continued to increase in temperature, the fuel rods melted and came in contact with residual water in the reactor and then leaked into the concrete containment building. It was the radioactive gases dissolved in the water that escaped and had to be slowly released that caused the situation to become a major news event.

The second nuclear accident was far more serious. It occurred near the small town of Chernobyl, in Ukraine, on April 26, 1986. The Chernobyl reactor used carbon as a neutron moderator. During an otherwise routine test run of the safety system of the reactor, the reactor temperature became too hot and the cooling water vaporized within the reactor, exploding the containment system. That ignited the carbon surrounding the fuel rods, ejecting the radioactive material directly into the atmosphere.

What conspired to initiate and then propagate this accident was operator error in the context of poor reactor design. The carbon-moderated reactor design used in the Chernobyl plant is intrinsically unstable because as the temperature of the core increases, the reactor nuclear fuel burn rate increases, establishing a positive feedback leading to irreversible runaway if not quickly controlled. This is why the U.S. and Western Europe employ water-moderated reactors—they have negative feedback. Second and equally serious, the Chernobyl plant did not have a concrete containment building so the explosion of the carbon medium surrounding the reactor was unconstrained. The UN estimates that 4000 excess deaths resulted from the Chernobyl nuclear accident. The area surrounding the Chernobyl plant remains contaminated.

Apart from nuclear waste as a safety issue, this then leaves us with the question: what are the issues of safety regarding nuclear power generation as we look to the future? Are there improvements in nuclear reactor design that could, through the basic physics of their design, markedly improve their intrinsic safety even in the advent of multiple human error? The answer to this important line of questioning is that yes, there are very significant changes to reactor design that can fundamentally change the safety issues associated with nuclear reactor operation.

One example is the pebble bed reactor design shown in Figure 13.4 that incorporates the uranium within a tray containing pebbles of pyrolytic graphite that can withstand very high temperatures. The pyrolytic graphite is capped with a silicon carbide ceramic heat shield that also easily withstands the maximum temperature that the reactor "bed" can reach even if all control of the reactor is inoperative. Thus the basic design of the reactor core—the bed of pebbles—is based upon the physics that limits the maximum attainable temperatures of the core. The key idea of this design is the intrinsic negative feedback built into the nuclear reaction itself. As the temperature of the pebble bed increases, the neutron velocity increases and at a threshold point—at temperatures well below the breakdown (melting) temperature of the bed materials—the neutrons are absorbed by the predominant uranium isotope, ^{238}U, which cannot sustain a chain reaction. Thus the negative feedback intrinsic to the nuclear fuel serves to *slow* the reaction as the temperature increases, resulting in the termination of increasing temperature well below the failure point of the reactor core pebble beds. Thus the reactor is designed so that, in the case of *all* the control and regulation systems failing—water flow, electrical, mechanical—the reactor will slow and reach a safe standby temperature.

There are also other cooling designs of conventional light water reactors that provide very significant margins of safety, such as the passive light water reactor design shown in Sidebar 13.3.

There is also, in any rational discussion of safety associated with primary energy generation, the question of the safety associated with each of the alternatives. One commonly accepted metric that can be applied to coal, lignite, peat, oil, gas, nuclear, biomass, hydro, and wind is the unit of deaths per GW·y (gigawatt-year) of primary energy generation. Using this unit, we first note that the death rates for primary energy generation vary greatly from country to country. For example, France, Japan, Belgium, Finland, Canada, the U.S., Sweden, Switzerland, and Russia (including Ukraine) have major nuclear power facilities. Yet the overwhelming

Sidebar 13.3 Passively Cooled Light Water Reactors

The Three Mile Island nuclear power plant accident in 1979 and the far more serious Chernobyl accident in 1986 left the nuclear power industry with a very serious public relations problem. In considering the potential role for nuclear power in the future, it is important to recognize the multiple examples of compounded human error in the case of Three Mile Island and the combination of human error and faulty design in the case of Chernobyl. The question is, can a Pressurized Light Water Reactor be designed such that the reactor can shut itself down independent of human error? The answer is that yes, reactors can be designed such that they do not require human intervention to fully shut down in the event of an accident. A class of reactors known as "passively stable light water reactors" constitutes an important example. This reactor design uses gravity fed water, stored in a reservoir, to inject cooling water into the reactor core encasement without the need for mechanical pumps. The water is injected if the vapor pressure of the core water exceeds a predetermined level. The schematic of the design is displayed here.

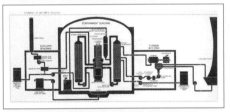

The containment shell is cooled by gravity fed injection pipes to keep the steel containment vessel well below the melt temperature of the metal. In addition to the passive cooling technique, these reactors are designed with far fewer pumps, valves, ducts and electrical subsystems than their predecessors. In fact one of the major strategic errors made by the US nuclear industry in the 1950s was not to refine and simplify reactor design before building large numbers of reactors and placing them in operation.

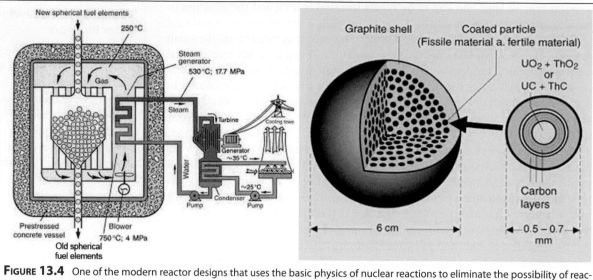

FIGURE 13.4 One of the modern reactor designs that uses the basic physics of nuclear reactions to eliminate the possibility of reactor run away, even with every possible human error in plant operation, is the pebble bed reactor shown in the left panel. The pebble design, shown in the right panel, places the uranium fuel within pyrolytic graphite pebbles that can withstand high temperatures without melting down. This is an important example of new reactor designs that are intrinsically safe.

event dictating the death rate from nuclear energy generation came from a single accident—Chernobyl. Another example is China; their death rate in coal mines per ton of delivered coal is 50 times that of most other nations. To summarize, in data taken from the European Union Project ExternE, Figure 13.5 presents the fatality rate in deaths per GW·y for the primary energy generating options globally.

Inspection of Figure 13.5 reveals that the extraction and distribution of petroleum is associated with the highest death rates—many from oil rigs, helicopters that go down at sea, refinery fires, pipeline fires, etc. In second place is coal mining followed by peat extraction and biomass. Of the *combustion* sources, gas has the lowest death rate per GW·y. Nuclear, hydro, and wind have the lowest death rates per unit of energy generated overall.

Nuclear Waste

Any discussion of nuclear waste is virtually certain to stir up intense sentiment. But it is also important to begin any discussion with the basic facts in the matter.

First, in the U.S., the volume of nuclear waste resulting from the generation of approximately 20% of our electric power by nuclear energy is about 1.2 liters per person per year. Most of this volume—90%—is low-level waste, 7% is intermediate-level waste, and 3% is high-level waste. It is the highly radioactive 3% that is typically stored at the reactor site in pools of water to absorb the heat generated in the radioactive decay of the fission products. In the high level waste, after 40 years, radioactive levels have dropped by three orders of magnitude. So, in steady state, we have the problem of storing radioactive waste that involves the low level nuclear waste (90% of the volume) and the intermediate level waste along with the depleted high level waste. This mixture of fission products, with half lives ranging from a few hundred years to 24,000 years, must be placed in a safe location. France dealt with this problem in the early 1950s, joining public support for, and public confidence in, a plan to bury the waste in inert canisters within geologically stable salt mines that had a very low probability of rupturing the canisters. In contrast, the U.S. moved quickly to build nuclear reactors without garnering public support and with no careful plan for nuclear waste storage. Now France generates 75% of its electricity use from nuclear without incident and with broad public support. The U.S. generates 20% of its electricity from nuclear power and is in political deadlock over the storage of nuclear waste. In recent years, the U.S. government has invested more than $100 billion in the development of a nuclear waste disposal site at Yucca Mountain, Nevada. But there is growing objection on the part of citizens of Nevada for using the state as a dumping ground for radioactive waste. In addition, it has become commonplace for the governors of states through which trains carrying nuclear waste must pass to ban the passage of those trains, essentially paralyzing the network of nuclear waste delivery to Nevada.

On the face of it, the prospect of nuclear waste storage in the U.S., given the need for reliable protection of radioactive waste for thousands of years, is unacceptable and a nonstarter in perpetuity. It is at an apparent impasse such as this that real numbers matter. While security for the nuclear waste for 10,000 years is the objective, it is important to quantify risk. For example, if we reduce the risk of leakage not to zero, but to 0.1%, we must then analyze the implications of a leakage with a probability of one chance in a thousand. Assuming conservatively that the radioactive level of the waste is 1000 times that of the original uranium ore, had it been left in the ground, the net risk (probability multiplied by danger) is 1000 × 0.001 = 1. Thus the risk is approximately the same as if the uranium had never been mined, but left in the ground.

But some important considerations emerge as we move forward in time. After approximately 300 years, the fission product's radioactivity will have dropped a factor of 10. Thus a probability of 1% that *all* the waste leaks out gives us the same radioactivity level as the unmined uranium. In fact, since we have assumed a probability of 1% for *all the waste to leak*, it is immediately clear that we could equally well accept a 100% probability that 1% of the waste would leak out. It is this perspective that places the actual risk from nuclear waste disposal in perspective.

But we develop another important perspective when we consider the risk associated with uranium ore that occurs naturally in the ground. In Colorado, a state endowed with a significant amount of uranium ore, the radioactivity in significant regions of the state is 20 times the legal limit established for Yucca Mountain. Importantly, the headwaters of the Colorado River run through and over these uranium rock deposits and this water is the drinking water source extending through the Western states to the cities of Los Angeles and San Diego. As we move forward to determine the wisdom of various choices for primary energy generation in the next decade, the next two decades and the next century, it is essential that we examine the numbers dispassionately and that we work toward a disciplined analysis of the problem.

Weapons Proliferation

The link that unifies the otherwise separate issues of nuclear power generation on the one hand and nuclear bombs on the other, is the issue of the fission products of nuclear power generation. As we have seen, the uranium used in a nuclear power generating plant is slightly enriched in ^{235}U over the naturally occurring 0.7% in standard uranium ore. Enrichment of ^{235}U is typically to about 3%, with the other 97% ^{238}U. Weapons grade uranium must be greater than 93% ^{235}U to achieve the required chain reaction. It is difficult technologically to achieve this degree of enrichment—it requires sophisticated technical infrastructure within an advanced national infrastructure.

In sharp contrast, the extraction of weapons grade plutonium, ^{239}Pu, is a straightforward matter of chemical separation in the *reprocessing* of spent nuclear fuel rods which contain considerable amounts of ^{239}Pu as a by-product of "burning" ^{235}U embedded in ^{238}U. The nuclear reaction sequence is reviewed in Sidebar 13.2 on the breeder reactor.

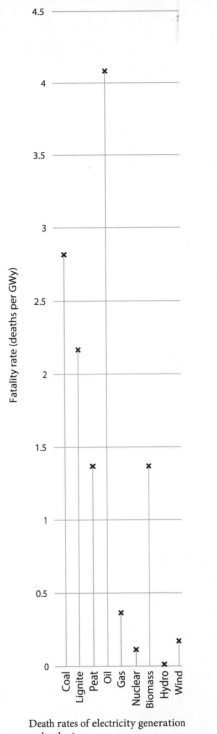

Death rates of electricity generation technologies.

✕: European Union estimates by the ExternE project.

FIGURE 13.5 A comparison of fatality rates (deaths per gigawatt-year of energy generation) for a number of candidate primary energy generation methods ranging from coal, to petroleum, to natural gas, to nuclear, to renewables. Note that the death rate per unit of energy generation is highest for petroleum, next highest for coal, and very low for nuclear, hydro, and wind.

The word "reprocessing" carries a specific meaning in the lexicon of nuclear energy discussions. It means quite literally the extraction of ^{239}Pu from the spent fuel rods; ^{239}Pu that can then be used to build a nuclear weapon. The other by-product of reprocessing, as Figure 13.4 shows, is the reconstituted blend of ^{238}U and ^{235}U that is then used in the fabrication of new fuel rods after enrichment back to 3% ^{235}U. While the first nuclear weapon—dropped on Japan August 6, 1945, was a uranium bomb, the second nuclear weapon dropped on Japan August 9, 1945, was a plutonium bomb. The key point from the perspective of nuclear proliferation today is that the continuing use of nuclear power requires the tracking of ^{239}Pu, because the amount of it proliferates with the use of uranium reactors for power generation. What is not generally recognized is that while the ^{239}Pu can be rather easily extracted in the reprocessing step of spent fuel rods, the *design* of a plutonium bomb requires remarkable sophistication because plutonium bombs can pre-detonate because of the ^{240}Pu isotope in the bomb material. This predetonation results in a very low yield from the bomb. To prevent this, a very precise delivery of the implosion shock to the nuclear material is required—and is rarely achieved. North Korea attempted a bomb test in 2006 using a plutonium bomb that fizzled, almost certainly a result of imprecise technology and predetonation of ^{240}Pu.

To summarize, the fuel enrichment for a uranium bomb is difficult; the bomb design is fairly easy. The fuel for a plutonium bomb is easy to acquire, the bomb design is difficult. Both North Korea and Iran are currently working to develop a plutonium weapon.

Nuclear Fusion

From virtually every perspective, nuclear fusion is the holy grail of clean energy generation. The fuel is abundant and inexpensive. Deuterium is naturally occurring in ordinary water and tritium can be produced from neutron bombardment of lithium or boron. There is no radioactive waste and no possibility of the reaction going out of control. The problem with nuclear fusion is that it is technically extremely difficult to achieve in a controlled manner because the deuterium and tritium nuclei repel each other, creating a *very* high potential energy barrier. In order to pass over this barrier, a temperature of 100 million degrees centigrade is required. The problem is made more difficult by the fact that at these temperatures, the pressure, $P = (nRT/V)$, becomes very large and the mixture (thermally) explodes. While sustained nuclear fusion has not been achieved, a very large international effort has been in progress for more than three decades in the quest for sustained fusion.

Two principal methods are under development. The first is to use very low density mixtures of deuterium and tritium so the pressure will not be uncontrollable and to use a magnetic field to confine the high temperature nuclei. This is called the "Tokamak" approach after the Russian experiment that first achieved—although *very* briefly—the threshold condition for fusion. A picture of the Tokamak machine built at Princeton University is shown in Figure 13.6. The nuclei are contained in the magnetic field that circumscribes the donut shaped interior of the reaction zone. A much larger machine is under construction by an international consortium in France. This is shown in Figure 13.7. Note the scale of the system by the human standing at the lower left of the figure. This project, named International Thermonuclear Experimental Reactor (ITER) is due for completion in 2016. The scale of the investment is a measure of the seriousness of the venture, but it is not clear whether a sustained nuclear fusion reaction can be achieved.

FIGURE 13.6 The interior of the Princeton Tokamak fusion reactor that uses a magnetic field to contain the low density mixtures of deuterium and tritium that are reacted to form ($^4_2$He), releasing energy in the process.

FIGURE 13.7 A cross sectional view of the International Thermonuclear Experimental Reactor (ITER) in France funded by an international consortium. The operating principle of this Tokamak machine is to magnetically contain the high temperature plasma containing deuterium and tritium. Operating temperatures of the plasma exceed 100 million Kelvin. Credit © ITER Organization, https://www.flickr.com/photos/oakridgelab/41783636452.

The second technique currently under test uses a very different approach—the use of lasers to direct very high intensity radiation into a very small volume containing a limited number of deuterium and tritium atoms. Thus the thermal explosion, $P = nRT/V$, is controlled by keeping the number of moles, n, *very* small. The objective is to see if the tritium-deuterium threshold can be reached under intense laser illumination and second, if it can, to produce sustained nuclear fusion using a very large number of small nuclear reactions to achieve the desired power output. A diagram of the laser system is displayed in Figure 13.8.

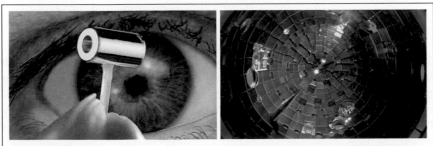

FIGURE 13.8 Another method to achieve high temperatures in the reaction of deuterium and tritium is to use multiple, convergent laser beams of the National Ignition Facility (NIF) at the Lawrence Livermore Laboratory in California. Multiple, high-powered laser beams are focused on a very small number of tritium and deuterium atoms within a very small reaction vessel shown in the left panel. The right-hand panel shows the multiple laser inputs that focus all beams into the center of the reaction vessel.

Misconceptions

Given the remarkable amount of misinformation surrounding nuclear power generation, we close this Framework section with a sampling of cases setting straight some common examples of these misconceptions.

First Misconception

Can a nuclear reactor explode like an atomic bomb? Isn't nuclear energy very dangerous because a reactor for power generation is just a nuclear bomb that is controlled by humans?

The answers to those questions are, fortunately, based entirely on the laws of physics, and do not depend upon human organizations nor human discipline.

The reason that a nuclear reactor will not, if it goes out of control, explode like an atomic bomb is that the large fraction of ^{238}U (~97%) in the core fuel means that the reaction will cease if the neutrons are not slowed, i.e. are not "moderated." (See Sidebar 13.1) If a moderated nuclear reactor runs out of control, the energy from the fission (the high temperatures produced) will build until the core structure blows apart. The structure blows apart just as it would if dynamite were detonated in the core structure, it does not blow apart because of a nuclear explosion. But this happens well before a nuclear reaction occurs that would release the binding energy of the nuclei contained in the ore. This results from the fact that atomic bombs use 93% enriched ^{235}U and nuclear power reactors use 3% enriched ^{235}U.

Thus if a power station is built with technology that prevents the core from reaching the threshold temperature for a thermal (non-nuclear) explosion such as a pebble bed reactor or a passive water reactor, no explosion will occur. If a reactor is built such that the reactor can reach a temperature for a thermal explosion given a chain of mechanical, electrical, and human failures, but it is built in a concrete containment structure, no nuclear explosion is possible and radiation from the thermal explosion will be contained.

Second Misconception

Because a nuclear power plant requires such a large amount of concrete and steel, building a nuclear power releases so much CO_2 that it would be better to simply produce energy by burning petroleum or coal.

Analysis of the energy required to build a power plant and the CO_2 released from the manufacturing and curing of concrete demonstrates that the construction of a 1 GW nuclear power plant releases 300,000 tons of CO_2—a very large number.

But with a typical reactor lifetime of 40 years, this translates to a carbon intensity resulting from the construction of the plant of

$$\text{carbon intensity} = \frac{\text{carbon release}}{(\text{years of operation})(\text{energy produced per year})}$$

$$= 1.4 \text{ g } CO_2/ \text{ kWh}$$

This is to be compared with the fossil fuel carbon intensity of 400 g CO_2/kWh. If we include all possible contributions to a *total carbon* intensity of nuclear power (transportation, construction, fuel processing, decommissions, etc.), recent estimates state that the total carbon intensity of nuclear power is approximately 4 g CO_2/kWh, 100 times less than the CO_2 output from a fossil fuel electrical power generating plant.

Third Misconception

If we switch to nuclear energy, wouldn't that contribute to global warming because of all the heat released directly into the atmosphere and the water systems?

Given what we already know about the ratio of global energy consumption by humans to the energy that circulates within the climate structure of the Earth, the answer is certainly no. But let's do the calculation explicitly for nuclear power.

Suppose we assume that one-third of all primary energy generation worldwide comes from nuclear (it is currently 6%). Assume further that the efficiency of energy generated from nuclear energy is 50% so if 30% of global energy (that is 5 TW in 2010 and 10 TW in 2050 in terms of power or 6×10^{13} kWh in 2050) is generated by nuclear, we must multiply this, the global use of energy per year, by a factor of two. Taking the average energy released to the environment by nuclear power production and use, this means that the energy added to the climate system would average about

$$\frac{6\times10^{13}\,\text{kWh/yr} + 12\times10^{13}\,\text{kWh/yr}}{2}$$

or approximately 9×10^{13} kWh/yr between now and 2050.

We can round this to 1×10^{14} kWh/yr of energy released by nuclear power generation to the environment. This is to be compared with the 1.7×10^{18} kWh/yr of energy circulating in the climate structure. Thus even with a significant overestimate of the fraction of total primary energy generation that comes from nuclear, the contribution from nuclear is less that one part in 15,000 of the energy circulating in the climate system.

Fourth Misconception

While nuclear power may be a key strategy for reducing carbon emission, it isn't possible to build nuclear power plants rapidly enough to make any real difference.

This question often arises because of the misinformation often circulated. For example, as David Mackay relates in his book "Sustainable Energy—Without the Hot Air," the *Oxford Research Group* (a UK think tank) released a report in 2005 stating that "For nuclear power to make a significant contribution to a reduction in global carbon emissions in the next two generations, the industry would have to construct nearly 3000 new reactors—or about one per week for 60 years. A civilian nuclear construction and supply program on this scale is a pipe dream, and completely unfeasible. The highest historic rate is 3.4 new reactors a year."

So let's check the numbers. We already know that we must increase total global power generation from 15 TW to approximately 35 TW by 2050. That is 20 TW over 40 years or 0.5 TW/year. If nuclear is to be a major contributor, it would grow to say 20% of global energy production, or 0.1 TW/year increase. For a nominal nuclear power plant of 1 GW, this is 100 GW/1 GW = 100 nuclear power plants constructed/year. For reference, let's examine the history of nuclear power plant construction. The primary expansion of nuclear power occurred in the 1970s and 1980s. During that period the primary builders of nuclear plants were France, U.S., Sweden, Finland, Switzerland, and Belgium. Figure 13.9 displays the increase in generated power from nuclear reactors as a function of time from 1970 to 2000. Production of new power plants reached 30 GW/yr (normal plant size 1 GW/plant) in the early 1980s and held that rate of plant production until the end of the decade.

Were nuclear power generation to be used broadly, rather than by a very limited number of countries, it is perfectly reasonable to expect a rate of power plant construction worldwide to reach three times that of the 1980s. The real issue with nuclear power is that of public perception and of public policy on a global basis. This requires a careful and dispassionate analysis of the numbers. When the Oxford Research Group stated that the "highest historic rate (of nuclear plant construction) is 3.4 new reactors a year," that number referred to just one country: France. It is always wise to compare global demand with global ability. It is quite common for organizations to distort numbers in an effort to promote a given agenda. Opinions will always vary, but every effort should be made to look at the actual numbers associated with policy judgements.

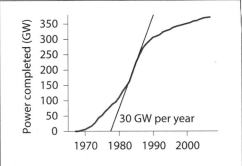

Graph of the total nuclear power in the world that was built since 1967 and that is still operational today. The world construction rate peaked at 30 GW of nuclear power per year in 1984.

FIGURE 13.9 The global build-up of nuclear reactor capability in terms of the increase in he *rate* of nuclear power capability is shown here as a function of calendar year. The *increase* in nuclear power plant capability reached a peak in the 1980s at about 30 GW per year of installed nuclear power. This was accomplished by a fairly small number of nations.

Chapter Core

Road Map for Chapter 13

Exploration of reactions associated with changes in nuclear structure began before science recognized that the structure of the atom was founded on the nucleus containing the proton and neutron that contained the preponderance of atomic mass, as deduced by the experiment of Rutherford as discussed in Chapter 2. This study of nuclear chemistry began with studies of radioactivity resulting from the decomposition of heavy nuclei into lighter ones.

In this chapter we investigate:
- The particles emitted by the nucleus leading to nuclear rearrangement.
- The character of subatomic particles.
- Nuclear fusion that combines light nuclei to form more massive nuclei, releasing energy.
- Nuclear stability that explores the binding energy of nucleons as a function of the number of nucleons in a nucleus.
- Nuclear fission that releases energy by splitting large nuclei into smaller ones.
- Radioactive dating that allows the determination of the age of artifacts by measuring the fraction of radioactive carbon-14, $^{14}_{6}C$, contained in the sample.

Road Map to Core Concepts	
1. Elementary Nuclear Particles and Reactions	
2. Nuclear Reactions: Fusion	
3. Nuclear Stability: Binding Energy	
4. Nuclear Reactions: Fission	
5. Radioactive Dating	

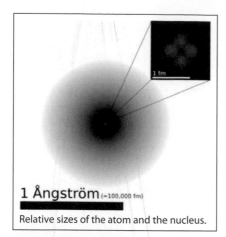

1 Ångström (=100,000 fm)

Relative sizes of the atom and the nucleus.

Marie Curie (1867–1934)

Pierre Curie (1859–1906)

Elementary Nuclear Particles and Reactions

The existence of the nucleus was proven for the first time by the experiments of Hans Geiger and Ernest Marsden in the laboratory of Ernest Rutherford, in 1909: they scattered α particles from a thin foil of gold and saw that while most particles were deflected very little, some were scattered back, which is only possible if the atoms in the thin foil contained a core (the nucleus) which was at least as massive as the bombarding particles. Subsequently, Rutherford provided the mathematical theory to explain this experimental result.

Nuclei contain two kinds of particles, **protons** and **neutrons**. These are collectively called nucleons. The proton has a positive charge, equal in magnitude to that of the electron. The neutron has no charge, and it has slightly larger mass than the proton. The size of the nucleus is about 10^{-15} m, about 5 orders of magnitude smaller than the atom, the latter being set by the orbits of electrons around the nucleus. On the other hand, the mass of the nucleus is about 4–5 orders of magnitude larger than the mass of electrons, since each nucleon is about 2000 times heavier than an electron.

There are three key players in radioactivity as first proposed by Marie Curie, who together with her husband Pierre Curie contributed greatly to nuclear physics, needed to describe emission of ionizing radiation by heavier elements:

- α particles, which are actually nuclei of the He atom;
- β particles, which are electrons that originate from the nuclei of atoms in nuclear decay processes;
- γ rays, which are quanta of electromagnetic radiation, that is, photons.

In discussing the chemistry of the nucleus, we need to identify an atom not only by the number of protons in its nucleus, which we have called the "atomic number," denoted by Z, but also by the number of neutrons in the nucleus, as well as the oxidation state. The basic notation in the following will be, as developed in Chapter 2:

$$_Z^A X^{+x}$$

where Z is the atomic number and A the mass number, which is equal to the sum of the number of neutrons and protons in the nucleus; as usual, the superscript on the right $+x$ is the oxidation state (equal to the number of electrons missing from the neutral atom, or equivalently, the excess positive charge of the ion). By analogy to the conservation of elements in a chemical reaction, in a nuclear reaction the mass number, atomic number and oxidation number are all conserved between the left-hand side (reactants) and right-hand side products of the reaction. In order to make this possible, we must assign the following mass, atomic, and oxidation numbers to the basic particles, the proton p, the neutron n, and the electron e:

$$_1^1 p^+: \quad \text{proton}$$

$$_0^1 n^0: \quad \text{neutron}$$

$$_{-1}^0 e^-: \quad \text{electron}$$

$$_1^0 e^+: \quad \text{positron}$$

We have also introduced a new particle, the positron, which is the so-called "antiparticle" of the electron: it has the opposite values for all the key quantities (charge, atomic number) of the electron.

When a particle and its antiparticle meet, they annihilate and produce photons, which carry the energy corresponding to the mass of the particle-antiparticle pair:

$$_{-1}^0 e^- + _1^0 e^+ \rightarrow 2\gamma + 1.02 \text{ MeV}$$

In the above equation we have used Einstein's equation that relates mass to energy:

$$E = mc^2$$

where c is the speed of light (3×10^8 m/sec). The mass of the electron is $m_e = 9.11 \times 10^{-31}$ kg, and the mass of the positron is the same (an antiparticle and a particle have exactly the same mass), which gives a total energy for the annihilation of the electron-positron pair:

$$2 \times 9.11 \times 10^{-31} \text{ kg} \times (3 \times 10^8 \text{ m/sec})^2 = 1.64 \times 10^{-13} \text{ J} = 1.022 \text{ MeV}$$

and we have also used the conversion to units of MeV = 10^6 eV (recall that 1 eV = 1.6×10^{-19} J). The energy scale of nuclear reactions is such that the MeV is a more convenient unit to describe them, as we will see in several examples below. In these units the mass of the electron (and the positron) is 0.511 MeV.

In order to be able to conserve both energy and momentum in the course of nuclear reactions, we must also introduce the neutrino ν and antineutrino $\bar{\nu}$, which carry no charge or mass or atomic number. The existence of the neutrino was hypothesized first by Wolfgang Pauli. Neutrinos interact so weakly with matter that they can go through the entire Earth without ever interacting with anything, so for the purposes of the present discussion we will simply ignore them.

With these definitions, the most basic nuclear reactions are:

$$_0^1 n^0 \rightarrow {}_1^1 p^+ + {}_{-1}^0 e^- + \bar{\nu}$$

$$_1^1 p^+ \rightarrow {}_0^1 n^0 + {}_1^0 e^+ + \nu$$

The mass of the neutron is 939.566 MeV, while the mass of the proton is 938.272 MeV. From this comparison, it is evident that the first reaction, the conversion of the neutron to a proton and an electron, would be exothermic, since the sum of the product masses (938.272 MeV + 0.511 MeV = 938.783 MeV) is smaller than the mass of the original particle. The rest of the energy difference (939.566 MeV – 938.783 MeV = 0.783 MeV) is carried away as kinetic energy of the products. In contrast, the second reaction should not happen spontaneously, because the mass of the products is larger than the mass of the original particle. Indeed, the proton is stable with an estimated half-life (more on the definition of half-life below) of 10^{35} years, which is much longer than the age of the universe, estimated to be about 14×10^9 years.

Albert Einstein (1879–1955)

Wolfgang Pauli (1900–1958)

Nuclear Reactions: Fusion

As we have seen already, the elementary nuclear particles are involved in reactions that change one into the other. Since larger nuclei are composed of neutrons and protons, more complicated reactions that transform nuclei are also possible. For a specific example, we examine the nuclear reactions that take place in the Sun, producing the enormous amount of energy that the Sun radiates. In these reactions, two hydrogen nuclei $_1^1 \text{H}^+$, that is, protons, come together to produce first a new element, the deuteron $_1^2 \text{D}^+$, whose nucleus consists of a proton and a neutron:

$$_1^1 \text{H}^+ + {}_1^1 \text{H}^+ \rightarrow {}_1^2 \text{D}^+ + {}_1^0 e^+ + \nu + 0.42 \text{ MeV}$$

In the next step, a deuteron nucleus and a hydrogen nucleus combine to produce a helium-3 nucleus, $_2^3 \text{He}^2$, which is not the normal kind of He nucleus, the latter having two protons and two neutrons and called helium-4:

$$_1^2 \text{D}^+ + {}_1^1 \text{H}^+ \rightarrow {}_2^3 \text{He}^{2+} + \gamma + 5.49 \text{ MeV}$$

In the last step, two helium-3 nuclei are combined to produce one helium-4 nucleus and two protons:

$$\,^{3}_{2}\text{He}^{2+} + \,^{3}_{2}\text{He}^{2+} \rightarrow \,^{4}_{2}\text{He}^{2+} + 2\,^{1}_{1}\text{H}^{+} + 12.86 \text{ MeV}$$

The helium-4 nucleus of the end reaction is very stable and the two hydrogen nuclei in the final product are available to participate in another reaction. This is called the proton-proton cycle, because it starts with two protons, which are regenerated at the end of the reaction so that it can start all over again. This is the dominant mechanism for energy generation in stars of the size of the Sun or smaller. The total energy produced in one cycle, starting with 6 protons and ending with one helium-4 nucleus and two protons is 26.72 MeV (this includes the energy of the photons produced by the annihilation of the positrons produced in the first step, with electrons).

All three steps in the proton-proton cycle combine smaller nuclei to form larger ones. This type of nuclear reaction is called **fusion**; reactions of the opposite type, producing smaller nuclei from larger ones, a process called **fission**, are also possible. We will examine both types of reactions in some detail.

Another fusion reaction that produces stable helium-4 nuclei from deuterium and tritium (whose nucleus consists of one proton and two neutrons) is displayed schematically in Figure 13.10 and here in equation form:

$$\,^{2}_{1}\text{D}^{+} + \,^{3}_{1}\text{T}^{+} \rightarrow \,^{5}_{2}\text{He}^{2+} \rightarrow \,^{4}_{2}\text{He}^{2+} + \,^{1}_{0}n + 17.6 \text{ MeV}$$

This particular reaction is interesting because the intermediate product of the fusion is the unstable helium-5 nucleus, which decays to the stable helium-4 nucleus by emitting a neutron and releasing energy ΔE carried away as kinetic energy of the products (the neutron has kinetic energy 14.1 MeV, the helium-4 nucleus 3.5 MeV). It is hoped that this energy can be harnessed for useful purposes in nuclear fusion reactors, but this has not been practically realized yet because of scientific and technological problems.

Another set of nuclear fusion reactions which take place in stars of size larger than the Sun is the so-called CNO-cycle, because it involves carbon, nitrogen, and oxygen nuclei. This set of reactions begins with a carbon-12 nucleus and 4 protons, and ends up with a carbon-12 nucleus and a helium-4 nucleus, releasing a substantial amount of energy:

$$\,^{12}_{6}\text{C} + \,^{1}_{1}\text{H} \rightarrow \,^{13}_{7}\text{N} + \gamma \qquad\quad + 1.95 \text{ MeV}$$
$$\,^{13}_{7}\text{N} \qquad \rightarrow \,^{13}_{6}\text{C} + \,^{0}_{1}e^{+} + \nu + 2.22 \text{ MeV}$$
$$\,^{13}_{6}\text{C} + \,^{1}_{1}\text{H} \rightarrow \,^{14}_{7}\text{N} + \gamma \qquad\quad + 7.54 \text{ MeV}$$
$$\,^{14}_{7}\text{N} + \,^{1}_{1}\text{H} \rightarrow \,^{15}_{8}\text{O} + \gamma \qquad\quad + 7.35 \text{ MeV}$$
$$\,^{15}_{8}\text{O} \qquad \rightarrow \,^{15}_{7}\text{N} + \,^{0}_{1}e^{+} + \nu + 2.75 \text{ MeV}$$
$$\,^{15}_{7}\text{N} + \,^{1}_{1}\text{H} \rightarrow \,^{12}_{6}\text{C} + \,^{4}_{2}\text{He} \qquad + 4.96 \text{ MeV}$$

Nuclear Stability: Binding Energy

If we simply think of the helium-4 nucleus as being formed by combining its elementary constituents, two protons and two neutrons, starting from those constituents as isolated particles,

$$2\,^{1}_{1}\text{H}^{+} + 2\,^{1}_{0}n^{+} \rightarrow \,^{4}_{2}\text{He}^{2+}$$

and count the total mass on each side of the reaction, we will find that there is a

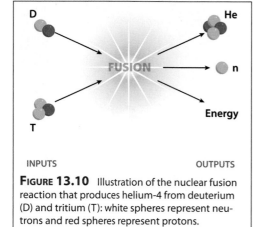

FIGURE 13.10 Illustration of the nuclear fusion reaction that produces helium-4 from deuterium (D) and tritium (T): white spheres represent neutrons and red spheres represent protons.

mass deficit on the right-hand side. The mass of each proton is 1.0073 u, where u is the atomic mass unit defined as:

$$1 \text{ u} = 1.6605402 \times 10^{-24} \text{ g}$$

while the mass of each neutron is 1.0087 u. The net mass of the helium-4 nucleus is 4.0015 u, so the deficit is

$$\Delta m = 2 (1.0073 + 1.0087) \text{ u} - 4.0015 \text{ u} = 0.0305 \text{ u}$$

This mass deficit has been converted into the binding energy that holds the nucleus together, because we would need to supply at least this amount of energy to the nucleus in order to break it apart into its constituents. Using Einstein's formula for relating mass to energy, we find that the binding energy for the helium-4 nucleus is

$$\Delta E = \Delta mc^2 = 4.54 \times 10^{-12} \text{ kg m}^2/\text{sec}^2 = 4.54 \times 10^{-12} \text{ J}$$

This means that formation of a single helium-4 nucleus releases this amount of energy, or formation of 1 mol of helium-4 would release 2.73×10^9 kJ/mol. This is an enormous amount of energy, compared to chemical reactions: in the latter, a typical chemical reaction produces about 100 kJ/mol. The difference is seven orders of magnitude: nuclear reactions produce between five and ten million times more energy than chemical reactions.

This impressive difference is due to the forces that bind the nuclei into stable entities. These forces are called nuclear forces. They are much stronger than the electromagnetic forces (the Coulomb force between charged particles), but their range is very limited: they act over very short distances, of approximately the size of the nucleus, which is 10^{-15} m, five orders of magnitude smaller than the size of atoms, which is 10^{-10} m. Nuclear forces have to be much stronger than electromagnetic forces because the protons in the nucleus strongly repel each other due to their positive charge. In fact, this strong repulsion is also the reason why neutrons are needed: neutrons provide the extra nuclear attraction to hold the protons within the confines of the nucleus. The larger the nucleus, the more neutrons are needed to counterbalance the strong electrostatic repulsion of the protons. For nuclei with sizes up to about 30 protons the number of neutrons and protons is approximately the same. For larger nuclei, the ratio of neutrons to protons is larger than 1.

The binding energy of nuclei per nucleon as a function of the mass number as displayed in Figure 13.11 increases almost steadily from the smallest nucleus (the hydrogen nucleus or proton in which the binding energy is naturally equal to zero) to the nucleus of iron-57, $^{57}_{28}\text{Fe}$. Beyond that, the binding energy decreases monotonically. Note that this is the same plot as Figure 13.2 except that it is inverted to show binding energy per nucleon as positive. Thus, the natural tendency of nuclei is to come together to form larger nuclei, which increases the binding energy per nucleon, thereby producing a more stable nucleus, up to iron-57; this process (fusion) was illustrated for the case of deuterium and tritium that combine to form helium-4. Beyond iron-57, the natural tendency would be

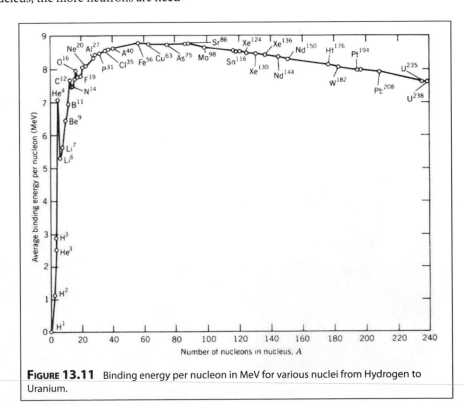

FIGURE 13.11 Binding energy per nucleon in MeV for various nuclei from Hydrogen to Uranium.

for large nuclei to fall apart into smaller fragments; this process (fission) is often triggered by bombardment of a nucleus that tends to be unstable by a neutron. The binding energy curve has some very pronounced spikes for small nuclei, namely helium-4, carbon-12, oxygen-16, and neon-20. These are the most stable nuclei, and they correspond to something like closed shells of nucleons, similar to what we found for the case of electrons in atoms.

Nuclear Reactions: Fission

Just like nuclei can come together to form larger ones, certain nuclei can break apart to smaller fragments, by what is called a nuclear *fission* reaction. The most famous of such reactions is one involving uranium nuclei. Enrico Fermi first hypothesized that elements beyond uranium-235, the so-called "transuranium" elements, might be produced by neutron bombardment of $^{235}_{92}U$, as displayed in Figure 13.12.

Enrico Fermi (1901–1954)

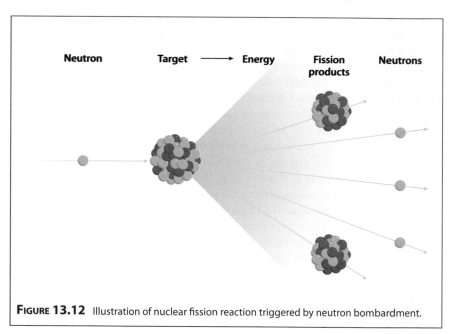

FIGURE 13.12 Illustration of nuclear fission reaction triggered by neutron bombardment.

The reaction hypothesized by Fermi actually happens, but it does not stop there; it continues to produce smaller nuclei and more neutrons as depicted in Figure 13.13 and in the equation:

$$^{235}_{92}U + {}^{1}_{0}n \rightarrow {}^{236}_{92}U \rightarrow {}^{141}_{56}Ba + {}^{92}_{36}Kr + 3\,{}^{1}_{0}n$$

This is an example of the breakup of the unstable uranium-236 nucleus into smaller fragments, a typical nuclear fission reaction.

The three neutrons produced from the fission reaction of uranium-236 can be used to induce more reactions, if there is enough mass ("critical mass") of uranium to absorb them. If this continues without inhibition, it leads to an explosive nuclear chain reaction. If, on the other hand, some other material is also present that partially absorbs the neutrons, then it is possible to control the reaction so that it is not explosive in nature. This is the basis for production of useful energy from nuclear fission reactions as is done in nuclear power stations. It is also the basis for the atomic bomb.

J. Robert Oppenheimer (1904–1967)

In a nuclear reactor, as discussed in the Framework, there are rods of other materials, called "control rods," which absorb the excess neutrons so that the chain reaction can be sustained in a controlled fashion. The heat produced from the nuclear reactions is used to turn water into steam, which then drives a steam turbine. The steam is subsequently condensed and the water is recycled through the reactor. Nuclear reactors are used to provide a large percentage of the electricity needs in many countries. Among large industrialized nations, France produces the largest fraction of its electricity from nuclear reactors, about 75%. In the United States, about 20% of the electricity is produced from nuclear power plants. The average for the world is about 6%.

If the nuclear reaction is not controlled, it produces a violent explosion, the atomic bomb as shown in Figure 13.13. The first atomic bomb was developed in the US during the Second World War, under the direction of the physicist J. Robert Oppenheimer.

The reaction in Figure 13.14 produces 3.2×10^{-11} J per uranium-235 nucleus. For 1 kg of uranium, this corresponds to 8.2×10^{10} kJ. To produce the same amount of energy from coal, we would need 3 tons! This comparison demonstrates the great efficiency in producing energy from nuclear reactions using a nuclear power plant as shown in Figure 13.15 and as discussed in the Framework section. The biggest disadvantage, assuming that technical solutions can be found to reduce the risk of accident, is that the products of the reaction are radioactive elements that need to be stored for a very long time until their radioactivity has diminished to a level that is not harmful. In some cases this can take centuries, but the issue must be placed in perspective as discussed in the Framework section.

FIGURE 13.13 Fission reaction: the atom bomb.

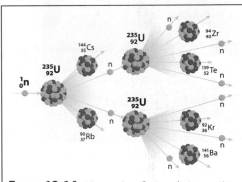

FIGURE 13.14 The nuclear fission chain reaction of uranium-235.

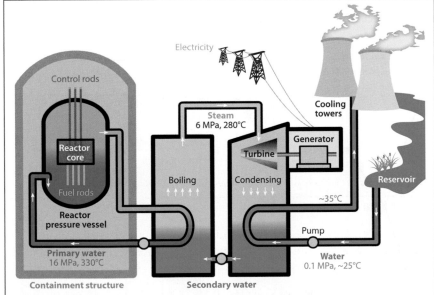

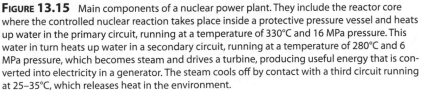

FIGURE 13.15 Main components of a nuclear power plant. They include the reactor core where the controlled nuclear reaction takes place inside a protective pressure vessel and heats up water in the primary circuit, running at a temperature of 330°C and 16 MPa pressure. This water in turn heats up water in a secondary circuit, running at a temperature of 280°C and 6 MPa pressure, which becomes steam and drives a turbine, producing useful energy that is converted into electricity in a generator. The steam cools off by contact with a third circuit running at 25–35°C, which releases heat in the environment.

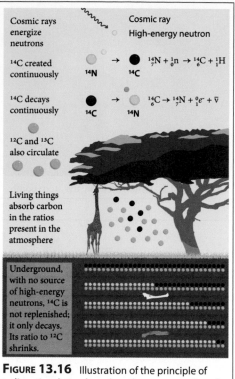

FIGURE 13.16 Illustration of the principle of radioactive dating based on the concentration of carbon-14.

Radioactive Dating

Nuclear reactions have found another important application. This is based on the existence of a radioactive form of carbon, carbon-14, which is produced in the upper atmosphere as shown in Figure 13.16 and in the equation:

$$^{14}_{7}N + ^{1}_{0}n \rightarrow ^{14}_{6}C + ^{1}_{1}H$$

In this reaction, high-energy neutrons in the upper atmosphere combine with the stable nitrogen-14 nuclei to form the radioactive carbon-14 nuclei. The carbon-14 nuclei enter the carbon reservoir at the Earth's surface, mixing with the stable carbon-12 nuclei in the ocean, atmosphere, soils, and tissues in living organisms. The carbon-14 nuclei are unstable and they decay to stable nitrogen-14 by beta-decay (emission of an electron and an antineutrino):

$$^{14}_{6}C \rightarrow ^{14}_{7}N + ^{0}_{-1}e^{-} + \overline{\nu}$$

Whatever decay of carbon-14 takes place in living organisms is replenished in steady-state with the carbon-12 from the atmosphere-biomass. At equilibrium with its environment, the rate of decay is 15.0 disintegrations per g per minute (A_0 = 15.0 dis/g·min). Since this is a spontaneous reaction, the decay rate is proportional to the amount present:

$$\frac{dN}{dt} = -\lambda N \Rightarrow N = N_0\, e^{-\lambda t}$$

where N_0 is the initial amount at time zero and the constant of proportionality λ is the rate of decay. The carbon-14 nuclei have a half-life of $t_{\frac{1}{2}}$ = 5730 years, that is, in

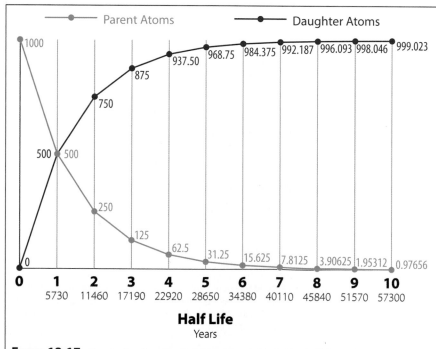

FIGURE 13.17 Decay of radioactive isotopes over a period of 10 half-lives: the blue line is the number of parent atoms (the ones undergoing decay), starting with 1000 parent atoms, while the red line is the number of daughter atoms (the products of the radioactive decay). The number of years under each value of half-life corresponds to actual time for the decay of radioactive carbon-14.

this time exactly half of the original number of nuclei have decayed and the other half remain in the radioactive state, as displayed in Figure 13.17.

The decay rate can be easily obtained from the half-life:

$$\frac{1}{2} = e^{-\lambda t_{1/2}} \Rightarrow \lambda = \frac{\ln(2)}{t_{1/2}} = 1.21 \times 10^{-4} \,/\mathrm{yr}$$

Now suppose that an archeological object such as that displayed in Figure 13.18 is measured to have a decay rate of $A_t = 7.9$ dis/g·min at the present time, which is much lower than the normal decay rate. What is the age of this object? The idea is that carbon-14 nuclei in the object are decaying, but they are not being replenished when the object is not in equilibrium with its natural environment (for example, it has been buried in the ancient site, and is not in contact with the atmosphere). For any object made of natural materials a long time ago, its content in carbon-14 is constantly decreasing and is not being replaced.

To find the age, we use the equation that relates the rate of decay (dN/dt) to the amount of radioactive nuclei (N), with the value of the measured activity A_t in place of N and the value of the original activity A_0 in place of N_0, because these activities are simply the number of radioactive carbon-14 nuclei at the two different times multiplied by their mass. This leads to, after taking the logarithm of both sides of the equation,

$$\ln\left[\frac{A_t}{A_0}\right] = -\lambda t$$

and using the values of A_t, A_0, λ, we find:

$$t = \frac{\ln\left[\dfrac{7.9}{15.0}\right]}{\lambda} = 5300 \text{ years}$$

FIGURE 13.18 Remains of a couple buried in embrace, found in Valdaro, Italy; radioactive dating showed they had lived during the Stone Age. (Credit: Dagmar Hollmann/Wikimedia Commons; license CC BY-SA 4.0.)

Standard Thermodynamic Values for Selected Substances

Note: All values at 298 K.

Substance or Ion	ΔH°_f (kJ/mol)	ΔG°_f (kJ/mol)	S° (J/mol·K)	Substance or Ion	ΔH°_f (kJ/mol)	ΔG°_f (kJ/mol)	S° (J/mol·K)
$e^-(g)$	0	0	20.87	**Calcium**			
Aluminum				$Ca(s)$	0	0	41.6
$Al(s)$	0	0	28.3	$Ca(g)$	192.6	158.9	154.78
$Al^{3+}(aq)$	−524.7	−481.2	−313	$Ca^{2+}(g)$	1934.1	—	—
$AlCl_3(s)$	−704.2	−628.9	110.7	$Ca^{2+}(aq)$	−542.96	−553.04	−55.2
$Al_2O_3(s)$	−1676	−1582	50.94	$CaF_2(s)$	−1215	−1162	68.87
Barium				$CaCl_2(s)$	−795.0	−750.2	114
$Ba(s)$	0	0	62.5	$CaCO_3(s)$	−1206.9	−1128.8	92.9
$Ba(g)$	175.6	144.8	170.28	$CaO(s)$	−635.1	−603.5	38.2
$Ba^{2+}(g)$	1649.9	—	—	$Ca(OH)_2(s)$	−986.09	−898.56	83.39
$Ba^{2+}(aq)$	−538.36	−560.7	13	$Ca_3(PO_4)_2(s)$	−4138	−3899	263
$BaCl_2(s)$	−806.06	−810.9	126	$CaSO_4(s)$	−1432.7	−1320.3	107
$BaCO_3(s)$	−1219	−1139	112	**Carbon**			
$BaO(s)$	−548.1	−520.4	72.07	$C(graphite)$	0	0	5.686
$BaSO_4(s)$	−1465	−1353	132	$C(diamond)$	1.896	2.866	2.439
Boron				$C(g)$	715.0	669.6	158.0
$B(\beta\text{-rhombo-}$ hedral)	0	0	5.87	$CO(g)$	−110.5	−137.2	197.5
				$CO_2(g)$	−393.5	−394.4	213.7
$BF_3(g)$	−1137.0	−1120.3	254.0	$CO_2(aq)$	−412.9	−386.2	121
$BCl_3(g)$	−403.8	−388.7	290.0	$CO_3^{2-}(aq)$	−676.26	−528.10	−53.1
$B_2H_6(g)$	35	86.6	232.0	$HCO_3^-(aq)$	−691.11	587.06	95.0
$B_2O_3(s)$	−1272	−1193	53.8	$CH_4(g)$	−74.87	−50.81	186.1
$H_3BO_3(s)$	−1094.3	−969.01	88.83	$C_2H_2(g)$	227	209	200.85
Bromine				$C_2H_4(g)$	52.47	68.36	219.22
$Br_2(l)$	0	0	152.23	$C_2H_6(g)$	−84.667	−32.89	229.5
$Br_2(g)$	30.91	3.13	245.38	$C_3H_8(g)$	−105	−24.5	269.9
$Br(g)$	111.9	82.40	174.90	$C_4H_{10}(g)$	−126	−16.7	310
$Br^-(g)$	−218.9	—	—	$C_6H_6(l)$	49.0	124.5	172.8
$Br^-(aq)$	−120.9	−102.82	80.71	$CH_3OH(g)$	−201.2	−161.9	238
$HBr(g)$	−36.3	−53.5	198.59	$CH_3OH(l)$	−238.6	−166.2	127
Cadmium				$HCHO(g)$	−116	−110	219
$Cd(s)$	0	0	51.5	$HCOO^-(aq)$	−410	−335	91.6
$Cd(g)$	112.8	78.20	167.64	$HCOOH(l)$	−409	−346	129.0
$Cd^{2+}(aq)$	−72.38	−77.74	−61.1	$HCOOH(aq)$	−410	−356	164
$CdS(s)$	−144	−141	−71	$C_2H_5OH(g)$	−235.1	−168.6	282.6

(continued)

Substance or Ion	ΔH_f° (kJ/mol)	ΔG_f° (kJ/mol)	S° (J/mol·K)	Substance or Ion	ΔH_f° (kJ/mol)	ΔG_f° (kJ/mol)	S° (J/mol·K)
$C_2H_5OH(l)$	−277.63	−174.8	161	**Copper**			
$CH_3CHO(g)$	−166	−133.7	266	$Cu(s)$	0	0	33.1
$CH_3COOH(l)$	−487.0	−392	160	$Cu(g)$	341.1	301.4	166.29
$C_6H_{12}O_6(s)$	−1273.3	−910.56	212.1	$Cu^+(aq)$	51.9	50.2	−26
$C_{12}H_{22}O_{11}(s)$	−2221.7	−1544.3	360.24	$Cu^{2+}(aq)$	64.39	64.98	−98.7
$CN^-(aq)$	151	166	118	$Cu_2O(s)$	−168.6	−146.0	93.1
$HCN(g)$	135	125	201.7	$CuO(s)$	−157.3	−130	42.63
$HCN(l)$	105	121	112.8	$Cu_2S(s)$	−79.5	−86.2	120.9
$HCN(aq)$	105	112	129	**Fluorine**			
$CS_2(g)$	117	66.9	237.79	$F_2(g)$	0	0	202.7
$CS_2(l)$	87.9	63.6	151.0	$F(g)$	78.9	61.8	158.64
$CH_3Cl(g)$	−83.7	−60.2	234	$F^-(g)$	−255.6	−262.5	145.47
$CH_2Cl_2(l)$	−117	−63.2	179	$F^-(aq)$	−329.1	−276.5	−9.6
$CHCl_3(l)$	−132	−71.5	203	$HF(g)$	−273	−275	173.67
$CCl_4(g)$	−96.0	−53.7	309.7	**Hydrogen**			
$CCl_4(l)$	−139	−68.6	214.4	$H_2(g)$	0	0	130.6
$COCl_2(g)$	−220	−206	283.74	$H(g)$	218.0	203.30	114.60
Cesium				$H^+(aq)$	0	0	0
$Cs(s)$	0	0	85.15	$H^+(g)$	1536.3	1517.1	108.83
$Cs(g)$	76.7	49.7	175.5	**Iodine**			
$Cs^+(g)$	458.5	427.1	169.72	$I_2(s)$	0	0	116.14
$Cs^+(aq)$	−248	−282.0	133	$I_2(g)$	62.442	19.38	260.58
$CsF(s)$	−554.7	−525.4	88	$I(g)$	106.8	70.21	180.67
$CsCl(s)$	−442.8	−414	101.18	$I^-(g)$	−194.7	—	—
$CsBr(s)$	−395	−383	121	$I^-(aq)$	−55.94	−51.67	109.4
$CsI(s)$	−337	−333	130	$HI(g)$	25.9	1.3	206.33
Chlorine				**Iron**			
$Cl_2(g)$	0	0	223.0	$Fe(s)$	0	0	27.3
$Cl(g)$	121.0	105.0	165.1	$Fe^{3+}(aq)$	−47.7	−10.5	−293
$Cl^-(g)$	−234	−240	153.25	$Fe^{2+}(aq)$	−87.9	−84.94	113
$Cl^-(aq)$	−167.46	−131.17	55.10	$FeCl_2(s)$	−341.8	−302.3	117.9
$HCl(g)$	−92.31	−95.30	186.79	$FeCl_3(s)$	−399.5	−334.1	142
$HCl(aq)$	−167.46	−131.17	55.06	$FeO(s)$	−272.0	−251.4	60.75
$ClO_2(g)$	102	120	256.7	$Fe_2O_3(s)$	−825.5	−743.6	87.400
$Cl_2O(g)$	80.3	97.9	266.1	$Fe_3O_4(s)$	−1121	−1018	145.3
Chromium				**Lead**			
$Cr(s)$	0	0	23.8	$Pb(s)$	0	0	64.785
$Cr^{3+}(aq)$	−1971	—	—	$Pb^{2+}(aq)$	1.6	−24.3	21
$CrO_4^{2-}(aq)$	−863.2	−706.3	38	$PbCl_2(s)$	−359	−314	136
$Cr_2O_7^{2-}(aq)$	−1461	−1257	214	$PbO(s)$	−218	−198	68.70

(*continued*)

A.2

Substance or Ion	ΔH_f° (kJ/mol)	ΔG_f° (kJ/mol)	S° (J/mol·K)	Substance or Ion	ΔH_f° (kJ/mol)	ΔG_f° (kJ/mol)	S° (J/mol·K)
$PbO_2(s)$	−276.6	−219.0	76.6	$N_2O_4(g)$	9.16	97.7	304.3
$PbS(s)$	−98.3	−96.7	91.3	$N_2O_5(g)$	11	118	346
$PbSO_4(s)$	−918.39	−811.24	147	$N_2O_5(s)$	−43.1	114	178
Lithium				$NH_3(g)$	−45.9	−16	193
$Li(s)$	0	0	29.10	$NH_3(aq)$	−80.83	26.7	110
$Li(g)$	161	128	138.67	$N_2H_4(l)$	50.63	149.2	121.2
$Li^+(g)$	687.163	649.989	132.91	$NO_3^-(aq)$	−206.57	−110.5	146
$Li^+(aq)$	−278.46	−293.8	14	$HNO_3(l)$	−173.23	−79.914	155.6
$LiF(s)$	−616.9	−588.7	35.66	$HNO_3(aq)$	−206.57	−110.5	146
$LiCl(s)$	−408	−384	59.30	$NF_3(g)$	−125	−83.3	260.6
$LiBr(s)$	−351	−342	74.1	$NOCl(g)$	51.71	66.07	261.6
$LiI(s)$	−270	−270	85.8	$NH_4Cl(s)$	−314.4	−203.0	94.6
Magnesium				**Oxygen**			
$Mg(s)$	0	0	32.69	$O_2(g)$	0	0	205.0
$Mg(g)$	150	115	148.55	$O(g)$	249.2	231.7	160.95
$Mg^{2+}(g)$	2351	—	—	$O_3(g)$	143	163	238.82
$Mg^{2+}(aq)$	−461.96	−456.01	118	$OH^-(aq)$	−229.94	−157.30	−10.54
$MgCl_2(s)$	−641.6	−592.1	89.630	$H_2O(g)$	−241.826	−228.60	188.72
$MgCO_3(s)$	−1112	−1028	65.86	$H_2O(l)$	−285.840	−237.192	69.940
$MgO(s)$	−601.2	−569.0	26.9	$H_2O_2(l)$	−187.8	−120.4	110
$Mg_3N_2(s)$	−461	−401	88	$H_2O_2(aq)$	−191.2	−134.1	144
Manganese				**Phosphorus**			
$Mn(s, \alpha)$	0	0	31.8	$P_4(s, white)$	0	0	41.1
$Mn^{2+}(aq)$	−219	−223	−84	$P(g)$	314.6	278.3	163.1
$MnO_2(s)$	−520.9	−466.1	53.1	$P(s, red)$	−17.6	−12.1	22.8
$MnO_4^-(aq)$	−518.4	−425.1	190	$P_2(g)$	144	104	218
Mercury				$P_4(g)$	58.9	24.5	280
$Hg(l)$	0	0	76.027	$PCl_3(g)$	−287	−268	312
$Hg(g)$	61.30	31.8	174.87	$PCl_3(l)$	−320	−272	217
$Hg^{2+}(aq)$	171	164.4	−32	$PCl_5(g)$	−402	−323	353
$Hg_2^{2+}(aq)$	172	153.6	84.5	$PCl_5(s)$	−443.5	—	—
$HgCl_2(s)$	−230	−184	144	$P_4O_{10}(s)$	−2984	−2698	229
$Hg_2Cl_2(s)$	−264.9	−210.66	196	$PO_4^{3-}(aq)$	−1266	−1013	−218
$HgO(s)$	−90.79	−58.50	70.27	$HPO_4^{2-}(aq)$	−1281	−1082	−36
Nitrogen				$H_2PO_4^-(aq)$	−1285	−1135	89.1
$N_2(g)$	0	0	191.5	$H_3PO_4(aq)$	−1277	−1019	228
$N(g)$	473	456	153.2	**Potassium**			
$N_2O(g)$	82.05	104.2	219.7	$K(s)$	0	0	64.672
$NO(g)$	90.29	86.60	210.65	$K(g)$	89.2	60.7	160.23
$NO_2(g)$	33.2	51	239.9	$K^+(g)$	514.197	481.202	154.47

(continued)

A.3

Substance or Ion	ΔH_f° (kJ/mol)	ΔG_f° (kJ/mol)	S° (J/mol·K)	Substance or Ion	ΔH_f° (kJ/mol)	ΔG_f° (kJ/mol)	S° (J/mol·K)
$K^+(aq)$	−251.2	−282.28	103	$Na_2CO_3(s)$	−1130.8	−1048.1	139
$KF(s)$	−568.6	−538.9	66.55	$NaHCO_3(s)$	−947.7	−851.9	102
$KCl(s)$	−436.7	−409.2	82.59	$NaI(s)$	−288	−285	98.5
$KBr(s)$	−394	−380	95.94	**Strontium**			
$KI(s)$	−328	−323	106.39	$Sr(s)$	0	0	54.4
$KOH(s)$	−424.8	−379.1	78.87	$Sr(g)$	164	110	164.54
$KClO_3(s)$	−397.7	−296.3	143.1	$Sr^{2+}(g)$	1784	—	—
$KClO_4(s)$	−432.75	−303.2	151.0	$Sr^{2+}(aq)$	−545.51	−557.3	−39
Rubidium				$SrCl_2(s)$	−828.4	−781.2	117
$Rb(s)$	0	0	69.5	$SrCO_3(s)$	−1218	−1138	97.1
$Rb(g)$	85.81	55.86	169.99	$SrO(s)$	−592.0	−562.4	55.5
$Rb^+(g)$	495.04	—	—	$SrSO_4(s)$	−1445	−1334	122
$Rb^+(aq)$	−246	−282.2	124	**Sulfur**			
$RbF(s)$	−549.28	—	—	$S(rhombic)$	0	0	31.9
$RbCl(s)$	−435.35	−407.8	95.90	$S(monoclinic)$	0.3	0.096	32.6
$RbBr(s)$	−389.2	−378.1	108.3	$S(g)$	279	239	168
$RbI(s)$	−328	−326	118.0	$S_2(g)$	129	80.1	228.1
Silicon				$S_8(g)$	101	49.1	430.211
$Si(s)$	0	0	18.0	$S^{2-}(aq)$	41.8	83.7	22
$SiF_4(g)$	−1614.9	−1572.7	282.4	$HS^-(aq)$	−17.7	12.6	61.1
$SiO_2(s)$	−910.9	−856.5	41.5	$H_2S(g)$	−20.2	−33	205.6
Silver				$H_2S(aq)$	−39	−27.4	122
$Ag(s)$	0	0	42.702	$SO_2(g)$	−296.8	−300.2	248.1
$Ag(g)$	289.2	250.4	172.892	$SO_3(g)$	−396	−371	256.66
$Ag^+(aq)$	105.9	77.111	73.93	$SO_4^{2-}(aq)$	−907.51	−741.99	17
$AgF(s)$	−203	−185	84	$HSO_4^-(aq)$	−885.75	−752.87	126.9
$AgCl(s)$	−127.03	−109.72	96.11	$H_2SO_4(l)$	−813.989	−690.059	156.90
$AgBr(s)$	−99.51	−95.939	107.1	$H_2SO_4(aq)$	−907.51	−741.99	17
$AgI(s)$	−62.38	−66.32	114	**Tin**			
$AgNO_3(s)$	−45.06	19.1	128.2	$Sn(white)$	0	0	51.5
$Ag_2S(s)$	−31.8	−40.3	146	$Sn(gray)$	3	4.6	44.8
Sodium				$SnCl_4(l)$	−545.2	−474.0	259
$Na(s)$	0	0	51.446	$SnO_2(s)$	−580.7	−519.7	52.3
$Na(g)$	107.76	77.299	153.61	**Zinc**			
$Na^+(g)$	609.839	574.877	147.85	$Zn(s)$	0	0	41.6
$Na^+(aq)$	−239.66	−261.87	60.2	$Zn(g)$	130.5	94.93	160.9
$NaF(s)$	−575.4	−545.1	51.21	$Zn^{2+}(aq)$	−152.4	−147.21	−106.5
$NaCl(s)$	−411.1	−384.0	72.12	$ZnO(s)$	−348.0	−318.2	43.9
$NaBr(s)$	−361	−349	86.82	$ZnS(s, zinc blende)$	−203	−198	57.7
$NaOH(s)$	−425.609	−379.53	64.454				

APPENDIX B

Equilibrium Constants for Selected Substances

Dissociation (Ionization) Constants (K_a) of Selected Acids

Name and Formula	Lewis Structure*	K_{a1}	K_{a2}	K_{a3}
Acetic acid CH_3COOH		1.8×10^{-5}		
Acetylsalicylic acid** $CH_3COOC_6H_4COOH$		3.6×10^{-4}		
Adipic acid $HOOC(CH_2)_4COOH$		3.8×10^{-5}	3.8×10^{-6}	
Arsenic acid H_3AsO_4		6×10^{-3}	1.1×10^{-7}	3×10^{-12}
Ascorbic acid $H_2C_6H_6O_6$		1.0×10^{-5}	5×10^{-12}	
Benzoic acid C_6H_5COOH		6.3×10^{-5}		
Carbonic acid H_2CO_3		4.5×10^{-7}	4.7×10^{-11}	
Chloroacetic acid $ClCH_2COOH$		1.4×10^{-3}		
Chlorous acid $HClO_2$		1.1×10^{-2}		

Note: All values at 298 K, except for acetylsalicylic acid.
*Acidic (ionizable) proton(s) shown in red. Structures shown have lowest formal charges.
**At 37°C in 0.15 M NaCl.

(*continued*)

Dissociation (Ionization) Constants (K_a) of Selected Acids (*continued*)

Name and Formula	Lewis Structure*	K_{a1}	K_{a2}	K_{a3}
Citric acid $HOOCCH_2C(OH)(COOH)CH_2COOH$		7.4×10^{-4}	1.7×10^{-5}	4.0×10^{-7}
Formic acid $HCOOH$		1.8×10^{-4}		
Glyceric acid $HOCH_2CH(OH)COOH$		2.9×10^{-4}		
Glycolic acid $HOCH_2COOH$		1.5×10^{-4}		
Glyoxylic acid $HC(O)COOH$		3.5×10^{-4}		
Hydrocyanic acid HCN		6.2×10^{-10}		
Hydrofluoric acid HF		6.8×10^{-4}		
Hydrosulfuric acid H_2S		9×10^{-8}	1×10^{-17}	
Hypobromous acid $HBrO$		2.3×10^{-9}		
Hypochlorous acid $HClO$		2.9×10^{-8}		
Hypoiodous acid HIO		2.3×10^{-11}		
Iodic acid HIO_3		1.6×10^{-1}		
Lactic acid $CH_3CH(OH)COOH$		1.4×10^{-4}		
Malic acid $HOOCCH=CHCOOH$		1.2×10^{-2}	4.7×10^{-7}	

(continued)

Dissociation (Ionization) Constants (K_a) of Selected Acids (*continued*)

Name and Formula	Lewis Structure*	K_{a1}	K_{a2}	K_{a3}
Malonic acid $HOOCCH_2COOH$		1.4×10^{-3}	2.0×10^{-6}	
Nitrous acid HNO_2		7.1×10^{-4}		
Oxalic acid $HOOCCOOH$		5.6×10^{-2}	5.4×10^{-5}	
Phenol acid C_6H_5OH		1.0×10^{-10}		
Phenylacetic acid $C_6H_5CH_2COOH$		4.9×10^{-5}		
Phosphoric acid H_3PO_4		7.2×10^{-3}	6.3×10^{-8}	4.2×10^{-13}
Phosphorous acid $HPO(OH)_2$		3×10^{-2}	1.7×10^{-7}	
Propanoic acid CH_3CH_2COOH		1.3×10^{-5}		
Pyruvic acid $CH_3C(O)COOH$		2.8×10^{-3}		
Succinic acid $HOOCCH_2CH_2COOH$		6.2×10^{-5}	2.3×10^{-6}	
Sulfuric acid H_2SO_4		Very large	1.0×10^{-2}	
Sulfurous acid H_2SO_3		1.4×10^{-2}	6.5×10^{-8}	

Dissociation (Ionization) Constants (K_b) of Selected Amine Bases

Name and Formula	Lewis Structure*	K_{b1}	K_{b2}
Ammonia NH_3		1.76×10^{-5}	
Aniline $C_6H_5NH_2$		4.0×10^{-10}	
Diethylamine $(CH_3CH_2)_2NH$		8.6×10^{-4}	
Dimethylamine $(CH_3)_2NH$		5.9×10^{-4}	
Ethanolamine $HOCH_2CH_2NH_2$		3.2×10^{-5}	
Ethylamine $CH_3CH_2NH_2$		4.3×10^{-4}	
Ethylenediamine $H_2NCH_2CH_2NH_2$		8.5×10^{-5}	7.1×10^{-8}
Methylamine CH_3NH_2		4.4×10^{-4}	
tert-Butylamine $(CH_3)_3CNH_2$		4.8×10^{-4}	
Piperidine $C_5H_{10}NH$		1.3×10^{-3}	
n-Propylamine $CH_3CH_2CH_2NH_2$		3.5×10^{-4}	

*Blue type indicates the basic nitrogen and its lone pair.

(*continued*)

Dissociation (Ionization) Constants (K_b) of Selected Amine Bases (*continued*)

Name and Formula	Lewis Structure*	K_{b1}	K_{b2}
Isopropylamine $(CH_3)_2CHNH_2$		4.7×10^{-4}	
1,3-Propylenediamine $H_2NCH_2CH_2CH_2NH_2$		3.1×10^{-4}	3.0×10^{-6}
Pyridine C_5H_5N		1.7×10^{-9}	
Triethylamine $(CH_3CH_2)_3N$		5.2×10^{-4}	
Trimethylamine $(CH_3)_3N$		6.3×10^{-5}	

Dissociation (Ionization) Constants (K_a) of Some Hydrated Metal Ions

Free Ion	Hydrated Ion	K_a
Fe^{3+}	$Fe(H_2O)_6^{3+}(aq)$	6×10^{-3}
Sn^{2+}	$Sn(H_2O)_6^{2+}(aq)$	4×10^{-4}
Cr^{3+}	$Cr(H_2O)_6^{3+}(aq)$	1×10^{-4}
Al^{3+}	$Al(H_2O)_6^{3+}(aq)$	1×10^{-5}
Cu^{2+}	$Cu(H_2O)_6^{2+}(aq)$	3×10^{-8}
Pb^{2+}	$Pb(H_2O)_6^{2+}(aq)$	3×10^{-8}
Zn^{2+}	$Zn(H_2O)_6^{2+}(aq)$	1×10^{-9}
Co^{2+}	$Co(H_2O)_6^{2+}(aq)$	2×10^{-10}
Ni^{2+}	$Ni(H_2O)_6^{2+}(aq)$	1×10^{-10}

Formation Constants (K_f) of Some Complex Ions

Complex Ion	K_f	Complex Ion	K_f
$Ag(CN)_2^-$	3.0×10^{20}	$Sn(OH)_3^-$	3×10^{25}
$Ag(NH_3)_2^+$	1.7×10^7	$Zn(CN)_4^{2-}$	4.2×10^{19}
$Ag(S_2O_3)_2^{3-}$	4.7×10^{13}	$Zn(NH_3)_4^{2+}$	7.8×10^8
AlF_6^{3-}	4×10^{19}	$Zn(OH)_4^{2-}$	3×10^{15}
$Al(OH)_4^-$	3×10^{33}		
$Be(OH)_4^{2-}$	4×10^{18}		
CdI_4^{2-}	1×10^6		
$Co(OH)_4^{2-}$	5×10^9		
$Cr(OH)_4^-$	8.0×10^{29}		
$Cu(NH_3)_4^{2+}$	5.6×10^{11}		
$Fe(CN)_6^{4-}$	3×10^{35}		
$Fe(CN)_6^{3-}$	4.0×10^{43}		
$Hg(CN)_4^{2-}$	9.3×10^{38}		
$Ni(NH_3)_6^{2+}$	2.0×10^8		
$Pb(OH)_3^-$	8×10^{13}		

Solubility-Product Constants (K_{sp}) of Slightly Soluble Ionic Compounds

Name, Formula	K_{sp}
Carbonates	
Barium carbonate, $BaCO_3$	2.0×10^{-9}
Cadmium carbonate, $CdCO_3$	1.8×10^{-14}
Calcium carbonate, $CaCO_3$	3.3×10^{-9}
Cobalt(II) carbonate, $CoCO_3$	1.0×10^{-10}
Copper(II) carbonate, $CuCO_3$	3×10^{-12}
Lead(II) carbonate, $PbCO_3$	7.4×10^{-14}
Magnesium carbonate, $MgCO_3$	3.5×10^{-8}
Mercury(I) carbonate, Hg_2CO_3	8.9×10^{-17}
Nickel(II) carbonate, $NiCO_3$	1.3×10^{-7}
Strontium carbonate, $SrCO_3$	5.4×10^{-10}
Zinc carbonate, $ZnCO_3$	1.0×10^{-10}
Chromates	
Barium chromate, $BaCrO_4$	2.1×10^{-10}
Calcium chromate, $CaCrO_4$	1×10^{-8}
Lead(II) chromate, $PbCrO_4$	2.3×10^{-13}
Silver chromate, Ag_2CrO_4	2.6×10^{-12}
Cyanides	
Mercury(I) cyanide, $Hg_2(CN)_2$	5×10^{-40}
Silver cyanide, $AgCN$	2.2×10^{-16}
Halides	
Fluorides	
Barium fluoride, BaF_2	1.5×10^{-6}
Calcium fluoride, CaF_2	3.2×10^{-11}
Lead(II) fluoride, PbF_2	3.6×10^{-8}
Magnesium fluoride, MgF_2	7.4×10^{-9}
Strontium fluoride, SrF_2	2.6×10^{-9}
Chlorides	
Copper(I) chloride, $CuCl$	1.9×10^{-7}
Lead(II) chloride, $PbCl_2$	1.7×10^{-5}
Silver chloride, $AgCl$	
Bromides	
Copper(I) bromide, $CuBr$	5×10^{-9}
Silver bromide, $AgBr$	5.0×10^{-13}
Iodides	
Copper(I) iodide, CuI	1×10^{-12}
Lead(II) iodide, PbI_2	7.9×10^{-9}
Mercury(I) iodide, Hg_2I_2	4.7×10^{-29}
Silver iodide, AgI	8.3×10^{-17}
Hydroxides	
Aluminum hydroxide, $Al(OH)_3$	3×10^{-34}
Cadmium hydroxide, $Cd(OH)_2$	7.2×10^{-15}

Name, Formula	K_{sp}
Calcium hydroxide, $Ca(OH)_2$	6.5×10^{-6}
Cobalt(II) hydroxide, $Co(OH)_2$	1.3×10^{-15}
Copper(II) hydroxide, $Cu(OH)_2$	2.2×10^{-20}
Iron(II) hydroxide, $Fe(OH)_2$	4.1×10^{-15}
Iron(III) hydroxide, $Fe(OH)_3$	1.6×10^{-39}
Magnesium hydroxide, $Mg(OH)_2$	6.3×10^{-10}
Manganese(II) hydroxide, $Mn(OH)_2$	1.6×10^{-13}
Nickel(II) hydroxide, $Ni(OH)_2$	6×10^{-16}
Zinc hydroxide, $Zn(OH)_2$	3×10^{-16}
Iodates	
Barium iodate, $Ba(IO_3)_2$	1.5×10^{-9}
Calcium iodate, $Ca(IO_3)_2$	7.1×10^{-7}
Lead(II) iodate, $Pb(IO_3)_2$	2.5×10^{-13}
Silver iodate, $AgIO_3$	3.1×10^{-8}
Strontium iodate, $Sr(IO_3)_2$	3.3×10^{-7}
Zinc iodate, $Zn(IO_3)_2$	3.9×10^{-6}
Oxalates	
Barium oxalate dihydrate, $BaC_2O_4 \cdot 2H_2O$	1.1×10^{-7}
Calcium oxalate monohydrate, $CaC_2O_4 \cdot H_2O$	2.3×10^{-9}
Strontium oxalate monohydrate, $SrC_2O_4 \cdot H_2O$	5.6×10^{-8}
Phosphates	
Calcium phosphate, $Ca_3(PO_4)_2$	1.2×10^{-29}
Magnesium phosphate, $Mg_3(PO_4)_2$	5.2×10^{-24}
Silver phosphate, Ag_3PO_4	2.6×10^{-18}
Sulfates	
Barium sulfate, $BaSO_4$	1.1×10^{-10}
Calcium sulfate, $CaSO_4$	2.4×10^{-5}
Lead(II) sulfate, $PbSO_4$	1.6×10^{-8}
Radium sulfate, $RaSO_4$	2×10^{-11}
Silver sulfate, Ag_2SO_4	1.5×10^{-5}
Strontium sulfate, $SrSO_4$	3.2×10^{-7}
Sulfides	
Cadmium sulfide, CdS	1.0×10^{-24}
Copper(II) sulfide, CuS	8×10^{-34}
Iron(II) sulfide, FeS	8×10^{-16}
Lead(II) sulfide, PbS	3×10^{-25}
Manganese(II) sulfide, MnS	3×10^{-11}
Mercury(II) sulfide, HgS	2×10^{-50}
Nickel(II) sulfide, NiS	3×10^{-16}
Silver sulfide, Ag_2S	8×10^{-48}
Tin(II) sulfide, SnS	1.3×10^{-23}
Zinc sulfide, ZnS	2.0×10^{-22}

Standard Electrode (Half-Cell) Potentials at 298 K

Half-Reaction	$E°$ (V)
$F_2(g) + 2e^- \rightleftharpoons 2F^-(aq)$	+2.87
$O_2(g) + 2H^+(aq) + 2e^- \rightleftharpoons O_2(g) + H_2O(l)$	+2.07
$Co^{3+}(aq) + e^- \rightleftharpoons Co^{2+}(aq)$	+1.82
$H_2O_2(aq) + 2H^+(aq) + 2e^- \rightleftharpoons 2H_2O(l)$	+1.77
$PbO_2(s) + 3H^+(aq) + HSO_4^-(aq) + 2e^- \rightleftharpoons PbSO_4(s) + 2H_2O(l)$	+1.70
$Ce^{4+}(aq) + e^- \rightleftharpoons Ce^{3+}(aq)$	+1.61
$MnO_4^-(aq) + 8H^+(aq) + 5e^- \rightleftharpoons Mn^{2+}(aq) + 4H_2O(l)$	+1.51
$Au^{3+}(aq) + 3e^- \rightleftharpoons Au(s)$	+1.50
$Cl_2(g) + 2e^- \rightleftharpoons 2Cl^-(aq)$	+1.36
$Cr_2O_7^{2-}(aq) + 14H^+(aq) + 6e^- \rightleftharpoons 2Cr^{3+}(aq) + 7H_2O(l)$	+1.33
$MnO_2(s) + 4H^+(aq) + 2e^- \rightleftharpoons Mn^{2+}(aq) + 2H_2O(l)$	+1.23
$O_2(g) + 4H^+(aq) + 4e^- \rightleftharpoons 2H_2O(l)$	+1.23
$Br_2(l) + 2e^- \rightleftharpoons 2Br^-(aq)$	+1.07
$NO_3^-(aq) + 4H^+(aq) + 3e^- \rightleftharpoons NO(g) + 2H_2O(l)$	+0.96
$2Hg^{2+}(aq) + 2e^- \rightleftharpoons Hg_2^{2+}(aq)$	+0.92
$Hg_2^{2+}(aq) + 2e^- \rightleftharpoons 2Hg(l)$	+0.85
$Ag^+(aq) + e^- \rightleftharpoons Ag(s)$	+0.80
$Fe^{3+}(aq) + e^- \rightleftharpoons Fe^{2+}(aq)$	+0.77
$O_2(g) + 2H^+(aq) + 2e^- \rightleftharpoons H_2O_2(aq)$	+0.68
$MnO_4^-(aq) + 2H_2O(l) + 3e^- \rightleftharpoons MnO_2(s) + 4OH^-(aq)$	+0.59
$I_2(s) + 2e^- \rightleftharpoons 2I^-(aq)$	+0.53
$O_2(g) + 2H_2O(l) + 4e^- \rightleftharpoons 4OH^-(aq)$	+0.40
$Cu^{2+}(aq) + 2e^- \rightleftharpoons Cu(s)$	+0.34
$AgCl(s) + e^- \rightleftharpoons Ag(s) + Cl^-(aq)$	+0.22
$SO_4^{2-}(aq) + 4H^+(aq) + 2e^- \rightleftharpoons SO_2(g) + 2H_2O(l)$	+.20
$Cu^{2+}(aq) + e^- \rightleftharpoons Cu^+(aq)$	+0.15
$Sn^{4+}(aq) + 2e^- \rightleftharpoons Sn^{2+}(aq)$	+0.13
$2H^+(aq) + 2e^- \rightleftharpoons H_2(g)$	**0.00**
$Pb^{2+}(aq) + 2e^- \rightleftharpoons Pb(s)$	−0.13
$Sn^{2+}(aq) + 2e^- \rightleftharpoons Sn(s)$	−0.14
$N_2(g) + 5H^+(aq) + 4e^- \rightleftharpoons N_2H_5^+(aq)$	−0.23
$Ni^{2+}(aq) + 2e^- \rightleftharpoons Ni(s)$	−0.25
$Co^{2+}(aq) + 2e^- \rightleftharpoons Co(s)$	−0.28
$PbSO_4(s) + H^+(aq) + 2e^- \rightleftharpoons Pb(s) + HSO_4^-(aq)$	−0.31
$Cd^{2+}(aq) + 2e^- \rightleftharpoons Cd(s)$	−0.40
$Fe^{2+}(aq) + 2e^- \rightleftharpoons Fe(s)$	−0.44
$Cr^{3+}(aq) + 3e^- \rightleftharpoons Cr(s)$	−0.74
$Zn^{2+}(aq) + 2e^- \rightleftharpoons Zn(s)$	−0.76
$2H_2O(l) + 2e^- \rightleftharpoons H_2(g) + 2OH^-(aq)$	−0.83
$Mn^{2+}(aq) + 2e^- \rightleftharpoons Mn(s)$	−1.18
$Al^{3+}(aq) + 3e^- \rightleftharpoons Al(s)$	−1.66
$Mg^{2+}(aq) + 2e^- \rightleftharpoons Mg(s)$	−2.37
$Na^+(aq) + e^- \rightleftharpoons Na(s)$	−2.71
$Ca^{2+}(aq) + 2e^- \rightleftharpoons Ca(s)$	−2.87
$Sr^{2+}(aq) + 2e^- \rightleftharpoons Sr(s)$	−2.89
$Ba^{2+}(aq) + 2e^- \rightleftharpoons Ba(s)$	−2.90
$K^+(aq) + e^- \rightleftharpoons K(s)$	−2.93
$Li^+(aq) + e^- \rightleftharpoons Li(s)$	−3.05

Note: Values given at 298 K. Reactions written as reductions. $E°$ value refers to all components in their standard states: for dissolved species in aqueous solutions, 1 M; for ideal gases, 1 atm pressure; for solids and liquids, the pure substance.

APPENDIX
D

Physical Constants

mass of the electron	m_e	$= 9.10939 \times 10^{-31}$ kg
mass of the neutron	m_n	$= 1.67493 \times 10^{-27}$ kg
mass of the proton	m_p	$= 1.67262 \times 10^{-27}$ kg
charge of the electron (or proton)	e	$= 1.60218 \times 10^{-19}$ C
atomic mass unit	amu	$= 1.66054 \times 10^{-27}$ kg
universal gas constant	R	$= 8.31447$ J/(mol·K)
		$= 8.20578 \times 10^{-2}$ (atm·L)/(mol·K)
Avogadro's number	N_A	$= 6.02214 \times 10^{23}$/mol
speed of light in a vacuum	c	$= 2.99792 \times 10^8$ m/s
standard acceleration of gravity	g	$= 9.80665$ m/s^2
Planck's constant	h	$= 6.62607 \times 10^{-34}$ J·s
Faraday constant	F	$= 9.64853 \times 10^4$ C/mol

SI Prefixes

a	f	p	n	µ	m	c	d	da	h	k	M	G	T
atto-	femto-	pico-	nano-	micro-	milli-	centi-	deci-	deka-	hecto-	kilo-	mega-	giga-	tera-
10^{-18}	10^{-15}	10^{-12}	10^{-9}	10^{-6}	10^{-3}	10^{-2}	10^{-1}	10^{1}	10^{2}	10^{3}	10^{6}	10^{9}	10^{12}

Common Conversions and Relationships

Length

SI unit: meter, m

1 km	$= 1000$ m
	$= 0.62$ mile (mi)
1 inch (in)	$= 2.54$ cm
1 m	$= 1.094$ yards (yd)
1 pm	$= 10^{-12}$ m $= 0.01$ Å

Mass

SI unit: kilogram, kg

1 kg	$= 10^3$ g
	$= 2.205$ lb
1 metric ton (t)	$= 10^3$ kg

Temperature

SI unit: kelvin, K

0 K	$= -273.15$
mp of H_2O	$= 0$°C (273.15 K)
bp of H_2O	$= 100$°C (373.15 K)
T (K)	$= T$ (°C) $+ 273.15$
T (°C)	$= [T$ (°F) $- 32]\frac{5}{9}$
T (°F)	$= \frac{9}{5}T$ (°C) $+ 32$

Volume

SI unit: cubic meter, m^3

1 dm^3	$= 10^{-3}$ m^3
	$= 1$ liter (L)
	$= 1.057$ quarts (qt)
1 cm^3	$= 1$ mL
1 m^3	$= 35.3$ ft^3

Energy

SI unit: joule, J

1 J	$= 1$ kg·m^2/s^2
	$= 1$ coulomb·volt (1 C·V)
1 cal	$= 4.184$ J
1 eV	$= 1.602 \times 10^{-19}$ J

Force

1 newton (N) $= 1$ kg·m·s^{-2}

Pressure

SI unit: pascal, Pa

1 Pa	$= 1$ N/m^2
	$= 1$ kg/m·s^2
1 atm	$= 1.01325 \times 10^5$ Pa
	$= 760$ torr
1 bar	$= 1 \times 10^5$ Pa

Math Relationships

$\pi = 3.1416$

volume of sphere	$= \frac{4}{3}\pi r^3$
volume of cylinder	$= \pi r^2 h$

The Elements (Atomic Numbers and Atomic Masses)

Name	Symbol	Atomic Number	Atomic Mass*	Name	Symbol	Atomic Number	Atomic Mass*
Actinium	Ac	89	(227)	Germanium	Ge	32	72.61
Aluminum	Al	13	26.98	Gold	Au	79	197.0
Americium	Am	95	(243)	Hafnium	Hf	72	178.5
Antimony	Sb	51	121.8	Hassium	Hs	108	(227)
Argon	Ar	18	39.95	Helium	He	2	4.003
Arsenic	As	33	74.92	Holmium	Ho	67	164.9
Astatine	At	85	(210)	Hydrogen	H	1	1.008
Barium	Ba	56	137.3	Indium	In	49	114.8
Berkelium	Bk	97	(247)	Iodine	I	53	126.9
Beryllium	Be	4	9.012	Iridium	Ir	77	192.2
Bismuth	Bi	83	209.0	Iron	Fe	26	55.85
Bohrium	Bh	107	(267)	Krypton	Kr	36	83.80
Boron	B	5	10.81	Lanthanum	La	57	138.9
Bromine	Br	35	79.90	Lawrencium	Lr	103	(257)
Cadmium	Cd	48	112.4	Lead	Pb	82	207.2
Calcium	Ca	20	40.08	Lithium	Li	3	6.941
Californium	Cf	98	(249)	Lutetium	Lu	71	175.0
Carbon	C	6	12.01	Magnesium	Mg	12	24.31
Cerium	Ce	58	140.1	Manganese	Mn	25	54.94
Cesium	Cs	55	132.9	Meitnerium	Mt	109	(268)
Chlorine	Cl	17	35.45	Mendelevium	Md	101	(256)
Chromium	Cr	24	52.00	Mercury	Hg	80	200.6
Cobalt	Co	27	58.93	Molybdenum	Mo	42	95.94
Copper	Cu	29	63.55	Neodymium	Nd	60	144.2
Curium	Cm	96	(247)	Neon	Ne	10	20.18
Darmstadtium	Ds	110	(281)	Neptunium	Np	93	(244)
Dubnium	Db	105	(262)	Nickel	Ni	28	58.70
Dysprosium	Dy	66	162.5	Niobium	Nb	41	92.91
Einsteinium	Es	99	(254)	Nitrogen	N	7	14.01
Erbium	Er	68	167.3	Nobelium	No	102	(253)
Europium	Eu	63	152.0	Osmium	Os	76	190.2
Fermium	Fm	100	(253)	Oxygen	O	8	16.00
Fluorine	F	9	19.00	Palladium	Pd	46	106.4
Francium	Fr	87	(223)	Phosphorus	P	15	30.97
Gadolinium	Gd	64	157.3	Platinum	Pt	78	195.1
Gallium	Ga	31	69.72	Plutonium	Pu	94	(242)

See https://www.nist.gov/pml/atomic-weights-and-isotopic-compositions-relative-atomic-masses

(continued)

Name	Symbol	Atomic Number	Atomic Mass*
Polonium	Po	84	(209)
Potassium	K	19	39.10
Praseodymium	Pr	59	140.9
Promethium	Pm	61	(145)
Protactinium	Pa	91	(231)
Radium	Ra	88	(226)
Radon	Rn	86	(222)
Rhenium	Re	75	186.2
Rhodium	Rh	45	102.9
Roentgenium	Rg	111	(272)
Rubidium	Rb	37	85.47
Ruthenium	Ru	44	101.1
Rutherfordium	Rf	104	(263)
Samarium	Sm	62	150.4
Scandium	Sc	21	44.96
Seaborgium	Sg	106	(266)
Selenium	Se	34	78.96
Silicon	Si	14	28.09
Silver	Ag	47	107.9
Sodium	Na	11	22.99
Strontium	Sr	38	87.62
Sulfur	S	16	32.07
Tantalum	Ta	73	180.9
Technetium	Tc	43	(98)
Tellurium	Te	52	127.6
Terbium	Tb	65	158.9
Thallium	Tl	81	204.4
Thorium	Th	90	232.0
Thulium	Tm	69	168.9
Tin	Sn	50	118.7
Titanium	Ti	22	47.88
Tungsten	W	74	183.9
Uranium	U	92	238.0
Vanadium	V	23	50.94
Xenon	Xe	54	131.3
Ytterbium	Yb	70	173.0
Yttrium	Y	39	88.91
Zinc	Zn	30	65.41
Zirconium	Zr	40	91.22

Periodic Table

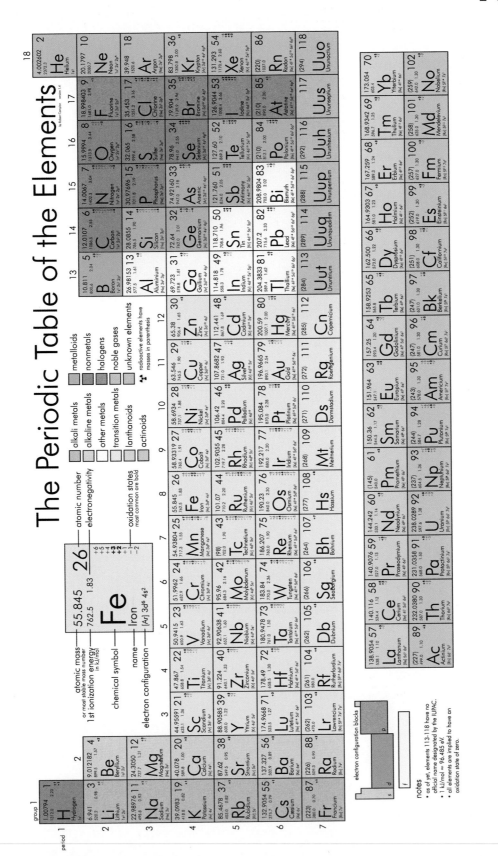

The Periodic Table of the Elements

Index